ENGR-
W9-BJL-158

Harnessing MicroStation/J™

Harnessing MicroStation/J™

G.V. Krishnan
James E. Taylor

Delmar Publishers

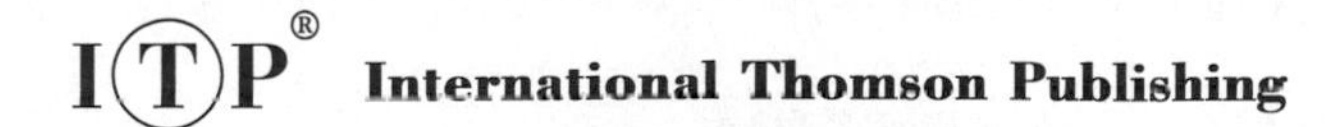

Albany • Bonn • Boston • Cincinnati • Detroit • London • Madrid
Melbourne • Mexico City • New York • Pacific Grove • Paris • San Francisco
Singapore • Tokyo • Toronto • Washington

NOTICE TO THE READER

Publisher does not warrant or guarantee any of the products described herein or perform any independent analysis in connection with any of the product information contained herein. Publisher does not assume, and expressly disclaims, any obligation to obtain and include information other than that provided to it by the manufacturer.

The reader is expressly warned to consider and adopt all safety precautions that might be indicated by the activities herein and to avoid all potential hazards. By following the instructions contained herein, the reader willingly assumes all risks in connection with such instructions.

The publisher makes no representation or warranties of any kind, including but not limited to, the warranties of fitness for particular purpose or merchantability, nor are any such representations implied with respect to the material set forth herein, and the publisher, takes no responsibility with respect to such material. The publisher shall not be liable for any special, consequential, or exemplary damages resulting, in whole or part, from the readers' use of, or reliance upon, this material.

Delmar Staff:
Publisher: Michael McDermott
Acquisitions Editor: Sandy Clark
Production Coordinator: Jennifer Gaines
Art and Design Coordinator: Mary Beth Vought
Editorial Assistant: Allyson Powell

COPYRIGHT © 1999
Delmar Publishers Inc.
an International Thomson Publishing Company
The ITP logo is a trademark under license.
Printed in Canada

Trademarks
Bentley and the "B" Bentley logo are registered trademarks. MicroStation/J is a trademark of Bentley Systems, Incorporated.
All other product names are acknowledged as trademarks of their respective owners.

Cover: Image courtesy of Brian Gassel/Thompson, Ventulett, Stainback & Associates Inc.

For more information, contact:

Delmar Publishers
3 Columbia Circle, Box 15-015
Albany, New York USA 12212-5015

International Thomson Publishing Europe
Berkshire House 168-173
High Holborn
London, WC1V 7AA
United Kingdom

Thomas Nelson Australia
102 Dodds Street
South Melbourne, Victoria 3205
Australia

Nelson Canada
1120 Birchmont Road
Scarborough, Ontario
Canada, M1K 5G4

International Thomson Publishing Southern Africa
Building 18, Constantia Park
240 Old Pretoria Road
P.O. Box 2459
Halfway House, 1685 South Africa

International Thomson Editores
Campos Eliseos 385, Piso 7
Colonia Polanco
11560 Mexico D. F. Mexico

International Thomson Publishing GmbH
Königswinterer Strasse 418
53227 Bonn Germany

International Thomson Publishing France
Tour Maine-Montparnesse
33, Avenue du Maine
75755 Paris Cedex 15, France

International Thomson Publishing – Japan
Hirakawacho Kyowa Building, 3F
2-2-1 Hirakaw-cho Chiyoda-ku
Tokyo 102 Japan

International Thomson Publishing Asia
60 Albert Street
#15-01 Albert Complex
Singapore 189969

All rights reserved. No part of this work covered by the copyright hereon may be reproduced or used in any form or by any means — graphic, electronic, or mechanical, including photocopying, recording, taping, or information storage and retrieval systems — without the written permission of the publisher.

1 2 3 4 5 6 7 8 9 10 XXX 04 03 02 01 00 99

Library of Congress Cataloging-in-Publication Data

Krishnan, G.V.
Harnessing MicroStation J / G.V. Krishnan, James E. Taylor.
p. cm.
ISBN: 0-7668-1248-0
1. MicroStation. 2. Computer-aided design–Computer programs.
I. Taylor, James E. II. Title.
TA174.K7455 1999
620'.0042'02855369–dc21

99-19000
CIP

Contents

DEDICATION

HATS OFF!

bhuvana
avinash
kavitha

dorothy

Introduction

Harnessing MicroStation/J gives you the necessary skills to start using a powerful CADD (Computer Aided Drafting and Design) program, MicroStation/J. We have created a comprehensive book providing information, references, instructions, and exercises for people of varied skill levels, disciplines, and requirements, for applying this powerful design/drafting software.

Now in its third edition, *Harnessing MicroStation/J* was written and updated as a comprehensive tool for the novice and the experienced MicroStation user, both in the classroom and on the job. The book includes new features introduced in both MicroStation SE and MicroStation/J.

Readers immediately gain a broad range of knowledge of the elementary CADD concepts necessary to complete a simple design. We do not believe the user should be asked to wade through all components of every tool or concept the first time that tool or concept is introduced. Therefore, we have set up the early chapters to cover and practice fundamentals as preparation for the advanced topics covered later in the book.

Harnessing MicroStation/J is intended to be both a classroom text and a desk reference. If you are already a user of earlier versions of MicroStation, you will see an in-depth explanation provided in the corresponding chapters for all new features. The new features in MicroStation SE and MicroStation/J give personal computer-based CADD even greater depth and breadth.

IN THIS BOOK

Chapter 1—Getting Started The beginning of this chapter describes the hardware you need to get started with MicroStation. The balance of the chapter explains how to start MicroStation/J; the salient features of dialogs, settings boxes, and MicroStation applications windows; input methods, design planes, and working units; saving changes and exiting the design file; and a summary of the enhancements in MicroStation SE and J.

Chapters 2, 3, and 4—Fundamentals These chapters introduce the basic element placement and manipulation tools needed to draw a moderately intricate design. All tools discussions are accompanied by examples. Ample exercises are designed to give students the chance to test their level of skill and understanding.

Chapter 5—AccuDraw and SmartLine An in-depth explanation is provided for AccuDraw and SmartLine, two powerful drawing tools.

Chapter 6—Manipulating Groups of Elements Introduces the Power Selector and Fence tools that allow you to manipulate groups of elements.

Chapter 7—Placing Text, Data Fields, and Tags Provides an in-depth explanation of various methods for placing text, data fields that serve as place holders for text to be placed later, and tags that allow you to add non-graphical information to the design.

Chapter 8—Element Modification This chapter introduces various tools for modifying the elements of your design. All tools are accompanied by examples.

Chapter 9—Measurement and Dimensioning Introduces tools that allow you to display the length, angle, or area of elements; place dimensions on elements; and customize the way dimensions are placed.

Chapter 10—Plotting This chapter introduces all the features related to plotting (creating paper copies of your design).

Chapter 11—Cells and Cell Libraries Introduces the powerful set of tools available in MicroStation for creating and placing symbols—called cells—and for storing them in Cell libraries. These tools permit you to group elements under a user-determined name and perform manipulations on the group as though they were a single element.

Chapter 12—Patterning Introduces the set of tools available in MicroStation to place repeating patterns to fill regions in a design, such as the hatch lines in a cross-section.

Chapter 13—Reference Files A powerful and timesaving feature of MicroStation is its ability to view other design files from an active design file (the one you currently have open for editing). MicroStation lets you display the contents of up to 255 other design files from the active design file. When you view a design file in this way, it is referred to as a Reference File. This chapter describes all the tools used to manipulate reference files.

Chapter 14—Special Features MicroStation provides some special features that, though less often used than the tools described in chapters 1 through 13, add power and versatility to the MicroStation tool set. This chapter introduces several such features.

Chapter 15—Internet Utilities MicroStation provides utilities that allow you to take part in collaborative engineering projects over the World Wide Web. With these utilities you can share design information over the Web and insert engineering data directly into your design from the Web.

Chapter 16—Customizing Introduces several tools for customizing MicroStation, such as creating multi-line definitions, custom line styles, and workspaces; creating and modifying tool boxes, tool frames, and function key menus; installing fonts; and using the archive utility.

Chapter 17—3D Design and Rendering Provides an overview of the tools and specific tools available for 3D design.

Appendices provide additional valuable information to the user.

Harnessing MicroStation/J provides a sequence suitable for learning, ample exercises, examples, review questions, and thorough coverage of the MicroStation program, that should make it a must for multiple courses in MicroStation, as well as self-learners, everyday operators on the job, and operators aspiring to customize MicroStation.

STYLE CONVENTIONS

In order to make this text easier for you to use, we have adopted certain conventions that are used throughout the book:

Convention	Example
Pull-down menu names appear with the first letter capitalized	Element pull-down menu
Tool box names appear with the first letter capitalized	Linear Elements tool box
A key icon indicates when you should respond by pressing a key on your keyboard	[Enter] Press the Enter or Return key [Shift] Press the Shift key [Ctrl] Press the Control key [F1] Press the F1 key
All tool names with an underlined letter can be invoked by holding down the Alt key and the appropriate letter	Element (Press [Alt] + l)
User input is indicated by boldface	**place line**
Instructions are indicated by italics and are enclosed in parentheses	Enter first point: *(Place data point or key-in coordinates)*

HOW TO INVOKE TOOLS

Throughout the book, instructions for invoking tools are summarized in tabular form similar to the example shown below:

Polygons tool box	Select the Place Block tool ...
Key-in Window	**place block orthogonal** (or **pl b o**) [Enter]

The left column tells you where the action will take place (such as, by selecting a pull-down menu), and the right column tells you specifically what the action is (such as, by selecting the Merge option from the File pull-down menu).

EXERCISE ICONS

A special icon is used to identify step-by-step Project Exercises. Exercises that give you practice with types of drawings that are often found in a particular engineering discipline are identified by icons that indicate the discipline. Exercises that are cross-discipline—that is, the skills used in the exercise are applicable for most or all disciplines—do not have a special icon designation. The following table presents all the exercise icons:

Type of Exercise	Icon	Type of Exercise	Icon
Project Exercises		Electrical	
Mechanical		Piping	
Architectural		Civil	

ACKNOWLEDGMENTS

This book was a team effort. We are very grateful to many people who worked very hard to help create this book. We are especially grateful to the following individuals at Delmar Publishers whose efforts made it possible to complete the project on time: Mr. Michael McDermott, Publisher; Ms. Sandy Clark, Acquisitions Editor; Mr. John Fisher, Developmental Editor; Ms. Jennifer Gaines, Production Coordinator; Ms. Mary Beth Vought, Art and Design Coordinator; and Ms. Allyson Powell, Editorial Assistant. The authors and Delmar Publishers would also like to acknowledge Mr. John Shanley and staff at Phoenix Creative Graphics, Production House.

In addition, the authors would like to acknowledge the following individuals who reviewed the previous edition of this book: Dennis C. Jackson, Malcolm A. Roberts, Jr., and Michael J. White. The authors and Delmar Publishers also gratefully acknowledge the thorough and thoughtful technical editing provided by David Newsom.

And, last but not least, special appreciation to Bentley Systems, Inc. for providing the MicroStation/J software and technical help whenever we needed and also special thanks to Applications Techniques, Inc. for providing the Pizzaz Plus software to capture screen layout.

chapter 1

Getting Started

The beginning of this chapter describes the hardware you need to get started with MicroStation J. If you need to set up the MicroStation J program on the computer and you are not familiar with the computer operating system (files, drives, directories, operating system commands, etc.), you may wish to review an Introduction to Windows operating system book, refer to the Installation Guide that came with the program, or consult the dealer from whom you purchased MicroStation. Once the computer is set up, you will have at your disposal a versatile design and drafting tool that continues to grow in power with each new version.

The balance of this chapter explains how to start MicroStation and gives an overview of the screen layout and the salient features of dialog and settings boxes. Detailed explanations and examples are provided for the concepts and commands throughout the chapters that follow.

At the end of the chapter, a list of enhancements for MicroStation SE and MicroStation J is provided. If you are already a user of MicroStation Version 95, check the list to see the new and improved features in MicroStation SE and MicroStation J. Throughout the book you will find in-depth explanations in the chapters corresponding to all the new features of MicroStation SE and MicroStation J. Features introduced in MicroStation SE and MicroStation J give personal computer–based CADD even greater depth and breadth.

HARDWARE CONFIGURATION

The configuration of your system is a combination of the hardware and software you have assembled to create your system. Countless PC configurations are available. The goal for a new computer user should be to assemble a PC workstation that will not block future software and hardware upgrades.

To use MicroStation, your computer system must meet certain minimum requirements. Following are the minimum recommended configurations for MicroStation J:

- Intel-compatible Pentium or compatible processor.
- Operating System: Windows NT version 3.51 or 4.0, Windows 95, or Windows 98.
- 32MB recommended, 16MB minimum.
- 200MB minimum hard disk (typical MicroStation J installation requires 120MB).
- Input device: mouse or tablet (tablet on Windows 98/95/NT requires vendor-supplied WINTAB driver).
- Supported graphics card (256 or more color card recommended for rendering).
- Dual screen graphics supported (Windows 98/95/NT and OS/2 require vendor-supplied drivers).

Processors

Select the one that best suits your workload. The faster and more powerful the processor, the better MicroStation performs. Thus, the processor of choice is generally the most technologically advanced one available. With the availability of fast Pentium-based systems under $2,000, a good platform for MicroStation can be attained from a number of sources. Several manufacturers of RISC processors are introducing PC-compatible systems that can now run most Windows applications.

Memory

To run MicroStation, your computer must be equipped with at least 16MB of RAM (random access memory). Depending on your MicroStation application, 32MB of RAM or more may be required for optimal performance. The availability of low-cost RAM has made this an easy upgrade for most MicroStation users.

Hard Drives

The hard disk is the personal computer's primary data storage device. A hard disk with at least 200MB capacity is required, but a larger capacity is advised. The hard disk accesses data at a rate of 3 milliseconds to 20 milliseconds. High performance is 3 milliseconds or faster, while 20 milliseconds is considered slow. Obviously, the faster the hard-disk drive, the more productive you will be with MicroStation.

Video Adapter and Display

As with any graphics program for PC CAD systems, MicroStation requires a video adapter capable of displaying graphics information. A video adapter is a printed circuit board that plugs into the central processing unit (CPU) and generates signals to drive a monitor. MicroStation supports a number of display options, ranging from low-priced monochrome setups to high-resolution color units. Some of the video display controllers can be used in combination, giving a two-screen display.

Input Devices

MicroStation supports several input device configurations. Data may be entered through the keyboard, a mouse, or a digitizing tablet with a cursor.

Keyboard

The keyboard is one of the primary input methods. It can be used to enter commands and responses.

Mouse

A mouse is used with the keyboard as a tracking device to move the crosshairs on the screen. MicroStation supports a two-button mouse and a three-button mouse (see Figure 1–1). As you move your pointing device around on a mouse pad or other suitable surface, the cursor will mimic your movements on the screen. It may take on the form of crosshairs when you are being prompted to select a point. Each of the buttons in the mouse is programmed to serve specific functions in MicroStation. See Table 1–1 for the specific functions that are programmed for the two- and three-button mouse.

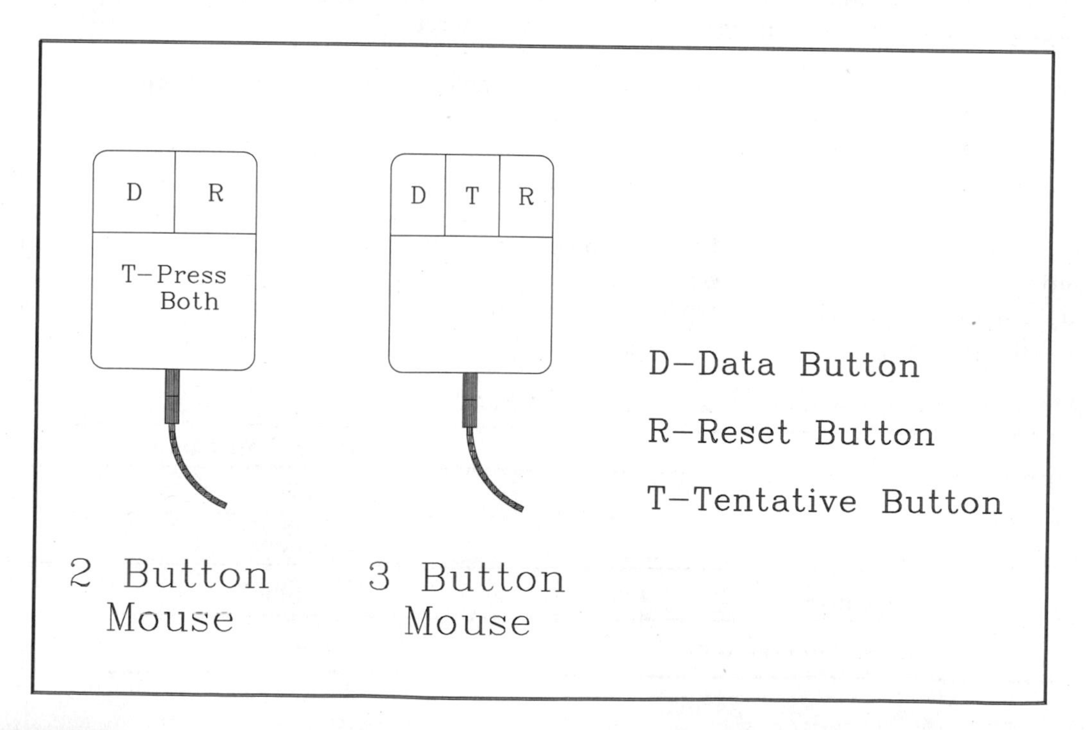

Figure 1–1 Pointing device—two-button mouse and three-button mouse

Table 1–1 Button Functions on Two- and Three-Button Mice

BUTTON FUNCTION	BUTTON POSITION	
	TWO-BUTTON MOUSE	THREE-BUTTON MOUSE
Data button	Left (first button)	Left (first button)
Reset button	Right (second button)	Right (third button)
Tentative button	Left and right simultaneously	Center (second button)

Digitizing Tablet

A digitizer is literally an electronic drawing board. An internal wiring system forms a grid of fine mesh, which corresponds to the coordinate points on the screen. The digitizer works with a puck. When the user moves the puck across the digitizer, crosshairs follow on the screen. The number of buttons on a puck depends on the manufacturer. A puck usually has four buttons, as shown in Figure 1–2; each of the buttons on the puck is programmed to serve specific functions in MicroStation. If there are more than four buttons on the puck, the first four are preprogrammed. For example, when you place a line, the points are selected by pressing the designated Data button in two locations. Although the user's attention is focused on the screen, the points are actually selected on the coordinates of the tablet. The coordinates from the tablet are then transmitted to the computer, which draws the image of a line on your screen.

Another advantage of a tablet is the ability to use a tablet menu for command selection. Menus have graphic representations of the commands and are taped to the digitizing tablet's surface. The menu is attached or activated with the key-in **AM=** followed by that menu's specific name. See Appendix B for a listing of the commands available in the table menu. The commands are chosen by placing the puck's crosshairs over the block that represents the command you want to use and pressing the designated command button. Another powerful feature of the tablet (not related to entering commands) is that it allows you to lay a map or other picture on the tablet and trace over it with the puck. See Table 1–2 for the specific functions that are programmed for the four-button puck. Detailed explanations of the functions are given later in the chapter, in the section on Input Methods.

Table 1–2 Button Functions on the Four-Button Puck

FUNCTION	FOUR-BUTTON PUCK
Data button	Top (yellow)
Reset button	Right (green)
Tentative button	Bottom (blue)
Command button	Left (white)

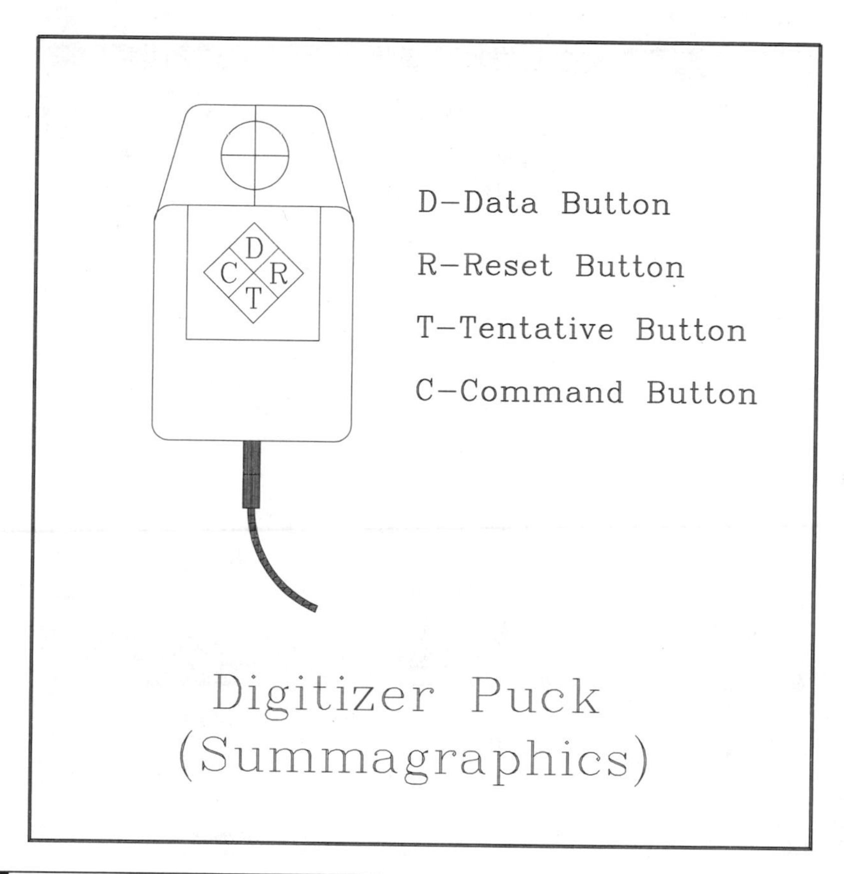

Figure 1–2 Pointing device—digitizer puck

STARTING MICROSTATION J

Click the Start button (on Windows 95, Windows 98, and Windows NT 4.0 operating systems), select the MicroStation J program group, and then select the MicroStation J program. MicroStation displays the MicroStation Manager dialog box, similar to Figure 1–3.

Note: All the screen captures shown in this textbook are taken from MicroStation J running in the Windows 98 operating system.

Before we start a new design file, let's discuss the important features of dialog boxes and settings boxes.

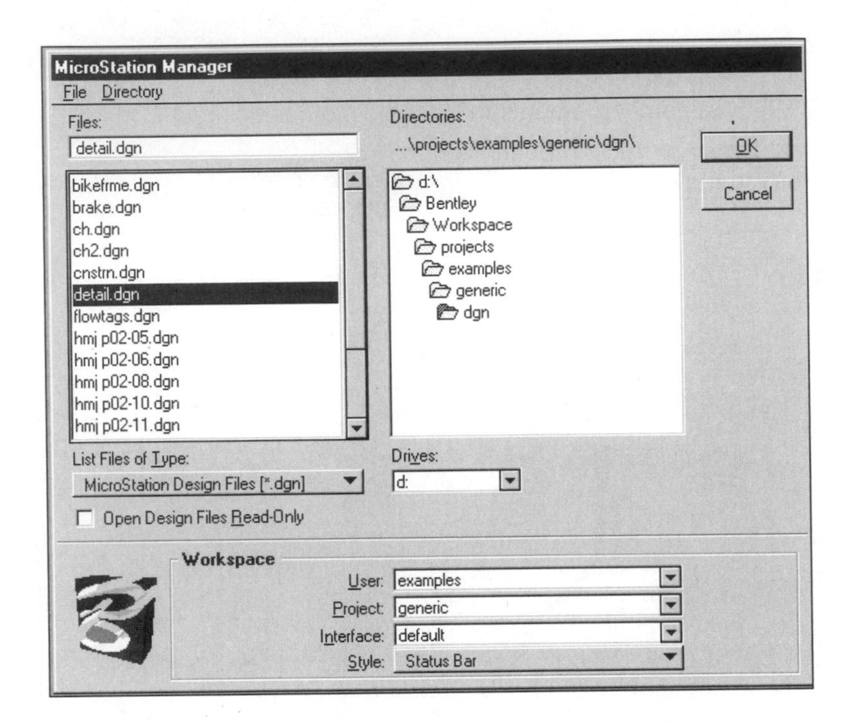

Figure 1–3 MicroStation Manager dialog box

USING DIALOG BOXES AND SETTINGS BOXES

A dialog box is a special type of window displayed by MicroStation. Dialog boxes were designed to permit the user to perform many actions easily within MicroStation. Dialog boxes force MicroStation to stop and focus on what is happening in that dialog box only. You cannot do anything else in MicroStation until you close the dialog box. The MicroStation Manager is a good example of this type of dialog box.

In addition to dialog boxes, MicroStation provides settings boxes. Several settings boxes can be left on the screen while you work in other areas of MicroStation. For instance, the Lock Toggles settings box (see Figure 1–4) can be left open as long as you need to use it. While it is open, you can turn the locks ON and OFF as you need and at the same time interact with other dialog boxes. To close the settings box, click the "X" symbol located in the top right corner of the settings box.

When you move the cursor onto a dialog box or settings box, the cursor changes to a pointer. You can use the arrow keys on your keyboard to make selections, but it is much easier with your pointing device. Another way to make selections is via the keyboard equivalents.

Title Bar and Menu Bar Item

MicroStation displays the title of the dialog box or settings box in the Title bar, as shown in Figure 1–5. Below the Title bar, MicroStation displays any available pull-down menus in the menu bar, as shown in Figure 1–5. In this case, two pull-down menus are available, File and Directory. Selecting from the list is a simple matter of moving the cursor down until the desired item is highlighted, then pressing the designated Pick (Data) button on the pointing device. If a menu item has an arrow to the right, it has a cascading submenu. To display the submenu, just click on the menu title. Menu items that include ellipses (...) display dialog boxes. To select these, just pick the menu item.

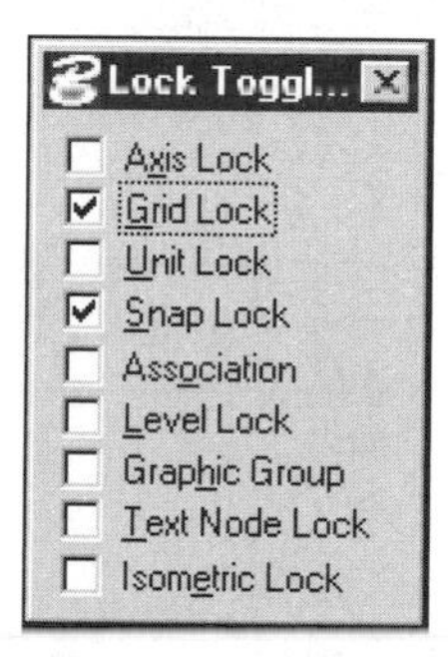

Figure 1–4 Lock Toggles settings box

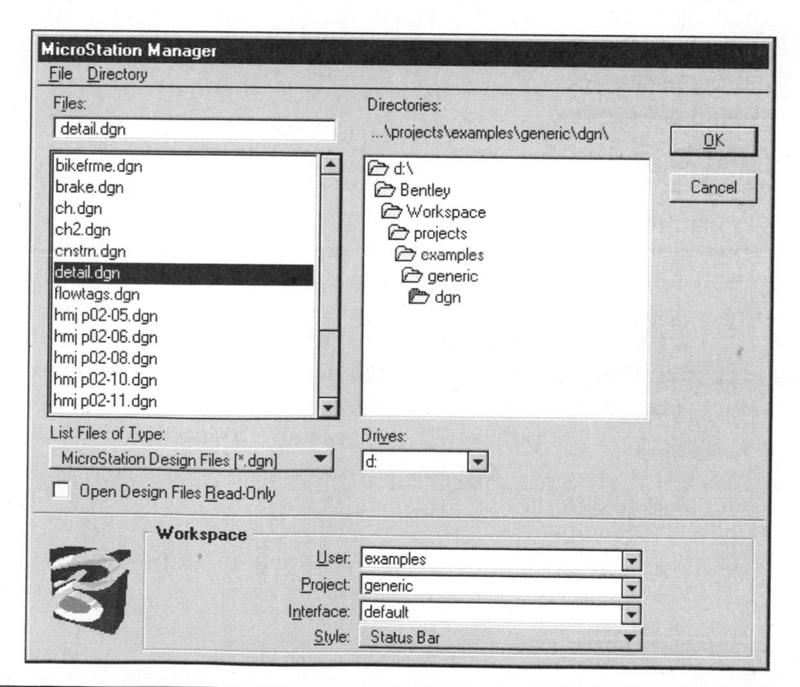

Figure 1–5 MicroStation Manager dialog box showing the Title bar and pull-down menus

Label Item

MicroStation displays the text (display only) to label the different parts of the dialog box or settings box, as shown in Figure 1–6.

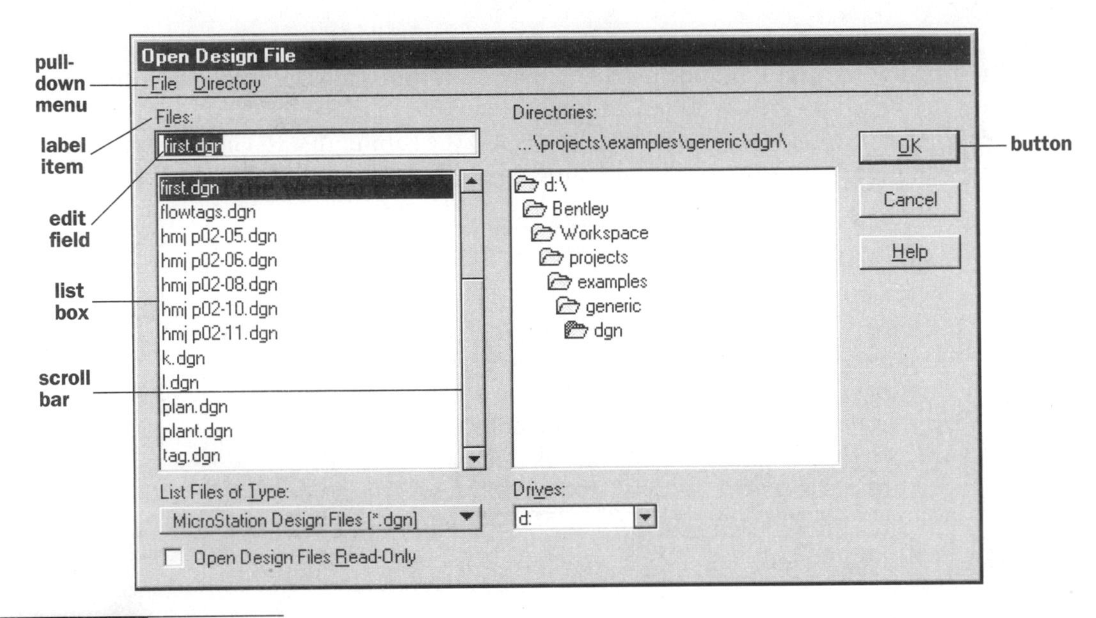

Figure 1–6 Dialog box

Edit Field

An edit field is an area that accepts one line of text entry. It normally is used to specify a name, such as a file name, including the drive and/or directory path or level name. Edit fields often make an alternative to selecting from a list of names when the desired name is not displayed in the list box. Once the correct text is keyed in, enter it by pressing Enter.

Moving the pointer into the edit field causes the text cursor to appear in a manner similar to the cursor in a word processor. If necessary, the text cursor, in combination with special editing keys, can help to facilitate changes to the text. You can see the text cursor and the pointer at the same time, making it possible to click the pointer on a character in the edit field and relocate the text cursor to that character. You can select a group of characters in the edit field to manipulate, for instance, or to delete, by highlighting the characters. This is accomplished by holding down the designated Pick button on your pointing device and dragging left or right. You can also use the right and left arrows on the keyboard to move the cursor right or left, respectively, across text, without affecting the text.

List Boxes and Scroll Bars

List boxes make it easy to view, select, and enter a name from a list of existing items, such as file names (see Figure 1–6). With the pointer, highlight the desired selection. The item, when clicked, will appear in the edit field. You can accept this item by clicking OK or by double-clicking on the item. List boxes are accompanied by scroll bars to facilitate moving long lists up and down in the list box. When you point and hold onto the slider box, you can move it up and down to cause the list to scroll. Pressing the up/down arrows causes the list to scroll up or down one item at a time.

Buttons

Actions are initiated immediately when one of the various buttons is clicked (OK or Cancel, for example). If a button (such as the OK button in Figure 1–6) is surrounded by a heavy line, it is the default button, and pressing Enter is the same as clicking that button. Buttons with action that is not acceptable will be disabled; they will appear grayed out. Buttons with ellipses (...) will cause that action's own dialog box (subdialog) to appear.

Toggle Buttons

A button that indicates an ON or OFF setting is also called a *check box* or *toggle button*. For instance, in Figure 1–7, the Axis, Grid, and Snap locks are set to ON and the remaining locks are set to OFF.

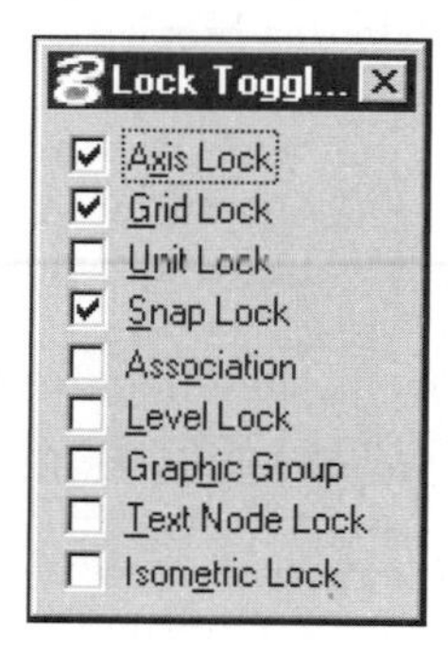

Figure 1–7 Toggle buttons

Option Button

A list of items is displayed when you click on the option button menu, and only one item may be selected from the list, as shown in Figure 1–8.

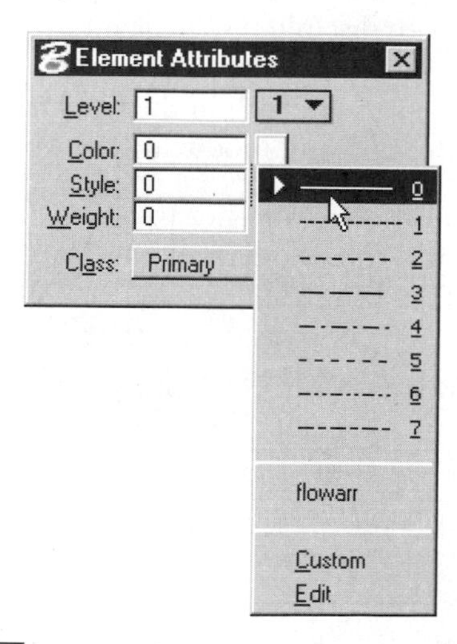

Figure 1–8 Option button menu

Let's get on to the business of starting a new design file in MicroStation.

BEGINNING A NEW DESIGN

To begin a new design, select the New... tool from the File pull-down menu, as shown in Figure 1–9.

The Create Design File dialog box opens, as shown in Figure 1–10.

Select the appropriate seed file (more about this in the section on Seed Files), enter a name for your new design in the Files: edit field, and click the OK button. The Create Design File dialog box is closed and control is passed to the MicroStation Manager. MicroStation by default highlights the name of the file you just created in the Files list box.

Before you click the OK button to open the newly created design file, make sure the appropriate User, Project, and Interface are selected from the option menu located at the bottom of the MicroStation Manager dialog box.

A user is a customized workspace drafting environment that permits the user to set up MicroStation for specific purposes. You can set up as many workspaces as you need. A workspace consists of "components" and "configuration files" for both the user and the project. By default, MicroStation selects the default workspace. If necessary, you can create or modify an existing workspace. Refer to Chapter 15 for a detailed description of creating or modifying workspaces.

The selection of the project sets the location and names of data files associated with a specific design project. Refer to Chapter 16 for setting up the project. By default, MicroStation selects the default in the Project option menu.

The selection of the interface sets a specific look and feel of MicroStation's tools and general on-screen operation. If necessary, you can change the selection of the interface from the Interface option menu. MicroStation comes with discipline-specific interfaces: civil, architecture, mechanical, drafting, and mapping. In addition, it has interfaces for previous versions (V4 and V5) and AutoCAD users. By default, MicroStation selects the default in the Interface option menu. Refer to Chapter 16 for a detailed explanation on creating and modifying the MicroStation interface.

The Style option menu allows you to select whether to use the older Command Window (V4 and V5) method of communicating with MicroStation or the Status bar. The default selection is the Status bar. Whichever style you select, MicroStation will remember it from session to session.

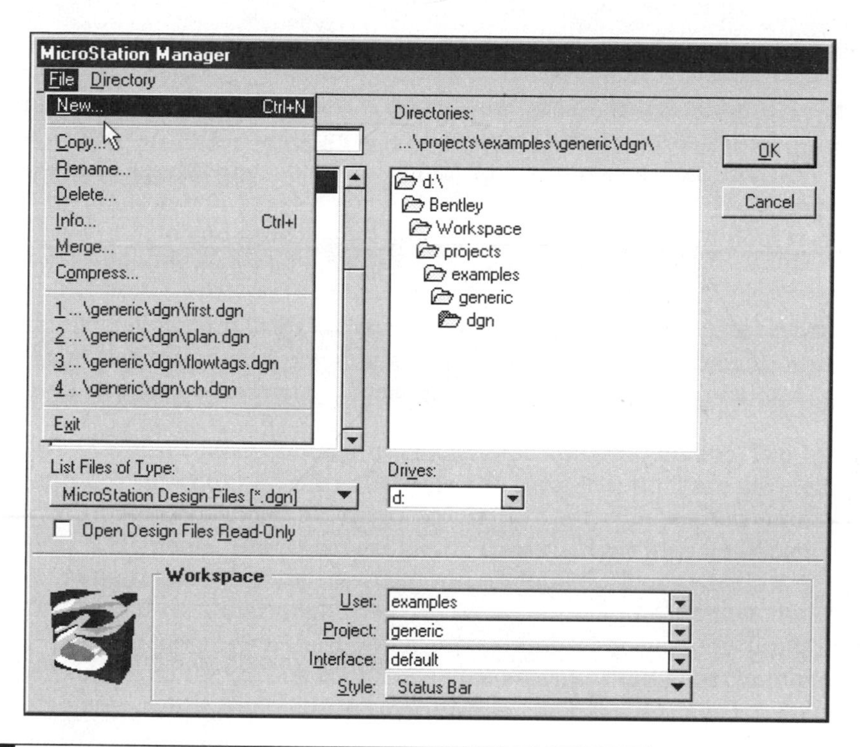

Figure 1–9 Invoking the New... tool from the File pull-down menu

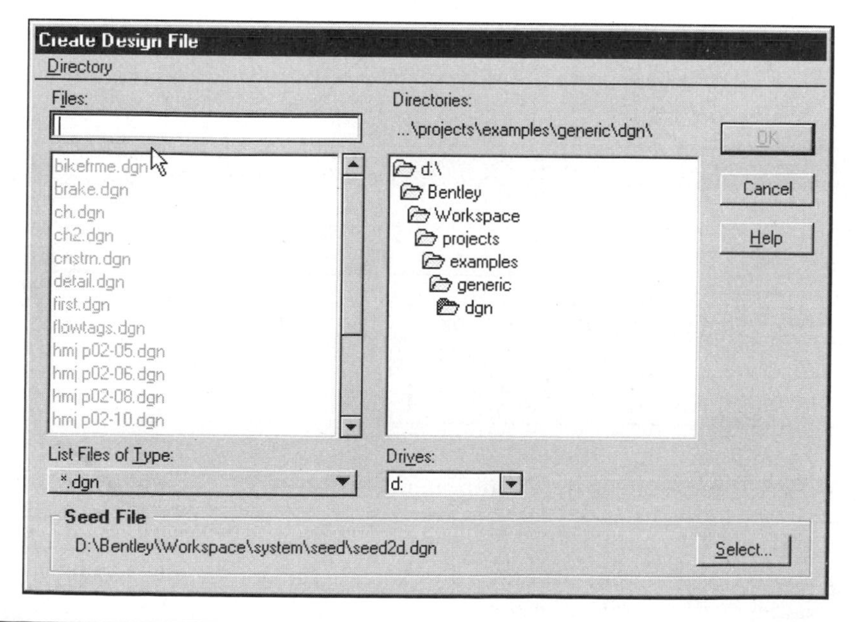

Figure 1–10 Create Design File dialog box

To open the new design file, click the OK button. Your screen will look similar to the one in Figure 1–11.

Note: If the design file name you key-in is the same as the name of an existing file name, MicroStation displays an Alert Box asking if you want to replace the existing file. Click the OK button to replace, or Cancel to reissue a new file name.

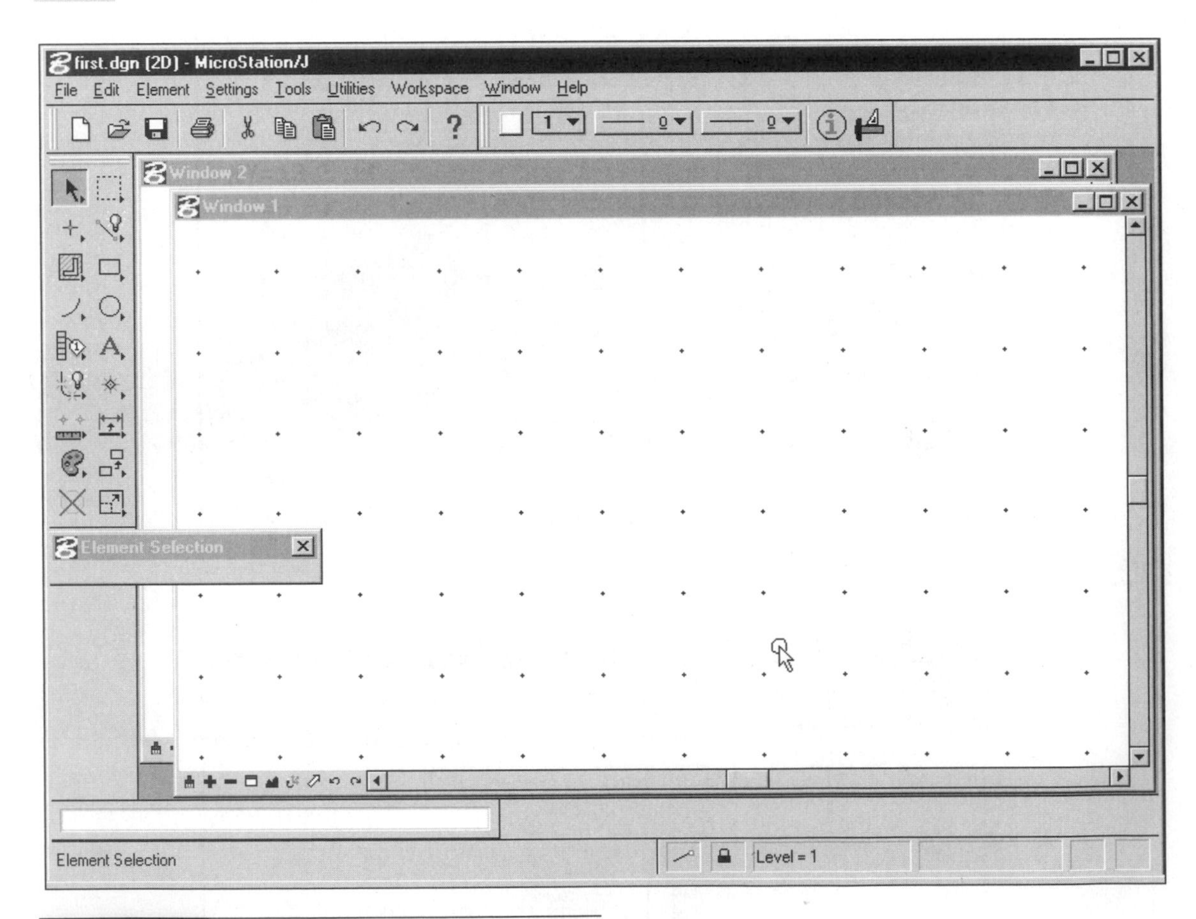

Figure 1–11 MicroStation application window

File Names

The name you enter in the Files: edit field will be the name of the file in which information about the design is stored. It must satisfy the requirements for file names as specified by the particular operating system on your computer.

File names and file extensions can contain up to 255 characters (Windows 95, Windows 98, and Windows 4.0). Names may be made up of combinations of uppercase and lowercase letters, numbers, the underscore (_), the hyphen (-), embedded spaces, and punctuation. Valid examples include:

this is my first design.dgn

first house.dgn

machine part one.dgn

PART_NO5.wrk

When MicroStation prompts for a design name, just type in the file name and MicroStation will append the extension .DGN by default. For instance, if you respond to the design name as FLOOR1, then MicroStation will create a file with the file spec FLOOR1.DGN. If you need to provide a different extension, key-in the extension with the file name.

As you progress through the lessons, note how various functions ask for names of files. If MicroStation performs the file processing, it usually adds the proper default extension.

The Path

If you want to create a new design file or edit a design file that is on a drive and/or folder other than the current drive/folder, you must furnish what is called the *path* to the design file as part of the file specification you enter. Specifying a path requires that you use the correct pathfinder symbols—the colon (:) and/or the backslash (\). The drive with a letter name (usually A through E) is identified as such by a colon, and the backslashes enclose the name of the directory where the design file is (or will be) located. Examples of path/key name combinations are as follows:

a:proj1	The file proj1 in the working directory on drive A.
b:\spec\elev	The file elev in the \spec directory on drive B.
\buil\john	The file john in the \buil directory.
ACME\doors	The file doors in the working directory's ACME subdirectory.
..\PIT\flange	The file flange in the parent directory's PIT subdirectory.

Note: Instead of specifying the path as part of the design file name, select the appropriate drive and folder from the list box and then select the design file from the files list box. MicroStation Manager displays the current drive and path in the dialog box.

Seed Files

Each time you use MicroStation's Create New File utility, a copy is made of an existing "prototype," or "seed," file. If necessary, you can customize the seed file. In other words, you can control the initial "makeup" of the file. To do so, open an existing seed file, make the necessary changes in the settings of the parameters, and place elements such as title block, and so on. Whenever you start a new design file, make sure to copy the appropriate seed file.

MicroStation programs come with several seed files. See Appendix F for a list of seed files and default working units. Depending on the discipline, use an appropriate seed file. For instance, if you plan to work on an architecture floor plan, then use the architecture seed file (SDARCH2D.DGN). When you open the Create Design file, MicroStation displays the name of the default seed file as shown in Figure 1–12. If necessary, you can change the default seed file by clicking the Select... button. MicroStation displays a list of available seed files as shown in Figure 1–13. Select the one you want from the list and click the OK button.

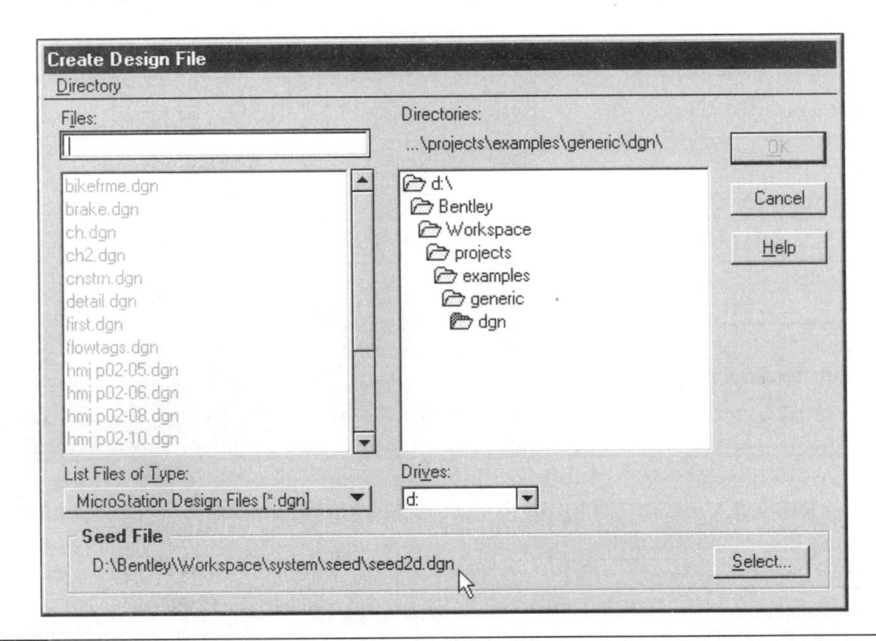

Figure 1–12 Create Design File dialog box displaying the name of the default seed file

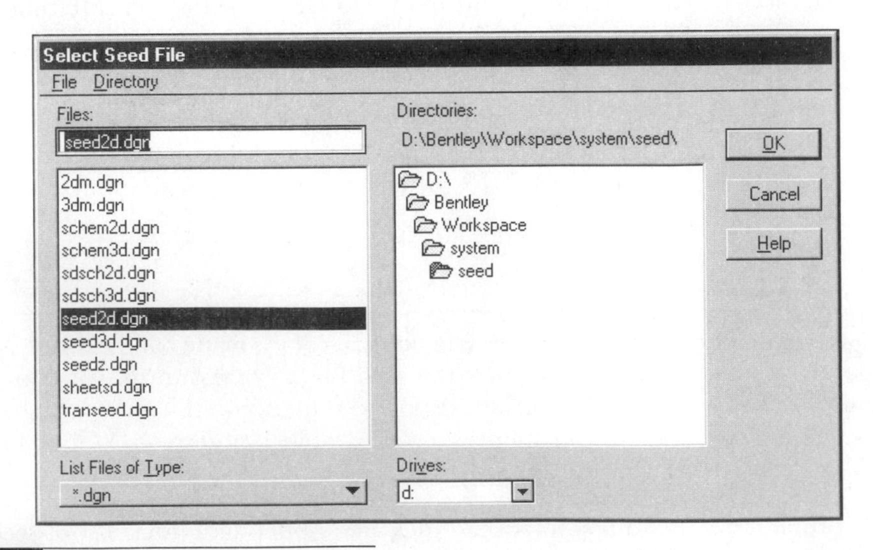

Figure 1–13 Select Seed File dialog box

OPENING AN EXISTING DESIGN FILE

Whenever you want to open an existing design file in MicroStation, simply click the name of the file in the Files: list box item of the MicroStation Manager dialog box, then click the OK button, or double-click the name of the file. If the design file is not in the current folder, change to the appropriate drive and folder from the Directories list box, then select the appropriate design file. In addition, MicroStation displays the names (including the path) of the last four design files opened in the File pull-down menu, as shown in Figure 1–14. If you need to open one of these four design files, click on the file name and MicroStation displays the design file.

Note: You can also create a new design file or open an existing design file by selecting the New... or Open... tool, respectively, from the File pull-down menu located in the MicroStation Application Window.

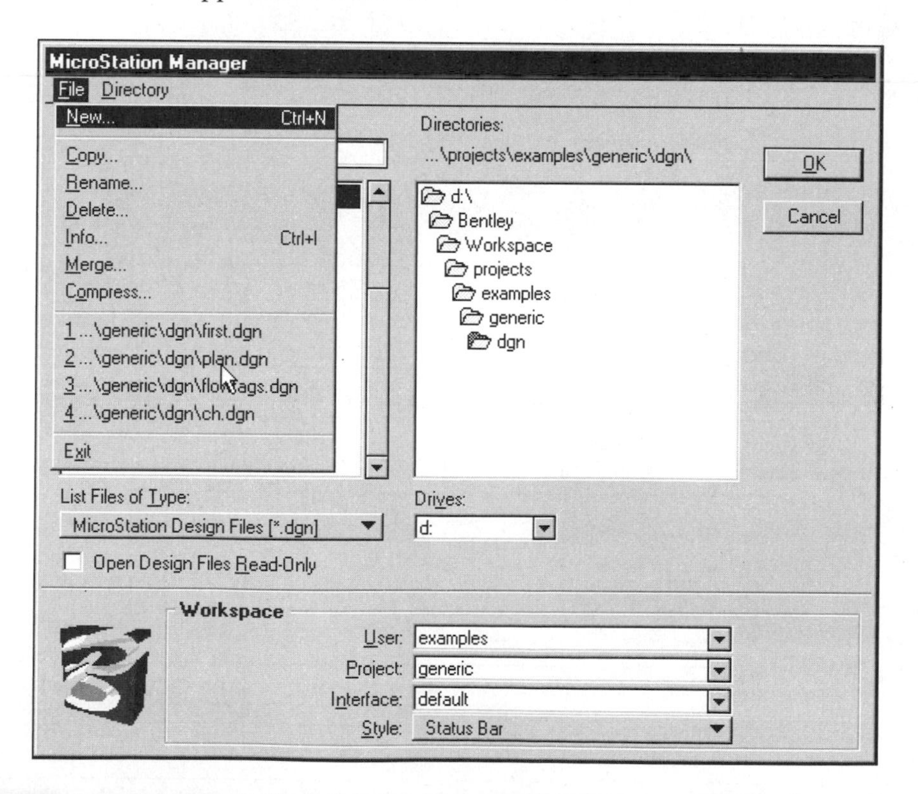

Figure 1–14 File Pull-down menu from the MicroStation Manager dialog box, displaying the names of the last four design files opened

Opening a Design in Read-Only Mode

MicroStation allows you to open a design in read-only mode by turning ON the toggle button for Open Design Files Read-Only, located in the MicroStation Manager dialog box. When the Read-

Only option is chosen, MicroStation opens the active design file in a read-only state and displays a disk icon with a large red X in the lower right corner of the application window. Any changes you make to the design will not be saved as part of the design file.

MICROSTATION APPLICATION WINDOW

The MicroStation application window consists of pull-down menus (also called *menu bars*), the Status bar, tool frames, tool boxes, the Key-in window, and view windows with the View Control bar (see Figure 1–15).

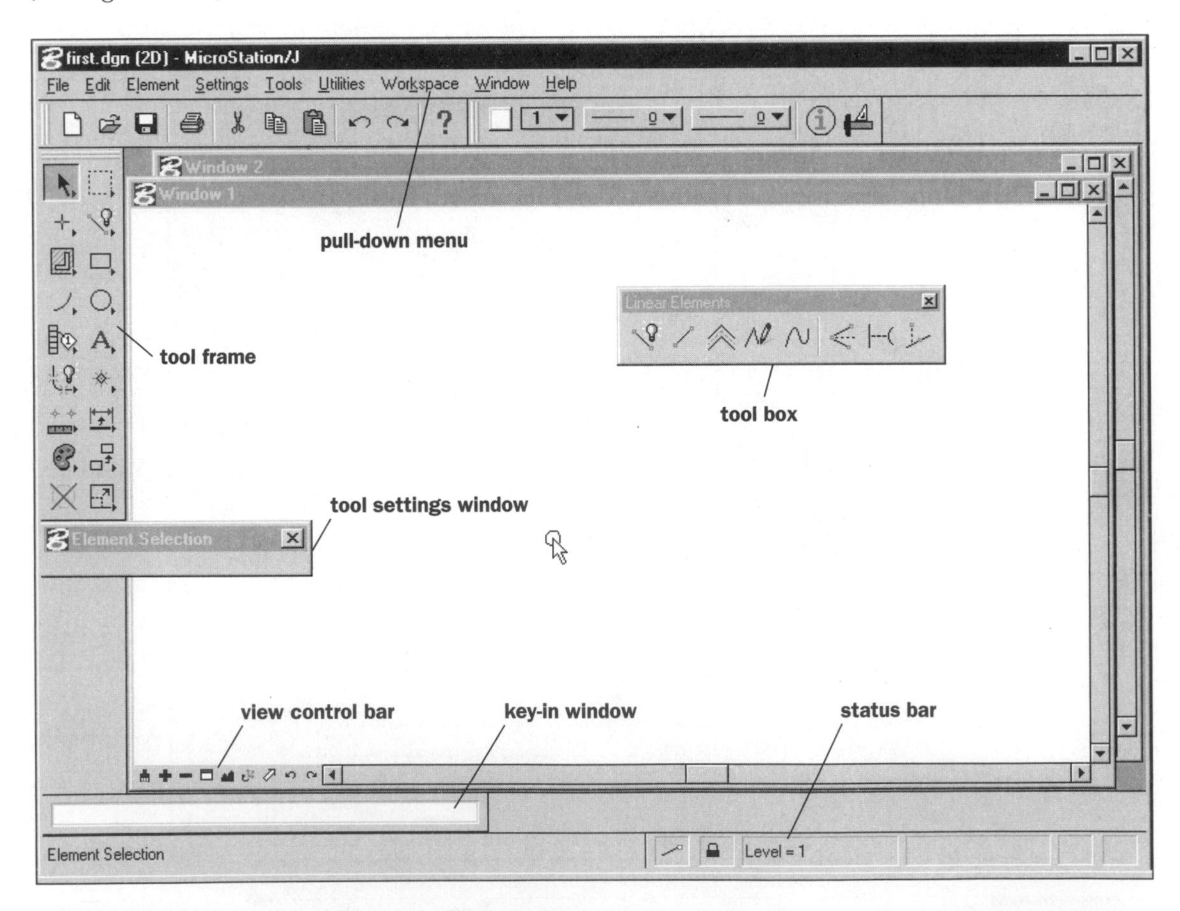

Figure 1–15 MicroStation application window

Pull-down Menus

The MicroStation Application Window has a set of pull-down menus, as shown in Figure 1–15. Several tool boxes, dialog boxes, and settings boxes are available from the pull-down menus. To select one of the pull-down menus, select the name of the pull-down menu. MicroStation displays the list of options available. Selecting from the list is a simple matter of moving the cursor down until the desired item is highlighted and then pressing the Data button on the pointing device. If a menu item has an arrow to the right, it has a cascading submenu. To display the submenu, just select the name of the submenu. Menu items that include ellipses (...) display dialog boxes. When a dialog box is displayed, no other action is allowed until that dialog box is dismissed or closed.

The Tools pull-down menu displays the list of available tool boxes in MicroStation. To select one of the available tool boxes, just select the name of the tool box and it will be displayed on the screen. Check marks are placed in the Tools menu to indicate open tool boxes. Choosing an item in the Tools pull-down menu toggles the state of the corresponding tool box. If the tool box is closed, it opens; if the tool box is open, it closes. You can place the tool box anywhere on the screen by dragging it with your pointing device. There is no limit to the number of the tool boxes that can be displayed on the screen.

To open multiple tool boxes at the same time, open the Tool Boxes dialog box from:

Pull-down menu	Tools > Tool Boxes...

MicroStation displays the Tool Boxes dialog box, similar to Figure 1–16.

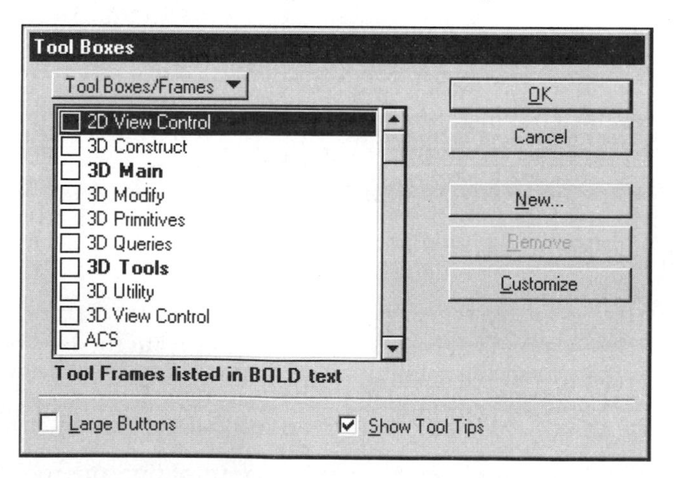

Figure 1–16 Tool Boxes dialog box

Select all the tool boxes to open by turning ON their check boxes, then click the OK button to close the dialog box. MicroStation displays the selected tool boxes.

If you are working with two monitors, you can drag the tool box to the second monitor. You can close the tool box by clicking on the "X" located in the top right corner of the tool box. When you open the tool box, it is displayed at the same location where it was previously open.

Tool Frames and Tool Boxes

Tool frames hold tool boxes of related tools and display the icon for the most recently invoked tool within each tool box. MicroStation provides six tool frames: Main, 3D Tools, Surface Modeling, Annotation, B-Spline Curves, and DD Design. The most important one is the Main tool frame, shown in Figure 1–17. The Main tool frame provides access to the majority of MicroStation's drawing tools, including measurement and dimensioning tools. Consider always leaving the Main tool frame open while drawing.

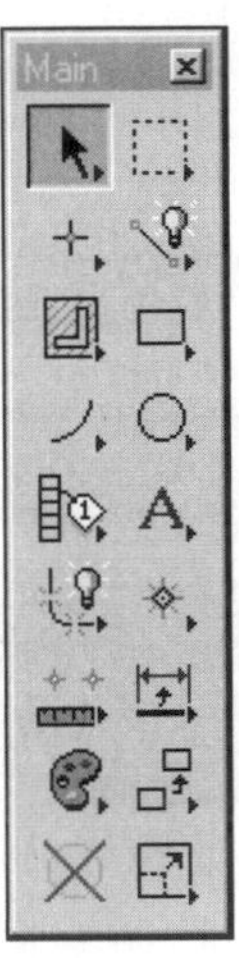

Figure 1–17 Main tool frame

Tool boxes consist of various tools. To access one of the tools from the tool box, click on the icon with the Data button and the appropriate tool is invoked. To access one of the tools from a tool box in the tool frame, click, hold, and drag the pointer to the appropriate icon and release the pointer. The selected tool is invoked. If necessary, you can tear the tool box from the tool frame. To do so, select the appropriate tool box by pressing and holding the tool box, then drag it to anywhere on the screen and release it. Figure 1–18 shows the Linear Elements tool box taken out from the Main tool frame.

Resizing the Tool Box

By clicking and dragging on the edge of the tool box you can change the shape of the tool box's window. Figure 1–19 shows the resizing of the Linear Elements tool box to three different layouts.

The tools contained within will move to fill the new window shape. The resized version of any given tool box will be remembered by MicroStation from session to session.

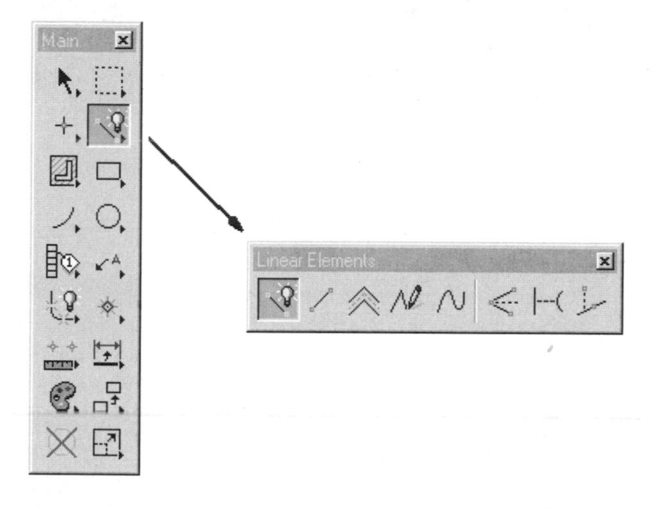

Figure 1–18 Linear Elements tool box taken out from the Main tool frame

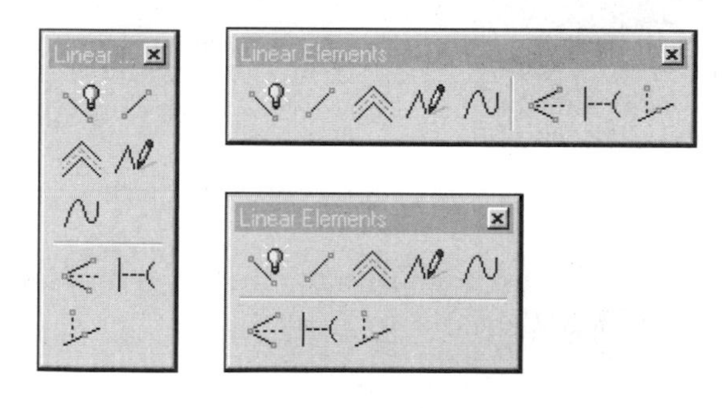

Figure 1–19 Linear Elements tool box resized to three different layouts

Docking a Tool Box

MicroStation allows you to dock a tool box to any edge of the MicroStation application window so it becomes part of the application window. Docking can happen on all four sides of the application window, and you can dock more than one tool box to each edge of the application window. Figure 1–20 shows tool boxes docked on four sides of the application window. When a tool box is docked to the top edge of the application window, it is most often referred to as a *tool bar.* For all practical purposes, a tool box and a tool bar are one and the same.

Note: To override the docking feature, hold Ctrl while dragging the tool box to the edge of the Application Window.

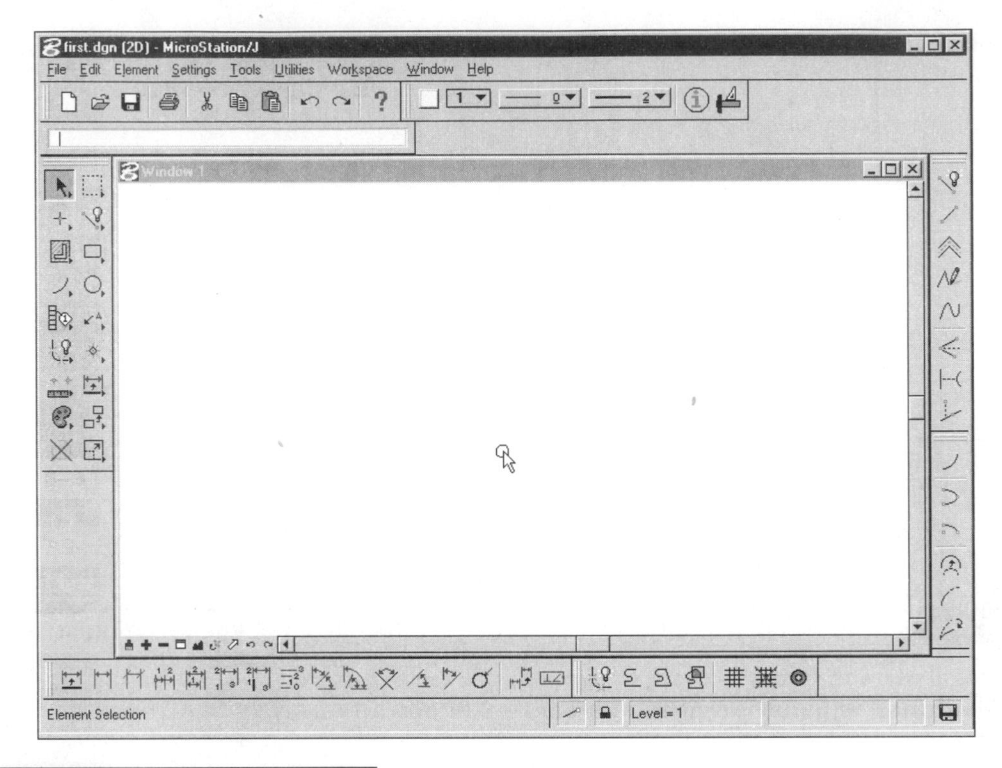

Figure 1–20 Docking of tool boxes

Undocking a Tool Box

To undock a tool box, grab part of the frame around the target tool box and drag it onto the main part of the MicroStation window. Once this is done, the tool box will return to its floating condition, or undragged.

Standard Tool Box (Also Called Standard Tool Bar)

By default, the Standard tool box (see Figure 1–21) is docked just below the pull-down menu. The Standard tool box contains icons that enable quick access to many commonly used File and Edit pull-down menu items. Table 1–3 lists the tools available in the Standard tool box and the corresponding items in the pull-down menus.

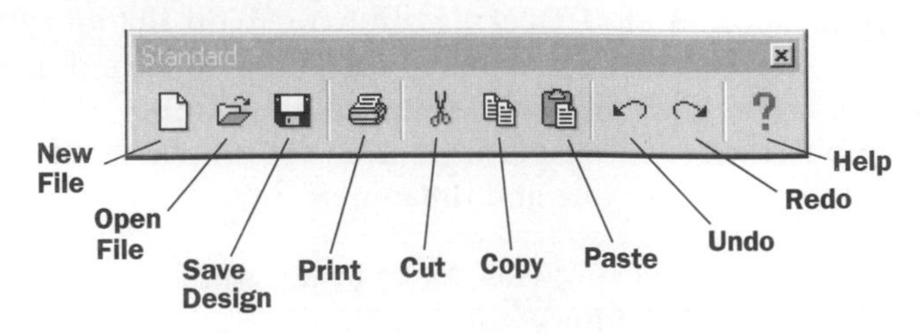

Figure 1–21 Standard tool box

Table 1–3 Tools Available in the Standard Tool Box

STANDARD TOOL BOX ITEM	PULL-DOWN MENU ITEM
New File	File > New...
Open File	File > Open...
Save Design	File > Save
Print	File > Print/Plot
Cut	Edit > Cut
Copy	Edit > Copy
Paste	Edit > Paste
Undo	Edit > Undo
Redo	Edit > Redo
Help	Help > Contents

Primary Tools Box (Also Called Primary Tool Bar)

By default, the Primary Tool box (see Figure 1–22) is docked along the side of the Standard tool box. The Primary tool box contains icons that provide access to the more frequently accessed element symbology settings, along with two tools. Table 1–4 lists the tools available in the Primary tool box.

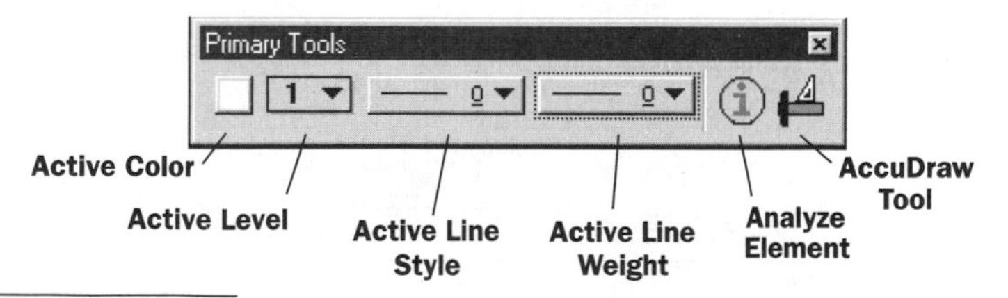

Figure 1–22 Primary tool box

Table 1–4 Tools Available in the Primary Tool Box

PRIMARY TOOL BOX ITEM	FUNCTION
Active Color	Sets the Active Color
Active Level	Sets the Active Level
Active Line Style	Sets the Active Line Style
Active Line Weight	Sets the Active Line Weight
Analyze Item	Reviews information about an element
Start AccuDraw	Starts the AccuDraw tools

Tool Tips

When you move the pointer over any icon, MicroStation displays a tool tip, as shown in Figure 1–23. In addition, MicroStation displays a brief description of the function of the command in the Status bar. This is very helpful, especially when you are unfamiliar with which icon goes with which tool.

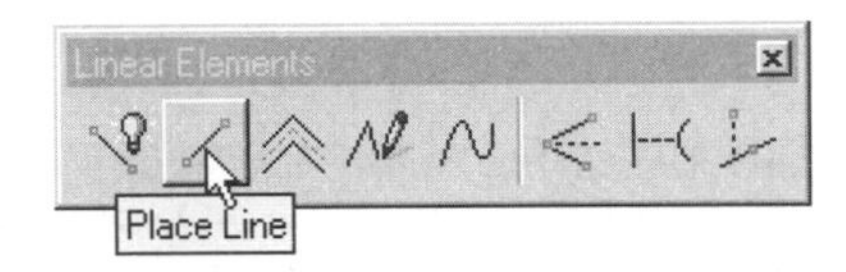

Figure 1–23 Invoking the Place Line tool (with the display of a tool tip) from the Linear Elements tool box

If necessary, you can disable the tool tips by turning off the option via the Help pull-down menu.

Tool Settings Window

Whenever you invoke a tool, MicroStation displays the controls required for adjusting the settings in the Tool Settings window. For example, if the Place Arc tool is selected, the Method, Radius, Start Angle, and Sweep Angle options are displayed in the Tool Settings window, as shown in Figure 1–24. If the Tool Settings window is closed, it opens automatically when a tool with settings is selected.

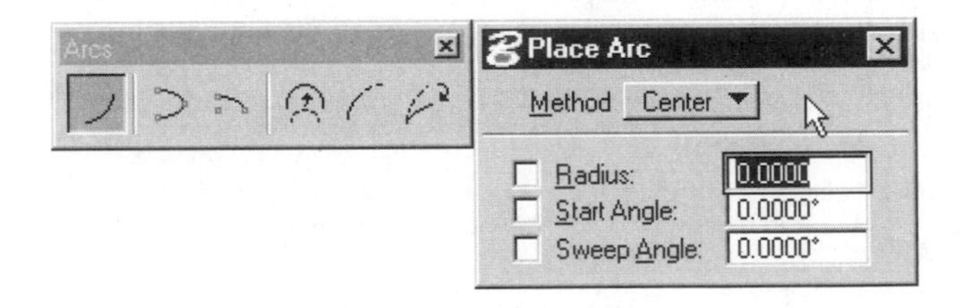

Figure 1–24 Place Arc tool box and Tool Settings window

Key-in Window

Key-ins are typed instructions entered into the Key-in window to control MicroStation. You can invoke any MicroStation command by typing the name of the command in full or in abbreviated form in the Key-in window. Open the Key-in window:

Pull-down menu	Utilities > Key-in

MicroStation displays the Key-in window, similar to Figure 1–25.

As you type, the characters are matched to keywords in the list box below the Key-in window and are automatically selected in the list box. If it selects the right key-in command, press the **Spacebar** to complete the key-in, then click the Key-in button or press Enter to enter the constructed key-in.

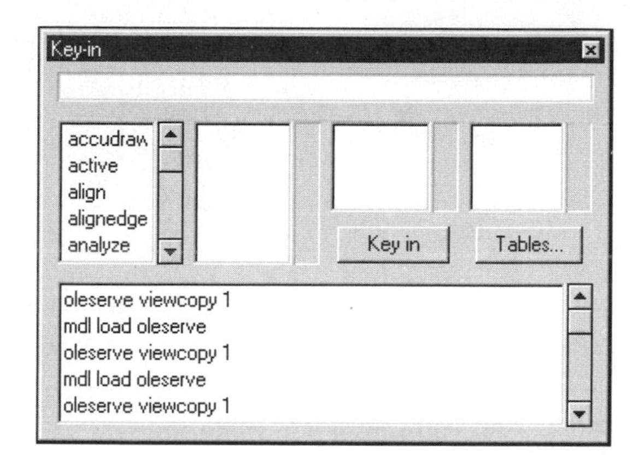

Figure 1–25 Key-in window

The list box in the Key-in window can also help you find and build key-ins. Scroll through the list of first words of key-ins in the left-most list box and select the keyword; it is then displayed in the key-in field. The subordinate, second-level keywords are shown in the key-in window's next list box. Select the desired keyword, and subsequently third-level keywords, if any, are shown. Select additional keywords, one per list box from left to right, until the desired key-in is constructed. To enter the constructed key-in, click the Key-in button or press Enter.

MicroStation stores submitted key-ins in a buffer so you can recall them and, if necessary, edit them. Press ↑ or ↓ on your keyboard repeatedly until the desired key-in text appears in the key-in field of the Key-in window. Make any necessary changes, if any. In addition, MicroStation lists the submitted key-ins in the list box at the bottom of the Key-in window, and you can select the desired key-in from that list box.

Status Bar

The Status bar, located at the bottom of the MicroStation Application Window, as shown in Figure 1–26, displays a variety of useful information, including prompts, messages, and the name of the selected tool.

The Status bar is divided into two sections, as shown in Figure 1–26.

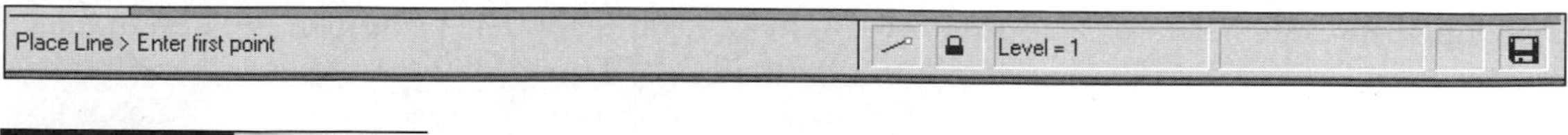

Figure 1–26 Status bar

Left-hand Section

The left-hand section of the Status bar shows the name of the selected tool, followed by either a "greater than" symbol (>) or a colon (:) and message text. The message text that follows ">" is the selected tool's prompts. For example, when you invoke the Place Line tool, MicroStation prompts on the left side of the Status bar as follows:

Place Line > Enter first point

The command programs guide you step by step as you perform an operation with a command. The text that follows the colon is a message that indicates a possible problem.

In addition, as you move the pointer on the tools in a tool box, the name of the selected tool and the associated message text are replaced with a description of the tool over which the pointer is located. This is intended as a form of online assistance.

Right-hand Section

The right-hand section of the Status bar consists of a series of fields. Following are the available fields, from left to right.

- The first field indicates the Snap Mode setting.
- The Locks icon in the second field allows you to open the Setting menu's Locks submenu. You can toggle the locks settings.
- The third field indicates the current level.
- The fourth field indicates the count of the selected elements. If this field is blank, no elements are selected.
- The fifth field indicates whether there is a fence in the design. If this field is blank, no fence is placed.
- The last field indicates whether changes to the active design file are unsaved. If the field is blank, there are no unsaved changes. If the field has a red icon with an "X" through it, the active design file is open for "read-only" access.

When you enter a tentative point or request quantitative information, the fields in this section to the right of the Snap Mode field are temporarily replaced with a single message field. To restore the fields, press the Reset button or click anywhere in the Status bar.

View Windows

MicroStation displays the elements you draw in the view windows. The portion of the design that is displaying in the view window is referred to as a *view.* With this part of the screen, all of the various commands you enter will construct your design. As you progress, you can Zoom In and Zoom Out to control the design's display. You can also move the scroll bar located both on the right side and at the bottom to pan the design. A view typically shows a portion of the design, but may show it in its entirety, as in Figure 1–27.

Eight view windows can be open (ON) at the same time, and all view windows are active. This lets you begin an operation in one view and complete it in another. You can move a view window by pressing and holding the cursor on the Title bar and dragging it to anywhere on the screen. You can resize the display window by clicking and dragging on its surrounding border, and shrink and expand the window by clicking the push buttons located at the top right corner of the window. To close the view window, click the "X" located at the top right corner of the view window.

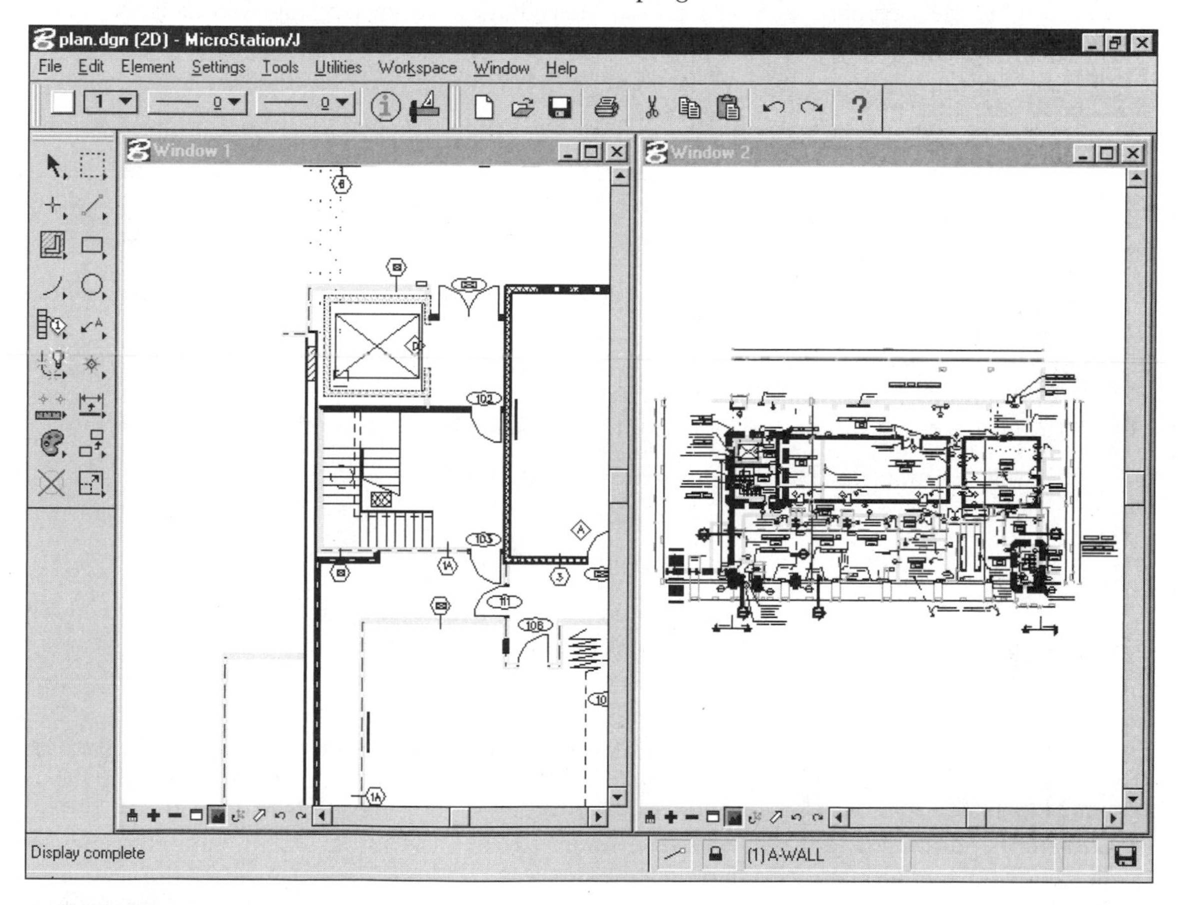

Figure 1–27 Two view windows displaying different portions of a design

MicroStation also provides a set of tools at the bottom left of the view window to control the view. View controls operate much like drawing tools; many even have "tool" settings. Detailed explanations of the View controls are provided in Chapter 3.

INPUT METHODS

Input method refers to the manner in which you tell MicroStation what tool to use and how to operate the tool. As mentioned earlier, the two most popular input devices are the mouse and the digitizing tablet, in addition to the keyboard.

Keyboard

To enter a command from the keyboard, simply key-in (type) the tool name in the Key-in window and click the Key-in button or press Enter. *Key-in* is the name given to the function of providing information via the keyboard to MicroStation. MicroStation key-in language is much like plain English. For example, keying-in **PLACE LINE** selects that line command; **DELETE ELEMENT** selects the delete command, and so on. See Appendix C for a list of the key-in commands available in MicroStation.

Pointing Devices

As mentioned earlier, MicroStation supports a two-button mouse, a three-button mouse, and a digitizer. Depending on your needs and hardware, select one of the three pointing devices.

Detailed explanations of the functions that are programmed to the pointing device buttons (mouse and digitizing tablet puck) follow.

Data Button

This is the most-used button on the mouse/puck. The Data button is used to do the following:

Select a tool from the pull-down menus and tool box.

Define location points in the design plane.

Identify elements that are to be manipulated.

In addition, it is used to accept tentative points, and generally tell the computer "yes" (accept) whenever it is prompted to do so. The Data button is also referred to as the *Identify button* or the *Accept button.*

Reset Button

The Reset button enables you to stop the current operation and reset MicroStation to the beginning of the current command sequence. For instance, when you are in the Place Line command, a series of lines can be drawn by using the Data button. When you are ready to stop the sequence, press the Reset button. MicroStation will stop the current operation and reset the Line sequence to the beginning. In addition, the Reset button also can reject a prompt, and generally tell the computer "no" (reject) whenever it is prompted to do so. The Reset button is also referred to as the *Reject button.*

Tentative Button

The tentative point is one of MicroStation's most powerful features. The Tentative button enables you to place a tentative (temporary) point on the screen. Once you are happy with the location of the point, accept it with the Data button. In other words, the tentative point lets you try a couple of places before actually selecting the final resting point for the data point. The Tentative button also can snap to elements at specific locations—for instance, the center and four quadrants of a circle, when the Snap lock is turned to ON. For a detailed explanation, see Chapter 3.

Command Button

The Command button is only available on the digitizer tablet puck and lets you choose commands from the tablet menu. To do so, look down the tablet menu, place the puck crosshairs in the box that represents the command you want to activate, and then press the Command button. The corresponding command is invoked.

See Tables 1–1 and 1–2 for the specific functions that are programmed for the two- and the three-button mouse and the four-button digitizer puck.

Cursor Menu

MicroStation has a cursor menu that can be made to appear at the location of the cursor by pressing the designated button on your pointing device. Two cursor menus are available, one for View Control tools and another for Snap Mode options.

To invoke the View Control tools (Figure 1–28), press Shift plus the Reset button. The menu includes all the available tools for controlling the display of the design.

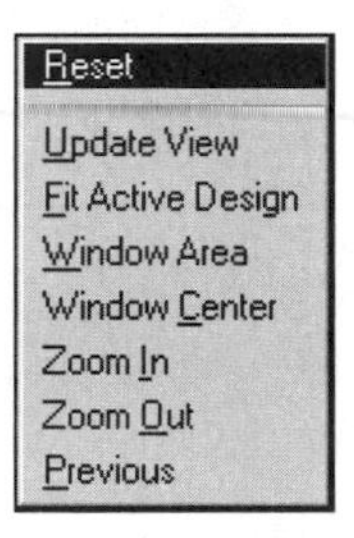

Figure 1–28 Cursor menu (Shift + Reset button)—View Control tools

To invoke the Snap Mode options (Figure 1–29), press Shift plus the Tentative button. The menu includes all the available Snap Mode options. The reason the Snap Mode options are in such ready access will become evident when you learn the significance of these functions.

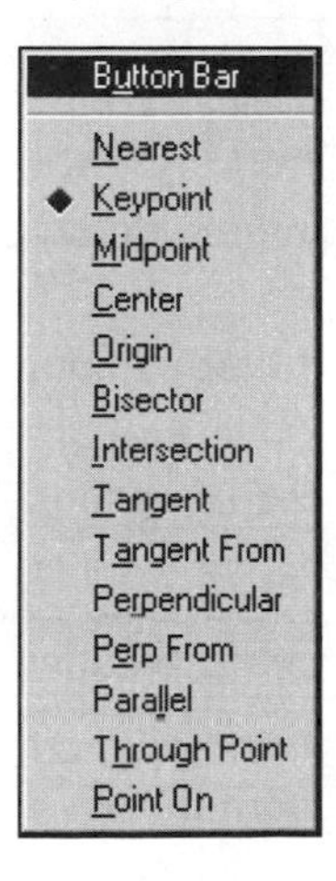

Figure 1–29 Cursor menu (Shift + Tentative button)—Snap Mode options

THE DESIGN PLANE

In conventional drafting, the drawing is normally done to a certain scale, such as 1' = 1"–0' or 1' = 1"–0'. But in MicroStation, you draw full scale: All lines, circles, and other elements are drawn and measured as full size. For example, if a part is 150 feet long, it is drawn as 150 feet actual size. When you are ready to plot the part, MicroStation scales the design to fit a given sheet size. Alternatively, you can specify a scale factor to plot on a given sheet size.

Whenever you start a new *two-dimensional* design, you get a design plane—the electronic equivalent of a sheet of paper on a drafting table. The *2D* design plane is a large, flat plane covered with an invisible matrix grid consisting of 4,294,967,296 (2^{32}) by 4,294,967,296 (2^{32}) coordinate intersections along the *X* and *Y* axes. The distance between two adjacent points is one positional unit or unit of resolution (UOR). The center of the design plane (2,147,483,648 by 2,147,483,648) is the global origin and is assigned coordinates (0,0), as shown in Figure 1–30. Any point to the right of the global origin has a positive *X* value; any point to the left has a negative *X* value. Any point above the global origin has a positive *Y* value; any point below has a negative *Y* value.

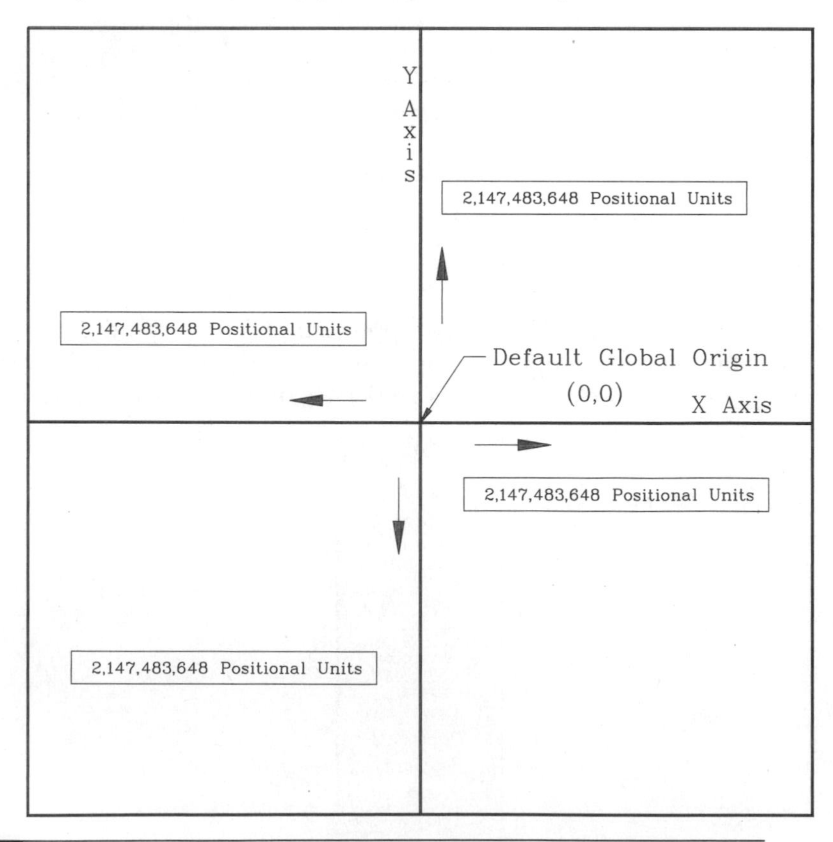

Figure 1–30 X and Y coordinates indicating the location of the global origin

If necessary, you can change the location or coordinates of the global origin. For example, an architect might want all coordinates to be positive in value, so he or she would set the global origin at the bottom left corner of the design plane.

WORKING UNITS

As mentioned earlier, a design plane is divided into positional units (2^{32} by 2^{32}), but, at the same time, you can draw in "real world" units such as feet and inches or meters and centimeters. These real-world units are called Working Units.

Working Units are comprised of Master or Major units (MU), Sub Units (SU), and working resolution or Positional Units (PU). The Master Unit is the largest unit being used in the design. The fractional parts of a Master Unit are called Sub Units. The number of Positional Units per Sub Unit is the working resolution. Working resolution determines both the precision with which elements are drawn and the working area of the design plane.

For example, you can draw a building floor plan specifying feet as Master Units, inches as Sub Units (12 parts make one Master Unit), and 1,600 Positional Units per Sub Unit. Adjacent data points can be entered as close as 1/1600 of an inch on a design plane stretching 223,696 feet (4,294,967,296 ÷ 1600 = 12) in each dimension. Enough room for a whole city!

The Working Units can be assigned any value necessary to your work. You can define a working area in terms of miles and quarters of a mile, meters and decimeters, inches and tenths of an inch, and so on. Working Units can be set to any real-world units. The number of Positional Units specified determines the value of the Sub Units.

When you wish to specify a distance, a radius, and so on, it is based on the Working Units. MicroStation has a standard syntax for doing this with the MU:SU:PU format. Each one of the three positions represents a value in relation to its respective unit. Whenever you key-in the value, make sure to use the colon (:) to separate Master Units, Sub Units, and Positional Units. Do *not* use the semicolon (;).

The following are the options available to key-in 3 1/8 feet when the Working Units are set as feet for Master Units, inches for Sub Units, and 1,600 Positional Units per Sub Unit:

3.125	*(In terms of feet, one need not specify SU and PU, since they are 0.)*
3:1.50	*(In terms of feet and inches.)*
3:1:800	*(In terms of feet, inches, Positional Units.)*
0:37.50	*(In terms of inches.)*
3:1 1/2	*(In terms of feet and inches in fractions.)*

Whenever you start a new design, you need not set the Working Units if you used the appropriate seed file to create the new design file. To draw an architectural floor plan, you copy SDARCH2D.DGN (architectural seed file); then MicroStation sets the Working Units to feet, inches, and 8,000 Positional Units. See Appendix F for a list of seed files and default Working Units available with MicroStation J.

If necessary, you can make changes to the current Working Units. To do so, open the Design File Settings dialog box:

Pull-down menu	Settings > Design File...

MicroStation displays the Design File Settings box, similar to Figure 1–31.

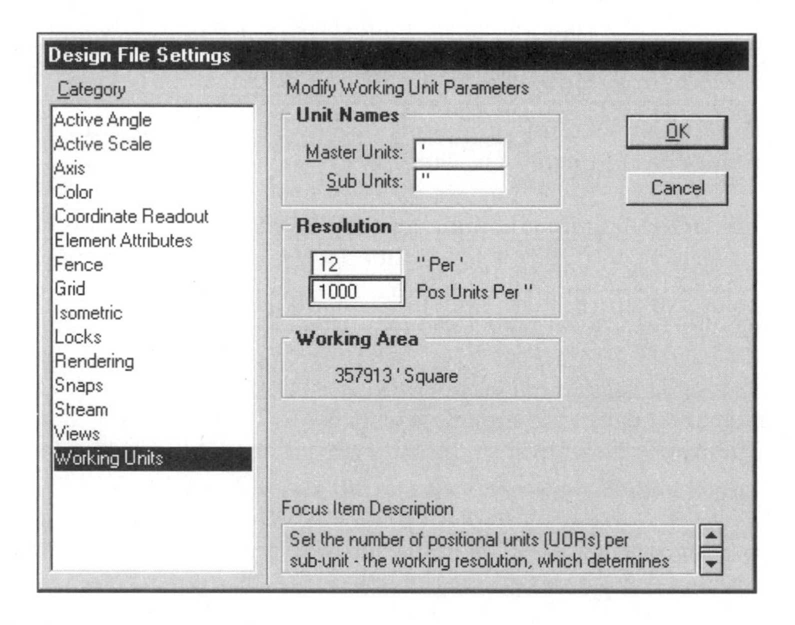

Figure 1–31 Design File Settings box

Select Working Units in the Category list box. MicroStation displays the appropriate controls needed to Modify the Working Units Parameters (see Figure 1–31).

Enter the appropriate unit names (two characters maximum length) in the Master Units edit box and Sub Units edit box located in the Units section of the dialog box. If necessary, change the number of Sub Units per Master Unit and the number of Positional Units per Sub Unit in the edit boxes provided in the Resolution section. Click the OK button to close the dialog box. MicroStation displays an Alert dialog box similar to the one shown in Figure 1–32. If you really want to make the changes, click the OK button to continue.

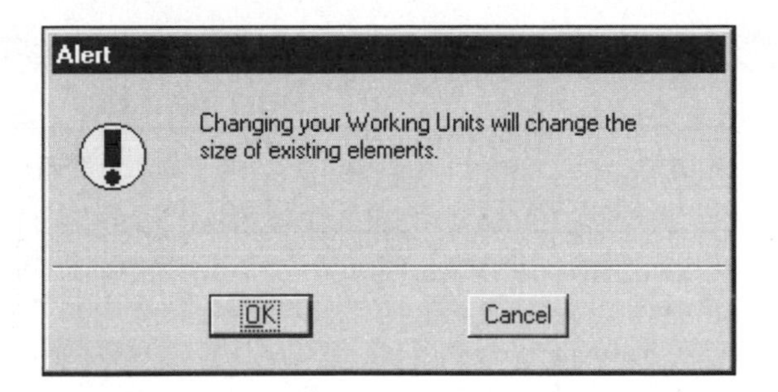

Figure 1–32 Alert dialog box

Make sure appropriate Working Units are set *before* you start drawing.

Note: Do not change the Working Units when you have already placed elements in the design. Changing the Working Units alters the size of existing elements.

SAVING CHANGES AND EXITING THE DESIGN FILE

Before we discuss how to place elements in the design file, let's discuss how to save the current design file. By default, MicroStation saves all elements in your design file as you draw them. There is no separate Save command. You can get out of your design file without doing the proper exit procedure and still not lose any work. Even if there is a power failure during a design session, you will get most of the design file back without significant damage.

If necessary, you can set the Immediately Save Design Changes toggle button to OFF. Then MicroStation provides a Save tool in the File pull-down menu to save the design file. No other automatic save feature is provided. Whenever you want to save the design file, you have to invoke the Save tool from the File pull-down menu.

To change the status of the toggle button for Immediately Save Design Changes, open the Preferences dialog box:

Pull-down menu	Workspace > Preferences...

MicroStation displays the Preference dialog box, similar to Figure 1–33.

Select the Operation option from the Category list box and set the toggle button to ON/OFF for Immediately Save Design Changes.

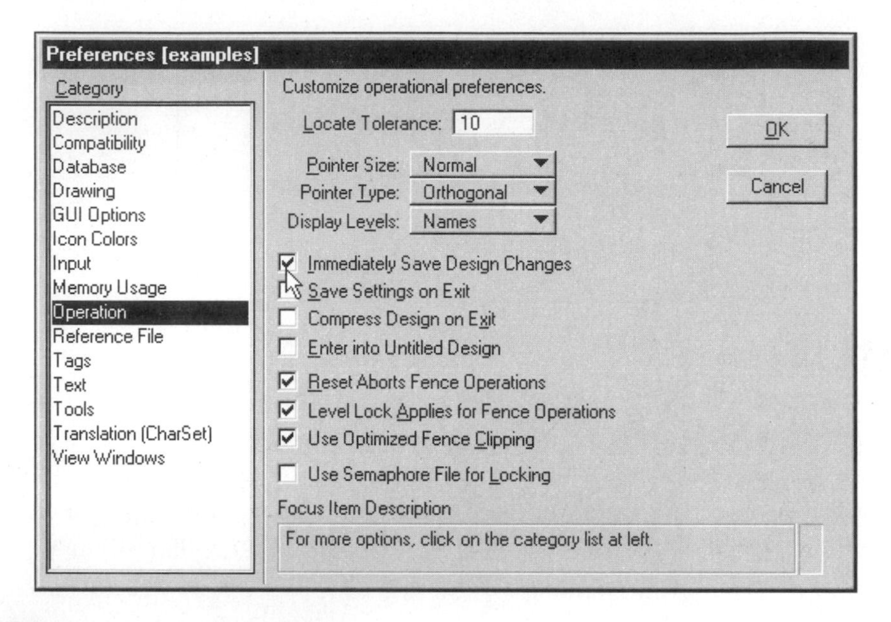

Figure 1–33 Preferences dialog box

To save the current design file to a different file name, invoke the Save As command:

Pull-down menu	File > Save As...

MicroStation displays the Save As dialog box. Select the folder where you want to save the design file, and key-in the name of the design file in the Files edit box. Click the OK button to save the file.

Saving the Design File Settings

To save the design file settings such as working units, grid spacing, or view settings between sessions, you must explicitly save the settings. To do so, invoke the Save Settings tool:

Pull-down menu	File > Save Settings
Key-in window	**file design** (or **fi**) Enter

MicroStation saves the current settings.

If you forget to save the settings, you will have to spend time adjusting the design and view settings to match what you had in place the last time you worked on the design.

To automatically save the settings on exiting the design file, set the toggle button for Save Settings on Exit to ON in the Preferences dialog box. By default it is set to OFF.

To change the status of the toggle button for Save Settings on Exit, open the Preferences dialog box:

Pull-down menu	Workspace > Preferences...

MicroStation displays the Preferences dialog box. Select the Operation option from the Category list box, and set the toggle button to ON/OFF for Save Settings on Exit.

Exiting the MicroStation Program

To exit the MicroStation program and return to the operating system, invoke the Exit tool:

Pull-down menu	File > Exit
Key-in window	**exit** (or **exi**) Enter

MicroStation exits the program and returns to the operating system.

If, instead, you prefer to return to the MicroStation Manager dialog box, invoke the Close tool:

Pull-down menu	File > Close
Key-in window	**close design** (or **clo d**) Enter

MicroStation returns to the MicroStation Manager dialog box.

GETTING HELP

When you are in a design file, MicroStation provides an online help facility available from the Help pull-down menu (see Figure 1–34). Online help is provided through the Help window, by specific topics, by searching for a text string within help topic names or help articles, or by browsing key-ins.

Contents

The Contents window lists the top-level topics, as shown in Figure 1–35. To see a list of more specific subtopics related to a topic in the list, select the topic. MicroStation displays a list of subtopics, and by selecting a subtopic, MicroStation displays the available help information on that topic. In addition, you can also search for a text string by clicking the Search button and MicroStation will display the Search settings box. Key-in the text string and click the Search button, and MicroStation will display the help available on the text string.

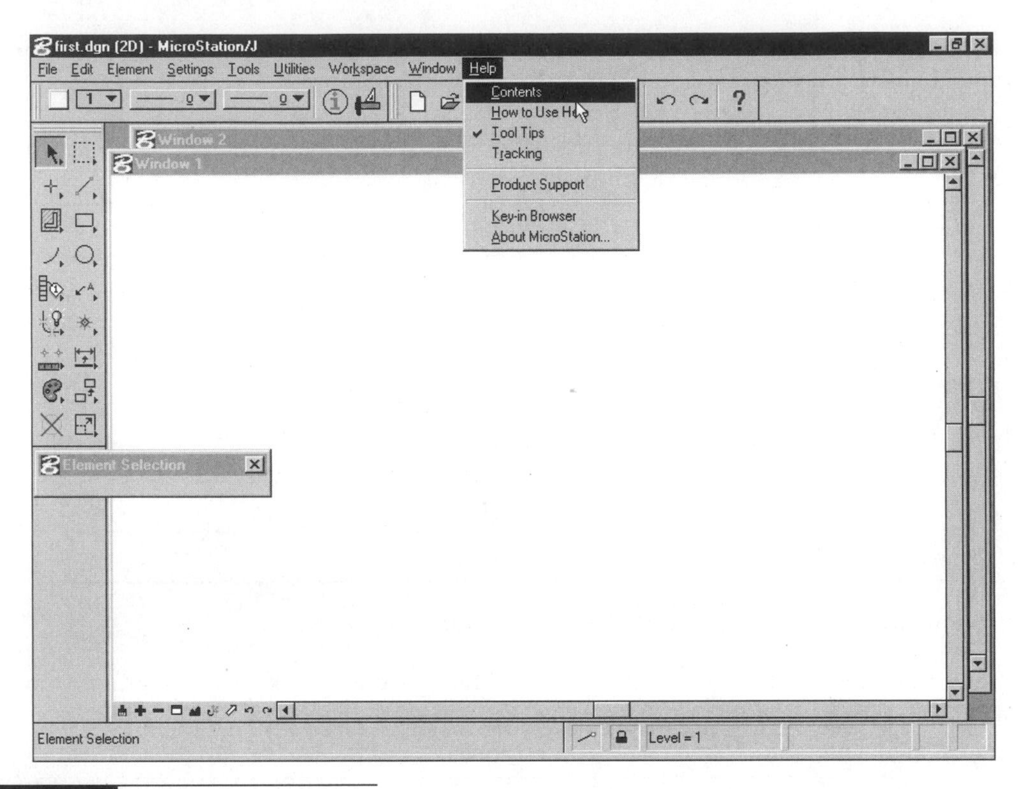

Figure 1–34 Pull-down menu Help

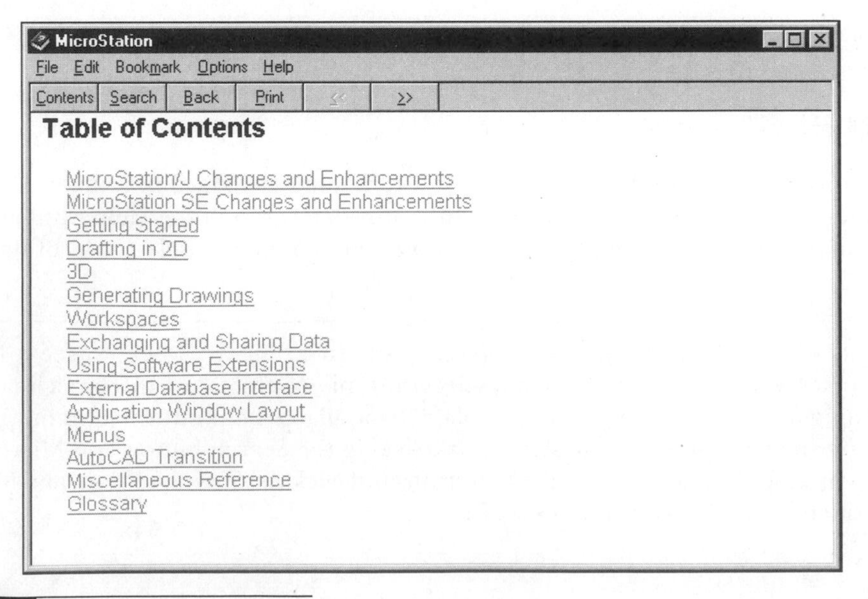

Figure 1–35 Help Contents window

The MicroStation Help window also has a Topic option menu displaying the current topic's parent, its parent's topic, and so on, up to the Contents. Selecting any one of the topics moves you up to that level in the Help document. You can also display the Help file's previous article and next article by clicking the Previous (<) and Next (>) buttons.

Search for Help On...

Selecting this option opens the Search settings box, as shown in Figure 1–36, which is used to search for a text string in the open Help file. Type the first few letters of the word you are looking for in the edit field and press Enter. MicroStation displays the Help information.

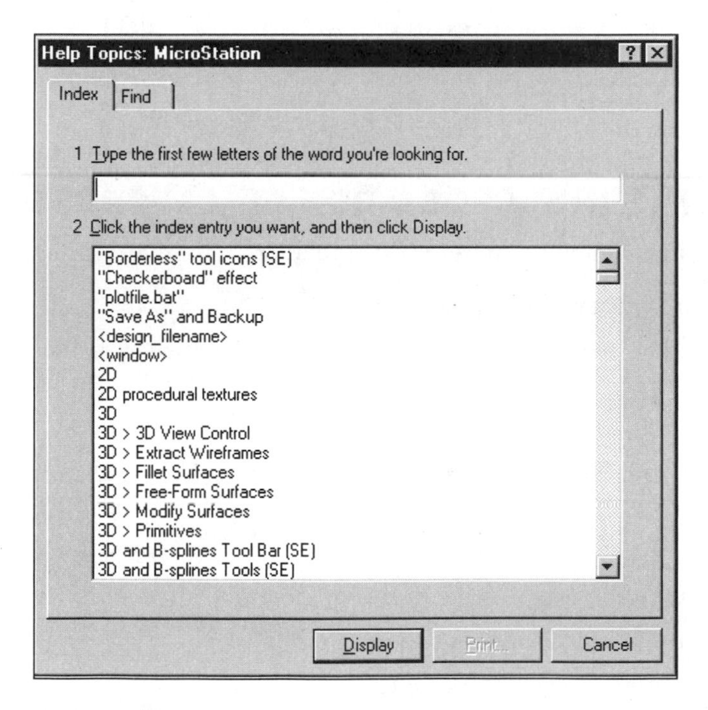

Figure 1–36 Search settings box

How to Use Help

This option opens the Help window that provides instructions for using online help.

Product Support

This option displays information about contacting MicroStation technical support.

Key-in Browser

This option opens the Key-in window, similar to the one shown in Figure 1–37. Refer to the section on the Key-in window, earlier in this chapter, for further details.

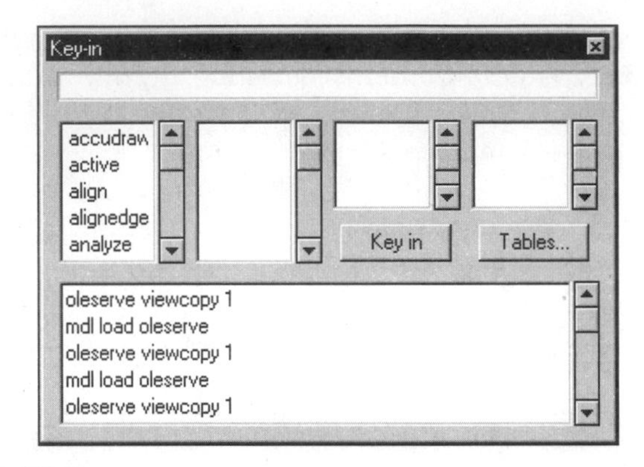

Figure 1–37 Key-in window

ENHANCEMENTS IN MICROSTATION SE / MICROSTATION J

Following are the enhancements that were added to MicroStation SE / MicroStation J.

- True Microsoft Windows Application
- New look and feel in User Interface—Microsoft Office–compliant (includes resizable tool boxes)
- Tool Consolidation
- Four additional types of fences—circular fence, existing element into a fence, setting fence based on a view's current extents, fence to encompass the entire design file
- Modify Element, with a new set of options
- Change Text Attributes—selectively changing individual attributes
- Place Note—multiple lines of text
- Productivity enhancements in Reference Files
- Printing and plotting enhancements, including easy-to-create pen tables
- AccuDraw—the most versatile and powerful new tool
- SmartLine—composite drawing tool for placing linear elements
- Flags—annotation in a design file, with reminders or suggestions for future changes
- Cell Selector—new utility that allows access to cells from multiple cell libraries without attaching each library to your design file
- Archive Utility—utility that allows you to combine all of the resources needed to support the design file
- Animation Producer—new set of tools for producing keyframe animation sequences
- Enhancements in customizing MicroStation tools
- MicroStation BASIC—New programming tool

chapter 2

Fundamentals I

Objectives

After completing this chapter, you will be able to:

- Draw lines, blocks, shapes, circles, and arcs.
- Drop blocks and shapes and delete elements.
- Use Precision Input.

PLACEMENT COMMANDS

MicroStation provides various tools for drawing objects. The primary drawing element is the line, and the Place Line tool enables you to draw series of lines. In addition, MicroStation provides tools for drawing objects such as blocks, shapes, circles, arcs, polygons, ellipses, multi-lines, and curves. This section explains in detail the various tools for drawing lines, blocks, shapes, circles, and arcs.

Place Line

Invoke the Place Line tool:

Linear Elements tool box	Select the Place Line tool (see Figure 2–1).
Key-in window	**place line** (or **pl l**) Enter

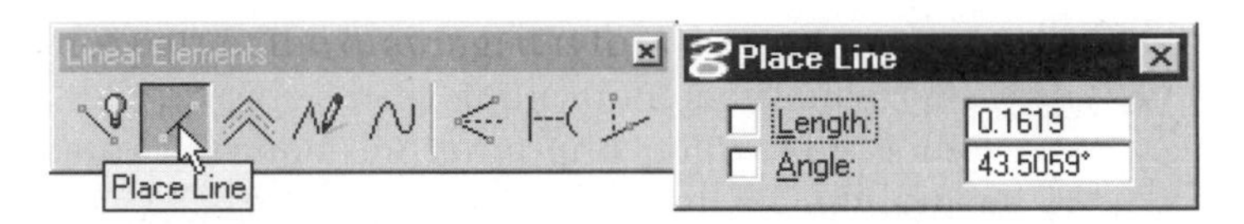

Figure 2–1 Invoking the Place Line tool from the Linear Elements tool box

MicroStation prompts:

Place Line > Enter first point

Specify the first point by providing a data point via your pointing device (mouse or puck) or by Precision Input (see the section on Precision Input for a more detailed explanation). After you specify the first point, MicroStation prompts:

Place Line > Enter end point

Specify the end of the line by placing a data point via your pointing device (mouse or puck) or by Precision Input. MicroStation repeats the prompt:

Place Line > Enter end point

Place a data point via the pointing device or by Precision Input to continue. To save time, the Place Line tool remains active and prompts for a new endpoint after each point you specify. When you have finished placing a series of lines, press the Reset button or invoke another tool to terminate the Place Line tool.

When placing data points with your pointing device to draw a series of lines, a rubber-band line is displayed between the starting point and the crosshairs. This helps you to see where the resulting line will go. In Figure 2–2 the dotted lines represent previous cursor positions. To specify the endpoint of the line, click the Data button. You can continue to place lines with the Place Line tool until you press the Reset button or select another tool.

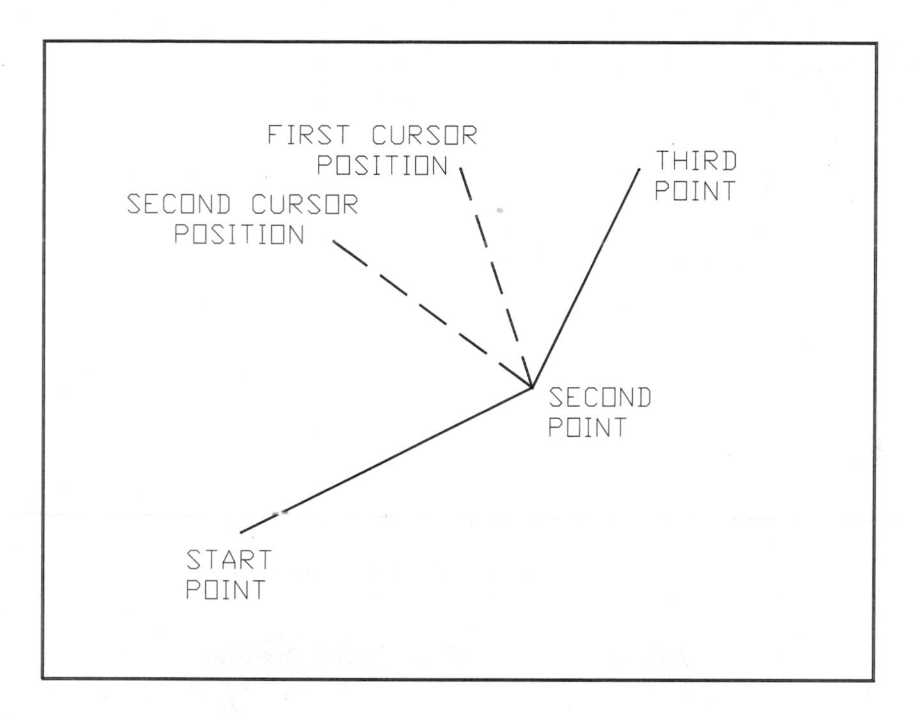

Figure 2–2 Placing data points with the cursor rather than with coordinates

Place Line to a Specified Length

To place a line to a specified length, select the Place Line tool in the Linear Elements tool box and turn ON the toggle button for Length: in the Tool Settings Window. Key-in the distance in MU:SU:PU in the Length: edit field. The prompts are similar to those for the Place Line tool, and you can place any number of line segments of specified length.

Place Line at an Angle

To place a line to a specified angle, select the Place Line tool in the Linear Elements tool box and turn ON the toggle button for Angle: in the Tool Settings Window. Key-in the angle in the Angle: edit field. The prompts are similar to those for the Place Line tool, and you can place any number of line segments of a specified angle.

If necessary, you can turn both of the toggle buttons ON for the Length: and Angle: edit fields; MicroStation allows you to place a line with a specific length and constrained angle.

Place Block

MicroStation allows you to draw a rectangular block by two different methods: orthogonal and rotated.

Orthogonal Method

The Place Block tool (orthogonal) allows you to place a rectangular block by selecting two points that define the diagonal corners of the shape. Place the two diagonal corners by specifying data points via your pointing device or by keying-in *two-dimensional (2D)* coordinates (see later discussion on "Precision Input").

Invoke the Place Block (orthogonal) tool:

Polygons tool box	Select the Place Block tool and Orthogonal from the Method: option menu located in the Tool Settings Window (see Figure 2–3).
Key-in window	**place block orthogonal** (or **pl b o**) Enter

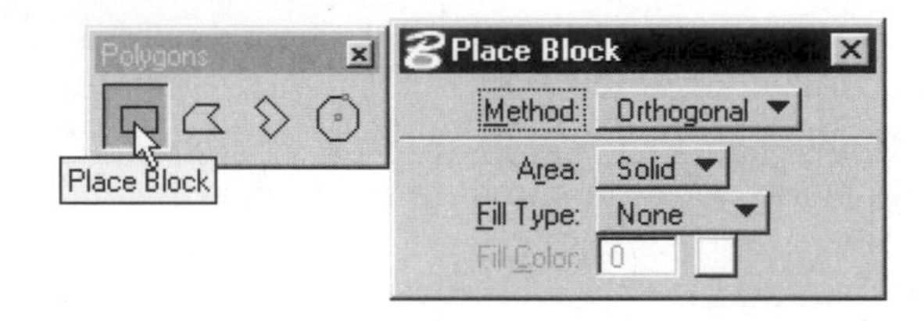

Figure 2–3 Invoking the Place Block (orthogonal) tool from the Polygons tool box

MicroStation prompts:

> Place Block > Enter first point *(Place a data point or key-in coordinates to define the start point of the block.)*
>
> Place Block > Enter opposite corner *(Place a data point or key-in coordinates to define the opposite corner of the block.)*

A block is a single element, and element manipulation tools such as Move, Copy, and Delete manipulate a block as one element. If necessary, you can make a block into individual line elements with the Drop Line String tool.

For example, the following command sequence shows placement of a block by placing two data points diagonally opposite to each other, as shown in Figure 2–4, using the Place Block (orthogonal) tool.

> Place Block > Enter first point *(Place a data point as shown in Figure 2–4.)*
>
> Place Block > Enter opposite corner *(Place a data point diagonally opposite to the first point.)*

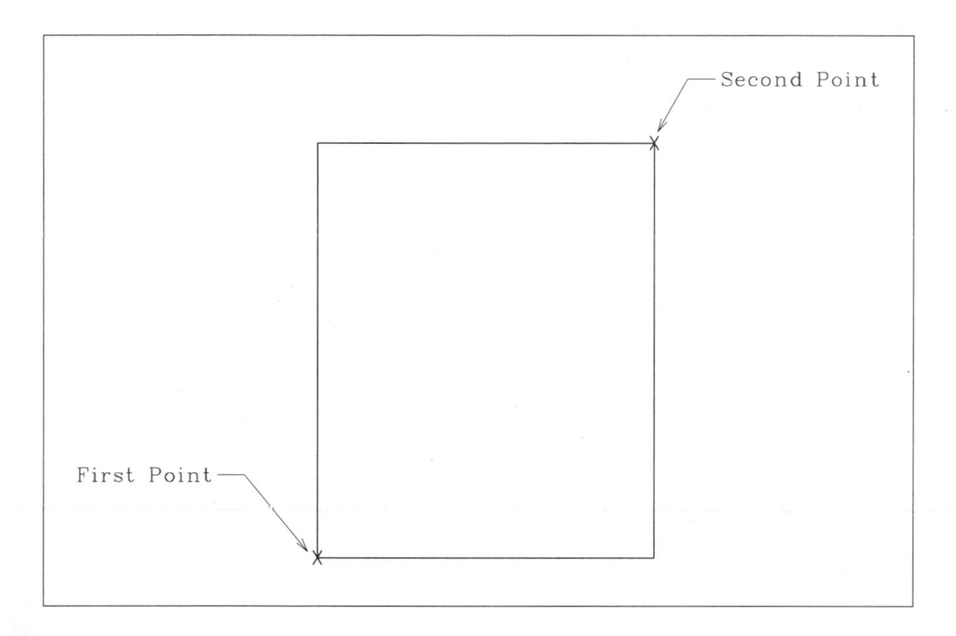

Figure 2–4 An example of placing an orthogonal block using the Place Block (orthogonal) tool

Rotated Method

The Place Block (rotated) tool allows you to place a rectangular block at any angle that is defined by the first two data points. The first data point defines the first corner of the block and the point that the block rotates around. The second data point defines the angle of the block, and the third data point, entered diagonally from the first, defines the opposite corner of the block.

Invoke the Place Block (rotated) tool:

Polygons tool box	Select the Place Block tool and Rotated from the Method: option menu located in the Tool Settings Window (see Figure 2–5).
Key-in window	**place block rotated** (or **pl b r**) Enter

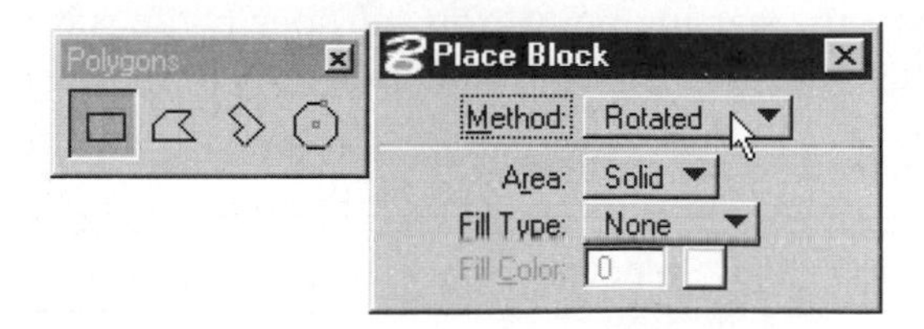

Figure 2–5 Invoking the Place Block (rotated) tool from the Polygons tool box

MicroStation prompts:

> Place Rotated Block > Enter first base point *(Place a data point or key-in coordinates to define the start point of the block.)*
>
> Place Rotated Block > Enter second base point *(Place a data point or key-in coordinates to define the angle of the block.)*
>
> Place Rotated Block > Enter diagonal point *(Place a data point or key-in coordinates to define the opposite corner of the block.)*

See Figure 2–6 for an example of placing a rotated block with the Place Block (rotated) tool by providing three data points.

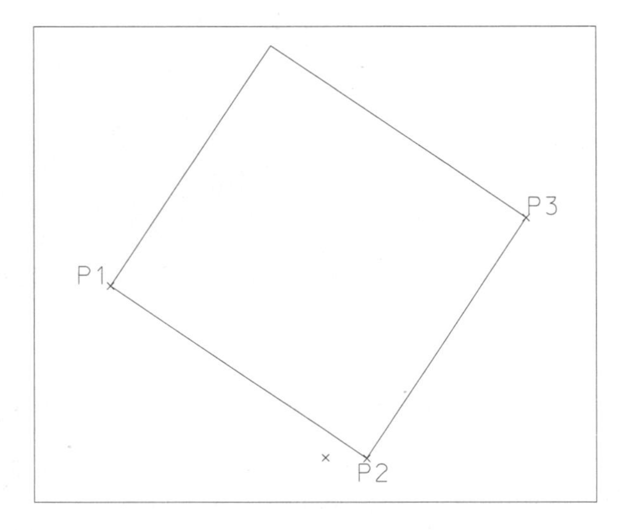

Figure 2–6 An example of placing a rotated block with the Place Block (rotated) tool

Similar to an orthogonal block, a rotated block is also a single element.

Place Shape

The Place Shape tool allows you to place a multisided shape defined by a series of data points (3 to 100) that indicates the vertices of the polygon. To complete the polygon shape, the last data point should be placed on top of the starting point. You can specify the starting point and subsequent points via Precision Input or by using your pointing device.

Invoke the Place Shape tool:

Polygons tool box	Select the Place Shape tool (see Figure 2–7).
Key-in window	**place shape** (or **pl sh**) Enter

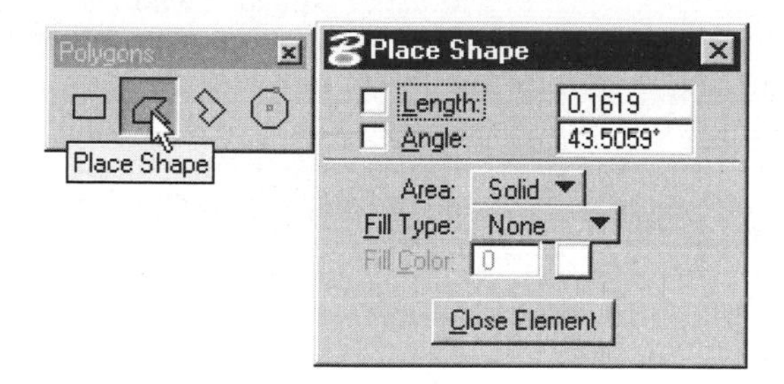

Figure 2–7 Invoking the Place Shape tool from the Polygons tool box

MicroStation prompts:

> Place Shape > Enter shape vertex *(Place a data point or key-in coordinates to define the starting point of the shape.)*
>
> Place Shape > Enter shape vertex *(Place a data point or key-in coordinates to define the vertex, or press Reset button to cancel.)*

Continue placing data points. To complete the polygon shape, place the last data point on top of the starting point or click the Close Element button located in the Tool Settings Window.

You can also draw a shape by constraining to Length and/or Angle by turning on the toggle button for Length: and for Angle:, appropriately located in the Tool Settings Window.

See Figure 2–8 for an example of placing a closed shape with the Place Shape tool by providing six data points.

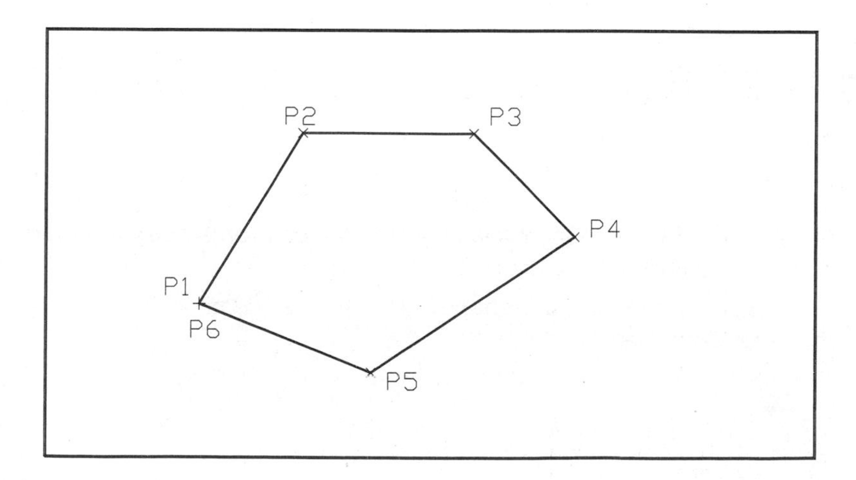

Figure 2–8 An example of placing a closed shape with the Place Shape tool

Similar to a block, a shape is also a single element. Element manipulation tools such as Move, Copy, and Delete manipulate the shape as one element. If necessary, you can make the shape into individual line elements with the Drop Line String tool.

Note: Area and Fill type options are explained in Chapter 12 on Patterning.

Place Orthogonal Shape

The Place Orthogonal Shape tool allows you to create a multisided shape that has adjacent sides at right angles. As with the Place Block (rotated) tool, the first two points define the vertices of the orthogonal shape. The additional points define the corners of the shape. To complete the polygon shape, the last data point should be placed on top of the starting point. You can specify the starting point and subsequent points with absolute or relative coordinates (see "Precision Input," later) or by using your pointing device.

Invoke the Place Orthogonal Shape tool:

Polygons tool box	Select the Place Orthogonal Shape tool (see Figure 2–9).
Key-in window	**place shape orthogonal** (or **pl sh o**) Enter

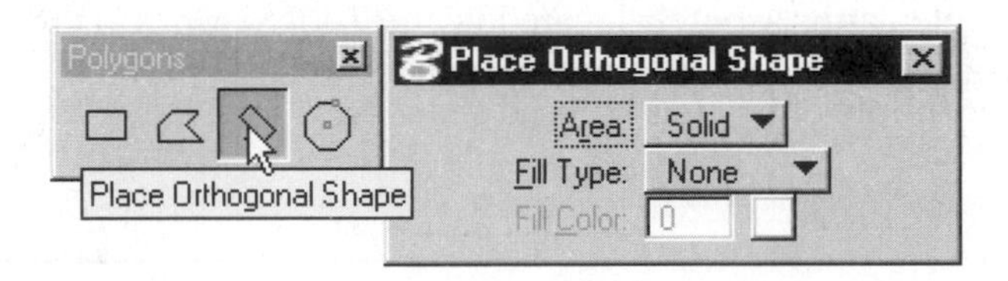

Figure 2–9 Invoking the Place Orthogonal Shape tool from the Polygons tool box

MicroStation prompts:

> Place Orthogonal Shape > Enter shape vertex *(Place a data point or key-in coordinates to define the start point of the shape.)*
>
> Place Orthogonal Shape > Enter shape vertex *(Place a data point or key-in coordinates to define the vertex.)*

MicroStation prompts for additional shape vertices. Continue placing data points. To complete the polygon shape, the last data point should be placed on top of the starting point.

Similar to a block, an orthogonal shape is also a single element. Element manipulation tools such as Move, Copy, and Delete manipulate the orthogonal shape as one element. If necessary, you can make the shape into individual line elements with the Drop Line String tool.

See Figure 2–10 for an example of placing an orthogonal shape with the Place Orthogonal Shape tool by providing nine data points.

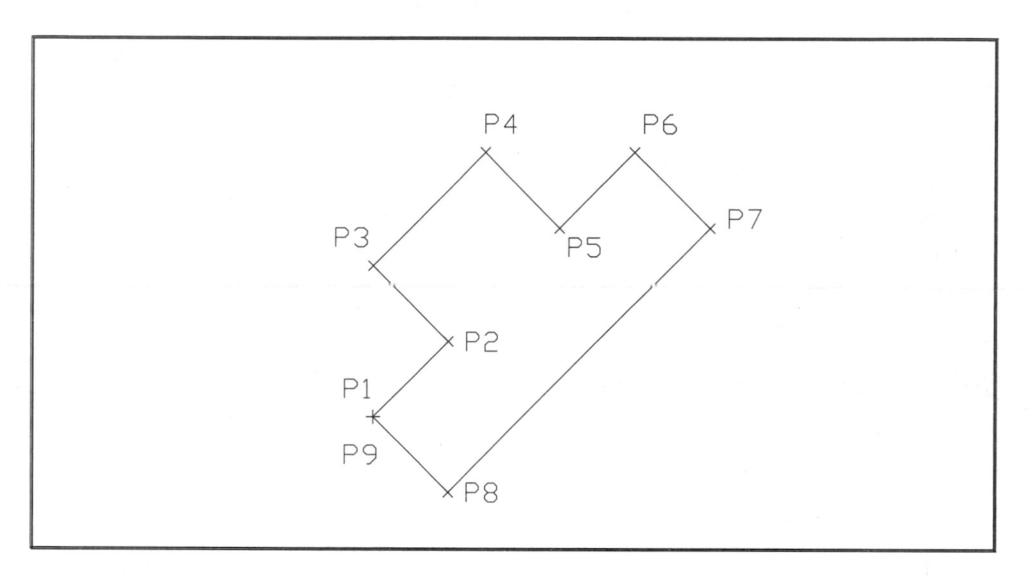

Figure 2–10 An example of placing an orthogonal shape with the Place Orthogonal Shape tool

Note: Area and Fill type options are explained in Chapter 12 on Patterning.

Place Circle

MicroStation offers several methods for drawing circles. These include Place Circle By Center, Place Circle By Edge, and Place Circle By Diameter. The appropriate method is selected from the Method: option menu located in the Tool Settings Window.

Place Circle By Center

With the Place Circle By Center tool, you can draw a circle by defining two points: the center point and a point on the circle.

Invoke the Place Circle By Center tool:

Ellipses tool box	Select the Place Circle tool and Center from the Method option menu located in the Tool Settings Window (see Figure 2–11).
Key-in window	**place circle center** (or **pl ci c**) Enter

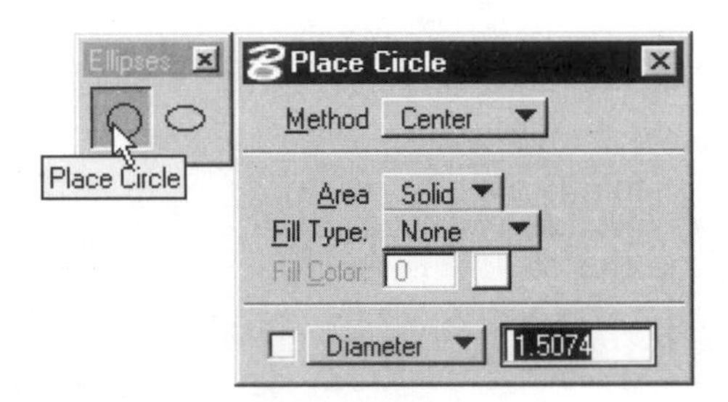

Figure 2–11 Invoking the Place Circle By Center tool from the Ellipses tool box

MicroStation prompts:

> Place Circle By Center > Identify Center Point *(Place a data point or key-in coordinates to define the center of the circle.)*
>
> Place Circle By Center > Identify Point on Circle *(Place a data point or key-in coordinates to define the edge of the circle.)*

Note: After you place the first data point, a dynamic image of the circle drags with the screen pointer.

To save time, the Place Circle By Center tool remains active and prompts for a new center point. When you are finished placing circles, invoke another tool to terminate the Place Circle By Center tool.

For example, the following command sequence shows placement of a circle with the Place Circle By Center tool (see Figure 2–12).

> Place Circle By Center > Identify Center Point *(place a data point to define the center point as shown in Figure 2–12)*
>
> Place Circle By Center > Identify Point on Circle *(place a data point to draw a circle)*

In the last example, MicroStation used the distance between the center point and the point given on the circle for the radius of the circle.

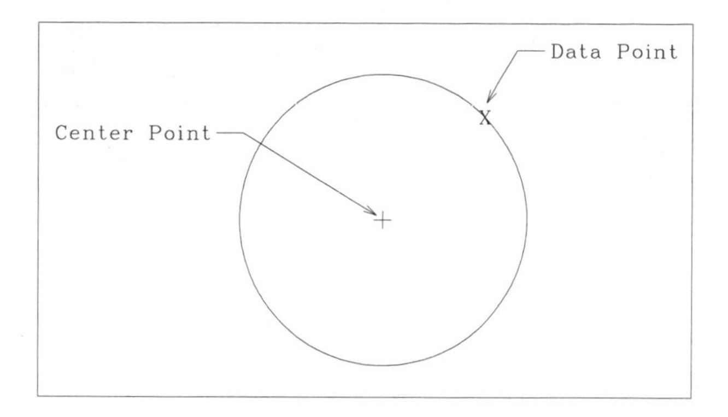

Figure 2–12 An example of placing a circle with the Place Circle By Center tool

You can also place a circle by its center by keying-in its diameter or radius. Select Diameter or Radius from the options menu located in the Tool Settings Window, turn on the toggle button, key-in the value in MU:SU:PU format, then press Enter or Tab. When the desired diameter/radius is entered using the MU:SU:PU format, a circle of that diameter/radius appears on the screen cursor. You will then be asked to identify the center point of the circle. Position your cursor where you want the center of the circle to be, and place a data point. Continue placing circles with this same diameter/radius, or press the Reset button to allow you to change the diameter/radius of the circle. If you do not wish to continue placing circles, invoke another tool to terminate the Place Circle By Center tool.

Place Circle By Edge

The Place Circle By Edge tool enables you to draw a circle by defining three data points on the circle.

Invoke the Place Circle By Edge tool:

Ellipses tool box	Select the Place Circle tool and Edge from the Method option menu located in the Tool Settings Window (see Figure 2–13).
Key-in window	**place circle edge** (or **pl ci e**) Enter

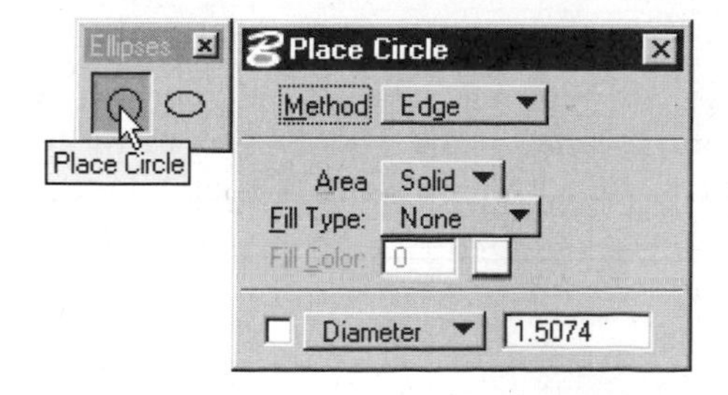

Figure 2–13 Invoking the Place Circle By Edge tool from the Ellipses tool box

MicroStation prompts:

> Place Circle By Edge > Identify Point on Circle *(Place a data point or key-in coordinates to define the first edge point of the circle.)*
>
> Place Circle By Edge > Identify Point on Circle *(Place a data point or key-in coordinates to define the second edge point of the circle.)*
>
> Place Circle By Edge > Identify Point on Circle *(Place a data point or key-in coordinates to define the third edge point of the circle.)*

For example, the following command sequence shows placement of a circle via the Place Circle By Edge tool (see Figure 2–14).

> Place Circle By Edge > Identify Point on Circle *(Place a data point to define the first point of the circle as shown in Figure 2–14.)*
>
> Place Circle By Edge > Identify Point on Circle *(Place a data point to define the second point of the circle.)*
>
> Place Circle By Edge > Identify Point on Circle *(Place a data point to define the third point of the circle.)*

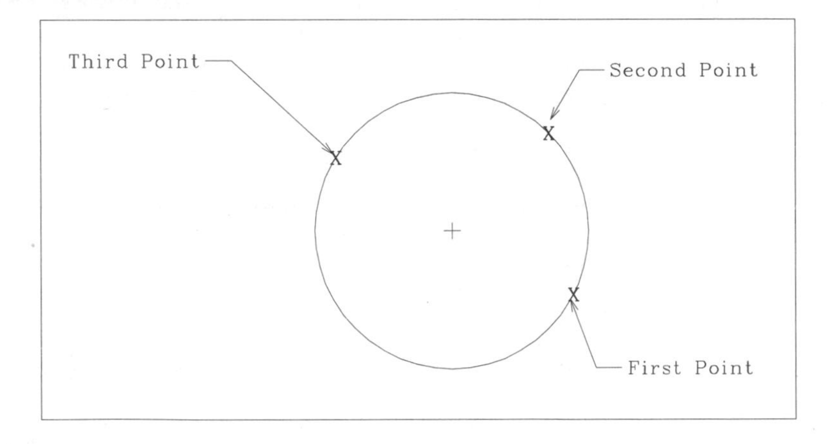

Figure 2–14 An example of placing a circle with the Place Circle By Edge tool

You can also place a circle by its edge by keying-in its diameter or radius. Select Diameter or Radius from the options menu located in the Tool Settings Window, turn on the toggle button, key-in the value in MU:SU:PU format, then press Enter or Tab. When the desired diameter/radius is entered in the MU:SU:PU format, MicroStation prompts for two data points, instead of three, to place a circle by edge.

Place Circle By Diameter

With the Place Circle By Diameter tool, you can draw a circle by defining two data points: two endpoints of the diameter.

Invoke the Place Circle By Diameter tool:

Ellipses tool box	Select the Place Circle tool and Diameter from the Method option menu located in the Tool Settings Window (see Figure 2–15).
Key-in window	**place circle diameter** (or **pl ci d**) Enter

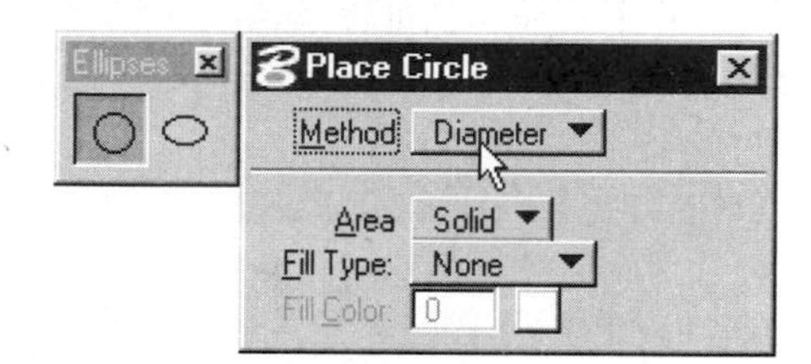

Figure 2–15 Invoking the Place Circle By Diameter tool from the Ellipses tool box

MicroStation prompts:

Place Circle By Diameter > Enter First Point on Diameter *(Place a data point or key-in coordinates to define the first endpoint of one of its diameters.)*

Place Circle By Diameter > Enter Second Point on Diameter *(Place a data point or key-in coordinates to define the second endpoint of one of its diameters.)*

For example, the following command sequence shows placement of a circle via the Place Circle By Diameter tool (see Figure 2–16).

Place Circle By Diameter > Enter First Point on Diameter *(Place a data point to define the first point to draw a circle as shown in Figure 2–16.)*

Place Circle By Diameter > Enter Second Point on Diameter *(Place a data point to define the second point to draw a circle.)*

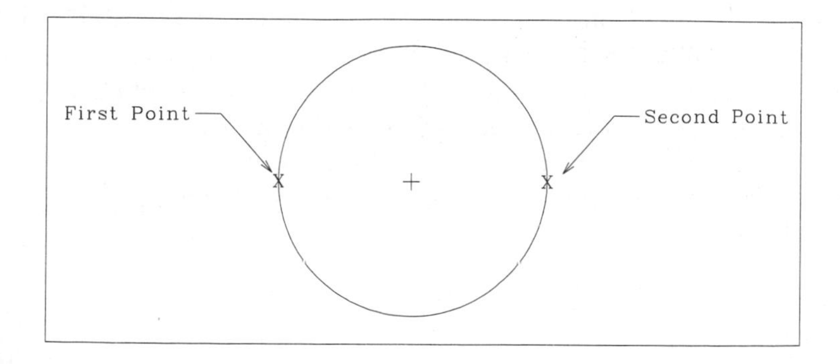

Figure 2–16 An example of placing a circle with the Place Circle By Diameter tool

Place Arc

Similar to placing circles, MicroStation offers two different methods for placing arcs: Place Arc By Center and Place Arc By Edge. The appropriate method is selected from the Method option menu located in the Tool Settings Window.

Place Arc By Center

The Place Arc By Center tool enables you to draw an arc defined by three points: the center point, the first arc endpoint, and the second arc endpoint.

Invoke the Place Arc By Center tool:

Arcs tool box	Select the Place Arc tool and Center from the Method: option menu located in the Tool Settings Window (see Figure 2–17).
Key-in window	**place arc center** (or **pl a c**) Enter

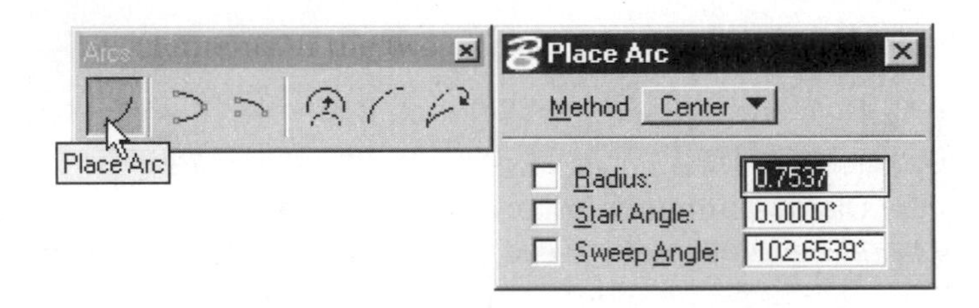

Figure 2–17 Invoking the Place Arc By Center tool from the Arcs tool box

MicroStation prompts:

Place Arc By Center > Identify First Arc Endpoint *(Place a data point or key-in coordinates to define the first arc endpoint.)*

Place Arc By Center > Identify Arc Center *(Place a data point or key-in coordinates to define the arc center.)*

Place Arc By Center > Identify Second Arc Endpoint *(Place a data point or key-in coordinates to define the second arc endpoint.)*

Note: After you place the first data point, a dynamic image of the arc drags with the screen pointer.

For example, the following command sequence shows placement of an arc with the Place Arc By Center tool (see Figure 2–18).

Place Arc By Center > Identify First Arc Endpoint *(Place a data point to define the first arc endpoint as shown in Figure 2–18.)*

Place Arc By Center > Identify Arc Center *(Place a data point to define the arc center.)*

Place Arc By Center > Identify Second Arc Endpoint *(Place a data point to define the arc endpoint.)*

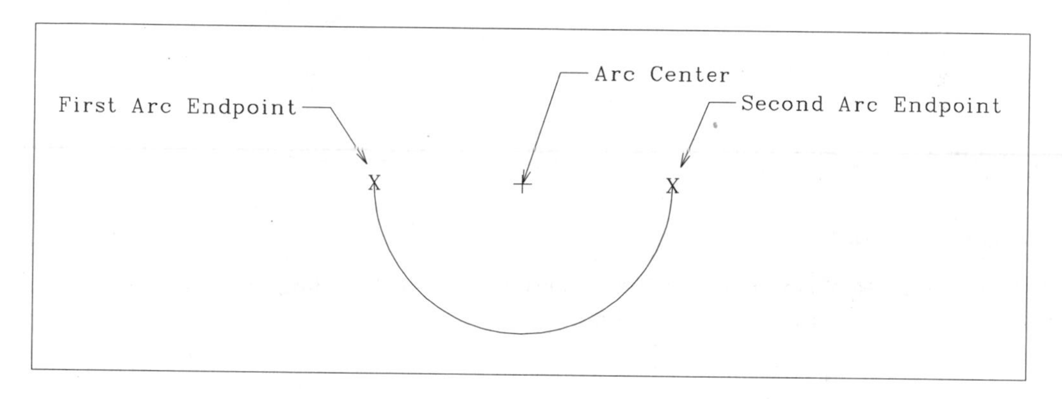

Figure 2–18 An example of placing an arc by the Place Arc By Center tool

You can also draw an arc by its center by keying-in its radius. To do so, turn on the toggle button for Radius:, located in the Tool Settings Window; key-in the appropriate value in MU:SU:PU format in the Radius: edit field; and press Enter or Tab. The prompts are similar to those for the Place Arc By Center tool, except the First Arc Endpoint and Second Arc Endpoint define the starting and ending directions of the arc.

Similarly, you can also constrain the Start Angle: and Sweep Angle: by keying-in appropriate angles in the respective edit fields. The MicroStation prompts depend on the number of constraints turned ON. For example, if Radius and Start Angle are preset, MicroStation prompts for the center point of the arc and the sweep angle; if Radius, Start Angle, and Sweep Angle are preset, MicroStation prompts only for the center of the arc.

Place Arc By Edge

The Place Arc By Edge tool allows you to draw an arc defined by three points on the arc.

Invoke the Place Arc By Edge tool:

Arcs tool box	Select the Place Arc tool and Edge from the Method: option menu located in the Tool Settings Window (see Figure 2–19).
Key-in window	**place arc edge** (or **pl a e**) Enter

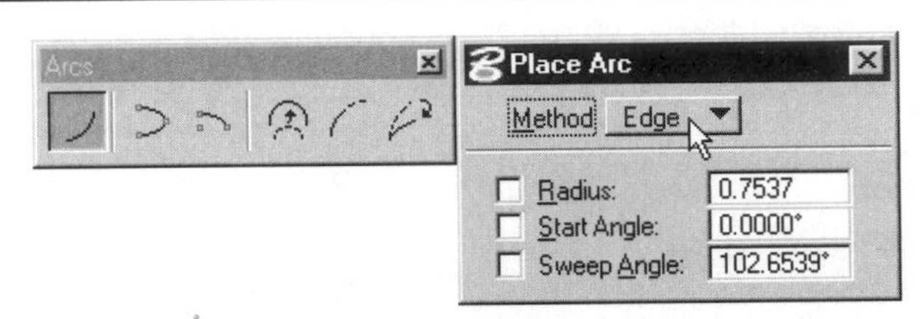

Figure 2–19 Invoking the Place Arc By Edge tool from the Arcs tool box

MicroStation prompts:

> Place Arc By Edge > Identify First Arc Endpoint *(Place a data point or key-in coordinates to define the first arc endpoint.)*
>
> Place Arc By Edge > Identify Point on Arc Radius *(Place a data point or key-in coordinates to define a point on the arc radius.)*
>
> Place Arc By Edge > Identify Second Arc Endpoint *(Place a data point or key-in coordinates to define the second arc endpoint.)*

For example, the following command sequence shows the placement of an arc with the Place Arc By Edge tool (see Figure 2–20).

> Place Arc By Edge > Identify First Arc Endpoint *(Place a data point to define the first arc endpoint as shown in Figure 2–20.)*
>
> Place Arc By Edge > Identify Point on Arc Radius *(Place a data point to define the arc radius.)*
>
> Place Arc By Edge > Identify Second Arc Endpoint *(Place a data point to define the second arc endpoint.)*

You can also draw an arc by its edge by keying-in the radius. To do so, turn on the toggle button for Radius:, located in the Tool Settings Window; key-in the appropriate value in MU:SU:PU format in the Radius: edit field; and press Enter or Tab. The prompts are similar to those for the Place Arc By Edge tool, except the First Arc Endpoint and Second Arc Endpoint define the starting and ending directions of the arc.

Similarly, you can also constrain the Start Angle: and Sweep Angle: by keying in appropriate angles in the respective edit fields. The MicroStation prompts depend on the number of constraints turned ON. For example, if Radius and Start Angle are preset, MicroStation prompts for the First Arc Endpoint and Second Arc Endpoint. If Radius, Start Angle, and Sweep Angle are preset, MicroStation prompts only for the First Arc Endpoint.

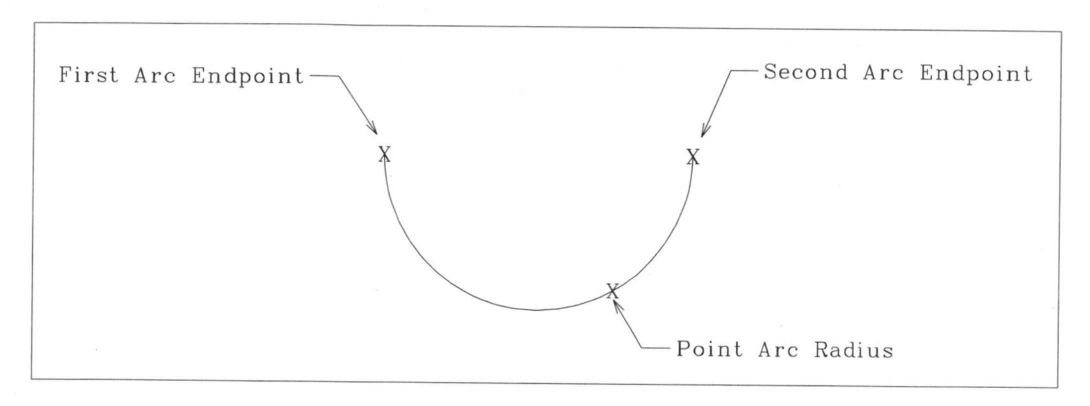

Figure 2–20 An example of placing an arc by the Place Arc By Edge tool

DELETE ELEMENT

MicroStation not only allows you to draw easily, it also allows you to manipulate the elements you have drawn. Of the many manipulation tools available, the Delete Element tool is probably the one you will use most often. Everyone makes mistakes, but MicroStation makes it easy to delete them.

Invoke the Delete Element tool:

Main tool box	Select the Delete Element tool (see Figure 2–21).
Key-in window	**delete** (or **del**) Enter

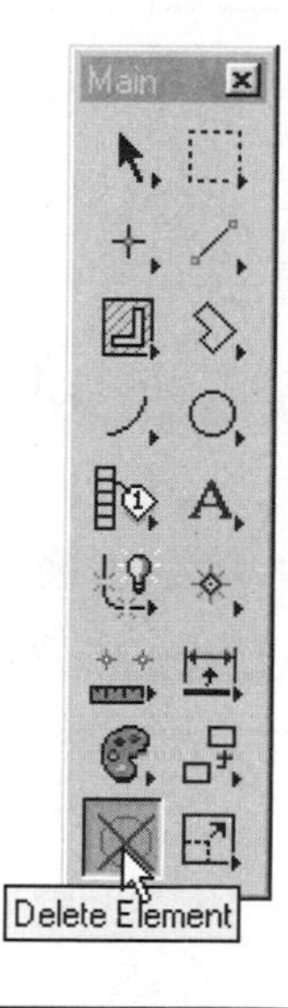

Figure 2–21 Invoking the Delete Element tool from the Main tool frame

MicroStation prompts:

> Delete Element > Identify element *(Identify the element to delete.)*
>
> Delete Element > Accept/Reject (select next input) *(Click the Accept button to delete the selected element, select another element to delete, or click the Reject button to terminate the command sequence.)*

If you change your mind about deleting an element, press the Reject button. If you need to delete additional elements, identify one after another and accept them. The only way to get out of the Delete Element tool is to invoke another tool.

Note: The Delete Element tool deletes only one element at a time. If you need to delete a group of elements, use the Fence Delete tool, explained in Chapter 6.

DROP LINE STRING/SHAPE STATUS

The Drop Line String/Shape Status tool causes blocks and shapes to separate into a series of connected individual line elements that can be manipulated as individual elements. Once a block or shape is dropped, it behaves as if it had been drawn with a Place Line tool.

Invoke the Drop Line String/Shape Status tool:

Drop tool box	Select the Drop Line String/Shape Status tool (see Figure 2–22).
Key-in window	**drop string** (or **dr st**) Enter

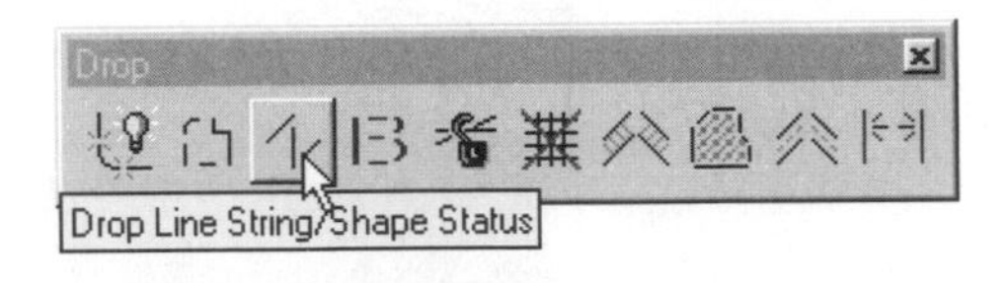

Figure 2–22 Invoking the Drop Line String/Shape tool from the Drop tool box

MicroStation prompts:

> Drop Line String/Shape Status > Identify element *(Identify the block or shape to be dropped.)*
>
> Drop Line String/Shape Status > Accept/Reject (select next input) *(Click the Accept button to accept, select another element to drop, or click the Reject button to reject.)*

PRECISION INPUT

MicroStation allows you to draw an object at its true size and then make the border, title block, and other non-object–associated features fit the object. The completed combination is reduced (or increased) to fit the plotted sheet size you require when you are plotting.

Drawing a not-to-scale schematic does not take advantage of MicroStation's full graphics and computing potential. But even though the symbols and distances between them have no relationship to any real-life dimensions, the sheet size, text size, line widths, and other visible characteristics of the drawing must be considered to give your schematic the readability you desire. Some planning, including sizing, needs to be applied to all drawings.

When MicroStation prompts for the location of a point, instead of providing the data point with your pointing device, you can use three Precision Input commands that enable you to place data points precisely. Each of the commands allows you to key-in by coordinate, including absolute rectangular coordinates, relative rectangular coordinates, and relative polar coordinates.

The rectangular coordinates system is based on specifying a point's location by giving its distances from two intersecting perpendicular axes for *two-dimensional (2D)* points or from three intersecting perpendicular planes for *three-dimensional (3D)* points. Each data point is measured along the *X* axis (horizontal) and *Y* axis (vertical) for *2D* design and along the *X* axis, *Y* axis, and *Z* axis (toward or away from the viewer) for *3D* design. The intersection of the axes, called the *origin* (XY=0,0), divides the coordinates into four quadrants for *2D* design, as shown in Figure 2–23.

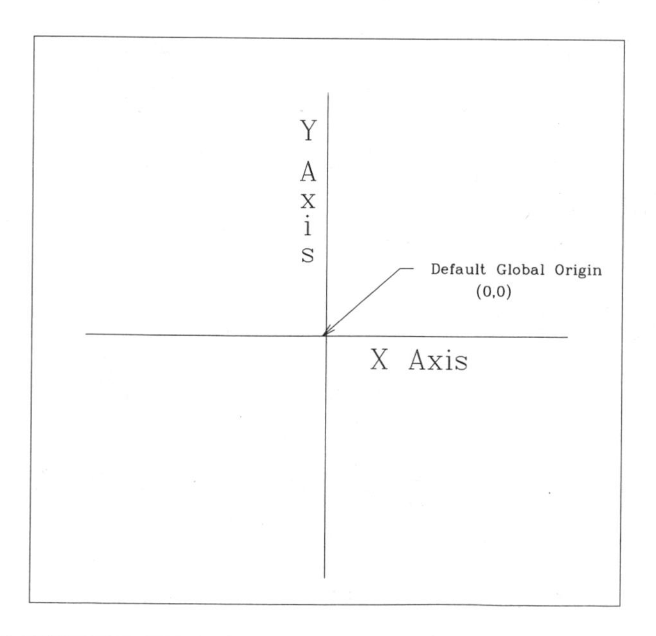

Figure 2–23 A 2D coordinate system

Absolute Rectangular Coordinates

Points are located by Absolute Rectangular Coordinates at an exact *X,Y* intersection on the design plane in relation to the Global Origin. By default, the Global Origin is located at the center of the design plane, as shown in Figure 2–24. The horizontal distance increases in the positive *X* direction from the origin, and the vertical distance increases in the positive *Y* direction from the origin. To enter an absolute coordinate, key-in:

XY=<X coordinate>**,**<Y coordinate> Enter

or

POINT ABSOLUTE <X coordinate>**,**<Y coordinate> Enter

The <X coordinate> and <Y coordinate> are the coordinates in MU:SU:PU in relation to the Global Origin. For example:

XY=2,4 Enter

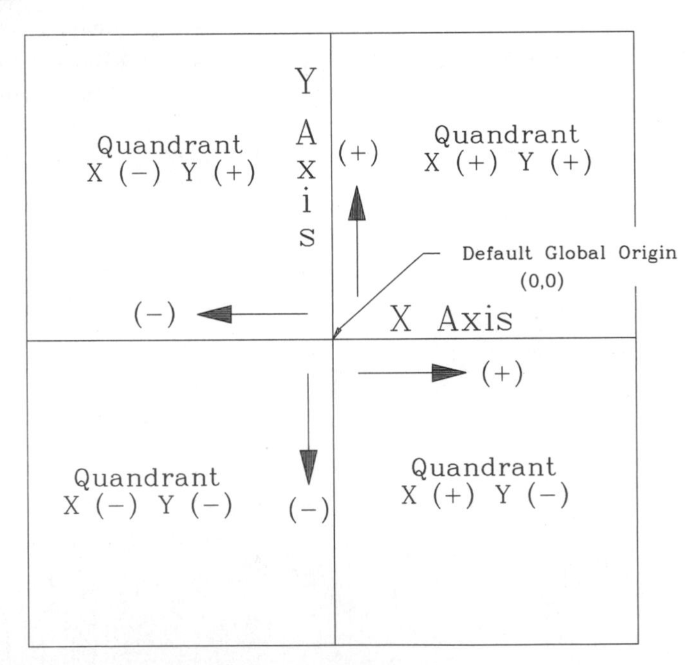

Figure 2–24 Showing Global Origin in a 2D design

The data point is located 2 master units from the origin along the *X* axis and 4 master units from the origin along the *Y* axis.

If necessary, you can relocate the Global Origin anywhere on or off the design plane. To do so, key-in **GO=MU:SU:PU,MU:SU:PU**, and MicroStation prompts:

Global Origin > Enter monument point *(Specify a point anywhere in the design plane to define the new global origin, or click the Reset button to automatically assign the coordinates to the lower-left corner of the design plane.)*

Relative Rectangular Coordinates

Points are located by Relative Rectangular Coordinates in relation to the last specified position or point in MU:SU:PU, rather than in relation to the origin. This is similar to specifying a point as an offset from the last point you entered. To enter a Relative Rectangular Coordinate, key-in:

DL=<X coordinate>**,**<Y coordinate> Enter

or

POINT DELTA <X coordinate>**,**<Y coordinate> Enter

The <X coordinate> and <Y coordinate> are the coordinates in relation to the last specified position or point. For example, if the last point specified was *X,Y*=4,4, the key-in:

DL=5,4 Enter

is equivalent to specifying the Absolute Rectangular Coordinates *X,Y*=9,8 (see Figure 2–25).

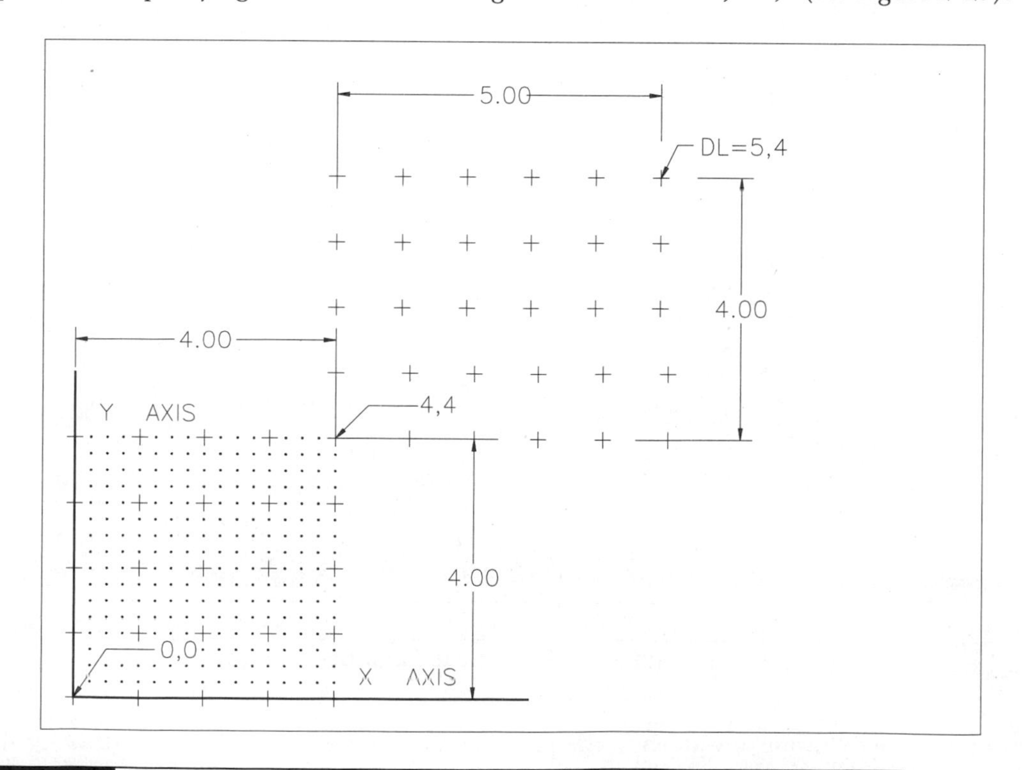

Figure 2–25 An example of placing a point by Relative Rectangular Coordinates

Relative Polar Coordinates

Relative Polar Coordinates are based on a distance from a fixed point at a given angle. In MicroStation, a Relative Polar Coordinate is determined by the distance and angle measured from the previous data point. By default, the angle is measured in a counterclockwise direction relative to the positive *X* axis. It is important to remember that points located by Relative Polar Coordinates are always positioned relative to the previous point, not to the Global Origin (0,0). To enter a Relative Rectangular Coordinate, key-in:

DI=<distance>**,**<angle> Enter

or

POINT DISTANCE <distance>**,**<angle> Enter

The <distance> and <angle> are specified in relation to the last specified position or point. The distance is specified in current working units (MU:SU:PU), and the direction is specified as an angle in current angular units relative to the *X* axis. For example, to specify a point at a distance of 6.4 Master Units from the previous point and at an angle of 39 degrees relative to the positive *X* axis (see Figure 2–26), key-in:

DI=6.4,39 Enter

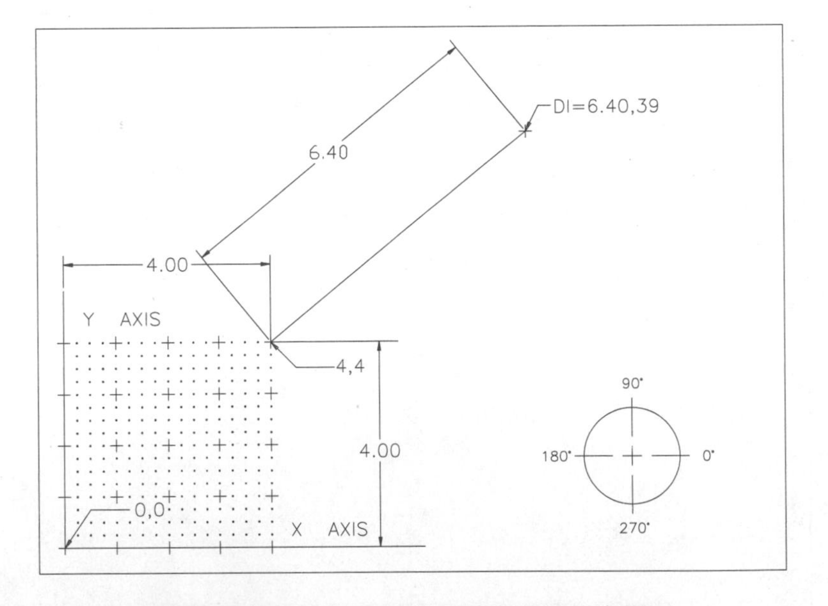

Figure 2–26 An example of placing a line by Relative Polar Coordinates

For example, the following key-ins show the placement of connected lines for the drawing shown in Figure 2–27 by the Place Line tool with absolute coordinates (see Figure 2–28):

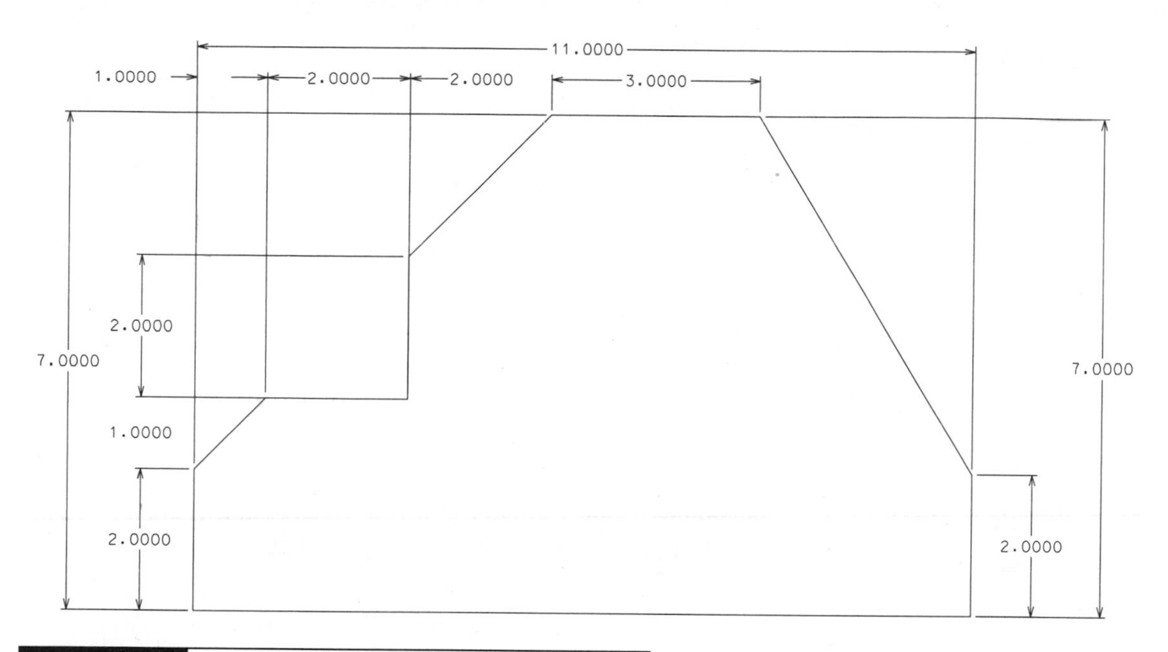

Figure 2–27 An example of placing connected lines

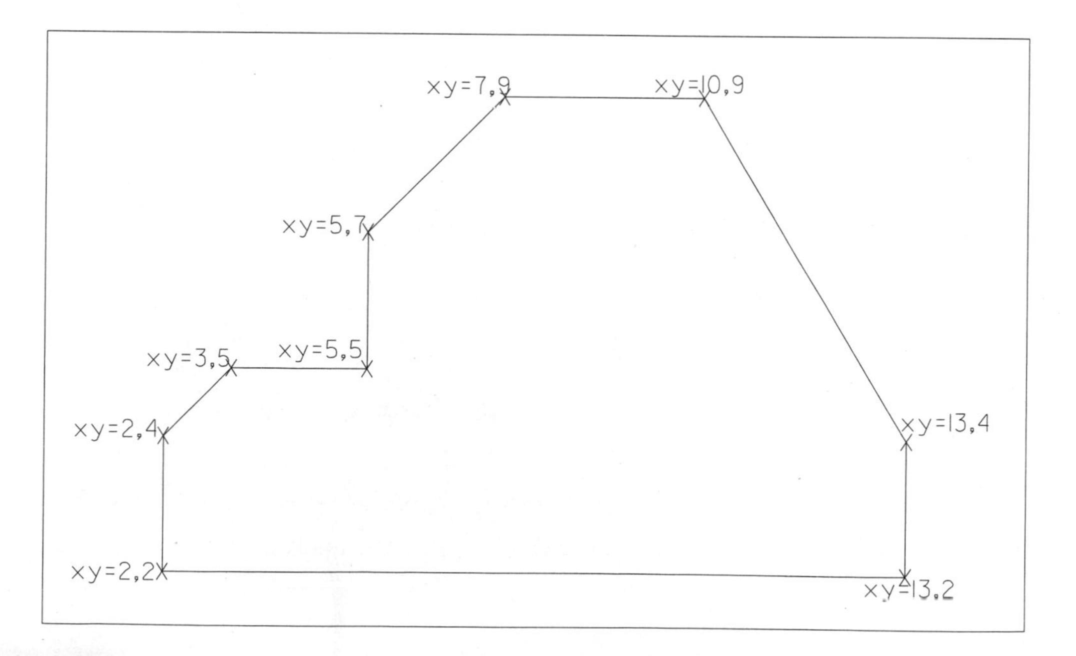

Figure 2–28 Placing connected lines using absolute coordinates

XY=2,2 Enter
XY=2,4 Enter
XY=3,5 Enter
XY=5,5 Enter
XY=5,7 Enter
XY=7,9 Enter
XY=10,9 Enter
XY=13,4 Enter
XY=13,2 Enter
XY=2,2 Enter
(click the Reset button to terminate the line sequence)

The following key-ins show the placement of connected lines for the drawing shown in Figure 2–27 by the Place Line tool with relative rectangular coordinates (see Figure 2–29).

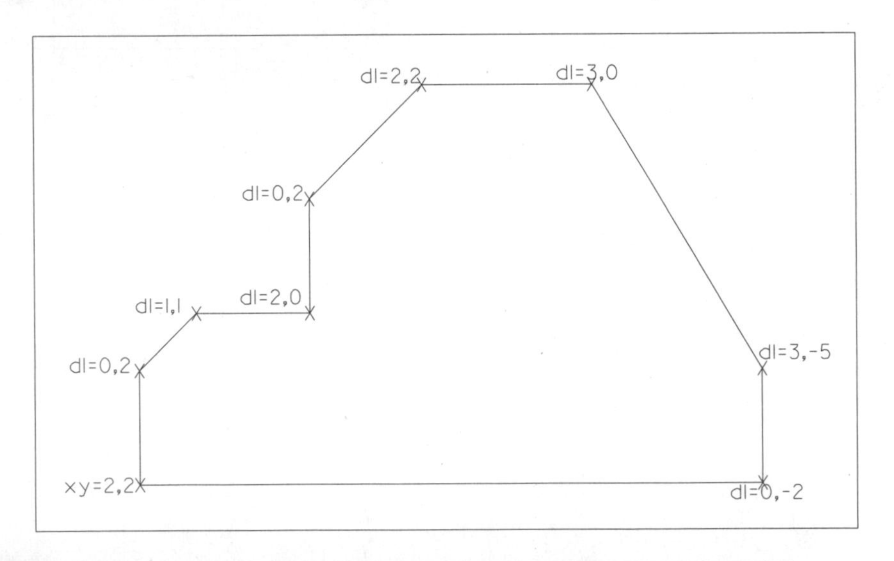

Figure 2–29 Placing connected lines using relative rectangular coordinates

XY=2,2 Enter
DL=0,2 Enter
DL=1,1 Enter
DL=2,0 Enter
DL=0,2 Enter
DL=2,2 Enter
DL=3,0 Enter

DL=3,-5 Enter
DL=0,-2 Enter
XY=2,2 Enter
(click the Reset button to terminate the line sequence)

The following key-ins show the placement of connected lines for the drawing shown in Figure 2–27 by the Place Line tool with relative polar coordinates and relative rectangular coordinates (see Figure 2–30):

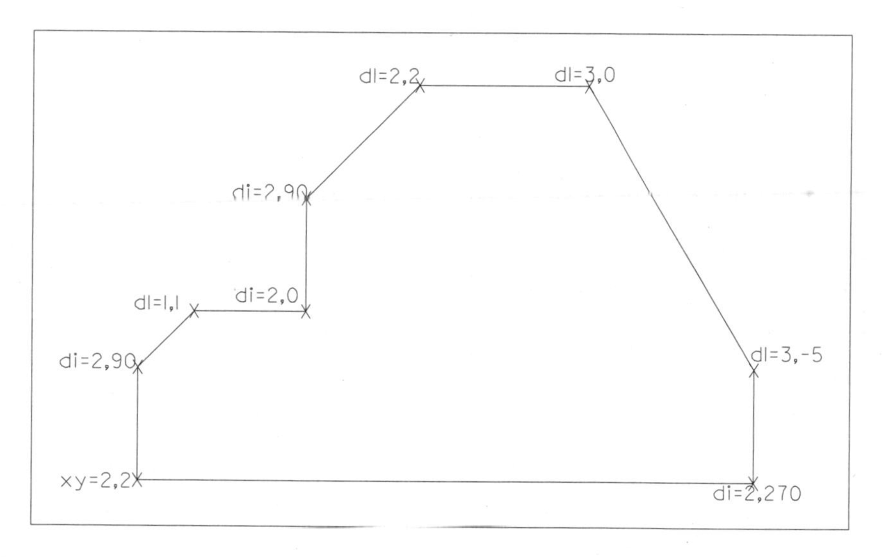

Figure 2–30 Placing connected lines using relative polar coordinates and relative rectangular coordinates

XY=2,2 Enter
DI=2,90 Enter
DL=1,1 Enter
DI=2,0 Enter
DI=2,90 Enter
DL=2,2 Enter
DL=3,0 Enter
DL=3,-5 Enter
DI=2,270 Enter
XY=2,2 Enter
(click the Reset button to terminate the line sequence)

REVIEW QUESTIONS

Write your answers in the spaces provided.

1. How many possible positions (positional units) are there in the *X* and *Y* directions of a *2D* design file?

   ____

2. Define Master Units: ______________________________

3. Define Sub Units: ______________________________

4. Define Positional Units: ______________________________

5. Name the tool that will make individual elements from a shape. ______________

6. Explain briefly the differences between the absolute rectangular coordinates and relative rectangular coordinates precision key-ins.

   ____

7. Name the three key-ins that are used in Precision Input for absolute rectangular coordinates, relative rectangular coordinates, and polar relative coordinates.

   ____

8. When MicroStation displays the information regarding the size of an element or coordinates, it does so in the following format:

   ________:________:________

9. If in a design file the Working Units are set up as inches, eighths, and 1600 positional units per eighth, what distance does 1:6:600 represent?

   ____

10. If in a design file the Working Units are set up as feet, inches, and 1600 positional units per inch, what Working Units expression is equivalent to 4'–0.3200"?

    ____

11. Name the three methods by which you can place circles in MicroStation.

    ____

12. If you wish to draw a circle by specifying three known points on the circle, invoke the ______________ tool.

13. The tool related to placing arcs is in the ______________ tool box.

14. The Place Arc Edge tool places an arc by identifying ___________ points on the arc.

15. Number of vertices allowed when you use the Place Shape tool range from ______________.

16. Number of data points MicroStation prompts to draw a rotated block when you use the Place Block Rotated tool is ____________.

17. Key-in to redefine the Global origin is ___________________.

18. What is the purpose of using the Save Settings command?

____

PROJECT EXERCISE

This project exercise provides step-by-step instructions for creating the design shown in Figure P2–1. The intent is to guide you in applying the concepts and tools presented in Chapters 1 and 2. (Note that these instructions are not necessarily the most efficient way to draw the objects. Your efficiency will improve as you learn more tools in later chapters.)

In this project, you'll learn how to do the following:

- Create a new design file.
- Set the working units.
- Draw the border using Precision Input.
- Draw the objects using the Place Line, Place Block, Place Circle, and Place Arc tools.

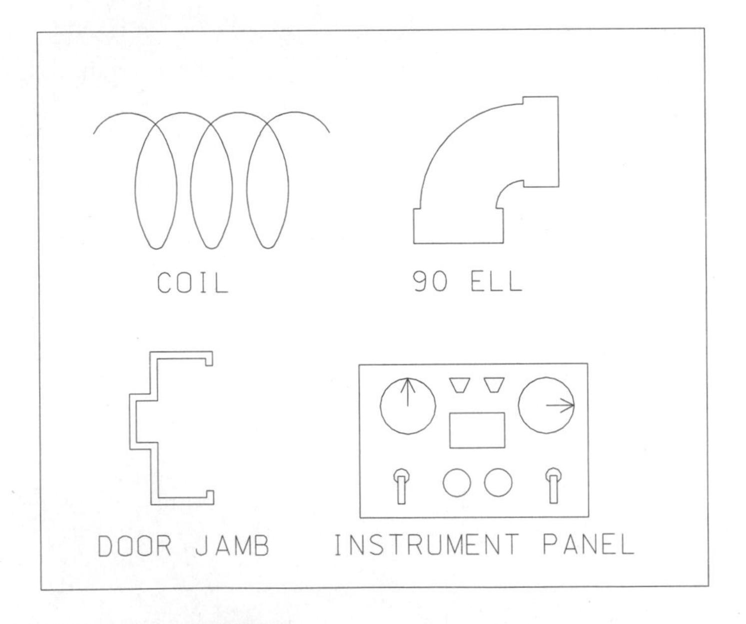

Figure P2–1 Completed project design

Note: Do not draw the dimensions. Dimensioning is introduced in a later chapter.

Note: As you complete each step in the project procedures, place a check mark by the step to help you keep up with where you are in the project.

Create a Design File

This procedure has you start MicroStation and create a design file for the project design.

STEP 1: Invoke the MicroStation program.

Example: Under Microsoft Windows 98, find the MicroStation program in the **Start** > Programs menu and select it.

STEP 2: In the MicroStation Manager dialog box, select the New... option from the File pull-down menu to open the Create New Design File dialog box.

STEP 3: If the "Seed File" area of the Create New Design File dialog box does not show SEED2D.DGN as the seed file, click the Select... button and select SEED2D.DGN as the seed file.

STEP 4: In the Create New Design File dialog box, type **CH2.DGN** as the name for the project design file in the Files edit field, then click the OK button to create the file, and close the dialog box.

STEP 5: In the MicroStation Manager dialog box, select the CH2.DGN file from the Files list box and click the OK button to load the file in MicroStation.

Set the Working Units and Draw a Border

This procedure presents the steps by which to do the following:

- Set the working unit's names to "IN" for the Master Units and "TH" for the Sub Units, and set the Resolution to "10" TH Per IN and "1000" Pos Units Per TH.
- Draw the border using the Place Block tool.

STEP 1: In the MicroStation application window, select the Design File... option from the Settings pull-down menu.

STEP 2: In the Design File Settings dialog box, select the Working Units category, and set the Working Units ratio as shown in Figure P2–2.

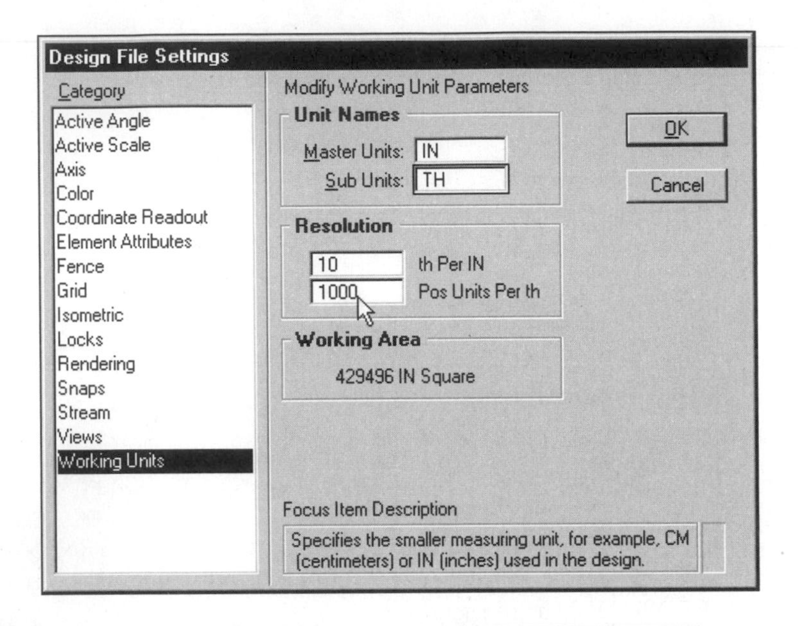

Figure P2–2 Design File Settings dialog box—Working Units ratio setup

STEP 3: Click the OK button to close the Design File Settings dialog box and keep the changes.

STEP 4: If an Alert box opens to tell you the Working Units are being changed, click its OK button to accept the changes.

STEP 5: Select Save Settings from the File pull-down menu to save the settings of the Working Units.

STEP 6: Invoke the Place Block tool from the Polygons tool box as shown in Figure P2–3.

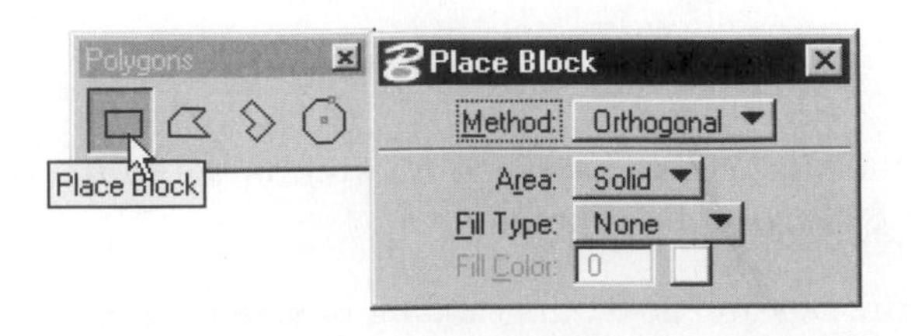

Figure P2–3 Invoking the Place Block tool from the Polygons tool box

MicroStation prompts:

Place Block > Enter first point *(Click in the Key-in window's input field, key-in* **XY=0,0** *as shown in Figure P2–4, and press* Enter*.)*

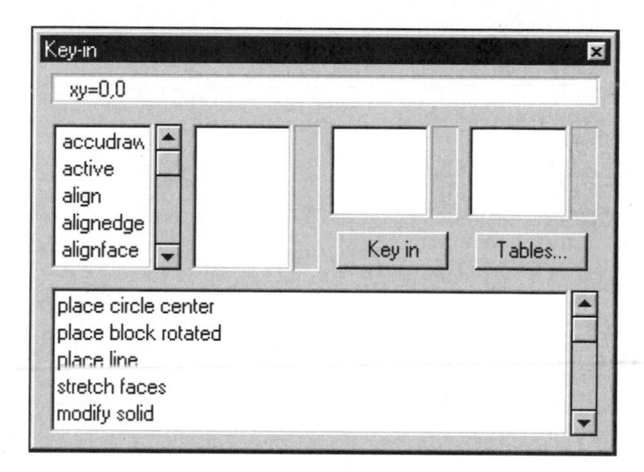

Figure P2–4 Key-in window

Note: If the Key-in window is not open, then select the Key-in option from the Utilities pull-down menu to open the Key-in window.

Place Block > Enter opposite corner *(Key-in* **XY=12,10** *in the Key-in window and press* Enter*.)*

STEP 7: Invoke the Fit View tool from the View Control bar (located in the lower-left corner of the view window) to display the complete border outline in the selected view. (Detailed explanation regarding the usage of the Fit View tool is provided in Chapter 3.)

STEP 8: Select Save Settings from the File pull-down menu to save the current settings.

If the procedure was executed correctly, a rectangle should be seen in the view, as shown in Figure P2–5.

Figure P2–5 Completed border outline

Draw the Door Jamb

This procedure describes the steps required to draw the door jamb shown in Figure P2–6. The door jamb is drawn with the Place Line tool using rectangular coordinates and polar coordinates.

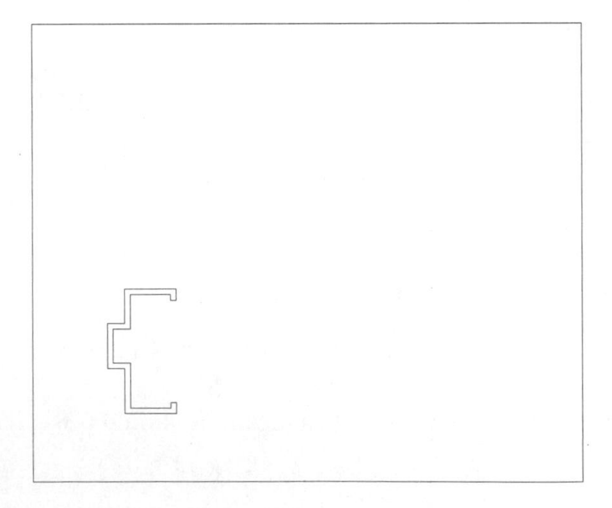

Figure P2–6 Door jamb

STEP 1: Invoke the Place Line tool from the Linear Elements tool box as shown in Figure P2–7.

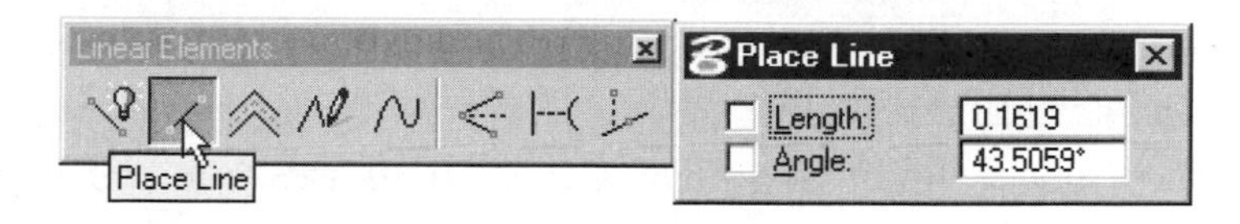

Figure P2–7 Invoking the Place Line tool from the Linear Elements tool box

MicroStation prompts:

Place Line > Enter first point *(Click in the Key-in window's input field, key-in* **XY=2,1.5**, *and press* Enter*.)*

Place Line > Enter end point *(Key-in* **XY=2,2.5** *in the Key-in window and press* Enter*.)*

Place Line > Enter end point *(Key-in* **XY=1.625,2.5** *in the Key-in window and press* Enter*.)*

Place Line > Enter end point *(Key-in* **XY=1.625,3.5** *in the Key-in window and press* Enter*.)*

Place Line > Enter end point *(Key-in* **XY=2,3.5** *in the Key-in window and press* Enter*.)*

Place Line > Enter end point *(Key-in* **XY=2,4.25** *in the Key-in window and press* Enter*.)*

Place Line > Enter end point *(Key-in* **XY=3.125,4.25** *in the Key-in window and press* Enter*.)*

Place Line > Enter end point *(Key-in* **XY=3.125,4** *in the Key-in window and press* Enter*.)*

Place Line > Enter end point *(Key-in* **XY=3,4** *in the Key-in window and press* Enter*.)*

Place Line > Enter end point *(Key-in* **DI=0.125,90** *in the Key-in window and press* Enter*.)*

Place Line > Enter end point *(Key-in* **DI=0.875,180** *in the Key-in window and press* Enter*.)*

Place Line > Enter end point *(Key-in* **DI=0.75,270** *in the Key-in window and press* Enter*.)*

Place Line > Enter end point *(Key-in* **DI=0.375,180** *in the Key-in window and press* Enter*.)*

Place Line > Enter end point *(Key-in* **DL=0,–0.75** *in the Key-in window and press* Enter*.)*

Place Line > Enter end point *(Key-in* **DL=0.375,0** *in the Key-in window and press* Enter*.)*

Place Line > Enter end point *(Key-in* **DL=0,–1** *in the Key-in window and press* Enter*.)*

Place Line > Enter end point *(Key-in* **DL=0.875,0** *in the Key-in window and press* Enter*.)*

Place Line > Enter end point *(Key-in* **DL=0,0.125** *in the Key-in window and press* Enter*.)*

Place Line > Enter end point *(Key-in* **DL=0.125,0** *in the Key-in window and press* Enter*.)*

Place Line > Enter end point *(Key-in* **DL=0,–.25** *in the Key-in window and press* Enter*.)*

Place Line > Enter end point *(Key-in* **DL=–1.125,0** *in the Key-in window and press* Enter*.)*

Place Line > Enter end point *(Click the Reset button to terminate the line sequence.)*

Draw the Instrument Panel

This procedure describes the steps required to draw the instrument panel shown in Figure P2–8. The instrument panel is drawn with the Place Line, Place Block, Place Circle, and Place Arc tools.

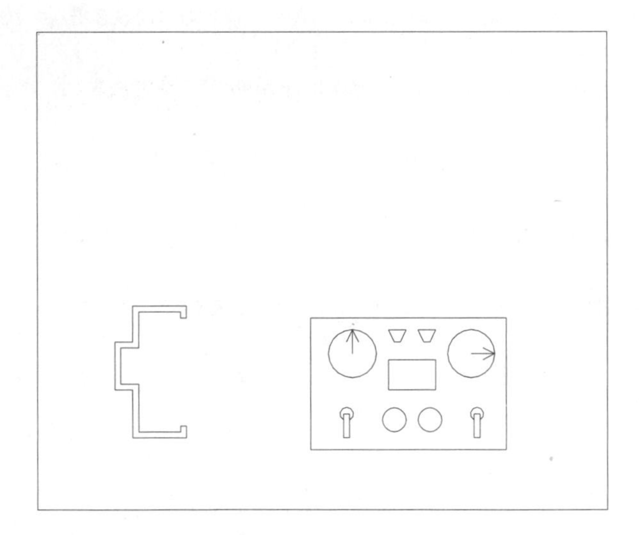

Figure P2–8 Door jamb and instrument panel

STEP 1: To draw the main outline of the instrument panel and the central rectangle block, invoke the Place Block tool from the Polygons tool box.

MicroStation prompts:

Place Block > Enter first point *(Click in the Key-in window's input field, key-in* **XY=5.75,1.25***, and press* Enter*.)*

Place Block > Enter opposite corner *(Key-in* **XY=9.875,4** *in the Key-in window and press* Enter*.)*

Place Block > Enter first point *(Click in the Key-in window's input field, key-in* **XY=7.375,2.5**, *and press* Enter*.)*

Place Block > Enter opposite corner *(Key-in* **XY=8.375,3.125** *in the Key-in window and press* Enter*.)*

STEP 2: To draw the trapezoidal shapes located in the top of the instrument panel, invoke the Place Line tool from the Linear Elements tool box.

MicroStation prompts:

Place Line > Enter first point *(Click in the Key-in window's input field, key-in* **XY=7.375,3.75**, *and press* Enter*.)*

Place Line > Enter end point *(Key-in* **DI=0.375,0** *in the Key-in window and press* Enter*.)*

Place Line > Enter end point *(Key-in* **XY=7.625,3.5** *in the Key-in window and press* Enter*.)*

Place Line > Enter end point *(Key-in* **DI=0.125,180** *in the Key-in window and press* Enter*.)*

Place Line > Enter end point *(Key-in* **XY=7.375,3.75** *in the Key-in window and press* Enter*.)*

Place Line > Enter end point *(Click the Reset button to terminate the line sequence.)*

Place Line > Enter first point *(Click in the Key-in window's input field, key-in* **XY=8,3.75**, *and press* Enter*.)*

Place Line > Enter end point *(Key-in* **DI=0.375,0** *in the Key-in window and press* Enter*.)*

Place Line > Enter end point *(Key-in* **XY=8.25,3.5** *in the Key-in window and press* Enter*.)*

Place Line > Enter end point *(Key-in* **DI=0.125,180** *in the Key-in window and press* Enter*.)*

Place Line > Enter end point *(Key-in* **XY=8,3.75** *in the Key-in window and press* Enter*.)*

Place Line > Enter end point *(Click the Reset button to terminate the line sequence.)*

STEP 3: To draw the circular gauges with the arrow pointers located in the top of the instrument panel, first invoke the Place Circle tool from the Ellipses tool box, as shown in Figure P2–9, to draw two circles. Select Center from the Method option menu in the Tool Settings Window, set the Radius to 0.5 master units, and turn ON the toggle button for radius.

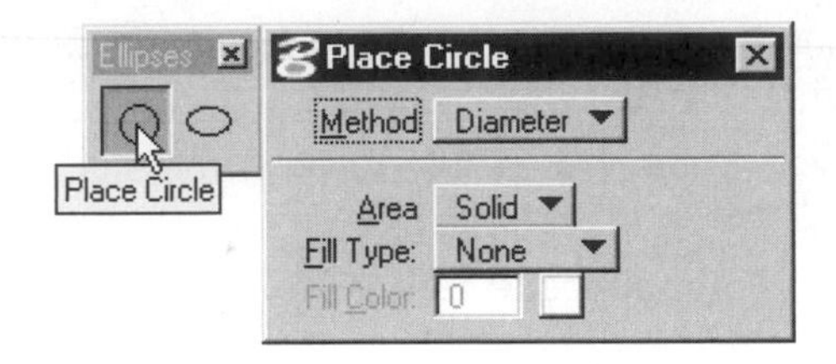

Figure P2–9 Invoking the Place Circle tool from the Ellipses tool box

MicroStation prompts:

Place Circle By Center > Identify Center Point *(Click in the Key-in window's input field, key-in* **XY=6.625,3.25**, *and press* Enter.*)*

Place Circle By Center > Identify Center Point *(Click in the Key-in window's input field, key-in* **XY=9.125,3.25**, *and press* Enter.*)*

And to draw the arrow pointers, invoke the Place Line tool from the Linear Elements tool box.

MicroStation prompts:

Place Line > Enter first point *(Click in the Key-in window's input field, key-in* **XY=6.5,3.5**, *and press* Enter.*)*

Place Line > Enter end point *(Key-in* **XY=6.625,3.75** *in the Key-in window and press* Enter.*)*

Place Line > Enter end point *(Key-in* **XY=6.75,3.5** *in the Key-in window and press* Enter.*)*

Place Line > Enter end point *(Click the Reset button to terminate the line sequence.)*

Place Line > Enter first point *(Click in the Key-in window's input field, key-in* **XY=6.625,3.25**, *and press* Enter.*)*

Place Line > Enter end point *(Key-in* **XY=6.625,3.75** *in the Key-in window and press* Enter.*)*

Place Line > Enter end point *(Click the Reset button to terminate the line sequence.)*

Place Line > Enter first point *(Click in the Key-in window's input field, key-in* **XY=9.375,3.375**, *and press* Enter.*)*

Place Line > Enter end point *(Key-in* **XY=9.625,3.25** *in the Key-in window and press* Enter.*)*

Place Line > Enter end point *(Key-in* **XY=9.375,3.125** *in the Key-in window and press* Enter.*)*

Place Line > Enter end point *(Click the Reset button to terminate the line sequence)*

Place Line > Enter first point *(Click in the Key-in window's input field, key-in* **XY=9.125,3.25**, *and press* Enter.*)*

Place Line > Enter end point *(Key-in* **XY=9.625,3.25** *in the Key-in window and press* Enter*.)*

Place Line > Enter end point *(Click the Reset button to terminate the line sequence.)*

STEP 4: To draw the two circles located in the bottom of the instrument panel, first invoke the Place Circle tool from the Ellipses tool box, select Center from the Method: option menu in the Tool Settings Window, set the Radius to 0.25 Master Units, and turn ON the toggle button for radius.

MicroStation prompts:

Place Circle By Center > Identify Center Point *(Click in the Key-in window's input field, key-in* **XY=7.5,1.875***, and press* Enter*.)*

Place Circle By Center > Identify Center Point *(Click in the Key-in window's input field, key-in* **XY=8.25,1.875***, and press* Enter *.)*

STEP 5: Invoke the Place Block tool from the Polygons tool box to draw the switches located in the bottom of the instrument panel.

MicroStation prompts:

Place Block > Enter first point *(Click in the Key-in window's input field, key-in* **XY=6.4375,1.5***, and press* Enter*.)*

Place Block > Enter opposite corner *(Key-in* **XY=6.5625,2** *in the Key-in window and press* Enter*.)*

Place Block > Enter first point *(Click in the Key-in window's input field, key-in* **XY=9.1875,1.5***, and press* Enter*.)*

Place Block > Enter opposite corner *(Key-in* **XY=9.3125,2** *in the Key-in window and press* Enter*.)*

STEP 6: Invoke the Place Arc tool from the Arcs tool box, and select Center from the Method option menu in the Tool Settings Window.

MicroStation prompts:

Place Arc By Center > Identify First Arc Endpoint *(Click in the Key-in window's input field, key-in* **XY=6.5625,1.875***, and press* Enter*.)*

Place Arc By Center > Identify Arc Center *(Click in the Key-in window's input field, key-in* **XY=6.5,2***, and press* Enter*.)*

Place Arc By Center > Identify Second Arc Endpoint *(Click in the Key-in window's input field, key-in* **XY=6.4375,1.875***, and press* Enter*.)*

Place Arc By Center > Identify First Arc Endpoint *(Click in the Key-in window's input field, key-in* **XY=9.3125,1.875***, and press* Enter*.)*

Place Arc By Center > Identify Arc Center *(Click in the Key-in window's input field, key-in* **XY=9.25,2**, *and press* Enter*.)*

Place Arc By Center > Identify Second Arc Endpoint *(Click in the Key-in window's input field, key-in* **XY=9.1875,1.875**, *and press* Enter*.)*

Draw the 90-degree Ell

This procedure describes the steps required to draw the 90-degree ell shown in Figure P2–10. The 90-degree ell is drawn with the Line and Arc tools.

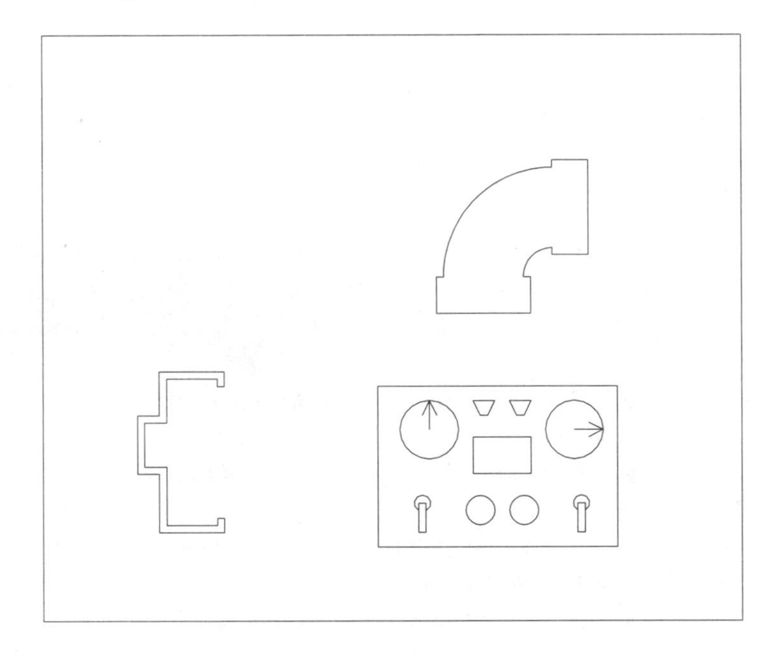

Figure P2–10 Door jamb, instrument panel, and 90-degree ell

STEP 1: To draw the lower lines of the 90-degree ell, invoke the Place Line tool from the Linear Elements tool box.

MicroStation prompts:

Place Line > Enter first point *(Click in the Key-in window's input field, key-in* **XY=8.25,5.875**, *and press* Enter*.)*

Place Line > Enter end point *(Key-in* **DI=0.125,0** *in the Key-in window and press* Enter*.)*

Place Line > Enter end point *(Key-in* **DI=0.625,270** *in the Key-in window and press* Enter*.)*

Place Line > Enter end point *(Key-in* **DI=1.625,180** *in the Key-in window and press* Enter*.)*

Place Line > Enter end point *(Key-in* **DI=0.625,90** *in the Key-in window and press* Enter*.)*

Place Line > Enter end point *(Key-in* **DI=0.125,0** *in the Key-in window and press* Enter*.)*

Place Line > Enter end point *(Click the Reset button to terminate the line sequence.)*

STEP 2: To draw the large outer arc, invoke the Place Arc tool from the Arcs tool box, and select Center from the Method option menu in the Tool Settings Window.

MicroStation prompts:

Place Arc By Center > Identify First Arc Endpoint *(Click in the Key-in window's input field, key-in* **XY=8.75,7.75**, *and press* Enter*.)*

Place Arc By Center > Identify Arc Center *(Click in the Key-in window's input field, key-in* **DL=0,-1.8750**, *and press* Enter*.)*

Place Arc By Center > Identify Second Arc Endpoint *(Click in the Key-in window's input field, key-in* **XY=6.875,5.875**, *and press* Enter*.)*

STEP 3: To draw the upper lines of the 90-degree ell, invoke the Place Line tool from the Draw tool box.

MicroStation prompts:

Place Line > Enter first point *(Click in the Key-in window's input field, key-in* **XY=8.75,7.75**, *and press* Enter*.)*

Place Line > Enter end point *(Key-in* **DI=0.125,90** *in the Key-in window and press* Enter*.)*

Place Line > Enter end point *(Key-in* **DI=0.625,0** *in the Key-in window and press* Enter*.)*

Place Line > Enter end point *(Key-in* **DI=1.625,270** *in the Key-in window and press* Enter*.)*

Place Line > Enter end point *(Key-in* **DI=0.625,180** *in the Key-in window and press* Enter*.)*

Place Line > Enter end point *(Key-in* **DI=0.125,90** *in the Key-in window and press* Enter*.)*

Place Line > Enter end point *(Click the Reset button to terminate the line sequence.)*

STEP 4: To draw the smaller outer arc, invoke the Place Arc tool from the Arcs tool box, and select Center from the Method option menu in the Tool Settings Window.

MicroStation prompts:

Place Arc By Center > Identify First Arc Endpoint *(Click in the Key-in window's input field, key-in* **XY=8.75,6.375***, and press* Enter*.)*

Place Arc By Center > Identify Arc Center *(Click in the Key-in window's input field, key-in* **DL=0,–0.5***, and press* Enter*.)*

Place Arc By Center > Identify Second Arc Endpoint *(Click in the Key-in window's input field, key-in* **XY=8.25,5.8750***, and press* Enter*.)*

Draw the Coil

This procedure describes the steps required to draw the coil shown in Figure P2–11. The coil is drawn with the Arc tool.

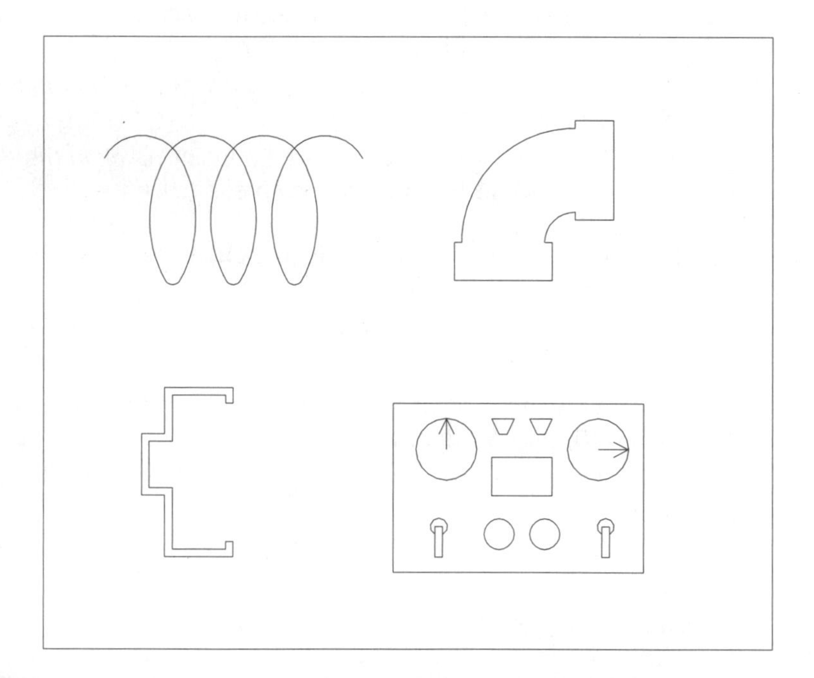

Figure P2–11 Door jamb, instrument panel, 90-degree ell, and coil

STEP 1: To draw the coil, invoke the Place Arc tool from the Arcs tool box, and select Edge from the Method option menu in the Tool Settings Window.

MicroStation prompts:

Place Arc By Edge > Identify First Arc Endpoint *(Click in the Key-in window's input field, key-in* **XY=1,7.25**, *and press* Enter*.)*

Place Arc By Edge > Identify Point on Arc Radius *(Click in the Key-in window's input field, key-in* **XY=1.625,7.625** *and press* Enter*.)*

Place Arc By Edge > Identify Second Arc Endpoint *(Click in the Key-in window's input field, key-in* **XY=2.25,7.25**, *and press* Enter*.)*

Place Arc By Edge > Identify First Arc Endpoint *(Click in the Key-in window's input field, key-in* **XY=2.25,7.25**, *and press* Enter*.)*

Place Arc By Edge > Identify Point on Arc Radius *(Click in the Key-in window's input field, key-in* **XY=2.5,6.25**, *and press* Enter*.)*

Place Arc By Edge > Identify Second Arc Endpoint *(Click in the Key-in window's input field, key-in* **XY=2.25,5.25**, *and press* Enter*.)*

Place Arc By Edge > Identify First Arc Endpoint *(Click in the Key-in window's input field, key-in* **XY=2.25,5.25**, *and press* Enter*.)*

Place Arc By Edge > Identify Point on Arc Radius *(Click in the Key-in window's input field, key-in* **XY=2.125,5.1750**, *and press* Enter*.)*

Place Arc By Edge > Identify Second Arc Endpoint *(Click in the Key-in window's input field, key-in* **XY=2,5.25**, *and press* Enter*.)*

Place Arc By Edge > Identify First Arc Endpoint *(Click in the Key-in window's input field, key-in* **XY=2,5.25**, *and press* Enter*.)*

Place Arc By Edge > Identify Point on Arc Radius *(Click in the Key-in window's input field, key-in* **XY=1.75,6.25**, *and press* Enter*.)*

Place Arc By Edge > Identify Second Arc Endpoint *(Click in the Key-in window's input field, key-in* **XY=2,7.25**, *and press* Enter*.)*

Place Arc By Edge > Identify First Arc Endpoint *(Click in the Key-in window's input field, key-in* **XY=2,7.25**, *and press* Enter*.)*

Place Arc By Edge > Identify Point on Arc Radius *(Click in the Key-in window's input field, key-in* **XY=2.6250,7.6250**, *and press* Enter*.)*

Place Arc By Edge > Identify Second Arc Endpoint *(Click in the Key-in window's input field, key-in* **XY=3.25,7.25**, *and press* Enter*.)*

Take a breath. This completes the first loop of the coil. To draw the second and third loops, refer to the following table for the coordinates.

First arc endpoint	Point on arc radius	Second arc endpoint
XY=3.25,7.25	XY=3.500,6.250	XY=3.25,5.25
XY=3.25,5.25	XY=3.125,5.175	XY=3.00,5.25
XY=3.00,5.25	XY=2.750,6.250	XY=3.00,7.25
XY=3.00,7.25	XY=3.625,7.625	XY=4.25,7.25
XY=4.25,7.25	XY=4.500,6.250	XY=4.25,5.25
XY=4.25,5.25	XY=4.125,5.175	XY=4.00,5.25
XY=4.00,5.25	XY=3.750,6.250	XY=4.00,7.25
XY=4.00,7.25	XY=4.625,7.625	XY=5.25,7.25

STEP 3: Invoke the Save settings from the File pull-down menu.

Congratulations! You have just successfully applied several MicroStation concepts in creating a design.

DRAWING EXERCISES 2–1 THROUGH 2–6

In Exercises 2–1 through 2–3, write down the coordinates necessary to draw the objects shown above the tables, then use the coordinates to draw the object. The coordinates are already entered in Exercise 2–1 as an example.

Note: Do not draw the dimensions or text.

Exercise 2–1

Draw the object using absolute coordinate key-ins (**XY=**<x,y>).

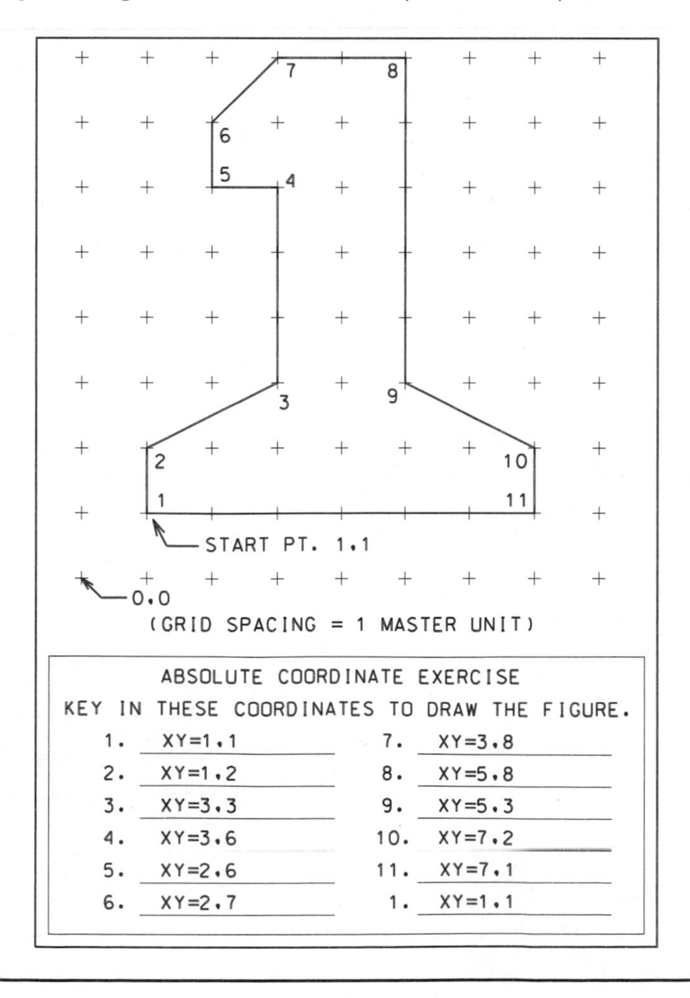

Exercise 2–2

Draw the object using relative coordinate key-ins (**DL=**<x,y>).

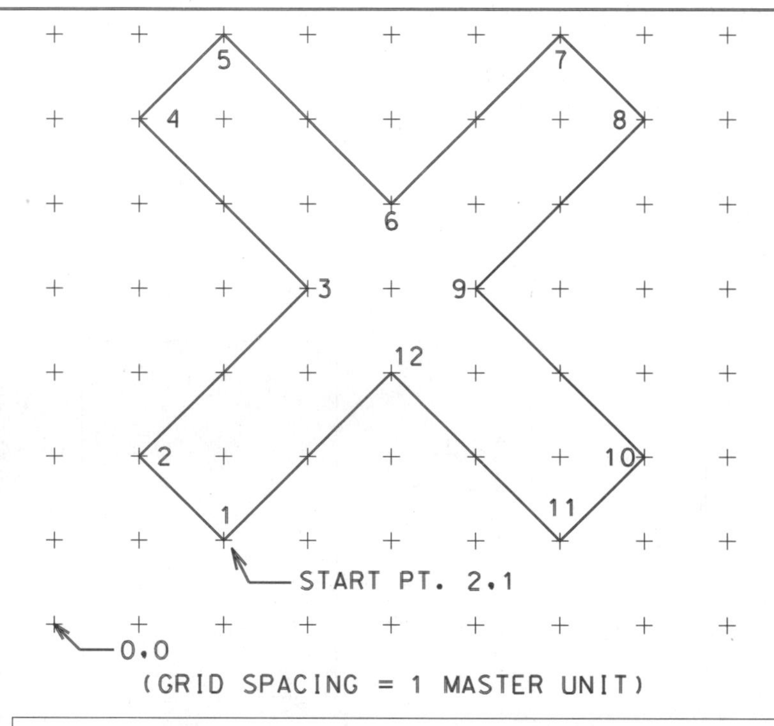

RELATIVE COORDINATE EXERCISE

1. ENTER THE COORDINATES IN THE TABLE BELOW.
2. KEY IN THE COORDINATES TO DRAW THE FIGURE.

1.	XY=2,1	7.	DL=
2.	DL=-1,1	8.	DL=
3.	DL=2,2	9.	DL=
4.	DL=	10.	DL=
5.	DL=	11.	DL=
6.	DL=	1.	DL=

Exercise 2–3

Draw the object using polar coordinate key-ins (**DI=<x,y>**).

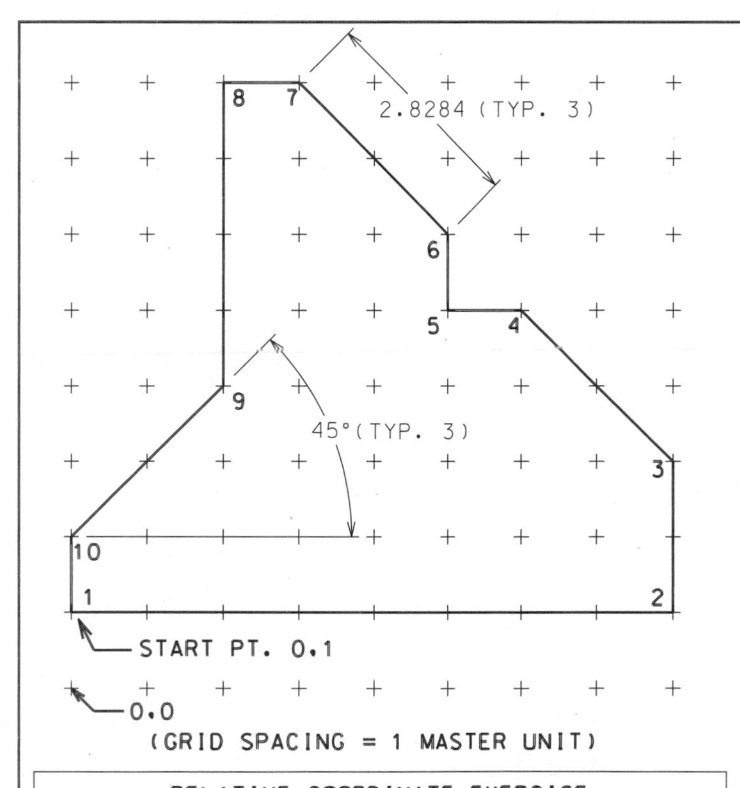

RELATIVE COORDINATE EXERCISE

1. ENTER THE COORDINATES IN THE TABLE BELOW.
2. KEY IN THE COORDINATES TO DRAW THE FIGURE.

1.	XY=0,1	7.	DI=
2.	DI=8,0	8.	DI=
3.	DI=2,90	9.	DI=
4.	DI=	10.	DI=
5.	DI=	11.	DI=
6.	DI=	1.	DI=

Exercise 2–4

Use the following table to set up the design file to draw the front elevation and select the best Precision Input method to use for each placement of the object.

Setting	**Value**
Seed File	SEED2D.DGN
Working Units	12 " Per ' and 8000 Pos Units Per "

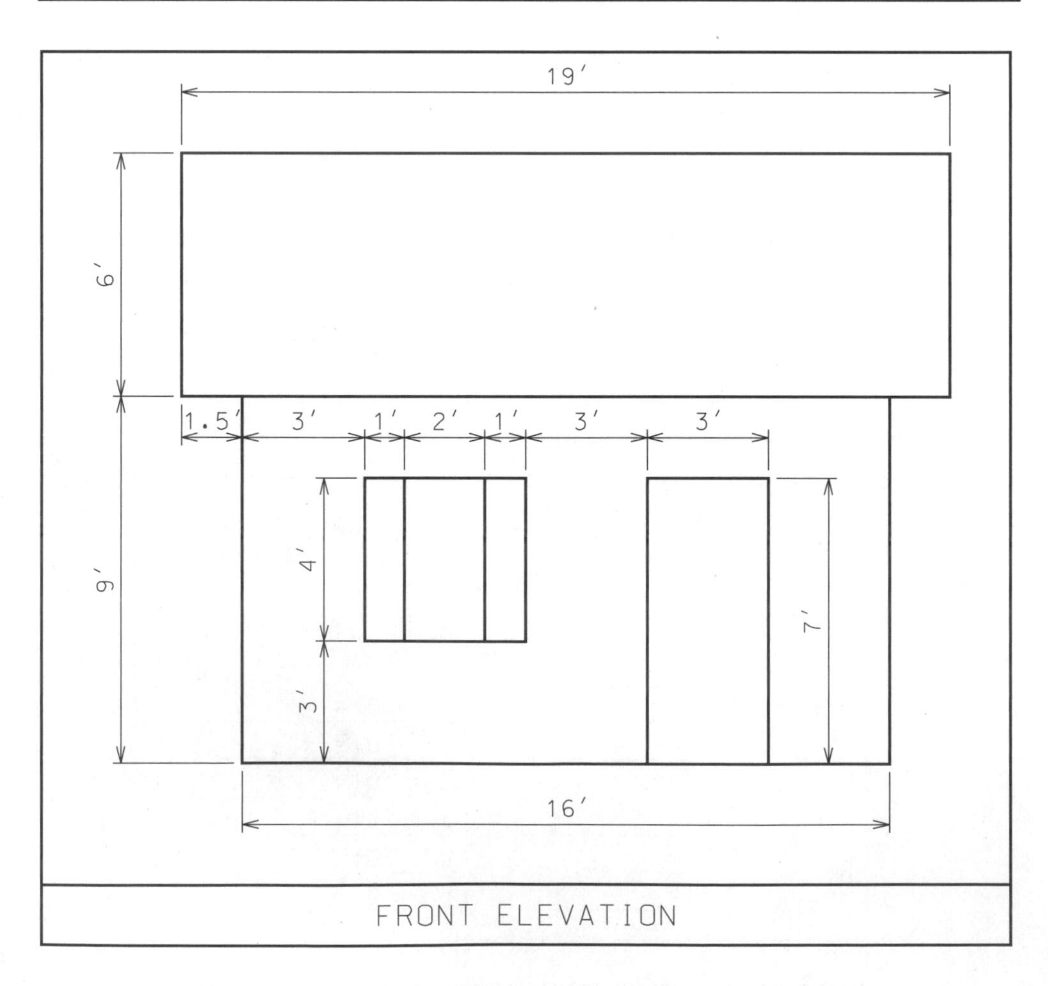

Exercise 2–5 and 2–6

Use the following table to set up the design files for Exercises 2–5 and Exercise 2–6 and select the best Precision Input method to use for each placement of the object.

Setting	Value
Seed File	SEED2D.DGN
Working Units	10 su Per mu and 1000 Pos Units Per su

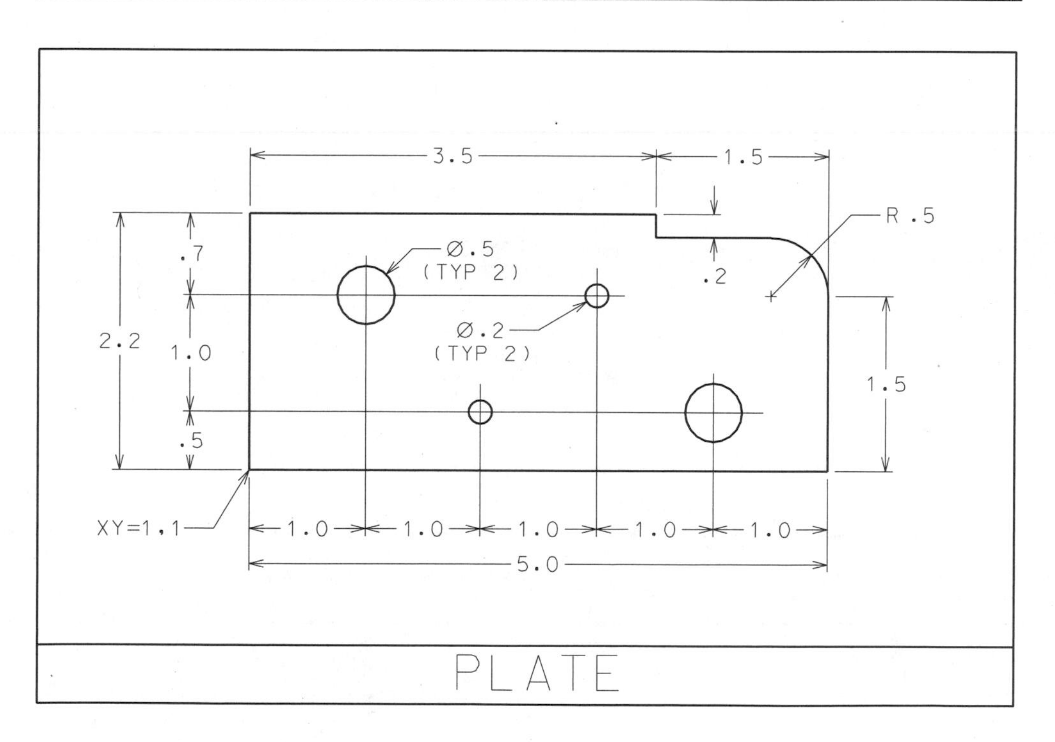

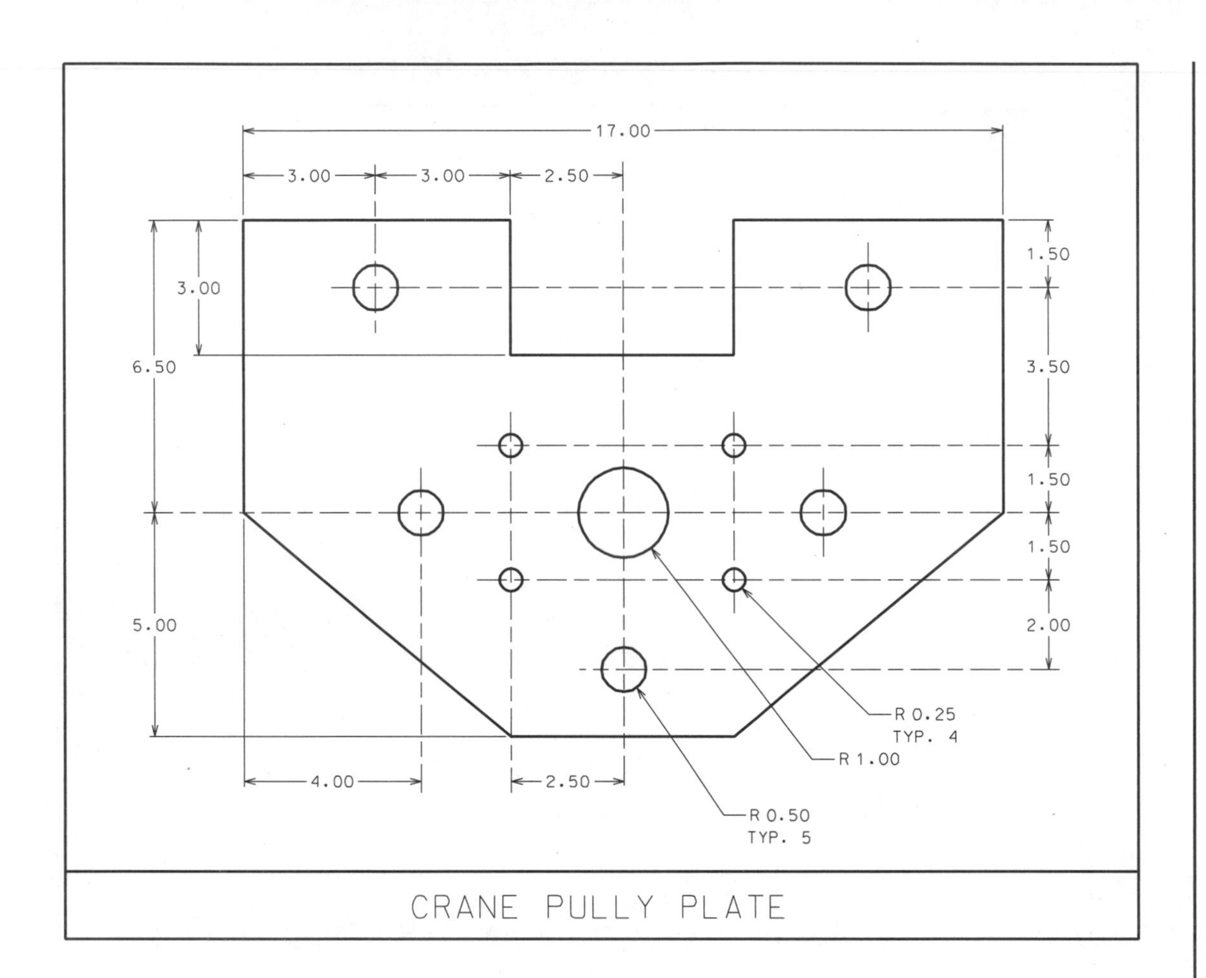
17.00
3.00
3.00
2.50
3.00
6.50
1.50
3.50
1.50
1.50
5.00
2.00
R 0.25
TYP. 4
R 1.00
4.00
2.50
R 0.50
TYP. 5
CRANE PULLY PLATE

chapter 3

Fundamentals II

Objectives

After completing this chapter, you will be able to do the following:

- Use drawing tools: grid, axis, units, and tentative snap
- Control and view levels
- Set element attributes
- Match element attributes
- Use View Control: zoom in, zoom out, window area, fit, pan, and update
- Use View windows and view attributes
- Use Undo and Redo tools

DRAWING TOOLS

MicroStation provides several different drawing tools to make your drafting and design layout easier.

The Grid System

Grids are a visual tool for measuring distances precisely and placing elements accurately. MicroStation displays a grid system that is similar to a sheet of graph paper. You can turn the Grid display to ON or OFF as needed, and the spacing can be changed at any time. There are two types of grid systems. The first is the Grid Reference, which appears on your screen as a cross; by default the spacing between crosses is set to one Master Unit (MU). The second is the Grid Unit (GU), which appears as a dot on the screen; by default the spacing between dots is set to one Sub Unit (SU). The grid is just a drawing tool, not part of the drawing; it is for visual reference only and is never plotted. Grids serve two purposes: they provide a visual indication of distances and, with Grid lock set to ON, they force all data points to start and end on a grid point. This is useful for keeping lines straight, ensuring that distances are exact, and making sure all elements meet.

Note: MicroStation overrides the Grid lock when you key-in the location of a point by precision input.

Grid Display

The Grid display (visual) can be set to ON or OFF. If it is set to ON, then you can see the Grid display on the screen. MicroStation, by default, displays a maximum of 90 dots and 46 crosses. You may not see any crosses or dots if you zoom out farther. When it is set to OFF, the grid is not displayed. You can change the status of the Grid display from the View Attributes setting box. To change the status, invoke the View Attributes settings box by selecting View Attributes from the Settings pull-down menu. The resulting settings box is similar to the one shown in Figure 3–1.

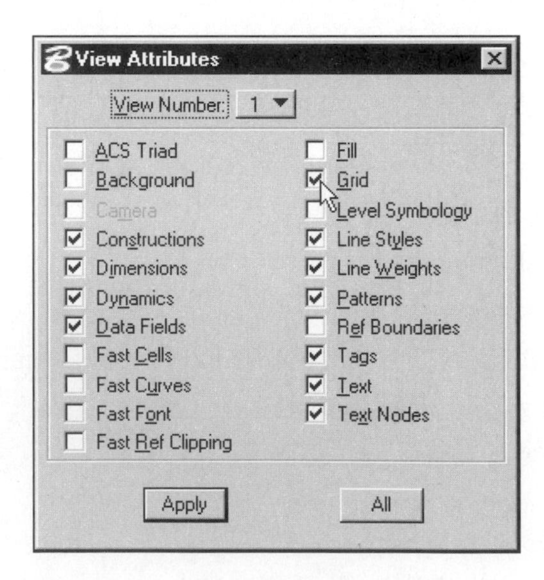

Figure 3–1 View Attributes settings box with the cursor position on the Grid toggle

Make any necessary change to the Grid display in the View Attributes settings box. If you only want the attribute change applied to a specific view, pick the number of the view you want in the View Number (default is the current working view) options menu at the top of the settings box, then click the Apply button. If you want to turn ON the Grid display on all the open view windows, click the ALL button. (For a detailed explanation of Views and the View Attributes settings box, see the View Windows and View Attributes section.)

Grid Spacing

When the Grid display is set to ON, you will notice a series of evenly spaced dots and crosses. The grid consists of a matrix of dots with the reference crosses falling at equally spaced intervals. The spacing between both the grid dots and the reference crosses may be changed at any time to suit your drawing needs by marking appropriate changes to the Grid settings.

To set up grid units:

Pull-down menu	Settings > Design File...

MicroStation displays a Design File Settings box similar to the one shown in Figure 3–2. Select Grid from the Category list, as shown in Figure 3–2, and MicroStation displays controls for adjusting the grid unit settings. In the settings box two edit fields are provided, one for Grid Master unit, the other for Grid Reference. The Grid Master unit defines the distance between the grid units (dots) and is specified in terms of MU:SU:PU. By default, this is set for one sub-unit. If necessary, you can override the default spacing by keying-in the spacing in the edit field in terms of MU:SU:PU.

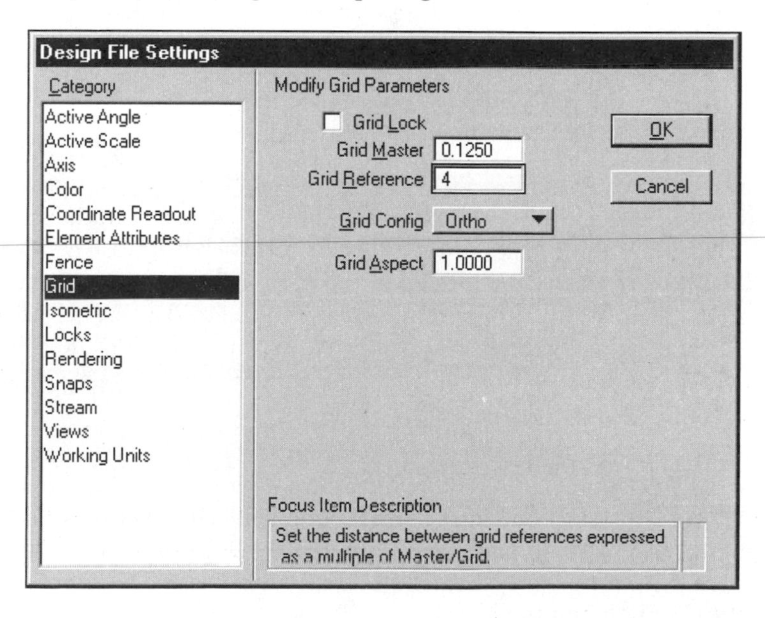

Figure 3–2 Design File Settings box

The Grid Reference is set to define the number (integer) of grid units between the grid reference (crosses). For example, let's say the Grid Master unit is set up to 0.125 inches and you would like to have the distance between the crossings to be 0.5 inches. Then the Grid Reference has to be set to 4 (0.5"/.125"), as shown in Figure 3–2.

Figure 3–3 shows the screen display for the grid settings.

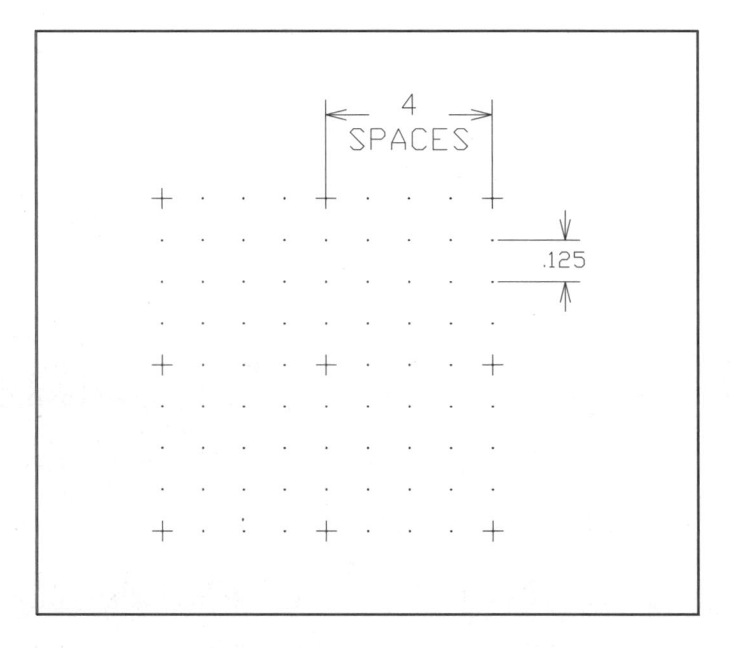

Figure 3–3 Screen display for the grid settings

You can also set the Grid Master unit and the Grid Reference by keying-in the spacing at the key-in window. To set the Grid Master unit, key-in **GU=<distance>** and press Enter. The <distance> has to be specified in MU:SU:PU. To set the Grid Reference, key-in **GR=<integer number>** and press Enter. The <integer number> is the number of grid units between the grid reference crosses. To keep the grid settings in effect for future editing sessions for the current design file, select Save Settings from the File pull-down menu.

Grid Configuration

MicroStation provides three choices to control the orientation of the Grid display: Orthogonal, Isometric, and Offset. The selection can be made from the Grid Settings box under the Grid Config options menu. The Orthogonal option aligns the grid points orthogonally (the default option). The Isometric option aligns the grid points isometrically. The Offset option offsets the rows by half the distance between the horizontal grid points.

Grid Aspect Ratio (*Y/X*)

The Grid Aspect Ratio edit field in the Grid Settings box allows you to set the ratio of vertical (*Y*) grid points to horizontal (*X*) grid points. The default is set at 1.000.

Grid Lock

The Grid Lock can be set to ON or OFF. When it is set to ON, MicroStation forces all the data points to the grid marks. You cannot place a data point in between the grid dots. By setting the Grid Lock to ON, you can enter points quickly, letting MicroStation ensure that they are placed precisely. You can always override the Grid Lock by keying-in absolute or relative coordinate points.

When Grid Lock is set to ON, you can identify the elements that were drawn on the grid. To identify an element that is not on the grid, simply set the Grid Lock to OFF and try again.

Note: The Grid Lock is effective regardless of the status of the Grid display. It still locks to grid points even if you cannot see the grid.

The Grid Lock can be set to ON or OFF (toggle) from the Grid Unit settings located in the Design File settings box. MicroStation displays the current status of the Grid Lock in the right-hand section of the Status bar. Click on the OK button to close the Design File Settings box after making the appropriate changes for the grid settings.

You can also toggle the Grid Lock: (1) from the settings box that appears by clicking the lock icon located on the right-hand section of the Status bar, (2) from the Settings pull-down menu's Locks submenu, or (3) from the Lock Toggles settings box, as shown in Figure 3–4.

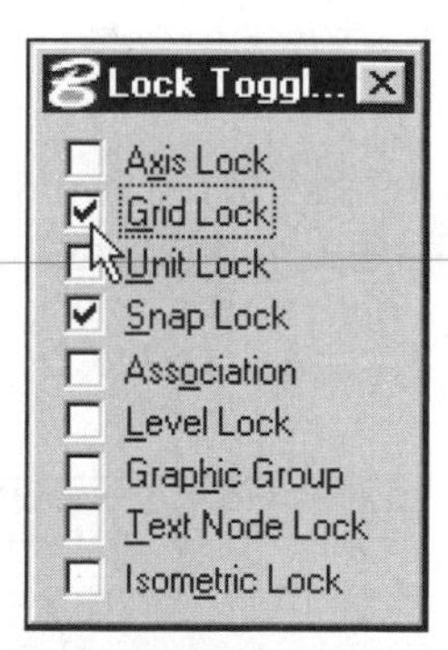

Figure 3–4 Lock Toggles settings box

To keep the grid settings (including the Grid Lock settings) in effect for future editing sessions for the current design file, choose Save Settings from the File pull-down menu.

Axis Lock

The Axis Lock forces each data point to lie at an angle (or multiples of that angle) from the previous data point. If necessary, you can change the Axis Start Angle, which is relative to the *X* axis, and the Axis Increment angle.

To set up Axis Lock settings:

Pull-down menu	Settings > Design File...

MicroStation displays the Design File Settings box. Select Axis from the Category list and MicroStation displays the controls for adjusting the Axis Lock settings. Key-in the appropriate Axis Start Angle in the Start Angle edit field and the Axis Increment angle in the Increment edit field. For example, whenever you want to draw horizontal or vertical lines, key-in an Axis Start Angle of 0 degrees and an Axis Increment angle of 90 degrees. Axis Start Angle and Axis Increment are only in effect when the Axis Lock is set to ON.

Snap Lock

The Snap Lock controls the placement of a tentative point at a specific point on an element, depending on the snap mode selected. Tentative snapping is a way of previewing a data point *before* it is actually entered in the design. Once a tentative point is placed in your design plane, a large cross appears to identify the tentative point. If it is snapped to an element, the element is highlighted and, in addition, MicroStation displays the absolute coordinates of the point selected in the left-side of the Status bar. If this is the point you wish to select, click the Accept button (same as the Data button) to confirm it. If, however, this is not the point you wish to select, move the cursor and click the tentative button again. This process of selecting another tentative point rejects the last tentative point and selects a new one highlighted by the large cross. Once you accept the tentative point with the Accept button, the large cross disappears. You may cancel the tentative point by clicking the Reset button.

A tentative point is placed by clicking the designated tentative button on your pointing device. For example, for a three-button mouse, click the middle button to place the tentative point; for a two-button mouse, press both buttons simultaneously to place a tentative point.

Snap Lock Mode

As mentioned earlier, the Snap Lock controls the placement of a tentative point at a specific point on an element, depending on the snap mode selected. If the Snap Lock toggle is set to ON, you can snap to a specific point on an element, depending on the snap mode selected. For example, if you set the snap mode to Center and turn ON the Snap Lock, the tentative button snaps to the center of circles, blocks, midpoints of lines, and segments of line strings. A tentative point can be placed while executing any MicroStation tool that requests a point, such as Line, Circle, Arc, Move, or Copy. If the Snap Lock toggle is set to OFF, tentative points do *not* snap to elements.

To set up Snap Lock settings:

Pull-down menu	Settings > Locks > Toggles
	Settings > Locks > Full

You can toggle the Snap Lock from either the Lock Toggles settings box or the Locks Full settings box, as shown in Figure 3–5.

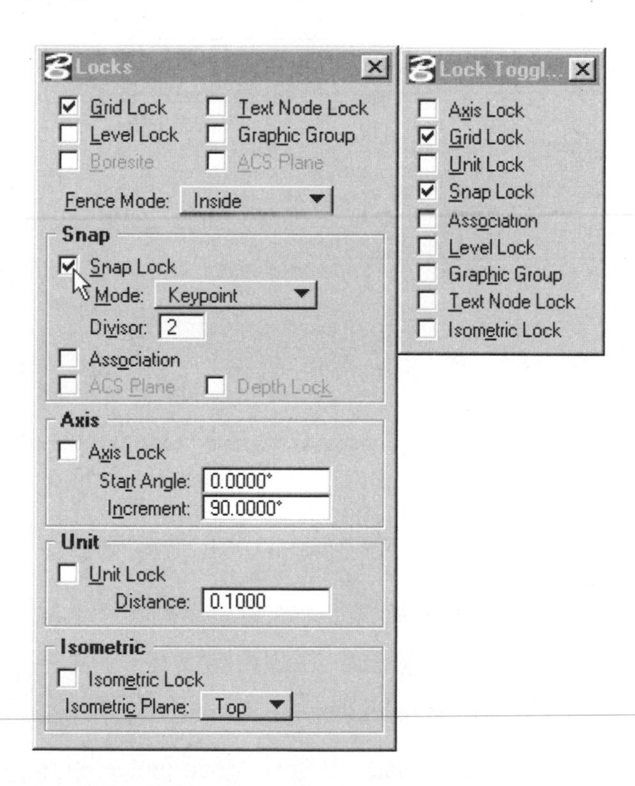

Figure 3–5 Locks Full setting box and Lock Toggles settings box

Selecting a Snap Mode

You can select an active (default) snap mode that always stays in effect, and, when you occasionally need a different one, you can select an override snap mode that applies only to the next tentative point.

To set snap active mode (default):

Pull-down menu	Settings > Locks > Full Select the desired snap mode from the Snap Mode option menu (see Figure 3–6).
Pull-down menu	Settings > Snaps > Button Bar Double-click the desired snap mode (see Figure 3–7).

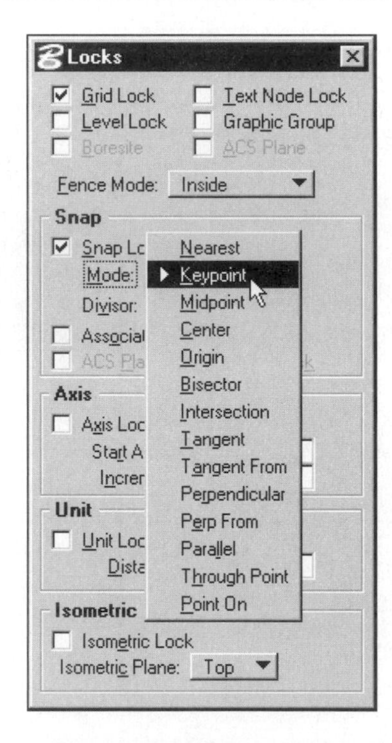

Figure 3–6 Locks settings box displaying the Snap Mode option menu

Figure 3–7 Snap mode button bar

You can also select the default snap mode from the pop-up Snap Mode menu. Hold down the Shift key and press the Tentative button—the pop-up Snap Mode appears as shown in Figure 3–8. While holding the Shift key, choose the desired Snap Mode by clicking on it.

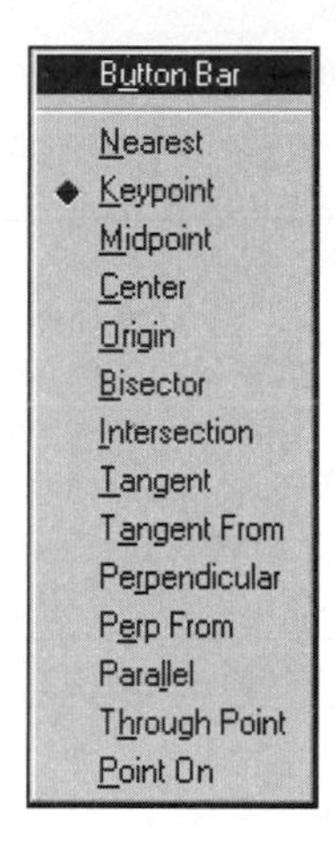

Figure 3–8 Pop-up Snap Mode menu

To set the Snap Mode override:

Pull-down menu	Settings > Snaps > Button Bar Then click the desired snap mode.
Pull-down menu	Settings > Snaps Then choose the desired Snap Mode override.

You can also select the override snap mode from the pop-up Snap Mode menu. Hold down the Shift key and press the Tentative button—the pop-up Snap Mode appears. Select the desired override snap mode from the Snap Mode menu.

MicroStation displays the snap mode icon that is in effect on the right-hand side of the Status bar. Also, MicroStation displays a diamond-shaped object to the left of the default snap active mode in the menu. If an override mode is selected, then a square appears to the left of the default snap active mode and a diamond shape object appears to the left of the current override snap mode.

Note: The number of snap modes included in the Snaps menu varies. The menu shows only the snap modes that are available for the active placement or manipulation command. Some commands do not allow all snap modes.

How to Use the Tentative Button

To use the Tentative button in placing and manipulating elements:

1. Select the placement or manipulation tool.
2. Select the snap mode you want (these modes are described later).

3. Point to the element you want to snap to, and click the Tentative button.
4. Click the Data button to accept the tentative point.
5. Continue using snap modes and tentative points as necessary to complete the placement or manipulation.

For example, if you want to start a line in the exact center of a block, select the Place Line tool, set the snap mode to Center, then click the Tentative button on the block. The block is highlighted and a large tentative cross appears at the exact center of the block. Place a data point to start the line at the block center (see Figure 3–9).

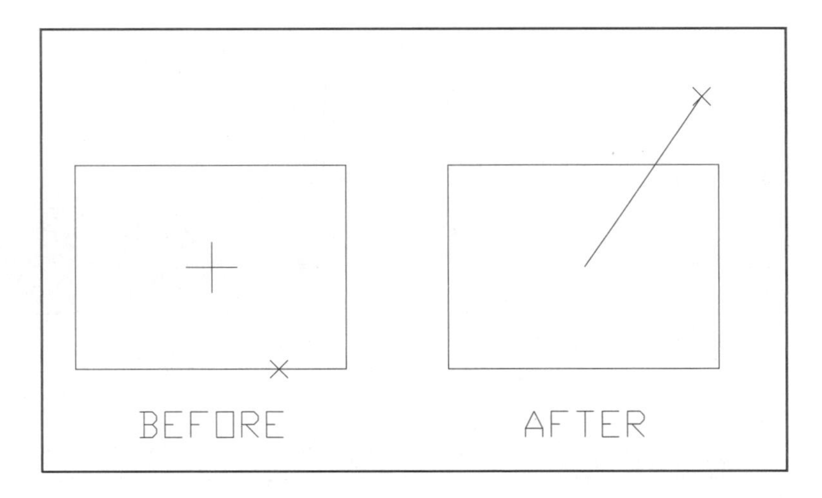

Figure 3–9 An example of a placement of the Tentative point

Keep in mind the following points when you use the tentative button.

- You can only snap to elements when Snap Lock is set to ON.
- If Grid Lock is set to ON and the element you are trying to snap to is placed between grid points, the Tentative button may snap to grid points rather than to the element you want. If that happens, set the Grid lock to OFF.
- When the Tentative button snaps to an element, the element is highlighted and the tentative cross appears at the snap to point. If the cross appears on the snap point but the element is not highlighted, you did not snap to the element; you may have snapped to a grid point close to the point you wanted.
- When you press the Tentative button, MicroStation starts searching for elements in the area immediately around the screen pointer. It selects elements in the order they were placed in the design. If the element it finds is not the one you want, just click the Tentative button again and the next element is found; there is no need to move the screen pointer location or to press the Reset button. If the Tentative button cycles through

all the elements in the area without finding the one you want, move the screen pointer closer to the element and press the Tentative button again. For example: You place a block, and then place a line starting very near one corner of the block. You need to snap to the end of the line for the next command, but the tentative point snaps to the corner of the block. Just click the Tentative button again, and it should snap to the end of the line. If the second snap also does not find the end of the line, move the screen pointer a little closer to it and snap again.

- You do not have to place the screen pointer exactly on the point of the element you want to snap to, just near it. In fact, to lessen the chance of snapping to the wrong element, it is best to move back along the element, away from other elements.

Types of Snap Modes

Following are the available snap modes.

Keypoint mode Keypoint mode allows tentative points to snap to predefined keypoints on elements. For a line, it is the endpoints of the line; for a circle, it is the center and four quadrants; for a block, it is the four corners, and so on. See Figure 3–10 for snap points for various element types. To snap to a key point on an element, position the cursor close to the key point (make sure the Snap Lock is set to ON and the keypoint mode is selected), and click the Tentative button. The tentative cross appears on the element's key point and the element highlights. If the tentative cross appears but the element does not highlight, you have not found the element's snap point. Press the Tentative button until the snap point is located, then press the Accept button.

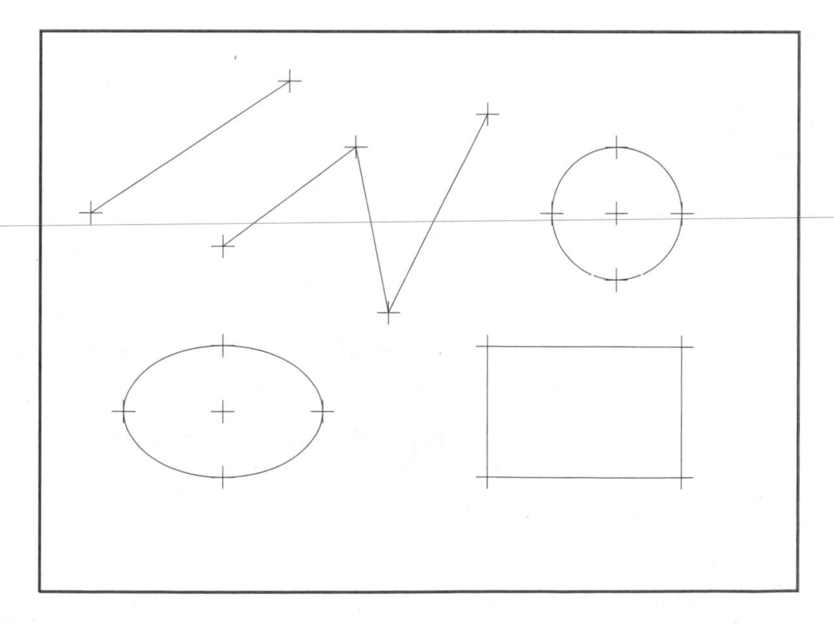

Figure 3–10 Keypoints for various element types

The *keypoint snap divisor* works with the keypoint snap mode and allows you to select additional snap points on an element by defining a value that divides the element into a specific number of divisions or parts. For example, setting the keypoint snap divisor to 5 divides an element into five equal divisions. Figure 3–11 shows the keypoint snaps for different keypoint snap divisor values. The keypoint snap divisor can be set in the Locks settings box (Full) by keying-in the value in the Divisor edit field. You can also set the value by keying-in at the key-in window: **KY=**<number of divisors> and press Enter.

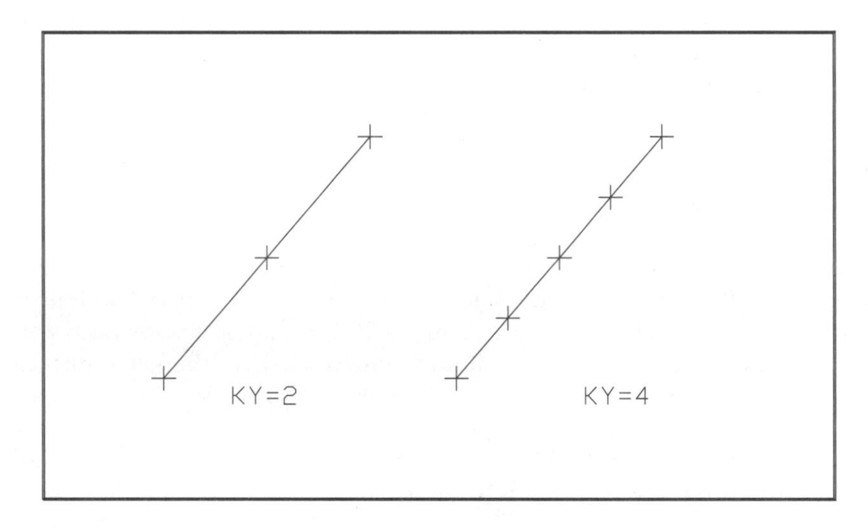

Figure 3–11 Example for various keypoint snap division

Nearest mode When active, the Nearest mode will place tentative points on any point of an element that is closest to the cursor. This rule remains the same among all element types, except as applied to text where the project point is the justification point. With the Nearest mode you are always certain that you can locate any point on any type of element. To pick a specific point on an element, position the cursor close to the point you want to select, ensure that the Snap Lock is set to ON, and that the Nearest mode is selected from the mode option menu. Press the Tentative button, and the tentative cross will appear at the closest point on the element as the element highlights. If the tentative cross appears but the element does not highlight, then you have not found the element. Press the Tentative button until a point is located, then press the Accept button.

Midpoint mode The Midpoint mode, when active, will place tentative points at the midpoint of an element or segment of a complex element (see Figure 3–12). The point position varies with different types of elements.

It bisects a line, arc, or partial ellipse.

It bisects the selected segment of a line string, block, multisided shape, or regular polygon.

It snaps to the 180-degree (9 o'clock) position of a circle or an ellipse.

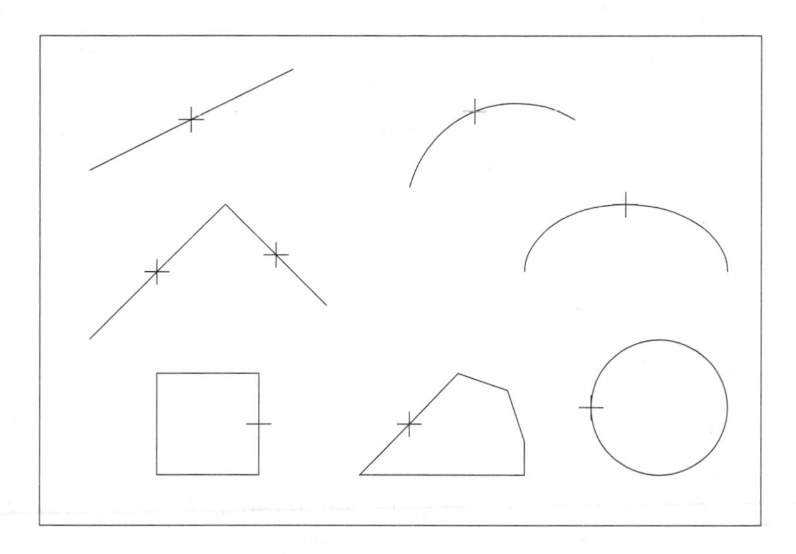

Figure 3–12 Midpoints for various element types

Center mode The Center mode causes tentative points to snap to the center of the space in the design occupied by an element (such as a circle, block, or arc); see Figure 3–13.

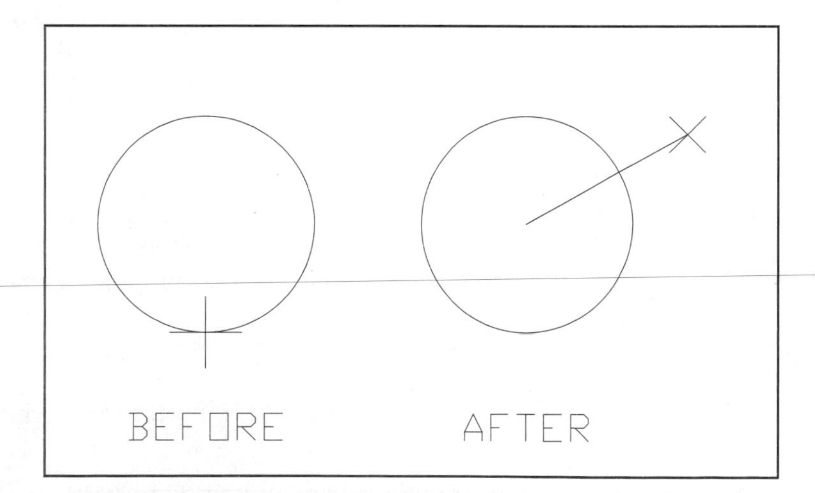

Figure 3–13 Example of snapping to the center of the circle

Intersection mode The Intersection mode causes tentative points to snap to the intersection of two elements. To find the intersection, snap to one of the intersecting elements. One or both of the elements may appear dashed while they are highlighted, but they return to normal appearance when you complete the command (see Figure 3–14).

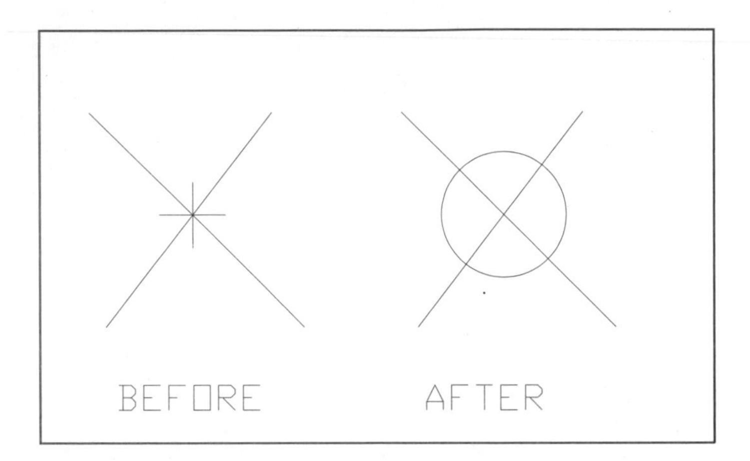

Figure 3–14 Example of snapping to the intersection of two elements in placing a circle

If the elements do not actually intersect, the tentative cross appears at the intersection of an imaginary extension of the two elements. If the two elements cannot be extended to an intersection, the message "Elements do not intersect" appears in the Status bar, and a tentative cross is not placed.

Through Point mode The Through Point mode causes tentative points to define a point on an existing element through which the element you are placing must pass (see Figure 3–15).

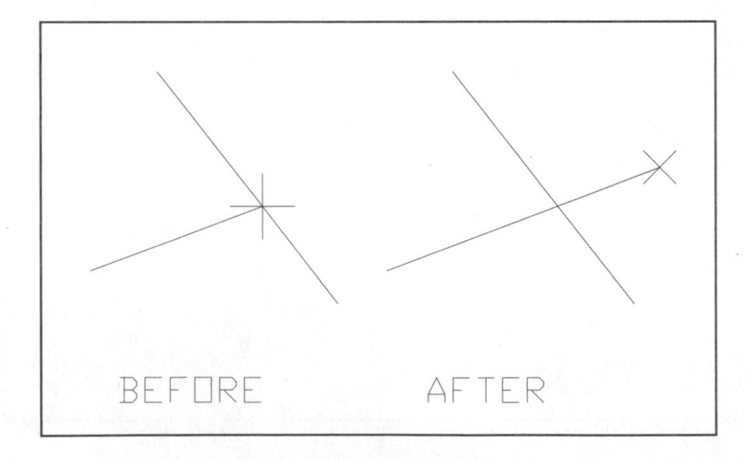

Figure 3–15 Example of snapping to a through point

Tangent mode The Tangent mode forces the element you are creating to be tangent to a nonlinear element (such as a circle, ellipse, or arc). The actual point of tangency varies, depending on how you place the element (see Figure 3–16).

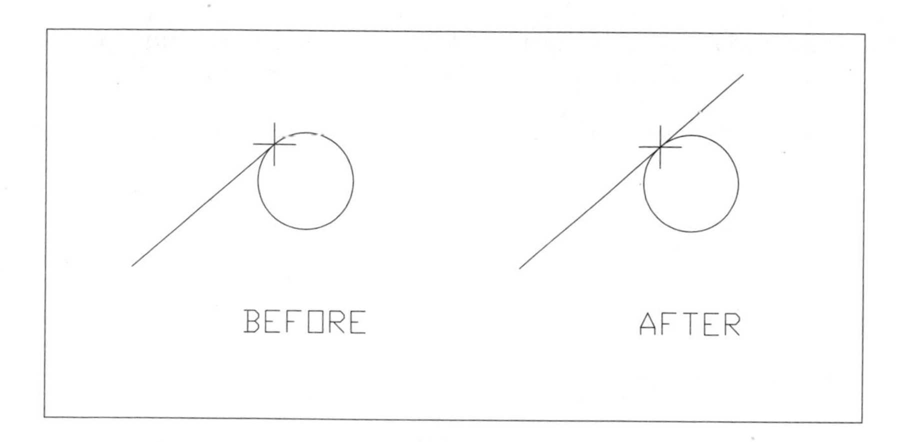

Figure 3–16 Example of snapping to a tangent point

Tangent From mode The Tangent From mode forces the element you are placing to be tangent to an existing nonlinear element (such as a circle, ellipse, or arc) at the point where you placed the tentative point (see Figure 3–17).

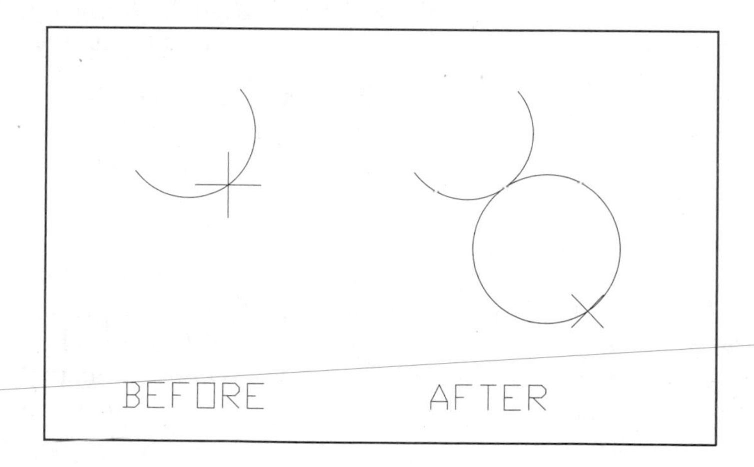

Figure 3–17 Example of snapping to a tangent on an existing nonlinear element

Origin mode The Origin mode snaps to the center of an arc, circle, origin of the text, or cell.

Bisector mode The Bisector mode sets the snap mode to bisect an element; the snap point varies with different types of elements.

Perpendicular mode The Perpendicular mode forces the element to be perpendicular to an existing element. The actual perpendicular point depends on the way the element is placed.

Perpendicular From mode The Perpendicular From mode forces the element to be perpendicular to an existing element at the point where you place the tentative point.

Parallel mode The Parallel mode forces the line or segment of the line string to be parallel to a linear element.

Point On mode The Point On mode snaps to the nearest element (after you have entered the first point of element placement) and constrains the next data point to lie on a closed element or anywhere on a linear element's line.

ELEMENT ATTRIBUTES

There are four important attributes associated with the placement of elements. The attributes include level, color, line style, and line weight.

Levels

MicroStation offers a way to group elements on levels in a manner similar to a designer drawing different parts of a design on separate transparent sheets. By stacking the transparent sheets one on top of another, the designer can see the complete drawing but can only draw on the top sheet. If the designer wants to show a customer only part of the design, he or she can remove from the stack the sheets that contain the parts of the design the customer does not need to see.

MicroStation supports the same functionality as the transparent sheets by providing you with 63 levels in each design file. For example, an architectural design in MicroStation might have the walls on one level, the dimensions on another level, electrical information on still another level, and so on. Separating parts of the design by level allows designers to turn on only the part they need to work on and to plot parts of the design separately.

You can only draw on one level at a time (the active level), but can turn ON or OFF the display of any number of levels, except the active level, in selected views. Elements on levels that are not displayed disappear from the view and cannot be plotted, but they are still in the design file. The same coordinate system and zoom factors apply to all levels. Levels are identified by numbers (1 through 63), and the right-hand side of the Status bar displays the number of the active level with LV=<level number>. If necessary, you can assign names to levels. For a more detailed explanation on naming levels, refer to Chapter 16.

When you manipulate an element, the resulting manipulation takes place on the same level on which the element was placed. For instance, a copy of an element goes on the same level as the original element, regardless of what level is currently active. The Change Element Attributes tool moves elements to different levels (discussed later).

Setting an Active Level

As just mentioned, you can have only one active level at any time, and with most tools that is the level on which new elements are placed. The level number to which the Active Level is set is shown in the Status bar and in the Primary Tools box.

To set the Active Level:

Primary Tools box	Select the appropriate level number from the Level option menu (see Figure 3–18).
Status bar	Click the Active Level field. The Set Active Level dialog box opens (see Figure 3–19). Select the appropriate level number and click on the OK button to close the dialog box.
Key-in window	**LV=**<level number> Enter

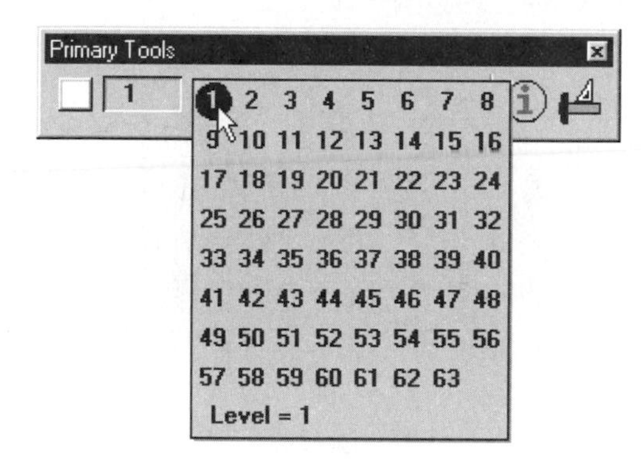

Figure 3–18 Primary Tools box—Level option menu

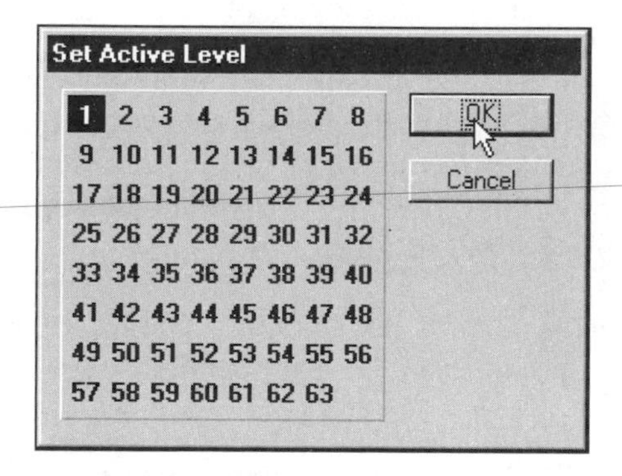

Figure 3–19 Set Active Level dialog box

Controlling the Display of Levels

The Level Display settings box controls the display of levels. You can control the display (ON or OFF) of one or more levels in one or more views.

To control the display of levels, open the View Levels settings box:

Pull-down menu	Settings > Level > Display
Key-in window	**ON=**<level numbers> Enter **OFF=**<level numbers> Enter

MicroStation displays the View Levels settings box as shown in Figure 3–20.

Figure 3–20 View Levels settings box

The settings box shows the number of each of the 63 total levels. The display of level numbers with a black background is set to ON, and the display of level numbers with a gray background is set to OFF. The View Number options menu at the top of the settings box displays the number of the view window to which the level display settings apply. The View Number menu also allows you to change the number of the view window displayed in the settings box (see the section on View Windows and View Attributes for a discussion on View setting).

The level number within a black circle is the current active level (the level on which all new elements are placed). The active level applies to all views.

To change the active level, double-click the level number you want to make active. Click once on a level number to toggle its display status between ON and OFF. To change the display status of a group of level numbers, drag the screen cursor across them while holding down the Data button on your pointing device.

Note: Display of the active level cannot be turned OFF.

The Apply button applies the current level display settings (ON or OFF) to the selected view window and applies the active level to all view windows. The All button applies the current level display settings (ON or OFF) and the active level to all views.

Note: The active level and the level display settings for each view remain in effect until you change them or exit from MicroStation. To keep them in effect for the next editing session, select Save Settings from the File pull-down menu.

Keeping Up with the Levels

Keeping up with what level everything is on can be confusing. To help overcome the confusion, MicroStation provides a level symbology table. Use the table to specify unique combinations of display color, weight, and style for each level. When level symbology is set to ON, all elements display using the symbology assigned to the level on which the elements are placed, rather than their true symbology. Chapter 14 describes the level symbology table in detail.

Element Color

The color for an element is very helpful in enabling you to differentiate between the elements on your design, especially when all the levels are displayed at one time. Before you place an element with a specific color, you have to select the active color.

To set the Active Color:

Primary Tools box	Select the appropriate color tile from the Color option menu (see Figure 3–21).
Key-in window	**CO=**<name of the color or color number anywhere from 0 to 255> Enter

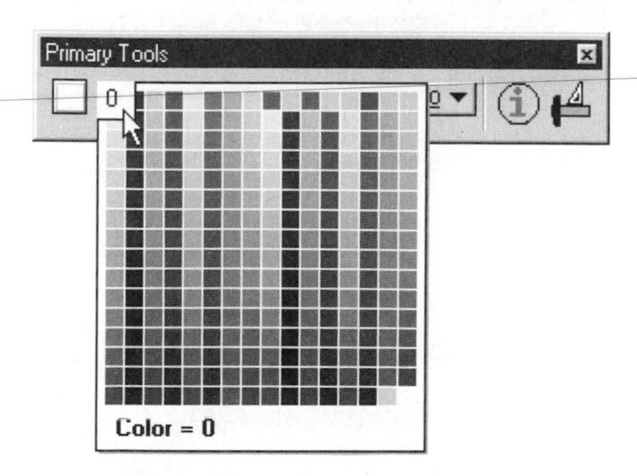

Figure 3–21 Primary Tools box—Color option menu

The actual colors shown depend on your monitor, graphics card, and what colors are defined in MicroStation's color table. Click on a color in the color palette to make it the active color.

Note: Setting up the active color does not affect elements that are already in the design plane, unless the Change Element Attributes tool (explained later in this chapter) is used.

To keep the active color in effect for future editing sessions for the current design file, select Save Settings from the File pull-down menu.

Element Line Style

Similar to color, MicroStation allows you to place elements with a specific line style (or line type). By default, MicroStation provides you with eight line styles (called internal line styles). In addition, MicroStation comes with numerous custom-made line styles. If necessary, you can change the custom-made line styles or add new ones. A detailed explanation for creating and modifying the custom-made line styles is given in Chapter 16.

Before you place an element with a specific line style, you have to make that line style an active line style.

To set the Active Line Style:

Primary Tools box	Select the appropriate Line Style from the Line Style option menu (see Figure 3–22).
Key-in window	**LC=**<name of the line style or line style number> Enter

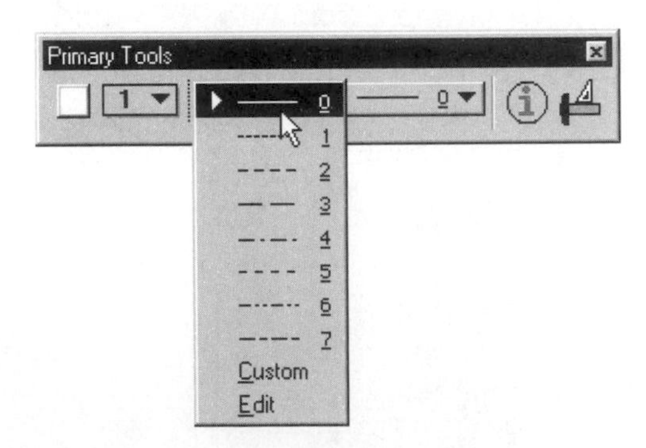

Figure 3–22 Primary Tools box—Line Style option menu

If you need to use one of the custom line styles, select Custom from the Line Style option menu, and MicroStation displays the Line Styles settings box, as shown in Figure 3–23. Select one of the available line styles from the list.

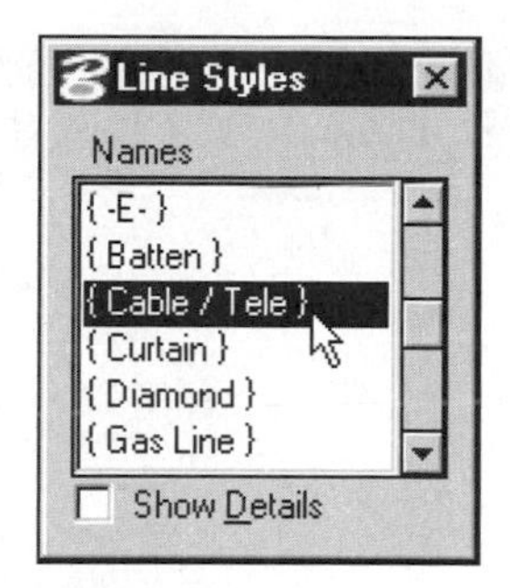

Figure 3–23 Listing of the customized line styles

Note: Selecting a new active line style does not affect elements that already exist in the design plane, unless the Change Element Attributes tool (explained later in the chapter) is used.

To keep the active line style in effect for future editing sessions for the current design file, select Save Settings from the File pull-down menu.

Element Line Weight

In MicroStation, weight refers to the width of the element. There are 32 line weights to choose from (numbered from 0 to 31), which is comparable to 32 different technical pens.

In drafting, the color, style, and even the weight (width) of the lines for creating elements in the design contribute to the "readability" or understanding of the design. For example, in a piping arrangement drawing, the line weight (width) for the pipe is the widest of all the lines on the drawing, to make the pipe stand out from the equipment, foundations, and supports.

To set the Active Element Line Weight:

Primary Tools box	Select the appropriate Line Weight from the Line Weight option menu (see Figure 3–24).
Key-in window	**WT=**<line weight number anywhere from 0 to 32> Enter

Note: Setting up the active line weight does not affect elements that already exist on the design plane, unless the Change Element Attributes tool (explained later in the chapter) is used.

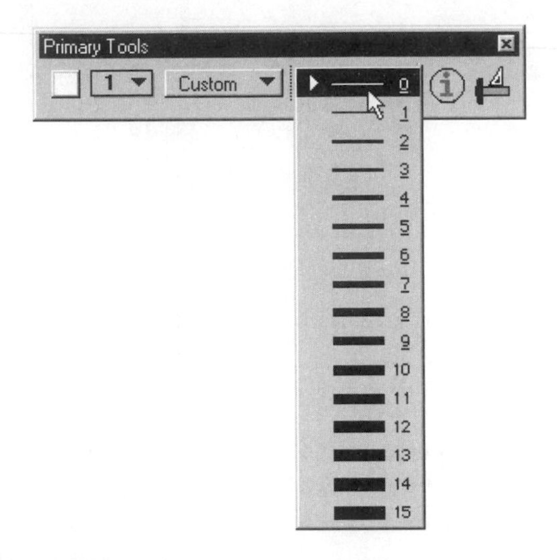

Figure 3–24 Primary Tools box—Line Weight option menu

To keep the active line weight in effect for future editing sessions for the current design file, select Save Settings from the File pull-down menu.

Element Attributes Settings Box

The Element Attributes settings box lets you control the attributes of elements, such as level, color, line style, and line weight. Instead of using four different tools (level, color, line style, and line weight) to set up the attributes, use the Element Attributes settings box to do the same thing in one place.

To set the Element Attributes:

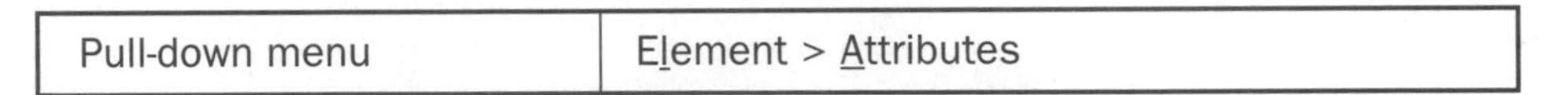

Pull-down menu	Element > Attributes

MicroStation displays the Element Attributes settings box as shown in Figure 3–25.

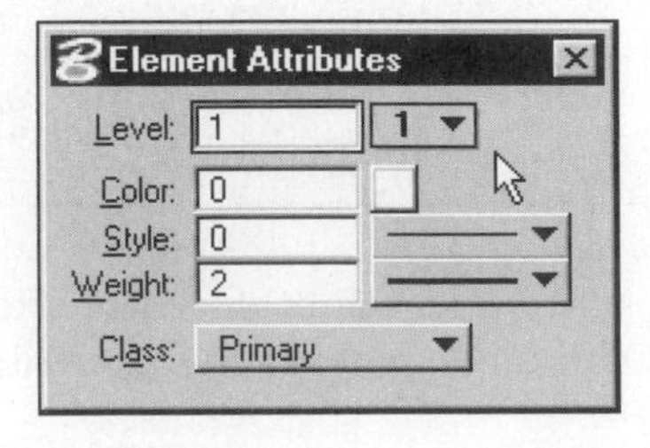

Figure 3–25 Element Attributes settings box

The Level edit field allows you to set the active level by keying-in the level number and pressing Enter or Tab.

The Color edit field allows you to set the current active color either by keying-in a numerical value in the Color edit field and pressing Enter or by choosing a color, without regard for numerical values, from the color palette. To open the color palette, which represents the active color table, click the colored button located next to the Color edit field and select the appropriate color from the palette.

The Style edit field allows you to set the current active line style either by keying-in a numerical value in the Style edit field and pressing Enter or by choosing a line style, without regard for numerical values, from the option menu located next to the Style edit field.

The Weight edit field allows you to set the current active line weight either by keying-in a numerical value in the Weight edit field and pressing Enter or by choosing a line weight, without regard for numerical values, from the option menu located next to the Weight edit field.

The Class option menu specifies the class of an element upon placement. Two options are available, Primary and Construction. Primary elements, the default option, are the elements that comprise the design. Construction elements are placed in the design plane, then employed as an aid to placing the primary elements. You create geometric constructions with fundamental entities such as lines, circles, and arcs to generate intersections, endpoints, centers, points of tangency, midpoints, and other useful data that might take a manual drafter considerable time to calculate or hand-measure on the board. From these you can create primary elements using intersections or other data generated from the construction elements. When the design is complete, display of the construction elements can be set to OFF from the View Attributes settings box (see the section on View Windows and View Attributes).

Note: If construction elements are displayed in a plotted view, they also plot. Set them to OFF before creating a plot of the finished drawing.

Change Element Attributes—Change Symbology

The Change Element Attributes tool enables you to change an element to the active element attributes (level, color, line style, line weight, and class).

Invoke the Change Element Attributes tool from:

Change Attributes tool box	Select the Change Element Attributes tool (see Figure 3–26).

Figure 3–26 Invoking the Change Element Attributes tool from the Change Attributes tool box

Turn on the toggle button to Level, Color, Style, Weight, or Class, and then make the necessary changes to the settings in the Tool Settings window. MicroStation prompts:

Change Element Attributes > Identify element *(Identify the element to make the necessary changes to the attributes.)*

Change Element Attributes > Accept/Reject (select next input) *(Click the Accept button to accept the changes for the selected element, select another element, or click the Reject button to reject the changes.)*

Match Element Attributes

In addition to being able to change the attributes of an element to the active settings, you can change the active attributes (level, color, style, and weight) to those that were in effect when an existing element was created, with the Match Element Attributes tool. This tool provides a quick way to return to placing elements with the same attributes as elements you placed earlier in the design.

Invoke the Match Element Attributes tool:

Change Attributes tool box	Select the Match Element Attributes tool (see Figure 3–27).

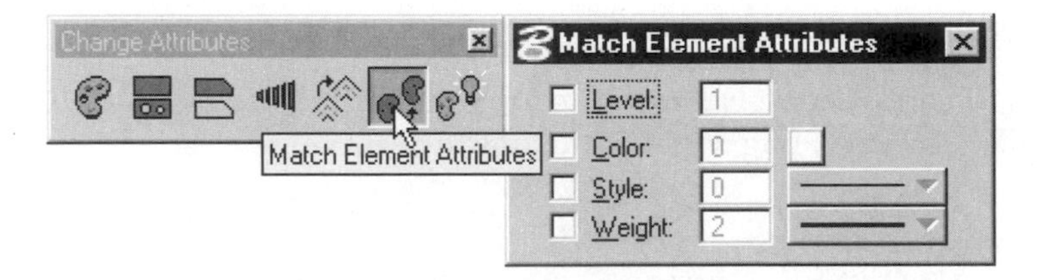

Figure 3–27 Invoking the Match Element Attributes tool from the Change Attributes tool box

Turn on the toggle button to the Level, Color, Style, or Weight to which you want the element to match in the Tool Settings window. MicroStation prompts:

Match Element Attributes > Identify element *(Identify the element to which you want to match the attributes.)*

Match Element Attributes > Accept/Reject (select next input) *(Click the Accept button to accept the changes for the selected element, select another element, or click the Reject button to reject the changes.)*

VIEW CONTROL

There are many ways to view a drawing in MicroStation. With the View Control tools you can select the portion of the drawing to be displayed. By letting you see your drawing in different ways, MicroStation gives you the means to draw more quickly, easily, and accurately.

The tools explained in this section are utility tools; they make your job easier and help you to draw more accurately.

The View Control tools can be selected from the View Control Bar located on the lower-left corner of the view window border, as shown in Figure 3—28; from the 2D View Control tool box (invoked from the Tools pull-down menu), as shown in Figure 3–29; or from the pop-up menu, as shown in Figure 3–30, that appears when you hold the Shift key and the Reset button at the same time. Selecting a View Control tool from the View Control Bar designates the view whose border contains the bar as the view on which to operate. If, instead, you select a View Control tool from the pop-up menu or from the 2D View Control tool box, then MicroStation prompts you to select a view window on which to operate.

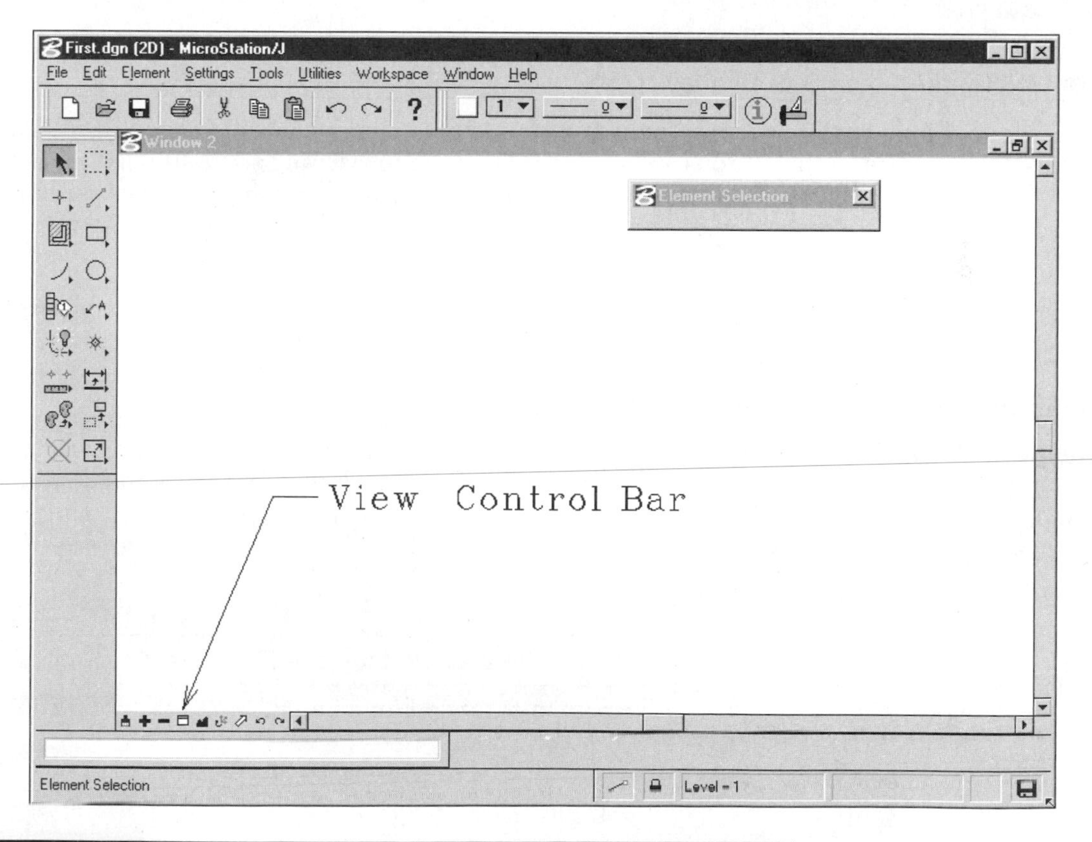

Figure 3–28 View Control Bar (lower left corner of the view window)

Figure 3–29 2D View Control tool box

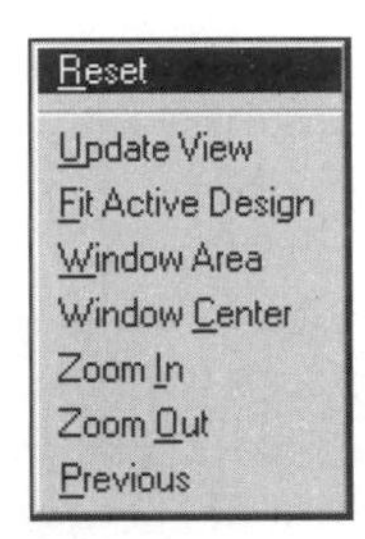

Figure 3–30 Pop-up menu

Controlling the Amount of Display

The amount of information that can be displayed in a view can be controlled in a way similar to using a zoom lens on a camera. You can increase or decrease the viewing area, although the actual size of the object remains constant. As you increase the visible size of the object, you view a smaller area of the drawing in greater detail. As you decrease the visible size of the object, you view a larger area. This ability gives greater accuracy and detail.

MicroStation provides three tools that control the amount of information that can be displayed on the screen view: Zoom In, Zoom Out, and Window Area.

Zoom In

The Zoom In tool increases the visible size of objects, allowing you to view a smaller area of the drawing in greater detail. The Zoom Ratio sets the factor by which the view is magnified, with the default value being set to 2. If necessary, you can change the Zoom Ratio between 1 and 50 by keying-in the appropriate value in the Tool Settings window.

Invoke the Zoom In tool:

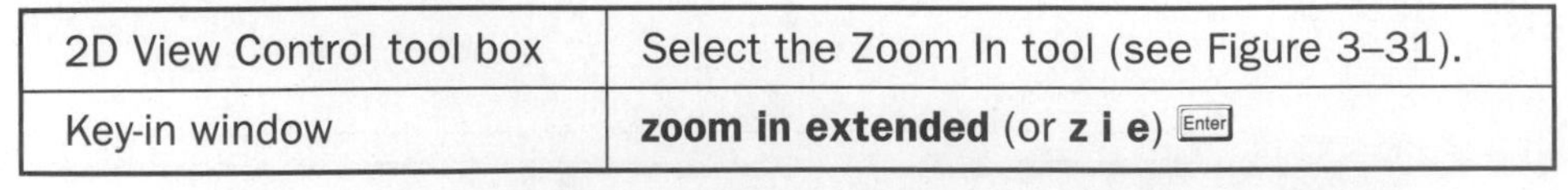

2D View Control tool box	Select the Zoom In tool (see Figure 3–31).
Key-in window	**zoom in extended** (or **z i e**) Enter

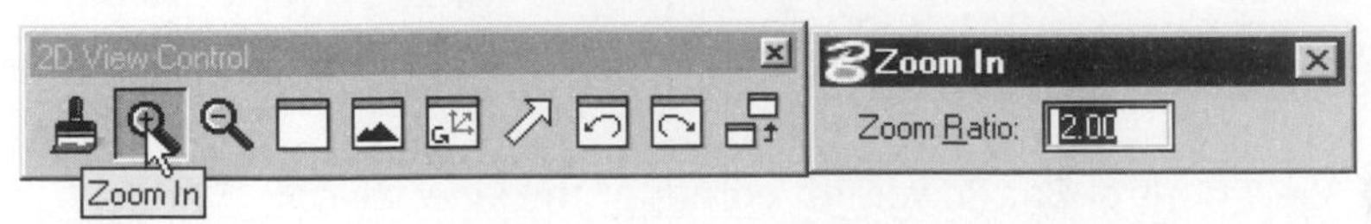

Figure 3–31 Invoking the Zoom In tool from the 2D View Control tool box

MicroStation prompts:

Zoom In > Enter zoom center point

When you move the pointer in the view, a rectangular box is displayed that indicates the new view boundary. Place a data point to define the center of the area of the view window to be displayed (see Figures 3–32a and 3–32b). You may continue defining data points to magnify further, or you may invoke a new tool.

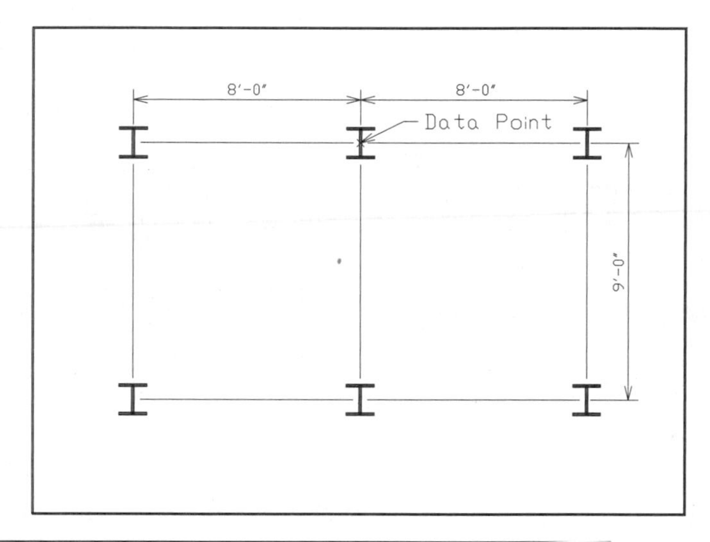

Figure 3–32a The design shown before the Zoom In tool is invoked

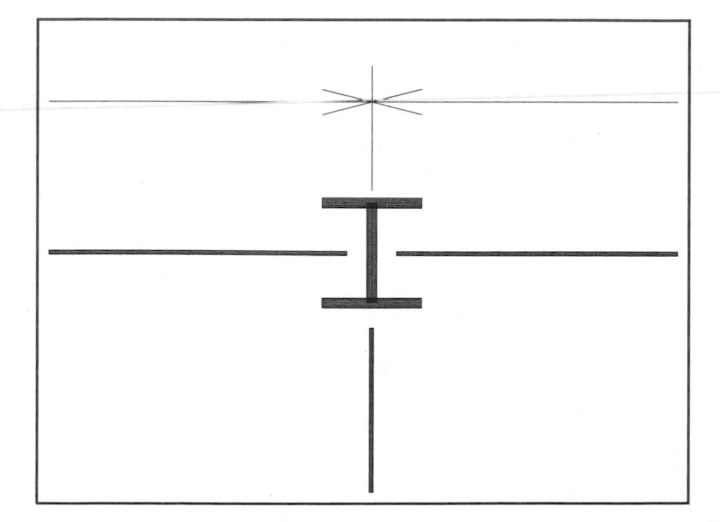

Figure 3–32b The design shown after the Zoom In tool is invoked

Zoom Out

The Zoom Out tool decreases the visible size of objects, allowing you to view a larger area of the drawing. The Zoom Ratio sets the factor by which the view magnification is decreased, and the default value is set to 2. If necessary, you can change the Zoom Ratio between 1 and 50 by keying-in the appropriate value in the Tool Settings window.

Invoke the Zoom Out tool:

2D View Control tool box	Select the Zoom Out tool (see Figure 3–33).
Key-in window	**zoom out extended** (or **z o e**) Enter

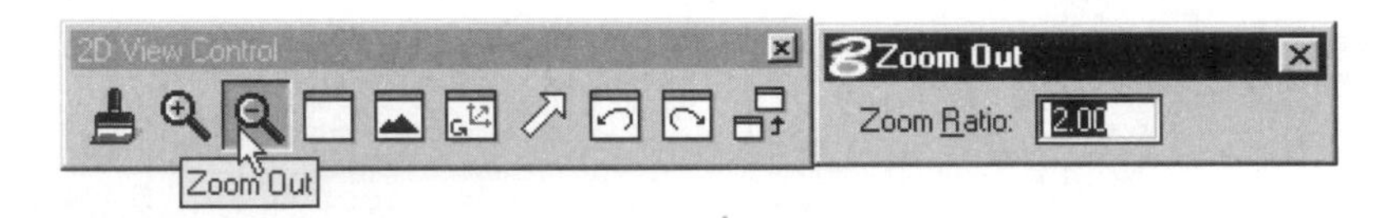

Figure 3–33 Invoking the Zoom Out tool from the 2D View Control tool box

MicroStation prompts:

Zoom Out > Enter zoom center point

Place a data point to define the center of the area of the view window with the decreased magnification (see Figures 3–34a and 3–34b). You may continue defining data points to decrease magnification further, or you may invoke a new tool.

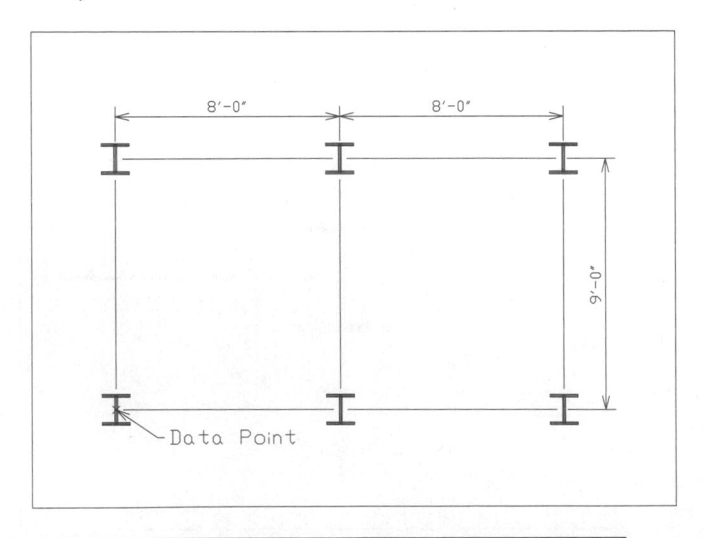

Figure 3–34a The design shown before the Zoom Out tool is invoked

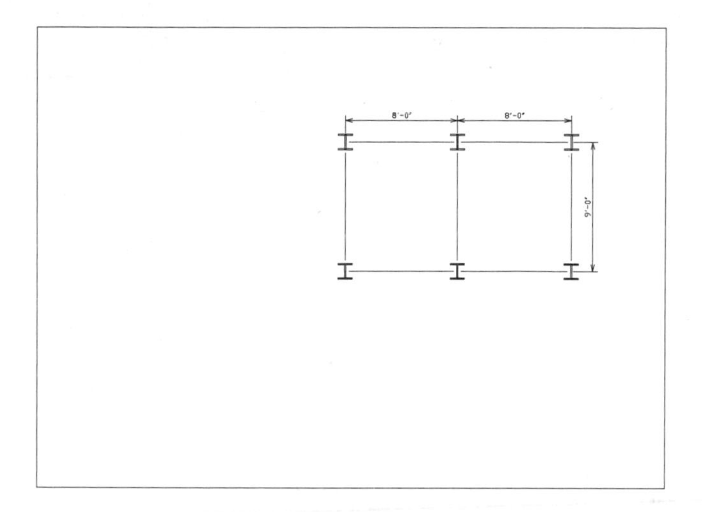

Figure 3–34b The design shown after the Zoom Out tool is invoked

Window Area

The Window Area tool allows you to specify an area of the design you wish to magnify by placing two opposite corner points of a rectangular window. The center of the area selected becomes the new display center, and the area inside the window is enlarged to fill the display as completely as possible. If necessary, you can change the destination window in the Tool Settings window.

Invoke the Window Area tool from:

2D View Control tool box	Select the Window Area tool (see Figure 3–35).
Key-in window	**window area extended** (or **w a e**) Enter

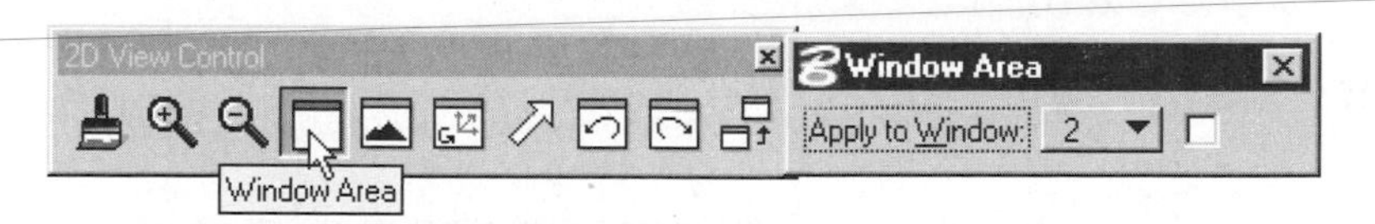

Figure 3–35 Invoking the Window Area tool from the 2D View Control tool box

MicroStation prompts:

Window Area > Define first corner point *(A full screen crosshair appears. Place a data point or key-in coordinates to define the first corner point.)*

Window Area > Define opposite corner point *(Place a data point or key-in coordinates to define the opposite corner point.)*

MicroStation updates the contents of the window in the destination window view (see Figures 3–36a and 3–36b). You may continue using the Window Area tool by defining another area and displaying in the window view, or you may select a new tool.

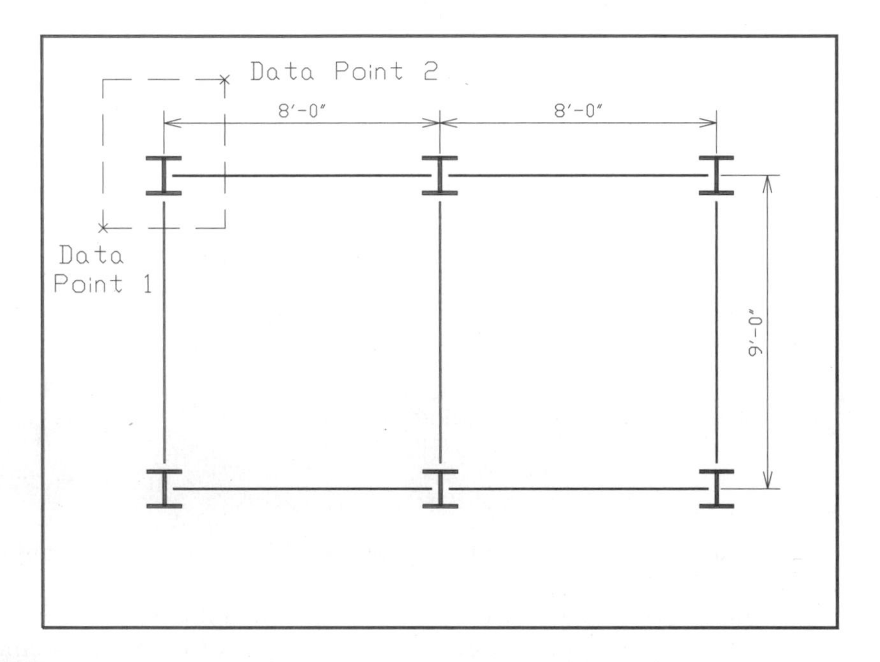

Figure 3–36a The design shown before the Window Area tool is invoked

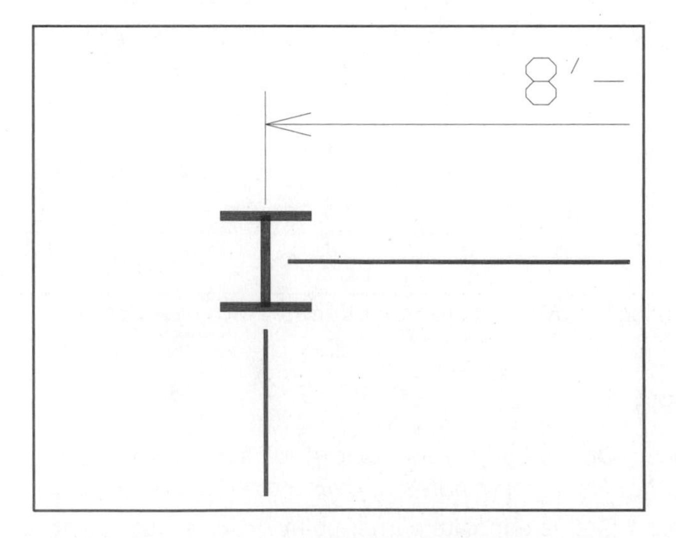

Figure 3–36b The design shown after the Window Area tool is invoked

Fit View

The Fit View tool lets you see the entire design. In a plan view, it zooms to show the entire design drawn on the design plane.

Invoke the Fit View tool from:

2D View Control tool box	Select the Fit View tool (see Figure 3–37).
Key-in window	**fit view extended** (or **fit v e**) Enter

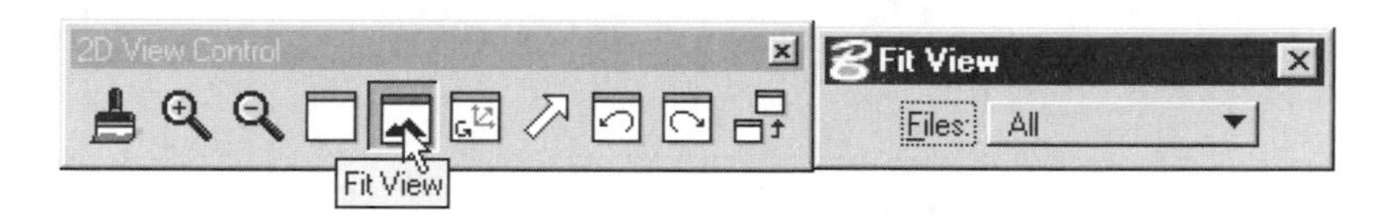

Figure 3–37 Invoking the Fit View tool from the 2D View Control tool box

MicroStation prompts:

> Fit View > Select view to fit *(Place a data point anywhere in the view to display the design.)*

MicroStation updates, showing all elements in the design plane (see Figures 3–38a and 3–38b).

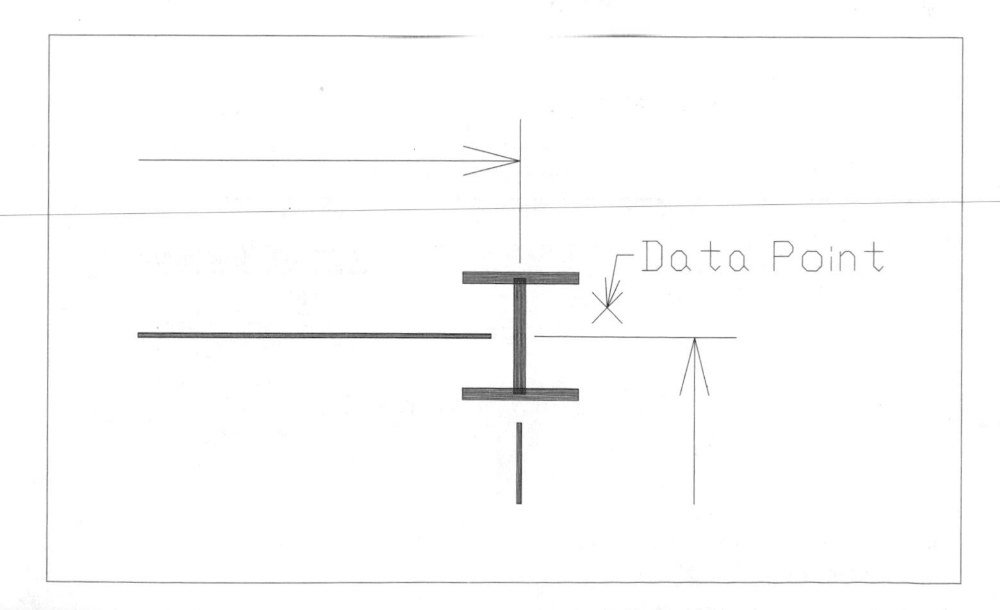

Figure 3–38a The design shown before the Fit View tool is invoked

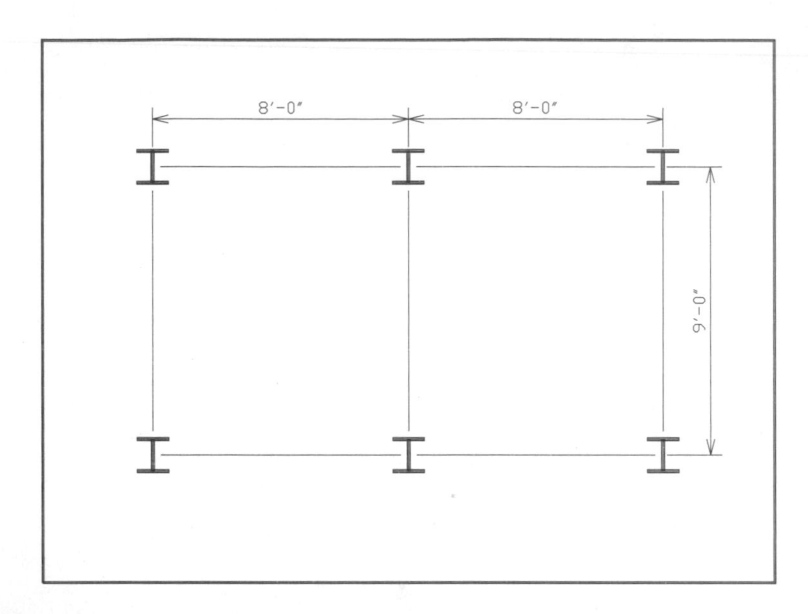

Figure 3–38b The design shown after the Fit View tool is invoked

Pan View

The Pan View tool lets you view a different portion of the design in the current view, without changing the magnification. You can move your viewing area to see details that are currently off-screen.

Invoke the Pan View tool:

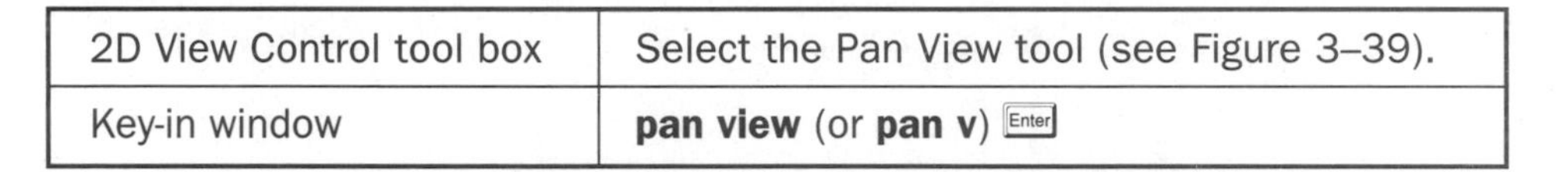

2D View Control tool box	Select the Pan View tool (see Figure 3–39).
Key-in window	**pan view** (or **pan v**) Enter

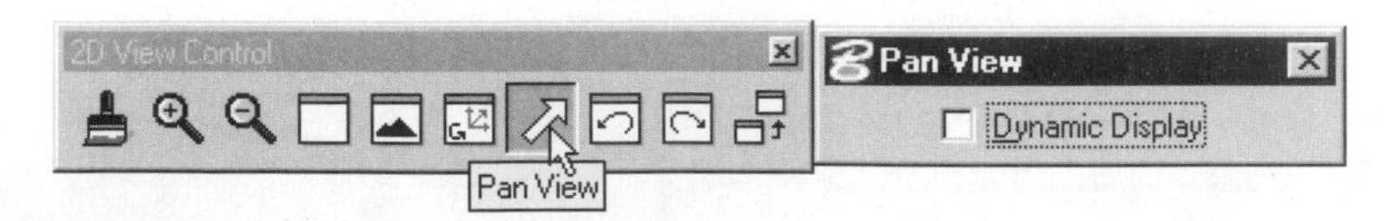

Figure 3–39 Invoking the Pan View tool from the 2D View Control tool box

MicroStation prompts:

Pan View > Select view *(Place a data point to select the view to pan and to define the origin for panning.)*

Pan View > Define amount of panning *(Place a data point to define the position where you want the origin to be displayed.)*

In addition, you can also perform dynamic panning: Hold the Shift key, press the Data button (the pointer location becomes the anchor point for panning), and drag the pointer in the direction to pan. When panning begins, the Shift key can be released. You can drag the pointer in any direction. Release the Data button to terminate the panning. The panning speed increases as the pointer is dragged farther away from the anchor point.

View Previous

The View Previous tool displays the last displayed view. You can restore back up to the previous six views.

Invoke the View Previous tool:

2D View Control tool box	Select the View Previous tool (see Figure 3–40).
Key-in window	**view previous** (or **vi p**) Enter

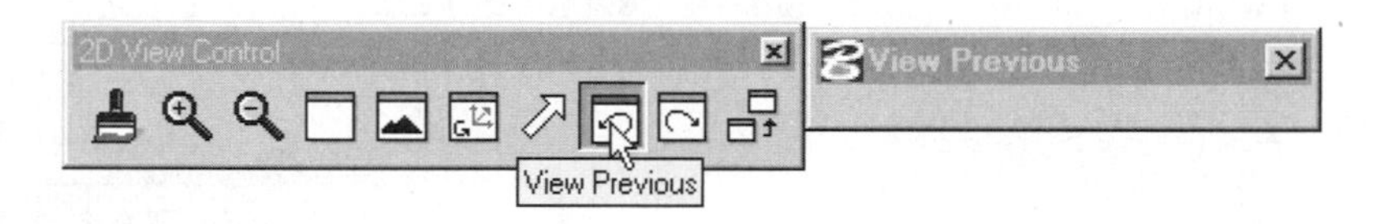

Figure 3–40 Invoking the View Previous tool from the 2D View Control tool box

MicroStation prompts:

View Previous > Select view *(Place a data point in a view to restore the previous view.)*

View Next

The View Next tool negates the view that was displayed by the View Previous tool.

Invoke the View Next tool:

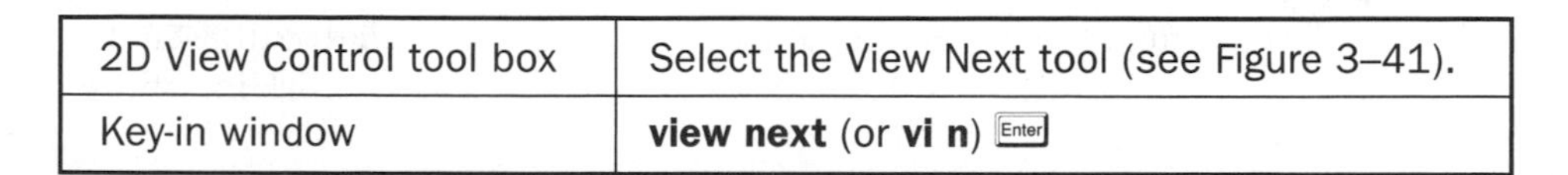

2D View Control tool box	Select the View Next tool (see Figure 3–41).
Key-in window	**view next** (or **vi n**) Enter

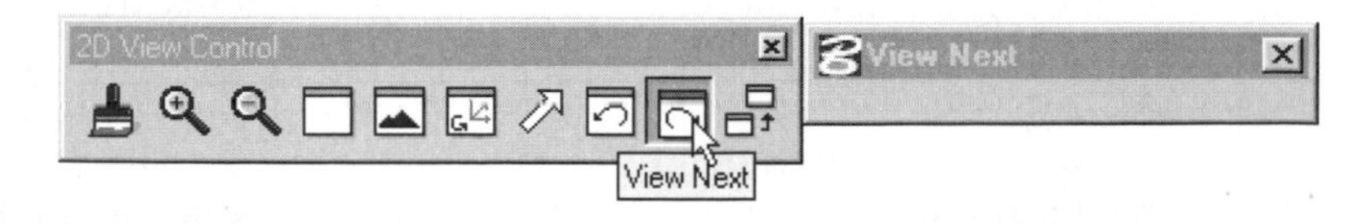

Figure 3–41 Invoking the View Next tool from the 2D View Control tool box

MicroStation prompts:

View Next > Select view *(Place a data point in a view to negate the view that was displayed by the View Previous tool.)*

Update View

The Update View tool instructs the computer to redraw the on-screen image. You can use this tool whenever you see an incomplete image of your design. If you delete an object on the display, there may be gaps in the outline of other elements it crossed or there may be grid dots that do not show up after deleting. If you update the display, the grid dots will be refreshed and all elements will be repainted.

Invoke the Update View tool:

2D View Control tool box	Select the Update View tool (see Figure 3–42).
Key-in window	**update view** (or **up**) Enter

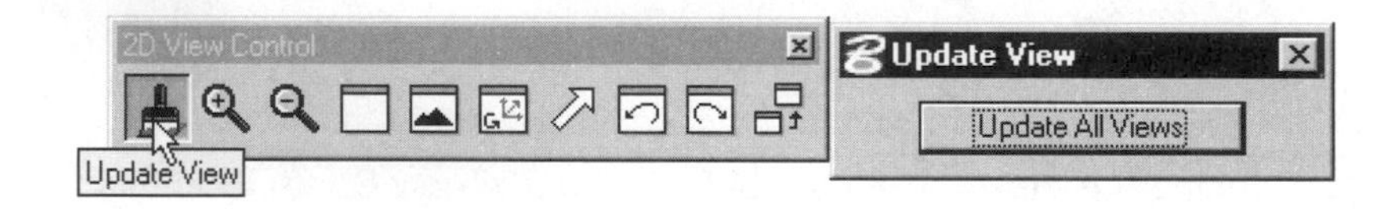

Figure 3–42 Invoking the Update View tool from the 2D View Control tool box

MicroStation prompts:

Update View > Select view *(Place a data point to update the view.)*

VIEW WINDOWS AND VIEW ATTRIBUTES

Thus far, you have been working in only one view window. That may have forced you to spend a lot of time using the view commands to set up the view for the areas you needed to draw in. MicroStation actually provides eight separate view windows (or cameras) that let you work in different parts of your design at the same time.

Each view window is identified by its view number (1–8). The view windows are similar to having eight zoom lens cameras that can be pointed at different parts of your design. For instance, in one view window you might display the entire drawing; in two other view windows you might be zoomed in close to widely separated design areas to show great detail (Figure 3–43). All eight view windows can be opened at the same time on your monitor or on either monitor of a two-monitor workstation.

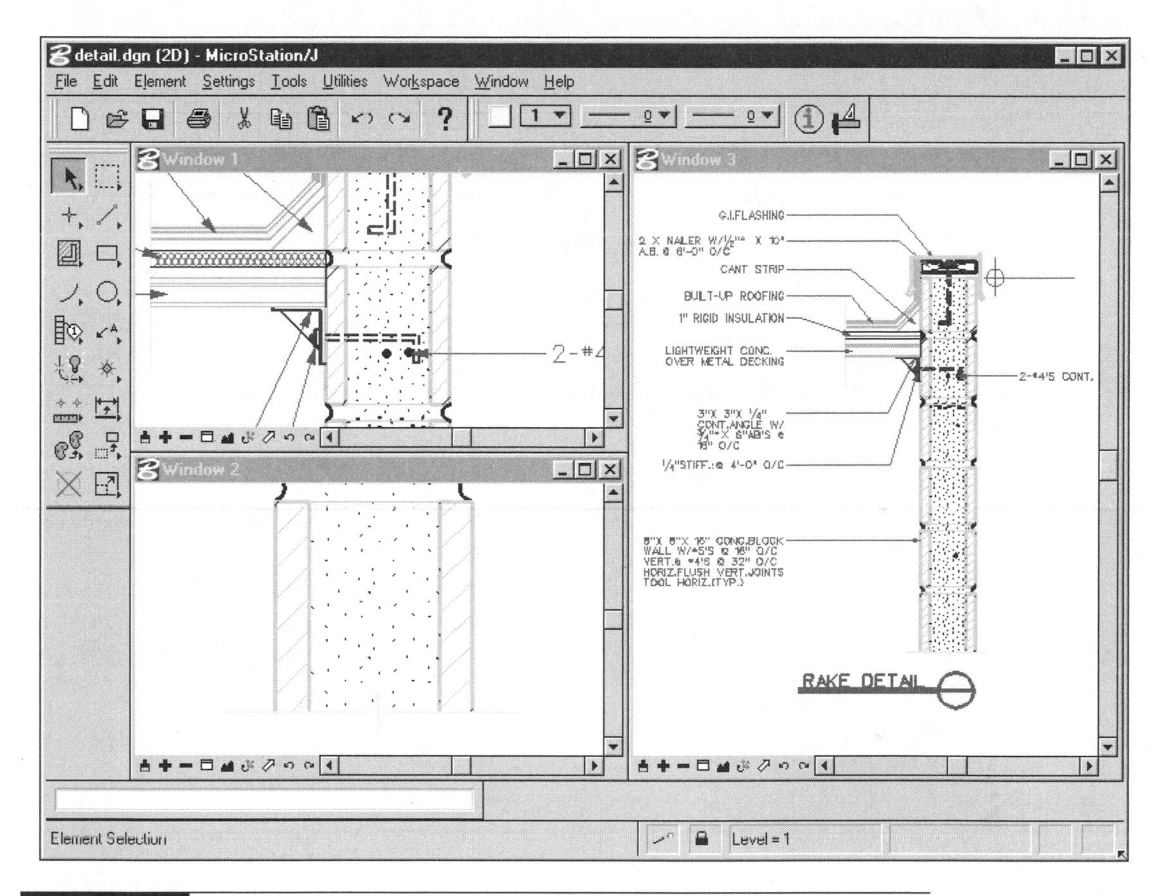

Figure 3–43 Example of three views showing different portions of a design.

The View Control tools you have already been introduced to (Zoom, Fit, Area, Center, and Update) work on any open view window. Let's look now at the View tools that help you position and size the eight view windows.

Opening and Closing View Windows

View windows are opened and closed either from the Open/Close submenu in the Windows pull-down menu, from the Open/Close dialog box, or by key-in commands.

To open or close one view window, select:

Pull-down menu	Window > Open/Close > View #
Key-in window	**view off** (or **vi off**) or **view on** (or **vi on**) Enter

To open or close several view windows quickly, select:

Pull-down menu	Window Open/Close, Dialog (see Figure 3–44).
Key-in window	**view off** (or **vi off**) or **view on** (or **vi on**) Enter

Figure 3–44 View of the Open/Close dialog box showing view windows 1 and 3 open

In addition to the Open/Close options just discussed, most operating systems provide a tool for closing a window in the window's "Control" menu. Figure 3–45 shows an example of such a window.

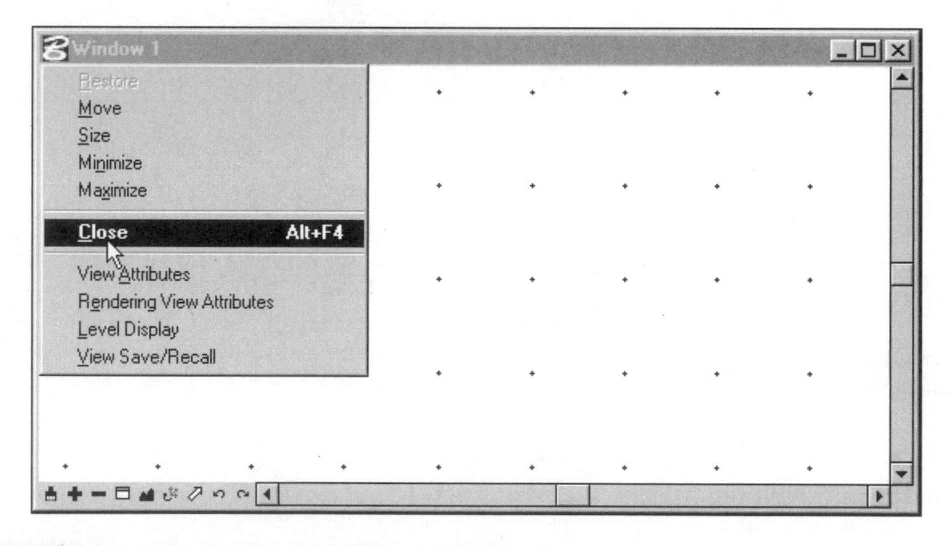

Figure 3–45 Window menu in Microsoft Windows 98 view window

Note: The view windows you open and close apply only to the current editing session. If you want the current arrangement of the view windows to be the same the next time you open the design file in MicroStation, select the Save Settings option from the File pull-down menu.

Arranging Open View Windows

The working area can become a little cluttered when several view windows are open, and MicroStation provides three housekeeping tools for cleaning up the clutter: Cascade, Tile, and Arrange. The tools are provided in the Window pull-down menu, as shown in Figure 3–46.

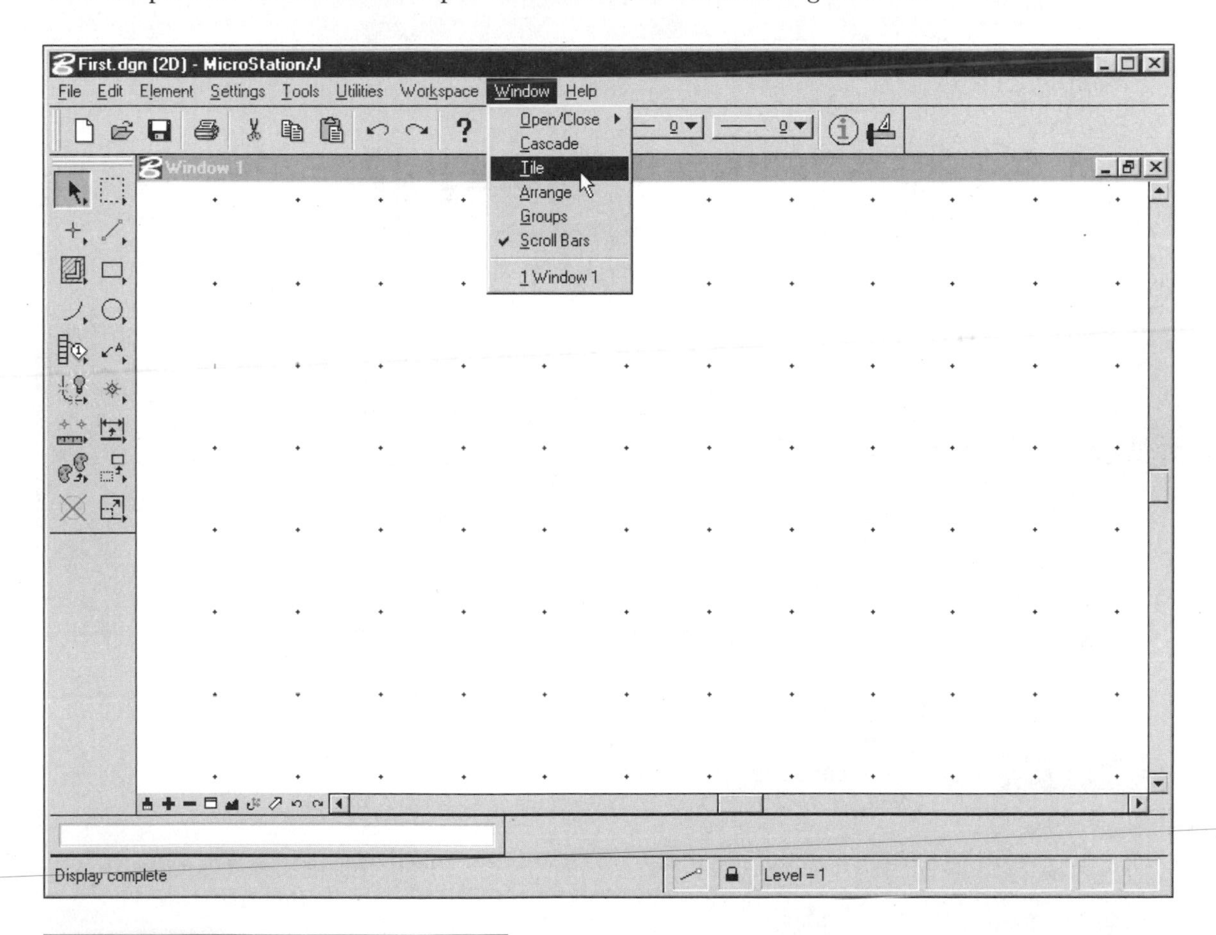

Figure 3-46 Window pull-down menu

Cascade

The Cascade tool stacks all open view windows in numerical order, with the lowest-numbered view window on top and the other view window title bars visible behind it, as shown in Figure 3–47.

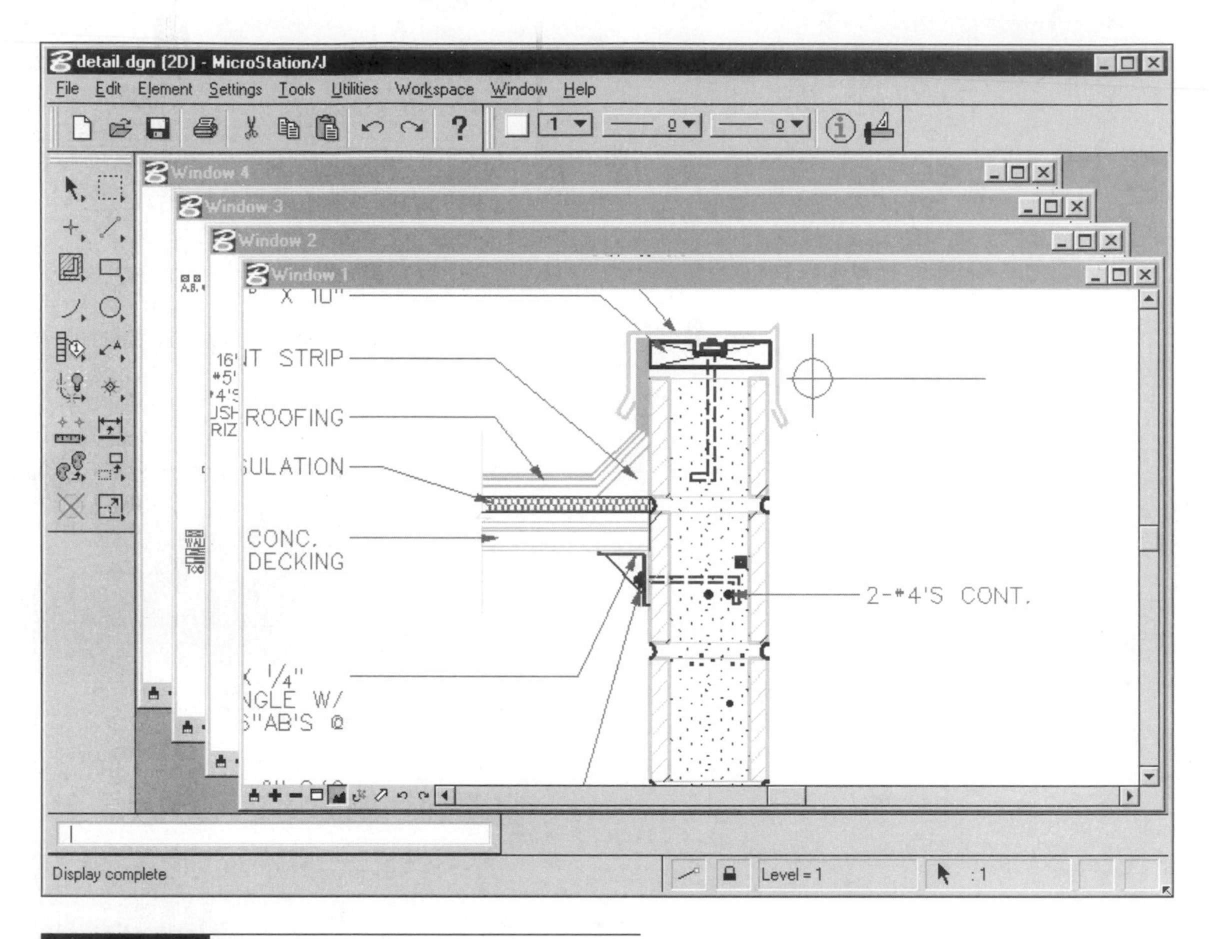

Figure 3–47 Example of cascaded view windows

Note: The view arrangement commands do not place any part of the views behind tool boxes that are attached to the side of the MicroStation workspace, as shown in Figures 3–47 and 3–48.

To cascade the open view windows, select the Cascade tool:

Pull-down menu	Window > Cascade
Key-in window	**window cascade** (or **w c**) Enter

The open view windows are cascaded (there are no MicroStation prompts). To work on a specific view, just click on the Title bar and it will pop up to the top.

Tile

The Tile tool arranges all open view windows side by side in a tiled fashion, with the lowest-numbered view window in the upper left, as shown in Figure 3–48.

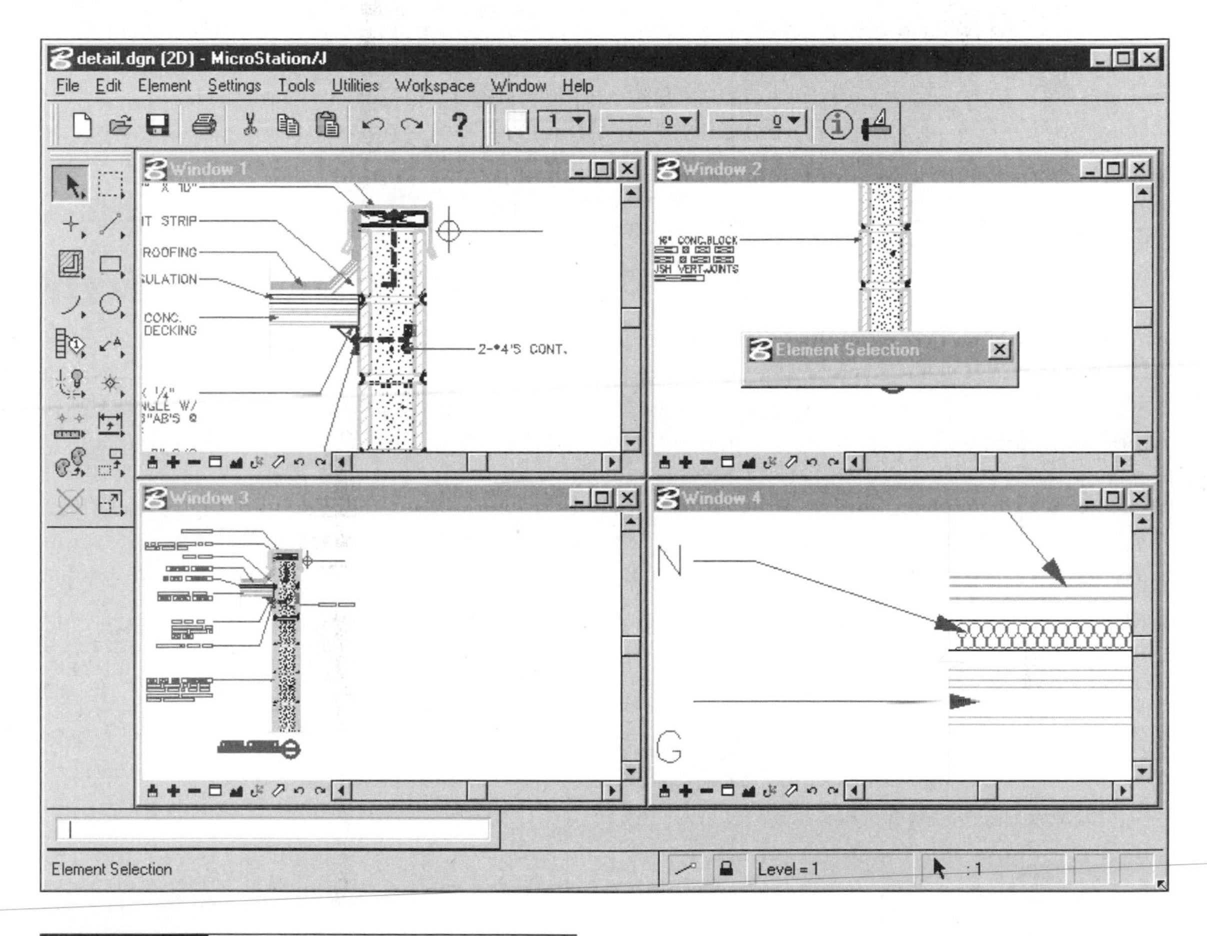

Figure 3–48 Example of tiled view windows

To tile the open view windows, select the Tile tool:

Pull-down menu	Window > Tile
Key-in window	**window tile** (or **w t**) Enter

The open view windows are tiled (there are no MicroStation prompts).

Arrange

The Arrange tool sizes and moves all open view windows as necessary to fill the MicroStation application window. The tool attempts to keep each view window as close to its original size and position as possible.

To arrange the open view windows, select the Arrange tool:

Pull-down menu	Window > Arrange
Key-in window	**window arrange** (or **w arr**) Enter

The open view windows are arranged to fill the MicroStation application window (there are no MicroStation prompts).

Arranging Individual View Windows

MicroStation provides a group of tools for controlling the size and position of individual view windows, in addition to the tools to open and close windows. You can move a view window to a new location in the workspace, resize it, minimize and maximize it, and pop it to the top when it is buried under a stack of other view windows.

Moving a View Window

To move a window to a new location in the MicroStation workspace:

1. Point to the window's title bar.
2. Press and hold the Data button.
3. Drag the window to the new location.
4. Release the Data button.

When you point to the title bar, the screen pointer changes to a different shape. As you drag the window, a dynamic of the window (or its outline) follows the screen pointer.

Resizing a View Window

To change the size of a window:

1. Point to the window border.
2. Press and hold the Data button.
3. Drag the border to a new position.
4. Release the Data button.

When you point to a window border, the screen pointer changes to a different shape. As you drag the window border, a dynamic of the window (or its outline) follows the screen pointer.

If you grab the border on a side, you can change the position of only that one border. If you grab the border on a corner, you can change the position of the two adjacent borders at the same time.

Minimizing and Maximizing a View Window

View windows contain a "minimize" button and a "maximize" button on the title bar (see Figure 3–49).

- Click the minimize button to reduce the view window to its minimum possible size.
- Click the maximize button to expand the view window to fill the MicroStation workspace and cover all other view windows.

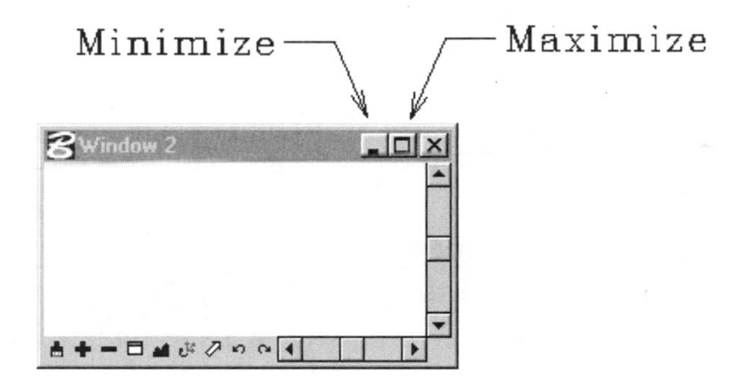

Figure 3-49 Minimize and Maximize buttons in Microsoft Windows 98

To return a view to the size it was before you clicked one of these buttons, click the button again. For example, if you maximized a view window, click its maximize button to return it to the size it was before being maximized.

Note: The appearance of the title bar buttons and of the window move and resize pointers varies among operating systems. For example, the resize pointer in Microsoft Windows 98 has a two-headed arrow pointing in the directions the window border can be dragged.

Finding a View Window

The tools to change the size and position of a view window can cause it to cover all or part of other view windows, thus creating a "stack" of windows. There are two ways to bring a buried window to the top of the view window stack:

- If you can see any part of the buried view window's title border or title bar, click on it.
- Open the Window pull-down menu, then click on the view name at the bottom of the pull-down menu. The names of all open view windows are displayed there.

Turning the View Window Scroll Bar ON and OFF

Each view window contains a horizontal and vertical scroll bar that can be used to position the view anywhere within the design plane. These bars can be set to OFF to obtain a little more design plane area display in each view window.

To turn view window slider bars ON and OFF, select:

Pull-down menu	Window > Scroll Bars

The command is a toggle switch that switches the state of the scroll bars in all open view windows every time you select it. The slider bars are turned ON or OFF immediately (there are no MicroStation prompts).

Note: The view control tools in the bottom left corner of the view windows also disappear when the scroll bars are set to OFF. To access the view control tools, open the View Control tool box from the Tools pull-down menu.

Creating and Using View Window Groups

MicroStation provides tools for creating and using named groups of windows in a design file. The groups can be opened and the view window control tools applied only to a selected group.

To open the Window Groups dialog box, select:

Pull-down menu	Window > Groups

The Window Groups dialog box opens and the Groups tools are available. Figure 3–50 presents an example of a Window Groups dialog box with two named groups; Table 3–1 describes the group commands.

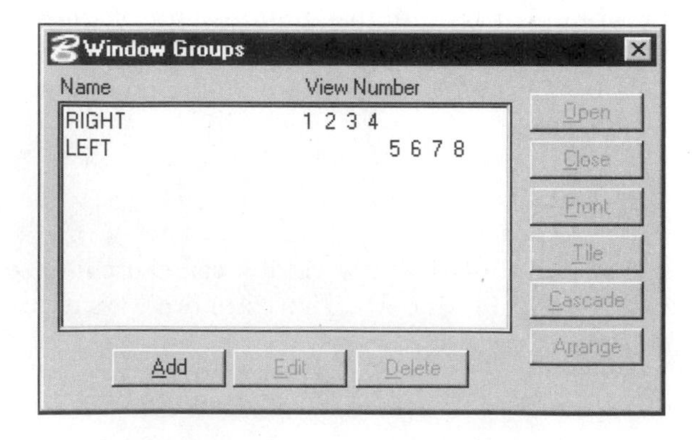

Figure 3–50 Window Groups dialog box

Table 3–1 Window Groups Dialog Box Parts

PART	DESCRIPTION
Names	Contains the group names and view window numbers in each group. Click on the group to select it for use.
Open	Opens the selected group's view windows. The views open at the same size and position as the last time they were opened, or, if this is the first opening in the editing session, with the same size and position as they were when settings were last saved.
Close	Closes all view windows assigned to the selected group.
Front	Brings the selected group's view windows to the front of other view windows.
Tile	Tiles only the selected group's view windows. Other open view windows are not tiled.
Cascade	Cascades only the selected group's view windows. Other open view windows are not cascaded.
Arrange	Arranges only the selected group's view windows to fill the MicroStation workspace. Other open view windows are not arranged.
Add	Opens the Edit Window Group dialog box, where you can create a new group name and assign view window numbers to it (see Figure 3–51).
Edit	Opens the Edit Window Group dialog box for the selected group, so you can edit the group name and assigned view window numbers.
Delete	Deletes the selected group.

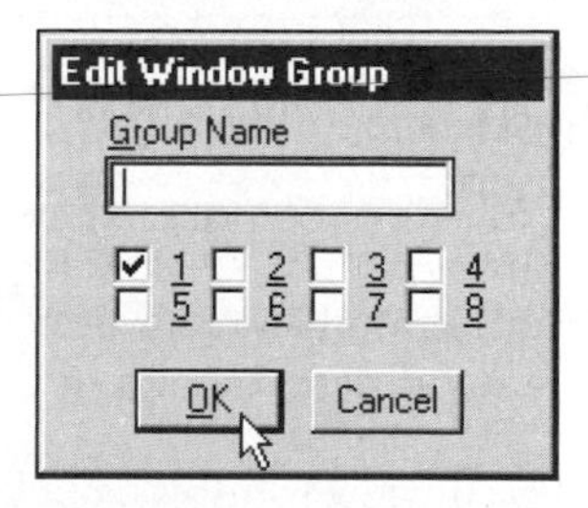

Figure 3–51 Edit Window Group dialog box

Note: New groups and changes to existing groups are saved automatically to the design file. It is not necessary to save settings to retain them.

Setting View Attributes

MicroStation allows a set of view attributes to be assigned to each of the open view windows. View attributes control the way elements and drawing aids appear in the view windows. Attributes are usually set to OFF to speed up view updating and to reduce clutter in a view. For example, if a view window contains a large amount of patterning, turning off the display of patterns can greatly reduce the update time.

To open the View Attributes dialog box, select:

Pull-down menu	Settings > View Attributes

MicroStation displays the View Attributes settings box as shown in Figure 3–52; each view attribute is explained briefly in Table 3–2.

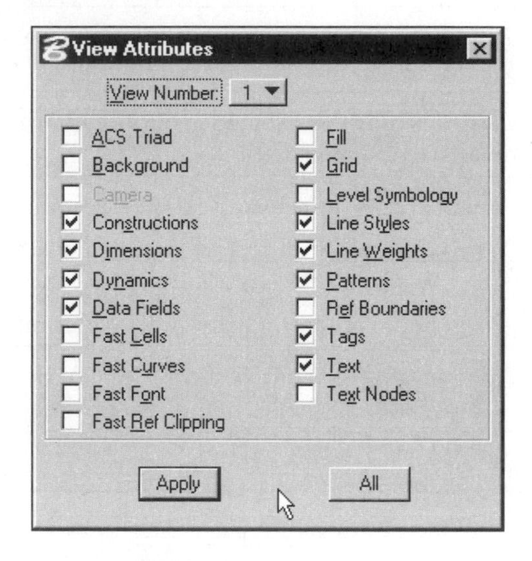

Figure 3–52 View Attributes settings box

Follow these steps to use the settings box to change view attributes:

1. Click the appropriate attribute toggle button ON or OFF as required.
2. If you want to set the attributes for all view windows, click the All button.
3. If you want to set the attributes for one view window, select its view window number in the View Number option menu and then click the Apply button.

Table 3–2 View Attributes

ATTRIBUTE	TURNS ON AND OFF THE DISPLAY O...
ASC Triad	The Auxiliary Coordinate System (ACS).
Background	The background image loaded with the Active Background command.
Camera	The 3D view camera.
Constructions	Elements placed with Construction Class mode active.
Dimensions	Dimension elements.
Dynamics	Dynamic updating of elements as they are placed in the design.
Data Fields	Data Field placeholder characters.
Fast Cells	The actual cells or a box indicating the area of the design occupied by cells.
Fast Curves	The actual curve string or straight line segments indicating the vertices.
Fast Font	The actual text font for each text element or MicroStation's fast font.
Fast Ref Clipping	The display of the largest block enclosing the reference file clipping boundaries.
Fill	The fill color in filled elements.
Grid	The grid (if the view is zoomed out far enough, the grid will be turned off even if this attribute is on).
Level Symbology	Elements according to the symbology table rather than the actual element symbology.
Line Styles	Elements with their actual line weights (when it is set to OFF, all elements are displayed with style 0, solid).
Line Weights	Elements with their actual line weights (when it is set to OFF, all elements are displayed at line weight 0).
Patterns	Pattern elements.
Ref Boundaries	The display of reference file clipping boundaries as dashed polygons.
Tags	The tag information for tagged elements.
Text	The display of text elements (when it is set to OFF, no text elements are displayed).
Text Nodes	Text nodes as small crosses with numeric identifiers.

Note: The View Attributes setting changes remain in effect until either they are changed again or the design file is closed. To keep them in effect for future editing sessions, select Save Settings from the File pull-down menu.

Saving Views

If you regularly work in several specific areas of a design, MicroStation provides a way for you to return to those areas quickly by saving the view setup and attributes under a user-defined name. To return to one of those areas, you provide the saved view name, then click in the view window you want set to have the saved view's setup.

Saving a View

To save a view setup, first align a view to display the area of the design you want to save, then set the view attributes you want to have in effect when you use the view.

Invoke the Saved Views settings:

Pull-down menu	Utilities > Saved Views

MicroStation displays the Saved Views settings box as shown in Figure 3–53.

Figure 3–53 Saved Views settings box

To save a view, first make sure the number of the view that contains the view setup is displayed in the View options menu in the Saved Views settings box. Then key-in a name and description in the Name: edit field and Description: edit field, respectively. Saved view names can be from one to six characters long and can consist of any combination of letters, numbers, dashes (-), periods (.), and underscores (_). Try to make your view name descriptive (to the extent that six characters allow).

Descriptions are optional and can be up to 27 characters long. Make it a habit always to use the description to explain what the saved view is set to. Your description will help other people who use your design file, and it will help you if you have not used the design file for a few weeks. Once you have provided a name and an optional description, click on the Save button. The view is saved, and the name and description appear in the Saved Views area of the Saved Views settings box. The saved view is now a permanent part of your design file.

You also can save a view by keying-in at the key-in window **sv=**<name>,<description> and pressing Enter. Replace <name> with the name you select for the saved view and <description> with a description of the saved view. MicroStation prompts:

> Select view *(Click the Data button in the view that contains the setup you want to save.)*

Attaching a Saved View

To restore a saved view, open the Saved Views settings box. Select the name of the view you want to restore from the list. Make sure the Dest View option menu (located at the bottom of the settings box) is displaying the number of the view to which you want to set the saved view. Click the Attach button.

You also can restore a view by keying-in at the key-in window field **vi=**<name>, and pressing Enter. Replace <name> with the name of the view to be restored. MicroStation prompts:

> Select view *(Click the Data button in the view where you want to restore the view.)*

Deleting a Saved View

To delete a saved view you no longer need, open the Saved Views settings box. Select the name of the view you want to delete from the list. Click on the Delete button, and the selected saved view is deleted.

You also can delete a view by keying-in at the key-in window **dv=**<name> and pressing Enter. Replace <name> with the name of the saved view to be deleted. The saved view is immediately deleted.

Note: If you have the Saved Views settings box displayed and you key-in **dv=**<name> to delete a saved view, the name and description of the deleted saved view may remain in the settings box Saved Views list, even though it has been deleted. To update the settings box, close and reopen it.

UNDO AND REDO

The Undo tool undoes the effects of the previous command or group of commands, depending on the option employed. The Redo tool is a one-time reversal of the effects of the previous Undo. Commands can be undone because all steps required for each command you use are stored in an Undo buffer in your computer's RAM. The Undo tool goes to that buffer to get the information necessary to put things back the way they were before the tool was invoked. The last command you executed is the first one undone, the next-to-last command is the next one undone, and so on.

Undo Tool

The Undo tool permits you to select the last command or a marked group of prior commands for undoing. To undo the last operation, invoke the Undo tool:

Standard tool box	Select the Undo tool (see Figure 3–54).
Key-in window	**undo** (or **und**) Enter

Figure 3–54 Invoking the Undo tool from the Standard tool box

To negate the last drawing operation, you can also select the Undo (action) option in the Edit pull-down menu. MicroStation displays the name of the last command operation that was performed in the Edit pull-down menu in place of (action). When you select the tool, MicroStation negates the last drawing operation.

Set Mark and Undo Mark

If you are at a point in the editing session at which you want to experiment but you want to be able to undo the experiment, you can place a mark in the design before you start.

To place a mark, invoke the Set Mark tool:

Pull-down menu	Edit > Set Mark
Key-in window	**mark** (or **mar**) Enter

To undo all the steps back to when the mark was placed, invoke the Undo Mark tool:

Pull-down menu	Edit > Undo Other > To Mark
Key-in window	**undo mark** (or **und m**) Enter

All the commands after the mark was placed are undone.

Undo All

The Undo All tool lets you negate all of the drawing operations recorded in the Undo buffer. Think twice before you invoke this tool.

To undo all the drawing operations recorded in the Undo buffer, invoke the Undo All tool:

Pull-down menu	Edit > Undo Other > All
Key-in window	**undo all** (or **und a**) Enter

MicroStation displays an alert box warning you that it will undo all of the drawing operations recorded in the Undo buffer. If you are not sure you want that to happen, just cancel the command.

Redo Tool

The Redo tool permits one reversal of a prior Undo command. It undoes the last undo. To undo the prior undo, the Redo tool should be invoked immediately after the Undo command. You can redo a series of negated operations by repeatedly choosing Redo.

To redo an undo, invoke the Redo tool:

Standard tool box	Select the Redo tool (see Figure 3–55).
Key-in window	**redo** Enter

Figure 3–55 Invoking the Redo tool from the Standard tool box

Things to Consider Before Undoing

Following are some points to consider before invoking the Undo or Redo tools.

- The Undo buffer resides in your computer's RAM; this buffer is limited in size. If you have issued more commands than the buffer can hold, the oldest commands can no longer be undone. For example, if the buffer can hold only information for 100 commands, you can undo only the last 100 commands. Compressing the design (from the File pull-down menu) clears the Undo buffer. No commands issued before the compress can be undone.

- Exiting from the design clears the Undo buffer. Commands issued in a previous editing session cannot be undone.
- The Undo commands back up through the Undo buffer. They are not always the best way to clean up a problem. For example, if five commands ago you placed a circle you want to get rid of, Undo forces you to undo the four commands issued after the circle placement to get to the circle. In this case, a better way to get rid of the circle is with the Delete Element tool.
- When you use one of the Undo commands, you are undoing commands, not elements. If the command manipulated multiple elements, Undo undoes the manipulation of *all* of those elements. For example, the Fence commands can manipulate hundreds of elements at one time. If you undo a Fence Contents Delete command, you get back all the elements that the fence deleted.

REVIEW QUESTIONS

Write your answers in the spaces provided.

1. How many possible positions (positional units) are there in the *X* and *Y* directions of a *2D* design file? ______
2. Define Master Units: ______
3. Define Sub Units: ______
4. Define Positional Units: ______
5. Name the tool that will make individual elements from a shape. ______
6. Explain briefly the differences between the absolute rectangular coordinates and relative rectangular coordinates precision key-ins. ______
7. Name the three key-ins that are used in precision input for absolute rectangular coordinates, relative rectangular coordinates, and polar relative coordinates. ______
8. Explain briefly the difference between the Zoom In and Zoom Out tools.

   ______
9. Which tool will get you closer to a portion of your design by a factor of 2?

   ______
10. Name the tool that will Rebuild, Refresh, or Update your views. ______
11. When MicroStation displays the information regarding the size of an element or coordinates, it does so in the following format:

    ______:______:______
12. Panning lets you view ______ .
13. To pan in a view, hold the Shift key and press the ______ button.
14. If in a design file the Working Units are set up as inches, eighths, and 1600 positional units per eighth, what distance does 1:6:600 represent?

    ______
15. If in a design file the Working Units are set up as feet, inches, and 1600 positional units per inch, what Working Units expression is equivalent to 4'-0.3200"?

    ______

16. What is the purpose of the Grid lock? ______________________________

17. The ______________________________ settings box allows you to control the display of the grid.

18. The Grid Master unit defines the distance between the ____________________ and is specified in terms of ________:________:________ .

19. The Grid Reference is set to define the ______________________________ .

20. To keep the grid settings in effect for future editing sessions for the current design file, invoke the ______________ tool.

21. The Aspect Ratio edit field in the Grid settings box allows you to set the

____ .

22. Name the three menu options that are available to select the Snap mode. ______________

23. Name four Snap modes available in MicroStation. ______________________________

24. Keypoint mode allows tentative points to snap to __________________ ______________.

25. The Midpoint mode snaps to the ________________________ position of a circle and an ellipse.

26. The Axis Lock forces each data point ______________________________.

27. List the four attributes associated with the placement of elements.

____

28. How many levels come with a new design file? ____________________

29. How many level(s) can be active at any time? ____________________

30. How many level(s) can you turn ON or OFF at any time in a specific view? __________

31. How many colors or shades of colors are available in the color palette? __________

32. What two-letter key-in makes a color active? __________

33. How many Line Styles (internal) are available in MicroStation? __________

34. In MicroStation, Line Weight refers to ______________________________ .

35. How many line weights are available in MicroStation ? ______________

36. Name the three methods by which you can place circles in MicroStation.

____

37. If you wish to draw a circle by specifying three known points on the circle, invoke the _________________________ tool.

38. The tools related to placing arcs are in the ______________ tool box.

39. The Place Arc Edge tool places an arc by identifying ___________ points on the arc.

PROJECT EXERCISE

This project exercise provides step-by-step instructions for creating the design shown in Figure P3–1. The intent is to guide you through applying the concepts and tools presented in Chapters 1, 2, and 3. (Note that these instructions are not the most efficient way to draw the objects. Your efficiency will improve as you learn more commands in later chapters.)

In this project, you will learn how to do the following:

- Create a new design file.
- Set the working units.
- Draw the border using precision input.
- Draw the objects using the Place Line, Place Block, Place Circle, and Place Arc tools using precision input and tentative snapping.

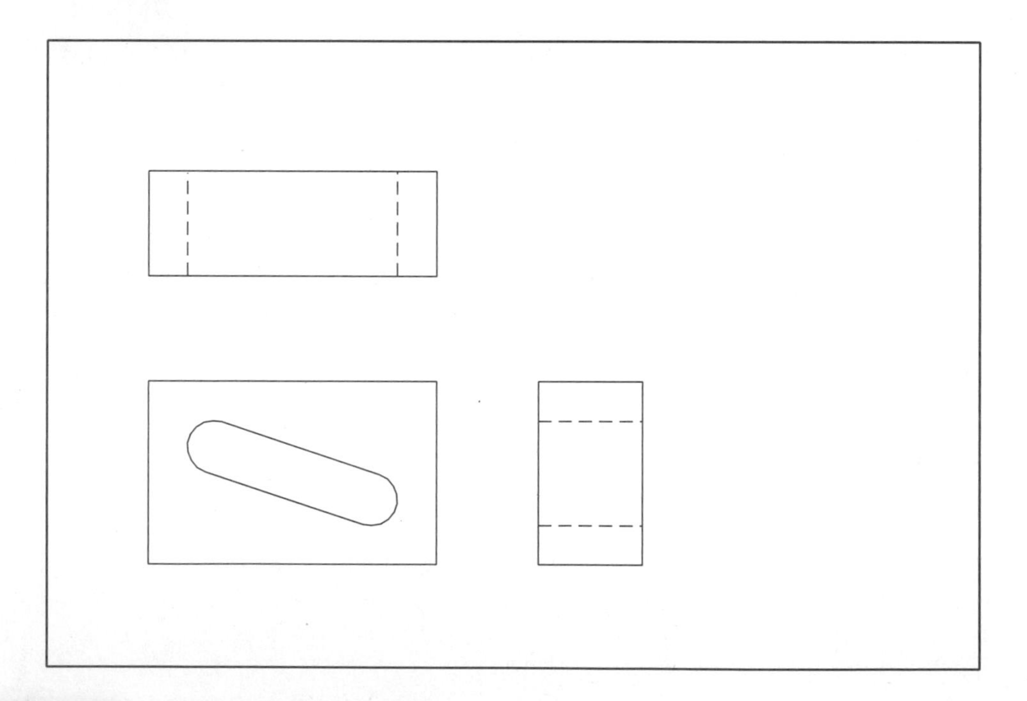

Figure P3–1 Completed project design

Note: Do not draw the dimensions. Dimensioning is introduced in a later chapter.

Note: As you complete each step in the project procedures, place a check mark by the step to help you keep up with where you are in the project.

Create a Design File

This procedure has you start MicroStation and create a design file for the project design.

STEP 1: Invoke the MicroStation program.

Example: Under Microsoft Windows 98, find the MicroStation program in the **Start** > Programs menu and select it.

STEP 2: In the MicroStation Manager dialog box, select the New... option from the File pull-down menu to open the Create New Design File dialog box.

STEP 3: If the "Seed File" area of the Create New Design File dialog box does not show seed2d.dgn as the seed file, click the Select button and select seed2d.dgn as the seed file.

Note: If you are working in Microsoft Windows, and MicroStation is loaded on the C drive, the path to seed2d.dgn should be:

C:\Bentley\Workspace\system\seed\seed2d.dgn

STEP 4: In the Create New Design File dialog box, type **CH3.DGN** as the name for the project design file in the Name edit field, then click on the OK button to create the file, and close the dialog box.

STEP 5: In the MicroStation Manager dialog box, select the CH3.DGN file from the files list box and click on the OK button to load the file in MicroStation.

Set the Working Units and Grid Spacing

This procedure presents the steps to do the following:

- Set the working unit ratios to 10 Sub Units (SUs) per Master Unit and 1000 Positional Units per Sub Unit.
- Set the Grid spacing to 0.1 and the grid reference to 10.
- Set the Element Attributes.

STEP 1: In the MicroStation application window, select the Design File... option from the Settings pull-down menu.

STEP 2: In the Design File dialog box, select the Working Units category, and set the Working Units ratios as shown in Figure P3–2.

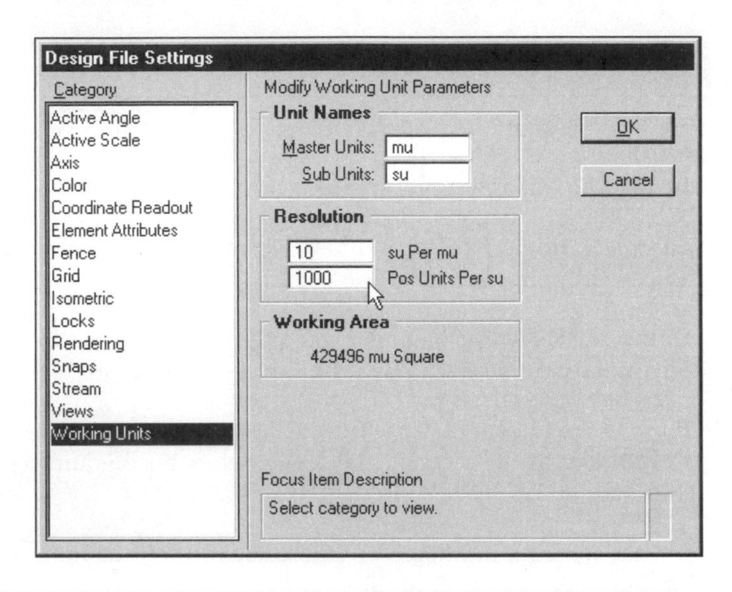

Figure P3–2 Design File Settings dialog box—Working Units ratio setup

STEP 3: In the Design File Settings dialog box, select the Grid category and set the Grid Master and Grid Reference, as shown in Figure P3–3.

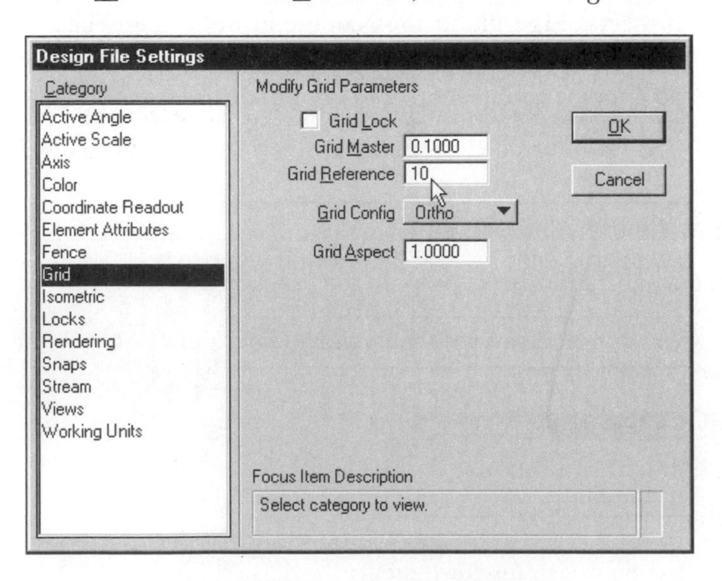

Figure P3–3 Design File Settings dialog box—Grid Units setup

STEP 4: Click on the OK button to close the Design File Settings dialog box and keep the changes.

STEP 5: If an Alert box opens to tell you the Working Units are being changed, click on the OK button to accept the changes.

STEP 6: Open the Element Attributes settings box by selecting Attributes from the Element pull-down menu. Set the Level to 10, Color to 1 (blue), Style to 0 (continuous), and Weight to 2, as shown in Figure P3–4.

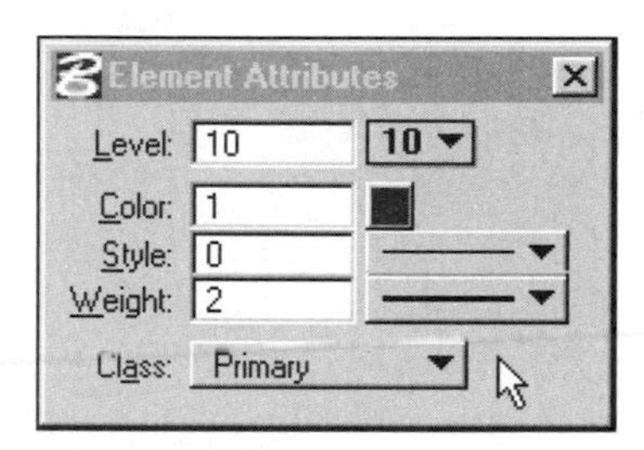

Figure P3–4 Element Attributes settings box

STEP 7: Select Save Settings from the File pull-down menu to save the working units, grid, and element attribute settings.

Draw the Border via Precision Input

This procedure presents the steps for employing the *XY* absolute coordinates to draw the border and for invoking the Fit View tool to see the complete border outline in the view window.

STEP 1: Open the Key-in window from the Utilities pull-down menu.

STEP 2: Invoke the Place Block tool from the Polygons tool box, and select the Orthogonal Method from the Tool Settings Window.

MicroStation prompts:

Place Block > Enter first point *(Click in the Key-in window's input field, key-in* **XY=0,0**, *and press* Enter.)

Place Block > Enter opposite corner *(Key-in* **XY=18,12** *in the Key-in window and press* Enter.)

STEP 3: Invoke the Fit View tool from the View Control Bar to display the complete border outline in the selected view.

STEP 4: Select Save settings from the File pull-down menu.

If the procedure was executed correctly, a rectangle should be seen in the view, as shown in Figure P3–5.

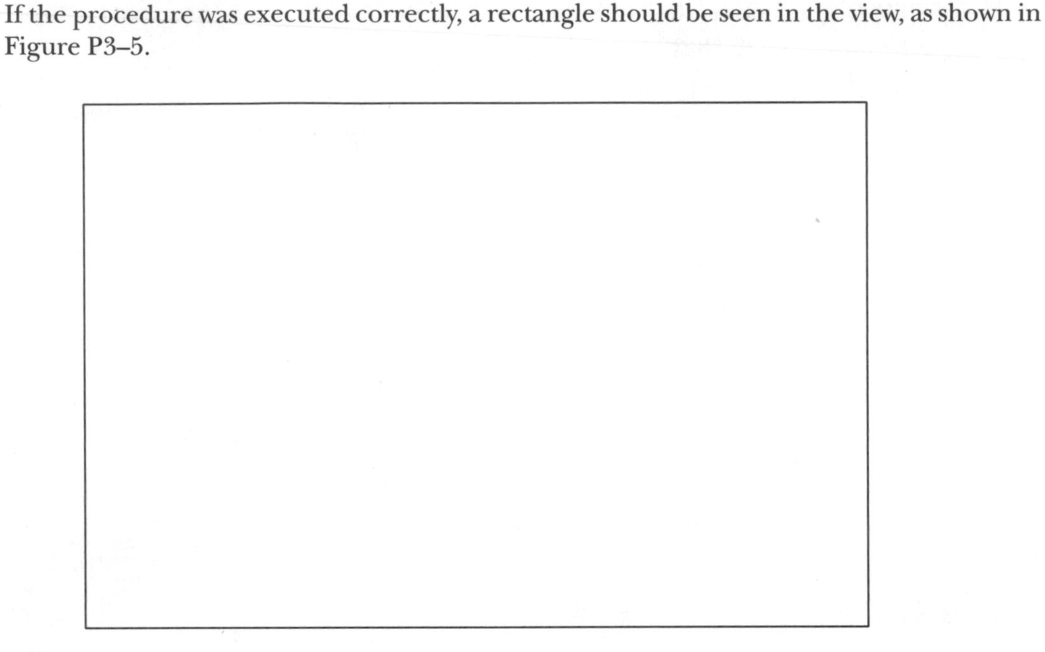

Figure P3–5 Completed border outline

Draw the Objects

STEP 1: Open the Element Attributes settings box by selecting Attributes from the Element pull-down menu. Set the Level to 12, Color to 3 (red), Style to 0 (continuous), and Weight to 1.

STEP 2: Draw two rectangles by invoking the Place Block tool from the Polygons tool box, and selecting the Orthogonal Method in the Tool Settings Window.

MicroStation prompts:

Place Block > Enter first point *(Click in the Key-in window's input field, key-in* **XY=2,7.5**, *and press* Enter*.)*

Place Block > Enter opposite corner *(Key-in* **XY=7.5,9.5** *in the Key-in window and press* Enter*.)*

Place Block > Enter first point *(Click in the Key-in window's input field, key-in* **XY=9.5,2**, *and press* Enter*.)*

Place Block > Enter opposite corner *(Key-in* **XY=11.5,5.5** *in the Key-in window and press* Enter*.)*

STEP 3: Draw another orthogonal block by using the appropriate tentative snaps.

MicroStation prompts:

Place Block > Enter first point *(Click Snaps icon in the Status bar, and select Intersection mode. Tentative snap to LINE 1 and LINE 4 as shown in Figure P3–6, then click the Data button to accept the Intersection snap.)*

Place Block > Enter opposite corner *(Click Snaps icon in the Status bar, and select Intersection mode. Tentative snap to LINE 2 and LINE 3 as shown in Figure P3–6, then click the Data button to accept the Intersection snap.)*

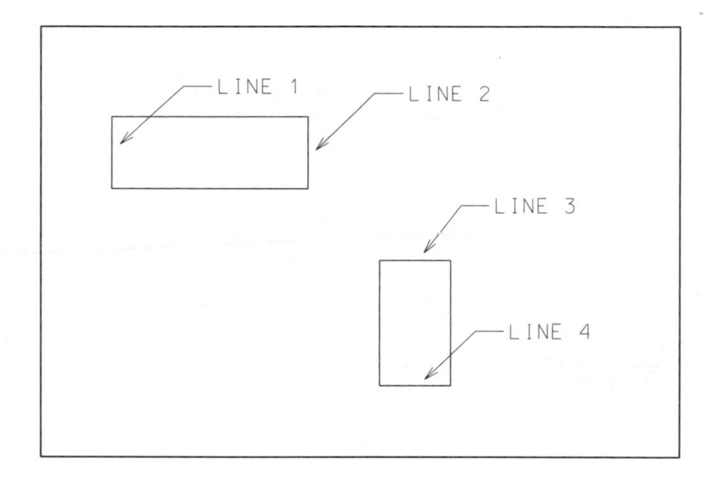

Figure P3–6 Identifying the lines whose intersections are used to draw a rectangle

After drawing the rectangle boxes, the drawing should look like Figure P3–7.

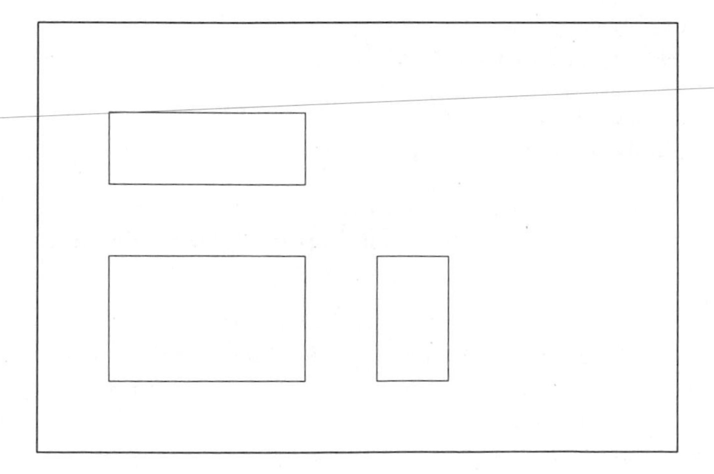

Figure P3–7 Completed drawing of rectangles

STEP 4: Open the Element Attributes settings box by selecting Attributes from the Element pull-down menu. Set the Level to 14, Color to 2 (green), Style to 0 (continuous), and Weight to 1.

STEP 5: Select Save settings from the File pull-down menu.

STEP 6: Invoke the Place Circle tool from the Ellipses tool box, and, from the Tool Settings window, select the Center Method and Radius from the Radius/Diameter options menu, turn ON the Radius toggle button, and enter 0.5 Master Units in the Radius edit field, as shown in Figure P3–8.

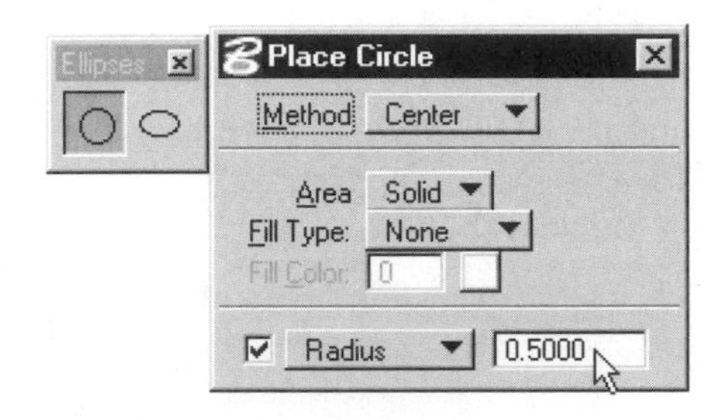

Figure P3–8 Place Circle Tool Settings window

STEP 7: Draw two circles.

MicroStation prompts:

Place Circle By Center > Identify Center Point *(Click in the Key-in window's input field, key-in* **XY=3.25,4.25**, *and press* Enter *.)*

Place Circle By Center > Identify Center Point *(Click in the Key-in window's input field, key-in* **XY=6.25,3.25**, *and press* Enter *.)*

STEP 8: Invoke the Place Line tool from the Linear Elements tool box and turn off the Length and Angle toggle buttons in the Tool Settings Window.

STEP 9: Draw a series of construction lines by using appropriate tentative snaps. These lines will be used later to aid in placing the hidden lines.

MicroStation prompts:

Place Line > Enter first point *(Click Snaps icon in the Status bar, and select keypoint mode. Tentative snap to QUA 1 as shown in Figure P3–9, then click the Data button to accept the Tentative Point.)*

Place Line > Enter endpoint (*Click Snaps icon in the Status bar, and select perpendicular mode., Tentative snap to LINE 5 as shown in Figure P3–9, then click the Data button to accept the Tentative Point.)*

Place Line > Enter endpoint *(Click the Reset button to terminate the line sequence.)*

Place Line > Enter first point *(Click Snaps icon in the Status bar, and select keypoint mode. Tentative snap to QUA 2 as shown in Figure P3–9, then click the Data button to accept the Tentative Point.)*

Place Line > Enter endpoint (*Click Snaps icon in the Status bar, and select perpendicular mode. Tentative snap to LINE 5 as shown in Figure P3–9, then click the Data button to accept the Tentative Point.)*

Place Line > Enter endpoint *(Click the Reset button to terminate the line sequence.)*

Place Line > Enter first point *(Click Snaps icon in the Status bar, and select keypoint mode., Tentative snap to QUA 3 as shown in Figure P3–9, then click the Data button to accept the Tentative Point.)*

Place Line > Enter endpoint (*Click Snaps icon in the Status bar, and select perpendicular mode. Tentative snap to LINE 6 as shown in Figure P3–9, then click the Data button to accept the Tentative Point.)*

Place Line > Enter endpoint *(Click the Reset button to terminate the line sequence.)*

Place Line > Enter first point *(Click Snaps icon in the Status bar, and select keypoint mode. Tentative snap to QUA 4 as shown in Figure P3–9, then click the Data button to accept the Tentative Point.)*

Place Line > Enter endpoint (*Click Snaps icon in the Status bar, and select perpendicular mode. Tentative snap to LINE 6 as shown in Figure P3–9, then click the Data button to accept the Tentative Point.)*

Place Line > Enter endpoint *(Click the Reset button to terminate the line sequence.)*

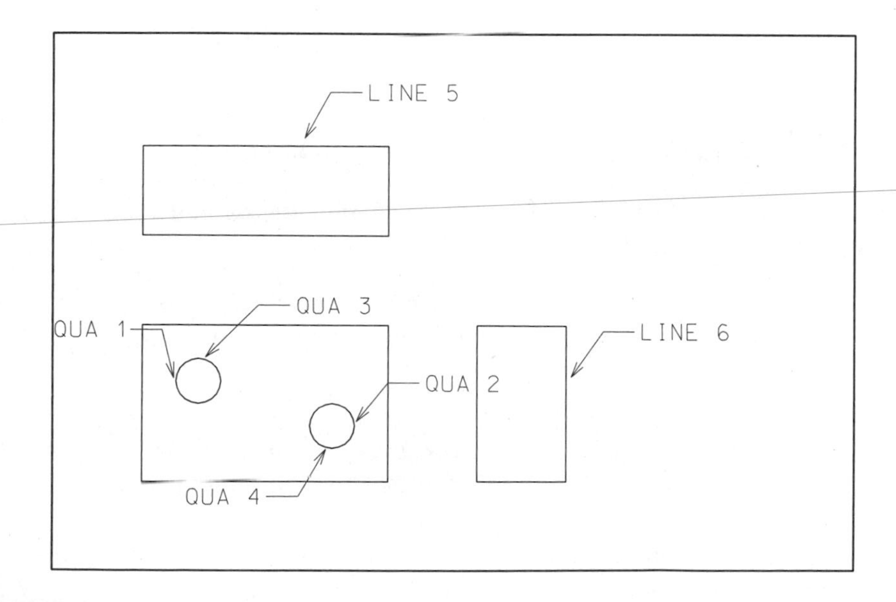

Figure P3–9 Identifying the points to draw additional lines

The drawing should look like Figure P3–10.

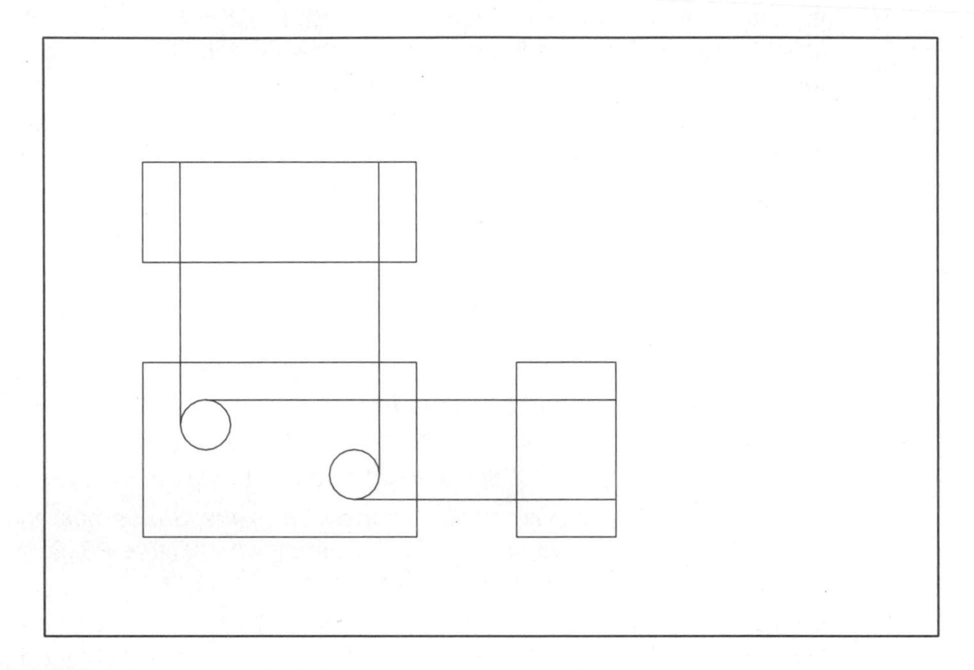

Figure P3–10 Completed drawing of construction lines

STEP 10: Open the Element Attributes settings box by selecting Attributes from the Element pull-down menu. Set the Level to 15, Color to 0 (white), Style to 2 (hidden), and Weight to 1.

STEP 11: Invoke Save settings from the File pull-down menu.

STEP 12: Set Intersection mode tentative snap as the default mode. Click the Snaps icon in the Status bar, then select Intersection mode while pressing Shift.

STEP 13: Invoke the Place Line tool from the Linear Elements tool box and draw the hidden lines by using intersection mode tentative snap.

MicroStation prompts:

Place Line > Enter first point *(Do Intersection Tentative snaps at POINT 1 as shown in Figure P3–11, then click the Data button to accept the Tentative Point.)*

Place Line > Enter endpoint *(Do Intersection Tentative snaps at POINT 2 as shown in Figure P3–11, then click the Data button to accept the Tentative Point.)*

Place Line> Enter endpoint *(Click the Reset button to terminate the line sequence.)*

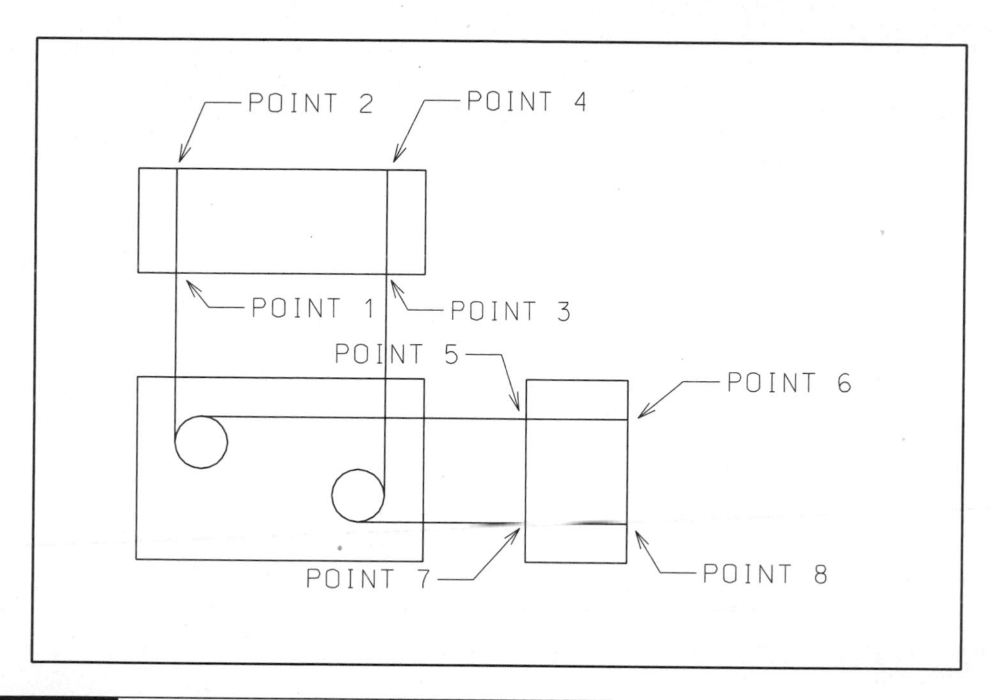

Figure P3–11 Identifying the points to draw object lines

STEP 14: Draw three additional lines using the Intersection snap method that was used in step 13 above.

From:	To:
POINT 3	POINT 4
POINT 5	POINT 6
POINT 7	POINT 8

STEP 15: Invoke the Delete Element tool from the Main tool frame to delete LINE 7, LINE 8, LINE 9, and LINE 10, as shown in Figure P3–12. Update the view window by invoking the Update View tool.

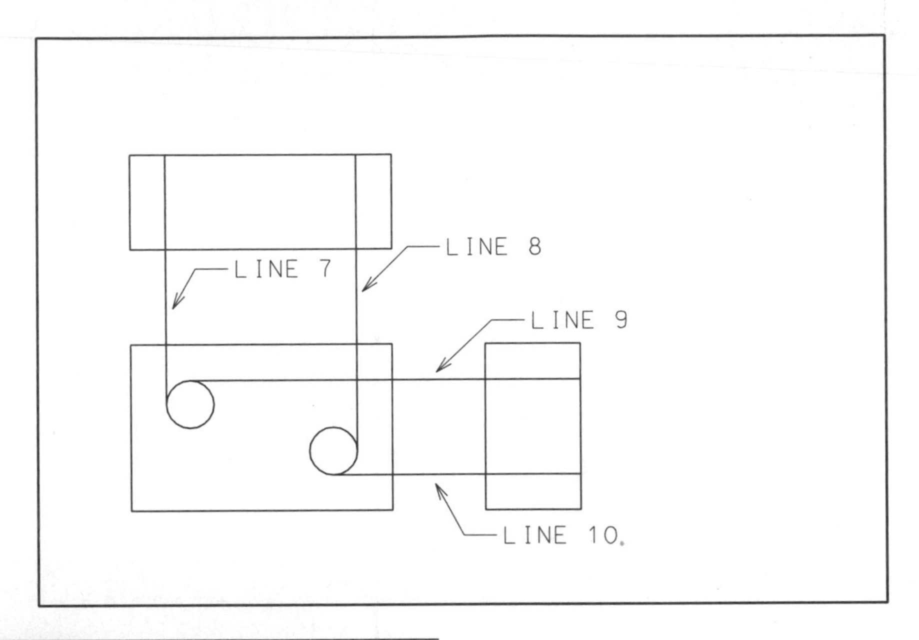

Figure P3–12 Identifying the lines to delete

After deleting the selected lines, the drawing should look like Figure P3–13.

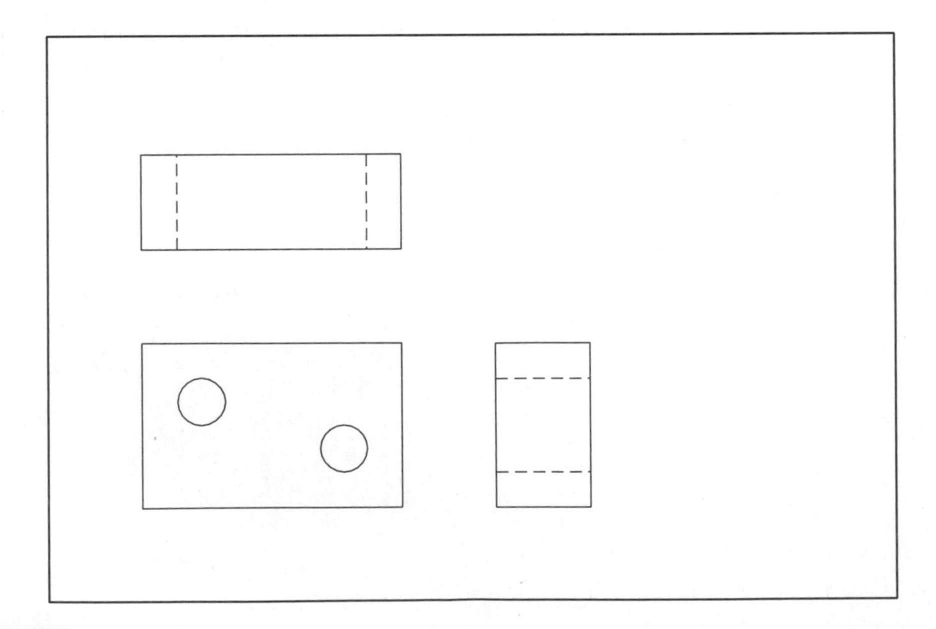

Figure P3–13 Completed drawing

STEP 16: Open the Element Attributes settings box by selecting Attributes from the Element pull-down menu. Set the Level to 12, Color to 3 (red), Style to 0 (continuous), and Weight to 1.

STEP 17: Invoke the Place Line tool from the Linear Elements tool box and draw lines tangent to the circles using appropriate tentative snaps.

MicroStation prompts:

Place Line > Enter first point *(Click Snaps icon in the Status bar, and select Tangent mode. Tentative snap in the upper half of the CIRCLE 1 as shown in Figure P3–14, then click the Data button to accept the Tentative Point.)*

Place Line > Enter endpoint (*Click Snaps icon in the Status bar, and select Tangent mode. Tentative snap in the upper half of the CIRCLE 2 as shown in Figure P3–14, then click the Data button to accept the Tentative Point.)*

Place Line > Enter endpoint *(Click the Reset button to terminate the line sequence.)*

Place Line > Enter first point *(Click Snaps icon in the Status bar, and select Tangent mode. Tentative snap in the lower half of the CIRCLE 1 as shown in Figure P3–14, then click the Data button to accept the Tentative Point.)*

Place Line > Enter endpoint (*Click Snaps icon in the Status bar, and select Tangent mode. Tentative snap in the lower half of the CIRCLE 2 as shown in Figure P3–14, then click the Data button to accept the Tentative Point.)*

Place Line > Enter endpoint *(Click the Reset button to terminate the line sequence.)*

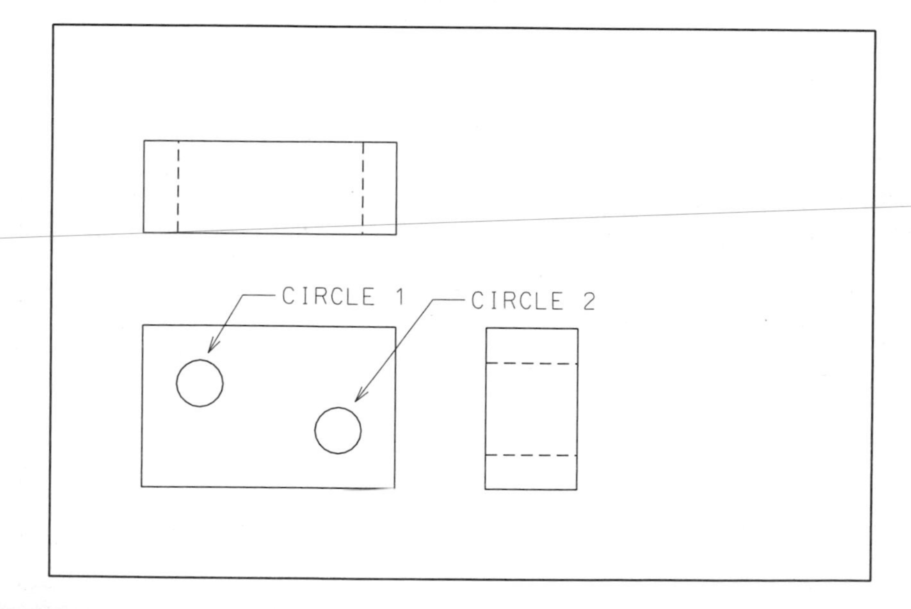

Figure P3–14 Identifying the tentative snap points to draw additional lines

STEP 18: Invoke the Place Arc tool from the Arcs tool box, and, in the Tool Settings Window, select the Center Method, and turn OFF the toggle buttons for Radius, Start Angle, and Sweep Angle.

STEP 19: Open the Snap Mode button bar and double-click the Keypoint mode to set it as the default tentative snap mode.

STEP 20: Draw two arcs to complete the cutout in the front view of the object.

MicroStation prompts:

Place Arc By Center > Identify First Arc Endpoint *(Tentative snap to the ENDPOINT 1 as shown in Figure P3–15, then click the Data button to accept the Tentative Point.)*

Place Arc By Center > Identify Arc Center *(Tentative snap to the center of the CIRCLE 1 as shown in Figure P3–15, then click the Data button to accept the Tentative Point.)*

Place Arc By Center > Identify Second Arc Endpoint *(Tentative snap to the ENDPOINT 2 as shown in Figure P3–15, then click the Data button to accept the Tentative Point.)*

Place Arc By Center > Identify First Arc Endpoint *(Tentative snap to the ENDPOINT 3 as shown in Figure P3–15, then click the Data button to accept the Tentative Point.)*

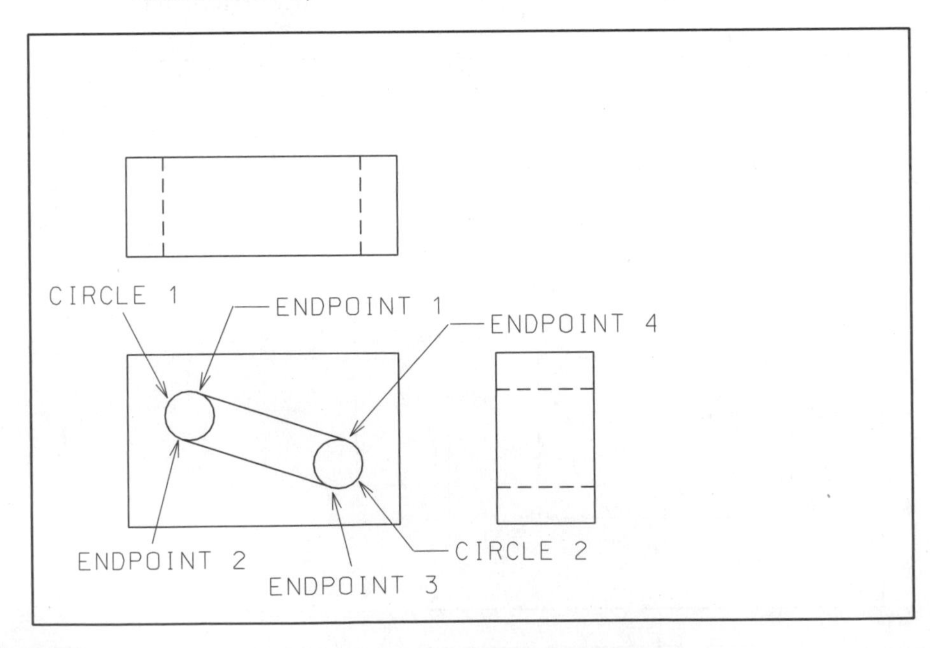

Figure P3–15 Identifying the tentative snap points to draw arcs

Place Arc By Center > Identify Arc Center *(Tentative snap to the center of the CIRCLE 2 as shown in Figure P3–15, then click the Data button to accept the Tentative Point.)*

Place Arc By Center > Identify Second Arc Endpoint *(Tentative snap to the ENDPOINT 4 as shown in Figure P3–15, then click the Data button to accept the Tentative Point.)*

STEP 21: From the Settings pull-down menu, open the Level submenu, then select Display to open the View Levels dialog box.

STEP 22: In the View Levels settings box, turn OFF level14, then click the Apply button.

STEP 23: Invoke the Save settings from the File pull-down menu.

The completed drawing should look like Figure P3–16.

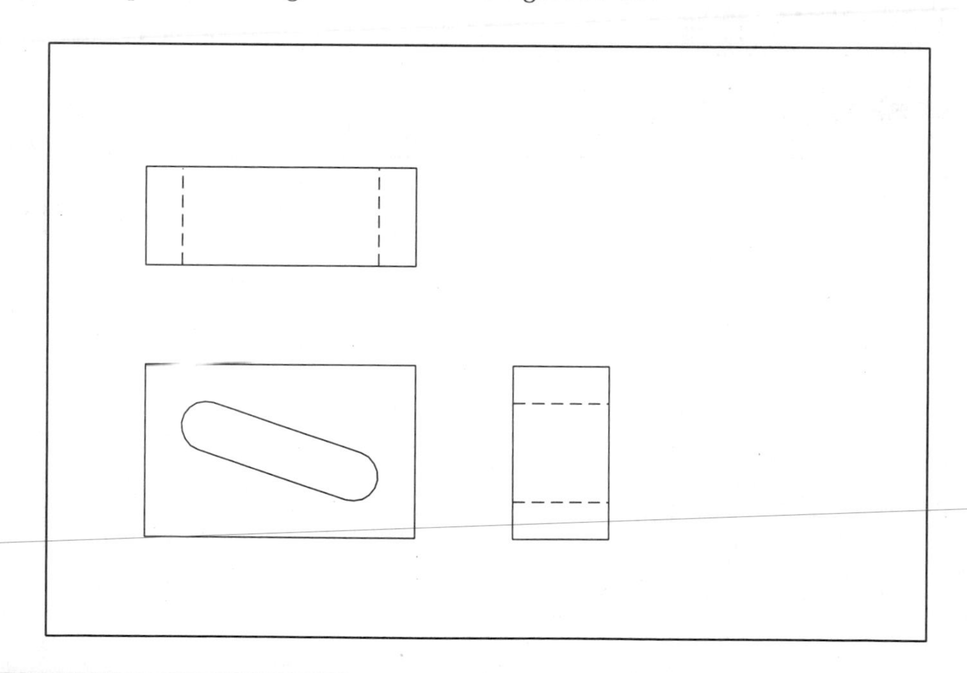

Figure P3–16 Completed drawing

DRAWING EXERCISES 3–1 THROUGH 3–6

Use the following table to set up the design files for Exercises 3–1 through 3–5.

SETTING	VALUE
Seed File	SEED2D.DGN
Working Units	10 th Per " and 1000 Pos Units Per th
Grid	Master = 0.25, Reference = 4, Grid Lock set to ON
Object Elements	Color = 0, Level = 1, Style = 0, Weight = 1
Hidden Lines	Color = 2, Level = 2, Style = 2, Weight = 1
Centerlines	Color = 3, Level = 3, Style = 4, Weight = 0

Exercise 3–1

Gasket

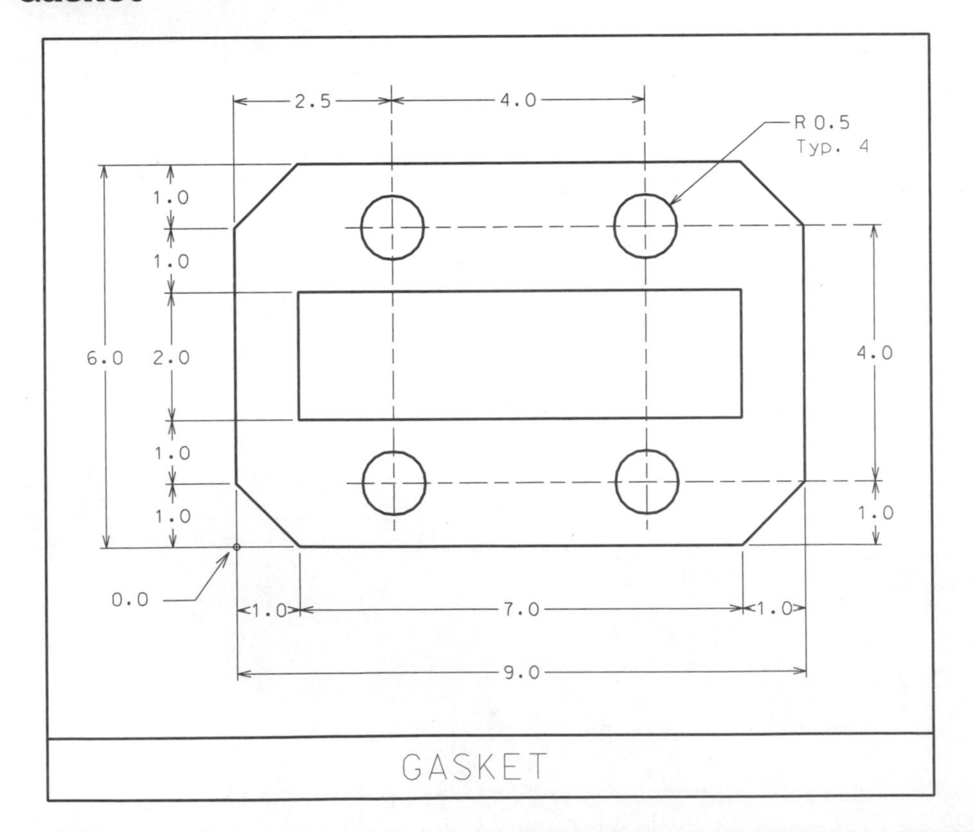

Exercise 3-2

Crane Pulley Plate

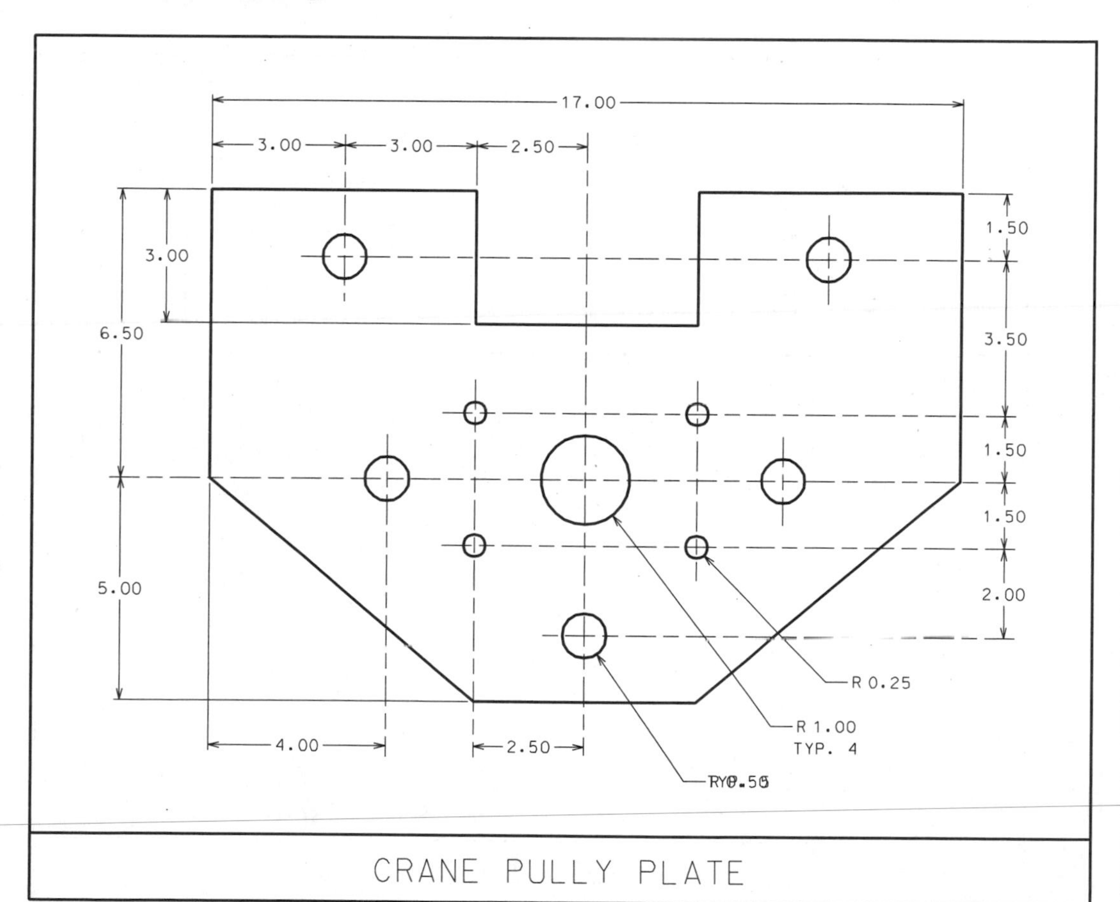

Exercise 3–3

Pipe Clamp

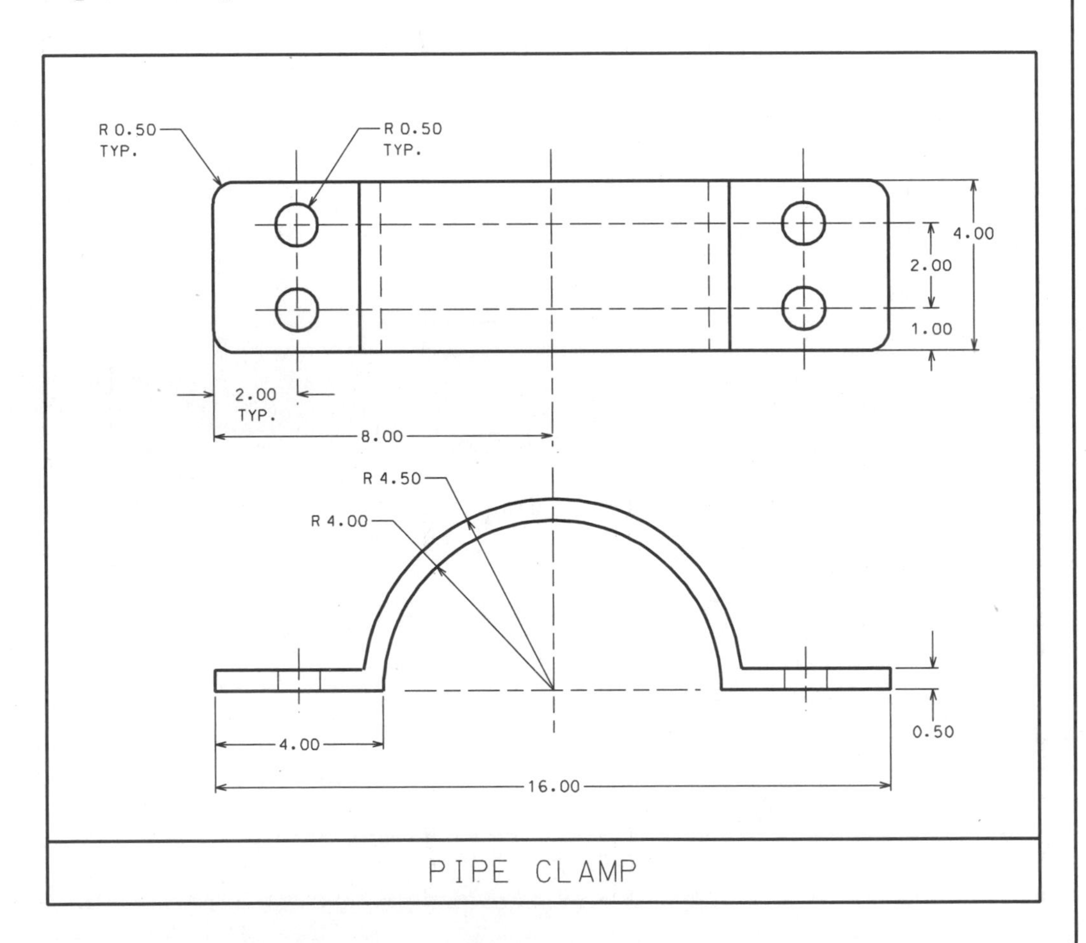

Exercise 3-4

Foundation Plan

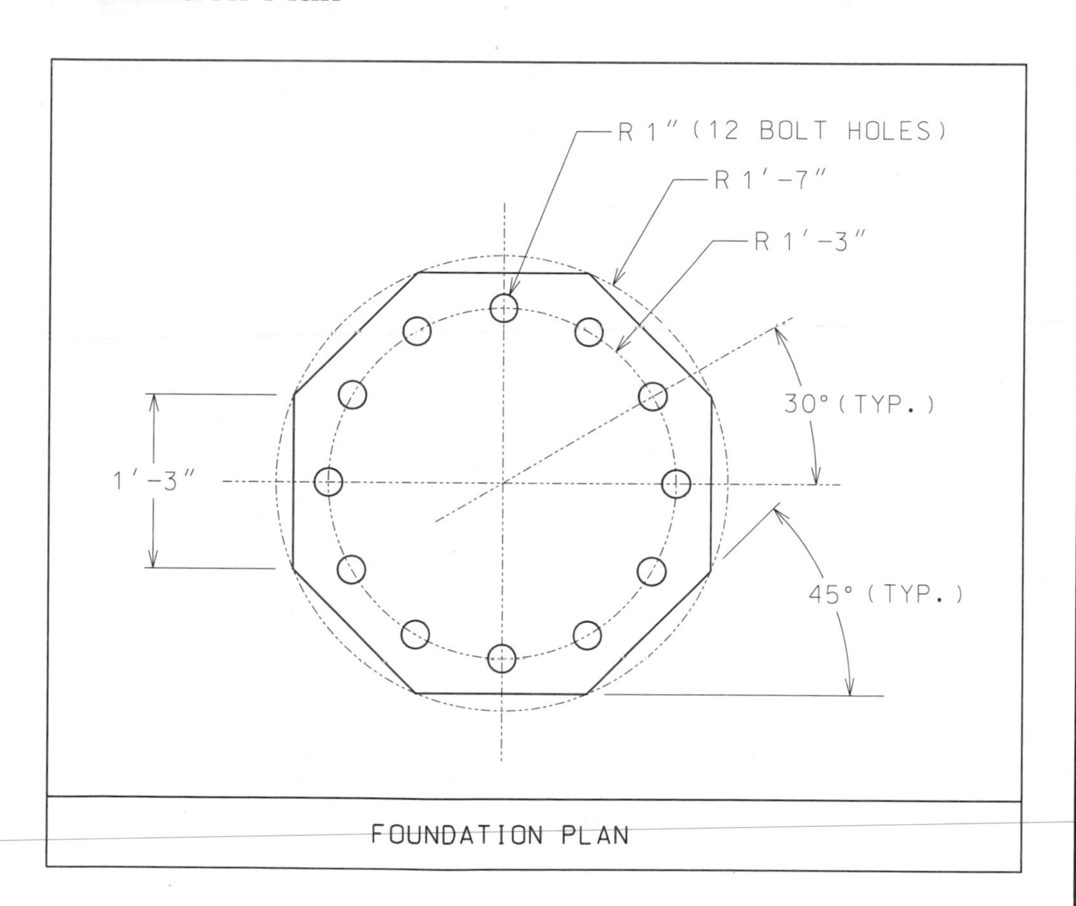

Exercise 3–5

Flow Diagram

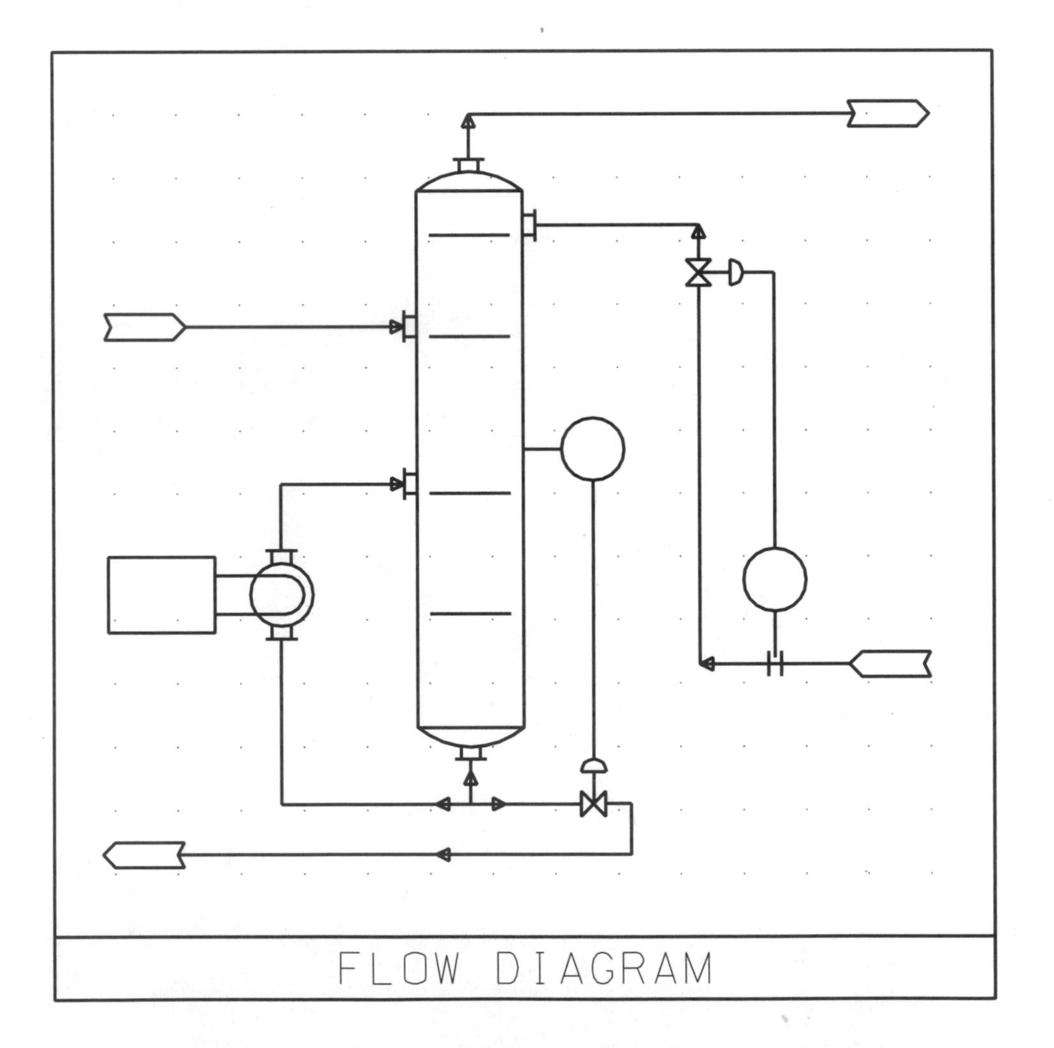

Use the following table to set up the design file for Exercises 3–6.

SETTING	VALUE
Seed File	SEED2D.DGN
Working Units	12 " Per ' and 8000 Pos Units Per in
Grid	Master = 0.25, Reference = 4, Grid Lock set to ON
Object Elements	Color = 0, Level = 1, Style = 0, Weight = 1
Hidden Lines	Color = 2, Level = 2, Style = 2, Weight = 1
Centerlines	Color = 3, Level = 3, Style = 4, Weight = 0

Exercise 3–6

Shop Floor Plan

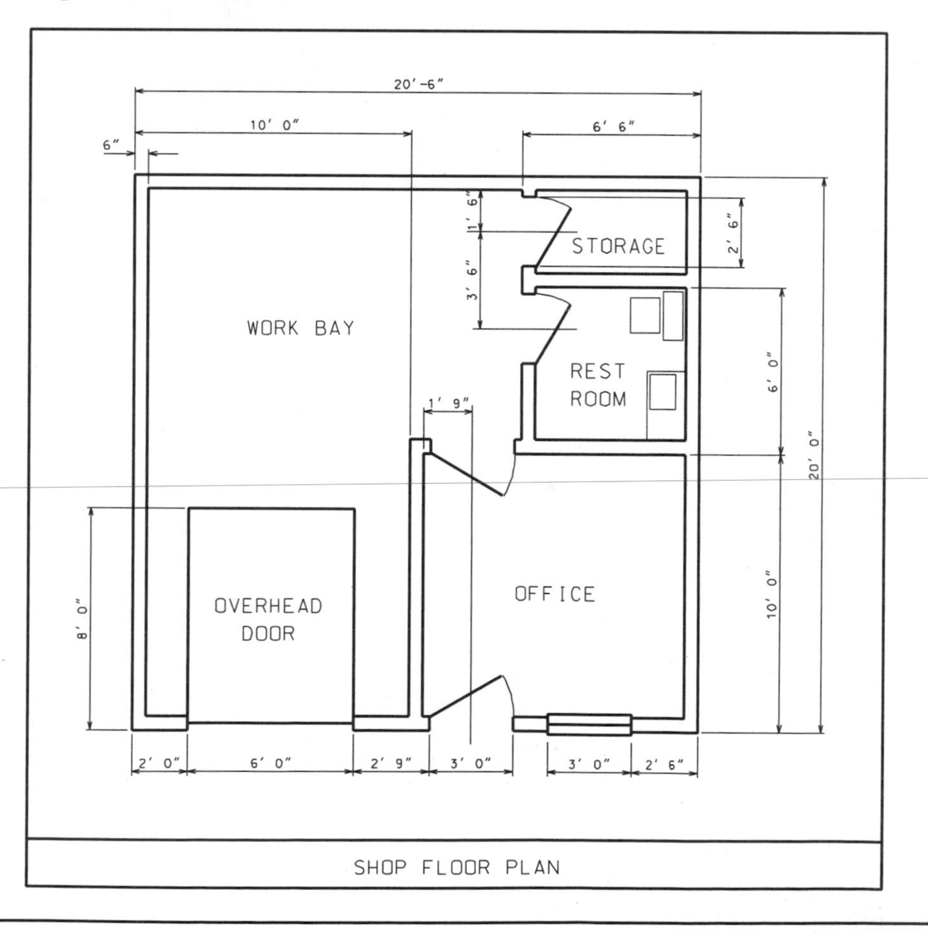

chapter 4

Fundamentals III

Objectives

After completing this chapter, you will be able to do the following:

- Draw ellipses, polygons, point curves, curve streams, and multi-lines
- Modify elements: fillets, chamfers, trim, and partial delete
- Manipulate elements: move, copy, move and copy parallel, scale original and copy, rotate original and copy, mirror original and copy, and array
- Place text: Set text parameters and place text by origin

PLACEMENT TOOLS

In this chapter, four more placement tools are explained: Ellipse, Curve (Place Point Curve and Place Curve Stream), and Multi-line tools. This adds to the placement tools already described in Chapter 2.

Place Ellipse

MicroStation offers two different methods for drawing an ellipse: Place Ellipse by Center and Edge and Place Ellipse by Edge Points. The appropriate method is selected from the Method option menu located in the Tool Settings window.

Place Ellipse By Center and Edge

The Place Ellipse By Center and Edge tool lets you draw an ellipse by defining three points: the center point, one end of the primary (major) axis, and one end of the secondary (minor) axis.

Invoke the Place Ellipse By Center and Edge tool:

Ellipses tool box	Select the Place Ellipse tool, then select Center from the Method option menu located in the Tool Settings window (see Figure 4–1).
Key-in window	**place ellipse center constrained** (or **pl el ce co**) Enter

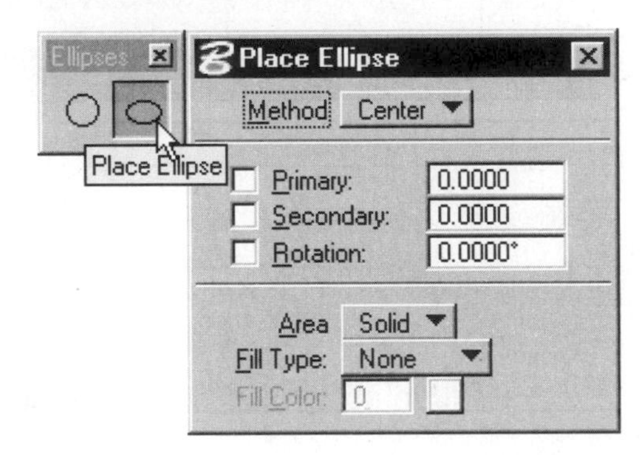

Figure 4–1 Invoke the Place Ellipse By Center and Edge tool from the Ellipses tool box

MicroStation prompts:

Place Ellipse By Center and Edge > Identify Ellipse Center *(Place a data point or key-in coordinates to define the center of the ellipse.)*

Place Ellipse By Center and Edge > Identify Ellipse Primary Radius *(Place a data point or key-in coordinates to define the one end of the primary axis.)*

Place Ellipse By Center and Edge > Identify Ellipse Secondary Radius *(Place a data point or key-in coordinates to define the one end of the secondary axis.)*

For example, the following tool sequence shows how to place an ellipse with the Place Ellipse by Center and Edge tool (see Figure 4–2).

Place Ellipse By Center and Edge > Identify Ellipse Center **XY=3,2** Enter

Place Ellipse By Center and Edge > Identify Ellipse Primary Radius **DL=2,0** Enter

Place Ellipse By Center and Edge > Identify Ellipse Secondary Radius **XY=3,3** Enter

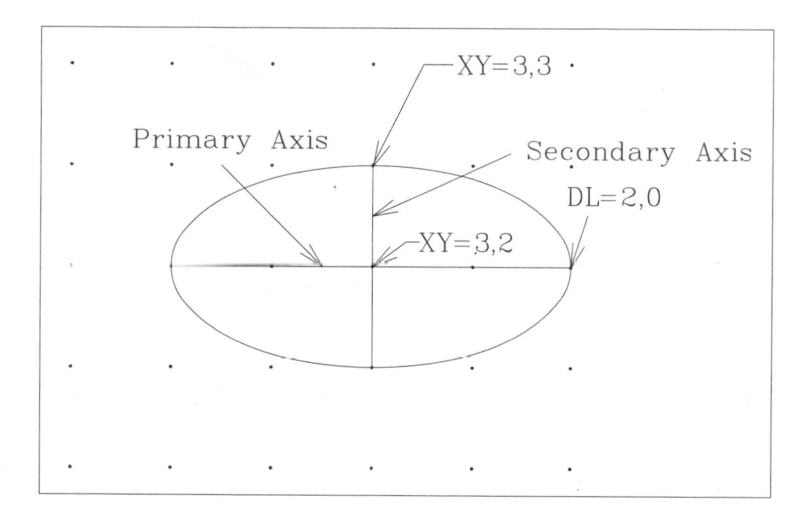

Figure 4–2 Example of placing an ellipse by means of the Place Ellipse by Center and Edge tool

You can also place an ellipse by its center and edge by keying-in the primary radius (constrained). To do so, turn ON the toggle button for Primary and key-in the value in MU:SU:PU format, then press Enter or Tab. MicroStation prompts you to identify the Ellipse Center and the Secondary Ellipse Radius. Similarly, you can also constrain the Secondary Radius and Rotation by turning ON the toggle buttons appropriately, keying-in the values in the edit fields, and pressing Enter or Tab.

MicroStation prompts depend on the number of constraints turned ON. For example, if the Primary and Secondary radius are preset, MicroStation prompts you to identify the center and rotation data points. If the Primary radius, Secondary radius, and Rotation are preset, MicroStation prompts you to identify the ellipse center point.

Place Ellipse By Edge Points

The Place Ellipse By Edge Points tool enables you to draw an ellipse by defining three points on the ellipse.

Invoke the Place Ellipse By Edge Points tool:

Ellipses tool box	Select the Place Ellipse tool, then select Edge from the Method option menu located in the Tool Settings window (see Figure 4–3).
Key-in window	**place ellipse edge constrained** (or **pl el ed co**) Enter

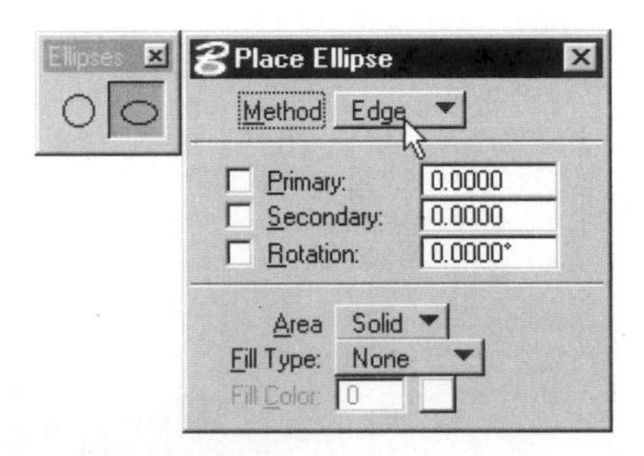

Figure 4–3 Invoke the Place Ellipse By Edge Points tool from the Ellipses tool box

MicroStation prompts:

Place Ellipse By Edge Points > Identify Point on Ellipse *(Place a data point or key-in coordinates to define the first point on the ellipse.)*

Place Ellipse By Edge Points > Identify Point on Ellipse *(Place a data point or key-in coordinates to define the second point on the ellipse.)*

Place Ellipse By Edge Points > Identify Point on Ellipse *(Place a data point or key-in coordinates to define the third point on the ellipse.)*

For example, the following tool sequence shows how to place an ellipse with the Place Ellipse by Edge Points tool (see Figure 4–4).

Place Ellipse By Edge Points > Identify Point on Ellipse **XY=1,2** Enter

Place Ellipse By Edge Points > Identify Point on Ellipse **XY=3,3** Enter

Place Ellipse By Edge Points > Identify Point on Ellipse **DL=4,0** Enter

You can also place an ellipse by edge points by keying-in the primary radius (constrained). To do so, turn ON the toggle button for Primary, key-in the value in MU:SU:PU format, and press Enter or Tab. MicroStation prompts you to identify the two edge points. Similarly, you can also constrain the Secondary radius and Rotation by turning ON the toggle button appropriately, keying-in the values in the edit fields, and pressing Enter or Tab.

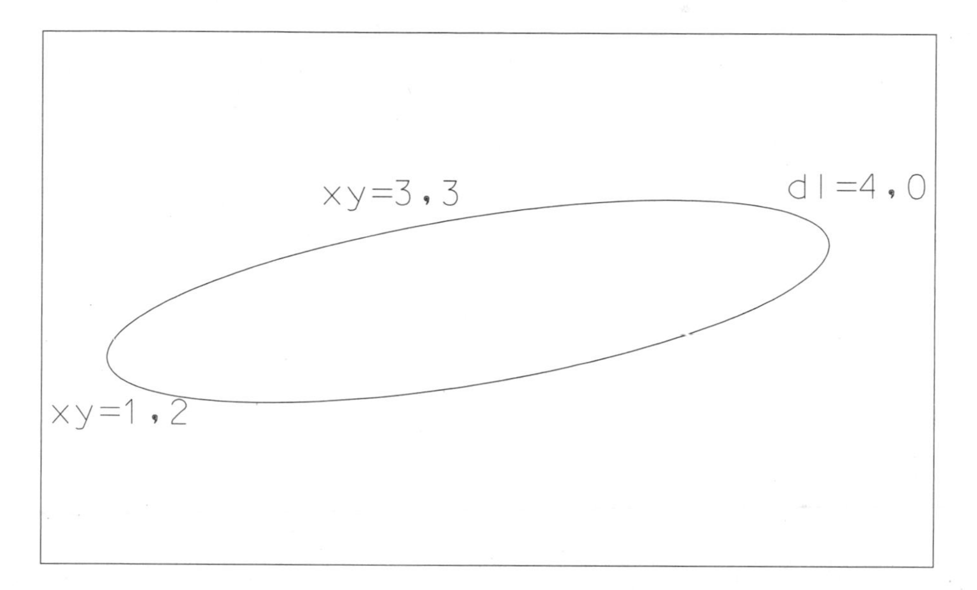

Figure 4–4 Example of placing an ellipse by means of the Place Ellipse By Edge Points tool

MicroStation prompts depend on the number of constraints turned ON. For example, if the Primary and Secondary radii are preset, MicroStation prompts you to identify an edge point and rotation data points. If the Primary radius, Secondary radius, and Rotation are preset, MicroStation prompts you to identify an edge point on the ellipse.

Note: Similar to placing arcs, you can also place half and quarter ellipses. The tools are located in the Arcs tool box.

Place Regular Polygon

You can place regular *two-dimensional* polygons (all edges are equal length, all vertex angles are equal) with the Place Polygon tool. The polygon can have from 3 to 100 sides. MicroStation provides three methods for placing regular polygons: Place Inscribed Polygon, Place Circumscribed Polygon, and Place Polygon by Edge. The appropriate method is selected from the Method option menu located in the Tool Settings window.

Place Inscribed Polygon

The Place Inscribed Polygon tool places a polygon of equal length for all sides inscribed inside an imaginary circle (see Figure 4–5) having the same diameter as the distance across opposite polygon corners.

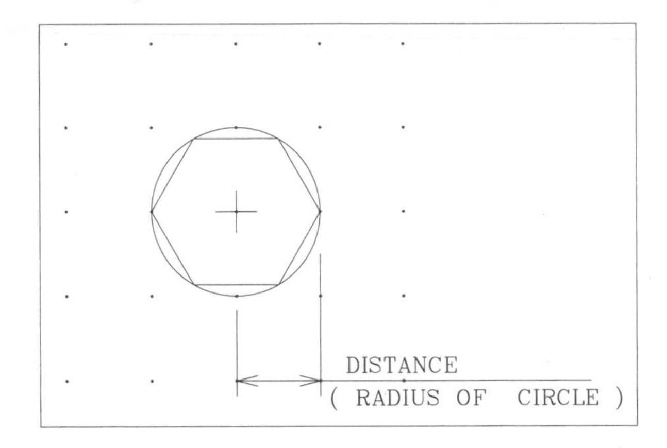

Figure 4–5 Example of displaying a polygon inscribed inside an imaginary circle

Invoke the Place Inscribed Polygon tool:

Polygons tool box	Select the Place Regular Polygon tool, then select Inscribed from the Method option menu located in the Tool Settings window (see Figure 4–6).
Key-in window	**place polygon inscribed** (or **pl pol in**) Enter

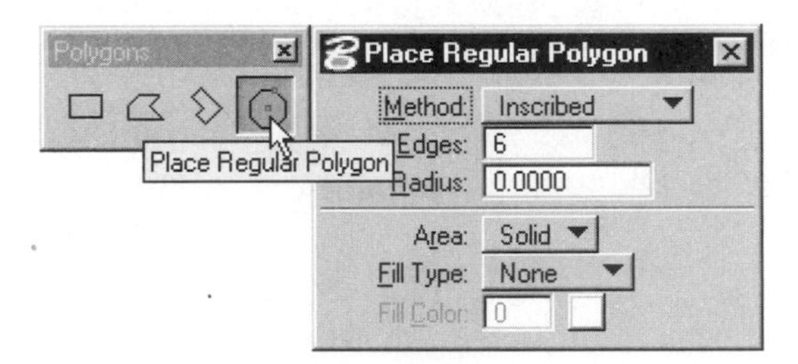

Figure 4–6 Invoke the Place Polygon Inscribed tool from the Polygons tool box

Key-in the number of sides of the polygon and the radius of the imaginary circle in the Edges and Radius edit fields, respectively, located in the Tool Settings window. If you set the Radius to 0, you can define the radius graphically with a data point or you can key-in coordinates.

MicroStation prompts:

Place Inscribed Polygon > Enter point on axis *(Place a data point or key-in coordinates to define the center of the polygon.)*

Place Inscribed Polygon > Enter first edge point *(Place a data point or key-in coordinates to define the radius of the imaginary circle, the polygon's rotation, and one vertex.)*

Place Circumscribed Polygon

The Place Circumscribed Polygon tool places a polygon circumscribed around the outside of an imaginary circle having the same diameter as the distance across the opposite polygon sides (see Figure 4–7).

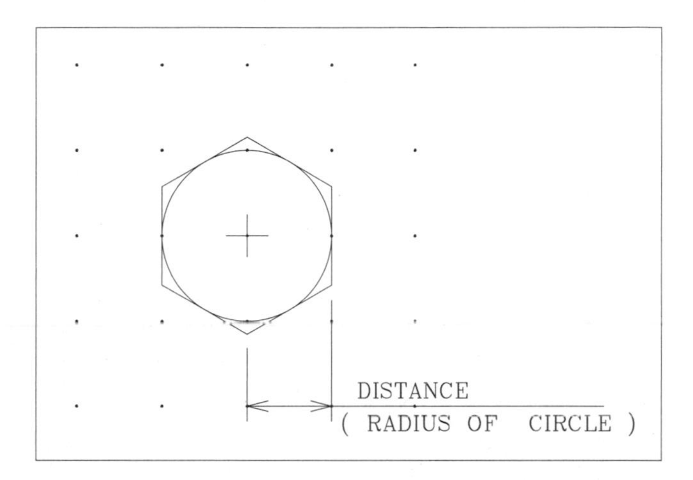

Figure 4–7 Example of displaying a polygon circumscribed around the outside of an imaginary circle

Invoke the Place Circumscribed Polygon tool:

Polygons tool box	Select the Place Regular Polygon tool, then select Circumscribed from the Method option menu located in the Tool Settings window (see Figure 4–8).
Key-in window	**place polygon circumscribed** (or **pl pol c**) Enter

Key-in the number of sides of the polygon and the radius of the imaginary circle in the Edges and Radius edit fields, respectively, located in the Tool Settings window. If you set the radius to 0, you can define the radius graphically with a data point or you can key-in coordinates.

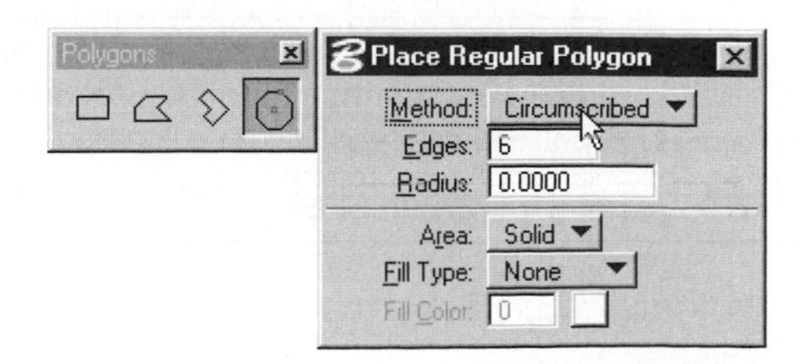

Figure 4–8 Invoke the Place Polygon Circumscribed tool from the Polygons tool box

MicroStation prompts:

Place Circumscribed Polygon > Enter point on axis *(Place a data point or key-in coordinates to define the center of the polygon.)*

Place Circumscribed Polygon > Enter radius or point on circle *(Place a data point or key-in coordinates to define the radius of the imaginary circle, the polygon's rotation, and one vertex.)*

Place Polygon By Edge

The Place Polygon By Edge tool allows you to place a polygon by defining two endpoints of a side of a polygon (see Figure 4–9).

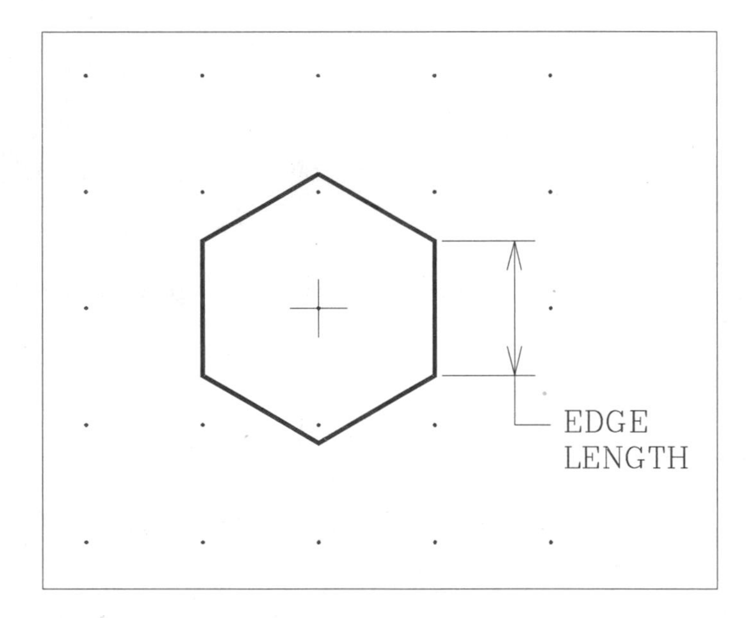

Figure 4–9 Example of a polygon placed by defining two endpoints of its side

Invoke the Place Polygon By Edge tool:

Polygons tool box	Select the Place Regular Polygon tool, then select Edge from the Method option menu located in the Tool Settings window (see Figure 4–10).
Key-in window	**place polygon edge** (or **pl pol ed**) Enter

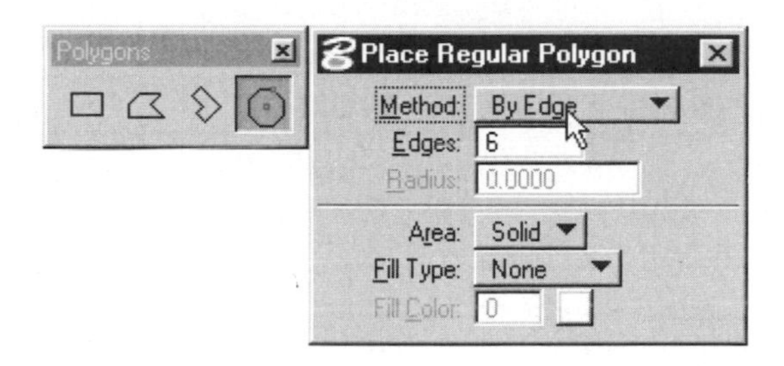

Figure 4–10 Invoke the Place Polygon by Edge tool from the Polygons tool box

Key-in the number of sides of the polygon in the Edges box, located in the Tool Settings window.

MicroStation prompts:

> Place Polygon by Edge > Enter first edge point *(Place a data point or key-in coordinates to define the vertex of the edge.)*
>
> Place Polygon by Edge > Enter next (CCW) edge point *(Place a data point or key-in coordinates to define the second edge point.)*

Place Point Curve

The Place Point Curve tool can place a *2D* (single-plane) curve element. This is accomplished by defining a series of data points the curve passes through. A curve element can have 3 to 97 vertices and is considered as one element. If more than 97 vertices are selected, MicroStation creates a complex chain consisting of one or more curved elements.

Invoke the Place Point Curve tool:

Linear Elements tool box	Select the Place Point or Stream Curve tool, then select Points from the Method option menu located in the Tool Settings window (see Figure 4–11).
Key-in window	**place curve point** (or **pl cu p**) Enter

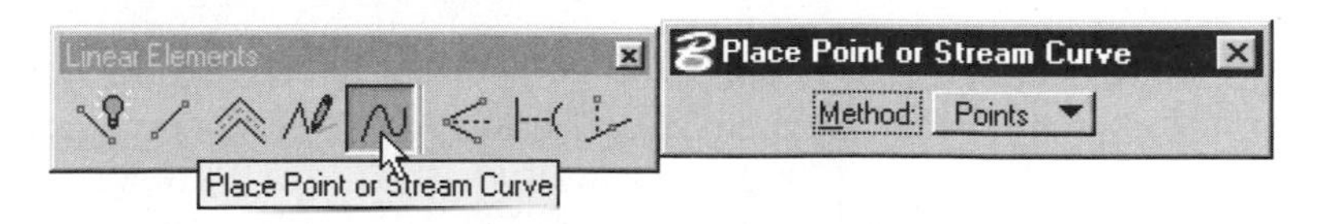

Figure 4–11 Invoke the Place Point Curve tool from the Linear Elements tool box

MicroStation prompts:

> Place Point Curve > Enter first point in curve string *(Place a data point or key-in coordinates to define the starting point of the curve.)*
>
> Place Point Curve > Enter point or RESET to complete *(Place a data point or key-in coordinates to define the next vertex, or click the Reset button to complete.)*

At least three points are required to describe a curved element. Once you are through defining curve data points or key-in coordinates, press the Reset button to terminate the tool sequence. See Figure 4–12 for an example of placing a curve using the Place Point Curve tool by providing five data points.

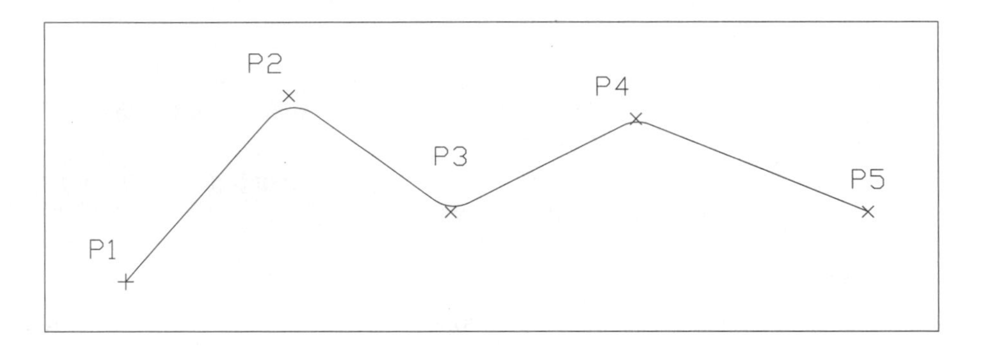

Figure 4–12 Example of placing an arc by the Place Point Curve tool

Place Stream Curve

The Place Stream Curve tool is used to place a curve stream that follows the movement of your cursor. As you move your input device, MicroStation records data points based on stream settings—active stream delta, active stream tolerance, active stream angle, and active stream area. A stream curve element can have 3 to 97 vertices. If more than 97 vertices are defined, MicroStation automatically creates a complex chain consisting of one or more curve elements.

Invoke the Place Curve Stream tool:

Linear Elements tool box	Select the Place Point or Stream Curve tool, then select Stream from the Method option menu located in the Tool Settings window (see Figure 4–13).
Key-in window	**place curve stream** (or **pl cu st**) Enter

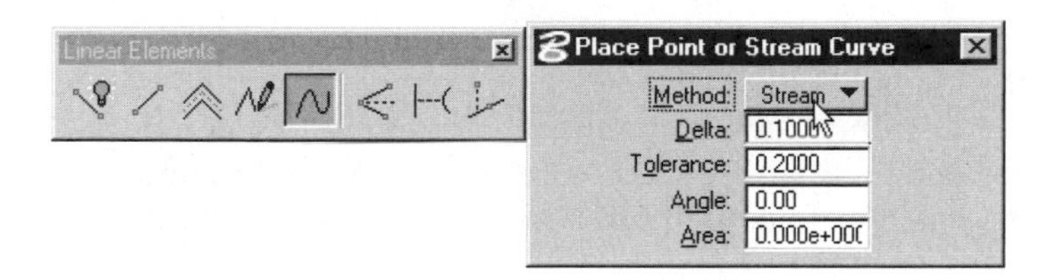

Figure 4–13 Invoke the Place Stream Curve tool from the Linear Elements tool box

MicroStation prompts:

> Place Stream Curve > Enter first point in curve string *(Place a data point or key-in coordinates to define the starting point of the curve stream.)*
>
> Place Stream Curve > Enter point or RESET to complete *(Move your cursor to define the curve stream. When you are finished, press the Reset button to complete the curve stream.)*

See Figure 4–14 for an example of placing a curve stream by means of the Place Curve Stream tool.

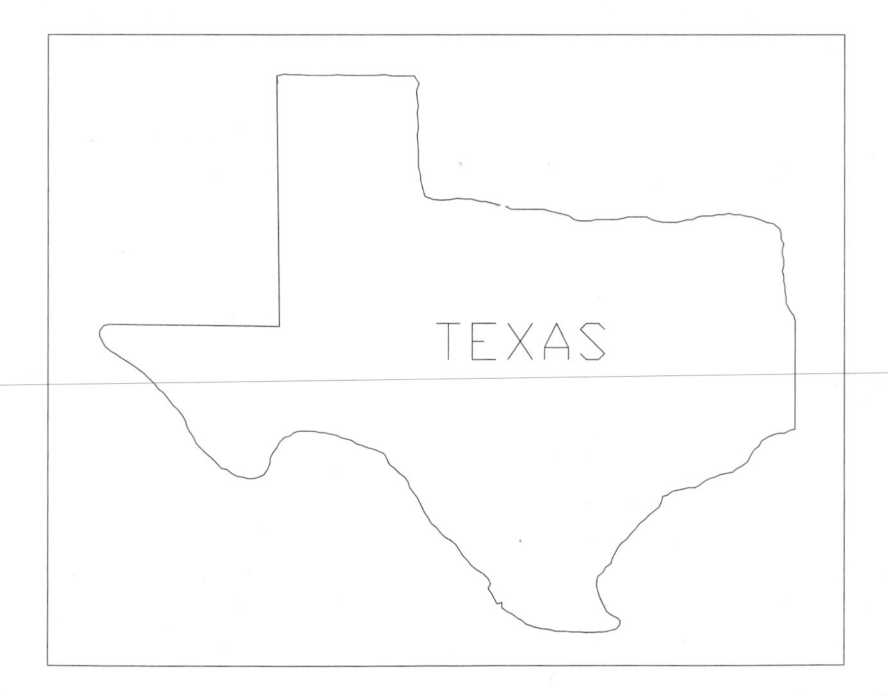

Figure 4–14 Example of placing a curve stream by means of the Place Curve Stream tool

Settings are provided with the Stream method to allow you to change the way the data points are placed: Delta sets the minimum distance in working units between points; Tolerance sets the maximum distance in working units between points; Angle sets the angle in degrees that the direction must change for the sampled point to be recorded as a data point; and Area sets the area that, when exceeded, causes the sampled point to be recorded as a data point.

Place Multi-line

With the Place Multi-line tool you can draw multiple parallel line segments, which are considered as one element. A multi-line can consist of as many as 16 separate, parallel lines of various line styles, weights, and colors, and the Place Multi-line tool allows you to draw a multi-line that is currently set as the active definition. If necessary, you can create or modify an existing multi-line definition with the help of the Multi-line settings box invoked from the Element pull-down menu. (Refer to Chapter 16 for a detailed description of creating or modifying an existing multi-line definition.)

Invoke the Place Multi-line tool:

Linear Elements tool box	Select the Place Multi-line tool (see Figure 4–15).
Key-in window	**place mline constrained** (or **pl m c**) Enter

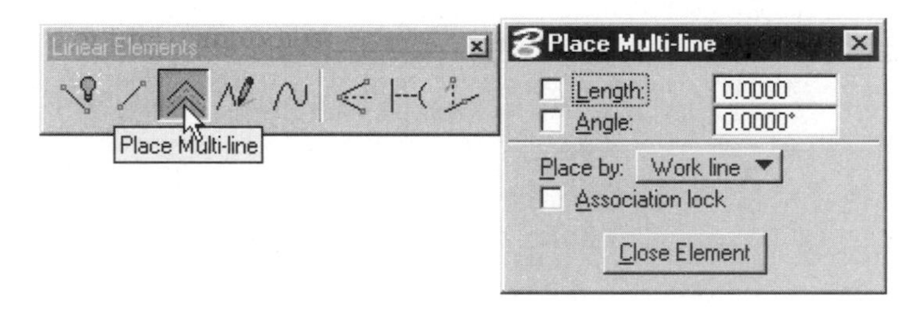

Figure 4–15 Invoke the Place Multi-line tool from the Linear Elements tool box

MicroStation prompts:

Place Multi-line > Enter first point *(Place a data point or key-in coordinates to define the starting point of the multi-line.)*

Place Multi-line > Enter vertex or Reset to complete *(Place a data point or key-in coordinates to define a vertex, or press the Reset button to complete.)*

Place Multi-line to a Specified Length

To place a multi-line to a specified length, select the Place Multi-line tool in the Linear Elements tool bar and turn ON the toggle button for Length in the Tool Settings window. Key-in the distance in MU:SU:PU in the Length edit field. The prompts are similar to those for the Place Multi-line tool, and you can place any number of line segments of specified length.

Place Multi-line to an Angle

To place a multi-line to a specified angle, select the Place Multi-line tool in the Linear Elements tool bar and turn ON the toggle button for Angle in the Tool Settings window. Key-in the angle in the Angle edit field. The prompts are similar to those for the Place Multi-line tool, and you can place any number of line segments of specified angle.

If necessary, you can turn both of the toggle buttons ON for Length and Angle, and MicroStation allows you to place a multi-line with a specified length and angle constrained.

MicroStation has a set of tools to edit Multi-lines called Multi-line Joints. See Chapter 8 for details about using Multi-line Joints.

ELEMENT MODIFICATION

MicroStation not only allows you to place elements easily, but also allows you to modify them as needed. This section discusses four important tools that will make your job easier: Fillet, Chamfer, Trim, and Delete part of element.

Construct Circular Fillet

The Construct Circular Fillet tool joins two elements (lines, line strings, circular arcs, circles, or shapes), two segments of a line string, or two sides of a shape with an arc of a specified radius. The arc will be placed tangent to the two elements it connects.

MicroStation constructs a circular fillet depending on the option selected from the Truncate option menu located in the Tools Settings window.

The None option places the fillet arc, but does not truncate the selected sides. The Both option places the fillet arc and at the same time truncates with the fillet at their point of tangency. The First option places the fillet arc and truncates the first side identified. See Figure 4–16 for examples of placing a fillet by the three different Truncate option methods.

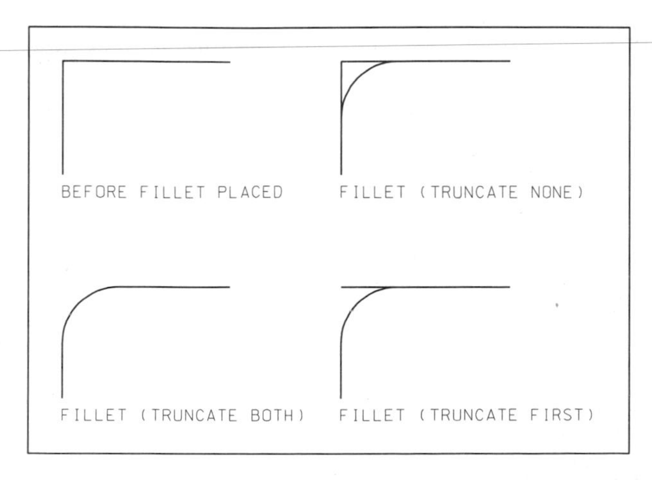

Figure 4–16 Examples of placing a fillet using the three truncation methods

Invoke the Construct Circular Fillet (no truncation) tool:

Modify tool box	Select the Construct Circular Fillet tool, key-in the Radius in MU:SU:PU in the Tool Settings window, then select None from the Truncate option menu located in the Tool Settings window (see Figure 4–17).
Key-in window	**fillet nomodify** (or **fill n**) Enter

Figure 4–17 Invoke the Circular Fillet (no truncation) tool from the Modify tool box

MicroStation prompts:

Circular Fillet (no truncation) > Select first segment *(Identify the first element or segment.)*

Circular Fillet (no truncation) > Select second segment *(Identify the second element or segment.)*

Circular Fillet (no truncation) > Accept-Initiate construction *(Click the Accept button to accept the placement of the fillet, or click the Reject button to reject the placement of the fillet.)*

Invoke the Construct Circular Fillet (Truncate Both) tool:

Modify tool box	Select the Construct Circular Fillet tool, key-in the Radius in MU:SU:PU in the Tool Settings window, then select Both from the Truncate option menu located in the Tool Settings window.
Key-in window	**fillet modify** (or **fill m**) Enter

MicroStation prompts:

Circular Fillet and Truncate Both > Select first segment *(Identify the first element or segment.)*

Circular Fillet and Truncate Both > Select second segment *(Identify the second element or segment.)*

Circular Fillet and Truncate Both > Accept-Initiate construction *(Click the Accept button to accept the placement of the fillet, or click the Reject button to reject the placement of the fillet.)*

Invoke the Construct Circular Fillet (Truncate Single) tool:

Modify tool box	Select the Construct Circular Fillet tool, key-in the Radius in MU:SU:PU in the Tool Settings window, and select First from the Truncate option menu located in the Tool Settings window.
Key-in window	**fillet single** (or **fill s**) Enter

MicroStation prompts:

Circular Fillet and Truncate Single > Select first segment *(Identify the first element or segment – the one that is to be truncated.)*

Circular Fillet and Truncate Single > Select second segment *(Identify the second element or segment.)*

Circular Fillet and Truncate Single > Accept-Initiate construction *(Click the Accept button to accept the placement of the fillet, or click the Reject button to reject the placement of the fillet.)*

Construct Chamfer

The Construct Chamfer tool is similar to the Construct Fillet tool, but it allows you to draw an angled corner instead of an arc. Its distance from the corner determines the size of the chamfer. If you want it to be a 45-degree chamfer, the two distances have to be the same. The Construct Chamfer tool can help you to construct a chamfer between two lines or between adjacent segments of a line string or shape.

Invoke the Construct Chamfer tool:

Modify tool box	Select the Construct Chamfer tool, and key-in the appropriate Distance 1 and Distance 2 in MU:SU:PU in the Tool Settings window (see Figure 4–18).
Key-in window	**chamfer** (or **ch**) Enter

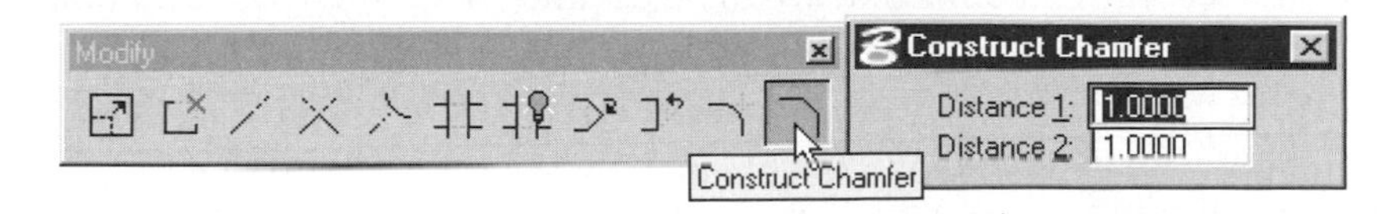

Figure 4–18 Invoke the Construct Chamfer tool from the Modify tool box

MicroStation prompts:

> Construct Chamfer > Select first chamfer segment *(Identify the first element or segment, as shown in Figure 4–19.)*
>
> Construct Chamfer > Select second chamfer segment *(Identify the second element or segment, as shown in Figure 4–19.)*
>
> Construct Chamfer > Accept-Initiate construction *(Click the Accept button to accept the placement of the chamfer, or click the Reject button to reject the placement of the chamfer.)*

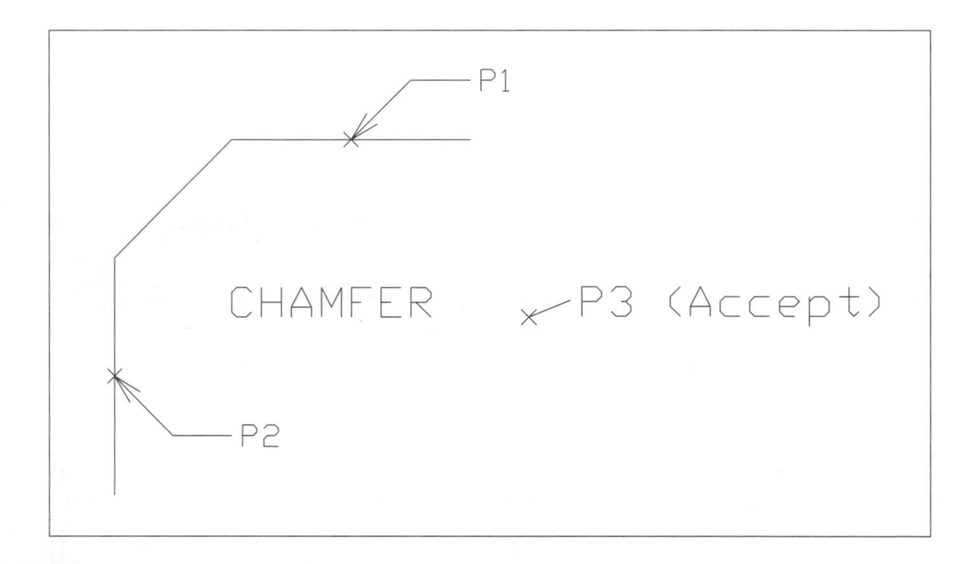

Figure 4–19 Example of placing the chamfer with the Construct Chamfer tool

Trim Elements

IntelliTrim combines into one tool the ability to Trim the part of selected elements that overlap a cutting element, Extend elements to a cutting element, or Cut elements into pieces. In addition, the tool provides a Quick mode that allows selecting one cutting element, and an Advanced mode that allows selecting one or more cutting elements.

- As you select the elements that are to be trimmed, extended, or cut, square guide posts appear at the point where the operation will take place (such as the point to which the elements will be extended).
- Elements can be extended to the elements, such as arcs, cell headers, complex shapes, complex strings, curves, ellipses, lines, line strings, shapes, and text nodes.
- Elements can be cut or trimmed, such as arcs, b-spline curves, complex shapes, complex strings, curves, ellipses, lines, line strings, and shapes.

- Elements that can be extended, such as b-spline curves, complex chains that end with a line or line string, lines, and line strings. Elements that cannot be extended will usually be deleted.

Quick Trim

When IntelliTrim is set to the Quick Mode and Trim Operation, one cutting element is selected, and temporary lines are drawn across all elements that are to be trimmed to the cutting element. The temporary lines define the part of the elements that are to be removed when the trimming takes place.

To execute a quick trim, invoke the IntelliTrim tool:

Modify tool box	Select the IntelliTrim Elements tool, then select the Quick Mode and the Trim Operation from the Tool Settings window (see Figure 4–20).
Key-in window	**trim multi** (or **tri m**) Enter (Then set the Mode to Quick and the Operation to Trim in the Tool Settings window.)

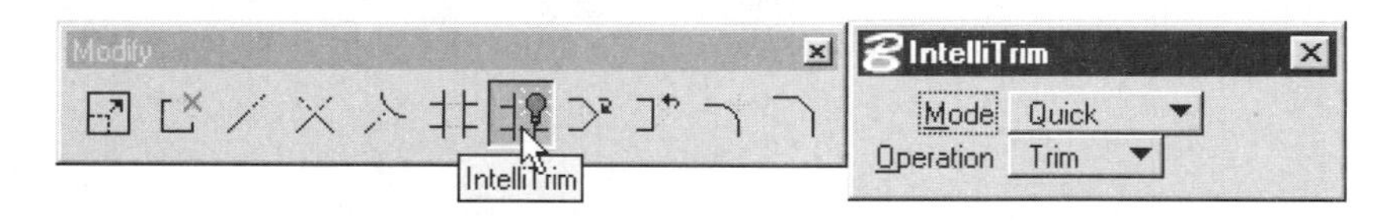

Figure 4–20 Invoke the IntelliTrim tool from the Modify tool box, then select the Quick Mode and the Trim Operation

MicroStation prompts:

IntelliTrim > Identify element *(Identify the element to define it as the cutting element.)*

IntelliTrim > Enter start point of the line *(Place a data point to identify the location of the start of the line that will define the part of the elements to be trimmed.)*

IntelliTrim > Enter endpoint of the line *(Place a data point to identify the location of the end of the temporary line that will define the part of the elements to be trimmed. The temporary line should pass across the part of the elements to be trimmed.)*

IntelliTrim > Enter start point of the line *(Continue identifying the start and endpoints of temporary lines until all elements to be trimmed are identified, then click the Reset button to complete the trimming operation.)*

For example, the following tool sequence shows how to use the IntelliTrim tool's Quick Mode to Trim the parts of a set of lines that are outside of an ellipse. The sequence of commands is illustrated in Figure 4–21.

IntelliTrim > Identify Element *(Identify the ellipse that will be the cutting element.)*

IntelliTrim > Enter start point of the line *(Place a data point on the left side of the ellipse, above the lines, to start the temporary line that will identify the lines to be trimmed on the left side of the ellipse.)*

IntelliTrim > Enter endpoint of the line *(Place a data point below the lines to identify the end of the temporary line so that it passes across each of the three lines on the left side of the ellipse.)*

IntelliTrim > Enter start point of the line *(Place a data point on the right side of the ellipse, above the lines, to start the temporary line that will identify the lines to be trimmed on the right side of the ellipse.)*

IntelliTrim > Enter endpoint of the line *(Place a data point to identify the end of the temporary line so that it passes across each of the three lines on the right side of the ellipse.)*

IntelliTrim > Enter start point of the line *(Click the Reset button to complete trimming the part of the lines outside of the ellipse.)*

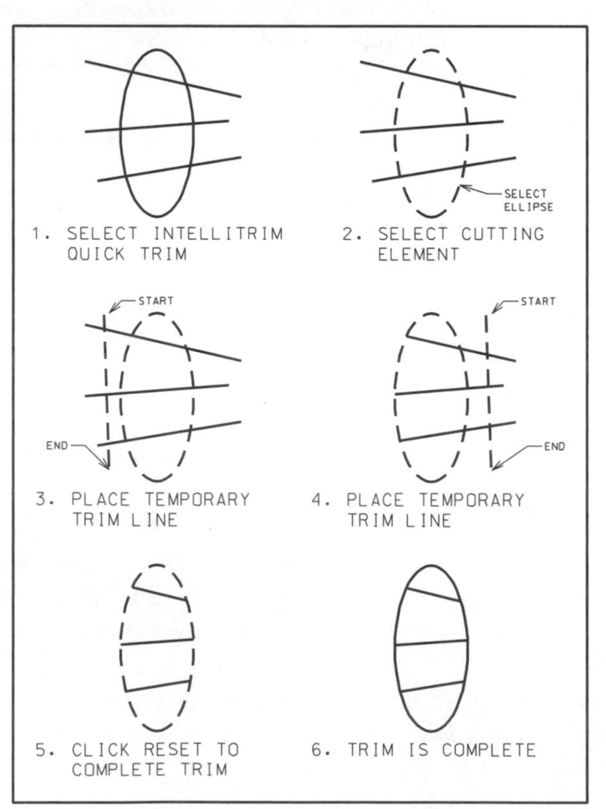

Figure 4–21 Example of trimming the parts of three lines that are outside of an ellipse

Quick Extend

When IntelliTrim is set to the Quick Mode and Extend Operation, one boundary element is selected, and temporary lines are drawn across all elements that are to be extended to the cutting element.

To execute a quick extend, invoke the IntelliTrim tool:

Modify tool box	Select the IntelliTrim Elements tool, then select the Quick Mode and Extend Operation from the Tool Settings window (see Figure 4–22).
Key-in window	**trim multi** (or **tri m**) Enter (Then set the Mode to Quick and the Operation to Extend in the Tool Settings window.)

Figure 4–22 Invoke the IntelliTrim tool from the Modify tool box, then select the Quick Mode and Extend Operation

MicroStation prompts:

IntelliTrim > Identify element *(Identify the element to define as the boundary element.)*

IntelliTrim > Enter start point of the line *(Place a data point to identify the location of the start of the line that will define elements to be extended.)*

IntelliTrim > Enter endpoint of the line *(Place a data point to identify the location of the end of the temporary line that will be extended. The temporary line should extend across the elements to be extended.)*

IntelliTrim > Enter start point of the line *(Continue identifying temporary lines until all elements to be extended are identified, then click the Reset button to complete the extension operation.)*

For example, the following tool sequence shows how to use the IntelliTrim tool's Quick Mode to Extend three lines to the left side of an ellipse. The sequence of commands is illustrated in Figure 4–23.

IntelliTrim > Identify Element *(Identify the ellipse that will be the boundary element.)*

IntelliTrim > Enter start point of the line *(Place a data point on the left side of the ellipse to start the temporary line that will identify elements to be extended.)*

IntelliTrim > Enter endpoint of the line *(Place a data point to identify the end of the temporary line so that it passes across each of the three lines to be extended.)*

IntelliTrim > Enter start point of the line *(Click the Reset button to complete trimming the part of the lines outside of the ellipse.)*

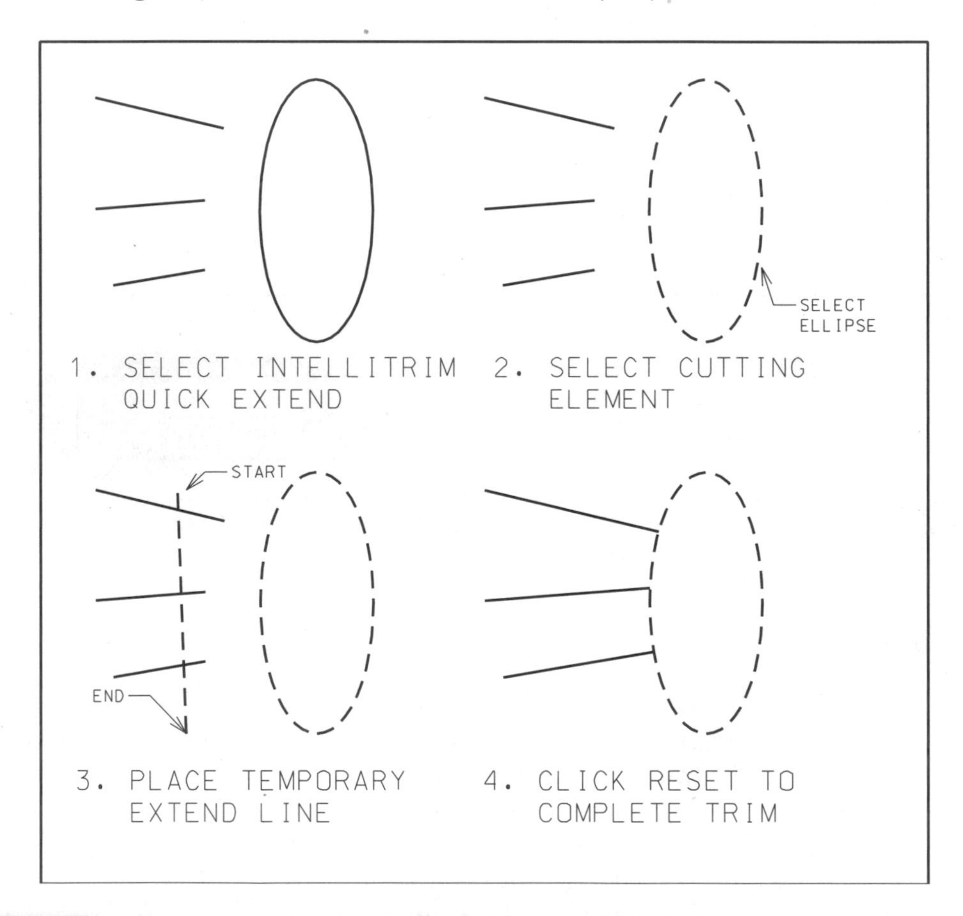
SELECT
ELLIPSE
1. SELECT INTELLITRIM
QUICK EXTEND
2. SELECT CUTTING
ELEMENT
START
END
3. PLACE TEMPORARY
EXTEND LINE
4. CLICK RESET TO
COMPLETE TRIM

Figure 4–23 Example of extending three lines that are outside of an ellipse

Quick Cut

When IntelliTrim is set to the Quick Mode and Cut Operation, temporary lines are drawn across all elements that are to be cut. The temporary lines define the position of the cuts on each element.

The cut elements are broken into separate elements whose endpoints are where temporary lines cross the original elements. Normally, you won't be able to tell that anything has changed until you select one of the new elements (such as by selecting one of them for deletion).

To execute a quick cut, invoke the IntelliTrim tool:

Modify tool box	Select the IntelliTrim Elements tool, then select the Quick Mode and the Cut Operation from the Tool Settings window (see Figure 4–24).
Key-in window	**trim multi** (or **tri m**) Enter (Then set the Mode to Quick and the Operation to Cut in the Tool Settings window.)

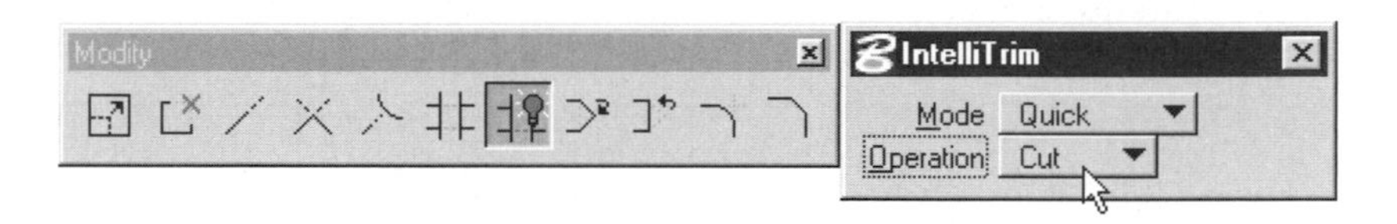

Figure 4–24 Invoke the IntelliTrim tool from the Modify tool box, then select the Quick Mode and the Cut Operation

MicroStation prompts:

> IntelliTrim > Enter start point of the line *(Place a data point to identify the location of the start of the line that will define the position of one of the cuts to be made on the elements it crosses.)*
>
> IntelliTrim > Enter endpoint of the line *(Place a data point to identify the location of the end of the temporary line that will define the cut points. The temporary line should cross the elements to be cut at the points where they are to be cut.)*
>
> IntelliTrim > Enter start point of the line *(Continue identifying temporary lines until all cut points are identified, then click the Reset button to complete the cutting operation.)*

For example, the following tool sequence shows how to use the IntelliTrim tool's Quick Mode to Cut two segments out of an ellipse. The sequence of commands is illustrated in Figure 4–25.

> IntelliTrim > Enter start point of the line *(Place a data point on the left side of the ellipse to start the temporary line that will identify the top of the segment to be cut out of the ellipse.)*
>
> IntelliTrim > Enter endpoint of the line *(Place a data point on the right side of the ellipse to identify the end of the temporary line that will identify the top of the segment to be cut out of the ellipse.)*
>
> IntelliTrim > Enter start point of the line *(Place a data point on the left side of the ellipse to start the temporary line that will identify the bottom of the segment to be cut out of the ellipse.)*

IntelliTrim > Enter endpoint of the line *(Place a data point on the right side of the ellipse to identify the end of the temporary line that will identify the bottom of the segment to be cut out of the ellipse.)*

IntelliTrim > Enter start point of the line *(Click the Reset button to complete making the cuts in the ellipse.)*

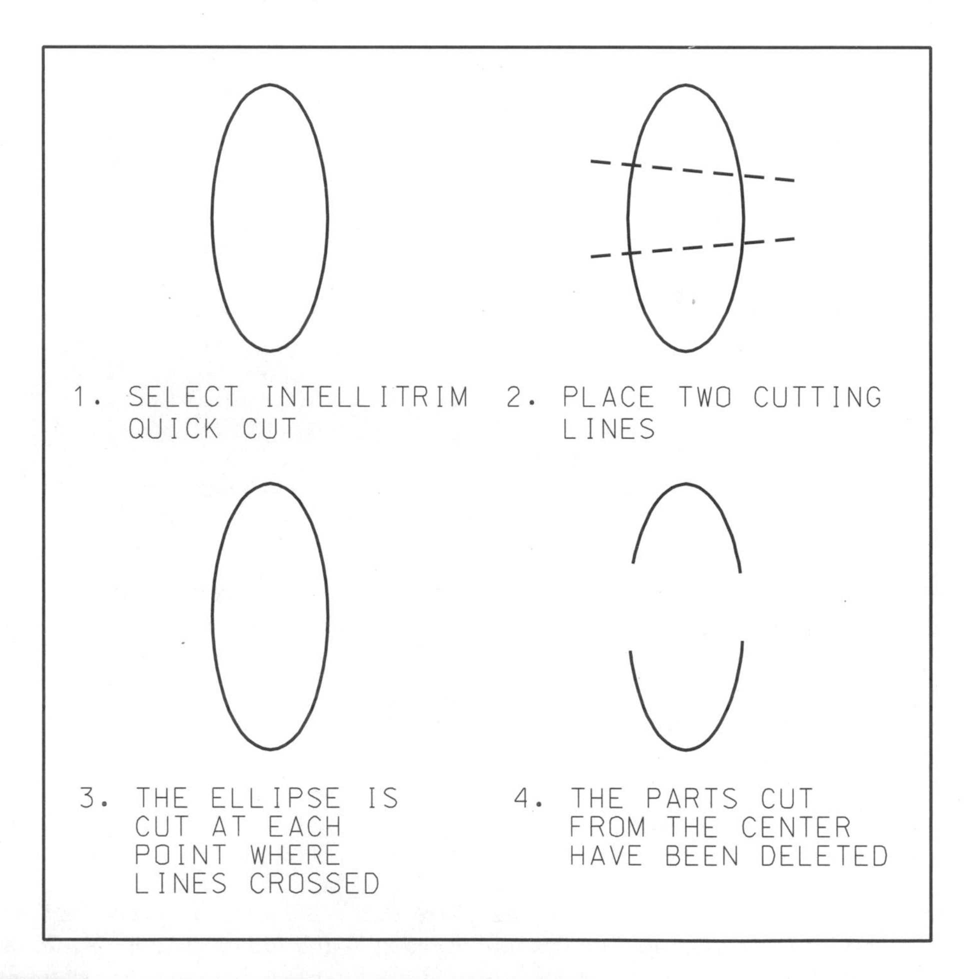
1. SELECT INTELLITRIM QUICK CUT
2. PLACE TWO CUTTING LINES
3. THE ELLIPSE IS CUT AT EACH POINT WHERE LINES CROSSED
4. THE PARTS CUT FROM THE CENTER HAVE BEEN DELETED

Figure 4–25 Example of making four cuts on an ellipse

Advanced Trim

When IntelliTrim is set to the Advanced Mode and Trim Operation, one or more cutting elements can be selected, and one or more elements can be trimmed to the cutting elements.

To execute an advanced trim, invoke the IntelliTrim tool:

Modify tool box	Select the IntelliTrim Elements tool, then select the Advanced Mode and the Trim Operation from the Tool Settings window, and, if necessary, select Cutting Elements radio button in the Tool Settings window (see Figure 4–26).
Key-in window	**trim multi** (or **tri m**) Enter (Then set the Mode to Advanced and the Operation to Trim in the Tool Settings window.)

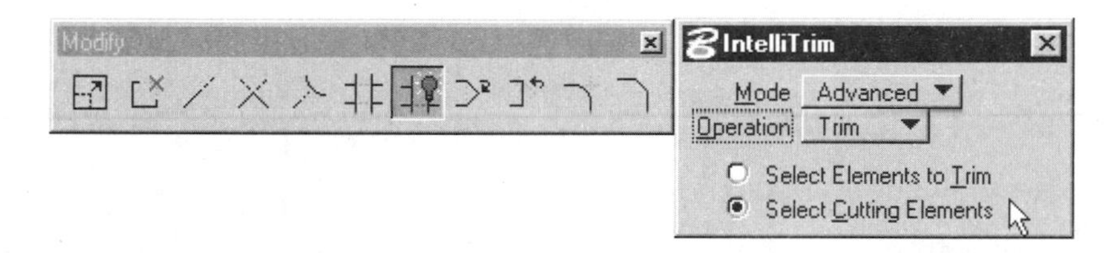

Figure 4–26 Invoke the IntelliTrim tool from the Modify tool box, then select the Advanced Mode and the Trim Operation

MicroStation prompts:

IntelliTrim > Identify cutting elements, reset to complete step *(Select each cutting element by clicking a data point on it, then click the Reset button.)*

IntelliTrim > Identify elements to trim, reset to complete step *(Select each element to be trimmed by clicking a data point on it, then click the Reset button.)*

IntelliTrim > Enter points near portions to keep, reset to complete command *(If the wrong part of any element has been trimmed, click a data point near it to switch the part that is trimmed. When the correct part of each element is shown trimmed, click the Reset button to make the trimming operation permanent.)*

For example, the following tool sequence shows how to use the IntelliTrim tool's Advanced Mode and Trim Operation to use two orthogonal blocks as cutting elements for trimming two shapes. The sequence of commands is illustrated in Figure 4–27.

IntelliTrim > Identify cutting elements, reset to complete step *(Select one of the orthogonal blocks by clicking a data point on it.)*

IntelliTrim > Identify cutting elements, reset to complete step *(Select the other orthogonal block by clicking a data point on it, then click the Reset button.)*

IntelliTrim > Identify elements to trim, reset to complete step *(Select one of the shapes by clicking a data point on it.)*

IntelliTrim > Identify elements to trim, reset to complete step *(Select the other shape by clicking a data point on it, then click the Reset button.)*

IntelliTrim > Enter points near portions to keep, reset to complete command *(If the wrong part of a shape was trimmed, click a data point next to it to switch the part to be trimmed. When the correct part of each shape is selected, click the Reset button to complete the trimming operation.)*

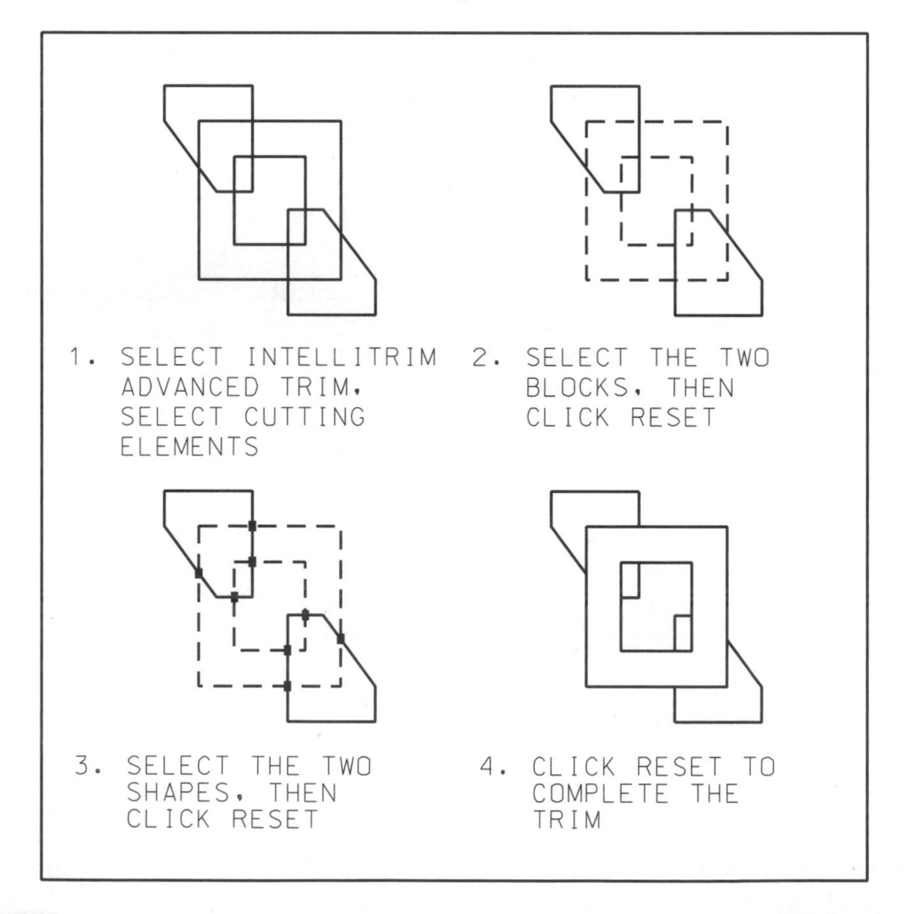

1. SELECT INTELLITRIM ADVANCED TRIM, SELECT CUTTING ELEMENTS
2. SELECT THE TWO BLOCKS, THEN CLICK RESET
3. SELECT THE TWO SHAPES, THEN CLICK RESET
4. CLICK RESET TO COMPLETE THE TRIM

Figure 4–27 Example of using two blocks as cutting elements for trimming two shapes

Advanced Extend

When IntelliTrim is set to the Advanced Mode and Extend Operation, one or more boundary elements can be selected, and one or more elements can be extended to the boundary elements.

To execute an advanced extend, invoke the IntelliTrim tool:

Modify tool box	Select the IntelliTrim Elements tool, then select the Advanced Mode and the Extend Operation from the Tool Settings window, and, if necessary, select Cutting Elements radio button in the Tool Settings window (see Figure 4–28).
Key-in window	**trim multi** (or **tri m**) Enter (Then set the Mode to Advanced and the Operation to Extend in the Tool Settings window.)

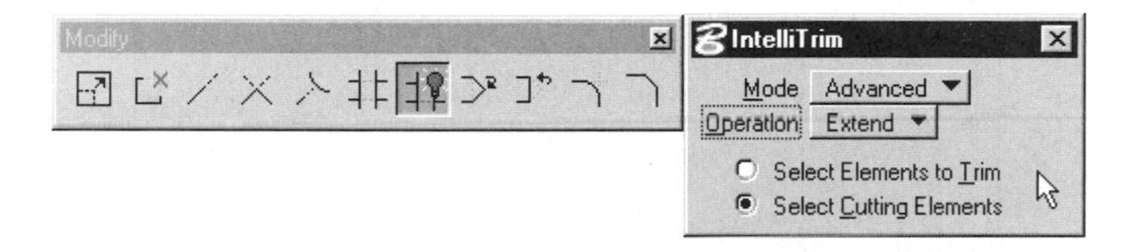

Figure 4–28 Invoke the IntelliTrim tool from the Modify tool box, then select the Advanced Mode and the Extend Operation

MicroStation prompts:

IntelliTrim > Identify cutting elements, reset to complete step *(Select each boundary element by clicking a data point on it, then click the Reset button.)*

IntelliTrim > Identify elements to extend, reset to complete step *(Select each element to be extended by clicking a data point on it, then click the Reset button.)*

IntelliTrim > Enter points near portions to keep, reset to complete command *(Click the Reset button to make the extend operation permanent.)*

For example, the following tool sequence shows how to use the IntelliTrim tool's Advanced Mode and Extend Operation to extend one line to one cutting block and another line to a second cutting block. The sequence of commands is illustrated in Figure 4–29.

IntelliTrim > Identify cutting elements, reset to complete step *(Select one of the orthogonal blocks by clicking a data point on it.)*

IntelliTrim > Identify cutting elements, reset to complete step *(Select the other orthogonal block by clicking a data point on it, then click the Reset button.)*

IntelliTrim > Identify elements to extend, reset to complete step *(Select one of the lines by clicking a data point on it.)*

IntelliTrim > Identify elements to extend, reset to complete step *(Select the other line by clicking a data point on it, then click the Reset button.)*

IntelliTrim > Enter points near portions to keep, reset to complete command *(Click the Reset button to complete the trimming operation.)*

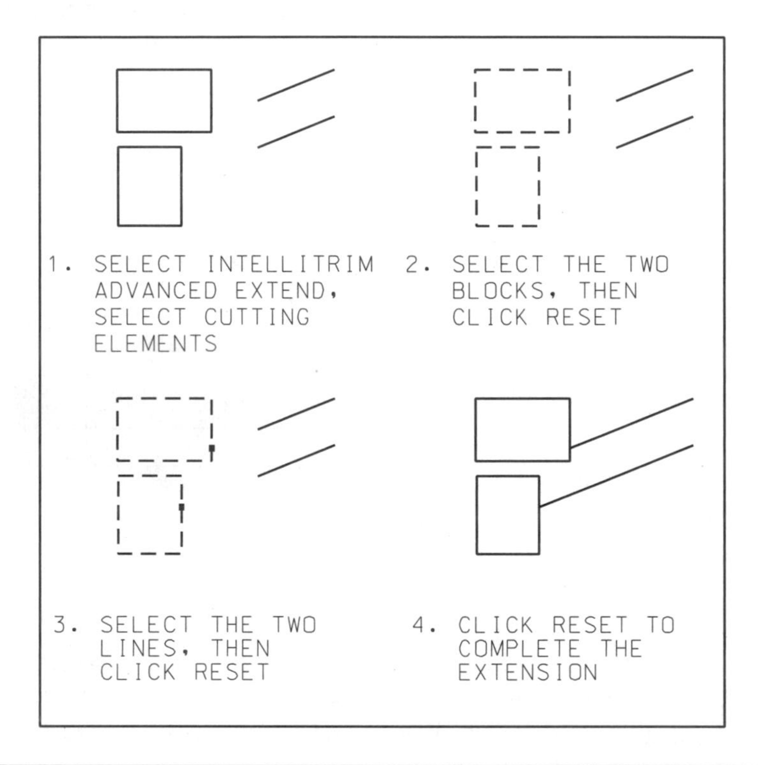

1. SELECT INTELLITRIM ADVANCED EXTEND, SELECT CUTTING ELEMENTS
2. SELECT THE TWO BLOCKS, THEN CLICK RESET
3. SELECT THE TWO LINES, THEN CLICK RESET
4. CLICK RESET TO COMPLETE THE EXTENSION

Figure 4–29 Example of using two blocks as cutting elements for extending two lines

Partial Delete

The Partial Delete tool allows you to delete part of an element. In the case of a line, line string, multi-line, curve, or arc, the Partial Delete tool removes part of the element, and the element is divided into two elements of the same type. A partially deleted ellipse or circle becomes an arc, and a shape becomes a line string.

Invoke the Partial Delete tool:

Modify tool box	Select the Partial Delete tool (see Figure 4–30.)
Key-in window	**delete partial** (or **del p**) Enter

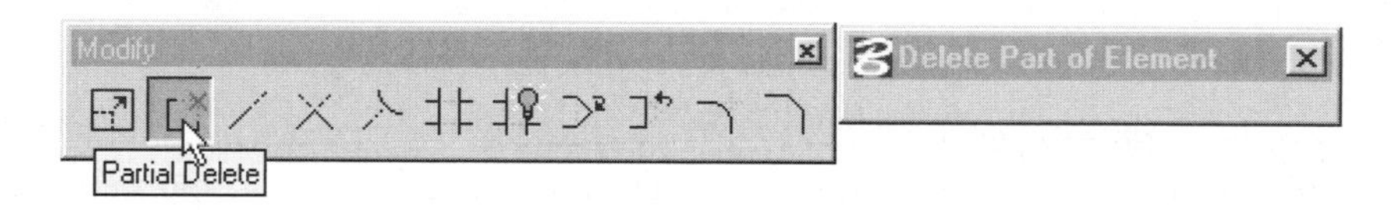

Figure 4–30 Invoke the Partial Delete tool from the Modify tool box

MicroStation prompts:

Delete Part of Element > Select start point for partial delete *(Identify the element where you want to start deleting partially.)*

Delete Part of Element > Select direction of partial delete *(This prompt appears only when you select a closed element, such as a circle, ellipse, or polygon. Move the pointer a short distance in the direction you want to cut, then click the data button.)*

Delete Part of Element > Select end pnt for partial delete *(Place a data point or key-in coordinates for the endpoint for partial delete.)*

See Figure 4–31 for examples of the Partial Delete tool.

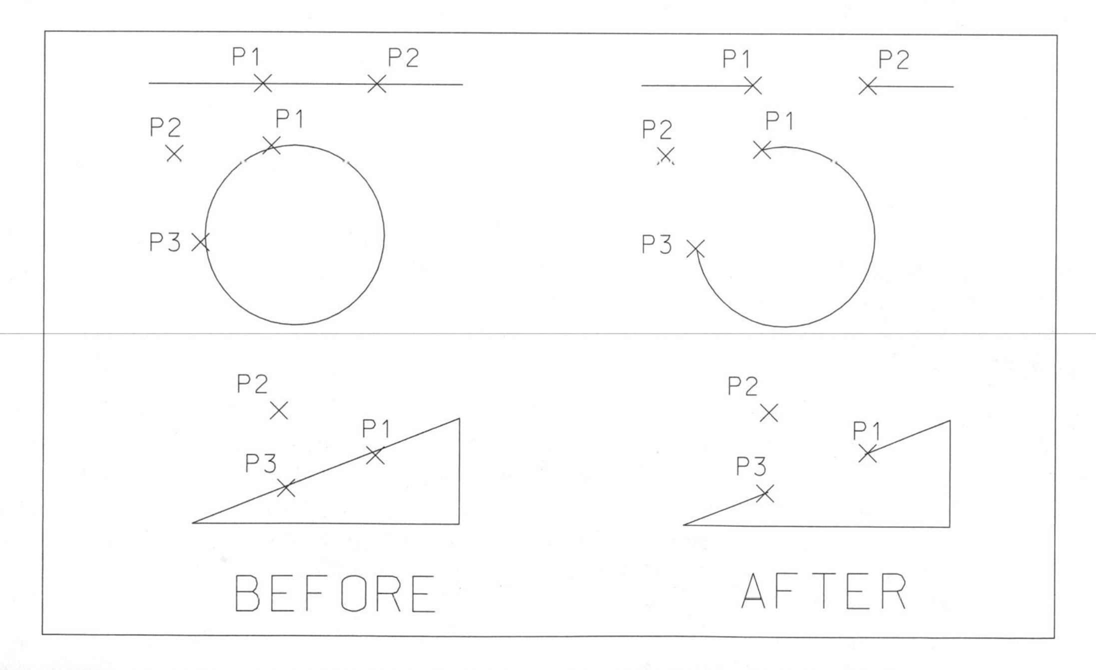

Figure 4–31 Examples of deleting part of an element with the Partial Delete tool

ELEMENT MANIPULATION

MicroStation not only allows you to draw entities easily, but also allows easy manipulation and modification of the objects you have drawn. To manipulate or modify an element is to make a change to one of its existing characteristics.

MicroStation offers two main categories of manipulation tools: single-element manipulation and multi-element manipulations. Single-element manipulation tools allow you to manipulate one element at a time, and multi-element manipulation tools manipulate groups of elements. This section discusses single-element manipulation tools. Multi-element manipulations are done with Element Selection tools and Fence manipulation tools (see Chapter 6).

After mastering the element manipulation tools and learning when and how to apply them, you will appreciate the power and capability of MicroStation. You can draw one element, then use the element manipulation tools to make copies quickly, saving you from having to draw each one separately. You will soon begin to plan ahead to utilize these powerful tools.

All the manipulation tools described here require you to identify the element to be manipulated, then to accept it. To identify an element, position the cursor until it touches the element and click the Data button. At this time the element highlights by changing color. If the highlighted element is the one you wanted to select, continue following the tool prompts shown in the prompt field. If the element that is highlighted is *not* the one you wanted to manipulate, click the Reject button to reject the element and try again.

Note: Be sure to check the status of your lock settings before you begin to modify your design. A good rule to follow is to turn OFF all of the locks that are not being used with the exception of the Snap Lock. It is very frustrating to try to select an element that does not lie on the grid when the Grid Lock is turned ON. The cursor bounces around from grid dot to grid dot, and it may be impossible to identify an element if it is not on the grid. It is simple to toggle the grid lock OFF quickly, identify the object, and then toggle the grid lock back ON again, if needed.

Copy Element

The Copy Element tool places a copy of the element at the specified displacement, leaving the original element intact. The copy is oriented and scaled the same as the original. You can make as many copies of the original as needed. Each resulting copy is completely independent of the original and can be manipulated and modified like any other element. The data point you enter to identify the element you want to copy becomes the (base) point on the element to which the cursor is attached. Select this point with care, and use the tentative snap if you need to snap to the element at a precise location (with the appropriate snap mode selected).

You can place any number of copies in your design file. Once you are through placing copies, click the Reset button to terminate the tool sequence. If necessary, you can select another element to copy, or you can invoke another tool to continue working on your design file.

Invoke the Copy Element tool:

Manipulate tool box	Select the Copy tool, then turn the Make Copy button ON in the Tool Settings window (see Figure 4–32).
Key-in window	**copy element** (or **cop el**) Enter

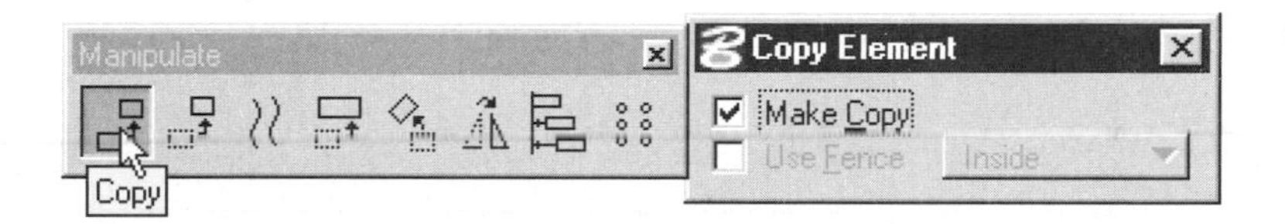

Figure 4–32 Invoke the Copy tool from the Manipulate tool box

MicroStation prompts:

Copy Element > Identify element *(Identify an element to copy.)*

Copy Element > Enter point to define distance and direction *(Provide the location of the copy by a data point or by keying-in coordinates.)*

Copy Element > Enter point to define distance and direction *(If necessary, copy to another location by providing a data point or by keying-in coordinates, and/or click the Reset button to terminate the tool sequence.)*

For example, the following tool sequence shows how to copy a line to the center of a circle using the Copy Element tool (see Figure 4–33).

Copy Element > Identify element *(Identify the line by snapping to the endpoint of the line.)*

Copy Element > Enter point to define distance and direction *(Snap to the center of the circle.)*

Copy Element > Enter point to define distance and direction *(Click the Reset button.)*

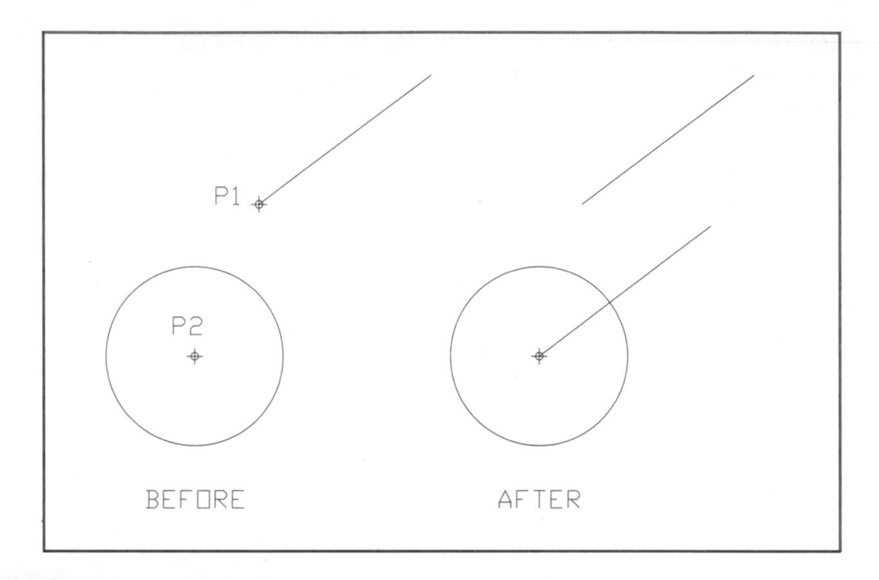

Figure 4–33 Example of copying an element by means of the Copy tool

Move Element

The Move Element tool allows you to move an element from one location to a new location without changing its orientation or size. The data point you enter to identify the element you want to move also becomes the (base) point on the element to which the cursor is attached. Select this point with care, and use the tentative snap if you need to snap to the element at a precise location (with the appropriate snap mode selected). Once you have moved the element to an appropriate location, click the Reset button to terminate the tool sequence. After clicking the Reset button, you can select another element to move, or you can invoke another tool to continue working on your design file.

Invoke the Move Element tool:

Manipulate tool box	Select the Move tool, then turn the Make Copy button OFF in the Tool Settings window (see Figure 4–34).
Key-in window	**move element** (or **mov e**) Enter

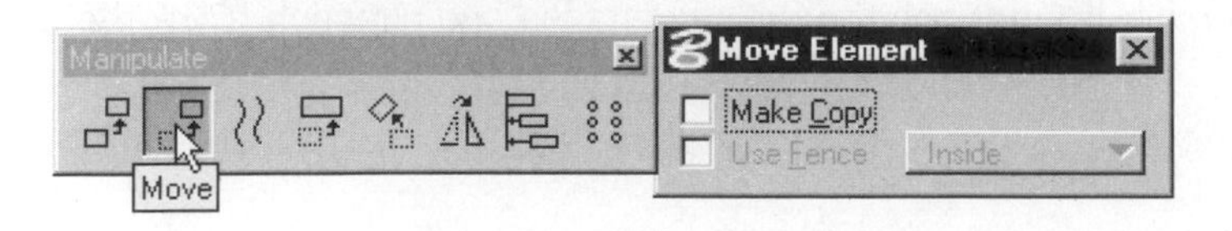

Figure 4–34 Invoke the Move tool from the Manipulate tool box

MicroStation prompts:

> Move Element > Identify element *(Identify an element to move.)*
>
> Move Element > Enter point to define distance and direction *(Reposition the element to its new location by providing a data point or keying-in coordinates.)*
>
> Move Element > Enter point to define distance and direction *(If necessary, move the element to another location by providing a data point or keying-in coordinates, and/or click the Reset button to terminate the tool sequence.)*

For example, the following tool sequence shows how to move a line to the center of a circle using the Move Element tool (see Figure 4–35).

> Move Element > Identify element *(Identify the line by snapping to the end point of the line.)*
>
> Move Element > Enter point to define distance and direction *(Snap to the center of the circle.)*
>
> Move Element > Enter point to define distance and direction *(Click the Reset button.)*

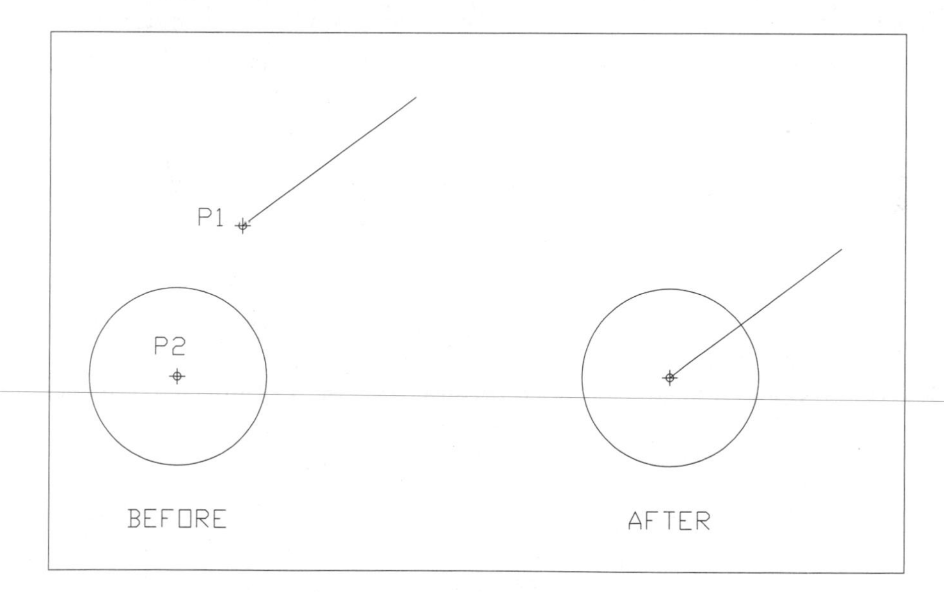

Figure 4–35 Example of moving an element with the Move tool

Move Parallel

The Move Parallel tool lets you move an element (such as a line, line string, multi-line, circles, curve, arc, ellipse, shape, complex chain, or complex shape) parallel to the original location of the element. The distance may be keyed in or defined by a data point.

Invoke the Move Parallel tool:

Manipulate tool box	Select the Move Parallel tool (see Figure 4–36).
Key-in window	**move parallel** (or **mov p**) Enter

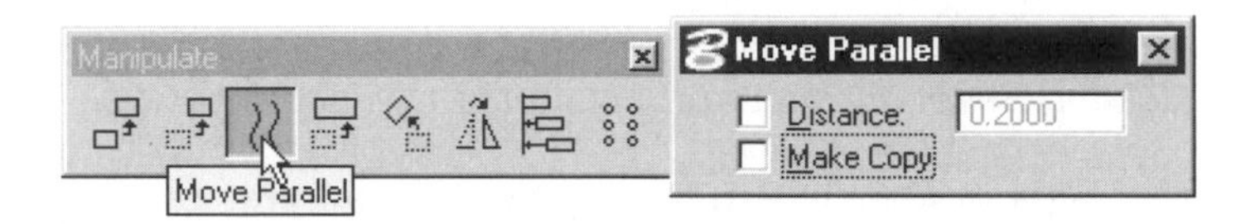

Figure 4–36 Invoke the Move Parallel tool from the Manipulate tool box

MicroStation prompts:

Move Parallel by Distance > Identify element *(Identify an element to move parallel.)*

Move Parallel by Distance > Accept/Reject (Select next input) *(Reposition the element to its new location by a data point or by keying-in coordinates.)*

Move Parallel by Distance > Accept/Reject (Select next input) *(If necessary, move it to another location by providing a data point or by keying-in coordinates, and/or click the Reset button to terminate the tool sequence.)*

For example, the following tool sequence shows how to move a line to another location parallel to the original location using the Move Parallel tool (see Figure 4–37).

Move Parallel by Distance > Identify element *(Identify the line.)*

Move Parallel by Distance > Accept/Reject (Select next input) *(Place a data point to move the element parallel to the original location.)*

Move Parallel by Distance > Accept/Reject (Select next input) *(Click the Reset button.)*

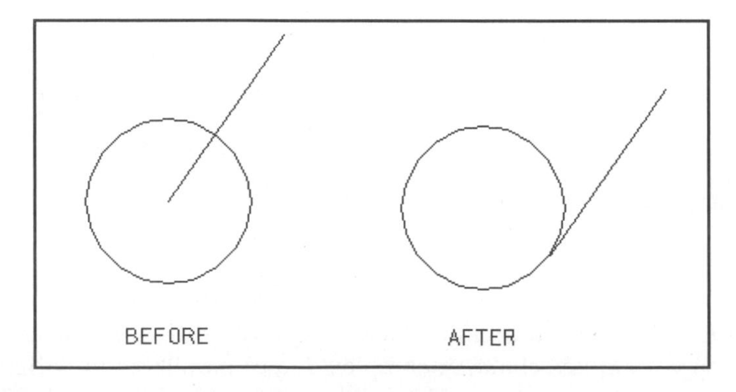

Figure 4–37 Example of moving an element parallel via the Move Parallel by Distance tool

You can also move an element parallel by specifying a distance. To do so, invoke the Move Parallel tool:

Manipulate tool box	Select the Move Parallel tool, key-in the distance in MU:SU:PU in the Distance edit field, and turn ON the Distance toggle button in the Tool Settings window (see Figure 4–38).
Key-in window	**move parallel keyin** (or **mov p k**) Enter

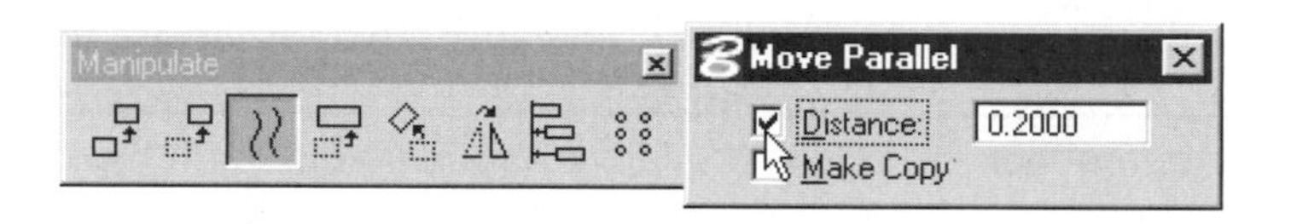

Figure 4–38 Invoke the Move Parallel Key-in tool from the Manipulate tool box

MicroStation prompts:

Move Parallel by Key-in > Identify element *(Identify an element to move parallel.)*

Move Parallel by Key-in > Accept/Reject (Select next input) *(Reposition the element to its new location by a data point.)*

Move Parallel by Key-in > Accept/Reject (Select next input) *(If necessary, move it to another location by providing a data point and/or click the Reset button to terminate the tool sequence.)*

Copy Parallel

Instead of moving the original element parallel, you can make a copy and then move the copy parallel to the original location of the element.

Invoke the Copy Parallel tool:

Manipulate tool box	Select the Move Parallel tool, then turn ON the Make Copy toggle in the Tool Settings window (see Figure 4–39).
Key-in window	**copy parallel** (or **cop p**) Enter

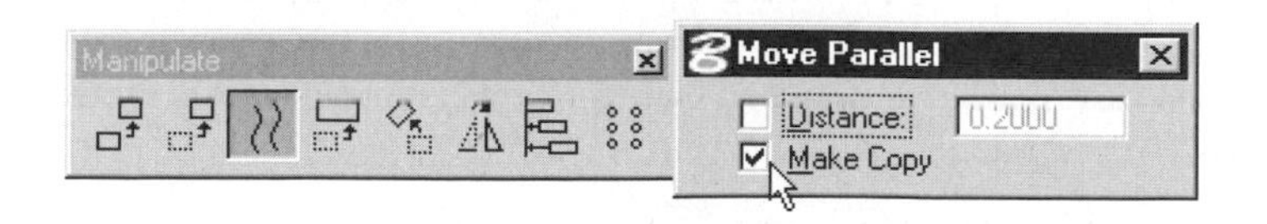

Figure 4–39 Invoke the Copy Parallel tool from the Manipulate tool box

MicroStation prompts:

Copy Parallel by Distance > Identify element *(Identify an element to copy parallel.)*

Copy Parallel by Distance > Accept/Reject (Select next input) *(Copy the element to its new location by a data point or by keying-in coordinates.)*

Copy Parallel by Distance > Accept/Reject > (Select next input) *(If necessary, copy it to another location by providing a data point or by keying-in coordinates, and/or click the Reset button to terminate the tool sequence.)*

You can also copy an element parallel by specifying a distance. To do so, invoke the Copy Parallel tool:

Manipulate tool box	Select the Move Parallel tool, key-in the distance in MU:SU:PU in the Distance edit field, and turn ON the toggle buttons for Distance and Copy in the Tool Settings window (see Figure 4–40).
Key-in window	**copy parallel keyin** (or **cop p k**) Enter

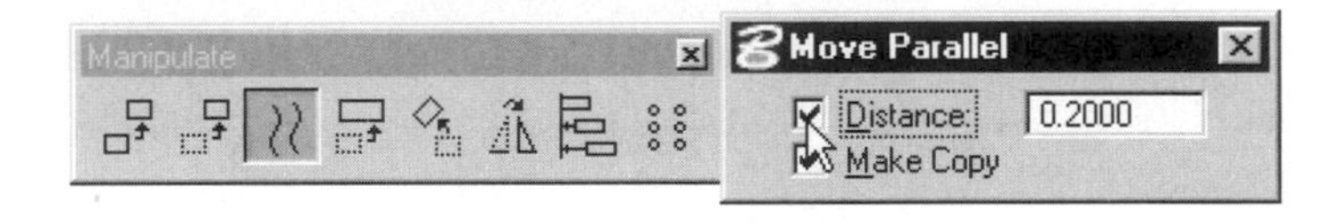

Figure 4–40 Invoke the Copy Parallel by Key-in tool from the Manipulate tool box

MicroStation prompts:

Copy Parallel by Key-in > Identify element *(Identify an element to copy parallel.)*

Copy Parallel by Key-in > Accept/Reject (Select next input) *(Copy the element to its new location by a data point.)*

Copy Parallel by Key-in > Accept/Reject (Select next input) *(If necessary, copy it to another location by providing a data point, and/or click the Reset button to terminate the tool sequence.)*

Scale Element Original

The Scale tool lets you increase or decrease the size of an existing element. If necessary, you can have a different scale factor for the *X* and *Y* axes. To enlarge an element, enter a scale factor greater than 1. For instance, a scale factor of 3 makes the selected element three times larger. To shrink an element, use a scale factor between 0 and 1. For instance, a scale factor of 0.75 shrinks the selected element to three-quarters of its current size.

MicroStation provides two methods by which you can scale an element: by setting an appropriate scale factor by key-in, and by specifying the scale factor graphically.

Scaling by Key-in

To scale an element by keying-in the scale factor, invoke the Scale Element tool:

Manipulate tool box	Select the Scale tool, then, in the Tool Settings window, select the Active Scale from the Method option menu and key-in the appropriate scale factors in the X Scale and Y Scale edit fields (see Figure 4–41).
Key-in window	**scale original** (or **sc o**) Enter

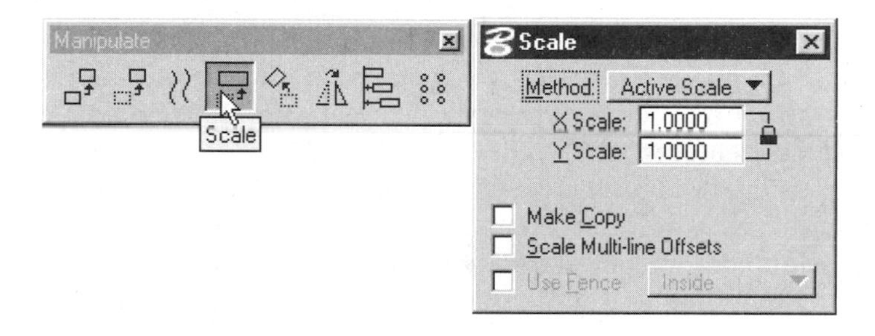

Figure 4–41 Invoke the Scale tool from the Manipulate tool box

MicroStation prompts:

Scale Element > Identify element *(Identify an element to scale.)*

Scale Element > Enter origin point (point to scale about) *(Reposition the scaled element to its new location by a data point or by keying-in coordinates.)*

Scale Element > Enter origin point (point to scale about) *(If necessary, scale it again by providing a data point or keying-in coordinates, and/or click the Reset button to terminate the tool sequence.)*

If you want the *X* Scale and *Y* Scale factors to be equal, close the small lock located to the right of the scale factor edit fields in the Tool Settings window. When the lock is closed you can key-in a value in either of the edit fields and the other field is automatically set equal to what you key-in. If the lock is open, you enter unequal *X* Scale and *Y* Scale factors.

For example, the following tool sequence shows how to scale an element to half its present size using the Scale Element tool (see Figure 4–42).

Scale Element > Identify element *(Identify the element to scale.)*

Scale Element > Enter origin point (point to scale about) *(Reposition the scaled element to its new location by a data point.)*

Scale Element > Enter origin point (point to scale about) *(Click the Reset button.)*

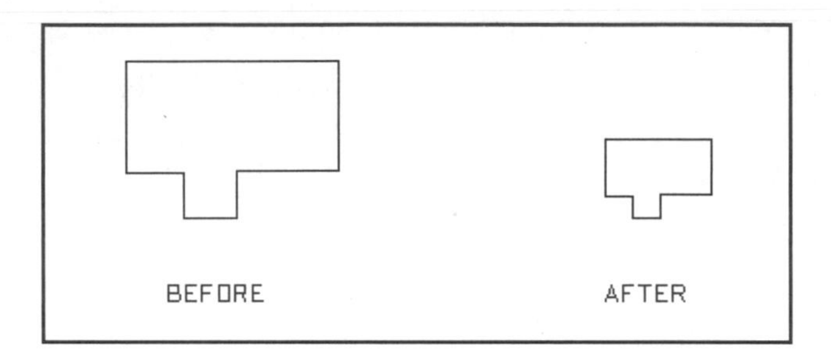

Figure 4–42 Example of scaling an element by means of the Scale tool

The scale factors can be negative numbers. A positive *X* and *Y* scale factor produces a scaled element that is the same orientation as the original element. A negative *X* and positive *Y* scale factor produces a scaled element that is a mirror (backward) image of the original element around the vertical axis. A positive *X* and negative *Y* produces a scaled element that is a mirror image around the horizontal axis. A negative *X* and *Y* produces a scaled element that is a mirror image around both a horizontal and vertical axis. Figure 4–43 shows the effect of positive and negative scale factors.

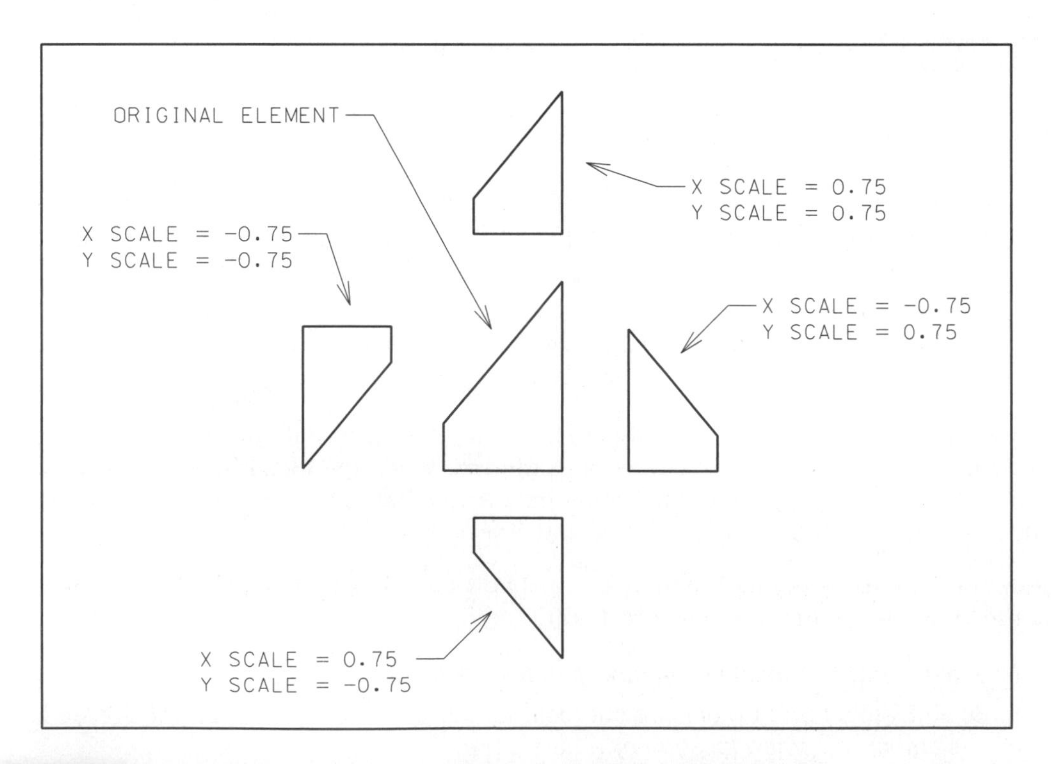

Figure 4–43 Example of the effect of positive and negative scale factors

Scaling Graphically

Scaling graphically involves providing three data points or keying-in their coordinates. The scale factors are computed by dividing the distance between the first and third points by the distance between the first and second points.

To scale an element graphically, invoke the Scale Element tool:

Manipulate tool box	Select the Scale tool, then select 3 points from the Method option menu (see Figure 4–44).
Key-in window	**scale points original** (or **sc p o**) Enter

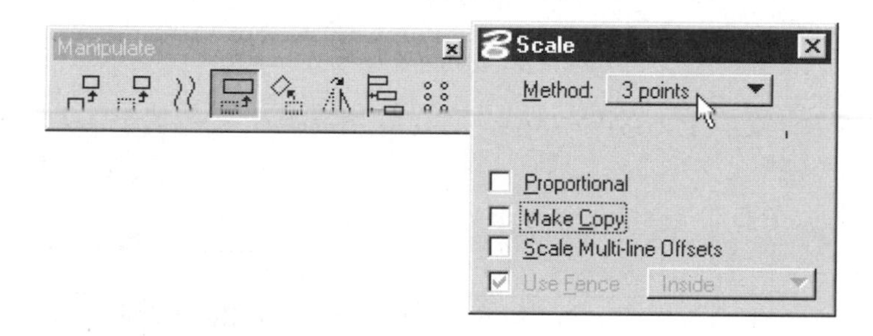

Figure 4–44 Invoke the Scale Element by 3 Points tool from the Manipulate tool box

MicroStation prompts:

> Scale Element by 3 Points > Identify element *(Identify an element to scale.)*
>
> Scale Element by 3 Points > Enter origin point (point to scale about) *(Place a data point or key-in coordinates to define the origin point.)*
>
> Scale Element by 3 Points > Enter reference point *(Place a data point or key-in coordinates to define the reference point.)*
>
> Scale Element by 3 Points > Enter point to define amount of scaling *(Place a data point or key-in coordinates to define the amount of scale.)*
>
> Scale Element by 3 Points > Enter point to define amount of scaling *(If necessary, scale it again by providing a data point or keying-in coordinates, and/or click the Reset button to terminate the tool sequence.)*

If necessary, to maintain the proportionality of the selected element you can turn on the Proportional toggle button in the Tool Settings window.

Scale Element Copy

Instead of scaling the original element, you can make a copy and then scale the copy. Similar to the Scale Element tool, MicroStation offers two methods by which to scale the copy: setting an appropriate scale factor by key-in, or specifying the scale factor graphically.

Invoke the Scale Element Copy tool:

Manipulate tool box	Select the Scale tool, then, in the Tool Settings window, select Active Scale or 3 Points from the Method option menu and turn ON the Make Copy toggle (see Figure 4–45).
Key-in window	**scale copy** (or **sc c**) Enter (By Active Scale) **scale points copy** (or **sc p c**) Enter (3 Point)

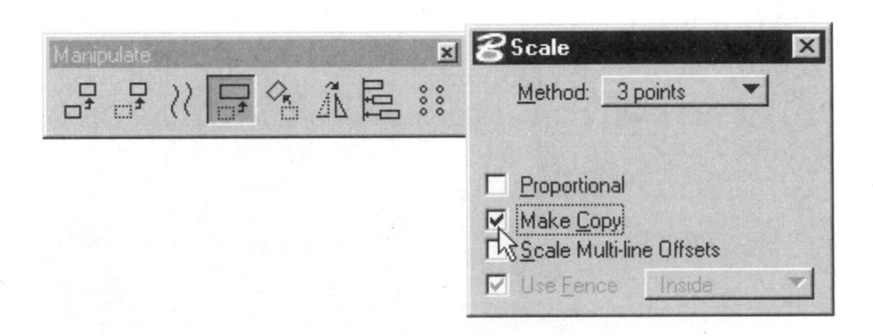

Figure 4–45 Invoke the Scale Element Make Copy tool from the Manipulate tool box

If the Active Scale method is selected, make sure to key-in appropriate scale factors in the *X* Scale and *Y* Scale edit fields. MicroStation prompts for Scale Element Copy are identical to the ones explained earlier for the Scale Element tool.

Rotate Element Original

The Rotate Element tool changes the orientation of an existing element by rotating it graphically about a specified pivot point. MicroStation provides three methods by which you can rotate an element: rotation by the active angle setting, rotation defined by two data points, and rotation defined by three data points.

Rotation by the Active Angle Setting

To rotate an element by the current active angle setting, invoke the Rotate Element tool:

Manipulate tool box	Select the Rotate tool, then select Active Angle from the Method option menu, and key-in the appropriate active angle in the edit field of the Tool Settings window (see Figure 4–46).
Key-in window	**rotate original** (or **ro o**) Enter.

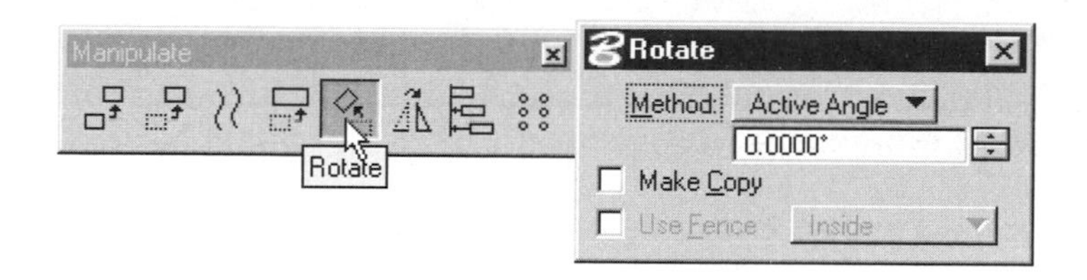

Figure 4–46 Invoke the Rotate tool from the Manipulate tool box

MicroStation prompts:

> Rotate Element > Identify element *(Identify an element to rotate.)*
>
> Rotate Element > Enter pivot point (point to rotate about) *(Reposition the rotated element to its new location by a data point or by keying-in coordinates.)*
>
> Rotate Element > Enter pivot point (point to rotate about) *(If necessary, rotate it again by providing a data point or keying-in coordinates, and/or click the Reset button to terminate the tool sequence.)*

For example, the following tool sequence shows how to rotate an element by 45 degrees from its present location using the Rotate Element tool (see Figure 4–47).

> Rotate Element > Identify element *(Identify the element to rotate.)*
>
> Rotate Element > Enter pivot point (point to rotate about) *(Reposition the rotated element to its new location by a data point.)*
>
> Rotate Element > Enter pivot point (point to rotate about) *(Click the Reset button.)*

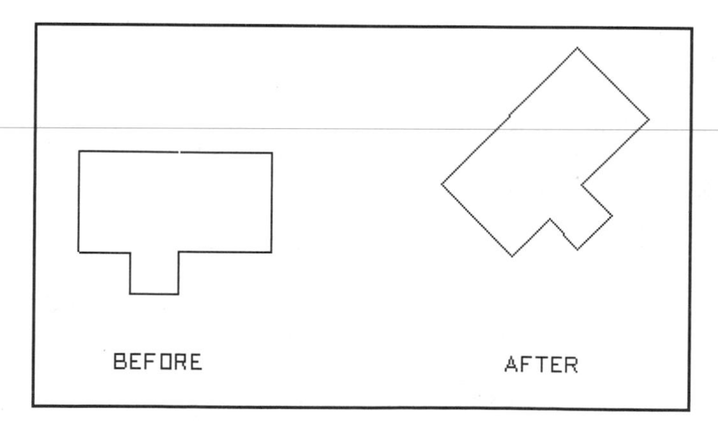

Figure 4–47 Example of rotating an element by means of the Rotate tool

Rotating by 2 Points

Rotating by 2 Points is defined by entering two data points or by keying-in coordinates. The angle of rotation is computed from the two data points.

To rotate an element by 2 points, invoke the Rotate Element tool:

Manipulate tool box	Select the Rotate tool, then select 2 Points from the Method option menu in the Tool Settings window (see Figure 4–48).
Key-in window	**spin original** (or **sp o**) Enter

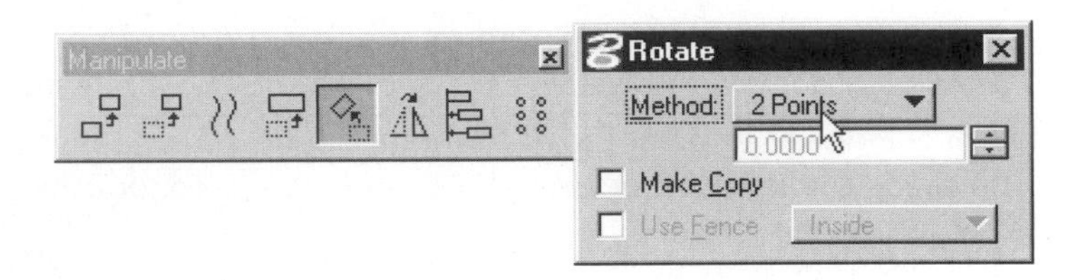

Figure 4–48 Invoke the Rotate Element by 2 Points tool from the Manipulate tool box

MicroStation prompts:

Spin Element > Identify element *(Identify an element to rotate.)*

Spin Element > Enter pivot point (point to rotate about) *(Place a data point or key-in coordinates to define the pivot point.)*

Spin Element > Enter point to define amount of rotation *(Place a data point or key-in coordinates to define the amount of rotation.)*

Spin Element > Enter point to define amount of rotation *(If necessary, rotate it again by providing a data point or keying-in coordinates, and/or click the Reset button to terminate the tool sequence.)*

Rotating by 3 Points

Rotating by 3 Points is defined by entering three data points or keying-in their coordinates. The angle of rotation is computed from the three data points.

To rotate an element by 3 points, invoke the Rotate Element tool:

Manipulate tool box	Select the Rotate tool, then select 3 Points from the Method option menu in the Tool Settings window (see Figure 4–49).
Key-in window	**rotate points original** (or **ro p o**) Enter

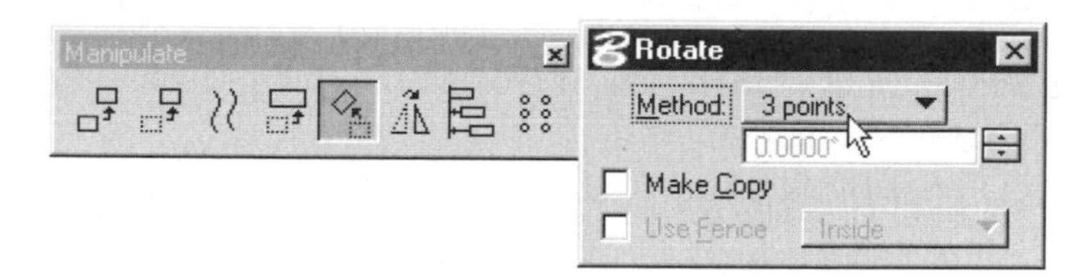

Figure 4–49 Invoke the Rotate Element by 3 Points tool from the Manipulate tool box

MicroStation prompts:

Rotate Element by 3 Points > Identify element *(Identify an element to rotate.)*

Rotate Element by 3 Points > Enter pivot point (point to rotate about) *(Place a data point or key-in coordinates to define the pivot point.)*

Rotate Element by 3 Points > Enter point to define start of rotation *(Place a data point or key-in coordinates to define the starting point of rotation.)*

Rotate Element by 3 Points > Enter point to define amount of rotation *(Place a data point or key-in coordinates to define the amount of rotation.)*

Rotate Element by 3 Points > Enter point to define amount of rotation *(If necessary, rotate it again by providing a data point or keying-in coordinates, and/or click the Reset button to terminate the tool sequence.)*

Rotate Element Copy

Instead of rotating the original element, you can make a copy and then rotate the copy. Similar to the Rotate Element tool, MicroStation offers three methods by which to scale the copy: rotation to the active angle, rotation defined by two data points, and rotation defined by three data points.

Invoke the Rotate Element Copy tool:

Manipulate tool box	Select the Rotate tool, then select either Active Angle, 2 Points, or 3 Points from the Method option menu, then turn ON the Make Copy toggle button (see Figure 4–50).
Key-in window	**rotate copy** (or **ro c**) Enter (Active Angle) **spin copy** (or **sp c**)) Enter (2 Point) **rotate points copy** (or **ro p c**)) Enter (3 Point)

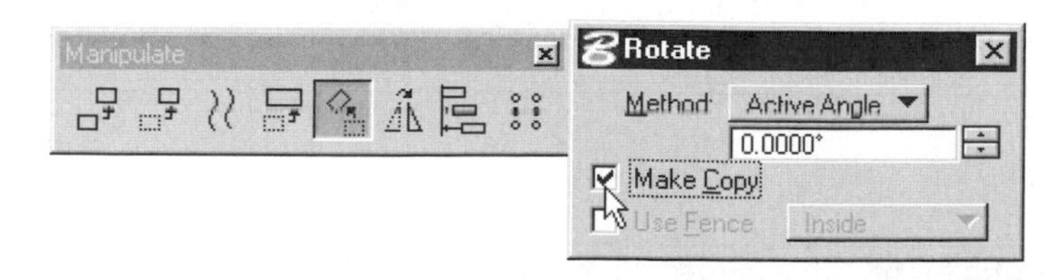

Figure 4–50 Invoke the Rotate Element Copy tool from the Manipulate tool box

If the Active Angle method is selected, make sure to key-in the appropriate active angle in the edit field located in the Tool Settings window. The MicroStation prompts for the Rotate Element Copy are identical to the ones just presented for the Rotate Element tool.

See Figure 4–51 for an example of rotating a copied element.

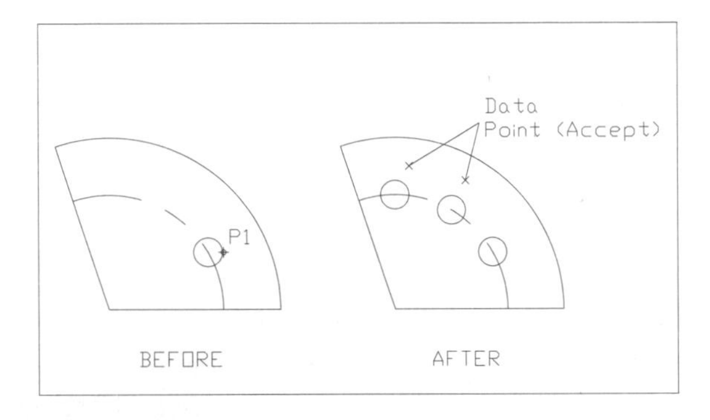

Figure 4–51 Example of rotating a copied element

Mirror Element Original

The Mirror Element tool creates a mirror (backward) image of an element. MicroStation provides three different methods by which you can mirror an element: along the *X* (horizontal) axis, along the *Y* (vertical) axis, and along a line defined by two data points (see Figure 4–52).

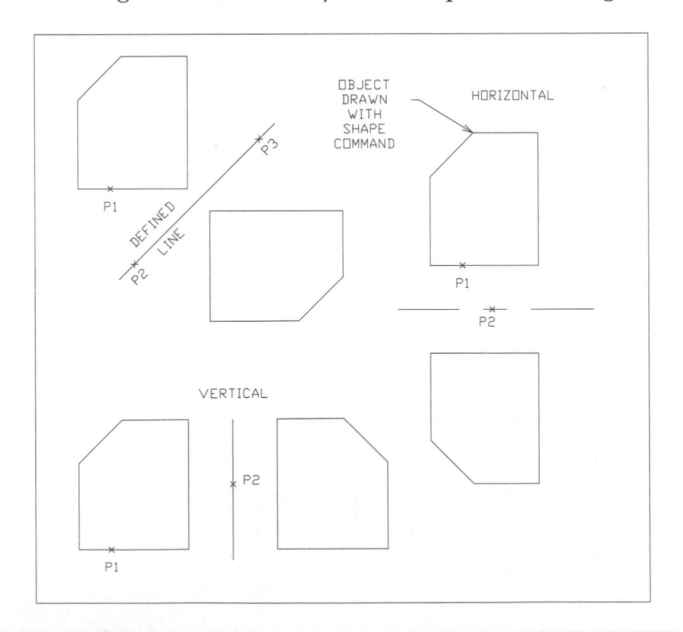

Figure 4–52 Examples of mirroring an element by three different methods

Mirror Image of an Element Along the *X* (Horizontal) Axis

To place a mirror image of an element along the horizontal axis, invoke the Mirror Element tool:

Manipulate tool box	Select the Mirror tool, then select Horizontal from the Mirror About option menu in the Tool Settings window (see Figure 4–53).
Key-in window	**mirror original horizontal** (or **mi o h**) Enter

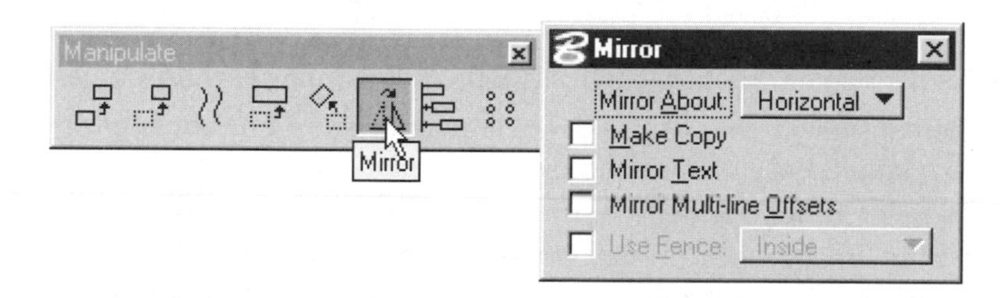

Figure 4–53 Invoke the Mirror Element tool from the Manipulate tool box

MicroStation prompts:

Mirror Element About Horizontal (Original) > Identify element *(Identify an element to mirror.)*

Mirror Element About Horizontal (Original) > Accept/Reject (Select next input) *(Place a data point or key-in coordinates to place a mirror image of the element.)*

Mirror Element About Horizontal (Original) > Accept/Reject (Select next input) *(If necessary, mirror it again by providing a data point or keying-in coordinates, or click the Reset button to terminate the tool sequence.)*

Mirror Image of an Element Along the *Y* (Vertical) Axis

To place a mirror image of an element along the vertical axis, invoke the Mirror Element tool:

Manipulate tool box	Select the Mirror tool, then select Vertical from the Mirror About option menu in the Tool Settings window.
Key-in window	**mirror original vertical** (or **mi o v**) Enter

The prompts are similar to those just presented for mirroring an element along the *X* axis.

Mirror Image of an Element Along a Line

To place a mirror image of an element along a line (defining two data points), invoke the Mirror Element tool:

Manipulate tool box	Select the Mirror tool then select Line from the Mirror About option menu.
Key-in window	**mirror original line** (or **mi o l**) Enter

MicroStation prompts:

Mirror Element About Line (Original) > Identify element *(Identify an element to mirror.)*

Mirror Element About Line (Original) > Enter 1st point on mirror line (or reject) *(Place a data point or key-in coordinates to place first point for mirror line.)*

Mirror Element About Line (Original) > Enter 2nd point on mirror line *(Place a data point or key-in coordinates to place second point for mirror line.)*

Mirror Element About Line (Original) > Enter 2nd point on mirror line *(If necessary, mirror it again by providing a data point or keying-in coordinates, or click the Reset button to terminate the tool sequence.)*

Mirror Element Copy

Instead of mirroring the original element, you can make a copy and then mirror the copy. Similar to the Mirror Element tool, MicroStation offers three methods by which to mirror the copy: along the *X* (horizontal) axis, along the *Y* (vertical) axis, and along a line defined by two data points.

Invoke the Mirror Element Copy tool:

Manipulate tool box	Select the Mirror tool, then select Horizontal, Vertical, or Line from the Method option menu and turn ON the Make Copy toggle button (see Figure 4–54).
Key-in window	**mirror copy horizontal** (or **mi c h**) Enter (Horizontal) **mirror copy vertical** (or **mi c v**)) Enter (Vertical) **mirror copy line** (or **mi c l**)) Enter (Line)

The MicroStation prompts for Mirror Element Copy are identical to those just presented for the Mirror Element tool.

See Figure 4–55 for an example of mirroring a copied element along a line.

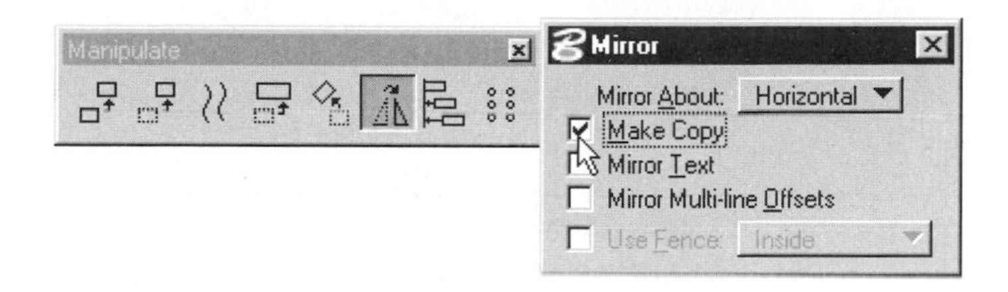

Figure 4–54 Invoke the Mirror Element Copy tool from the Manipulate tool box

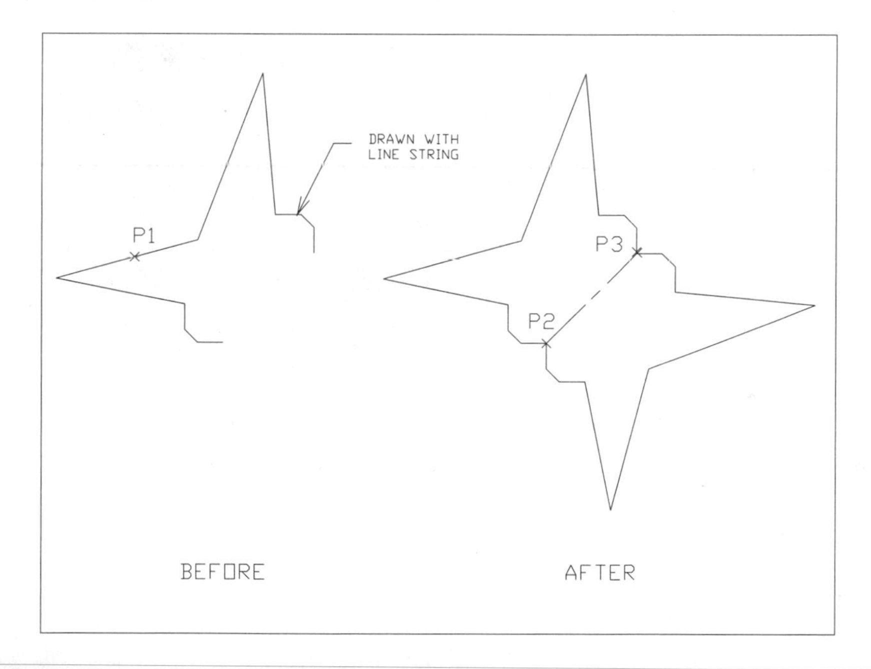

Figure 4–55 Example of mirroring a copied element

Construct Array

The Construct Array tool makes multiple copies of a selected element in either rectangular or polar arrays. In a rectangular array, you place copies in rows and columns by specifying the number of rows, the number of columns, and the spacing between rows and columns (row spacing and column spacing may differ). The whole rectangular array can be rotated to a selected angle. In the polar array, you place copies in a circular fashion by specifying the number of copies, the angle between two adjacent copies (delta angle), and whether or not the element will be rotated as it is copied.

Rectangular Array

To place multiple copies of an element by rows and columns, invoke the Construct Array (Rectangular) tool:

Manipulate tool box	Select the Construct Array tool, then select Rectangular from the Array Type option menu (see Figure 4–56).
Key-in window	**array rectangular** (or **ar r**) Enter

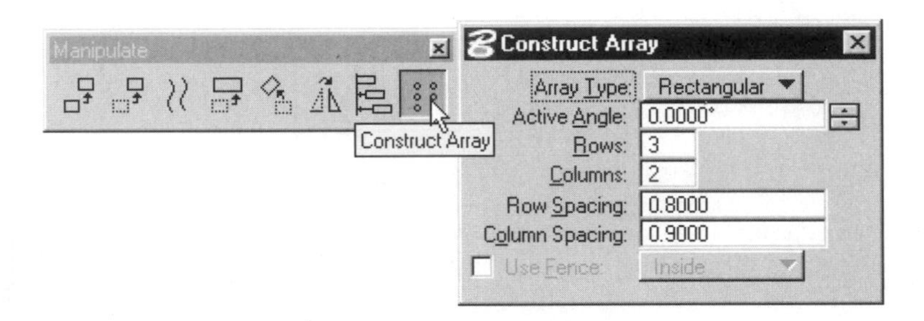

Figure 4–56 Invoke the Construct Array (Rectangular) tool from the Manipulate tool box

In the Tool Settings window, key-in: the degrees of rotation in the Active Angle edit field; the number of rows and columns in the Rows and Columns edit fields, respectively; and the distance between rows and the distance between columns in the Row Spacing and Column Spacing edit fields, respectively.

MicroStation prompts:

Rectangular Array > Identify element *(Identify an element to array.)*

Rectangular Array > Accept/Reject (Select the next input) *(Click the Accept button to place copies, or click the Reject button to disregard the selection.)*

Note: Any combination of a whole number of rows and a whole number of columns may be entered (except both 1 row and 1 column, which would not create any copies). MicroStation includes the original element in the number you enter. A positive distance for the column and row spacing causes the elements to array toward the right and upward. A negative distance for the column and row spacing causes the elements to array toward the left and downward.

For example, the following tool sequence constructs a six-row by eight-column array with rows that are 1.5 Master Units apart and columns that are 2.75 Master Units apart (see Figure 4–57).

Rectangular Array > Identify element *(Identify the element.)*

Rectangular Array > Accept/Reject (Select next input) *(Click the Accept button to place the copies.)*

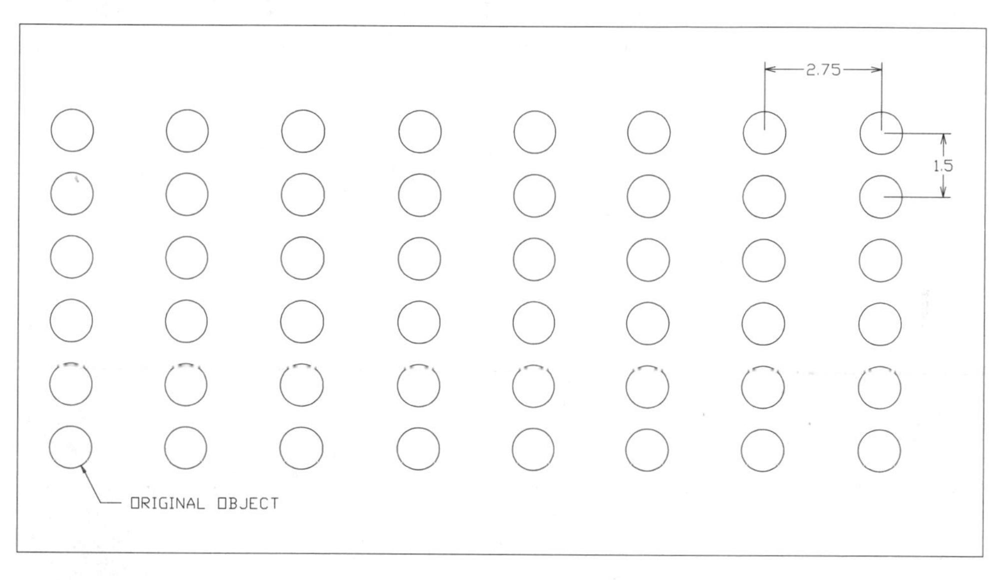

Figure 4–57 Example of placing a rectangular array

Polar Array

To place multiple copies of an element in a circular fashion, invoke the Construct Array (Polar) tool:

Manipulate tool box	Select the Construct Array tool, then select Polar from the Array Type option menu in the Tool Settings window (see Figure 4–58).
Key-in window	**array polar** (or **ar p**) Enter

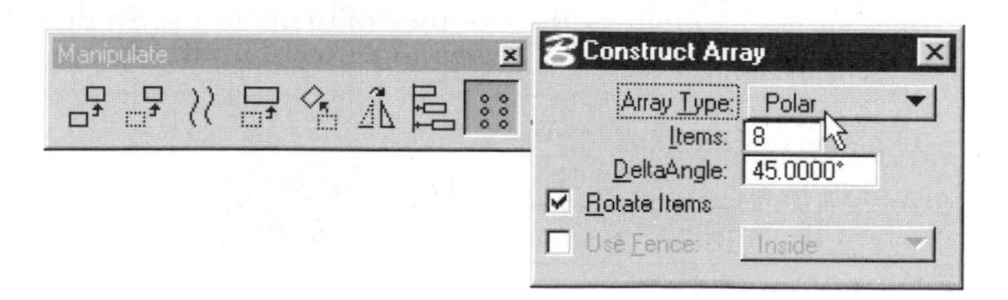

Figure 4–58 Invoke the Construct Array (Polar) tool from the Manipulate tool box

In the Tool Settings window, key-in the required number of copies of the selected element in the Items edit field, and specify the angle between adjacent items in the Delta Angle edit field. To rotate the elements as they are copied, toggle the Rotate Items button to ON. Figure 4–59 shows the difference between rotating and not rotating the elements as they are copied. Once you set all the necessary parameters, MicroStation prompts:

Polar Array > Identify element *(Identify an element to array.)*

Polar Array > Accept, select center/Reject *(Specify the center point for the array by pressing the Data button or by keying-in coordinates, or click the Reject button to disregard the selection.)*

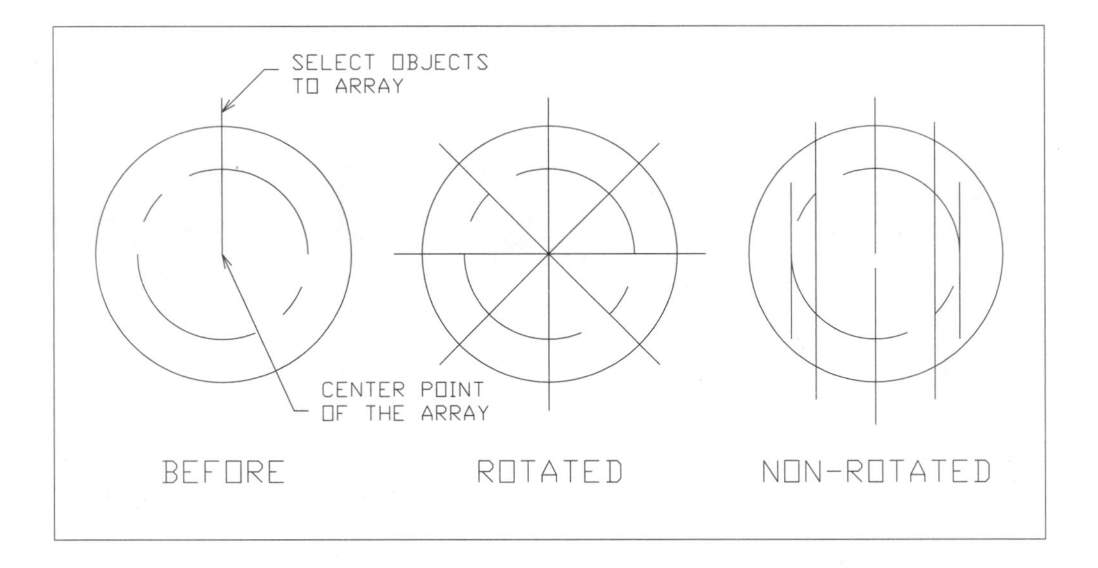

Figure 4–59 Example of rotating and not rotating elements as they are copied

Note: Key-in a whole number for the number of items to be copied, and MicroStation includes the original element in the number of array items. In other words, if you request seven items, your array will consist of the original element and six copies.

The following tool sequence shows an example (see Figure 4–60) of using the Construct Array tool to place a polar array when the number of items is set to 8 and the delta angle is set to 45 degrees. MicroStation prompts:

Polar Array > Identify element *(Identify the shape.)*

Polar Array > Accept, select center/Reject *(Place a tentative point at the center of the circle and accept it.)*

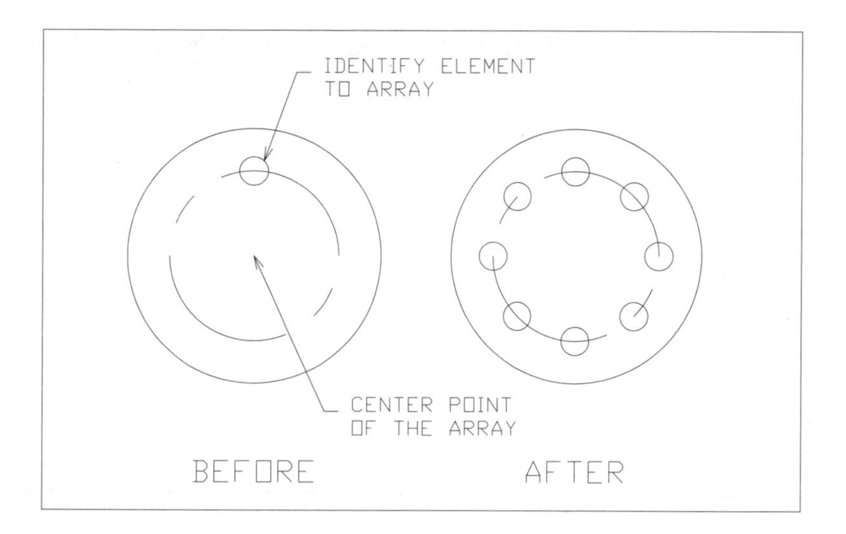

Figure 4–60 Example of placing a polar array

TEXT PLACEMENT

You have learned how to draw the geometric shapes that make up your design. Now it is time to learn how to annotate your design. When you draw by hand on paper, adding descriptions of the design components and the necessary shop and fabrication notes is a time-consuming, tedious process. MicroStation provides several text placement tools that greatly reduce the time and tedium of text placement.

The text placement procedure includes setting up the text parameters (size, line spacing, style, etc.), selecting a placement tool, typing your text, and then placing it in the design. Each string of text you place is a single element to which all of the manipulation tools can be applied.

Note: If you do not know how to type, you can place text quickly and easily after a period of learning the keyboard and developing typing skills. If you create designs that require a lot of text entry, it may be worth your time to learn the proper technique. There are several computer programs that can help you teach yourself to type, and almost all colleges offer typing classes. If you have no time to learn proper typing, there is no need to worry—many "two-finger" typists place text in their designs productively.

Text Parameters

Before you can place text in your design, you have to make sure the text parameters, such as the font, text size, line spacing, and justification, are set up appropriately.

The Text settings box allows you to set the text parameters. Invoke the Text Settings box:

Pull-down menu	Element > Text (see Figure 4–61).

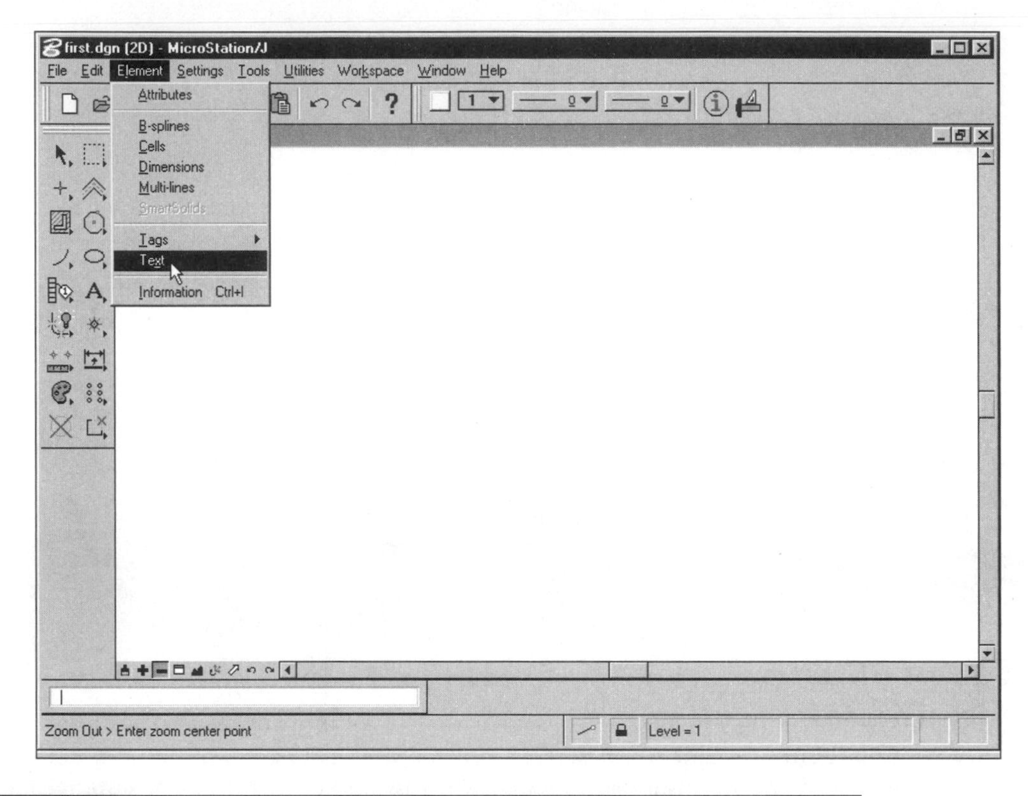

Figure 4–61 Invoke the Text settings box from the Element pull-down menu

MicroStation displays the Text settings box as shown in Figure 4–62.

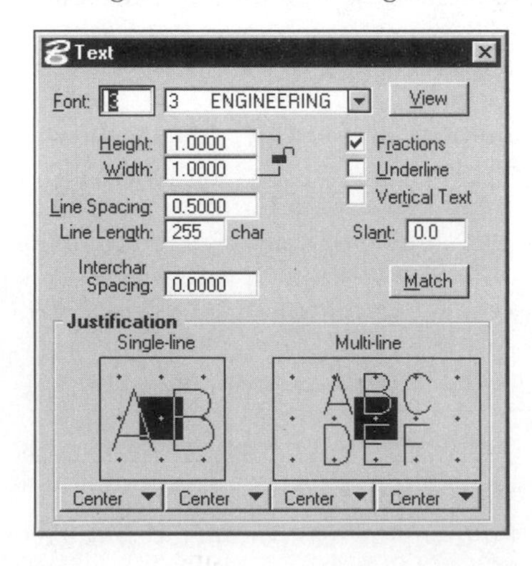

Figure 4–62 Text settings box

The parameters that can be set in the Text Settings box are: font (style), text size (height and width), line spacing (space between adjacent lines of text), line length, intercharacter spacing, single character fractions, underline and vertical text, text slant, and justification. Text color, weight, level, and angle are set in other settings boxes.

Font

Before you start placing text, you must decide what style (font) you want to use. Do you want fancy text, italic text, block text, or some other font? Numbers identifies text fonts, and MicroStation can support up to 255 different fonts. To find out what fonts are loaded in your copy of MicroStation, click the View button in the Text settings box. MicroStation displays the Fonts settings box listing the available fonts (see Figure 4–63).

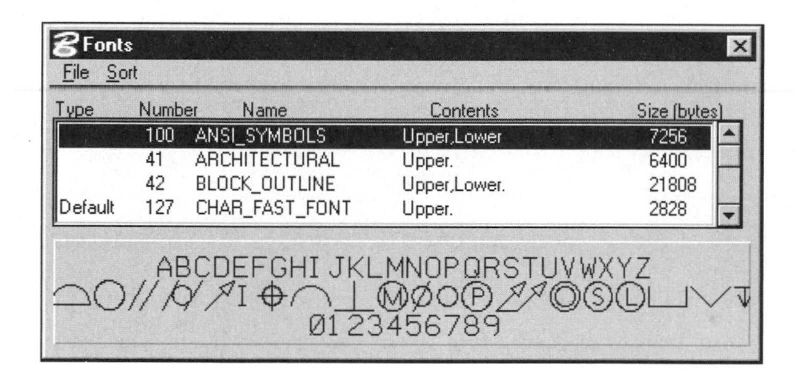

Figure 4–63 Fonts settings box

The top half of the Fonts settings box lists all the fonts loaded in MicroStation. Each line in the list area describes one font. Use the scroll bar to view all of the available fonts.

To see what a font looks like, click on the font's description line. An example of the font you click on appears in the bottom half of the settings box (refer to Figure 4–63). Some fonts lack lowercase letters, and some have no single-character fractions. If a font does not include one type of character, that type will *not* show up in the font example. The font description in the upper half of the Fonts settings box also tells you what types of characters the font contains.

When you select a symbol font, the letters you type produce symbols rather than the letters. For example, font 102 contains uppercase and lowercase letters that produce different symbols (such as arrowheads) rather than letters.

To select a font, click on the font's description. MicroStation displays an example in the lower half of the Fonts settings box. Click in the lower half of the window to make the displayed font the active font. You also can select the font by keying-in **FT=<#>** or name of the font (where <#> is the number of the font) at the key-in window, and pressing Enter. The selected font number becomes the active font.

After you select a font number, MicroStation displays your selection in the Status bar and in the Font fields of the Text settings box. The font number you select remains the active font until you select another font number or exit MicroStation. To keep the font number active for the next time you edit the design, select Save Settings from the File pull-down menu.

Text Size

After you select a text font, you must tell MicroStation what size you want the text to be, both height and width, specified in working units (MU:SU:PU).

If you are drawing an unscaled schematic, or if you are going to plot your design full size, selecting a text size is simple—just enter the size you want your text to be when you plot it.

If you are drawing a design that must be scaled when plotted, selecting a text size is a little more complicated. As mentioned earlier, in MicroStation you draw objects in their full size (real-world size), and tell MicroStation what scale to use when it plots the design to paper. MicroStation scales down everything in the design to fit the size of paper you choose for plotting, including the text. Therefore you must scale up your text by the *inverse* of the plot scale so it will be the correct size when you plot.

For example, if you are creating a design that will be plotted at 1" = 10', and you want your text size to be 0.1 inch, your text height in the design must be 1 foot. If 1 inch of plotter paper equals 10 foot, then 0.1 inch of plotter paper equals 1 foot.

Let's put that into a formula:

Text height in design = (design units ÷ plotter units) × plotted text size

Now let's try the formula for providing 1/8-in. text when we plot at 1/8" = 1'. Our design units are 1 foot, our plotter units are 1/8 inch, and we want our plotted text size to be 1/8-inch:

Text height in design = (1" ÷ 1/8") × 1/8" = 1'.

Thus, we need to place 1-foot tall text in the design.

To specify the text size, key-in the text height and the text width in the Height and Width edit fields, respectively.

There is a small lock symbol to the right of the text Height and Width fields. If you want your text height and width to be equal, click on the lock symbol to close the lock. When the lock is closed, you can key-in a value in either of the size fields and the other will automatically be set equal to what you type. If the lock is open, you must enter each field separately.

You can also set the text size in the Key-in window by using one or more of the following key-ins:

TX= <size> *(to set both the text height and the text width with one key-in)*

TH= <size> *(to set only the text height)*

TW= <size> *(to set only the text width)*

In each key-in tool, replace <size> with the text size (in working units) and press Enter.

Once you set the text size, it remains active until you either change it or exit MicroStation. Select Save Settings from the File pull-down menu to save the settings for the next time you load your design file in MicroStation.

Underline and Vertical Text

To place text with a line below it as shown in Figure 4–64, turn ON the toggle button for Underline in the Text settings box. Similarly, to place text vertically, as shown in Figure 4–64, turn ON the toggle button for Vertical text in the Text settings box.

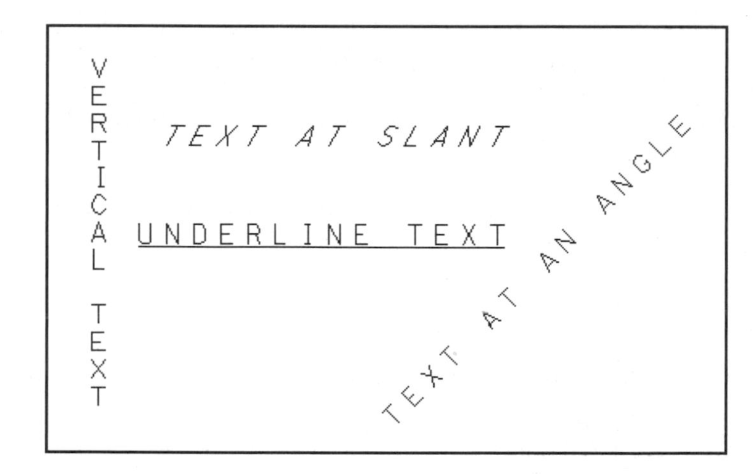

Figure 4–64 Examples of placing underlined text, vertical text, slanted text, and text at an angle

Slant

To place text at a slant or angle as shown in Figure 4–64, key-in the slant or angle in the Slant edit field in the Text Attributes settings box. The slant or angle can be anywhere from –89 degrees to 89 degrees.

Line Spacing and Line Length

When you place text, it becomes one element in your design. While you are typing the text, you can press the Enter to create multi-line text that is treated as one element. If you plan to enter multi-line text, you must tell MicroStation how much space to leave between the text lines and the maximum number of characters you want on one line.

There are no firm rules for setting line spacing. But if you set it to a value less than half the text height, the lines may appear too close together when plotted.

For the majority of text work, the maximum number of characters per line is not important; just leave it set at the default value of 255 characters (the maximum it can be). If you try to type more characters in one line of text than the maximum allows, the text will wrap to a new line at the maximum number of characters. (It wraps even if you are in the middle of a word.)

To specify the space between adjacent lines of text and the maximum characters allowed on one line, key-in the appropriate values in the Line Spacing edit field (in MU:SU:PU) and the Line Length edit field, respectively.

You can also key-in the Line Spacing and Line Length in the Key-in window with the following key-ins:

LS = <space> *(to set space between lines)*

LL = <charc> *(to set the maximum characters per line)*

For <space>, key-in the line spacing in working units; for <charc>, key-in the maximum number of characters per line; then press Enter.

Once you set the Line Spacing and Line Length, these settings remain active until you change them or exit MicroStation. Selecting Save Settings from the File pull-down menu saves the settings for the next time you load your design file in MicroStation.

Intercharacter Spacing

To specify the spacing between two characters within a text string, key-in the value in MU:SU:PU in the Interchar Spacing edit field in the Text settings box.

Justification

To place the text you typed, first you have to define a data point in your design. Before you do so, you have to tell MicroStation where to place the text in relation to that data point. That relationship is called the justification. The Text settings box provides an excellent visual aid to setting up the justification (see Figure 4–65).

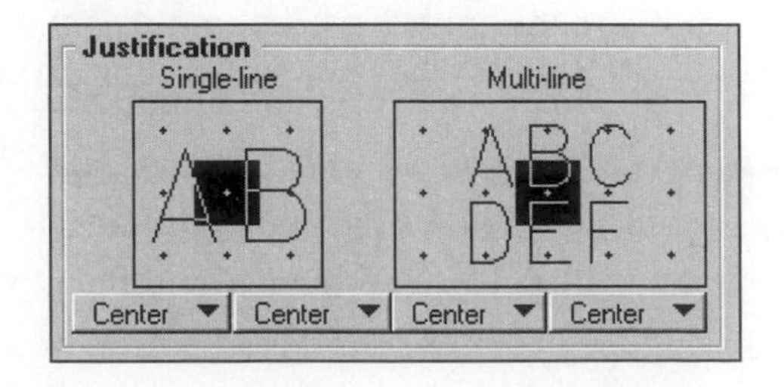

Figure 4–65 Text settings box showing the justification options

In the window are two pictures of large text over a grid of dots. The text is displayed in the currently active font. The dark square in each picture shows the relation of the Data button to the text. The text String picture on the left defines the justification when you place a single line of text. The multi-line text picture defines the justification when you place text that is on more than one line.

To set the justification, click on one of the grid points in the justification pictures, or select the justification from the sets of option menus below the pictures. Once set, the justification remains active until you either change it or exit MicroStation. Select Save Settings from the File pull-down menu to save the active justification for the next editing session.

The Multi-line Text justification also determines which side of the text will be smooth. Additional options available for justification in the multi-line text picture are Left Margin and Right Margin. These two justifications employ the Line Length setting that was discussed earlier. When you select a Left Margin justification, the right edge of the multi-string text is placed equivalent to the number of characters of Line Length from the data point. If a Right Margin justification is chosen, the left edge of the multi-string text is placed equivalent to the number of characters of Line Length from the data point.

Note: A common mistake of inexperienced MicroStation users (and occasionally of experienced users) is forgetting that the outside settings set the Multi-line Text justification to margin. They click these thinking they are selecting left or right justification. If the line length is set to 255 characters, the results can be startling when the text is placed.

Angle

Set the appropriate active angle to place the text string at an angle. This can be set by keying-in **AA=**<angle> in the key-in window, or in the Tool Settings window when the Place Text tool is active. The default Active Angle is 0 degrees.

Color, Weight, and Level

Set the appropriate color, weight, and level to place the text. This can be done by selecting Attributes from the Element pull-down menu, or from the Primary toolbar.

Text View Attributes

There is one last thing to check before you start placing text in your design file. Make sure the Text view attribute is set to ON. Select View Attributes from the Settings pull-down menu.

If the Text view attribute is set to ON, all text that is placed in the design will appear in the view; if it is set to OFF, all text disappears from the view. Updates may be completed faster when no text is displayed, but you must be careful not to use the space occupied by the text.

If the Fast Font view attribute is set to ON, all text is displayed in font 127, regardless of the font that was used to place it. Font 127 is a simple font that updates more quickly than other fonts. Text size is affected by font. So if you turn Fast Font to ON, the text may appear to take up more room than it does with its true font.

If the Text Nodes view attribute is set to ON, you will see a cross and a number placed at the data point of multi-line text strings. For the majority of your work, keep this view attribute set to OFF.

Detailed explanation is provided for View Attributes in Chapter 3 and for Text Nodes in Chapter 7.

Place Text Tools

Following are the tools MicroStation provides for placing text.

- The Place Text By Origin tool places text at a data point you define.
- The Place Fitted Text tool scales the text to fill the space between two data points.
- The Place Text Above Element tool places text above a line you have identified.
- The Place Text Below Element tool places text below a line you have identified.
- The Place Text on Element tool places the text on the identified line and removes the portion of the line where the text is placed.
- The Place Text Along Element tool places text along a curved element.
- The Place Note tool places text at the end of a line and arrowhead.

When you select one of these text placement tools, the Text Editor box is displayed, as shown in Figure 4–66. This box provides a place to type the text and some helpful text editing commands.

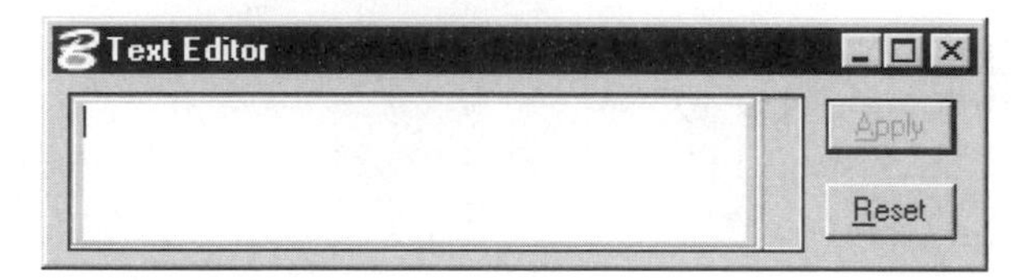

Figure 4–66 Text Editor

If necessary, you can resize the Text Editor box so you can see more of what you are typing. Point to the box border, press the data button, and drag it to the new size.

To type text in the box, place the screen cursor in the box and click the Data button. When you see a text cursor similar to the cursor in a word processor, you may start typing.

For multi-line text, press Enter at the place where you want the new line to start and continue typing. If you do not press Enter, the text will wrap to a new line when you reach the right end of the Text Editor box, but all the text will be on one line when you place it in your design. You can place multi-line text only when you press Enter.

The key-ins described in Table 4–1 position the text cursor within the text in the Text Editor box.

Table 4–1 Positioning the Text Cursor

PRESS:	TO MOVE THE TEXT CURSOR:
←	Left one character
→	Right one character
Ctrl + ←	Left one word
Ctrl + →	Right one word
Home	To the beginning of the current text line
End	To the end of the current text line
↑	Up to the previous line of text
↓	Down to the next line of text
Page Up	Straight up into the first text line
Page Down	Straight down into the last text line
Page Up + Home	Up to the beginning of the first text line
Page Up + End	Down to the end of the last text line

The key-ins described in Table 4–2 delete characters from the text in the Text Editor box.

Table 4–2 Keys that Delete Text

PRESS:	TO DELETE:
Backspace	The character to the left of the text cursor
Delete	The character to the right of the text cursor
Shift + Backspace	All characters from the text cursor to the beginning of the word
Alt + Delete	All characters from the text cursor to the end of the word
Ctrl + Delete	All characters from the text cursor to the end of the current line
Reset button in Text Editor box	All characters in the Text Editor box

The key-ins described in Table 4–3 select or deselect text in the Text Editor box. Selected text is shown with a dark background. Selected text can be moved, copied, or deleted.

Table 4–3 Selecting Text with Key-Ins

PRESS:	TO SELECT (OR DESELECT IF ALREADY SELECTED):
Shift + ←	The character to the left of the text cursor
Shift + →	The character to the right of the text cursor
Ctrl + Shift + ←	The characters from the text cursor to the left end of a word
Ctrl + Shift + →	The characters from the text cursor to the right end of a word
Ctrl + A	To select all text in the Text Editor box
← or →	To deselect all previously selected text

The pointing device actions described in Table 4–4 select or deselect text in the Text Editor box.

Table 4–4 Selecting Text with the Pointing Device

POINTING DEVICE ACTION	RESULT
Press the Data button and drag the screen cursor across the text	Selects all the text you drag across
Double-click the Data button	Selects the word the cursor is in
Hold down Shift + Data button and drag across the text	Adds more text to the text already selected
Click the Data button in an area where there is no text	Deselects all previously selected text

The actions required to replace, delete, and copy previously selected text are shown in Table 4–5.

Table 4–5 Replacing, Deleting, and Copying Selected Text

ACTION	RESULT
Start typing characters	Replace the selected text with the text you type
Press BackSpace	Delete all the selected text
Press Delete	Delete all the selected text
Press Ctrl + Insert	Copy the selected text to a buffer
Press Shift + Insert	Paste the previously copied or deleted text at the text cursor position

In the next section, Place Text By Origin is explained. The remaining text placement tools are explained in Chapter 7.

Place Text By Origin

The Place Text tool places the text at the data point you define via the active text parameters (font, size, line spacing, line length, and justification), the active color, the active line weight, and the active angle.

You also can place multi-line text by pressing Enter while typing the text in the Text Editor box. Remember that in the Text settings box there are separate justification fields for text strings (all the text in one line) and for multi-line text.

To place text by origin, invoke the Place Text tool:

Text tool box	Select the Place Text tool, then select By Origin from the Method option menu in the Tool Settings window (see Figure 4–67).
Key-in window	**place text** (or **pl tex**) Enter

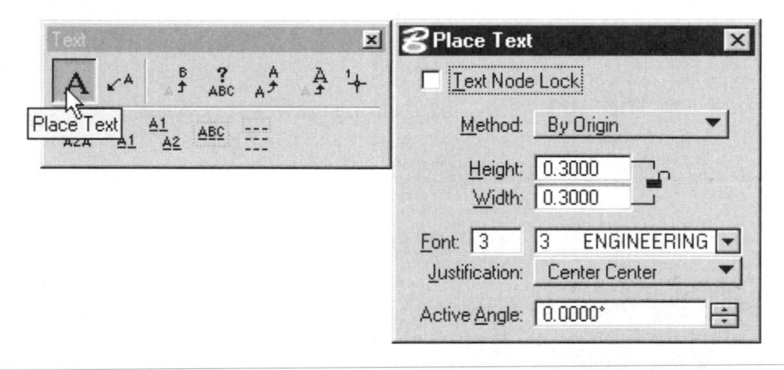

Figure 4–67 Invoke the Place Text (By Origin) tool from the Text tool box

MicroStation prompts:

> Place Text > Enter Text *(Type the appropriate text string in the Text Editor, then place a data point in the design to indicate the text justification point.)*
>
> Place Text > Enter more chars or position text *(Continue placing copies of the text in the design, change the text in the Text Editor box before continuing, or select another tool.)*

Each copy of the text you place becomes a single element that can be manipulated like any other element. The only key point in a text string or multi-line element is the placement point. As you place the text, you can change any of the text or text attributes.

REVIEW QUESTIONS

Write your answers in the spaces provided.

1. The Place Polygon tool places polygons that can have a maximum of ____________ sides.
2. The Place Point Curve tool is used ____.
3. The Place Stream Curve tool is used ____.
4. The Multi-line tool allows you to place up to ________ separate lines of various ____________, ____________, and ___________ with a single tool.
5. The Place Fillet tool joins two lines, adjacent segments of a line string, arcs, or circles with an _____________ of a specified radius.
6. Name the three methods by which you can control the removal of extension lines when placing the fillet.

   ____

7. The Chamfer tool allows you to draw a ________________ instead of an arc.
8. The purpose of the Trim tool is ____.
9. What is the name of the tool that will delete part of an element? ___________________
10. Name the two categories of manipulation tools available in MicroStation.

    ____

11. The Copy tool is similar to the Move tool, but it ________________.
12. Name at least three element manipulation tools available in MicroStation.

    ____

13. To rotate an element, the key-in tool is _______________.
14. The Array tool can make multiple copies of a selected element in either __________________ or __________________ arrays.
15. List the four parameters you have to specify for a rectangular array.

    ____

    ____

16. Explain briefly the functions of the Undo and Redo tools.

____

____

17. The Text settings box is invoked from the menu ________________.

18. Name three text parameters that can be changed from the Text settings box.

19. You can change the current font by keying-in ______________________.

20. List the two key-ins that can change the text height and text width.

____

21. Name the text parameter that controls the distance between two lines of text in placing multi-line text.

____

22. With the current working units set to MU=1 in., SU=4 qt., and PU=1000, how would you set your text size to one-eighth of an inch?

____

23. If a design is to be plotted at a scale of 1/2 inch equals 1 foot, what should be the text size in the design to plot at 1/8 inch? (Note: Working Units are set to feet, inches, and 1600 positional units per inch.)

____

PROJECT EXERCISE

This project exercise provides step-by-step instructions for creating the design shown in Figure P4–1. The intent is to guide you in applying the concepts and tools presented in Chapters 1 through 4. (Note that the instructions are not necessarily the most efficient way to draw the objects. Your efficiency will improve as you learn more tools in later chapters.)

This project introduces the use of the following tools:

- Placement: Line String, Arc, Circle
- Manipulation: Mirror, Copy Parallel, Fillet, Array

Notes: The dimensions are not part of this project. They are included in Figure P4–1 only to show the size of the design.

As you complete each step in the project procedures, place a check mark by the step to help you keep up with where you are in the project.

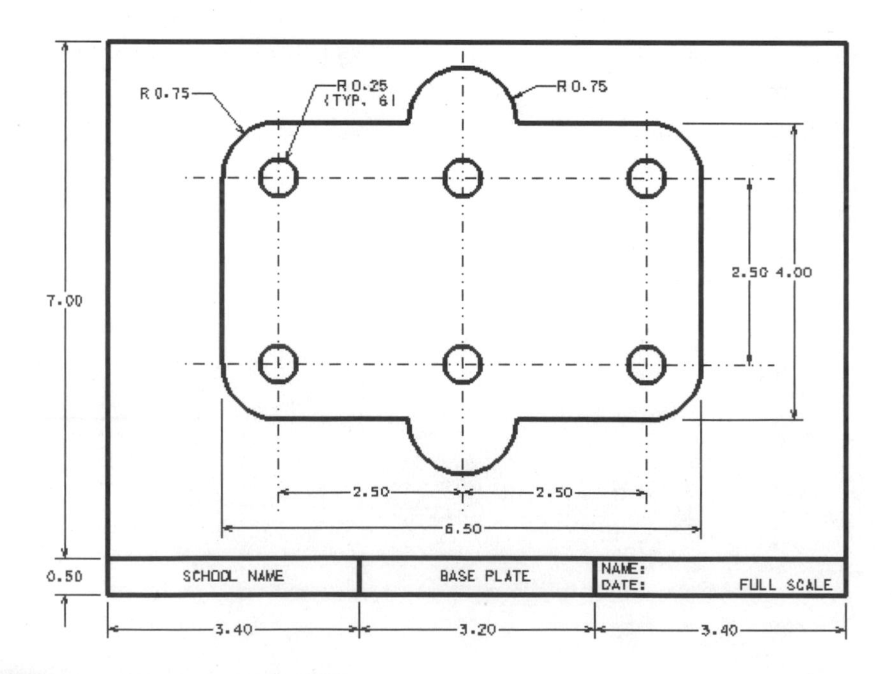

Figure P4–1 Completed project design

Prepare the Design File

This procedure has you start MicroStation, create a design file, and enter the initial settings.

STEP1: Invoke MicroStation by the normal technique for the operating system on your workstation.

STEP 2: Create a new design file named CH4.DGN using the SEED2D.DGN seed file.

STEP 3: In the Design File dialog box, set the Working Unit ratios to 1:10:1000 (see Figure P4–2).

STEP 4: Set the Grid Master to 0.1 and the Grid Reference to 10, and turn OFF the Grid Lock (see Figure P4–3).

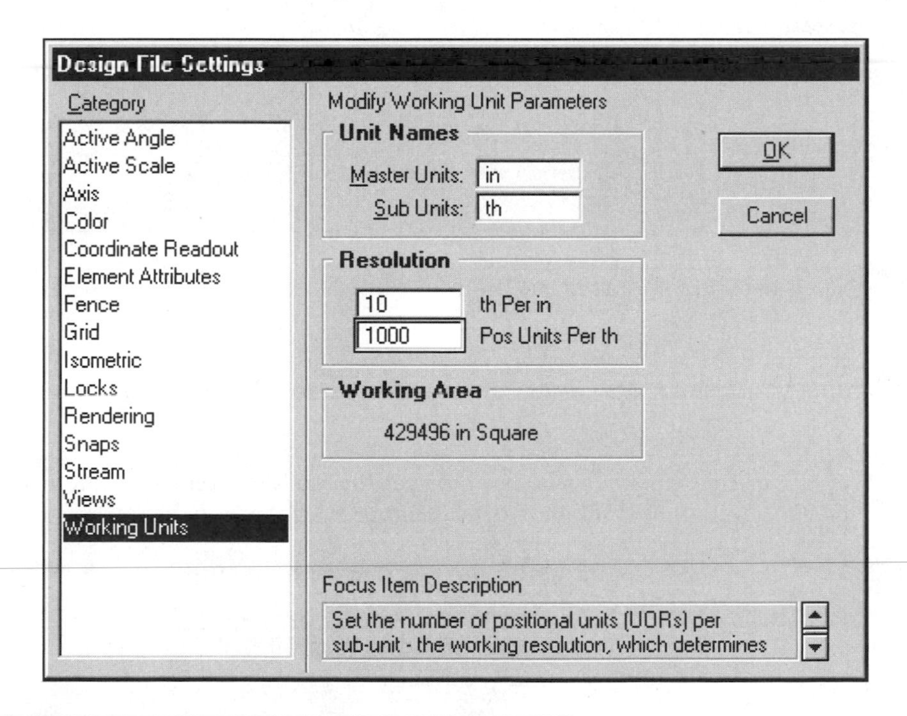

Figure P4–2 Set the Working Unit ratios as shown here

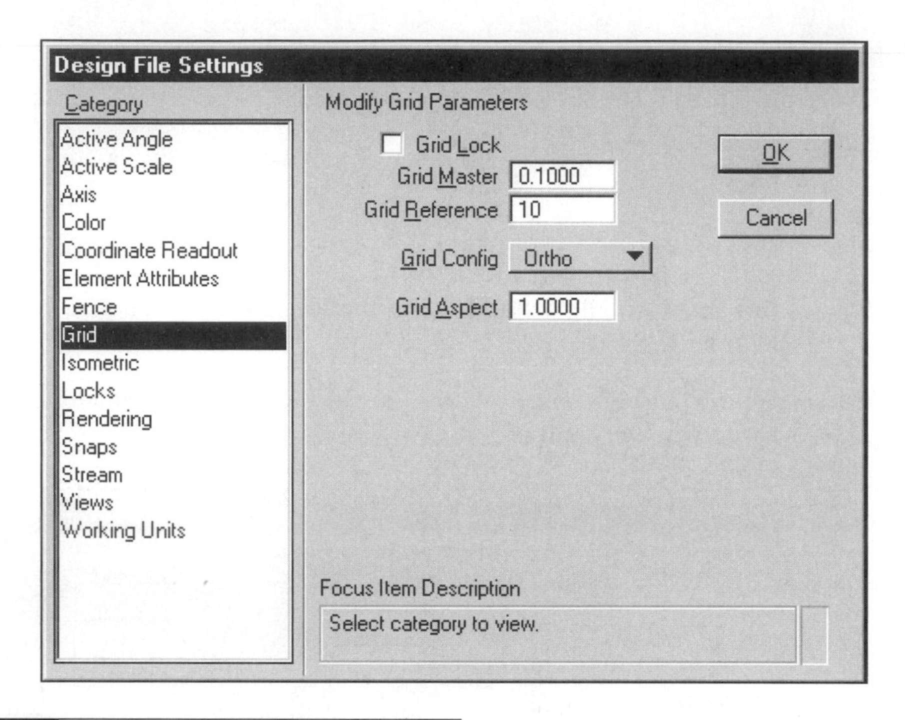

Figure P4–3 Set the Grid Units as shown here

STEP 5: Click the Snaps icon in the Status bar, and select Keypoint mode while pressing Shift.

STEP 6: In the Primary Tools tool box, set the Active Level to 10, the Color to blue, and the Line Weight to 2 (see Figure P4–4).

STEP 7: Select Save settings from the File pull-down menu.

Figure P4–4 Set the element attributes as shown here

Draw the Border and Title Block

This procedure presents the steps for drawing the border and title block, as shown in Figure P4–5.

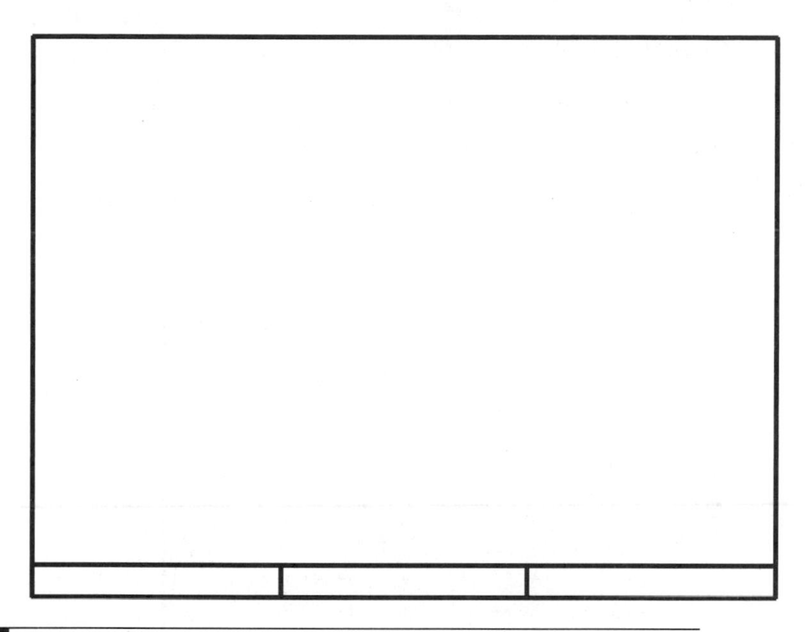

Figure P4–5 Border and title block before the title block text is entered

STEP 1: Create the border by drawing a block 12 inches wide by 9 inches tall, with the lower left corner at **XY=0,0**.

STEP 2: Fit the view.

STEP 3: Create the title block area by drawing a horizontal line across the width of the block and one-half inch above the bottom of the block.

STEP 4: Divide the title block into three equal areas by drawing two vertical lines.

STEP 5: Select Save Settings from the File pull-down menu.

Compare your completed border to the one shown in Figure P4–5.

Fill in the Title Block Text

This procedure has you place text in the title block, as shown in Figure P4–6.

SCHOOL NAME | BASE PLATE | NAME: DATE: SCALE: FULL SCALE

Figure P4–6 Filled-in title block

STEP 1: Change the Line Weight to 0.

STEP 2: Open the Text settings box from the Element pull-down menu, and set the text parameters as shown in Figure P4–7.

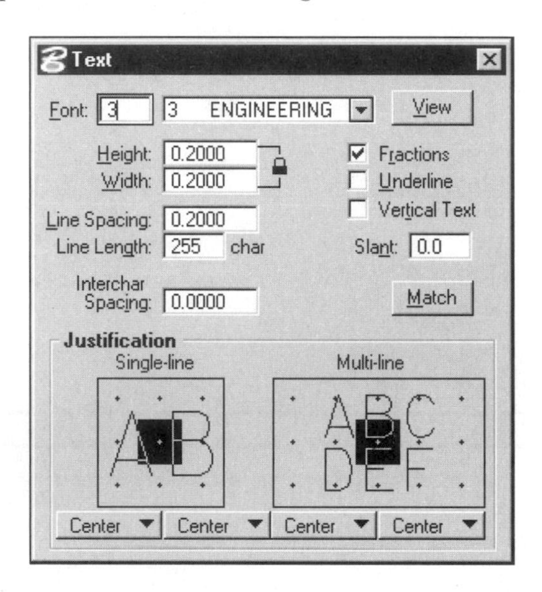

Figure P4–7 Enter the text parameters shown here

STEP 3: Select Save Settings from the File pull-down menu.

STEP 4: Invoke the Place Text tool from the Text tool box, then select By Origin from the Tool Settings window.

MicroStation prompts:

Place Text > Enter Text *(In the Text Editor window type a school or company name, then place the text centered in the left title block area.)*

Place Text > Enter Text *(Click the Text Editor Reset button, type* **BASE PLATE**, *then place the text centered in the center title block area.)*

STEP 5: In the Text settings box, set the Height and Width to 0.125, the Line Spacing to 0.1, and both justifications to Left Top.

STEP 6: In the View Attributes settings box, set the Text Node view attribute to OFF.

STEP 7: Use either the Zoom In or Window Area tool to zoom in close to the right title block area.

STEP 8: In the right title block area, place the text strings shown in Figure P4–6. Insert your name to the right of "NAME:" and today's date to the right of "DATE:".

STEP 9: Fit the view, and select Save Settings from the File pull-down menu.

Compare your completed title block to the one shown in Figure P4–6.

Draw the Center Lines

This procedure describes the steps required to draw one horizontal centerline and one vertical centerline, then, with the Copy Parallel tool, create the additional centerlines, as shown in Figure P4–8.

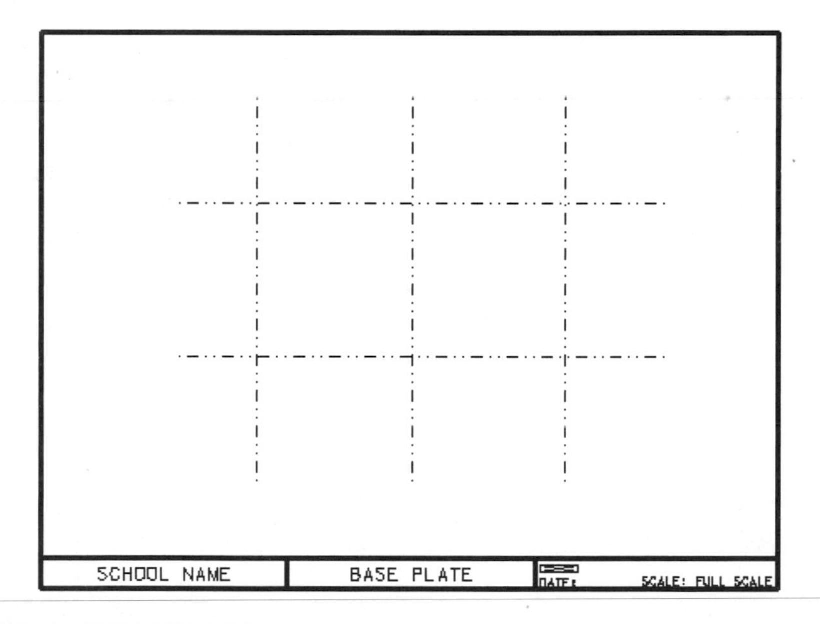

Figure P4–8 Completed centerlines

STEP 1: Set the Active Level to 2, the Color to green, the Line Weight to 0, and the Line Style to 6. Then select Save Settings from the File pull-down menu.

STEP 2: Place the top horizontal centerline 8 inches long starting at **XY = 2.25,6.25**.

STEP 3: Place the left vertical centerline 6.25 inches long starting at **XY =3.5,1.75**.

STEP 4: Invoke the Move Parallel tool from the Manipulate tool box, set the Make Copy and Distance toggle buttons to ON, and key-in **2.5** in the Distance edit field in the Tool Settings window.

MicroStation prompts:

> Copy Parallel by Key-in > Identify element *(Select the horizontal centerline.)*
>
> Copy Parallel by Key-in > Accept/Reject (select next input) *(Click the Data button below the horizontal line, then click the Reset button to release the line.)*

STEP 5: Make two parallel copies of the vertical line, each at a distance of 2.5 inches.

Compare your completed centerlines to Figure P4–8.

Draw Part of the Base Plate Outline

This procedure draws the left half of the base plate outline using the Place Line, Fillet, and Arc tools, as shown in Figure P4–9.

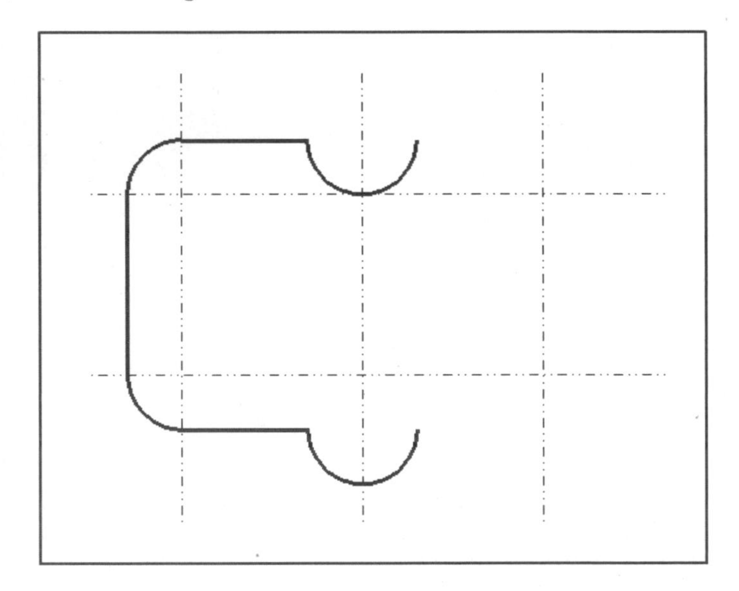

Figure P4–9 Result of drawing the left half of the base plate outline

STEP 1: Set the Active Level to 1, the Color to 0, the Line Weight to 2, and the Line Style to 0, then select Save Settings from the File pull-down menu.

STEP 2: Place three lines using these precision key-ins (see Figure P4–10):

- XY=5.25,3
- DI=2.5,180
- DI=4,90
- DI=2.5,0

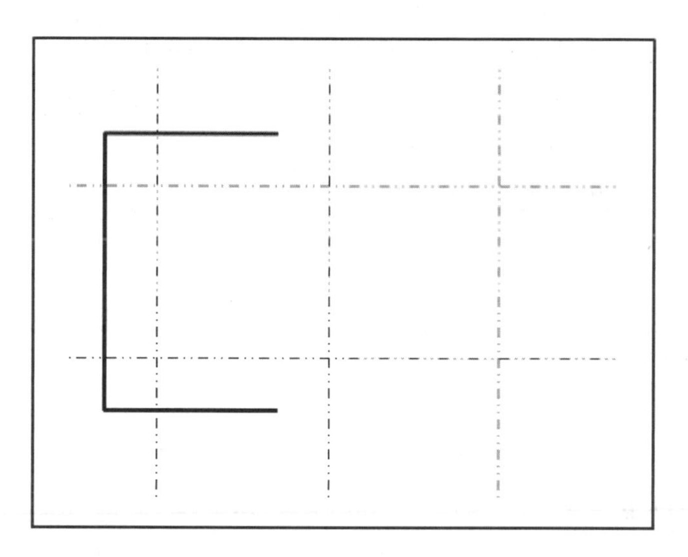

Figure P4–10 View after placing three lines in Step 2

STEP 3: Invoke the Construct Circular Fillet tool from the Modify tool box, then set the Radius to 0.75 and the Truncate option to Both in the Tool Settings window.

MicroStation prompts:

Circular Fillet and Truncate Both > Select first segment *(Select the bottom horizontal line that was drawn in Step 2.)*

Circular Fillet and Truncate Both > Select second segment *(Select the vertical line.)*

Circular Fillet and Truncate Both > Accept-Initiate construction *(Click the Data button in space to place the fillet.)*

STEP 4: Place a 0.75-inch fillet at the intersection of the vertical line and the top horizontal line.

STEP 5: Invoke Place Arc from the Arcs tool box, select the Edge option from the Method option menu, and set the Radius to 0.75, the Start Angle to 180, and the Sweep Angle to 180 in the Tool Settings window.

MicroStation prompts:

Place Arc By Edge > Identify First Arc Endpoint *(Keypoint Snap to the right end of the bottom horizontal line and place a data point.)*

Place Arc By Edge > Identify First Arc Endpoint *(Keypoint Snap to the right end of the top horizontal line and place a data point. Then click the Reset button.)*

Compare your completed left half to Figure P4–9.

Note: The next procedure fixes the rotation of the top arc.

Complete the Base Plate Outline

This procedure uses the Mirror and Copy tools to complete the outline of the base plate, as shown in Figure P4–11.

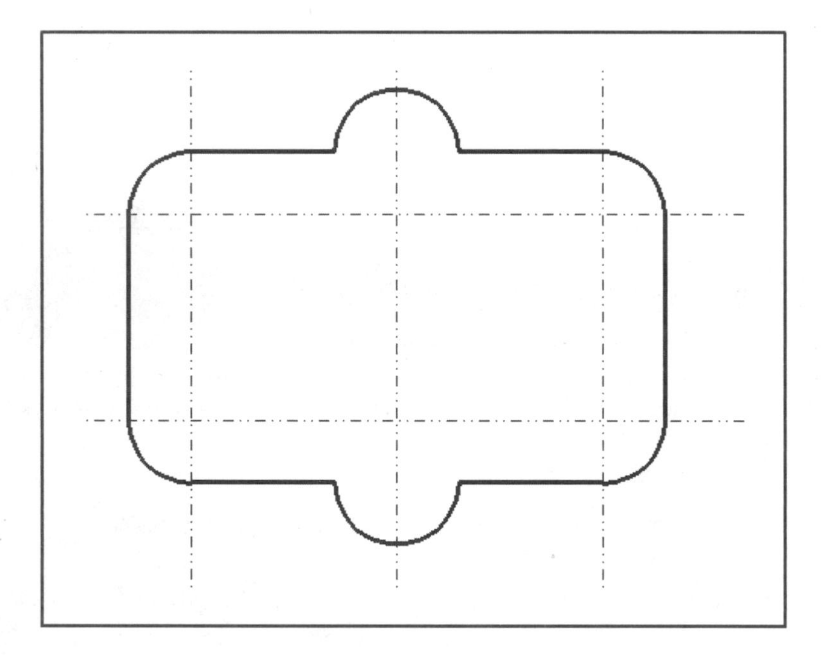

Figure P4–11 Completed base plate outline

STEP 1: Invoke the Mirror tool from the Manipulate tool box, then set Mirror About to Horizontal and set the toggle button for Make Copy to OFF in the Tool Settings window.

MicroStation prompts:

Mirror Element About Horizontal (Original) > Identify element *(Select the top arc.)*

Mirror Element About Horizontal (Original) > Accept/Reject (select next input) *(Keypoint snap to the right end of the top horizontal line, place a data point, then click the Reset button to release the arc.)*

STEP 2: Change the Mirror About option to Vertical, and set the toggle button for Make Copy to ON.

MicroStation prompts:

> Mirror Element About Vertical (Copy) > Identify element *(Select the top fillet.)*
>
> Mirror Element About Vertical (Copy) > Accept/Reject (select next input) *(Keypoint snap to the center vertical center line, place a data point, then click the Reset button to release the fillet.)*

STEP 3: Mirror copy the bottom fillet to the right side of the base plate outline.

STEP 4: Mirror copy the left vertical base plate line to the right side of the outline.

STEP 5: Invoke the Copy tool from the Manipulate tool box.

MicroStation prompts:

> Copy Element > Identify element *(Keypoint Snap to the left end of the top horizontal line in the base plate outline and click the Data button.)*
>
> Copy Element > Accept/Reject (select next input) *(Keypoint Snap to the right end of the top arc and click the Data button.)*
>
> Copy Element > Accept/Reject (select next input) *(Keypoint Snap to the right end of the bottom arc, click the Data button, then click the Reset button.)*

Compare your design to Figure P4–11.

Place the Bolt Hole Circles

This procedure uses the Place Circle and Array tools to place the bolt hole circles in the design, as shown in Figure P4–12.

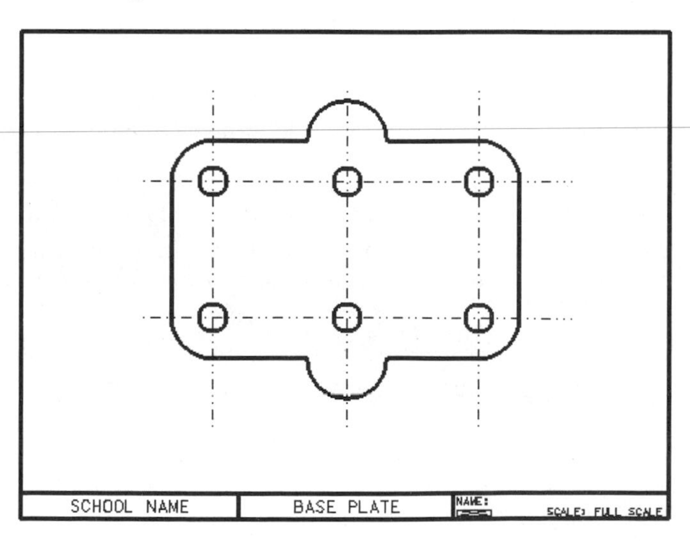

Figure P4–12 Design after placing the bolt hole circles

STEP 1: Click the Snaps icon in the Status bar, and select Intersection mode while pressing Shift.

STEP 2: Invoke the Place Circle tool from the Ellipse tool box, select the Center Method, and key-in 0.25 in the Radius edit field in the Tool Settings window.

MicroStation prompts:

Place Circle by Center > Identify Center Point *(Tentative snap at the intersection of the vertical centerline and the lower horizontal centerline, and click the Accept button to place the circle as shown in Figure P4–13. Then click the Reset button.)*

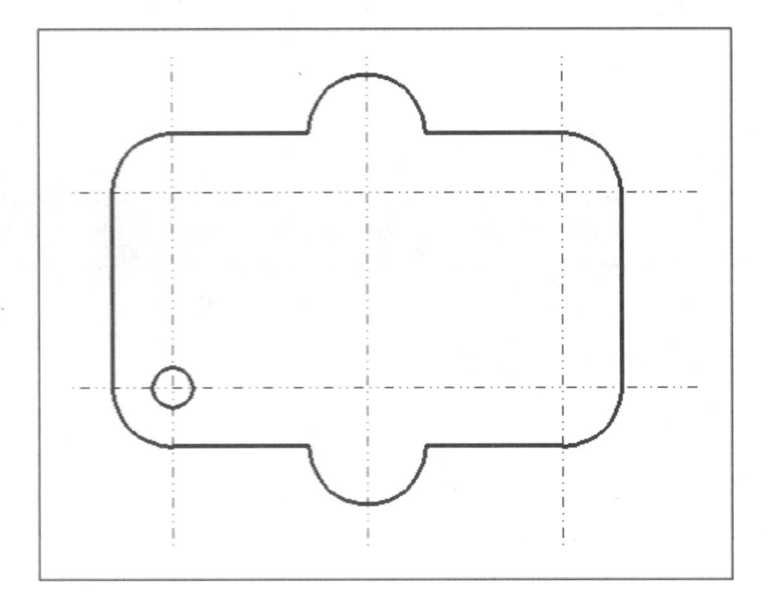

Figure P4–13 A circle is placed on the lower left center intersection

STEP 3: Invoke the Construct Array tool from the Manipulate tool box, set the Array Type to Rectangular, the Active Angle to 0, Rows to 2, Columns to 3, the Row Spacing to 2.5, and the Column Spacing to 2.5 in the Tool Settings window.

MicroStation prompts:

Rectangular Array > Identify element *(Select the circle.)*

Rectangular Array > Accept/Reject (select next input) *(Click the Data button in space to initiate construction of the array.)*

STEP 4: Select Save Settings from the File pull-down menu.

Compare your design to Figure P4–12.

DRAWING EXERCISES 4–1 THROUGH 4–10

Use the following table to set up the design files for Exercises 4–1 through 4–3.

SETTING	VALUE
Seed File	SEED2D.DGN
Working Units	12 " Per ' and 8000 Pos Units Per "
Grid	Grid Master = .25, Grid Reference = 4 Grid Lock ON
Object Elements	Color = 0, Level = 1, Style = 0, Weight = 1
Hidden Lines	Color = 0, Level = 1, Style = 3, Weight = 1
Centerlines	Color = 3, Level = 2, Style = 6, Weight = 0

Exercise 4–1

Machine Part

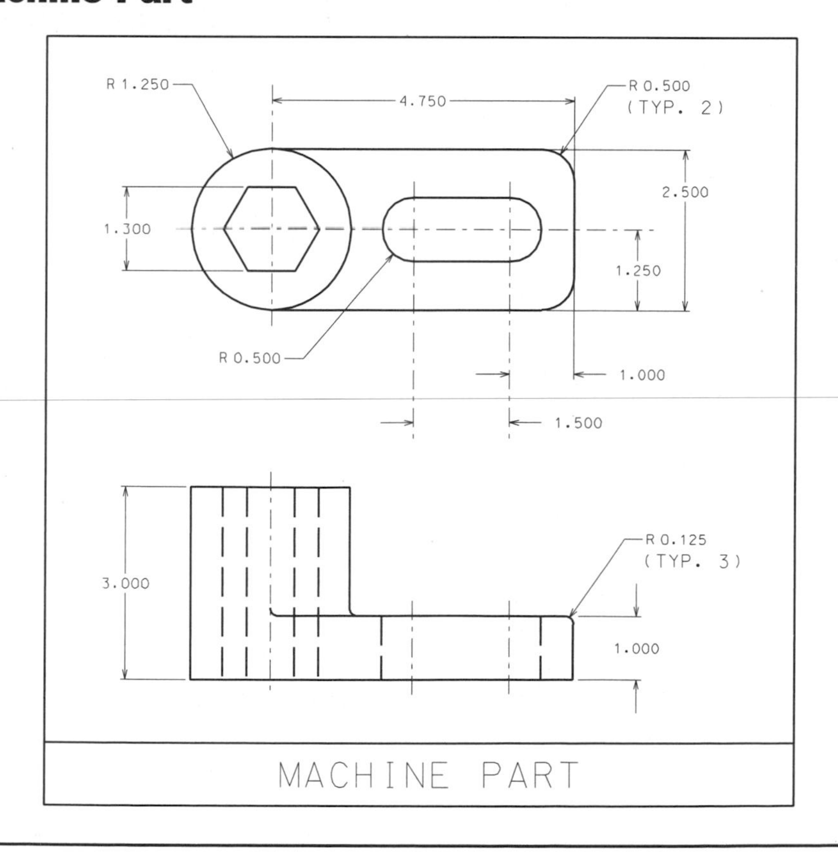

Exercise 4–2

Plate

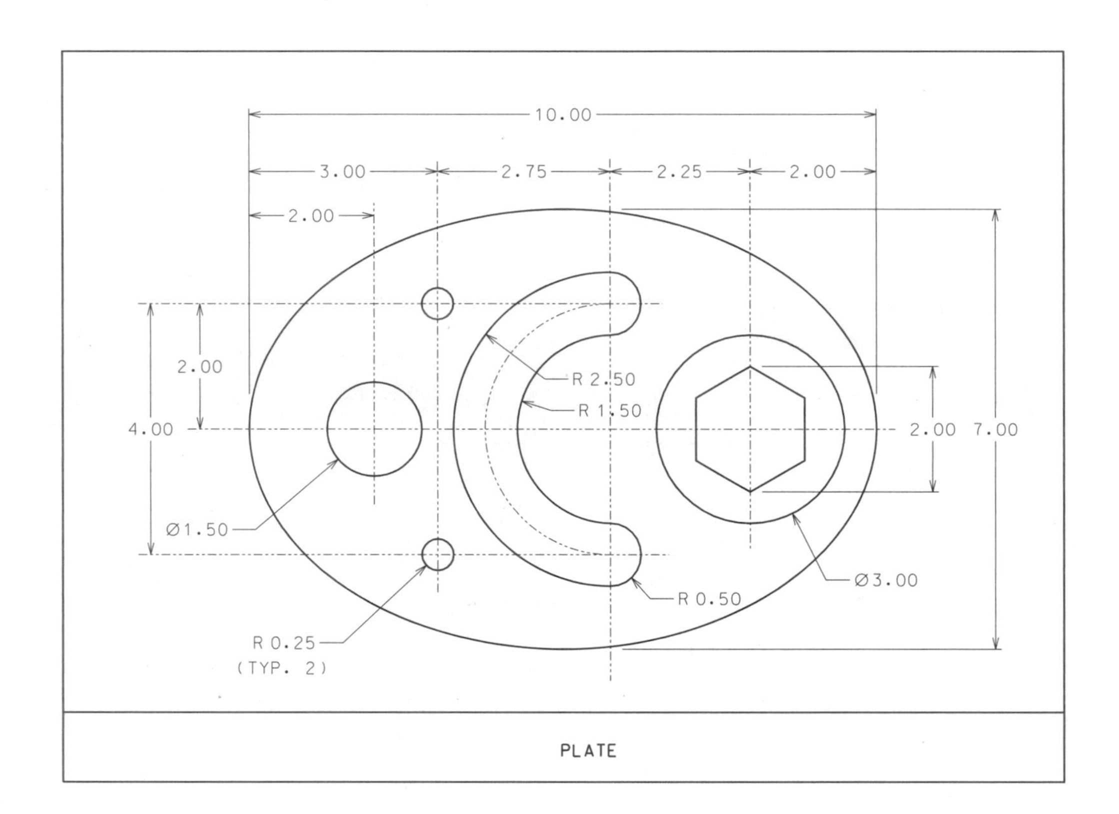

Exercise 4–3

Machine Part

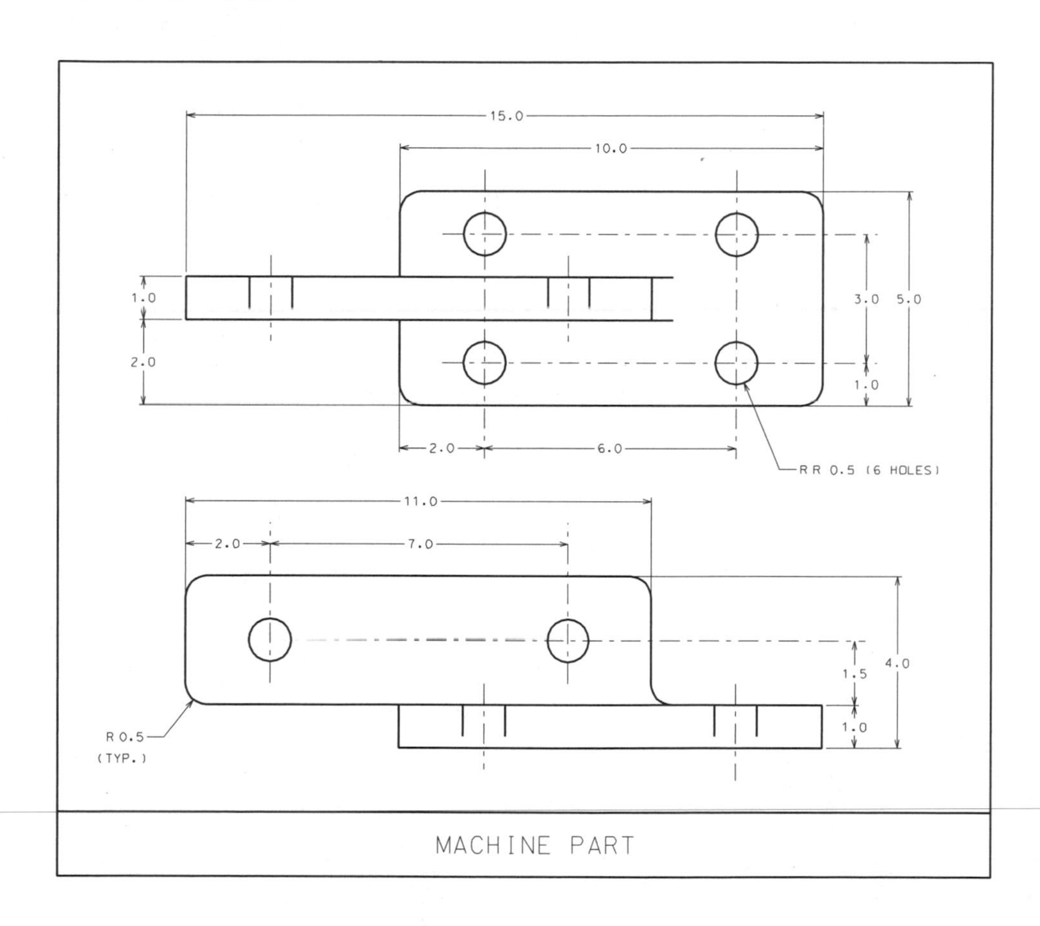

Use the following table to set up the design files for Exercises 4–4 and 4–5.

SETTING	VALUE
Seed File	SEED2D.DGN
Working Units	12 " per ' and 8000 PU per "
Grid	Grid Master = .25, Grid Reference = 4, Grid Lock OFF
Object Elements	Color = 0, Level = 1, Style = 0, Weight = 1
Grid Lines	Color = 3, Level = 2, Style = 6, Weight = 0

Exercise 4–4

Topography Map

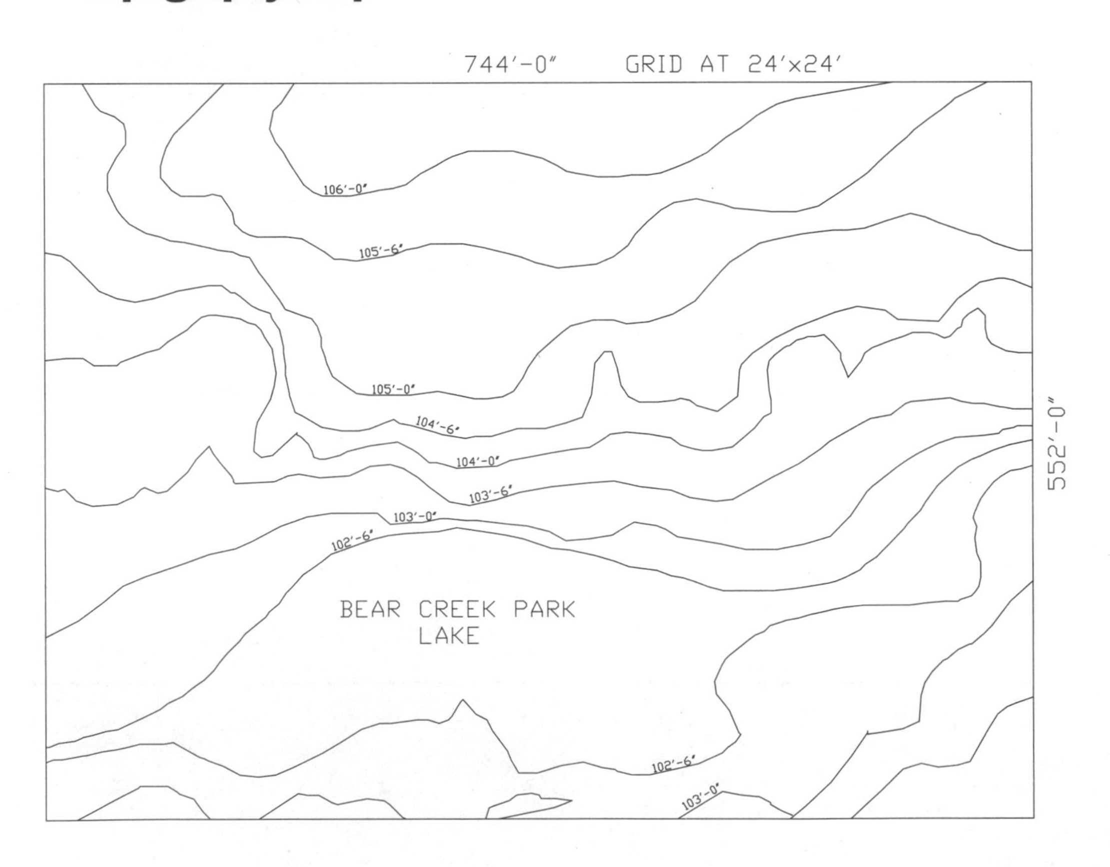

Exercise 4–5

Front Elevation

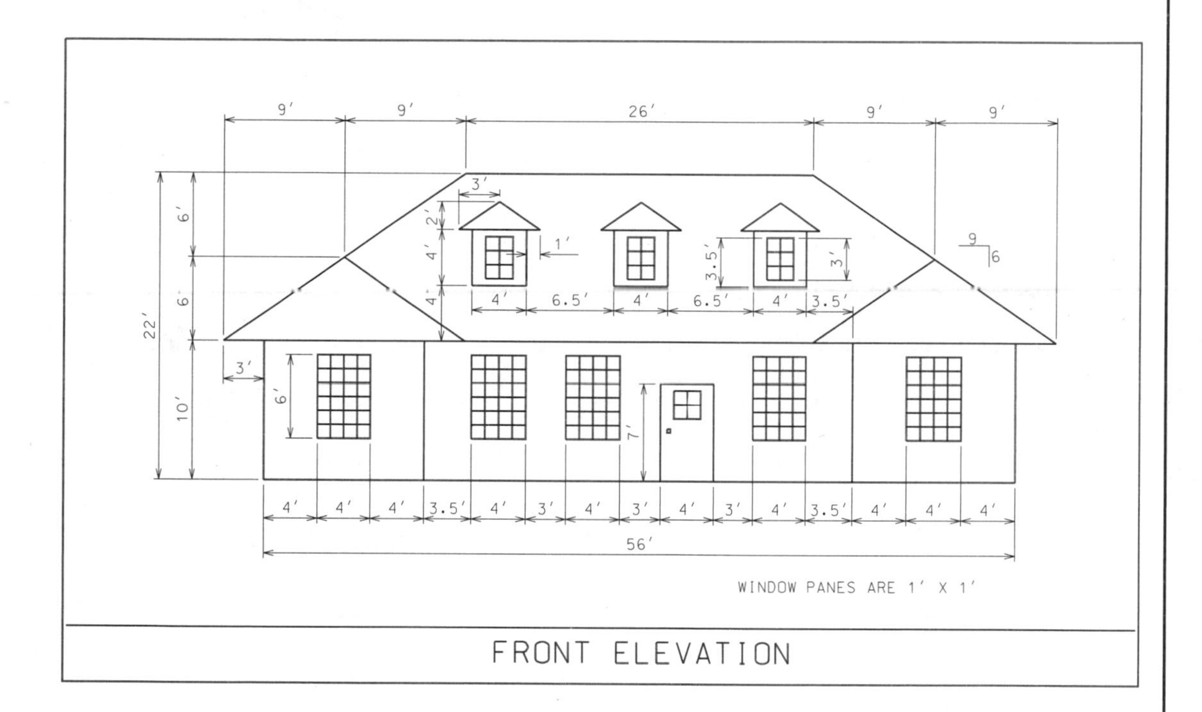

Use the following table to set up the design files for Exercises 4–6 through 4–9.

SETTING	VALUE
Seed File	SEED2D.DGN
Working Units	MU = IN, SU = 10 TH, PU = 1000
Grid	Grid Master = .1, Grid Reference = 10, Grid Lock ON
Object Elements	Color = 0, Level = 1, Style = 0, Weight = 1
Center Lines	Color = 3, Level = 2, Style = 6, Weight = 0

For Figures 4–6 and 4–7, set the Font to 3 and the text Height and Width to 0.125".

Exercise 4–6

Organization Chart

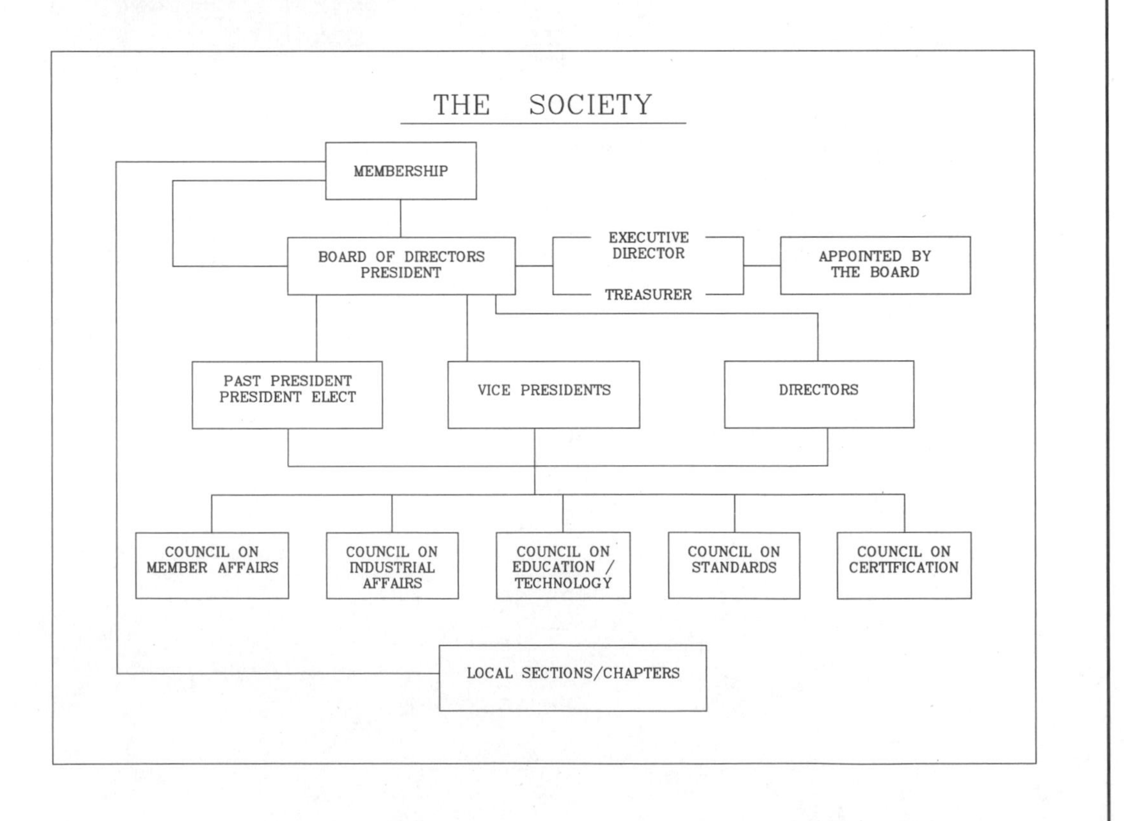

Exercise 4–7

Schematic Diagram

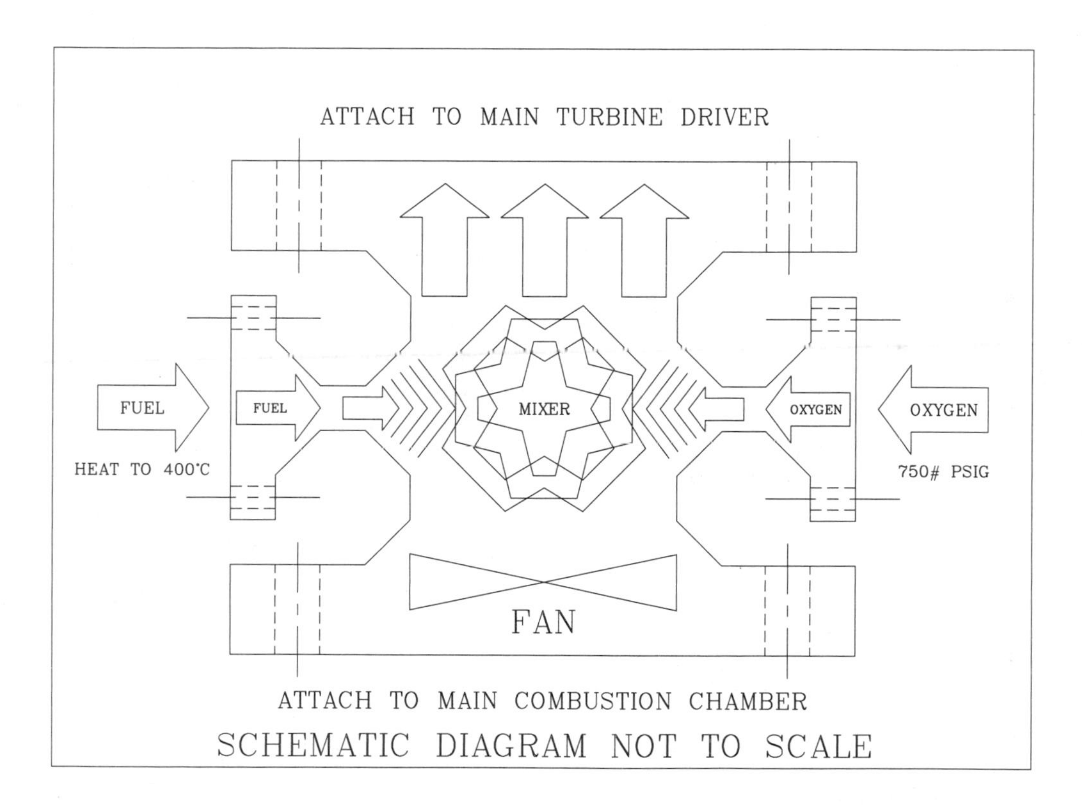

Exercise 4–8

Flange Gasket

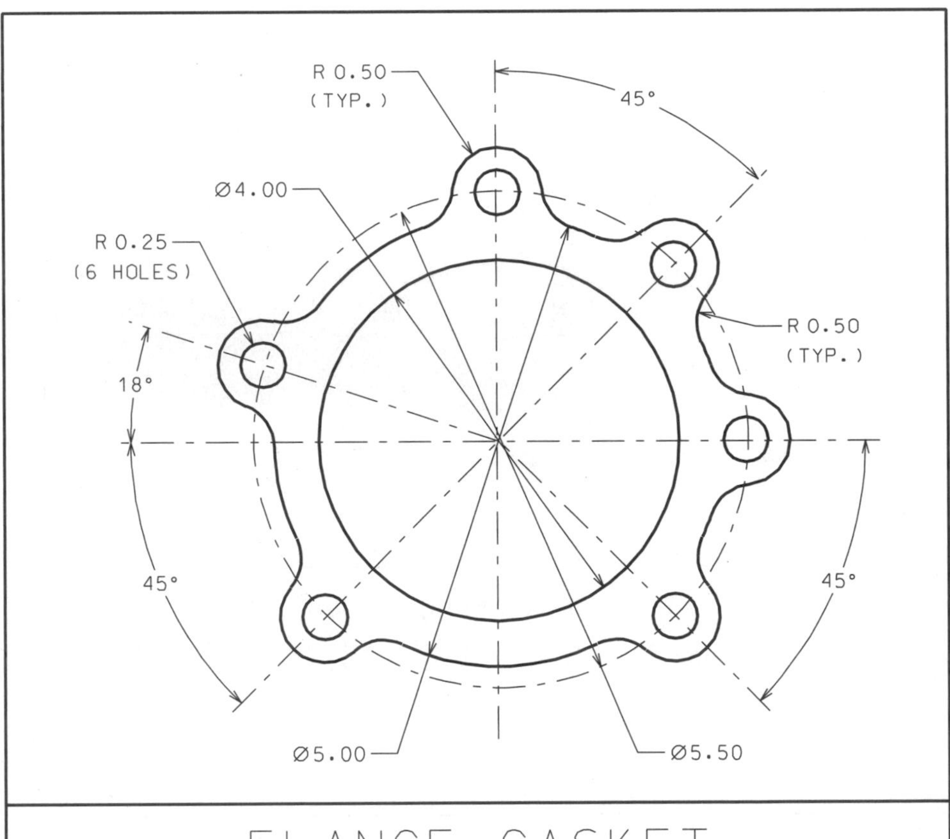

Exercise 4–9

Machine Parts

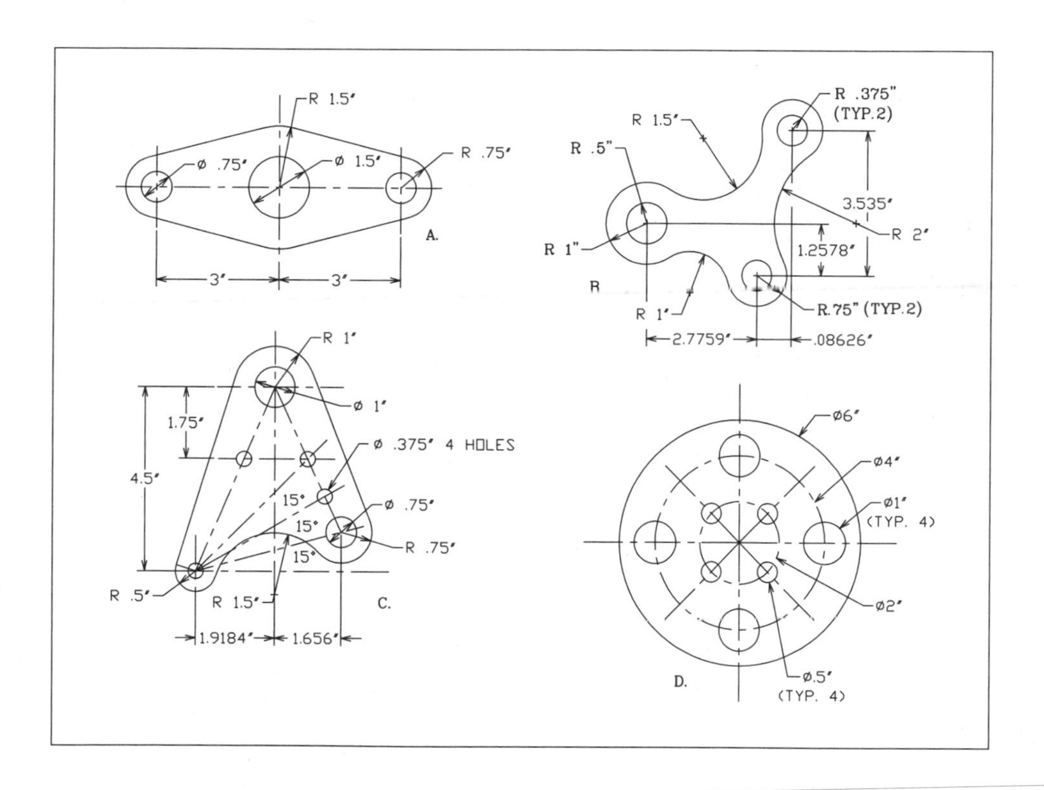

SETTING	VALUE
Seed File	SEED2D.DGN
Working Units	1 SU Per mm and 10000 Pos Units Per SU
Grid	Grid Master = 1, Grid Reference = 10, Grid Lock OFF
Object Elements	Color = 0, Level = 1, Style = 0, Weight = 1
Hidden Elements	Color = 0, Level = 2, Style = 3, Weight = 1
Center Elements	Color = 1, Level = 3, Style = 6, Weight = 0
Text	Font = 3, Text Height, Width, and Spacing = 3

Exercise 4–10

Spring

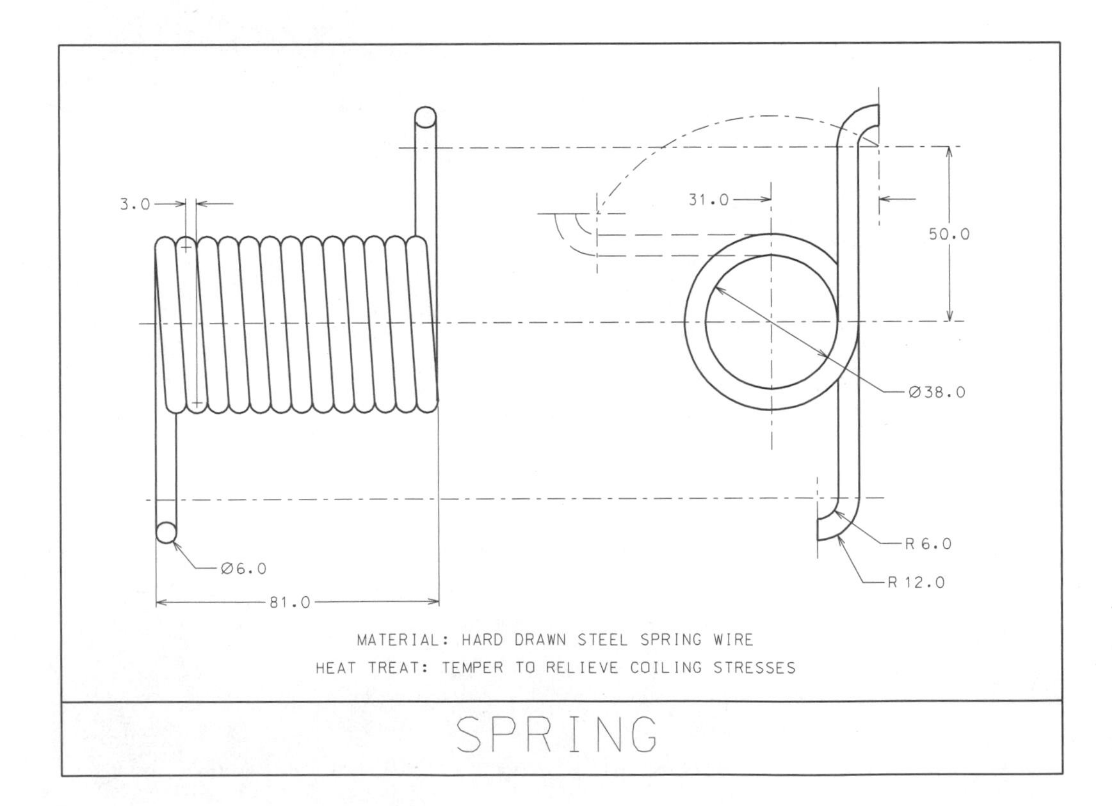

chapter 5

AccuDraw and SmartLine

Objectives

After completing this chapter, you will be able to do the following:

- Set up AccuDraw.
- Use AccuDraw to place elements with fewer data points and less typing.
- Use SmartLine to draw complex models quickly with one tool.

GET TO KNOW ACCUDRAW

AccuDraw is a powerful MicroStation feature that increases your drawing productivity by tracking what you did and attempting to anticipate what you will do next. It includes a drawing compass, an input window, settings windows with sub-windows, and a set of key-in shortcuts (see Figure 5–1).

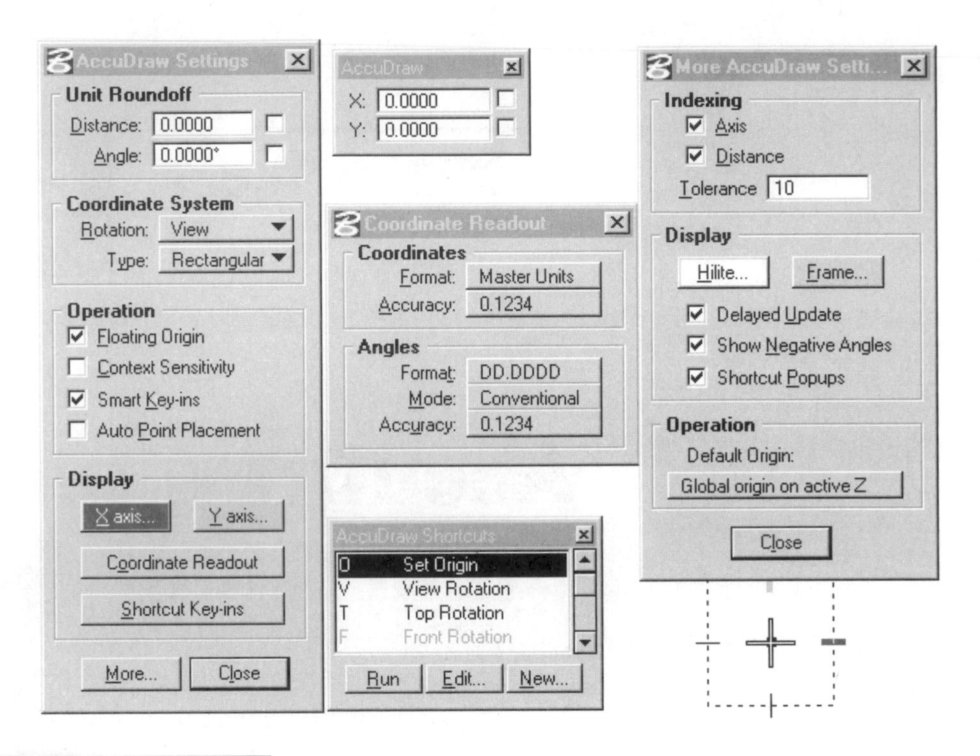

Figure 5–1 AccuDraw tools

Starting AccuDraw

AccuDraw is not active the first time MicroStation is activated, but once you start AccuDraw it opens every time MicroStation is called.

Invoke AccuDraw:

Primary tool box	Select the Start AccuDraw tool (see Figure 5–2).
Key-in window	**accudraw activate** (or **a a**) Enter

MicroStation displays:

Start AccuDraw point input tool

Figure 5–2 Invoking the AccuDraw tool from the Primary tool box

The AccuDraw window opens either as a floating window (see Figure 5–3a) or docked at the edge of the MicroStation workspace (see Figure 5–3b). After it is activated, it enhances all drawing tools.

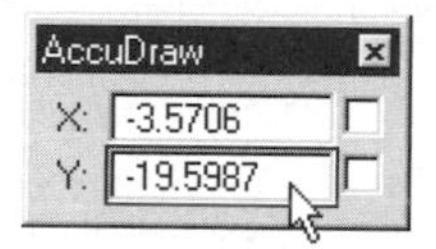

Figure 5–3a AccuDraw window shown floating in the View window

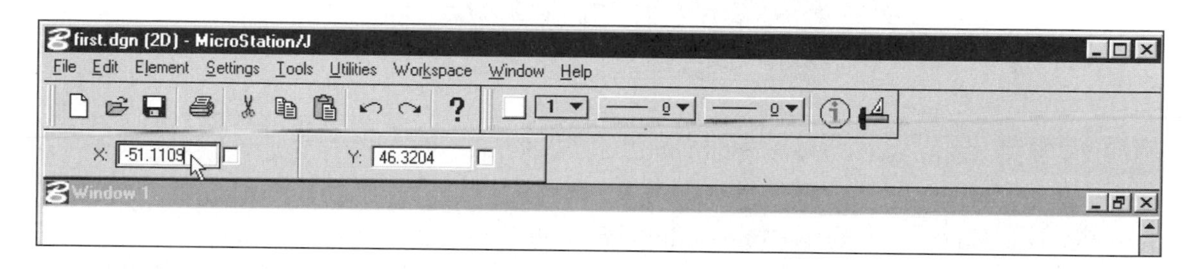

Figure 5–3b AccuDraw window docked at the top of the MicroStation application window

Key-In Shortcuts

When AccuDraw is started, a set of one- and two-character shortcuts becomes available to you. Shortcuts control AccuDraw actions and select AccuDraw features.

When focus is on the AccuDraw window, MicroStation checks all characters you key-in to see if they are AccuDraw shortcuts. When AccuDraw is active, focus usually defaults to the AccuDraw window.

Some of the shortcuts will be introduced throughout this chapter. Instructions for opening a window from which you can view descriptions of all shortcut definitions, as well as a table listing all shortcuts, is provided at the end of this chapter.

The AccuDraw Compass

When AccuDraw is active, the AccuDraw compass appears whenever you place a data point in the design. This compass is the center of the AccuDraw drawing plane and is your main focus for input.

Coordinate System

The AccuDraw drawing plane includes polar and rectangular coordinate systems for locating points. The systems are like those provided by the precision key-in tools, except all offsets in the AccuDraw coordinate systems are from the AccuDraw compass origin, not the design plane's origin point. Figures 5–4a and 5–4b show the appearance of the compass for each coordinate system.

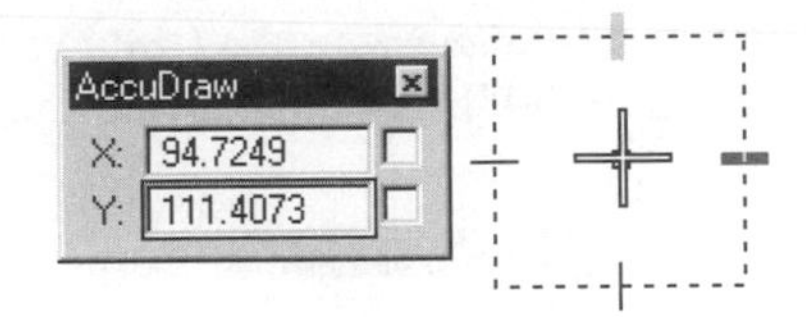

Figure 5–4a Compass with the rectangular coordinate system active

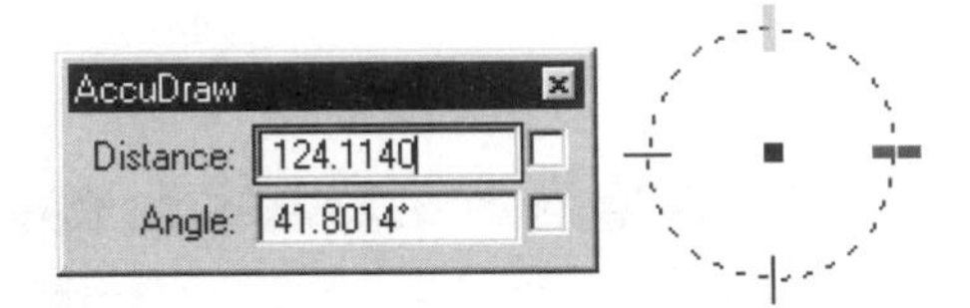

Figure 5–4b Compass with the polar coordinate system active

To switch coordinate systems, use one of these methods:

Keyboard shortcut	**Spacebar** (when focus is on the AccuDraw window)
Key-in window	**accudraw mode** (or **a m**) Enter toggles between modes

The coordinate system is switched (there are no MicroStation prompts).

Orthogonal Axes

The center of the AccuDraw design plan is indicated by a dot in the center of the compass. The orientation of the AccuDraw drawing plane *X* axis and *Y* axis is indicated by short tick marks (lines) crossing the compass rectangle or circle.

To aid in distinguishing the two axes, the positive *X* axis tick mark is a red line and the positive *Y* axis tick mark is a green line. These colors can be changed from the AccuDraw Settings box (discussed later in this chapter).

Compass Movement

The compass moves to and centers on each data point you enter. If you are drawing lines or line segments, the compass also rotates to align its *X* axis parallel to the angle of the last line segment drawn with the positive *X* axis direction pointing in the direction the line was drawn. Figure 9–5 shows an example of the Rectangular coordinate system compass at the end of a line rotated 45 degrees. There are options to turn off both of these features in the AccuDraw Settings box.

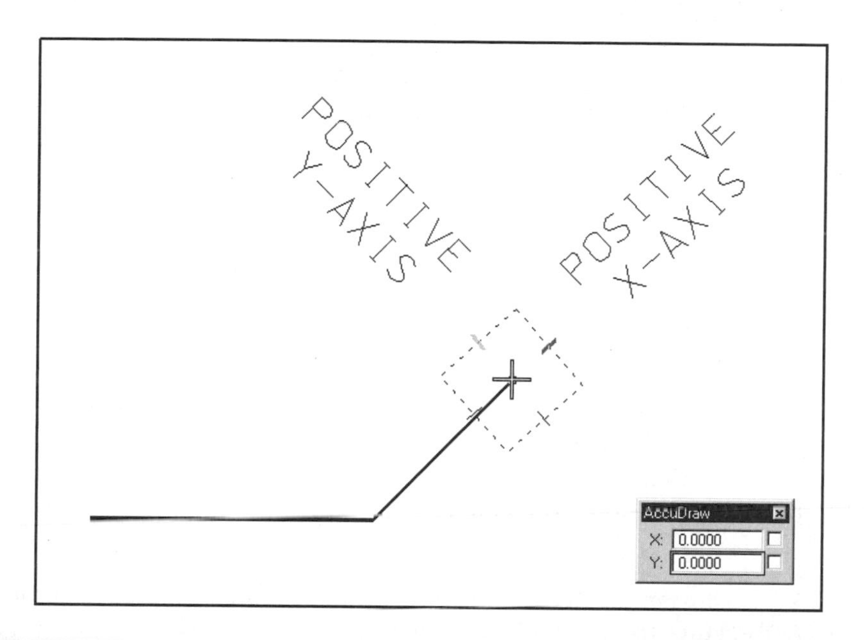

Figure 5–5 Example of the AccuDraw compass rotated to match the rotation of the last line placed in the design

Shortcuts are provided to change the position and rotation of the compass:

- **O** (not zero) moves the compass from its current location to the position of the screen cursor. This is useful for starting element placement at a specific point on an existing element. Snap to the element, then press the **O** key to put the compass at the tentative point.
- **T** realigns the compass parallel with the design's *X* axis (the red positive *X* axis tick mark will point to the right).

The AccuDraw settings box has options to lock the compass at its current location so it no longer floats to the last data point. There is also a setting that locks the compass at its current rotation.

Compass Indexing

As you move the cursor away from the center of the compass, AccuDraw tracks where the cursor is in relation to the compass *X* and *Y* axes. When the cursor is almost lined up with the *X* axis or *Y* axis, AccuDraw indexes (snaps) to the axis. Figure 5–6 shows an example of the cursor indexed to the *X* axis while placing a circle by center (as indicated by the horizontal line extending from the center of the compass).

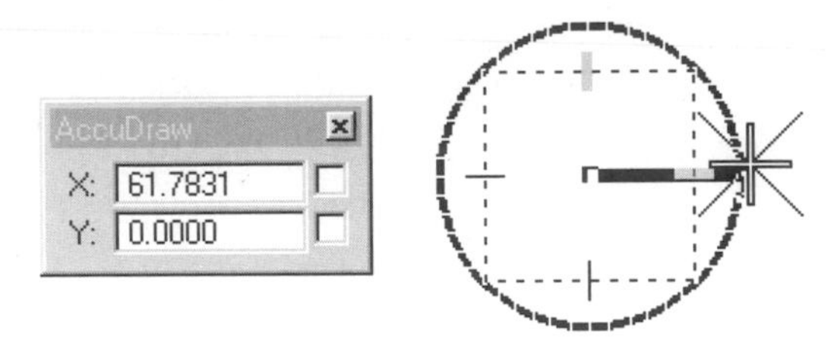

Figure 5-6 The AccuDraw compass indexed to the X axis

The AccuDraw Window

The AccuDraw window provides a place to type *X* and *Y* axis offsets in the Rectangular coordinate system and Distance and Angle offsets in the Polar coordinate system (see Figures 5–4a and 5–4b). This window almost completely eliminates the need to type precision input codes such as **DL=** and **DI=**.

As you move the screen cursor, the AccuDraw window's input fields are updated automatically with the cursor's position relative to the AccuDraw origin.

Selecting Fields

AccuDraw usually has focus on the AccuDraw window edit field you are most likely to type in next, so there is often no need to select the field first. Field focus is indicated by a change in the appearance of the field. If the wrong edit field is in focus, just press the Tab key to move the focus to the edit field you need to type a value in.

For example, in Microsoft Windows 98, the field in focus is indicated by a rectangle around the field and a blinking cursor in the field (see Figure 5–7). Highlight methods vary among operating systems.

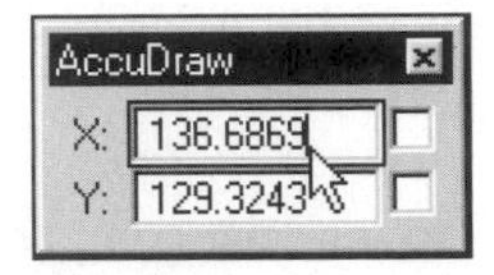

Figure 5-7 The AccuDraw window with the X axis in focus

When you type in a distance or an angle, it appears in the field that is in focus, and that field is locked until the next data point is placed. The toggle button to the right of the edit field is turned on when the field is locked.

For example, if the compass is in rectangular mode and the dynamic image is indexed to the positive *X* axis indicator, the *Y* edit field in the window contains 0 and the *X* edit field has focus. If you type a number on the keyboard, it is placed in the *X* edit field and the field is locked.

Accepting Field Contents

When the AccuDraw window fields contain the correct offset from the compass origin, click the Data button to place the data point at the offset values. When you place the data point, the AccuDraw edit fields are unlocked.

Locking Fields

As mentioned earlier, each edit field in the AccuDraw window has a lock toggle button that locks the field when you type something in it. The field remains locked until you place the next data point. You can also lock or unlock the fields by either clicking on the lock or by typing a shortcut key. The lock shortcut keys are as follows:

- **X** locks and unlocks the *X* edit field when the Rectangular coordinate system is active.
- **Y** locks and unlocks the *Y* edit field when the Rectangular coordinate system is active.
- **D** locks and unlocks the Distance edit field when the Polar coordinate system is active.
- **A** locks and unlocks the Angle edit field when the Polar coordinate system is active.

Negative Distances

The direction of the screen cursor from the compass origin indicates the direction of your input in the drawing plane, so there is usually no need to type the negative sign for distances that are to the left of or down from the AccuDraw origin. Move the screen cursor near the area where the next data point will be placed to establish the correct sign.

Recall Previous Values

AccuDraw remembers the values previously entered in the AccuDraw window, and it takes the previous distance as a hint for the next distance.

Previous Distance

AccuDraw remembers the linear distance between the last two data points. If the cursor is indexed to either Rectangular coordinate axis, a tick mark at the screen cursor indicates when the linear distance from the last data point to the current cursor location is equal to the linear distance between the last two data points. For Polar coordinates, the tick mark will appear at any angle of rotation. Figure 5–8 shows the tick mark at the cursor position for rectangular coordinates.

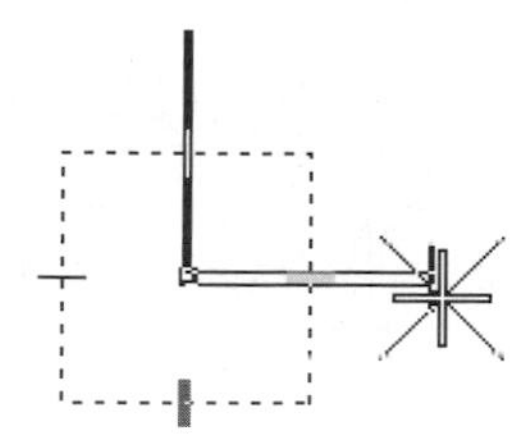

Figure 5–8 Current distance is equal to previous distance, as indicated by tick mark at cursor position

Recalling Previous Values

Each time you place a data point, AccuDraw remembers the offsets used to place that point. All distance values are stored in a buffer and all angle values are stored in a separate buffer. Press the Page Up key to load the last value used in the edit field that is in focus. Press the Page Up key again and the next-to-last value is loaded. Each time you press the key a saved value is loaded in the edit field.

All distance values are stored in the same buffer, so they can be applied to any coordinate distance field (*X* axis, *Y* axis, or Polar Distance).

Popup Calculator

MicroStation provides a calculator that works with AccuDraw to allow you to do calculations using the values in the AccuDraw window fields. To use the calculator on the value of the field with focus:

1. Type a symbol for the mathematical operation you want: + (add), – (subtract), * (multiply), or / (divide).
2. Type a value to complete the calculation.
3. Place a data point to accept the calculation, or click the Esc key to reject the calculation.

When you type one of the mathematical symbols, the calculation opens in a separate window below the AccuDraw field with focus. The calculation result is shown at the bottom of the calculator window. Figure 5–9 shows an example of the calculator being applied to the *X* input field.

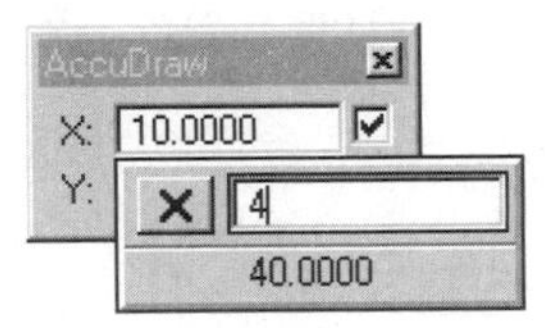

Figure 5–9 The calculator window

Replace Current Value

To perform a calculation that does not use the value in the AccuDraw field with focus, type an equal sign (=) to start the calculation. When you start the calculation with an equal sign, the calculation results are stored in the AccuDraw field rather than in the calculator window (see Figure 5–10).

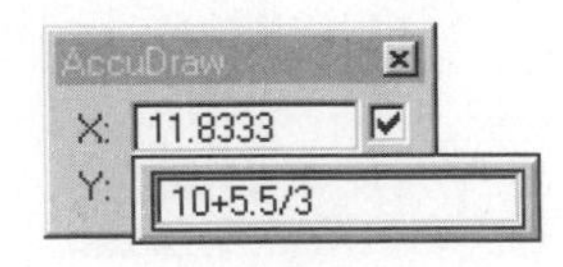

Figure 5–10 The calculator window when you start by typing an equal sign (=)

Complex Calculations

Parenthesis can be used to create complex calculations. For example; the calculation, "=10*(2+3)" places 50 in the field. The two and three are added, then the result is multiplied by ten (see Figure 5–11).

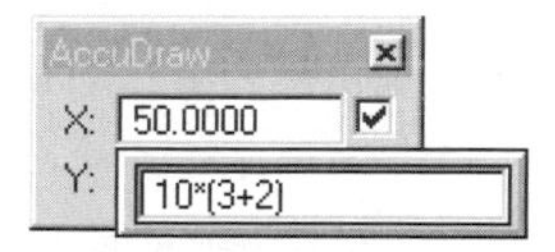

Figure 5–11 The use of parentheses in a calculation

Using Smart Lock

Smart Lock allows you to constrain the next data point to the nearest axis. To use Smart Lock, move the cursor near to the axis and direction you want the next data point constrained to, then press the Enter Key. If the rectangular compass is active, either the *X* or *Y* axis will be locked. If the polar compass is active, the Angle will be locked at 0, 90, 180, or 270 degrees.

To turn off Smart Lock and unconstrain the next data point, press the Enter key again. The Enter key is a toggle that turns Smart Lock on and off.

CHANGE ACCUDRAW SETTINGS

AccuDraw provides four settings boxes that can be used to change the appearance of the compass and the way AccuDraw works.

AccuDraw Settings Box

The AccuDraw Settings box allows you to set round-off values, control the compass, set the color of the compass axis tick marks on the compass, and open the other AccuDraw settings boxes.

Invoke the AccuDraw Settings box:

Pull-down menu	Settings > AccuDraw
Keyboard shortcut	**GS** (with focus on the AccuDraw window)
Key-in window	**accudraw dialog settings** (or **a d se**) Enter

MicroStation opens the AccuDraw Settings box, as shown in Figure 5–12.

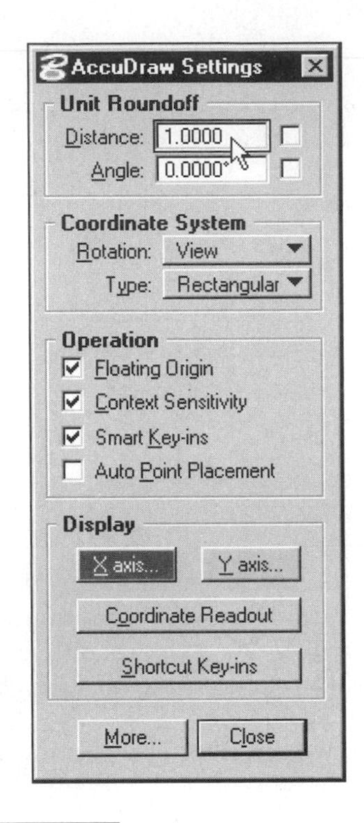

Figure 5-12 The AccuDraw Settings box

Unit Roundoff Options

The Unit Roundoff options allow you to set and lock AccuDraw to a distance and angle roundoff value. When a unit roundoff lock is set, values are forced to the roundoff value or multiples of it as you move the screen cursor.

The Distance Roundoff Lock limits rectangular and polar distances to the roundoff value, in working units (MU:SU:PU) or multiples of it. The distance is always calculated from the current compass location (not from a grid point).

For example, set the Distance roundoff value to 5 and lock the value to force all distances (rectangular *X* and *Y*, and polar Distance) to increase or decrease by increments of 5 (–10, –5, 0, 5, 10, etc.) as you drag the cursor in the design.

The Angle Roundoff Lock limits polar angles to the roundoff value or multiples of it.

For example, set the Angle roundoff value to 30 degrees to force all angles to increments of 30 degrees (0, 30, 60, 90, etc.) as you drag the cursor in the design.

Note: Values keyed into AccuDraw window edit fields override the roundoff lock.

Coordinate System Options

The Coordinate System options control the compass angle and type.

The Rotation options menu allows you to align the compass to the following settings:

- Top—Aligns the compass to the top view in *three dimensions,* or the current view axes in *two dimensions* (same as View).
- Front—Aligns the compass to the front view in *3D* only.
- Side—Aligns the compass to the side view in *3D* only.
- View—Aligns the compass to the current view axes.
- Auxiliary—Aligns the compass to the last defined auxiliary coordinate system.
- Context—A temporary orientation affected by several factors, including the RQ keyboard shortcut.

The Type options menu allows you to switch the compass between the Polar and Rectangular coordinate systems (the same as pressing the Spacebar when focus is on the AccuDraw window).

Operation Options

The Operation options control the compass action and other features.

- When the Floating Origin toggle button is set to ON the compass moves to the latest data point.
- When the Context Sensivity toggle button is set to ON the compass rotates, when placing lines or line segments, to match the angle formed by the last placed line segment. This button is usually OFF.
- When the Smart Key-ins toggle button is ON, the AccuDraw focus goes to the field that is not locked, when you lock a field. For example, in the Rectangular coordinate system, lock the *Y* axis and focus jumps to the *X* axis. If the button is set to OFF, focus will stay on the locked axis, and you must use the tab key to jump to the other field.
- When the Auto Point Placement toggle button is set to ON, you can type a value in an AccuDraw window edit field, then press Enter to apply the value to the design. If this button is set to OFF, you must click the Data button to apply the value to the design.

Display Options

The Display options allow you to change the color of the positive *X* and *Y* axis marks on the compass. Figure 5–13 shows a typical Modify Axis Color settings box.

To change the color:

1. Select the desired color from the Modify Axis Color settings box.
2. Click on the OK button to close the settings box.

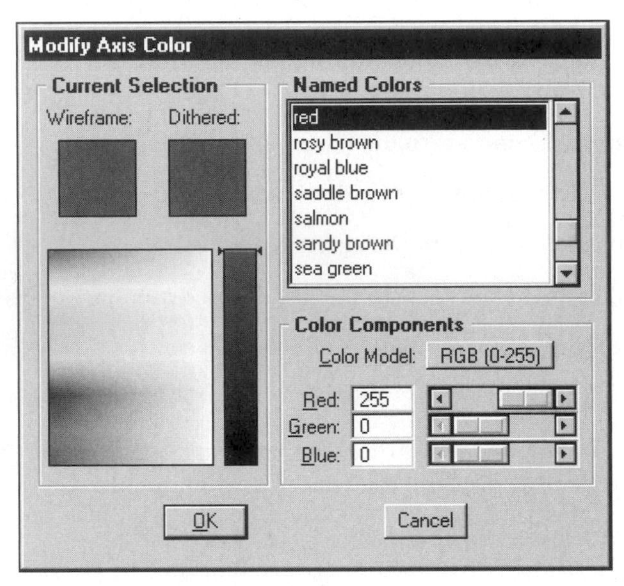

Figure 5–13 Modify Axis Color settings box

The Coordinate Readout, Shortcut Key-ins, and More buttons open the settings boxes that are discussed later in this chapter.

The Close button closes the AccuDraw Settings box.

More AccuDraw Settings Box

The More AccuDraw Settings box provides more options for modifying the appearance and operation of the compass.

Invoke the More AccuDraw Settings box:

AccuDraw Settings box	Click the More button.
Keyboard shortcut	**GM** (with focus on the AccuDraw window)
Key-in window	**accudraw dialog moresettings** (or **a d mo**) Enter

MicroStation opens the More AccuDraw Settings box, as shown in Figure 5–14.

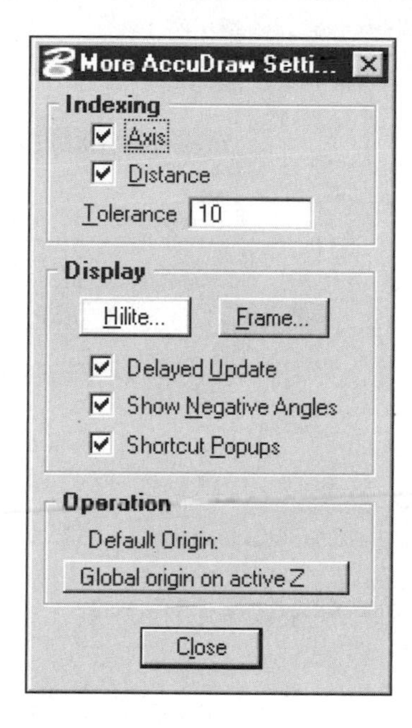

Figure 5–14 The More AccuDraw Settings box

Indexing Options

The Indexing options allow you to change the way indexing works.

- The Axis toggle button turns axis indexing On and Off. When it is Off, AccuDraw will not "grab" the dynamic image when it is at the compass 0, 90, 180, and 270 degree positions.
- The Distance toggle button turns distance indexing On and Off. When it is Off, AccuDraw will not indicate when the line you are about to draw is the same length as the previously drawn line.
- The Tolerance edit field code allows you to control how accurately AccuDraw calculates the indexing point.

Display Options

The Display options set the compass color and control items that appear in the AccuDraw window.

- The Hilite and Frame buttons allow you to set the color of the compass center point and outline. When you click one of the buttons, the Modify Axis Color settings box appears, as shown in Figure 5–15.

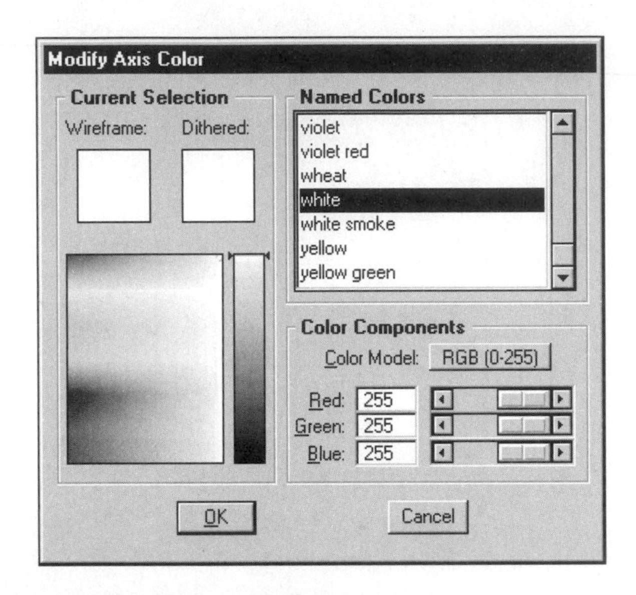

Figure 5–15 The Modify Axis Color settings box

When the Delayed Update toggle button is OFF, the values in the AccuDraw window always display the current location of the cursor as you move it around. When the button is ON, the fields are not updated until you hold the cursor still for about a second.

The Show Negative Angles toggle button controls the way the angle of rotation is presented in the Polar coordinate compass. If the button is:

- OFF, the Angle edit field always displays the counterclockwise rotation value, even if you move the screen cursor in a clockwise direction. For example, move the cursor 45 degrees clockwise and the edit field displays 315 degrees of counterclockwise rotation.
- ON, clockwise movement of the cursor will cause negative angles increasing from zero to be displayed in the edit field. For example, move the cursor 45 degrees clockwise and the edit field displays –45 degrees of clockwise rotation.

The Shortcut Popups toggle button controls the display of confirmation messages below the AccuDraw window when you press a keyboard shortcut. If the button is:

- OFF, no messages are displayed.
- ON, the name of the one-character shortcut appears below the AccuDraw window until you release the shortcut key, or a list of shortcuts starting with the first key you press appears for two-character shortcuts (see Figure 5–16).

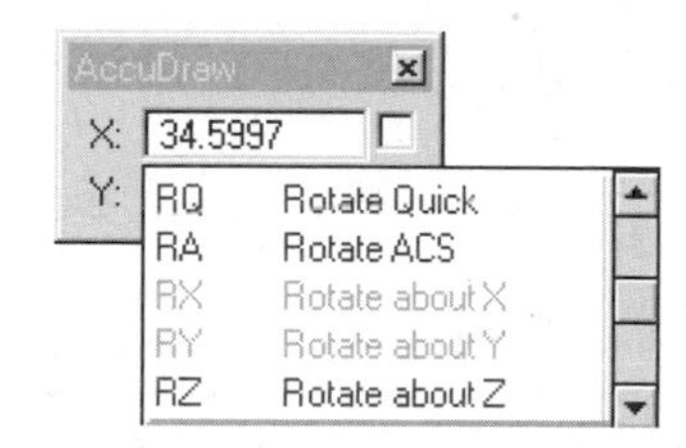

Figure 5–16 The list of two-letter shortcuts that appears when the first letter of the shortcut is pressed

Operation Options

The Default Origin options control values displayed in the AccuDraw window when the compass is not in the view (when no placement or manipulation tool is in use).

The Close button closes the More AccuDraw Settings box.

AccuDraw Shortcuts

The AccuDraw Shortcuts box displays all defined AccuDraw keyboard shortcuts, and allows you to edit and create shortcuts (editing and creating shortcuts are not discussed in this book).

Invoke the Shortcuts settings box:

AccuDraw Settings box	Click the Shortcut Key-ins button.
Keyboard shortcut	**?** (with focus on the AccuDraw window)
Key-in window	**accudraw dialog shortcuts** (or **a d sh**) Enter

MicroStation opens the AccuDraw Shortcuts settings box, as shown in Figure 5–17.

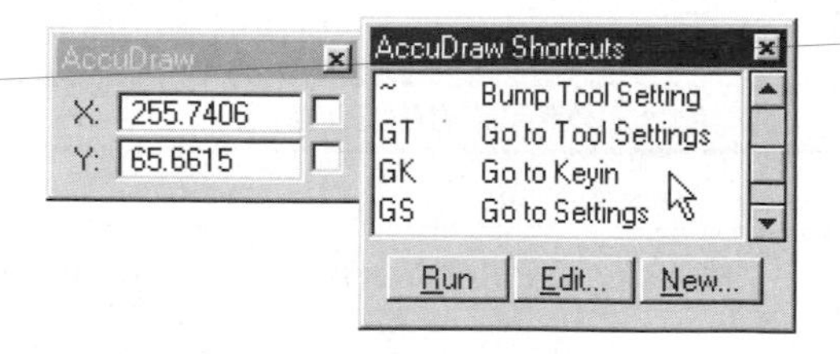

Figure 5–17 The AccuDraw Shortcuts settings box.

As has been discussed earlier in this chapter, AccuDraw includes one- and two-letter keyboard shortcuts that allow you to change the action AccuDraw is about to perform. A shortcut is invoked by typing the shortcut while focus is in the AccuDraw window. Table 5–1 lists the shortcuts available in *2D* designs (additional shortcuts are available for *3D* designs).

Note: AccuDraw supports up to 400 keyboard shortcuts.

Table 5–1 AccuDraw Shortcuts for *2D* design.

KEY-IN	ACCUDRAW DIRECTIVE
?	Open the AccuDraw Shortcuts window.
~	Bumps an item in the tool settings dialog box. For example; it will change to the next option in the Method menu.
Spacebar	Toggle between Rectangular and Polar coordinate compass.
M	Open or set focus to the Data Point Key-in window that can be used for precision input (*XY*=, DL=, DI=, etc.).
O	Move the origin point to the current screen cursor location.
P	Key-in window appears to key-in single data point.
Q	Close the AccuDraw tool.
G + A	Get a saved ACS (Auxiliary Coordinate System)
G + K	Open, or move focus, to the Key-in window (same as choosing Key-in from the Utility menu).
G + M	Open or set focus to the AccuDraw More Settings dialog box.
G +	Go to the Settings dialog box.
G + T	Open or set focus to the AccuDraw Tool Settings.
W + A	Save the drawing plane alignment as an ACS.
LOCKS	**ACCUDRAW DIRECTIVE**
Enter	Toggle Smart Lock ON and OFF.
X	Toggle the Rectangular coordinate *X* value lock ON and OFF.
Y	Toggle the Rectangular coordinate *Y* value lock ON and OFF.
D	Toggle the Polar coordinate Distance value lock ON and OFF.
A	Toggle the Polar coordinate Angle value lock ON and OFF.
L	Toggle the Index lock ON and OFF. If the Index lock is OFF, the only way to index the cursor position to an axis is to use Smart Lock.

ROTATE THE DRAWING PLANE	ACCUDRAW DIRECTIVE
R	Rotate the compass between its current rotation and the drawing's top view. It is a toggle switch that alternates between rotating the compass between the two positions.
T	Rotate the drawing plane to align with the top view.
V	Rotate the drawing plane to align with the view (its normal rotation).
R + Q	Temporarily rotate the drawing plane about the compass origin point. The lock is turned OFF after the next data point.
R + A	Permanently rotate the drawing plane. It stays active after the current tool terminates.
R + Z	Rotate the drawing plane 90 degrees about its *Z* axis. (In a *2D* drawing, the *Z* axis is perpendicular to the drawing plane.)

SNAP MODES	ACCUDRAW DIRECTIVE
C	Activate Center snap mode.
I	Activate Intersect snap mode.
N	Activate Nearest snap mode.
K	Open the Keypoint Snap divisor window so the snap divisor can be set.

Closing AccuDraw

You can close AccuDraw, but after you learn to use it you will probably seldom want to. There are three ways to close it:

AccuDraw input window title bar (if floating in View window)	Click the Close (X) button.
AccuDraw shortcut	**Q** (when focus is on the AccuDraw window)
Key-in window	**accudraw quit** (or **a q**) Enter

The AccuDraw window closes and the next data point does not have a compass around it.

WORKING WITH ACCUDRAW

You've seen the parts of AccuDraw and how to change the way it functions. Now let's look at how we can apply it to various tools available in MicroStation.

Example of Simple Placement Task

This example shows how AccuDraw can be used to draw the arrowhead shown in Figure 5–18 (this is a simplified illustration of using the AccuDraw compass and window, and is not a lesson in drawing arrowheads).

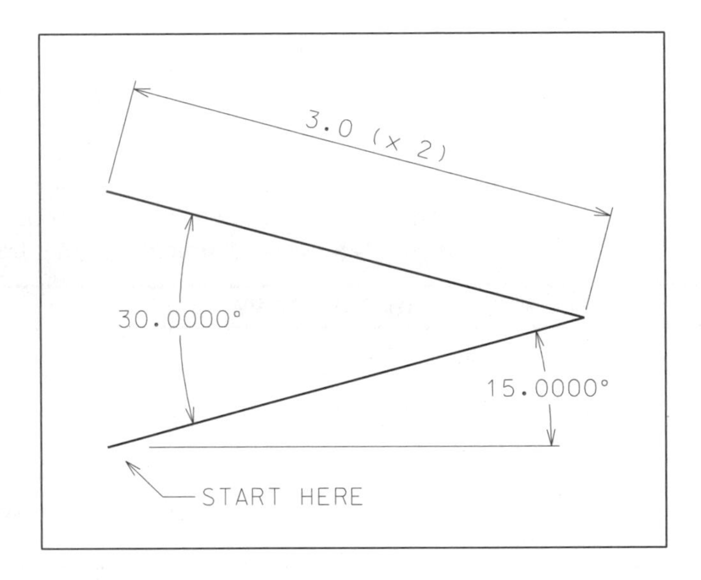

Figure 5–18 Simple Arrowhead drawn with AccuDraw

1. Open the AccuDraw Settings box and turn on the Floating Origin and Context Sensivity toggle buttons.
2. Invoke AccuDraw and the Place Line tool.
3. Place a data point to define the lower left end of the arrowhead.
4. If the AccuDraw Rectangular coordinates compass appears, click on the Spacebar to switch to the Polar coordinates compass.
5. Drag the cursor a short distance to the right and upwards.
6. If the AccuDraw window is focused on the Angle edit field, click Tab to focus on the Distance edit field.
7. Type 3 to place, and lock the line length in the Distance edit field.
8. Press Tab to move focus to the Angle edit field, then type 15 to place, and lock the line rotation angle in the Angle edit field.
9. Click the data button to draw the bottom half of the arrowhead.

10. Drag the cursor to the left and up until the length index tick mark appears at the screen cursor, then press X to lock that length in the Distance edit field (the length should be 3).
11. If the AccuDraw window is not focused on the Angle edit field, click [Tab] to focus on it.
12. Type 30 to place, and lock the angle of the second line in the Angle edit field.
13. Click the data button to draw the top half of the arrowhead.

Moving the Compass Origin

In almost all uses of AccuDraw, the compass origin is on the previous data point, but it can be moved without placing a data point. To relocate the compass origin, key-in the shortcut **O** (*not* zero). The compass origin relocation depends on the current location.

- If the compass is currently not visible, it appears at the last data point location.
- If the compass is currently visible, it relocates to the current pointer location.
- If there is an active tentative point, the compass jumps to the tentative point.

Using Tentative Points with AccuDraw

Tentative points can be used with the compass to place elements in precise relationships to other elements, just as was done with precision key-ins. For example, to start a line 2 Master Units (MUs) to the right of the corner of an existing element, do the following:

1. Invoke the Place Line tool.
2. Snap to the corner of the existing element from which the offset is to be measured.
3. Key-in the shortcut **O** to release the compass origin and move it to the tentative point.
4. If the Polar coordinate system is not active, press the **Spacebar** to switch to it. (This example uses Polar coordinates, but it can be done in Rectangular coordinates as well.)
5. In the AccuDraw window, key-in **2** in the Distance field and **0** in the Angle field.
6. Click the Data button, and the first point of the line is placed 2 units to the right of the tentative point, as shown in Figure 5–19.
7. Finish drawing the line.

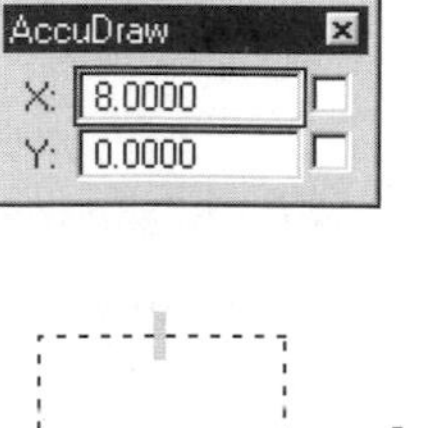

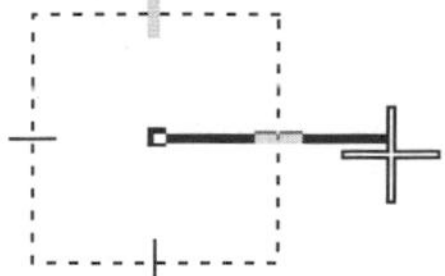

Figure 5–19 Example of starting a line at an offset from a tentative point

Note: AccuDraw provides keyboard shortcuts for selecting some of the tentative snap modes, which are listed in Table 5–1.

Rotating the Drawing Plane

In *2D* designs, the AccuDraw drawing plane can be rotated about the *Z* axis (which is perpendicular to the *2D* drawing plane) any time the compass is visible. To rotate the plane, click the first shortcut letter, **R**, and the Shortcuts window appears. Select the type of rotation from the shortcut list of **R** options.

The **RQ** and **RA** shortcuts allow you to define the rotation dynamically by dragging the cursor and clicking the Data button. **RQ** only locks the compass at the selected rotation for the next data point, after which the compass is released. **RA** locks the compass permanently at the selected rotation.

Each time the **RZ** shortcut is selected, the plane is rotated 90 degrees counterclockwise about the *Z* axis.

Each time the **B** shortcut is selected, the plane is rotated between the compass' current rotation and the drawing's Top rotation.

Note: The compass rotates automatically to the same angle as the previously placed line segment when the context sensitivity toggle button is set to ON. These rotate shortcuts allow you to override that rotation.

Placing Elements with AccuDraw Active

AccuDraw generally enhances placement tools by automating some steps and guiding the user. Here are two examples of using AccuDraw with placement tools.

Ellipse

With AccuDraw, the first two data points for the Ellipse still define the primary axis, as before, but the third data point is locked automatically on the secondary axis. You can simply type in the radius or drag the cursor to where it should be.

For example, place an ellipse with a major axis 8 MUs long and rotated 30 degrees, and a minor axis 4 MUs long:

1. Invoke the Place Ellipse tool.
2. Place the first data point to define one end of the major axis.
3. If the rectangular compass is active, switch to the polar compass.
4. In the AccuDraw window, set the Distance to 4 and the Angle to 30.
5. Click the Data button to define the major ellipse axis.
6. In the AccuDraw window, set the Distance to 2.
7. Click the Data button to complete the ellipse by defining its minor axis.

Block

Fewer steps are required to place a rotated block with specific dimensions when using AccuDraw. Place the first data point, then use Polar coordinates to specify the rotation angle and the length of the block at that angle. Place the data point, and the compass axes rotate to the angle. Switch to Rectangular coordinates and enter the *Y* axis length to complete the block, then place the final data point.

For example, place a 3 x 5 block rotated 15 degrees:

1. Invoke the Place Rotated Block tool.
2. Place the first data point to define the lower left corner of the rotated block.
3. If the rectangular compass is active, switch to the polar compass.
4. In the AccuDraw window, set the Distance field to 3 and the Angle to 15.
5. Click the Data button to define the bottom edge of the rotated block (the compass switches to rectangular coordinates).
6. In the AccuDraw window, set the *X* distance to 3 and the *Y* distance to 5.
7. Click the Data button to complete the rotated block.

Manipulating Elements with AccuDraw Active

AccuDraw also enhances manipulation of elements. Here is an example of placing two copies of the lower left block shown in Figure 5–20.

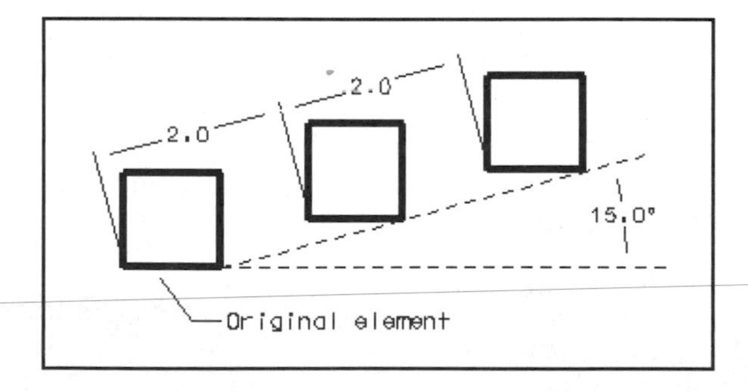

Figure 5–20 Example of using AccuDraw with the Copy Element tool

If AccuDraw is not active, invoke it.

1. Invoke Copy Element from the Manipulate tool box.
2. Open the AccuDraw Settings window, set the compass to Polar, set and lock the Distance roundoff to 2 and the Angle roundoff value to 15.
3. Identify the element by doing a Keypoint snap to the lower left corner of the block to be copied, then click the Data button.

4. Drag the cursor until the AccuDraw window shows an Angle of 15 and a Distance of 2.
5. Click the Data button to place the copy.
6. Repeat steps 4 and 5 for the second copy.

Using the AccuDraw tool appropriately you will see a significant increase in productivity.

PLACE SMARTLINE TOOL

The SmartLine tool places a chain of connected line and arc segments as individual elements, or as a line string, shape, complex chain, or complex shape. The vertexes between segments can be a sharp point, a tangent arc (rounded), or a chamfer. The tool settings (see Table 5–2) can be changed on the fly to allow any combination of segments and vertices.

Invoke SmartLine:

Linear Elements tool bar	Select the Place SmartLine tool (see Figure 5–21) and make the initial tool settings in the Tool settings window (see Table 5–2).
Key-in window	**place smartline** (or **pl sm**) Enter

MicroStation prompts:

Place SmartLine > Enter first vertex *(Place the first data point to start the SmartLine.)*

Place SmartLine > Enter next vertex or reset to complete *(Place the remaining data points. Change the Segment, Vertex Type, and Rounding Radius or Chamfer Offset settings as required between points.)*

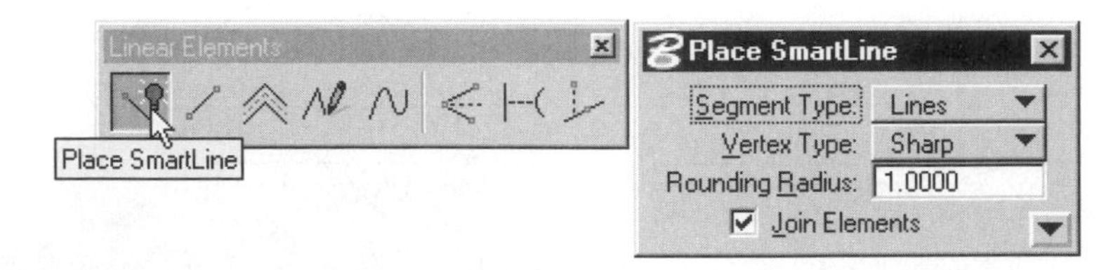

Figure 5–21 Place SmartLine tool icon in the Linear Elements tool box

You can complete the tool sequence by clicking the Reset button to create an open element, or you can place a Tentative snap to the first data point and click the Accept button to create a closed element.

Note: If the Join Elements check box in the Tool settings window is turned ON, the segments between the vertices are joined and the completed SmartLine is one element. If it is set to OFF, each segment is a separate element.

Table 5–2 The SmartLine Tool Settings

SETTING	EFFECT
Segment Type	Set the element placed with each data point after the first one. Select one of these types: Line Arc (Placed by defining the center and end points. The arc will sweep clockwise or counterclockwise depending on which way you drag the cursor when defining the end point.)
Vertex Type	Set the shape of each vertex to one of these types: Sharp Rounded (a tangent arc) Chamfered
Rounding Radius	Enter the rounded vertex radius in working units (MU:SU:PU). **Note:** The Rounding Radius prompt appears only when the Vertex Type is Rounded.
Chamfer Offset	Enter the offset of each end of the chamfer from the vertex point. Each chamfer offset is equal. **Note:** The "Chamfer Offset" prompt appears only when the Vertex Type is Chamfer.
Join Elements	*Toggle button turned ON:* The segments between the vertices are joined and the completed SmartLine is one element.
	Toggle button turned OFF: Each segment is a separate element.
The following setting appears in the Settings window only if Join Elements is checked and you have a tentative point snapped to the first data point of the SmartLine.	
Closed Element	*Set to ON:* Creates a closed element (shape or complex shape) when the tentative snap is accepted.
	Set to OFF: Accepting the tentative snap does not create a closed element.

Table 5–2 The SmartLine Tool Settings (continued)

SETTING	EFFECT
The following settings appear in the Settings window only if Closed Element is turned ON and you have a tentative point snapped to the first data point of the SmartLine.	
Area	Sets the active area of the closed element to Solid or Hole. (The reason for these settings is discussed in Chapter 11 on Patterning.)
Fill Type	The Fill Type options control the closed element's fill: None (no fill) Opaque (filled with active color) Outlined (filled with fill color)
Fill Color	Sets the fill color for the closed element (unless the Fill Type is set to None): If Fill Type is Opaque, the active color is selected. If Fill Type is Outlined, the fill color can be selected here.

Placing the Segments

As you are placing a SmartLine, you can switch between placing a line and placing an arc.

- To place a line, define the two segment end points.
- To place an arc, define the arc center and sweep end point.
- To change the direction of an arc, swing the cursor in the desired sweep direction.

Using SmartLine with AccuDraw

The SmartLine tool was designed for use with AccuDraw active. When SmartLine is employed with AccuDraw, the AccuDraw drawing plane automatically does the following:

- Moves to each new data point.
- Rotates to align with each newly defined segment, which makes it easier to define new segments tangent or perpendicular to the aligned segment.
- Switches to polar coordinates when defining an arc segment.

Example of Drawing with SmartLine and AccuDraw

This example uses SmartLine with AccuDraw to draw the object shown in Figure 5–22. The letters in the figure point to the locations of all data points.

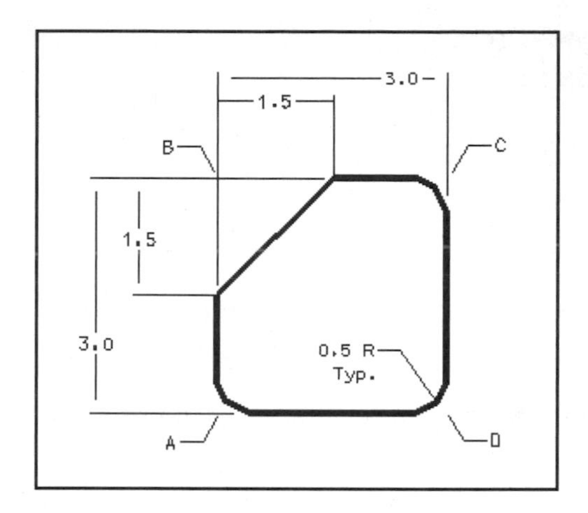

Figure 5–22 Example of a drawing made with the SmartLine tool

1. If AccuDraw is not active, invoke the AccuDraw tool.
2. Open the AccuDraw Settings window.
3. Set and lock the Distance roundoff value to 1.5 and the Angle roundoff value to 90.
4. Set the compass to Polar.
5. Invoke SmartLine from the Lines tool box.
6. Set SmartLine to draw Lines, and place a 1.5-inch Chamfer Offset chamfer at the vertices.
7. Define a point to start the object (Point A).
8. Slide the cursor up until the AccuDraw window Distance field displays 3, then click the Data button (Point B).
9. Slide the cursor to the right until the AccuDraw window Distance field displays 3, then click the Data button (Point C).
10. Set SmartLine Rounding Radius to 0.5 inch.
11. Slide the cursor down until the AccuDraw window Distance field displays 3, then click the Data button (Point D).
12. Slide the cursor to the left until the AccuDraw window Distance field displays 3 and touches the first point. Click the Data button (Point A).

MicroStation creates the object as a complex shape if the Join Elements toggle button is set to ON. If it is set to OFF, the object created consists of five separate lines and three separate arcs.

MicroStation provides various options to draw lines and arcs with the SmartLine tool. See Table 5–2 for details on the various options.

REVIEW QUESTIONS

Write your answers in the spaces provided.

1. How do you activate AccuDraw?
2. Name the two coordinate systems you can use with AccuDraw.
3. Name three settings you can adjust in the AccuDraw settings box.
4. What is the purpose of rounding off distances and angles in the AccuDraw settings box?
5. How do you recall the previous values by using AccuDraw?
6. What is the shortcut key-in that will move the compass origin from the previous data point?
7. Explain briefly how you can use the Tentative points with AccuDraw.
8. Explain briefly the benefits of manipulating elements with AccuDraw active.
9. Explain the difference between the Place Line tool and the Place SmartLine tool.
10. Name the two segment types you can use with the SmartLine tool.

PROJECT EXERCISE

This project exercise provides step-by-step instructions for creating the design shown in Figure P5–1. The intent is to guide you in applying AccuDraw and SmartLine.

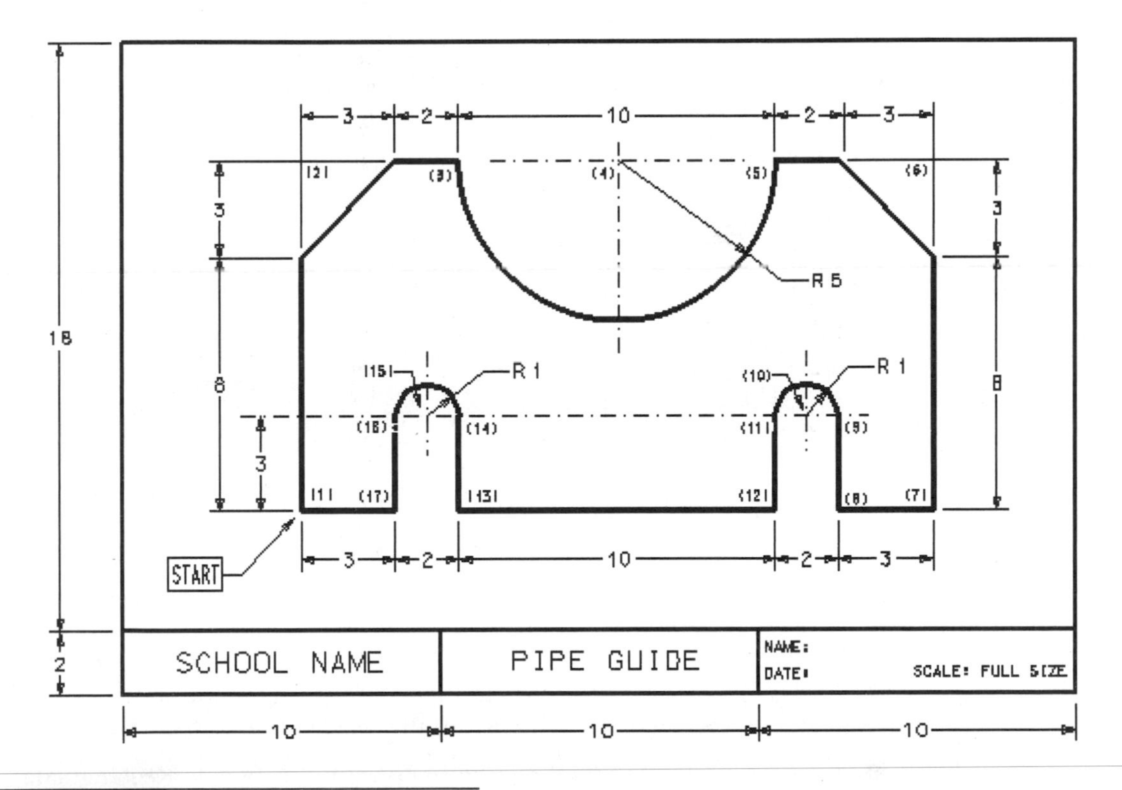

Figure P5–1 Completed project design

Note: The dimensions are not part of this project. They are included in Figure P5–1 only to show the size of the design.

Prepare the Design File

Note: As you complete each step in the project procedures, place a check mark by the step to help you keep up with where you are in the project.

In this procedure you start MicroStation, create a design file, and enter the initial settings.

STEP 1: Invoke MicroStation by the normal technique for the operating system on your workstation.

STEP 2: Create a new design file named CH5.DGN using the SEED2D.DGN seed file.

STEP 3: In the Design File dialog box:

- Set the Working Units names to "IN" for the Master Units and "TH" for the Sub Units; and set the Resolution to "10" Th Per In and "1000" Pos Units per Th.
- Set the Grid Master to 0.1 and the Grid Reference to 10, and turn the Grid lock OFF.

STEP 4: Invoke the AccuDraw tool from the Primary Tool bar (see Figure P5–2).

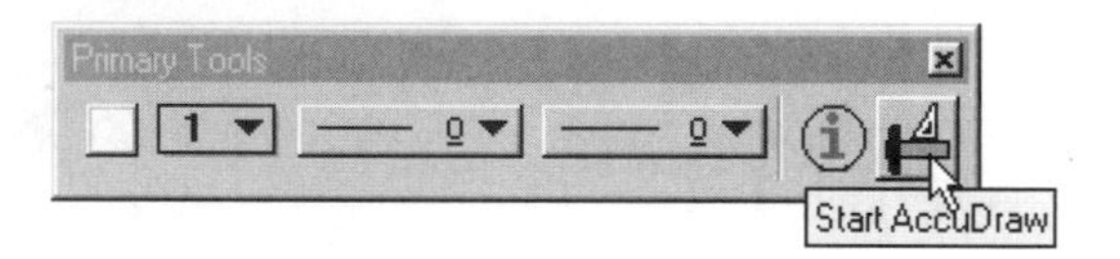

Figure P5–2 Invoking the AccuDraw tool

STEP 5: Open the AccuDraw settings box from the Settings pull-down menu and adjust the settings as follows:

- *Unit Roundoff:* Set the Distance to 1.000 and the toggle button to ON. Then set the Angle to 90.000 degrees and the toggle button ON.
- *Coordinate System:* Set the Rotation to Top and the Type to Rectangular.
- *Operation:* Set the toggle buttons for Floating Origin and Smart Key-ins to ON. Set the toggle button for Context Sensitivity and Auto Point Placement to OFF.
- *Display:* If you have trouble seeing the colors red and green, change the *X* axis and *Y* axis colors.

Make sure the settings are properly adjusted by referring to Figure P5–3. Click the Close button to close the settings box.

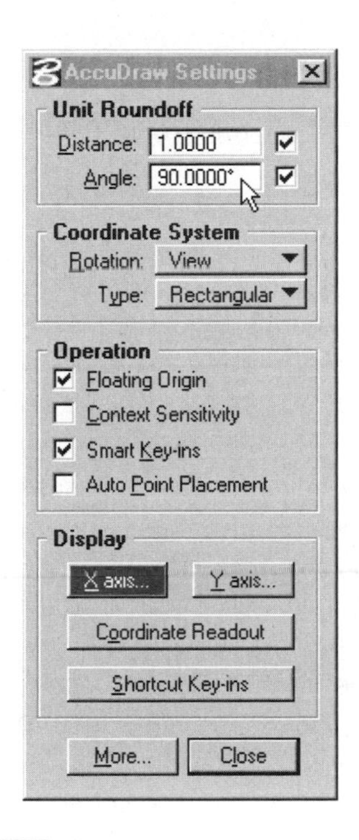

Figure P5–3 AccuDraw settings box

STEP 6: Invoke Save settings from the File pull-down menu.

Draw the Border and Title Block

Draw the border and title block as shown in Figure P5–1, employing AccuDraw to aid in element placement.

STEP 1: Invoke the Place Block tool from the Polygons tool box.

MicroStation prompts:

Place Block > Enter first point *(Click in the AccuDraw window's "X:" edit field and key-in* ***0****. Click in the "Y:" edit field and key-in* ***0****. Then press* Enter*.)*

Place Block > Enter opposite corner *(Key-in* ***30*** *in the AccuDraw "X:" edit field and* ***20*** *in the "Y:" edit field. Then click the Data button to place the upper right corner of the block.)*

STEP 2: Fit the view window.

STEP 3: Invoke the Place Line tool from the Linear Elements tool box.

MicroStation prompts:

Place Line > Enter first point *(Keypoint snap to the lower left corner of the block, type **O** to release the AccuDraw origin, and type **X** to lock the AccuDraw X axis at 0.0000. Drag the drawing pointer vertically upward until the AccuDraw "Y:" field is equal to 2.0000. Then click the Data button to start the line.)*

Place Line > Enter end point *(Type **Y** to lock the AccuDraw Y axis at 0.0000. Then drag the drawing pointer to the right until "X:" equals 30.0000, and click the Data button to complete the line. Click the Reset button.)*

Place Line > Enter first point *(Keypoint snap to the lower left corner of the block, type **O** to release the AccuDraw origin, and type **Y** to lock the AccuDraw Y axis at 0.0000. Drag the drawing pointer to the right until the AccuDraw "X:" field is equal to 10.0000, then click the Data button to start the line.)*

Place Line > Enter end point *(Type **X** to lock the AccuDraw X axis at 0.0000. Then drag the drawing pointer vertically upward until "Y:" equals 2.0000, and click the Data button to complete the line. Click the Reset button.)*

Place Line > Enter first point *(Keypoint snap to the lower left corner of the block, type **O** to release the AccuDraw origin, and type **Y** to lock the AccuDraw Y axis at 0.0000. Drag the drawing pointer to the right until the AccuDraw "X:" field is equal to 20.0000, then click the Data button to start the line.)*

Place Line > Enter end point *(Type **X** to lock the AccuDraw X axis at 0.0000. Then drag the drawing pointer vertically upward until "Y:" equals 2.0000, and click the Data button to complete the line. Click the Reset button.)*

STEP 4: Place the text in the title block using font 3, 0.6 for the large text size, and 0.3 for the small text size:

- Replace "SCHOOL NAME" with your school or company name, or make up a name.
- Place your name to the right of "NAME."
- Place today's date to the right of "DATE."

Draw the Design

Draw the pipe guide shown in Figure P5–1 using AccuDraw and SmartLine.

STEP 1: Invoke Place SmartLine from the Linear Elements tool box, as shown in Figure P5–4. In the Tool Settings window, set the Vertex Type to Chamfered, and key-in **3** in the Chamfer Offset edit field.

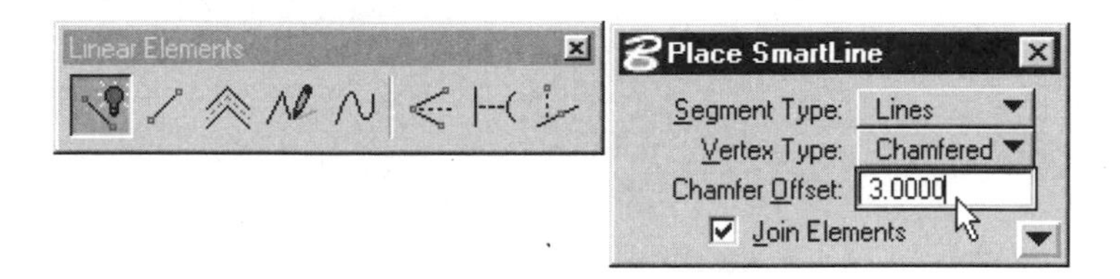

Figure P5–4 Invoke Place SmartLine and set the Chamfer Offset to 3

Note: Numbers in parentheses have been added to the following MicroStation prompts and to Figure P5–1, to help you keep up with where you are in the procedure of drawing the pipe guide. Those numbers do **not** appear in the MicroStation prompt on the screen.

MicroStation prompts:

(1) Place SmartLine > Enter first vertex *(Place the first data point approximately at the START point shown in Figure P5–1.)*

(2) Place SmartLine > Enter the next vertex or reset to complete *(Type **X** to lock the AccuDraw X axis at 0.0000, and drag the drawing pointer straight up until the AccuDraw rectangular coordinate "Y:" field equals 11.0000, as shown in Figure P5–5. Then click the Data button.)*

(3) Place SmartLine > Enter the next vertex or reset to complete *(Type **Y** to lock the AccuDraw Y axis at 0.0000, drag the screen pointer straight to the right until the AccuDraw rectangular coordinates "X:" field equals 5.0000, then click the Data button.)*

(4) Place SmartLine > Enter the next vertex or reset to complete *(Change the SmartLine Segment Type to Arcs, as shown in Figure P5–6.)*

Place SmartLine > Enter arc center *(Type **Y** to lock the AccuDraw Y axis at 0.0000, drag the drawing pointer straight to the right until the AccuDraw coordinate "X:" field equals 5.0000, then click the Data button to place the arc center.)*

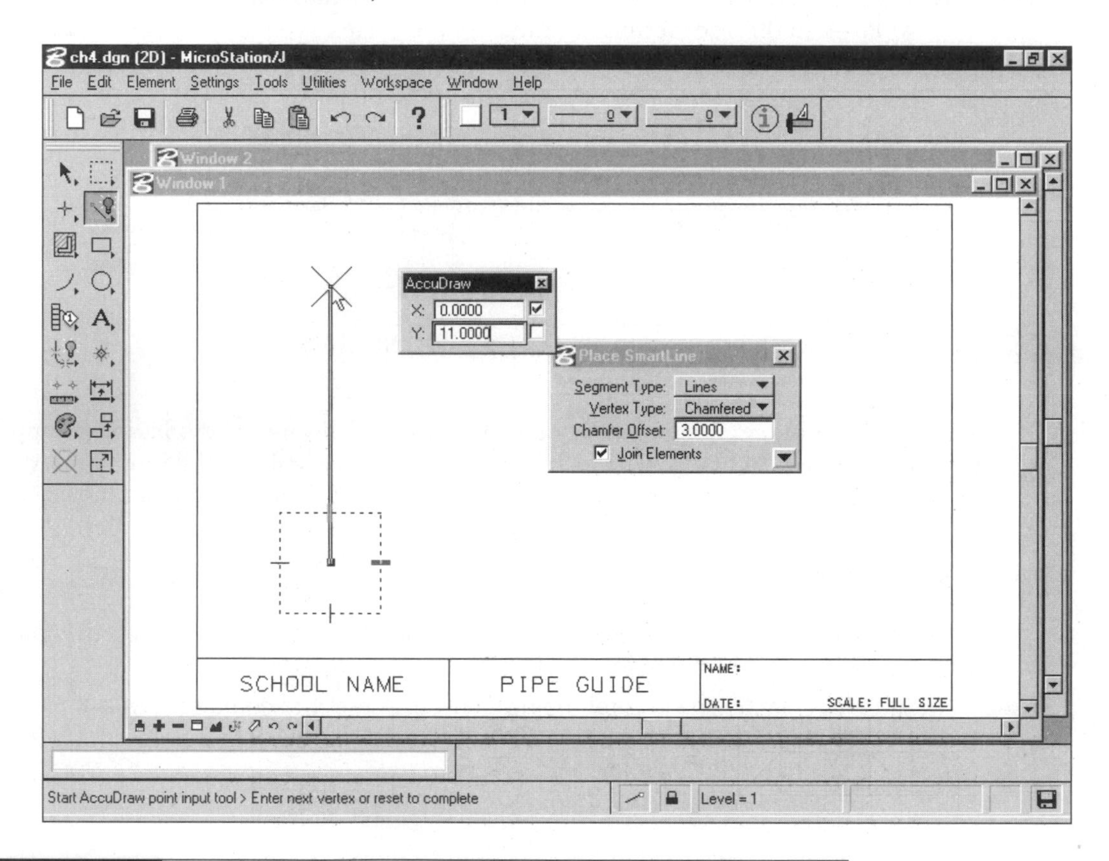

Figure P5-5 AccuDraw rectangular coordinate values for data point (2)

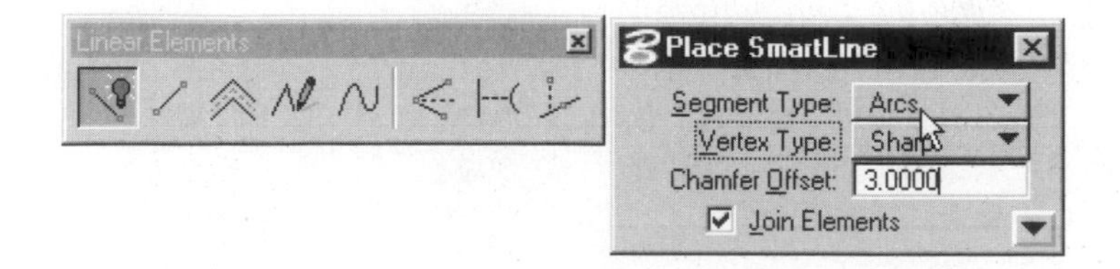

Figure P5-6 Place SmartLine Tool Settings window showing the change made to Arcs

(5) Place SmartLine > Define the sweep angle *(Type **Y** to lock the AccuDraw Y axis at 0.0000, drag the drawing pointer straight to the right until the AccuDraw coordinate "X:" field equals 5.0000, and click the Data button to complete the arc.)*

(6) Place SmartLine > Enter arc center *(Change the SmartLine Segment Type to Lines.)*

Place SmartLine > Enter the next vertex or reset to complete *(Type **Y** to lock the AccuDraw Y axis at 0.0000, drag the drawing pointer to the right until the AccuDraw coordinate "X:" field equals 5.0000, and click the Data button.)*

(7) Place SmartLine > Enter the next vertex or reset to complete *(Type **X** to lock the AccuDraw X axis at 0.0000, drag the drawing pointer down vertically until the AccuDraw coordinate "Y:" field equals –11.0000, and click the Data button.)*

(8) Place SmartLine > Enter the next vertex or reset to complete *(Change the SmartLine Vertex Type to Sharp. Type **Y** to lock the AccuDraw Y axis at 0.0000, drag the drawing pointer to the left until the AccuDraw coordinate "X:" field equals –3.0000, and click the Data button.)*

(9) Place SmartLine > Enter the next vertex or reset to complete *(Type **X** to lock the AccuDraw X axis at 0.0000, drag the drawing pointer vertically up until the AccuDraw coordinate "Y:" field equals 3.0000, and click the Data button.)*

(10) Place SmartLine > Enter the next vertex or reset to complete *(Change the SmartLine Segment Type to Arcs.)*

Place SmartLine > Enter arc center *(Type **Y** to lock the AccuDraw Y axis at 0.0000, drag the drawing pointer to left until the AccuDraw coordinate "X:" equals –1.0000, and click the Data button.)*

(11) Place SmartLine > Define sweep angle *(Type **Y** to lock the AccuDraw Y axis at 0.0000, drag the drawing pointer to left until the AccuDraw coordinate "X:" equals –1.0000, and click the Data button to complete the arc.)*

(12) Place SmartLine > Enter arc center *(Change the SmartLine Segment Type to Lines.)*

Place SmartLine > Enter the next vertex or reset to complete *(Complete the remainder of the pipe guide as shown in Figure P5–1.)*

STEP 2: Invoke the Save settings from the File pull-down menu.

DRAWING EXERCISES 5–1 THROUGH 5–5

Use the following table to set up the design files for Exercises 5–1 through 5–3.

Setting	Value
Seed File	SEED2D.DGN
Working Units	10 TH Per IN and 1000 Pos Units Per TH
Grid	Master = .25, Reference = 4, Grid Lock ON

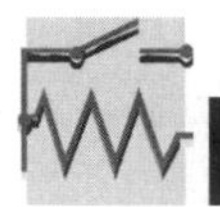

Exercise 5–1

Input-output card

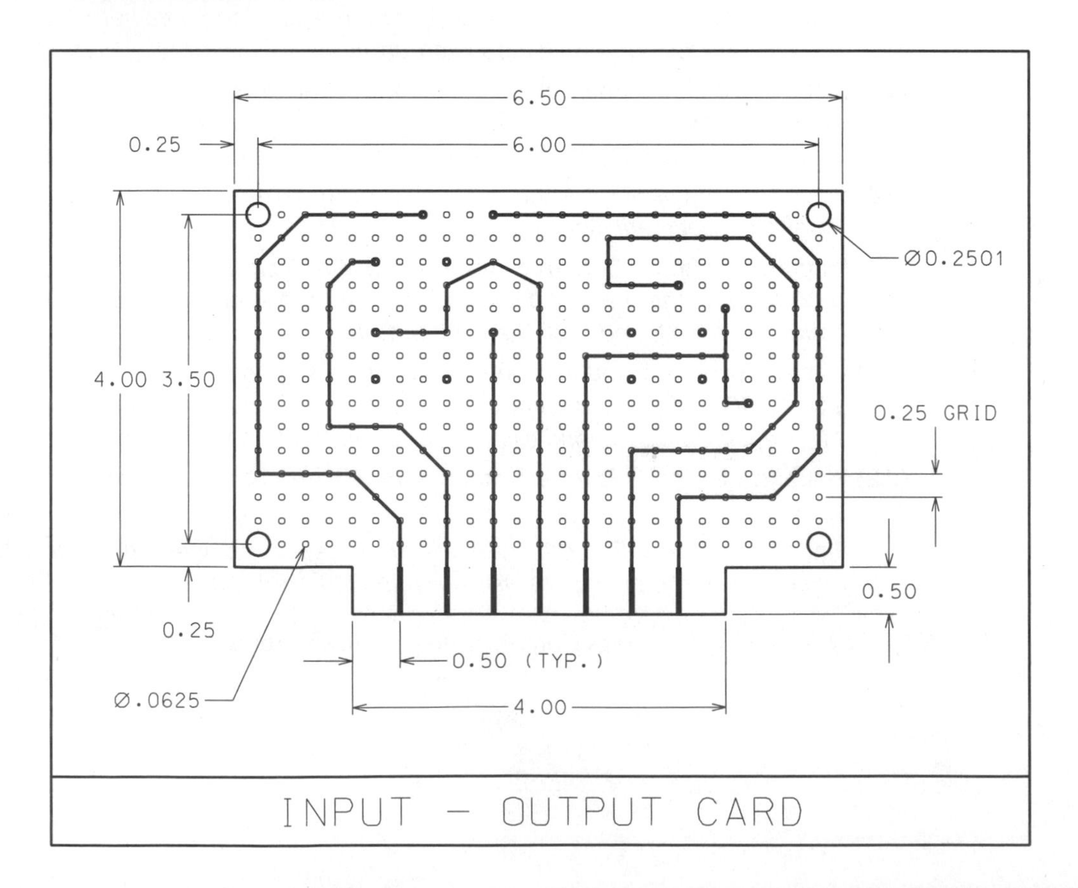

Exercise 5–2

Machine part

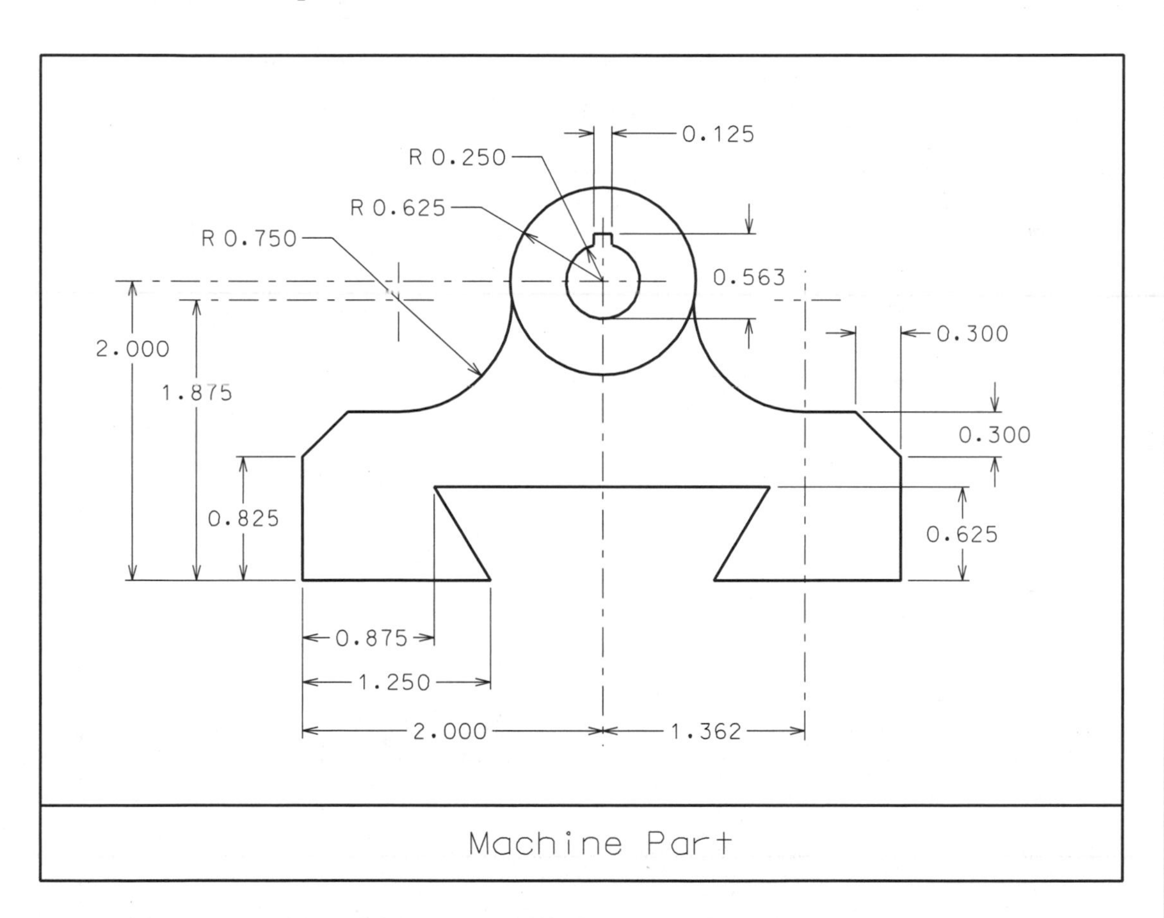

Exercise 5–3

Machine part

Use the following table to set up the design files for Exercises 5–4 through 5–5.

Setting	Value
Seed File	SEED2D.DGN
Working Units	12" Per ' and 8000 Pos Units per "
Grid	Master = .25, Reference = 4, Grid Lock ON

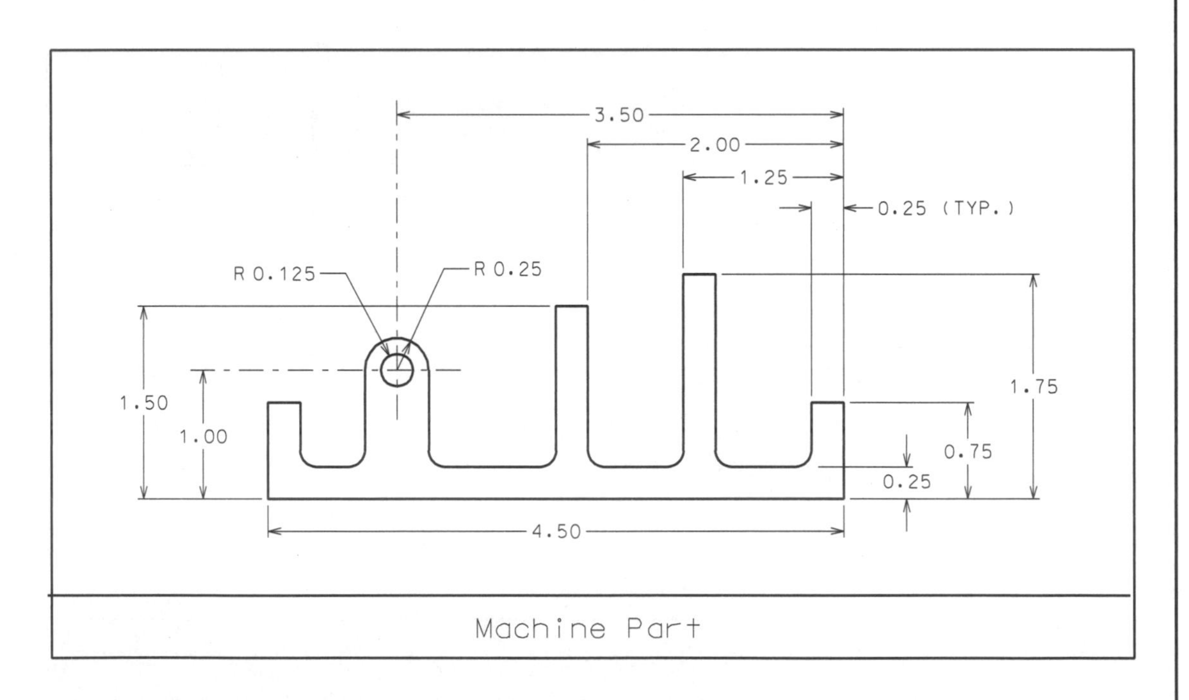

Exercise 5-4

Plot plan

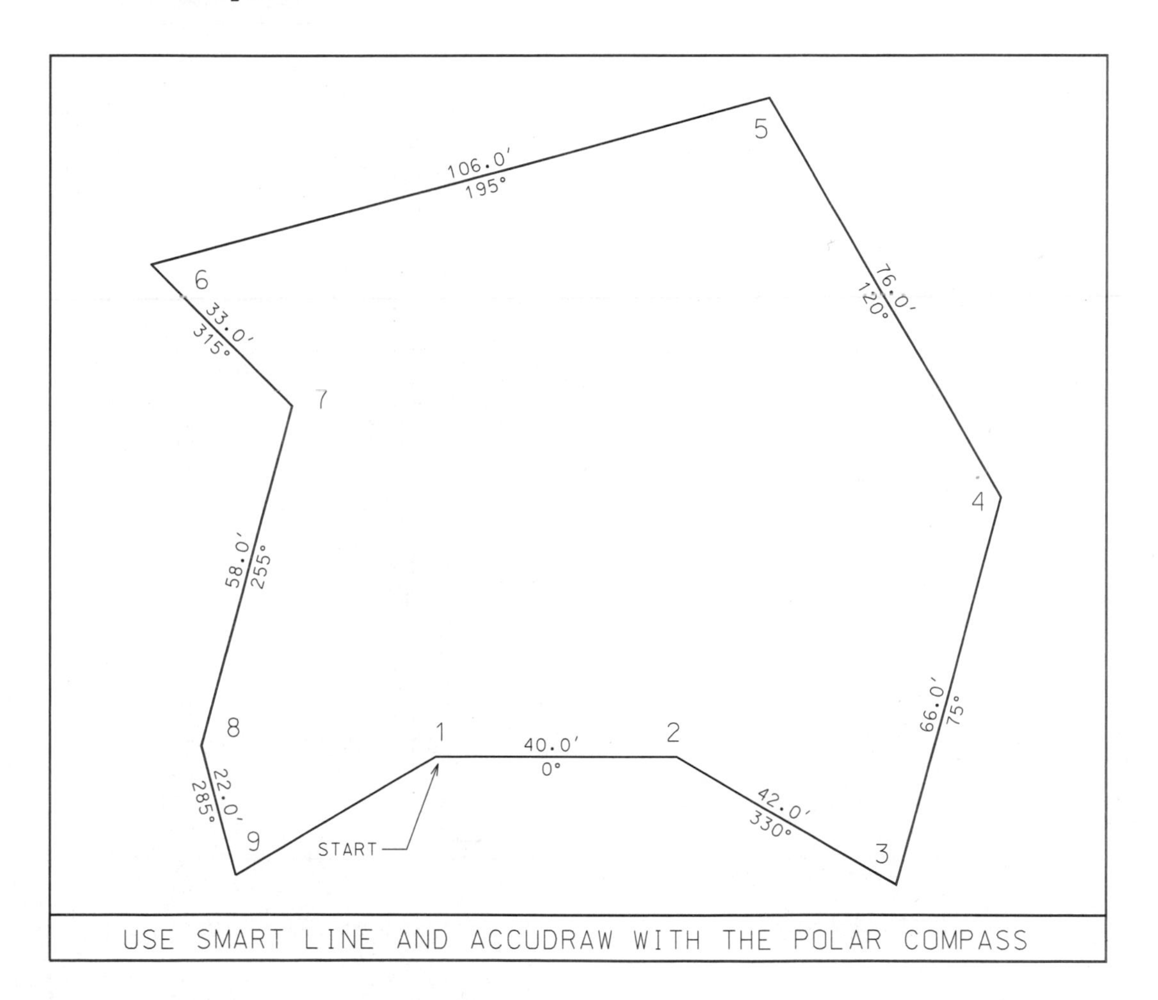

Exercise 5–5

Master bath floor plan

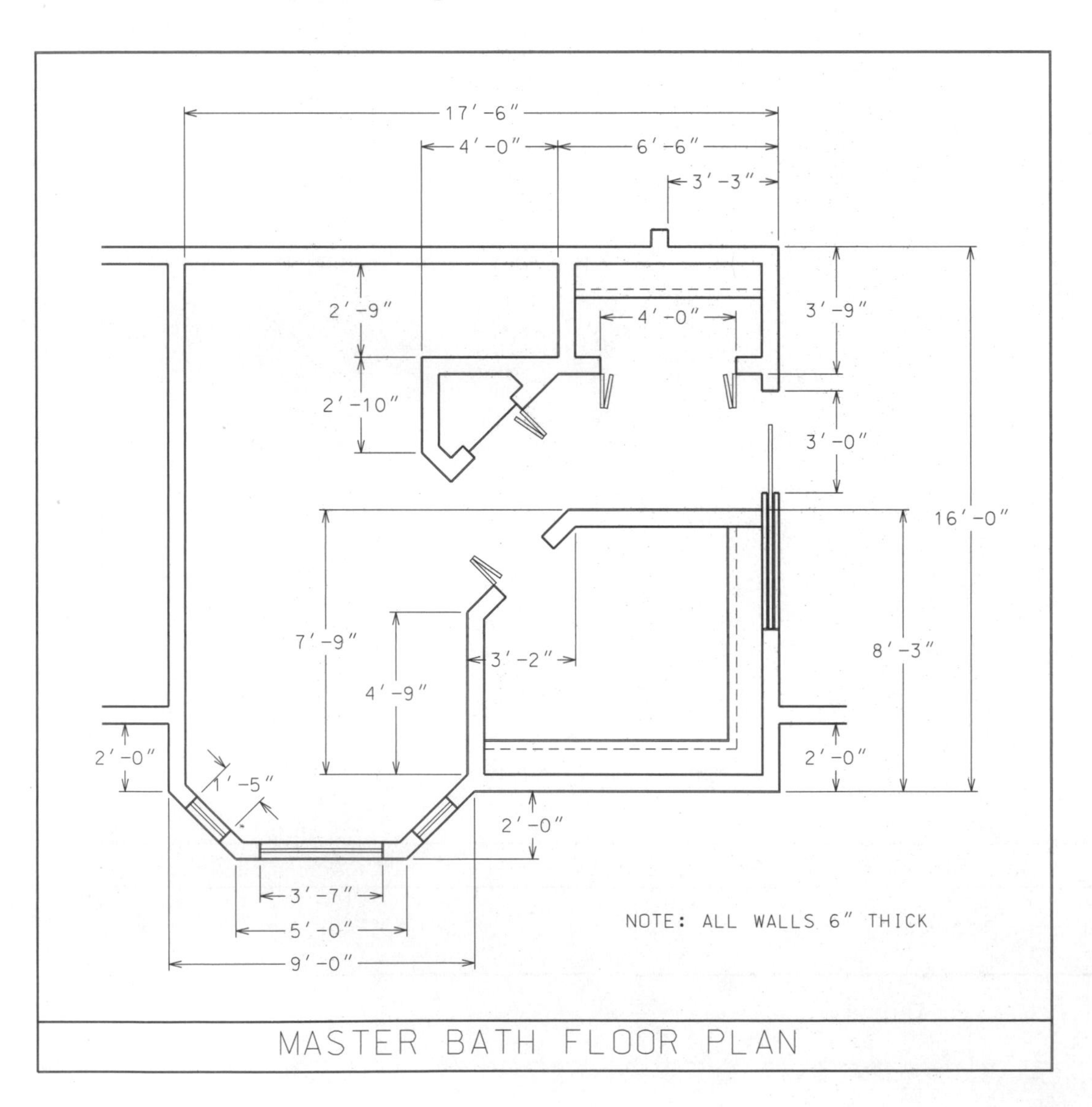

chapter 6

Manipulating a Group of Elements

Objectives

After completing this chapter, you will be able to do the following:

- Select elements with the PowerSelector tool and manipulate them
- Place fences and manipulate fence contents

ELEMENT SELECTION

While you were practicing the element manipulation tools described in the preceding chapters, did little squares occasionally appear on the corners of one of your elements? Did they make the tools act differently from the way the book said they would? This chapter turns those "handles" from a nuisance into a useful feature by showing you how the PowerSelector and Element Selector tools provide a powerful new way to manipulate elements.

Selecting Elements with the PowerSelector Tool

In the previous chapters you first invoked an element manipulation tool, and then identified the element you wanted to manipulate. With the PowerSelector Tool, you select or deselect one element, several elements, or all elements in the design. You can then manipulate all elements in the "selection set."

When elements are selected, "handles" appear on the elements. See Figure 6–1 for examples of the handles on different element types.

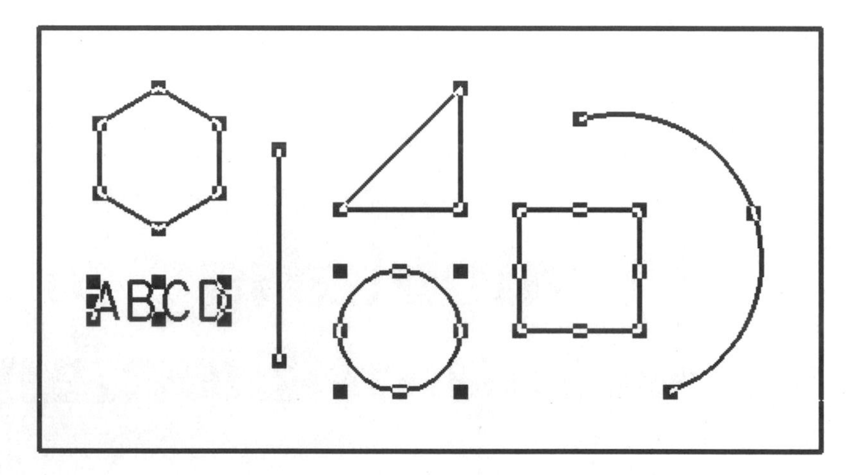

Figure 6–1 Examples of the handles on different element types

Invoke the PowerSelector tool:

Element Selection tool box	Select the PowerSelector tool, then pick a selection Method and Mode from the Tool Settings window (see Figure 6–2).
Key-in window	**powerselector** (or **pow**) Enter

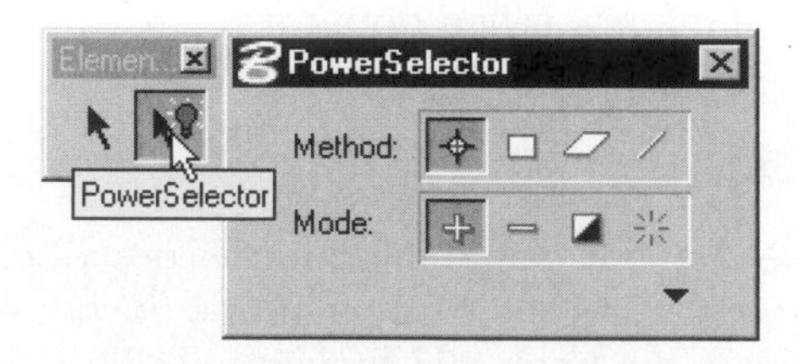

Figure 6–2 Invoking the PowerSelector tool from the Main tool frame

The PowerSelector action and prompt depend on the position of the cursor and the Method and Mode combination selected. If the cursor is on one of the buttons in the Tool Settings window, the prompt displays the name of the button and the keyboard shortcuts that will activate the button's function. If the cursor is on the View Window, the prompt tells you what selection Method and Mode combination is active.

The Method buttons control the way elements are selected or deselected. Table 6–1 describes the four Methods. The first three Mode buttons determine if the Method buttons select or deselect elements. The fourth Mode button toggles between selecting all elements and deselecting all elements. Table 6–2 describes the four modes.

Table 6–1 The PowerSelector Methods

Method	Action
Individual	Select or deselect individual elements by clicking on them or by dragging a block around them. Activate this Method by clicking on the first Method button from the left, or by pressing either the **Q** or **U** key.
Block	Select or deselect groups of elements by dragging a block around them. Activate this Method by clicking on the second Method button, or by pressing either the **W** or **I** key.
Shape	Select or deselect elements by defining a multi-sided, closed shape that can have up to 101 vertices. Activate this Method by clicking on the third Method button, or by pressing either the **E** or **O** key.
Line	Select or deselect elements by drawing a line. All elements the line touches are included. Activate this Method by clicking on the fourth Method button, or by pressing either the **R** or **P** key.

Table 6–2 The PowerSelector Modes

Mode	Action
Add	Causes the active PowerSelector Method to add elements to the current selection set. Activate this Mode by clicking the first Mode button from the left, or by pressing either the **A** or **J** key.
Subtract	Causes the active PowerSelector Method to remove elements from the current selection set. Activate this Mode by clicking the second Mode button, or by pressing either the **S** or **K** key.

Table 6–2 *(continued)*

Mode	Action
Invert	Causes the active PowerSelector Method to add previously unselected elements to the selection, and to remove previously selected elements from the selection set. Activate this Mode by clicking the third Mode button, or by pressing either the **D** or **L** key.
Select All or Clear	Is a toggle button that works independently of the active PowerSelector Method. If there are any selected elements, this button Clears all element selections. The result is no selected elements in the design. If there are no selected elements, this button Selects All elements in the design. Activate this Mode by clicking the fourth Mode button, or by pressing either the **F** or **Semicolon (;)** key.

Select Elements

The following discussion provides an example of selecting elements using the Block Method.

Invoke the PowerSelector tool:

Main tool frame	Select the PowerSelector tool, then select the Block Method and the Invert Mode from the Tool Settings window (see Figure 6–3).
Key-in window	**powerselector** (or **pow**) Enter (Activate PowerSelector) **powerselector area block** (or **pow ar b**) Enter (Selects Block Method) **powerselector mode invert** (or **pow m i**) Enter (Selects Invert Mode)

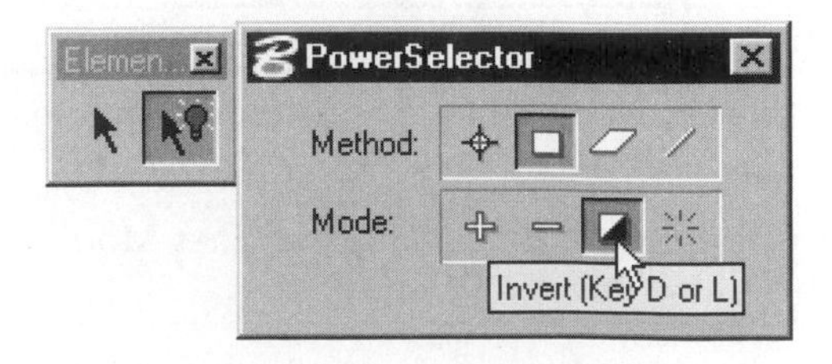

Figure 6–3 Invoking the PowerSelector tool and setting the options to select elements

MicroStation prompts (when you move the cursor off of the Tool Settings window):

> PowerSelector > Place Shape for elements to invert in set *(Place a data point to define one corner of the selection block, as shown in Figure 6–4.)*
>
> PowerSelector > Place Shape for elements to invert in set *(Place a data point to define the diagonally opposite corner of the selection block. The elements are selected, as shown in Figure 6–5.)*

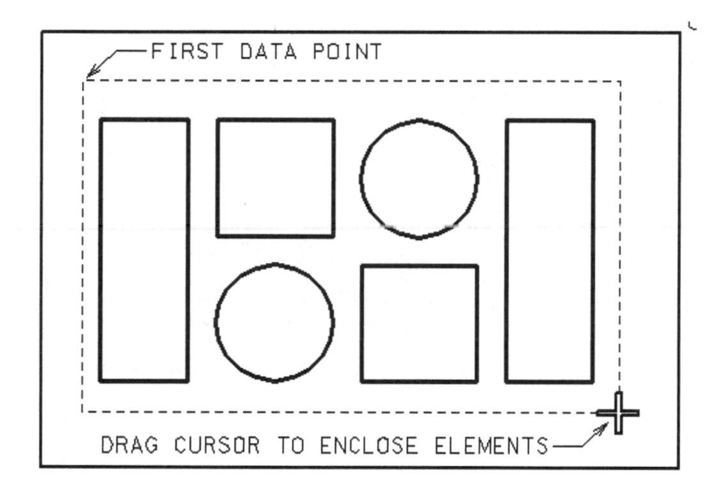

Figure 6–4 Define one corner of the selection block

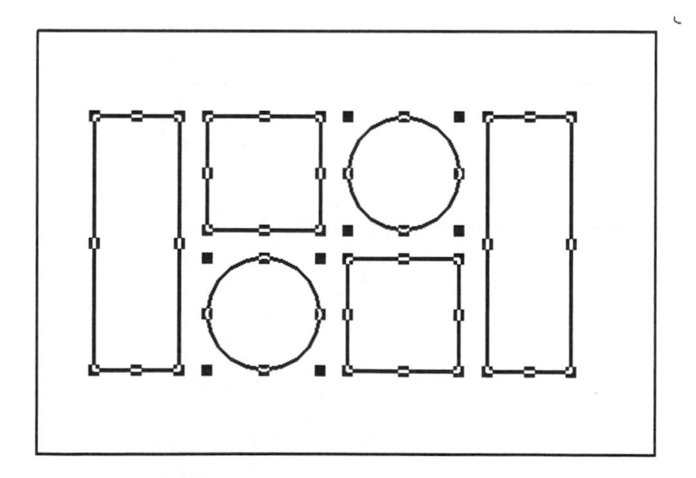

Figure 6–5 Elements inside the block are selected.

Remove Elements from a Selection Set

The following discussion provides an example removing selected elements from a selection set. The example removes the selected circles from the set that was selected in the previous example.

Invoke the PowerSelector tool from:

Tool Selection Window	Select the Individual Method from the Tool Settings window (see Figure 6–6).
Key-in window	**powerselector area individual** (or **pow ar i**) Enter (Only works when PowerSelector tool is active.)

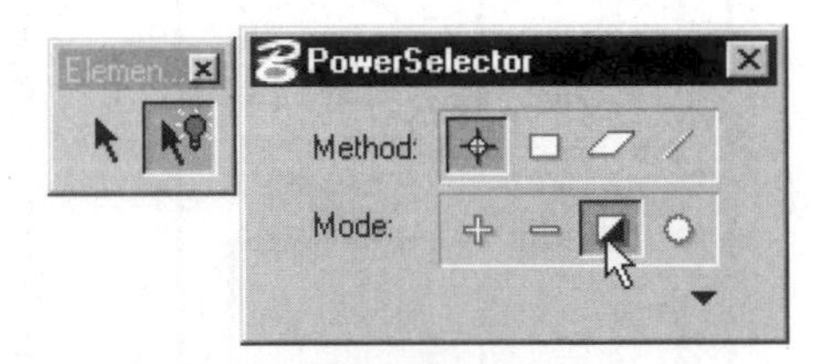

Figure 6–6 Selecting the PowerSelector tool's Individual Method in the Tool Settings window

MicroStation prompts (when you move the cursor off of the Tool Settings window):

PowerSelector > Identify element to invert in set *(Place a data point on one of the circles to remove it from the selection set.)*

PowerSelector > Identify element to invert in set *(Place a data point on the other circle to remove it from the selection set. The handles are removed from the circles, but the handles remain on the other elements, as shown in Figure 6–7.)*

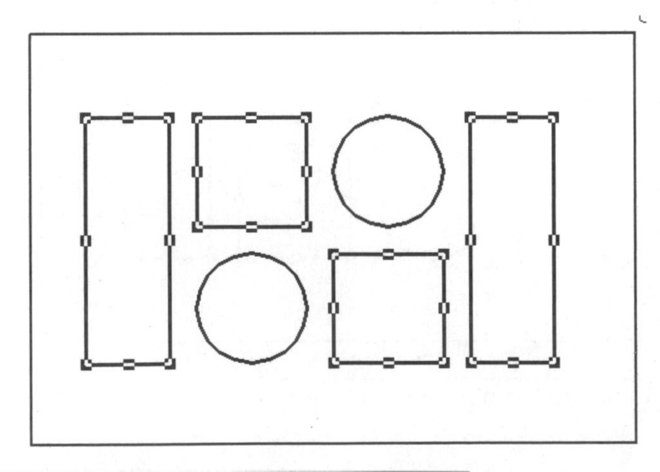

Figure 6–7 Circles are removed from the selection group

Select Elements with the Element Selection Tool

The Element Selection tool box also contains a tool named Element Selection. The new PowerSelector tool almost makes the Element Selection tool obsolete, but there are a few situations where it is still useful.

The Element Selection tool is similar to the Individual Method and Invert Mode of the PowerSelector tool, in that you can select an individual element by clicking on it, or several elements by dragging a block around them. If you click on, or drag a block around previously selected elements, those elements are unselected.

The action of the Element Selection tool is different than PowerSelector when you want to add additional elements to the selection set or remove elements from the set. To add elements, hold down the Ctrl key while you select the additional elements. To remove elements from the selection, hold down the Ctrl key while you click on, or drag, a block around each element to be removed from the set. To remove all elements from the set, click the Data button without holding down the Control key, and without touching any elements.

Invoke the Element Selection tool:

Element Selection tool box	Select the Element Selection tool (see Figure 6–8).
Key-in window	**choose element** (or **cho e**) Enter

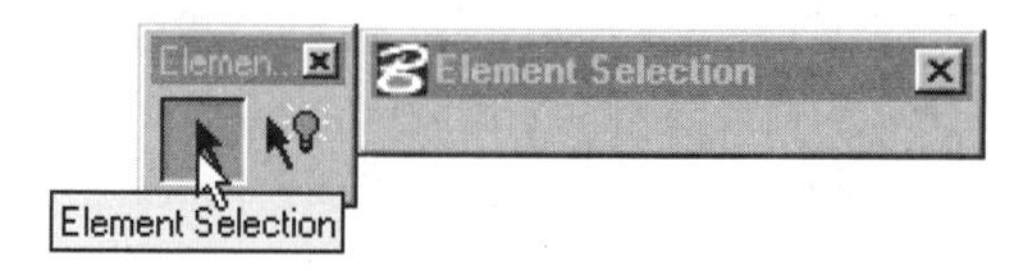

Figure 6–8 Invoking the Element Selection tool from the Element Selection tool box

The screen pointer changes to an arrow with a circle, similar to the one shown in Figure 6–9, and MicroStation prompts:

> Element Selection *(Select elements using the methods discussed in the previous paragraphs.)*

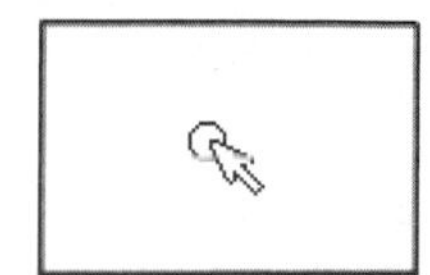

Figure 6–9 Screen pointer shape when the Element Selection tool is active

Consolidate Elements into a Group

The MicroStation Group option consolidates all selected elements (those that have selection handles) into a permanent group that acts like a single element when manipulated. The group has handles on its boundary, as shown in Figure 6–10.

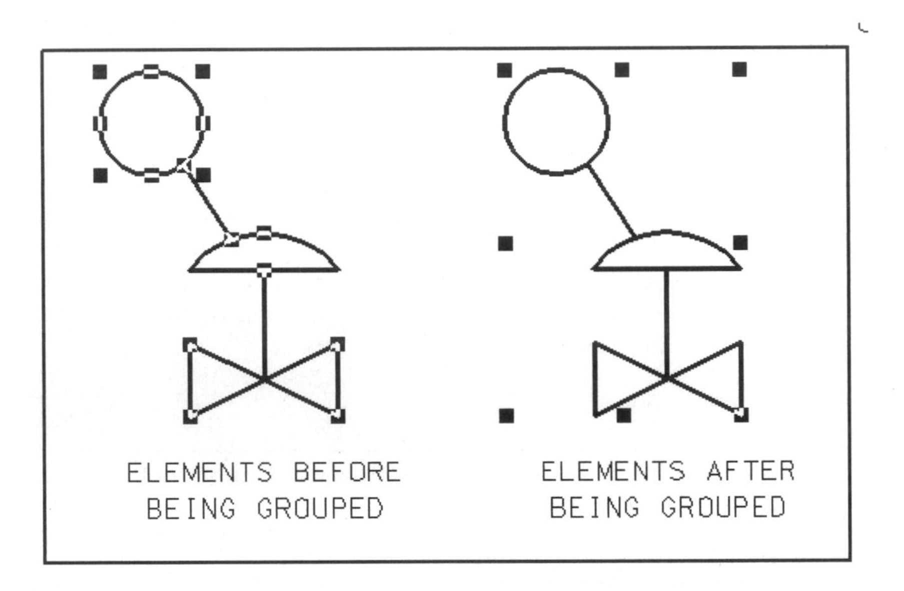

Figure 6–10 Example of elements before and after being grouped

To create a group, select the elements, then invoke the Group tool:

Pull-down menu	Edit > Group
Key-in window	**group selection** (or **gr s**) Enter

The selected elements are immediately consolidated into a group with one set of handles on the group boundary (there are no MicroStation prompts).

Note: The Group tool is dimmed in the Edit pull-down menu if no elements are selected.

Ungroup Consolidated Elements

The MicroStation Ungroup option permanently ungroups all consolidated elements that have handles. The ungrouped elements can be manipulated separately. When consolidated elements are ungrouped, the group boundary handles are replaced by sets of handles on each element.

To ungroup a group, select the group (or groups) of elements, then invoke the Ungroup tool:

Pull-down menu	Edit > Ungroup
Key-in window	**ungroup** (or **ung**) Enter

The selected group (or groups) immediately return into separate elements (there are no MicroStation prompts).

Note: The Ungroup tool is dimmed in the Edit pull-down menu if no elements are selected.

Locking Selected Elements

The MicroStation Lock option locks selected elements to prevent them from being manipulated. If you attempt to select a locked element for manipulation, MicroStation ignores the element and displays the message "Element not found" in the Status bar. Locking is an easy way to protect completed parts of a design from accidental changes.

To lock elements, select the elements, then invoke the Lock tool:

Pull-down menu	Edit > Lock
Key-in window	**change lock** (or **chan lo**) Enter

The selected elements are immediately locked and protected from manipulation (there are no MicroStation prompts).

Note: The Lock tool is dimmed in the Edit pull-down menu if no elements are selected.

Unlock Selected Elements

Locked elements can be unlocked with the MicroStation Unlock option. After unlocking, the elements can be manipulated.

To unlock elements, select the group (or groups) of, then invoke the Unlock tool from:

Pull-down menu	Edit > Unlock
Key-in window	**change unlock** (or **chan u**) Enter

The selected elements are immediately unlocked (there are no MicroStation prompts).

Note: The Unlock tool is dimmed in the Edit pull-down menu if no elements are selected.

Drag Selected Elements to a New Position

When the Element Selection tool is active, selected elements can be dragged (moved) to a new location in the design plane. You do not need to select the Move tool to move them.

1. Select the element or elements to be moved.
2. If PowerSelector is active, switch to the Element Selection tool.
3. Press and hold the Data button anywhere on one of the selected element outlines (but *not* on a handle).
4. Drag the elements to the new location.
5. Release the Data button to place the elements at the new location.

The elements are placed at the new location and removed from the original location, as shown in Figure 6–11 (there are no MicroStation prompts).

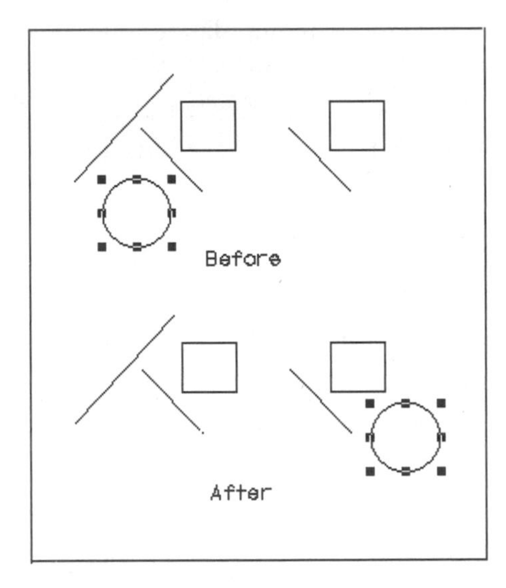

Figure 6–11 Example of dragging a selected element to a new location

Note: The PowerSelector tool cannot be active when you want to drag elements to a new location, because it will capture the Data button press, and assume you are adding elements to the selection set or removing them from the set.

Drag an Element Handle to Change Its Shape

When the Element Selection tool is active, the geometric shape of an element can be changed by dragging one of the element's handles to a new location in the design—you do not need to use the Modify Element tool.

1. Select the element to be modified.
2. If PowerSelector is active, switch to the Element Selection tool.
3. Press and hold the Data button on the element handle to be modified.
4. Drag the handle to the new shape.
5. Release the Data button to complete the modification.

The element's shape is changed (there are no MicroStation prompts).

See Figure 6–12 for an example of modifying an element's shape.

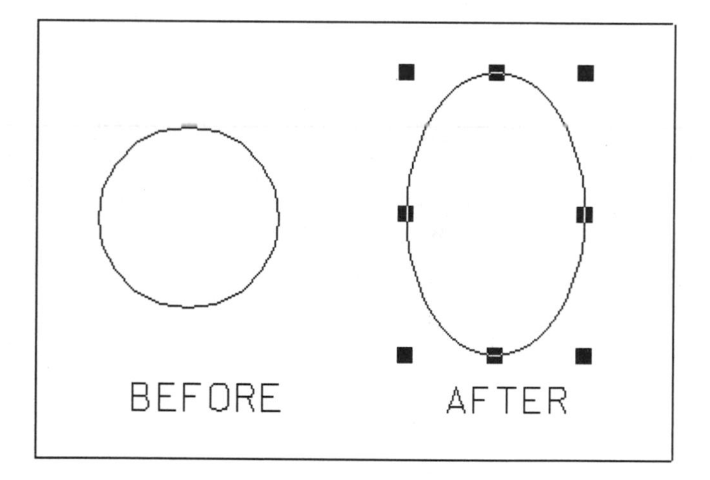

Figure 6–12 Example of modifying an element's shape

Notes: The PowerSelector tool cannot be active when you want to drag elements to a new location, because it will capture the Data button press, and assume you are adding elements to the selection set or removing them from the set.

You can only modify the geometric shape of one element by dragging its handle. If you want to modify the shape of several elements proportionally, first select and group them.

Deleting Selected Elements

All selected elements can quickly be deleted by pressing ~DelKey~ on the computer keyboard or by selecting the Delete tool in the Main tool frame. To delete elements:

1. Select the elements to be deleted.
2. Click [Delete] or select the Delete tool from the Main tool frame.

All selected elements are deleted (there are no MicroStation prompts).

Note: On some systems, the Backspace Key will also delete selected elements.

Change the Attributes of Selected Elements

The element attributes (color, level, style, and weight) of selected elements can be changed simply by selecting the desired attribute in the Primary tool box. For example, to change the attributes of a group of elements:

1. Select the elements whose attributes are to be changed.
2. From the Primary toolbox, select the desired attributes (e.g., change the line weight from 2 to 4 and the color from green to red).

The selected elements are immediately changed to the new attribute each time an attribute is changed (there are no MicroStation prompts).

Note: This only works when you select new attributes in the Primary toolbar. If you use the Element Attributes dialog box to change attributes, the attributes of the selected elements are not changed.

Manipulation Tools that Recognize Selected Elements

Several MicroStation manipulation tools work with selected elements. The tools that recognize selected elements are the Array, Copy, Delete, Mirror, Move, Rotate, Scale, and Change Element attributes.

These manipulation tools affect all selected elements as if they were one element, and the tools exit after completing the requested change.

To manipulate the elements, first select them, then invoke the appropriate manipulation tool. The prompts are slightly different from the prompts that you see when you manipulate individual elements. Always read the MicroStation tool prompts in the Status bar. For example: the Move and Copy tools still require two data points, but they define the relative distance to move or copy the selected elements. In other words, the new location of the elements has the same relationship to the second data point as the original elements had to the first data point (see Figure 6–13).

Note: If a tool cannot work with selected elements, the element handles disappear when the tool is selected.

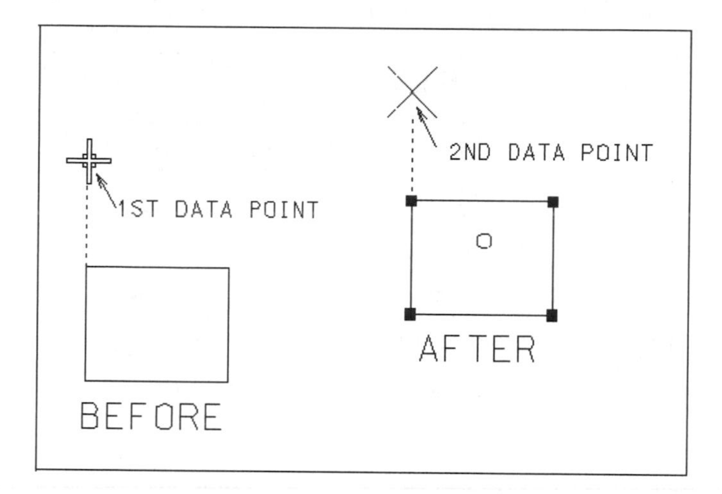

Figure 6–13 Example of using the Move tool with a selected element

FENCE MANIPULATION

The Fence manipulation tools provide another way to manipulate sets of elements. A fence is placed around the elements to be manipulated, and then the fence contents tools can manipulate all elements in the fence. Only one fence at a time can be placed in the design plane, and it remains active until either a new fence is placed or the design file is closed.

The Fence tool box is opened from the Main tool frame, as shown in Figure 6–14.

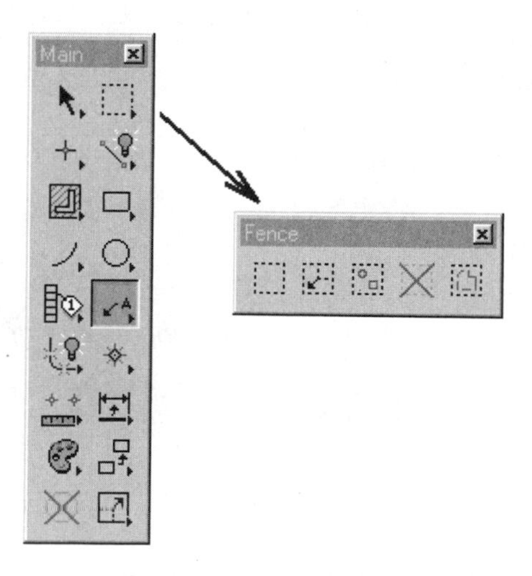

Figure 6–14 Fence tool box location in the Main tool frame

Placing a Fence

Six types of fences can be placed in the design:

- A block defined by diagonally opposite data points
- A shape defined by a series of vertex data points
- A circle defined by center and edge data points
- An existing closed element
- A fence defined by the contents of the selected View
- A fence defined by the contents of the active design file

Fence Block

The Block Fence Type places the fence as an orthogonal block.

Invoke the Place Fence Block tool:

Fence tool box	Select the Place Fence tool, then select Block from the Fence Type option menu in the Tool settings Window (see Figure 6–15).
Key-in window	**place fence block** (or **pl f b**) Enter

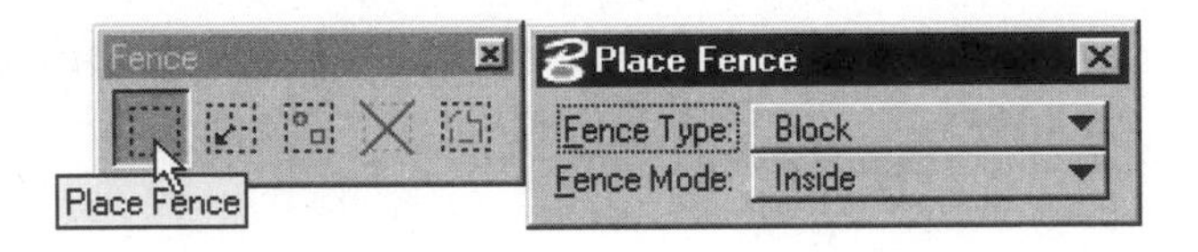

Figure 6–15 Invoking the Place Fence Block tool from the Fence tool box

MicroStation prompts:

Place Fence Block > Enter first point *(Define one corner of the block in the design plane.)*

Place Fence Block > Enter opposite corner *(Define the diagonally opposite corner of the block in the design plane.)*

Shape Fence

The Shape Fence Type places the fence as a closed multi-sided shape that can have up to 101 vertexes.

Invoke the Place Fence Shape tool:

Fence tool box	Select the Place Fence tool, then select Shape from the Fence Type option menu in the Tool settings Window (see Figure 6–16).
Key-in window	**place fence shape** (or **pl f s**) Enter

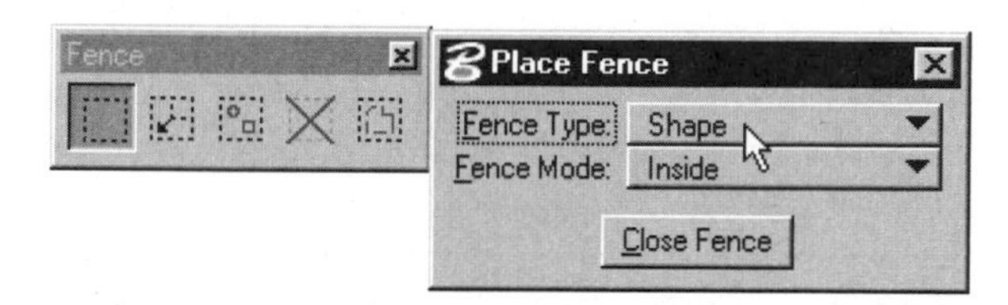

Figure 6–16 Invoking the Place Fence Shape tool from the Fence tool box

MicroStation prompts:

> Place Fence Shape > Enter Fence Points *(Define the location of each fence vertex in the design plane.)*

To complete the fence shape, either place the last vertex data point on top of the first fence data point or click the Close Shape button in the Tool Settings window, as shown in Figure 6–16.

Fence Circle

The Circle Fence Type places the fence as a circle that you place by defining the center of a point on the circumference of the circle.

Invoke the Place Fence Circle tool:

Fence tool box	Select the Place Fence tool, then select Circle from the Fence Type option menu in the Tool settings Window (see Figure 6–17)
Key-in window	**place fence circle** (or **pl f c**) Enter

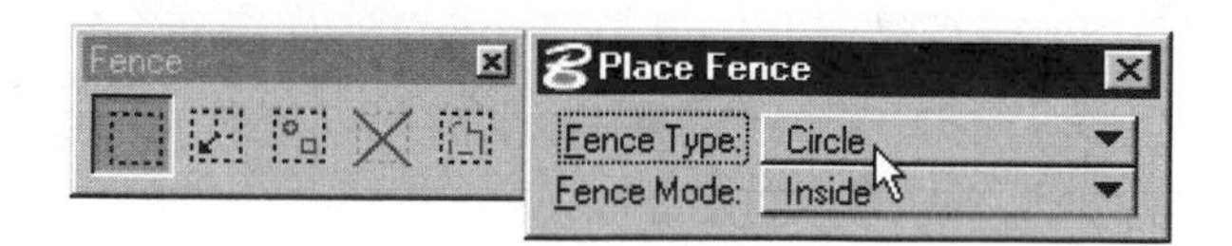

Figure 6–17 Invoking the Place Fence Circle tool from the Fence tool box

MicroStation prompts:

Place Fence Circle > Enter circle center *(Define the location of the fence center in the design plane.)*

Place Fence Circle > Enter edge point *(Define a data point on the circumference of the fence circle.)*

Fence from Element

The Element Fence Type places the fence on top of a selected closed element. A closed element is a circle, ellipse, shape, or complex shape.

Invoke the Place Fence From Element tool:

Fence tool box	Select the Place Fence tool, then select Element from the Fence Type option menu in the Tool settings Window (see Figure 6–18).
Key-in window	**place fence from shape** (or **pl f e**) Enter

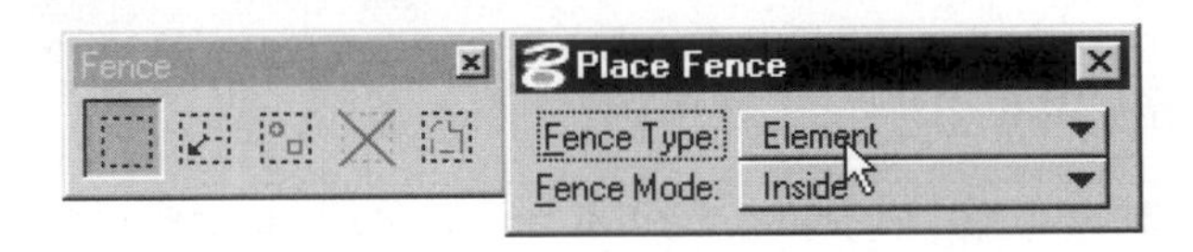

Figure 6–18 Invoking the Place Fence Element tool from the Fence tool box

MicroStation prompts:

Create Fence From Element > Identify element *(Select the closed element whose outline will define the fence.)*

Create Fence From Element > (Accept/Reject) Shape Element *(Click anywhere in the design plane to accept the element.)*

Fence From View

The From View Fence Type places a fence block that is the size of the selected View Window.

Invoke the Place Fence From View tool:

Fence tool box	Select the Place Fence tool, then select From View from the Fence Type option menu in the Tool settings Window (see Figure 6–19)
Key-in window	**place fence view** (or **pl f v**) Enter

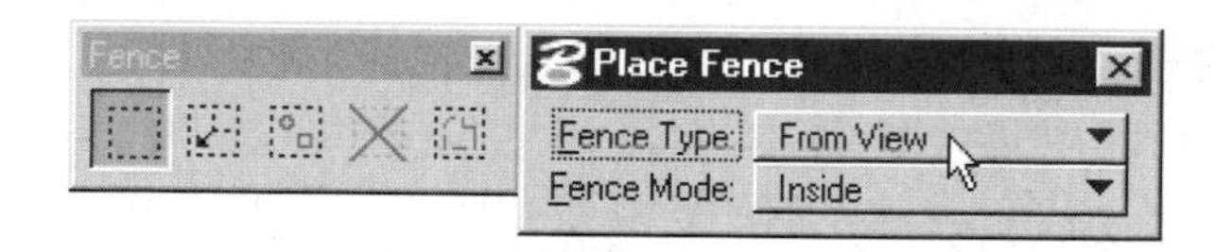

Figure 6–19 Invoking the Place Fence From View tool from the Fence tool box

MicroStation prompts:

Create Fence From View > Select View *(Place a data point anywhere on the view to place a fence in that view.)*

MicroStation places a fence block the exact size of the view window. The fence outline appears along the edges of the view window.

Fence From Active Design File

The From Design File Fence Type places a fence block that encloses all elements in the design file.

Invoke the Place Fence From Active Design File tool:

Fence tool box	Select the Place Fence tool, then select From Design File from the Fence Type option menu in the Tool settings Window (see Figure 6–20).
Key-in window	**place fence active** (or **pl f a**) Enter

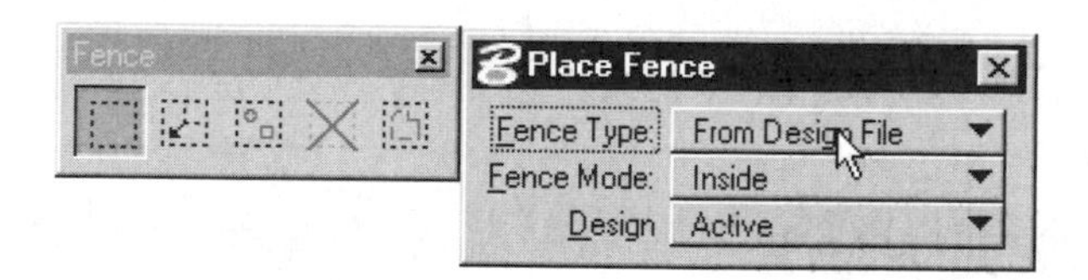

Figure 6–20 Invoking the Place Fence From Active Design File tool from the Fence tool box

MicroStation prompts:

Create Fence From Active Design File > Select View *(Click anywhere in the view window to place the fence.)*

Create Fence From Active Design File > Fence placed - <Reset> to place again. *(Proceed with using the fence contents manipulation commands.)*

Fence Selection Mode

Before you manipulate the contents of a fence, select the Fence Mode to control which elements are actually manipulated by the fence. For example, you can elect to manipulate only elements that are completely inside the fence, or only elements that are completely outside the fence.

Six Fence Mode options are available from the Tool Settings window when a Fence tool is active (see Figure 6–21). The Fence Mode options are also available in the Settings>Locks>Full dialog box.

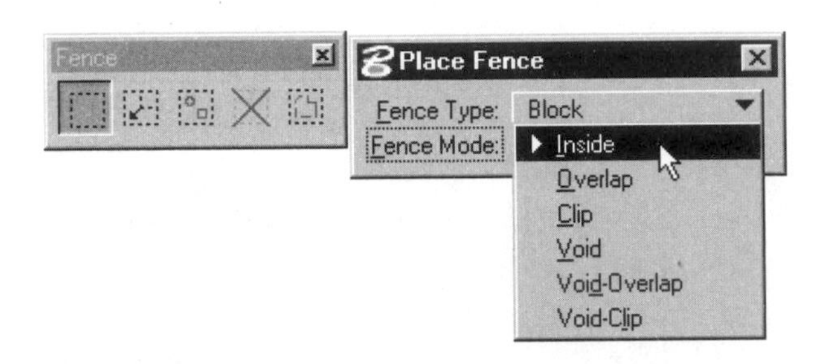

Figure 6–21 Fence Mode options in the Tool Settings window

Following are descriptions of each of the six fence modes.

Inside Mode

Inside Fence Mode limits manipulation to the elements completely inside the fence. For example, the Delete Fence Contents tool deletes circles A, B, and D, shown in Figure 6–22a.

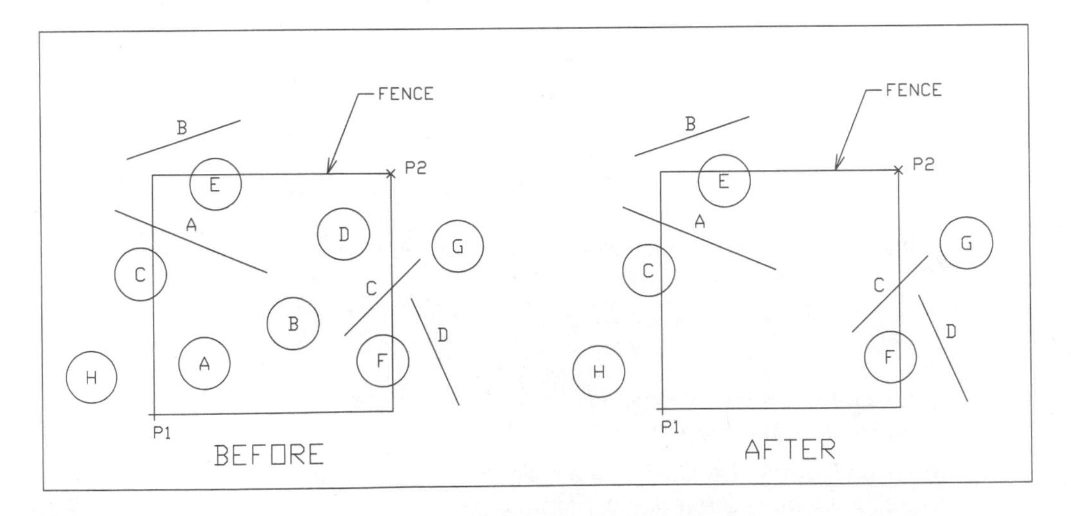

Figure 6–22a Example of deleting the fence contents with Inside mode selected

Overlap Mode

Overlap Fence Mode limits manipulation to the elements that are inside and overlapping the fence. For example, the Delete Fence Contents tool deletes circles A, B, C, D, E, and F, and lines A and C, shown in Figure 6–22b.

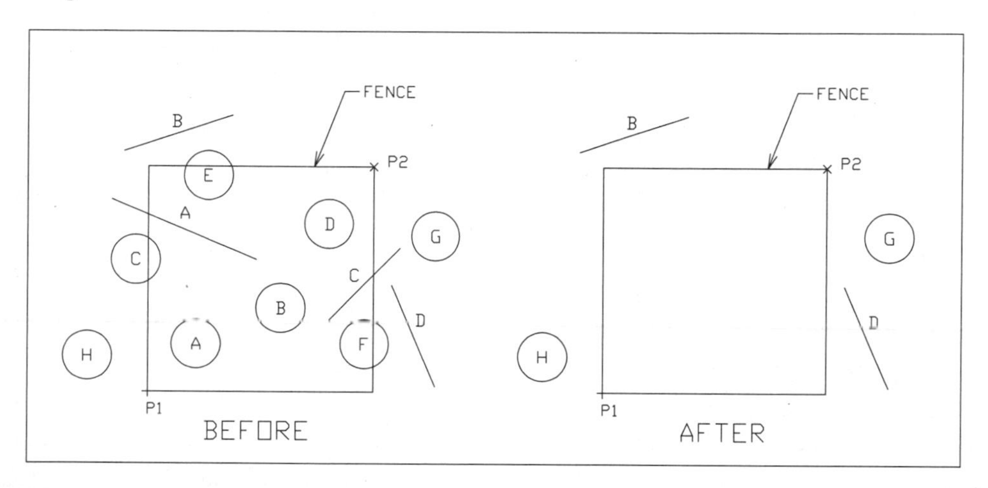

Figure 6–22b Example of deleting the fence contents with Overlap mode selected

Clip Mode

Clip Fence Mode limits manipulation to elements inside the fence and the inside part of elements overlapping the fence. Elements overlapping the fence are clipped at the fence boundary. For example, the Delete Fence Contents tool deletes circles A, B, and D, part of circles C, E, and F, and parts of lines A and C, shown in Figure 6–22c.

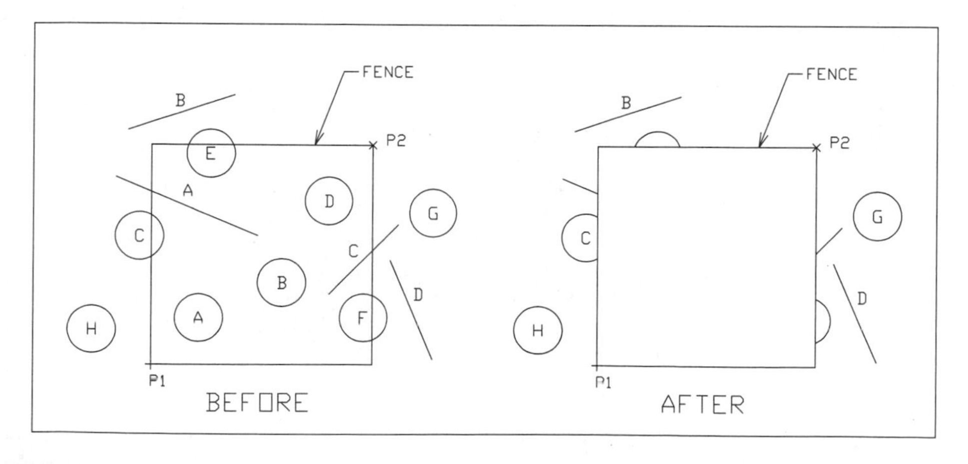

Figure 6–22c Example of deleting the fence contents with Clip mode selected

Void Mode

Void Fence Mode limits manipulation to elements completely outside the fence. For example, the Delete Fence Contents tool deletes circles G and H and lines B and D, shown in Figure 6–22d.

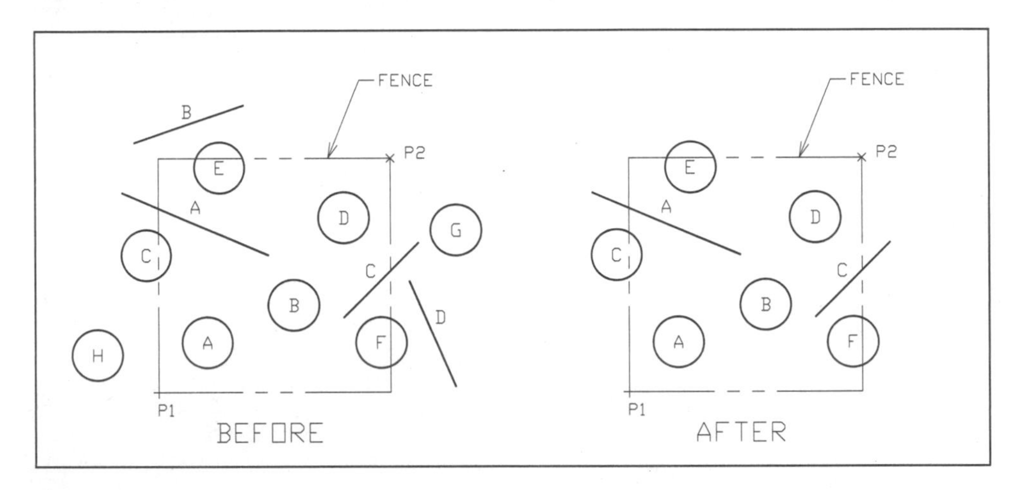

Figure 6–22d Example of deleting the fence contents with Void mode selected

Void-Overlap Mode

Void-Overlap Fence Mode limits manipulation to elements outside and overlapping the fence. For example, the Delete Fence Contents tool deletes circles C, E, F, G, and H, and lines A, B, C, and D, as shown in Figure 6–22e.

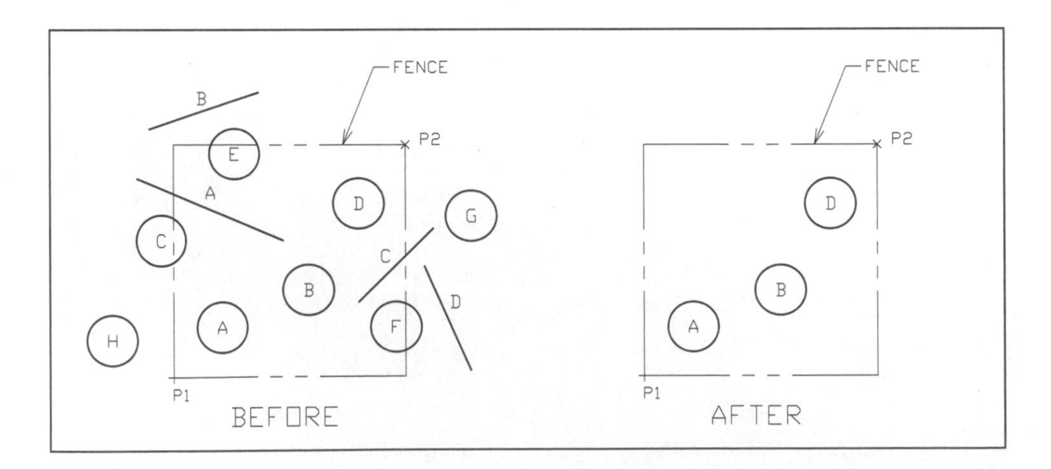

Figure 6–22e Example of deleting the fence contents with Void-Overlap mode selected

Void-Clip Mode

Void-Clip Fence Mode limits manipulation to elements outside the fence and the parts of the overlapping elements that are outside the fence. Elements are clipped at the fence boundary. For example, the Delete Fence Contents tool deletes circles G and H, lines B and D, parts of circles C, E, and F, and parts of lines A and C, as shown in Figure 6–22f.

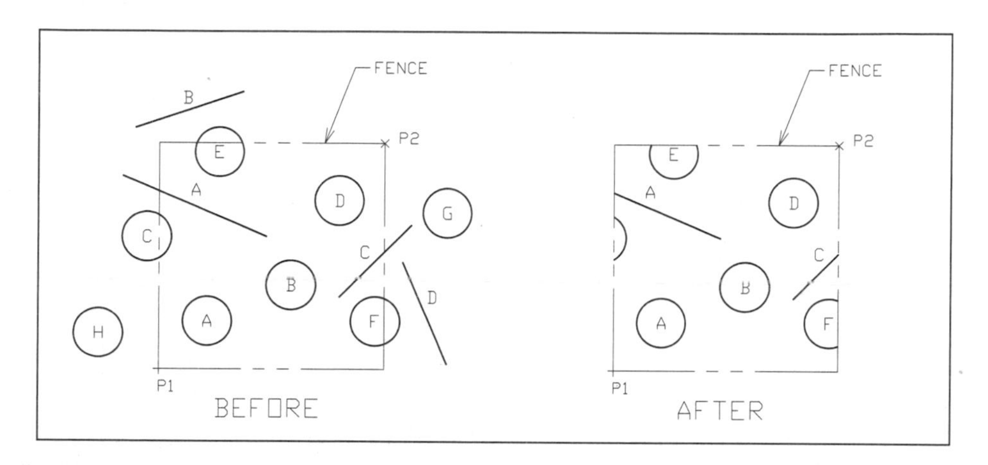

Figure 6–22f Example of deleting the fence contents with Void-Clip mode selected

Modifying a Fence's Shape or Location

You just finished placing a complicated fence shape and there, sitting outside the fence, is an element that should be inside the fence. There is no need to place the fence again; the Modify Fence tool can modify a fence vertex or move the fence to a new location.

Modifying Fence Shape

To modify a fence, invoke the Modify Fence Vertex tool:

Fence tool box	Select the Modify Fence tool, then select Vertex from the Modify Mode option menu in the Tool Settings window (see Figure 6–23).
Key-in window	**modify fence** (or **modi f**) Enter

Figure 6–23 Invoking the Modify Fence Vertex tool from the Fence tool box

MicroStation prompts:

> Modify Fence Vertex > Identify vertex *(Click the Data button on the fence outline near the vertex to be modified, drag the vertex to the new position, and click the Data button again. Click the Reset button to complete the modification.)*

See Figure 6–24 for an example of modifying a fence.

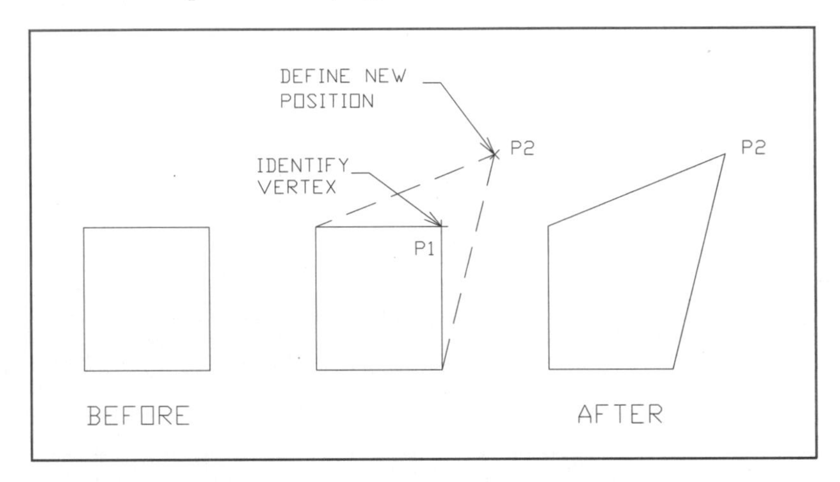

Figure 6–24 Example of modifying a fence vertex

> **Note:** After the first data point is placed, a dynamic image of the fence drags with the screen pointer.

Moving a Fence

To move a fence to a new location in the design plane, invoke the Move Fence tool:

Fence tool box	Select the Modify Fence tool, then select Position from the Modify Mode option menu in the Tool Settings window (see Figure 6–25).
Key-in window	**move fence** (or **mov f**) Enter

Figure 6–25 Invoking the Modify Fence tool (Move Position) from the Fence tool box

MicroStation prompts:

> Move Fence Block/Shape > Define origin *(Click the Data button in the design plane to identify the relative starting position of the move.)*
>
> Move Fence Block/Shape > Define distance *(Click the Data button in the design plane to identify the relative position to which the fence is to be moved. Click the Reset button to complete the move.)*

The fence is moved a distance equal to the distance between the two data points, and the relationship of the final fence position to data point two is the same as the original fence position was to data point one.

Note: The Modify Fence tool modifies only the shape or position of the fence, not the contents of the fence. The Modify Fence Contents tools are used to modify the elements contained by the fence.

Manipulate Fence Contents

After you place a fence and select the appropriate fence selection mode, you are ready to manipulate the contents of the fence. There is a fence contents manipulation equivalent for each of the element manipulation tools discussed in the previous chapters. The only difference is that you do not have to select the elements to manipulate them—the fence does that for you.

The Copy, Move, Scale Rotate, Mirror, and Array tools in the Manipulate tool box can be switched between element manipulation and fence contents manipulation. The Use Fence toggle button in the Tool Settings window determines which type of manipulation is done. If the button is set to ON, the fence contents are manipulated; if it is set to OFF, individual elements are manipulated. Figure 6–26 shows the Manipulate tool box with the Copy tool selected and the Use Fence toggle button set to ON in the Tool Settings window.

Similarly, when you invoke the Change Element Attributes tool from the Change Attributes tool box, a toggle button in the Tool Settings window can switch between element and fence manipulation.

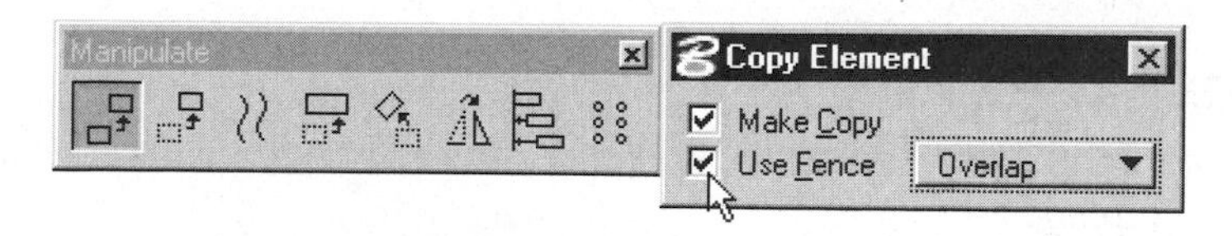

Figure 6–26 Invoking the Copy Element tool from the Manipulate tool box with the Use Fence toggle button set to ON

In addition, MicroStation provides three fence contents manipulation tools:

- The Manipulate Fence Contents tool, invoked from the Fence tool box, has an Operation: option menu in the Tool Settings window from which you can select the type of manipulation you need (see Figure 6–27).

- The Delete Fence Contents tool, invoked from the Fence tool box, deletes all elements within the contents of the fence.
- The Drop Fence Contents tool, invoked from the Fence tool box, drops all complex elements within the contents of the fence.

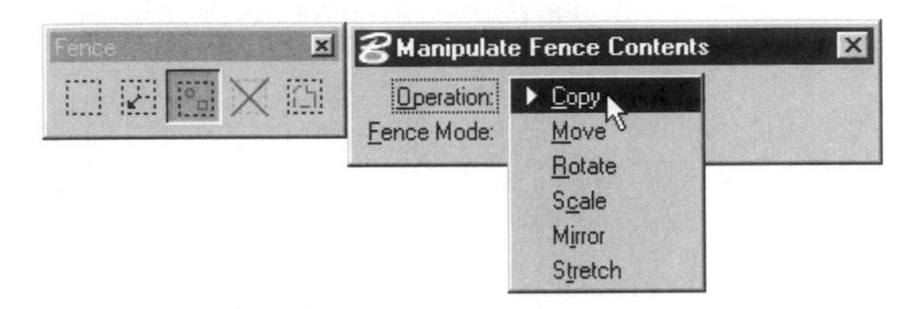

Figure 6–27 Manipulate Fence Contents tool Operation: option menu

For example, to move the contents of the fence to a new location in the design plane, select the Manipulate Fence Contents tool:

Fence tool box	Select the Manipulate Fence Contents tool, then select Move from the Operation menu in the Tool Settings Tool box (see Figure 6–28).
Key-in window	**fence move** (or **f mo**) Enter

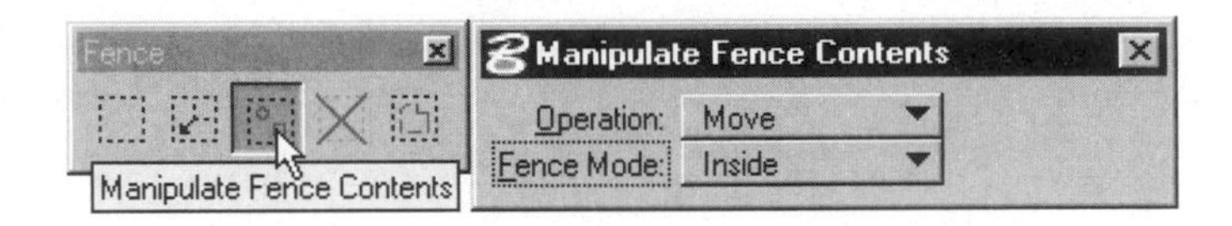

Figure 6–28 Invoking the Manipulate Fence Contents Move tool from the Fence tool box

MicroStation prompts:

Move Fence Contents > Define origin *(Locate the starting position of the move in the design plane.)*

Move Fence Contents > Define distance *(Locate the final destination in the design plane.)*

Stretch Fence Contents

The Fence Stretch tool allows you to stretch the contents of a fence. There is no equivalent element manipulation tool for this command.

- Line, Line String, Multi-line, Curve String, Shape, Polygon, Arc, and Cell (see Chapter 10) elements that overlap the fence are stretched.
- Circle and Ellipse elements that overlap the fence are ignored.
- Elements completely inside the fence are moved.

The Stretch tool only checks to see if an inside mode (Inside, Overlap, or Clip) is active, or if a void mode (Void, Void-Overlap, or Void-Clip) is active. If an inside mode is active, the element vertexes inside the fence are stretched. If a void mode is active, the element vertexes outside the fence are stretched.

See Figure 6–29 for an example of stretching the contents of a fence when one of the Inside Modes is active.

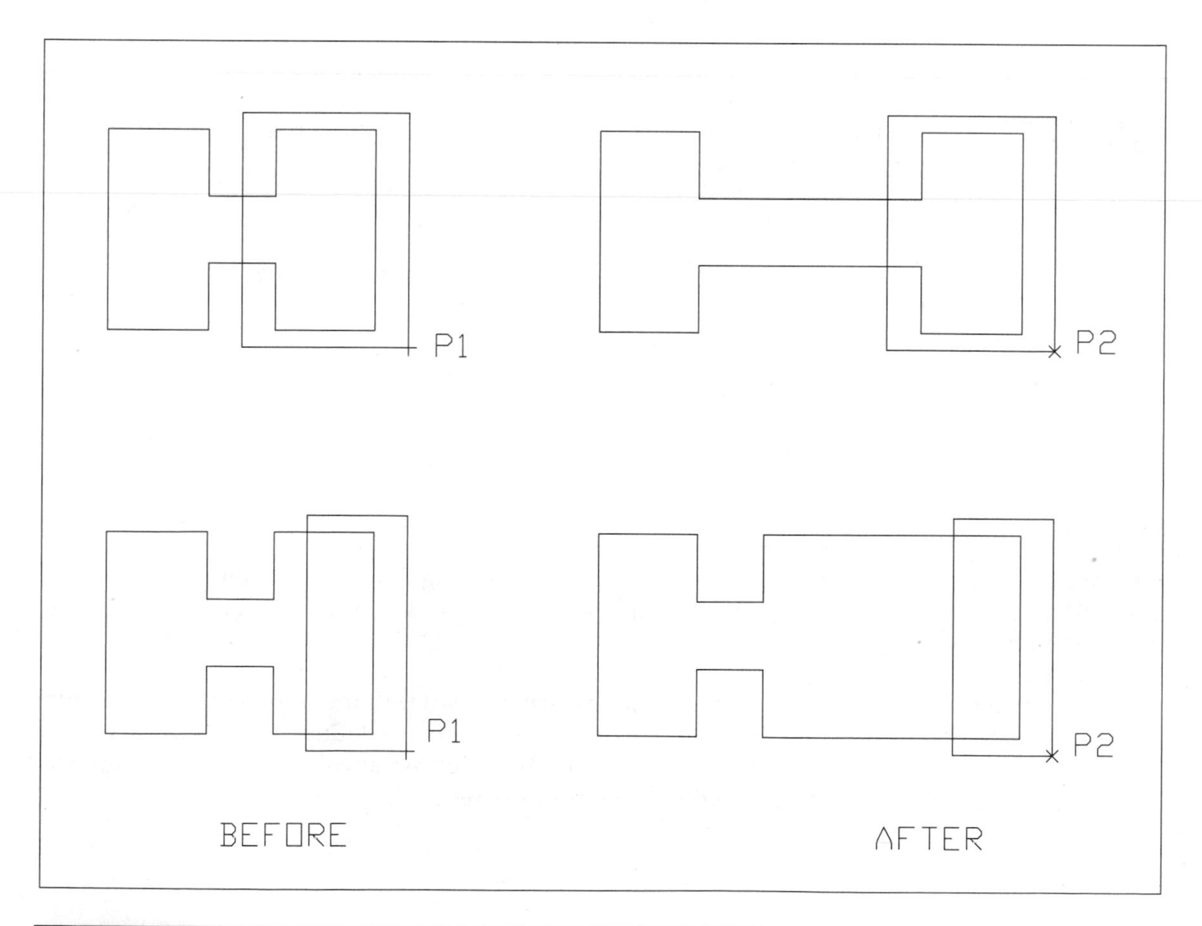

Figure 6–29 Example of stretching the contents of a fence

To stretch a group of elements, place a fence that overlaps the elements you want to stretch, then select the Fence Stretch tool:

Fence tool box	Select the Manipulate Fence Contents tool and Stretch from the Operations option menu (see Figure 6–30).
Key-in window	**fence stretch** (or **f st**) Enter

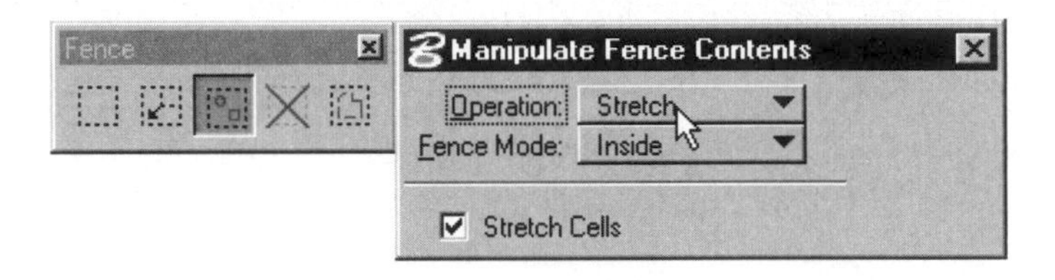

Figure 6–30 Invoking the Manipulate Fence Contents Stretch tool from the Fence tool box

MicroStation prompts:

Fence Stretch > Define origin *(Locate a relative point in the design plane to start the stretch.)*

Fence Stretch > Define distance *(Locate a relative point in the design plane to end the stretch.)*

The new fence location has the same relationship to the second data point as the original fence location did to the first data point.

Remove a Fence

To remove a fence, invoke the Place Fence tool and the existing fence will be removed. If you do not want to place another fence, select another tool and continue working. There is no separate tool to remove a fence.

Note: Always remove the fence after you are finished with it to protect from accidental fence contents manipulation. For example, if a fence is defined and you accidentally select the tool to delete the fence contents while thinking you selected the delete element tool, you could delete the contents of the fence.

REVIEW QUESTIONS

Write your answers in the spaces provided.

1. The PowerSelector tool, ____________ Method, and ____________ Mode allow you to select all elements by dragging a block around them.
2. The PowerSelector tool Mode that allows selecting and unselecting elements at one time is the ____________ mode.
3. The PowerSelector tool's Individual Method and Add Mode can be turned on by pressing two keyboard shortcuts. What two keys will turn them on? ____________________.
4. You can only change the shape of an element by dragging one of its handles when the ______________________ tool is active.
5. Briefly explain the purpose of locking individual elements.
   ____
   ____
6. List the six fence selection modes available in MicroStation.
   ____
   ____
7. Explain briefly the difference between the Overlap and Void-Overlap mode.
   ____
   ____
8. The Fence Stretch tool will stretch an arc. TRUE or FALSE
9. The element manipulation tools will manipulate the fence contents when the __________________ button is turned ON.
10. How do you remove a Fence?
    ____
    ____

PROJECT EXERCISE

This project exercise provides step-by-step instructions for creating the design shown in Figure P6–1. The intent is to guide you in applying Element Selection and Fence manipulations.

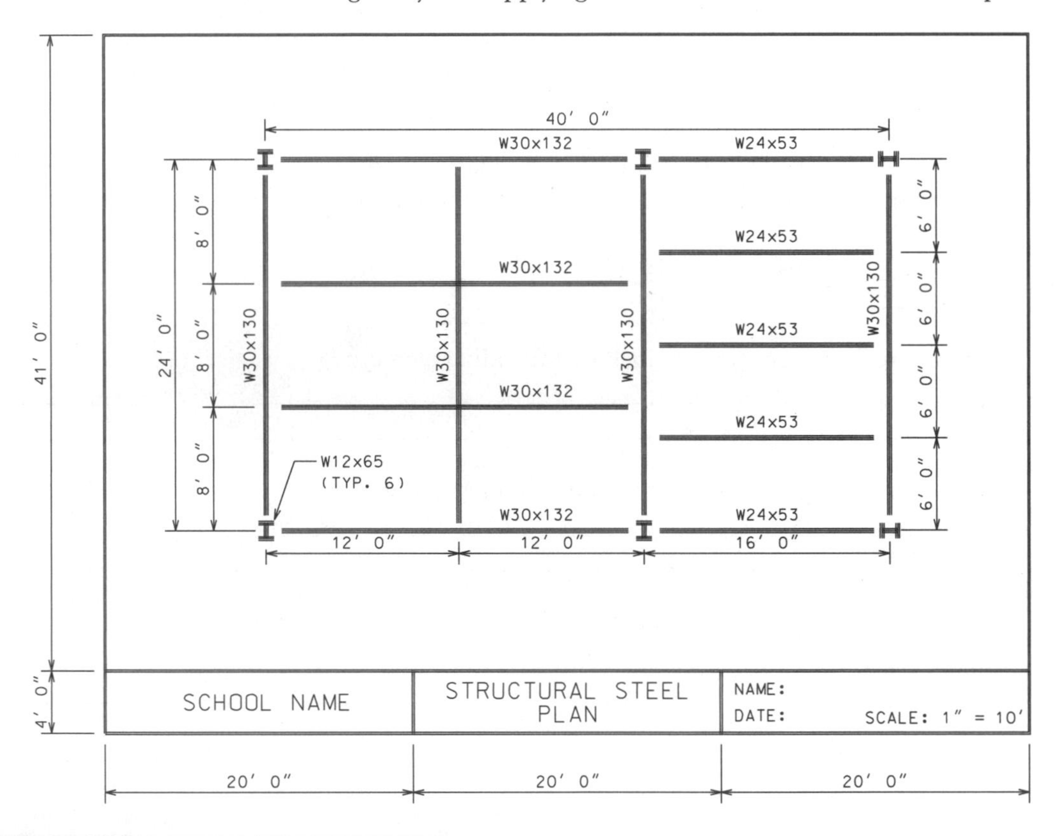

Figure P6–1 Completed project design

Note: The text and dimensions placed on the structure and members are not part of this project. They are included in Figure P6–1 as an aid to drawing the design.

Prepare the Design File

This procedure starts MicroStation, creates a design file, and enters the initial settings.

Note: As you complete each step in the project procedures, place a check mark by the step to help you keep up with where you are in the project.

STEP 1: Invoke MicroStation by the normal technique for the operating system on your workstation.

STEP 2: Create a new design file named CH6.DGN using the SEED2D.DGN seed file.

STEP 3: In the Design File dialog box:

- Set the Working Unit names to ' for the Master Units and " for the Sub Units; and set the Resolution to 12" Per ' and 8000 Pos Units per ".
- Set the Grid Master to 0.5, the Grid Reference to 2, and turn the Grid lock ON.

STEP 4: Invoke the AccuDraw from the Primary Tool box.

STEP 5: Open the AccuDraw settings box from the Settings pull-down menu, and adjust the values as follows:

- *Unit Roundoff Distance:* Set to 0.5 and turn the toggle button ON.
- *Unit Roundoff Angle:* Set to 90.000 and turn the toggle button ON.
- *Coordinate System:* Set the Rotation to Top and the Type to Rectangular.
- *Operation:* Set the toggle buttons for Floating Origin and Smart Key-ins to ON; and for Context Sensitivity and Auto Point Placement, set them to OFF.

STEP 6: Using Figure P6–1 as a guide, draw the boarder and title block on level 10.

- Replace “SCHOOL NAME” with your school or company name, or make up a name.
- Place your name to the right of “NAME.”
- Place today’s date to the right of “DATE.”

Draw the First I-beam

This procedure describes the steps required to draw the I-beam shown in Figure P6–2.

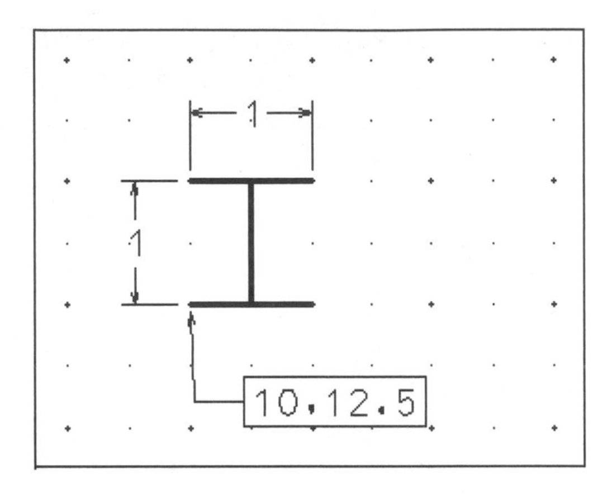

Figure P6–2 Draw the first column

STEP 1: If View Window 2 is not open, select the Open/Close submenu from the Window pull-down menu, and turn on View Window 2.

STEP 2: From the Window pull-down menu, select the Cascade option.

STEP 3: Fit View Window 1.

STEP 4: Set the Active Level to 1, the Line Weight to 2, and the Color to green.

STEP 5: Invoke Save Settings from the File pull-down menu.

STEP 6: Invoke the Place Line tool from the Linear Elements tool box.

MicroStation prompts:

Place Line > Enter first point *(Keypoint snap to the lower left corner of the border block, type* **O** *to release the AccuDraw origin, and drag the cursor so the* X *axis is set to 10.0000 and the* Y *axis is set to 12.5. Click the Data button to locate the start of the bottom I-beam line.)*

Place Line > Enter endpoint *(Drag the cursor so the* X *axis is set to 1.0000 and the* Y *axis is set to 0.0000. Click the Data button to complete the line.)*

Place Line > Enter endpoint *(Click the Reset button.)*

STEP 7: In View Window 1, invoke the Window Area tool, then, in the Tool Settings window, set the Apply to Window option to 2.

MicroStation prompts:

Window Area > Define first corner point *(Place a data point about 2 feet above and to the left of the I-beam line that was just completed.)*

Window Area > Define opposite corner point *(Drag the dynamic rectangle below and to the right of the line, then place a data point to place the view area in View Window 2.)*

STEP 8: Invoke Save Settings from the File pull-down menu.

STEP 9: In View Window 2, use Center Snap to place a vertical 1'-long line centered above the line you just drew, then place a 1'-long top horizontal line centered above the vertical line, as shown in Figure P6–2.

Select and Group the I-beam Lines

This procedure groups the three lines forming the I-beam so they can be manipulated as one element.

STEP 1: Invoke the PowerSelector tool from the Main tool frame, then select the Individual Method and Invert Mode in the Tool Settings window.

MicroStation prompts:

PowerSelector > Place Shape for elements to add to set *(Position the cursor above and to the left of the I-beam, then press and hold down the Data button while you drag the Selection rectangle around the I-beam. Release the Data button to select the three lines, as shown in Figure P6–3.)*

STEP 2: Invoke the Group option from the Edit pull-down menu. MicroStation creates a group of the three selected lines. Place a Data point on the I-beam to remove the handles.

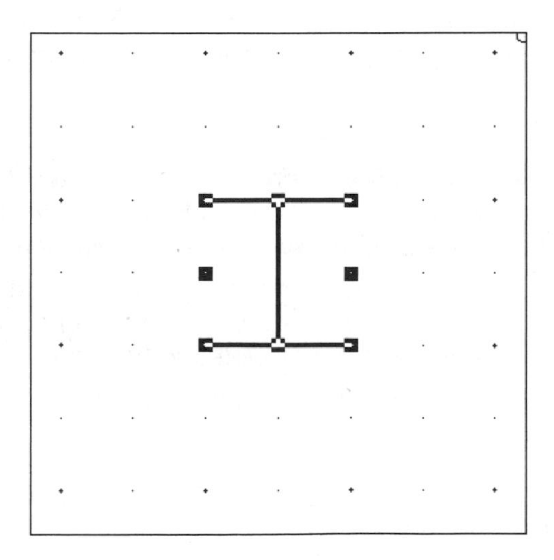

Figure P6–3 I-beam handles after element selection

Create the Two Rows of Columns

This procedure uses the Copy Element, Rotate Copy, Place Fence Block, and Copy Fence Contents tools with AccuDraw to create the two rows of three I-beams each, as shown in Figure P6–4.

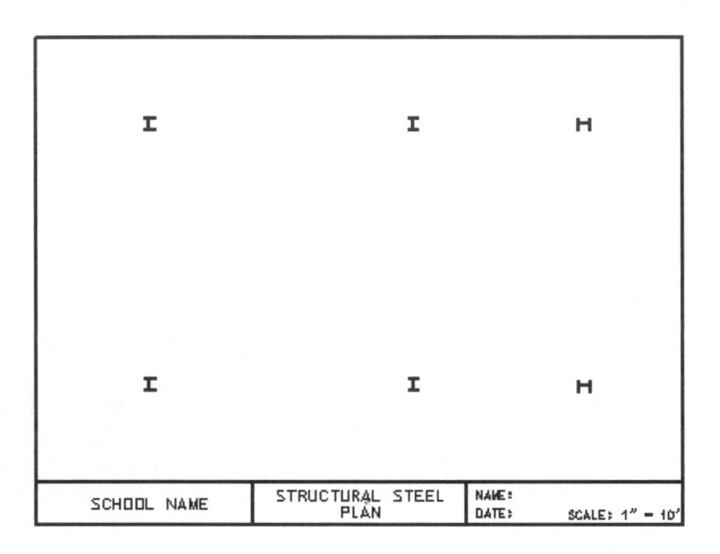

Figure P6–4 Two rows of I-beams

STEP 1: Click the title bar of View Window 1 to return focus to it.

STEP 2: Invoke the Copy Element tool from the Manipulate tool box, then turn the Make Copy button ON in the Tool Settings window.

MicroStation prompts:

Copy Element > Identify element *(Select the I-beam, type* **Y** *to lock the AccuDraw Y axis at 0.0000, then drag the manipulation pointer right to AccuDraw coordinate X = 24. Click the Data button to make the first copy in the bottom row.)*

Copy Element > Accept/Reject (select next input) *(Type* **Y** *to lock the AccuDraw Y axis at 0.0000, then drag the manipulation pointer to the right to X = 16. Click the Data button to complete the bottom row, as shown in Figure P6–5.)*

Copy Element > Accept/Reject (select next input) *(Click the Reset button to terminate the tool sequence.)*

STEP 3: In View Window 1, define a small Window Area, to be placed in View Window 2, around the right-most I-beam.

STEP 4: Click the title bar or border of View Window 2 to return focus to it.

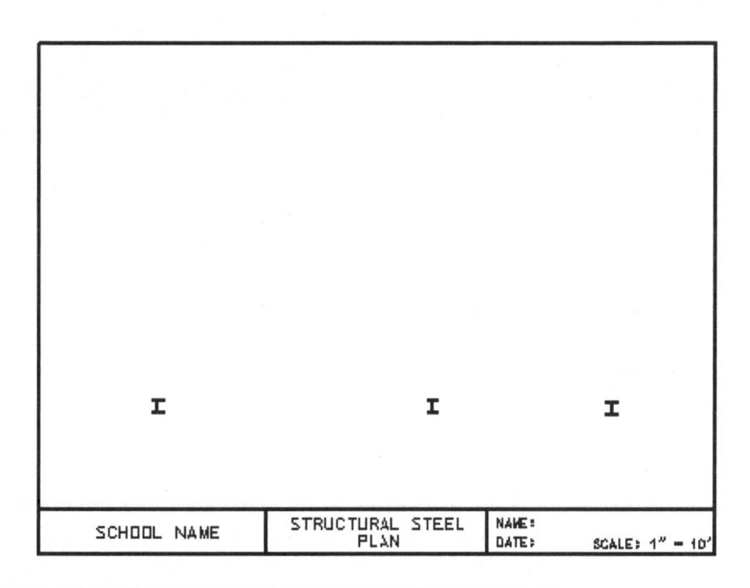

Figure P6–5 Bottom row after the I-beam is copied two times

STEP 5: Invoke the Rotate tool from the Manipulate tool box, then, in the Tool Settings window, set the Method to Active Angle, set the Active Angle to 90, and set the toggle button for Make Copy to OFF.

MicroStation prompts:

Rotate Element > Identify element *(Identify the I-beam.)*

Rotate Element > Enter pivot point (point to rotate about) *(Click the Data button in the center of the I-beam's vertical line to pivot the I-beam about its center point, then click the Reset button.)*

STEP 6: Click the title bar or border of View Window 1 to return focus to it.

STEP 7: Invoke the Place Fence tool from the Fence tool box, then, in the Tool Settings window, set the Fence Type to Block and the Fence Mode to Inside.

MicroStation prompts:

Place a Fence Block > Enter first point *(Place a data point above and to the left of the left-most I-beam.)*

Place a Fence Block > Enter opposite corner *(Drag the dynamic fence image around the three I-beams, then place a data point to complete the fence.)*

STEP 8: Invoke the Copy tool from the Manipulate tool box, then, in the Tool Settings window, set the toggle buttons for Make Copy and Use Fence to ON.

MicroStation prompts:

Copy Fence Contents > Enter first point *(Place a data point somewhere near the bottom of the view.)*

Copy Fence Contents > Enter point to define distance and direction *(Type* **X** *to lock the AccuDraw X axis at 0.0000, then drag the manipulation pointer up to Y = 24. Click the Data button to create the top I-beam row, as shown in Figure P6–4.)*

Copy Fence Contents > Enter point to define distance and direction *(Click the Reset button to terminate the tool sequence.)*

STEP 9: Invoke the Place Fence tool again to remove the fence.

STEP 10: Invoke Save Settings from the File pull-down menu.

Draw the Outside Structural Members

This procedure places a Block element for the outside structural members, then uses the Partial Delete tool to cut away the parts of the Block that overlap the I-beams, as shown in Figure P6–6.

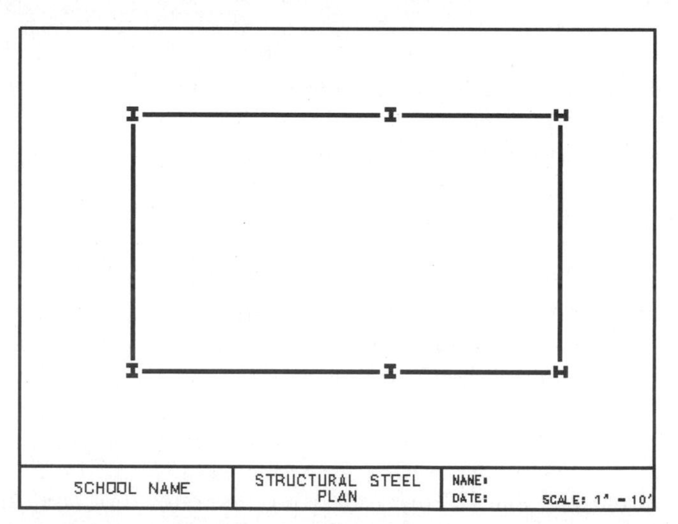

Figure P6–6 I-beams and outside structural members

STEP 1: Set the Active Level to 2, the Line Weight to 2, and the Color to blue.

STEP 2: In View Window 1, place a Block element with its lower left corner in the center of the lower left I-beam and its upper-right corner in the center of the upper-right I-beam, as shown in Figure P6–7.

STEP 3: In View Window 1, define a small Window Area, to be placed in View Window 2, around the lower-right I-beam.

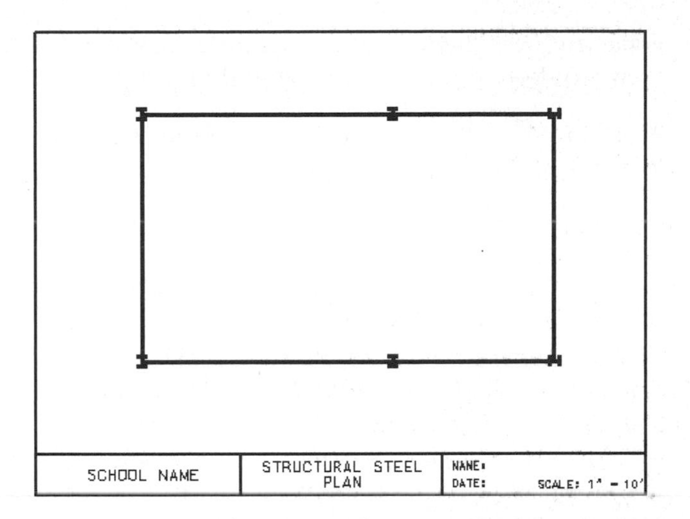

Figure P6–7 Result of placing a block for the outside structural members

STEP 4: Focus on View Window 2, then invoke the Partial Delete tool from the Modify tool box.

MicroStation prompts:

> Delete Part of Element > Select start point for partial delete *(Select the block one Grid point to the left of the I-beam in View Window 2.)*
>
> Delete Part of Element > Select direction of partial delete *(Drag the manipulation pointer a short distance toward the I-beam, and click the Data button.)*
>
> Delete Part of Element > Select end point of partial delete *(Drag the manipulation point to one Grid point above the I-beam, and place a data point to complete the partial delete, as shown in Figure P6–8.)*

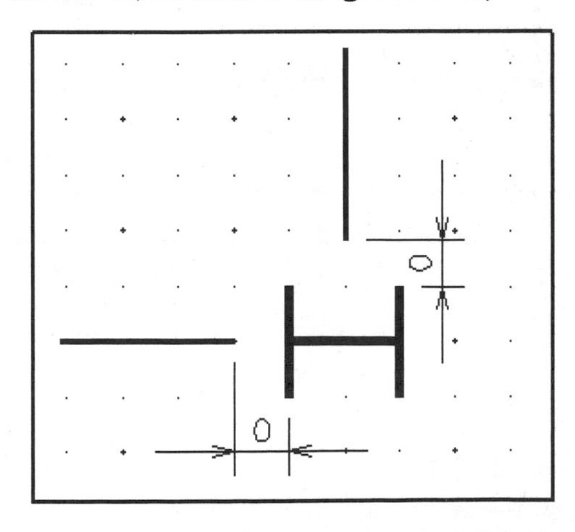

Figure P6–8 Amount of block to delete partially over each I-beam

STEP 5: Focus on View Window 1, then define a small Window Area, to be placed in View Window 2, around the upper-right I-beam.

STEP 6: Focus on View Window 2, then invoke the Delete part of the Element tool from the Modify tool box.

MicroStation prompts:

Delete Part of Element > Select start point for partial delete *(Select the line one Grid point below the I-beam.)*

Delete Part of Element > Select end point of partial delete *(Drag the manipulation pointer to one Grid point to the left of the I-beam, and place a data point to complete the partial delete.)*

STEP 7: Repeat Steps 5 and 6 for the other four I-beams.

Draw the Interior Structural Members

This procedure uses the Move Parallel and Extend Element to Intersection tools to draw the interior structural Members as shown in Figure P6–9.

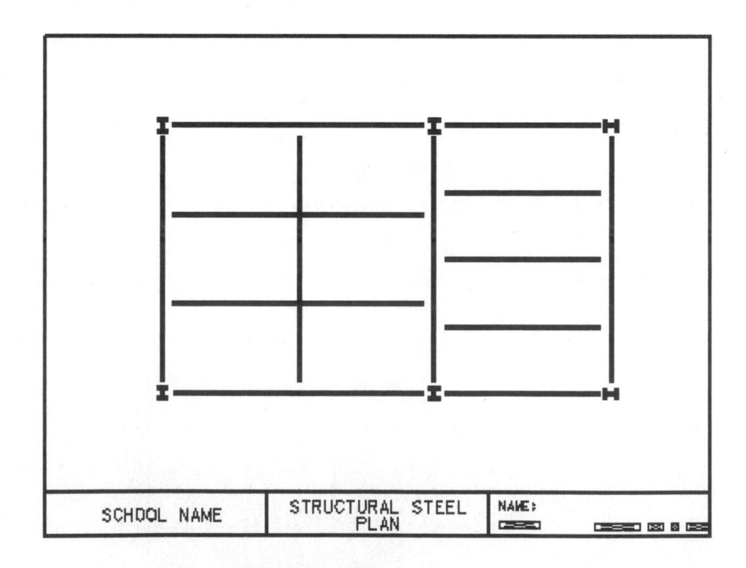

Figure P6–9 Completed interior structure members

STEP 1: Focus on View Window 1.

STEP 2: Invoke the Move Parallel tool from the Manipulate tool box, then, in the Tool Settings window, turn the toggle buttons for Distance and Make Copy to ON, and key-in **12** in the Distance edit field.

MicroStation prompts:

Copy Parallel by Key-in > Identify element *(Select the left vertical line.)*

Copy Parallel by Key-in > Accept/Reject (select next input) *(Move the manipulation pointer to the right of the element, and click the Data button two times to place two parallel copies of the line, as shown in Figure P6–10. Click the Reset button.)*

Copy Parallel by Key-in > Accept/Reject (select next input) *(Click the Reset button.)*

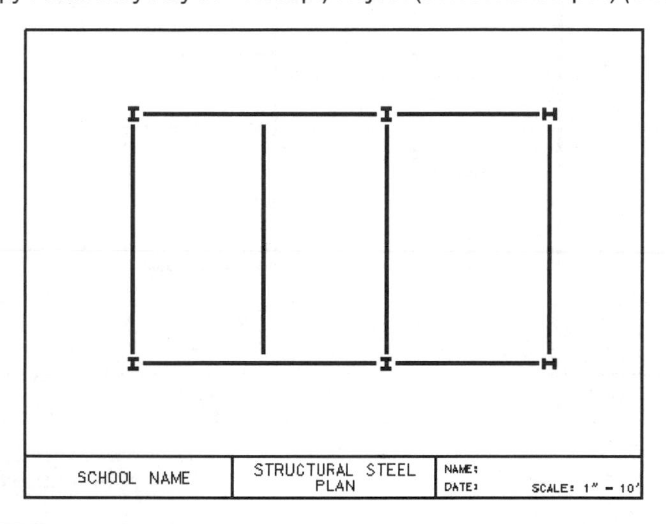

Figure P6–10 Place two parallel copies of the left vertical line, 12' apart

STEP 3: Make two parallel copies of the top left horizontal line, each 8' apart below the line, as shown in Figure P6–11.

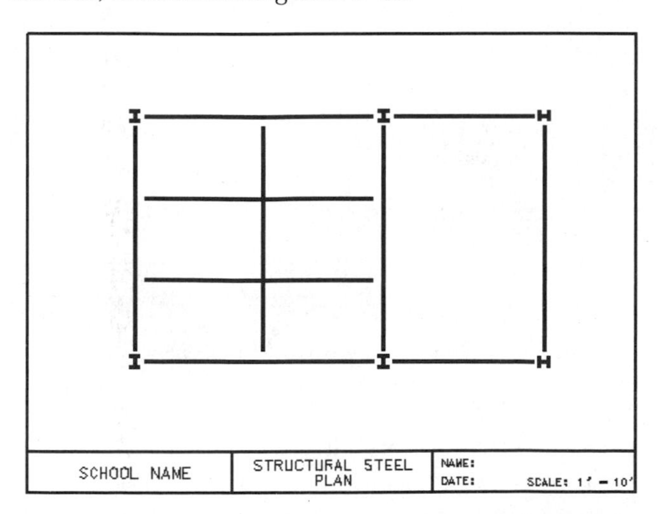

Figure P6–11 Place two parallel copies of the top left horizontal line, 8' apart below the line

STEP 4: Make three parallel copies of the top right horizontal line, each 6' apart below the line, as shown in Figure P6–12.

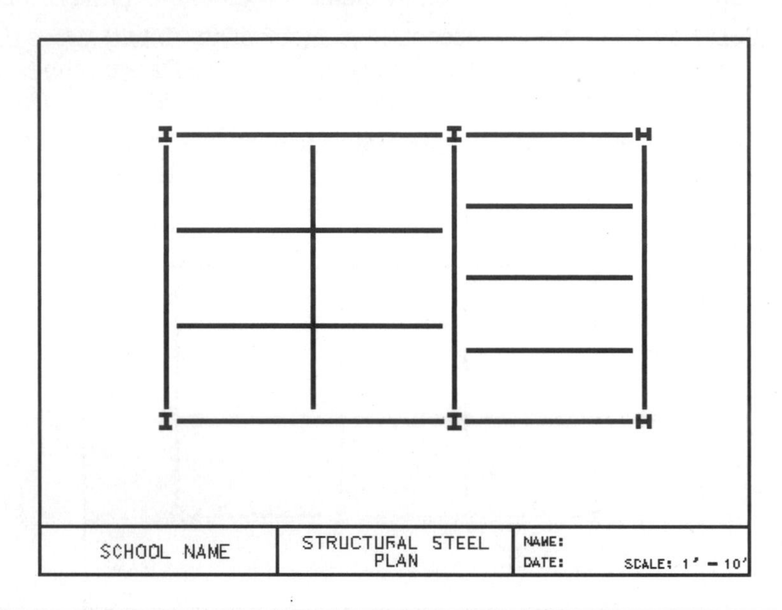

Figure P6–12 Place three parallel copies of the top right horizontal line, each 6' apart below the line

STEP 5: Invoke Save Settings from the File pull-down menu.

STEP 6: Refer to the Chapter 9 Project Exercise for placing text and dimensioning.

DRAWING EXERCISES 6–1 THROUGH 6–5

Use the following table to set up the design files for Exercises 6–1 through 6–3.

SETTING	VALUE
Seed File	SEED2D.DGN
Working Units	10 TH Per IN and 1000 Pos Units Per TH
Grid	Master = .1, Reference = 10, GRID Lock ON

Exercise 6–1

Flange gasket

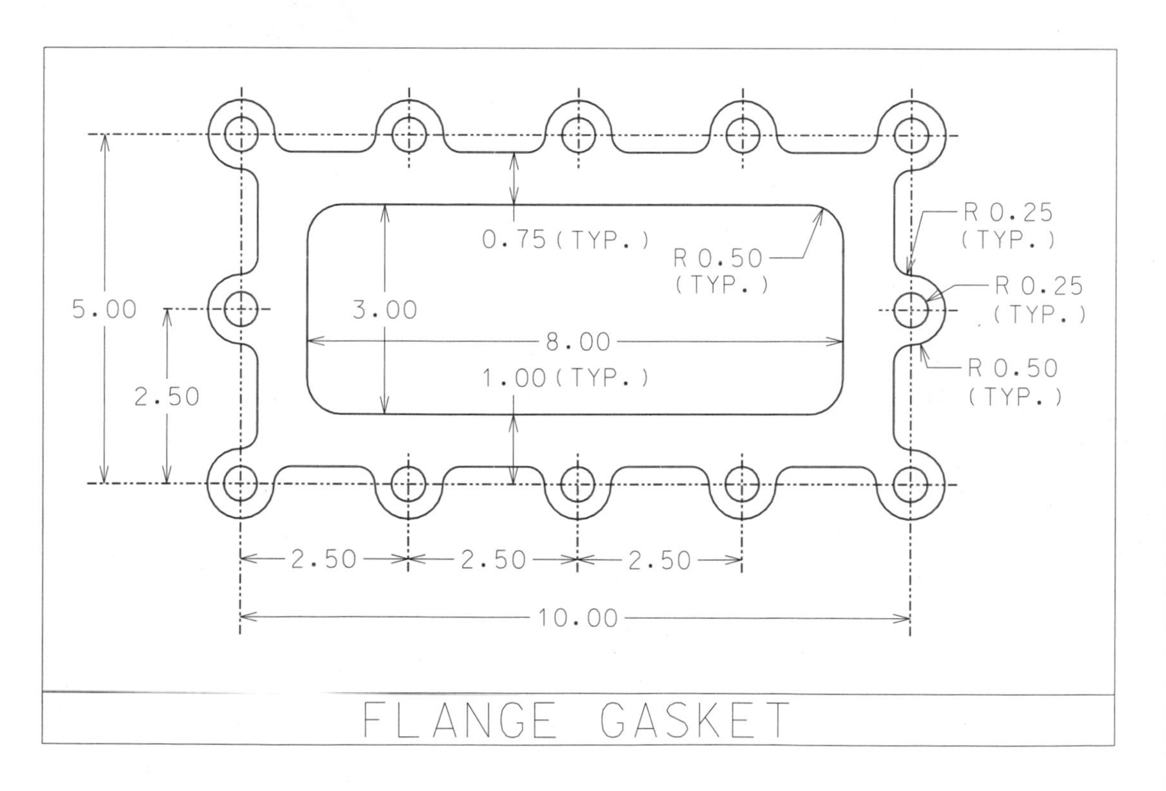

Exercise 6–2

Machine part

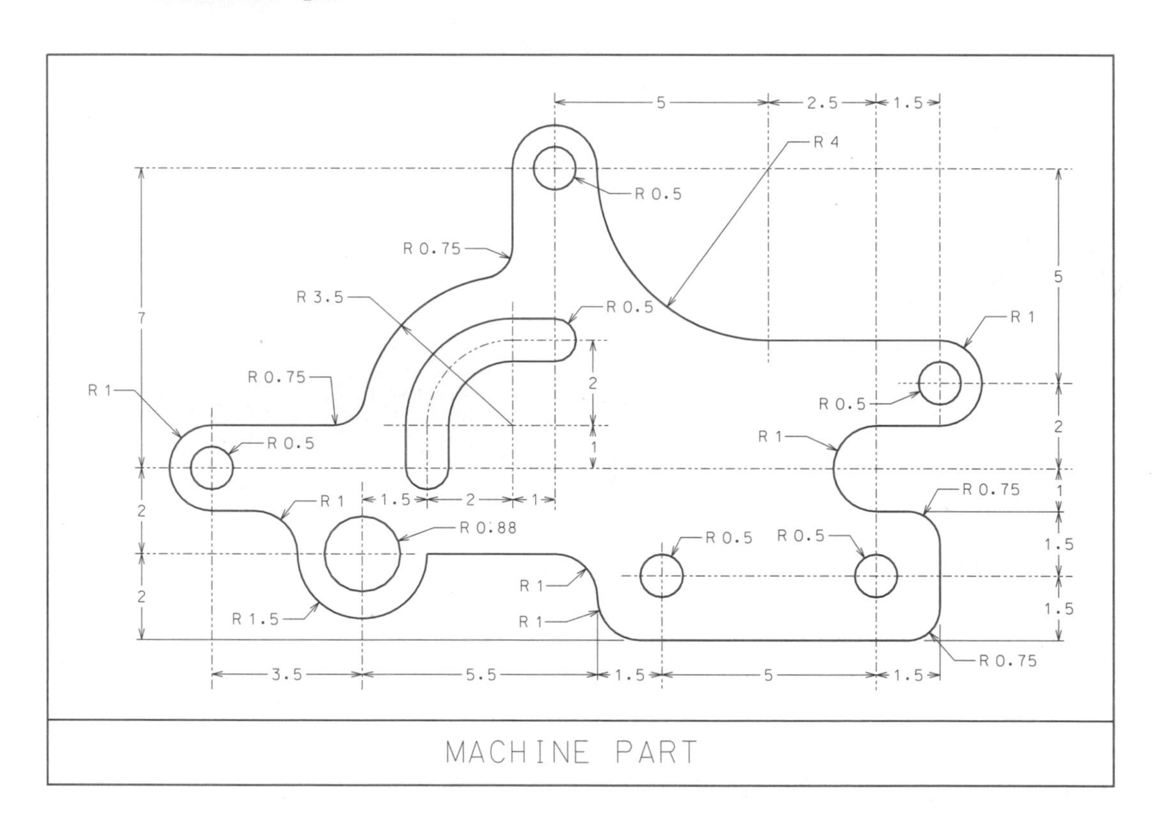

Exercise 6–3

Rotary pressure joint

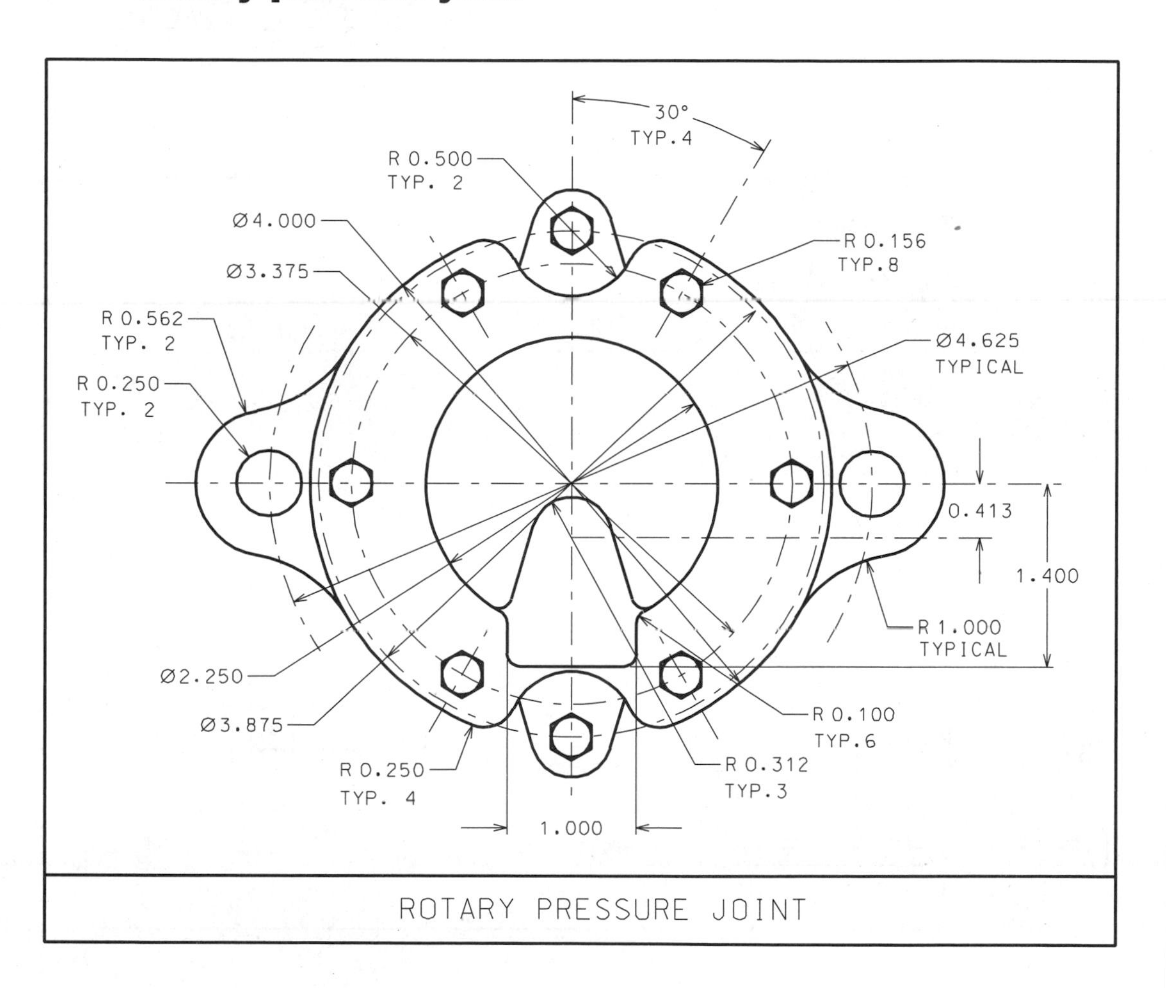

Use the following table to set up the design files for Exercises 6–4 and 6–5.

SETTING	VALUE
Seed File	SEED2D.DGN
Working Units	12"s Per ' and 8000 Pos Units Per "
Grid	Master = 0.5, Reference = 24, GRID Lock ON

Exercise 6–4

Leaded glass design

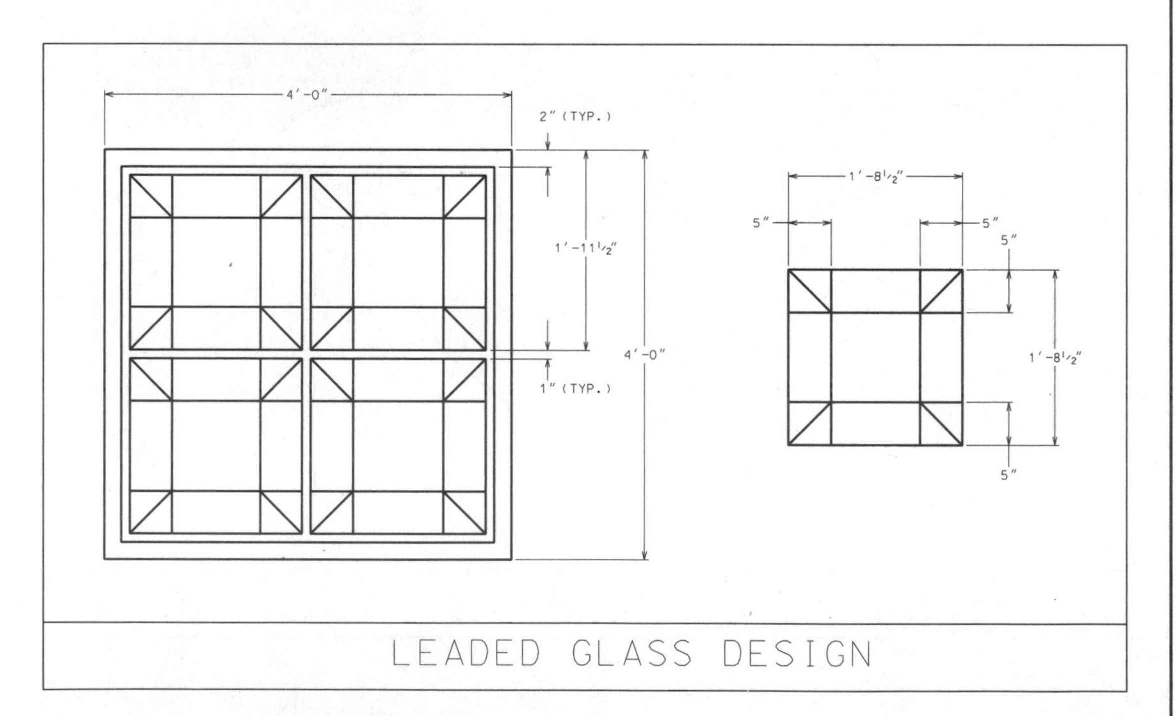

Exercise 6–5

Custom Doors

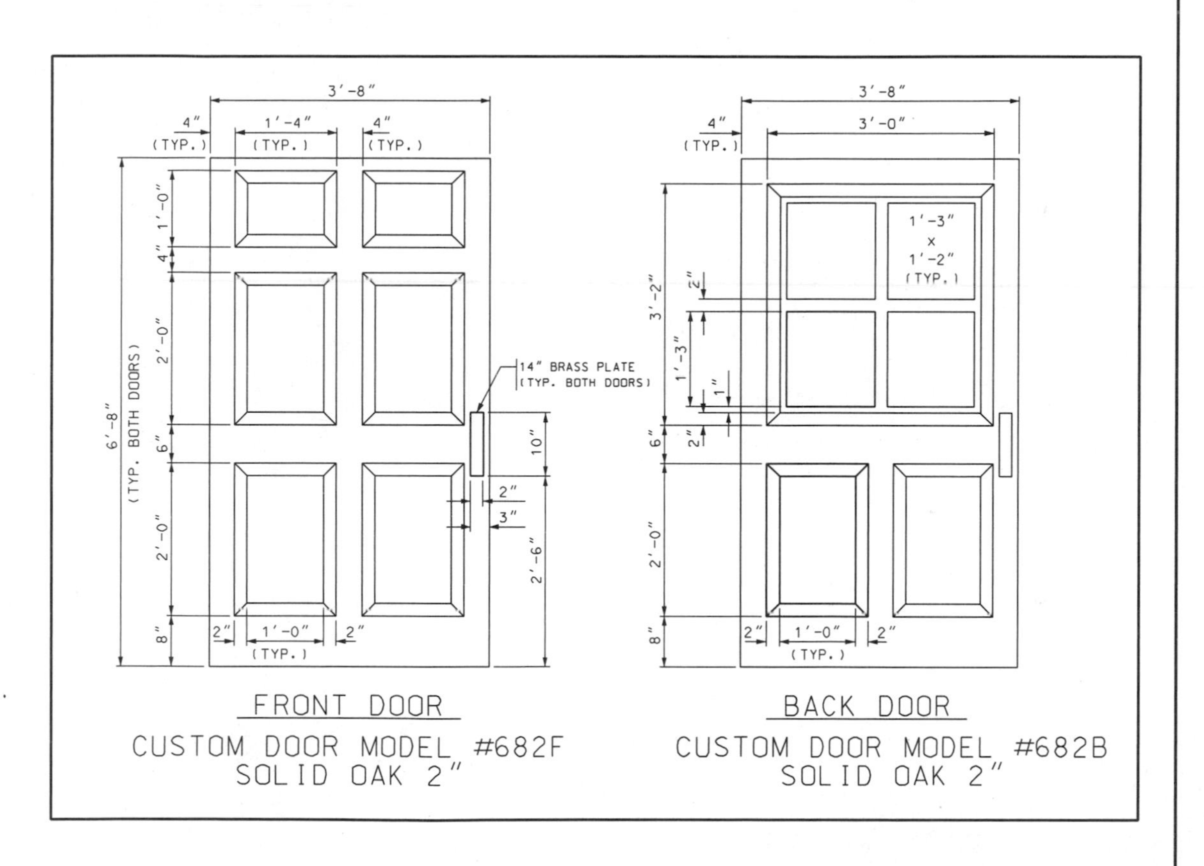

chapter 7

Placing Text, Data Fields and Tags

Objectives

After completing this chapter, you will be able to do the following:

- Place single-character fractions
- Use several tools to place text elements
- Import text from other computer applications
- Edit the content of text elements
- Manipulate the attributes of text elements
- Place notes in the design
- Create and use "fill-in-the-blanks" Text Node and Data Field elements
- Place and manage Tags

PLACE TEXT

Chapter 4 discussed setting Text Attributes and placing Text by Origin. In this section we discuss additional text placement tools available in MicroStation.

Place a Natural Fraction in One-Character Position

Natural fractions are several characters long, which can take up a lot of space in the design. For example, 9/16 is four characters long. To reduce the space required for such fractions, MicroStation adds one-character natural fractions to several fonts and provides a switch in the Text settings window for turning on these one-character fonts.

Figure 7–1 shows an example of natural fractions placed both as separate characters and as one-character fractions.

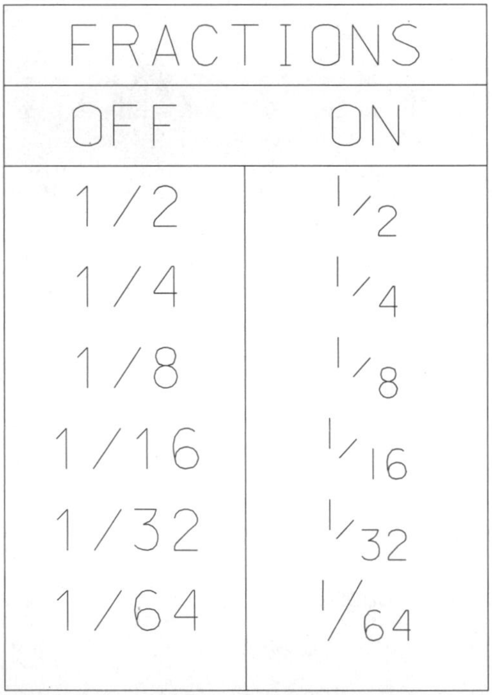

FRACTIONS	
OFF	ON
1/2	½
1/4	¼
1/8	⅛
1/16	1/16
1/32	1/32
1/64	1/64

Figure 7–1 Examples of natural fractions

Invoke one-character natural fractions placement mode:

Pull-down menu	Element > Text

MicroStation displays the Text settings box. Turn the Fractions toggle button to ON, as shown in Figure 7–2.

Natural fractions placed in text strings *after* the button is turned ON are placed as one-character fractions. Existing multicharacter natural fractions are not changed. Select Save Settings from the File pull-down menu to make the setting permanent.

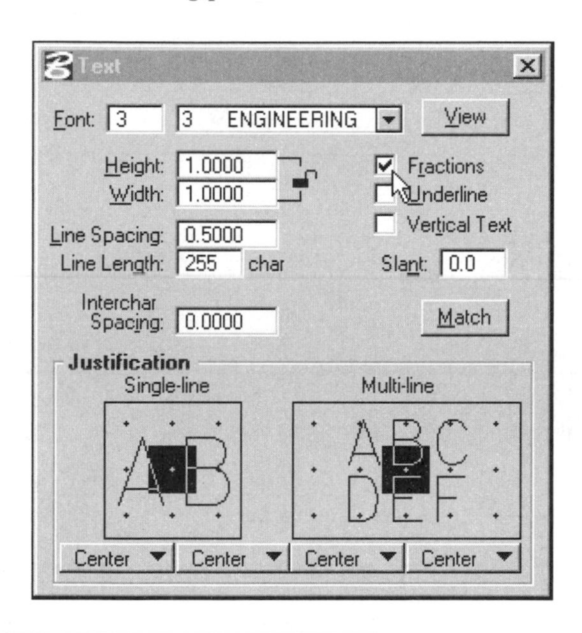

Figure 7–2 Fractions button in the Text settings window

Things to Keep in Mind About Natural Fractions

- If a font does not include natural fractions, or does not contain the particular fractions you enter, those fractions are placed as separate characters even when the Fraction button is ON.
- To find out which fonts support one-character natural fractions, click the View window in the Text Attributes settings box to open the Fonts dialog box. If the word "Fractions" is in the Contents column of a font, the font supports one-character fractions.
- If you include a natural fraction in a string of text, there must be a space character before and after the fraction in order for MicroStation to recognize it as a natural fraction.
- One-character fractions take up slightly more space than single characters, so you may need to insert extra space characters to keep the fraction from running into the characters before and after it.

Place Fitted Text

The Place Fitted Text tool scales and rotates the text you type in the Text Editor between two data points, as shown in Figure 7–3.

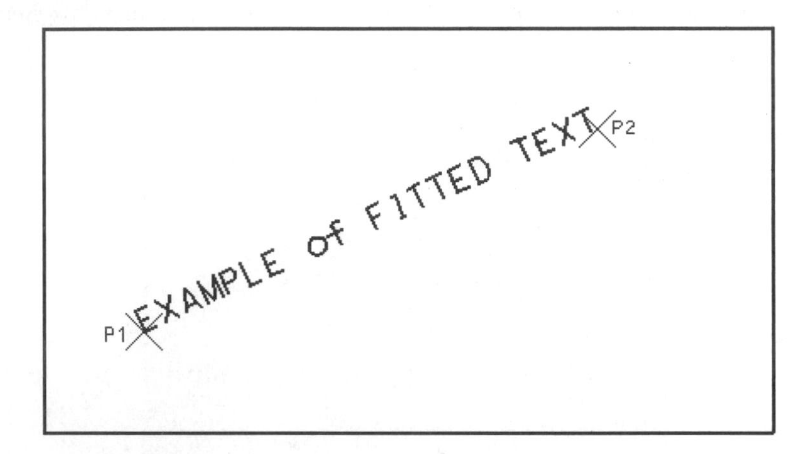

Figure 7–3 Example of placing fitted text

Invoke the Place Fitted Text tool:

Text tool box	Select the Place Text tool, then select Fitted from the Method option menu in the Tool Settings window (see Figure 7–4).
Key-in window	**place text fitted** (or **pl tex f**) Enter

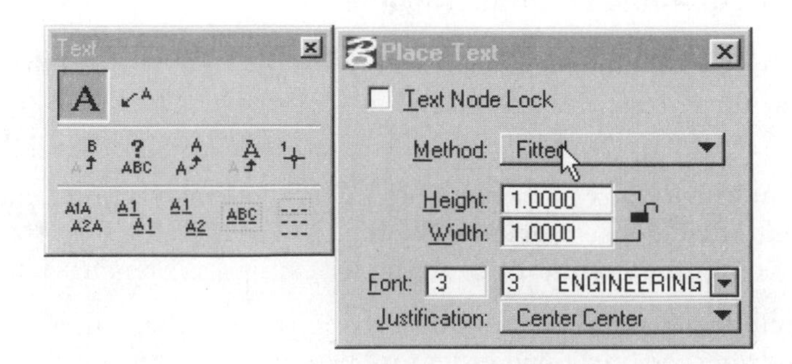

Figure 7–4 Invoking the Place Text (Fitted) tool from the Text tool box

MicroStation prompts:

> Place Fitted Text > Enter text *(Type the text in the Text Editor window, then define the starting point for placing the text.)*
>
> Place Fitted Text > Define endpoint of text *(Define the end point of the text.)*
>
> Place Fitted Text > Enter more chars or position text *(Either place more fitted copies of the text string, enter a new text string, or select another tool.)*

Things to Keep in Mind About Fitted Text

- The only Text settings used by the tool are the active font and text justification.
- Top, Center, and Bottom text justification determine where the text lines up in relation to an imaginary line between the two data points. The text in Figure 7–3 was placed with one of the Bottom justifications.
- If you attempt to insert a line break in the Fitted Text string, MicroStation will ignore it and keep the text in one line.
- Fitted Text strings are normal text elements and can be manipulated like any other element.

Place Text Above, Below, or On an Element

MicroStation provides tools for placing a text string above, below, or on a line or a segment of a linear element, as shown in Figure 7–5.

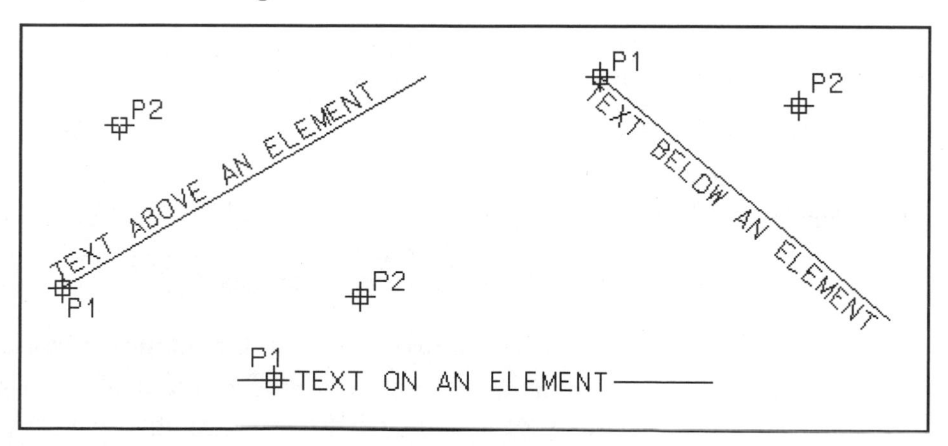

Figure 7–5 Examples of text placed above, below, and on a line

Invoke the tools:

Text tool box	Select the Place Text tool, then select Above Element, Below Element, or On Element from the Method option menu in the Tool Settings window (see Figure 7–6).
Key-in window	**place text above** (or **pla tex a**) Enter **place text below** (or **pla tex b**) Enter **place text on** (or **pla tex o**) Enter

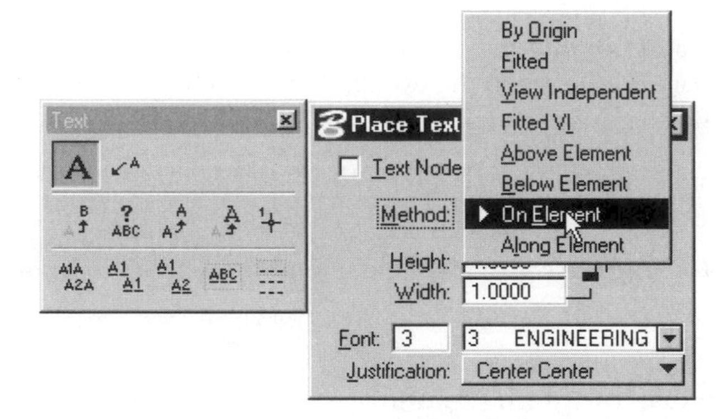

Figure 7–6 Above, Below, and On Element options in the Place Text settings window

MicroStation prompts are identical for all of the three options (except for the tool name). Following are the MicroStation prompts when the Above option is selected from the Method option menu.

Place Text Above Element > Enter text *(Type the text in the Text Editor window.)*

Place Text Above Element > Identify element *(Identify the element where the text is to be placed. A dynamic image of the text appears above the element.)*

Place Text Above Element > Accept/Reject (Select next input) *(Click the Data button to accept the placement, or click the Reset button to reject it.)*

Things to Keep in Mind About Placing Text Above, Below, or On a Linear Element

- Left, Center, and Right text justification determine where the text lines up in relation to the point on the line or segment where the element was identified. The examples in Figure 7–5 were all placed with one of the Left text justification settings.
- If you insert a line break in the text string, MicroStation accepts it but places only the first line of the multi-line text string. For example, if you type **Pump** Enter **3B**, only "Pump" will be placed.

- The space between the element and the text string for the Above and Below tools is equal to the text Line Spacing attribute. Select Text from the Element pull-down menu to change the Line Spacing.
- When text is placed on an element, a hole is cut in the element and the text is placed in the hole. If the element is a line or line string, two separate, unrelated line or line string elements result from the insertion. If the element is closed, it changes to an open element (for example, a circle becomes an arc) after the insertion.
- Once placed, the text string has no relation to the element it was placed above, below, or on. For example, if the text string placed on an element is deleted, the two element pieces do not rejoin.
- Text strings placed in this way are normal text elements that can be manipulated like any other element.
- If the Text Editor window contains text when you invoke one of these tools, MicroStation skips the first prompt and asks you to select the element.
- If you attempt to select an element when the Text Editor window is empty, MicroStation prompts you to "Enter characters first," and the element is not selected.

Place Text Along an Element

The Place Text Along Element tool places text above or below an element. The text follows the contour of an arc, a circle, or a curve, and can bend around the vertex of a linear multi-segment element. See Figure 7–7 for examples of placing text along elements.

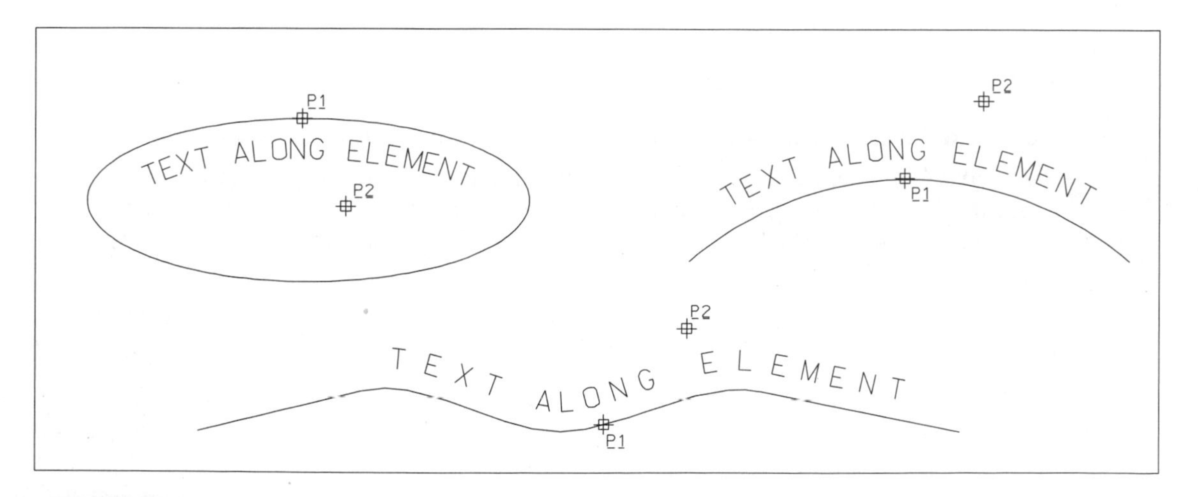

Figure 7–7 Examples of text placed along elements

Text elements are linear, so to make text follow the contour of a curving element, the Place Text Along tool places each character as a separate element. To compensate for tight curves, settings are provided to allow increasing the space between each character (Interchar Spacing) and the space between the text and the element (Line Spacing) in working units (MU:SU:PU).

Invoke the Place Text Along Element tool:

Text tool box	Select the Place Text tool, then select Along Element from the Method option menu in the Tool Settings window (see Figure 7–8).
Key-in window	**place text along** (or **pl tex al**) Enter

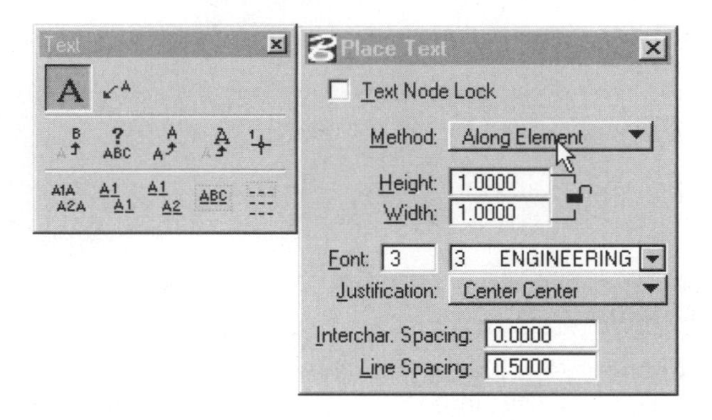

Figure 7–8 Invoking the Place Text (Along Element) tool from the Text tool box

MicroStation prompts:

Place Text Along Element > Enter text *(Type the text in the Text Editor window.)*

Place Text Along Element > Identify element, text location *(Define the point on the element where the text is to be placed. Dynamic images of the text appear both above and below the element, as shown in Figure 7–9, and you must select the one to place.)*

Place Text Along Element > Accept, select text above/below *(Click the Data button on the side of the element where you want the text placed, or click the Reset button to reject it.)*

The text on the side of the element you selected is placed, and the dynamic image on the other side disappears.

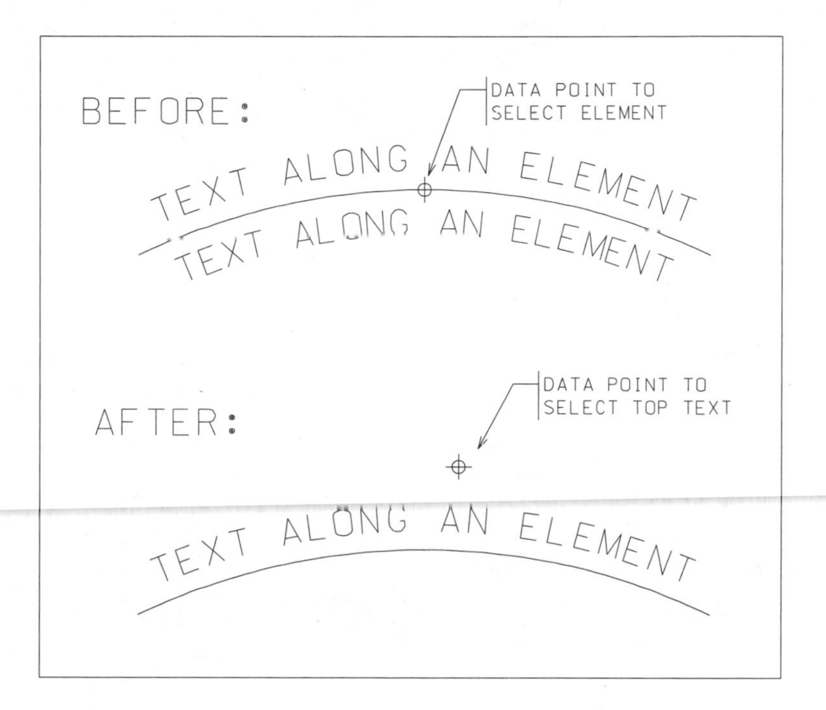

Figure 7–9 Example of placing text along an element

Things to Keep in Mind About Placing Text Along an Element

- Left, Center, and Right text justification determine where the text lines up in relation to the point where the element was identified. The examples in Figure 7–7 were placed with Center justification.
- If you insert a line break in the text string, MicroStation accepts it but only places the first line of the multi-line text string.
- Once placed, the text string has no relation to the element it was placed along. For example, if you move the element the text was placed along, the text does not move.
- Each character placed in this method is a normal text element and can be manipulated like any other element.
- If the Text Editor window contains text when you invoke the tool, MicroStation skips the first prompt and asks you to select the element.
- If you attempt to select an element when the Text Editor window is empty, MicroStation prompts, "Enter characters first," and the element is not selected.

Placing Notes

The Place Note tool allows you to place single and multi-line notes in the design. In addition, for multi-line notes MicroStation provides an option to draw a box around the notes.

Place Single-line Note

The Single-line Note tool places a single-line note at the end of a leader.

Invoke the Place Note tool:

Text tool box	Select the Place Note tool, then select Single-line from the Type option menu in the Tool Settings window (see Figure 7–10).
Key-in window	**place note** (or **pl not**) Enter

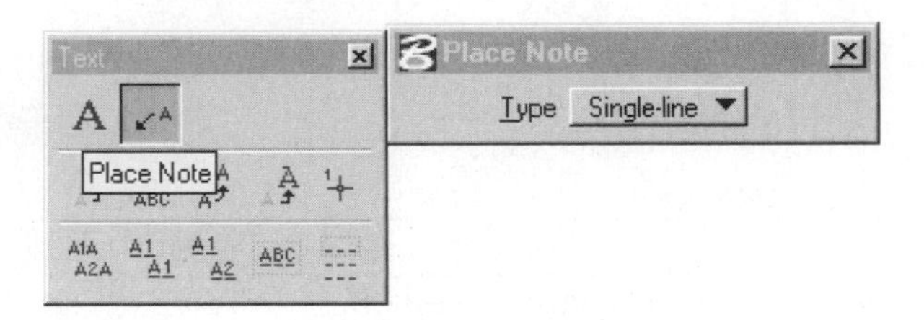

Figure 7–10 Invoking the Place Note tool from the Text tool box

MicroStation prompts:

> Place Single-line Note > Define start point *(Define the starting point where the leader arrowhead is to be placed.)*
>
> Place Single-line Note > Define next point or <Reset> to abort *(Define the point where the single-line note text is to be placed.)*

The text is placed with the active font and active text size settings. MicroStation provides three different methods by which to place text at the end of the leader line: In-Line, Above, and Horizontal. You can select one of the available methods from the Dimension Settings window.

Invoke the Dimension Settings window:

Pull-down menu	Element > Dimensions

MicroStation displays the Dimension Settings box. Select the Text category from the options list on the left side of the dialog box, and MicroStation displays the available options, as shown in Figure 7–11.

In the Orientation menu, select one the following: In Line, Above, or Horizontal. Table 7–1 describes the orientation options, and Figure 7–12 provides examples of each orientation.

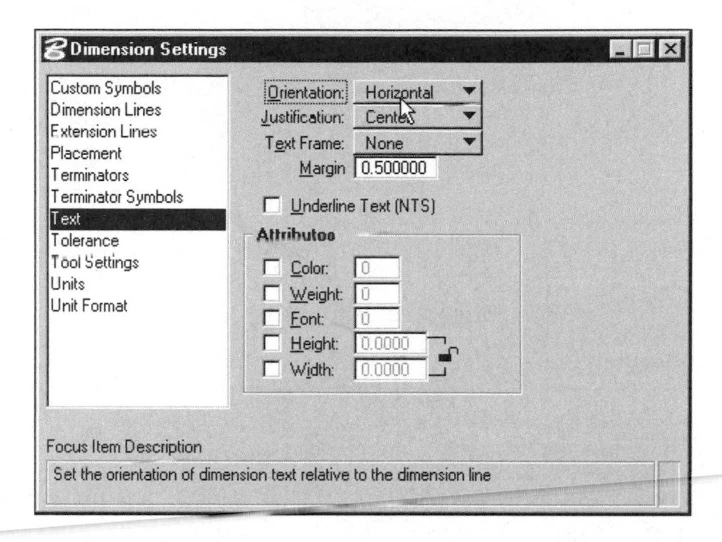

Figure 7–11 Dimension Settings box

Table 7–1 The Dimension Text Orientation Options

SETTING	PLACES THE NOTE TEXT AT . . .
In-Line	The end of the dimension line and at the same rotation as the line.
Above	Above the dimension line and at the same rotation as the line.
Horizontal	The end of a horizontal piece of leader line that is attached to the dimension line.

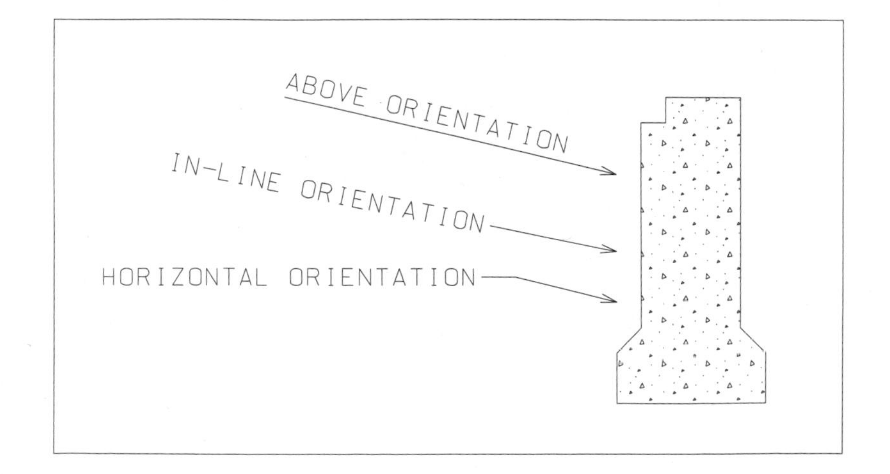

Figure 7–12 Examples of Dimension Text Orientation on Single-line Note placement orientation

Place Multi-line Note

The Multi-line Note tool places a multi-line note at the end of a leader line. Note text is placed with the active Text Size settings.

Invoke the Place Note tool:

Text tool box	Select the Place Note tool, then select Multi-line from the Type option menu in the Tool Settings window (refer to Figure 7–10).
Key-in window	**place note multi** (or **pl not m**) Enter

MicroStation prompts:

> Place Multi-line Note > Define start point *(Define the starting point where the leader arrowhead is to be placed.)*
>
> Place Multi-line Note > Define next point or <Reset> to abort *(Define the point where the multi-line note text is to be placed.)*

MicroStation provides several options in the Tool Settings window for controlling the placement of multi-line notes. Each option is described in Table 7–2, and Figure 7–13 shows examples of Multi-line notes placed with the various options active.

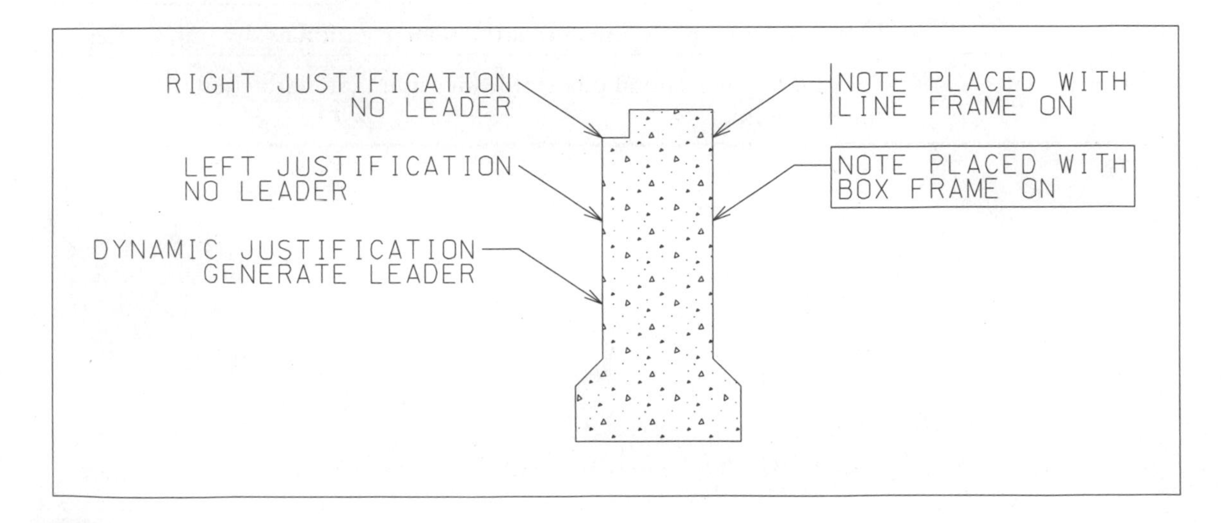

Figure 7–13 Examples of multi-line notes

Table 7–2 Multi-line Note Tool Settings

SETTING	EFFECT
Font	Selects one of the available fonts
Text Frame	Provides options to frame the note text: None—no frame is placed Box—draws a box around the text Line—places a line next to the smooth margin of the text
Justification	Controls the alignment of the note text: Left—left margin of text is smooth Right—right margin of text is smooth Dynamic—which margin is smooth switches sides so the smooth margin is always next to the end of the dimension line
Generate Leader	If ON, a short horizontal leader line is added to the end of the dimension line next to the note text
Associate Lock	If ON, a tentative snap to an element before placing the first placement point causes the note to be associated with the element If an element with an associated note is moved or scaled, the note's dimension line remains attached to the same place on the element after the moving or scaling is completed, but the note text does not move

Import Text

The Import Text tool allows you to import text from a file created by another computer application.

The file to be imported must contain only unformatted (ASCII) text. Thus, the other application's text formatting cannot be imported. For example, the Microsoft Word for Windows word processing application has an export option in its Save As option that creates an unformatted text file with line breaks.

The way the tool imports the text depends on the number of characters and lines in the text file. The break point is 128 lines or 2048 characters. If the number of lines or characters in the file is:

- *Less than the break point,* the text is imported as one multi-line text element (a Text Node). All text settings apply to the element, and a dynamic image of the text follows the cursor until you define the placement point. The relationship of the text to the placement point is determined by the active text justification.

- *Greater than the break point*, the text is placed in a "Graphic Group," with each line of text placed as a separate text element. All text settings apply except Text Justification, and there is no dynamic image of the text before you define the placement point. The placement point is *always* the upper left-hand corner of the top line of text.

Invoke the Import Text tool:

Pull-down menu	File > Import > Text (see Figure 7–14)
Key-in window	**include** (or **in**) Enter

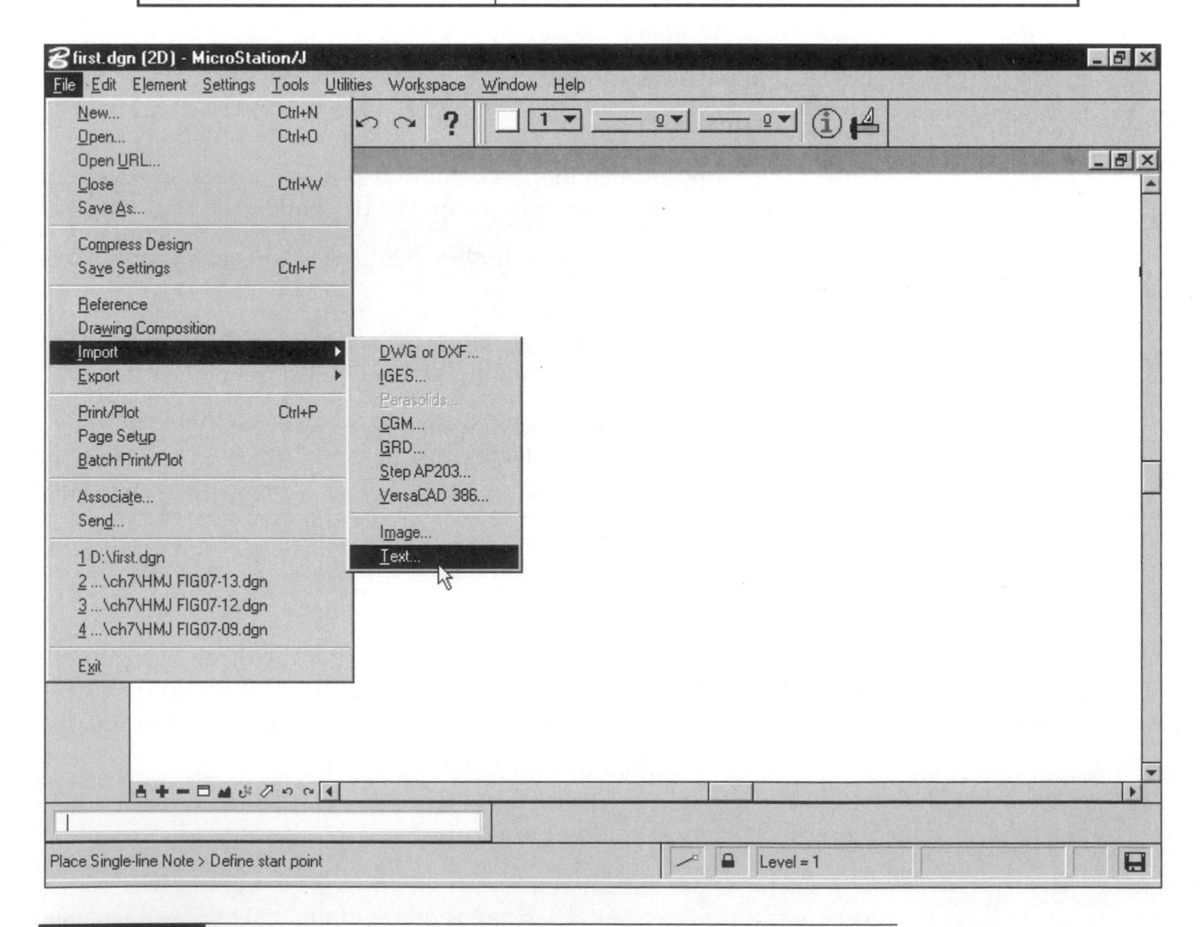

Figure 7–14 Invoking the Import Text tool from the File pull-down menu

MicroStation displays the Include Text File dialog box, as shown in Figure 7–15. Select the appropriate file that contains the text to be imported, then click on the OK button.

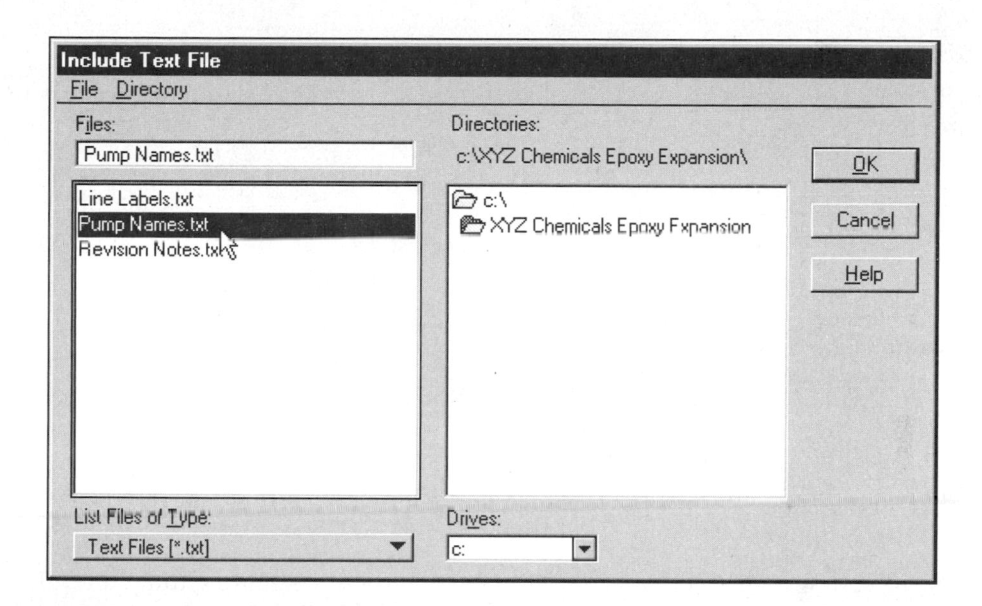

Figure 7–15 Include Text File dialog box

If the text is being placed as a Text Node, MicroStation prompts:

> Import Text File > Enter text node origin *(Define the point in the drawing plane where the text is to be placed.)*

If the text is being placed as a graphic group, MicroStation prompts:

> Import Text File > Identify upper left of text block *(Define the point in the drawing plane where the upper left corner of the text is to be placed.)*

Handling Imported Tabs

By default, each tab character in the imported text is replaced with space characters. MicroStation allows you to change the number of spaces to be used in place of tabs before the text is imported. Set the number of spaces:

Key-in window	**TB=<#>** Enter *(Replace <#> with the number of space characters to use—for example: TB=3.)*

MicroStation responds with a message in the right side of the Status bar:

> Tab interval = <#>

where <#> is replaced with the number you entered.

Text Attribute Settings

The text file can also contain MicroStation element and text attribute setting key-ins to control the way the text appears when placed.

The Rules for Adding Attribute Setting Tools to a Text File are:

- Standard MicroStation attribute key-ins are used.
- Each key-in must be preceded by a period.
- The key-in must be the only thing on the line.
- The settings act on the text that follows them in the file.
- The key-ins are not placed in the text string.
- If an attribute setting is not included in the file, the drawing's current active setting applies.
- The settings in the imported file become the drawing's active settings after the text is imported.

Table 7–3 lists useful setting key-ins, Table 7–4 lists two additional text control settings that are not element settings, and Figure 7–16 shows an example of using key-ins in an imported text file.

Note: An imported file that contains text attribute setting key-ins is always placed as a series of one-line text strings, and the placement point is the upper left corner of the top line of text. Each text element is part of a graphic group.

Table 7–3 Element Attribute Key-ins

KEY-IN	SETS THE ACTIVE . . .
.AA=	Angle degrees
.CO=	Color number
.FT=	Font number
.LS=	Line Spacing in working units (MU:SU:PU)
.LV=	Level number
.TH=	Text Height in working units (MU:SU:PU)
.TW=	Text Width in working units (MU:SU:PU)
.TX=	Text Size (sets height and width equal) in working units (MU:SU:PU)
.WT=	Weight number

Table 7–4 Imported Text Control Settings

SETTING	EFFECT ON THE IMPORTED TEXT
.Indent #	Indents each following line of text with "#" number of spaces
.Newgg	Ends the current graphic group and starts a new one for the following text strings

```
.FT=7
.CO=3
.TH=1:5
.TW=1
This text is placed using font number 7 and color number 3. It is
1:5 working units high and 1 working unit wide.
.FT=0
.CO=0
.WT=2
.TX=:8
This text is placed using font number 0, color number 0, and weight of 2.
It is :8 working units high and wide.
```

Figure 7–16 Example of placing Element Attributes in an imported text file

TEXT MANIPULATION TOOLS

The manipulation tools (such as Move, Copy, and Rotate, among others) manipulate the text element but not the text itself. Changes to the text in the element are handled by a set of text manipulation tools that include editing the text, setting the active text settings to match an existing text element, changing a text element's settings to match the current active text settings, copying and incrementing numbers in text elements, and displaying the text settings used to place a text element.

Edit Text Elements

The Edit Text tool provides a way to change the text in an existing text element.

Invoke the Edit Text tool:

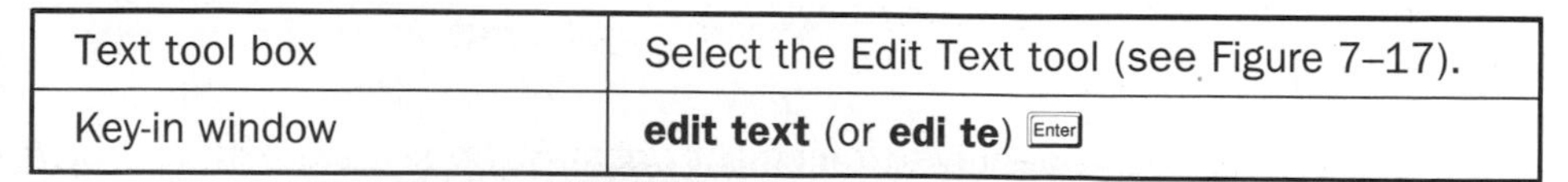

Text tool box	Select the Edit Text tool (see Figure 7–17).
Key-in window	**edit text** (or **edi te**) Enter

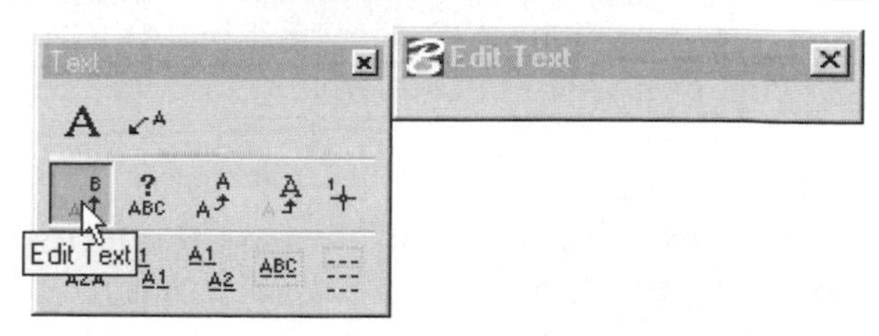

Figure 7–17 Invoking the Edit Text tool from the Text tool box

MicroStation prompts:

> Edit Text > Identify element *(Select the text element to be edited.)*
>
> Edit Text > Accept/Reject (Select next input) *(Click the Data button again to accept the text, or click the Reset button to reject it.)*

MicroStation displays the text string in the Text Editor window. Make the required changes to the text, then click the Text Editor window's Apply button to place the edited text in the design. Refer to the discussion of "Place Text Tools" in Chapter 4 for notes on using the Text Editor window.

Match Text Attributes

The Match Text Attributes tool sets the active text attributes to match those of the text element that is already placed in the design. The tool changes the active font, text size, line spacing, and text justification to the settings of the selected text element, and all text placed afterwards uses the new active settings.

Invoke the Match Text Attributes tool:

Text tool box	Select the Match Text Attributes tool (see Figure 7–18).
Key-in window	**match text** (or **mat t**) Enter

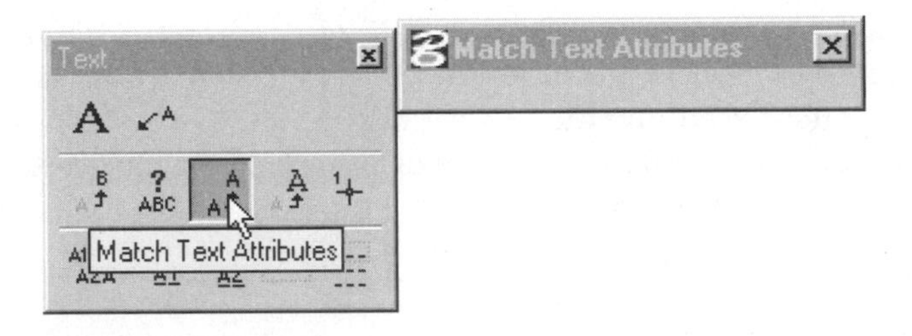

Figure 7–18 Invoking the Match Text Attributes tool from the Text tool box

MicroStation prompts:

> Match Text Attributes > Identify text element *(Identify the text element to match.)*
>
> Match Text Attributes > Accept/Reject (Select next input) *(Click the Data button to set the active settings to selected text element, or click the Reset button to reject it.)*

MicroStation sets the new active text attributes and displays them in the right-hand side of the Status bar. To make these changes permanent, select Save Settings from the File pull-down menu.

Note: The Match Text Attributes tool can also be invoked by clicking the Match button in the Text settings dialog box.

Change Text Attributes

The Change Text Attributes tool changes the attributes of an existing text element from the settings used to place it to the current active settings. For example, if the selected text element's text height is 1:5 Working Units and the active text height is 3:0 Working Units, then with the Change Text Attributes tool you can change the text element's text height to 3:0 Working Units.

Invoke the Change Text Attributes tool:

Text tool box	Select the Change Text Attributes tool (see Figure 7–19), then, in the Tool Settings window, turn ON the toggle button for each attribute that needs to be changed, and select or type the required attributes in the option menus and edit fields of the attributes you turn on.
Key-in window	**modify text** (or **modi te**) Enter

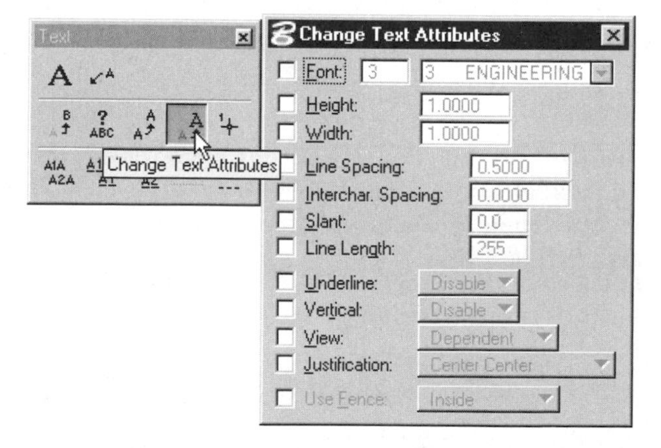

Figure 7–19 Invoking the Change Text Attributes tool from the Text tool box

MicroStation prompts:

Change Text Attributes > Identify text *(Select text element.)*

Change Text Attributes > Accept/Reject (Select next input) *(Click the Data button to change the attributes of the selected text element, or click the Reset button to reject it. This data point can also select another text element at the same time it accepts the current element.)*

Display Text Attributes

Display Text Attributes is an information-only tool that displays the attributes that were used to place an existing text element.

Invoke the Display Text Attributes tool:

Text tool box	Select the Display Text Attributes tool (see Figure 7–20).
Key-in window	**identify text** (or **i t**) Enter

Figure 7–20 Invoking the Display Text Attributes tool from the Text tool box

MicroStation prompts:

Display Text Attributes > Identify text *(Identify the text element.)*

MicroStation displays the attributes in the left side of the Status bar. If desired, you can select another text element. The text attributes displayed in the Status bar are different for one-line text elements and multi-line text elements (Text Nodes):

- *For one-line text elements,* the displayed attributes include the text height and width, the level the element is on, and the font number.
- *For multi-line text elements,* the displayed attributes include the Text Node number, the maximum characters per line, the line spacing, the level the element is on, and the font number.

Copy and Increment Text

Annotating a series of objects with an incremented identification (such as P100, P101, P102) would be a tedious job without the Copy/Increment Text tool. This tool copies and increments numbers in text strings. To make incremented copies, just select the element to be copied and incremented, then place data points at each location where an incremented copy is to be placed.

A Tag Increment setting in the Tool Settings window allows you to set a positive or negative increment value. For example, an increment value of 10 causes each copy to be 10 greater than the previous one. A value of –10 causes each new copy to be 10 less than the previous one.

Only the numeric portion of a text string is incremented, and, if the string contains more than one numeric portion separated by nonnumeric characters, only the right-most numeric portion is incremented. For example, only the 30 in the string P100-30 will be incremented (P100-31, P100-32, P100-33, and so on).

To place a series of incremented text strings, first place the starting text string, then invoke Copy/Increment Text:

Text tool box	Select the Copy/Increment Text tool, then, optionally, set the Tag Increment value in the Tool Settings window (see Figure 7–21).
Key-in window	**increment text** (or **incr t**) Enter

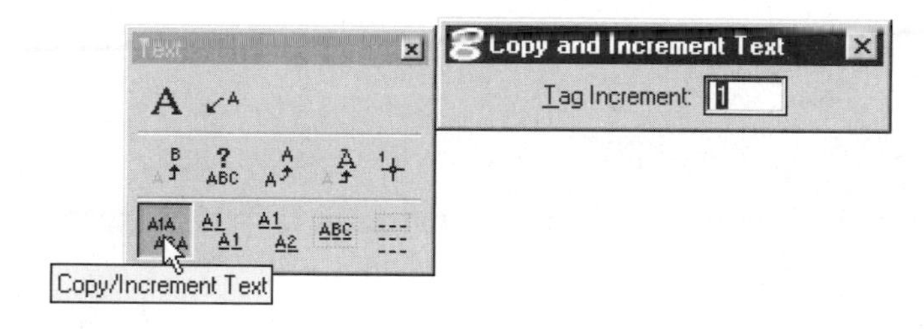

Figure 7–21 Invoking the Copy/Increment Text tool from the Text tool box

MicroStation prompts:

Copy and Increment Text > Identify element *(Identify the text string to be copied and incremented.)*

Copy and Increment Text > Accept/Reject (Select next input) *(Define the location of each incremented copy, or press the Reset button to reject the copy.)*

Note: The Copy/Increment tool only accepts single-line text strings that contain numbers. If you attempt to select a string that does not contain numbers, or a multi-line string, the message "Element not found" appears in the Status Bar, and the element is not accepted.

TEXT NODES

The Text Node tool provides a way to reserve space in a design where text is to be placed later. Once a Text Node is placed, it saves the active element and text attribute settings. When text is added to the node at a later time, the text takes on those settings.

Nodes are most often used in "fill-in-the-blank" forms that can be inserted in a design and filled in with information specific to the design. A common example is a title block form that has all the required fields held with Text Nodes. Use of the form provides a standard title block layout for all designs.

View Text Nodes

The visual indication of a text node is a unique identification number and a cross indicating the node origin point. Figure 7–22 shows examples of the way empty and filled-in Text Nodes appear in a design.

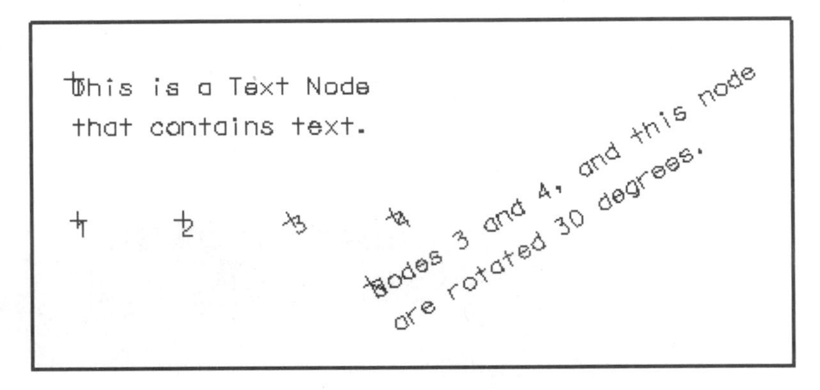

Figure 7–22 Examples of Text Node indicators

The Text Node view attribute controls the display and plotting of Text Node indicators for selected views. You can change the display of Text Node view attribute from the View Attributes settings box (invoked from the Settings pull-down menu). The Text Node toggle button is at the bottom of the right column in the View Attributes window, as shown in Figure 7–23.

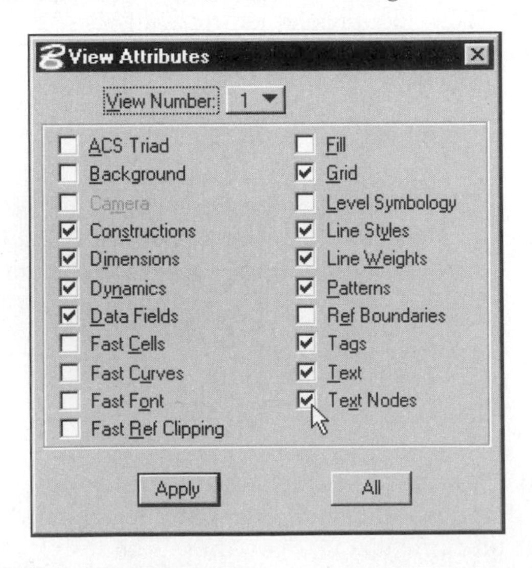

Figure 7–23 The View Attributes settings box

Note: When a multi-line text string is placed in the design, a Text Node is assigned to the string. If the Text Node view attribute is ON, the node cross and number will appear with the text element at the justification point.

Text Node Attributes

The Text Node attributes, such as text size, font, and spacing, are set in the Text settings box and are similar to those for the Place Text tool, except for text justification. The text justification is set in the Multi-line Text Justification area of the Text settings box.

Place Text Nodes

To place text nodes, invoke the Place Text Node tool:

Text tool box	Select the Place Text Node tool, then, optionally, set the Active Angle (to control the angle at which the node is placed) in the Tool Settings window (see Figure 7–24).
Key-in window	**place node** (or **pla n**) Enter

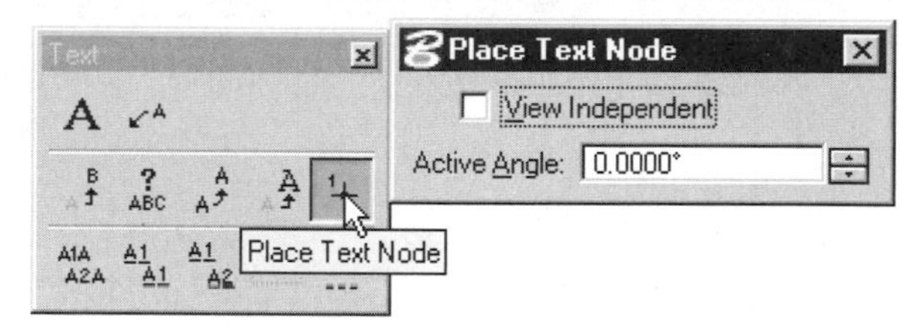

Figure 7–24 Invoking the Place Text Node tool from the Text tool box

MicroStation prompts:

Place Text Node > Enter text node origin *(Define the origin point for each node that is to be placed.)*

Notes: If the Text Node view attribute is turned OFF, nothing appears to happen when the Place Text Node tool is invoked to place empty nodes. Turn ON the Text Node view attribute to see the results of placing empty nodes.

The Place Text Node settings include a View Independent option that applies only to three-dimensional designs. If the toggle button is set to ON, MicroStation prompts "Place View Independent Text Node," but there is no difference in the way nodes are placed in a two-dimensional design.

Fill In Text Nodes

Text can only be placed on a text node when the Text Node Lock is ON. You can turn ON the lock from the Tool Settings window when you invoke the Place Text By Origin tool.

Invoke the Place Text tool:

Text tool box	Select the Place Text tool, then select By Origin from the Method option menu, and turn ON the Text Node Lock toggle button in the Tool Settings window (see Figure 7–25).
Key-in window	**place text** (or **pl te**) Enter

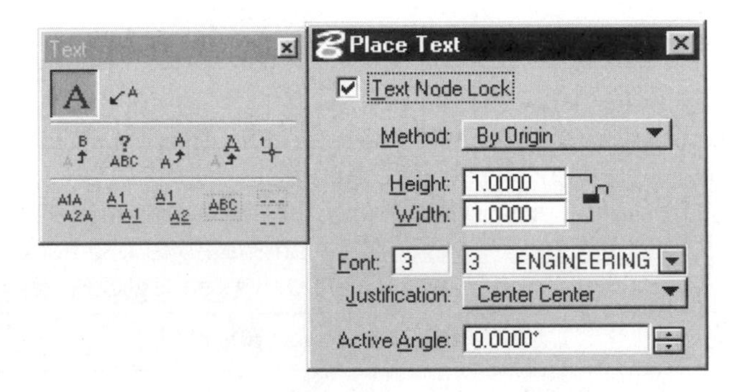

Figure 7–25 Invoking the Place Text By Origin tool with the Text Node Lock set to ON

MicroStation prompts:

Place Text > Enter text *(Enter the text in the Text Editor window, then select the Text Node the text is to be placed on.)*

Place Text *(Click a second data button to accept the previous node. This click can also select another node to place the same text on. You can also enter more text in the Text Editor window before clicking again, or click the Reset button to clear the Text Editor window.)*

The text is placed using the element and text attributes in effect when the node was created.

Note: If the selection or acceptance points are placed in an empty space or on a node that already contains text, MicroStation displays the message "Text node not found" in the Status bar. When the Text Node Lock is set to ON, you can only place text on empty Text Nodes.

DATA FIELDS

Data Fields or Enter Data Fields are similar to Text Nodes in that they create placeholders for text that will be filled in later. Data Fields, though, are more powerful than Text Nodes because there are tools available to automate filling in the fields.

A common use for Data Fields is to provide placeholders for descriptive text in cells (symbols). For example, a control valve cell might contain Data Fields for the valve type, size, and identification code. Cells are discussed in Chapter 11.

Data Field Character

Data Fields are created by typing a contiguous string of underscores in ordinary single-line and multi-line text strings. Any of the text placement tools can create the fields (At Origin, Fitted, Above, Below, On, and Along). For example, "_____" is a five-character Enter Data field.

When you fill in text, you can have one text character per underscore, so when creating the field you must anticipate the number of text characters that are to be placed in the field. The Edit Text tool can be used to add or remove underscores in Data Fields that have already been placed in the design.

A text string can contain more than one data field. For example, the string, "Pump ____ - __" contains two data fields. If you insert a line break in a string of underscores, there will be a separate data field on each line.

Note: The underscore is the reserved character used to create fields for each Data Field character. You can change the reserved character from the MicroStation Preferences settings box (Preferences are discussed in Chapter 16).

Data Field View Attributes

The View Attributes settings box includes a Data Field toggle button for turning ON and OFF the display of the Data Field underscores for a selected view, as shown in Figure 7–26.

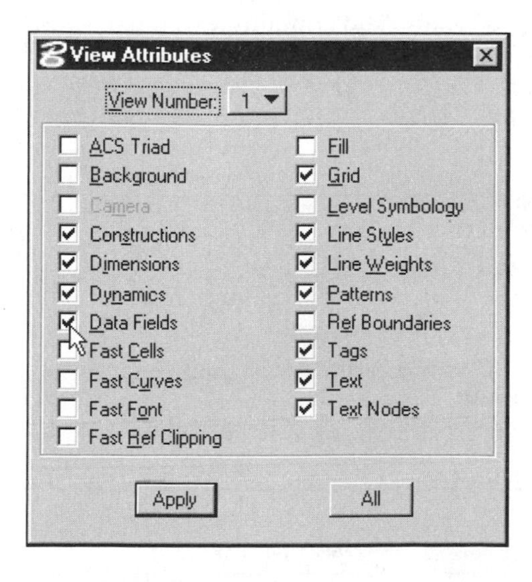

Figure 7–26 View Attributes settings box

When the Data Field View Attributes toggle button is:

- OFF—the underscores disappear from the selected view. If a field has been filled in, the fill-in text is still visible.
- ON—the underscores are visible in the selected view, and they plot. If the data fields contain characters, the underscores appear at the bottom of each fill-in character.

Set Justification for Data Field Contents

The Data Field contents can be justified Left, Right, or Center within the field when there are fewer characters in the field than underscores. Data Field content justification is different than text string justification, and is applied to the Data Field only *after* the text string is placed.

Invoke the Data Field justification tool:

Key-in window	**justify left** (or **ju l**) Enter **justify center** (or **ju c**) Enter **justify right** (or **ju r**) Enter

The MicroStation prompt includes the selected justification. The MicroStation prompt shown here appears when Center justification is selected:

Center Justify Enter_Data Field > Identify element *(Click on the Data Field to be justified.)*

No acceptance is required for this tool. The field is justified as soon as it is selected.

Note: If the justified Data Field contains text, the position of that text does not change. If you replace the existing text after changing the justification, the new text takes on the new justification.

Fill In Data Fields

MicroStation provides two tools to fill in text in Data Fields. The Fill In Single Enter-Data Field tool allows you to place text by identifying a specific Data Field. The Auto Fill In Enter_Data Fields tool prompts you to select a specific view, then MicroStation finds, and lets you fill in, each empty Data Field in the view in the order they were created.

Fill In Single Enter_Data Field Tool

Invoke the Fill In Single Enter_Data Field tool:

Text tool box	Select the Fill In Single Enter_Data Field tool (see Figure 7–27).
Key-in window	**edit single** (or **edi s**) Enter

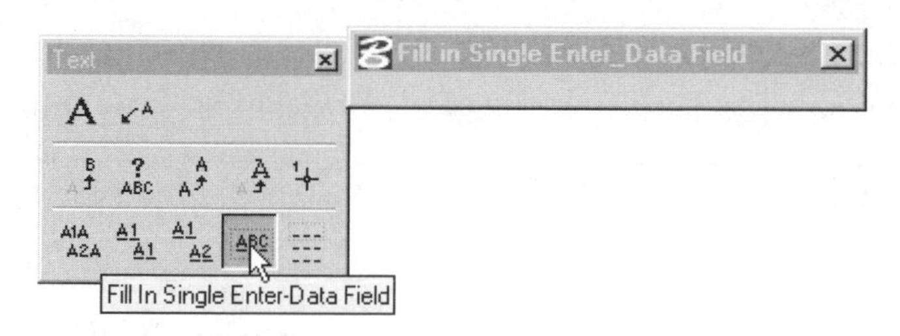

Figure 7–27 Invoking the Fill In Single Enter_Data Field tool from the Text tool box

MicroStation prompts:

> Fill in Single Enter_Data Field > Identify element *(Identify the Data Field.)*

The Text Editor window opens. Type the text in the Text Editor window and press Enter. MicroStation places the text in the selected Data Field. You can continue by identifying additional Data Fields.

Auto Fill In Enter_Data Fields Tool

Invoke the Auto Fill In Enter_Data Fields tool:

Text tool box	Select the Auto Fill In Enter_Data Fields tool (see Figure 7–28).
Key-in window	**edit auto** (or **edi au**) Enter

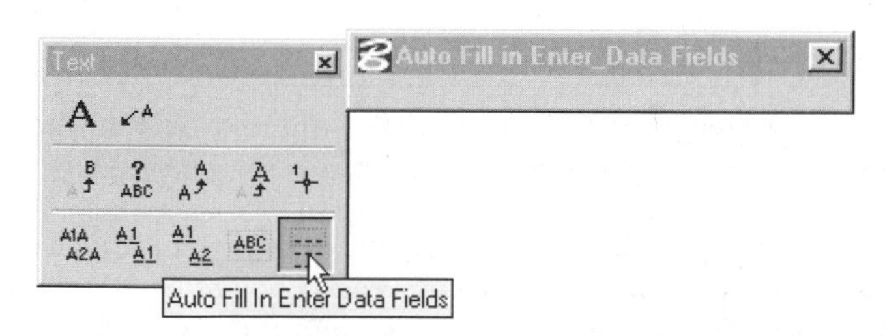

Figure 7–28 Invoking the Auto Fill In Enter_Data Fields tool from the Text tool box

MicroStation prompts:

> Auto Fill in Enter_Data Fields > Select view *(Click the Data button anywhere in the view containing the fields to be filled in.)*
>
> Auto Fill in Enter_Data Fields > <CR> to fill in or DATA for next field *(Type the text in the Text Editor window and press Enter, or click the Data button to skip the field.)*

The tool continues through the view selecting empty Data Fields in the order they were created. It skips Data Fields that already contain text.

Copy Data Fields

To copy the contents of one Data Field to another Data Field, invoke the Copy Enter–Data Field tool:

Text tool box	Select the Copy Enter_Data Field tool (see Figure 7–29).
Key-in window	**copy ed** (or **cop e**) Enter

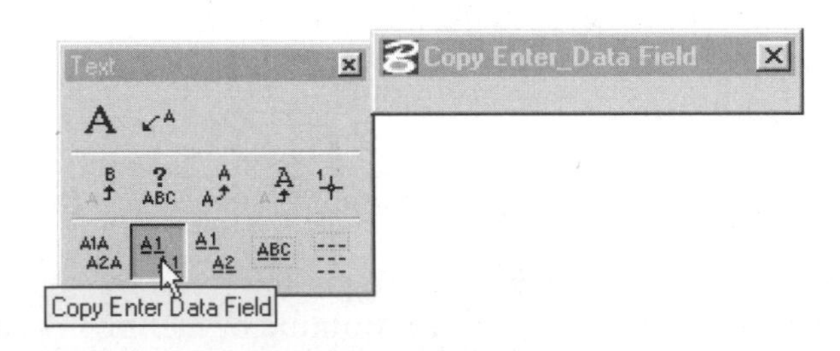

Figure 7–29 Invoking the Copy Enter_Data Field tool from the Text tool box

MicroStation prompts:

Copy Enter_Data Field > Select enter data field to copy *(Select the Data Field containing the text to be copied, then click in each Data Field to which the text is to be copied.)*

Copy and Increment Data Fields

The Copy and Increment Enter_Data Field tool copies the text from a filled-in Data Field, then increments the numeric portion of the text and places it in the empty Data Field you select.

A Tag Increment setting in the Tool Settings window allows setting a positive or negative increment value. For example, an increment value of 10 causes each copy to be 10 greater than the previous one; a value of –10 causes each new copy to be 10 less than the previous one.

To copy the contents of one Data Field to another Data Field and increment the numeric portion of the copy, invoke the Copy/Increment Enter-Data Field tool:

Text tool box	Select the Copy/Increment Enter_Data Field tool (see Figure 7–30), and, optionally, set the Tag Increment value in the Tool Settings window.
Key-in window	**increment ed** (or **incr e**) Enter

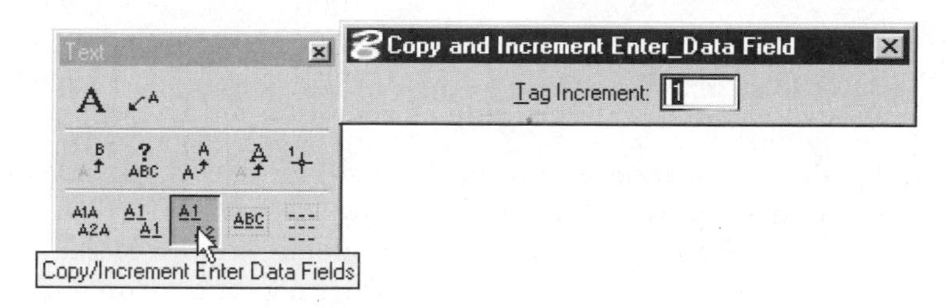

Figure 7–30 Invoking the Copy and Increment Enter_Data Field tool from the Text tool box

MicroStation prompts:

> Copy and Increment Enter_Data Field > Select enter data field to copy *(Select the Data Field containing the text to be copied, then click in each Data Field to which the text is to be copied and incremented.)*

Edit Text in a Data Field

The number of underscore characters in an existing Data Field can be changed by editing the text string in the Text Editor window.

In the Text Editor window the underscores are represented by spaces enclosed in pairs of angle brackets. For example, the string "PUMP __-__" will appear as "Pump << >>-<< >>" in the Text Editor window.

- To *shorten a Data Field*, remove spaces from between the angle brackets.
- To *lengthen a Data Field*, insert spaces (or underscores) between the angle brackets.
- To *delete a Data Field* completely, delete the angle brackets and the spaces between them.
- To *insert a new Data Field:*

1. Position the cursor at the insertion point in the text string.
2. Type a pair of left angle brackets (<<).
3. Type the spaces (or underscores) to define the length of the Data Field.
4. Type a pair of right angle brackets (>>).

TAGS

Engineering drawings have long served to convey more than just how a model looks. Drawings must tell builders and fabricators how actually to construct the design. This nongraphical information includes such things as the construction material, how many to make, colors, where to obtain materials, and what finishes to apply to surfaces. When models were created on paper, painstaking work was required to extract lists of this information from the drawings. A major innovation of CAD models is the ability to automate the creation of such lists.

MicroStation provides this automation by attaching "Tags" to objects. Any element, or element group, can be tagged with descriptive information, and tag reports can be requested. For example, each electrical fixture in an architectural floor plan can be tagged with its rating, order number, price, and project name. An estimator could extract a fixture tag report from the design and insert the resulting data in a spreadsheet to obtain the total project cost for electrical fixtures. A purchasing agent could use the tag reports from several projects to order fixtures and take advantage of quantity discounts. Receiving clerks could employ the order numbers and project names to route the received fixtures to the correct projects.

MicroStation's Tag tools are helpful when the tagging requirements are fairly simple and when the project must import or export drawings from other CAD packages that store nongraphical data inside their design files. For complex tagging requirements, MicroStation supports connections to databases (which is beyond the scope of this textbook).

Tags can be placed on any element in a design file. Figure 7–31 shows an example of tags in design files. The tags are assigned to a small point (actually a short line) in each tract of land in a plot plan. The points are given just to provide an element to hook the tag to. Each "Tract" tag set includes the tract identification number, the purchase status, and the tract size.

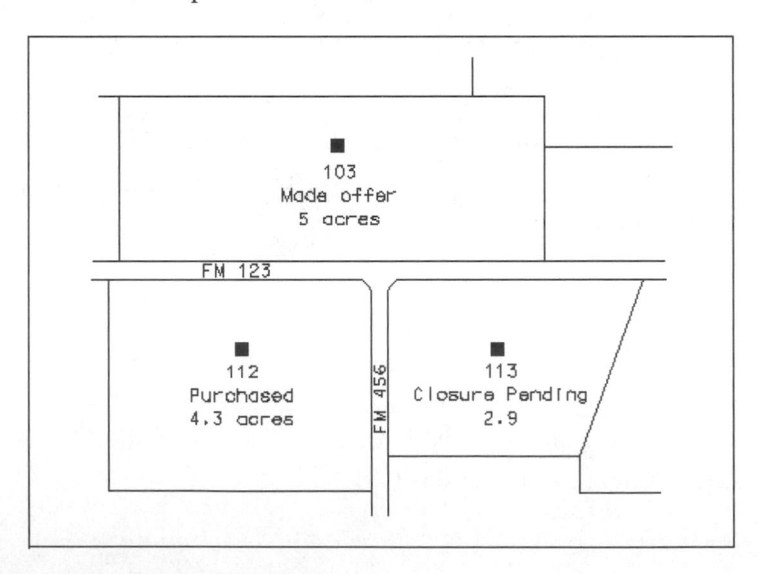

Figure 7–31 Tags displayed in a plot plan

Tag Terms

Adding tags to a design requires an understanding of tag terminology. Table 7–5 defines the important tagging terms.

Table 7–5 Tagging Terms

TERM	DEFINITION	EXAMPLES
Tag set	A set of associated tags. For each tag, it provides the tag name, display attributes, data type, and default value.	Separate tag sets for doors, windows, and electrical fixtures.
Tag	Nongraphical attributes that may be attached to graphical elements.	Part number, size, material of construction, vendor, price.
Tag report template	A file that specifies the tag set and the set's member tags to include on each line of the report. One tag set per template.	For the fixtures set, report the part number, rating, price, and project name of each tagged element.
Tag report	A list of all tags based on a tag report template.	F300-2, 220V, $300, New ABC, Inc. building.
Tag set library	Files containing tag set definitions for use in multiple design files.	A library of architectural tag sets.

Note: If you delete or move an element with attached tags, the tags are deleted or moved as well.

Create a Tag Set and Tags

The first step in creating nongraphical tags in a design is to create the tag set and define the tags in the set.

To create a tag set and to define the tags in a set, open the Tag Sets settings box:

Pull-down menu	Element > Tags > Define
Key-in window	**mdl load tags define** (or **md l tags define**) Enter

MicroStation displays the Tag Sets settings box as shown in Figure 7–32. All defined tag sets are displayed on the left side of the window, and the tag names for the selected tag set are displayed on the right side of the window. In the pictured example, the design has one tag set (doors) with four tags (desc, id, price, and size). Under the tag set and tag names areas of the settings box are buttons for creating and maintaining the sets and tags.

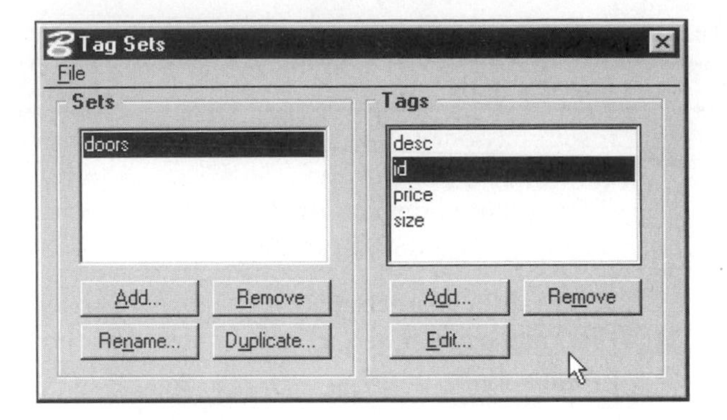

Figure 7–32 Tag Sets settings box

To create a new tag set name, click the Add... button in the Tag Sets area of the settings box; MicroStation displays the Tag Set Name dialog box, as shown in Figure 7–33. Key-in the Tag Set name in the Name field of the Tag Set Name dialog box, and click on the OK button to create the new Tag Set.

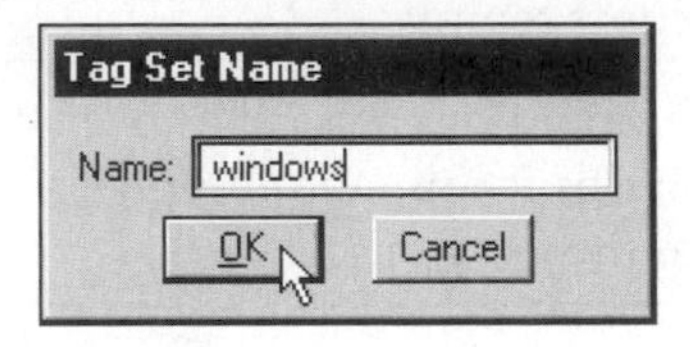

Figure 7–33 Tag Set Name dialog box

To create a new tag under a specific tag set, first highlight the Tag Set from the available Tag Sets list, and click the Add... button from the right side of the settings box under the Tags names. MicroStation displays the Define Tag dialog box, as shown in Figure 7–34.

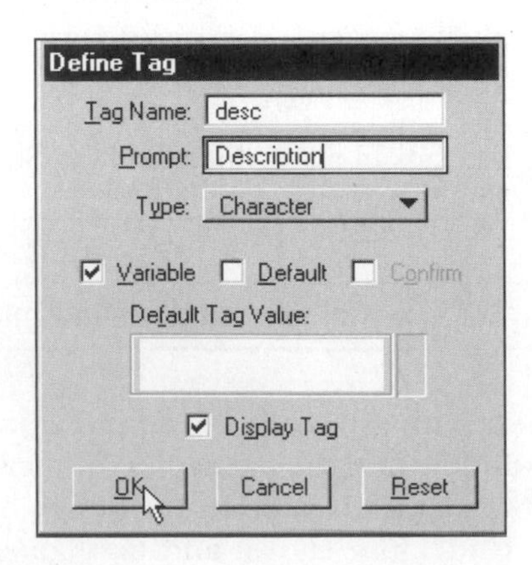

Figure 7–34 Define Tag dialog box

Key-in the appropriate information in the fields provided in the Define Tag dialog box. Refer to Table 7–6 for a detailed explanation of the available fields in the Define Tag dialog box.

Table 7–6 Tag Attributes

ATTRIBUTE	DESCRIPTION
Tag Name	The name of the tag.
Tag Prompt	A 32-character-maximum text string that will serve to tell the user what the tag is for when it is assigned to an element.
Tag Data Type	Tags are one of three types (determines what type of information is placed in the tag): Character—a text string Integer—a whole number Real—a number with a fractional part
Variable	A toggle switch that, when OFF, prevents the tag value from being changed with the Edit Tags tool (discussed later). If ON, the value can be edited.
Default	A toggle switch that, when OFF, uses the default tag value and prevents the tag value from being changed with the Edit Tags tool. If ON, the toggle switch uses the default but allows editing of the value.
Tag Default Value	A default tag value that is initially assigned to a tag when the tag set is assigned to an element. It can be overridden.
Display Tag	Controls how the tags are displayed in the views and what can be done to them.

Maintain Tag Set Definitions

The Tag Sets settings box provides options for maintaining existing tag set definitions. The options include:

- *Remove*—remove (delete) the selected Tag Set (with all its tags) or Tag. A confirmation window opens, and you initiate the removal by clicking on the OK button.
- *Rename*—change a tag set's name. It opens the Tag Set Name window, in which you type the new Tag Set name.
- *Duplicate*—create a duplicate copy of a Tag set. It opens the Tag Set Name window, in which you type the name to use for the duplicate Tag Set.
- *Edit*—edit the attributes of the selected Tag. It opens the Define Tag window for you to edit the tag attributes.

Attach Tags to Elements

To assign a tag to an element, invoke the Attach Tags tool:

Tags tool box	Select the Attach Tags tool, then select a Tag Sets name from the Tool Settings window (see Figure 7–35).

Figure 7–35 Invoking the Attach Tags tool from the Tags tool box

MicroStation prompts:

Attach Tags > Identify element *(Identify the element to which tags are to be attached.)*

Attach Tags > Accept/Reject (Select next input) *(Click the Accept button to accept the selected element, and, if required, select the next element to which tags will be attached after the current one is completed.)*

When the element is accepted, the Attach Tags window opens. Figure 7–36 provides an example of the window displaying the tag names for a "doors" tag set; Table 7–7 describes each field.

Table 7–7 The Fields in the Attach Tags Window

FIELD	DESCRIPTION
Name	The name of each tag in the set.
Value	The default value, if any, of each tag in the set.
Display	A toggle switch that can be used to turn ON or OFF the display of each tag.
Prompt	The input prompt for the selected tag (in Figure 7–36, the prompt for the "desc" tag is for "Description").
Value Field	Next to the prompt is an input field where values are entered for the tags. If a tag has a default value, the value appears here when the tag is selected.

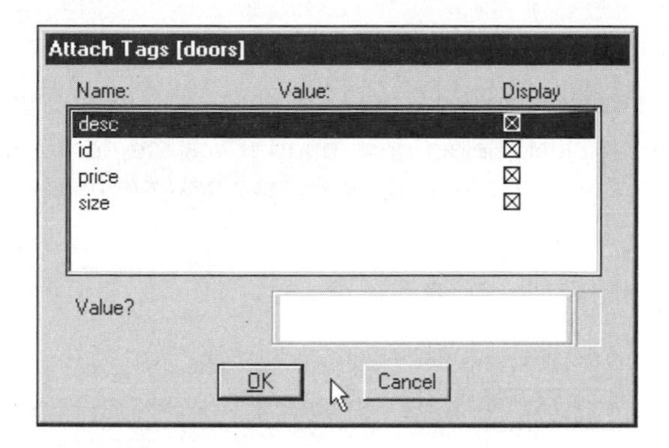

Figure 7–36 Typical Attach Tags window

Filling In Tag Values

To fill in the tag values:

1. Optionally, turn the Display button ON or OFF for each of the tags.
2. Select a tag from the window by clicking on its name.
3. Type its value in the entry field, unless the tag has an acceptable default value.
4. Repeat steps 2 and 3 for each tag.
5. Click on the OK button to close the window and to place the tag values, or Cancel to discard the changes.

If one or more of the tags are to be displayed, MicroStation prompts:

Attach Tags > Place Tag *(Select the location for placing the tags in the design.)*

The tag values are placed using the active element and text attribute settings. If no tags are to be displayed, there is no prompt for placing the tags.

Edit Attached Tags

To make changes to the tag values attached to an element, invoke the Edit Tags tool:

Tags tool box	Select the Edit Tags tool (see Figure 7–37).
Key-in window	**edit tags** (or **edi t**) Enter

Figure 7–37 Invoking the Edit Tags tool from the Tags tool box

MicroStation prompts:

> Edit Tags > Identify element *(Select the element containing tags to be edited.)*
>
> Edit Tags > Accept/Reject (Select next input) *(Click the Accept button to accept the selected element, and, if required, select the next element containing tags to be edited.)*

The Edit Tags window opens. Figure 7–38 provides an example of the window displaying the tag names for a "doors" tag set.

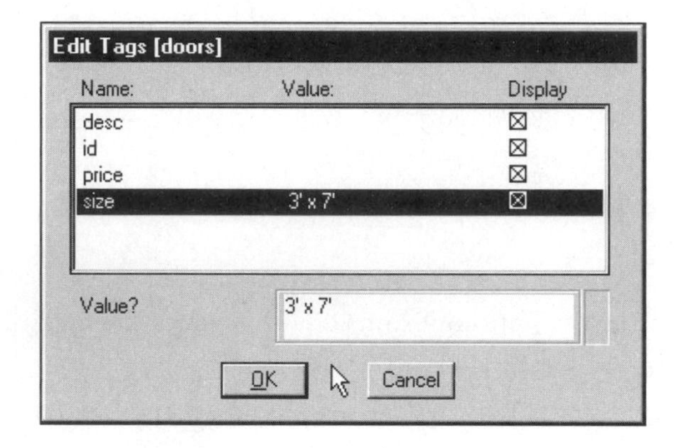

Figure 7–38 Typical Edit Tags window

Editing Tag Values

To edit a tag value:

1. Select the tag.
2. Optionally, turn the Display button ON or OFF for the tag.
3. Edit its value in the entry field.
4. Repeat steps 1 through 3 for each tag that must be changed (edited).
5. Click on the OK button to close the window and to place the edited tag values, or Cancel to discard the changes and leave the tag values unchanged.

Note: Tags that were created with the Variable option OFF cannot be edited.

Review Attached Tags

The Review Tags tool allows you to select an element and view the tags attached to it. No changes can be made to the tags with this tool.

Invoke the Review Tags tool:

Tags tool box	Select the Review Tags tool (see Figure 7–39).
Key-in window	**review tags** (or **rev t**) Enter

Figure 7–39 Invoking the Review Tags tool from the Tags tool box.

MicroStation prompts:

Review Tags > Identify element *(Identify the element containing tags to be reviewed.)*

Review Tags > Accept/Reject (Select next input) *(Click the Accept button to accept the selected element, and, if required, select the next element containing tags to be reviewed, or click the Reset button to reject the element.)*

Review Tags > Select Tag to review *(Select the Tag to review.)*

If the selected element has more than one tag set attached:

1. A Review Tags window opens showing the names of the attached tag sets, as in Figure 7–40.
2. Select the tag set to review, and click on the OK button.
3. Another Review Tags window opens showing the tags for the selected tag set, as in Figure 7–41.

If the selected element has only one tag set attached, a Review Tags window opens showing the tags for the attached tag set.

Figure 7–40 Review Tags window showing attached tag sets

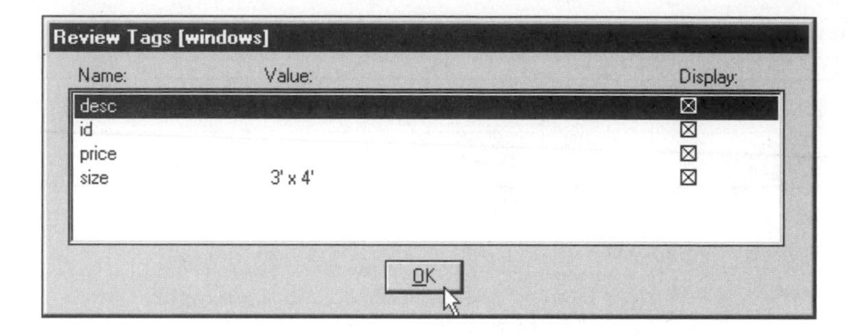

Figure 7–41 Review Tags window showing tags in the attached tag set

Change Tags

The Change Tags tool allows you to change the values assigned to tags in the design. For example, you can change every "Door Material" tag in the design that has the value "Pine," to "White Oak."

Invoke the Change Tags tool:

Tags tool box	Select the Change Tags tool, then, select required change options in the Tool Settings window (see Figure 7–42).
Key-in window	**change tags** (or **chan t**) Enter

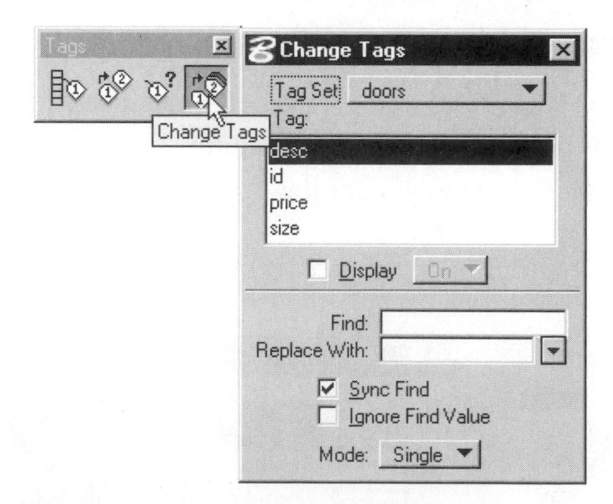

Figure 7–42 Invoking the Change Tags tool from the Tags tool box

When you invoke the Change Tags tool, the Tool Settings window contains several fields for controlling the way the tag changes are handled. Table 7–8 describes each option. MicroStation presents a different prompt for each Change Tags Mode.

Table 7–8 The Fields in the Change Tags Window

FIELD	DESCRIPTION
Tag Set	An options menu that allows you to select the Tag Set that contains the Tag you need to change.
Tag	Lists all Tags for the selected Tag Set. You pick the Tag to be changed from this list.
Display	If the Display toggle button is on, you can turn ON or OFF the display of the selected Tag in the design. If the display status is changed, it will be initiated when the change is completed.
Find	Provides an edit field in which you can type the value you want to change.
Replace With	Provides an edit field in which you can type new value to be assigned to the selected Tag.
Sync Find	Provides a toggle button that enables and disables the function of the Find field.
Ignore Find Value	Provides control over the way the tool handles the contents of the Find edit field. If this toggle button is ON, the contents of the Find edit field are ignored and all occurrences of the selected Tag are changed to the contents of the Replace With edit field. If this button is OFF, only occurrences of the selected Tags that contain the text in the Find edit field are changed to the contents of the Replace With edit field.
Mode	This option menu controls how the tool searches for occurrences of the selected Tag. If the Mode is Single, each Tag must be selected before its value can be changed. If the Mode is Fence, all occurrences of the selected Tag inside the fence are changed (a fence must be defined before this Mode can be used). If the Mode is All, all occurrences of the selected Tag in the design are changed to the new value.

For example, all Material Tags in the Doors Tag Set that have a value of "Pine" need to be changed to "White Oak." The following steps explain how to do it.

1. Invoke Change Tags from the Tags tool box.
2. In the Tool Settings window, enter the following settings:

   - In the Tag Set menu, select the Doors Tag Set.
   - In the Tag list, select the Material Tag.

- Turn the Display toggle button ON and select On from the Display menu, so you can see all occurrences of the Material Tag in the design after the tool action is completed.
- In the Find edit field, type **Pine**.
- In the Replace With edit field, type **White Oak**.
- Turn on the Sync Find toggle button to enable searching for Material Tags with values equal to the contents of the Find edit field.
- Turn off the Ignore Find Value toggle button to make the tool use the contents of the Find edit field for the search.
- Set the Mode to All to cause the tool to find all occurrences of the Materials Tag in the design (MicroStation prompts, "Change Tag > Accept/Reject entire design file").

3. Click a data button anywhere in any open View Window to initiate the Tag Change.

Report Tags

As mentioned earlier, MicroStation creates a tag report that lists the nongraphical information required to fabricate the model created in the project's design files.

Creating reports is a two-step process. First, a tags template is created to control what is included in each type of report. Second, a report is generated based on the tag template.

Generating a Tags Template

A tag template defines the data columns in a tag report. Each column is either a tag from one tag set or an element attribute. To create a tag template, open the Generate Template settings box:

Pull-down menu	Element > Tags > Generate Templates
Key-in window	**mdl load tags template** (or **md l tags template**) Enter

MicroStation displays the General Templates settings box, as shown in Figure 7–43; Table 7–9 describes the fields in the settings box.

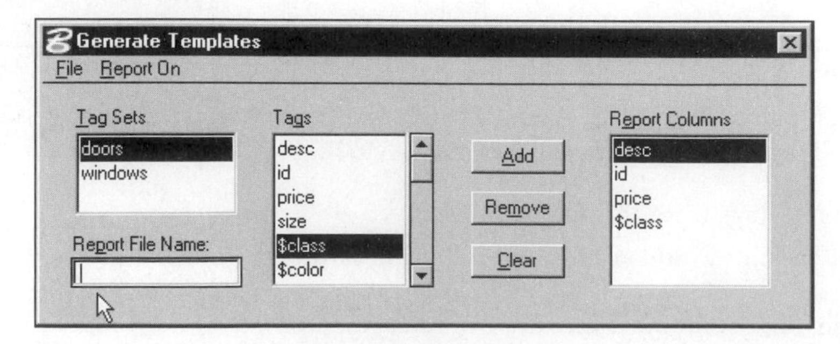

Figure 7–43 Typical Generate Templates settings box

Table 7–9 Generate Templates Window Fields

FIELD	DESCRIPTION
Tag Sets	Lists the tag sets attached to the design file. Select the one for which the template is to be created.
Tags	Lists all tags in the selected tag set and all element attributes that can be included in the report. The names in the list that start with a dollar sign ($) are element attributes.
Report Columns	Lists the tags and element attributes that have been selected for inclusion in the report. Each name will be the name of a column in the report, and the order in the window determines the column order in the report.
Report File Name	Key-in an eight-character-maximum report file name, and, optionally, a three-character-maximum file extension. This name will be used for the report files generated from this template.
Report On Menu	Select the type of elements to include in the report: *Tagged elements*—Include only tagged elements in the report. *All elements*—Include all elements, both tagged and untagged. If all elements are included but no element attribute columns are included, the untagged elements will show up in the report as empty rows.
File Menu	*Open*—Opens the Open Template dialog box, from which you can select an existing template file to open. *Save*—Saves the template information to the same file that was previously opened or saved as. *Save As*—Opens the Save Template As dialog box, from which you can save the template information to any directory path with a file name that you supply in the window. Use this tool to save new templates. Both the Open and Save Template As windows display the default template files directory path the first time they are opened.

Following are the steps to create a new template in the Generate Templates settings box.

1. Key-in the file name for the report in the Report File Name: edit.
2. For each Tag or element attribute to include in the report, select the name in the Tags column and click on the Add button.

3. If you added a name by mistake, select it in the Report Columns list and click the Remove button.
4. When all report columns have been created, select the Save option from the File menu to save the new template using the name currently in the Report File Name field.

Generating a Tags Report

Tag reports list tags and element data information based on Tags Templates. To generate a tag report, open the Generate Reports settings box:

Pull-down menu	Element > Tags > Generate Reports....
Key-in window	**mdl load tags report** (or **md l tags report**) Enter

MicroStation displays the Generate Reports settings box, as shown in Figure 7–44; Table 7–10 describes the fields in the settings box.

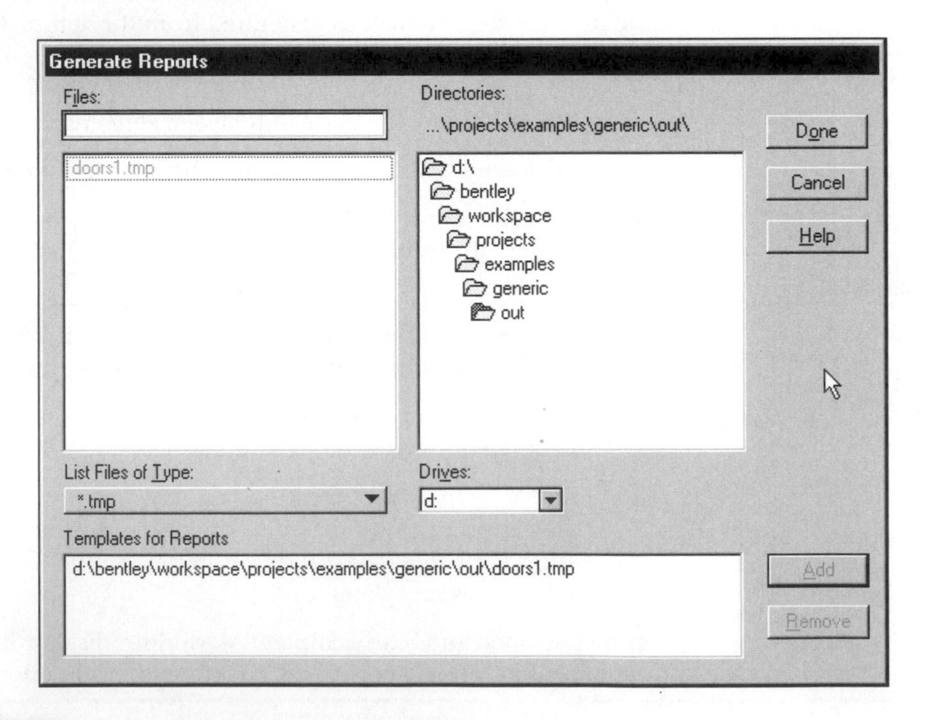

Figure 7–44 Typical Generate Reports window

Table 7–10 Generate Reports Window Fields

FIELD	DESCRIPTION
Files	Lists the existing tag template files in the directory path shown in Directories.
Directories	Lists the directory path to the templates, and is used to change the path. It opens initially displaying the MicroStation default templates path.
List Files of Type	Provides an option menu for selecting the type of template files to display. It has two options:
	*.tmp—List files with the extension "tmp" (the default tem plates file extension).
	All files [*.*]—List all files in the directory path.
Drives	Provides a menu for selecting the letter of the disk drive containing the template files.
Templates for Reports	Displays the templates that have been selected for the report. Separate report files are generated for each template in this list.

Following are the steps to generate a report from the Generate Templates settings box:

1. If required, change the directory path and file type to display the required templates.
2. For each report to be generated, click on the required template file name in the Files list, then click on the Add button. The file specification for each selected template appears in the Templates for Reports list.
3. If a mistake was made in selecting a template, select it in the Templates for Reports list, then click the Remove button to remove it. The Remove button is then dimmed, unless a template has been selected.
4. After all required templates have been selected, click the Done button to generate the reports.

Tag reports are created using the template's file name and the "rpt" extension. The reports are stored in MicroStation's default reports path: BENTLEY/PROGRAM/MICROSTATION/IMGMNGR/OUT. The folder under which this path is found varies, depending on how MicroStation was installed and the computer's operating system. The complete path is defined in the MS_TAGREPORTS configuration variable.

Accessing the Reports

Tag reports are in ASCII files (also called "flat files") that can be accessed in several ways. For example:

- View and print a report with one of the operating system's text viewers, such as NotePad in Microsoft Windows.
- Import a report into another application, such as Excel, the Microsoft Windows spreadsheet application.

Tag Set Libraries

MicroStation provides a tool to export the tag sets created in a design file for insertion in any design file. To export the tag sets from the current design file, open the Define Tags window. The tools to create and use tag set libraries are all in the Define Tags window. Open the Tag Sets settings box:

Pull-down menu	Element > Tags > Define.
Key-in window	**mdl load tags define** (or **md l tags define**) Enter

MicroStation displays the Tag Sets settings box.

Creating a Tag Set Library

Following are the steps to create a tag set library.

1. Select the tag set from the Sets list box.
2. Invoke the Create Set Library... option from Export submenu located in the pull-down menu File, as shown in Figure 7–45. MicroStation displays the Export Tag Library dialog box.
3. Key-in the library file name in the Export Tag Library dialog box, and click on the OK button. MicroStation creates a tag set library file with the default file extension "tlb" and stores it in the default reports path BENTLEY/PROGRAM/MICROSTATION/IMGMNGR/OUT.

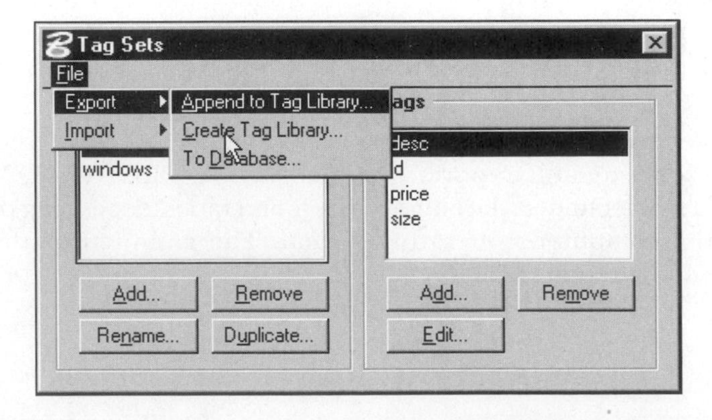

Figure 7–45 Invoking the Create option to open the Export Tag Library dialog box

Appending a Tag Set to an Existing Library

In addition to being able to create a new tag set library, MicroStation also provides a tool to append a tag set to an existing library. Following are the steps to append a tag set to an existing library.

1. From the Sets list box located in the Tag Sets settings box, select the tag set to append to an existing library.
2. Invoke the Append... option from the Export submenu located in the pull-down menu File. MicroStation displays the Export Tag Library dialog box.
3. Select the library file name to which the tag set is to be appended, and click on the OK button. MicroStation appends the selected tag set to the library.
5. Repeat steps 2 and 3 for each tag set to be appended to a library.

Copying a Tag Set from a Selected Library

Following are the steps to copy tag sets from a selected library into the current design.

1. Invoke the Import... option from the pull-down menu File located in the Tag Sets settings box, and MicroStation displays the Open Tag Library.
2. Select the library file, and MicroStation displays the Import Sets dialog box. Select the tag sets to copy into the current design, and click on the OK button. MicroStation copies the selected tag sets into the current Tag Sets lists.
3. If more tag sets need to be copied for another library, go back to step 1.

REVIEW QUESTIONS

Write your answers in the spaces provided.

1. To place natural fractions in the one-character position, turn ON the ________________ toggle button.
2. The Place Fitted Text tool fits the text between two ____________________.
3. When you place a text string above a line with the Place Text Above tool, the distance between the line and the text is controlled by the ________________.
4. Explain briefly why you might use Intercharacter Spacing attribute in placing text.
   ____
   ____
5. Under what circumstance might you use the Match Text Attributes tool?
   ____
   ____
6. What would you key-in to set the Tag Increment to 5? ____________________________
7. To determine the text attributes of an existing text element in a design file, invoke the ________________ tool.
8. Explain briefly the purpose of placing nodes in a design file.
   ____
   ____
9. Explain briefly the purpose of defining tags.
   ____
   ____
10. List the steps involved in generating a template and report files.
    ____
    ____

PROJECT EXERCISE

This project exercise provides step-by-step instructions for creating the circuit board drilling diagram and the hole locations table shown in Figure P7–1. The project provides practice for placing and filling in Text Nodes and Data Fields, and for placing multi-line text elements.

Note: Dimensions are shown in some of the instruction pictures in this project. They are included as an aid to drawing the design, and are not to be drawn as part of this project.

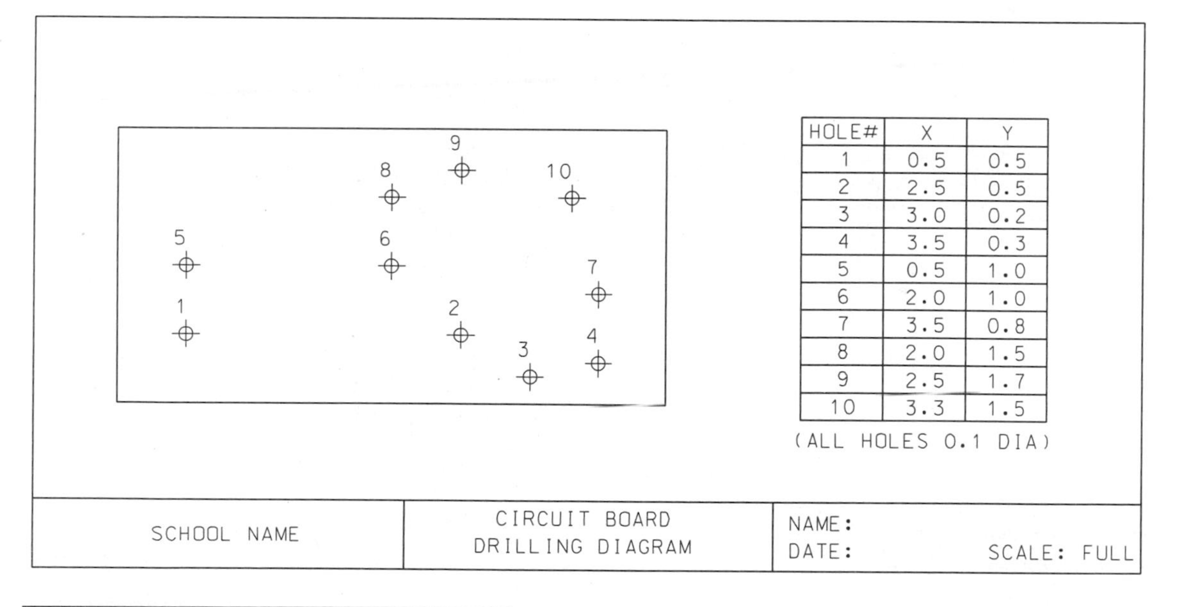

HOLE#	X	Y
1	0.5	0.5
2	2.5	0.5
3	3.0	0.2
4	3.5	0.3
5	0.5	1.0
6	2.0	1.0
7	3.5	0.8
8	2.0	1.5
9	2.5	1.7
10	3.3	1.5

(ALL HOLES 0.1 DIA)

Figure P7–1 Completed project design

Prepare the Design File and Draw the Border

This procedure starts MicroStation, creates a design file, enters the initial settings, and draws the border and title block.

Note: As you complete each step in the project procedures, place a check mark by the step to help you keep up with where you are in the project.

STEP 1: Invoke MicroStation, and create a new design file named CH7.DGN using the SEED2D.DGN seed file.

STEP 2: In the Design File dialog box, set the following:

- Working Unit ratios to 10 TH Per IN and 1000 Pos Units Per TH.

STEP 3: If focus is not on View Window 1, focus on it by clicking a Data button on its title bar or border.

STEP 4: Invoke the Text settings box from the Element pull-down menu, and set the settings shown in Figure P7–2.

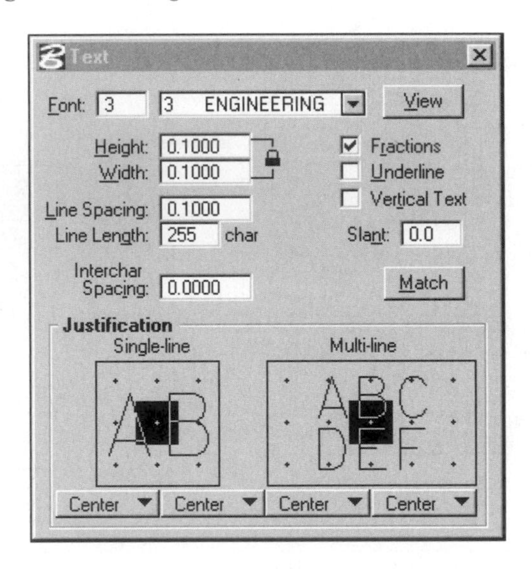

Figure P7–2 Initial Text settings

STEP 5: Turn on AccuDraw.

STEP 6: Open the AccuDraw settings box from the Settings pull-down menu, and set up AccuDraw as follows:

- Lock the Unit Roundoff Distance to 0.1 inch.
- Set the Coordinate system Rotation to Top.
- Set the Coordinate System Type to Rectangular.
- Turn ON the Floating Origin, Context Sensitivity, and Smart Key-ins locks.
- Turn OFF the Auto Point Placement lock.

STEP 7: Set the Active Level to 10, the Color to white (0), and the Line Weight to 1.

STEP 8: Using Figure P7–3 as a guide, draw the border and title block.

STEP 9: Fit View Window 1.

STEP 10: Invoke Save Settings from the File pull-down menu.

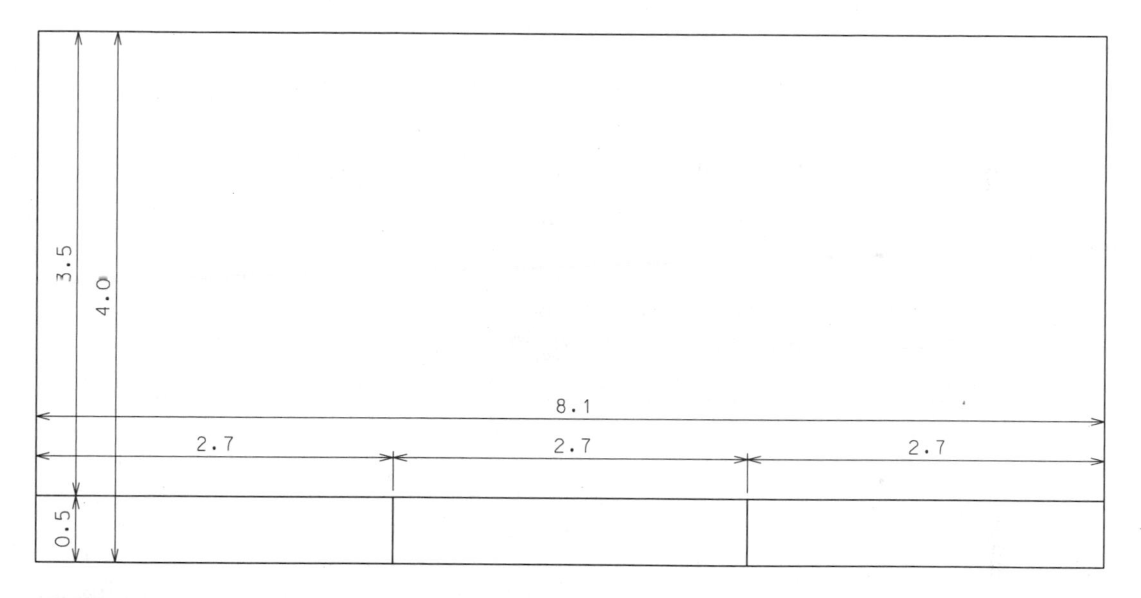

Figure P7–3 Dimensions for drawing the border and title block

Enter the Title Block Text

This procedure places Text Nodes in the title block area, then places text on each Text Node.

STEP 1: Select View Attributes from the Settings pull-down menu.

STEP 2: In the View Attributes dialog box, turn ON the Text Node attribute, then Apply the settings to View Window 1.

STEP 3: Invoke the Place Text Node tool from the Text tool box, then, in the Tool Settings window, set the Active Angle to 0.

STEP 4: Keypoint snap to the lower-left corner of the border, then press the **O** key to center the AccuDraw compass on the keypoint.

STEP 5: Place the first text node at an offset from the compass of X = 1.35 and Y = 0.25.

STEP 6: Place the second text node at an offset from the previous node of X = 2.7 and Y = 0.

STEP 7: Select the Text option from the Elements pull-down menu.

STEP 8: In the Text dialog box, change the Multi-Line justification to Left Center.

STEP 9: Place the third text node at an offset from the previous node of X = 1.45 and Y = 0.1

STEP 10: Place the fourth text node at an offset from the previous node of X = 0.0 and Y = –0.2

STEP 11: In the Text dialog box, change the Multi-Line justification to Right Center.

STEP 12: Place the fourth text node at an offset from the previous node of X = 2.5 and Y = 0.0

The title block area of the design should now be similar to Figure P7–4 (note that the node numbers will be different than those shown in the figure).

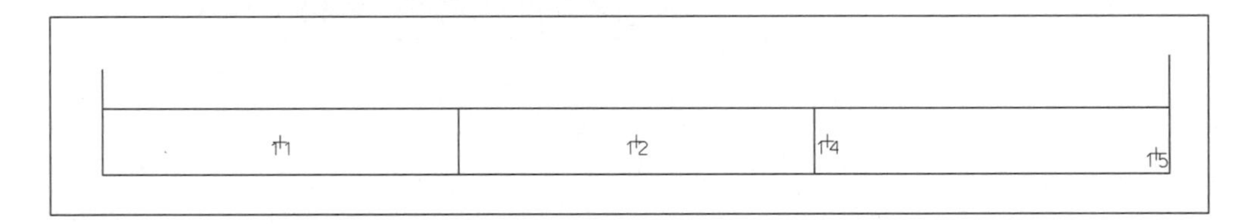

Figure P7–4 The Text Nodes in the title block area

STEP 13: Invoke the Place Text tool from the Text tool box, then, in the Tool Settings window, select the By Origin Method and turn ON the Text Node Lock.

STEP 14: Type your school name in the Text Editor window (or make up a name).

STEP 15: After you type the text, click the Data button on the left-most Text Node to tentatively place the school name on the node, then click the Data button a second time to initiate actually placing the text on the node.

STEP 16: Click the Reset button to drop the image of the text from the cursor and clear the Text Editor window.

STEP 17: Using Figure P7–5 as a guide, place text on the other Text Nodes in the title block.

- Place your name to the right of "NAME:"
- Place today's date to the right of "DATE:"

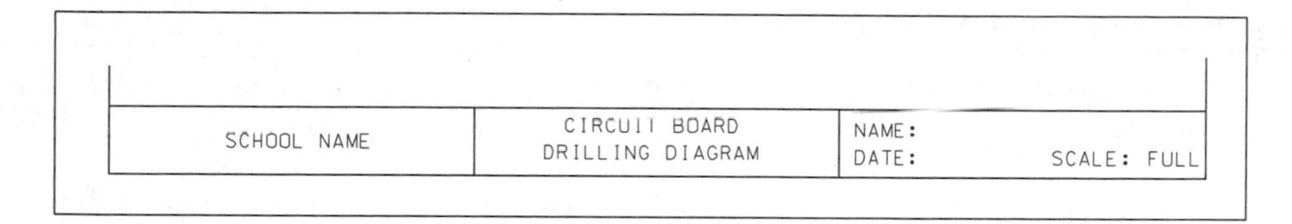

Figure P7–5 Title block text (shown with the Text Node view attribute OFF)

STEP 18: In the View Attributes dialog box, turn the Text Node attribute OFF, then Apply the settings to View Window 1.

STEP 19: Invoke Save Settings from the File pull-down menu.

Draw the Circuit Board

This procedure draws the circuit board.

STEP 1: Set the Active Level to 1, the Line Weight to 1, and the Color to blue (1), then invoke Save Settings from the File pull-down menu.

STEP 2: Using Figure P7–6 as a guide, draw the circuit board.

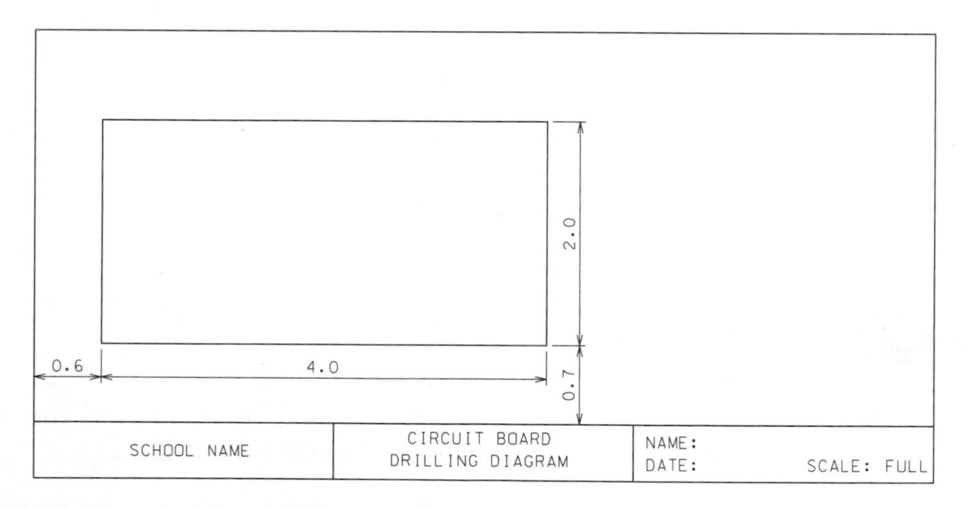

Figure P7–6 Circuit board dimensions

STEP 3: If View Window 2 is not open, open it and focus on it.

STEP 4: In View Window 2, turn OFF the display of level 10 (to hide the border and title block), then Fit the view.

STEP 5: Invoke Save Settings from the File pull-down menu.

Draw the Symbols for the Circuit Board Holes

This procedure draws the symbol for circuit board hole number 1, then copies it to place the other nine hole symbols.

STEP 1: Invoke the Place Circle tool from the Ellipses tool box, then, from the Tool Settings window, select the Center Method, and lock the Diameter to 0.1 inch.

STEP 2: Keypoint snap to the lower-left corner of the circuit board, then press the **O** key to center the AccuDraw compass on the keypoint.

STEP 3: Place the circle center at an offset from the compass of X = 0.5 and Y = 0.5.

STEP 4: Draw two lines through the center of the circle, as shown in Figure P7–7.

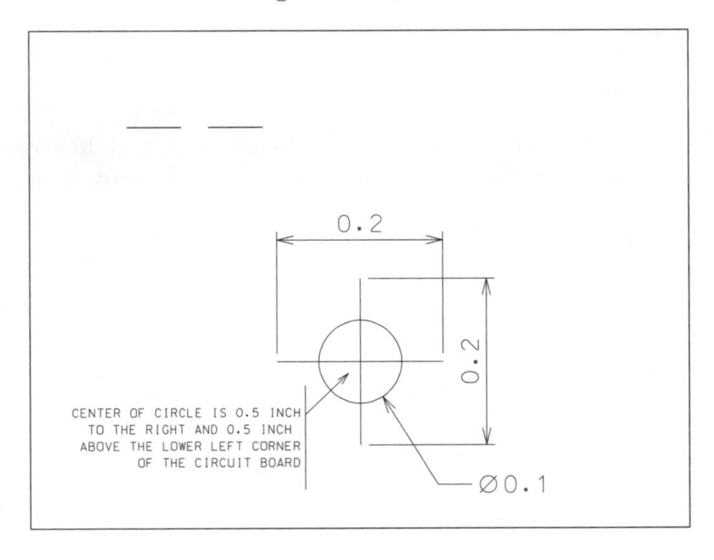

Figure P7–7 Dimensions of the first circuit board hole symbol

STEP 5: Invoke the Place Text tool from the Text tool box, then, in the Tool Settings window, select By Origin Method, set the Text Lock to OFF and set the Active Angle to 0.

STEP 6: Type two underscores ("__") in the Text Editor window (to create an Enter Data Field) and place the text string above and to the left of the circle, as shown in Figure P7–7.

STEP 7: If the Key-in window is closed, open it.

STEP 8: In the Key-in window, type **justify right** Enter, then select and accept the Enter Data Field to set right text justification within the field.

STEP 9: Use the PowerSelector tool to select the circle, two lines, and Enter Data Field that form the hole symbol.

STEP 10: Select Group from the Edit pull-down menu to group the selected elements.

STEP 11: Invoke the Copy command from the Manipulations tool box and select the hole symbol group you just created.

STEP 12: Keypoint snap to the lower left corner of the circuit board block, then press the **O** key to center the AccuDraw compass on the keypoint.

STEP 13: In the AccuDraw settings box, turn OFF the Floating Origin lock to keep the compass on the corner of the circuit board block while you place copies of the hole symbol.

STEP 14: Place hole symbol 2, at an offset from the AccuDraw compass of X = 2.5 and Y = 0.5.

STEP 15: Using the hole positions table in Figure P7–8 as a guide, place the other eight hold symbols on the circuit board. Figure P7–9 shows the completed circuit board.

STEP 16: In the AccuDraw settings box, turn ON the Floating Origin.

STEP 17: Invoke Save Settings from the File pull-down menu.

HOLE#	X	Y
1	0.5	0.5
2	2.5	0.5
3	3.0	0.2
4	3.5	0.3
5	0.5	1.0
6	2.0	1.0
7	3.5	0.8
8	2.0	1.5
9	2.5	1.7
10	3.3	1.5

(ALL HOLES 0.1 DIA)

Figure P7–8 The hole positions table

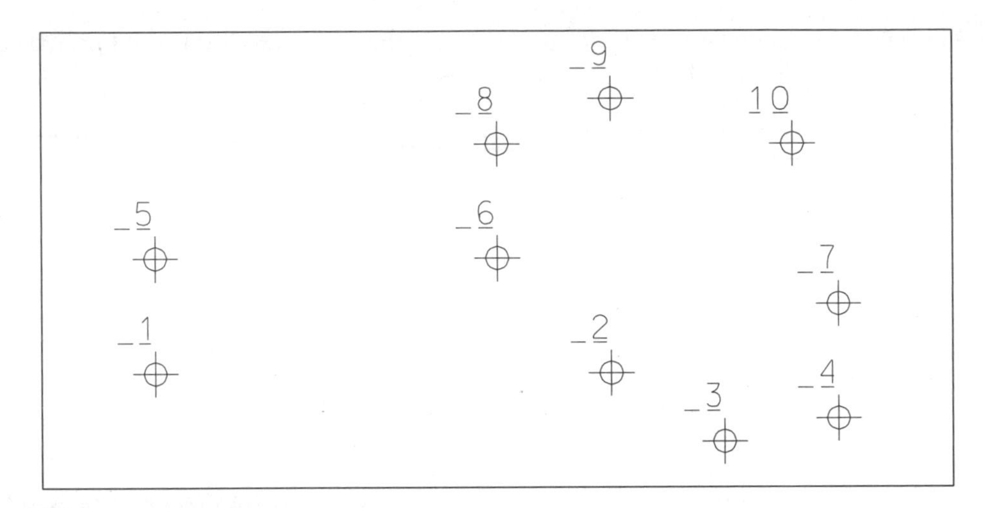

Figure P7–9 The hole symbols in the circuit board

Fill in the Hole Symbol Data Fields

This procedure places numbers in each of the hole symbol Enter Data Fields.

STEP 1: Invoke the Fill in Single Enter_Data Field tool from the Text tool box.

STEP 2: Select the first Enter Data Field order (refer to Figure P7–9 for the numeric order), type the number **1** in the Text Editor window, then press Enter.

STEP 3: Invoke the Copy and Increment Enter_Data Fields tool from the Text tool box, then, in the Tool Settings window, set the Tag Increment to **1**.

STEP 4: Select the Enter Data Field in which you just placed the number **1** to tell the tool what number to increment.

STEP 5: Select each of the other hole symbol Enter Data Fields in numeric order, then click one extra time to accept the last field.

STEP 6: Invoke Save Settings from the File pull-down menu.

Draw the Hole Locations Table

This procedure draws the hole locations table in View Window 3 and uses multi-line text strings to place the text in the table.

STEP 1: Turn on View Window 3, then Fit the view, and set focus to it.

STEP 2: Using Figure P7–10 as a guide, draw the table outline block and grid lines.

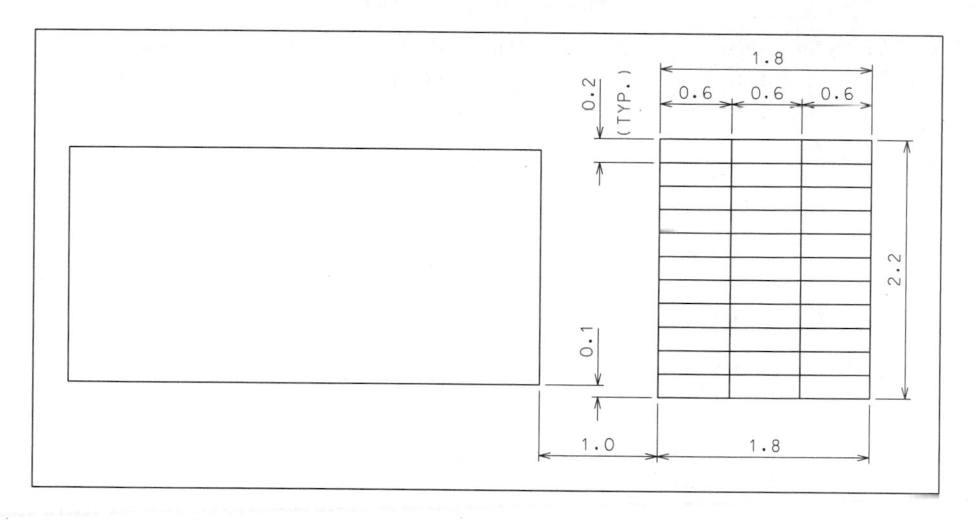

Figure P7–10 The hole locations table dimensions

STEP 3: Use the Window Area tool to fill the view with the hole locations table.

STEP 4: Invoke Save Settings from the File pull-down menu.

STEP 5: Invoke the Place Text tool, then, in the Tool Settings window, select the By Origin Method and set the Justification to Center.

STEP 6: In the Text Editor window, type in all the text for the left table column, as shown in Figure P7–11.

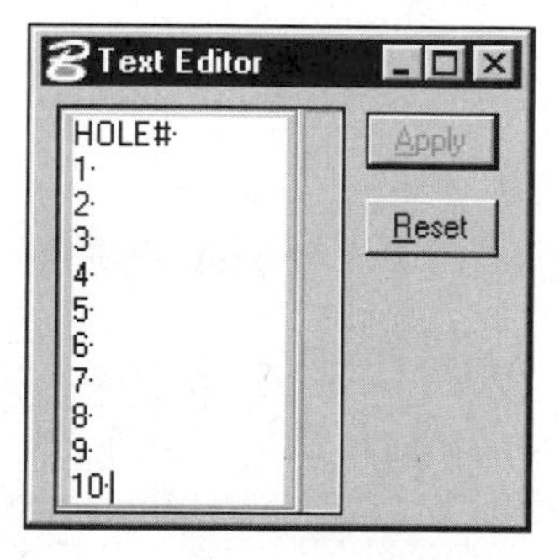

Figure P7–11 The Text Editor window containing the text for the first table column

STEP 7: Drag the multi-line text string into the left table column, center it within the column, click the Data button to place it, then click the Reset button to clear the Text Editor.

STEP 8: Using Figure 7–8 as a guide, place multi-line text strings for the *X* and *Y* table columns.

STEP 9: Place the single-line text string below the table.

Prepare to Close the Design

This procedure checks the design for mistakes, compresses it, and saves the current settings.

STEP 1: Set focus to View Window 1, by clicking on its title bar or border.

STEP 2: Fit View Window 1.

STEP 3: Compare your design to Figure P7–1 and, if necessary, correct mistakes.

STEP 4: Select Compress Design from the File pull-down menu.

STEP 5: Invoke Save Settings from the File pull-down menu.

DRAWING EXERCISES 7–1 THROUGH 7–5

Use the following table to set up the design files for Exercises 7–1 and 7–2.

SETTING	VALUE
Seed File	SEED2D.DGN
Working Units	10 TH Per IN and 1000 Pos Units Per TH
Grid	Grid Master = 0.1, Grid Reference = 10, Grid Lock ON

Exercise 7–1

Textbook logo

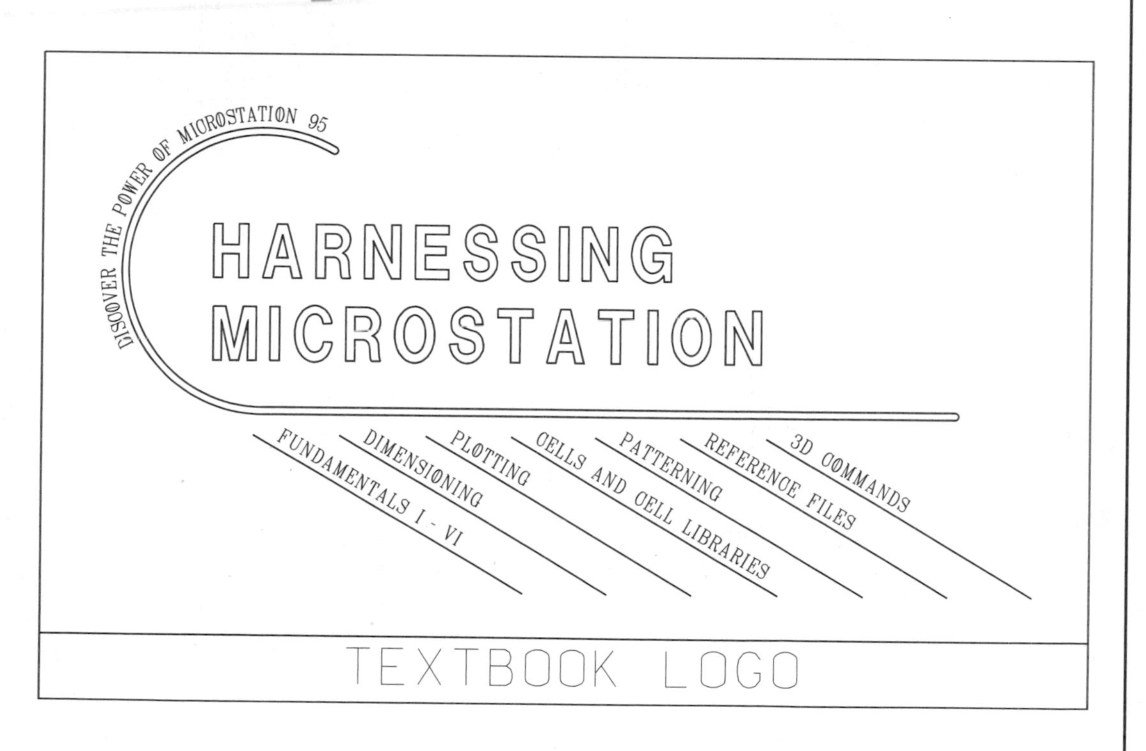

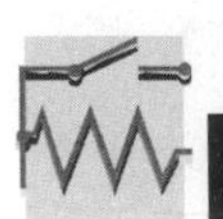

Exercise 7-2

Terminal strip

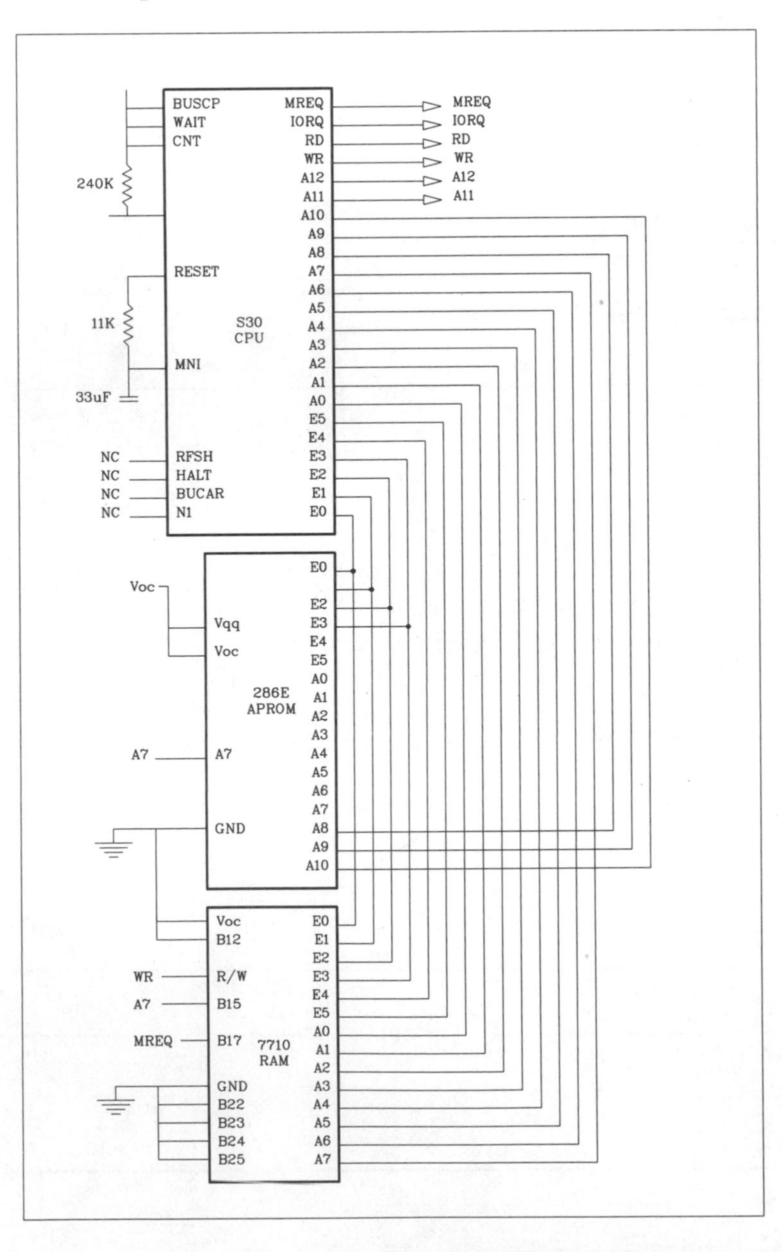

Use the following table to set up the design files for Exercises 7–3 through 7–5.

SETTING	VALUE
Seed File	SEED2D.DGN
Working Units	12" Per ' and 12000 Pos Units Per "
Grid	Grid Master = .5, Grid Reference = 24, Grid Lock ON

In Exercise 7–3, make your best estimate to determine the dimensions of the building section.

Exercise 7–3

Building section

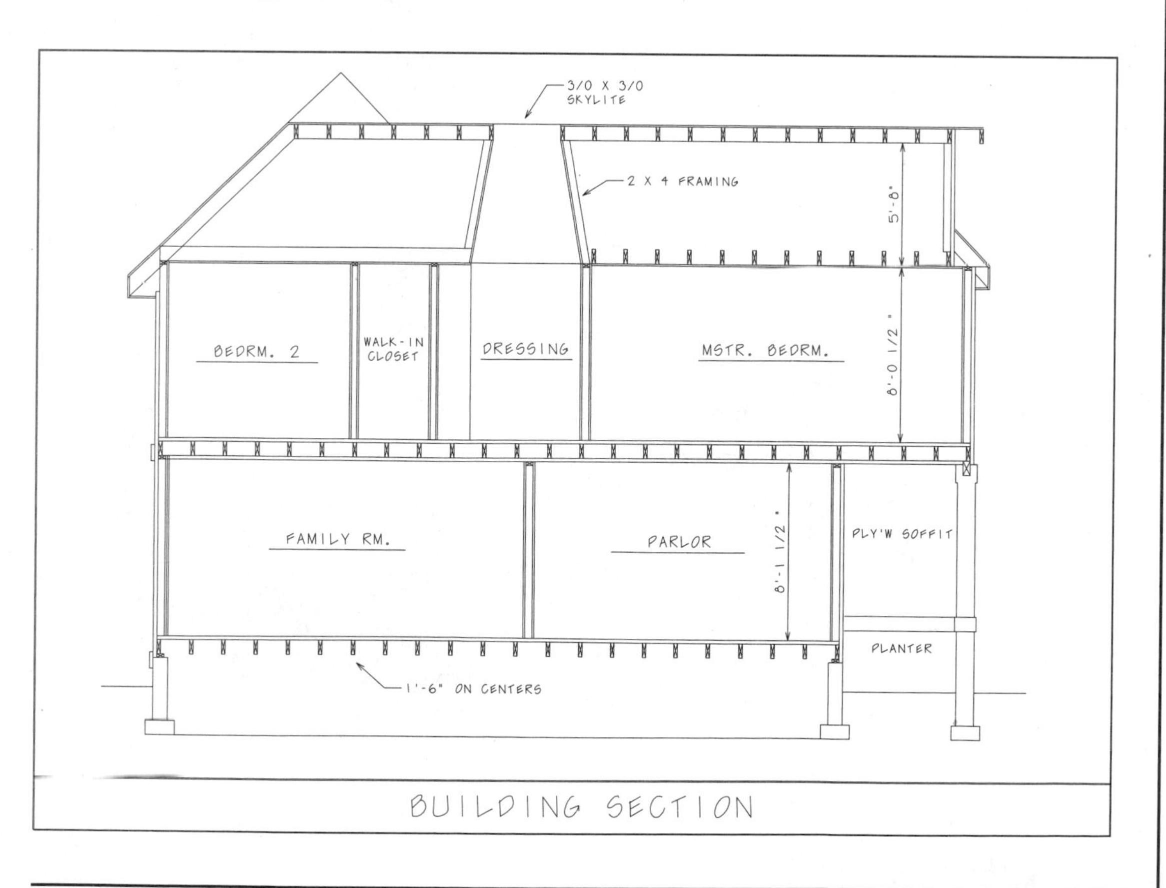

For Exercise 7–4, draw the structural plan and place the descriptive text. Do *not* draw the dimensions.

Exercise 7–4

Structural steel plan

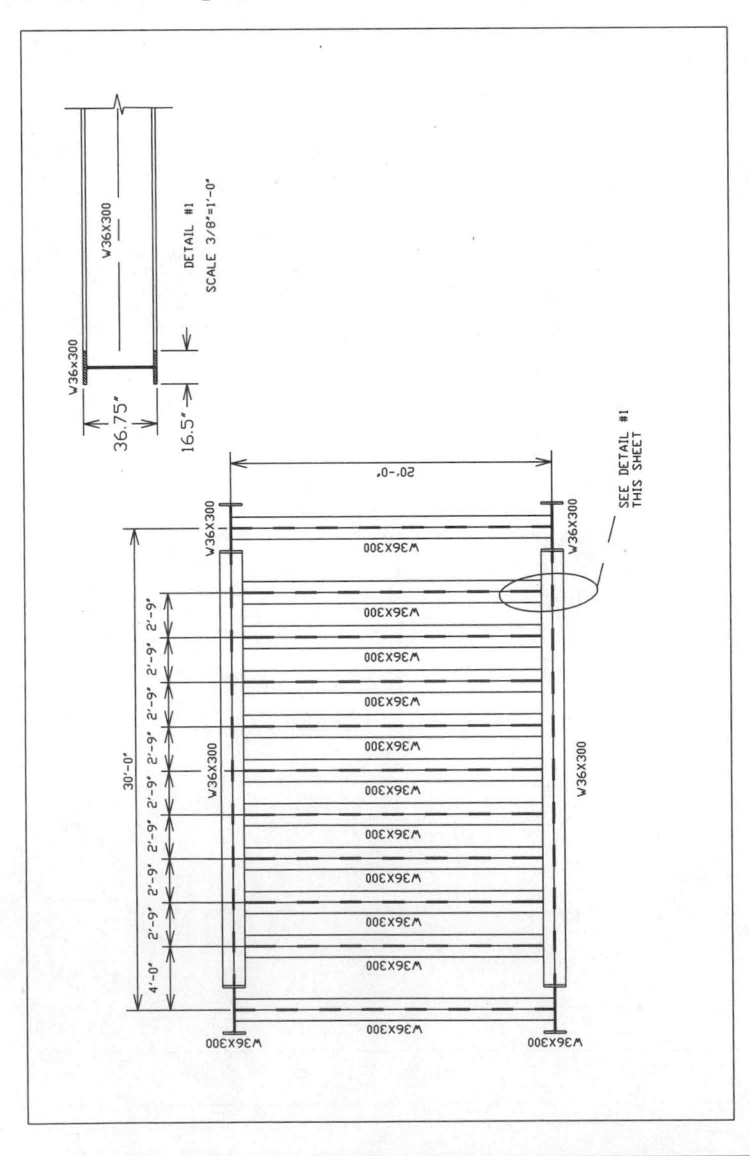

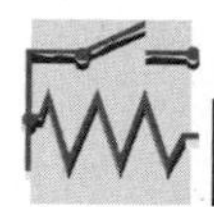

Exercise 7–5

Site plan

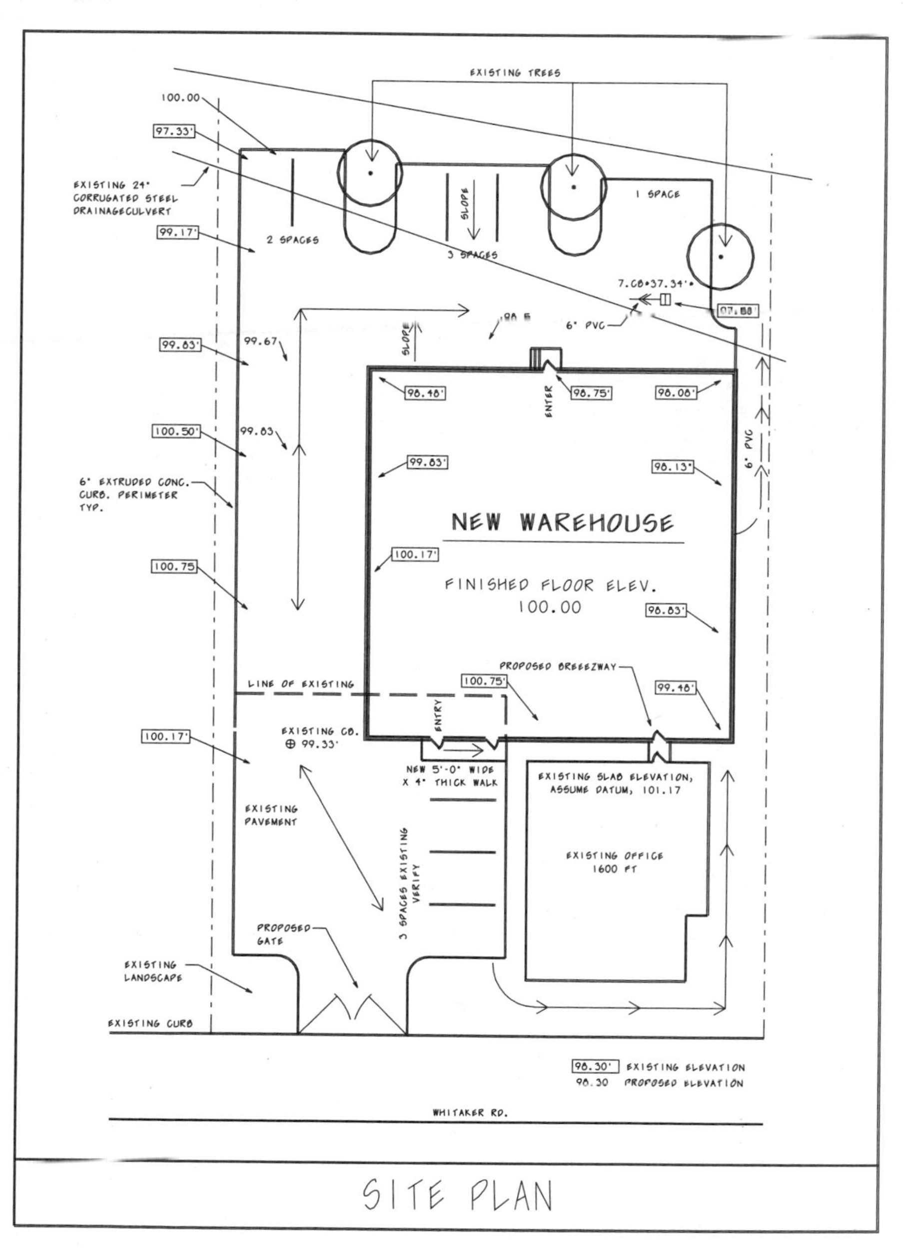

chapter 8

Element Modification

Objectives

After completing this chapter, you will be able to:

- Extend Elements
- Modify vertices and arcs
- Modify elements with AccuDraw
- Create complex shapes and chains
- Create multi-line profiles
- Modify multi-line joints

ELEMENT MODIFICATION—EXTENDING LINES

MicroStation allows you not only to place elements easily, but also to modify them as needed. Three tools that are helpful for cleaning up the intersections of elements are available in

MicroStation. The tools—Extend Element, Extend Elements to Intersection, and Extend Element to Intersection—are available from the Modify Element tool box.

Extend Element

The Extend Line functions to extend or shorten a line, line string, or multi-line via a graphically defined length (with a data point) or via a keyed-in distance. Figure 8–1 shows examples of extending lines.

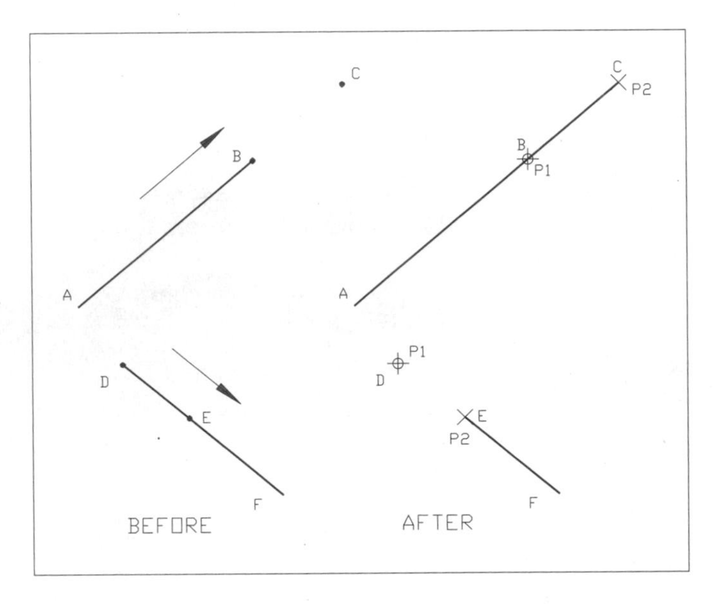

Figure 8–1 Examples of extending and shortening an element graphically

Extend Element Graphically

To extend an element graphically, invoke the Extend Element tool:

Modify tool box	Select the Extend Line tool (see Figure 8–2).
Key-in window	**extend line** (or **ext l**) Enter

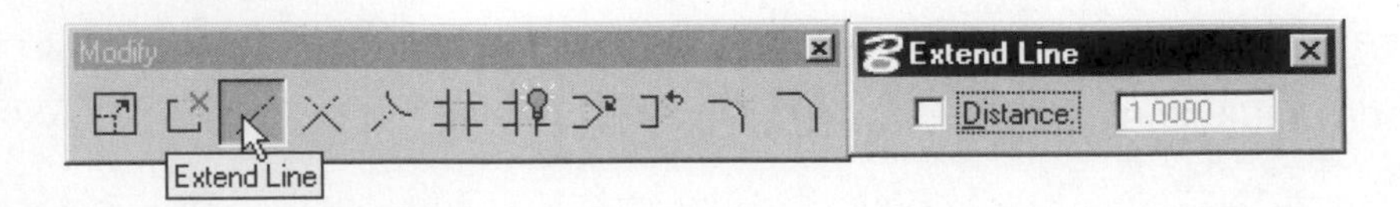

Figure 8–2 Invoking the Extend Line tool from the Modify Element tool box

MicroStation prompts:

Extend Line > Identify element *(Identify the element near the end to be extended or shortened.)*

Extend Line > Accept or Reject (Select next input) *(Drag the element to the new length and click the Data button to accept, or click the Reject button to disregard the modification.)*

Extend Line by Key-in

To extend an element by keying-in the distance, invoke the Extend Element tool:

Modify tool box	Select the Extend Element tool, then turn on the Distance toggle button and key-in the distance in the Tool Settings window (see Figure 8–3).
Key-in window	**extend line keyin** (or **ext l k**) Enter

Figure 8–3 Invoking the Extend Element tool via key-in from the Modify Element tool box

MicroStation prompts:

Extend Line by Key-in > Identify element *(Identify the element near the end to be extended or shortened.)*

Extend Line by Key-in > Accept/Reject (Select next input) *(Click the Data button again anywhere in the design plane to accept the extension.)*

Note: To shorten the element, key-in a negative value in the Distance edit field.

Extend Elements to Intersection

Two elements can be extended or shortened to create a clean intersection between the two. Elements that can be extended to a common intersection with each other are lines, line strings, arcs, half ellipses, and quarter ellipses. Figure 8–4 shows several examples of possible extensions to an intersection.

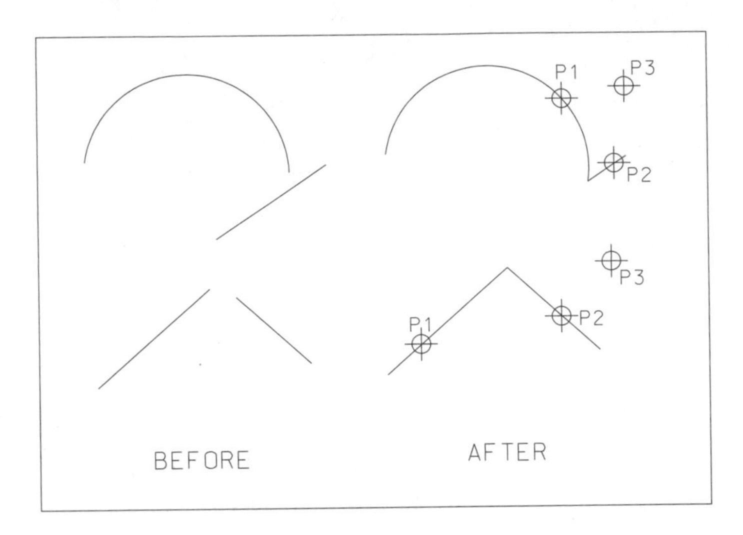

Figure 8–4 Examples of extending elements to a common intersection

To extend two elements to their common intersection, invoke the Extend two Elements to Intersection tool:

Modify tool box	Select the Extend Elements to Intersection tool (see Figure 8–5).
Key-in window	**extend line 2** (or **ext l 2**) Enter

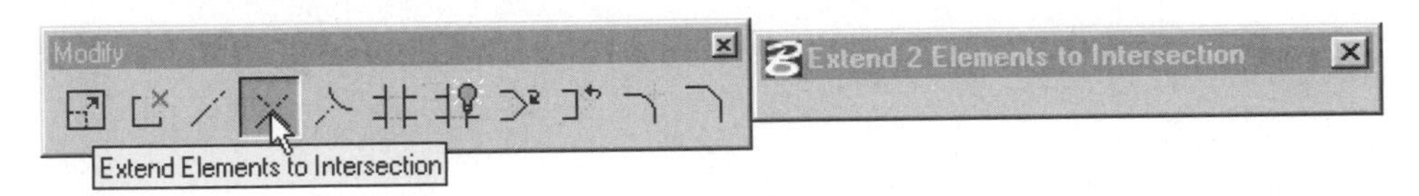

Figure 8–5 Invoking the Extend Elements to Intersection tool from the Modify Element tool box

MicroStation prompts:

> Extend 2 Elements to Intersection > Select first element to extension *(Identify one of the two elements.)*
>
> Extend 2 Elements to Intersection > Select element for intersection *(Identify the second element.)*
>
> Extend 2 Elements to Intersection > Accept/Initiate Intersection *(Place a data point anywhere in the view to initiate the intersection.)*

Dynamic update shows the intersection as soon as you select the second element, but the intersection is not actually created until you accept it by clicking the Data button a third time.

Note: If an element overlaps the intersection, select it on the part you want to keep. The part of the element beyond the intersection is deleted. If dynamic update shows the wrong part of the element deleted, click the Reset button to back up and try again.

Extend Element to Intersection

The Extend Element to Intersection tool serves to change the endpoint of the first selected line to extend to the second selected line, line string, shape, circle, or arc. Elements that can be extended are lines, line strings, arcs, half ellipses, and quarter ellipses. Figure 8–6 shows several examples of possible extensions to intersection.

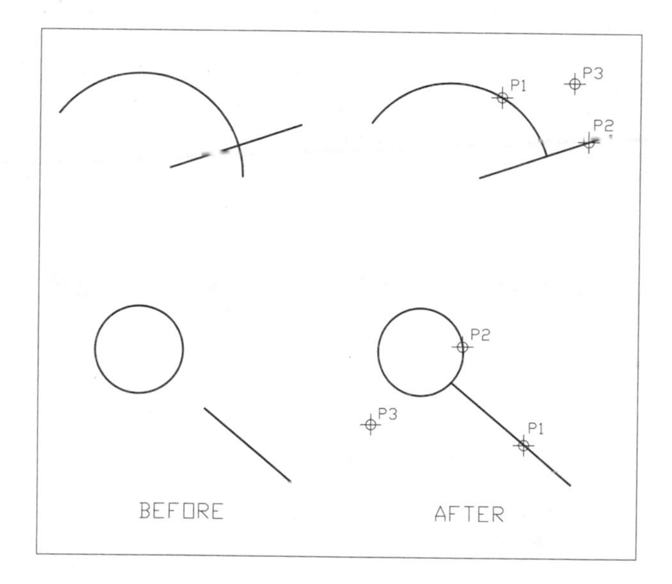

Figure 8–6 Examples of extending an element to an intersection

To extend an element to its intersection with another element, invoke the Extend Element to Intersection tool:

Modify tool box	Select the Extend Element to Intersection tool (see Figure 8–7).
Key-in window	**extend line intersection** (or **ext l in**) Enter

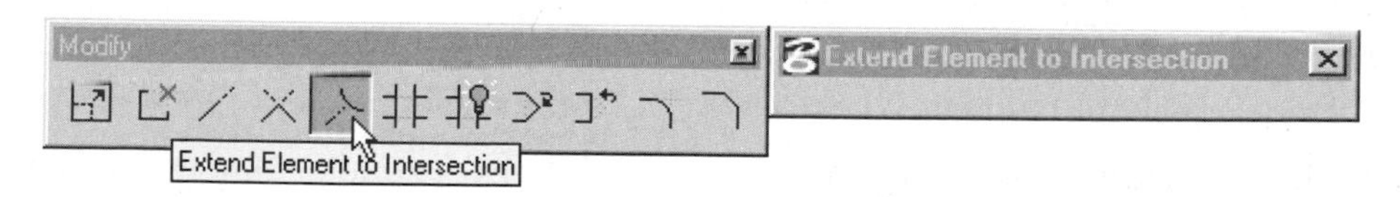

Figure 8–7 Invoking the Extend Element to Intersection tool from the Modify Element tool box

MicroStation prompts:

> Extend Element to Intersection > Select first element for extension *(Identify the element to extend.)*
>
> Extend Element to Intersection > Select element for intersection *(Identify the element to which the first element will be extended.)*
>
> Extend Element to Intersection > Accept/Initiate Intersection *(Place a data point anywhere in the view to initiate the intersection.)*

Dynamic update shows the intersection as soon as you select the second element, but the intersection is not actually created until you accept it by clicking the Data button a third time.

Note: If the element to be extended overlaps the intersection, select it on the part you want to keep. The part of the element beyond the intersection is deleted. If dynamic update shows the wrong part of the element deleted, click the Reset button to back up and try again.

ELEMENT MODIFICATION—MODIFYING VERTICES

Several tools are available to modify the geometric shape of elements by moving, deleting, or inserting vertices. For example, you can change the size of a block by grabbing and moving one of the vertices of the block, or you can turn the block into a triangle by deleting one of the vertices.

Modify Element

The Modify Element tool can modify the geometric shape of any type of element except text elements. Here are the types of modifications it can make:

- Move a vertex or segment of a line, line string, multi-line, curve, B-spline control polygon, shape, complex chain, or complex shape
- Scale a block about the opposite vertex
- Modify rounded segments of complex chains and complex shapes created with the Place SmartLine tool while preserving their tangency
- Change rounded segments of complex chains and complex shapes to sharp, and vice versa
- Scale a circular arc while maintaining its sweep angle (use the Modify Arc Angle tool to change the sweep angle of an arc)
- Change a circle's radius or the length of one axis of an ellipse (if the ellipse axes are made equal, the ellipse becomes a circle and only the radius can be modified after that)
- Move dimension text or modify the extension line length of a dimension element

Typical element modifications are shown in Figure 8–8.

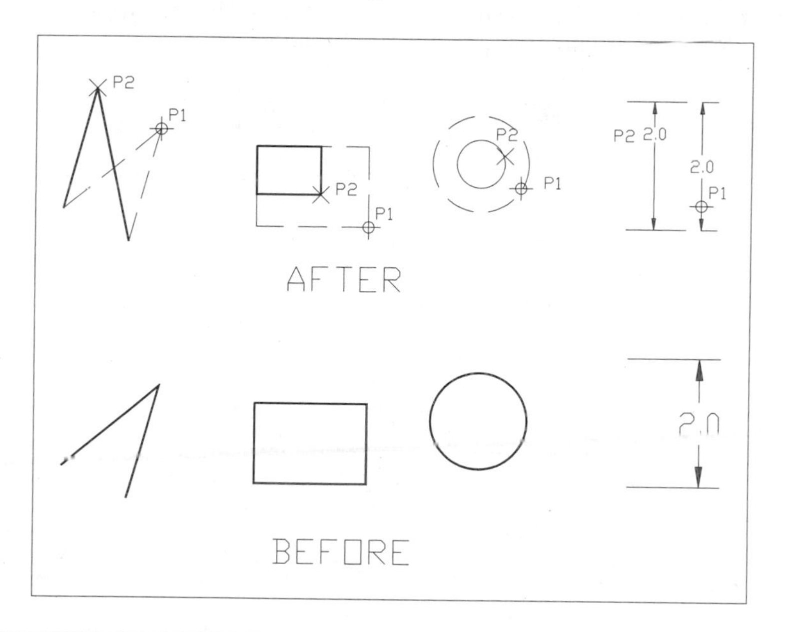

Figure 8–8 Examples of element modifications

Invoke the Modify Element tool:

Modify tool box	Select the Modify Element tool (see Figure 8–9).
Key-in window	**modify element** (or **modi e**) Enter

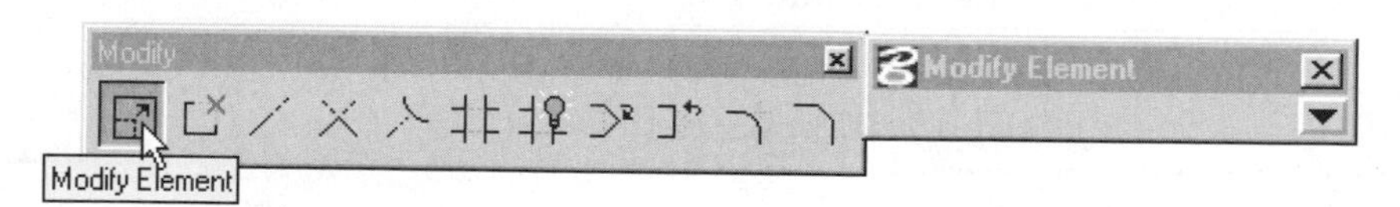

Figure 8–9 Invoking the Modify Element tool from Modify tool box

MicroStation prompts:

Modify Element > Identify element *(Identify the element to be modified.)*

Modify Element > Accept/Reject (Select next input) *(Move the selection point to the desired new location and place a data point to complete the modification, or click the Reset button to deselect the element.)*

When a vertex is selected for modification, the Tool Settings window presents options for modifying the shape of the vertex. Figure 8–10 shows a typical Modify Element tool settings window when a vertex is selected. The settings are described in Table 8–1.

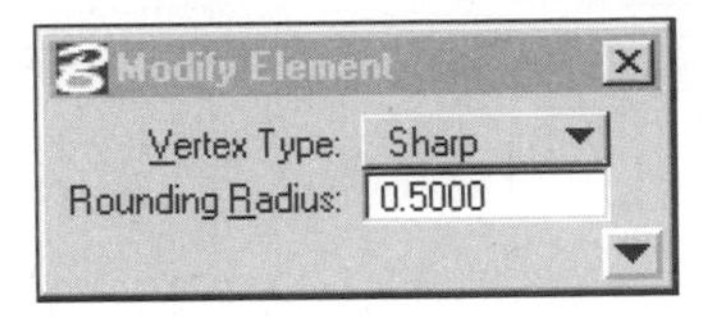

Figure 8–10 Modify Element tool settings window

Table 8–1 Modify Element Tool Settings for Vertices

SETTING	EFFECT OF SETTING
Vertex Type	Set the shape of each vertex to one of these types: Sharp Rounded (a Fillet) Chamfered
Rounding Radius	Enter the rounded vertex radius in working units (MU:SU:PU). **Note:** The "Rounding Radius" prompt appears only when the Vertex Type is "Rounded."
Chamfer Offset	Enter the offset of each end of the chamfer from the vertex point. Each offset is equal. **Note:** The "Chamfer Offset" prompt appears only when the Vertex Type is "Chamfer."
Orthogonal	If an orthogonal vertex is identified, turn this ON to maintain the orthogonal shape of the vertex.

Note: If you select a segment near its center, the segment is moved. If you select it near a vertex, that vertex is moved.

Using Modify Element and AccuDraw Together

Turn on AccuDraw before invoking the Modify Element tool to benefit from the extra drawing aids AccuDraw provides. These aids make Modify Element a more efficient tool. For example, the angle can be locked for a line to make it easy to adjust only the line length, or the length can be locked to make it easy to change only the rotation angle.

Delete Vertex

The Delete Vertex tool removes a vertex from a shape, line string, or curve string. Figure 8–11 shows an example of deleting a vertex from a line string.

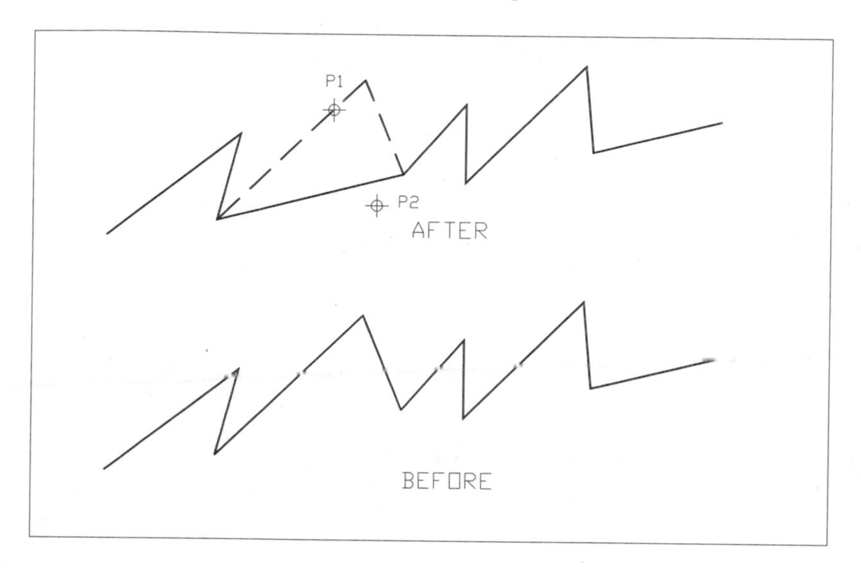

Figure 8–11 Example of deleting a vertex from a line string

To delete a vertex from an element, invoke the Delete Vertex tool:

Modify tool box	Select the Delete Vertex tool (see Figure 8–12).
Key-in window	**Delete Vertex** (or **del v**) Enter

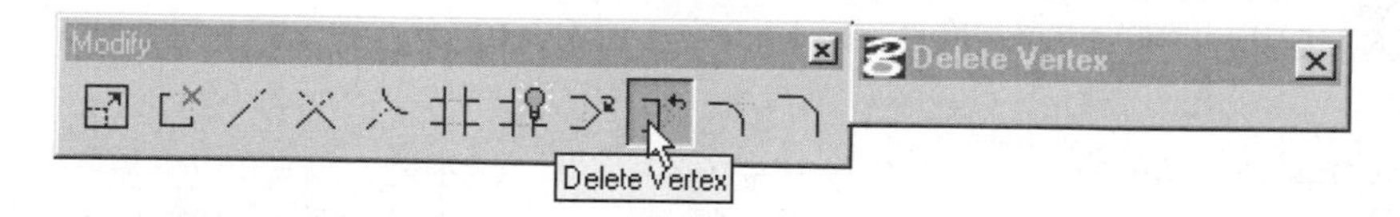

Figure 8–12 Invoking the Delete Vertex tool from the Modify Element tool box

MicroStation prompts:

Delete Vertex > Identify element *(Identify the element near the vertex you want to delete.)*

Delete Vertex > Accept/Reject (Select next input) *(Click the Data button to accept the deleted vertex, or click the Reject button to disregard the modification.)*

When you select the vertex to delete, dynamic update shows the element without the vertex, but it is not actually removed until you click the Data button again. The second data point can also select another vertex to delete.

Note: If the element has only the minimum number of vertices required to define that type of element, you cannot delete a vertex from it. The tool indicates it is deleting the vertex, but nothing is deleted. For example, a minimum of three vertices is required to define a shape.

Insert Vertex

The Insert Vertex tool inserts a new vertex into a shape, line string, or curve string. Figure 8–13 shows an example of inserting a vertex for a line string.

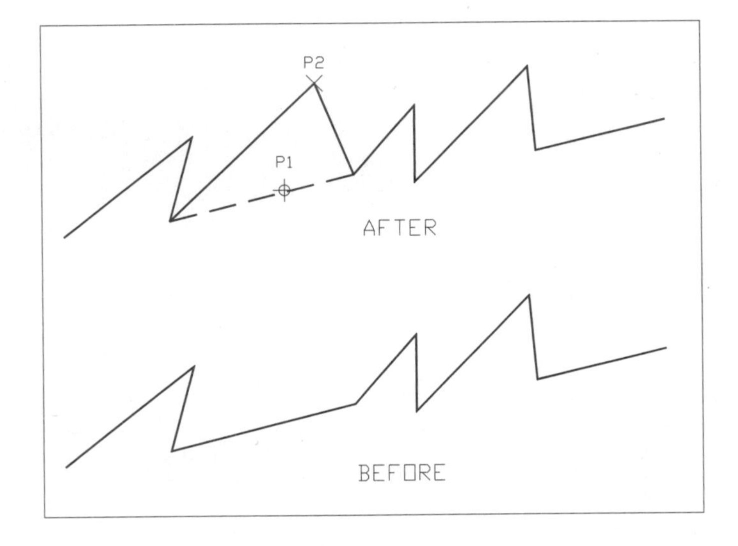

Figure 8–13 Example of inserting a vertex

To insert a vertex from an element, invoke the Insert Vertex tool:

Modify tool box	Select the Insert Vertex tool (see Figure 8–14).
Key-in window	**insert vertex** (or **ins v**) Enter

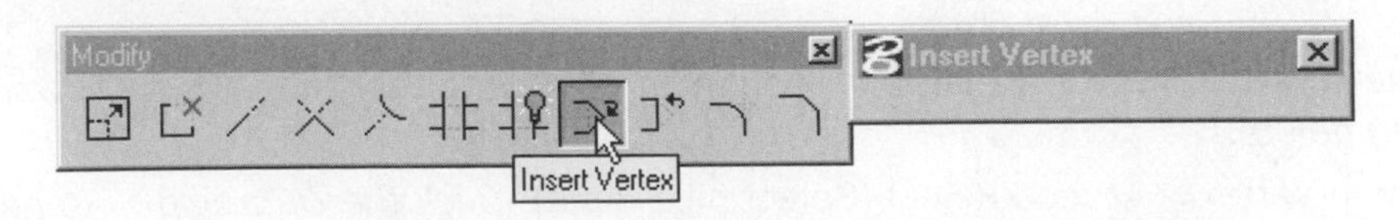

Figure 8–14 Invoking the Insert Vertex tool from the Modify Element tool box

MicroStation prompts:

> Insert Vertex > Identify element *(Identify the element at the point where you want the vertex inserted.)*
>
> Insert Vertex > Accept/Reject (Select next input) *(Drag the new vertex to where you want it in the design plane and click the Data button to insert it, or click the Reject button to reject your selection.)*

When you select the element at the point where you want the new vertex inserted, a dynamic image of the new vertex follows the screen pointer until you place the second data point where you want the vertex located. The second data point causes the new vertex to be inserted, and dynamic update continues dragging the new vertices. Click the Reset button or select another MicroStation tool once you are through with the modification.

ELEMENT MODIFICATION—MODIFYING ARCS

After you place an arc, you can modify its radius, sweep angle, and axis. The tools are available in the Arcs tool box, or you can key-in the tools at the key-in window.

Modify Arc Radius

The Modify Arc Radius tool changes the length of the radius of the selected arc. Figure 8–15 shows examples of modifying an arc radius.

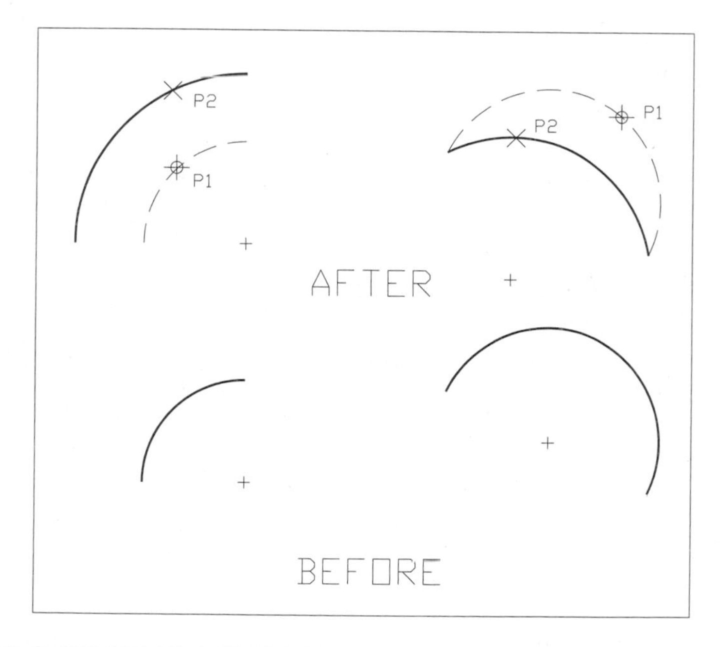

Figure 8–15 Examples of modifying an arc radius

To modify the radius of an arc, invoke the Modify Arc Radius tool:

Arcs tool box	Select the Modify Arc Radius tool (see Figure 8–16).
Key-in window	**modify arc radius** (or **modi a r**) Enter

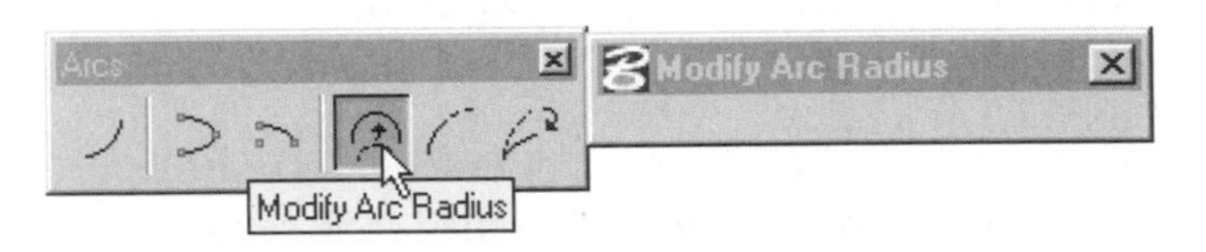

Figure 8–16 Invoking the Modify Arc Radius tool from the Arcs tool box

MicroStation prompts:

> Modify Arc Radius > Identify element *(Identify the arc to modify.)*
> Modify Arc Radius > Accept/Reject (Select next input) *(Reposition the arc and click the Data button to place it, or click the Reject button to reject the modification.)*

Dynamic update shows the arc following the screen pointer after you select the arc. The arc is actually modified once you place the second data point, after which dynamic update continues to drag the arc. Click the Reset button or select another MicroStation tool once you are through with your modifications.

Modify Arc Angle

The Modify Arc Angle tool increases or decreases the sweep angle of the selected arc. Figure 8–17 shows examples of modifying an arc angle.

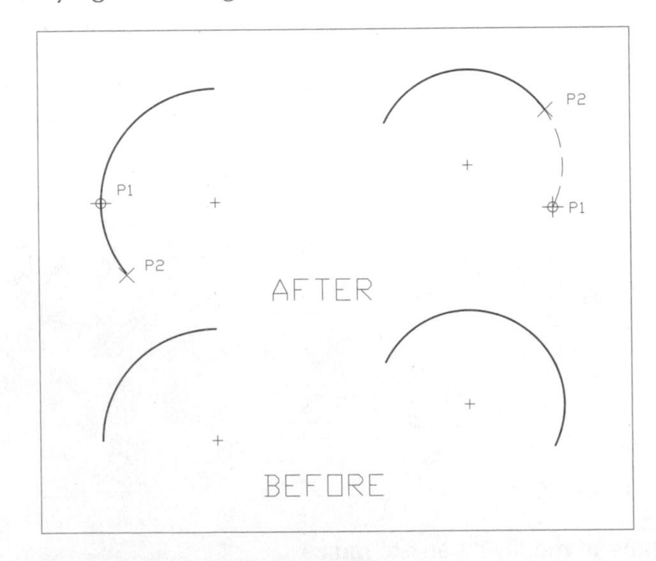

Figure 8–17 Examples of modifying an arc angle

To modify the sweep angle of an arc, invoke the Modify Arc Angle tool:

Arcs tool box	Select the Modify Arc Angle tool (see Figure 8–18).
Key-in window	**modify arc angle** (or **modi a a**) Enter

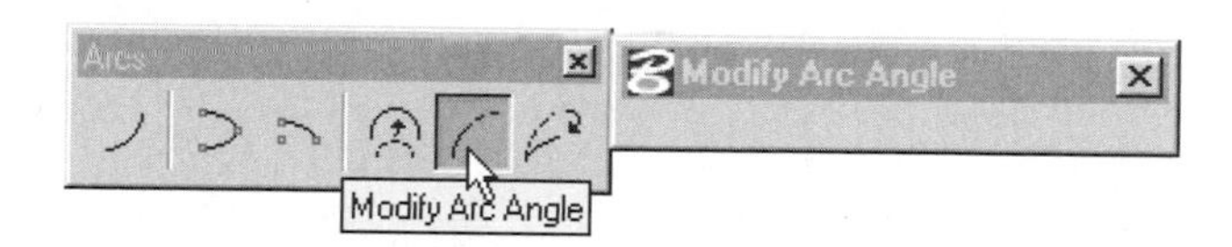

Figure 8–18 Invoking the Modify Arc Angle tool from the Arcs tool box

MicroStation prompts:

Modify Arc Angle > Identify element *(Identify the arc near the end whose sweep angle you want to modify.)*

Modify Arc Angle > Accept/Reject (Select next input) *(Reposition the end of the arc and click the Data button to place it, or click the Reject button to reject the modification.)*

Dynamic update shows the arc following the screen pointer after you select the arc. The arc sweep angle is actually modified once you place the second data point, after which dynamic update continues to drag the arc. Click the Reset button or select another MicroStation tool once you are through with the modification.

Note: If you drag the sweep angle around until the arc appears to be a circle, it is still an arc. Arcs that look like circles can be confusing later when using tools like patterning. If the arc should have been a circle, place a circle instead and delete the arc.

Modify Arc Axis

The Modify Arc Axis tool changes the major or minor axis radius of the selected arc. Figure 8–19 shows an example of modifying an arc axis.

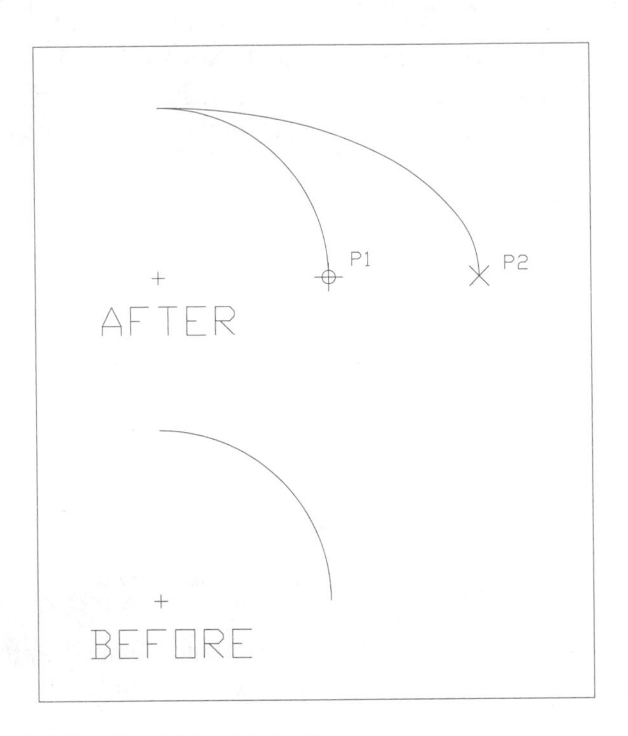

Figure 8–19 Example of modifying an arc axis

To modify an arc axis, invoke the Modify Arc Axis tool:

Arcs tool box	Select the Modify Arc Axis tool (see Figure 8–20).
Key-in window	**modify arc axis** (or **modi a ax**) Enter

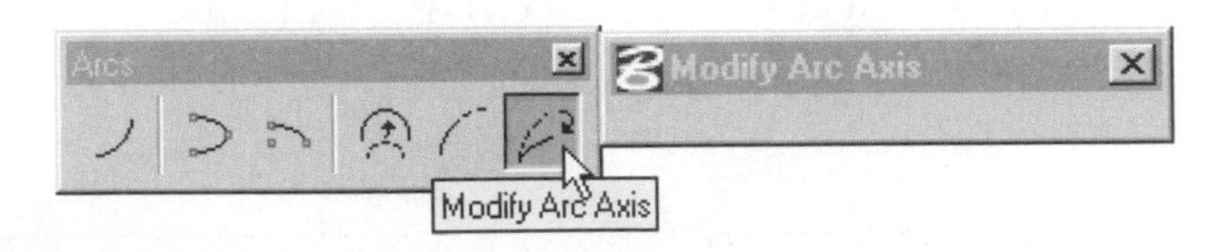

Figure 8–20 Invoking the Modify Arc Axis tool from the Arcs tool box

MicroStation prompts:

> Modify Arc Axis > Identify element *(Identify the arc whose axis has to be modified.)*
>
> Modify Arc Axis > Accept/Reject (Select next input) *(Reposition the axis of the arc and click the Data button to place it, or click the Reject button to reject the modification.)*

Dynamic update shows the arc following the screen pointer after you select the arc. The arc is actually modified once you place the second data point, after which dynamic update continues to drag the arc. Click the Reset button or select another MicroStation tool once you are through with the modification.

Note: After an arc's axis has been changed, the Modify Arc Radius tool can no longer be used on the element.

Create Complex Chais and Shapes

The Create Complex Chain and Create Complex Shape tools turn groups of connected elements into one complex element. A complex shape is a closed element (you could say it "holds water"), and a complex chain is an open element (the "water" can flow out between the two ends of the chain). The element manipulation tools treat the elements in a complex group as one element.

When you create a complex chain or shape from separate elements, the elements take on the current active element attributes (all the available settings in the Element Attributes settings box), and any gaps between the elements are closed. You can key-in the Maximum allowable gap in the Max Gap edit field. Figure 8–21 shows a group of individual elements before and after being turned into a complex shape.

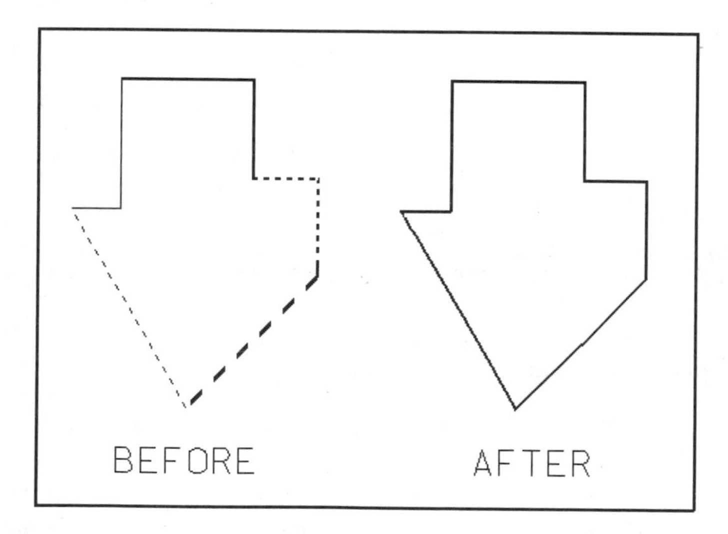

Figure 8–21 Example of a shape being turned into a complex shape

You can create a complex chain or shape manually by selecting each element to be included, or automatically by letting MicroStation find each element. If you want the elements that make up the complex chain or shape to be individual elements again, you can drop them with the Drop Complex tool (the dropped elements keep the attributes of the complex shape).

A quick way to check to see if you really created a complex group from the elements is to apply one of the element manipulation tools to it. If the elements are complex, dynamic update shows an image of all elements following the screen pointer, and the element type in the Status bar tells you it is either a Complex Chain or a Complex Shape.

Note: It is easy to make a mistake in creating complex chains and shapes, and only experience will make you competent. If you goof while creating a complex chain or shape, undo it and try again.

Create a Complex Chain Manually

When you create a complex chain manually, you must select and accept, in order, each of the elements to be included in the chain. Any gaps between the elements are closed, and the complex chain takes on the current active element attributes.

To create a complex chain manually, invoke the Create Complex Chain tool:

Groups tool box	Select the Create Complex Chain tool, then select Manual from the Method option menu in the Tool Settings window (see Figure 8–22).
Key-in window	**create chain** (or **cr ch**) Enter

Figure 8–22 Invoking the Create Complex Chain tool from the Groups tool box

MicroStation prompts:

Create Complex Chain > Identify element *(Identify the first element to include in the complex chain.)*

Create Complex Chain > Accept/Reject (Select next input) *(Identify the next element to include in the complex chain and continue selecting elements in order until all elements are selected. When all elements are selected and accepted, click the Reset button to create the complex chain.)*

Figure 8–23 provides an example of creating a complex chain manually.

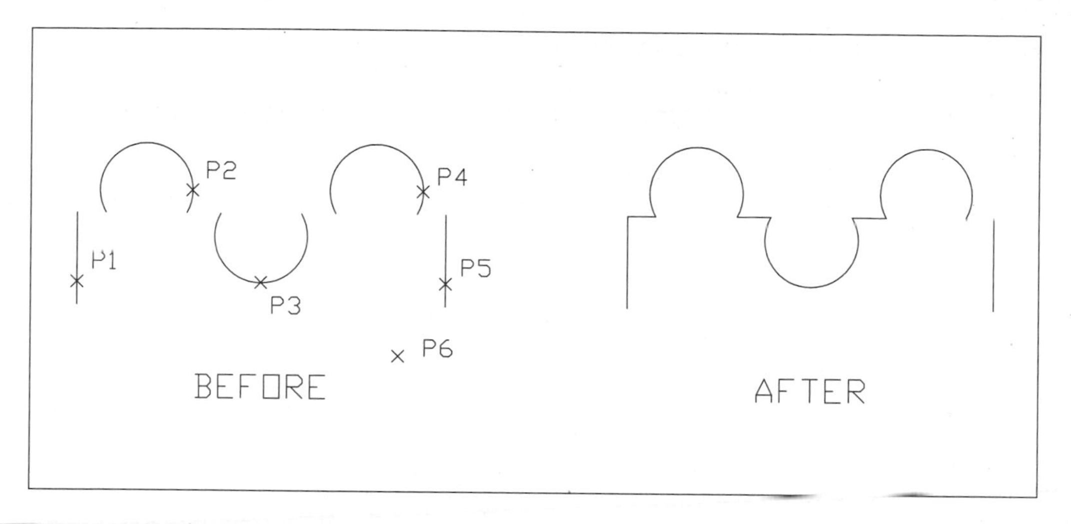

Figure 8–23 Example of creating a complex chain manually

Note: Sometimes when you update a view, the lines closing the gaps between elements in the complex chain may disappear. They are still there—update the view again, and they should reappear.

Create a Complex Chain Automatically

To create a complex chain automatically, start by selecting and accepting the first element in the chain. After that, MicroStation finds and highlights more elements, in series, and you must accept or reject each one.

The automatic version of the tool also allows you to specify a maximum gap that tells MicroStation how far away, in working units, from the end of the previous element it can search for another element. If the tolerance is set to zero, the next element must touch the last selected one before MicroStation finds it.

If there are two or more possible elements at a junction, MicroStation tells you that there is a fork in the path and selects one of the possible elements. You can either accept the selected element or reject it and have MicroStation highlight another possible element in the fork.

Note: If the complex chain contains many elements, and there are not very many forks, the automatic method is probably faster than the manual method. If there are many fork points, creating the chain manually may go faster.

To create a complex chain automatically, invoke the Create Complex Chain command:

Groups tool box	Select the Create Complex Chain tool, then select Automatic from the Method option menu in the Tool Settings window (see Figure 8–24).
Key-in window	**create chain automatic** (or **cre ch a**) Enter

Figure 8–24 Invoking the Create Complex Chain Automatic tool from the Groups tool box

MicroStation prompts:

Automatic Create Complex Chain > Identify element *(Identify the first element to include in the complex chain.)*

Automatic Create Complex Chain > Accept/Reject (Select next input) *(Move the screen pointer in the direction you want the search to go and click the Data button to accept the first element, or click the Reject button to reject it and start over.)*

If there are no forks in the path from the previous element, MicroStation prompts:

Automatic Create Complex Chain > Accept Chain Element *(Click the Data button to accept the element and continue the search, or click the Reset button to complete the chain with the previous element.)*

If there is a fork in the path from the previous element, MicroStation prompts:

Automatic Create Complex Chain > FORK—Accept or reject to See Alternate *(To accept the fork element MicroStation selected, click the Data button; or click the Reject button to disregard the current selection and select another fork element.)*

The process continues until MicroStation cannot find another element to add or until you reject a selection when there is no fork in the path. You *cannot* end a search at a fork point. Figure 8–25 shows an example of creating a complex chain automatically.

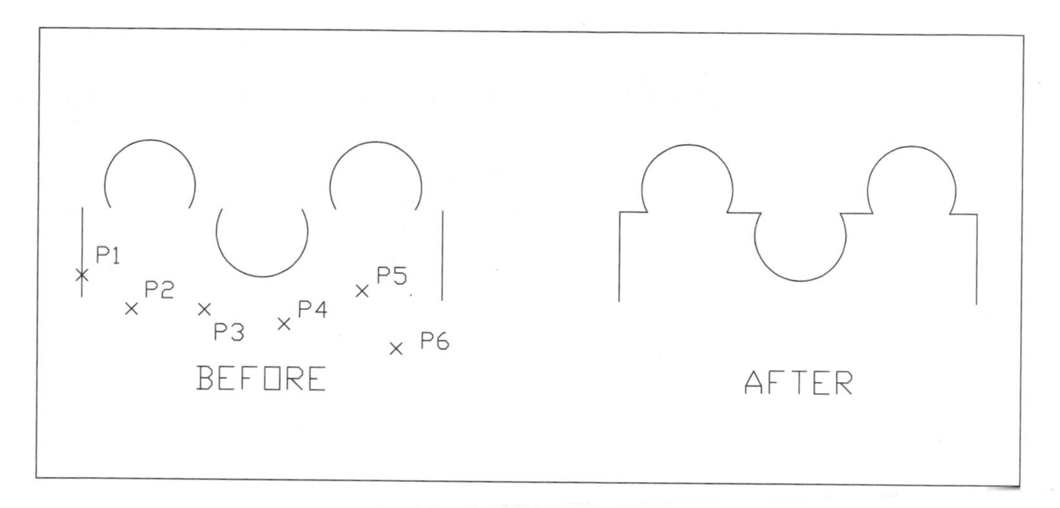

Figure 8–25 Example of creating a complex chain automatically

Create a Complex Shape Manually

To create a complex shape manually, you must select and accept, in order, each of the elements to be included in the shape. Any gaps between the elements are closed, and the complex chain takes on the current active element attributes.

To create a complex shape manually, invoke the Create Complex Shape tool:

Groups tool box	Select the Create Complex Shape tool, then select Manual from the Method option menu Tool Settings window (see Figure 8–26).
Key-in window	**create shape** (or **cr s**) Enter

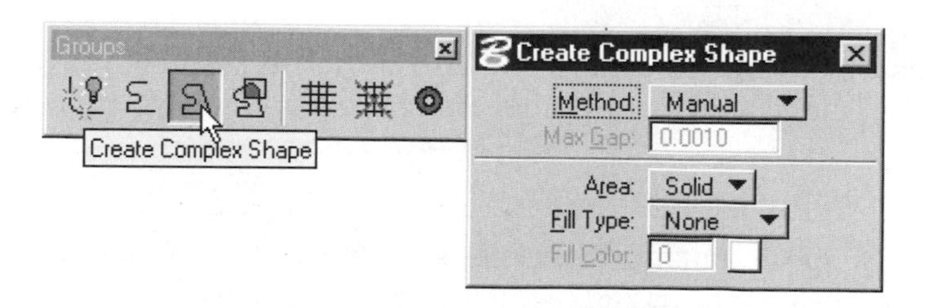

Figure 8–26 Invoking the Create Complex Shape tool from the Groups tool box

MicroStation prompts:

> Create Complex Shape > Identify element *(Identify the first element to include in the complex shape.)*
>
> Create Complex Shape > Accept/Reject (Select next input) *(Identify the next element to include in the complex shape and continue selecting elements in order until all elements are selected. When all elements are selected and the shape appears closed, click the Reset button to create the complex shape.)*

Figure 8–27 provides an example of how to create a complex chain manually.

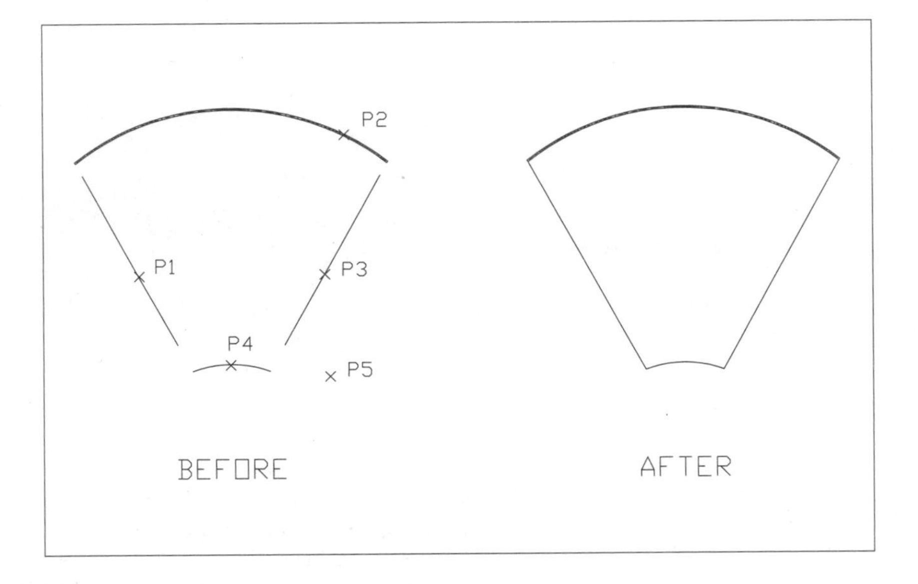

Figure 8–27 Example of creating a complex shape manually

If the end of the last element is touching the start of the first element, you can click the Data button back on top of the first element to close the shape without clicking the Reset button.

Create a Complex Shape Automatically

To create a complex shape automatically, start by selecting and accepting the first element in the shape. After that, MicroStation finds and highlights more elements, in series, and you must accept or reject each one.

If there are two or more possible elements at a junction, MicroStation tells you there is a fork in the path and picks one of the possible elements. You can either accept or reject the element and have MicroStation highlight another possible element.

The automatic version of the tool also allows you to specify a maximum gap that tells MicroStation how far away, in working units, from the end of the previous element it can search for another element. If the tolerance is set to zero, the next element must touch the last selected one before MicroStation finds it.

Note: If the complex shape contains many elements, and there are not very many forks, the automatic method is probably faster than the manual method. If there are many fork points, creating the shape manually may go faster.

To create a complex shape automatically, invoke the Create Complex Shape tool:

Groups tool box	Select the Create Complex Shape tool, then select Automatic from the Method option menu, and if necessary, key-in the gap distance in the Max Gap edit field in the Tool Settings window (see Figure 8–28).
Key-in window	**create shape automatic** (or **cr s a**) Enter

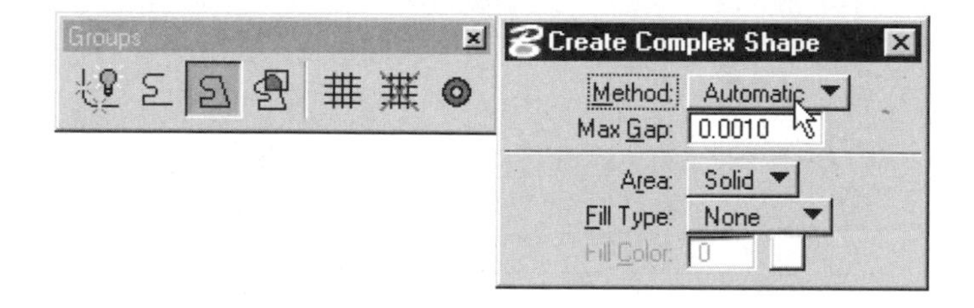

Figure 8–28 Invoking the Create Complex Shape Automatic tool from the Groups tool box

MicroStation prompts:

Automatic Create Complex Shape > Identify element *(Identify the first element to include in the complex shape.)*

Automatic Create Complex Shape > Accept/Reject (Select next input) *(Move the screen pointer in the direction you want the search to go and click the Data button to accept the first element, or click the Reject button to reject it and start over.)*

If there are no forks in the path from the previous element, MicroStation prompts:

Automatic Create Complex Shape > Accept chain Element *(Click the Data button to accept the element and continue the search, or click the Reset button to complete the chain with the previous element.)*

If there is a fork in the path from the previous element, MicroStation prompts:

> Automatic Create Complex Chain > FORK—Accept or reject to See Alternate *(To accept the fork element MicroStation selected, click the Data button; click the Reject button to disregard the current selection and select another fork element.)*

The process continues until MicroStation cannot find another element to add or until you reject a selection when there is no fork in the path. You *cannot* end a search at a fork point. If you press the Reset button when another element is highlighted, that element is not used in the shape. Figure 8–29 shows an example of creating a complex shape automatically.

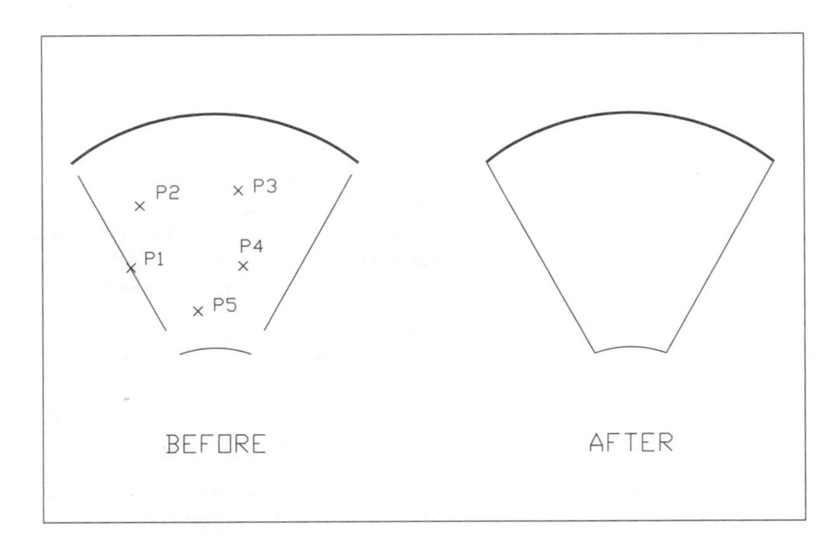

Figure 8–29 Example of creating a complex shape automatically

Create a Region

The Create Region tool creates a complex shape, similar to Complex Shape tools. You can create a complex shape from either of the following:

- The union, intersection, or difference between two or more closed elements
- A region bounded by elements that have endpoints that are closed together by the Maximum Gap

A Keep Original button controls the handling of the original elements. If the button is set to OFF, the original elements are deleted and only the complex region remains. If the button is set to ON, the original elements remain in the design.

Create a Complex Shape from Element Intersection

The Intersection option allows you to create a complex shape from a composite area formed from the area that is common to two closed elements.

To create a complex shape from the intersection of two overlapping circles as shown in Figure 8–30, invoke the Create Region tool:

Groups tool box	Select the Create Region tool, then, in the Tool Settings window, select Intersection from the Method option menu, and, if you want to keep the original elements, turn on the Keep Original toggle button (see Figure 8–31).
Key-in window	**create region intersection** (or **cr r in**) Enter

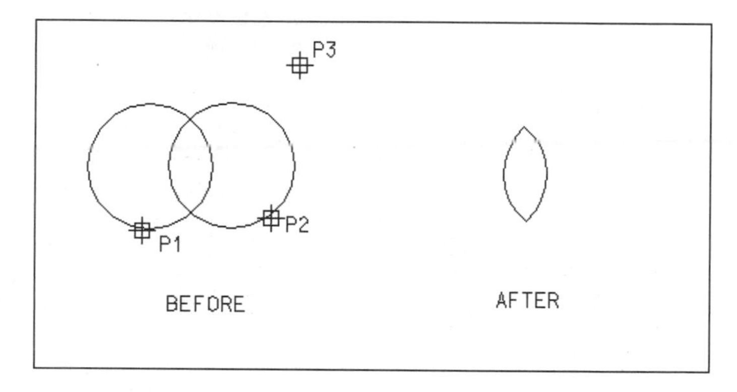

Figure 8–30 Example of creating a complex shape with the Intersection option

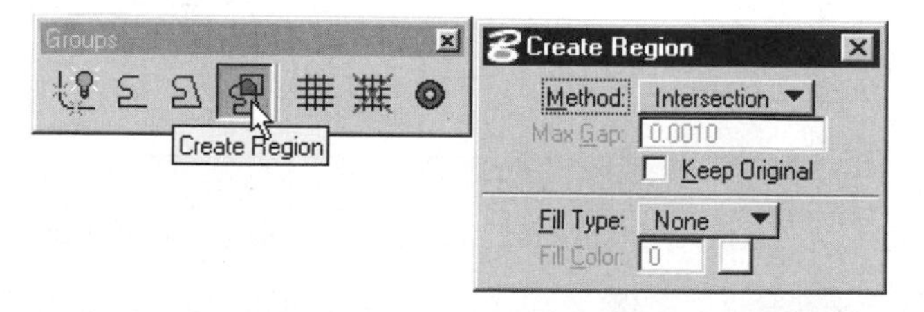

Figure 8–31 Invoking the Create Region From Element Intersection tool from the Create Region tool box

MicroStation prompts:

Create Region From Element Intersection > Identify element *(Identify one of the two circles.)*

Create Region From Element Intersection > Accept/Reject (Select next input) *(Identify the second circle.)*

Create Region From Element Intersection > Accept/Reject (Select next input) *(Click the Data button again to accept the selection of the second circle.)*

Create Region From Element Intersection > Identify additional/Reset to complete *(Click the Reset button to create the complex shape.)*

Create a Complex Shape from Element Union

The Union option allows you to create a complex shape from a composite area formed in such a way that there is no duplication between two closed elements. The total resulting area can be equal to or less than the sum of the areas in the original closed elements.

To create a complex shape from the union of two overlapping circles as shown in Figure 8–32, invoke the Create Region tool:

Groups tool box	Select the Create Region tool, then, in the Tool Settings window, select Union from the Method option menu, and, if you want to keep the original elements, turn on the Keep Original toggle button (see Figure 8–33).
Key-in window	**create region union** (or **cr r u**) Enter

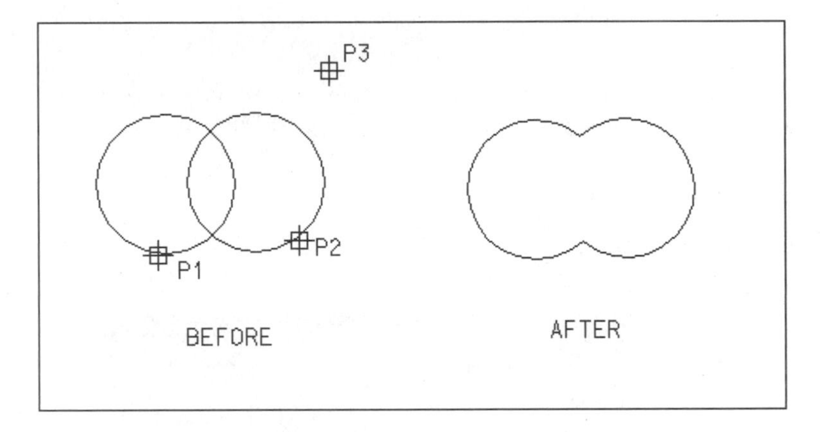

Figure 8–32 Example of creating a complex shape with the Union option

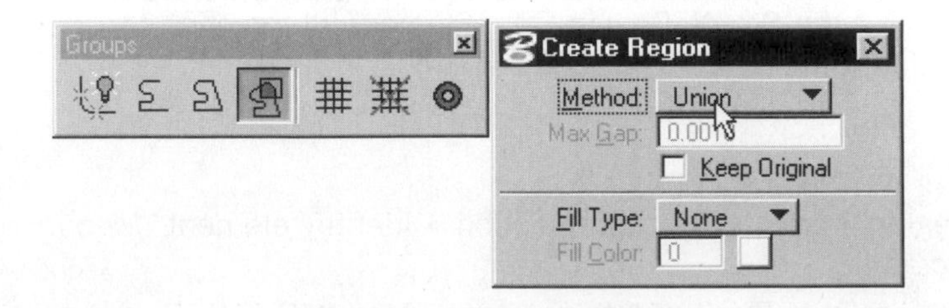

Figure 8–33 Invoking the Create Region From Element Union tool from the Create Region tool box

MicroStation prompts:

> Create Region From Element Union > Identify element *(Identify one of the two circles.)*
>
> Create Region From Element Union > Accept/Reject (Select next input) *(Identify the second circle.)*
>
> Create Region From Element Union > Accept/Reject (Select next input) *(Click the Data button again to accept the selection of the second circle.)*
>
> Create Region From Element Union > Identify additional/Reset to complete *(Click the Reset button to create the complex shape.)*

Create a Complex Shape from Element Difference

The Difference option allows you to create a complex shape from a closed element after removing from it any area it has in common with a second element.

To create a complex shape from the difference of two overlapping circles as shown in Figure 8–34, invoke the Create Region tool:

Groups tool box	Select the Create Region tool, then, in the Tool Settings window, select Difference from the Method option menu, and, if you want to keep the original elements, turn on the Keep Original button (see Figure 8–35).
Key-in window	**create region difference** (or **cr r d**) Enter

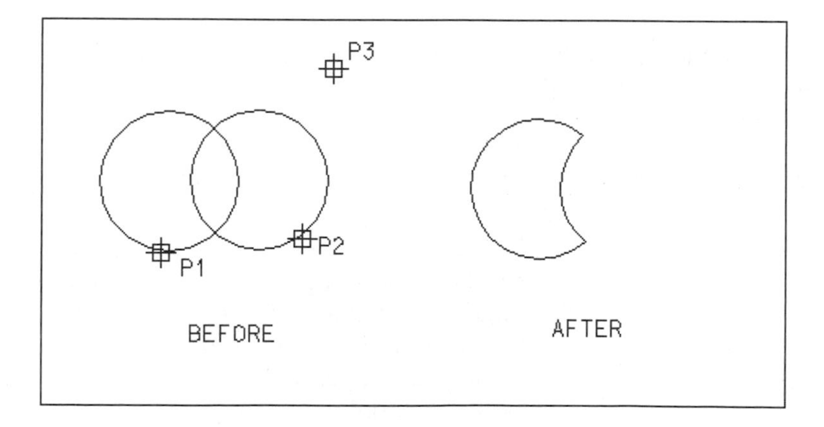

Figure 8–34 Example of creating a complex shape with the Difference option

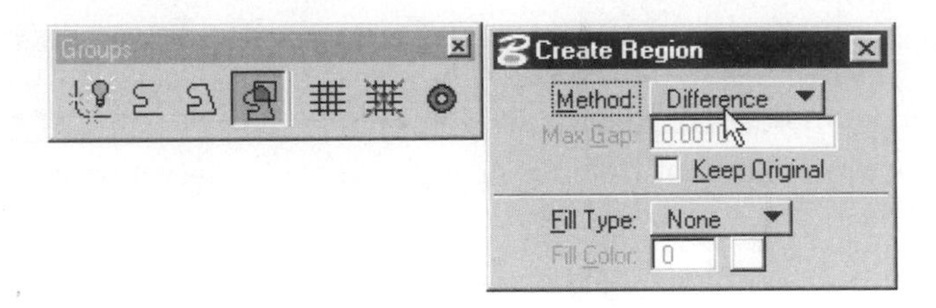

Figure 8–35 Invoking the Create Region From Element Difference tool from the Create Region tool box

MicroStation prompts:

> Create Region From Element Difference > Identify element *(Identify the circle on the left.)*
>
> Create Region From Element Difference > Accept/Reject (Select next input) *(Identify the circle on the right.)*
>
> Create Region From Element Difference > Accept/Reject (Select next input) *(Click the Data button again to accept the selection of the second circle.)*
>
> Create Region From Element Difference > Identify additional/Reset to complete *(Click the Reset button to create the complex shape.)*

Create a Complex Shape from an Enclosed Area

The Flood option allows you to create a complex shape made up of one or more elements. MicroStation prompts you to pick a point inside the closed area. When you place the first data point inside the area, MicroStation searches for the elements that enclose the area, highlights pieces of the elements as it finds them, and then creates the complex shape.

You can specify a maximum gap between elements. If the gap is zero, all elements must touch. If the gap is greater than zero, the enclosed area will only be found if all gaps between elements are less than the gap value.

To create a complex shape from an enclosed area as shown in Figure 8–36, invoke the Create Region tool:

Groups tool box	Select the Create Region tool, then, in the Tool Settings window, select Flood from the Method option menu. If necessary, turn ON the Keep Original toggle button to keep the original elements and enter a value in the Max Gap field to specify how far apart the elements can be (see Figure 8–37).
Key-in window	**create region flood** (or **cr r f**) Enter

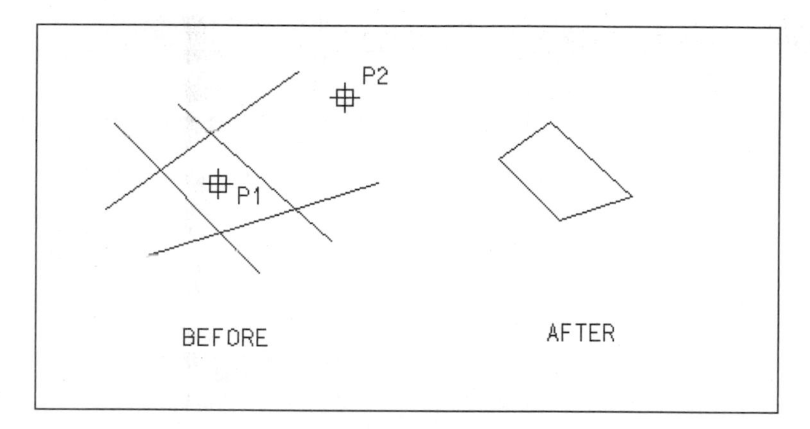

Figure 8–36 Example of creating a complex shape with the Flood option

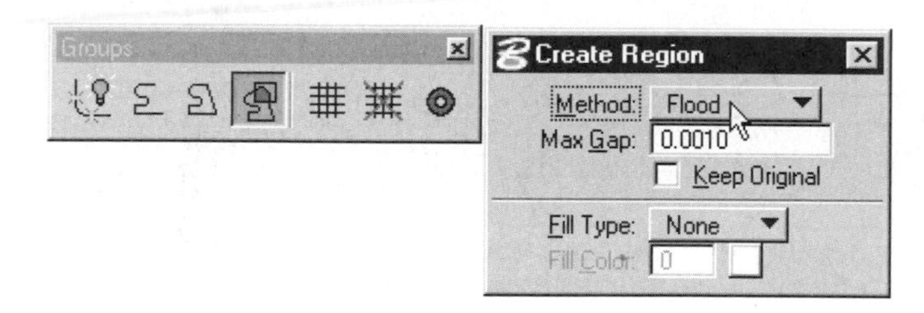

Figure 8–37 Invoking the Create Region Flood option from the Create Region tool box

MicroStation prompts:

Create Region From Area Enclosing Point > Enter data point *(Place a data point inside the enclosed area.)*

Create Region From Area Enclosing Point > Accept-Create a complex shape *(Click the Data button to accept the complex shape.)*

Note: If the tool does not find an enclosing area, it displays the following message in the status bar: "Error – No enclosing region found."

DROP COMPLEX CHAINS AND SHAPES

If you want to return the elements in a complex chain or shape to individual elements, you can drop their complex status. The Drop Complex Status tool drops an individual complex group. The Fence Drop Complex Status tool drops all complex groups within the boundary of a fence.

Dropped complex elements return to being individual elements, but they keep the element attributes (color, weight, etc.) of the complex shape. The elements do not return to their original attributes.

Drop a Complex Chain or Shape

To drop a complex chain or shape, invoke the Drop Element tool:

Groups tool box	Select the Drop Element tool, then, in the Tool Settings window, turn ON the Complex toggle button, and turn OFF the other toggle buttons (see Figure 8–38).
Key-in window	**drop element** (or **dr e**) Enter

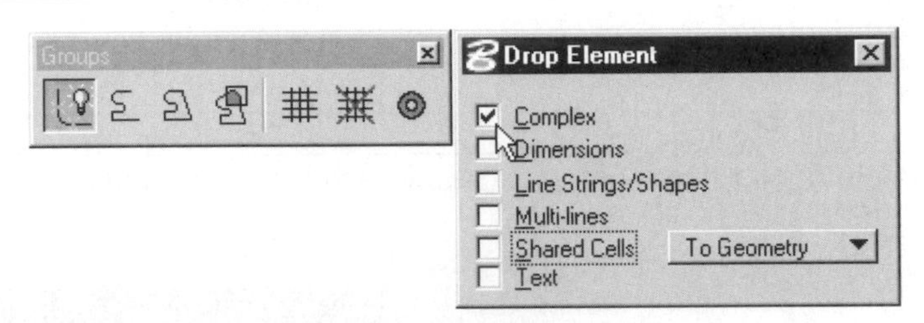

Figure 8–38 Invoking the Drop Element tool from the Groups tool box

MicroStation prompts:

Drop Complex Status > Identify element *(Identify the complex chain or shape to drop into individual elements.)*

Drop Complex Status > Accept/Reject (select next input) *(Click the Data button to accept the change in status from complex element into individual elements, or click the Reject button to reject the change in the status.)*

Note: If there were gaps between the elements that make up the complex shape or chain, the gaps will return after the complex shape or chain is dropped. You may not see the gaps until you repaint the view window.

Drop Several Complex Chains or Shapes

The Drop Complex Status of Fence Contents tool breaks all the complex elements enclosed in a fence into separate elements. Before you invoke the tool, place a fence that encloses all the complex elements you want to drop.

Invoke the Drop Fence Contents tool:

Fence tool box	Select the Drop Fence Contents tool, then select the desired Fence Mode from the Tool Settings window (see Figure 8–39).
Key-in window	**fence drop** (or **fe dr**) Enter

Figure 8–39 Invoking the Drop Fence Contents tool from the Fence tool box

MicroStation prompts:

> Drop Complex Status of Fence Contents > Accept/Reject *(Click the Data button to accept the change in status from complex element into individual elements, or click the Reject button to disregard the change in the status.)*

DEFINE MULTI-LINES

In Chapter 4 we learned how to place multi-lines using the active multi-line profile (usually three parallel lines with a dashed center line). Now we are going to learn how to customize the active multi-line profiles.

MicroStation provides the Multi-lines settings box for defining a new multi-line profile. Table 8–2 describes each multi-line component that can be defined in the settings box, and Figure 8–40 shows examples of typical multi-line profiles.

Table 8–2 The Multi-line components

Component	Description
Line	The parallel lines placed by the multi-line. You can define from one to sixteen lines components and set unique attributes for each one (color, weight, style, level).
Start Cap	An element that can be placed across the start of the multi-line to close it. The cap can be an arc or line completely across the start end, or arc across each pair of line components. You can also specify the attributes of the start cap.
End Cap	An element that can be placed across the end of the multi-line to close it. The cap can be an arc or line completely across the start end, or arc across each pair of line components. You can also specify the attributes of the start cap.
Joint	A line that can be placed at each multi-line joint (vertex). You can specify the attributes of the joint.

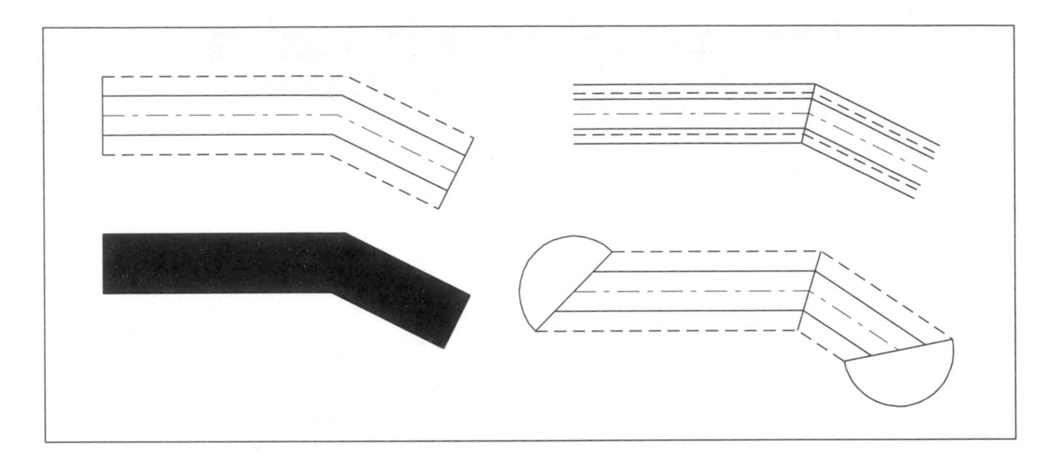

Figure 8–40 Examples of typical multi-line definitions

Invoke the Multi-Lines settings box:

Pull-down menu	Element > Multi-lines
Key-in window	**dialog multiline open** (or **di mu o**) Enter

MicroStation displays the Multi-lines settings box as shown in Figure 8–41.

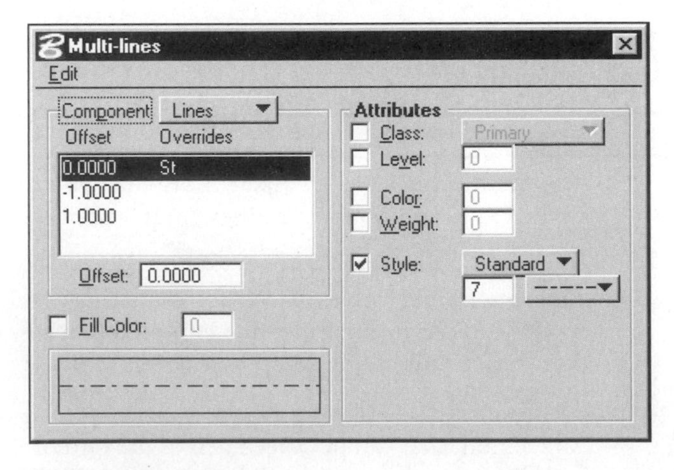

Figure 8–41 Multi-lines settings box

Note: The menu options displayed in the Multi-lines settings box depend on what component option is selected.

Define Line Components

To edit the line components in the active multi-line profile, select Lines from the Component options menu in the Multi-lines settings box. When you add a new line component to a multi-line definition, you must specify the line's offset from the placement point (data point in the design) and its attributes.

The multi-line placement point is called the "working line" and is the zero offset point in the set of parallel line components. The working line may not be an actual line. For example, if three lines are defined with offsets of –2, 0, and +2 respectively, the line with 0 offset is the working line. If two lines are defined with offsets of –2 and +2 respectively, the working line is at 0 offset (half-way in between the two parallel lines), but there is no line at the 0 offset position.

The Multi-line settings box provides several options for viewing and maintaining the line components of a multi-line (see Figure 8–41):

- The Offset/Overrides list box shows the offset (in working units) of each defined line component and the line attributes that are being controlled from this settings box. It also allows you to select the line you need to edit or delete.
- The current definition area at the bottom left of the settings box provides a graphic picture of the current multi-line definition.
- The Edit pull-down menu in the settings box's menu bar, provides options to Insert, Delete, and Duplicate line components.
- The Offset edit field provides a place to type in the offset of the selected line component from the working line.
- The Attributes area provides options for setting the selected line component's attributes.

Add a New Line

To add a line to the active multi-line definition, invoke the Insert option:

Pull-down menu (Multi-lines settings box)	Edit > Insert

MicroStation adds the new line above the selected line in the Components list box.

Delete a Line

To delete a component line from the active multi-line definition, select the component line from the Components list box, then invoke the Delete option:

Pull-down menu (Multi-lines settings box)	Edit > Delete

MicroStation deletes the selected component line from the multi-line definition.

Create a Duplicate of an Existing Line

To add a component line with the same attributes as the currently selected, invoke the Duplicate option:

Pull-down menu (Multi-lines settings box)	Edit > Duplicate

MicroStation adds a component line with the same attributes as the selected line.

Set a Line's Offset

The Offset edit field sets the distance in Working Units (MU:SU:PU) from the working line to the selected component line. A positive offset places the line above the working line, and a negative offset places it below the working line. A zero offset places the line on the working line.

To enter the offset, select the line from the Offset/Overrides list, then type the offset in the Offset edit field and press the Tab key. Each line offset should be unique. If two lines have the same offset, they will be placed one on top of the other.

Set a Line's Attributes

When you create a new line component, it defaults to using the current active attributes for the design. You can override the active attributes by turning on the toggle buttons in the Attributes area of the settings box. When an attribute button is on, that attribute is saved with the multi-line definition.

To set the attributes for an individual component line, select the component line in the Offset/Overrides list box, then set the appropriate attribute options that are located in the top right side of the Multi-lines settings box. If the toggle button to the left of an attribute option is:

- ON, the line will always be placed using that attribute setting.
- OFF, the line will always be placed using the design attribute value in effect when the multi-line line is placed. When the button is OFF, the options are dimmed and cannot be accessed.

Figure 8–42 shows attributes being set as part of the multi-line definition.

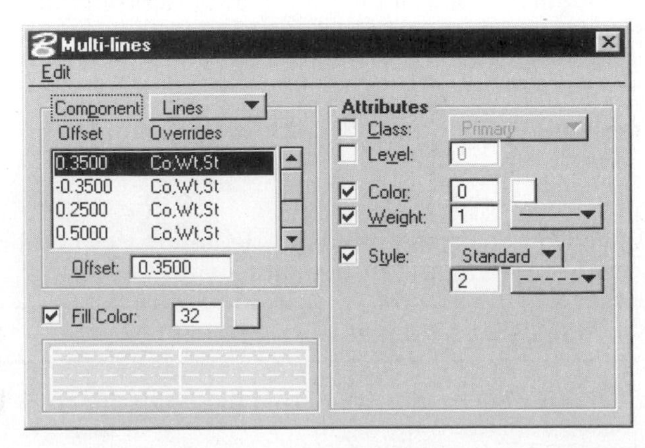

Figure 8–42 Example of line component attributes set as part of the multi-line definition

Control Multi-line Fill

The Fill color toggle button controls the color of the background area within the multi-line's parallel lines.

Turn the button ON and enter a color number in the edit field to fill the multi-line with the selected color when it is placed. Turn it OFF to have an unfilled multi-line.

You will only see filled multi-lines if the Fill view attribute is turned ON (Settings > View Attributes).

Start Cap or End Cap Components

The Start Cap and End Cap options specify the appearance of the start and end of the multi-line. To define the start and end caps, select either Start Cap or End Cap from the Component options menu. Figure 8–40 shows examples of different start and end cap arrangements. Figure 8–43 shows the dialog box when the Start Cap option is selected.

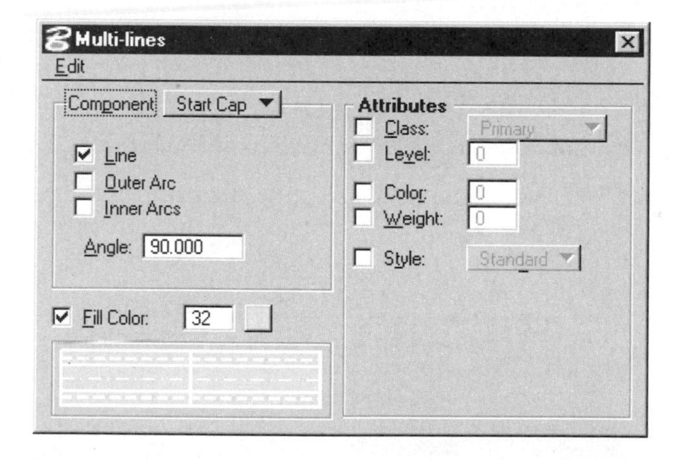

Figure 8–43 Start Cap selection displaying the controls for specifying the appearance of the start cap

The following Start Cap and End Cap options are available in the settings box:

- The Line toggle button controls the display of the cap as a line.
- The Outer Arc toggle button controls the display of the cap as an arc connecting the outermost component lines.
- The Inner Arcs toggle button controls the display of the cap, with arcs connecting pairs of inside component lines. If there is an even number of inside component lines, all are connected; if there is an odd number of inside component lines (three or more), the middle line is not connected.
- The Angle edit field sets the cap's line angle in degrees.

To set the attributes for a cap, set the appropriate attribute options that are located in the top right side of the Multi-lines settings box. If the toggle button to the left of an attribute option is:

- ON, the cap will always be placed using that attribute setting.
- OFF, the cap will always be placed using the design attribute value in effect when the multi-line line is placed. When the button is OFF, the options are dimmed and cannot be accessed.

Joints Component

To display a joint line at each multi-line vertex, select Joints from the Component options menu. One of the multi-line examples in Figure 8–40 has joints displayed. Figure 8–44 shows the Joints options in the settings box.

The Display Joints toggle button controls the placement of joint lines in multi-line elements. If it is set to ON, the joint lines are displayed at vertices.

To set the attributes for displayed joints, set the appropriate attribute options that are located in the top right side of the Multi-lines settings box. If the toggle button to the left of an attribute option is:

- ON, the joints will always be placed using that attribute setting.
- OFF, the joints will always be placed using the design attribute value in effect when the multi-line line is placed. When the button is OFF, the options are dimmed and cannot be accessed.

Note: Manage Groups setting box allows you to save the newly created multi-line definition. Refer to Chapter 14 for a detailed explanation of the various options available.

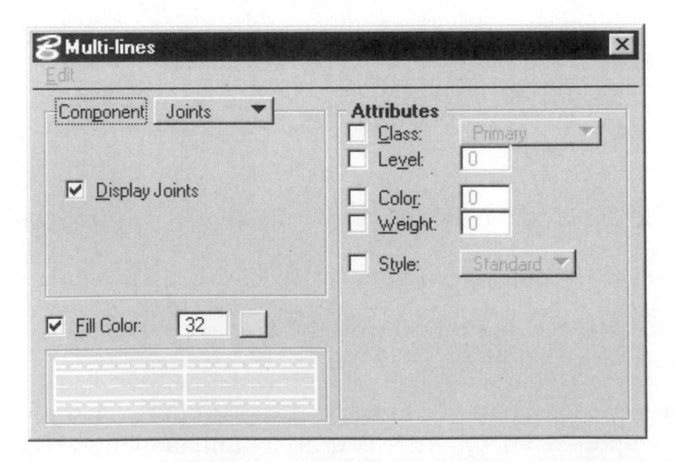

Figure 8–44 The Joints option selection displaying the controls for specifying the appearance of joints

MODIFY MULTI-LINE JOINTS

When multi-lines cross over other multi-lines, the intersection formed by the crossing needs to be cleaned up. The Multi-Line Joints tool box provides several tools for cleaning up such intersections, and for cutting holes in multi-lines.

Construct Closed Cross Joint

The Construct Closed Cross Joint tool cuts all lines that make up the first multi-line selected at the point where it crosses the second multi-line, as shown in Figure 8–45.

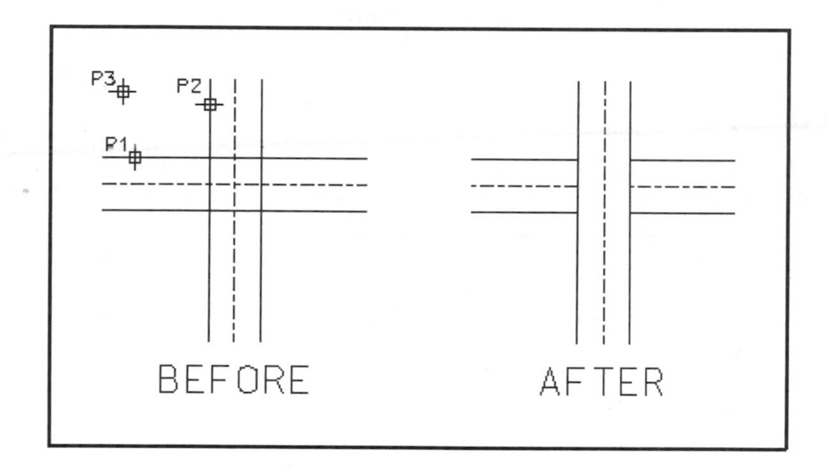

Figure 8–45 Example of using a Construct Closed Cross Joint tool

Invoke the Construct Closed Cross Joint tool:

Multi-line Joints tool box	Select the Construct Closed Cross Joint tool (see Figure 8–46).
Key-in window	**join cross closed** (or **jo cr c**) Enter

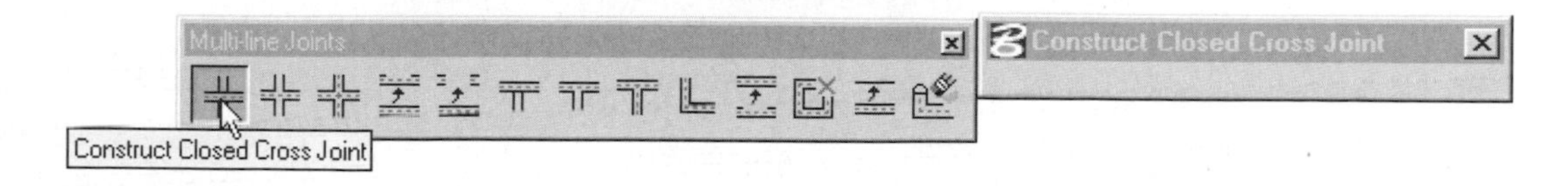

Figure 8–46 Invoking the Construct Closed Cross Joint tool from the Multi-line Joints tool box

MicroStation prompts:

> Construct Closed Cross Joint > Identify element *(Identify the multi-line P1 as shown in Figure 8–45.)*
>
> Construct Closed Cross Joint > Identify element *(Identify the multi-line P2 as shown in Figure 8–45.)*
>
> Construct Closed Cross Joint > Identify element *(Click the Data button anywhere in the view to initiate cleaning up of the intersection, or click the Reject button to reject the change.)*

Note: For each Multi-Line Joint tool, dynamic update shows the intersection cleaned up after the second data point, but it does not become permanent until you provide the third data point.

Construct Open Cross Joint

The Construct Open Cross Joint tool cuts all lines that make up the first multi-line you select and cuts only the outside line of the second multi-line, as shown in Figure 8–47.

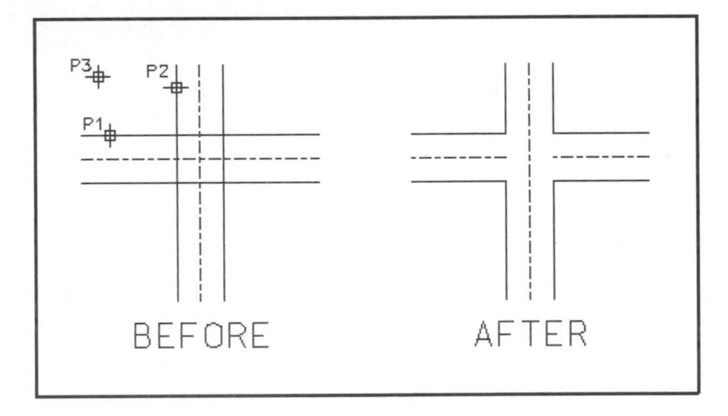

Figure 8–47 Example of using a Construct Open Cross Joint tool

Invoke the Construct Open Cross Joint tool:

Multi-line Joints tool box	Select the Construct Open Cross Joint tool (see Figure 8–48).
Key-in window	**join cross open** (or **jo cr o**) Enter

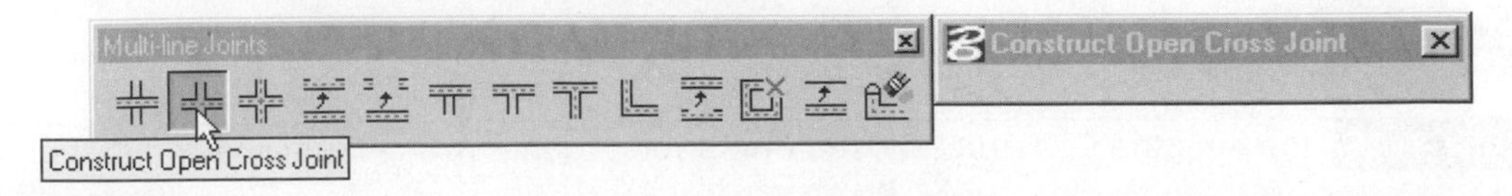

Figure 8–48 Invoking the Construct Open Cross Joint tool from the Multi-line Joints tool box

MicroStation prompts:

> Construct Open Cross Joint > Identify element *(Identify the multi-line P1 as shown in Figure 8–47.)*
>
> Construct Open Cross Joint > Identify element *(Identify the multi-line P2 as shown in Figure 8–47.)*
>
> Construct Open Cross Joint > Identify element *(Click the Data button anywhere in the view to initiate cleaning up of the intersection, or click the Reject button to reject the change.)*

Construct Merged Cross Joint

The Construct Merged Cross Joint tool cuts all lines that make up each of the intersecting multi-line you select, except the center lines, as shown in Figure 8–49. If there are no center lines, all lines in each multi-line are cut.

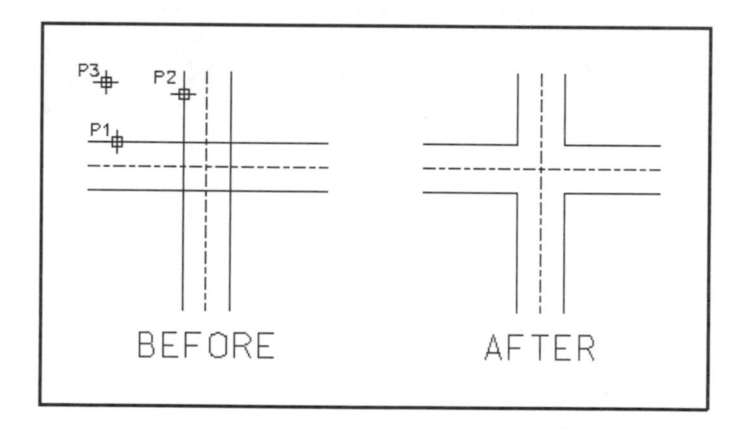

Figure 8–49 Example of using a Construct Merged Cross Joint tool

Invoke the Construct Merged Cross Joint tool:

Multi-line Joints tool box	Select the Construct Merged Cross Joint tool (see Figure 8–50).
Key-in window	**join cross merge** (or **jo cr m**) Enter

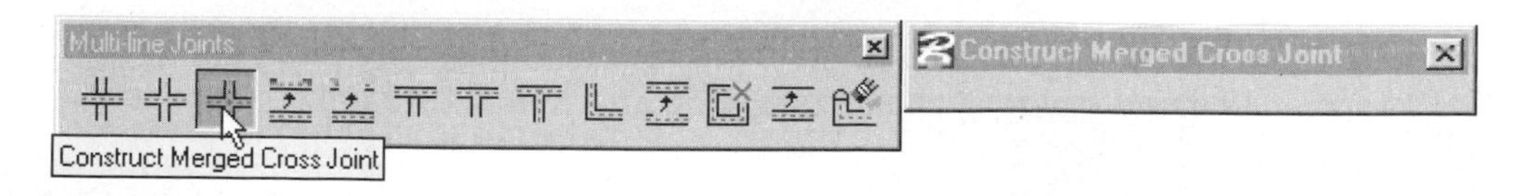

Figure 8–50 Invoking the Construct Merged Cross Joint tool from the Multi-line Joints tool box

MicroStation prompts:

> Construct Merged Cross Joint > Identify element *(Identify the multi-line P1 as shown in Figure 8–49.)*
>
> Construct Merged Cross Joint > Identify element *(Identify the multi-line P2 as shown in Figure 8–49.)*
>
> Construct Merged Cross Joint > Identify element *(Click the Data button anywhere in the view to initiate cleaning up of the intersection, or click the Reject button to reject the change.)*

Construct Closed Tee Joint

The Construct Closed Tee Joint tool extends or shortens the first multi-line you identify to its intersection with the second multi-line. The first multi-line ends at the near side of the intersecting multi-line, which is left intact, as shown in Figure 8–51.

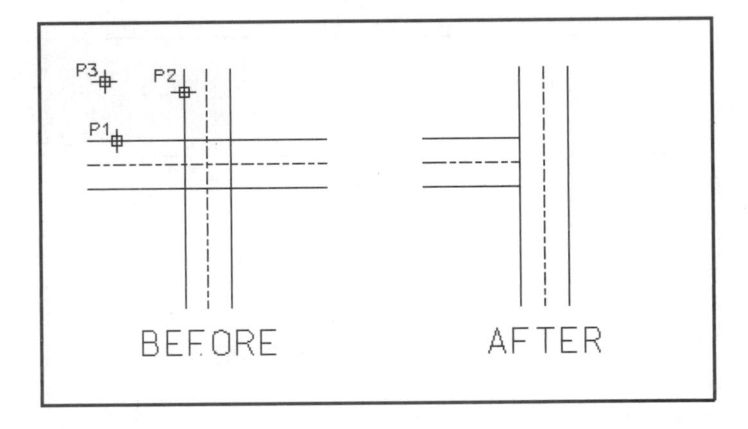

Figure 8–51 Example of using a Construct Closed Tee Joint tool

Invoke the Construct Closed Tee Joint tool:

Multi-line Joints tool box	Select the Construct Closed Tee Joint tool (see Figure 8–52).
Key-in window	**join tee closed** (or **jo t c**) Enter

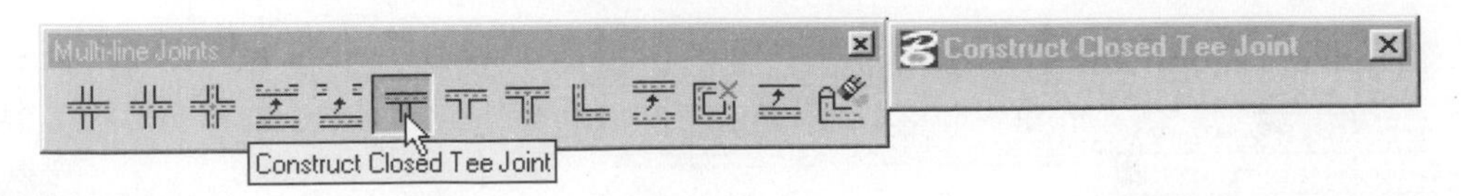

Figure 8–52 Invoking the Construct Closed Tee Joint tool from the Multi-line Joints tool box

MicroStation prompts:

> Construct Closed Tee Joint > Identify element *(Identify the multi-line P1 as shown in Figure 8–51.)*
>
> Construct Closed Tee Joint > Identify element *(Identify the multi-line P2 as shown in Figure 8–51.)*
>
> Construct Closed Tee Joint > Identify element *(Click the Data button anywhere in the view to initiate cleaning up of the intersection, or click the Reject button to reject the change.)*

Construct Open Tee Joint

The Construct Open Tee Joint tool is similar to the Closed Tee Joint tool, except it leaves an open end at the intersecting multi-line, as shown in Figure 8–53.

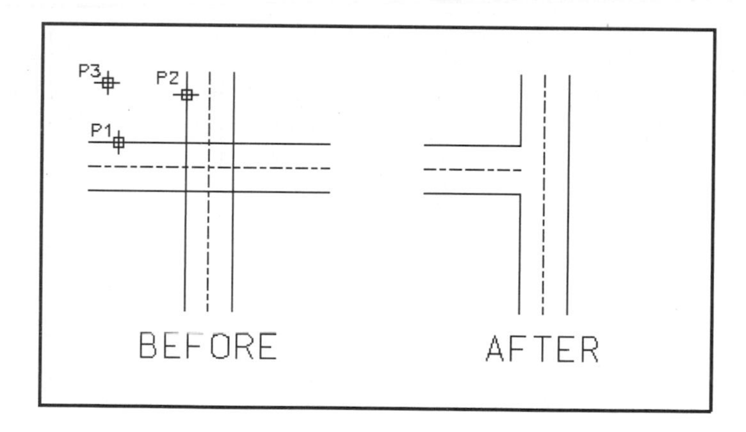

Figure 8–53 Example of using a Construct Open Tee Cross Joint tool

Invoke the Construct Open Tee Joint tool:

Multi-line Joints tool box	Select the Construct Open Tee Joint tool (see Figure 8–54).
Key-in window	**join tee open** (or **jo t o**) Enter

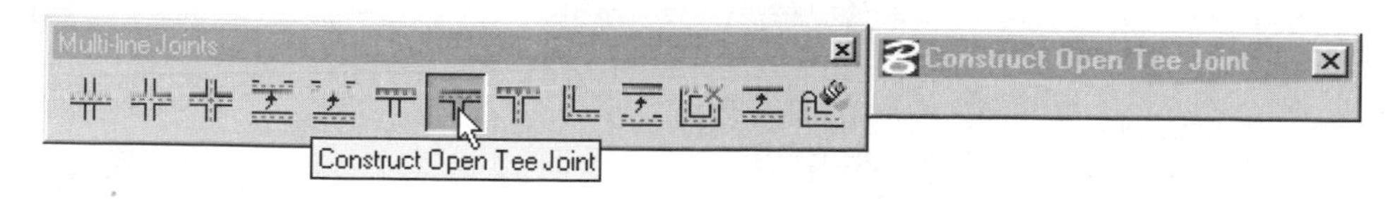

Figure 8–54 Invoking the Construct Open Tee Joint tool from the Multi-line Joints tool box

MicroStation prompts:

Construct Open Tee Joint > Identify element *(Identify the multi-line P1 as shown in Figure 8–53.)*

Construct Open Tee Joint > Identify element *(Identify the multi-line P2 as shown in Figure 8–53.)*

Construct Open Tee Joint > Identify element *(Click the Data button anywhere in the view to initiate cleaning up of the intersection, or click the Reject button to reject the change.)*

Construct Merged Tee Joint

The Construct Merged Tee Joint tool is similar to the Open Tee Joint tool, except the center line of the first multi-line is extended to the center line of the intersecting multi-line, as shown in Figure 8–55.

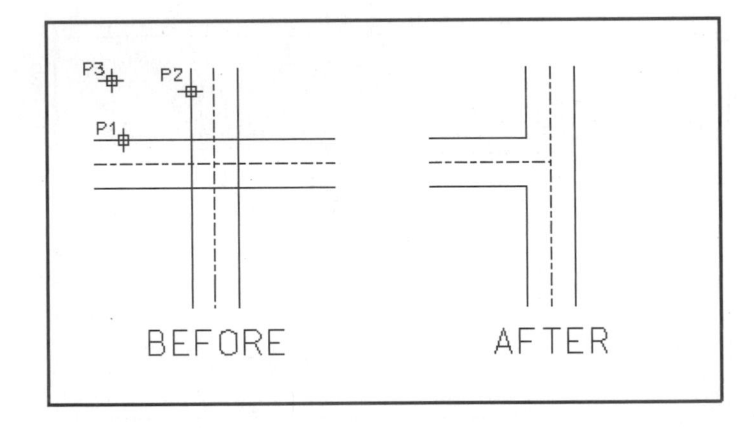

Figure 8–55 Example of using a Construct Merged Tee Joint tool

Invoke the Construct Merged Tee Joint tool:

Multi-line Joints tool box	Select the Construct Merged Tee Joint tool (see Figure 8–56).
Key-in window	**join tee merge** (or **jo t m**) Enter

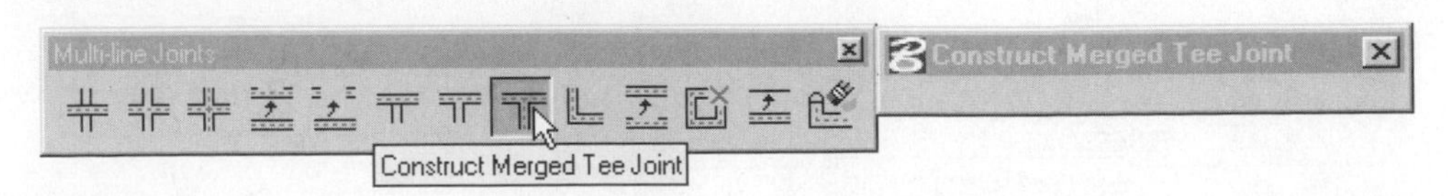

Figure 8–56 Invoking the Construct Merged Tee Joint tool from the Multi-line Joints tool box

MicroStation prompts:

> Construct Merged Tee Joint > Identify element *(Identify the multi-line P1 as shown in Figure 8–55.)*
>
> Construct Merged Tee Joint > Identify element *(Identify the multi-line P2 as shown in Figure 8–55.)*
>
> Construct Merged Tee Joint > Identify element *(Click the Data button anywhere in the view to initiate cleaning up of the intersection, or click the Reject button to reject the change.)*

Construct Corner Joint

The Construct Corner Joint tool lengthens or shortens each of the two multi-lines you select as necessary to create a clean intersection, as shown in Figure 8–57.

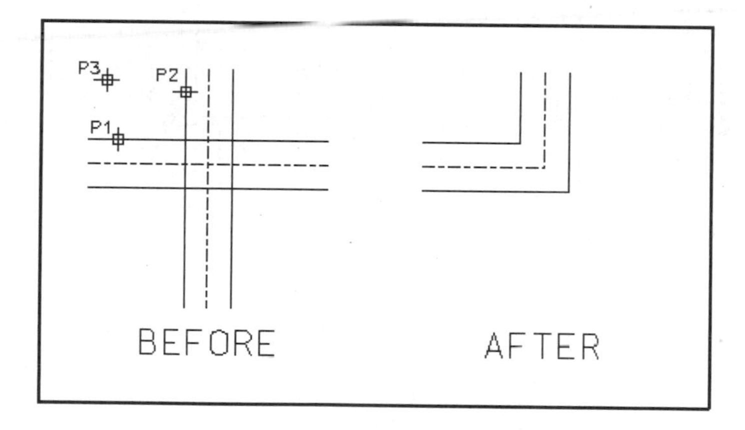

Figure 8–57 Example of using a Construct Corner Joint tool

Invoke the Construct Corner Joint tool:

Multi-line Joints tool box	Select the Construct Corner Joint tool (see Figure 8–58).
Key-in window	**join corner** (or **jo c**) Enter

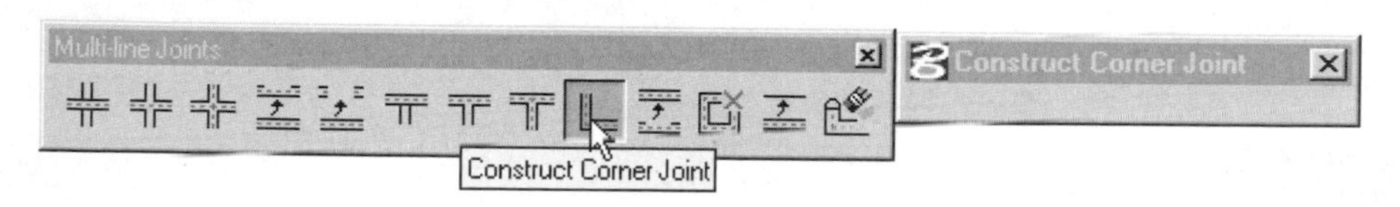

Figure 8–58 Invoking the Construct Corner Joint tool from the Multi-line Joints tool box

MicroStation prompts:

Construct Joint > Identify element *(Identify the multi-line P1 as shown in Figure 8–57.)*

Construct Joint > Identify element *(Identify the multi-line P2 as shown in Figure 8–57.)*

Construct Joint > Identify element *(Click the Data button anywhere in the view to initiate cleaning up of the intersection, or click the Reject button to reject the change.)*

Cut Single Component Line

The Cut Single Component Line tool cuts a hole in the line you select in a multi-line from the first data point to the second data point, as shown in Figure 8–59.

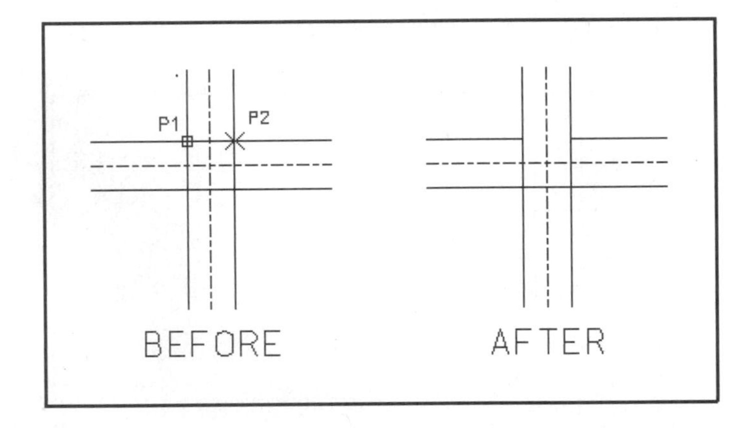

Figure 8–59 Example of using a Cut Single Component Line tool

Invoke the Cut Single Component Line tool:

Multi-line Joints tool box	Select the Cut Single Component Line tool (see Figure 8–60.)
Key-in window	**cut single** (or **cu s**) Enter

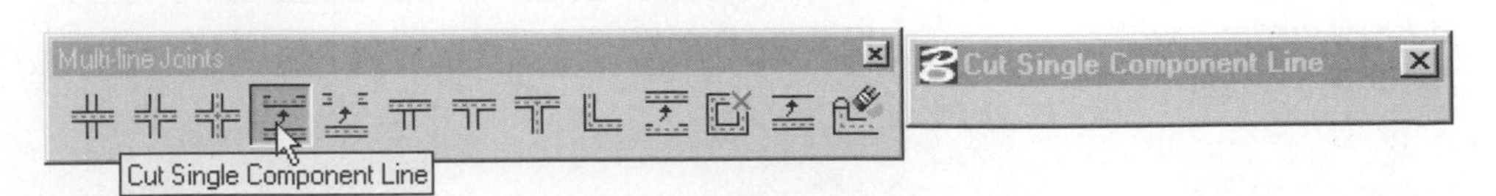

Figure 8–60 Invoking the Cut Single Component Line tool from the Multi-line Joints tool box

MicroStation prompts:

> Cut Single Component Line > Identify element *(Identify the multi-line at P1 as shown in Figure 8–59.)*
>
> Cut Single Component Line *(Identify the multi-line at P2 as shown in Figure 8–59 to remove the portion of the line, or click the Reset button to reject the change.)*

Cut All Component Lines

The Cut All Component Lines tool cuts a hole in the multi-line you select from the first data point to the second data point, as shown in Figure 8–61. The multi-line is still an element after this tool cuts a hole in it.

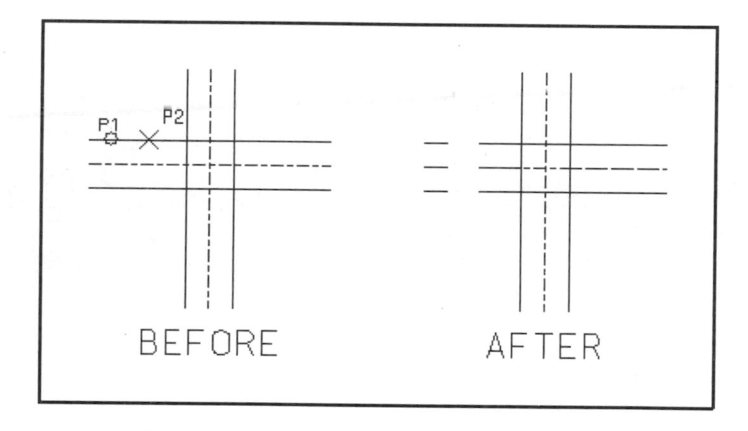

Figure 8–61 Example of using a Cut All Component Lines tool

Invoke the Cut All Component Lines tool:

Multi-line Joints tool box	Select the Cut All Component Lines tool (see Figure 8–62.)
Key-in window	**cut all** (or **cu a**) Enter

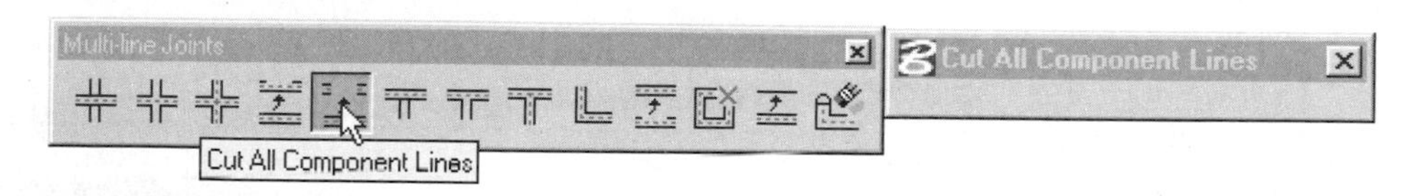

Figure 8–62 Invoking the Cut All Component Lines tool from the Multi-line Joints tool box

MicroStation prompts:

> Cut All Component Lines > Identify element *(Identify the multi-line at P1 as shown in Figure 8–61.)*
>
> Cut All Component Lines *(Identify the multi-line at P2 as shown in Figure 8–61 to remove the portion of the multi-line, or click the Reset button to reject the change.)*

Uncut Component Lines

The Uncut Component Lines tool provides a special undo tool for multi-lines. With it you can undo a cut in one line of a multi-line. Identify one end of the cut with a Keypoint snap, then click the Data button to accept the snap, and click the Data button a second time to initiate removing the cut, as shown in Figure 8–63.

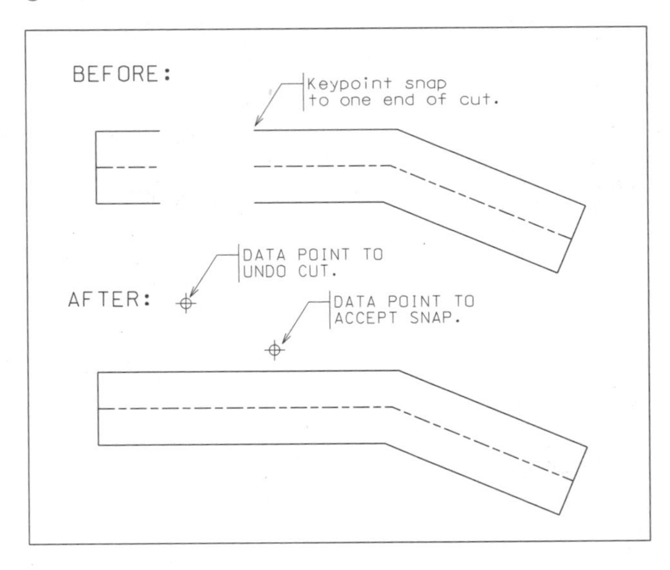

Figure 8–63 Example of using an Uncut Component Lines tool

Invoke the Uncut Component Lines tool:

Multi-line Joints tool box	Select the Uncut Component Lines tool (see Figure 8–64).
Key-in window	**uncut** (or **un**) Enter

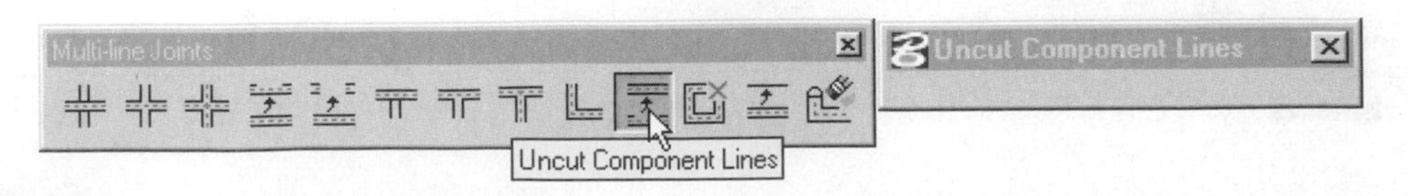

Figure 8–64 Invoking the Uncut Component Lines tool from the Multi-line Joints tool box

MicroStation prompts:

> Uncut Component Lines > Identify element *(Keypoint snap to one end of the cut, as shown in Figure 8–63.)*
>
> Uncut Component Lines *(Place a data point to accept the Keypoint snap, or click the Reset button to reject the change.)*
>
> Uncut Component Lines *(Place a second data point to undo the cut, or click the Reset button to reject the change.)*

Multi-line Partial Delete

The Multi-line Partial Delete tool enables you to reduce the length of a multi-line element by selecting one end of it, or break it into two separate elements by cutting a hole in it. This tool also allows you to control cap placement on each end of the hold, using one of the following four options in the Cap Mode option menu:

- None: No caps are created, similar to the Cut All Component Lines tool.
- Current: Uses the start cap and end cap definitions that were in effect when the multi-line was originally placed.
- Active: Uses the active multi-line profile start cap and end cap definitions.
- Joint: Place a 90-degree joint line on each end of the hole.

Invoke the Multi-line Partial Delete tool:

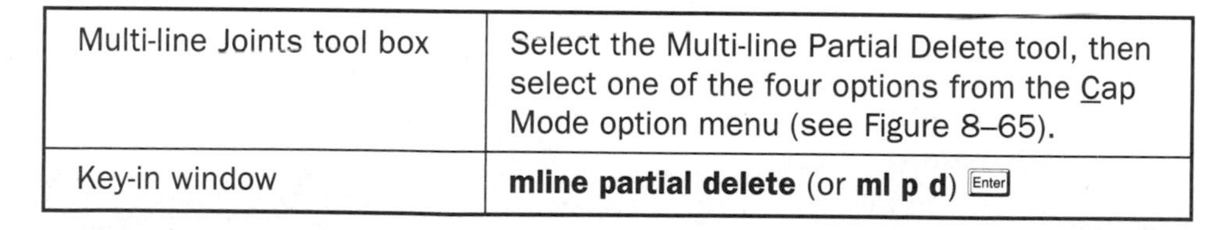

Multi-line Joints tool box	Select the Multi-line Partial Delete tool, then select one of the four options from the Cap Mode option menu (see Figure 8–65).
Key-in window	**mline partial delete** (or **ml p d**) Enter

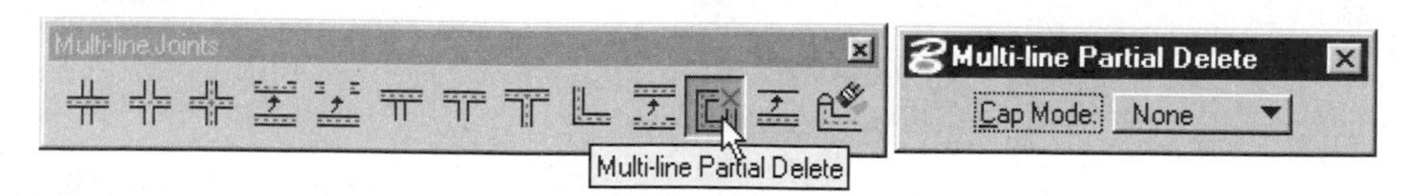

Figure 8–65 Invoking the Multi-line Partial Delete tool from the Multi-line Joints tool box

MicroStation prompts:

> Multi-line Partial Delete > Identify multi-line at start of delete *(Identify the multi-line at one end of the part to delete.)*
>
> Multi-line Partial Delete > Define length of delete *(Place a data point to define the length of the delete, or click the Reset button to reject the change.)*

Move Multi-line Profile

The Move Multi-line Profile tool will move an individual component line of a multi-line, or reposition the working line of a multi-line without moving any component lines.

Move Component

The Component Move option allows you to move a selected line component in a multi-line element. The line can only be moved in parallel to the other lines in the multi-line element.

To move a line component, invoke the Move Multi-line Profile:

Multi-line Joints tool box	Select the Move Multi-line Profile tool, then select Component from the Move option menu in the Tool settings Window (see Figure 8–66).
Key-in window	**mline edit profile** (or **ml e p**) Enter

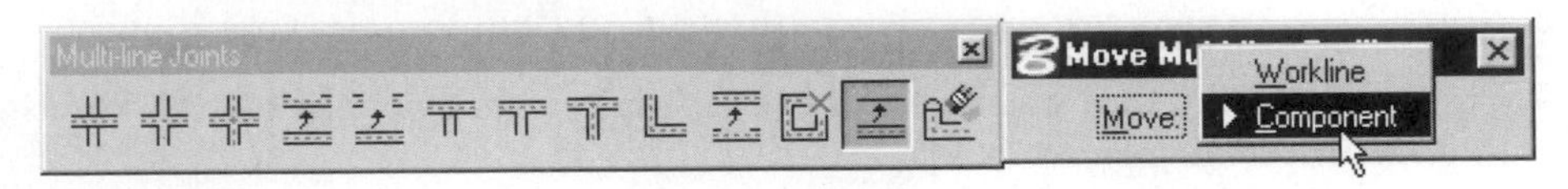

Figure 8–66 Invoking the Move Multi-line Profile tool from the Multi-line Joints tool box

MicroStation prompts:

> Move Multi-line Profile > Identify multi-line component to move *(Identify the component to move.)*
>
> Move Multi-line Profile > Define component position (reset to reject element) *(Place a Data point to reposition the component, or click the Reset button to reject the move.)*

Move Working Line

The Work line Move option allows you move the location of the insertion point (working line) in a multi-line element. This option has the effect of changing the zero Offset position. You usually will not see any change in the multi-line after using this tool.

The result of a working line move becomes apparent when you do element manipulations on the multi-line element. For example, if you use the Modify Element tool to move one end of the multi-line element, it will pivot about the new working line position.

To move the working line, invoke the Move Multi-line Profile:

Multi-line Joints tool box	Select the Move Multi-line Profile tool, then select Workline from the Move option menu in the Tool settings Window (see Figure 8–66).
Key-in window	**mline edit profile** (or **ml e p**) Enter

MicroStation prompts:

> Move Multi-line Profile > Identify multi-line profile *(Identify the multi-line.)*
>
> Move Multi-line Profile > Define new workline (reset to reject element) *(Place a Data point to reposition the working line, or click the Reset button to reject the move.)*

Edit Multi-line Cap

The Edit Multi-line Cap tool changes the end cap of a multi-line. MicroStation provides four options under the Cap Mode option menu:

- None: Removes any end caps. The effect is the same as with the Cut All Component Lines tool.
- Current: Does not change the end cap; enabled only when Adjust Angle is turned ON.
- Active: Uses the active multi-line definitions for the end cap.
- Joint: Uses the identified multi-line's joint definition instead of the end cap definition; ensures the end cap will always be 90 degrees.

Invoke the Edit Multi-line Cap tool:

Multi-line Joints tool box	Select the Edit Multi-line Cap tool, then select one of the four options from the Cap Mode option menu in the Tool Settings window (see Figure 8–67).
Key-in window	**mline edit cap** (or **ml e c**) Enter

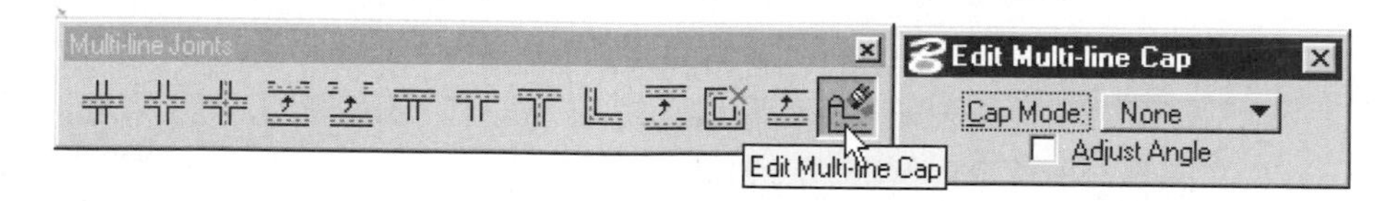

Figure 8–67 Invoking the Edit Multi-line Cap tool from the Multi-line Joints tool box

MicroStation prompts:

> Edit Multi-line Cap > Identify multi-line near the end cap to modify *(Identify the multi-line near the end cap.)*
>
> Edit Multi-line Cap > Data to change end cap (reset to reject element) *(Place a data point to change the end cap, or click the Reject button to reject the change.)*

REVIEW QUESTIONS

Write your answers in the spaces provided.

1. Explain briefly the options available with the Extend Line tool.

   ____

   ____

2. List the element types that can be modified by means of the Modify Element tool.

   ____

   ____

3. List the tools available to modify an arc.

   ____

   ____

4. Explain the difference between creating a chain manually and creating one automatically.

   ____

   ____

5. To drop a complex chain, invoke the ______________ tool.

6. Explain the difference between the Construct Closed Cross Joint tool and the Construct Merged Cross Joint tool.

   ____

   ____

7. Explain the difference between the Construct Closed Tee Joint tool and the Construct Merged Tee Joint tool.

   ____

   ____

8. Give the steps involved in moving the Multi-line profile.

   ____

   ____

PROJECT EXERCISE

This project exercise provides step-by-step instructions for creating the Utility–Storage Floor Plan shown in Figure P8–1. The intent is to guide you in applying the Multi-line setting, placement, and joints tools.

Note: The dimensions are not part of this project. They are included in Figure P8–1 as an aid to drawing the design.

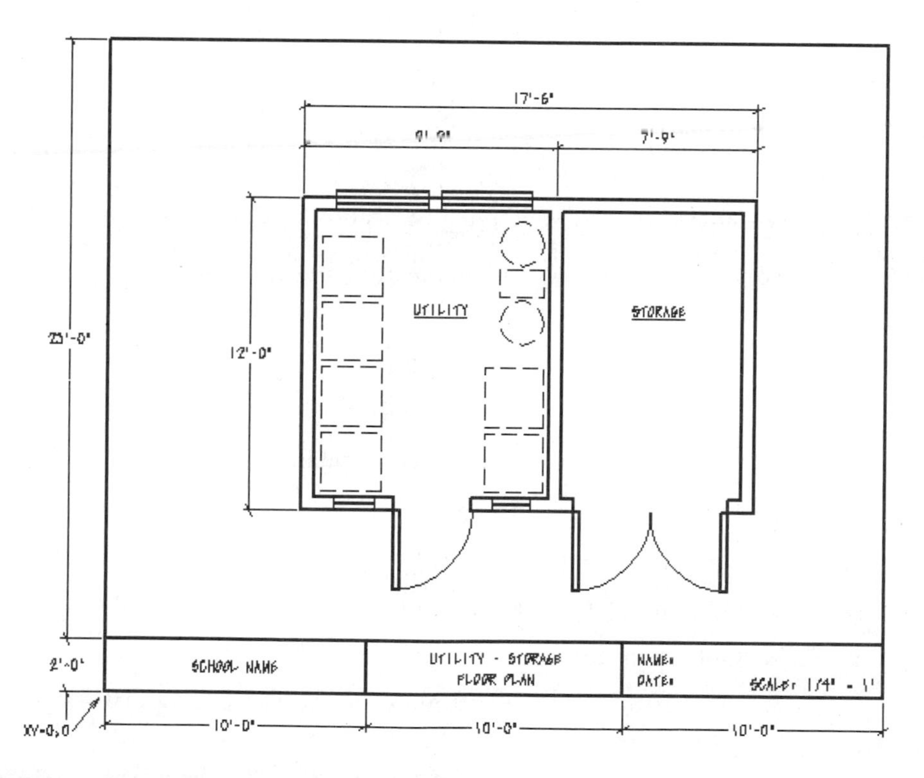

Figure P8–1 Completed project design

Prepare the Design File

This procedure starts MicroStation, creates a design file, and enters the initial settings.

Note: As you complete each step in the project procedures, place a check mark by the step to help you keep up with where you are in the project.

STEP 1: Invoke MicroStation, and create a new design file named CH8.DGN using the SEED2D.DGN seed file.

STEP 2: In the Design File dialog box set the following:

- Working Unit ratios to 12" Per ' and 8000 Pos Units Per "
- Grid Master to 0.5, Grid Reference to 2, and Grid Lock ON.

STEP 3: Invoke the Text settings box from the pull-down menu Element, and adjust the settings as follows:

- Font = 41 - Architectural
- Text Height = 0.4
- Text Width = 0.3
- Line Spacing = 0.4
- Single-line and Multi-line justifications = Center Center

STEP 4: Turn on AccuDraw, and make the following settings in the AccuDraw settings box:

- Unit Roundoff Distance = 0:1
- Coordinate System Rotation = Top
- Coordinate System Type = Rectangular
- Operation Floating Origin = ON
- Operation Context Sensitivity = ON
- Operation Smart Key-ins = ON
- Operation Auto Point Placement = OFF
- Open Coordinate Readout settings box, set Coordinates Format to Sub Units

STEP 5: Select Save Settings from the File pull-down menu.

Draw the Border and Title Block

This procedure draws the border and title block, then places the required text in the title block.

STEP 1: Using Figure P8–1 as a guide, draw the 30' x 25' border and title block on level 10. Place the lower-left corner of the border at XY=0,0.

STEP 2: Fit the view window.

STEP 3: Place the title block text:

- Replace "SCHOOL NAME" with your school or company name, or make up a name.
- Place your name to the right of "NAME:"
- Place today's date to the right of "DATE:"

STEP 4: Select Save Settings from the File pull-down menu.

Create a Multi-line Profile

This procedure sets up the multi-line element to be used to draw the walls of the utility–storage floor plan.

STEP 1: Open the Multi-lines settings box by selecting Multi-lines from the Element pull-down menu.

STEP 2: In the Multi-lines dialog box, select Lines from the Component options menu.

STEP 3: If the Offset/Overrides list box contains more than two line components, delete all but two of them. To delete each extra line:

- Select a line in the Offset/Overrides list box.
- Select Delete from the dialog box's Edit pull-down menu.

STEP 4: Select the first line in the Offset/Overrides list box, then set the Offset to 0.0000 (the working line), and set the Attributes as shown in Figure P8–2.

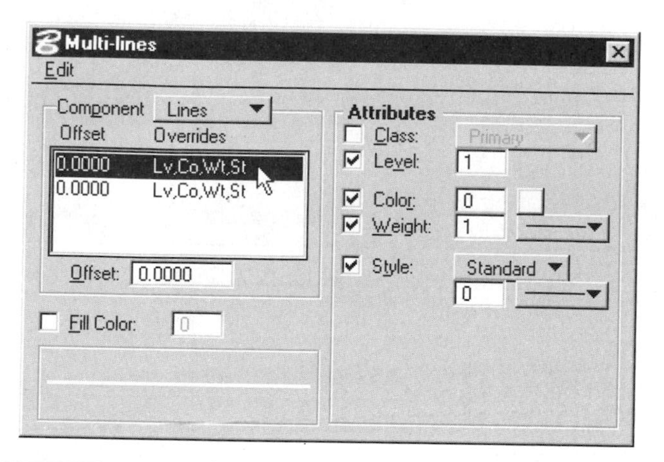

Figure P8–2 Settings for the working line in the wall's Multi-line profile

STEP 5: Select the second line in the Component list box and set the Attributes as shown in Figure P8–3.

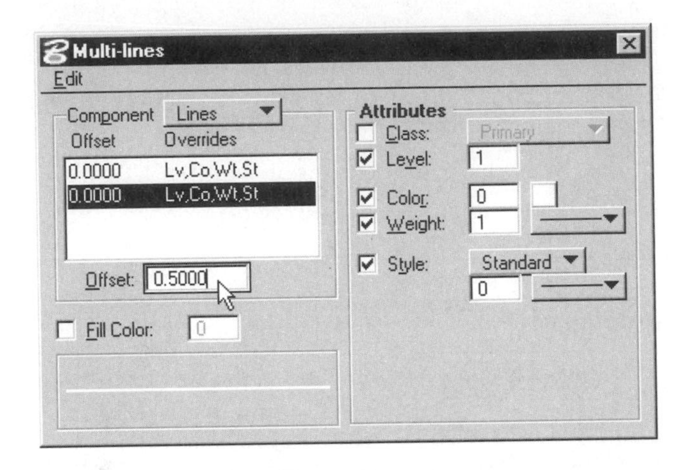

Figure P8–3 Setting the Attributes for the second line in the wall's Multi-line profile

STEP 6: Select Start Cap from the Component option menu, then set the cap type and Attributes as shown in Figure P8–4.

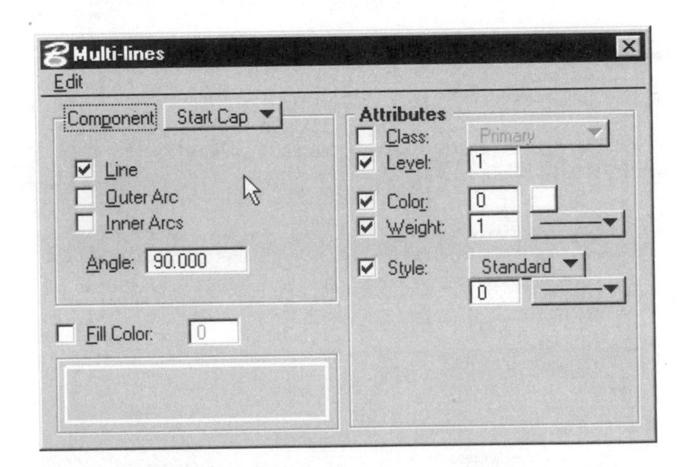

Figure P8–4 Start Cap settings for the wall's Multi-line profile

STEP 7: Select End Cap from the Component option menu, then set the cap type and Attributes as shown in Figure P8–5.

STEP 8: Select Joints from the Component option menu, then turn off the Display Joints toggle button.

STEP 9: Close the Multi-lines settings box.

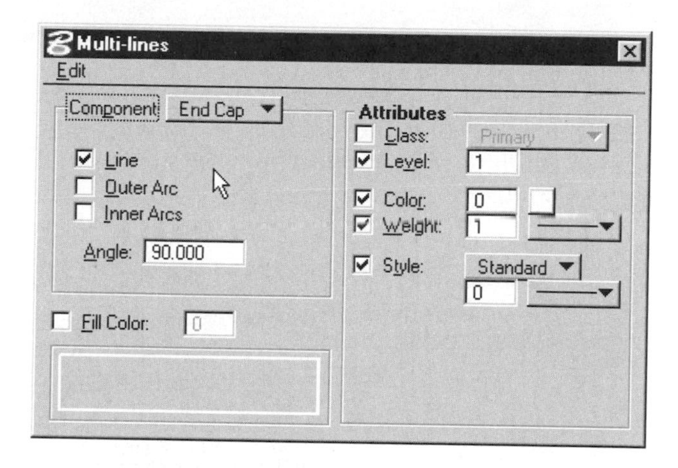

Figure P8–5 End Cap settings for the wall's Multi-line profile

STEP 10: Select Save Settings from the File pull-down menu.

Draw the Walls

This procedure uses the Multi-line tool and AccuDraw to draw the utility–storage floor plan walls, as shown in Figure P8–6.

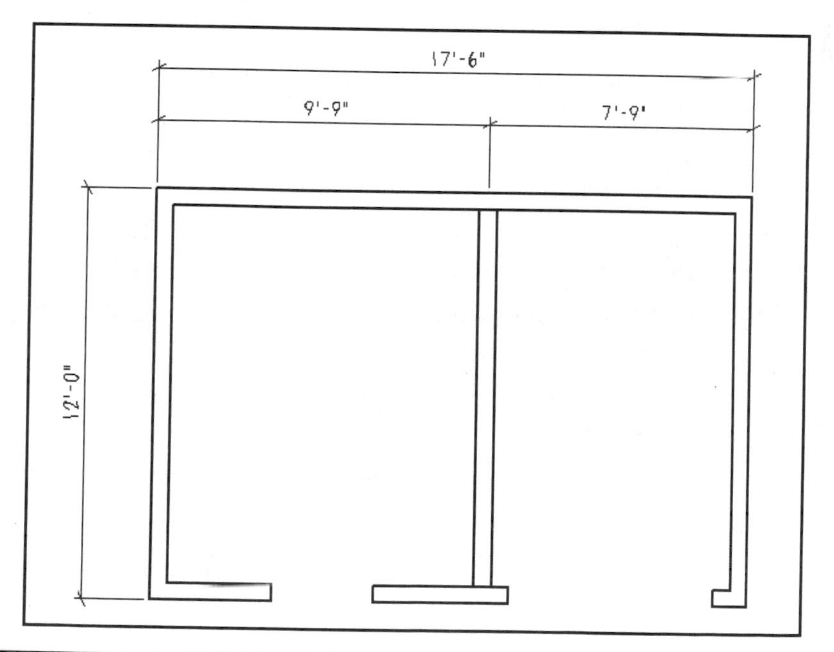

Figure P8–6 The walls before holes are cut for the windows

STEP 1: Set the Active Level to 1 and the Line Weight to 2, then select Save Settings from the File pull-down menu.

STEP 2: Invoke the Place Multi-line tool from the Linear Elements tool box, then, in the Tool Settings window, select Workline from the Place by options menu and turn off the Length and Angle toggle buttons, as shown in Figure P8–7.

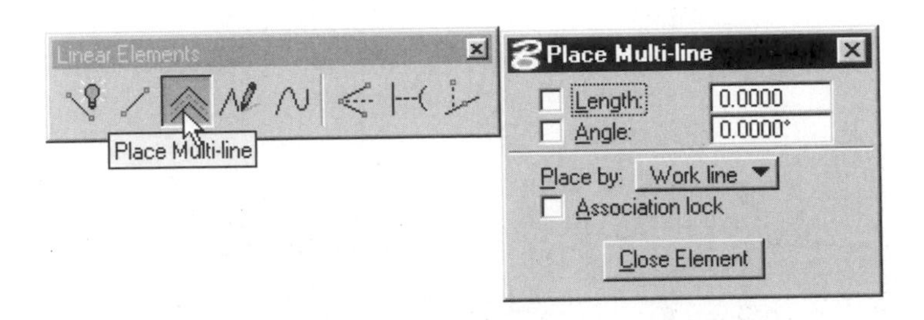

Figure P8–7 Invoking the Place Multi-line tool

MicroStation prompts:

Place Multi-line > Enter first point *(Keypoint snap to the lower-left corner of the border, press the* ***0*** *key to position the AccuDraw compass at the Keypoint, then place a data point at this offset from the compass:* X = *24 and* Y = *7.)*

Place Multi-line > Enter vertex or Reset to complete *(Place the next data point at this offset from the compass:* X = *1 and* Y = *0.)*

Place Multi-line > Enter vertex or Reset to complete *(Place the next data point at this offset from the compass:* X = *0 and* Y = *12.)*

Place Multi-line > Enter vertex or Reset to complete *(Place the next data point at this offset from the compass:* X = *0 and* Y = *17.5.)*

Place Multi-line > Enter vertex or Reset to complete *(Place the next data point at this offset from the compass:* X = *0 and* Y = *12.)*

Place Multi-line > Enter vertex or Reset to complete *(Place the next data point at this offset from the compass:* X = *0 and* Y = *3.5.)*

Place Multi-line > Enter vertex or Reset to complete *(Click the Reset button.)*

The completed Multi-line element is shown in Figure P8–8.

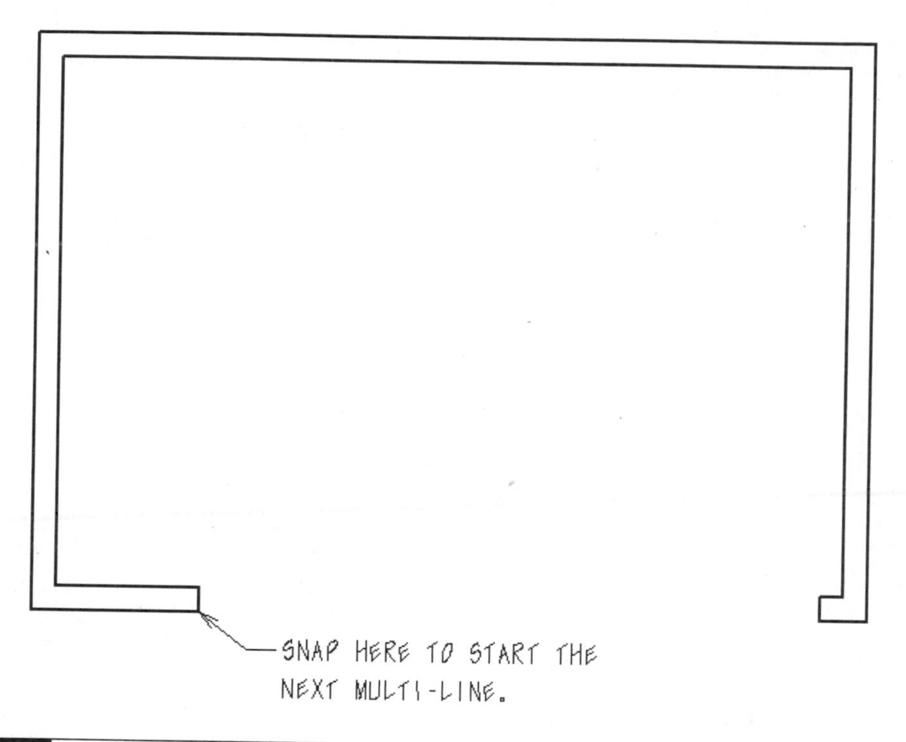
SNAP HERE TO START THE
NEXT MULTI-LINE.

Figure P8–8 First Multi-line element

STEP 3: With the Multi-line tool still active, start the next short piece of multi-line outer wall by Keypoint snapping to the lower right corner of the bottom left horizontal wall, as shown by the note in Figure P8–8.

STEP 4: Press the **O** key to position the AccuDraw compass on the Keypoint.

STEP 5: Start the new Multi-Line at this offset from the compass: $X = 3$ and $Y = 0$.

STEP 6: Place the next data point at this offset from the compass: $X = 4$ and $Y = 0$.

STEP 7: Click the Reset button to complete the Multi-line.

STEP 8: Use the Window Area tool to move in close to the walls you just completed, then click the Reset button to return to placing Multi-line elements.

The completed Multi-line element is shown in Figure P8–9.

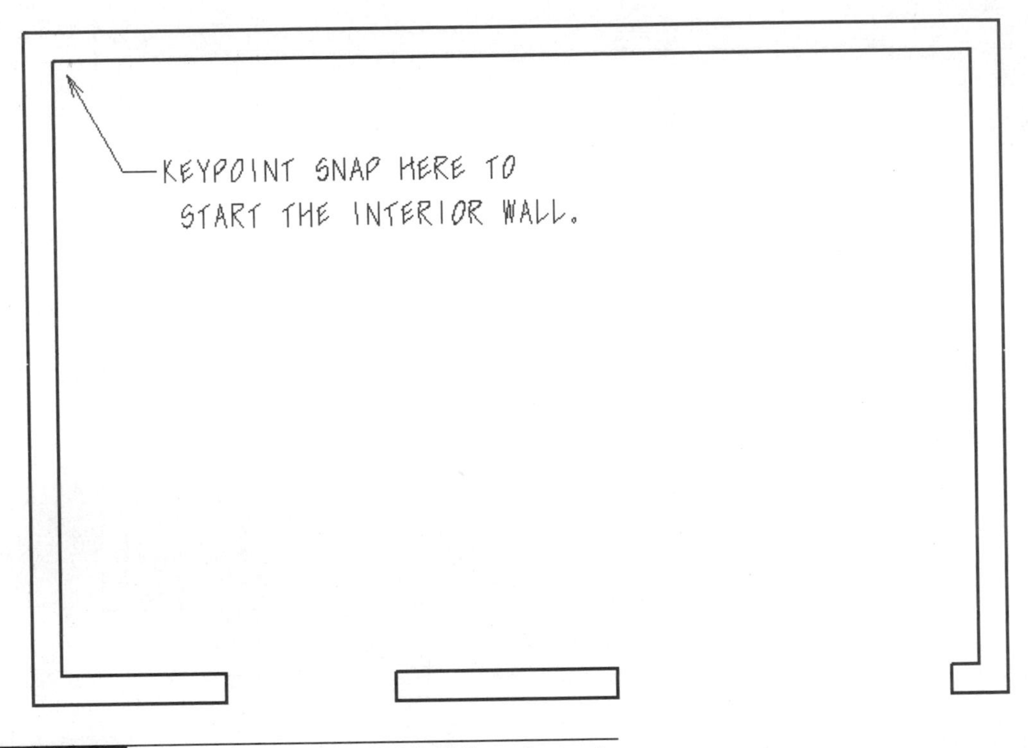
KEYPOINT SNAP HERE TO
START THE INTERIOR WALL.

Figure P8–9 Completed outer wall Multi-line elements

STEP 9: Draw the interior wall by Keypoint snapping to the inside of the top-left corner of the outer wall, as shown by the note in Figure P8–9.

STEP 10: Press the **O** key to position the AccuDraw compass on the Keypoint.

STEP 11: Start the new Multi-Line at this offset from the compass: $X = 9$ and $Y = 0$.

STEP 12: Place the next data point at this offset from the compass: $X = 0$ and $Y = -11$.

STEP 13: Click the Reset button to complete the Multi-line.

The completed Multi-line element is shown in Figure P8–6.

Cut Holes for Placing Windows in the Outer Wall

This procedure uses the Multi-line Joints tools to clean up the inner and outer wall intersections and to cut four holes in the utility room wall for windows, as shown in Figure P8–10.

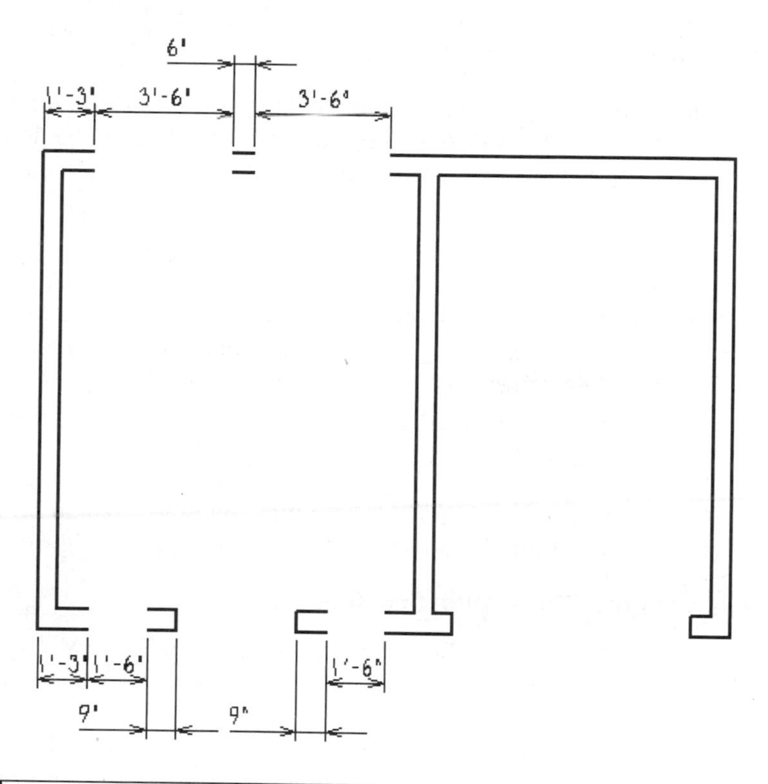

Figure P8–10 The walls after cutting holes for windows and wall unions

STEP 1: Open the Multi-line Joints tool box from the Tools pull-down menu.

STEP 2: To create the joint at the top of the interior wall, invoke the Construct Open Tee Joint tool from the Multi-line Joints tool box.

MicroStation prompts:

> Construct Open Tee Joint > Identify element *(Select the interior Multi-line element near its top, select the exterior Multi-line element, then click the Data button in space to complete the joint.)*

STEP 3: Repeat step 2 for the joint at the bottom of the interior wall.

STEP 4: To create the window openings next to the utility room door, invoke the Cut All Component Lines tool from the Multi-line Joints tool box.

MicroStation prompts:

Cut All Component Lines > Identify element *(Keypoint snap to the lower-left outside corner of the outer wall, then press the* ***O*** *key to position the AccuDraw compass on the Keypoint, and start the cut at this offset from the compass:* X = *1.25* *and* Y = *0.)*

Cut All Component Lines *(Complete the cut at this offset from the compass:* X = *1.5 and* Y = *0.)*

Cut All Component Lines > Identify element *(Keypoint snap to the lower-left corner of the wall on the right side of the utility room door, then press the* ***O*** *key. Start the cut at this offset from the compass:* X = *.75 and* Y = *0.)*

Cut All Component Lines *(Complete the cut at this offset from the compass:* X = *1.5 and* Y = *0.)*

STEP 5: Use the Cut All Component Lines tool to cut two 3.5'-wide window holes in the top wall of the utility room. Use the dimensions in Figure P8–10 as a guide.

Draw the Windows with the Place Multi-line Tool

This procedure changes the Multi-line settings, then uses the Place Multi-line tool to place windows in the opening created in the utility room exterior walls, as shown in Figure P8–11.

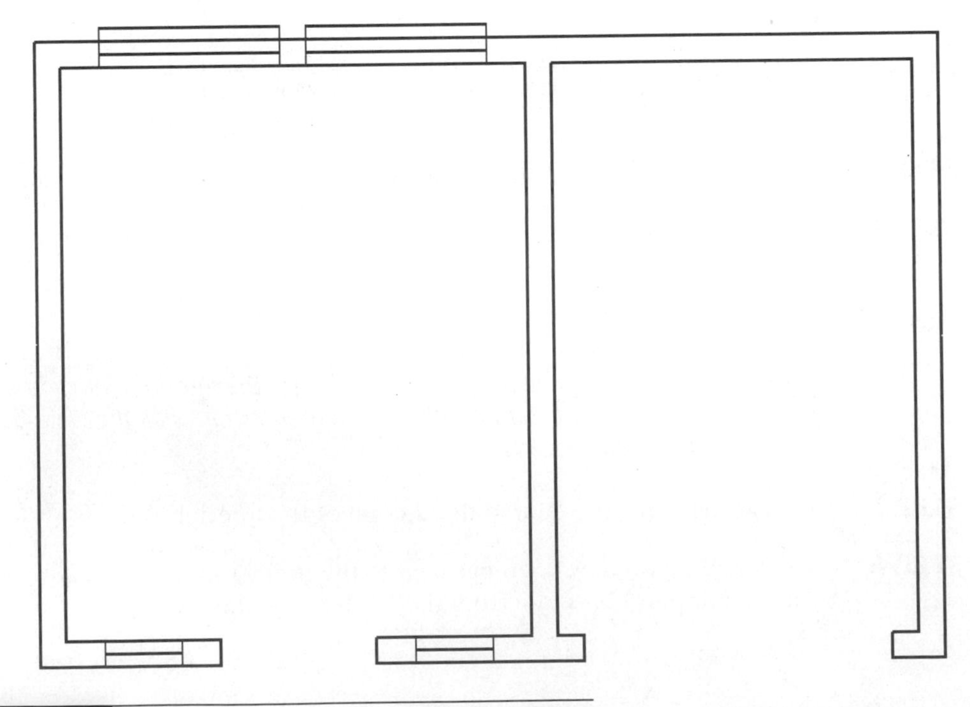

Figure P8–11 Utility room walls with windows inserted

STEP 1: Select Multi-lines from the Element pull-down menu.

STEP 2: In the Multi-lines settings box, Insert a Line component with an Offset of 0.25, and set the new line to the same attribute values as the other two line components.

STEP 3: Invoke the Place Multi-line tool from the Linear Element tool box, and place Multi-line elements in the spaces on each side of the utility room door.

STEP 4: In the Multi-lines settings box, Insert a fourth Line component with an Offset of 0.75 and set the new line to the same attribute values as the other three line components.

STEP 5: Place Multi-line elements in the spaces cut into the top utility room wall.

Complete the Floor Plan

This procedure places the equipment in the utility room and the room names in each room, as shown in Figure P8–12.

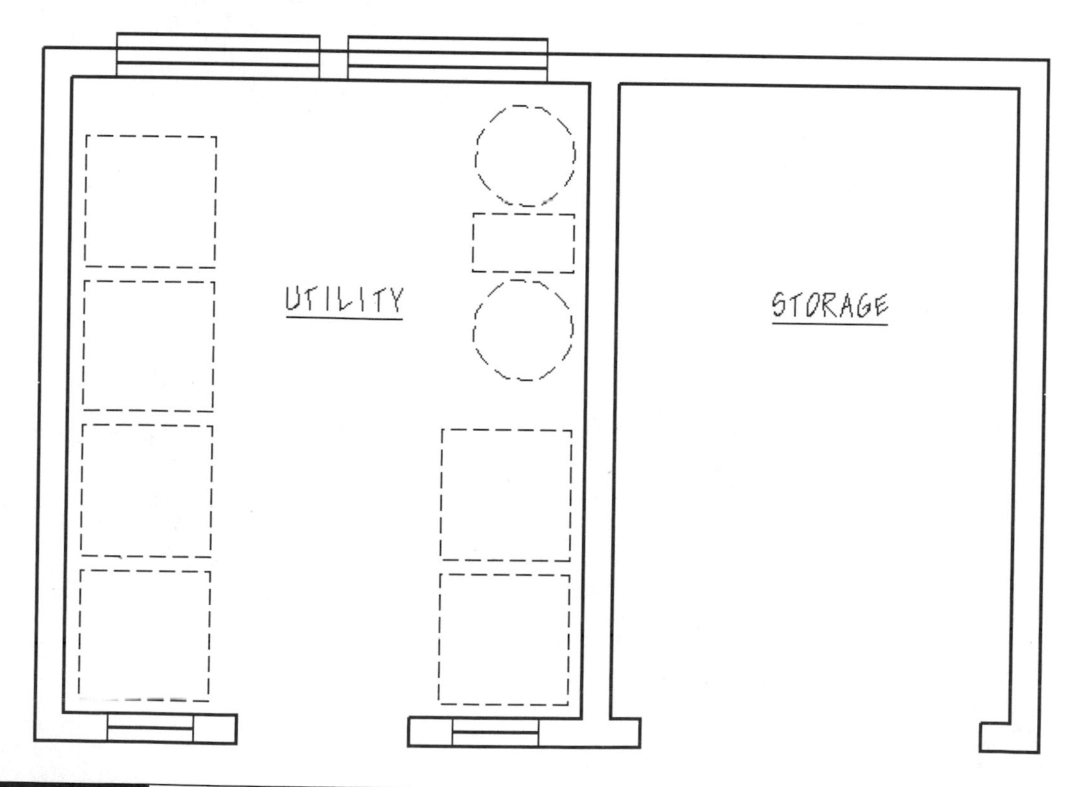

Figure P8–12 Completed floor plan

STEP 1: Place the equipment symbols in the utility room with the following dimensions:

- 6 each, 2.25' x 2.25' blocks
- 1 each, 1.0' x 1.75' block
- 2 each, 0.875-radius circles

STEP 2: Place the room names with the same text font and size as the title block text.

STEP 3: Invoke the Fit View tool to fit the view in the window view.

STEP 4: Compare your design to Figure P8–1 and, if necessary, correct mistakes.

STEP 5: Compress the design and save the design settings.

DRAWING EXERCISES 8–1 THROUGH 8–5

Use the following table to set up the design files for Exercises 8–1 through 8–3.

SETTING	VALUE
Seed File	SEED2D.DGN
Working Units	10 TH Per IN AND 10000 Pos Units Per IN
Grid	Master = 0.1, Grid Reference = 10

Exercise 8–1

Machine part

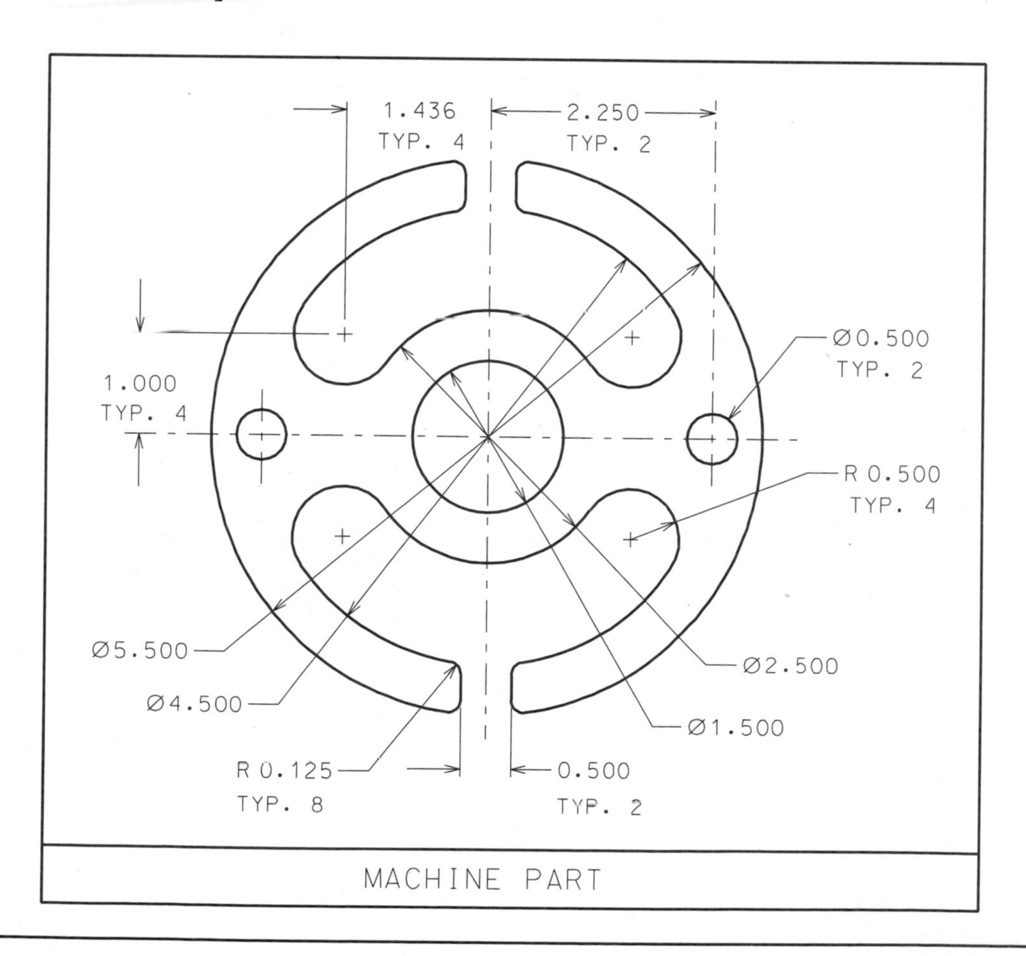

Exercise 8-2

Bracket set

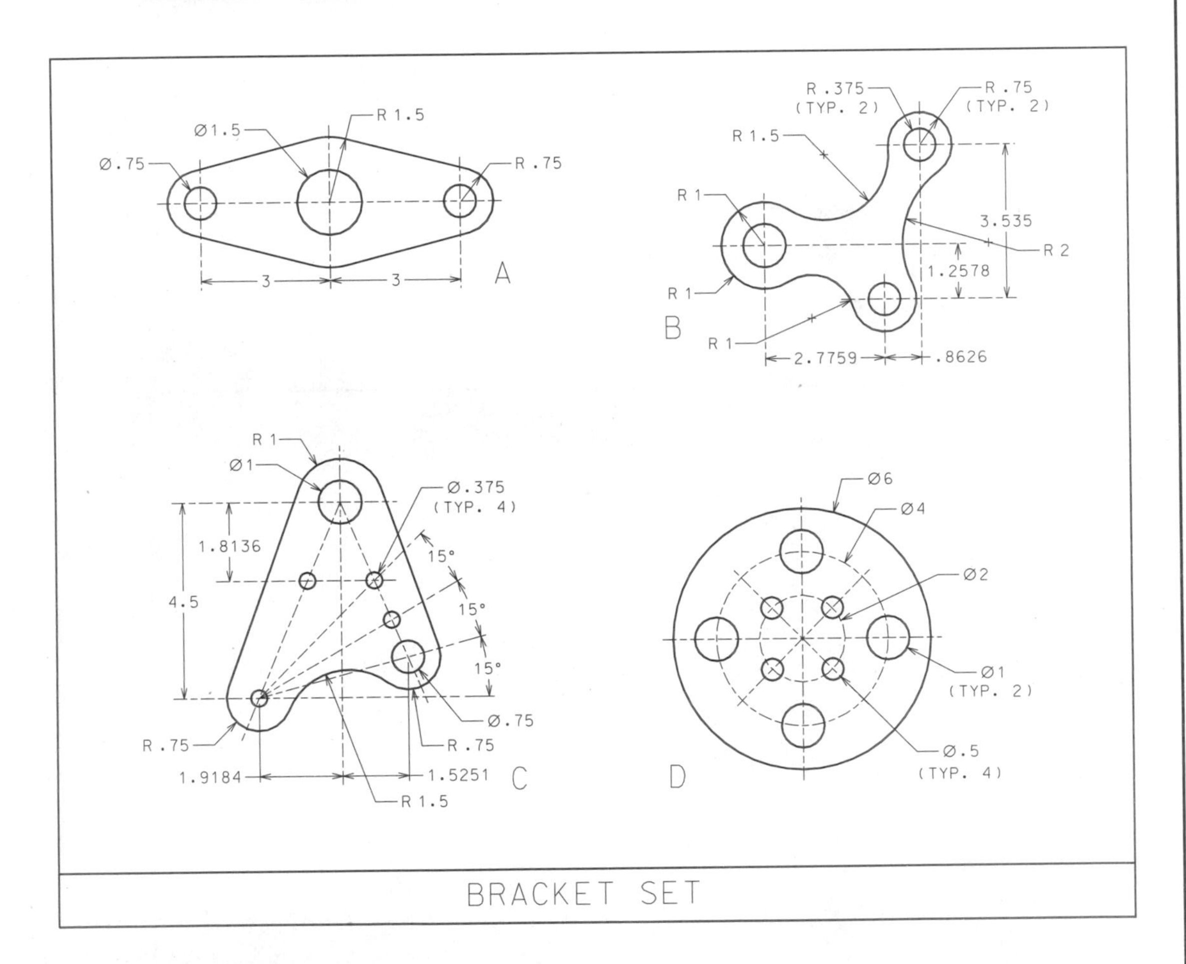

Exercise 8–3

Mounting bracket

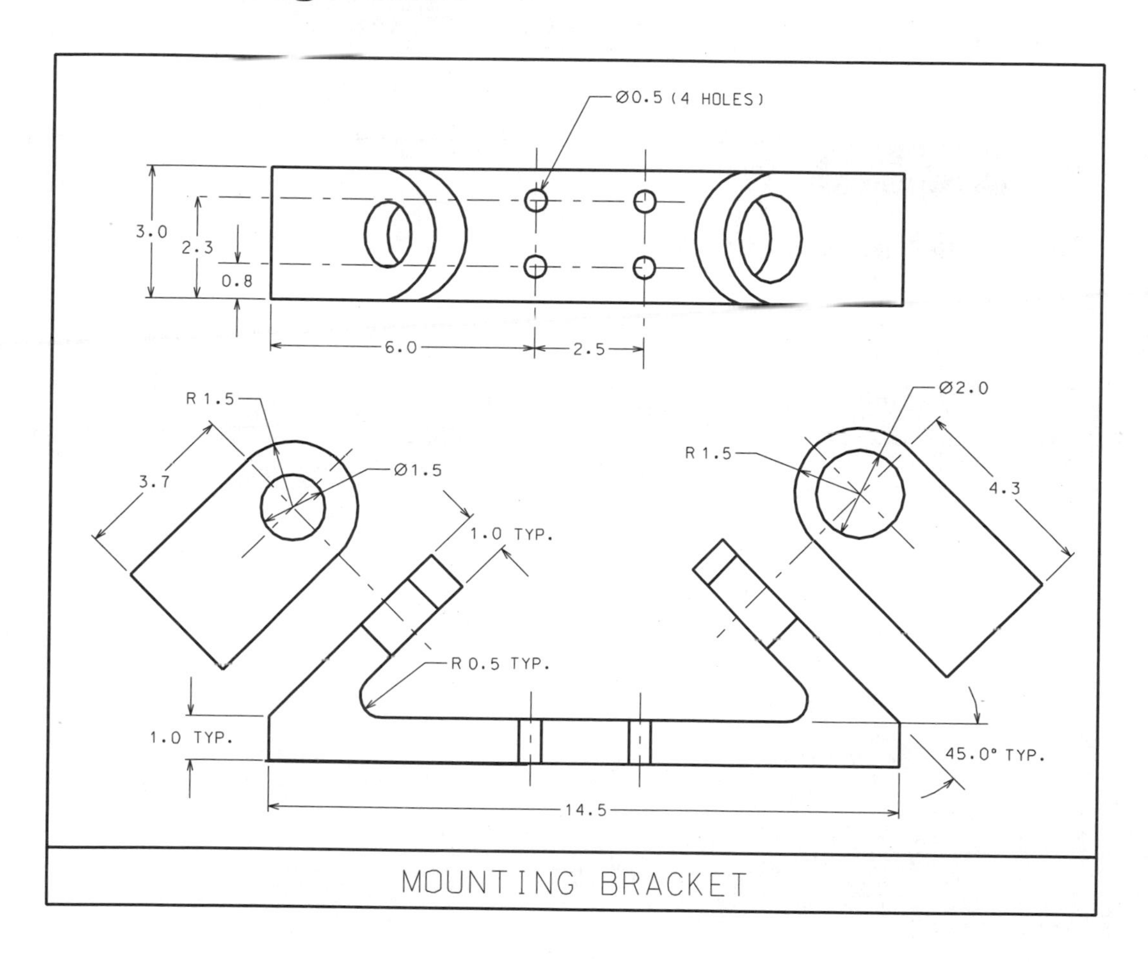

Use the following table to set up the design files for Exercises 8–4 and 8–5.

SETTING	VALUE
Seed File	SEED2D.DGN
Working Units	12" Per ' and 12000 Pos Units Per "
Grid	Master = 0.5, Grid Reference = 24

Exercise 8–4

Small business park plan

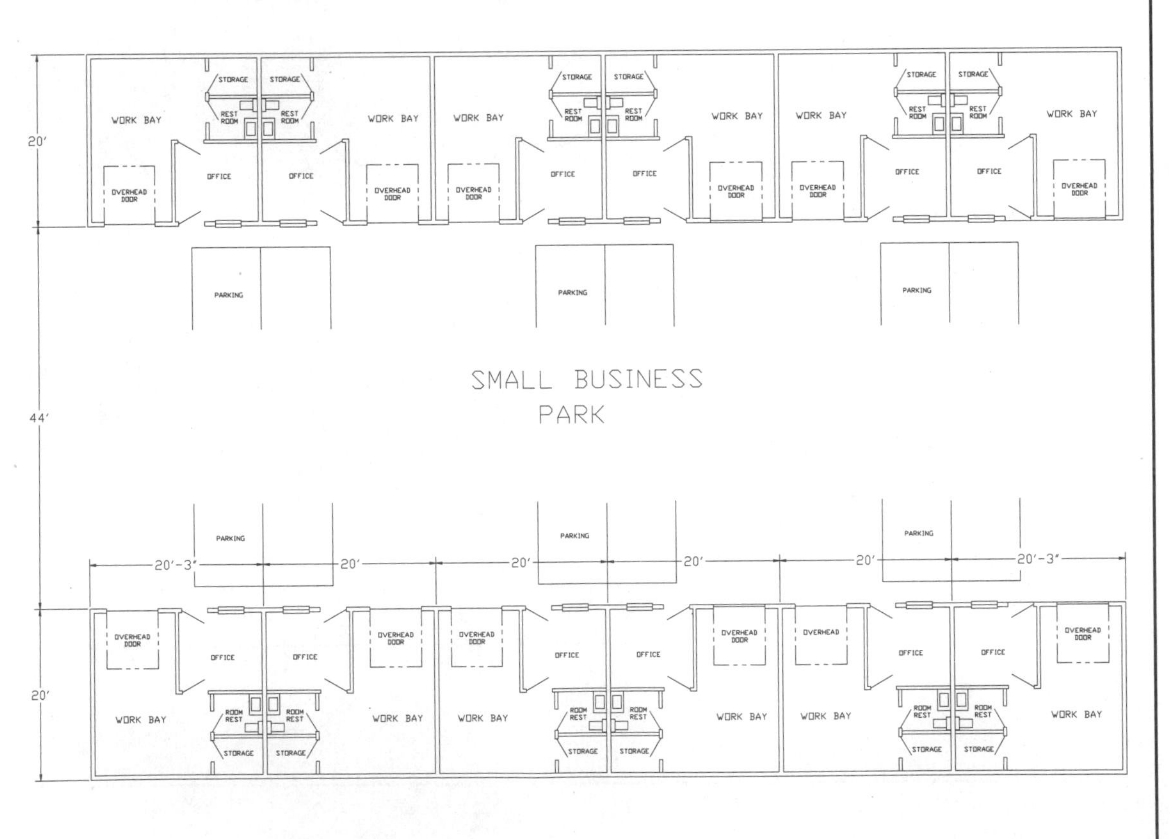

Exxercise 8–5

Self-storage warehouse plan

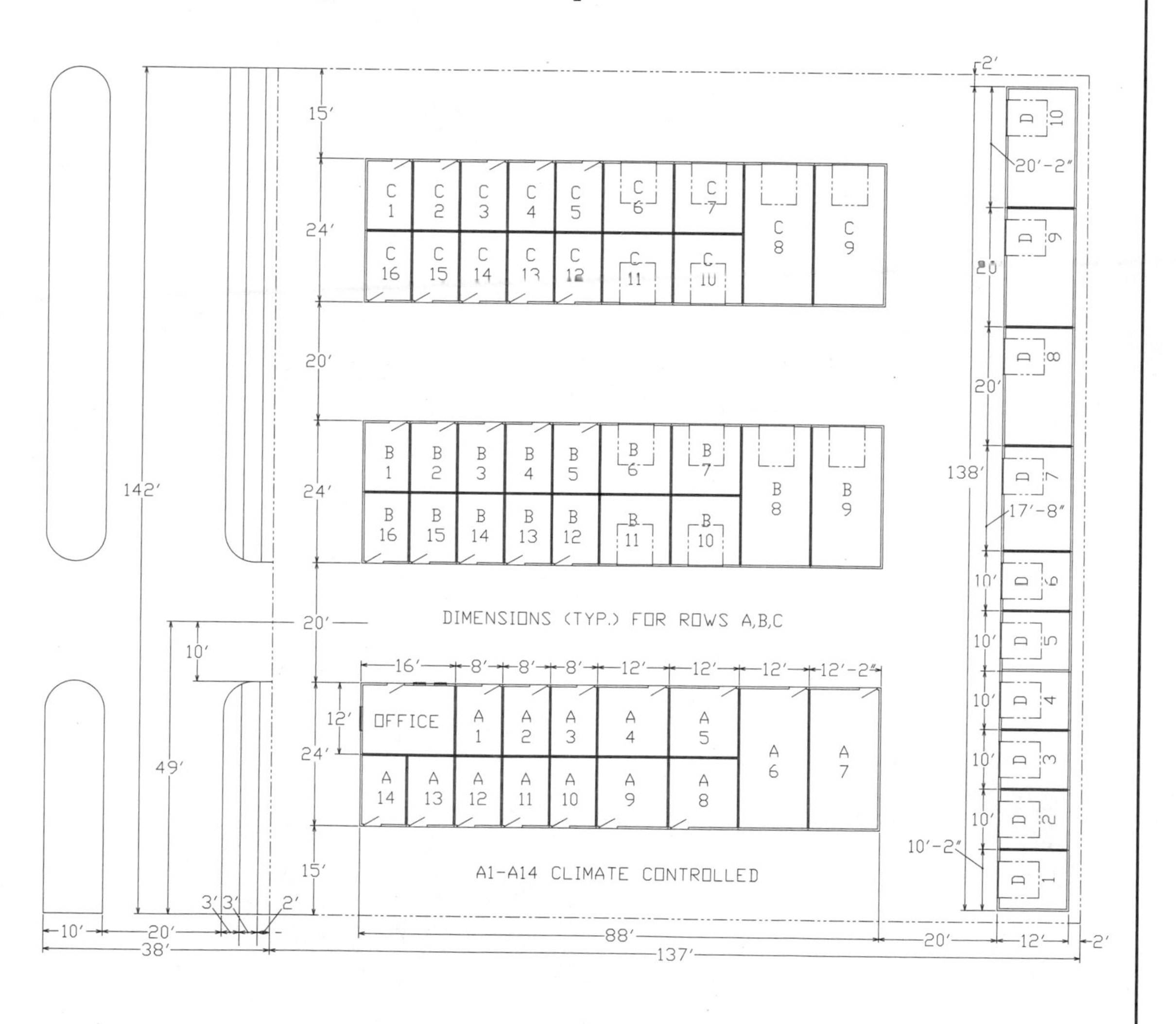

chapter 9

Measurement and Dimensioning

Objectives

After completing this chapter, you will be able to do the following:

- Use the measurement tools, such as Measure Distance, Measure Radius, Measure Angle, Measure Length, and Measure Area.
- Use the dimensioning tools for linear, angular, and radial measurement.
- Adjust the dimension settings.

MEASUREMENT TOOLS

"Is that line really 12 feet long?" "What's the radius of that circle?" "What is the surface area of that foundation?" MicroStation can answer these questions with the measurement tools.

The measurement tools do nothing to your design. They just display distances, areas, and angles in the Status bar.

All measurement tools are available from the Measure tool box, shown in Figure 9–1.

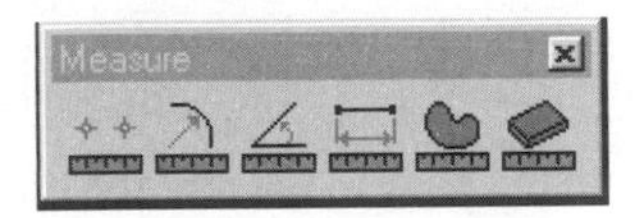

Figure 9–1 Measure tool box

Measure Distance

MicroStation provides four distance measurement options. Distance options include measuring the distance between points you define, the distance along an element between points you define, the perpendicular distance from an element, and the minimum distance between two elements.

Measure Distance Between Points

This tool measures the cumulative straight-line distance from the first data point, through successive data points, to the last data point you define.

Invoke the Measure Distance Between Points tool:

Measure tool box	Select the Measure Distance tool, then select Between Points from the Distance option menu in the Tool Settings window (see Figure 9–2).
Key-in window	**measure distance points** (or **me dist po**) Enter

Figure 9–2 Invoking the Measure Distance Between Points tool from the Measure tool box

MicroStation prompts:

> Measure Distance Between Points > Enter start point *(Place a data point to start the measurement.)*
>
> Measure Distance Between Points > Define distance to measure *(Place a data point to define the distance to measure and continue placing data points to define additional measurement segments. Click the Reset button to terminate the tool sequence.)*

As you place each data point, the cumulative linear distance between points is displayed in the Status bar.

Measure Distance Along Element

This tool measures the cumulative distance along an element, from the data point that selects it through successive data points on the element to the last data point you define.

Invoke the Measure Distance Along Element tool:

Measure tool box	Select the Measure Distance tool, then select Along Element from the Distance option menu in the Tool Settings window (see Figure 9–3).
Key-in window	**measure distance along** (or **me dist a**) Enter

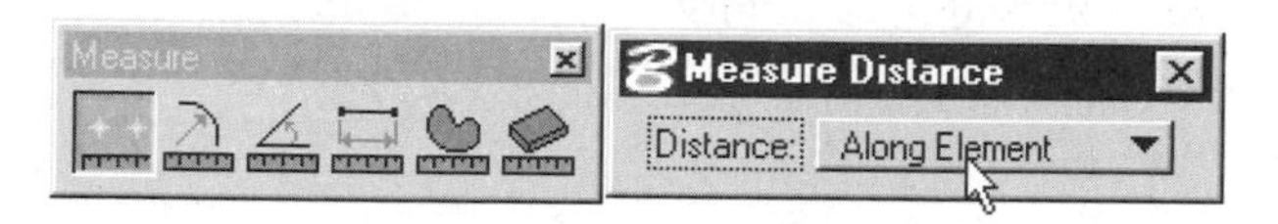

Figure 9–3 Invoking the Measure Distance Along Element tool from the Measure tool box

MicroStation prompts:

Measure Distance Along Element > Identify Element @ first point *(Place a data point on the element to start the measurement.)*

Measure Distance Along Element > Enter end point *(Place a data point on the element to end the measurement.)*

Measure Distance Along Element > Measure more points/Reset to select *(Continue placing data points on the element to obtain the cumulative measurement from the first data point through the succeeding points, or click the Reset button to terminate the measurement.)*

As you place each data point after the first one, the cumulative distance along the element is displayed in the Status bar.

Measure Distance Perpendicular From Element

This tool measures the perpendicular distance from a point to an element.

Invoke the Measure Distance Perpendicular From Element tool:

Measure tool box	Select the Measure Distance tool, then select Perpendicular from the Distance option menu in the Tool Settings window (see Figure 9–4).
Key-in window	**measure distance perpendicular** (or **me dist p**) Enter

Figure 9–4 Invoking the Measure Distance Perpendicular From Element tool from the Measure tool box

MicroStation prompts:

> Measure Distance Perpendicular From Element > Enter start point *(Identify the element from which to measure the perpendicular distance.)*
>
> Measure Distance Perpendicular From Element > Enter end point *(Place a data point to measure the perpendicular distance to the element.)*
>
> Measure Distance Perpendicular From Element > Measure more points/Reset to select *(Continue placing data points to obtain additional perpendicular measurements from the element, or click the Reset button to terminate the measurement.)*

A line from the element to the cursor shows the location of the calculation. The line is a temporary image that disappears when you click the Reset button or select another tool.

Note: If you place the measurement end point beyond the end of a linear element (such as a line or a box), the perpendicular is calculated from an imaginary extension of the measured element.

Measure Minimum Distance Between Elements

This tool measures the minimum distance between two elements.

Invoke the Measure Minimum Distance Between Elements tool:

Measure tool box	Select the Measure Distance tool, then select Minimum Between from the Distance option menu in the Tool Settings window (see Figure 9–5).
Key-in window	**measure distance minimum** (or **me dist m**) Enter

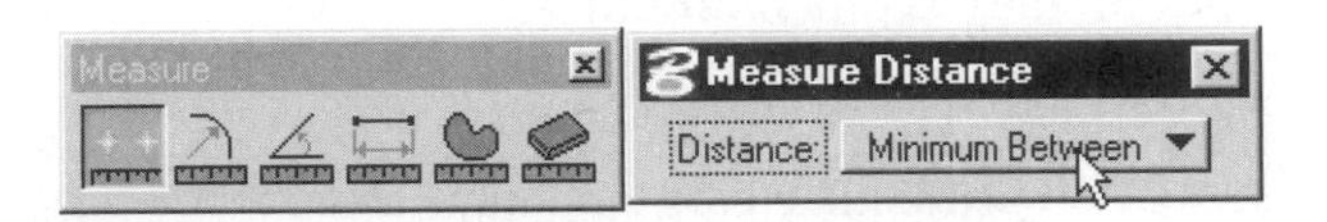

Figure 9–5 Invoking the Measure Minimum Distance Between Elements tool from the Measure tool box

MicroStation prompts:

> Measure Minimum Distance Between Elements > Identify first element *(Identify the first element.)*
>
> Measure Minimum Distance Between Elements > Accept, Identify 2nd element/Reject *(Identify the second element, or click the Reject button to start all over again.)*
>
> Measure Minimum Distance Between Elements > Accept, Initiate min dist calculation *(Click the Data button to initiate the minimum distance calculation.)*

After you place the third data point, a line appears in the design to indicate where the minimum distance is, and the minimum distance is displayed in the Status bar. The line is a temporary image that disappears when you click the Reset button or select another tool.

Measure Radius

The Measure Radius tool displays the radius of arcs, circles, partial ellipses, and ellipses in the Status bar.

Invoke the Measure Radius tool:

Measure tool box	Select the Measure Radius tool (see Figure 9–6).
Key-in window	**measure radius** (or **me r**) Enter

Figure 9–6 Invoking the Measure Radius tool from the Measure tool box

MicroStation prompts:

> Measure Radius > Identify element *(Identify the element to measure the radius.)*
>
> Measure Radius > Accept, Initiate Measurement *(Click the Data button to accept the element and initiate the measurement.)*

After the second data point is placed, the element's radius is displayed, in the current Working Units, in the Status bar. If the element you are measuring is an ellipse or a partial ellipse, the major axis and minor axis radii are displayed.

Measure Angle

The Measure Angle Between Lines tool measures the minimum angle formed by two elements.

Invoke the Measure Angle Between Lines tool:

Measure tool box	Select the Measure Angle tool (see Figure 9–7).
Key-in window	**measure angle** (or **me a**) Enter

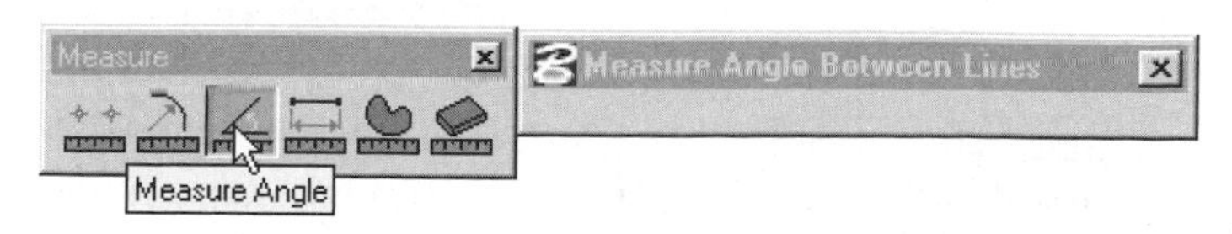

Figure 9–7 Invoking the Measure Angle tool from the Measure tool box

MicroStation prompts:

> Measure Angle Between Lines > Identify first element *(Identify the first element.)*
>
> Measure Angle Between Lines > Accept, Identify 2nd element/Reject *(Identify the second element to measure the angle.)*
>
> Measure Angle Between Lines > Accept, Initiate Measurement *(Click the Data button to accept the element and initiate the measurement.)*

After you accept the elements, the angle between the two elements is displayed in the Status bar.

Measure Length

The Measure Length tool measures the total length of an open element or the length of the perimeter of a closed element.

When you invoke the Measure Length tool, a Tolerance (%) field appears in the Tool Settings window. Tolerance sets the maximum allowable percentage of the distance between the true curve and the approximation for measurement purposes. A low value produces a very accurate measurement but may take a long time to calculate. The default value is sufficient in most cases.

Invoke the Measure Length tool:

Measure tool box	Select the Measure Length tool (see Figure 9–8).
Key-in window	**measure length** (or **me l**) Enter

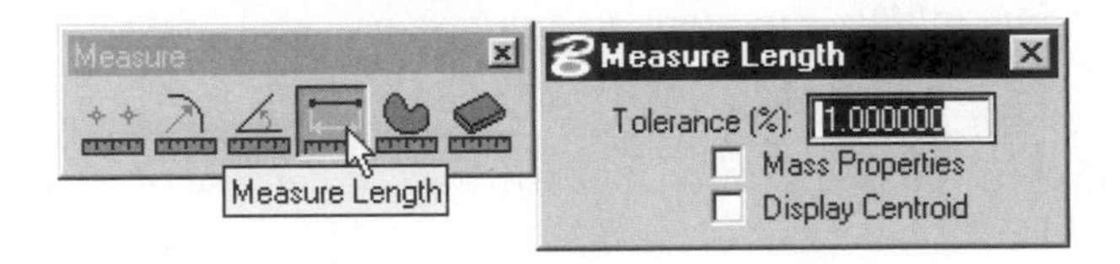

Figure 9–8 Invoking the Measure Length tool from the Measure tool box

MicroStation prompts:

> Measure Length > Identify element *(Identify the element to measure.)*
>
> Measure Length > Accept, Initiate Measurement *(Click the Data button to accept the element and initiate the measurement.)*

After the second data point is placed, the total length of the element, or element perimeter, is displayed in the Status bar.

You can use this tool to measure the cumulative length of several elements by first employing the Element Selection tool to select all the elements you want to include in the measurement. After selecting the elements, select Measure Length from the Measure tool box, and the cumulative length of all selected elements appears in the Status bar.

Measure Area

MicroStation provides seven different ways to measure areas. Area options include measuring the area of a closed element; a fence; the intersection, union, or difference of two overlapping closed elements; a group of intersecting elements; or a group of points.

When you select the Measure Area tool, a Tolerance (%) field appears in the Tool Settings window. Tolerance sets the maximum allowable percentage of the distance between the true curve and the approximation for area calculation purposes. A low value produces a very accurate area but may take a long time to calculate. The default value is sufficient in most cases.

Measure Area

The Measure Area tool measures the area of a closed element, such as a circle, ellipse, shape, or block.

Invoke the Measure Area tool:

Measure tool box	Select the Measure Area tool, then select Element from the Method option menu in the Tool Settings window (see Figure 9–9).
Key-in window	**measure area element** (or **me ar e**) Enter

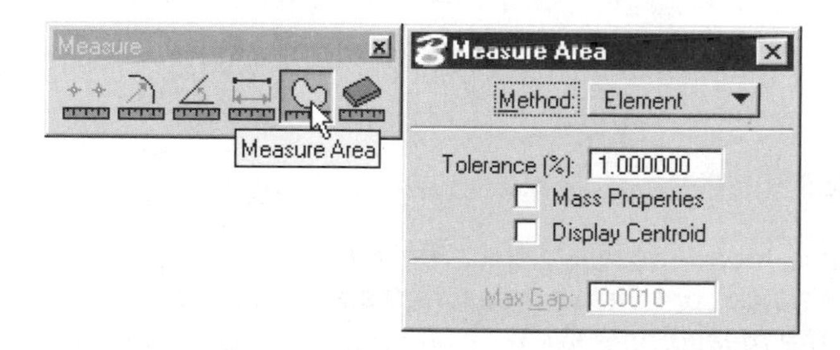

Figure 9–9 Invoking the Measure Area tool from the Measure tool box.

MicroStation prompts:

> Measure Area > Identify element *(Identify the closed element to measure the area.)*
>
> Measure Area > Accept, Initiate Measurement *(Click the Data button to accept the element and initiate the measurement.)*

After the second data point is selected, the element's area and perimeter length are displayed in the Status bar.

You can use this tool to measure the cumulative area of several closed elements by first employing the Element Selection tool to select all the elements you want to include in the area measurement. After selecting the elements, select Measure Area from the Measure tool box, and the cumulative area of all selected elements then appears in the Status bar.

Measure Area Fence

This tool measures the area enclosed by a fence.

Place a fence and then invoke the Measure Fence Area tool:

Measure tool box	Select the Measure Area tool, then select Fence from the Method option menu in the Tool Settings window (see Figure 9–10).
Key-in window	**measure area fence** (or **me ar f**) Enter

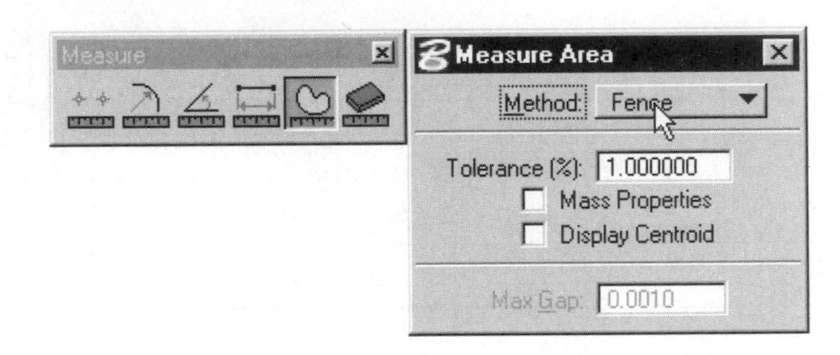

Figure 9–10 Invoking the Measure Fence Area tool from the Measure tool box

MicroStation prompts:

Measure Fence Area > Accept/Reject Fence Contents *(Click the Data button to accept the fence contents to measure the area, or click the Reject button to disregard the measurement.)*

MicroStation displays the area of the fence in the Status bar.

Measure Area Intersection, Union, or Difference

These options measure areas formed by intersecting closed elements. The intersection option allows you to determine the area that is common to two closed elements. The union option allows you to determine the area in such a way that there is no duplication between two closed elements.

The difference option allows you to determine the area that is formed from a closed element after removing from it any area that it has in common with the other selected element.

Invoke the Measure Element Union (or Difference, or Intersection) Area tools:

Measure tool box	Select the Measure Area tool, then select Union, Difference, or Intersection from the Method option menu in the Tool Settings window (see Figure 9–11).
Key-in window	**measure area union** | **difference** | **intersection** (or **me ar u** | **d** | **i**) Enter

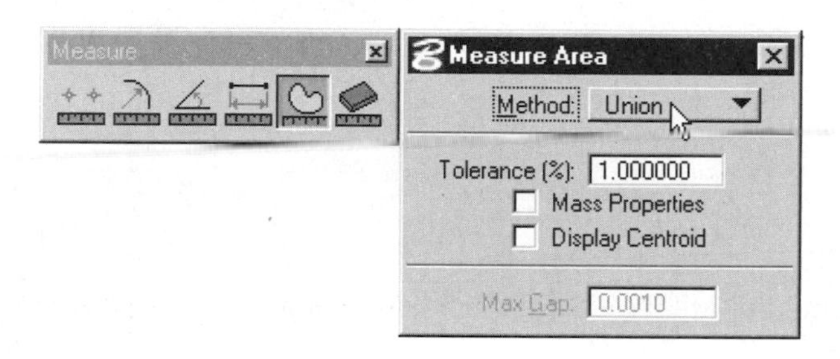

Figure 9–11 Invoking the Measure Element Union Area tool from Measure tool box

MicroStation prompts:

Measure Element Union Area > Identify the element *(Identify the first element.)*

Measure Element Union Area > Accept/Reject (Select next input) *(Identify the second element.)*

Measure Element Union Area > Accept/Reject (Select next input) *(Click the Data button to accept the element, or identify another element.)*

Measure Element Union Area > Identify Additional/Reset to terminate *(Identify additional elements, or click the Reset button to terminate and initiate the measurement.)*

After you select all the elements and place the last data point in space, MicroStation displays an image of only the part of the elements that is included in the type of area you select. Click the reset Button to cause the elements to reappear, the elements reappear in their entirety, and MicroStation displays the area and perimeter length in the Status bar.

Measure Area Flood

This measures the area enclosed by a group of elements. Invoke the Measure Area Enclosing Point tool:

Measure tool box	Select the Measure Area tool, then select Flood from the Method option menu in the Tool Settings window (see Figure 9–12).
Key-in window	**measure area flood** (or **me ar fl**) Enter

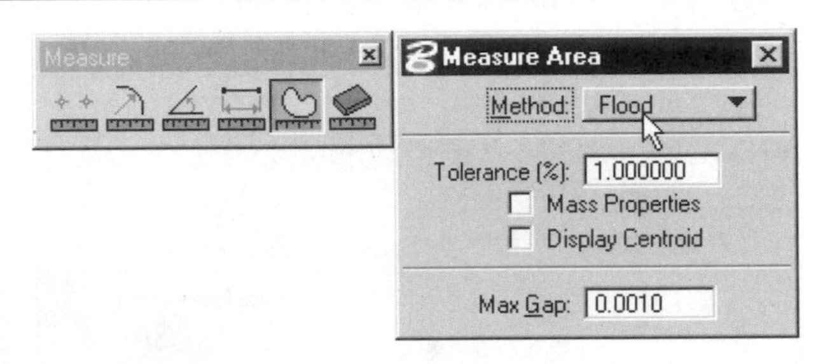

Figure 9–12 Invoking the Measure Area Enclosing Point tool from the Measure tool box

MicroStation prompts:

> Measure Area Enclosing Point > Enter data point inside area *(Click the Data button inside the area enclosed by the elements.)*
>
> Measure Area Enclosing Point > Accept, Initiate Measurement *(Click the Data button to accept and initiate measurement.)*

After you click the Data button, a small spinner will appear in the Status bar. The spinner spins to indicate that MicroStation is determining the area enclosed by the elements. As MicroStation traces the area, it highlights the elements. When the area has been determined, the spinner stops spinning and the area and perimeter length appear in the Status bar.

Note: The Max Gap setting controls the space that is allowed between the elements that define the flood area. If the gap is zero, the elements must be touching.

Measure Area Points

This tool measures the area formed by a set of data points you enter. It assumes the perimeter of the area is formed by straight lines between the data points. An image of the area is displayed as you enter the data points. The image is temporary and disappears when you update the screen or select another tool.

Invoke the Measure Area Defined By Points tool:

Measure tool box	Select the Measure Area tool, then select Points from the Method option menu in the Tool Settings window (see Figure 9–13).
Key-in window	**measure area points** (or **me ar p**) Enter

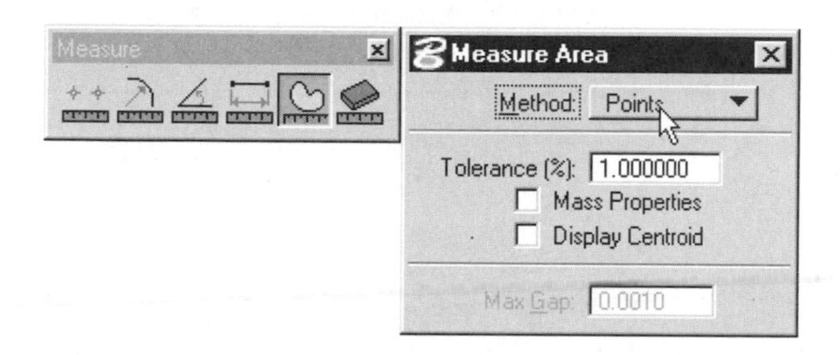

Figure 9–13 Invoking the Measure Area Defined By Points tool from the Measure tool box

MicroStation prompts:

> Measure Area Defined By Points > Enter shape vertex *(Place data points to define the vertices of the area to be measured. When the area is completely defined, click the Reset button to initiate area measurement.)*

As you are entering data points, a closed dynamic image of the area appears on the screen. When you press the Reset button, the area and the perimeter length appear in the Status bar.

DIMENSIONING

MicroStation's dimensioning features provide an excellent way to add dimensional information to your design, such as lengths, widths, angles, tolerances, and clearances.

Dimensioning of any drawing is generally one of the last steps in manual drawing; however, it does not need to be the last step in your MicroStation drawing. If you place the dimensions and find out later they must be changed because the size of the objects they are related to have changed, MicroStation allows you to stretch or extend the objects and have the dimensions change automatically to the new size. MicroStation provides three basic types of dimensions: linear, angular, and radial dimensioning. Figure 9–14 shows examples of these three basic types of dimensions.

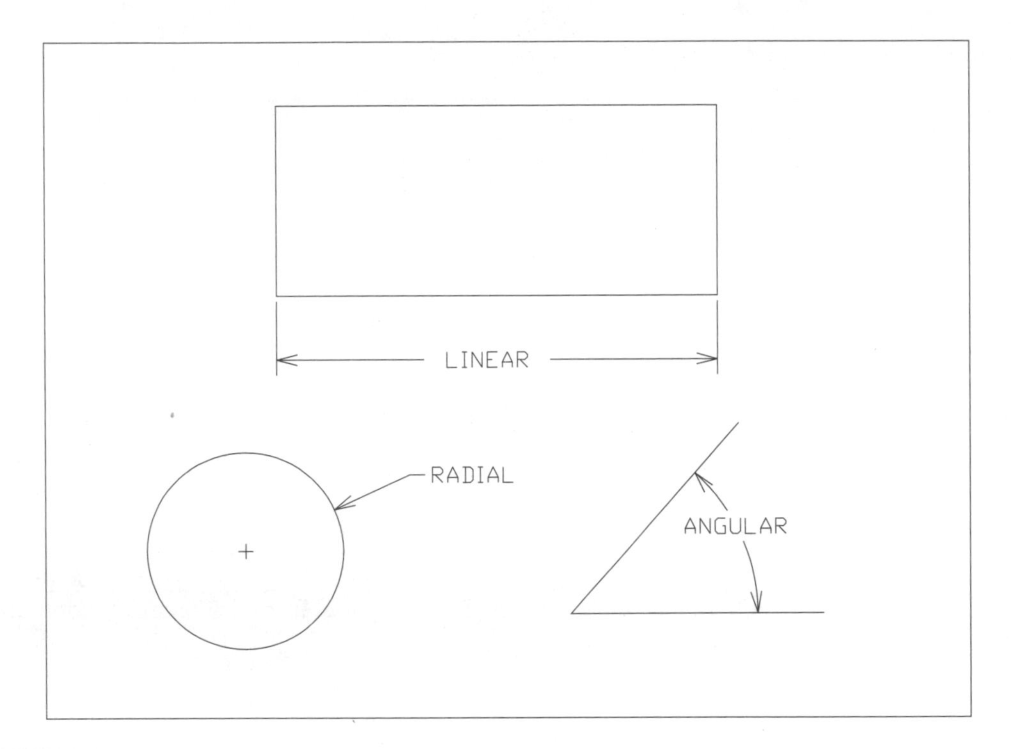

Figure 9–14 Examples of the three basic types of dimensions

All the available dimensioning tools in MicroStation are found in the Dimension tool box, as shown in Figure 9–15.

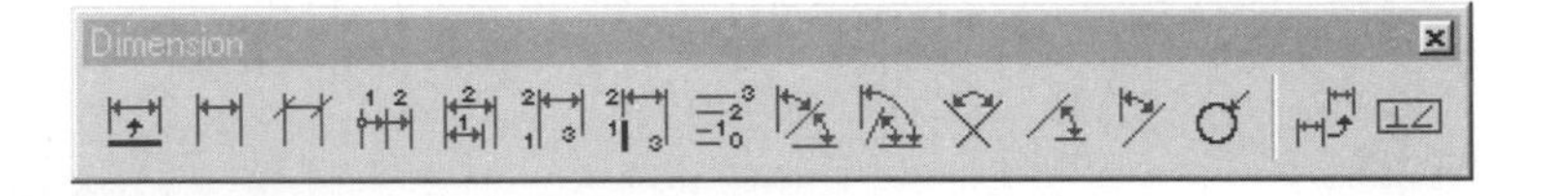

Figure 9–15 Dimension tool box

Dimensioning Terminology

The following terms occur commonly in the MicroStation dimensioning procedures.

Dimension Line

This is a line with markers at each end (arrows, dots, tick marks, etc.). The dimensioning text is located along this line; you may place it above the line or in a break in the dimension line. Usually, the dimension line is inside the measured area. If there is insufficient space, MicroStation places the dimensions and draws two short lines outside the measured area with arrows pointing inward.

Extension Lines

The extension lines (also called *witness lines*) are the lines that extend from the object to the dimension line. Extension lines normally are drawn perpendicular to the dimension line. (Several options that are associated with this element will be reviewed later in this chapter.) Also, you can suppress one or both of the extension lines.

Terminators

Terminators are placed at one or both ends of the dimension line, depending on the type of dimension line placed. MicroStation allows you to use arrows, tick marks, or arbitrary symbols of your own choosing for the terminators. You can also adjust the size of the terminator.

Dimension Text

This is a text string that usually indicates the actual measurement. You can accept the default measurement computed automatically by MicroStation, or change it by supplying your own text.

Figure 9–16 shows the different components of a typical dimension.

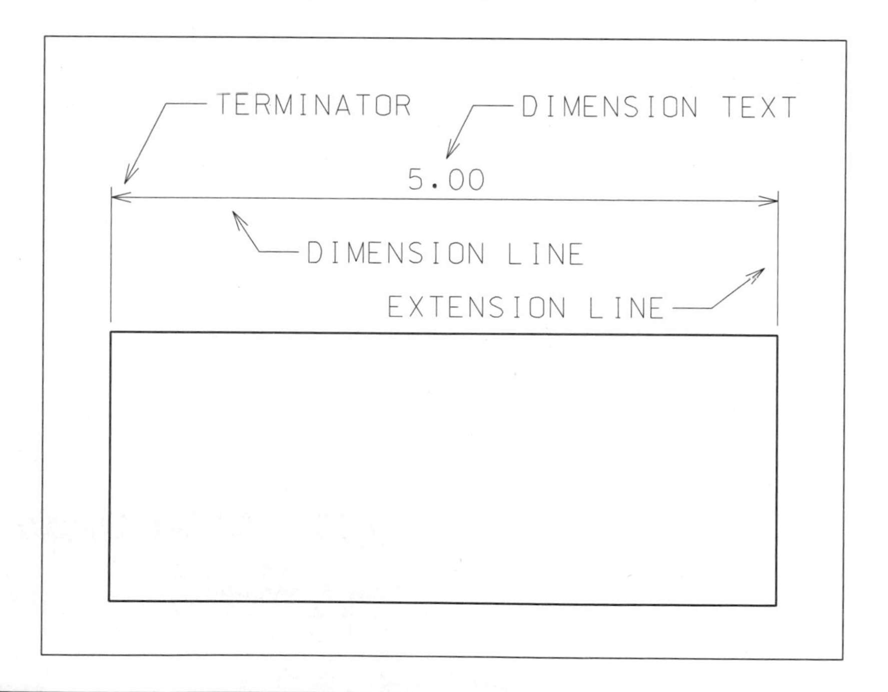

Figure 9–16 Different components of a typical dimension

Leader

The leader line is a line from text to an object on the design, as shown in Figure 9–17. For some dimensioning, the text may not fit next to the object it describes; hence, it is customary to place the text nearby and draw a leader from it to the object.

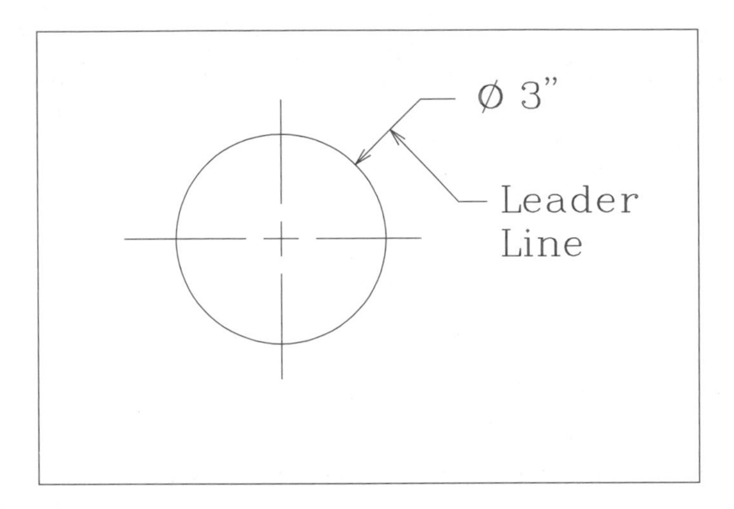

Figure 9–17 Example of placing a leader line

Associative Dimensioning

Dimensions can be placed as either associative dimensions or normal (nonassociative) dimensions. Associative dimensioning links dimension elements to the objects being dimensioned. An association point does not have its own coordinates, but is positioned by the coordinates of the point with which it is associated. When you change the size, shape, or position of an element, MicroStation modifies the associated dimension automatically to reflect the change. It also draws the dimension entity at its new location, size, and rotation.

To place associative dimensions, turn ON the Association Lock in the Tool Settings window when you invoke one of the dimensioning tools, as shown in Figure 9–18. If you place a dimension when the Association Lock is OFF, the dimension will not associate with the object dimensioned, and if the object is modified, the dimension is not changed.

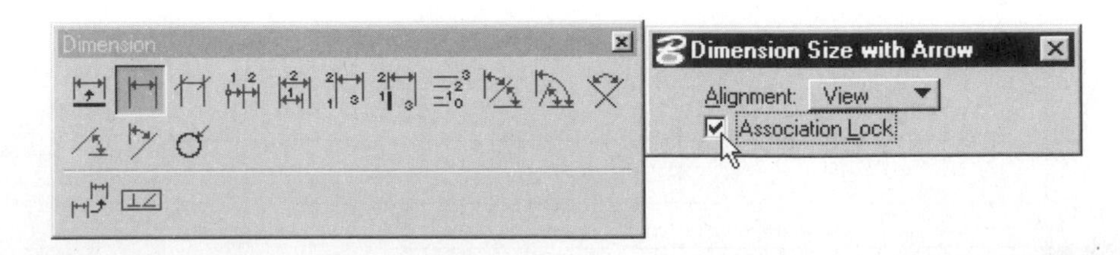

Figure 9–18 Displaying a toggle for the Association Lock in the Tool Settings window

Placing associative dimensions can significantly reduce the size of a design file that has many dimensions, since a dimension element is usually smaller than its corresponding individual elements.

Alignment Controls

The alignment controls the orientation of linear dimensions. View, Drawing, True, and Arbitrary are the available options. The options are selected from the Alignment option menu in the Tool Settings window when you invoke one of the linear dimensioning tools.

- The View option aligns linear dimensions parallel to the view *X* or *Y* axis. This is useful when dimensioning *three dimensional* reference files with dimensions parallel to the viewing plane.
- The Drawing option aligns linear dimensions parallel to the design plane *X* or *Y* axis.
- The True option aligns linear dimensions parallel to the element being dimensioned. The extension lines are constrained to be at right angles to the dimension line.
- The Arbitrary option places linear dimensions parallel to the element being dimensioned. The extension lines are not constrained to be at right angles to the dimension line. This is useful when dimensioning elements in *2D* isometric drawings. The Iso Lock toggle button must be set to ON.

Figure 9–19 shows examples of placing linear dimensions with different alignment controls.

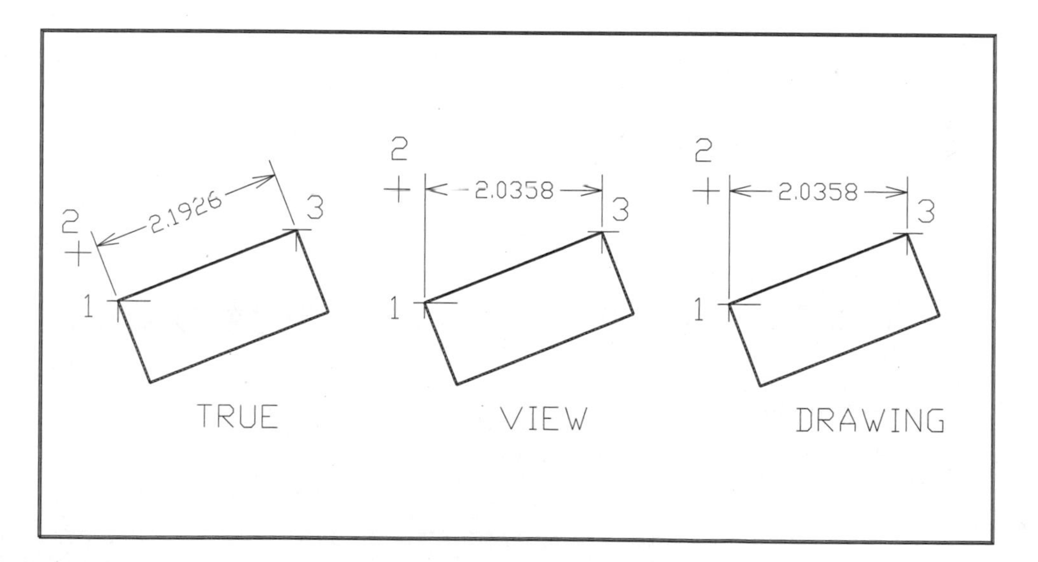

Figure 9–19 Examples of placing linear dimensions with different alignment controls

Linear Dimensioning

Linear dimensioning tools allow you to dimension such linear elements as lines and line strings. Following are the tools available to dimension the linear elements.

- Dimension Size with Arrows—This tool allows you to dimension the linear distance between two points (length).
- Dimension Size with Stroke—This is similar to the Dimension Size with Arrow tool, except the terminators are set to strokes instead of arrows.
- Dimension Location—With this tool you can dimension linear distances from an origin (datum), with the dimensions placed in line (chained).
- Dimension Location (Stacked)—This tool also allows you to dimension linear distances from an origin (datum), but with the dimensions stacked.
- Dimension Size Perpendicular to Points—With this tool you can dimension the linear distance between two points. The first two data points entered define the dimension's *Y* axis.
- Dimension Size Perpendicular to Line—This tool dimensions the linear distance perpendicular from an element to another element or point. The dimension's *Y* axis is defined by the element identified.

Dimension Size with Arrows

To place linear dimension with arrows, invoke the Dimension Size with Arrows tool:

Dimension tool box	Select the Dimension Size with Arrows tool. If necessary, turn ON Association Lock and select an Alignment in the Tool Settings window (see Figure 9–20).
Key-in window	**dimension size arrow** (or **dim si a**) Enter

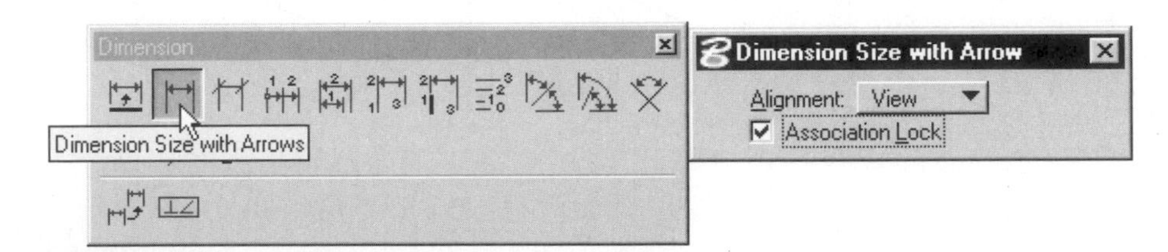

Figure 9–20 Invoking the Dimension Size with Arrow tool from the Dimension tool box.

MicroStation prompts:

Dimension Size with Arrow > Select start of dimension *(Place a data point as shown in Figure 9–21 (point 1) to define the starting point of the dimension.)*

Dimension Size with Arrow > Define length of extension line *(Place a data point as shown in Figure 9–21 (point 2) to define the length of the extension line.)*

Dimension Size with Arrow > Select dimension endpoint *(Place a data point as shown in Figure 9–21 (point 3) to define the end point of the dimension.)*

Dimension Size with Arrow > Select dimension endpoint *(Continue placing data points as shown in Figure 9–21 (points 4 and 5) to continue linear dimensioning in the same direction, or click the Reset button to change the direction, or click the Reset button twice to start all over again.)*

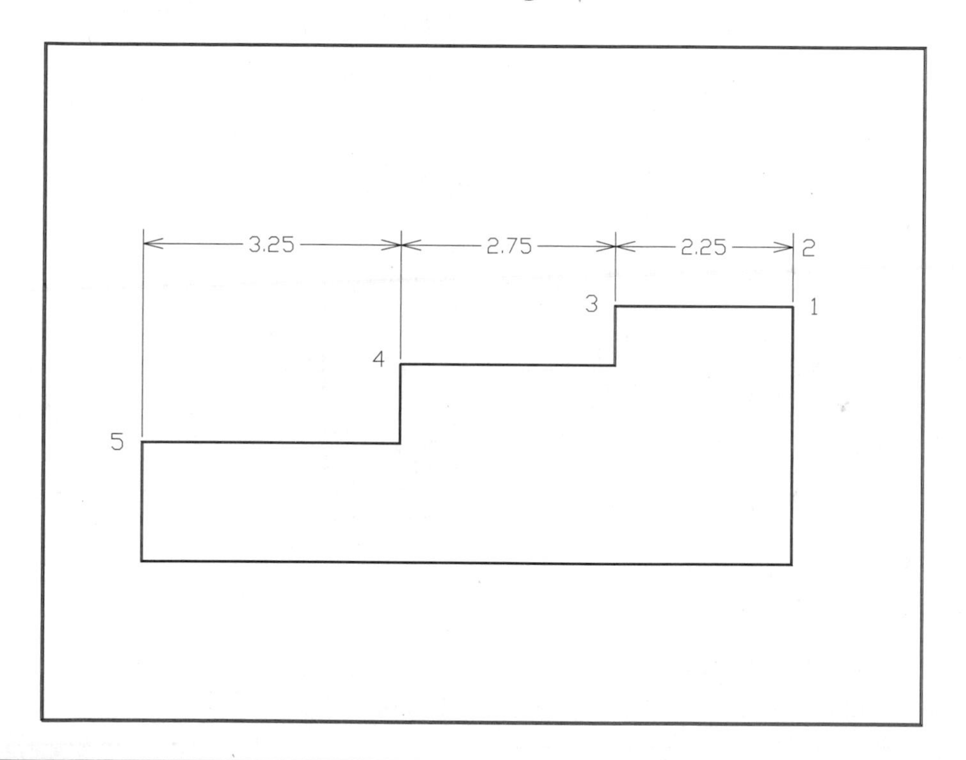

Figure 9–21 Example of placing linear dimensioning with the Dimension Size with Arrows tool

Note: Use the appropriate Snap Lock when placing the data points for the starting point and end point of the dimension line.

Edit Dimension Text

After you place dimensions, the dimension text can be edited using the Edit Text tool in the Text tool box. To edit a piece of dimension text, select the Edit Text tool, then select and accept the dimension text to be edited. MicroStation displays the Dimension Text dialog box similar to the one shown in Figure 9–22.

The asterisk (*) in the Primary Text edit field indicates the current default dimension text string.

- If you need to change the default dimension string, delete the asterisk (*) and key-in the new dimension text string.
- If you need to add any prefix and/or suffix text string to the default dimension text string, keep the asterisk (*) and type in the appropriate text string in the Text edit field.

Click on the OK button to keep the changes, or click on the Cancel button to disregard the changes, and MicroStation closes the Dimension Text dialog box.

The Secondary Text edit field allows you to edit the dimension text of the secondary dimension when you are placing dual dimensions. Primary and Secondary dimensioning is discussed later in this chapter.

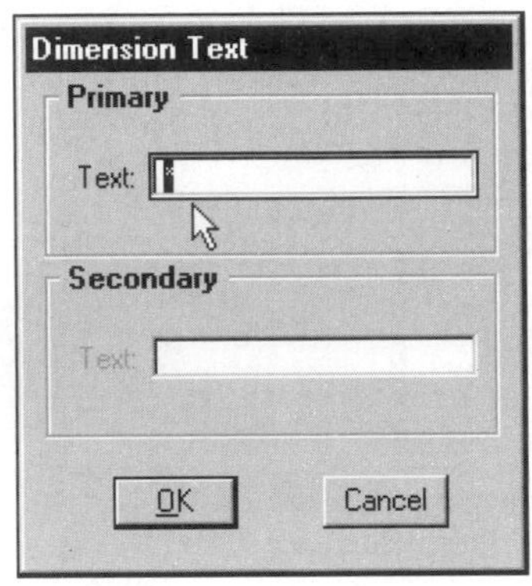

Figure 9–22 Dimension Text dialog box

Dimension Size with Stroke

To place linear dimensions with strokes, invoke the Dimension Size Stroke tool:

Dimension tool box	Select the Dimension Size Stroke tool. If necessary, turn ON Association Lock and select an Alignment option in the Tool Settings window (see Figure 9–23).
Key-in window	**dimension size stroke** (or **dim si s**) Enter

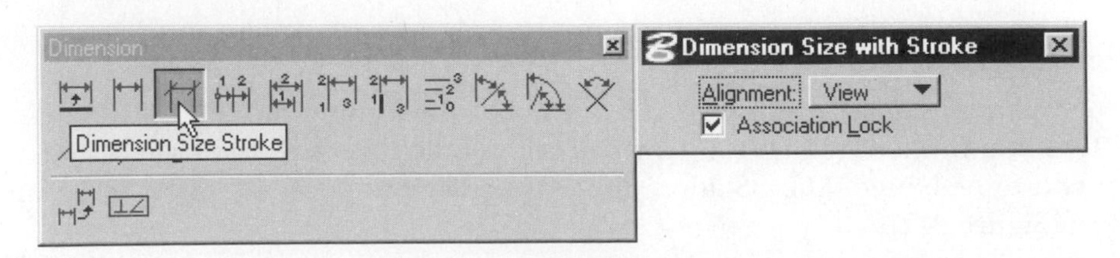

Figure 9–23 Invoking the Dimension Size Stroke tool from the Dimension tool box

MicroStation prompts are similar to those for the Dimension Size with Arrows tool. Instead of placing arrows at the end of the dimension line, MicroStation places strokes, as shown in Figure 9–24.

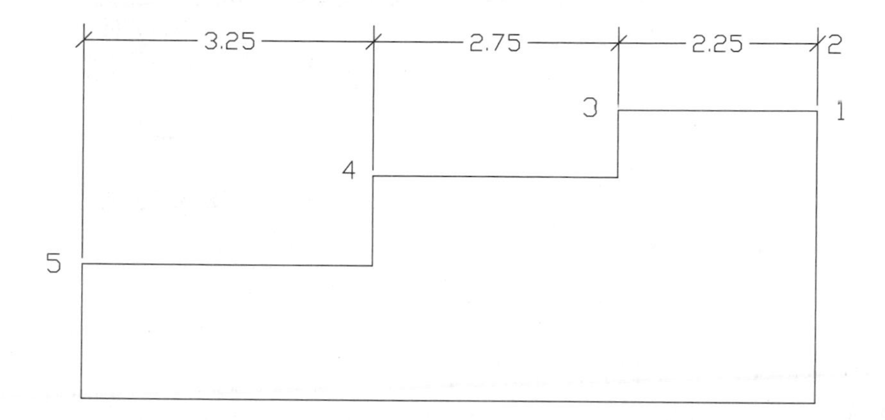

Figure 9–24 Example of placing linear dimensioning with the Dimension Size Stroke tool

Dimension Location

The Dimension Location tool enables you to dimension linear distances from an origin (datum), as shown in Figure 9–25. The dimensions are placed in a line (chain). All dimensions are measured on an element originating from a common surface, centerline, or center plane. The Dimension Location tool is commonly used in mechanical drafting.

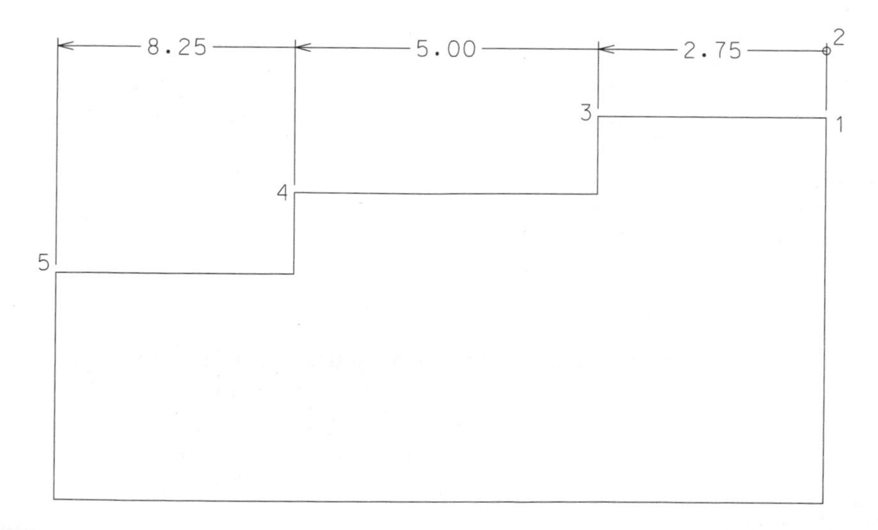

Figure 9–25 Example of placing linear dimensioning from an origin (datum) with the Dimension Location tool

To place linear dimensions in a chain, invoke the Dimension Location tool:

Dimension tool box	Select the Dimension Location tool. If necessary, turn ON Association Lock and select an Alignment option in the Tool Settings window (see Figure 9–26).
Key-in window	**dimension location single** (or **dim lo s**) Enter

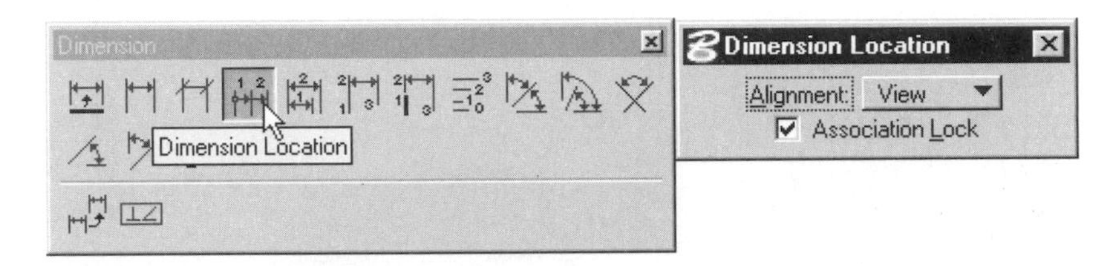

Figure 9–26 Invoking the Dimension Location tool from the Dimension tool box

MicroStation prompts:

> Dimension Location > Select start of dimension *(Place a data point to define the origin.)*
>
> Dimension Location > Define length of extension line *(Place a data point to define the length of the extension line.)*
>
> Dimension Location > Select dimension endpoint *(Place a data point to define the end point of the dimension. If necessary, press* Enter *to edit the dimension text.)*
>
> Dimension Location > Select dimension endpoint *(Continue placing data points to continue linear dimensioning in the same direction, or click the Reset button to change the direction, or click the Reset button twice to start all over again.)*

Note: Use the appropriate Snap Lock when placing the data points for the starting point and end point of the dimension line.

Dimension Location (Stacked)

With the Dimension Location (Stacked) tool you can dimension linear distances from an origin (datum), as shown in Figure 9–27. The dimensions are stacked. All dimensions are measured on an element originating from a common surface, centerline, or center plane. The Dimension Location (Stacked) tool is commonly used in mechanical drafting because all dimensions are independent, even though they are taken from a common datum. If necessary, you can change the stack offset distance; see the later section on Dimension Settings.

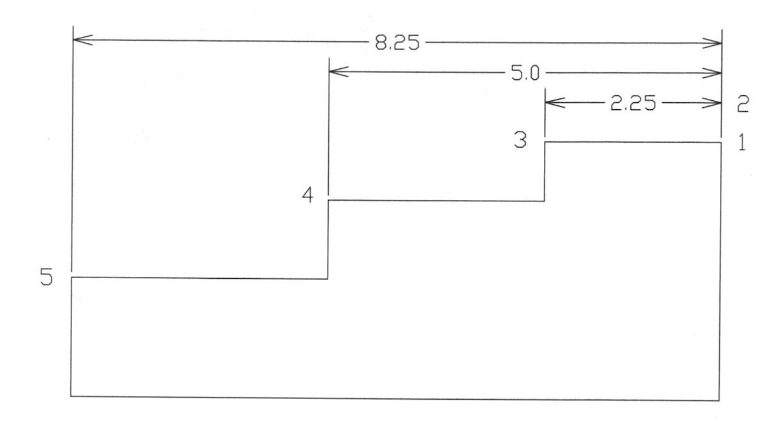

Figure 9–27 Example of placing linear dimensioning (stacked) from an origin (datum) via the Dimension Location (Stacked) tool

To place linear dimensions in a stack, invoke the Dimension Location (Stacked) tool:

Dimension tool box	Select the Dimension Location (Stacked) tool. If necessary, turn ON Association Lock and select an Alignment option in the Tool Settings window (see Figure 9–28).
Key-in window	**dimension location stacked** (or **dim lo st**) Enter

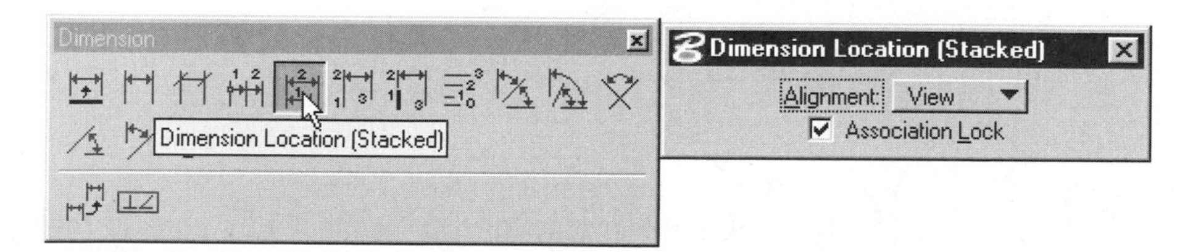

Figure 9–28 Invoking the Dimension Location (Stacked) tool from the Dimension tool box

MicroStation prompts:

Dimension Location (Stacked) > Select start of dimension *(Place a data point to define the origin.)*

Dimension Location (Stacked) > Define length of extension line *(Place a data point to define the length of the extension line.)*

Dimension Location (Stacked) > Select dimension endpoint *(Place a data point to define the end point of the dimension line. If necessary, press* Enter *to edit the dimension text.)*

Dimension Location (Stacked) > Select dimension endpoint *(Continue placing data points to continue linear dimensioning in the same direction, or click the Reset button to change the direction, or click the Reset button twice to start all over again.)*

Note: Use the appropriate Snap Lock when placing the data points for the starting point and end point of the dimension line.

Dimension Size Perpendicular to Points

The Dimension Size Perpendicular to Points tool can dimension the linear distance between two points. The first two data points entered define the dimension's *Y* axis, as shown in Figure 9–29.

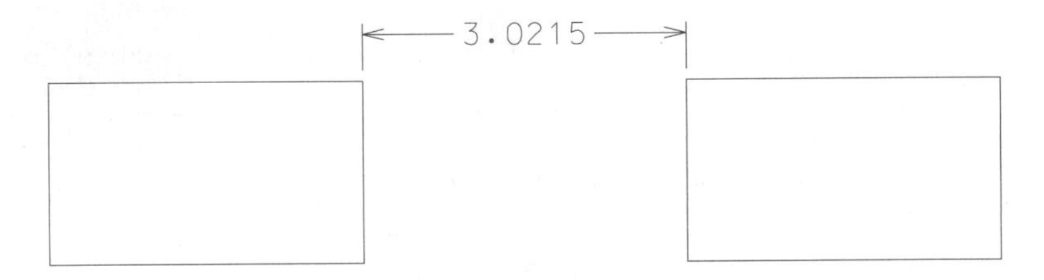

Figure 9–29 Example of placing linear dimensioning with the Dimension Size Perpendicular to Points tool

To place linear dimensions between two points, invoke the Dimension Size Perpendicular to Points tool:

Dimension tool box	Select the Dimension Size Perpendicular to Points tool. If necessary, turn ON Association Lock in the Tool Settings window (see Figure 9–30).
Key-in window	**dimension size perpendicular to points** (or **dim si p p**) Enter

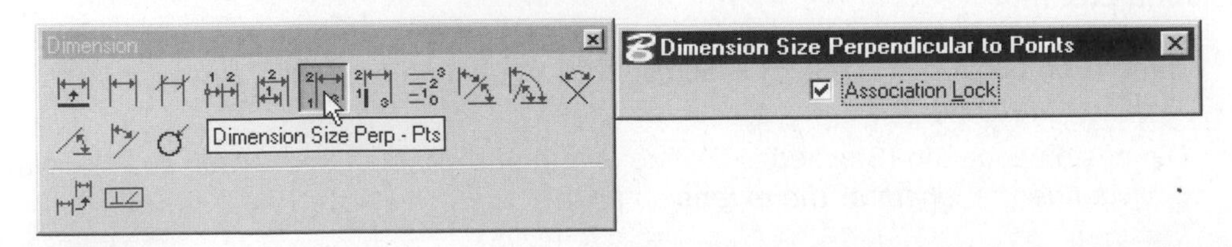

Figure 9–30 Invoking the Dimension Size Perpendicular to Points tool from the Dimension tool box

MicroStation prompts:

Dimension Size Perpendicular to Points > Select base of first dimension line *(Place a data point to define the base point of the first dimension line.)*

Dimension Size Perpendicular to Points > Select end of extension line *(Place a data point to define the length of the extension line.)*

Dimension Size Perpendicular to Points > Select dimension endpoint *(Place a data point to define the end point of the dimension line.)*

Note: Use the appropriate Snap Lock when placing the data points for the starting point and end point of the dimension line.

Dimension Size Perpendicular to Line

The Dimension Size Perpendicular to Line tool dimensions the linear distance perpendicular from an element to another element. The dimension's *Y* axis is defined by the element identified.

To place a linear dimension perpendicular from one element to another, invoke the Dimension Size Perpendicular to Line tool:

Dimension tool box	Select the Dimension Size Perpendicular to Line tool. If necessary, turn ON Association Lock in the Tool Settings window (see Figure 9–31).
Key-in window	**dimension size perpendicular to line** (or **dim si p l**) Enter

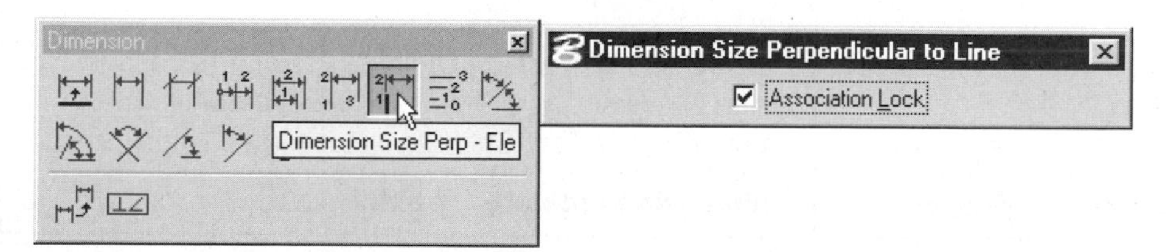

Figure 9–31 Invoking the Dimension Size Perpendicular to Line tool from the Dimension tool box

MicroStation prompts:

Dimension Size Perpendicular to Line > Select base of first dimension line *(Place a data point to define the base point of the first dimension line.)*

Dimension Size Perpendicular to Line > Select end of extension line *(Place a data point to define the length of the extension line.)*

Dimension Size Perpendicular to Line > Select dimension endpoint *(Place a data point to define the end point of the dimension line.)*

Note: Use the appropriate Snap Lock when placing the data points for the starting point and end point of the dimension line.

Dimension Ordinates

Ordinate dimensioning is common in mechanical designs. It labels distances along an axis from a point of origin on the axis along which the distances are measured. See Figure 9–32 for an example of ordinate dimensioning.

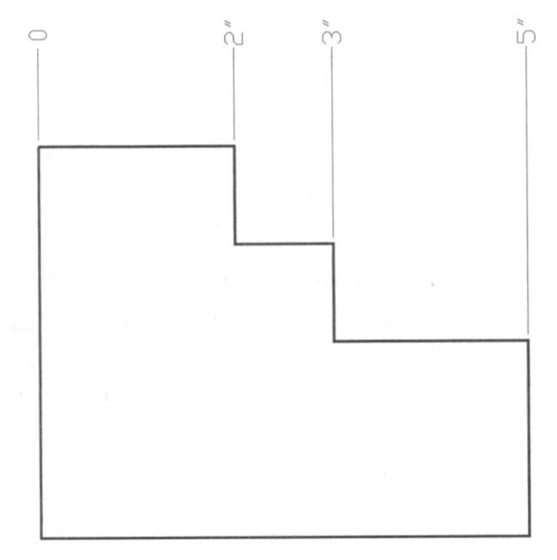

Figure 9–32 Example of ordinate dimensioning

To place ordinate dimensions, invoke the Dimension Ordinates tool:

Dimension tool box	Select the Dimension Ordinates tool. If necessary, turn ON Association Lock and select an Alignment option in the Tool Settings window (see Figure 9–33).
Key-in window	**dimension ordinate** (or **dim o**) Enter

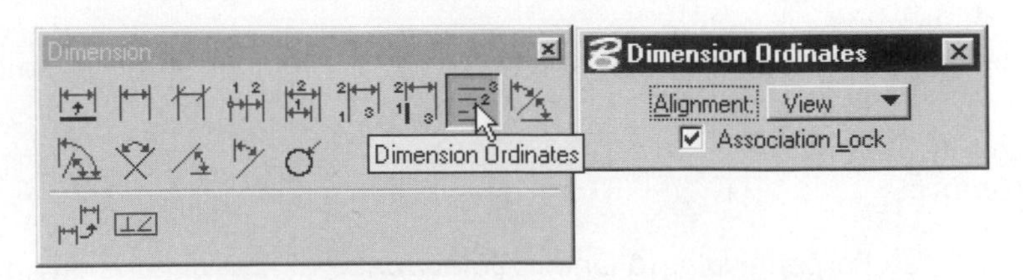

Figure 9–33 Invoking the Dimension Ordinates tool from the Dimension tool box

MicroStation prompts:

> Dimension Ordinates > Select ordinate origin *(Place a data point from which all ordinate labels are to be measured.)*
>
> Dimension Ordinates > Select ordinate direction *(Place a data point to indicate the rotation of the ordinate axis.)*
>
> Dimension Ordinates > Select dimension endpoint *(Place a data point to define the end point of the dimension line. If necessary, press Enter to edit the dimension text.)*
>
> Dimension Ordinates > Select start of dimension *(Place data points at each place where you want an ordinate dimension placed. These points define the base of the extension line. After placing all the points, click the Reset button to terminate the tool sequence.)*

Note: To prevent the text for a higher ordinate from overlapping the text for a lower ordinate, turn ON the Stack Dimensions toggle button in the Dimension Settings box—Tool Settings category.

Angular Dimensioning

The angular dimensioning tools create dimensions for the angle between two non-parallel lines, using the conventions that conform to the current dimension variable settings. "Angle" is defined by *Webster's Ninth New Collegiate Dictionary* as "a measure of an angle or of the amount of turning necessary to bring one line or plane into coincidence with or parallel to another." Following are the five different tools available to create angular dimensions.

- Dimension Angle Size—Each dimension (except the first) is computed from the end point of the previous dimension.
- Dimension Angle Location—Each dimension is computed from the dimension origin (datum).
- Dimension Angle Between Lines—To dimension the angle between two lines, two segments of a line string, or two sides of a shape.
- Dimension Angle from *X* axis—To dimension the angle between a line, a side of a shape, or a segment of a line string and the view *X* axis.
- Dimension Angle from *Y* axis—To dimension the angle between a line, a side of a shape, or a segment of a line string and the view *Y* axis.

Dimension Angle Size

To place angular dimensioning with the Dimension Angle Size tool, invoke the tool:

Dimension tool box	Select the Dimension Angle Size tool. If necessary, turn ON Association Lock in the Tool Settings window (see Figure 9–34).
Key-in window	**dimension angle size** (or **dim a s**) Enter

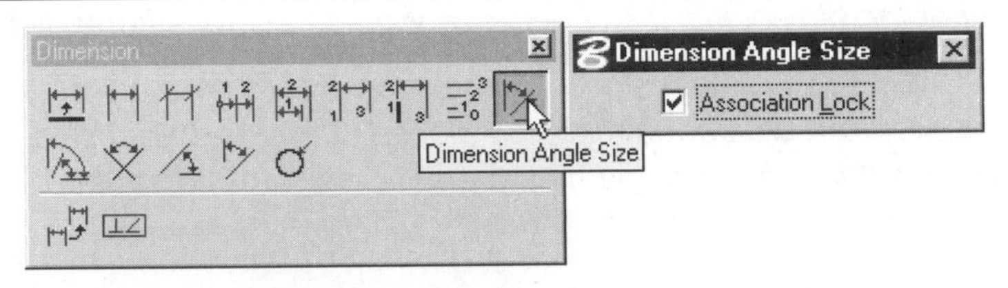

Figure 9–34 Invoking the Dimension Angle Size tool from the Dimension tool box

MicroStation prompts:

Dimension Angle Size > Select start of dimension *(Place a data point as shown in Figure 9–35 (point P1) to define the start of the dimension, which is measured counterclockwise from this point.)*

Dimension Angle Size > Define length of extension line *(Place a data point as shown in Figure 9–35 (point P2) to define the length of the extension line.)*

Dimension Angle Size > Enter point on axis *(Place a data point as shown in Figure 9–35 (point P3) to define the vertex of the angle.)*

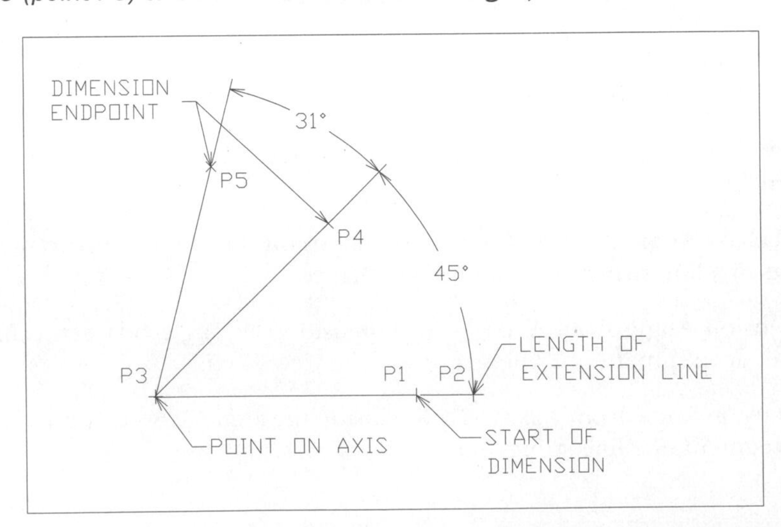

Figure 9–35 Example of placing angular dimensioning via the Dimension Angle Size tool

Dimension Angle Size > Select dimension endpoint *(Place a data point as shown in Figure 9–35 (point P4) to define the end point of the dimension line. If necessary, press* Enter *to edit the dimension text.)*

Dimension Angle Size > Select dimension endpoint *(Continue placing data points for angular dimensioning, and click the Reset button to complete the tool sequence.)*

Dimension Angle Location

To place angular dimensioning with the Dimension Angle Location tool, invoke the tool:

Dimension tool box	Select the Dimension Angle Location tool. If necessary, turn ON Association Lock in the Tool Settings window (see Figure 9–36).
Key-in window	**dimension angle location** (or **dim a lo**) Enter

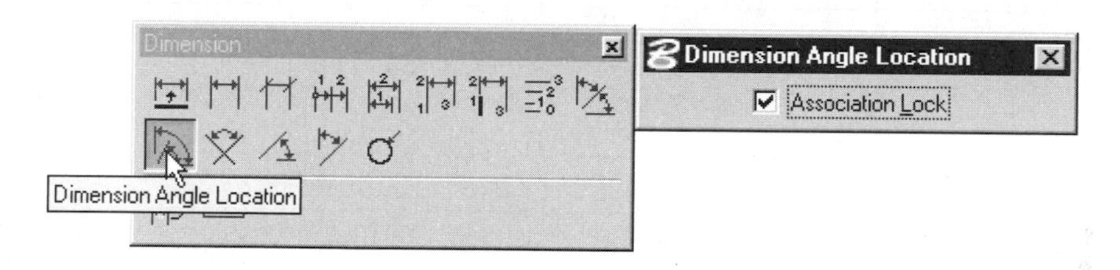

Figure 9–36 Invoking the Dimension Angle Location tool from the Dimension tool box

MicroStation prompts:

Dimension Angle Location > Select start of dimension *(Place a data point as shown in Figure 9–37 (point P1) to define the start of the dimension, which is measured counterclockwise from this point.)*

Dimension Angle Location > Define length of extension line *(Place a data point as shown in Figure 9–37 (point P2) to define the length of the extension line.)*

Dimension Angle Location > Enter point on axis *(Place a data point as shown in Figure 9–37 (point P3) to define the vertex of the angle.)*

Dimension Angle Location > Select dimension endpoint *(Place a data point as shown in Figure 9–37 (point P4) to define the end point of the dimension. Press* Enter *to edit the dimension text.)*

Dimension Angle Location > Select dimension endpoint *(Continue placing data points for angular dimensioning, and click the Reset button to complete the tool sequence.)*

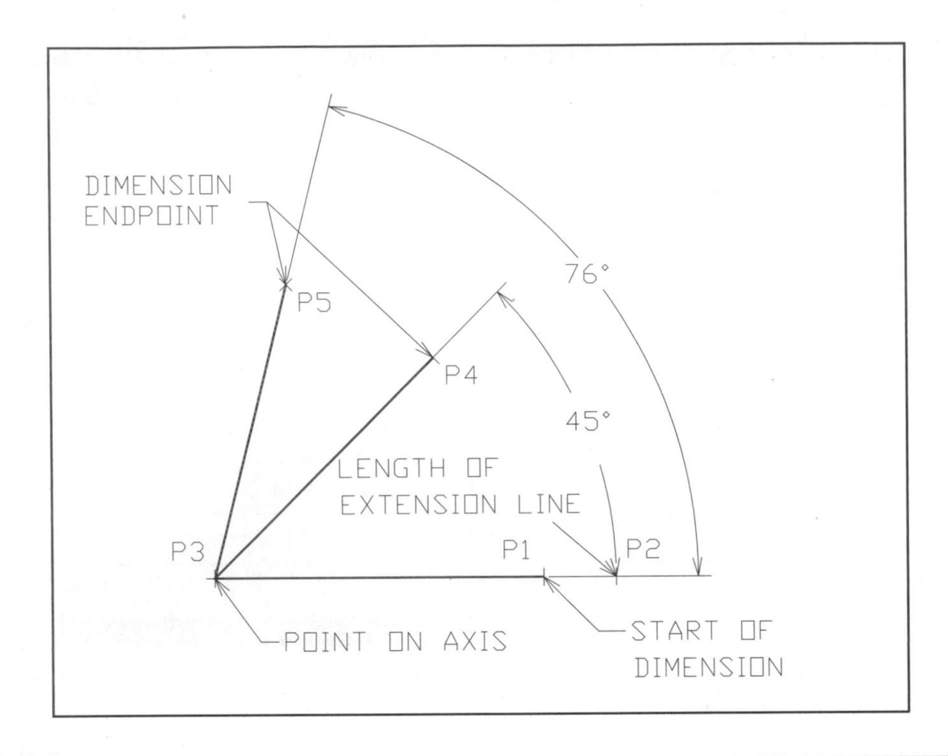

Figure 9–37 Example of placing angular dimensioning via the Dimension Angle Location tool

Dimension Angle Between Lines

To place angular dimensioning with the Dimension Angle Between Lines tool, invoke the tool:

Dimension tool box	Select the Dimension Angle Between Lines tool. If necessary, turn ON Association Lock in the Tool Settings window (see Figure 9–38).
Key-in window	**dimension angle lines** (or **dim a l**) Enter

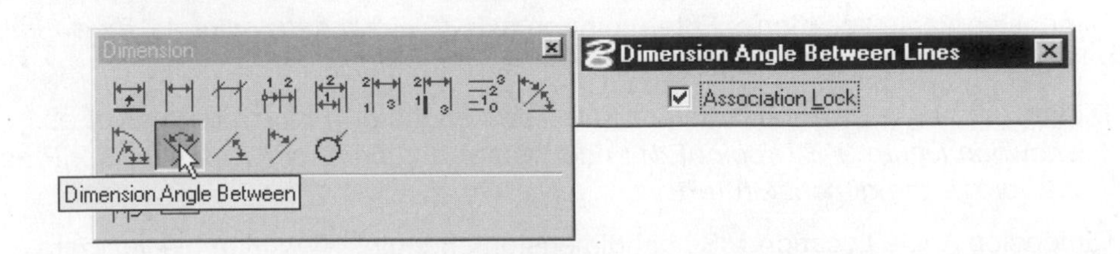

Figure 9–38 Invoking the Dimension Angle Between Lines tool from the Dimension tool box

MicroStation prompts:

> Dimension Angle Between Lines > Select first line *(Identify the first line or segment, as shown in Figure 9–39 (point P1).)*
>
> Dimension Angle Between Lines > Select second line *(Identify the second line or segment, as shown in Figure 9–39 (point P2).)*
>
> Dimension Angle Between Lines *(Place a data point to define the location of the dimension line, as shown in Figure 9–39 (point P3).)*

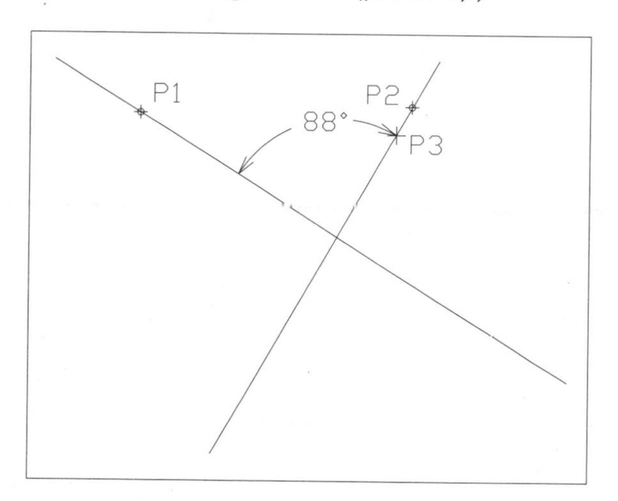

Figure 9–39 Example of placing angular dimensioning with the Dimension Angle Between Lines tool

Dimension Angle from *X*-Axis

To place angular dimensioning with the Dimension Angle from *X*-axis tool, invoke the tool:

Dimension tool box	Select the Dimension Angle from *X*-axis tool. If necessary, turn ON Association Lock in the Tool Settings window (see Figure 9–40).
Key-in window	**dimension angle x** (or **dim a x**) Enter

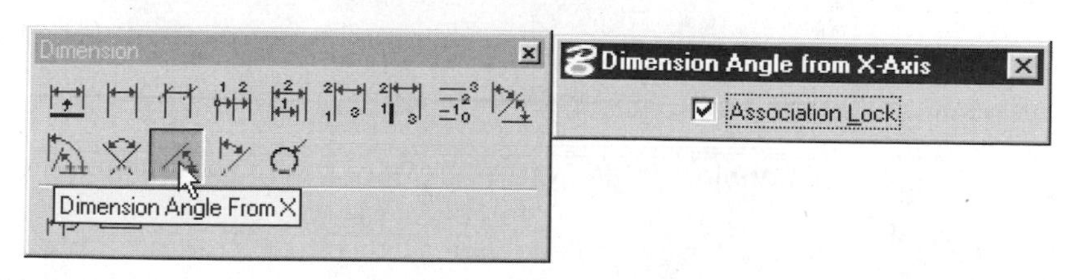

Figure 9–40 Invoking the Dimension Angle from X-Axis tool from the Dimension tool box

MicroStation prompts:

> Dimension Angle from *X* Axis > Identify element *(Identify the element, as shown in Figure 9–41 (point P1).)*
>
> Dimension Angle from *X* Axis > Accept, define dimension axis *(Place a data point, as shown in Figure 9–41 (point P2), to specify the location and direction of the dimension.)*

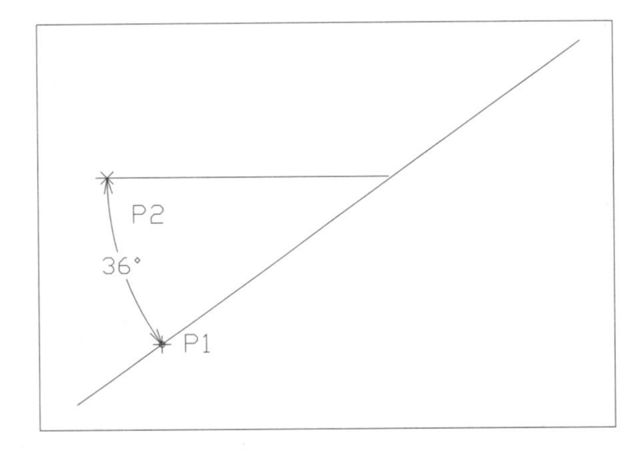

Figure 9–41 Example of placing angular dimensioning with the Dimension Angle from X Axis tool

Dimension Angle from *Y*-Axis

To place angular dimensioning with the Dimension Angle from *Y*-axis tool, invoke the tool:

Dimension tool box	Select the Dimension Angle from *Y*-axis tool. If necessary, turn ON Association Lock in the Tool Settings window (see Figure 9–42).
Key-in window	**dimension angle y** (or **dim a y**) Enter

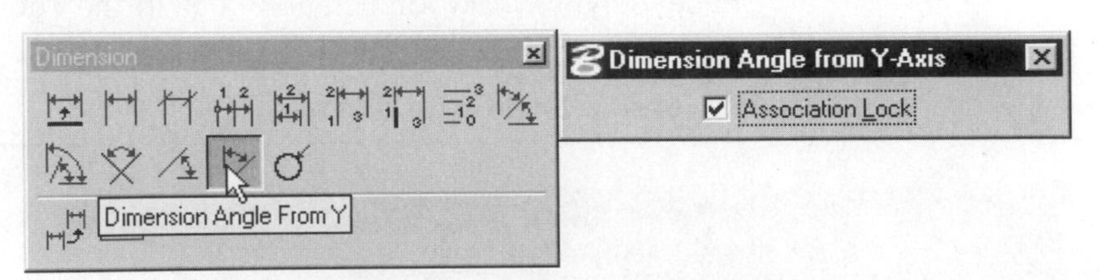

Figure 9–42 Invoking the Dimension Angle from Y-Axis tool from the Dimension tool box

MicroStation prompts:

> Dimension Angle from *Y* Axis > Identify element *(Identify the element.)*

Dimension Angle from *Y* Axis > Accept, define dimension axis *(Place a data point to specify the location and direction of the dimension.)*

Dimension Arc Size

The Dimension Arc Size tool lets you dimension a circle or circular arc. Each dimension is computed from the end point of the previous dimension, except the first one, similar to the Dimension Angle Size tool.

To dimension arcs with the Dimension Arc Size tool, invoke the tool:

Key-in window	**dimension arc size** (or **dim ar s**) Enter

MicroStation prompts:

Dimension Arc Size > Select start of dimension *(Place a data point, as shown in Figure 9–43 (point P1), to define the origin point. The dimension is measured counterclockwise from this point. This point must select an arc, circle, or ellipse.)*

Dimension Arc Size > Define length of extension line *(Place a data point, as shown in Figure 9–43 (point P2), to define the length of the extension line.)*

Dimension Arc Size > Select dimension endpoint *(Place a data point, as shown in Figure 9–43 (point P3), to define the dimension end point. If necessary, press* Enter *to edit the dimension text.)*

Dimension Arc Size > Select dimension endpoint *(Continue placing data points to continue angular dimensioning, and click the Reset button to terminate the tool sequence.)*

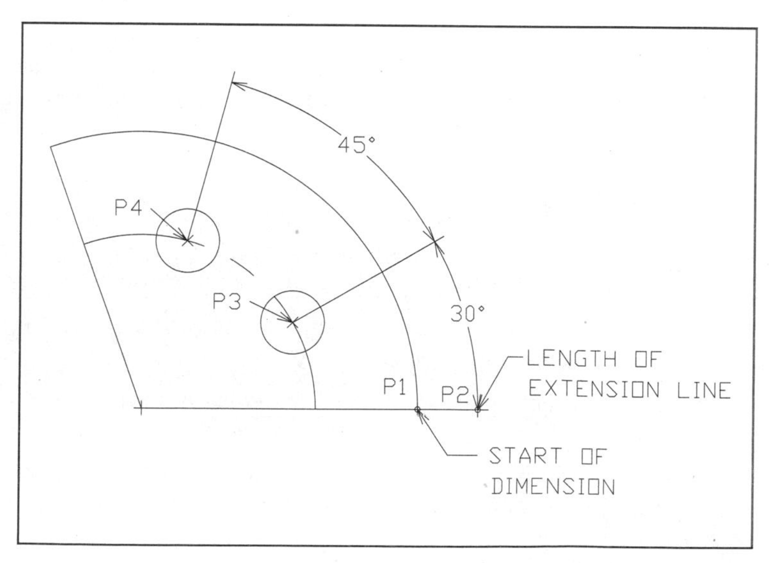

Figure 9–43 Example of placing arc dimensioning with the Dimension Arc Size tool

Dimension Arc Location

The Dimension Arc Location tool enables you to dimension a circle or circular arc. Each dimension is computed from the dimension origin (datum), as shown in Figure 9–44.

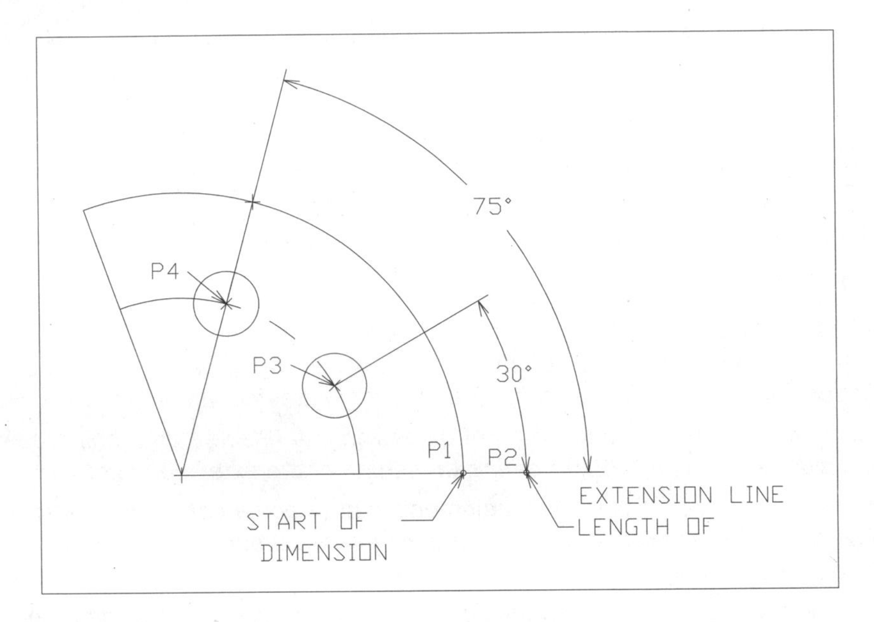

Figure 9–44 Example of placing arc dimensioning via the Dimension Arc Location tool

To dimension arcs with the Dimension Arc Location tool, invoke the tool:

Key-in window	**dimension arc location** (or **dim ar l**) Enter

MicroStation prompts:

Dimension Arc Location > Select start of dimension *(Place a data point to define the origin point. The dimension is measured counterclockwise from this point. This point must select an arc, circle, or ellipse.)*

Dimension Arc Location > Define length of extension line *(Place a data point to define the length of the extension line.)*

Dimension Arc Size > Select dimension endpoint *(Place a data point to define the dimension end point. If necessary, press* Enter *to edit the dimension text.)*

Dimension Arc Size > Select dimension endpoint *(Continue placing data points to continue angular dimensioning, and click the Reset button to terminate the tool sequence.)*

Dimension Radial

The Dimension Radial feature provides tools to create dimensions for the radius or diameter of a circle or arc and to place a center mark. Following are the tools available for radial dimensioning:

- Radius—To dimension the radius of a circle or a circular arc.
- Radius Extended—Identical to the Radius tool, except the leader line continues across the center of the circle, with terminators that point outward.
- Diameter—To dimension the diameter of a circle or a circular arc.
- Diameter Extended—Identical to the Diameter tool, except the leader line continues across the center of the circle, with terminators that point outward.
- Center Mark—To place a center mark at the center of a circle or circular arc.
- Dimension Diameter Parallel—To dimension the diameter of a circle using extension lines and a dimension line.

Dimension Radius

To place a radial dimension with the Dimension Radius tool, invoke the tool:

Dimension tool box	Select the Dimension Radial tool, then, in the Tool Settings window, select Radius from the Mode option menu, and if necessary, turn Association Lock and select an Alignment option (see Figure 9–45).
Key-in window	**dimension radius** (or **dim radiu**) Enter

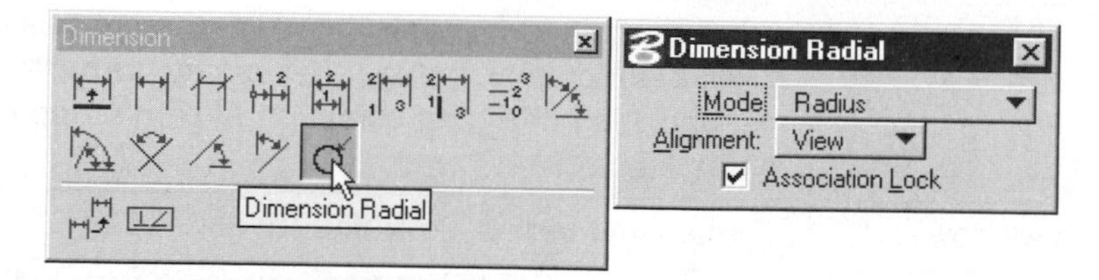

Figure 9–45 Invoking the Dimension Radial tool from the Dimension tool box

MicroStation prompts:

Dimension Radius > Identify element *(Identify a circle or arc, as shown in Figure 9–46.)*

Dimension Radius > Select dimension endpoint *(Place a data point inside or outside the circle or arc to place the dimension line, as shown in Figure 9–46.)*

Note: The placement of the dimension text (horizontal or in-line) is set by selecting the Text Orientation in the Dimension Settings dialog box.

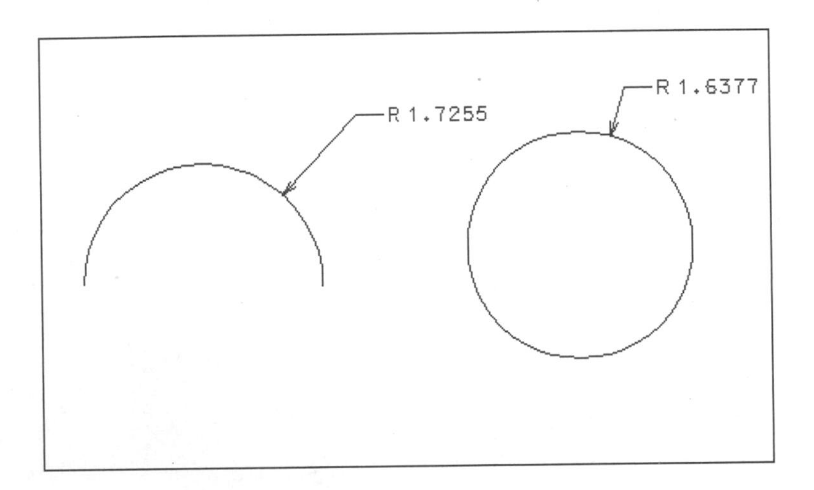

Figure 9–46 Examples of placing radial dimensions with the Dimension Radius tool

Dimension Radius (Extended Leader)

To place radial dimensioning with the Dimension Radius (Extended Leader) tool, invoke the tool:

Dimension tool box	Select the Dimension Radial tool, then, in the Tool Settings window, select Radius Extended from the Mode option menu, and if necessary, turn ON Association Lock and select an Alignment option. (see Figure 9–47).
Key-in window	**dimension radius extended** (or **dim radiu e**) Enter

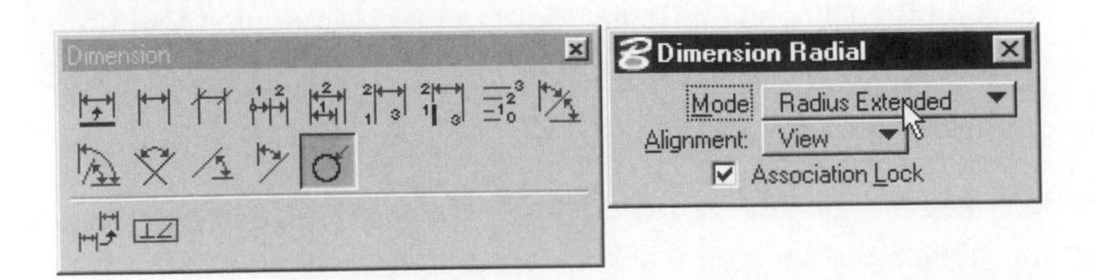

Figure 9–47 Invoking the Dimension Radial (Radius Extended) tool from the Dimension tool box

MicroStation prompts:

Dimension Radius (Extended Leader) > Identify element *(Identify a circle or arc, as shown in Figure 9–48.)*

Dimension Radius (Extended Leader) > Select dimension endpoint *(Place a data point inside or outside the circle or arc to place the dimension line as shown in Figure 9–48.)*

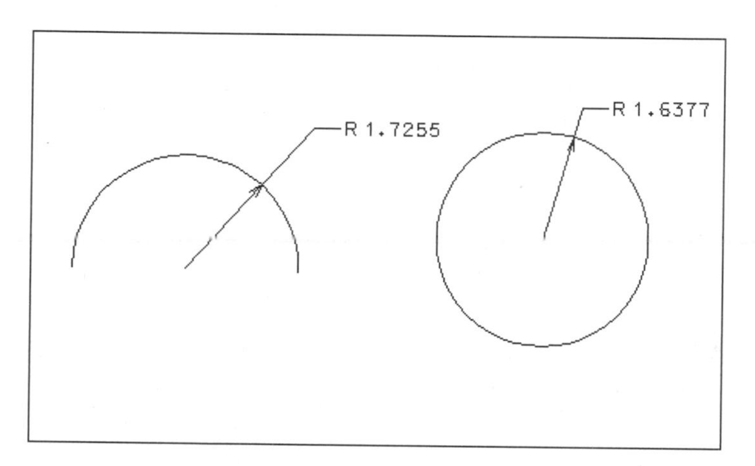

Figure 9–48 Examples of placing radial dimensions with the Dimension Radius (Extended Leader) tool

Dimension Diameter

To place diameter dimensioning with the Dimension Diameter tool, invoke the tool:

Dimension tool box	Select the Dimension Radial tool, then in the Tool Settings window, select Diameter from the Mode option menu, and if necessary, turn ON Association Lock and select an Alignment option (see Figure 9–49).
Key-in window	**dimension diameter** (or **dim d**) Enter

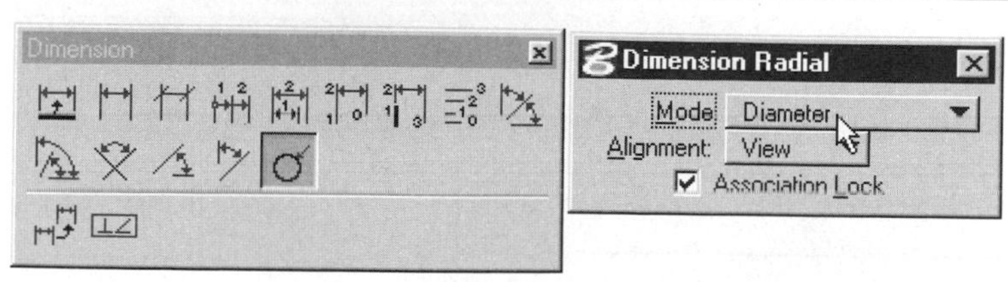

Figure 9–49 Invoking the Dimension Radial (Diameter) tool from the Dimension tool box

MicroStation prompts:

> Dimension Diameter > Identify element *(Identify a circle or arc, as shown in Figure 9–50.)*
>
> Dimension Diameter > Select dimension endpoint *(Place a data point inside or outside the circle or arc to place the dimension line, as shown in Figure 9–50.)*

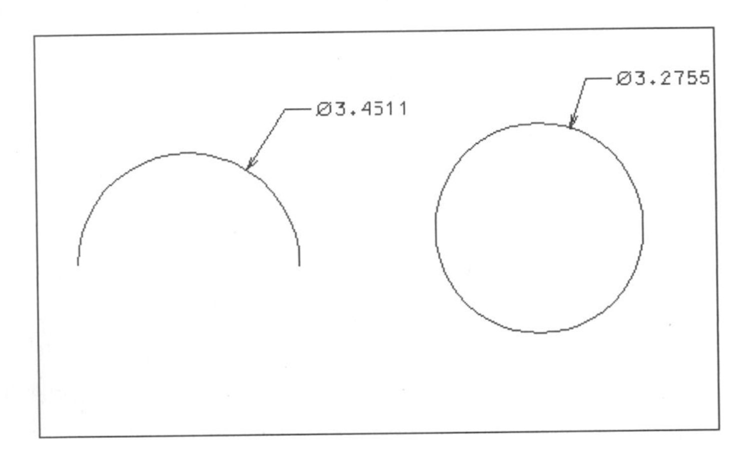

Figure 9–50 Examples of placing diameter dimensions with the Dimension Diameter tool

Dimension Diameter (Extended Leader)

To place diameter dimensioning with the Dimension Diameter (Extended Leader) tool, invoke the tool:

Dimension tool box	Select the Dimension Radial tool, then in the Tool Settings window, select Diameter Extended from the Mode option menu, and if necessary, turn ON Association Lock and select an Alignment option (see Figure 9–51).
Key-in window	**dimension diameter extended** (or **dim d e**) Enter

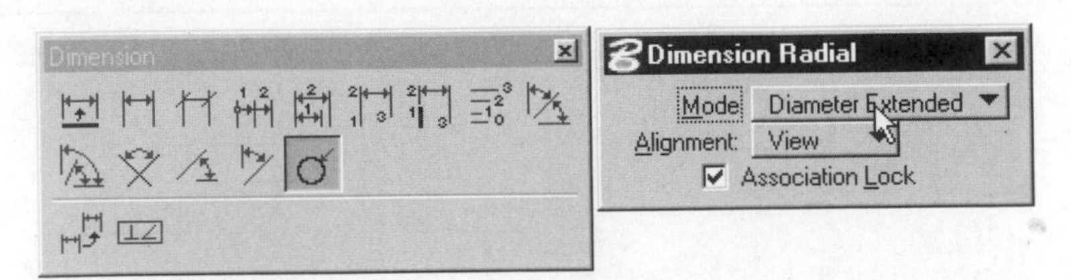

Figure 9–51 Invoking the Dimension Radial (Diameter Extended) tool from the Dimension tool box

MicroStation prompts:

Dimension Diameter (Extended Leader) > Identify element *(Identify a circle or arc, as shown in Figure 9–52.)*

Dimension Diameter (Extended Leader) > Select dimension endpoint *(Place a data point inside or outside the circle or arc to place the dimension line, as shown in Figure 9–52.)*

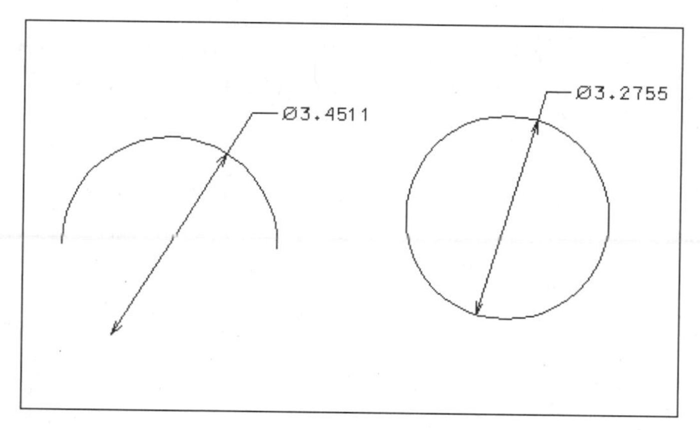

Figure 9–52 Examples of placing diameter dimensions via the Dimension Diameter Extended tool

Place Dimension Diameter Parallel

The Place Dimension Diameter Parallel places a dimension on a circle using extension lines and a dimension line, as shown in Figure 9–53.

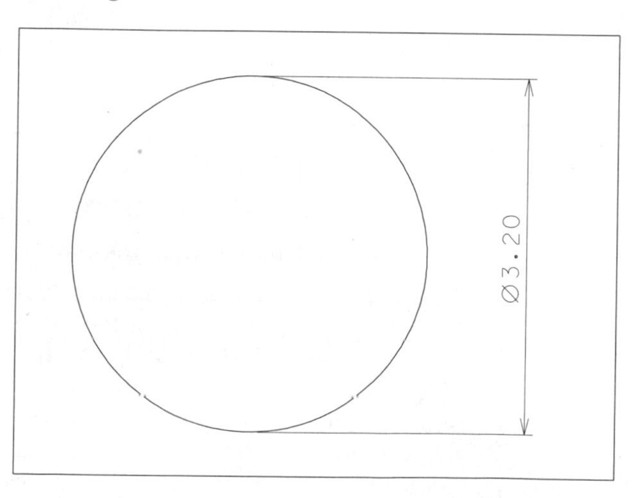

Figure 9–53 Example of placing a Dimension Diameter Parallel on a circle

To place a Dimension Diameter Parallel, invoke the tool:

Key-in window	**dimension diameter parallel** (or **dim d p**) Enter

MicroStation prompts:

Dimension Diameter Parallel > Identify element *(Identify the circle to be dimensioned.)*

Dimension Diameter Parallel > Select dimension end point *(Drag the dimension line to the desired position and click the Data button to complete the dimension.)*

Place Center Mark

The Place Center Mark tool can place a center mark at the center of a circle or circular arc, as shown in Figure 9–54.

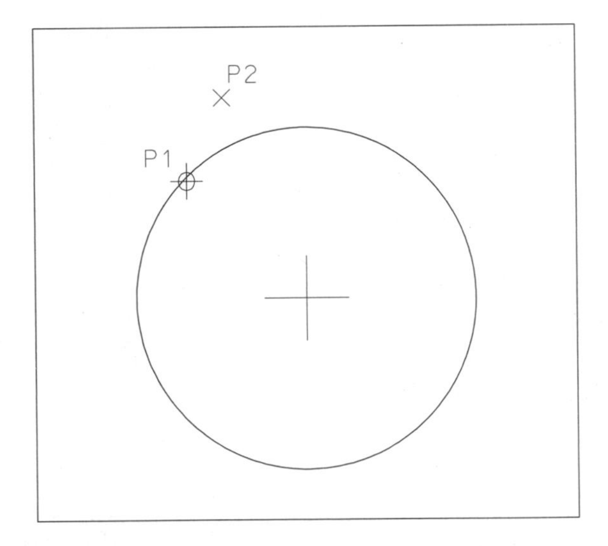

Figure 9–54 Example of placing a center mark by the Place Center Mark tool

To place a center mark with the Place Center Mark tool, invoke the tool:

Dimension tool box	Select the Dimension Radial tool, then in the Tool Settings window, select Center Mark from the Mode option menu, and if necessary, turn ON Association Lock, an Alignment option, and key-in the appropriate size of the center mark in the Center Size edit field (see Figure 9–55).
Key-in window	**dimension center mark** (or **dim c m**) Enter

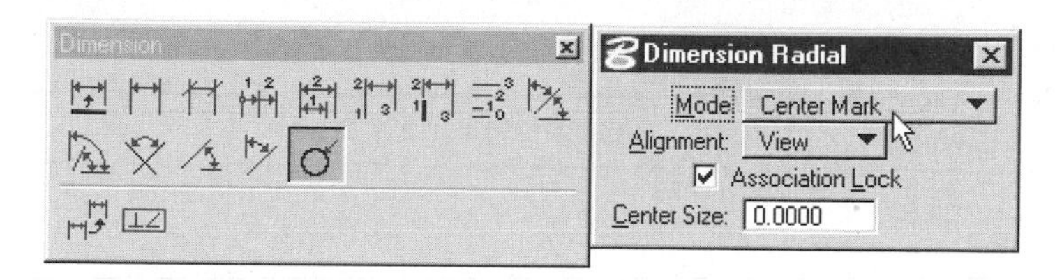

Figure 9–55 Invoking the Dimension Radial (Center Mark) tool from the Dimension tool box

MicroStation prompts:

Place Center Mark > Identify element *(Identify a circle or arc to place a center mark on.)*

Place Center Mark > Accept (next input) *(Click the Accept button, or identify another circle or arc to place the center mark.)*

Note: If the Center Size: text is set to 0.0000, then it applies the text size set in the Text settings box.

Label Line

The Label Line tool places the line length above the line you select and the line rotation angle below the line. You can place a label on a line, line string, block, closed shape, or multi-line.

MicroStation determines which side of the element is the top; what the angle of rotation is depends on how you drew it. The angle of the line or line segment, for example, is measured as a counterclockwise rotation from the first data point to the second. Exercise care in deciding how to draw elements you plan to label with the Label Line tool, or you may not get the angle of rotation you expected. See Figure 9–56 for examples of line labels.

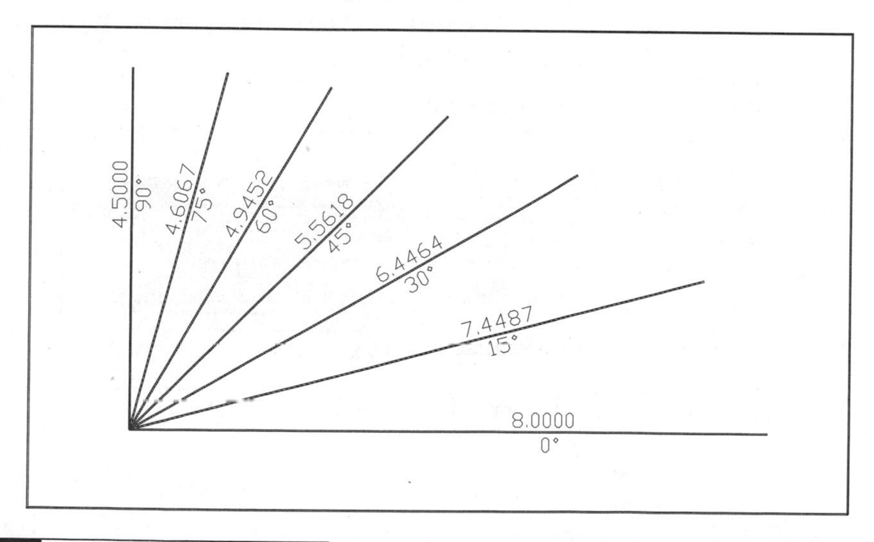

Figure 9–56 Examples of line labels

Invoke the Label Line tool:

Key-in window	**Label Line** (or **l l**) Enter

MicroStation prompts:

Label Line > Identify element *(Identify the element on which you want to place the label.)*

Label Line > Accept/Reject (Select next input) *(Click the Data button to accept, or click Reset to disregard the dimensioning.)*

Dimension Element

The Dimension Element provides a fast way to dimension an element. Simply identify the element and MicroStation selects the type of dimensioning tool it thinks is best. For linear elements it selects one of the linear dimensioning tools; for circles, arcs, and ellipses it selects one of the radial dimensioning tools.

When you invoke the Dimension Element tool, the tool name Dimension Element appears in the Status bar. As soon as you select an element, the name of the dimensioning tool MicroStation intends to use appears in the Status bar. If you don't want that selected dimension tool, press Enter and MicroStation switches to another dimensioning tool. Keep pressing Enter until you find the tool you want to use. You can also press the Next button in the Tool Settings window to change dimensioning tools.

Invoke the Dimension Element tool:

Dimension tool box	Select the Dimension Element tool. If necessary, turn ON Association Lock and select an Alignment option (see Figure 9–57).
Key-in window	**dimension element** (or **dim e**) Enter

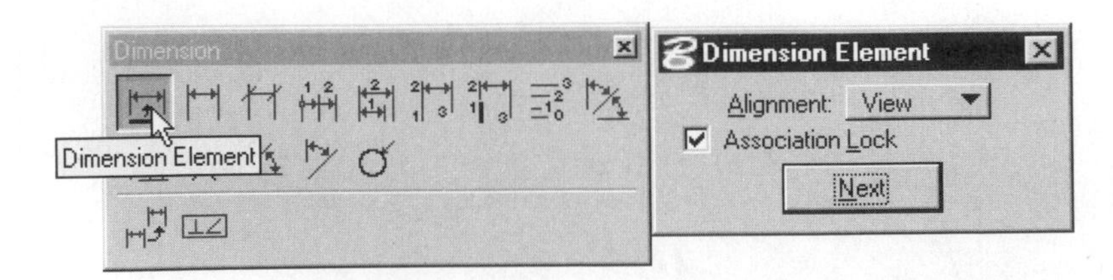

Figure 9–57 Invoking the Dimension Element tool from the Dimension tool box

MicroStation prompts:

> Dimension Element > Select element to dimension *(Identify the element or segment of the block or line string element.)*
>
> Dimension Element > Accept (Press Return to switch tool) *(Place a data point to indicate where you want the dimension placed, or press* Enter *as many times as necessary to find the dimensioning tool you want to use and place a data point for the location of the dimension element.)*

For most dimension tools, the second data point indicates the length and direction of the extension line. See Figure 9–58 for examples of placing dimensions with the Dimension Element tool.

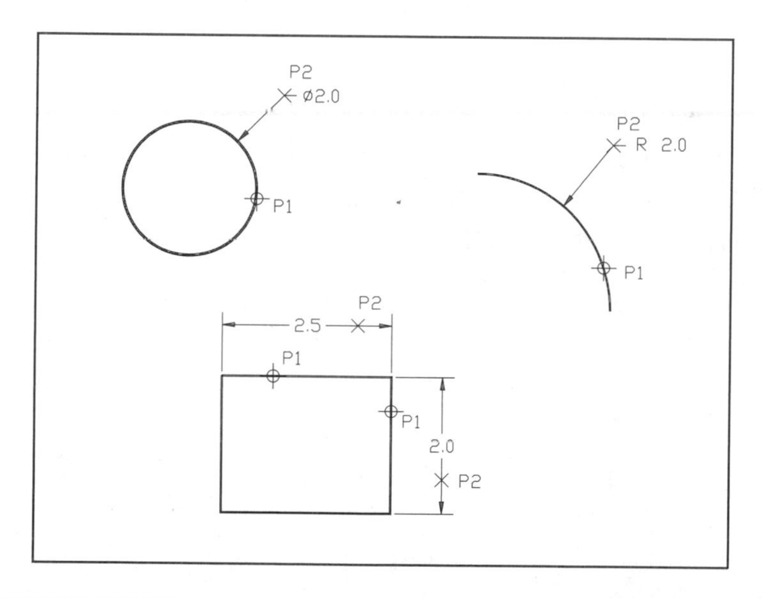

Figure 9–58 Examples of placing dimensioning with the Dimension Element tool

Geometric Tolerance

The Geometric Tolerance settings box (see Figure 9–59) helps you to build feature control frames with geometric tolerance symbols. The feature control frames are used with the Place Note tool and Place Text tool. The pull-down menu Fonts available in the settings box allows you to select one of the two fonts, 100—Ansi symbols, and 101—Feature Control Symbols. When a specific font is chosen, the buttons in the settings box reflect the availability of the symbols in the selected font. Whenever you want to add the symbols, click on the buttons as part of placing the text to the Place Note tool and Place Text tool.

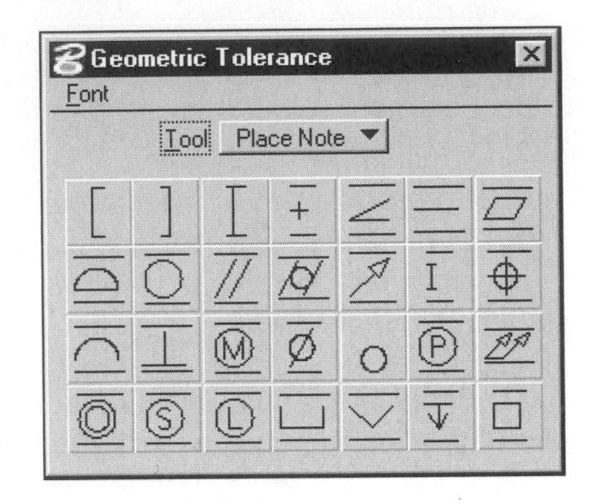

Figure 9–59 Geometric Tolerance settings box

The left bracket ([) and right bracket (]) form the ends of compartments; the vertical line (|) separates compartments.

Invoke the Geometric Tolerance settings box:

Dimension tool box	Select the Geometric Tolerance tool (see Figure 9–60).
Key-in window	**mdl load geomtol** Enter

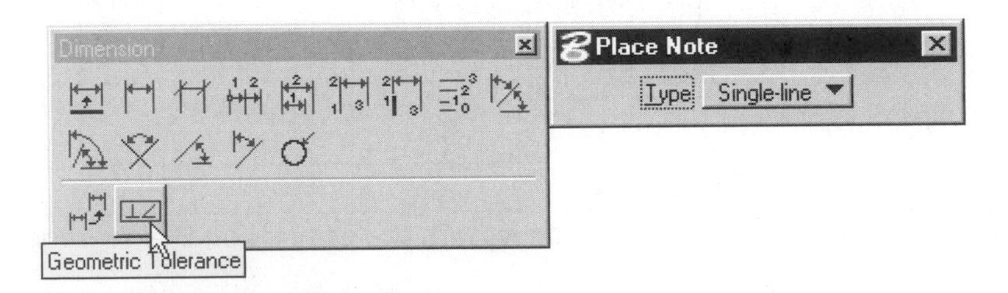

Figure 9–60 Invoking the Geometric Tolerance tool from the Dimension tool box

Select one of the two tools from the Geometric Tolerance settings box, Place Note or Place Text. MicroStation sets up the appropriate tool settings in the Tool Settings window. Follow the prompts to place text.

DIMENSION SETTINGS

MicroStation provides dimension settings options that allow you to customize dimension style and accuracy. The settings are flexible enough to accommodate users in all engineering disciplines.

The entire set of dimensioning settings can be saved with the design so they will be the active settings in later design sessions.

To change the active dimension settings, invoke the Dimension Settings box:

Pull-down menu	Element > Dimensions.
Key-in window	**dialog dimsettings** (or **di dims**) Enter

MicroStation displays the Dimension Settings box, as shown in Figure 9–61, which serves to control the settings for dimensioning.

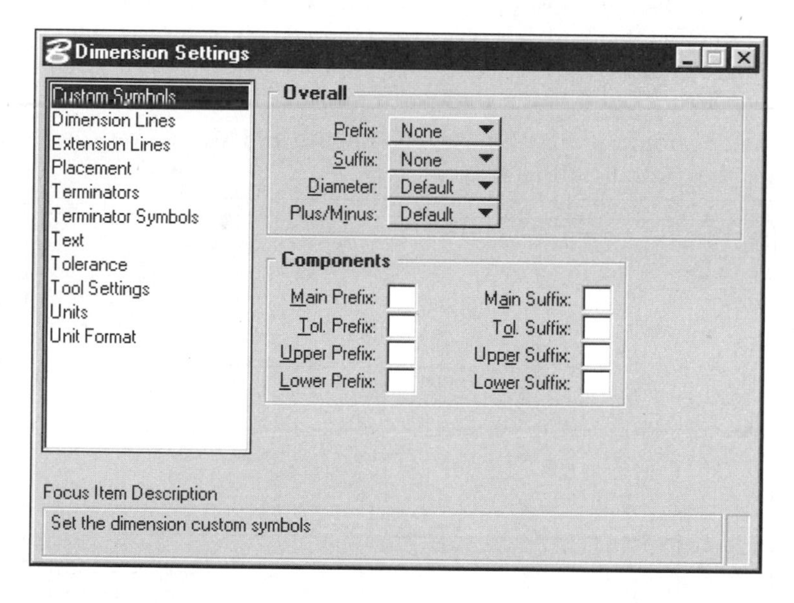

Figure 9–61 Dimension Settings box

MicroStation lists the available categories in alphabetical order on the left side of the settings box. Selecting a category causes the appropriate options for the category to be displayed to the right of the category list. The "Focus Item Description" area at the bottom of the window describes the purpose of the option with focus (the one you last clicked on).

Following are the categories and associated options available in the Dimension Settings box.

Custom Symbols

The Custom Symbols category, as shown in Figure 9–61, provides options for inserting symbols (characters from symbol fonts or cells) in the different parts of the dimension text.

Overall Options

The Overall options allow you to place an overall dimension text prefix and suffix, and to replace the default diameter and plus/minus symbols.

There are three options each for placing the Overall Prefix and Suffix:

- None—Do not insert a suffix or prefix.
- Symbol—When this option is selected, Char and Font fields appear in the settings box. Type the number of a font in the Font field, and the character to use in the Char field. That character will be used as the suffix or prefix.
- Cell—When this option is selected, a Name field appears in the settings box. Type the name of a cell in the field, it is placed as a shared cell (refer to Chapter 11 for a detailed discussion of Cells).

Figure 9–62 shows an example of dual dimensions (English and Metric) with an "X" for the Overall Prefix and a "Y" for the Overall Suffix.

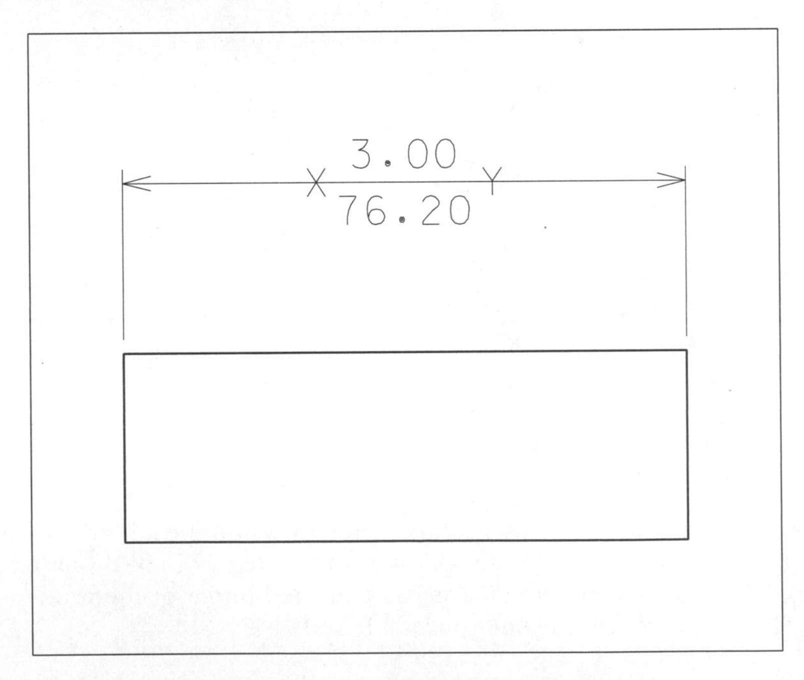

Figure 9–62 Example of Overall Prefix and Suffix in a linear dimension

There are two options each for selecting the Overall Diameter and Plus/Minus:

- Default—Use the default symbol provided by MicroStation.

- Symbol—When this option is selected for the Diameter, Char and Font fields appear in the settings box. Type the number of a font in the Font field, and the character to use in the Char field. That character will be used as the diameter symbol. When this option is selected for the Plus/Minus, an edit field appears in which you can type the number of a symbol from "0" to "9". (0 is the default symbol.)

Components Options

The Components options allow you to specify characters to use the prefixes and suffixes for parts of the dimension text. These options only allow you to specify the character, but not the font.

- Main Prefix and Main Suffix are the same as the Overview Prefix and Suffix.
- Tol. Prefix and Tol. Suffix place characters before and after the tolerance part of the dimension text, when tolerance is turned on. Figure 9–63 shows an example of placing an "X" before the tolerance part of a dimension and a "Y" after. Tolerance settings are discussed later in this chapter.
- Upper Prefix and Upper Suffix place characters before and after the upper text of a dual dimension (such as English above the dimension line and Metric below).
- Lower Prefix and Lower Suffix place characters before and after the lower text of a dual dimension.

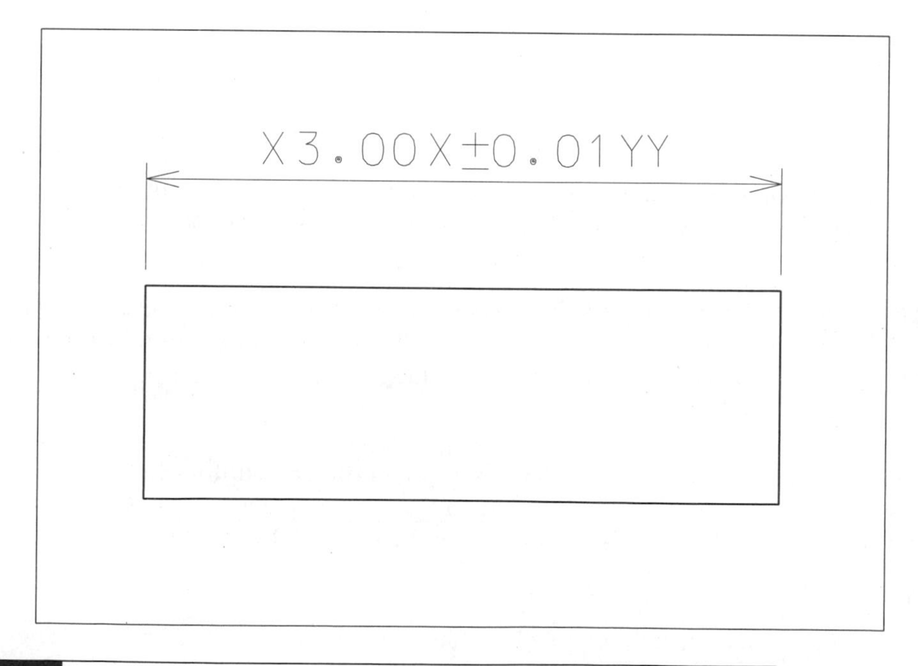

Figure 9–63 Example of Tolerance Prefix and Suffix in a linear dimension

Dimension Lines

The Dimension Lines category, as shown in Figure 9–64, provides options for changing the appearance of dimension lines.

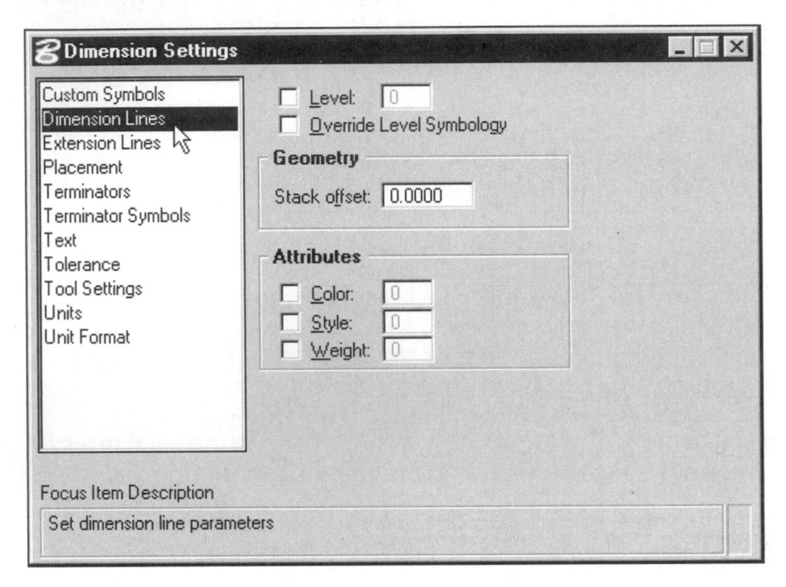

Figure 9–64 Dimension Settings box, Dimension Lines category

Dimension Level

The Level toggle button and edit field allow you to set the level that dimensions are placed on. When the toggle button is on, dimensions are placed on the level whose number is in the edit box, regardless of what the design's active level is. The complete dimensions are placed on the selected level (not just the dimension lines).

The Override Level Symbology toggle switch, when turned on, shows dimensions using their actual attributes (color, weight, and style), rather than the attributes set by the Level Symbology (refer to Chapter 14 for a detailed description of Level Symbology).

Geometry Option

The Stack Offset edit field controls the spacing between dimension lines in stacked dimensions. Enter values in Working Units (MU:SU:PU). When the number is zero, MicroStation sets the space between the stacked dimension lines equal to twice the text height.

Attributes

The Attributes options allow you to set the dimension line Color, Style, and Weight. When you turn on an attribute's toggle button, an options menu for the attribute appears to the right of the attribute name. Select the desired attribute value from the options menu, or type number in the provided edit field.

When the toggle buttons are on, the dimension lines are placed using the attributes defined in this settings box, not the design's active attributes.

Extension Lines

The Extension Lines category, as shown in Figure 9–65, provides options for changing the appearance of extension lines.

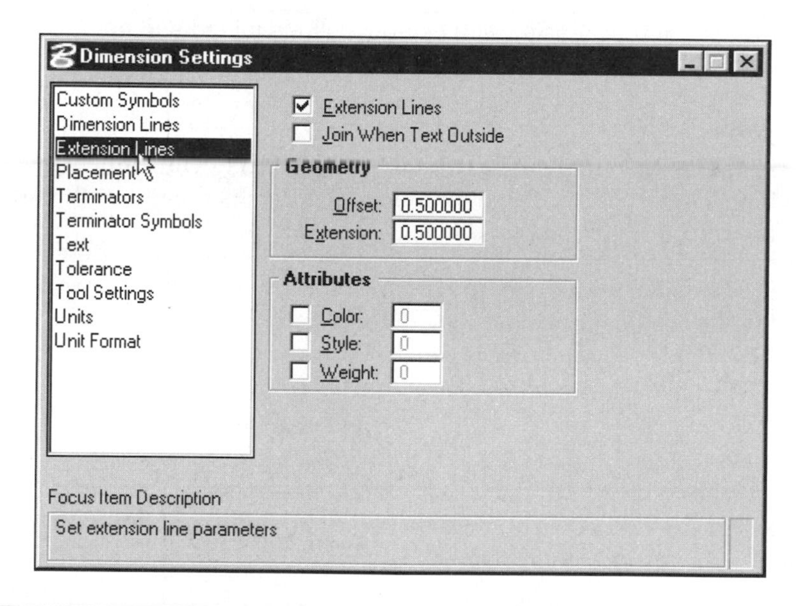

Figure 9–65 Dimension Settings box, Extension Lines category

Extension Line Creation

Two options at the top of the settings box control the creation of extension lines in a dimension:

- If the Extension Lines toggle button is ON, the extension lines are drawn with the dimension. If the button is OFF, the dimension is placed in the same way, but the extension lines are not drawn.
- If the Join When Text Outside toggle button is ON, the dimension line is drawn between the two extension lines, even when the space between the extension lines is so small that the dimension text must be placed outside of the extension line pair. If the button is OFF, no connecting dimension line will be drawn in tight dimensions. Figure 9–66 shows an example of a tight dimension with this button ON and with it OFF.

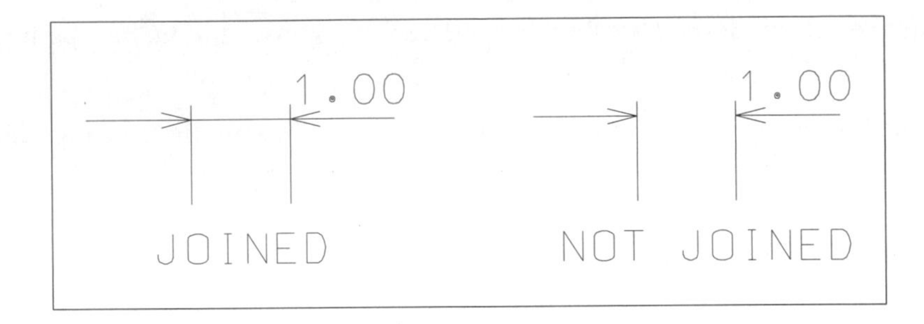

Figure 9–66 Example of a dimension with extension lines joined and not joined

Geometry Options

The Offset edit field allows you to set the gap between the object being dimensioned and the start of the extension line. The gap is calculated by multiplying the text height by the value in this field. For example, if the value is 1.0, the offset is equal to the text height.

The Extension edit field allows you to set the distance the extension line extends beyond the end dimension line. The extension is calculated by multiplying the text height by the value in this field. For example, if the value is 0.5, the offset is equal to half the text height.

Attributes Options

The Attributes options allow you to set the extension line Color, Style, and Weight. When you turn on an attribute's toggle button, an options menu for the attribute appears to the right of the attribute name. Select the desired attribute value from the options menu, or type number in the provided edit field.

When the toggle buttons are on, the extension lines are placed using the attributes defined in this settings box, not the design's active attributes.

Placement

The Placement category, shown in Figure 9–67, provides options for controlling how dimensions are aligned in relation to the dimensioned element, where the dimension text is placed on the dimension line, and how reference file elements are dimensioned.

The Alignment menu provides options that control how the dimension is placed in relation to the dimensioned object:

- View—Aligns linear dimensions parallel to the view *X* or *Y* axis.
- Drawing—Aligns linear dimensions parallel to the design plane *X* or *Y* axis.
- True—Aligns linear dimensions parallel to the element being dimensioned.
- Arbitrary—Aligns linear dimensions parallel to the element being dimensioned. The extension lines are not constrained to be at right angles to the dimension line.

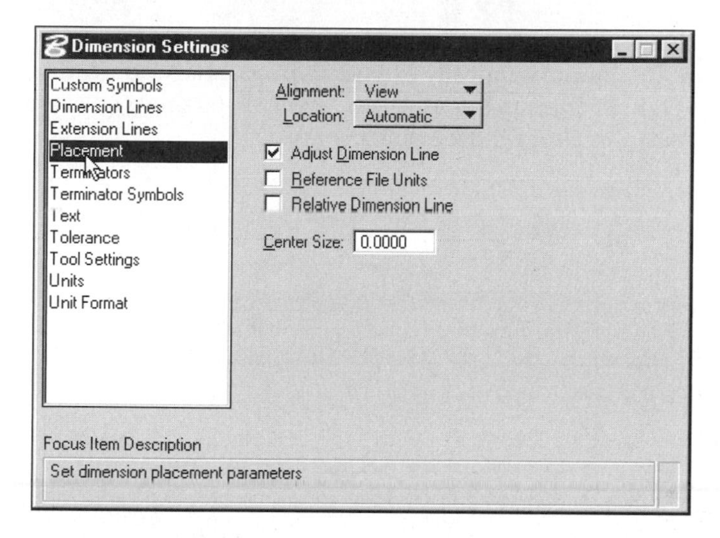

Figure 9–67 Dimension Settings box, Placement category

Figure 9–68 shows examples of placing linear dimensions using different alignment controls.

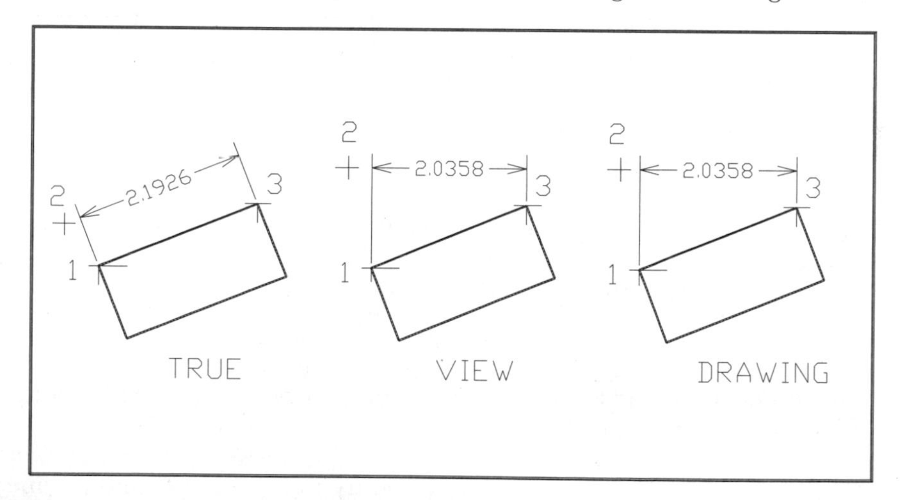

Figure 9–68 Examples of placing linear dimensions via different alignment controls

The Location menu provides options for controlling where the dimension text is placed:

- Automatic—Automatically places the dimension text according to the justification setting (left, center, or right on the dimension line). Justification is discussed later in this chapter
- Semi-Auto—Places the dimension according to the justification setting if the text fits between the extension lines. If not, you must specify the position of the text.
- Manual—You must specify the dimension text position for all dimensions.

If the Adjust Dimension Line toggle button is ON, the dimension line and text are dynamically moved and extension lines are dynamically extended, if necessary, to prevent overlaying existing dimension text. If it is OFF, no adjustments are made. Figure 9–69 shows an example of a dynamically adjusted center dimension in a set of three dimensions.

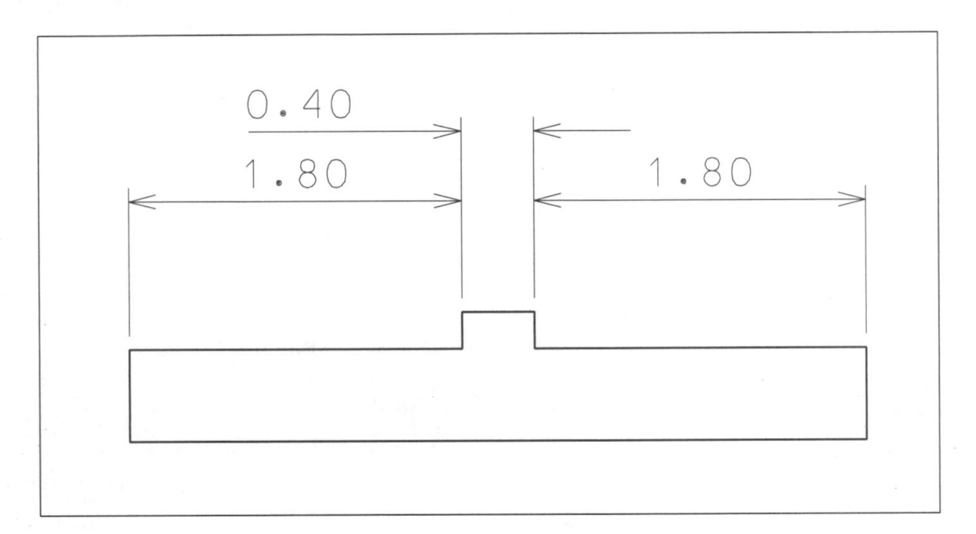

Figure 9–69 Example of dynamically adjusted dimension

If the Reference File Units toggle button is:

- ON, the Working Units of the reference file are used when dimensions are placed on elements in the reference file to allow for reference file scaling. (Refer to Chapter 13 for a detailed discussion of Reference Files.)
- OFF, the active design file's Working Units are always used.

The Relative Dimension Line toggle button comes into play when the element a dimension is associated with is modified. If the button is:

- ON, the dimension will be moved as necessary to keep the extension line the same length.
- OFF, the dimension stays in the same position and the length of the extension line is varied as necessary to maintain the dimension's relationship to the element.

The Center Size edit field allows you to set the default size, in Working Units, for the center mark that is placed from the Dimension Radial tool. The number you enter is half the overall height and width of the center mark. If the field contains a zero, the overall center mark height and width is equal to the text height.

Terminators

The Terminators category, as shown in Figure 9–70, provides options that control the placement of dimension terminators and their appearance after placement.

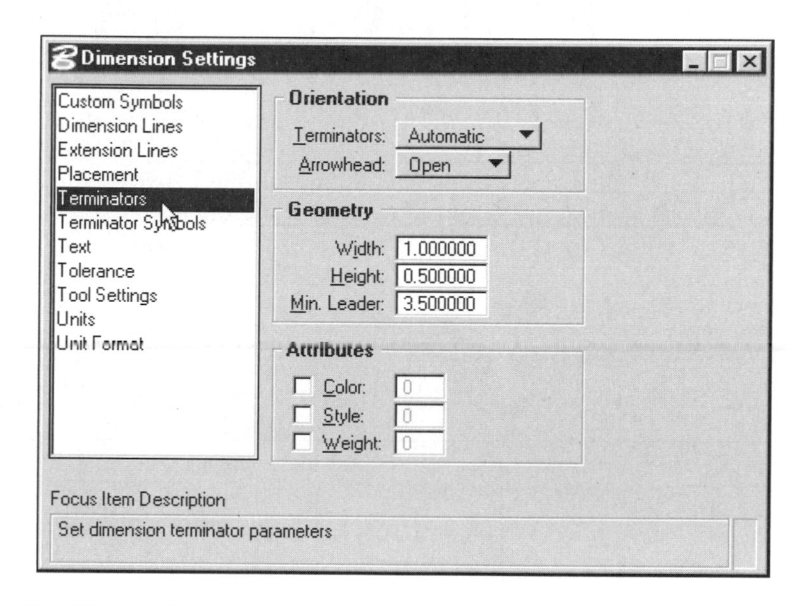

Figure 9–70 Dimension Settings box, Terminators category

Orientation Options

The terminator orientation options control where the terminators are placed and the style of terminator arrowhead. The default dimension terminator is an arrowhead, but, as we will learn in the Terminator Symbols category, other types of terminators can be used.

The Terminator options menu controls the placement of the dimension terminators:

- Automatic—If MicroStation decides there is enough room between the extension lines, it will place the terminators and a dimension line between the extension lines. If it decides there is not enough room, it places the terminators and dimension lines outside the extensions (there will not be a dimension line between the extension lines).
- Inside—MicroStation always places the terminators and a dimension line between the extension lines.
- Outside—MicroStation always places the terminators and dimension lines outside the extension lines (there will not be a dimension line between the extension lines).
- Reversed—MicroStation always places the terminators outside the extension lines and places a dimension line both inside and outside the extension lines.

The Arrowhead options menu controls the style of arrowhead used for the dimension terminators:

- Open—The arrowheads look like angle brackets.
- Closed—The arrowheads look like triangles.
- Filled—the arrowheads look like triangles and are filled with the outline color.

Figure 9–71 shows examples of each terminator placement position and each style of arrowhead.

Note: To make the filled arrowheads actually appear filled, turn on the Fill View Attribute for the view window you are working in. The View Attributes settings box is opened from the Settings pull-down menu.

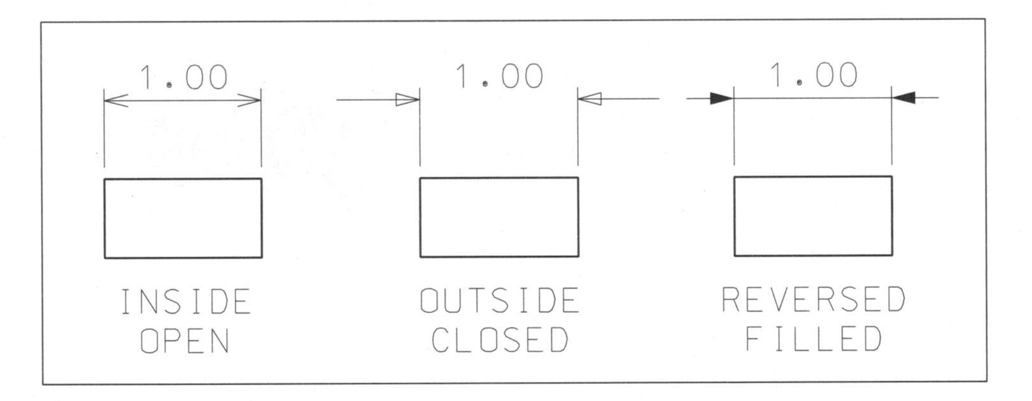

Figure 9–71 Examples of terminator placement positions and styles

Geometry Options

Three Geometry edit fields control the size of arrowhead terminators. Each size is the product of the dimension text height multiplied by the number you enter in the edit field. For example, if the text size is one inch and you enter a 2.0 in the width field, the arrowhead width will be two inches.

- Width—Allows you to set the width of the arrowhead from tip to back (parallel to the dimension line).
- Height—Allows you to set the height of the arrow head at its widest part (at right angles to the dimension line).
- Min. Leader—Allows you to control the minimum length of the dimension line the arrowhead is placed on. For example, if the arrowheads are placed Outside the extension lines, the short pieces of dimension lines on the outside of the extension lines are the minimum leader value in length.

Attributes Options

The Attributes options allow you to set the terminator Color, Style, and Weight. When you turn ON an attribute's toggle button, an options menu for the attribute appears to the right of the attribute name. Select the desired attribute value from the options menu, or type a number in the provided edit field.

When the toggle buttons are ON, the terminators are placed using the attributes defined in this settings box, not the design's active attributes.

Terminator Symbols

The Terminator Symbols category, as shown in Figure 9–72, provides controls to specify alternate symbols (characters from symbol fonts or cells) for each of the four types of dimension terminators.

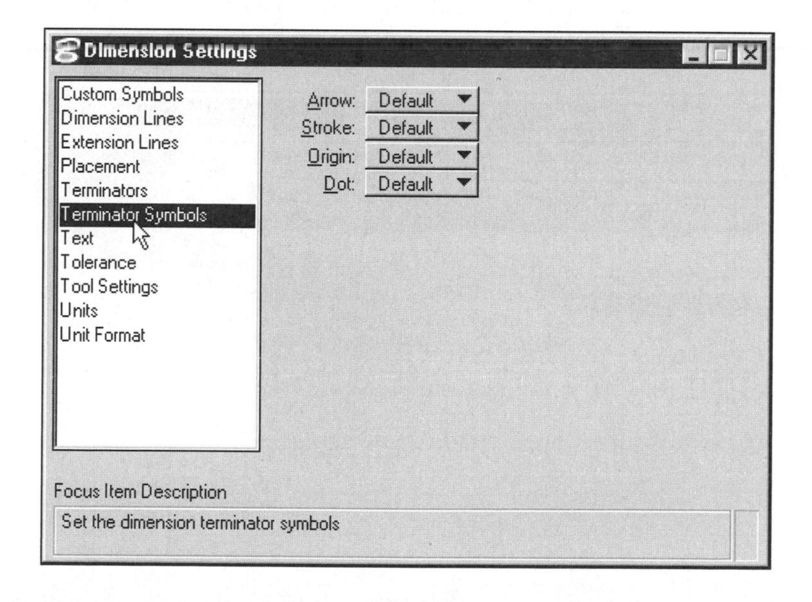

Figure 9–72 Dimension Settings box, Terminator Symbols category

The three types of terminators are:

- Arrow—The terminator placed by most of the dimensioning tools.
- Stroke—The terminator placed by the Dimension Size Stroke tool.
- Origin—The start of the dimension symbol placed by the Dimension Location Stacked tool.
- Dot—The dimension dot symbol.

Each type of terminator has an options menu that provides the following options:

- Default—Use the default symbol provided by MicroStation.
- Symbol—When this option is selected, Char and Font fields appear in the settings box. Type the number of a font in the Font field, and the character to use in the Char field. That character will be used as the terminator.
- Cell—When this option is selected, a Name field appears in the settings box. Type the name of a cell in the field, and it is placed as a shared cell (refer to Chapter 11 for a detailed discussion of Cells).

Text

The Text category, as shown in Figure 9–73, provides options for controlling the placement and appearance of dimension text.

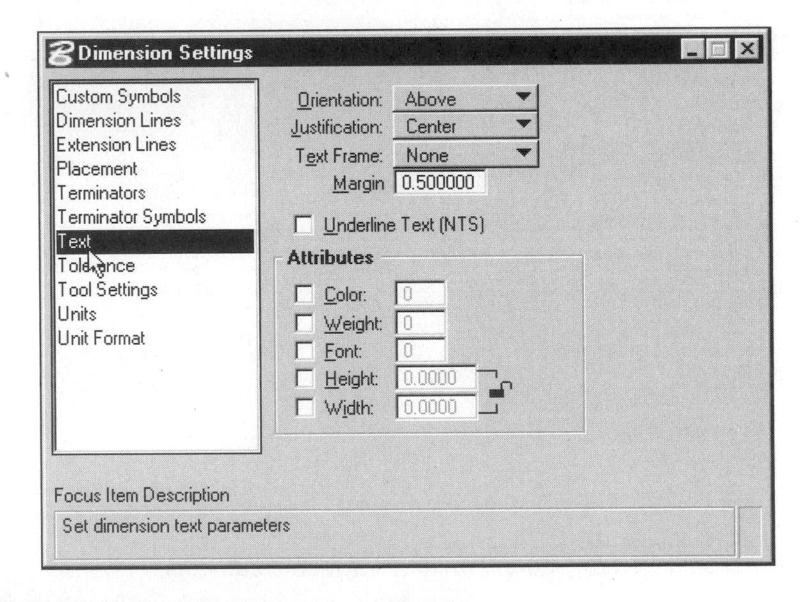

Figure 9–73 Dimension Settings box, Text category

Placement Options

The Orientation menu provides options to control placement of the dimension text relative to the dimension line:

- In Line—Places text on the dimension line.
- Above—Places text above the dimension line.
- Horizontal—Places text horizontally.

Figure 9–74 shows examples of placing linear dimension with different orientation modes.

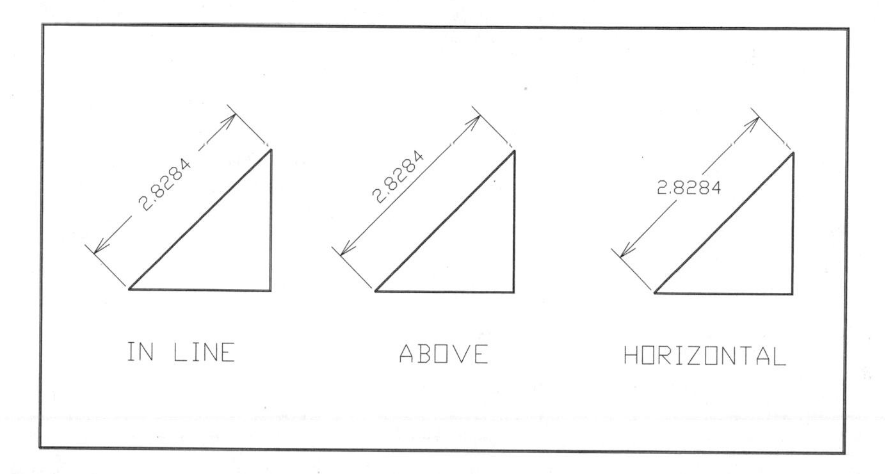

Figure 9–74 Examples of placing linear dimensioning in different Orientation modes.

The Justification menu provides options to control where on the dimension line the text is placed when the text placement location is set for Automatic or Semi-Auto:

- Left—Place the text at the left end of the dimension line.
- Center—Place the text at the center of the dimension line.
- Right—Place the text at the right end of the dimension line.

Figure 9–75 shows examples of placing linear dimensions with different justification modes.

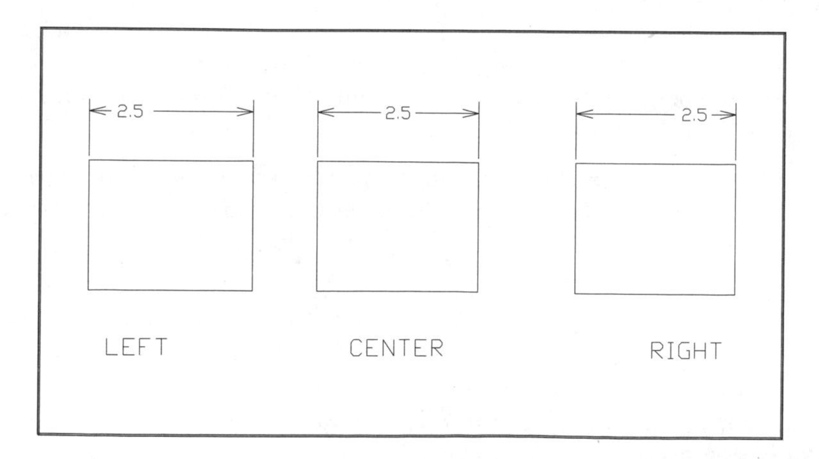

Figure 9–75 Examples of placing linear dimensioning in different Justification modes

The Text Frame menu provides options for placing a frame around the dimension text:

None—No frame.

Box—Box frame.

Capsule—Capsule frame.

Figure 9–76 shows examples of placing linear dimensions with different Text frame modes.

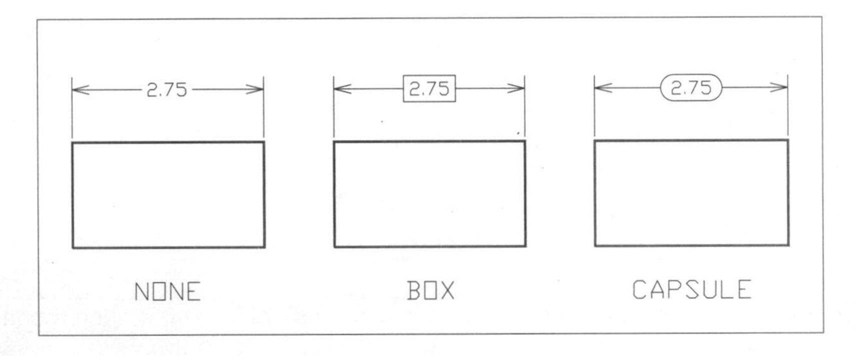

Figure 9–76 Examples of placing linear dimensioning in different Text Frame modes

The Margin edit field allows you to control the gap between each end of the dimension text and the dimension line when the text is placed in-line or at the end of the leader in a radial dimension. The gap is the product of the number you enter in the field multiplied by the dimension text height.

If the Underline Text (NTS) toggle button is ON, all dimension text is placed with a line under the characters. If it is OFF, no underline is placed.

Attributes Options

The Attributes options allow you to set the dimension text Color, Weight, Height, and Width. When you turn on the color or weight toggle button, an options menu for the attribute appears to the right of the attribute name. Select the desired attribute value from the options menu, or type a number in the provided edit field.

When you turn on the height or width toggle button, you can enter the height or width, in working units, in the associated edit fields. If you want to force the height and width to always be equal, click on the lock symbol to close it before entering one of the values.

When the toggle buttons are ON, the terminators are placed using the attributes defined in this settings box, not the design's active attributes.

Tolerance

The Tolerance category, as shown in Figure 9–77, provides options for placing dimension tolerance values in the dimension text.

Tolerance values are added to the dimensions when the Tolerance Generation toggle button is turned ON.

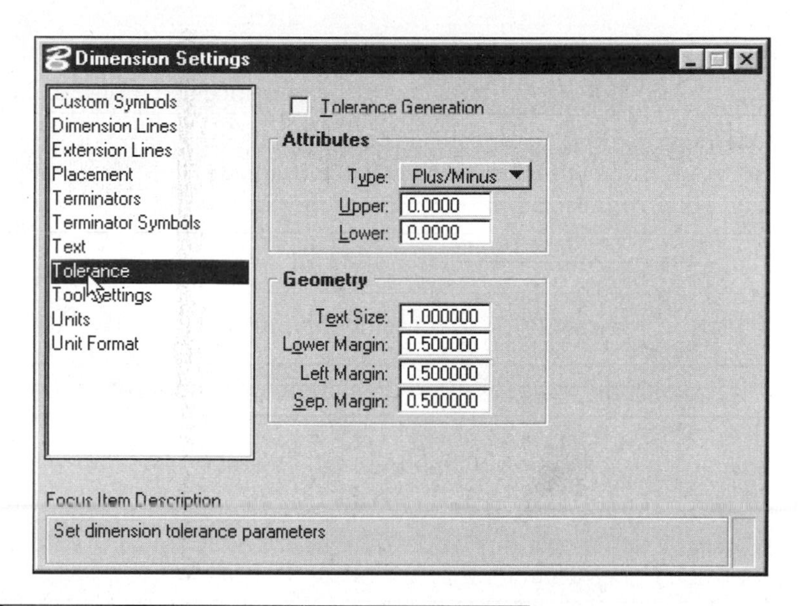

Figure 9–77 Dimension Settings box, Tolerance category

Attributes Options

The Type menu provides options for displaying tolerance values as a Plus/Minus addition to the dimension value or as a dimension Limit.

The Upper and Lower edit fields allow you to enter the tolerance range maximum and minimum values in working units (MU:SU:MU).

Figure 9–78 shows examples of each tolerance type with the upper and lower tolerance values equal to 0.002.

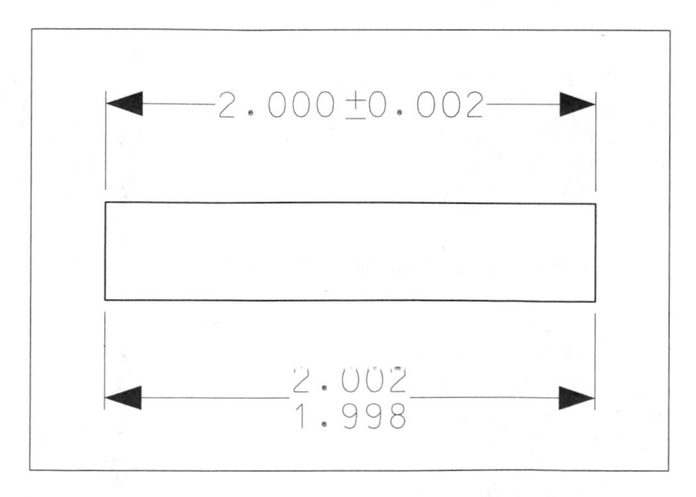

Figure 9–78 Examples of Plus/Minus and Limit tolerance attributes

Geometry Options

The Geometry options control the size and position of the tolerance text relative to the dimension text.

- Text Size—Sets the tolerance text size, specified as a multiple of the dimension text Height and Width.
- Lower Margin—Sets the space, specified as a multiple of text size, between the dimension line and the bottom of the dimension text.
- Left Margin—Sets the horizontal space, specified as a multiple of text size, between tolerance text and dimension text.
- Sep. Margin—Sets the vertical space, specified as a multiple of text size, between tolerance values.

Tool Settings

The Tool Settings category, as shown in Figure 9–79, provides options to customize the settings associated with individual dimensioning tools.

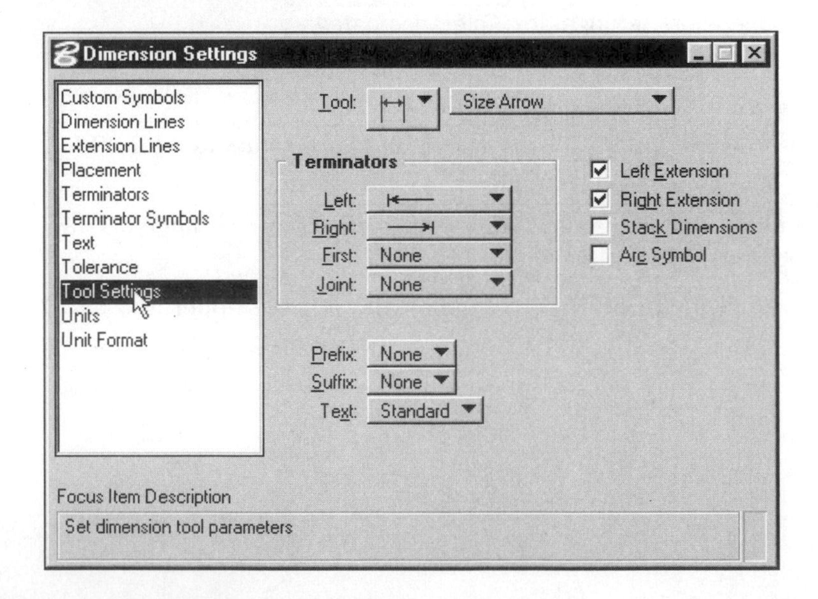

Figure 9–79 Dimension Settings box, Tool Settings category

The two Tool menus allow you to select the dimension tool you want to customize. The menu on the left contains pictures of each tool, and the menu on the right contains the names of each tool. Select a tool from either menu.

Each tool has its own set of customization options that appear in the settings box when you select the tool from a Tool menu.

Units

The Units category, as shown in Figure 9–80, provides options for placing two different styles of dimension text, Mechanical and AEC (Architectural, Engineering, Construction). The Units setting also allows you to choose to display one dimension value or a Primary and Secondary dimension value on each dimension. For example, the Primary dimension value might be in feet and the secondary in meters.

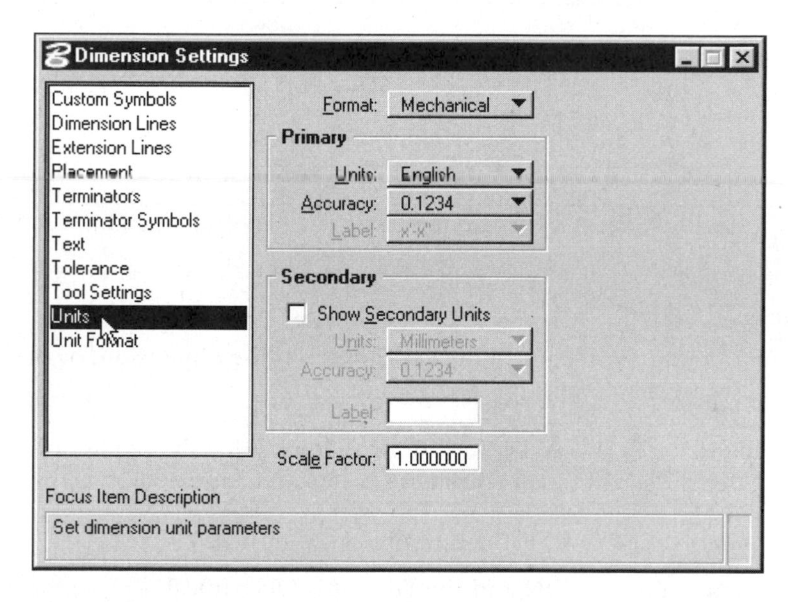

Figure 9–80 Dimension Settings box, Units category

Format Option

The Format menu allows you to select either of the two dimensioning styles:

- Mechanical—Places dimension text using master units without a units name, and allows displaying the fractional part of the measurement as a decimal or stacked fraction (for English units only).
- AEC—Provides options for placing the dimension text using master units and sub-units or master units only, and for adding the master and sub-unit names to the dimension.

Figure 9–81 shows an example of a nine-foot long object dimensioned using the Mechanical and AEC styles.

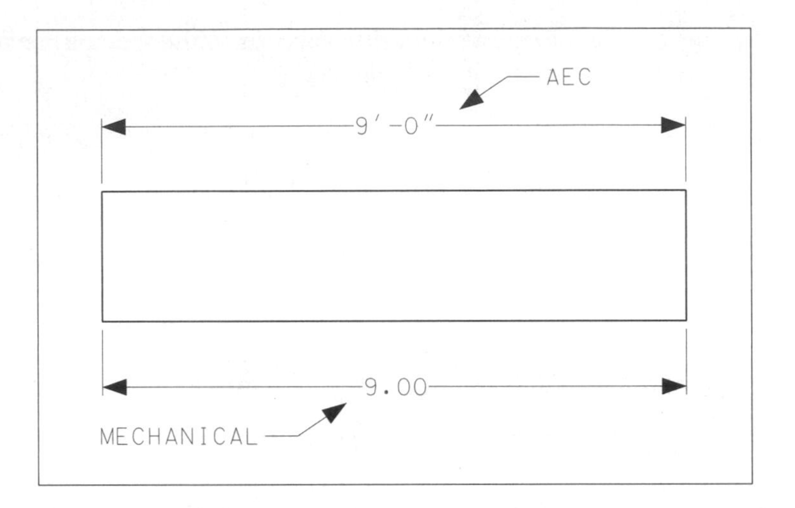

Figure 9–81 Example of object dimensioned using the AEC and Mechanical styles

Primary Options

The three Primary option menus control the appearance of the dimension, or the first dimension when primary and secondary dimensions are placed.

The Units menu allows you to choose between dimensioning using the English or Metric units. The actual options vary depending on which Format is selected and if the Secondary units are shown.

- If Secondary units are OFF, the options are English and Metric.
- If Secondary units are ON and the Format is Mechanical, the options are Inches and Millimeters.
- If Secondary units are ON and the Format is AEC, the options are Feet and Meters.

The Accuracy menu allows you to select the dimensioning accuracy from zero to eight decimal places, or from ½" to 1/64" (English units only).

The Label menu allows you to select the style of AEC dimension (it's not available for Mechanical).

Note: MicroStation uses the design's working units and working units names for the Primary dimension. Select the measurement unit that matches the design's working units. For example, if the working units are feet and inches, select an English unit for the Primary dimension.

Secondary Options

To add secondary dimension text to each dimension, turn ON the Show Secondary Units toggle button. These use the three Secondary options to control the dimension appearance.

The Units menu allows you to choose the measurement units for the secondary dimension. The options vary depending on the Primary Units selection:

- If the Primary Units are Feet or Inches, the Secondary options will be Millimeters, Centimeters, or Meters.
- If the Primary Units are Meters or Millimeters, the Secondary options will be Inches or Feet.

The Accuracy menu offers the same options as the Primary Accuracy menu.

The Label edit field allows you to type a one- or two-character label that will be placed with the secondary dimension text. When you select a Secondary Unit, a default label is placed in this field, but you can change it.

Scale Factor Option

The Scale Factor edit field allows you to enter a dimension scaling value to compensate for a design that was not drawn true size. For example:

- If the design is half its true size, enter a Scale Factor of 2 to double all dimensions.
- If the design is twice its true size, enter a Scale Factor of 0.5 to cut all the dimensions in half.

Unit Format

The Unit Format category, shown in Figure 9–82, provides options to change the display format of dimension text.

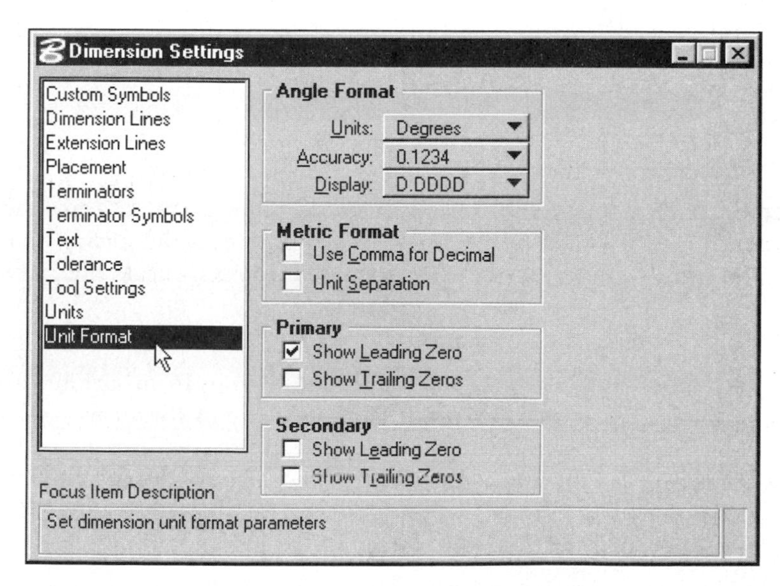

Figure 9–82 Dimension Settings box, Unit Format category

Angle Format Options

The Angle Format Options control the dimension units used for the angular dimensions placed by the Dimension Angle Size, Location, Between, From *X*, and From *Y* tools.

- The Units menu allows you to place angle dimensions using Length in working units, or Degrees.
- If the Units are in Degrees, the Accuracy menu allows you to set the dimension accuracy from zero to eight decimal places.
- If the Units are in Degrees, the Display menu allows you to display the angle as decimal degrees (D.DDDD); degrees, minutes, and seconds (DD°MM'-SS"); or Centesimal degrees (C.CCCC).

Note: The Centesimal system measures a right angle by dividing it into 100 equal parts. Units are referred to as "grades," "grads," or "centesimal degrees."

Metric Format

The two Metric Format toggle buttons allow you to set the unit separator symbols to match the usage standards in Europe.

- The Use Comma for Decimal toggle button, when ON, causes a switch in the use of the comma and period in numbers (e.g., the comma is used to separate the whole and fractional part of a number).
- The Unit Separation toggle button, when ON, causes a space to be placed after the thousands and millions place when dimensions are in the metric format.

Primary and Secondary Options

The Primary and Secondary toggle buttons allow you to Show Leading Zero and Show Trailing Zeros in the dimension text. Figure 9–83 shows an example of a dimension that is set to three decimal places of accuracy, with the zero display toggle buttons ON and OFF.

Match Dimension Settings

The steps required to set up dimensioning involve several settings from various categories and are time consuming. After setting everything up, it is all too easy to forget to save it as part of the current design file by using the Save Settings option. If you lose the dimension settings but have placed dimension elements with appropriate settings in your design, the Match Dimension tool can set the current settings by matching them to the settings in effect when the dimensions were placed.

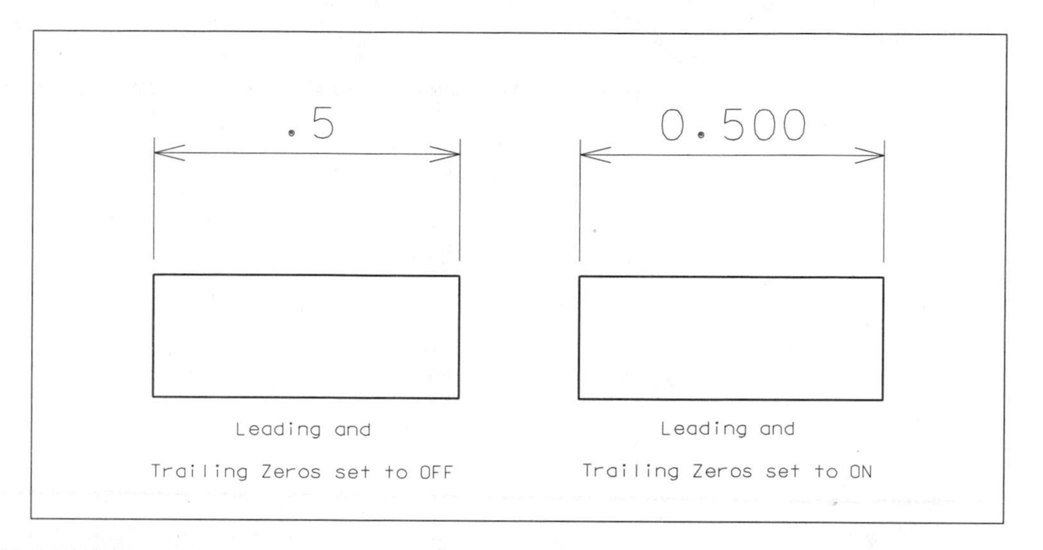

Figure 9–83 Dimension example with leading and trailing zero display ON and OFF.

Invoke the Match Dimension tool:

Pull-down menu	Tools > Match (Then select the Match Dimension Settings tool from the Match tool box (see Figure 9–84).)
Key-in window	**Match Dimension** (or **mat d**) Enter

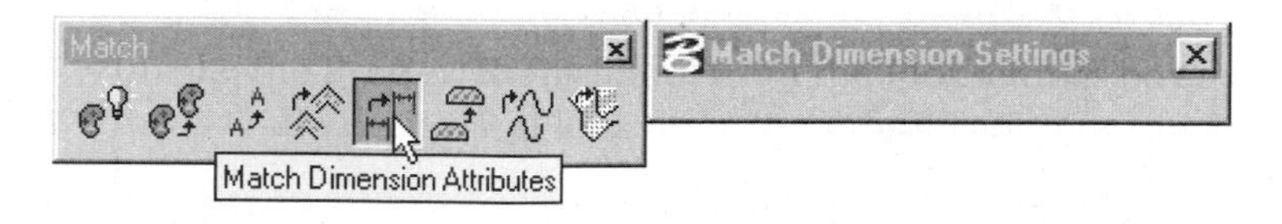

Figure 9–84 Invoking the Match Dimension Settings tool from the Match tool box

MicroStation prompts:

Match Dimension Settings > Identify element *(Identify the dimension element whose dimension settings you want to match.)*

Match Dimension Settings > Accept/Reject (Select next input) *(Click the Data button to set the selected dimension element settings as the current dimension settings, or click the Reject button to reject the settings.)*

Update Dimension

The Update Dimension tool can change a dimension element to the active dimension attributes.

Invoke the Update Dimension tool:

Dimension tool box	Select the Update Dimension tool (see Figure 9–85).
Key-in window	**Change Dimension** (or **chan d**) Enter

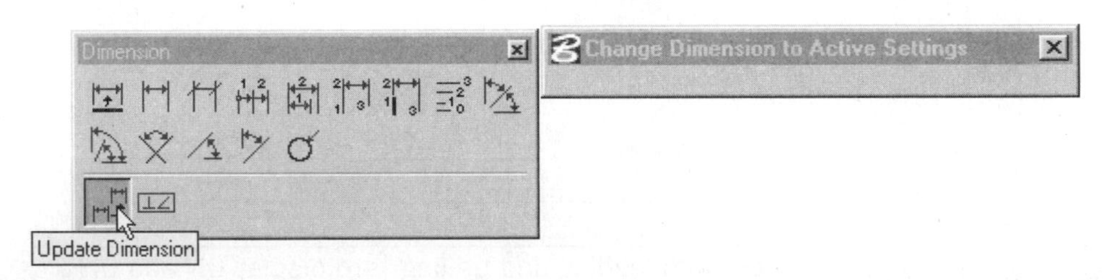

Figure 9–85 Invoking the Update Dimension tool from the Dimension tool box

MicroStation prompts:

Change Dimension to Active Settings > Identify element *(Identify the dimension element to set it to the active dimension attributes.)*

Match Dimension Settings > Accept/Reject (Select next input) *(Click the Data button to update the dimension attributes to the selected element, or click the Reject button to disregard the selection of the element.)*

REVIEW QUESTIONS

Write your answers in the spaces provided.

1. The dimension line is a ______________________________ .

2. The extension lines are ______________________________ .

3. The leader line is a ______________________________ .

4. Associative dimensioning links ______________________________.

5. To place associative dimensions, the Association Lock must be ______________ .

6. What are the three options available for orientation of dimension text relative to the dimension line?

   ____

7. The justification field setting for dimension text does <u>not</u> apply when the __________ placement location mode is selected.

8. Name the two options that are available with dimension text length format.

   ____

9. What are the three types of linear dimensioning available in MicroStation?

   ____

10. The linear dimensioning tools are provided in the __________ tool box.

11. Name the two types of dimensioning included in circular dimensioning.

    ____

12. The Dimension Angle Size tool is used to ______________________________ .

13. The Place Center Mark tool serves to ______________________________ .

14. List the four options available with the Measure Distance tool.

    ____

15. The Measure Radius tool provides information on such elements as:

    ____

16. List the seven area options available with the Measure Area tool.

____

____

17. The Measure Area Flood option measures the area enclosed _______________

18. The Geometric Tolerance settings box is used to build _______________

19. Ordinate dimensions are used to label _______________ .

20. The Label Line tool places the _______________ and _______________ .

PROJECT EXERCISE

This project exercise provides step-by-step instructions for placing dimensions on the Chapter 4 project design, as shown in Figure P9–1. The intent of this project is to guide you in applying the Dimension settings and dimension placement tools.

Note: If you have not drawn this design, refer to the instructions at the end of Chapter 4.

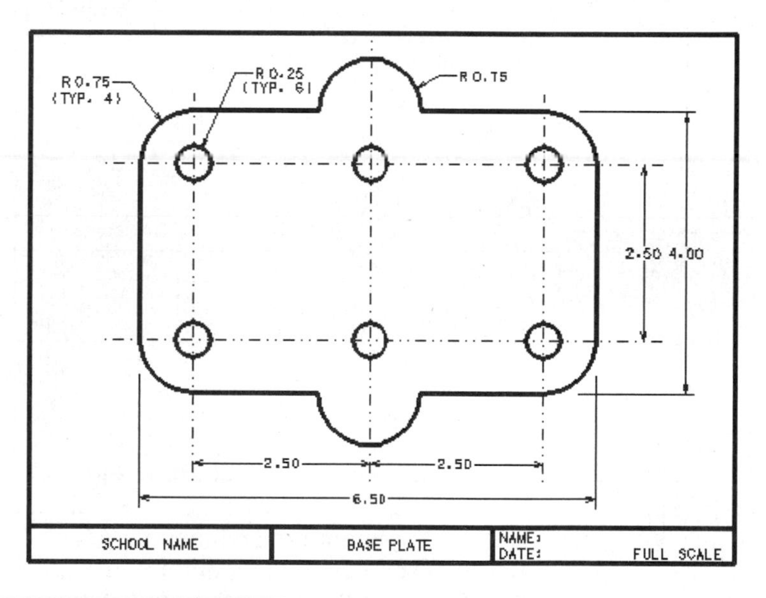

Figure P9–1 Chapter 4 project design

Set Up Dimensioning for the Chapter 4 Project

This procedure loads the Chapter 4 project design file in MicroStation, then enters the Dimension settings.

Note: As you complete each step in the project procedures, place a check mark by the step to help you keep up with where you are in the project.

STEP 1: Invoke MicroStation using the normal technique for the operating system on your workstation, and open the Chapter 4 project design file named CH4.DGN.

STEP 2: If necessary, invoke the Fit View tool to fit the view.

STEP 3: Open the Dimension Settings box by selecting Dimensions from the Element pull-down menu. MicroStation displays the dimension settings similar to Figure P9–2.

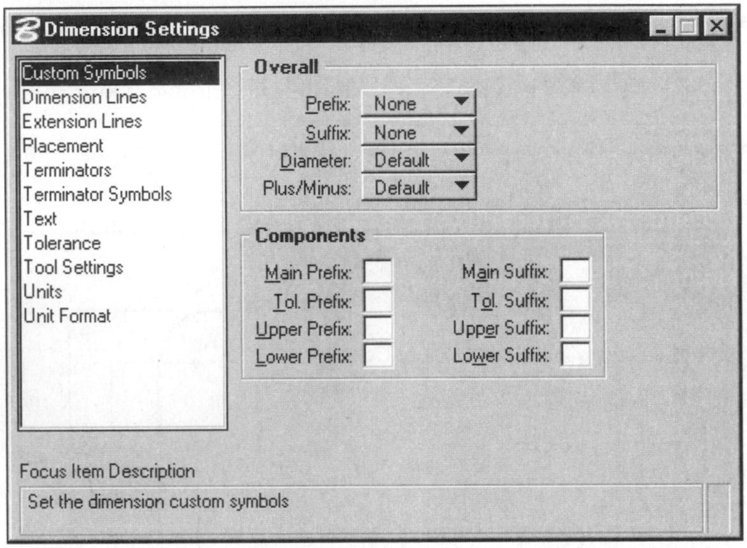

Figure P9–2 Dimension Settings box

STEP 4: Set the dimension settings as follows in the Dimension Settings box:

Placement	Set the Alignment option menu to View. Set the Location option menu to Automatic. Set the Adjust Dimension Line toggle button to ON.
Text	Set the Orientation option menu to Horizontal. Set the Justification option menu to Center. Set the Text Frame to None. Set the Margin to 0.50000. Set the Text Height and Width to 0.1250.
Units	Set the Format option menu to Mechanical. Set the Primary Units option menu to English. Set the Primary Accuracy to 0.12.
Units Format	Set the Angle Units option menu to Degrees. Set the Angle Accuracy to 0.12. Set the Angle Display option menu to D.DDDD Set the Show Trailing Zero toggle button to ON.

STEP 5: Select Save Settings from the File pull-down menu.

Note: No dimension tolerance settings are required, because tolerance dimensioning is not used on this design.

Place the Radial Dimensions

This procedure places the radial dimensions and adds text below the dimension on the circle, as shown in Figure P9–3.

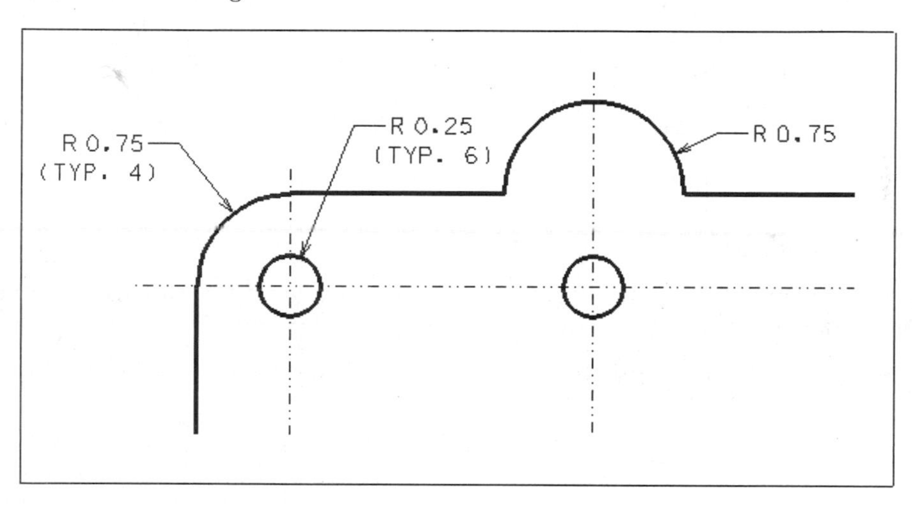

Figure P9–3 Completed radial dimensions

STEP 1: Invoke the Dimension Radial tool from the Dimension tool box. Then in the Tool Settings window, set the Mode to Radius and the Alignment to View, and turn ON Association Lock (see Figure P9–4).

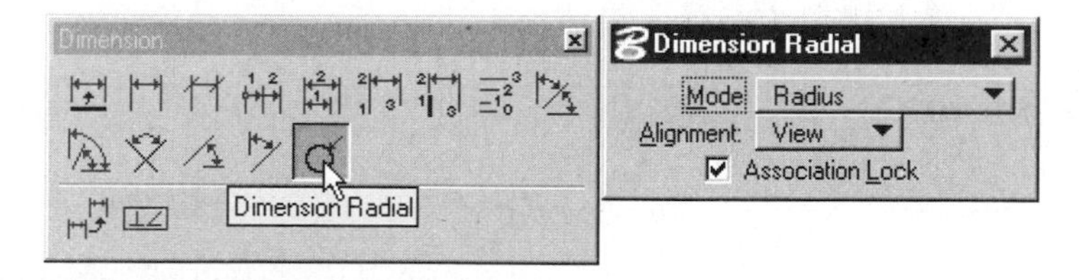

Figure P9–4 Invoking the Dimension Radial tool from the Dimension tool box

MicroStation prompts:

Dimension Radius > Identify element *(Select the top, left fillet.)*

Dimension Radius > Select dimension endpoint *(Drag the image of the dimension text to the location where you want to place it, then place a data point to complete the dimension.)*

STEP 2: Place a radial dimension on the top-left circle.

STEP 3: Place a radial dimension on the top arc.

STEP 4: Invoke the Place Text at Origin tool from the Text tool box, then in the Tool Settings window, set the text Height and Width to 0.125.

MicroStation prompts:

> Place Text > Enter text *(In the Text Editor window, type the string (TYP. 4). Place the text centered under the fillet's dimension text, as shown in Figure P9–3.)*

STEP 5: Repeat the procedure of step 4 to place (TYP. 6) below the circle's dimension text.

STEP 6: Select Save Settings from the File pull-down menu.

Place Outer Linear Dimensions

This procedure places the linear overall length and width dimensions, as shown in Figure P9–5.

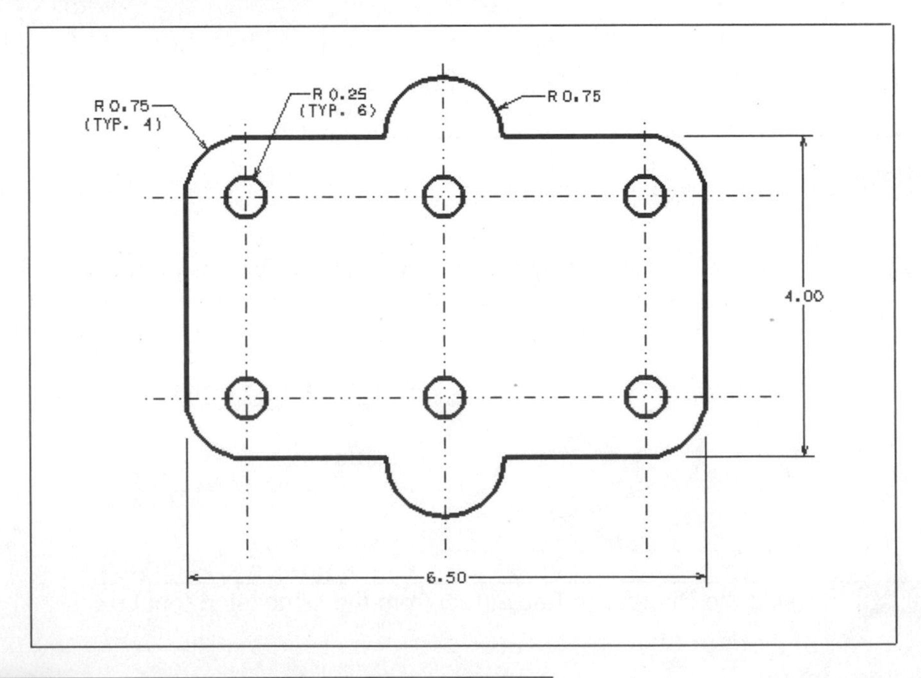

Figure P9–5 Completed length and width dimensions

STEP 1: To place the height dimension, invoke the Dimension Size with Arrows tool from the Dimension tool box. Then in the Tool Settings window, set the Alignment to View and turn ON Association Lock (see Figure P9–6).

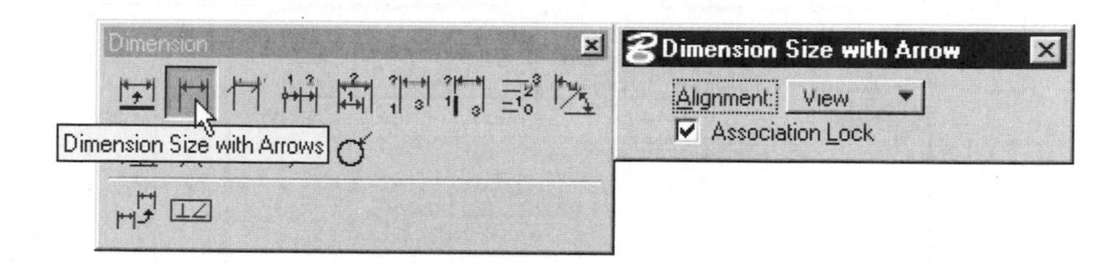

Figure P9–6 Invoking the Dimension Size with Arrows tool from the Dimension tool box

MicroStation prompts:

Dimension Size with Arrow > Select start of dimension *(Keypoint snap at the joining point between the horizontal line and the top fillet, and place a data point to define the starting point of the dimension.)*

Dimension Size with Arrow > Define length of extension line *(Drag the placement pointer about 1.5 Master Units straight to the right, then place a data point to define the length of the extension line.)*

Dimension Size with Arrow > Select dimension endpoint *(Keypoint snap at the joining point between the horizontal line and the bottom fillet, and place a data point to complete the height dimension.)*

STEP 2: Click the Reset button twice, then dimension the 6.5 horizontal length of the base plate.

Place Centerline Dimensions

This procedure turns OFF placement of extension lines, then dimensions the center-lines, as shown in Figure P9–7.

STEP 1: If the centerlines are not long enough to hold the dimensions, extend them with the Extend Line tool.

STEP 2: Open the Dimension settings box by selecting Dimensions from the Element pull-down menu. MicroStation displays the dimension settings.

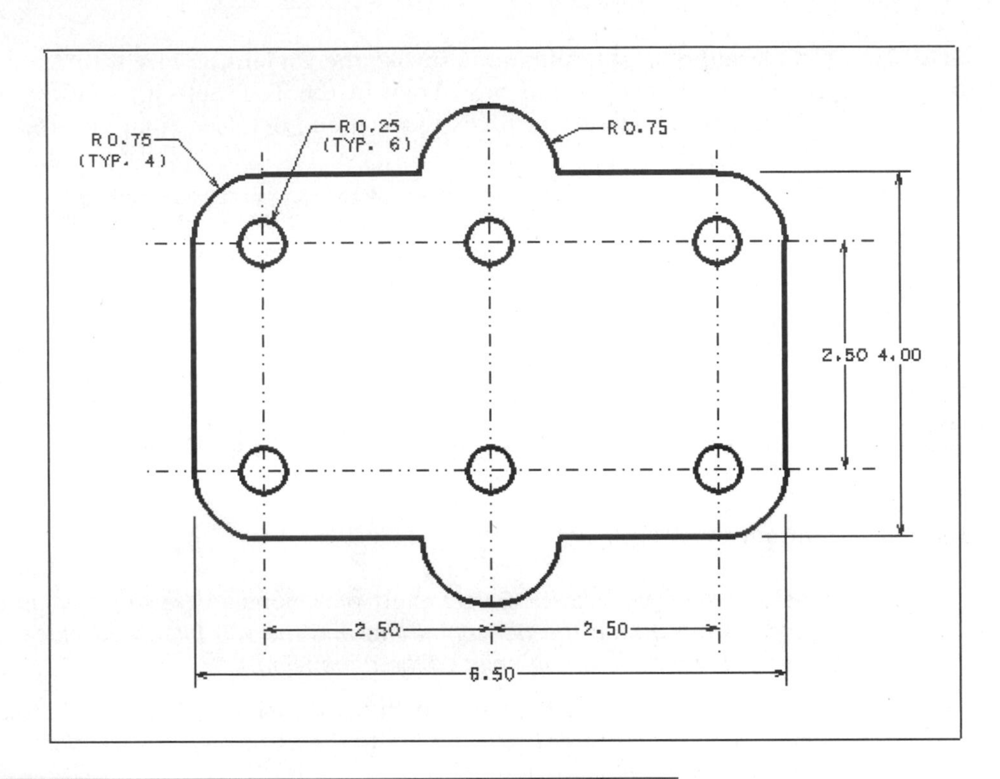

Figure P9–7 Completed overall length and width dimensions

STEP 3: In the Dimension Settings box, select the Extension Lines option, then set the toggle button to OFF for the Extension Lines. The settings box should now match the settings shown in Figure P9–8.

STEP 4: To place the vertical dimension, invoke the Dimension Size with Arrow tool from the Dimension tool box.

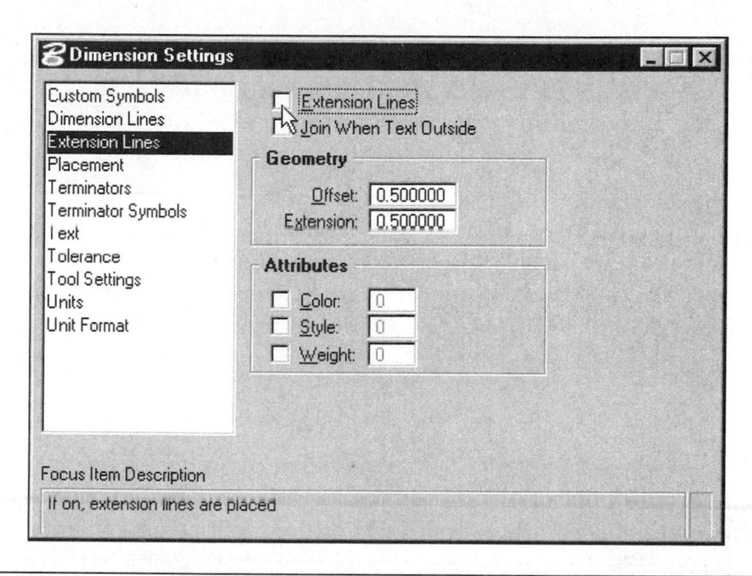

Figure P9–8 Dimension Settings box with the Extension Lines option toggled to OFF

MicroStation prompts:

Dimension Size with Arrow > Select start of dimension *(Keypoint snap to the right end of the top centerline, then place a data point.)*

Dimension Size with Arrow > Define length of extension line *(Drag the placement pointer about .25 Master Units to the right, then place a data point.)*

Define Size with Arrow > Select dimension endpoint *(Keypoint snap to the right end of the lower horizontal centerline.)*

STEP 5: Click the Reset button twice, then place the horizontal centerline dimensions.

STEP 6: Select Save Settings from the File pull-down menu.

DRAWING EXERCISES 9–1 THROUGH 9–5

Exercise 9–1

Use Mechanical dimensioning to place dimensions on the flange gasket created in Exercise 4–8 on page 238.

Exercise 9–2

Use Mechanical dimensioning to place dimensions on the machine part created in Exercise 5–2 on page 275.

Exercise 9–3

Tolerance dimensioning

Use the following table to set up the design file for Exercise 9–3, then draw and dimension the object shown in the figure.

SETTING	VALUE
Seed File	SEED2D.DGN
Working Units	10 IN Per TH and 1000 Pos Units Per TH
Grid	Grid Master = 0.1, Grid Reference = 10

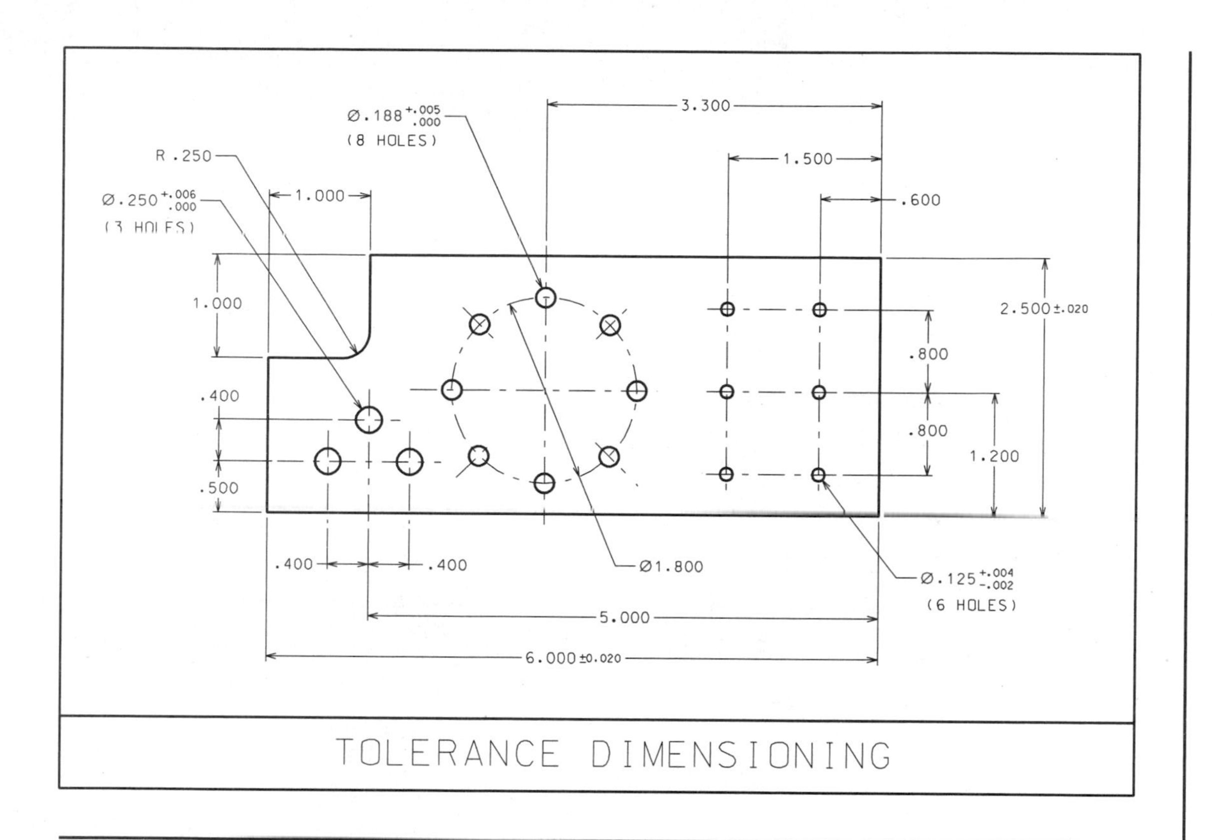

Exercise 9–4

Use AEC dimensioning to place dimensions on the master bathroom floor plan created in Exercise 5–5 on page 278.

Exercise 9–5

Use AEC dimensioning to place dimensions on the custom doors created in Exercise 6–5 on page 321.

chapter 10

Plotting

One task has not changed much in the transition from board drafting to CAD, and that is obtaining a hard copy. The term *hard copy* describes a tangible reproduction of a screen image. The hard copy is usually a reproducible medium from which prints are made, and it can take many forms, including slides, videotape, prints, and plots. This chapter describes the most common process for getting a hard copy: plotting.

Objectives

After completing this chapter, you will know how to do the following:

- How the plotting process works.
- What components are involved in the process.
- How to create a plot file.
- How to create a hard copy.
- How to use the BatchPlot utility.

OVERVIEW OF THE PLOTTING PROCESS

In manual drafting, if you need your design to be done in two different scales, you physically have to draw the design for different scales. In CAD, on the other hand, with minor modifications you can plot the same design in different scale factors on different-size paper.

To plot a design with MicroStation you carry out a three-step process:

1. Set up the view to be plotted or place a fence around the part of the design to be plotted.
2. Use the Plot settings box to preview the plot and create a plot file.
3. Send the plot file to the plotter.

MicroStation lets you get a hard copy of your design file by creating a "plot file" and sending the file to a plotting device to create a hard copy.

The plot file describes all the elements in the plot area in a language the plotting device can understand, and provides commands to control the plotting device. It is separate from your design file, and contains the design as it existed when the plot file was created. If you make changes to the design after creating the plot file and want to plot the new design, you must create a new plot file.

MicroStation stores plot files in the directory path contained in the MS_PLTFILES configuration variable. By default, that path is <disk>:\Bentley\Workspace\projects\examples\generic\out. Replace <disk> with the letter of the disk that contains the MicroStation program. (See Chapter 16 for a detailed explanation of configuration variables.)

Plotting devices (printers and plotters) put the information contained in the plot file on the hard copy page. MicroStation supports many types and models of plotting devices. There are electrostatic plotters that provide only shades of gray and more expensive models that plot in color. Pen plotters use ink pens contained in a movable rack. A mechanical control mechanism selects pens and moves them across the page under program control.

MicroStation provides a plotter driver file for each supported plotting device. The information contained in this file (combined with the information you supply through the Plot settings box) tells MicroStation how to create the plot file and send it to the plotting device.

The default path where the plotter driver files are located is <device>:\Bentley\Program \microstation\plotdrv\. Replace <device> with the letter of the drive containing the MicroStation program.

The plotter driver file name usually consists of the device model number plus .PLT as the extension. MicroStation provides two plotter driver files for most plotting devices, one for English measurement units and one for metric units. See Table 10–1 for the list of the supported plotters and corresponding sample plotter driver files supplied with MicroStation J.

The plotter driver file specifies the following:

- Plotter model
- Number of pens the plotter can use
- Resolution and units of distance on the plotter
- Pen change criteria
- Name, size, offset, and number for all paper sizes
- Stroking tolerance for arcs and circles
- Border around the plot and information about the border comment
- Pen speeds, accelerations, and force, where applicable
- Pen-to-element color or weight mapping
- Spacing between multiple strokes on a weighted line
- Number of strokes generated for each line weight
- Definitions for user-defined line styles (for plotting only)
- Method by which plots are generated
- Actions to be taken at plot's start and end and on pen changes

Table 10–1 Supported Plotters and Corresponding Sample Plotter Driver Configuration Files

SUPPORTED PLOTTER	PLOTTER DRIVER FILE
Calcomp 907	cal104x.plt, cal524xx.plt, cal906.plt, cal907.plt, cal906.plt, ver8536.plt, ver8524.plt
DMPL (Houston Instrument)	hidmp40.plt, hidmp56.plt, hidmp52.plt, ioline.plt
EPSP	epson8.plt, epson8h.plt, epson24.plt, epscripc.plt, epscripm.plt
HP-GL	drftpro.plt, mutoh500.plt, drftmstr.plt, hp7470a.plt, hp7550a.plt, hp7580b.plt, hp7585b.plt, hp7475a.plt, hp7440a.plt
HP-GL/2	hpgl2.plt, hpljet3.plt, hpljet4.plt, hpljet4v.plt, hpdjet.plt, hpxl300.plt, hp650c.plt, drftprop.plt, novajet2.plt
PCL	hpljet.plt, hlp200c.plt, hppc.plt
PostScript	pscript.plt, pscriptc.plt
SVF	svfhires.plt, svflores.plt

Plotter driver files can be edited with any text editor. For more information on the contents of these files, and changes you can make to them, consult the *MicroStation User Guide.*

Note: If you modify a sample plotter driver file, it is a good idea to retain the original file and to save the modified file as a new file with a different name.

CREATING PLOT FILES

To create a plot file, invoke the Print/Plot tool:

Pull-down menu	File > Print/Plot
Key-in window	**plot** (or **plo**) Enter

MicroStation displays the Plot settings box, as shown in Figure 10–1.

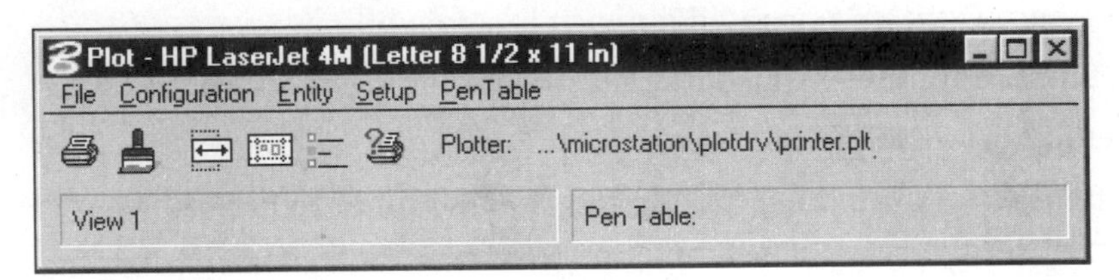

Figure 10–1 Plot settings box

MicroStation displays the name of the selected plotter driver file name on the right side of the Plot settings box.

If necessary, you can change the plotter driver file by selecting:

Plot settings box tool bar	Plotter Driver (see Figure 10–2)
Pull-down menu (Plot settings box)	Setup > Driver...

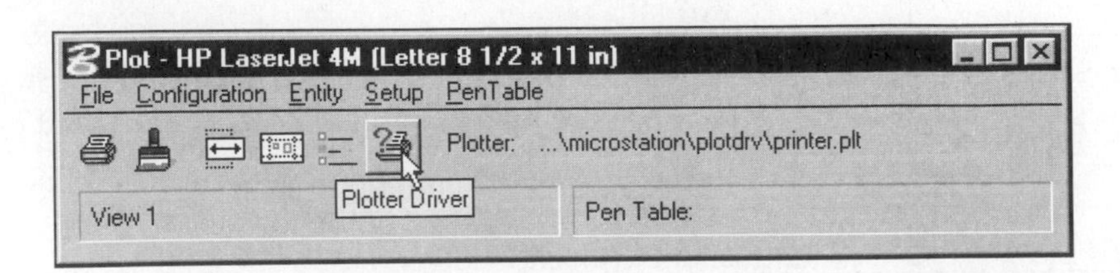

Figure 10–2 Invoking the Plotter Driver from the Plot settings box tool bar

MicroStation displays the Select Plotter Driver File dialog box, as shown in Figure 10–3. Select the appropriate driver file and click on the OK button.

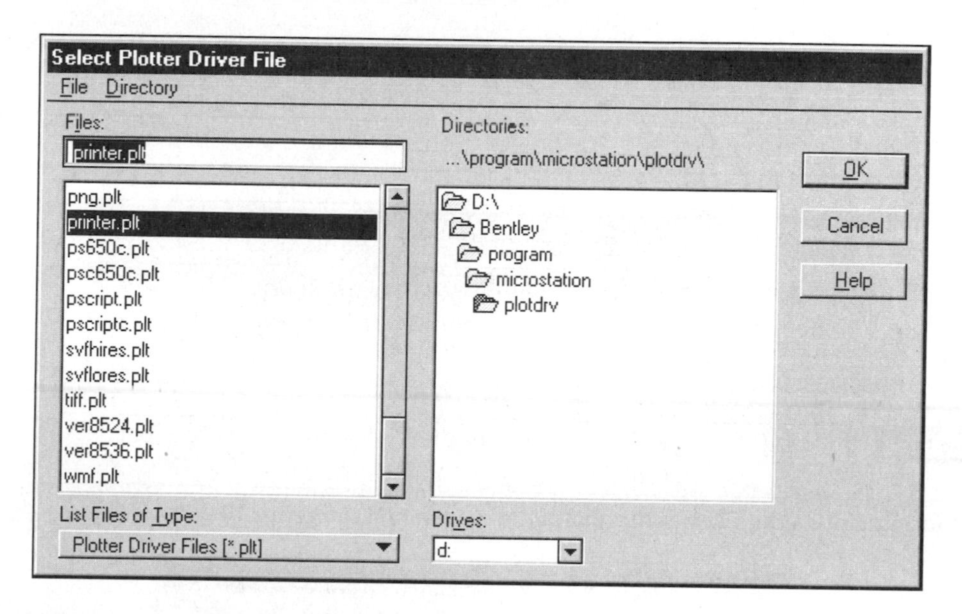

Figure 10–3 Select Plotter Driver File dialog box

If you are using Microsoft Windows and select printer.plt as the plotter driver file, MicroStation plots with the default printer selected in the Windows Printer Manager.

Note: The initial default plotter driver file is specified by the MS_PLTR configuration variable. If you change the plotter driver file, MicroStation automatically updates the MS_PLTR configuration variable with the new device name.

Selecting the Area of the Design to Plot

By default, MicroStation selects View 1 as the view to plot. If necessary, you can change the view to plot by invoking the command:

Pull-down menu (Plot settings box)	Entity > View > 1 to 8

MicroStation displays the selected view number in the left corner of the settings box. To plot from the view containing the fence, first place the fence to include the elements to plot. Then select the Fence option in the Entity pull-down menu in the Plot settings box. MicroStation displays the view number with the word "Fence" in the left corner of the settings box.

Page Setup

To set the page for plotting, invoke the Page Setup tool:

Plot settings box tool bar	Select the Page Setup tool (see Figure 10–4).
Pull-down menu (Plot settings box)	Setup > Page...

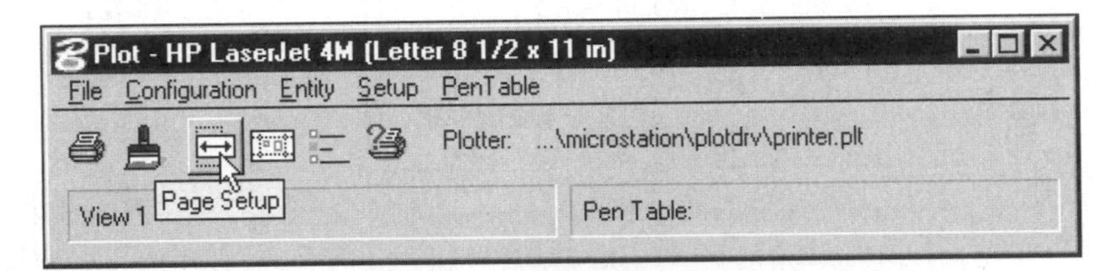

Figure 10–4 Invoking the Page Setup tool from the Plot settings box tool bar

MicroStation displays the Page Setup dialog box, as shown in Figure 10–5.

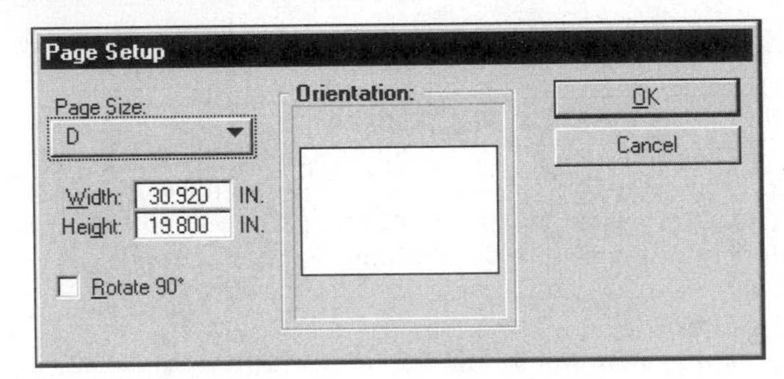

Figure 10–5 Page Setup dialog box

Select the desired page size from the Page Size option menu. The Rotate 90° toggle button determines whether or not the plot is rotated 90 degrees on the page. Click on the OK button to close the Page Setup dialog box.

Plot Layout

The Plot Layout settings box allows you to adjust the margins and scale of the plot. To set the plot layout, invoke the Plot Layout tool:

Plot settings box tool bar	Select the Plot Layout tool (see Figure 10–6).
Pull-down menu (Plot settings box)	Setup > Layout...

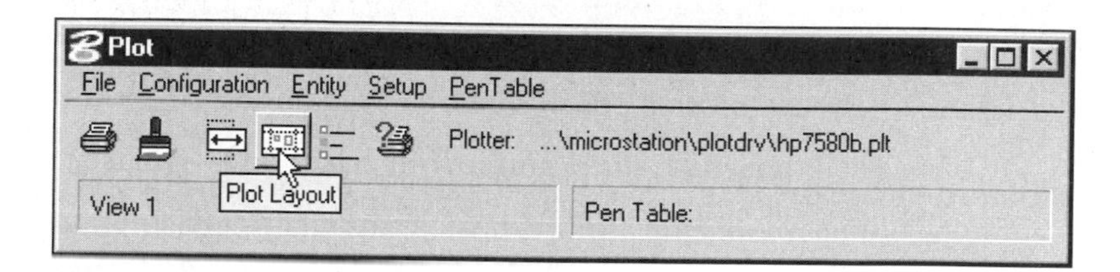

Figure 10–6 Invoking the Plot Layout tool from the Plot settings box tool bar

MicroStation displays the Plot Layout dialog box, as shown in Figure 10–7.

Plot Layout
Margins:
Left Margin: 0.000 IN.
Bottom Margin: 0.024 IN.
Plot Width: 30.920 IN.
Plot Height: 19.752 IN.
Scale to 100 % of normal
Scale to 1.1676 ft:in / IN.
Automatic:
Center to Page
Maximize
Page Layout:
Form Name: D
Width: 30.920 IN.
Height: 19.800 IN.
OK
Cancel

Figure 10–7 Plot Layout dialog box

The black rectangle in the Page Layout area represents the selected page for the selected plotter. The blue rectangle inside it represents the plot area. If Fence is chosen, the blue rectangle represents the smallest rectangular area that encloses the fenced area.

The Left Margin and Bottom Margin edit fields set the left and bottom edges of the printable area of the paper to the origin of the plot. To center the plot area, turn ON the toggle button for Center to Page, and MicroStation automatically displays the new offset settings for Left Margin and Bottom Margin.

The Plot Width and Plot Height edit fields display the size of the plot for the selected page. The "Scale to" edit fields display the percentage of normal size and the ratio of Master Units to plotter units. If you make changes in one of the four fields, MicroStation reflects the corresponding changes in the remaining three edit fields. For example, if you set the design to be plotted to 50% of normal, then MicroStation automatically displays the scale in terms of the ratio of Master Units to plotter units and the appropriate Plot Width and Plot Height.

Turn ON the toggle button for Maximize to fit the selected view or fenced area into as much of the printable area as possible. The plot is centered either horizontally or vertically if it does not fit the printable area exactly.

Click on the OK button to save the changes and close the Plot Layout dialog box.

Plot Options

The Plot Options settings box allows you to adjust the appearance of the plot. Invoke the Plot Options settings:

Plot settings box tool bar	Select the Plot Options tool (see Figure 10–8).
Pull-down menu (Plot settings box)	Setup > Options...

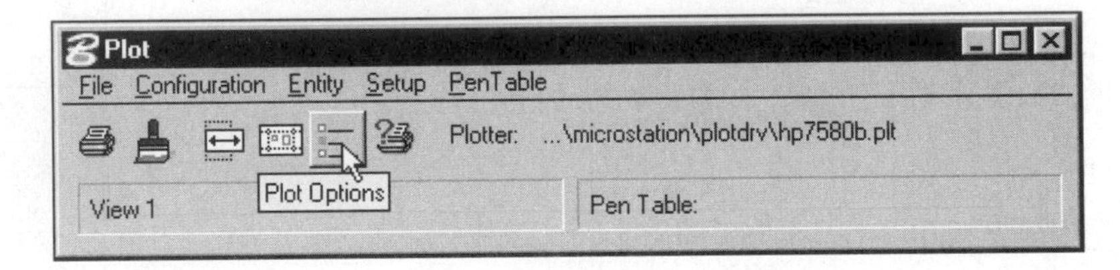

Figure 10–8 Invoking the Plot Options tool from the Plot settings box tool bar

MicroStation displays the Plot Options dialog box, as shown in Figure 10–9.

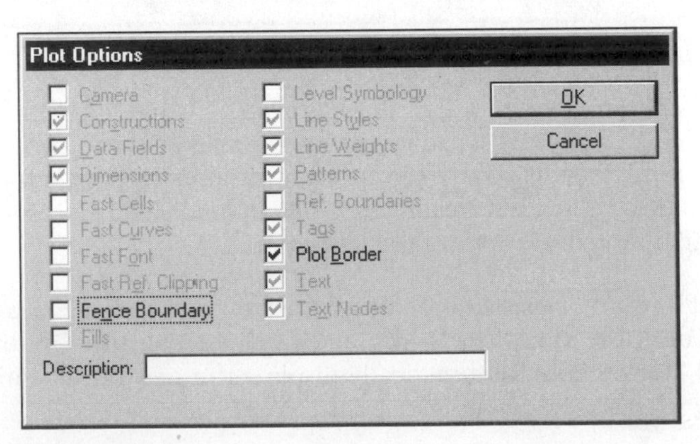

Figure 10–9 Plot Options dialog box

Toggle the check boxes to turn ON or OFF the appearance options. If the element to plot is a view window or a fence, the View Attributes settings box determines most aspects of the plot's appearance, and the corresponding options in the Plot Options settings box are dimmed (disabled). If necessary, make the changes in the View Attributes settings box.

If necessary, you can add a comment to the plot by typing the appropriate information in the Description edit field. Click on the OK button to save the changes and close the Plot Options dialog box.

Previewing the Plot

To display a plot preview image, invoke the Preview tool:

Plot settings box tool bar	Select the Preview Refresh tool (see Figure 10–10).
Pull-down menu (Plot settings box)	File > Preview

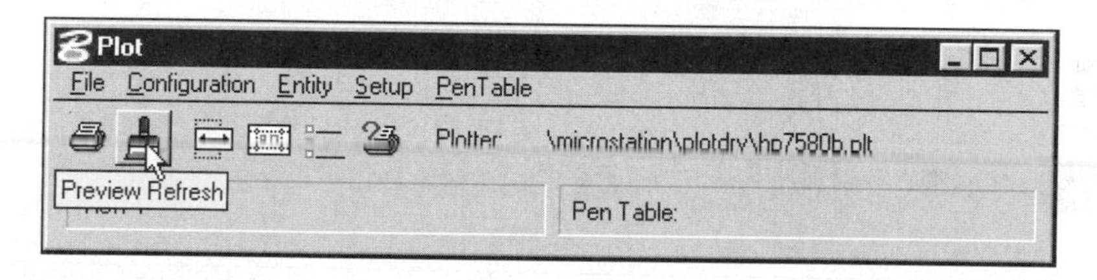

Figure 10–10 Invoking the Preview Refresh tool from the Plot settings box tool bar

MicroStation expands the Plot settings box to reveal the preview image, as shown in Figure 10–11, and changes the title of the settings box from "Plot" to "Plot Preview."

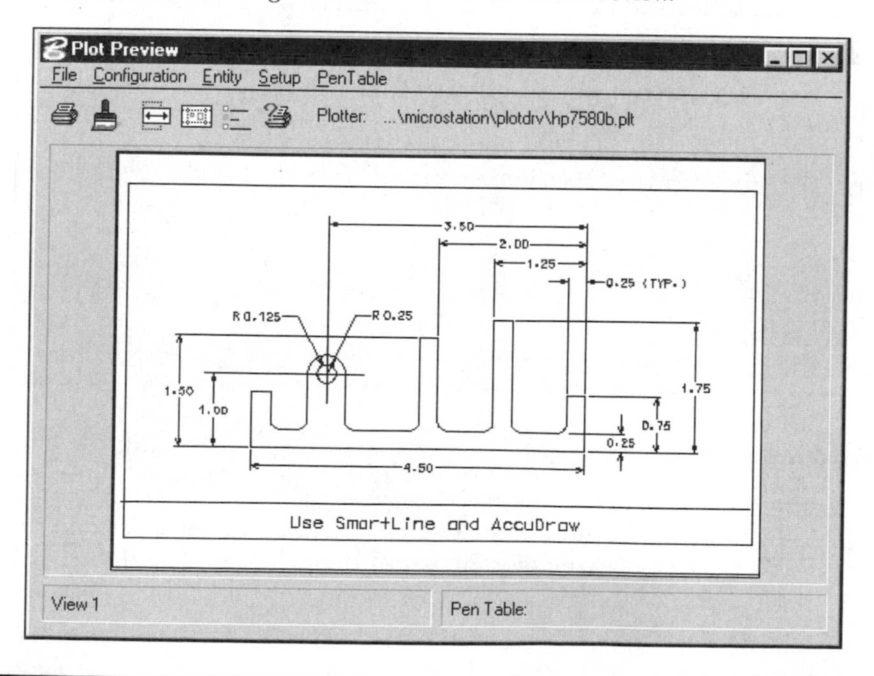

Figure 10–11 Plot Preview settings box

Creating a Plot File

To create a plot file, invoke the Plot tool:

Plot settings box tool bar	Select the Plot tool (see Figure 10–12).
Pull-down menu (Plot settings box)	File > Plot

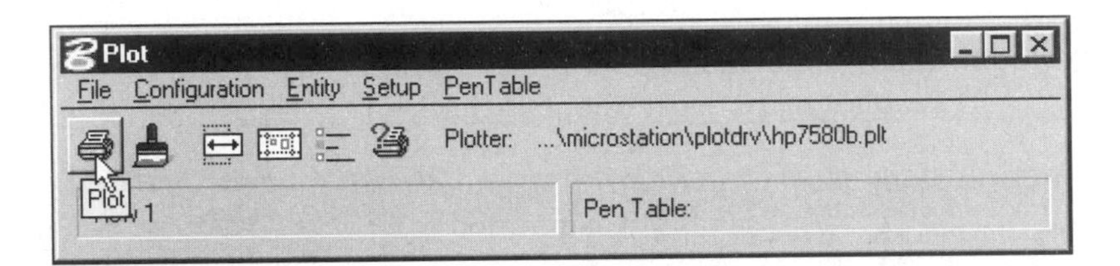

Figure 10–12 Invoking the Plot tool from the Plot settings box tool bar

MicroStation displays the Save Plot As dialog box, as shown in Figure 10–13.

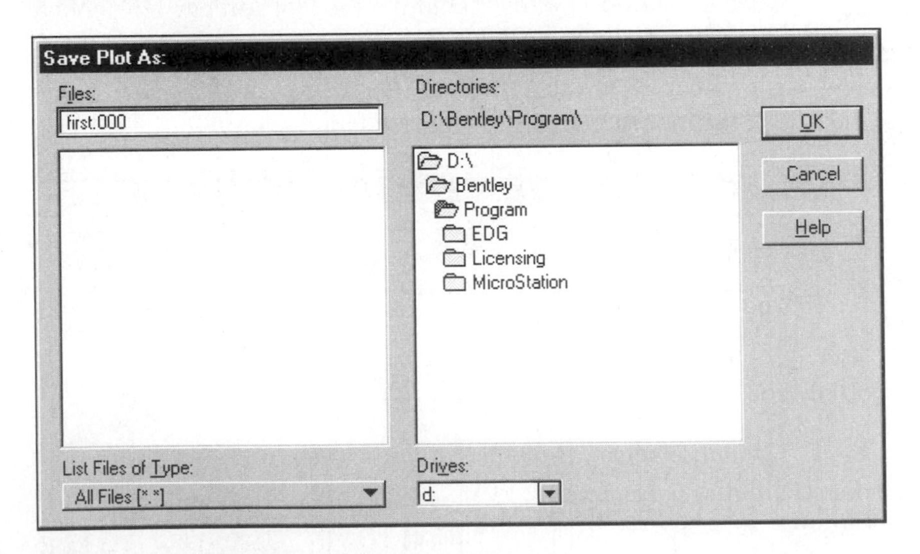

Figure 10–13 Save Plot As dialog box

The default plot file name is the same as the design file name, and the .000 extension is added to the file name. If necessary, change the plot file name, then click on the OK button to create the plot file.

By default, the plot file is saved in the <disk>:Bentley\Workspace\projects\examples\generic\out\ directory. Replace <disk> with the letter of the disk that contains the MicroStation program.

If the selected file name already exists, an Alert window opens. Click on the OK button to overwrite the file, or click the Cancel button to return to the Save File As dialog box and enter a new file name.

Note: If you have selected Printer.plt as the plotter driver, then MicroStation plots directly to the plotter configured in the Windows operating system.

GENERATING THE PLOT FROM THE PLOT FILE

The PLOTUTIL utility program sends a plot file to a plotter. You can run the utility on another PC without MicroStation. Make sure to copy the PLOTUTIL utility and the configured plotter driver file to the PC from which you want to plot the design files.

Invoke the **PLOTUTIL** program. MicroStation prompts for the file name to plot. Type the name of the plot file name and press Enter, and the design file is plotted.

You can also use PLOTFILE.BAT batch file to plot from the plot file. The batch file sets the environment variables used by PLOTUTIL to locate the plotter driver file (MS_PLTR) and the plot files (MS_PLTFILES). If necessary, you can change the settings of the environment variables.

PLOT CONFIGURATION FILES

Plot configuration files let you save the plot information specific to a design file. Plot configuration files are a way to streamline repetitive plotting tasks. Information saved in a plot configuration file includes the following:

- Plotting area
- Plot option settings
- Fence location
- Displayed levels
- Page size, margin, and scale
- Pen table if attached

Before you create a plot configuration file, set the appropriate controls in the Plot settings box. To create a plot configuration file, open the Plot Configuration File dialog box:

Pull-down menu (Plot settings box)	Configuration > New...

MicroStation displays the Plot Configuration File dialog box, as shown in Figure 10–14.

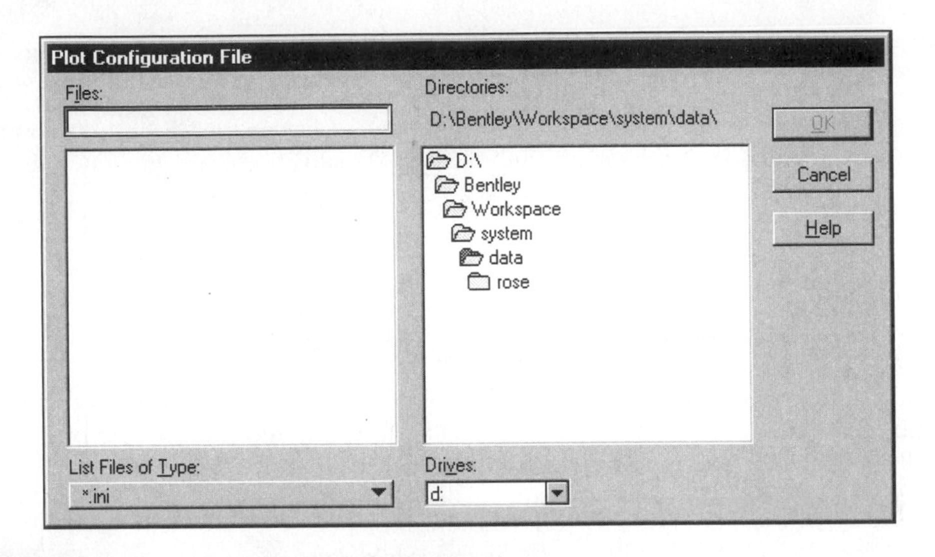

Figure 10–14 Plot Configuration File dialog box

Key-in the name of the configuration file in the Files edit field and click on the OK button. MicroStation saves the configuration to the given file name with the .INI extension, and by default it is saved in the <disk>:Bentley\Workspace\system\data\ directory.

To open an existing configuration file, invoke the Open tool from the Configuration pull-down menu in the Plot settings box. MicroStation displays the Plot Configuration File dialog box. Select the appropriate configuration file and click on the OK button. MicroStation makes the necessary changes to the plot settings.

PEN TABLES

The pen table is a data structure that allows you to modify the appearance of a plot without modifying the design file, by performing one or more of the following at plot-creation time:

- Changing the appearance of elements
- Determining the plotting order of the active design file and its attached reference files
- Specifying text string substitutions

A pen table is stored in a pen table file. The pen table consists of sections that are tested against each element in the design. When a match is found, the output options are applied to the element. The modified element is then converted into plot data, which in turn is written to the plot file. At no time are the elements of the design file or its reference files modified.

Creating a Pen Table

To create a pen table, invoke the New tool:

Pull-down menu (Plot settings box)	PenTable > New...

MicroStation displays the Create New Pen Table file dialog box. Key-in the pen table file name and click on the OK button. MicroStation displays the Modify Pen Table settings box, as shown in Figure 10–15.

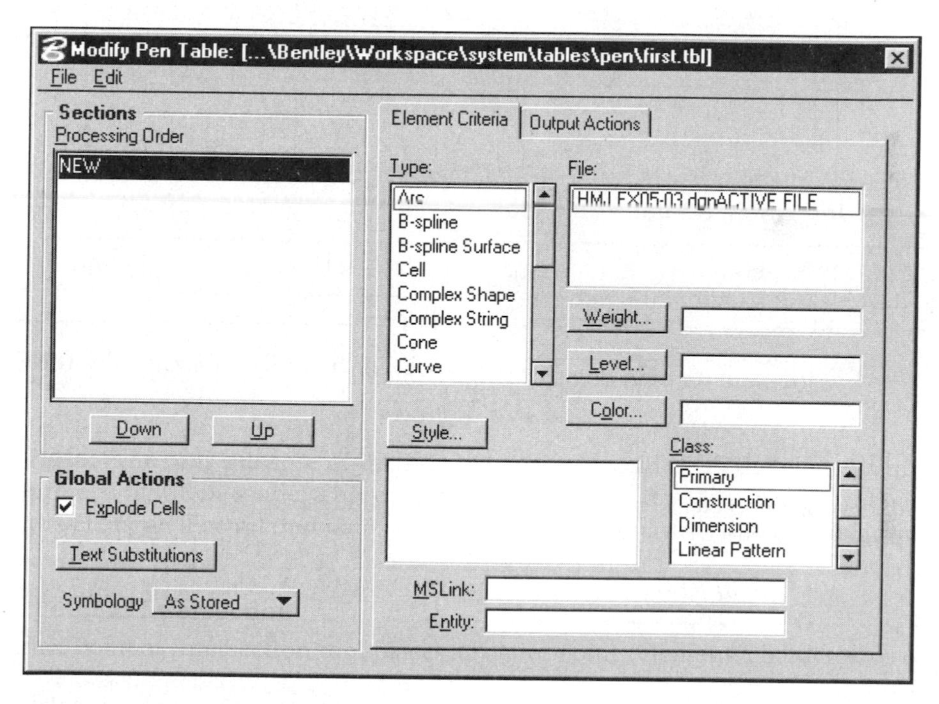

Figure 10–15 Modify Pen Table settings box

By default, MicroStation adds the section called NEW in the Sections list box. You can either rename the section or insert a new one and delete the NEW section.

Renaming a Pen Table Section

To rename a section, first select the name of the section in the list box, then invoke the Rename tool:

Pull-down menu (Modify Pen Table settings box)	Edit > Rename Section...

MicroStation displays the Rename Section dialog box. Key-in the new name and click the OK button to rename the section.

Inserting a New Pen Table Section

Above an Existing Section

To insert a new section above an existing section, first select the name of the section in the list box, then invoke the Insert New Section Above tool:

Pull-down menu (Modify Pen Table settings box)	Edit > Insert New Section Above...

MicroStation displays the Insert Section dialog box. Key-in the new name and click the OK button to insert the new section.

Below an Existing Section

To insert a new section below an existing section, first select the name of the section in the list box, then invoke the Insert New Section Below tool:

Pull-down menu (Modify Pen Table settings box)	Edit > Insert New Section Below...

MicroStation displays the Insert Section dialog box. Key-in the new name and click the OK button to insert the new section.

Deleting a Pen Table Section

To delete a section, first select the name of the section in the list box, then invoke the Delete Section command:

Pull-down menu (Modify Pen Table settings box)	Edit > Delete Section

MicroStation deletes the selected section from the list box.

If necessary, you can change the section's position in the processing order. First select the section in the list box, then click Down or Up to change the processing order.

Modifying a Pen Table Section

To modify a pen table section, follow these numbered steps.

STEP 1: Highlight the name of the section to modify from the Sections list box.

STEP 2: *Optional:* Set the toggle button for Explode Cells.

ON Each cell component element is evaluated independently against the element criteria; the defined output action applies to each element of the cell.

OFF Each cell header is evaluated against the element criteria; the defined output action applies to each element of the cell.

STEP 3: *Optional:* To substitute text in the design with alternate text for plotting, click the Text Substitutions button. MicroStation displays the Text Substitutions settings box, as shown in Figure 10–16.

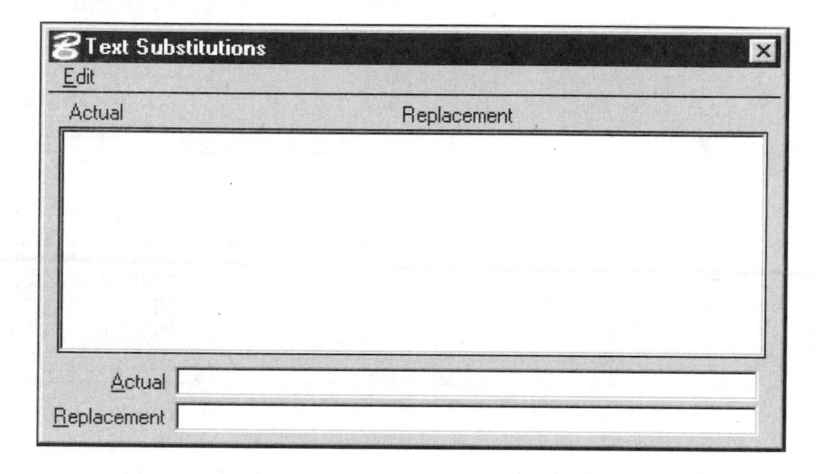

Figure 10–16 Text Substitutions settings box

To insert a text substitution entry, invoke the Insert New tool:

Pull-down menu (Text Substitutions settings box)	Edit > Insert New

An entry labeled "Original" appears in the list box and in the Actual edit field. Replace "Original" with the string in the design to be replaced for plotting purposes. Type the replacement text string in the Replacement edit field and press Enter.

Note: The defined text string substitutions apply universally to all text elements in the plot and only to exact matches of the specified strings.

In addition, you can also replace a text string in a design with a file name, current date, or time.

To replace a text string in a design with a file name, date, or time, select options from the Edit pull-down menu (Text Substitutions settings box) as listed in Table 10–2.

Table 10–2 Replacing a Text String with a File Name, Date, or Time

EDIT MENU ITEM	ACTUAL STRING IN THE DESIGN*	REPLACEMENT STRING FOR PLOTTING	EFFECT
Insert Abbreviated Filename	$FILENAME$	_FILEA_	Replaces the actual text string with the file name of the active design file. The replacement file name is truncated to the size of the actual string.
Insert Filename	$FILE$	_FILE_	Replaces the actual text string with the file name of the active design file. No truncation.
Insert Date	$DATE$	_DATE_	Replaces the actual text string with the current date.
Insert Time	$TIME$	_TIME_	Replaces the actual text string with the current time.

* The actual text string is shown with the dollar sign character ($) as the delimiter character just to differentiate it from normal text. It is *not* necessary to have the delimiter character as part of the text string in the design file. You can replace any text string in the design with the file name, date, or time.

To delete a text substitution entry, first highlight the text string substitution, then invoke the Delete tool:

Pull-down menu (Text Substitutions settings box)	Edit > Delete

The selected text string substitution is deleted from the settings box.

Note: You can see the substitutions to the text string by clicking the Preview Refresh icon in the Plot Preview settings box.

STEP 4: To set the element criteria, first select the Element Criteria tab (shown in Figure 10–17), located in the top right side of the Modify Pen settings box.

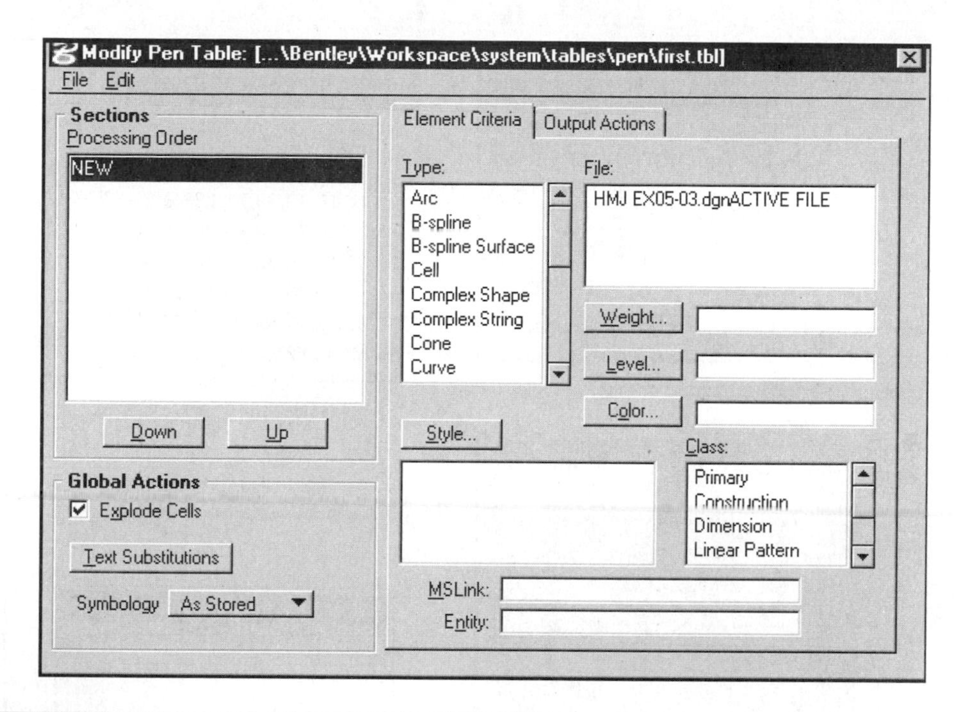

Figure 10–17 Selecting the Element Criteria tab in the Modify Pen settings box

Then select element types from the Type list box, a file name from the File list box, and the class or classes of elements to include in the selection criteria. To select multiple items from the list box, hold down Ctrl and then select the items.

Note: The File list box lists the name of the active design file and all the reference files attached to the active design file.

You can also select or deselect all the element types, files, and classes by selecting the appropriate tool from the Edit pull-down menu in the Pen settings box.

In addition, you can set the selection criteria based on weight, level, color, or style by keying-in the appropriate values in the edit fields or by clicking the appropriate button in the Modify Pen settings box. MicroStation displays the appropriate dialog boxes, shown in Figure 10–18. Use the controls in the dialog box to make the selections, and click the OK button to close the dialog box.

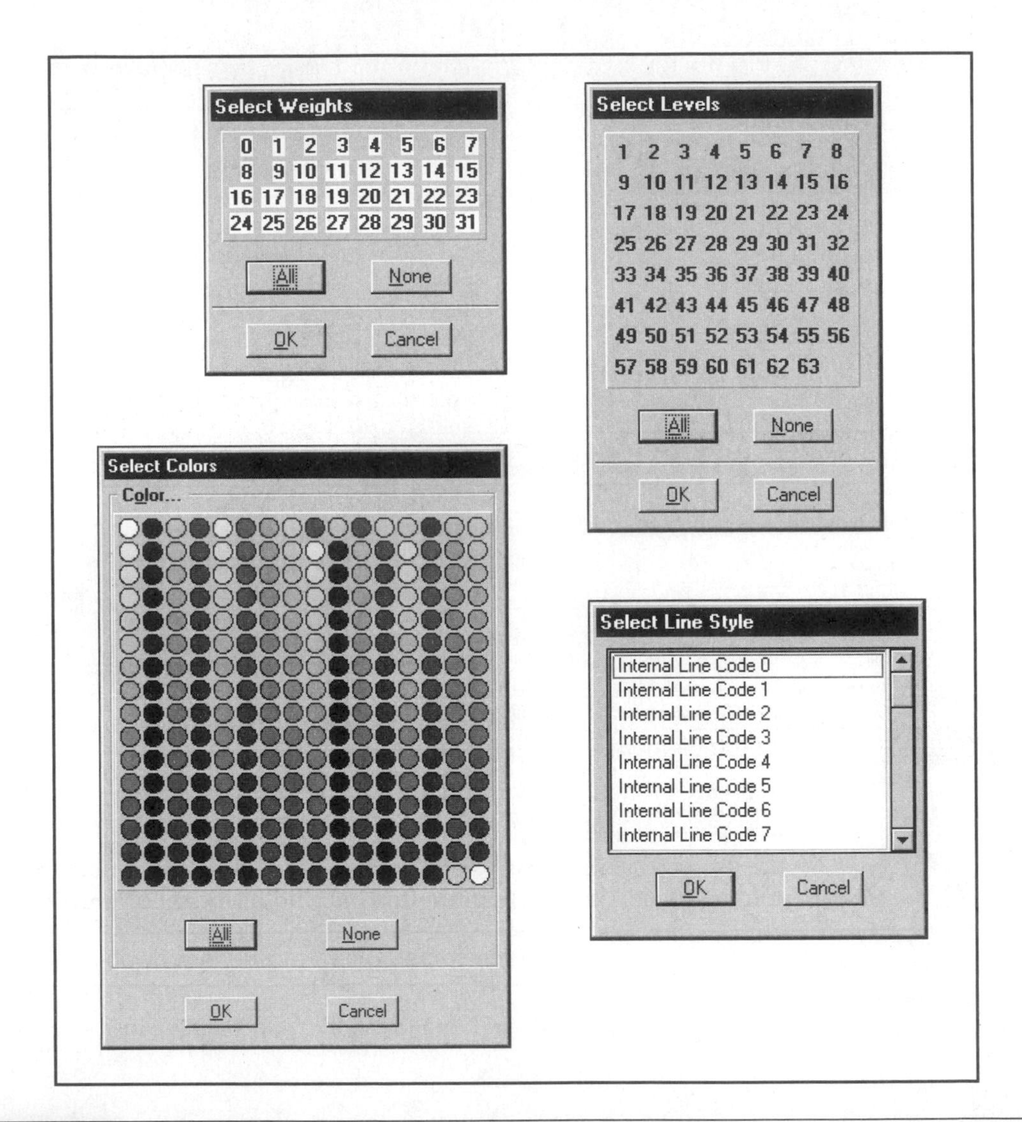

Figure 10–18 Select Weights, Select Levels, Select Colors, and Select Line Style dialog boxes

STEP 5: To set the output actions, first select the Output Actions tab (shown in Figure 10–19), located in the top right side of the Modify Pen settings box.

Select one of the available options from the Master Control option menu. Table 10–3 explains the available Master Control options.

Note: Do not prioritize elements unless it is significant to the plot, since prioritized elements require additional processing time and memory.

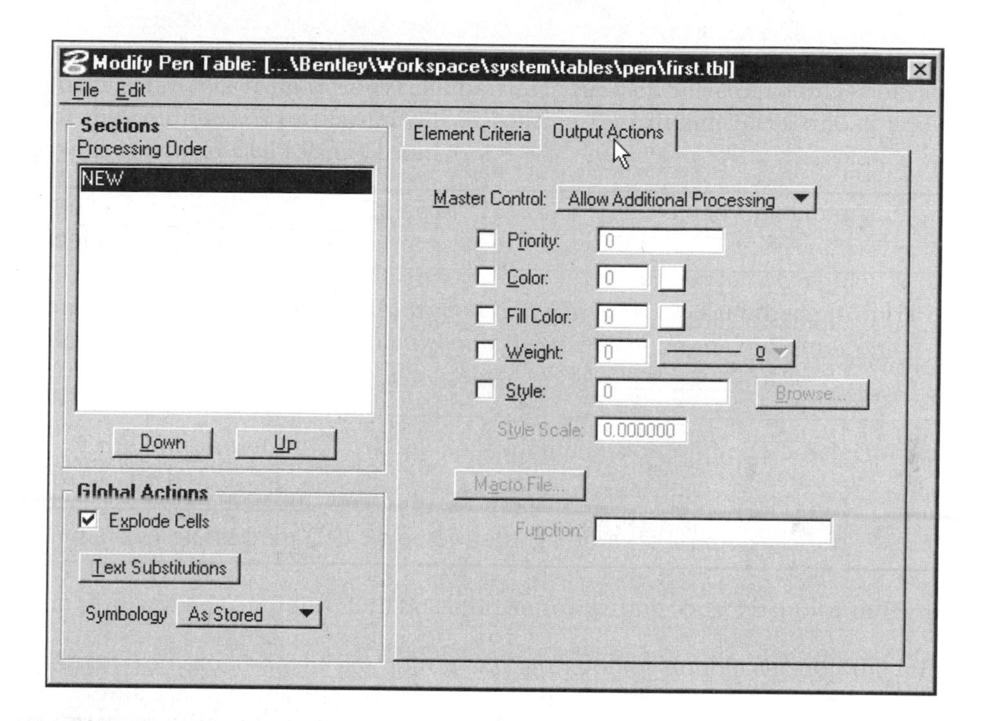

Figure 10–19 Selecting the Output Actions tab in the Modify Pen settings box

Table 10–3 Master Control Menu Options

MASTER CONTROL OPTION	EFFECT
Allow Additional Processing (default)	The output actions are applied to the selection criteria, and any loaded MDL applications that you desire to process the element are invoked.
No Additional Processing	The output actions are applied to the selection criteria, and no MDL applications that you desire to process the element are invoked.
Don't Display Element	The elements that satisfy the selection criteria are not plotted, and no MDL applications that you desire to process the element are invoked.
Call BASIC Macro Function	The output actions are applied to the selection criteria, and additional processing is performed by the designated function in the designated BASIC macro.

Optional: Turn ON the toggle button for Priority to set the priority. Key-in the desired priority value (range: –2147483648 to 2147483647) in the Priority edit field. Elements with a lower priority value are plotted before elements with a higher priority value. Unprioritized elements are always plotted before all prioritized elements.

Optional: Turn ON the toggle buttons to override the Color, Weight, Level, or Style appropriately for the elements that satisfy the selection criteria. Key-in the values in the edit fields or choose the desired attributes from the pop-up palette. If the element is to be plotted with a custom line style, key-in the desired line style scale factor in the Style Scale field.

STEP 6: Before you plot, click the Preview Refresh icon in the Plot Preview settings box. MicroStation displays all the elements that are set to plot in the Preview box, with appropriate changes as per the settings of the Output Actions. Once you are satisfied with the changes, create the plot file.

STEP 7: To save the pen table, invoke the Save command:

Pull-down menu (Modify Pen Table settings box)	File > Save

MicroStation saves the modifications made to the existing pen table.

To save the modifications to a different pen table file, invoke the Save As command:

Pull-down menu (Modify Pen Table settings box)	File > Save As

MicroStation displays the Create Pen Table File dialog box. Key-in the name of the file to save the settings, and click the OK button.

STEP 8: To disable pen table processing, unload the pen table. To unload the pen table, invoke the Unload tool:

Pull-down menu (Modify Pen Table settings box)	File > Exit/Unload

MicroStation unloads the pen table.

You can also unload the pen table from the Pen Table pull-down menu located in the Plot settings box.

BATCH PLOTTING

MicroStation provides a utility program called BatchPlot to plot sets of design files. This utility program allows plotting of multiple design files. You can compose and re-use job sets that identify design files to be plotted and the specifications that describe how they should be plotted. You can also plot individual files or subsets of the files in large job sets for spot-checking.

To invoke the BatchPlot program, select:

Pull-down menu	File > Batch Print/Plot

MicroStation displays the BatchPlot dialog box, as shown in Figure 10–20.

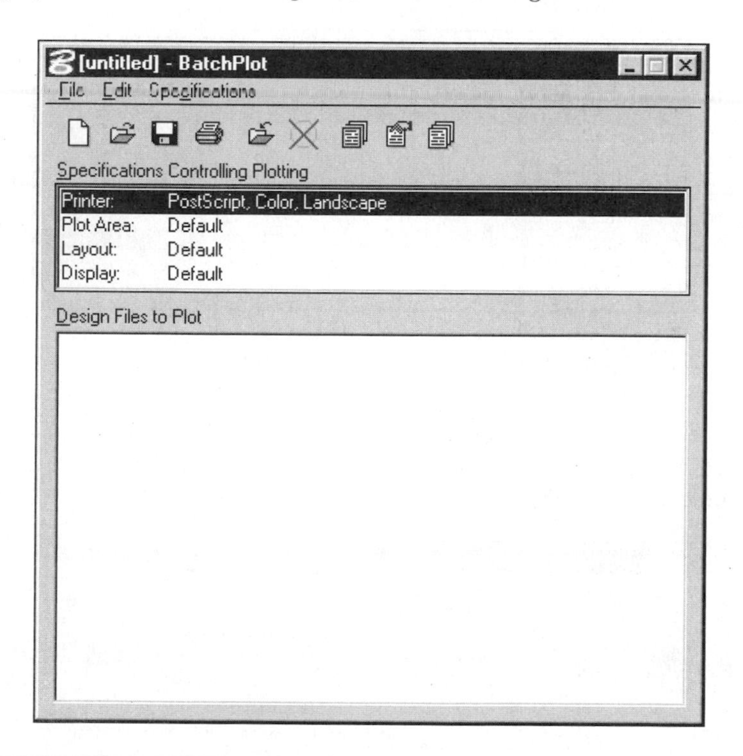

Figure 10–20 BatchPlot dialog box

Setting Plot Specifications

Plot specification is a named group of instructions describing how to perform a certain step in the plotting process. The BatchPlot utility program provides four specification types: Printer, Plot Area, Layout, and Display. The Printer specification type describes the printer, paper size, and post-processing options. The Plot Area specification type selects the portions of the design file to plot.

The Layout specification type places a representation of the given plot area on the paper at the specified size and position. And the Display specification allows you to select a pen table and setting the attributes for plotting.

Printer Specification

The Printer specification type describes the printer, paper size, and post-processing options. The utility allows you to create a new printer specification, and modify or delete an existing specification. To create a new, modify or delete a printer specification, open the BatchPlot Specification Manager:

Pull-down menu (BatchPlot settings box)	Specifications > Manage

MicroStation displays BatchPlot Specification Manager as shown in Figure 10–21.

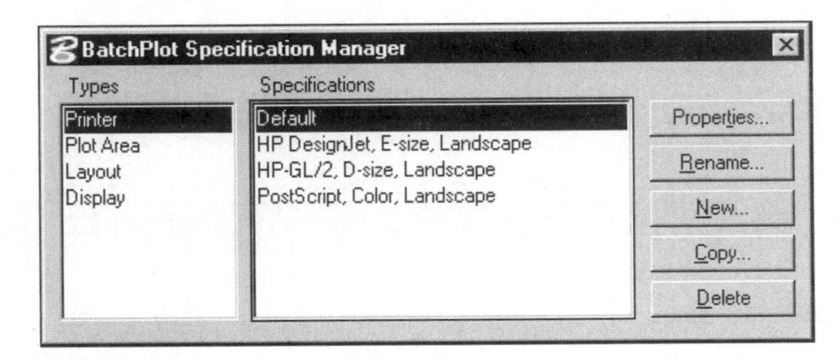

Figure 10–21 BatchPlot Specification Manager dialog box

To change the properties of an existing printer specification, select the appropriate specification from the Specifications list box, then click the Properties... button. MicroStation displays a Printer Properties dialog box similar to Figure 10–22.

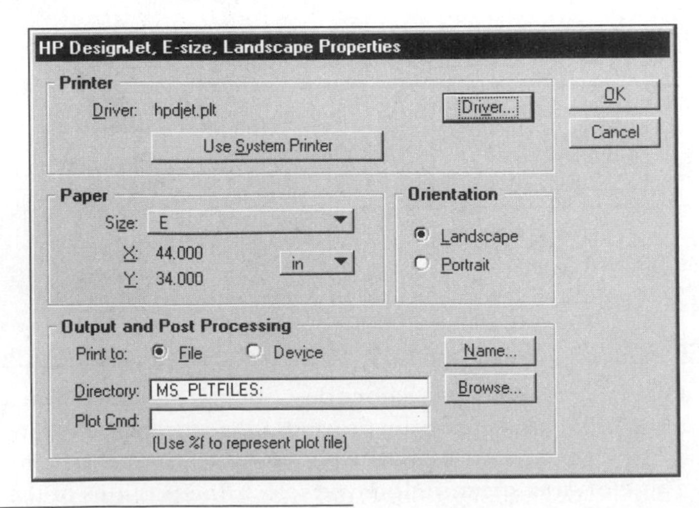

Figure 10–22 Printer Specification dialog box

The Printer specification dialog box allows you to select a specific printer driver, paper size, orientation, and output and post processing settings. Click the OK button to accept the changes and close the dialog box.

To create a new printer specification, click the New... button in the BatchPlot Specification Manager. MicroStation displays the New Printer Specification Name dialog box. Key-in the name in the Name: edit field and click the OK button. If necessary, you can change the properties of the selected Printer specifications. MicroStation lists the newly created printer specification in the Specifications list box.

To rename a new printer specification, first select the specification you want to rename, then click the Rename... button in the BatchPlot Specification Manager. MicroStation displays Rename Printer Specification dialog box. Key-in the new name in the Name: edit field and click the OK button. MicroStation lists the renamed printer specification in the Specifications list box.

To create a printer specification from an existing one, first select the specification from which you want to copy, then click the Copy... button in the BatchPlot Specification Manager. MicroStation displays the New Printer Specification Name dialog box. Key-in the name in the Name: edit field and click the OK button. If necessary, you can change the properties of the newly created Printer specifications. MicroStation lists the newly created printer specification in the Specifications list box.

To delete a printer specification, first select the specification you want to delete from the Specifications list box, then click the Delete button in the BatchPlot Specification Manager. MicroStation deletes the selected Printer Specification from the Specifications list box.

To change the current default selection of the printer specification, invoke the Select... tool:

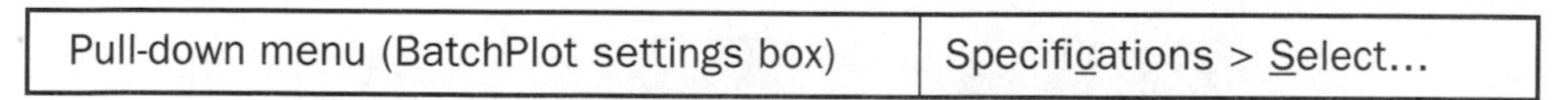

Pull-down menu (BatchPlot settings box)	Specifications > Select...

MicroStation displays Select Printer Specification dialog box shown in Figure 10–23.

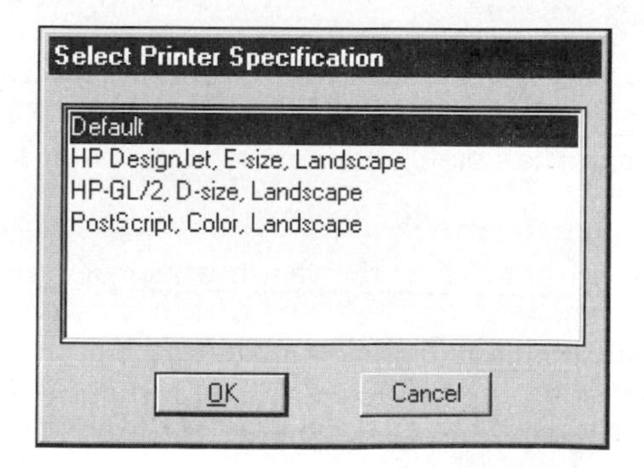

Figure 10–23 Select Printer Specification dialog box

Select the appropriate printer specification from the list box and click the OK button.

If necessary, you can also change the properties of the selected printer specification by double-clicking the name of the specification from the Specifications list box in the BatchPlot dialog box. MicroStation displays the Printer Properties dialog box, similar to Figure 10–22. Make the necessary changes and click the OK button to accept the changes and close the dialog box.

Plot Area

The Plot Area specification type allows you to select the portions of the design file to plot. The utility allows you to create a new Plot Area specification, and modify or delete an existing specification. To create a new, modify or delete a Plot Area specification, open the BatchPlot Specification Manager:

Pull-down menu (BatchPlot settings box)	Specifications > Manage

MicroStation displays the BatchPlot Specification Manager, shown in Figure 10–21. To change the properties of an existing Plot Area specification, first select the Plot Area from the Types list box, then select the appropriate specification from the Specifications list box and click the Properties... button. MicroStation displays the Plot Area Properties dialog box, similar to Figure 10–24.

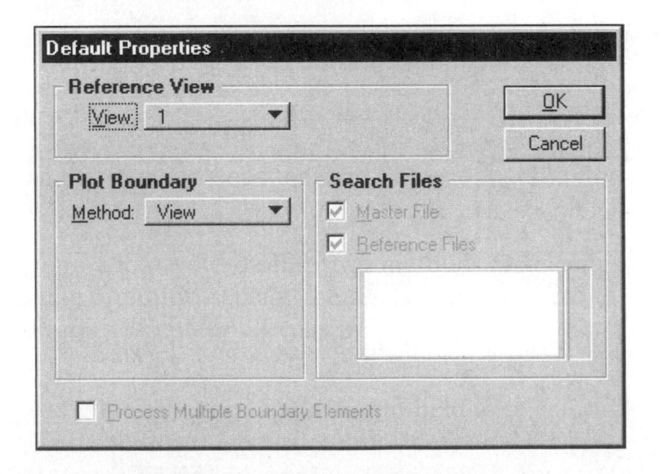

Figure 10–24 Plot Area Properties dialog box.

The View: options menu allows you to select the view number or saved view to plot.

The Plot Boundary section of the dialog box allows you to select the boundary-defining elements (similar to placing a fence by snapping to the vertices of a shape). The option menu in the Plot Boundary section includes three options: View, Shape, and Cell. The View selection plots to the extent of the view window. The Shape selection plots an area bounded by a particular shape. Specify the attributes of the shape in the appropriate fields in the Boundary section of the dialog box. The

Cell selection plots an area bounded by a cell. Specify the name of the cell in the Name: edit field in the Boundary section of the dialog box.

The Master File and Reference Files settings set the limit for the search for the boundary-defining shape or cell. By default, the utility searches each master file and all of its reference files to find the boundary-defining shape or cell. If necessary, you can restrict the search to specific reference files by typing their logical names or filenames in the Reference Files field.

The Process Multiple Boundary Elements control whether to generate the plots for each boundary element found in the design file. Toggle this box ON if you want to plot for each boundary element found and OFF to generate a plot for the first boundary element found only.

Click the OK button to accept the changes and close the dialog box.

Similar to Printer specification, you can also create a new Plot Area specification, rename an existing one, or delete one.

To change the current selection of the Plot Area specification, invoke the Select... tool:

Pull-down menu (BatchPlot settings box)	Specifications > Select...

MicroStation displays the Select Plot Area Specification dialog box, shown in Figure 10–25.

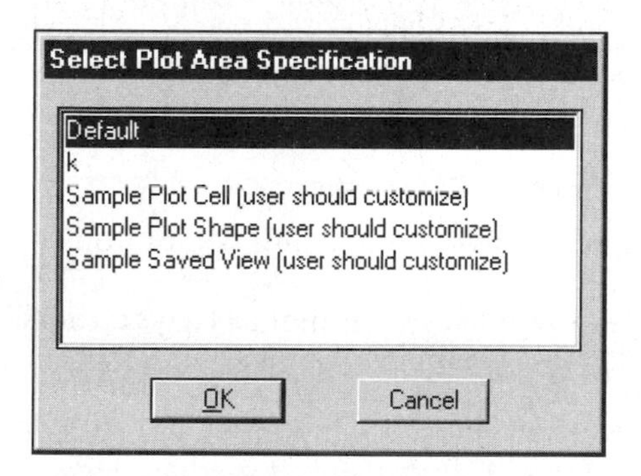

Figure 10–25 Select Plot Area Specification dialog box

Select the appropriate Plot Area specification from the list box and click the OK button.

If necessary, you can also change the properties of the selected Plot Area specification by double-clicking the name of the specification from the Specifications list box in the BatchPlot dialog box. MicroStation displays the Plot Area Properties dialog box, similar to Figure 10–24. Make the necessary changes and click the OK button to accept the changes and close the dialog box.

Layout

The Layout specification type describes how the utility program determines the size and position of each plot. The utility allows you to create a new Layout specification, and modify or delete an existing specification. To create a new, or modify or delete a Layout specification, open the BatchPlot Specification Manager:

Pull-down menu (BatchPlot settings box)	Specifications > Manage

MicroStation displays the BatchPlot Specification Manager, as shown in Figure 10–21. To change the properties of an existing Layout specification, first select the Layout from the Types list box, then select the appropriate specification from the Specifications list box, and click the Properties... button. MicroStation displays the Layout Properties dialog box, similar to Figure 10–26.

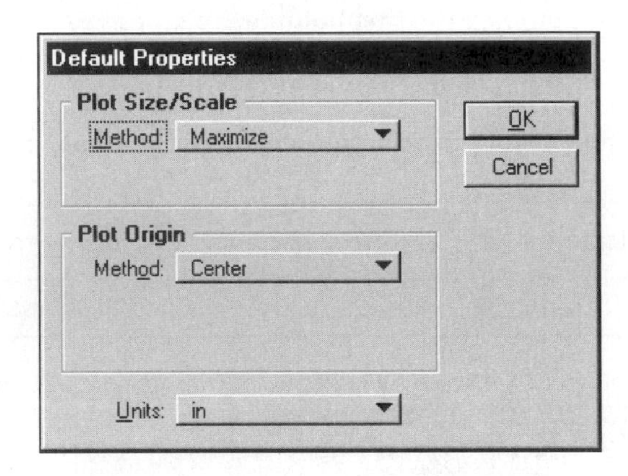

Figure 10–26 Layout Properties dialog box

The Plot Size/Scale Method: menu provides five plot size options:

- The Maximize selection makes each plot as large as possible, given the paper size and orientation in the job set's printer specification.
- The Scale selection allows you to specify a scale factor in terms of master units in the design file to physical units of the output media.
- The % of Maximum Size selection allows you to specify an integer value between 10 and 100 percent of its maximum possible size.
- The X Size selection allows you to specify an explicit X size (width) for the plot.
- The Y Size selection allows you to specify an explicit Y size (height) for the plot.

The Plot Origin Method: menu provides two options:

- The Center selection centers each plot on the output media.
- The Manual Offset selection allows you to specify explicit *X* and *Y* offsets for the plot. The offsets provided are relative to the media's lower-left margin.

Click the OK button to accept the changes and close the dialog box.

Similar to Printer and Plot Area specifications, you can create a new Layout specification, rename from an existing one, or delete one.

To change the current selection of the Layout specification, invoke the Select... tool:

Pull-down menu (BatchPlot settings box)	Specifications > Select...

MicroStation displays the Select Layout Specification dialog boxes shown in Figure 10–27.

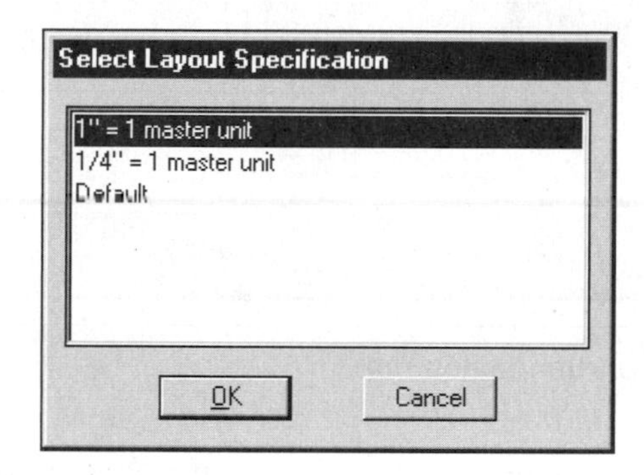

Figure 10–27 Select Layout Specification dialog box

Select the appropriate Layout specification from the list box and click the OK button.

If necessary, you can also change the properties of the selected Layout specification by double-clicking the name of the specification from the Specifications list box in the BatchPlot dialog box. MicroStation displays the Layout Properties dialog box, similar to Figure 10–26. Make the necessary changes and click the OK button to accept the changes and close the dialog box.

Display

The display specification controls the appearance of printed elements. You can control the plotting equivalents of the view attributes, and, in addition, you can specify a pen table that will resymbolize the plot. To create a new, modify or delete a Display specification, open the BatchPlot Specification Manager:

Pull-down menu (BatchPlot settings box)	Specifications > Manage

MicroStation displays the BatchPlot Specification Manager shown in Figure 10–21. To change the properties of an existing Display specification, first select Display from the Types list box, then select the appropriate specification from the Specifications list box, and click the Properties... button. MicroStation displays the Display Properties dialog box similar to Figure 10–28.

Default Properties

Options

Camera:	As-is	Level Symbology:	As-is
Constructions:	As-is	Line Styles:	As-is
Data Fields:	As-is	Line Weights:	As-is
Dimensions:	As-is	Patterns:	As-is
Fast Cells:	As-is	Plot Border:	As-is
Fast Curves:	As-is	Ref Boundaries:	As-is
Fast Font:	As-is	Tags:	As-is
Fast Ref Clipping:	As-is	Text:	As-is
Fence Boundary:	As-is	Text Nodes:	As-is
Fill:	As-is		

OK

Cancel

Pen Table

Filename:

Browse...

Figure 10–28 Display Properties dialog box

The options provided are similar to the View Attributes setting options. Make necessary changes to the display options. In addition, you can also specify the name of the pen table file you want to apply for plotting the selected design file. Click the OK button to accept the changes and close the dialog box.

Similar to Printer, Plot Area, and Layout specifications, you can create a new Display specification, rename an existing one, or delete one.

To change the current selection of the Display specification, invoke the Select… tool:

Pull-down menu (BatchPlot settings box)	Specifications > Select...

MicroStation displays the Select Display Specification dialog box shown in Figure 10–29.

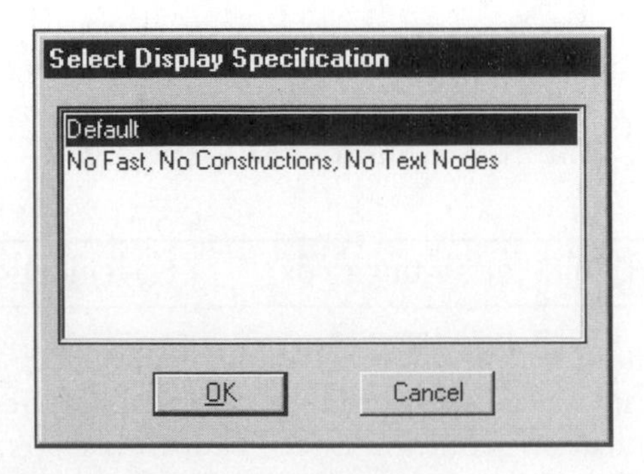

Figure 10–29 Select Display Specification dialog box

Select the appropriate Display specification from the list box and click the OK button.

If necessary, you can also change the properties of the selected Display specification by double-clicking the name of the specification from the Specifications list box in the BatchPlot dialog box. MicroStation displays the Display Properties dialog box, similar to Figure 10–28. Make the necessary changes and click the OK button to accept the changes and close the dialog box.

Design Files to Plot

To select design files to plot, invoke the Add Files tool:

Pull-down menu (BatchPlot settings box)	Edit > Add Files...

MicroStation displays the Select Design Files to Add dialog box shown in Figure 10–30.

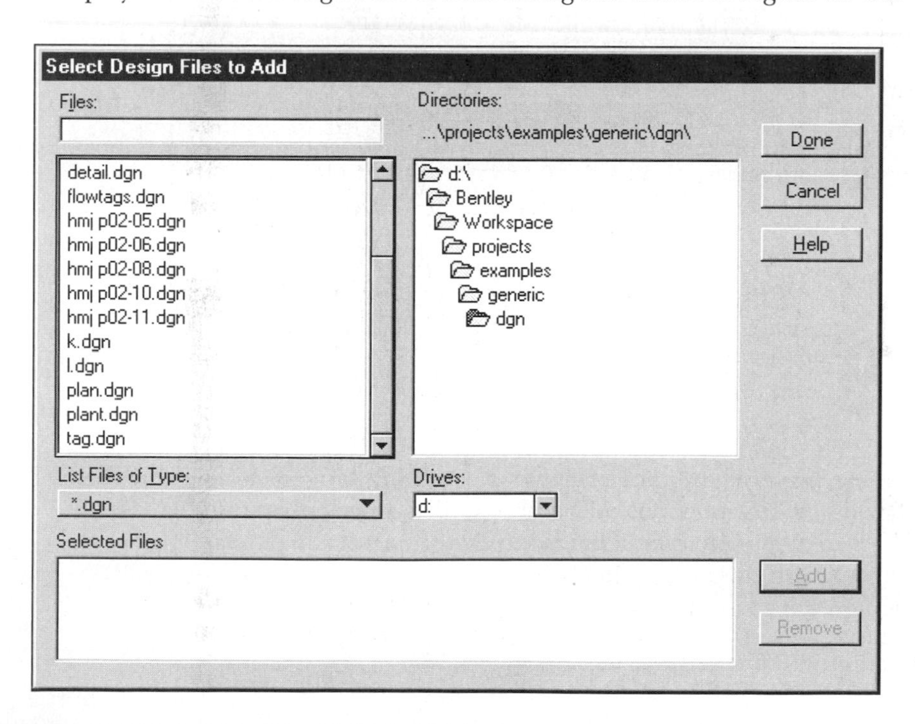

Figure 10–30 Select Design Files to Add dialog box

Select the design files from the appropriate directory and then click the Add button. Selected files are added to the Design Files to Plot list box. Click Done button when the selection is complete. The selected files are plotted in the order in which they appear in the list box. If necessary, you can rearrange the files by highlighting their names and invoking one of the modify tools available in the Edit pull-down menu.

To save the current selection of the design files set to plot, invoke Save tool:

Pull-down menu (BatchPlot settings box)	File > Save

MicroStation displays the Save Job Set File dialog box. Key-in the name of the file in the Files: edit field to which you want to save the current job set. Click the OK button to save the current job set and close the dialog box.

To plot the current selection of the design files, invoke the Print tool:

Pull-down menu (BatchPlot settings box)	File > Print

MicroStation displays Print dialog box similar to Figure 10–31.

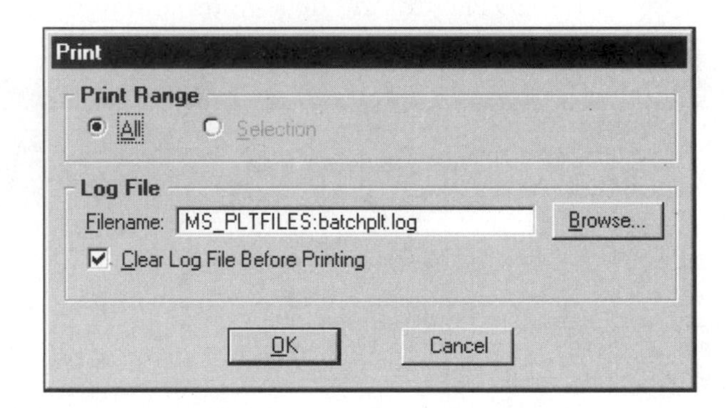

Figure 10–31 Print dialog box

Select one of the two radio buttons available in the Print Range section of the dialog box. The All option allows MicroStation to plot all the design files in the current job set. The Selection option plots only the design files that are selected explicitly in the current job set. Click the OK button to plot the selection and close the dialog box.

PROJECT EXERCISE

This project exercise provides step-by-step instructions for plotting the Chapter 6 Project Exercise design, shown in Figure P10–1.

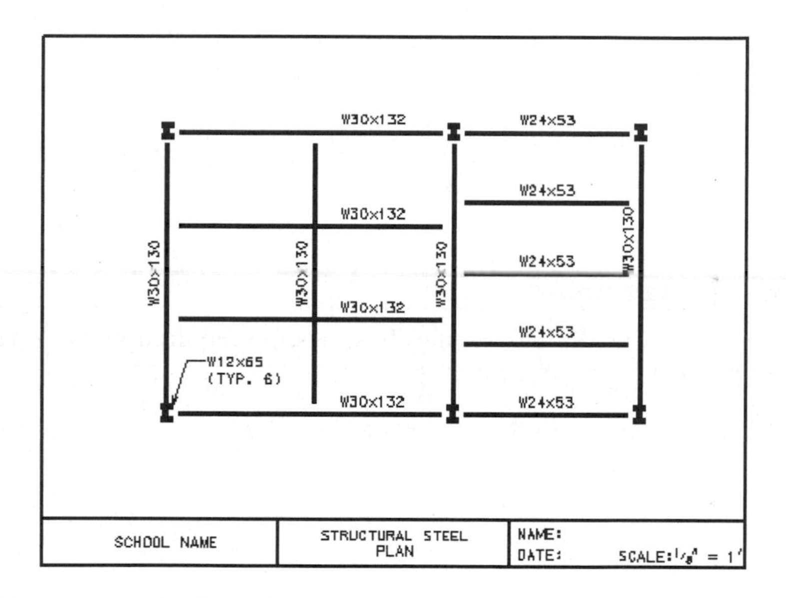

Figure P10–1 Completed project design

Note: As you complete each step in the project procedures, place a check mark by the step to help you keep up with where you are in the project.

Set Up the Design for Plotting

This procedure prepares the structural steel plan design for plotting.

STEP 1: Start MicroStation and open the design file named CH6.DGN.

Note: If you did not do the Chapter 6 project, refer to Chapter 6 for instructions.

STEP 2: Invoke the Fit View tool to fit the view window.

STEP 3: Set the default Snap mode to Keypoint.

STEP 4: Invoke the Place Fence tool from the Fence tool box, and, in the Tool Settings window, select the Block option.

MicroStation prompts:

> Place Fence Block > Enter first point *(Keypoint snap to one corner of the border block, then place a data point.)*
>
> Place Fence Block > Enter opposite corner *(Keypoint snap to the diagonally opposite corner, then place a data point to complete placing the Fence Block.)*

Note: A successful Keypoint snap is indicated by the tentative cross appearing on the corner and the entire border block switching to the highlight color.

Set Up the Plotting Parameters

This procedure opens the Plot settings box, sets the required plotting parameters, and plots the fence contents.

STEP 1: Invoke the Print/Plot command from the File pull-down menu. MicroStation displays the Plot settings box.

STEP 2: Select the appropriate plotter driver by clicking the Plotter Driver tool in the Plot settings box. (For this exercise we have selected hp7585b.plt).

STEP 3: Select the Fence option from the Entity pull-down menu in the Plot settings box.

STEP 4: Invoke the Page Setup tool from the Plot settings tool box, and set the Page Size as shown in Figure P10–2. Click the OK button to close the dialog box.

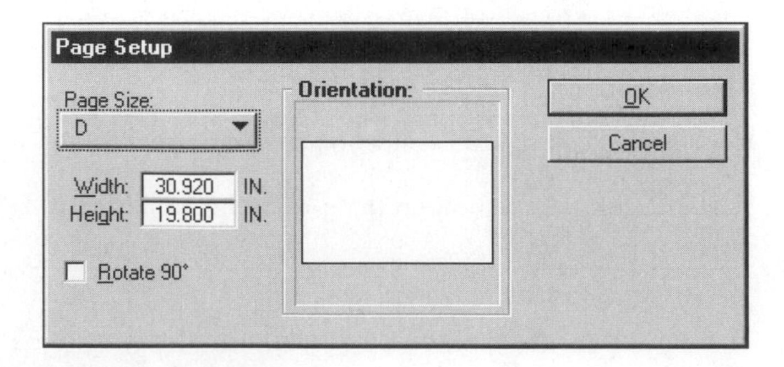

Figure P10–2 Page Setup dialog box

STEP 5: Invoke the Plot Layout tool from the Plot settings tool box, and set the Plot Layout settings as shown in Figure P10–3. Click the OK button to close the dialog box.

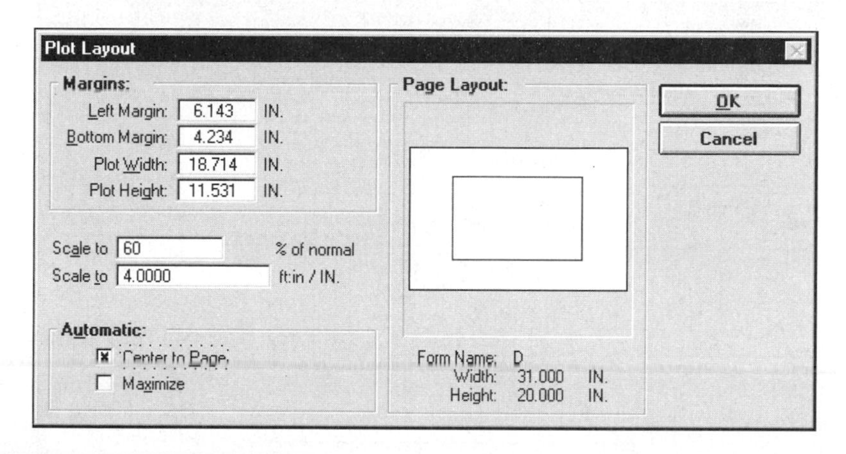

Figure P10–3 Plot Layout dialog box

Note: The Scale to: edit field says 4.0000 ft:in/IN and is equivilent to ¼" = 1'–0" on a Size-D paper.

STEP 6: Invoke the Plot Options tool from the Plot settings tool box and set the toggle button for Plot Border to OFF, as shown in Figure P10–4. Click the OK button to close the dialog box.

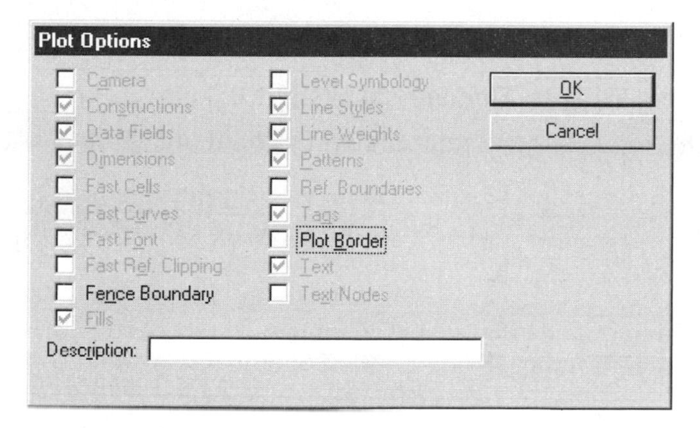

Figure P10–4 Plot Options dialog box

STEP 7: Invoke the Preview Refresh command from the Plot settings box tool bar to preview the way the design will look on the page (see Figure P10–5).

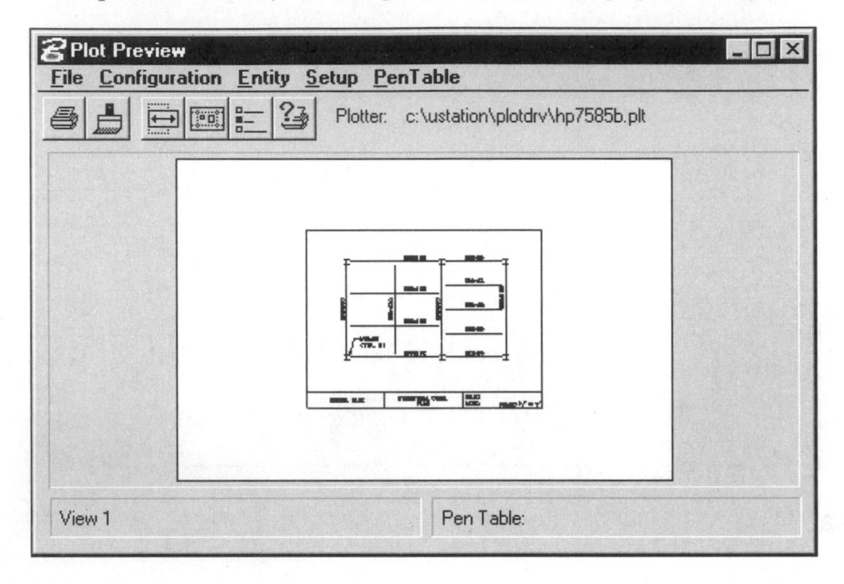

Figure P10–5 The Plot settings box with Plot Preview turned on

STEP 8: Submit the plot to the printer/plotter by invoking the Plot tool from the Plot settings box tool bar.

Note: Leave the Plot settings box open for the next procedure.

Plot the Design Again With All Elements at Weight 0

This procedure uses the Pen Table option to plot the same design with all elements at weight 0.

STEP 1: Invoke the New... tool from the PenTable pull-down menu in the Plot settings box.

STEP 2: In the Create Pen Table File window, key-in **CH10.TBL** in the Files edit field, then click the OK button. MicroStation displays the Modify Pen Table settings box, similar to Figure P10–6.

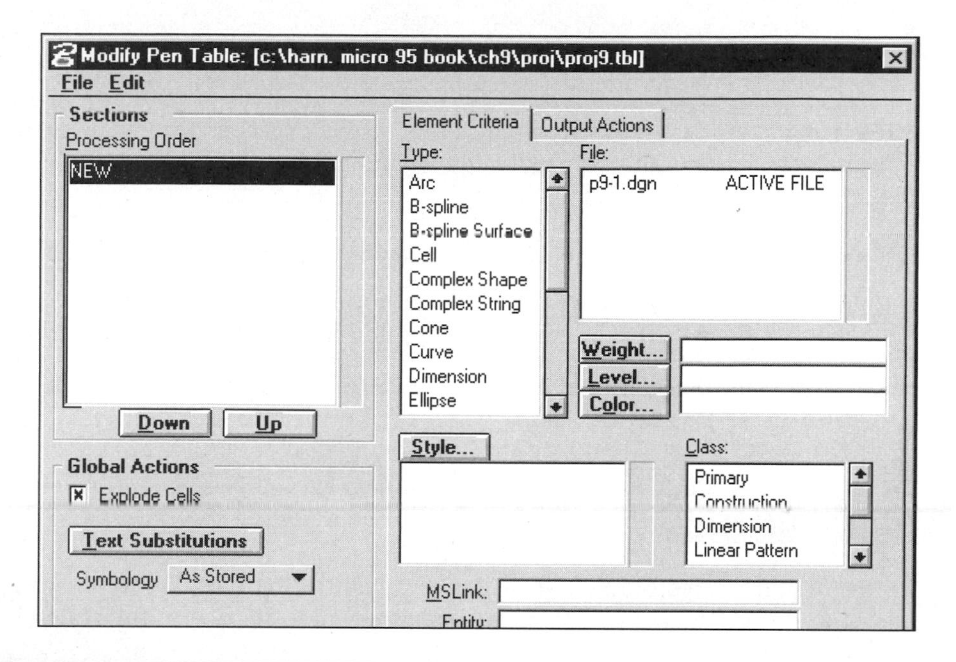

Figure P10–6 Modify Pen Table settings box

STEP 3: Select the active design file in the File: list box and click the Output Actions tab. Set the toggle button for Weight to ON and set the Line Weight to 0.

STEP 4: Invoke the Save tool from the File pull-down menu in the Modify Pen Table settings box.

STEP 5: Invoke the Preview Refresh tool from the Plot settings box tool bar to preview the way the design will look after the modification in the Pen Table settings box.

Note: This weight setting is for plotting only. The weight of the elements in the design is unchanged.

STEP 6: Invoke the Plot tool from the Plot settings box tool bar to plot the design with all elements at weight 0.

DRAWING EXERCISES 10–1 THROUGH 10–5

The following exercises have you create plots of the exercises that were created in earlier chapters. The requested plot scales are intended to create plots on letter-size (8.5" x 11") paper.

- If the printer/plotter you have access to can handle larger sheet sizes, plot at larger scales.
- If the printer/plotter you have access to cannot plot at the recommended scales, reduce the scale by half. For example, if it cannot create a full-scale plot, create a half-scale plot.

Exercise 10–1

Open the design created in Exercise 4–5 on page 235, fit the view, size the view window to minimize the space outside the design area, then plot the view at full scale (1.0000 IN:TH/IN).

Exercise 10–2

Open the design created in Exercise 5–2 on page 275, place a fence by snapping to diagonally opposite corners of the design border, then plot the contents of the fence at full scale (1.0000 IN:TH/IN).

Exercise 10–3

Open the design created in Exercise 9–3 on page 524-25, place a fence by snapping to diagonally opposite corners of the design border, then plot the contents of the fence at full scale (1.0000 IN:TH/IN).

Exercise 10–4

Open the design created in Exercise 5–5 on page 278, place a fence by snapping to diagonally opposite corners of the design border, then plot the contents of the fence at ¼" = 1', (4.0000'."/IN).

Exercise 10–5

Open the design created in Exercise 6–5 on page 321, place a fence by snapping to diagonally opposite corners of the design border, then plot the contents of the fence at ½" = 1', (2.0000'/IN).

chapter 11

Cells and Cell Libraries

Objectives

After completing this chapter, you will be able to do the following:

- Create cell libraries
- Attach cell libraries
- Create cells
- Select active cells
- Place cells
- Place line terminators
- Place point elements, characters, and cells
- Maintain cells and cell libraries
- Place and maintain shared cells
- Use and modify cells from cell selector

CELLS

Cells are like the variously shaped cutouts in a manual drafting template. You draw standard symbols on paper by tracing the outline of the symbol's cutout in the template. In MicroStation you do the same thing by placing a copy of a cell in your design file. Figure 11–1 shows some common uses of cells in various engineering disciplines.

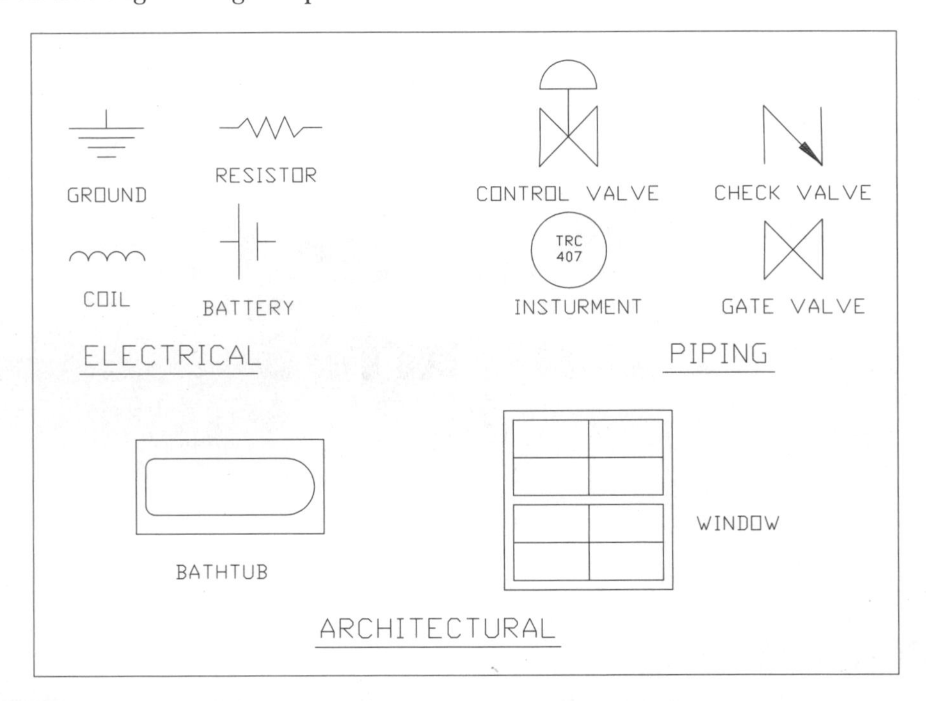

Figure 11–1 Common uses of cells in various engineering disciplines

Even though a cell contains separate elements, the copy you place in a design file acts like a single element when manipulated with tools such as Delete, Rotate, Array, and Mirror. When placing cells, you can change the scale and/or rotation angle of the original object(s). Cells save time by eliminating the need to draw the same thing more than once, and they also promote standardization.

CELL LIBRARIES

If cells are like the holes in a plastic template, cell libraries are the template. A cell library is a file that holds cells. Most engineering companies that use MicroStation have several cell libraries to provide standard symbols for all of their design files. Some companies also create sets of cells stored in libraries that they offer for sale.

You can access any of the cells from the cell library and place copies of them in your current design file. If you create a new cell, it is automatically placed in the library that is attached to your design file.

You can also place a cell that is not in the attached cell library by keying-in the name of the cell. MicroStation searches for the cell in the cell library list specified by the Cell Library List configuration variable (MS_CELLIST). (Refer to Chapter 16 for how to set up the configuration variables.) Cell libraries are searched in their order in the list. If wild-card characters are involved, cell libraries are searched in alphabetical order.

Although there is no limit to the number of cells you can store in a library, you don't want to end up with a very large, hard-to-manage library. It is advisable to create separate cell libraries for specific disciplines, such as electrical fixtures, plumbing, and HVAC.

Creating a New Cell Library

To start the process of creating a new cell library, invoke the Cell Library settings box:

Pull-down menu	Element > Cells (see Figure 11–2).
Key-in window	**dialog cellmaintenance** (or **di ce**) Enter

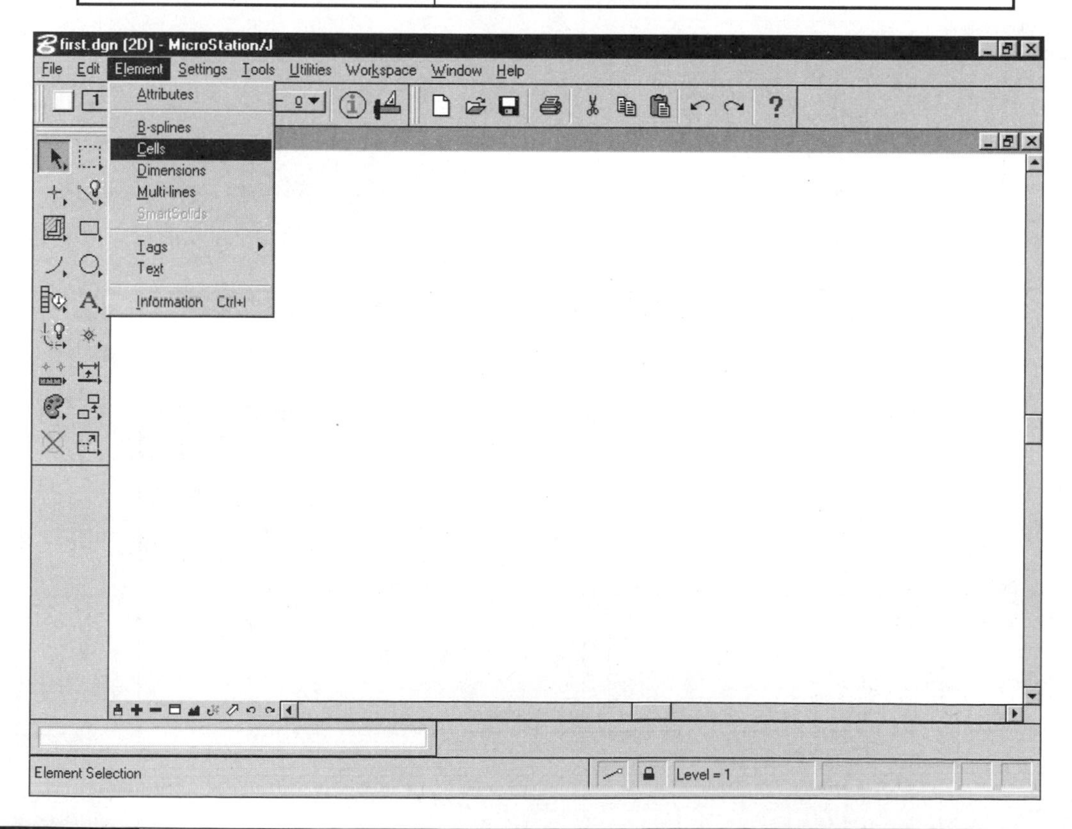

Figure 11–2 Invoking the Cell Library settings box from the Element pull-down menu.

A settings box appears similar to the one shown in Figure 11–3.

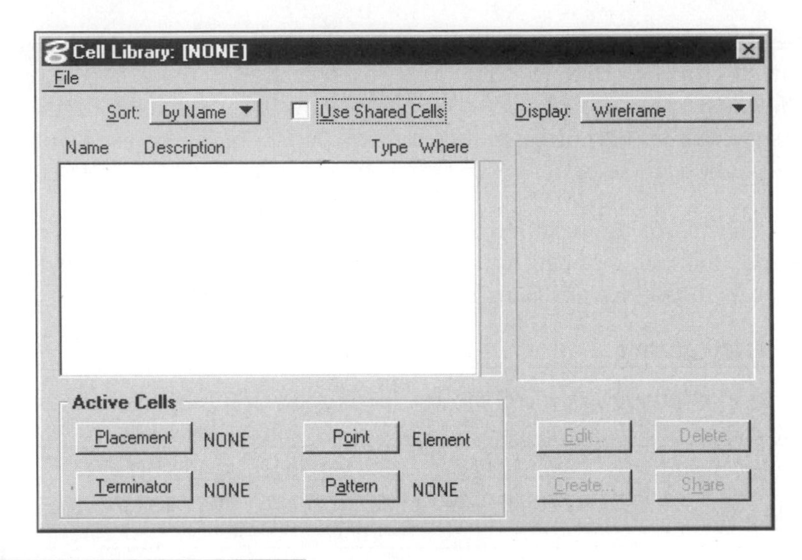

Figure 11–3 Cell Library settings box

To create a new cell library file, open the Create Cell Library dialog box:

Pull-down menu in the Cell Library settings box	File > New

MicroStation displays the Create Cell Library dialog box, as shown in Figure 11–4.

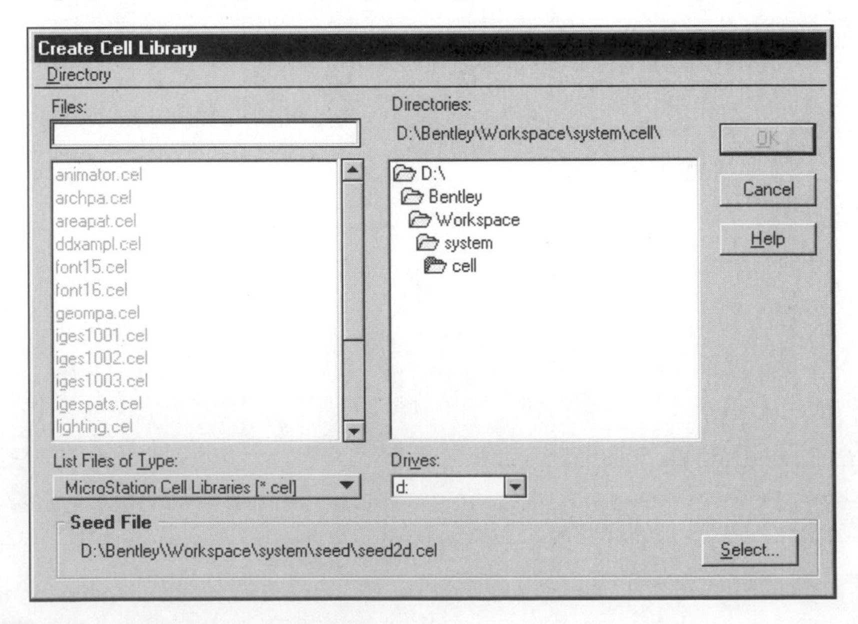

Figure 11–4 Create Cell Library dialog box

Following are the steps to create a new cell library file.

1. Check the Seed File specification at the bottom of the Create Cell Library dialog box:

   - SEED2D.CEL for *2-dimensional* designs.
   - SEED3D.CEL for *3-dimensional* designs.

2. If necessary, you can change the seed file. Click the Select button to open the dialog box for selecting a seed file, and select the appropriate seed file.
3. Key-in a name for the new library in the Name edit field. Do *not* type a period or extension. MicroStation appends the ".CEL" extension to the file name.
4. Click the OK button to create the new cell library and to close the Create Cell Library dialog box.

Note: Before you close the dialog box, make sure the cell library file is created in the appropriate directory. If it isn't, find and select the appropriate directory from the Directories list before you key-in the file name.

When you create a new cell library, MicroStation automatically attaches it to your design file, acknowledges the attachment with a message in the status bar , and places the filename in the Cell Library dialog box title bar.

Your cell library is now available for use. You can start creating cells to store in the library.

Attaching an Existing Cell Library

To attach an existing cell library file to the active design file, invoke the Cell Library settings box:

Pull-down menu	Element > Cells

Open the Attach Cell Library dialog box:

Pull-down menu in the Cell Library settings window	File > Attach
Key-in window	**rc=**<name of the library> Enter

MicroStation displays the Attach Cell Library dialog box. Select the appropriate cell library from the file list, and click the OK button to attach the library and close the Attach Cell Library dialog box.

MicroStation acknowledges the attachment with a message in the status bar. Once the cell library is attached, the Cell List in the Cell Library settings box displays the names and descriptions of the cells in the library, similar to the one shown in Figure 11–5.

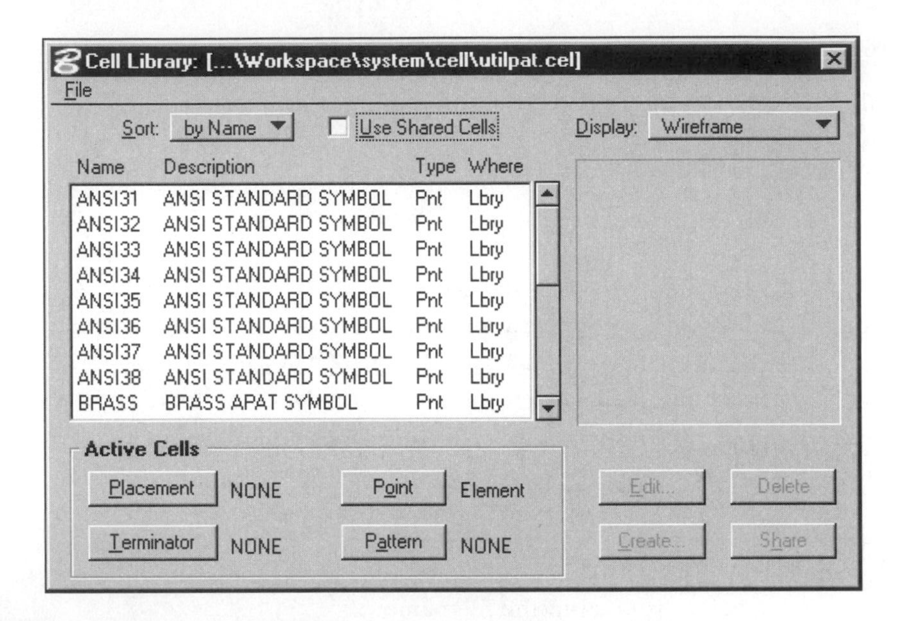

Figure 11-5 Listing of the cell names and descriptions in the Cell Library settings box

Things to Remember about Cell Library Attachment

- The attachment is permanent as long as MicroStation can find the library file.
- You can attach only one cell library at a time. If you attach another library, the first library is detached.
- When the library is attached, all the library's cells are available to you for placement, and you can create new cells in the library.

Note: The new cells you create are stored only in the cell library that is attached to the current design file, not in the cell libraries that are attached through the environmental variable.

CREATING YOUR OWN CELLS

When you need to place copies of a symbol for which no cell currently exists, you can create your own cell, store it in the attached library, and then use it in any design file.

Before You Start

Here are a few things to consider before starting to draw the elements that will make up your cell.

Working Units

The Working Units in the design file you use to create the cell should be the same as the Working Units of the design files in which you plan to place copies of the cell. If the Working Units are different, you may have to scale the cell every time you place it.

Cell Elements

There are no restrictions on the elements that make up the cell. All element types can be placed in a cell; they can be drawn on any level; and they can be any color, weight, or style. The elements you draw do not become a cell. Placing a copy of your elements into the attached cell library creates the cell. After you create your cell, you may delete the original elements from the design file.

Cell Rotation

Always draw the object you are going to make into a cell with 0 degrees of rotation. That means it should be upright and facing to the right (see Figure 11–6). Sticking with zero rotation for cells makes it easier to understand what happens to the cell when it is rotated.

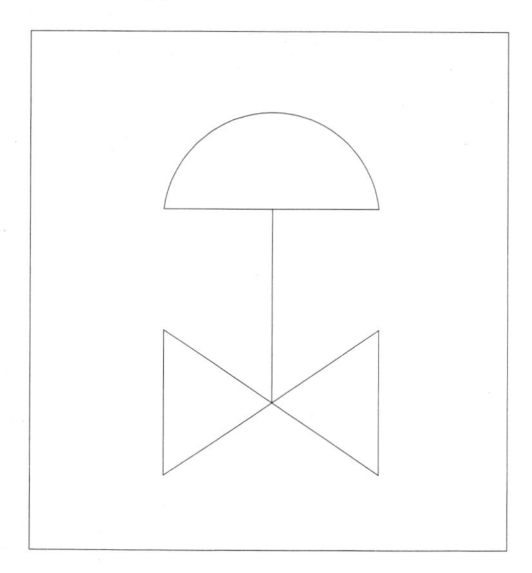

Figure 11–6 Example of a symbol drawn upright and facing to the right

Cell Origin

Defining the cell's origin point in the design places cells. The cell's origin point is created during the process of creating the cell. Consider how the cell will be used when deciding exactly where in the cell elements to create the origin. If the cell is to be placed connected to other elements, place the origin point at the connection point. For instance, if the cell is a control valve, place the origin where the valve is to be attached to the pipeline, as shown in Figure 11–7.

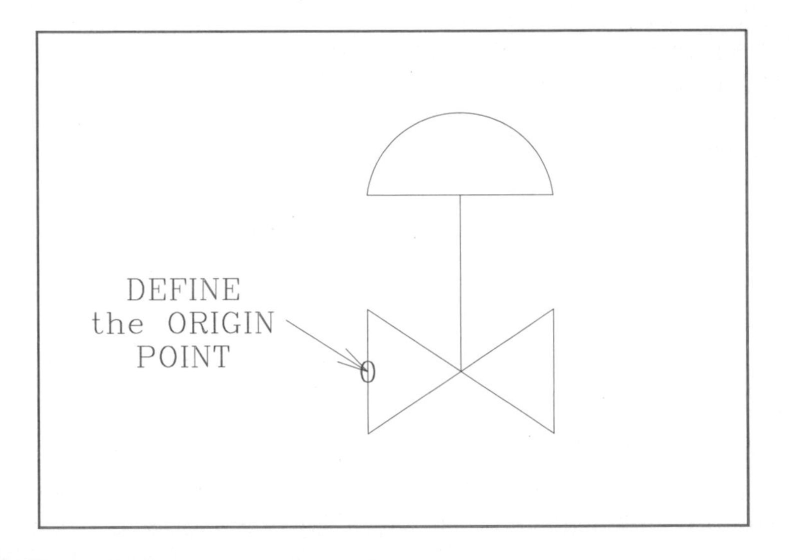

Figure 11–7 Example of defining a cell origin

Steps for Creating a Cell

Following are the steps to create a new cell.

STEP 1: Draw the elements.

STEP 2: Group the elements by placing a fence around them, then set the fence mode to Inside.

STEP 3: Define the cell's origin point by invoking the Define Cell Origin tool:

Cells tool box	Select the Define Cell Origin tool (see Figure 11–8).
Key-in window	**define cell origin** (or **d c o**) Enter

MicroStation prompts:

Define Cell Origin *(Place a data point where the origin is to be located. Snap, if necessary, to place the point precisely.)*

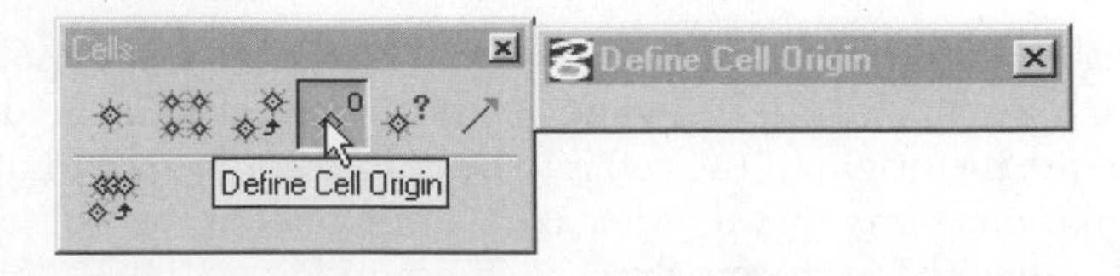

Figure 11–8 Invoking the Define Cell Origin tool from the Cells tool box

Note: If you accidentally set the origin point in the wrong place, just place another one. The cell will be created using the last point. The letter "O" appears at each origin placement point. The "O" is not an element; it only provides a visual indication that an origin has been placed.

STEP 4: Click the Create button located in the bottom right corner of the Cell Library settings box. MicroStation displays the Create New Cell dialog box, similar to the one shown in Figure 11–9.

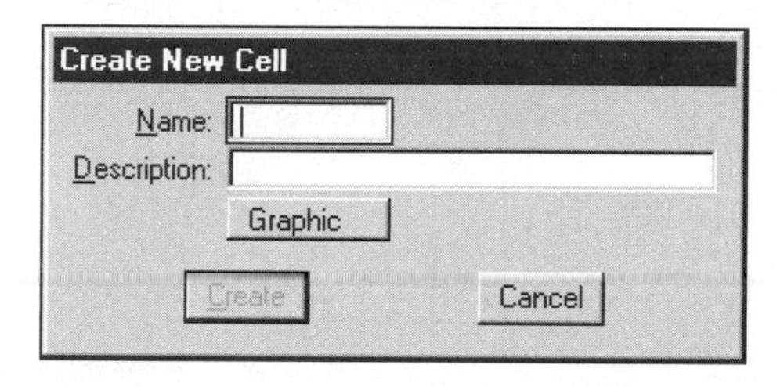

Figure 11–9 Create New Cell dialog box

STEP 5: Click in the Name field in the Create New Cell dialog box, and enter a name for your cell (don't press Enter). You must provide a one- to six-character alphanumeric name for your cell. The name can contain only letters, numbers, and underscores (_). Help other people who will have to use your cells by creating descriptive cell names (as best as you can in six characters).

Good Examples: PNP is a good name for a cell that provides a PNP transistor. If the cell provides a globe valve symbol, a good name might be GLBVAL.

Bad Examples: CELL-1 is a legal cell name but does not provide any information about the intended use of the cell. GLOBE_VALVE is an unacceptable cell name because it contains too many characters.

STEP 6: Click on the Description field and enter a 1- to 27-character description for your cell. Take the time to provide a description for every cell you create. If your cell libraries contain lots of cells, the descriptions will help other users figure out the purpose you intended for the cell. It will also help you when you go back to the cell library after being away from it for a few weeks.

STEP 7: Select the appropriate cell type from the option menu. Cells that can be placed in a design come in two types—Graphic and Point. The default cell type is Graphic. In the next section we will discuss the difference between the two types and how to specify which type to create.

STEP 8: Click the Create button.

When you complete the procedure, the fenced elements are copied to the attached cell library, and your new cell appears in the cells list area of the Cell Library settings box (see Figure 11–10). In addition to the cell name, the cells list displays the cell description, type, and location. The cell is available for placement in the current design file and in any other design files to which your cell library is attached.

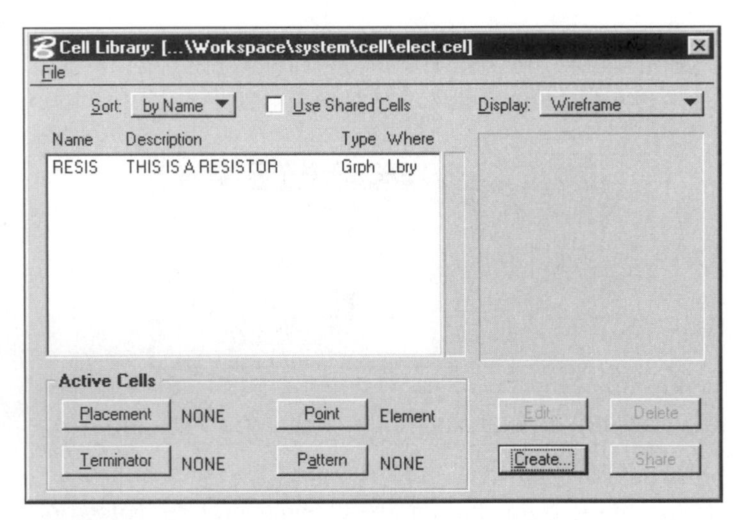

Figure 11–10 Displaying the name and description of the new cell in the Cell Library settings box

Refer to Figure 11–11 to review the steps for the creation of a cell.

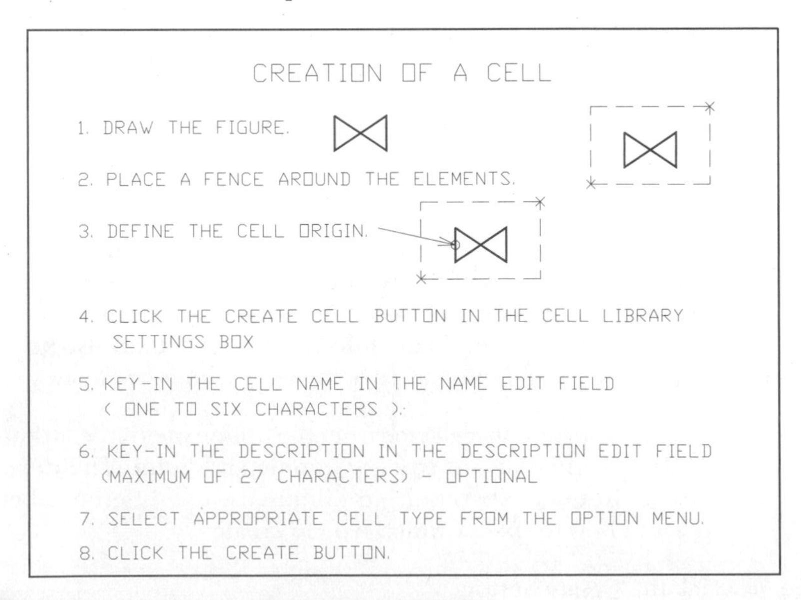

CREATION OF A CELL
1. DRAW THE FIGURE.
2. PLACE A FENCE AROUND THE ELEMENTS.
3. DEFINE THE CELL ORIGIN.
4. CLICK THE CREATE CELL BUTTON IN THE CELL LIBRARY SETTINGS BOX
5. KEY-IN THE CELL NAME IN THE NAME EDIT FIELD (ONE TO SIX CHARACTERS).
6. KEY-IN THE DESCRIPTION IN THE DESCRIPTION EDIT FIELD (MAXIMUM OF 27 CHARACTERS) - OPTIONAL
7. SELECT APPROPRIATE CELL TYPE FROM THE OPTION MENU.
8. CLICK THE CREATE BUTTON.

Figure 11–11 Review of the steps for creating a cell

Graphic and Point Cell Types

As just mentioned, two cell types, Graphic and Point, can be placed in your design. The default cell type is Graphic (also called Normal). Here are the differences between Graphic and Point cells.

A Graphic cell, when placed in a design file:

- Keeps the symbology (color, weight, style, and levels) of its elements.
- Remembers the levels on which its elements were drawn. (MicroStation provides two methods for placing the cells—Absolute and Relative modes, discussed in detail later in the chapter.)
- Retains the keypoints of each cell element.

A Point cell, when placed in a design file:

- Takes on the current active symbology settings (color, weight, and style).
- Places all cell elements on the current active level, regardless of what level they were drawn on.
- Has only one keypoint—the cell's origin point. The individual cell elements do not have keypoints.

Selecting the Cell Type

You can select either Graphic or Point cell type from the Create New Cell dialog box option menu when you provide the cell's name and description. The option menu has four cell types. The other two, Menu and Tutorial, do not create cells that can be placed in your design file. They are for customizing the way you use MicroStation and are beyond the scope of this book.

ACTIVE CELLS

To place a copy of a cell in your current design file, make the cell the active placement cell, then select a cell placement tool from the Cells tool box. To select the cell, click on the cell name from the cell list provided in the Cell Library settings box, then click one of the four buttons provided under Active Cell. The four types of active cells give you the freedom to place the cell under varying conditions in your design file.

- The active Placement cell is used with a group of tools that place cells at a data point.
- The active line Terminator cell is used with a tool that places a cell on the end of an element.
- The active Point cell is used with a group of tools that place cells in geometric relation to other elements.
- The active Pattern cell is used with a group of tools that create patterns, such as running bond brick.

Active placement, active terminator, and active point cells are explained in detail in this chapter. Active pattern cells are explained in Chapter 12.

Note: If you double-click on a cell in the Cell Library settings box, the name of the cell is paced next to the Placement button, and the Place Active Cell tool is invoked. (The tool is discussed later.)

Alternate Methods

You also can use key-in to select an active cell for each type of cell placement. In the key-in window, type:

ac=<name> *and press* Enter *to select the active placement cell.*

lt=<name> *and press* Enter *to select the active terminator cell.*

pt=<name> *and press* Enter *to select the active point cell.*

ap=<name> *and press* Enter *to select the active pattern cell.*

In each tool, replace <name> with the name of the cell you want to use.

Example: ac=coil Enter

Several cell placement tools provide a cell name field in the Tool Settings window. You can also select the active cell by typing its name in this field.

Active Placement Cell

Four tools are provided to place the active placement cell at data points in your design file.

- Place Active Cell—Places copies of the active placement cell at data points you specify. The cell is placed at the active angle and active scale.
- Place Active Cell (Interactive)—Places copies of the active placement cell interactively and allows you to specify the cell rotation and scale graphically.
- Select and Place Active Cell—Lets you select a cell already placed in your design file to be the active placement cell, then places copies at the data points you specify.
- Place Active Cell Matrix—Places a rectangular matrix of the active placement cell, with the lower left corner of the matrix at the data point you specify.

Place Active Cell

Invoke the Place Active Cell tool:

Cells tool box	Select the Place Active Cell tool(see Figure 11–12).
Key-in window	**place cell icon** (or **pl ce i**) Enter

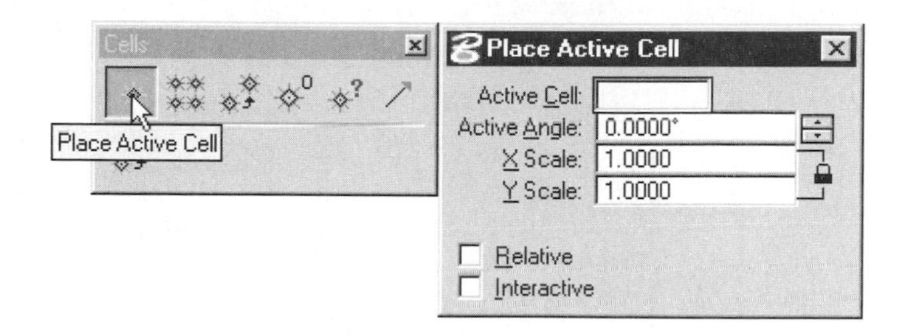

Figure 11–12 Invoking the Place Active Cell tool from the Cells tool box

MicroStation prompts:

> Place Active Cell > Enter cell origin *(Define the origin point at each location where a copy of the cell is to be placed.)*

Each time you place a data point, a copy of the cell is placed such that its origin is at your data point. While this tool is active, the screen cursor drags a dynamic image of the cell. The cell's origin point is at the screen cursor position.

Note: If you do not have an active cell and invoke the Place Active Cell tool, MicroStation displays the message *No Active Cell* in the Status bar. You must declare an active cell before you can use this tool (discussed earlier in this section).

The Place Active Cell tool places the active cell at the current Active Angle, X Scale, and Y Scale. Change the angle and scale any time while you are using the tool, and the next cell copy you place will use your new angle and scale settings. This feature allows you to place the active cell rotated and scaled all in one operation.

Change the angle and scale by keying-in the values in the Tool Settings box. In addition to the edit boxes provided in the Tool Settings window when you invoke the Place Active Cell tool, two toggle buttons are provided. One is for relative placement of the cell, the other for interactive placement of the cell.

Placing Graphic Cell in Absolute or Relative Mode

MicroStation allows you to place graphic cells in Absolute or Relative mode when you use the Place Active Cell or Select and Place Active Cell tools. (The Select and Place Active Cell tool is explained later in this chapter.) The default placement mode for each tool is Absolute. To place cells in Relative mode, turn ON the Relative toggle button in the Tool Settings window.

In **Absolute** placement mode, the graphic cell's elements are placed on the levels on which they were drawn originally, regardless of the active level setting. For example, if a graphic cell contains elements on levels 1 and 3, they are placed in the design file on levels 1 and 3 (see Figure 11–13a).

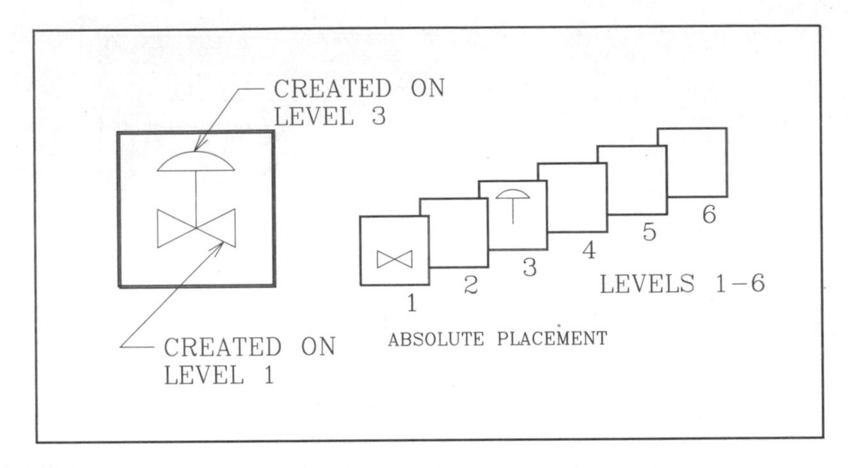

Figure 11–13a Example of placing a cell by Absolute placement mode

In **Relative** placement mode, the graphic cell element on the lowest level is placed on the active level and all other cell levels are shifted by the same amount as the lowest level.

For example, if the graphic cell with elements on levels 1 and 3 is placed in Relative mode and the active level is 4, the cell elements are placed as follows:

- Cell elements on level 1 are moved to level 4.
- Cell elements on level 3 are moved to level 6 (4 + 2 = 6) (see Figure 11–13b).

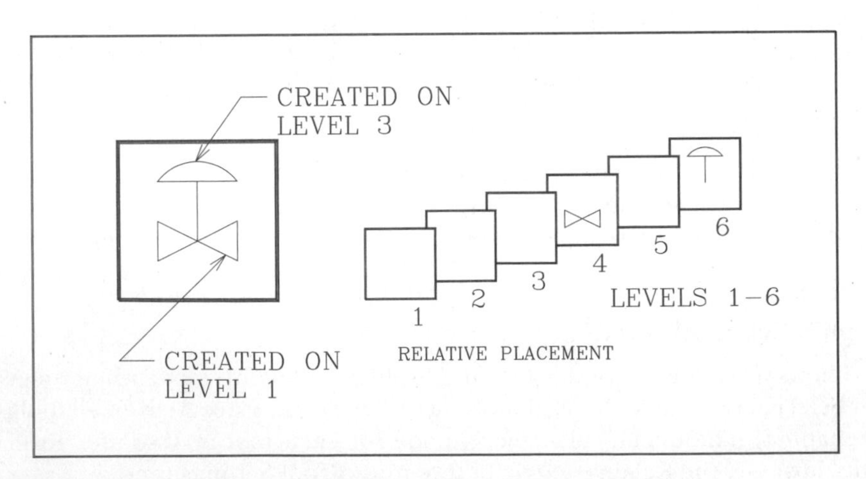

Figure 11–13b Example of placing a cell by Relative placement mode

If you want to place a graphic cell by Relative mode, make sure the toggle button is turned ON for Relative mode.

Interactive Placement

Interactive placement is helpful when you need to align your cell with existing elements but you don't know what angle and scale to make it. You define the angle and scale graphically as you place the cell.

To place the active cell interactively, turn ON the Interactive toggle button in the Tool Settings window. When a cell is placed interactively, MicroStation prompts:

> Place Active Cell (Interactive) > Enter cell origin *(Define the origin point of the cell.)*
>
> Place Active Cell (Interactive) > Enter scale or corner point *(Either place a data point to scale the cell or key-in the scale factor in the key-in window.)*
>
> Place Active Cell (Interactive) > Enter rotation angle or point *(Either place a data point to rotate the cell or key-in the rotation angle in the key-in window.)*

Select and Place Cell

Often, when working in an existing design, you will need to place additional copies of a cell that was placed earlier but is no longer your active placement cell. The Select and Place Cell tool allows you to select that cell for placement simply by clicking on a copy of the cell in the design plane. The cell you click becomes the active placement cell, and dynamic update shows it at the screen pointer position. Additional data points place copies of the cell.

Invoke the Select and Place Cell tool:

Cells tool box	Select the Select and Place Cell tool (see Figure 11–14).
Key-in window	**select cell icon** (or **se c i**) Enter

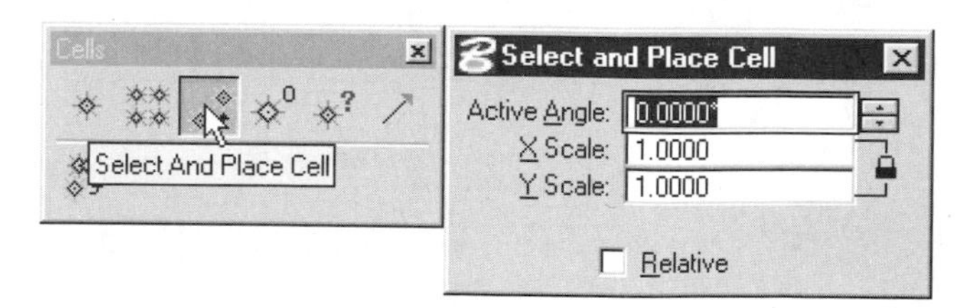

Figure 11–14 Invoking the Select and Place Cell tool from the Cells tool box

MicroStation prompts:

> Select and Place Cell > Identify element *(Select the cell to be copied.)*
>
> Select and Place Cell > Accept/Reject (Select next input) *(Accept the cell by placing a data point at the location where the first copy is to be placed.)*
>
> Select and Place Cell > Enter cell origin *(Place a data point at the location of each additional copy of the cell.)*

The Select and Place Cell tool places the active cell at the active angle and scale, just as the Place Active Cell tool did. Set the angle and scale in the Tool Settings window.

Note: Similar to the Place Active Cell tool, you can place the cell in Relative mode by toggling the Relative button to ON.

When you click on a cell in your design file, MicroStation checks to make sure it is actually a cell. When you click the second time to indicate where you want to place a copy of the cell, MicroStation checks to see if the cell is in the attached cell library. If the cell is not in the attached library, MicroStation displays the message *Cell not found* in the Status bar and waits for you to select another cell.

Place Active Cell Matrix

Do you need to place two rectangular rows of electronic components in a circuit diagram? The Place Active Cell Matrix tool can do it for you. With this tool, you can place the active cell quickly in a rectangular matrix whose parameters you define (see Figure 11–15).

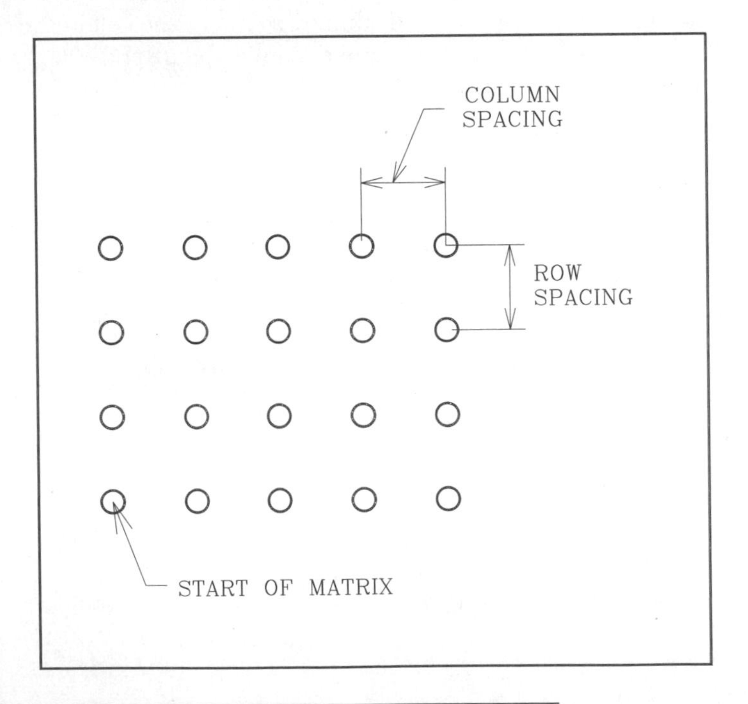

Figure 11–15 Example of placing a cell in a rectangular array

The Place Active Cell Matrix tool requires five parameters:

- Active Cell—The active cell name

- Rows—The number of rows in the matrix
- Columns—The number of columns in the matrix
- Row Spacing—The space, in Working Units, between the rows
- Column Spacing—The space, in Working Units, between the columns

The row and column spacing is from origin point to origin point. It is not the distance between the cells.

Invoke the Place Active Cell Matrix tool:

Cells tool box	Select the Place Active Cell Matrix tool (see Figure 11–16), then set the matrix parameters in the Tool Settings window.
Key-in window	**matrix cell** (or **matr c**) Enter

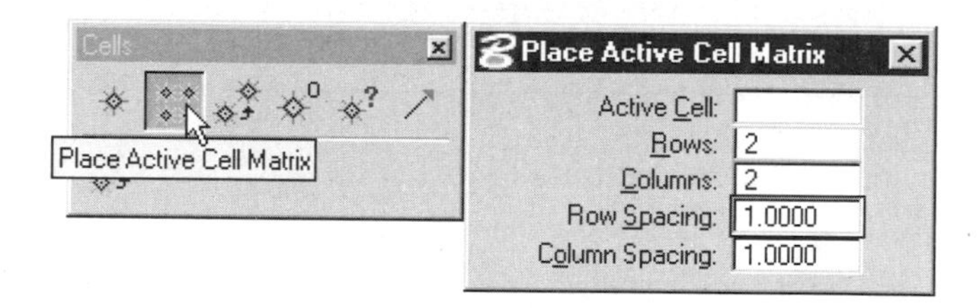

Figure 11–16 Invoking the Place Active Cell Matrix tool from the Cells tool box

MicroStation prompts:

Place Active Cell Matrix > Enter lower left corner of matrix *(Place a data point to define the origin location of the cell in the lower left corner.)*

The data point you place to start the matrix designates the position of the origin of the lower left cell in the matrix (see Figure 11–15).

The Place Active Cell Matrix tool places each cell in the matrix at the active angle and active scale. The edit fields available in the Tool Settings window for this tool do not include angle and scale fields. You have to use the settings boxes available from the Settings pull-down menu or key-in the active angle and active scale.

Place Active Line Terminator

There will be occasions when you want to place a cell (such as an arrowhead) at the end of a line and have it rotated to match the line's rotation. Invoking the Place Active Line Terminator tool can do this. It places the active terminator cell at the end of the element you select and automatically rotates it to match the rotation of the element at the point of connection. See Figure 11–17 for examples of placing line terminators.

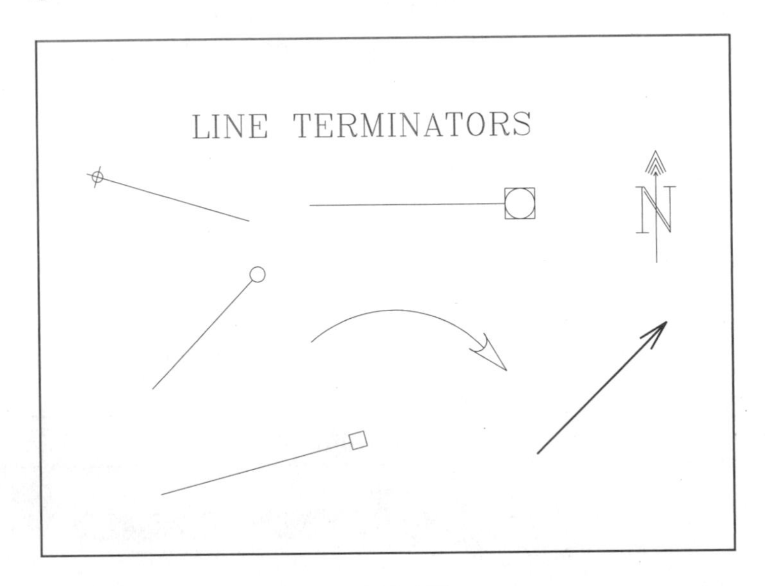

Figure 11–17 Examples of placing line terminators

Invoke the Place Active Line Terminator tool:

Cells tool box	Select the Place Active Line Terminator tool (see Figure 11–18), and, optionally, enter the Terminator cell name in the Tool Settings window.
Key-in window	**place terminator** (or **pl t**) Enter

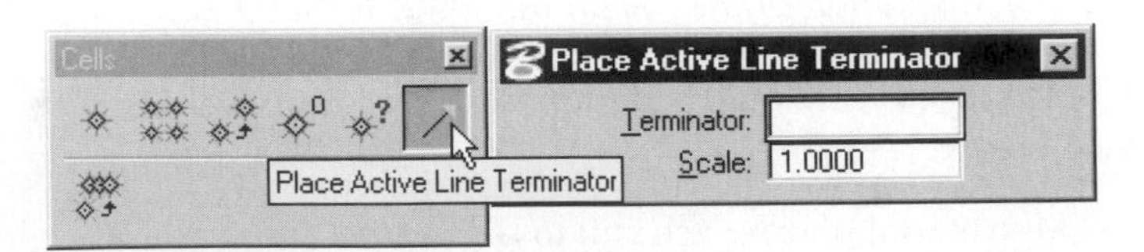

Figure 11–18 Invoking the Place Active Line Terminator tool from the Cells tool box

MicroStation prompts:

Place Active Line Terminator > Identify element *(Select the element near the end where the terminator is to be placed.)*

Place Active Line Terminator > Accept/Reject (Select next input) *(Click the Data button again to place the terminator, and, optionally, select another element to place a terminator.)*

If the second data point does not identify another element, MicroStation displays the message *Element not found* in the Status bar. Ignore the message.

Notes: If you do not have an active cell and invoke the Place Active Line Terminator tool, MicroStation displays the message *No Active Cell* in the Status bar.

Beginning users of MicroStation often select the element to place a terminator by first snapping to the element with the Tentative button. That is not necessary, because MicroStation finds the end of the element automatically. Just place a data point near the end of the element on which you want to place the terminator.

Terminator Scale

The Place Active Line Terminator tool has its own Scale factor. If the scale factor is:

- Greater than 1, the cell is scaled up
- Equal to 1, the cell is placed at its true size
- Less than 1, the cell is scaled down

The scale is set in the Tool Settings window. You can change the terminator scale as often as necessary while placing terminator cells. Each time you change it, the next terminator cell you place is scaled to your new factor.

Note: The Place Active Line Terminator tool does not use the active angle. It rotates the cell as necessary to match the rotation of the element it terminates.

Point Placement

MicroStation provides six point tools that place a dot, character, or cell in your design file.

- Place Active Point—Places a single point at the data point you specify.
- Construct Points Between Data Points—Places a set of equally spaced points between two data points.
- Project Point Onto Element—Places a point on an element at the point on the element nearest to a data point.
- Construct Point at Intersection—Places a point at the intersection of two elements.

- Construct Points Along Element—Places a set of equally spaced points along an element between two data points on the element.
- Construct Point at @Dist Along Element—Places one point at a predefined distance along an element from a data point.

Types of Points

Before you use any of the point tools, you must set the type of point you want to place. An options menu allows you to select the type of point; and two edit fields allow you to supply required information, as shown in Figure 11–19. The available point types are as follows.

- *Element*—Places dots (0-length lines). If you want to make these dots more noticeable, increase the active line weight before placing them.
- *Character*—Places a text character in the currently active font. If a symbol font is active, the character point will be a symbol. Font characters are placed at the active angle of rotation. Key-in the appropriate character in the Character edit field located in the Tool Settings window.
- *Cell*—Places the currently active point cell as the point. Point cells are placed at the active angle and active scale. Key-in the appropriate cell name in the Cell edit field located in the Tool Settings window.

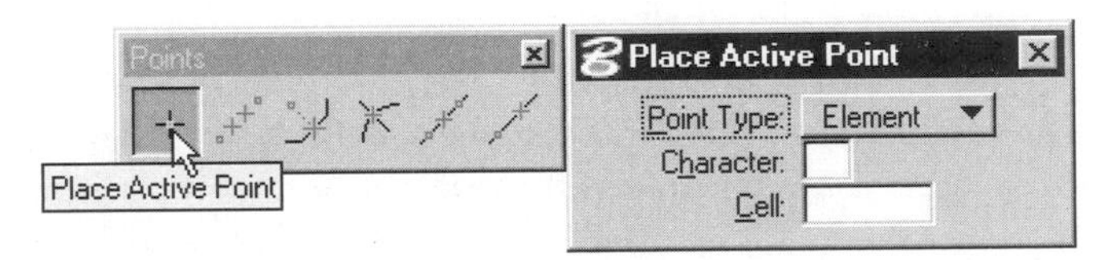

Figure 11–19 Invoking the Place Active Point tool from the Points tool box

You can also select a cell as the active point from the Cell Library settings box. To do so:

1. Open the Cell Library settings box from the pull-down menu Element.
2. Find the cell you want in the cell list and highlight it by clicking with the Data button.
3. Click on the Point button (located on the bottom left of the settings box).

Note: Don't confuse using a cell as the active point with point and graphic cells. You can make either a point or a graphic cell the active point cell. The Point tools place graphic cells in the Absolute mode only.

Place Active Point

Invoke the Place Active Point tool:

Points tool box	Select the Place Active Point tool (see Figure 11–19).
Key-in window	**place point** (or **pl po**) Enter

MicroStation prompts:

Place Active Point > Enter point origin *(Place a data point.)*

You can place any number of points in one tool sequence. Press the Reset button to terminate the tool sequence.

Construct Points Between Data Points

The Construct Points Between Data Points tool will place a specified number of points (element, character, or cell) between two data points. The number of points placed includes the two placed on your data points.

Invoke the Construct Points Between Data Points tool:

Points tool box	Select the Construct Points Between Data Points tool (see Figure 11–20), then in the Tool Settings window, enter the number of Points.
Key-in window	**construct points between** (or **constru po b**) Enter

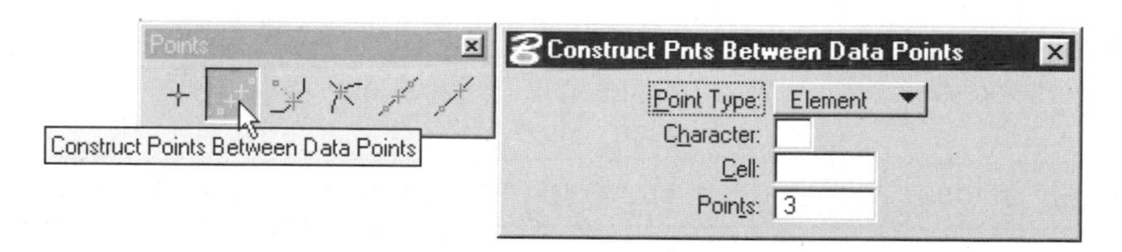

Figure 11–20 Invoking the Construct Points Between Data Points tool from the Points tool box

MicroStation prompts:

Construct Pnts Between Data Points > Enter first point *(Place a data point to define the location of the first point in the series.)*

Construct Pnts Between Data Points > Enter endpoint *(Place a data point to define the location of the last point in the series.)*

After you place the first set of points, you can continue placing additional sets. Each set will use the last data point of the previous set as its starting point. To start over with a first data point again, click the Reset button on your pointing device. See Figure 11–21 for an example of placing 10 points between two data points.

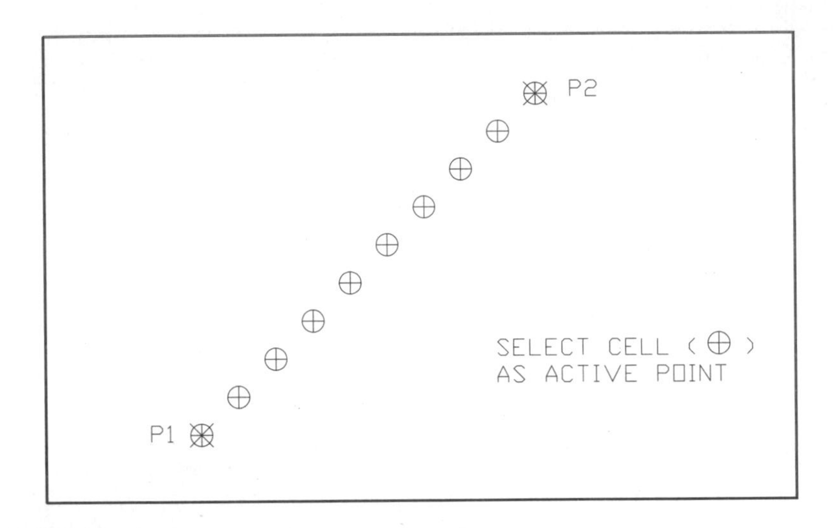

Figure 11–21 Example of placing 10 points between two data points

Project Point Onto Element

The Project Point Onto Element tool places the active point (element, character, or cell) on the selected element at the point projected from the acceptance data point.

Invoke the Project Point Onto Element tool:

Points tool box	Select the Project Point Onto Element tool (see Figure 11–22).
Key-in window	**construct point project** (or **constru po p**) Enter

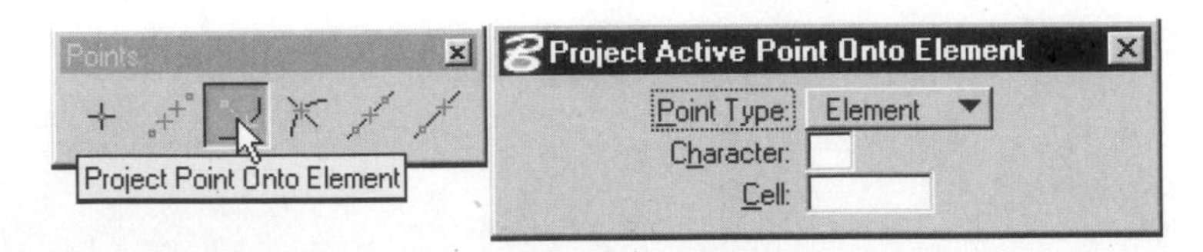

Figure 11–22 Invoking the Project Point Onto Element tool from the Points tool box

MicroStation prompts:

> Construct Active Point Onto Element > Identify element *(Select the element.)*
>
> Construct Active Point Onto Element > Accept/Reject (Select next input) *(Select the point in the design from which to project the point, and, optionally, select the next element to which a point is to be projected.)*

If your accept Data button does not identify another element, MicroStation displays the message *Element not found* in the Status bar. Ignore the message. See Figure 11–23 for an example of how to place an active point projected onto an element.

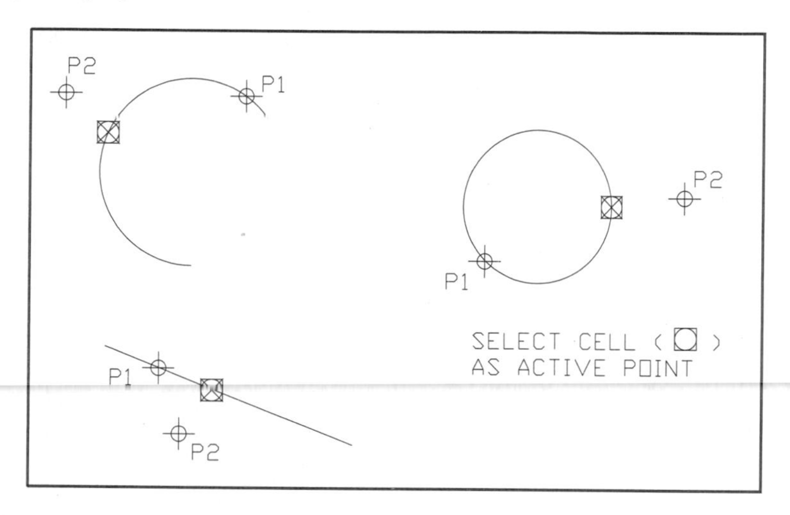

Figure 11–23 Example of placing an active point projected onto an element

Note: The only purpose of the first data point is to identify the element to which the projection will be made. There is no need to use the Tentative button when identifying the element. The second data point is the one that may need precise placement.

Construct Point at Intersection

The Construct Point at Intersection tool places an active point (element, character, or cell) at the intersection of two elements.

Invoke the Construct Point at Intersection tool:

Points tool box	Select the Construct Point at Intersection tool (see Figure 11–24).
Key-in window	**construct point intersection** (or **constru po i**) Enter

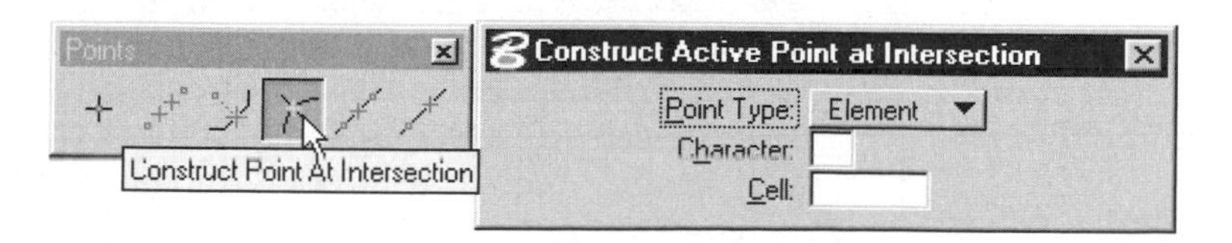

Figure 11–24 Invoking the Construct Point at Intersection tool from the Points tool box

MicroStation prompts:

> Construct Active Point at Intersection > Select element for intersection *(Select one of the elements.)*
>
> Construct Active Point at Intersection > Select element for intersection *(Select the other element.)*
>
> Construct Active Point at Intersection > Accept - Initiate intersection *(Place a data point to place the point.)*

The acceptance data point only initiates placement of the point; it does not identify another element. See Figure 11–25 for an example of how to place an active point at an intersection of two elements.

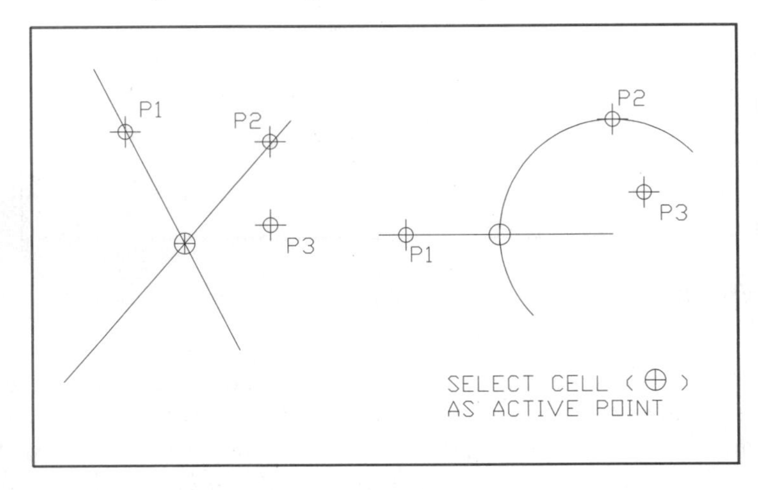

Figure 11–25 Example of placing an active point at an intersection of two elements

Note: If the two elements intersect more than once (such as a line passing through a circle), identify the two elements close to the intersection where you want the point to be placed. There is no need to use the Tentative button. Just place the Data button close to the intersection.

Construct Points Along Element

The Construct Points Along Element tool places a set of active points (elements, characters, or cells) equally spaced along an element between two data points on the element.

Invoke the Construct Points Along Element tool:

Points tool box	Select the Construct Points Along Element tool (see Figure 11–26), then in the Tool Settings window, enter the number of Points.
Key-in window	**construct point along** (or **constru po a**) Enter

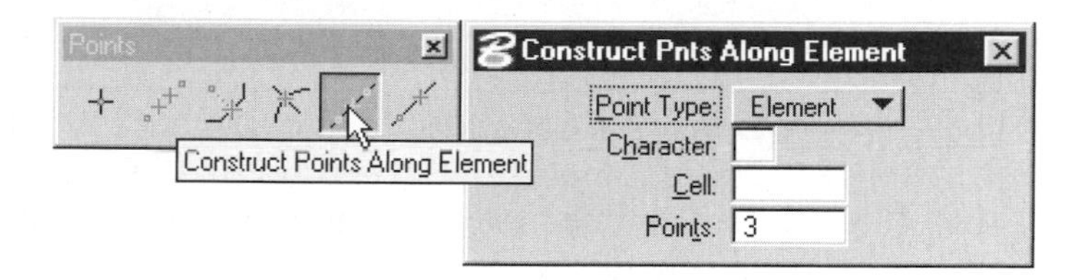

Figure 11–26 Invoking the Construct Points Along Element tool from the Points tool box

MicroStation prompts:

> Construct Pnts Along Element > Enter first point *(Select the element at the location where the first point is to be placed.)*
>
> Construct Pnts Along Element > Enter endpoint *(Define the location on the element where the last point is to be placed, and the points are placed.)*

You can continue placing points along elements, and, if necessary, you can also change the active point at any time while placing them. See Figure 11–27 for an example of how to place 10 points along an element between two data points.

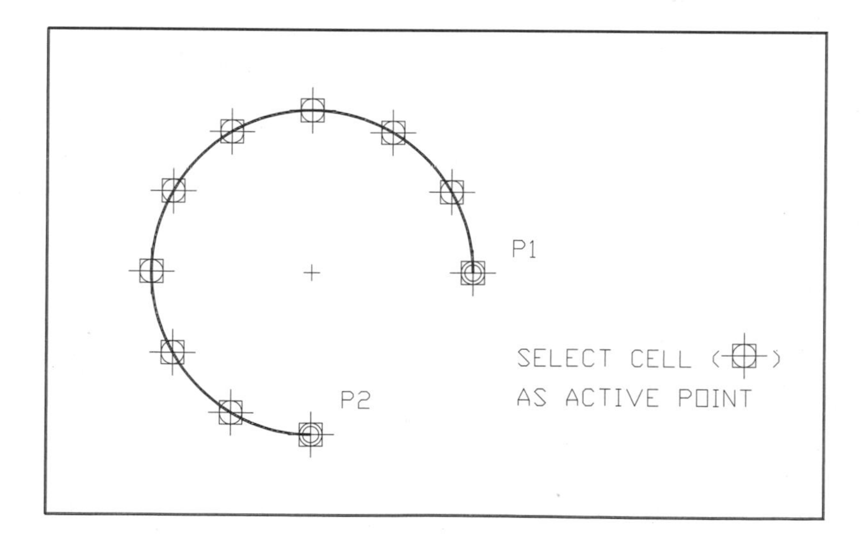

Figure 11–27 Example of placing 10 points along an element between two data points

Construct Point at @ Distance Along Element

The Construct Point at @ Distance Along Element tool places the active point (element, character, or cell) at a keyed-in distance along an element from the data point that identified the element.

Invoke the Construct Point @ Distance Along Element tool:

Points tool box	Select the Construct Point @ Distance Along Element tool (see Figure 11–28), then in the Tool Settings window, enter the Distance in working units.
Key-in window	**construct point distance** (or **constru po d**) Enter

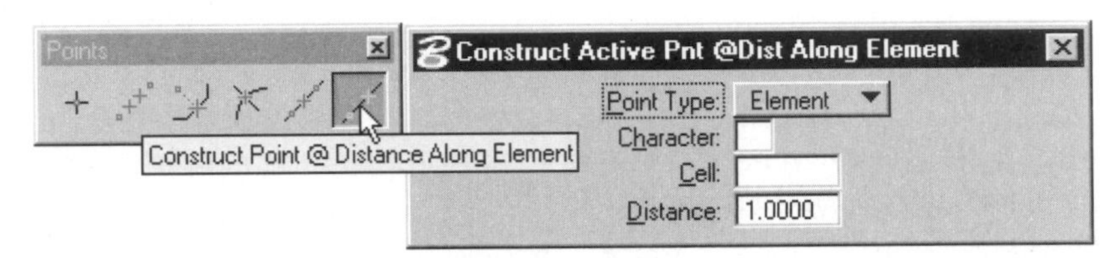

Figure 11–28 Invoking the Construct Point at @ Distance Along Element tool from the Points tool box

MicroStation prompts:

> Construct Active Pnt @Dist Along Element > Identify element *(Identify the element.)*
> Construct Active Pnt @Dist Along Element > Accept/Reject (Select next input) *(Place a data point to accept the construction, and, optionally, select the next element for construction.)*

You can continue selecting elements along which you want to place a point and you can change the distance at any time while using the tool. See Figure 11–29 for an example of how to place a point at a specified distance along an element.

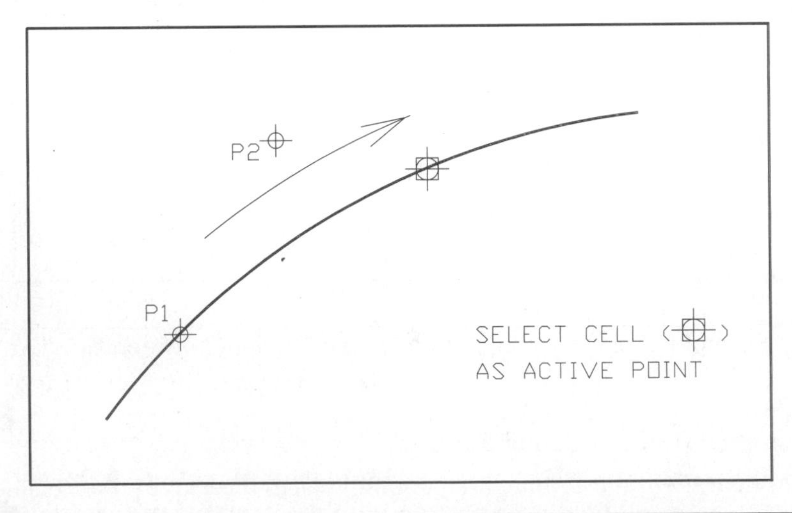

Figure 11–29 Example of placing a point at a specified distance along an element

CELL SELECTOR

As mentioned earlier, MicroStation allows you to place cells by invoking one of the available Place Cell tools and selecting the appropriate cell from the Cell Library settings box, or by typing the name of the cell in the Tool Settings window. In addition, you can also place cells by selecting the appropriate cell from the Cell Selector dialog box. The Cell Selector dialog box shows thumbnail representations of the cells stored in a cell library. The Cell Selector contains copies of the cells from the library.

Opening the Cell Selector Settings Box

Open the Cell Selector settings box:

Pull-down menu	Utilities > Cell Selector

MicroStation displays a Cell Selector settings box, as shown in Figure 11–30. If a cell library is already attached to the current design file, MicroStation displays thumbnail representations of the cells stored in the library, as shown in Figure 11–30.

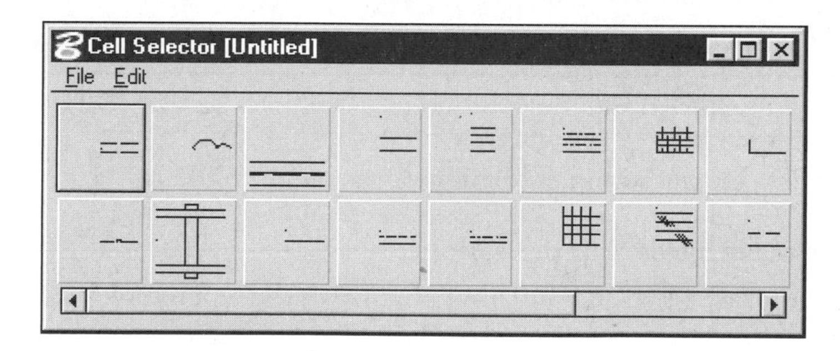

Figure 11–30 Cell Selector settings box

If you have not already attached a cell library to the current design file when you open the Cell Selector settings box, MicroStation opens the Select Cell Library dialog box. Select the cell library file you want to attach to the current design file from the Files list box, and click the OK button to close the dialog box. MicroStation displays thumbnail representations of the cells stored in the attached library.

Clicking a cell thumbnail button in the Cell Selector settings box has the effect of activating the associated cell and selecting the Place Active Cell tool.

Customizing the Button Configuration

The settings box is resizable and can be customized with buttons for cells from different cell libraries. Each button can be set to display either the associated cell thumbnail representation or the cell name, and it can be associated with a key-in, which is activated when the button is clicked.

MicroStation allows you to save the button configuration to a cell selector file with a file extension of .CSF. The cell selector files contain each button's configuration information and the size and number of buttons in the configuration.

By default, MicroStation loads the button configuration set by the system variable MS_CELLSELECTOR. If the system variable is not defined, then the cells from the attached library are loaded in the Cell Selector settings box. If neither condition is met, MicroStation displays a dialog box for selecting a cell library when you open the Cell Selector settings box.

Assigning a File Name

To assign a file name to the current button configuration, invoke the New tool:

Pull-down menu from Cell Selector settings box	File > New

MicroStation displays the Define Cell Selector File dialog box. Key-in the name of the file to assign a file name to the current button configuration.

Saving the Current Configuration

To save the current button configuration, invoke the Save tool:

Pull-down menu from Cell Selector settings box	File > Save

MicroStation saves the current button configuration to the given file name.

Opening the Previously Saved Configuration

To open the previously saved button configuration, invoke the Open tool:

Pull-down menu from Cell Selector settings box	File > Open...

MicroStation displays the Select Cell Selector File dialog box. Select the appropriate file name from the Files list box to open the previously saved cell selector configuration file.

Saving the Current Configuration to a New Cell Selector File

To save the current button configuration to a new cell selector file, invoke the Save As tool:

Pull-down menu from Cell Selector settings box	File > Save As...

MicroStation displays the Save Cell Selector File dialog box. Key-in the name of the file to save the current button configuration.

Adding Buttons from Other Cell Libraries

As mentioned earlier, the Cell Selector settings box displays thumbnail representations of the cells stored in a cell library. MicroStation allows you to open additional cell libraries from the Cell

Selector settings box and load the buttons with additional cells. In other words, MicroStation allows you to display the cells from various cell libraries in the Cell Selector settings box, and you can select any of the cell buttons to place a cell in your current design file. To load additional cell libraries, invoke the Load Cell Library tool:

Pull-down menu from Cell Selector settings box	File > Load Cell Library...

MicroStation displays the Select Cell Library to Load dialog box. Select the appropriate file name from the Files list box to load the cell library into the current button configuration.

Editing Button Settings

To edit one of the button settings, click the appropriate button in the Cell Selector settings box, and then invoke the button tool:

Pull-down menu from Cell Selector settings box	Edit > Button

MicroStation displays the Configure Cell Selector Button settings box, as shown in Figure 11–31.

Note: You can also edit the button configuration by double-clicking the button in the Cell Selector settings box. MicroStation displays the Configure Cell Selector Button settings box.

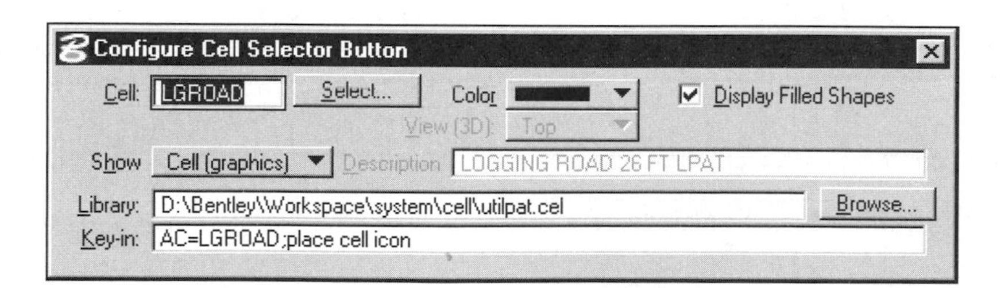

Figure 11–31 Configure Cell Selector Button settings box

- The Cell edit field displays the cell name of the selected button. If necessary, you can change the cell name by selecting a different name from the Select Cell settings dialog box displayed when you click the Select... button.
- The Color option menu allows you to select a different color for the display of the thumbnail representation of the cell.
- The toggle button for the Show menu provides options that allow you to control the display on the button. You can select Cell Graphics to display a picture of the cell, Cell Name, Description, or Cell and Name, to display a picture of the cell in the button and the cell name below the button.

- The cell Description edit field is available only when the Description option is selected in the Show menu. It allows you to edit the cell description that appears in the button.
- The Display Filled Shapes toggle button controls the display of the filled shapes on the button.
- The Library edit field displays the name of the cell library associated with the selected button. You can change the cell library by selecting a different name from the Select Cell Library dialog box displayed when you click the Browse... button.
- The Key-in edit field shows the key-in associated with the cell to be inserted. You can enter multiple key-ins separated by a semi-colon (;). This field can contain up to 511 characters.

Inserting a New Button

To insert a new button after the button that has the focus, invoke the Insert tool:

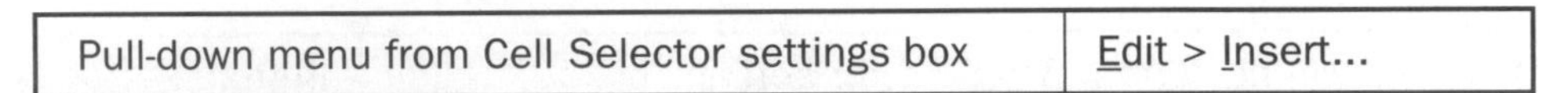

Pull-down menu from Cell Selector settings box	Edit > Insert...

MicroStation displays the Define Button settings box, as shown in Figure 11–32.

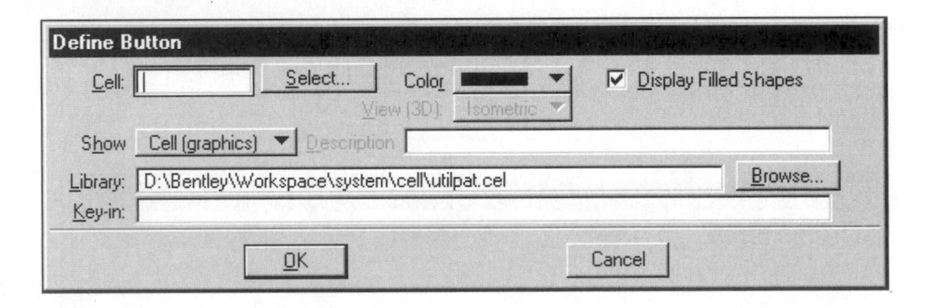

Figure 11–32 Define Button settings box

Key-in or select the appropriate cell library and the cell to define the selected button. Type the associated key-in in the key-in edit field. Click the OK button to configure the new button, or click the Cancel button to close the dialog box without configuring the new button.

Deleting Button Configurations

To delete the configuration of a single button, click the appropriate button in the Cell Selector settings box, and then invoke the Delete tool:

Pull-down menu from Cell Selector settings box	Edit > Delete

MicroStation deletes the configuration of the selected button.

To clear the configuration for all buttons, invoke the Clear Configuration tool:

Pull-down menu from Cell Selector settings box	Edit > Clear Configuration

MicroStation clears the configuration for all buttons in the Cell Selector settings box.

Changing Button and Gap Size

To change the size of the buttons in pixels, invoke the Button Size tool:

Pull-down menu from Cell Selector settings box	Edit > Button Size...

MicroStation displays the Define Button Size dialog box, as shown in Figure 11–33.

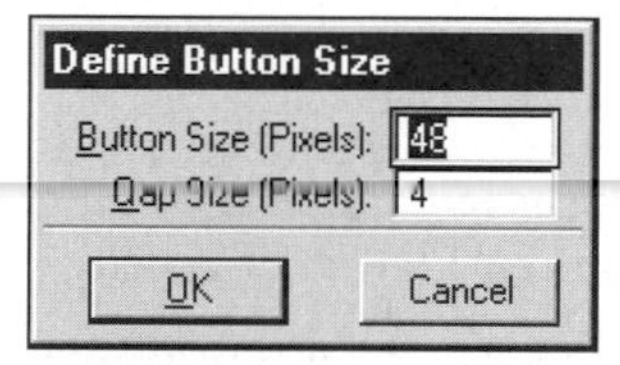

Figure 11–33 Define Button Size dialog box

The Button Size and Gap Size edit fields set the size (in pixels) of the button and of the gap between the buttons, respectively.

Standardizing New Button Configurations

MicroStation allows you to set the defaults for new button configuration. To set the defaults for new button configuration, invoke the Define Defaults tool:

Pull-down menu from Cell Selector settings box	Edit > Defaults

MicroStation displays the Define Defaults settings box, as shown in Figure 11–34.

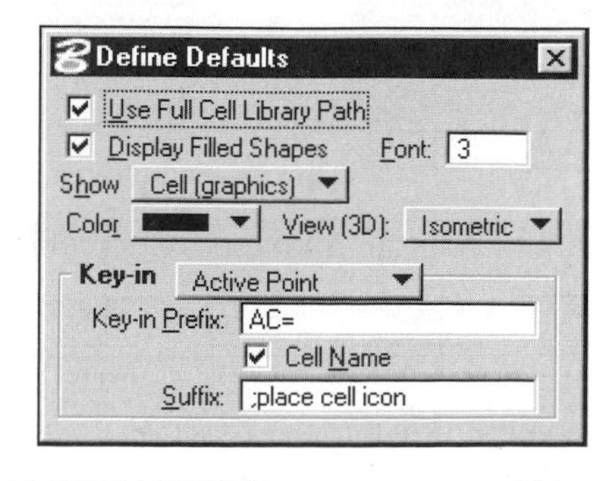

Figure 11–34 Define Defaults settings box

Set the appropriate default settings in the Define Defaults settings box and close the settings box. MicroStation uses all the settings for new button configurations.

CELL HOUSEKEEPING

You have seen several tools that place copies of cells in your design file. Now let's look at additional tools that affect cells already placed in your design file. The available tools include the following:

- Identify Cell—Displays the name and other related information of the selected cell.
- Replace Cell—Replaces a cell in the design file with another cell from the currently attached cell library.
- Drop Complex Status—Breaks up a cell into its individual elements. The elements lose their identities as cells.
- Drop Fence Contents—Breaks up all cells contained in a fence to their individual elements. The elements lose their identities as cells.
- Fast Cells View—Speeds up view updates by displaying only a box showing the location of all cells in the view, rather than the cell elements.

Identify Cell

The Identify Cell tool displays the name of a selected cell. It is useful when you want to use a cell already placed in your design file as a terminator cell or point cell but you don't know the cell's name. The information is displayed in the Status bar.

To determine the name of a cell in the active design, invoke the Identify Cell tool:

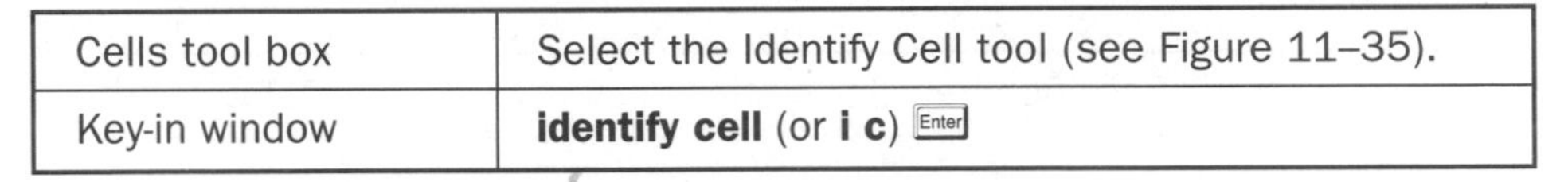

Cells tool box	Select the Identify Cell tool (see Figure 11–35).
Key-in window	**identify cell** (or **i c**) Enter

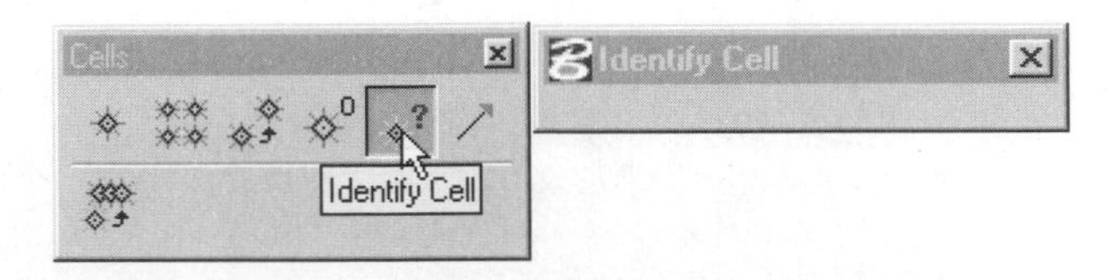

Figure 11–35 Invoking the Identify Cell tool from the Cells tool box

MicroStation prompts:

Identify Cell > Identify element *(Identify the cell.)*

Identify Cell > Accept/Reject (Select next input) *(Place a data point to accept the cell, and, optionally, to select the next cell.)*

The name of the cell appears in the Status bar after the second data point.

Replace Cells

The Replace Cells tool updates cells in the design file with the cell of the same name from the attached cell library, or replaces cells in the design with other cells from the library. It is useful when the design of a cell is changed and there are copies of the old cell in your design file, or when design changes require different cells (such as a different type of valve).

Update Cells

The Update Method replaces selected cells in the active design with cells of the same name that are in the attached cell library. Cells can be selected individually for updating, or groups of fenced cells can be updated. If you want to update a group of elements, place a fence around them before invoking this tool.

To update cells, invoke the Replace Cells tool:

Cells tool box	Select the Replace Cell tool, then in the Tool Settings window, select the Update Method, and, if you are updating a group of fenced cells, turn ON the Use Fence toggle button and set the required Fence Mode (see Figure 11–36).
Key-in window	**replace cells extended** (or **rep cells e**) Enter

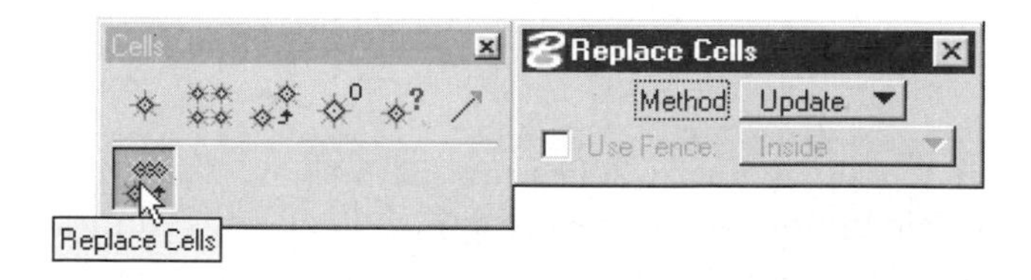

Figure 11–36 Invoking the Replace Cells tool and Update Method from the Cells tool box

If the Use Fence button is OFF, MicroStation prompts:

> Replace Cell > Identify Cell *(Identify the cell to be updated.)*
>
> Replace Cell > Accept/Reject *(Place a data point to initiate the update, and, optionally, to select the next cell.)*

The name of the cell appears in the Status bar after the first data point. The second data point also can identify another cell to be replaced.

If you need to replace a shared cell (discussed later in this chapter), identify one of them and MicroStation will replace all the instances of the shared cell with the same name.

If the Use Fence button is ON, MicroStation prompts:

Replace Cell > Accept/Reject Fence *(Click the data button anywhere in the design.)*

Note: The replaced cell may shift position in your design file. That happens when the new cell's origin point was not defined in the same relationship to the cell elements as the old cell's origin.

Replace Cells

The Replace Method replaces selected cells in the design with the active placement cell or a cell that you select in the design. You can replace a selected cell, a group of fenced cells, or all cells with the same name as the one you select. If you want to update a group of elements, place a fence around them before invoking this tool.

When you select the Replace Method, several options appear in the Tool Settings window (see Figure 11–37):

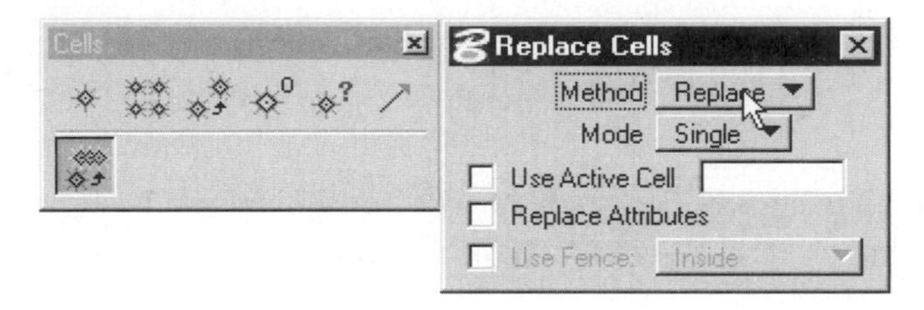

Figure 11–37 Invoking the Replace Cells tool and Replace Method from the Cells tool box

- The Mode menu allows you to select two methods of selecting cells for replacement. Select the Single option to replace the selected cell (or group of fenced cells). Select the Global option to replace all cells in the design that have the same name as the selected cell.
- Use the Active Cell toggle button and its edit field to select the replacement cell. If the button is ON, the cell in the edit field replaces the selected cell (you can type a cell name in the field). If the button is OFF, MicroStation prompts you to select a replacement cell from the design.
- The Replace Attributes toggle button determines how user data assigned to the cells (such as tags) will be handled. If the button is ON, the replacement cell's user data is assigned the replaced cell. If the button is OFF, the replaced cell's user data is retained.
- If a fence is defined in the design, the Use Fence Contents toggle button can be turned ON to replace groups of cells. If Single Mode is in effect, all cells grouped by the fence

are replaced. If Global Mode is in effect, all cells grouped by the fence are replaced and all ungrouped cells with the same name as grouped cells are replaced. The Fence Mode (Inside, Overlap, etc.) controls what cells are in the fence group.

To replace cells, invoke the Replace Cells tool:

Cells tool box	Select the Replace Cell tool, then select the desired options in the Tool Settings window.
Key-in window	**replace cells extended** (or **rep cells e**) Enter

If the Use Fence and Use Active Cell buttons are both OFF, MicroStation prompts:

Replace Cell > Identify Cell *(Identify the cell to be replaced.)*

Replace Cell > Identify Replacement Cell *(Identify the cell that will replace the cell you just selected.)*

Replace Cell > Accept/Reject Replacement Cell *(Click anywhere in the design to initiate the replacement.)*

If the Use Fence button is OFF and the Use Active Cell button is ON, MicroStation prompts:

Replace Cell > Identify Cell *(Identify the cell to be replaced.)*

Replace Cell > Accept/Reject *(Click anywhere in the design to initiate replacing the selected cell with the cell in the Use Active Cell edit field.)*

If the Use Fence button is ON and the Use Active Cell button is OFF, MicroStation prompts:

Replace Cell > Identify Replacement Cell *(Identify the cell that will replace the cells grouped by the fence.)*

Replace Cell > Accept/Reject Replacement Cell *(Click anywhere in the design to initiate the replacement.)*

If the Use Fence button is ON and the Use Active Cell button is ON, MicroStation prompts:

Replace Cell > Accept/Reject Fence *(Click anywhere in the design to initiate the replacement.)*

Drop Complex Status

The cells placed in your design file are "complex shapes" that act like one element when manipulated. If you need to change the shape of a cell in your design file, you must first "drop" the cell to break it into separate elements. The Drop Element tool does that for you. A dropped cell loses its identity as a cell and becomes separate, unrelated elements.

To drop a cell, first invoke the Drop Element tool:

Groups tool box	Select the Drop Element tool, and then turn ON the Complex toggle button (see Figure 11–38).
Key-in window	**drop element** (or **dr e**) Enter

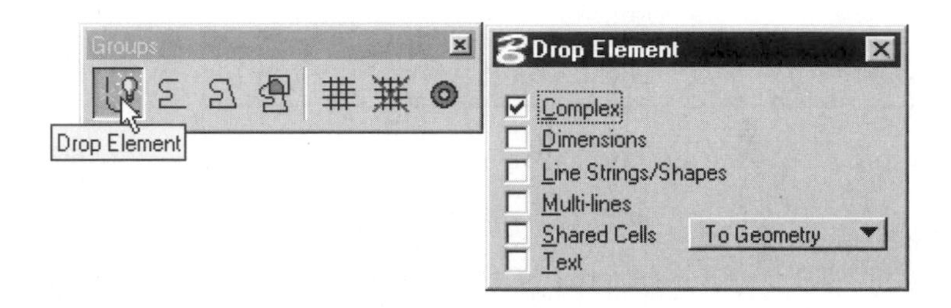

Figure 11–38 Invoking the Drop Element tool from the Groups tool box

MicroStation prompts:

Drop Complex Status > Identify element *(Identify the cell to be dropped.)*

Drop Complex Status > Accept/Reject (select next input) *(Place a data point to drop the cell, and, optionally, to select the next cell to be dropped.)*

Drop Fence Contents

The Drop Fence Contents tool breaks all the cells enclosed in a fence into separate elements. Before you select this tool, group the cells to be dropped by placing a fence and selecting the appropriate fence mode.

Invoke the Drop Fence Contents tool:

Fence tool box	Select the Drop Fence Contents tool (see Figure 11–39).
Key-in window	**fence drop complex** (or **f dr c**) Enter

Figure 11–39 Invoking the Drop Fence Contents tool from the Fence tool box

MicroStation prompts:

Drop Complex Status of Fence Contents > Accept/Reject fence contents *(Place a data point to drop all cells grouped by the fence.)*

Note: If you create a new cell that contains copies of existing cells, the new cell is "nested." Nested cells contain pointers to the other cells they contain, rather than the actual elements of the other cells. This saves space but has the potential to create problems later. If you delete a cell from the cell library, and that cell is nested in other cells, you also delete part of all the cells that refer to it.

Fast Cells View

Numerous cells in a view may cause view updates to be completed too slowly. If you see that happening, you can speed up updating by turning on the Fast Cells View attribute. When that attribute is ON for a view, a box is displayed at each cell location, rather than at the cell elements. The box is the same size as the cell's range.

To turn on the Fast Cells View attribute:

1. Open the View Attributes settings box from the Settings pull-down menu.
2. Turn ON the Fast Cells toggle button.
3. Check the View Number menu, and, if necessary, change it to the number of the view where you want the Fast Cells View attribute to be ON.
4. Click the Apply button to turn on the attribute for the selected view.
5. If you no longer need the View Attributes settings box, close it.

The Fast Cells View attribute stays ON for the selected view until you turn it OFF or exit from MicroStation. To make it permanent, select Save Settings from the File pull-down menu.

Note: Turn OFF Fast Cells View Attribute before you plot, if you want the cells to appear in the plot.

LIBRARY HOUSEKEEPING

You've seen how to place cells in your design file and take care of them. Now let's look at some housekeeping tools that help you take care of the cells in the attached cell library. The discussion includes explanations of how to do the following:

- Edit a cell's name and description
- Delete a cell from the Cell Library

- Compress the attached Cell Library
- Create a new version of a cell

All these tools affect the cells in the attached library, not the cells in your design file.

Edit a Cell's Name and Description

The Edit Cell Information dialog box, which is available from the Cell Library settings box, allows you to change a cell's name and description. If you need to create a new version of a cell, and you want to keep the old one around, rename it before creating the new version. If the person who designed the cell failed to provide a description, you can provide one to help other users of the cell library figure out what is in it. Following is the step-by-step procedure for renaming a cell and changing the cell description.

1. If the Cell Library settings box is not already open, select Cells from the Element pull-down menu to open it.
2. Select the cell you want to edit from the Cells list.
3. Click the Edit button to open the Edit Cell Information dialog box (see Figure 11–40).
4. Make the necessary changes to the cell name and description.
5. Click the Modify button to make the necessary changes and close the Edit Cell Information dialog box.

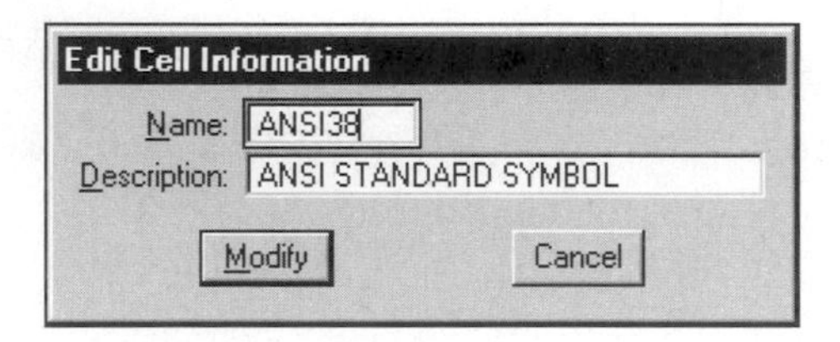

Figure 11–40 Edit Cell Information dialog box

Note: Changing the cell name and description in the library does not affect the cells already in the design file (or any other design file). They keep their old names.

Alternate Method

To rename a cell from the key-in window, type **CR=**<old>, <new>, and press Enter. Replace <old> with the cell's current name and <new> with the new cell name. The key-in only changes the cell names. You cannot use a key-in to change descriptions.

Delete a Cell from the Library

If a cell becomes obsolete, or if it was not drawn correctly, it can be deleted from the cell library using the Delete button in the Cell Library settings box. When you ask to delete a cell, MicroStation opens an Alert dialog box to ask you if you really want to delete the cell. Following are the step-by-step procedures for deleting a cell from the Cell Library.

1. If the Cell Library settings box is not already open, select Cells from the Element pull-down menu to open it.
2. Select the cell you want to delete from the Cells list.
3. Click the Delete button.
4. The Alert dialog box opens to ask you if you really want to delete the cell.
5. If you really want to delete it, click the OK button. If you selected the wrong cell or changed your mind, click the Cancel button.

Note: This procedure deletes a cell from the cell library. It does not delete copies of the cell already placed in the design file (or any other design file).

Alternate Method

To delete a cell from the key-in window, type **CD=<name>**, and press Enter. Replace <name> with the cell's name.

When you use the key-in to delete a cell, MicroStation does not open the Alert window to ask you if you really want to delete the cell.

Note: The Undo tool will not undo the deleting of a cell from the attached library.

Compress the Attached Cell Library

Deleting a cell from a cell library does not really delete it. The cell is marked as deleted and is no longer available, but its elements still take up space in the cell library file on the disk. To get rid of the no-longer-usable cell elements, you must compress the cell library with the Compress Library option in the Cell Library settings box's File pull-down menu. Following are the step-by-step procedures to compress the Cell Library.

1. If the Cell Library settings box is not already open, select Cells from the Element pull-down menu to open it.
2. From the File pull-down menu in the Cell Library settings box, select Compress.

The attached cell library is compressed, and the Status bar indicates that the Cell Library is compressed.

Create a New Version of a Cell

Occasionally the need will arise to replace a cell with an updated version of that cell. The geometric layout of the object represented by the cell may have changed, or you may have discovered that a mistake was made when the cell was originally drawn. The cell elements cannot be edited in the library. You must create a new cell and place it in the library.

Follow this procedure to replace a cell in the library with a new version of the cell:

1. Place a copy of the cell in a design file that has the same Working Units as the design file in which the cell was created.
2. Drop the cell.
3. Delete the old cell from the library, or rename it.
4. Make the required changes to the cell elements in the design.
5. If possible, place the cell origin at the same place as in the old cell.
6. Create the new cell from the modified elements.
7. If you deleted the old cell, compress the library.
8. If you had already placed copies of the cell in your design files, invoke the Replace Cell tool to update those copies with the new version of the cell.

SHARED CELLS

Thus far, each time you placed a cell, a separate copy of the cell's elements was placed in the design file. That uses up a lot of disk space in a design file that contains many cells.

Shared cells can help with the disk space problem. Each time you place a shared cell, the placement refers back to the shared copy rather than placing more elements in your design file. No matter how many copies of a shared cell you place, only one copy is actually in your design file.

When you declare a cell to be shared, MicroStation places a copy of it in your design file. You can have the shared cell's elements stored in your design file even though no copies of the cell have ever been placed in the design. Later, when you use the cell placement tools, the placements refer to the locally stored cell elements rather than to the copy in the library. Shared cells can be placed even when no cell library is attached.

Let's look how you can do the following:

- Turn on the shared cell feature
- Determine which cells are shared
- Declare a cell to be shared
- Place shared cells
- Turn a shared cell into an unshared cell
- Delete the shared cell copy from your design file

Turn on the Shared Cell Feature

The Use Shared Cells toggle button in the Cell Library settings box (see Figure 11–41) turns the shared cell feature ON or OFF. Each time you click the button, you toggle between using shared cells and not using them.

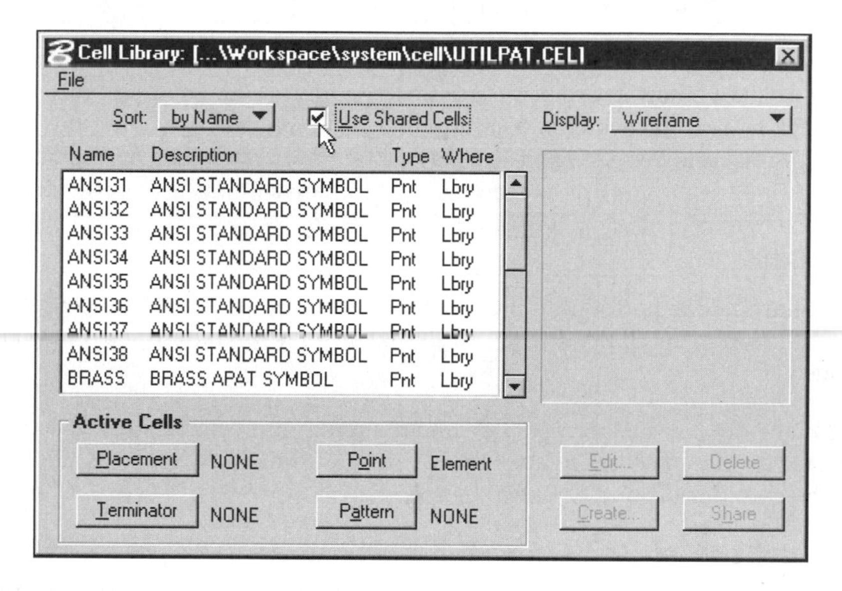

Figure 11–41 Displaying the status of the toggle button for the Use Shared Cells in the Cell Library settings box

Determine Which Cells Are Shared

When the shared cells feature is ON, all shared cells in your design file show up in the Cells List in the Cell Library settings box (see Figure 11–41). Shared cells are indicated by "Shrd" in the Where column.

If the shared cell is also in the currently attached cell library, the list shows the shared copy in your design file—not the one in the library.

Each time you select a shared cell in your design file, the Status bar will tell you it is a shared cell.

Declare a Cell to Be Shared

Use the Share button located in the bottom right in the Cell Library settings box to place copies of the shared cells in your design file. Following are the step-by-step procedures:

1. If the Cell Library settings box is not already open, select Cells from the Element pull down menu to open it.
2. If the shared cells feature is not currently active, click on the Use Shared Cells button to turn it ON.

3. Select the cell you want to be shared from the Cells List.
4. Click the Share button.

When you click the Share button, "Shrd" appears in the Where column of the selected cell, and a copy of the cell is placed in your current design file.

This procedure stores a shared copy of the cell in your design file but does not actually place the cell in the design. This is a handy way to create a seed file for a group of design files that will use the same set of cells. The procedure includes creating a new design file, attaching the cell library, and declaring all required cells shared. Then you can create new design files by copying the new seed file. Each copied file contains a copy of the elements that make up the shared cells.

Place Shared Cells

When the Use Shared Cells button in the Cell Library settings box is ON, all cell placement tools place shared cells. Placing a cell makes the cell shared, even if you have not declared the cell to be shared previously.

The tools that place the active placement, terminator, and point cells all work the same for shared cells as they do for unshared cells. The Replace Cell tool automatically replaces all the copies of the shared cell when you identify one of them.

Turn a Shared Cell Into an Unshared Cell

The Drop Element tool provides a Shared Cells toggle button for turning shared cells either into unshared cells or individual elements in the design. The menu associated with the Shared Cells toggle button controls the way the shared cell is dropped. The To Geometry option drops the cells all the way back to individual elements in the design, and the To Normal Cell option removes the shared status (each shared cell is replaced with a local copy of the cell).

Invoke the Drop Element tool:

Groups tool box	Select the Drop Element tool, then in the Tool Settings window, turn ON the Shared Cells toggle button, turn OFF the other toggle buttons, and select the way you want to drop the shared cells (see Figure 11–42).
Key-in window	**drop element** (or **dr e**) Enter

MicroStation prompts:

Drop Element > Identify element *(Select the shared cell to be converted.)*

Drop Element > Accept/Reject (select next input) *(Place a second data point to convert the selected cell, and, optionally, select the next cell to convert.)*

If no cell is selected with the second data point, the error message *Element not found* appears in the Status bar. Ignore it.

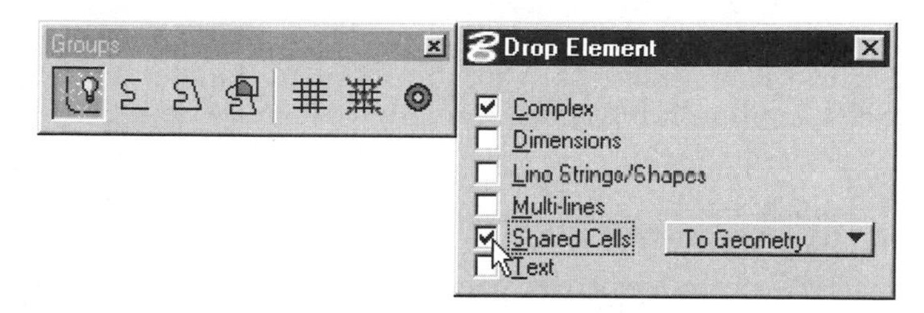

Figure 11–42 The Drop Element tool set to drop a shared cell

Delete the Shared Cell Copy from Your Design File

The placements of shared cells can be deleted like any other element, but the actual shared cell copy takes a little more work to delete. It is done in the Cell Library settings box via the Delete button.

Here is the procedure for deleting the shared cell elements from your design file.

1. Either delete all placements of the shared cell or make them into unshared cells (see earlier).
2. If the Cell Library settings box is not already open, select Cells from the Element pull-down menu to open it.
3. If the Use Shared Cells toggle button is OFF, turn it ON.
4. Select the shared cell you want to delete from the Cells List and make sure the cell has "Shrd" in the Where column.
5. Click the Delete button (located in the bottom right of the Cell Library settings box).
6. The Alert dialog box opens to ask you if you really want to delete the cell. Click the OK button to delete the shared cell, or the Cancel button to cancel the operation.

This procedure deletes only the copy of the shared cell elements in your design file. It does not delete anything from the attached cell library.

Note: If there are still any placements of the shared cell in your design file when you try to delete the shared cell, an Alert dialog box appears with a message stating that you cannot delete the cell. In this case the Alert dialog box's Cancel and OK buttons only close the box—click on either one of them.

DIMENSION-DRIVEN CELLS

MicroStation allows you to create a special type of a cell called a dimension-driven cell. When you place the dimension-driven cell, you can dynamically change relationships between the elements that were defined when it was created. For a detailed explanation of creating dimension-driven cells, refer to Chapter 14.

REVIEW QUESTIONS

Write your answers in the spaces provided.

1. Explain briefly the difference between a cell and cell library.
   ____
2. How many cell libraries can you attach at one time to a design file? ____________
3. List the steps involved in creating a cell.
   ____
   ____
4. What is the alternate key-in **AC=** used for? ____________
5. What is the file that has an extension of .CEL? ____________
6. What does it mean to place a cell Absolute?
   ____
7. What does it mean to place a cell Relative?
   ____
8. Explain briefly the differences between a graphic cell and a point cell.
   ____
   ____
9. How many cells can you store in a library? ____________
10. What is the purpose of defining an active cell as a terminator?
    ____
11. What is the purpose of turning ON the Fast Cells View attribute?
    ____
12. Explain briefly the benefits of declaring a cell as a shared cell.
    ____
    ____

PROJECT EXERCISE

This project exercise provides step-by-step instructions for creating the design shown in Figure P11–1. The intent is to guide you in creating and using cells.

Note: The dimensions included in several figures in this project are presented only as an aid to completing the design. They are *not* to be drawn.

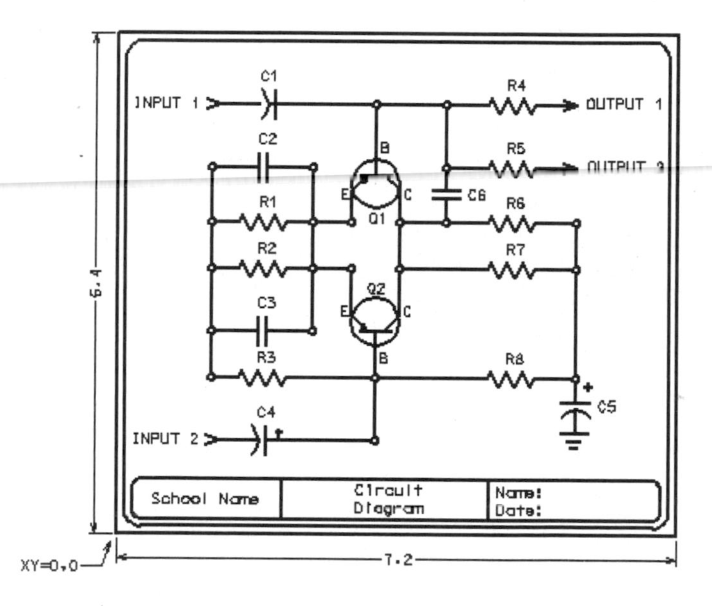

Figure P11–1 Completed Project design

Prepare the Design File

This procedure starts MicroStation, creates a design file, and enters the initial settings.

Note: As you complete each step in the project procedures, place a check mark by the step to help you keep up with where you are in the project.

STEP 1: Invoke MicroStation using the normal technique for the operating system on your workstation.

STEP 2: Create a new design file named CH11.DGN using the SEED2D.DGN seed file.

In the Design File dialog box set the:

- Working units Master Units name to IN (for inches), the Sub Units name to TH (for tenths of an inch), and the resolution ratios to 10 and 1000.
- Grid Master to 0.1, Grid Reference to 10, and Grid Lock to ON.

STEP 3: Set the Active Level to 1 and Line Weight to 1.

STEP 4: Invoke the Save Settings tool from the File pull-down menu.

Create the Arrowhead Cell

This procedure creates the arrowhead shown in Figure P11–2, creates a cell library, and creates a cell from a copy of the arrowhead elements.

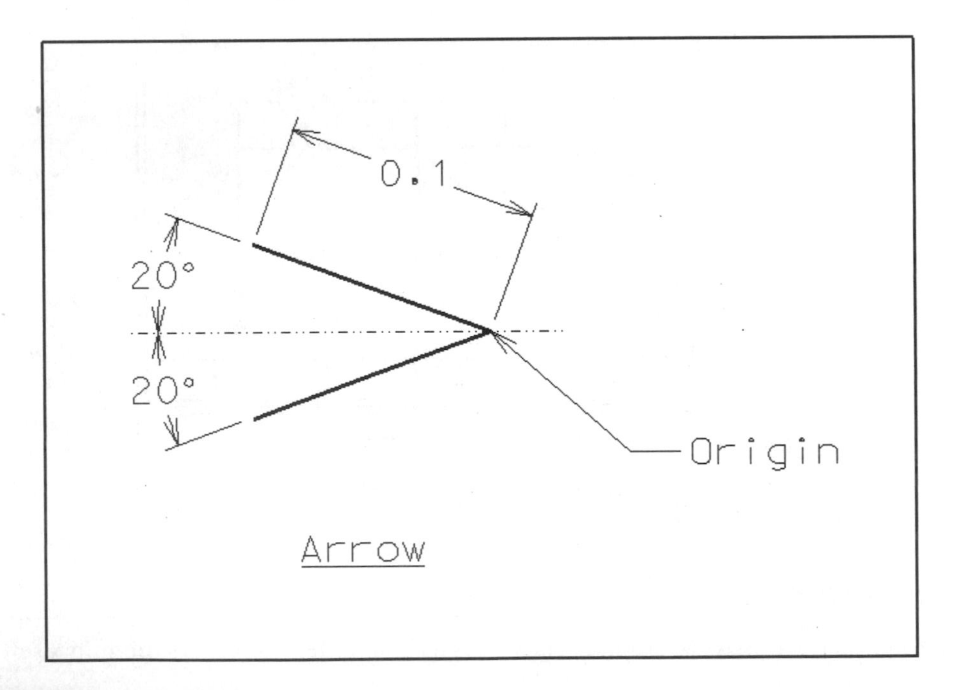

Figure P11–2 Arrowhead cell

STEP 1: Align a view window to display an area about 1 inch by 1 inch.

STEP 2: Invoke the Place Line tool from the Linear Elements tool box, and, in the Tool Settings window, set the Length to 0.1 inch, the Angle to 160, and the Length and Angle toggle buttons to ON.

MicroStation prompts:

> Place Line > Enter first point *(Place a data point to place the 0.1-inch line at 160 degrees rotation.)*
>
> Place Line > Enter first point *(In the Tool Settings window, change the Angle to 200 degrees, then place the line at the right end of the previous line.)*
>
> Place Line > Enter first point *(Click the Reset button, and, in the Tool Settings window, set the toggle buttons for the Length and Angle locks to OFF.)*

STEP 3: Place a Fence Block around the two lines that form the arrowhead.

STEP 4: Invoke the Define Cell Origin tool from the Cells tool box.

MicroStation prompts:

> Define Cell Origin > Define origin *(Place a data point on the arrow's point, as shown in Figure P11–2.)*

STEP 5: Open the Cells settings box by selecting the Cells from the Element pull-down menu. MicroStation displays the Cells settings box, as shown in Figure P11–3.

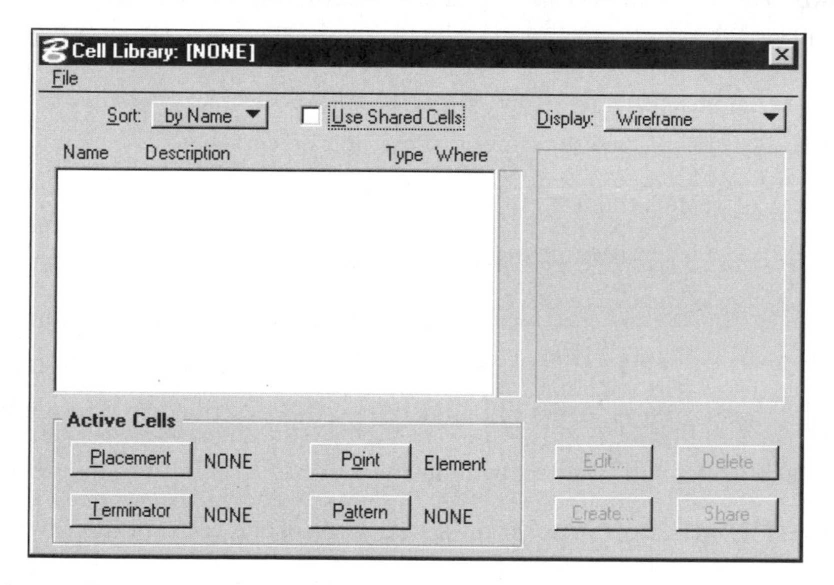

Figure P11–3 Cell Library settings box

STEP 6: Select the New option from the File pull-down menu in the Cell Library settings box. MicroStation opens the Create Cell Library dialog box.

STEP 7: In the Create Cell Library dialog box's Files window, key-in the file name **PROJ11**, and then click the OK button to create and attach the cell library.

STEP 8: In the Cell Library settings box, click the Create button to open the Create New Cell dialog box.

STEP 9: In the Create New Cell dialog box, do the following:

- Key-in **arrow** in the Name field.
- Key-in **Arrowhead symbol** in the Description field.
- Select the Point option menu.
- Click the Create button to create the cell.

STEP 10: Delete the two lines and remove the fence.

Create Additional Cells

This procedure draws the objects shown in Figure P11–4 and creates cells from them.

Note: Draw only the heavy-line-weight electronic symbols shown in Figure P11–4. Do *not* place the text or dimensions. The text and dimensions are shown only as an aid to creating the symbols.

STEP 1: Draw the resistor symbol (upper left object).

STEP 2: Create a cell from the resistor elements:

a. Place a Fence Block around the resistor.

b. Define a cell origin at the left end of the resistor, as indicated in Figure P11–4.

c. Create a Point cell named "res" and described as "Resistor Symbol."

STEP 3: Delete the resistor elements from the design file, and update the view.

STEP 4: Draw all elements in the PNP transistor symbol (bottom row, right symbol), except the arrowhead.

STEP 5: To place the arrowhead in the PNP transistor symbol, invoke the Place Active Line Terminator tool from the Cells tool box, and, in the Tool Settings window, key-in **arrow** in the Terminator field and **1** in the Scale field.

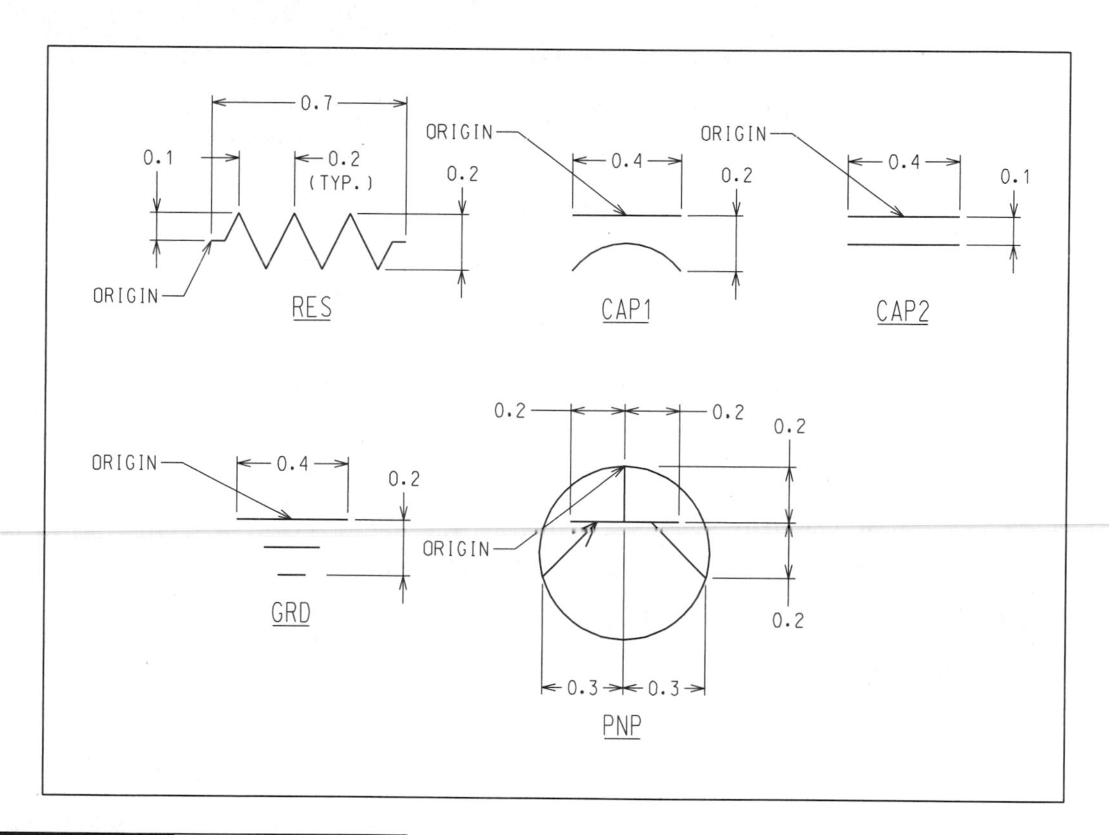

Figure P11–4 Electronic symbols

MicroStation prompts:

> Place Active Line Terminator > Identify element *(Identify the line close to the endpoint of the line where you want to place the arrow.)*
>
> Place Active Line Terminator > Accept/Reject (Select next input) *(Click the Data button in space to place the arrowhead on the line.)*

STEP 6: Create a point cell from the transistor elements using "PNP" as the cell name and "PNP Transistor symbol" for the description.

STEP 7: Draw and create point cells of the other three electronic symbols shown in Figure P11–4. Place the origin at the location indicated in the figure, and use the underlined text for the cell names. Here are the names and descriptions of each cell:

- CAP1 Capacitor symbol 1
- CAP2 Capacitor symbol 2
- GRD Ground symbol

STEP 8: Delete all elements from the design file.

Create the Circuit Diagram

This procedure uses the electronic symbol cells to create the diagram shown in Figure P11–1.

STEP 1: With Figure P11–1 as a guide, draw the border and title block on level 10.

- Replace “SCHOOL NAME” with your school or company name, or make up a name.
- Place your name to the right of “NAME.”
- Place today’s date to the right of “DATE.”

STEP 2: Set up the view window:

- Fit the view.
- Set the Active Level to 2.
- Set the Line Weight to 1 and the Color to green.

STEP 3: Invoke the Save Settings tool from the pull-down menu File to save the settings.

Note: The following steps place a transistor symbol cell, then draw outward from the transistor. There are several equally productive ways to draw the diagram.

STEP 4: Invoke the Place Active Cell tool from the Cells tool box, and, in the Tool Settings window, key-in **pnp** in the Active Cell edit field, **0** in the Active Angle edit field, and **1** in the X and Y Scale edit fields. Set the toggle buttons for Relative and Interactive to OFF.

MicroStation prompts:

Place Active Cell > Enter cell origin *(Place a data point to place the cell approximately at the same location as shown for the top PNP transistor symbol in Figure P11–1.)*

Place Active Cell > Enter cell origin *(Click the Reset button to terminate the tool sequence.)*

STEP 5: Use the Mirror Element About Horizontal (Copy) tool to create a mirror image copy of the just-placed PNP cell below the original, as shown in Figure P11–5.

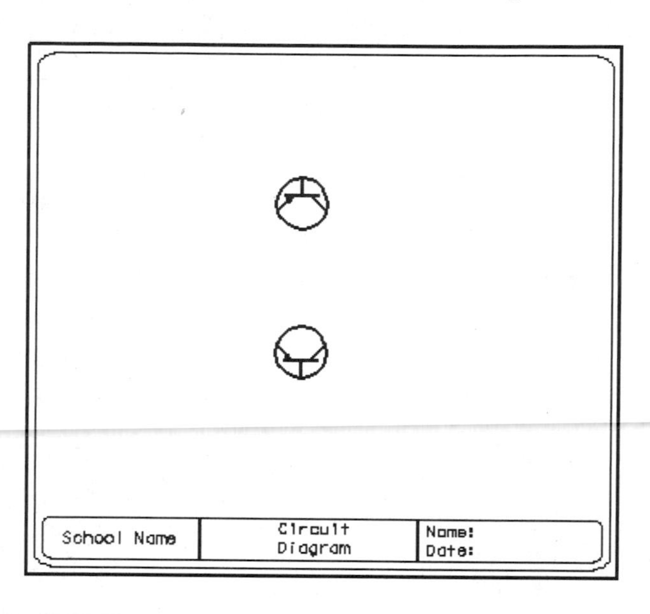

Figure P11–5 Two copies of the PNP cell

STEP 6: Continue drawing lines and placing cells by referring to Figure P11–1.

Note: Most of the CAP1 and CAP2 cells must be placed rotated. To rotate them, key-in **270** (or **–90**) in the Tool Setting window's Active Angle field before placing the cell.

STEP 7: Place circles on the line intersections with the Place Circle By Center tool, with the Radius field set to 0.05.

STEP 8: Remove the lines from inside the intersection circles with the IntelliTrim tool's Advanced Mode. Use the circles as the Cutting Elements.

STEP 9: Invoke the Save Settings tool from the pull-down menu File to save the settings.

DRAWING EXERCISES 11–1 THROUGH 11–5

Use the following table to set up the design files for all Chapter 11 exercises.

SETTING	VALUE
Seed File	SEED2D.DGN
Working Units	10 TH Per IN and 1000 Pos Units Per TH
Grid	Grid Master = 0.1, Grid Reference = 10, Grid Lock ON

Exercises 11–1 through 11–3

Create the piping flowsheet symbols shown in Figure E11–1, and make each into a cell. Use the cells to create the flow diagrams in Exercises 11–1 through 11–3.

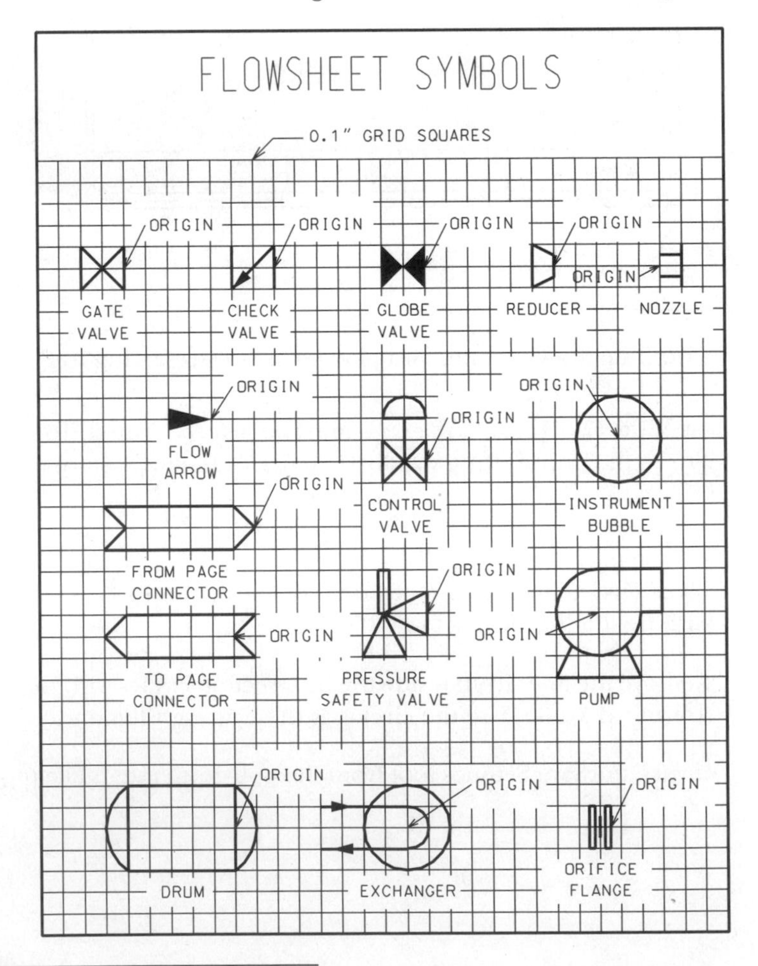

Figure E11–1 Piping flowsheet symbols

Exercise 11–1

Process flow diagram

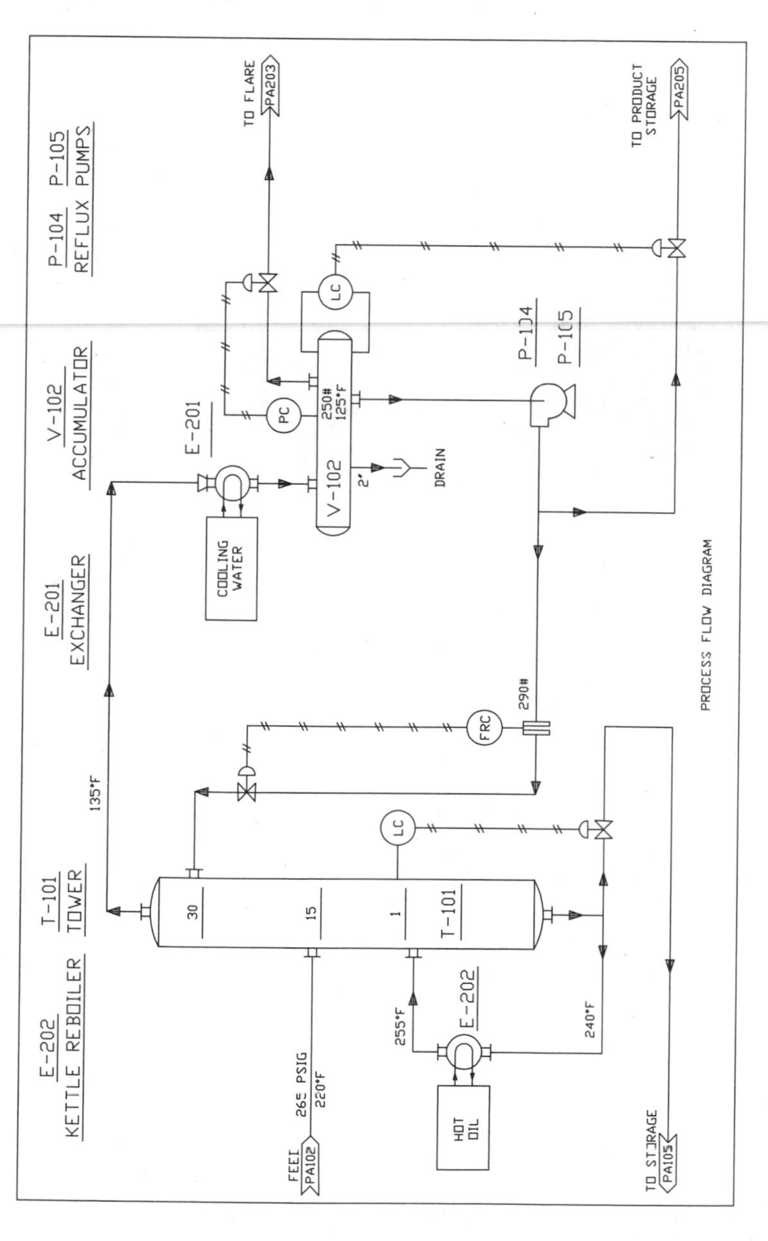

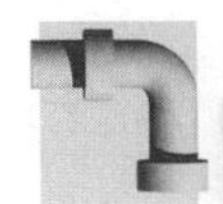

Exercise 11–2

Reactor and exchanger flow diagram

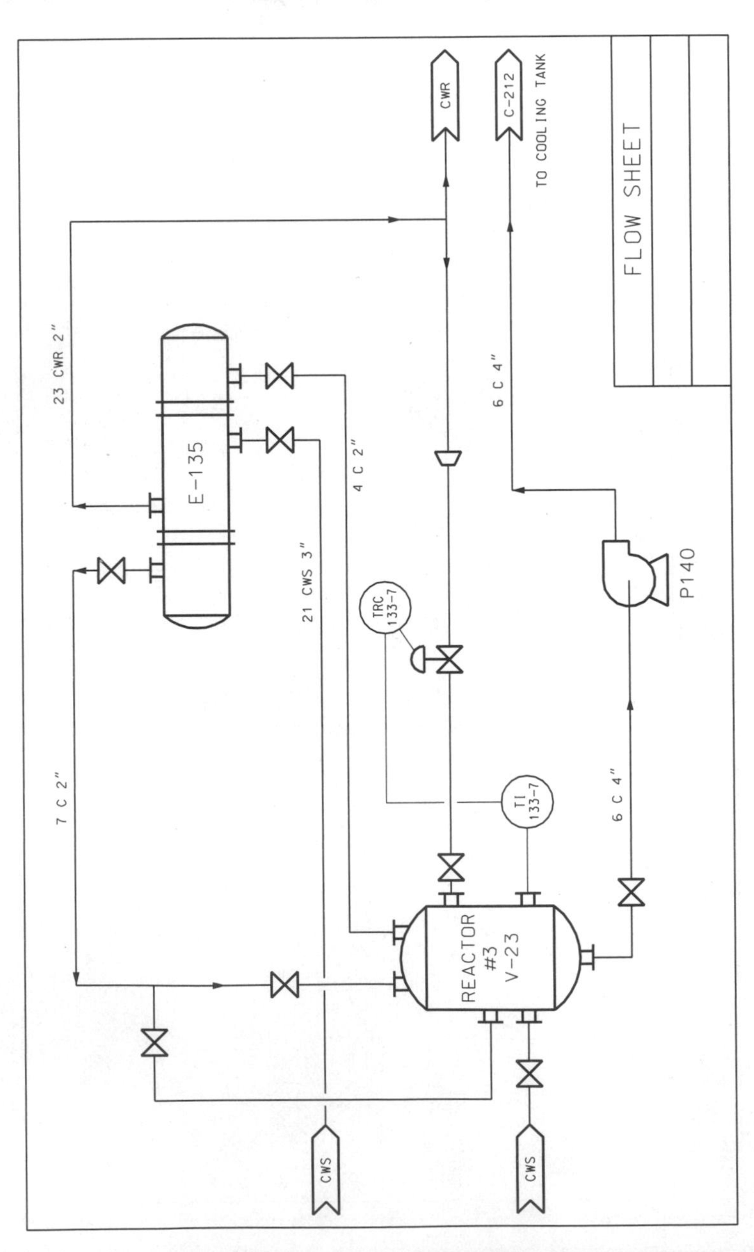

Exercise 11-3

Wastewater control panel

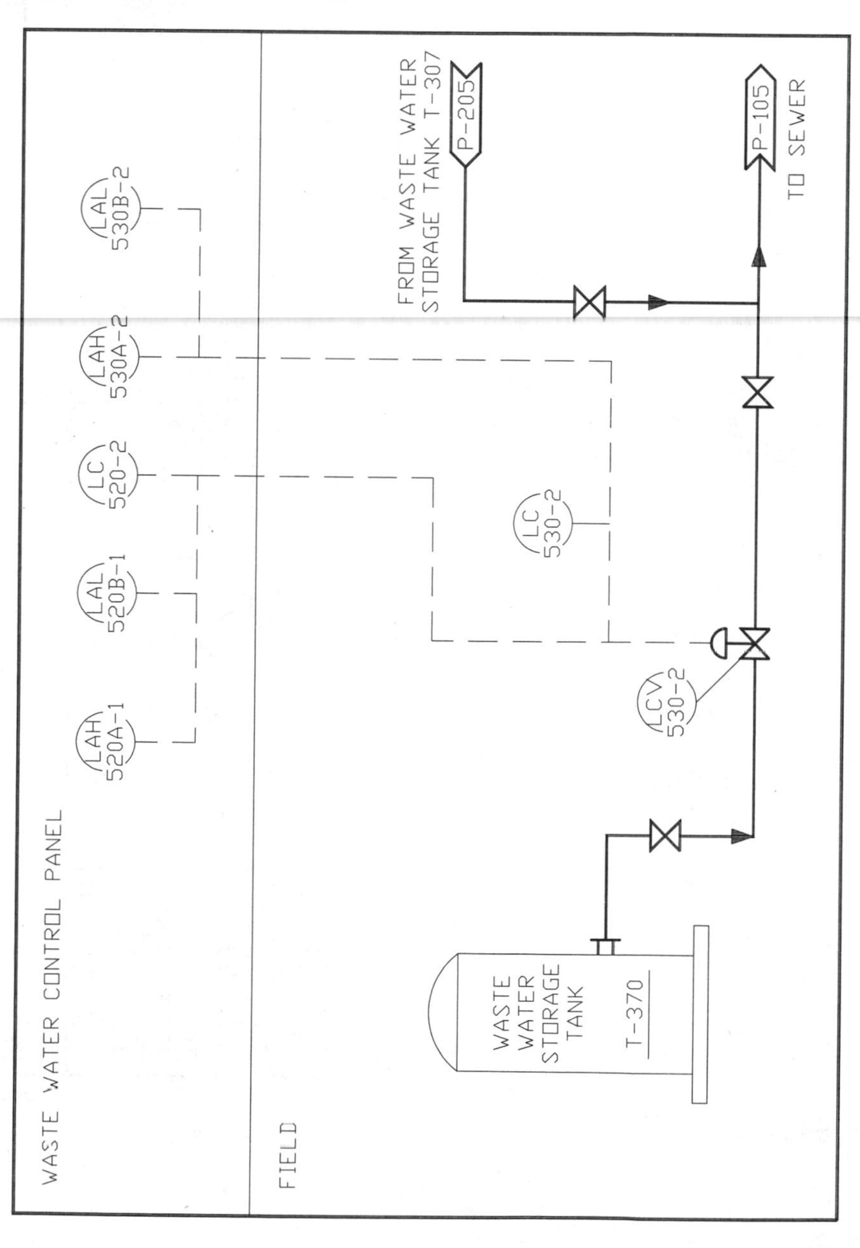

Exercises 11–4 and 11–5

Create the electrical symbols shown in Figure E11–2, and make each into a cell. Use the cells to create the diagrams in Exercises 11–4 and 11–5.

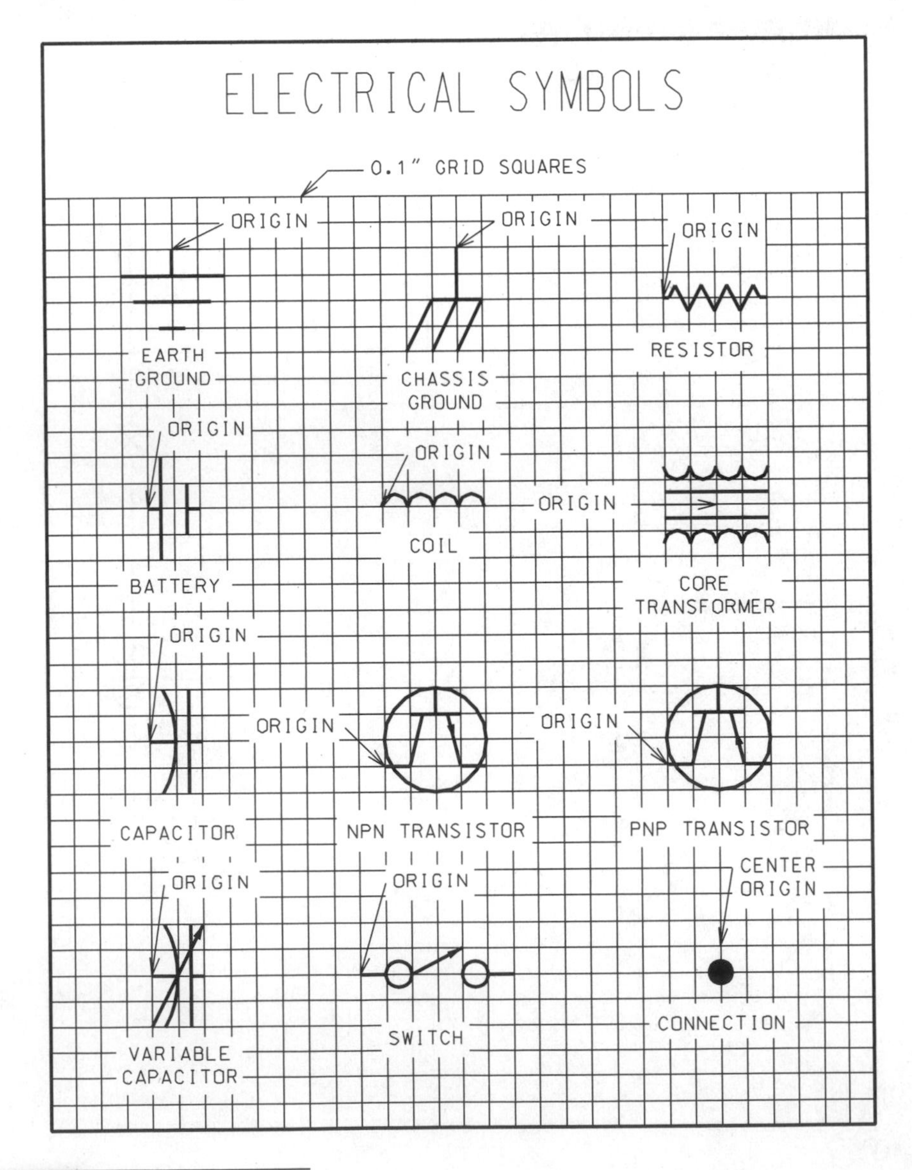

Figure E11–2 Electrical symbols

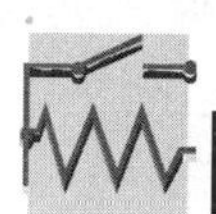

Exercise 11–4

Electrical diagram

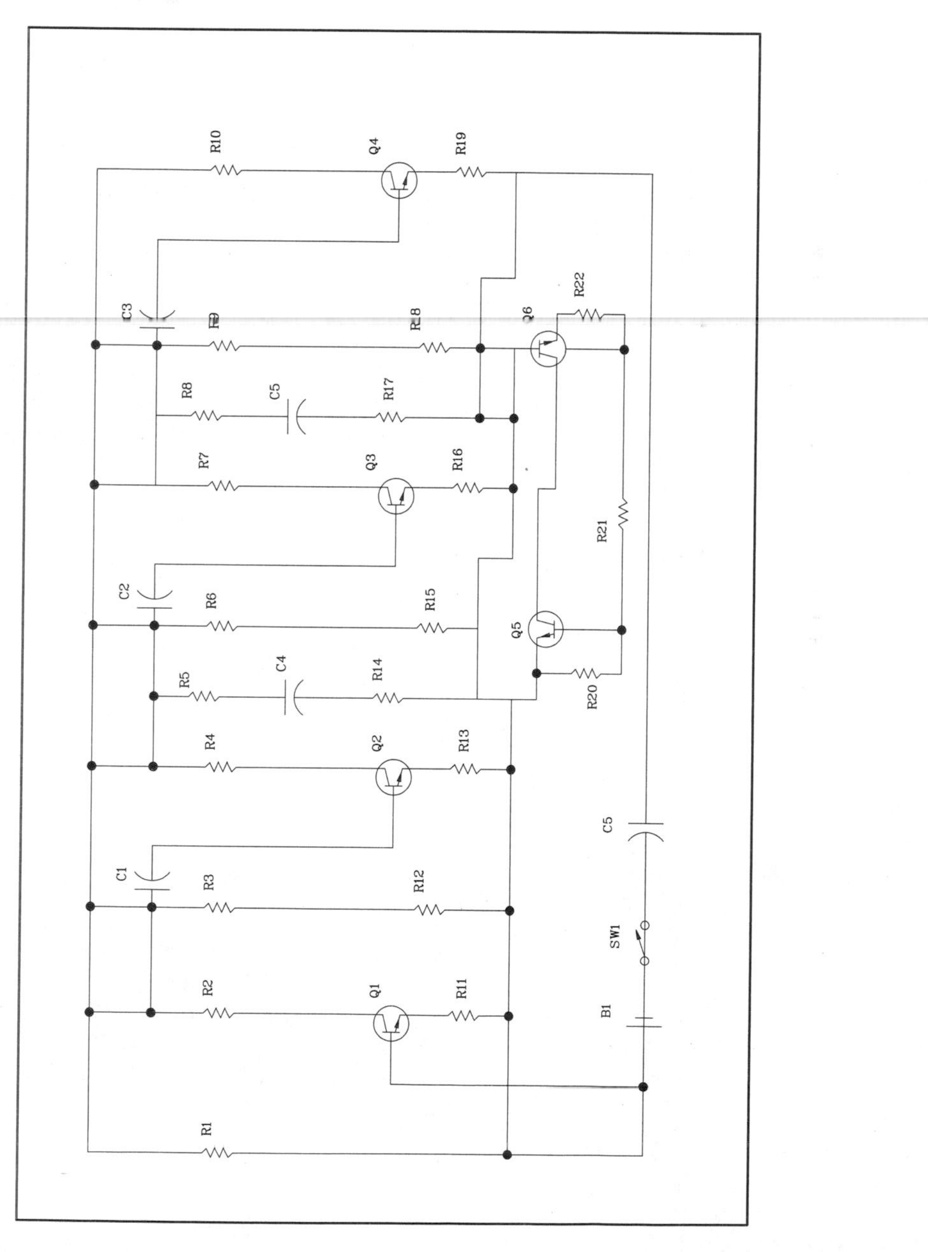

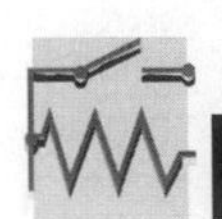

Exercise 11–5

Electrical diagram

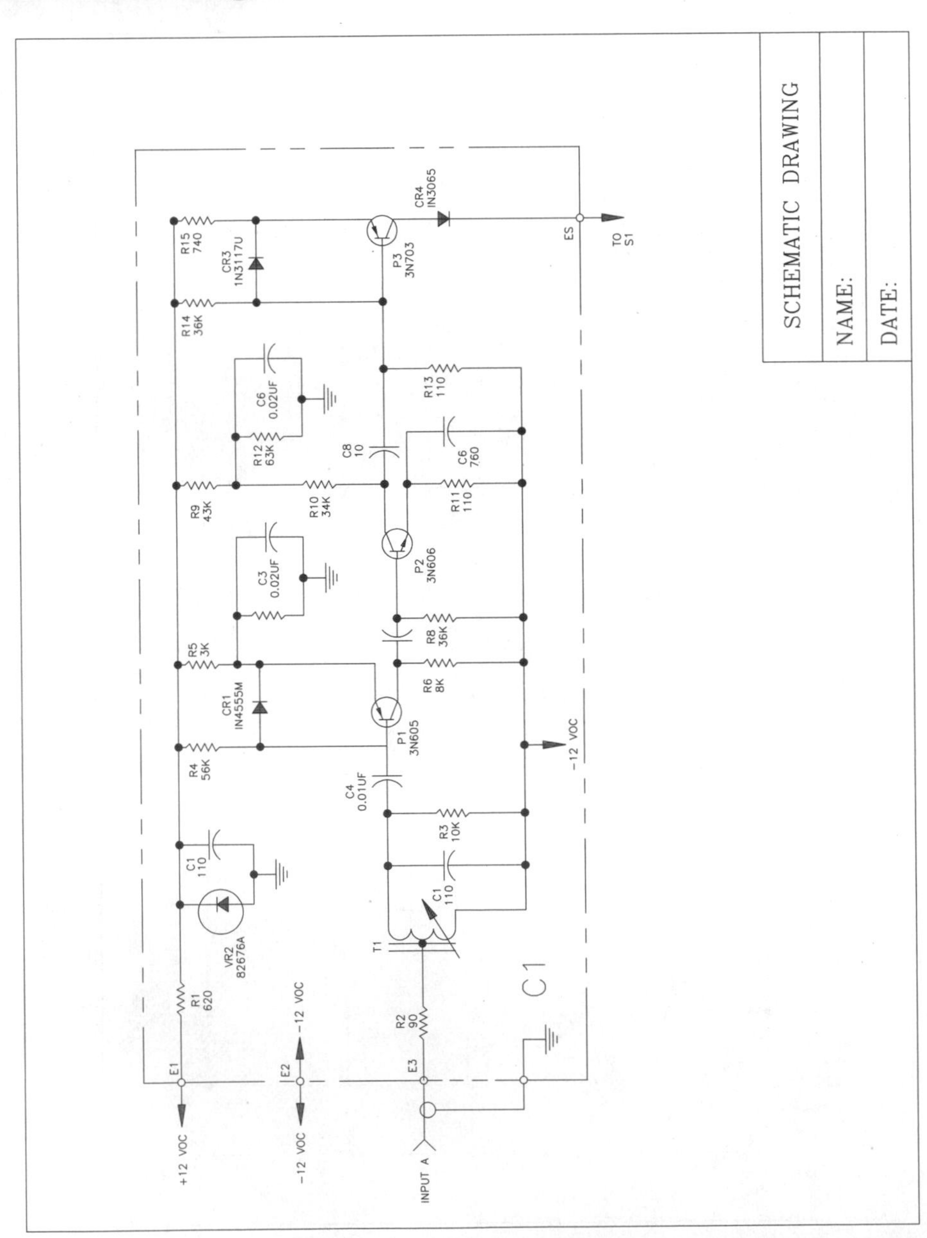

chapter 12

Patterning

Patterns are used in drawings for several reasons. Cutaways (cross sections) are hatched to help the viewer differentiate among components of an assembly and to indicate what the material is made of. Patterns on surfaces depict material and add to the readability of the drawing. In general, patterns help communicate information about design. Because drawing patterns is a repetitive task, it is an ideal computer-aided drafting application.

Objectives

After completing this chapter, you will be able to do the following:

- Control the display of patterns in view windows
- Place hatching, crosshatching, and area patterns via seven placement methods
- Manipulate patterns
- Fill elements

CONTROLLING THE VIEW OF PATTERNS

Patterning can place so many elements in a design that view updates may start taking an unacceptable time to complete on slow workstations. To overcome that, MicroStation provides the pattern view attribute to turn the display of pattern elements ON and OFF.

The Patterns view attribute should be ON when placing new patterns so the placement results can be seen. When work with patterns is completed, the Patterns view attribute can be turned OFF to speed up view updates. Figure 12–1 shows a crosshatch pattern with the view attribute ON and with it OFF.

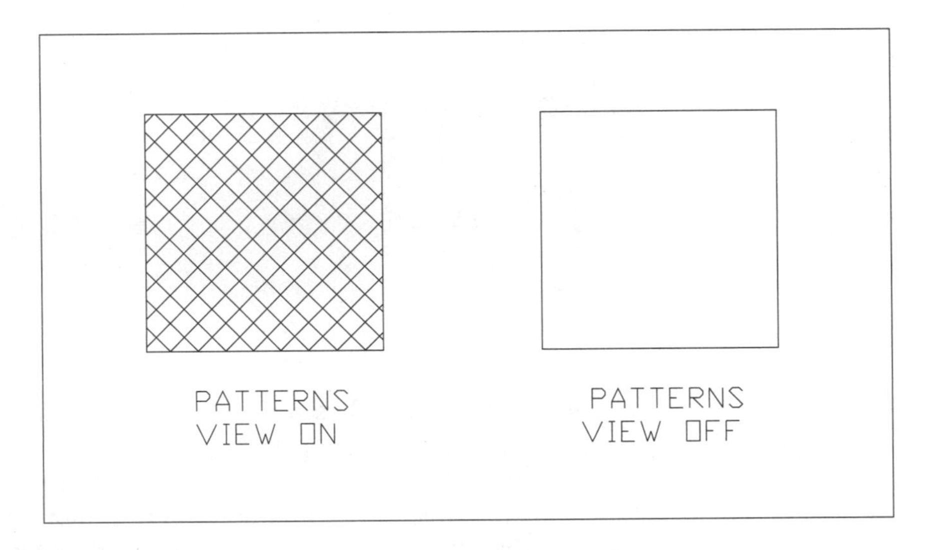

Figure 12–1 Example of turning the Patterns view attribute ON and OFF

To change the Patterns view status, invoke the View Attributes settings box:

Settings pull-down menu	Select View Attributes
Key-in window	**dialog viewsettings** (or **di views**) Enter

MicroStation displays the View Attributes settings box, as shown in Figure 12–2. Set the Patterns view attribute button ON or OFF, select a View Number from the top of the settings box, then click the Apply button, or click the All button to set the attributes for all open view windows.

Note: If you place patterns in a view where the Patterns view attribute is turned OFF, the patterns are placed but they do not show up on the screen. To avoid the confusion that that can cause, set the Patterns view ON before placing patterns.

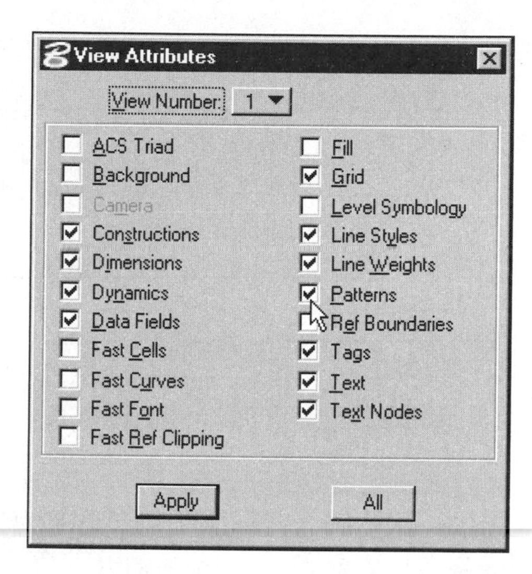

Figure 12–2 View Attributes settings box with the Patterns view attribute turned ON

PATTERNING COMMANDS

MicroStation provides three tools for placing patterns in a design file:

- **Hatch Area** places a set of parallel lines in the pattern area by invoking the Hatch Area tool from the Patterns tool box, as shown in Figure 12–3a.

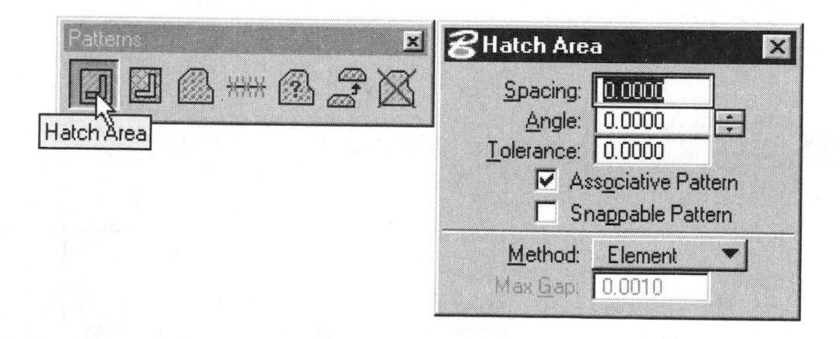

Figure 12–3a Hatch Area icon and settings box

- **Crosshatch Area** places two sets of parallel lines in the pattern area by invoking the Crosshatch Area tool from the Patterns tool box, as shown in Figure 12–3b.

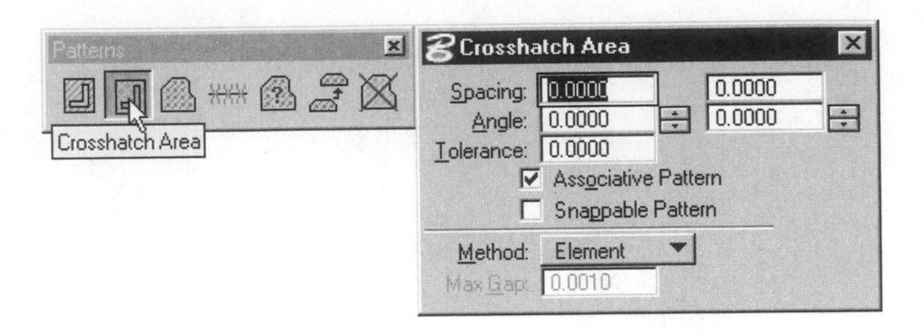

Figure 12–3b Crosshatch Area icon and settings box

- **Pattern Area** fills the pattern area with tiled copies of the active pattern cell by invoking the Pattern Area tool from the Patterns tool box, as shown in Figure 12–3c.

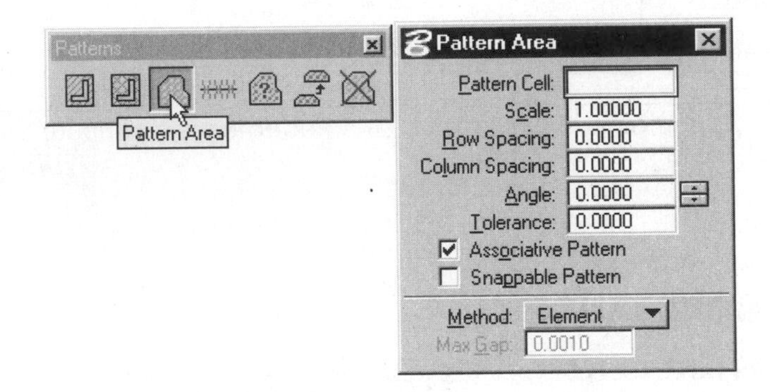

Figure 12–3c Pattern Area icon and settings box

Examples of the patterns placed by each tool are shown in Figure 12–4.

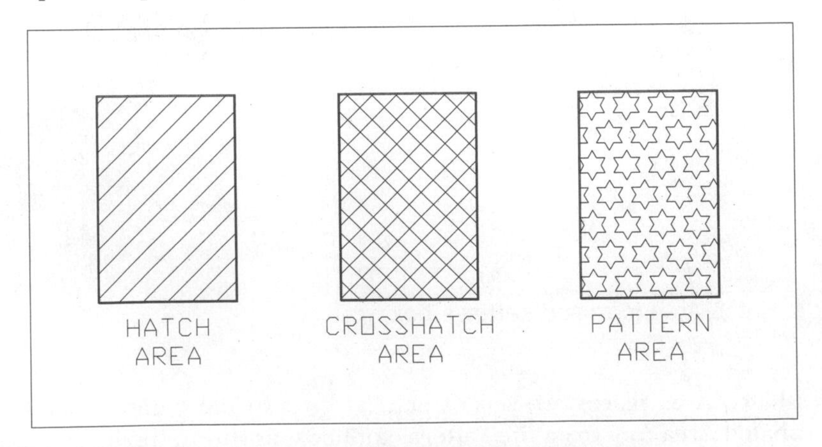

Figure 12–4 Examples of hatching, crosshatching, and patterning

TOOL SETTINGS FOR PATTERNING

The following discussion describes the tool settings for the three patterning tools.

Spacing

Spacing sets the space in Working Units (MU:SU:PU) between patterning elements.

- Hatching—One field to set the space between the lines
- Crosshatching—Two fields to set the space between each set of lines
- Pattern Area—Two fields to set the space between cell rows and columns

If the Hatch space or both of the Crosshatch spaces are zero, an error is displayed in the Status bar and no pattern is placed. Zero spaces are okay for the Pattern Area tool.

If the second crosshatch line space is set to zero, it will be placed equal to the first line spacing.

Angle

Angle sets the counterclockwise rotation angle (from the positive *X* axis direction) for the patterning elements.

- Hatching—One field to set the angle of the lines
- Crosshatching—Two fields to set the angle of each set of lines
- Pattern Area—One field to set the angle of the rows of cells

If the first crosshatch line angle is greater than zero and the second is equal to zero, the second set is placed at right angles to the first set. For example, if the first set's angle is 45 degrees and the second is zero degrees, the first set is placed at 45 degrees and the second set at 135 degrees (45+90).

Tolerance

Tolerance sets the variance between the true section curve and the approximation when a curved element is patterned. The curve is approximated by a series of straight-line segments, and a low tolerance number increases the accuracy of the approximation by reducing the length of each segment. The low tolerance number also means a larger design file and slower pattern placement.

Associative Pattern

Associative Pattern is a toggle button that, when ON, associates the pattern elements with the element they pattern. Changes to an associated element also affect the pattern. For example, if the element is stretched, the pattern expands to fill the new size. Figure 12–5 shows an example of pattern association.

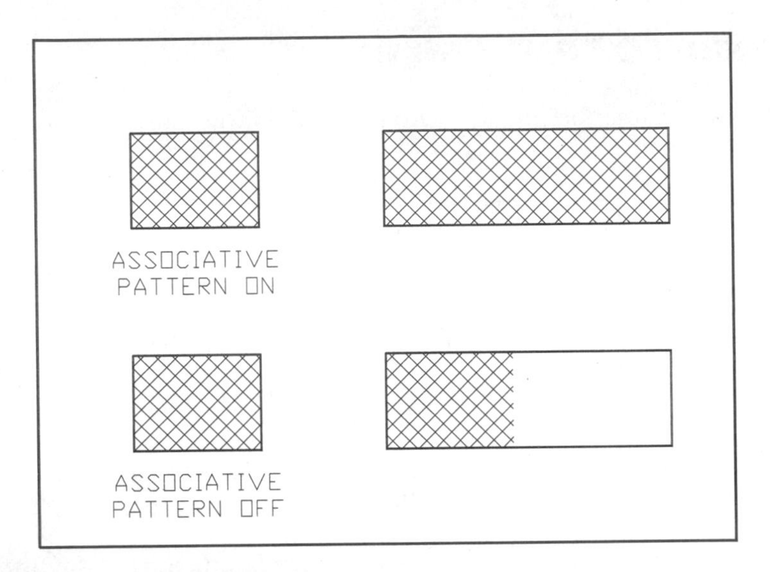

Figure 12-5 Example of the associative pattern setting

Snappable Pattern

Snappable pattern is a toggle button that, when ON, allows snapping to the pattern elements after they are placed. If it is OFF, tentative points cannot snap to the pattern elements.

Pattern Cell

The Pattern Cell edit field is used by the Pattern Area tool to provide the name of the cell that will fill the pattern area. If no cell name is provided, the Pattern Area tool displays an error message in the Status bar and no pattern is placed.

Scale

The Scale edit field is used by the Pattern Area tool to scale the cell placements. A number less than 1 reduces the size of each placement; a number greater than 1 increases the size.

Note: If the cell to be used for the pattern area was created with different Working Unit ratios than the design file, the cell may need to be scaled. If you are not sure of the correct scale, make the cell the active placement cell, then select the Place Active Cell tool. The dynamic image of the cell shows you what size it is in your design. Adjust the cell placement scale until the dynamic image is the correct size, then set the Pattern Area scale to the same value.

Method

The Method menu provides options to control how the patterned area is determined. The specific options are discussed later in the chapter.

Max Gap

The Max Gap edit field allows you to control the maximum space in Working Units (MU:SU:PU) between the enclosing elements in Flood patterning mode (discussed later in this chapter).

PATTERNING METHODS

Seven Methods are provided in the Tool Settings window for defining the area to be patterned, as shown in Figure 12–6.

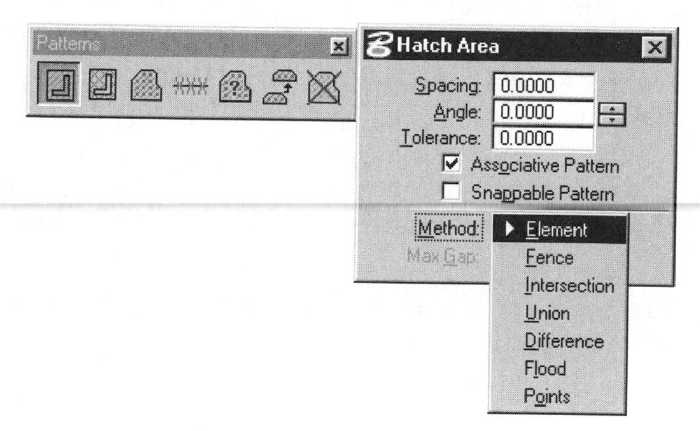

Figure 12–6 Patterns Method drop-down menu

The following list describes the patterning area methods; Figure 12–7 shows examples of each method.

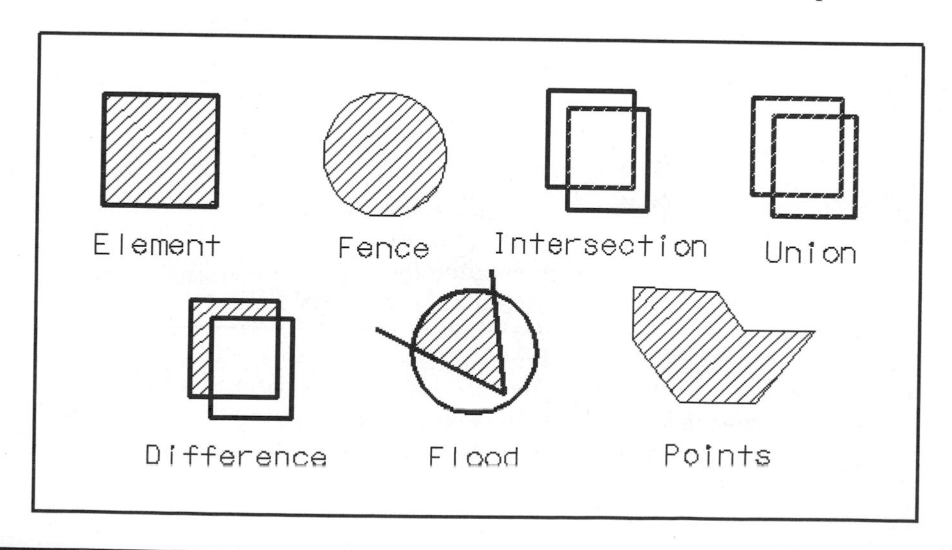

Figure 12–7 Examples of the seven area patterning methods

- **Element**—Patterns a selected closed element, such as a Shape or Ellipse, or between components of a multi-line.
- **Fence**—Patterns the area enclosed by a fence.
- **Intersection**—Patterns the area common to two or more overlapping elements.
- **Union**—Patterns two or more selected elements as if they were one element. The elements do not have to overlap.
- **Difference**—Patterns the part of the first selected closed element that does not overlap the other selected closed elements. If the first element is completely inside the second element, no pattern is placed.
- **Flood**—Patterns the area enclosed by a set of separate elements surrounding the placement data point. If there is a gap between enclosing elements that is greater than the Max Gap tool setting, the area is not patterned and an error message appears in the Status bar. If the area is complicated, the tool may take a long time to decide if the area is truly enclosed.
- **Points**—Patterns a temporary closed shape created by a set of data points placed after the tool is invoked.

PLACING PATTERNS

Invoke the pattern placement tools from:

Patterns tool box	Select the required pattern placement tool: • Hatch Area (see Figure 12–3a) • Crosshatch Area (see Figure 12–3b) • Pattern Area (see Figure 12–3c) Optionally, modify the pattern settings in the Tool Settings window.
Key-in window	• **hatch** (or **h**) Enter • **crosshatch** (or **cro**) Enter • **pattern area** (or **pat a**) Enter

Note: The placement prompts depend on the area definition method selected. The following discussion shows the prompts for each area method and uses the Hatch Area tool as an example. The prompts are the same for Crosshatch and Pattern Area.

Element

When the patterning area method is Element, MicroStation prompts:

> Hatch Area > Identify element *(Select the closed element to be patterned as shown in Figure 12–8.)*
>
> Hatch Area > Accept @pattern intersection point *(Click the Data button again to accept the element and initiate pattern placement.)*

Note: For each patterning area method, the acceptance point defines a point through which pattern elements pass—one line for Hatch Area, the intersection of two lines for Cross-hatch Area, and the origin point of one of the pattern cells for Pattern Area. If the acceptance point is not within the patterned area, the intersection will be on the edge of the area at the closest extension of the acceptance point.

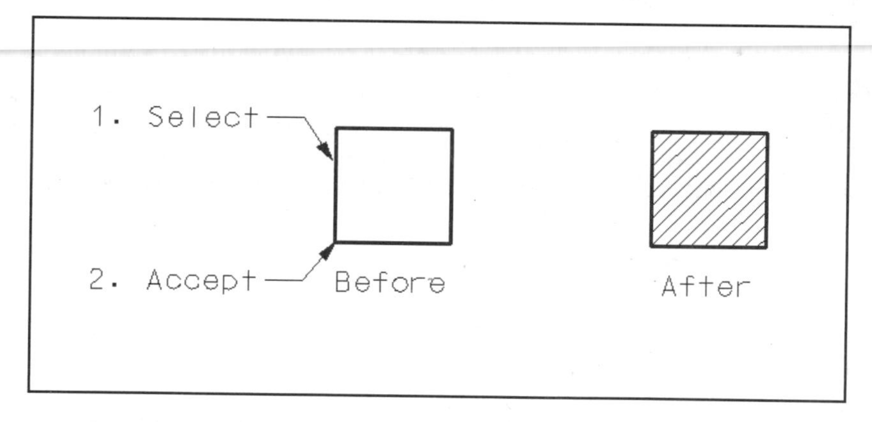

Figure 12-8 Example of patterning a closed element

Fence

When the patterning area method is Fence, MicroStation prompts:

> Hatch Fence > Accept/Reject fence contents *(Place a data point to define an intersection point for the pattern, as shown in Figure 12–9, and initiate patterning.)*

The example in Figure 12–9 is a circular fence. The fence must be placed before it can be patterned. The fence is not part of the pattern and removing it does not affect the pattern.

If the Associative pattern toggle button is ON, there will be an outline around the pattern in the shape of the fence. If the button is OFF, there won't be an outline.

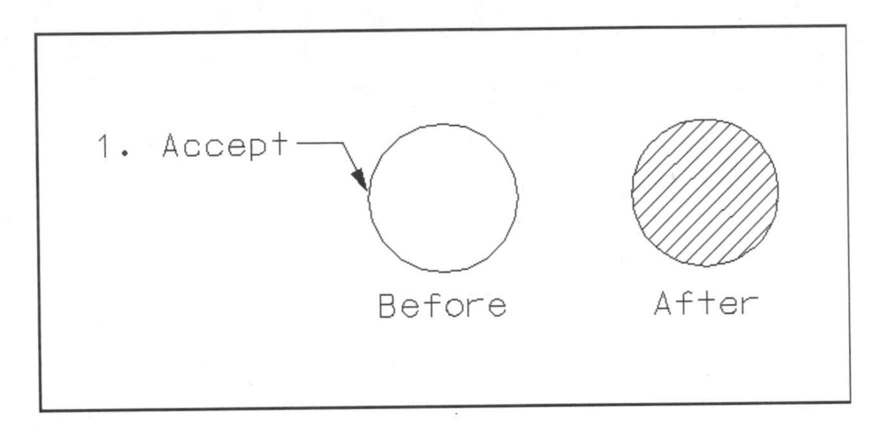

Figure 12–9 Example of patterning a fence

Intersection

When the patterning area method is Intersection, MicroStation prompts:

> Hatch Element Intersection > Identify element *(Select the first element, as shown in Figure 12–10.)*
>
> Hatch Element Intersection > Accept/Reject (Select next input) *(Select all the other elements that make up the intersection, then click the Data button in space.)*
>
> Hatch Element Intersection > Identify additional/Reset to complete *(Click the Reset button to initiate the patterning.)*

As each element is accepted, the dynamic image changes the appearance of the selected elements to show only the intersection area. The elements are not affected, and reappear after the tool is completed.

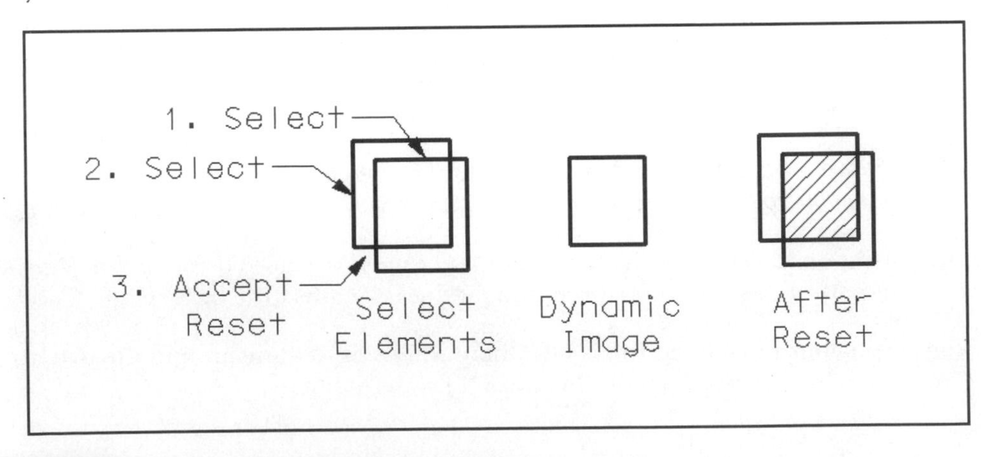

Figure 12–10 Example of intersection patterning

Union

When the patterning area method is Union, MicroStation prompts:

> Hatch Element Union > Identify element *(Select the first element, as shown in Figure 12–11.)*
>
> Hatch Element Intersection > Accept/Reject (Select next input) *(Select all the other elements that make up the union, then click the Data button in space.)*
>
> Hatch Element Union > Identify additional/Reset to complete *(Click the Reset button to initiate the patterning.)*

As each element is accepted, the dynamic image changes the appearance of the selected elements to show the union area with interior segments removed. The elements are not affected, and reappear after the tool is completed.

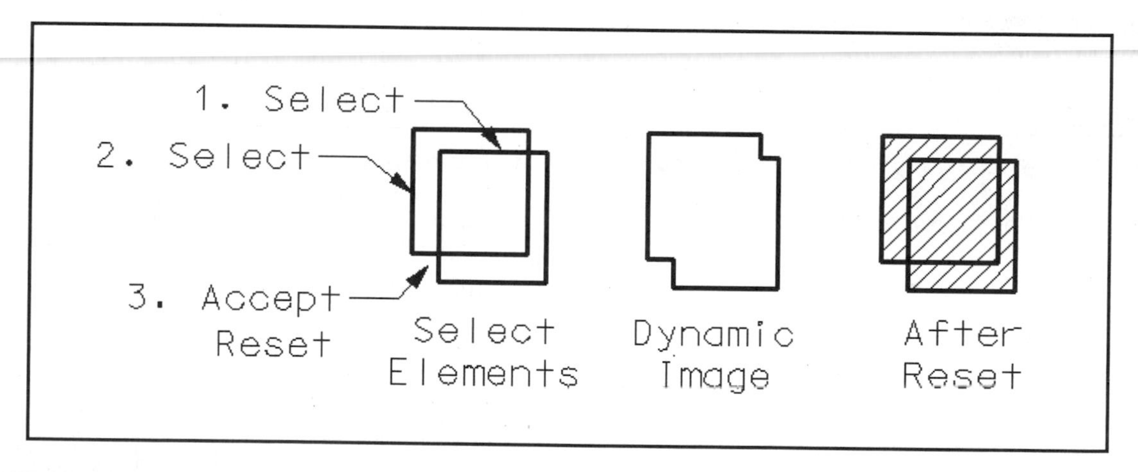

Figure 12–11 Example of union patterning

Difference

When the patterning area method is Difference, MicroStation prompts:

> Hatch Element Difference > Identify element *(Select the first element, as shown in Figure 12–12.)*
>
> Hatch Element Difference > Accept/Reject (Select next input) *(Select all the other elements that make up the difference, then click the Data button in space.)*
>
> Hatch Element Difference > Identify additional/Reset to complete *(Click the Reset button to initiate the patterning.)*

As each element is accepted, the dynamic image changes the appearance of the selected elements to show only the difference area. The elements are not affected, and reappear after the tool is completed.

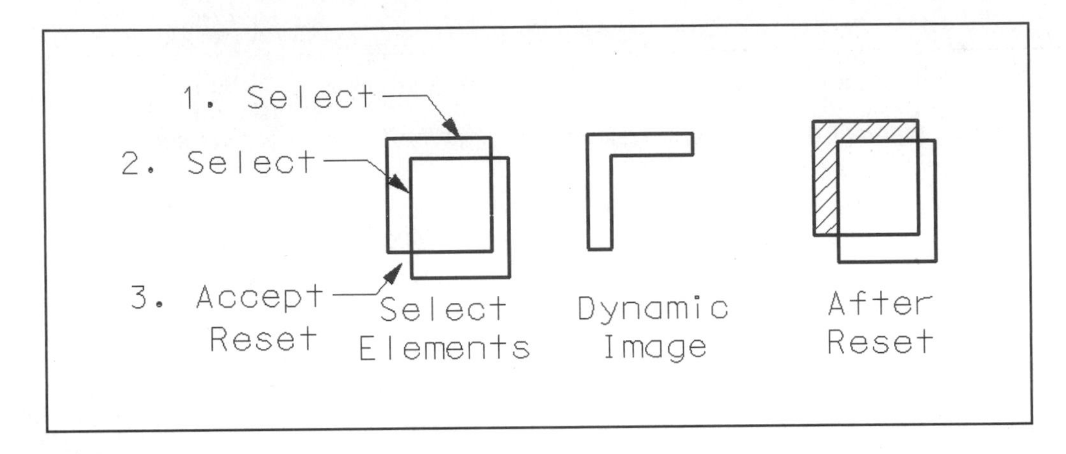

Figure 12–12 Example of difference patterning

Flood

When the patterning area method is Flood, MicroStation prompts:

> Hatch Area Enclosing Point > Enter data point inside area *(Place a data point inside the area to be patterned, as shown in Figure 12–13.)*
>
> Hatch Area Enclosing Point > Accept @pattern intersection point *(Click the Data button again to initiate patterning.)*

After the first data point is placed, a dynamic image is drawn around the perimeter of the flood area.

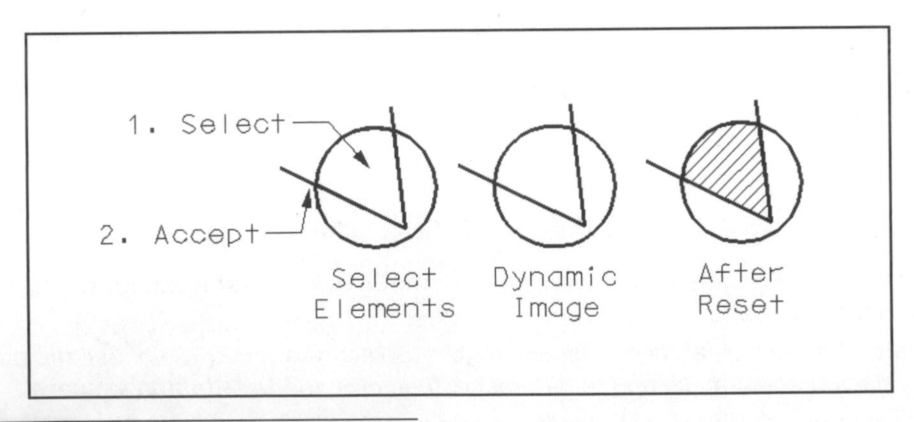

Figure 12–13 Example of flood patterning

Points

When the patterning area method is Points, MicroStation prompts:

Hatch Area Defined By Points > Enter shape vertex *(Place the first three vertex points of the shape, as shown in Figure 12–14.)*

Hatch Area Defined By Points > Enter point or Reset to complete *(Continue entering shape vertex points, or click the Reset button to complete the shape and initiate patterning.)*

Hatch Area Defined By Points > Identify additional/Reset to complete *(Click the Reset button to initiate the patterning.)*

After the third point, the dynamic image shows the shape of the pattern area if the Reset button were to be clicked.

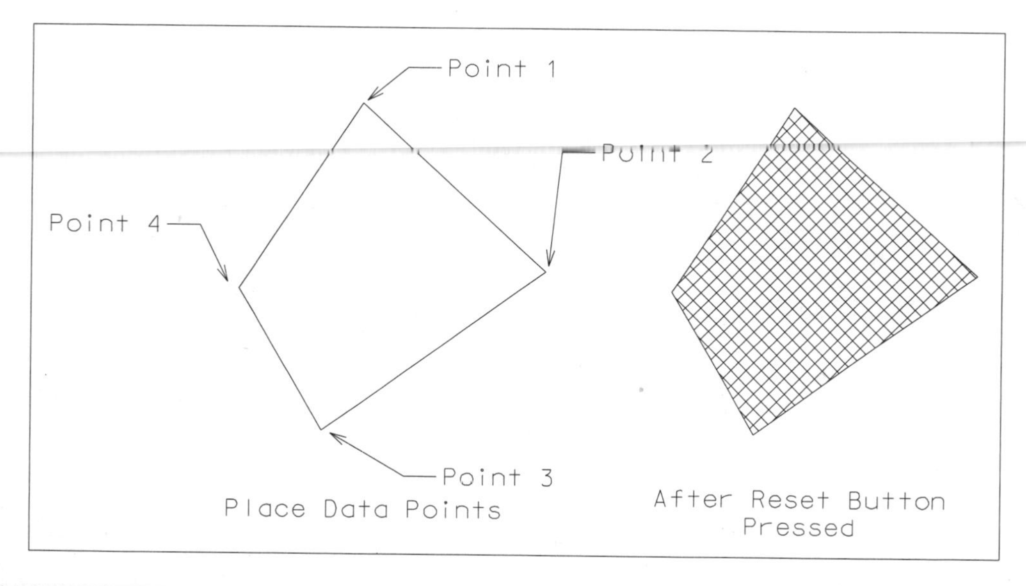

Figure 12–14 Example of patterning an area enclosed by points

LEAVING HOLES IN PATTERNS

Often the area to be patterned contains elements that should not be patterned. The Difference method can be used to leave holes in patterns, but that method can be rather tedious when there are lots of elements that should be holes, such as windows in an elevation of a building. An easier way to handle it would be to create closed elements that cannot be patterned. MicroStation provides the Solid and Hole Area option for closed elements (circle, block, shape, etc.) to allow you to do that. Solid elements can be patterned, but Hole elements cannot.

For example, Figure 12–15 shows a brick pattern on a wall that contains two windows. The windows were placed as blocks with the Hole Area option selected. The wall was placed as a block with Solid Area selected.

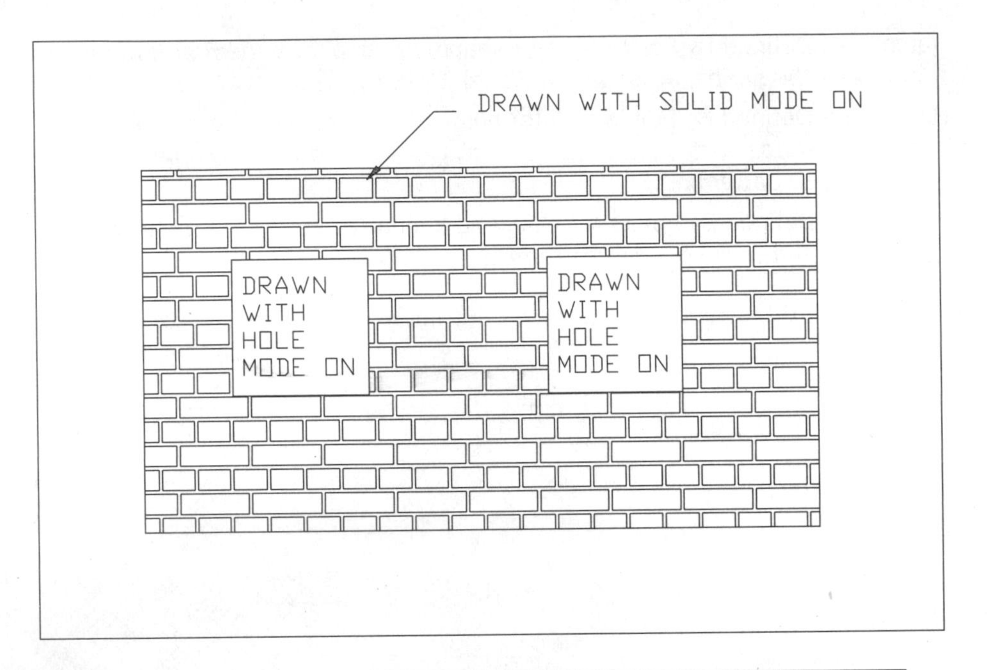

Figure 12–15 Example of the effect of Hole elements within a patterned element

Text placed with Hole Area mode set to ON also cannot be patterned, as shown in Figure 12–16.

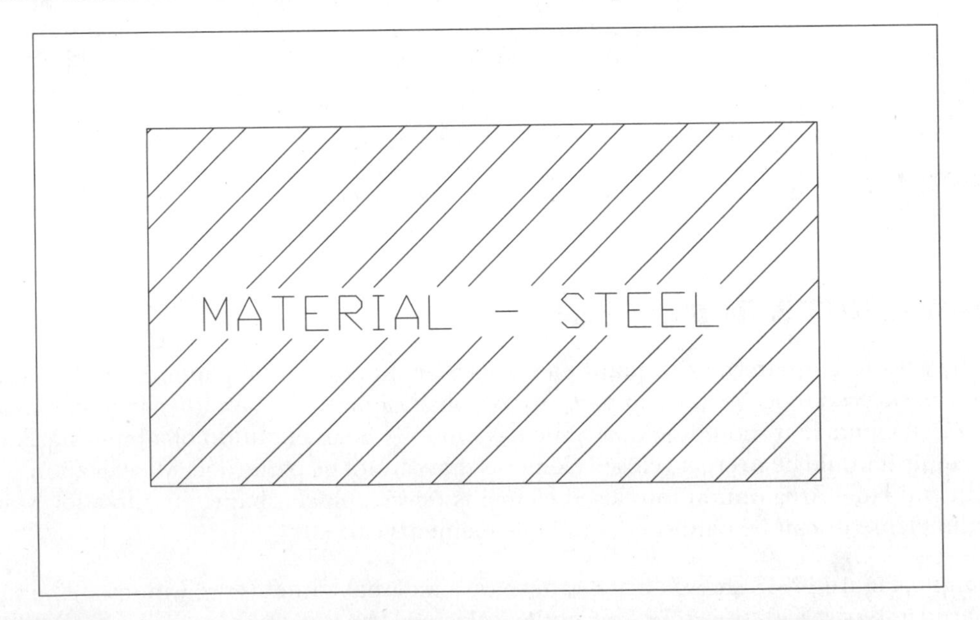

Figure 12–16 Example of patterning over a text string placed in Hole Area mode

The Tool Settings window for each tool that places closed elements includes a menu for setting the Area mode. An example is shown in Figure 12–17. The Area option you select becomes the active setting, and it is applied to all closed elements you create after that.

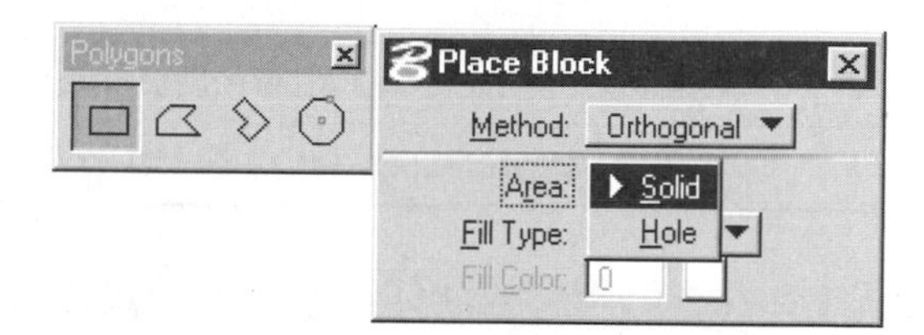

Figure 12–17 Example of the Area mode tool setting for a closed element placement tool

Note: Only the Element and Fence patterning methods recognize Hole elements. The other patterning methods (Intersection, Union, Difference, Flood, and Points) ignore the Hole setting and pattern over Hold elements. The Associative Lock toggle button must also be OFF for Hole elements to be recognized.

Changing an Element's Area

If an element was placed via the wrong Area mode, you can change it to the correct area with the Change Element to Active Area tool. Invoke the tool from:

Change Attributes tool box	Select the Change Element to Active Area tool, then select the required Area (Solid or Hole) in the Tool Settings window (see Figure 12–18).
Key-in window	**change area** (or **Chan a**) Enter

Figure 12–18 Invoking the Change Element to Active Area tool from the Change Attributes tool box

MicroStation prompts:

Change Element to Active Area > Identify element *(Identify the element.)*

Change Element to Active Area > Accept/Reject (Select next input) *(Click the Data button to initiate the area change, or click Reset to reject the selected element. Optionally, also select the next element to be changed.)*

DELETING PATTERNS

A special delete tool for patterns deletes all pattern elements but not the element that contains the pattern. Invoke the Delete Pattern tool:

Patterns tool box	Select the Delete Pattern tool (see Figure 12–19).
Key-in window	**delete pattern** (or **del pat**) Enter

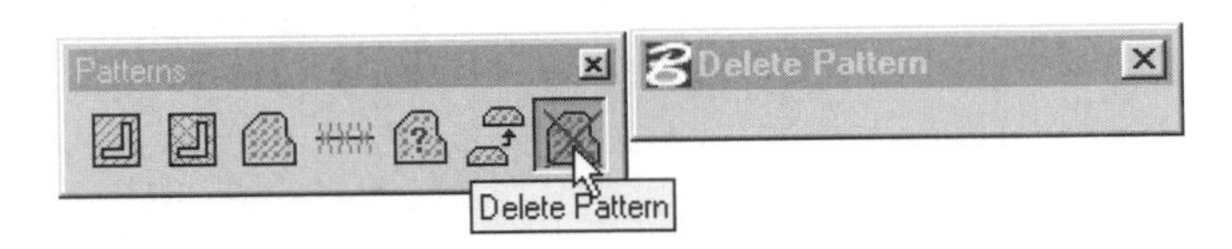

Figure 12–19 Invoking the Delete Pattern tool from the Patterns tool box

MicroStation prompts:

Delete Pattern > Identify element *(Identify the pattern to be deleted, then click the Data button again to accept and delete the pattern.)*

Note: Patterns can contain a large number of elements. After deleting patterns, you may want to select the Compress Design tool from the pull-down menu File to remove the deleted elements from the design file. Remember though, that when you compress the design, you also clear the Undo buffer.

MATCHING PATTERN ATTRIBUTES

If additional patterns need to be placed with the same patterning attribute settings as existing patterns, the Match Active Pattern tool provides a fast way to set the attributes. The tool sets the active patterning attributes to match those of a selected pattern.

Invoke the Match Active Pattern tool:

Patterns tool box	Select the Match Active Pattern tool (see Figure 12–20).
Key-in window	**match pattern** (or **mat pat**) Enter

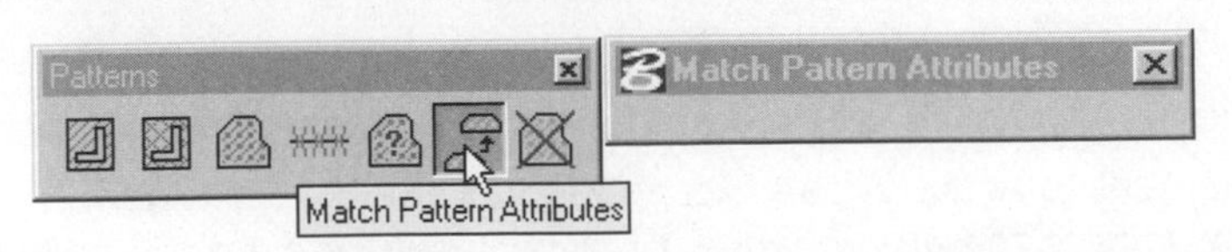

Figure 12–20 Invoking the Match Pattern Attributes tool from the Patterns tool box

MicroStation prompts:

> Match Pattern Attributes > Identify element *(Identify the element containing the pattern to be matched.)*
>
> Match Pattern Attributes > Accept/Reject (Select next input) *(Click the Data button to accept the element or the Reject button to reject it.)*

After the second data point is placed, the active patterning elements match those of the selected pattern, and the active settings appear in the Status bar.

Note: The Match Pattern Attributes tool is also available in the Match tool box that is opened from the Tools pull-down menu.

FILLING AN ELEMENT

The Tool Settings window for tools that place closed elements (Circle, Block, Shape, etc.) include a Fill Type menu that allows you fill the interior of the element with color. The Fill Type menu includes three settings:

- **None**—The element is transparent.
- **Opaque**—The element is filled with the active (outline) color.
- **Outlined**—The fill color of the element is controlled by the Fill Color attribute (which can be different from the element outline color).

In addition to the Fill Type menu, a Fill Color menu is provided for setting the color in the following manner:

- Fill and outline when the Opaque Fill Type is active.
- Fill only when the Outlined Fill Type is active.

Figure 12–21 shows a typical Tool Settings window, with the Fill Type menu open; Figure 12–22 shows examples of opaque and outlined fill types.

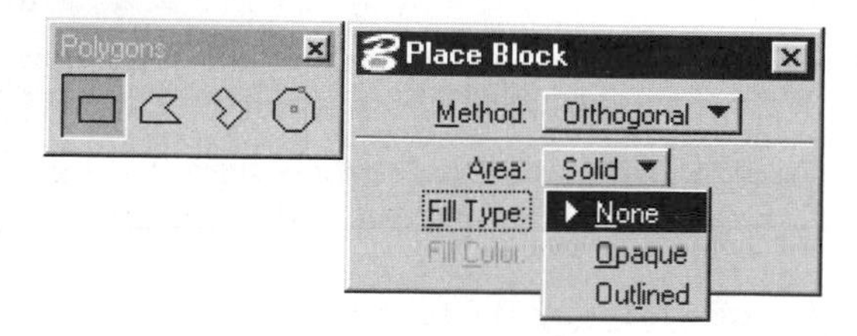

Figure 12–21 Fill Type menu in the Tool Settings window

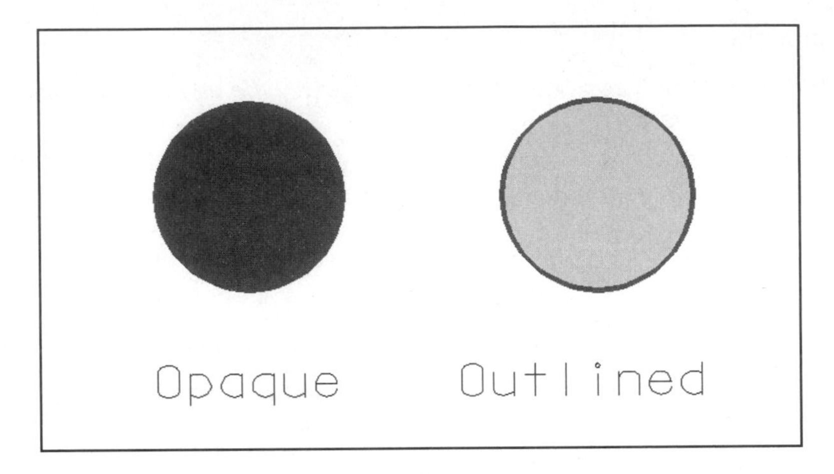

Figure 12–22 Examples of Opaque and Outlined Fill Types

To place filled closed elements, just select the desired placement tool, then select the desired Fill Type and Fill Color before placing the element. The Fill settings you select become the active setting, and they are applied to all closed elements you create after that.

Area Fill View Attribute

Before placing filled closed elements, make sure the Fill View attribute is turned ON for the view window being used to place the elements. If the attribute is set to OFF, filled elements appear to be transparent, and they plot that way.

To turn ON the Fill View attribute, invoke the View Attributes window:

Settings pull-down menu	Select View Attributes
Key-in window	**dialog viewsettings** (or **di views**) Enter

Set the Fill switch ON or OFF as required (see Figure 12–23), then:

- Click the Apply button to apply the view settings to the View Number shown at the top of the window, or
- Click the All button to apply the view settings to all open view windows.

The applied view attributes stay in effect until they are changed or the file is closed. To make the settings permanent, select Save Settings from the File pull-down menu.

Note: If a plotter does not print the fill area of elements, it may be because of a setting in the plotter's configuration file. Consult the MicroStation technical documentation for information on plotter configuration.

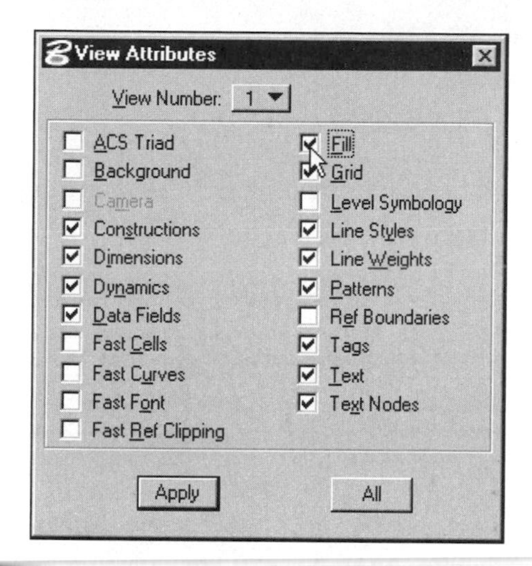

Figure 12–23 The Fill attribute in the View Attributes settings box

Changing the Fill Type of an Existing Element

The Change Element to Active Fill Type tool changes the fill type of existing elements. Invoke the tool:

Change Attributes tool box	Select the Change Element to Active Fill Type tool, and set the desired Fill Type and Fill Color in the Tool Settings window (see Figure 12–24).
Key-in window	**dialog change fill** (or **chan f**) Enter

Figure 12–24 Invoking the Change Element to Active Fill Type tool from the Change Attributes tool box

MicroStation prompts:

> Change Element to Active Fill Type > Identify element *(Select the element to be changed.)*
>
> Change Element to Active Fill Type > Accept/Reject (Select next input) *(Click the Data button again to initiate the Fill Type change and, optionally, select the next element to change; or click the Reset button to reject the element.)*

If the second data point is not on an element, the error message *Element not found* appears in the Status bar. Ignore this message.

REVIEW QUESTIONS

Write your answers in the spaces provided.

1. List the three tools MicroStation provides for area patterning.

   ____

2. Explain briefly the difference between Hatch Area and Crosshatch Area patterning.

   ____

   ____

3. List the methods that are available to set the area patterning.

   ____

4. Explain with illustrations the differences between the Intersection, Union, and Difference options in area patterning.

   ____

   ____

5. Explain the difference between the Element and Points options in area patterning.

   ____

   ____

6. What is the purpose of providing a second data point in hatching a closed element?

   ____

7. The Pattern Area tool places copies of the Active Patterning ____________ in the area you select.

8. What is the purpose of turning ON the Associative Pattern toggle button?

   ____

9. What is the purpose of placing elements in a Hole Area mode?

   ____

10. Explain briefly the purpose of invoking the Match Pattern Attributes tool.

    ____

    ____

11. Explain briefly the three options available for area fill placement.

    ____

    ____

12. Explain the steps involved in switching an existing element between area fill modes.

    ____

    ____

PROJECT EXERCISE

This project exercise provides step-by-step instructions for creating the design shown in Figure P12–1. The intent is to guide you in placing patterns.

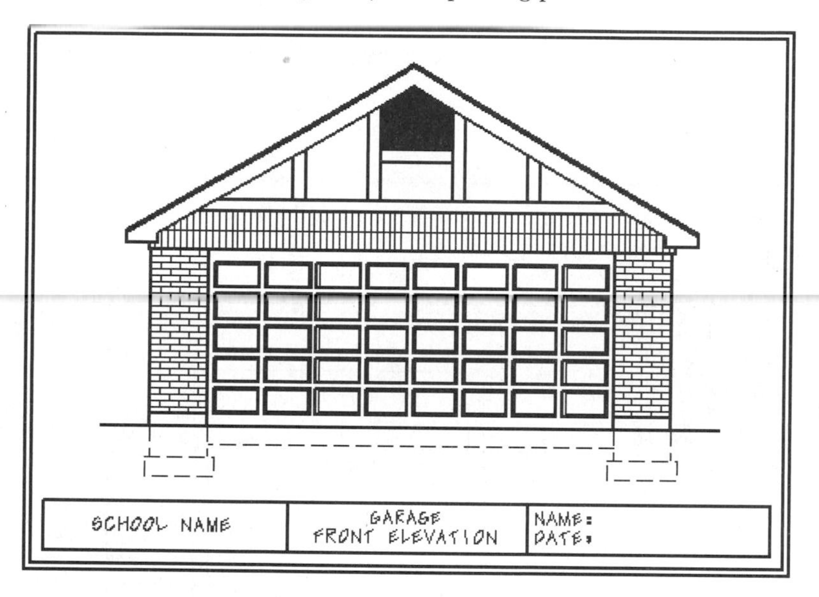

Figure P12–1 Completed project design

Prepare the Design File

This procedure starts MicroStation, creates a design file, and enters the initial settings.

Note: As you complete each step in the project procedures, place a check mark by the step to help you keep up with where you are in the project.

STEP 1: Invoke MicroStation by the normal technique for the operating system on your workstation.

STEP 2: Create a new design file named CH12.DGN using the SEED2D.DGN seed file.

In the Design File dialog box:

- Set the Working Units Master Units name to ' (for feet), the Sub Units name to " (for inches), and the resolution ratios to 12 and 8000.
- Set the Grid Master to 0.25, Grid Reference to 4, and Grid Lock to ON.

STEP 3: Invoke the Save Settings from the File pull-down menu to save the settings.

STEP 4: Draw the border and title block on level 10 (the outer border block is 30' x 6" wide by 21' x 6" high):

- Replace "SCHOOL NAME" with your school or company name, or make up a name.
- Place your name to the right of "NAME."
- Place today's date to the right of "DATE."

Create Cells

This procedure creates the garage door panel cell shown in Figure P12–2 and the brick patterning cell shown in Figure P12–3.

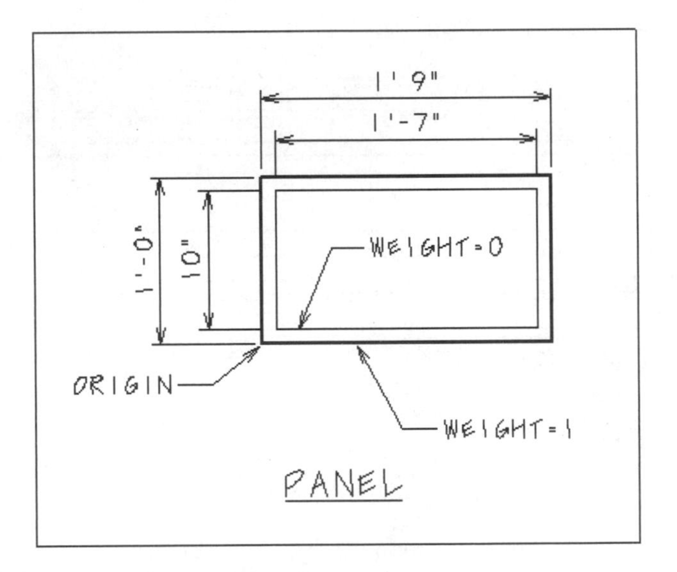

Figure P12–2 Garage door panel cell construction

STEP 1: Align a view window to display an area about 3' x 3'.

STEP 2: Create a new cell library file named CH12.CEL.

STEP 3: Use Figure P12–2 as a guide to draw the two rectangles making up the garage door panel. Do not draw any of the dimensions or text.

STEP 4: Create a cell from the two rectangles named *panel,* with the origin defined at the lower left corner of the outer rectangle.

STEP 5: Delete the two rectangles.

STEP 6: Use Figure P12–3 as a guide to draw the brick pattern.

Note: The brick pattern lines in Figure P12–3 are shown heavy to make them easier to distinguish from the dimensions. Draw them at Line Weight 0.

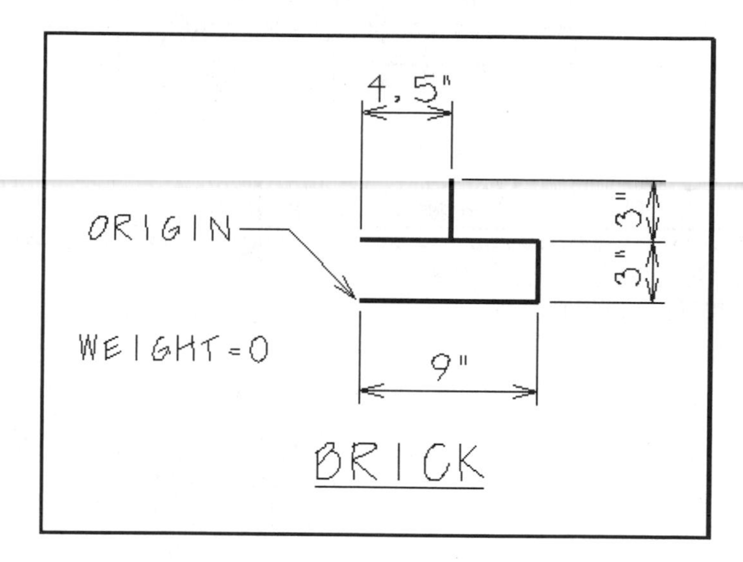

Figure P12–3 Brick pattern cell construction

STEP 7: Create a cell from the brick elements named *brick,* with the origin defined at the lower left corner. Do *not* draw any of the dimensions or text.

STEP 8: Delete the brick elements.

STEP 9: Fit the view and save the design settings.

Draw the Garage

This procedure creates the garage front elevation, as shown in Figure P12–4.

Note: If a dimension is missing, estimate the proper placement.

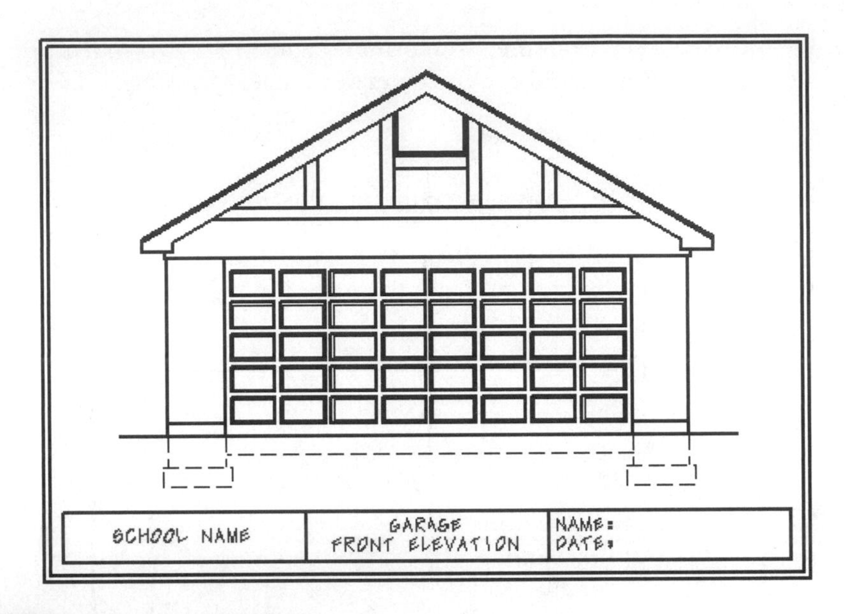

Figure P12–4 Garage front elevation

STEP 1: Set the Line Weight to 1, Active Level to 2, Color to Green.

STEP 2: Using Figure P12–5 as a guide, draw the garage wall and door. Use the Place Block tool to create the outline of the brick wall on each side of the garage door.

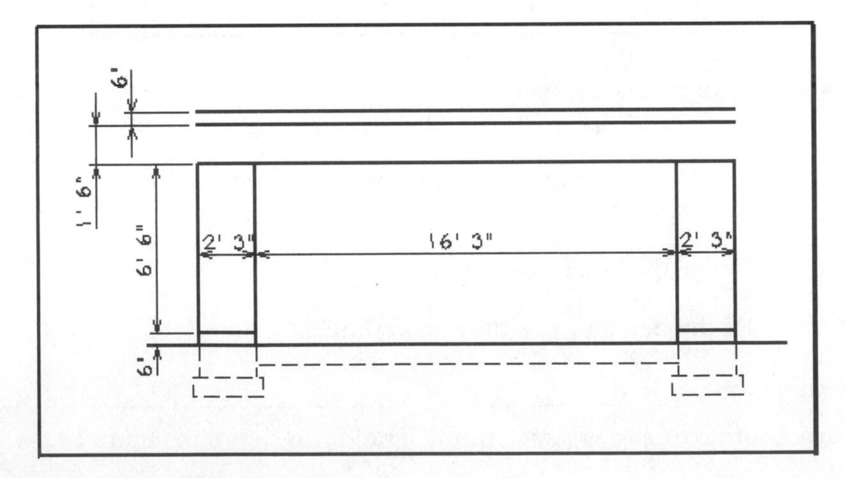

Figure P12–5 Garage wall and door detail

STEP 3: With Figure P12–6 as a guide, draw the garage roof.

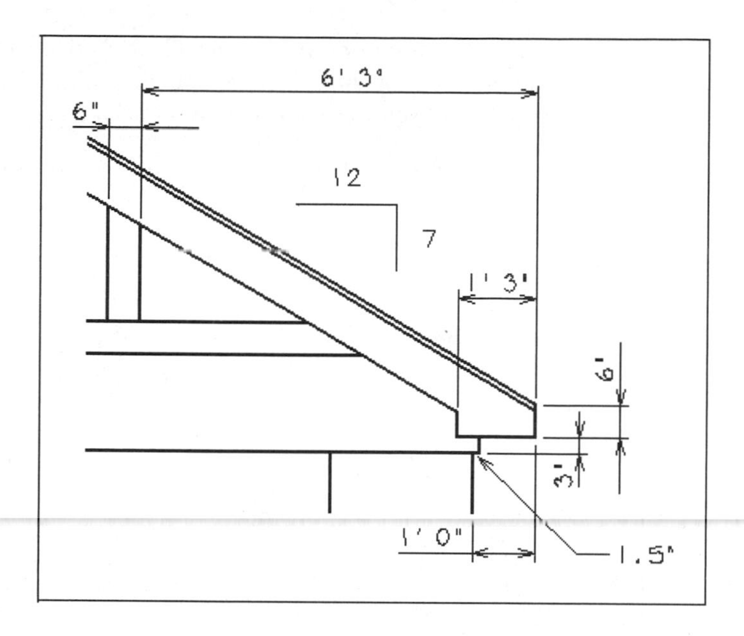

Figure P12–6 Garage roof detail

STEP 4: With Figure P12–7 as a guide, draw the garage attic vent.

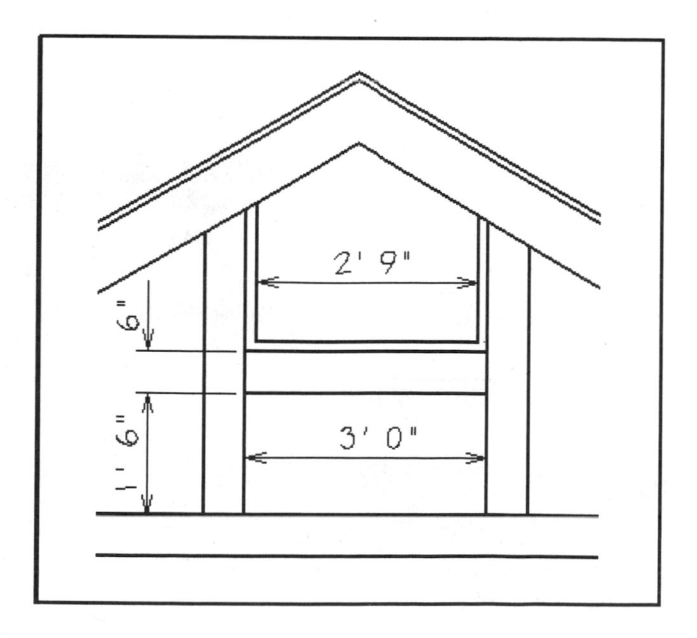

Figure P12–7 Garage vent detail

STEP 5: Using Figure P12–8 as a guide, invoke the Place Active Cell Matrix tool to place a rectangular array of the panel cell in the garage door. Figure P12–9 shows the required tool settings for the array.

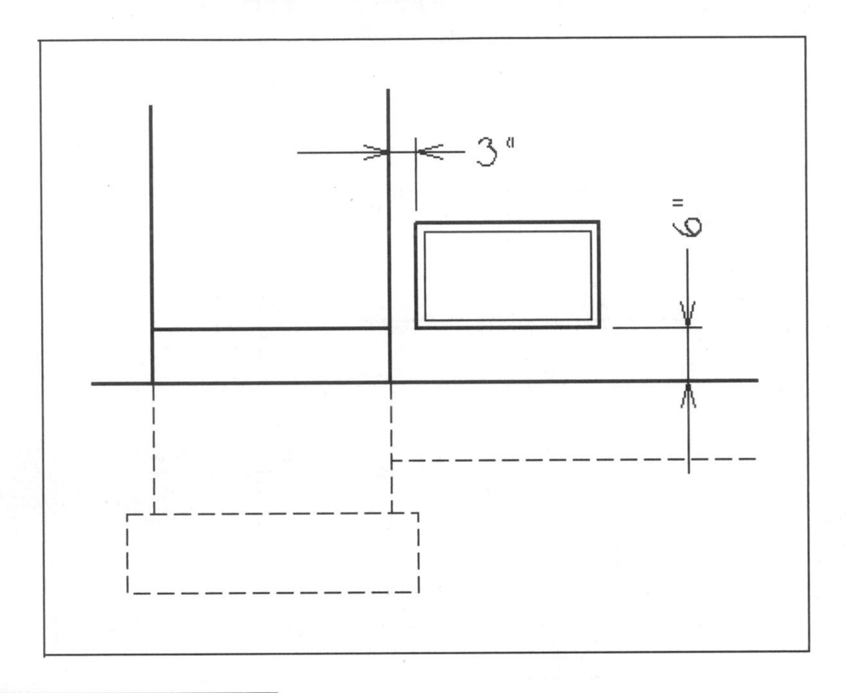

Figure P12–8 Garage door detail

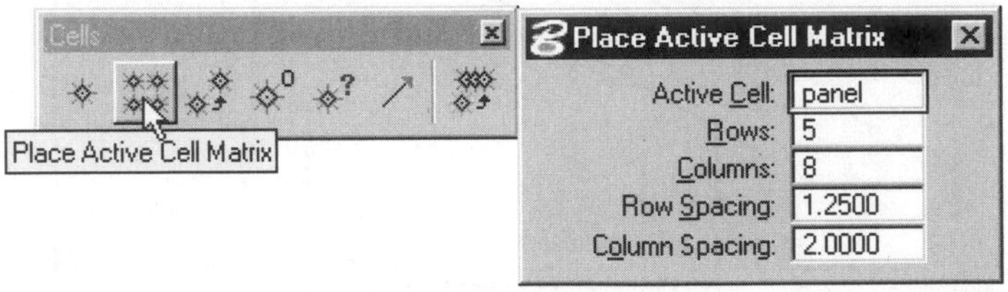

Cells

Place Active Cell Matrix

Place Active Cell Matrix

Active Cell: panel

Rows: 5

Columns: 8

Row Spacing: 1.2500

Column Spacing: 2.0000

Figure P12–9 Place Active Cell Matrix tool settings

Place the Brick Patterns

This procedure uses the Pattern Area tool to place the brick pattern on each side of the garage door, the Crosshatch Area tool to place the brick pattern above the garage door, and the Hatch Area tool to place the louvers in the attic vent.

STEP 1: Set the Active Level to 4.

STEP 2: Use the Window Area tool to fill the view with the rectangle on the right side of the garage door.

STEP 3: Invoke the Pattern Area tool from the Patterns tool box, and, in the Tool Settings window, adjust the tool settings as shown in Figure P12–10.

Note: If you did not use a closed element for the brick walls, set the patterning Method to Flood.

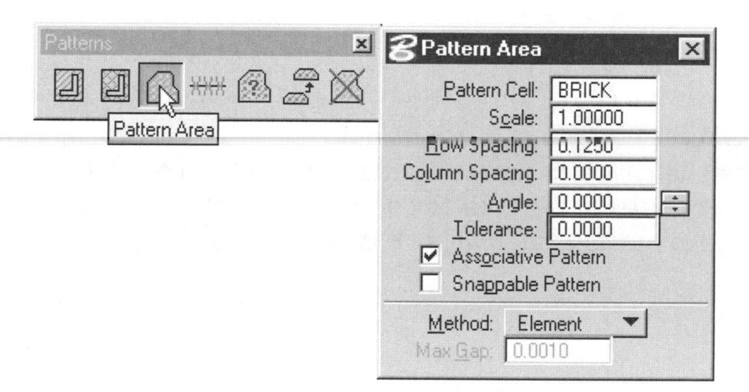

Figure P12–10 Pattern Area tool settings

MicroStation prompts:

Pattern Area > Identify element *(Select the block element to be patterned on the right side of the garage door.)*

Pattern Area > Accept @pattern intersection point *(Keypoint snap to the lower left corner of the block element to be patterned, then place a data point to initiate patterning.)*

STEP 4: Arrange the view window to display the block element on the left side of the garage door, and use the Pattern Area tool to place a brick pattern in it.

STEP 5: Arrange the view window to display all of the area above the garage door that is to be filled with two vertical brick rows.

STEP 6: Invoke the Crosshatch Area tool from the Patterns tool box, and, in the Tool Settings window, make the tool settings shown in Figure P12–11.

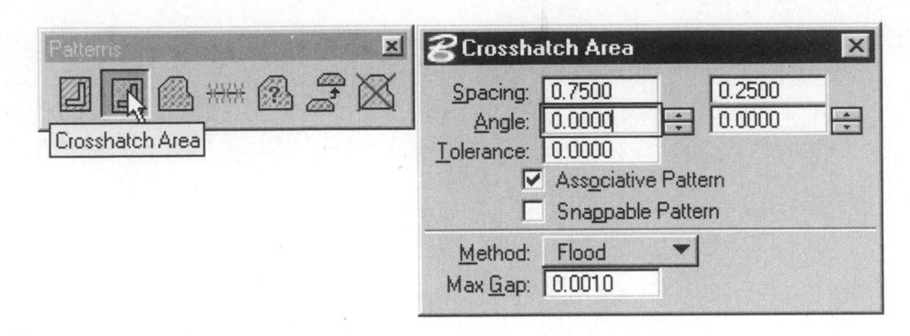

Figure P12–11 Required Crosshatch Area tool settings

MicroStation prompts:

Crosshatch Area > Enter data point inside area *(Click inside the brick area above the garage door.)*

Crosshatch Area > Accept @pattern intersection point *(Keypoint snap to the lower left corner of the area to be crosshatched, as shown in Figure P12–12, then place a data point to initiate crosshatching.)*

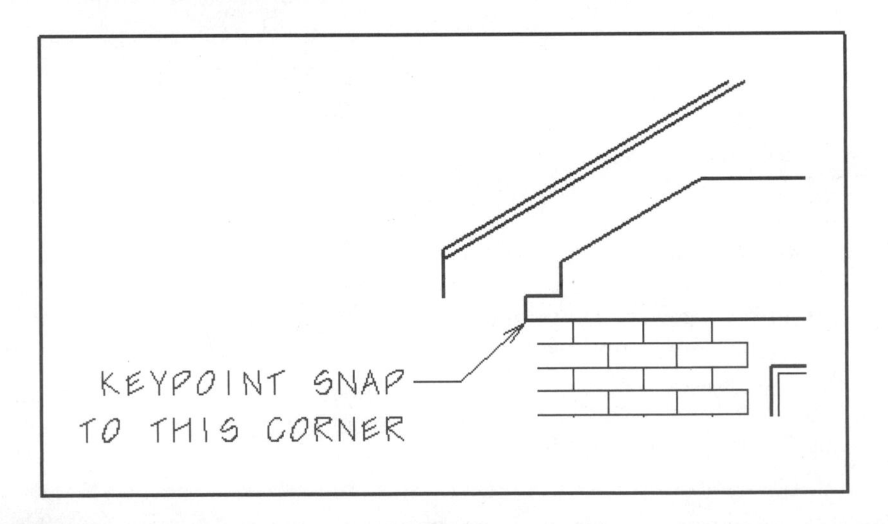

Figure P12–12 Acceptance point for crosshatching the area above the garage door

STEP 7: Arrange the view window to fill the view with the attic vent area.

STEP 8: Invoke the Hatch Area tool from the Patterns tool box, and, in the Tool Settings window, make the tool settings shown in Figure P12–13.

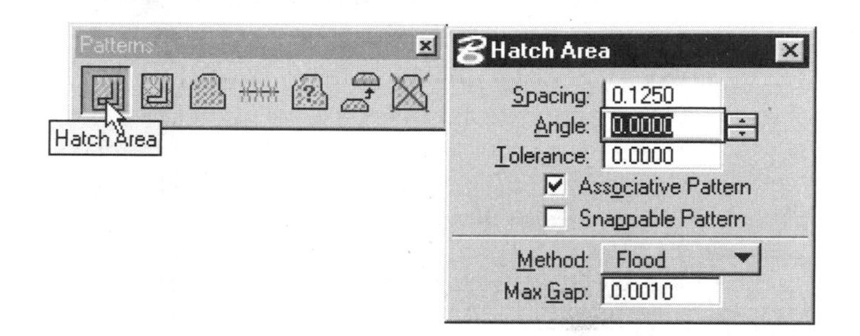

Figure P12–13 Required Hatch Area tool settings

MicroStation prompts:

Hatch Area > Enter data point inside area *(Click inside the vent area.)*

Hatch Area > Accept @pattern intersection point *(Keypoint snap to the lower left corner of the area to be hatched, then place a data point to initiate hatching.)*

STEP 9: Invoke the Fit View tool to fit the view.

STEP 10: Invoke the Save Settings tool from the File pull-down menu to save the settings.

DRAWING EXERCISES 12–1 THROUGH 12–5

Use the following table to set up the design files for Exercises 12–1 through 12–3.

SETTING	VALUE
Seed File	SEED2D.DGN
Working Units	10 TH Per IN and 1000 Pos Units Per TH
Grid	Grid Master = 0.1, Grid Reference = 10, Grid Lock to ON

Exercise 12–1

Nozzle

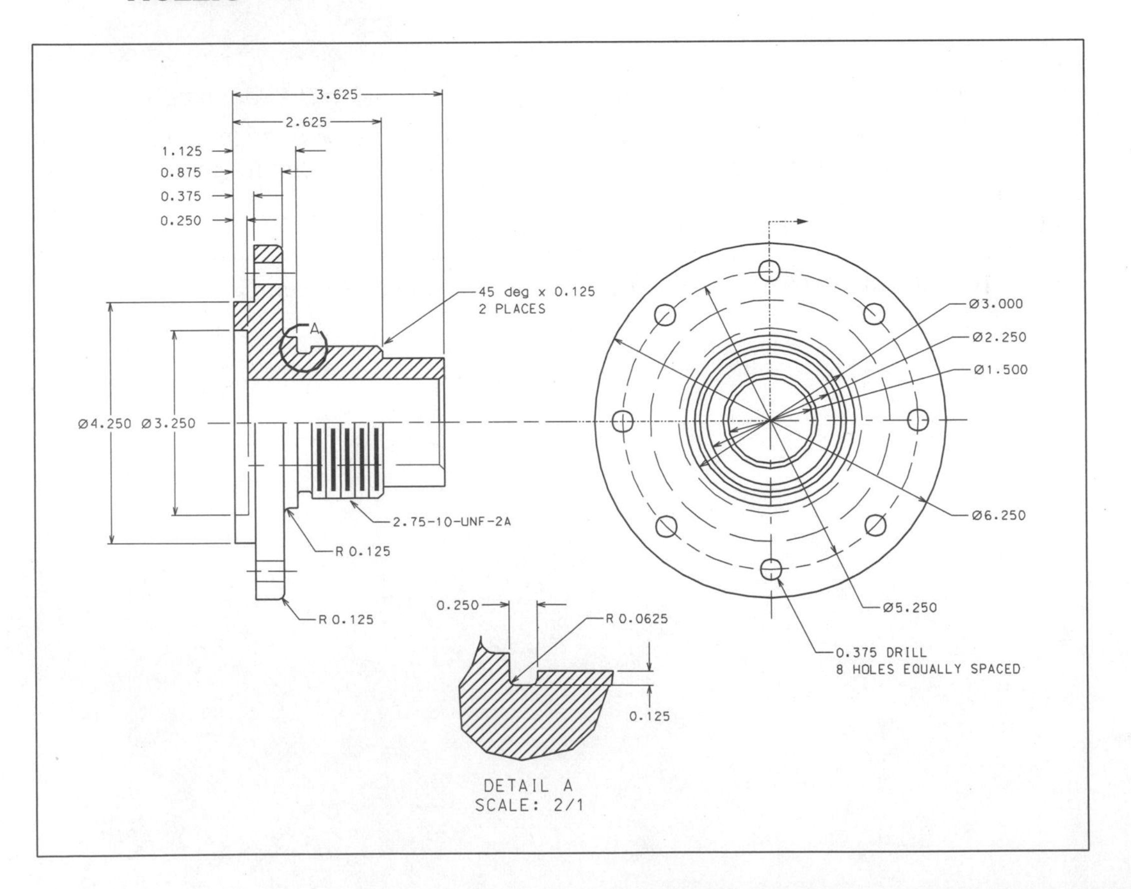

Exercise 12–2

Beam compass pointer

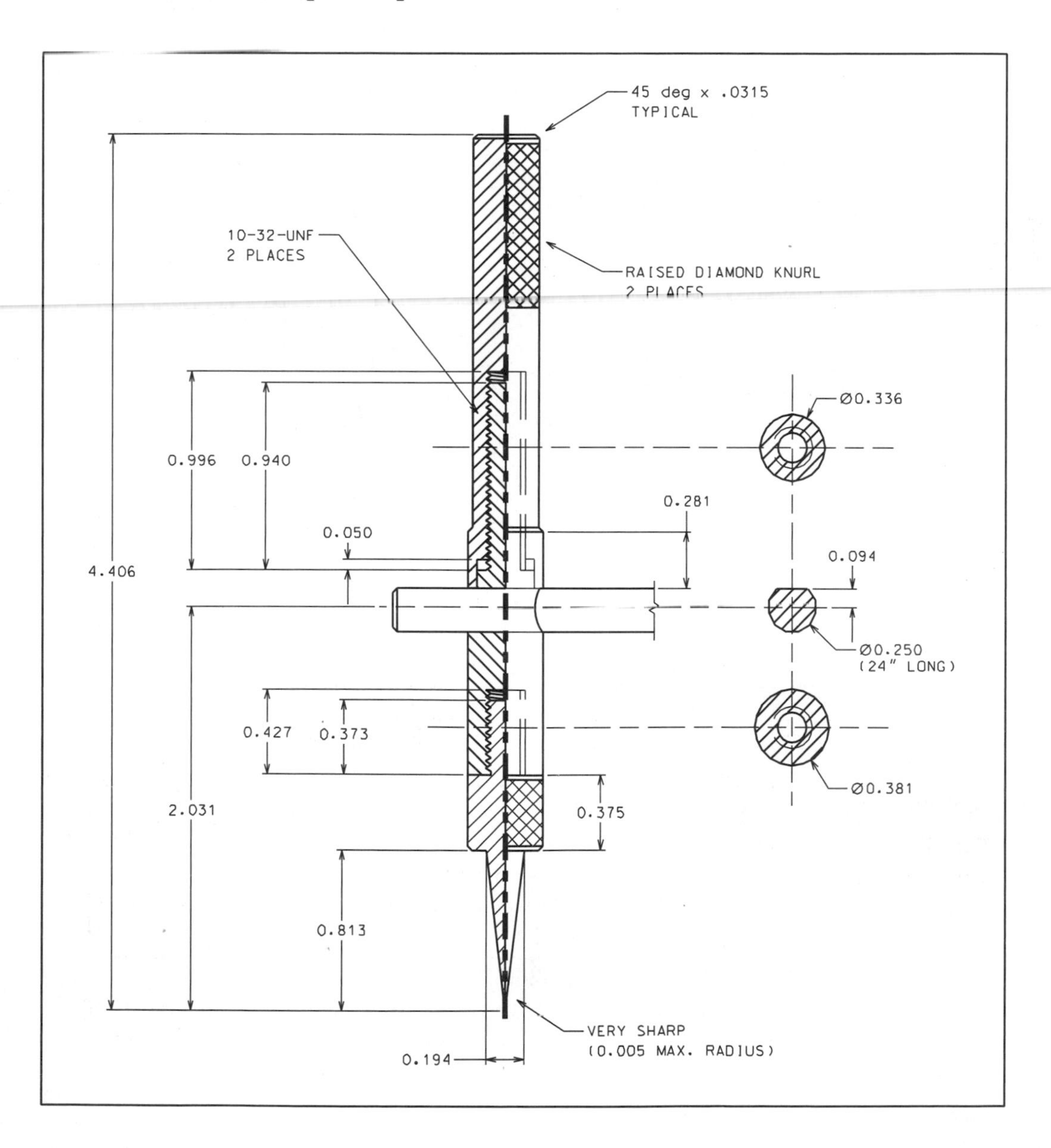

Exercise 12–3

Backplate cast aluminum

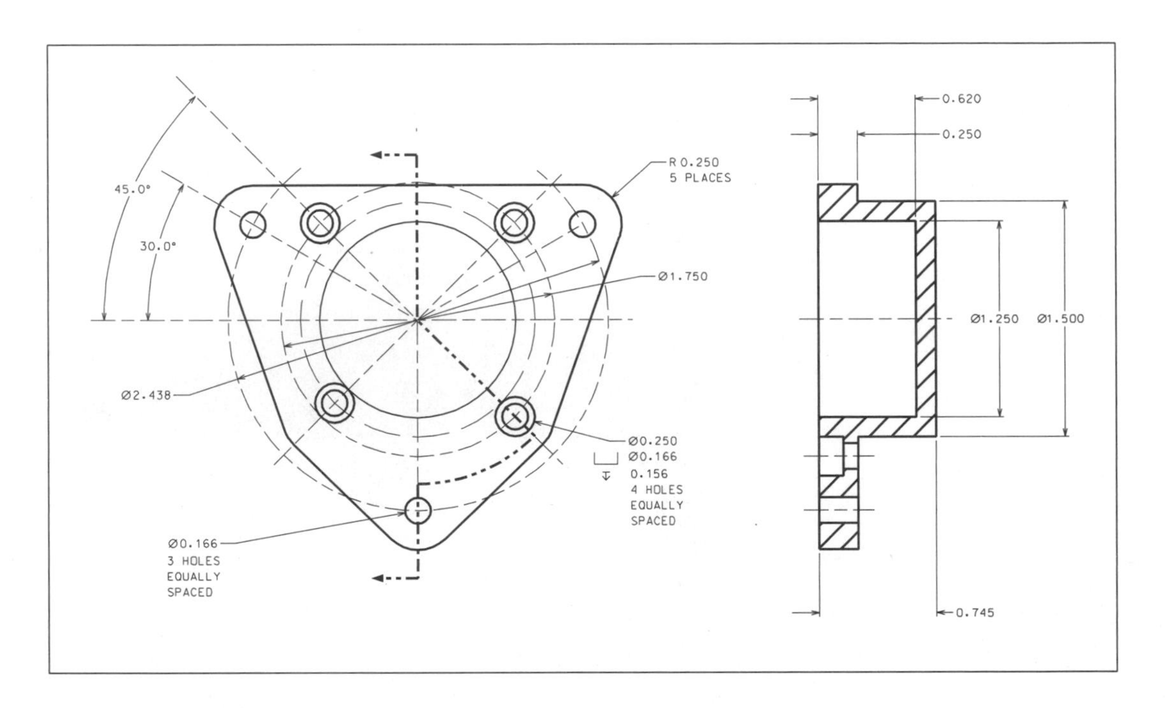

Use the following table to set up the design files for Exercises 12–4 and 12–5.

SETTING	VALUE
Seed File	SEED2D.DGN
Working Units	12" Per ' and 8000 Pos Units Per "
Grid	Grid Master = 0.5, Grid Reference = 24, GRID Lock to ON

Note: The cells used for patterning in the next two exercises were taken from a cell library furnished with MicroStation. The default path to the library in Microsoft Windows installations is: \Bentley\Workspace\System\Cell\Archpa.cel

Exercise 12–4

Parade Stand

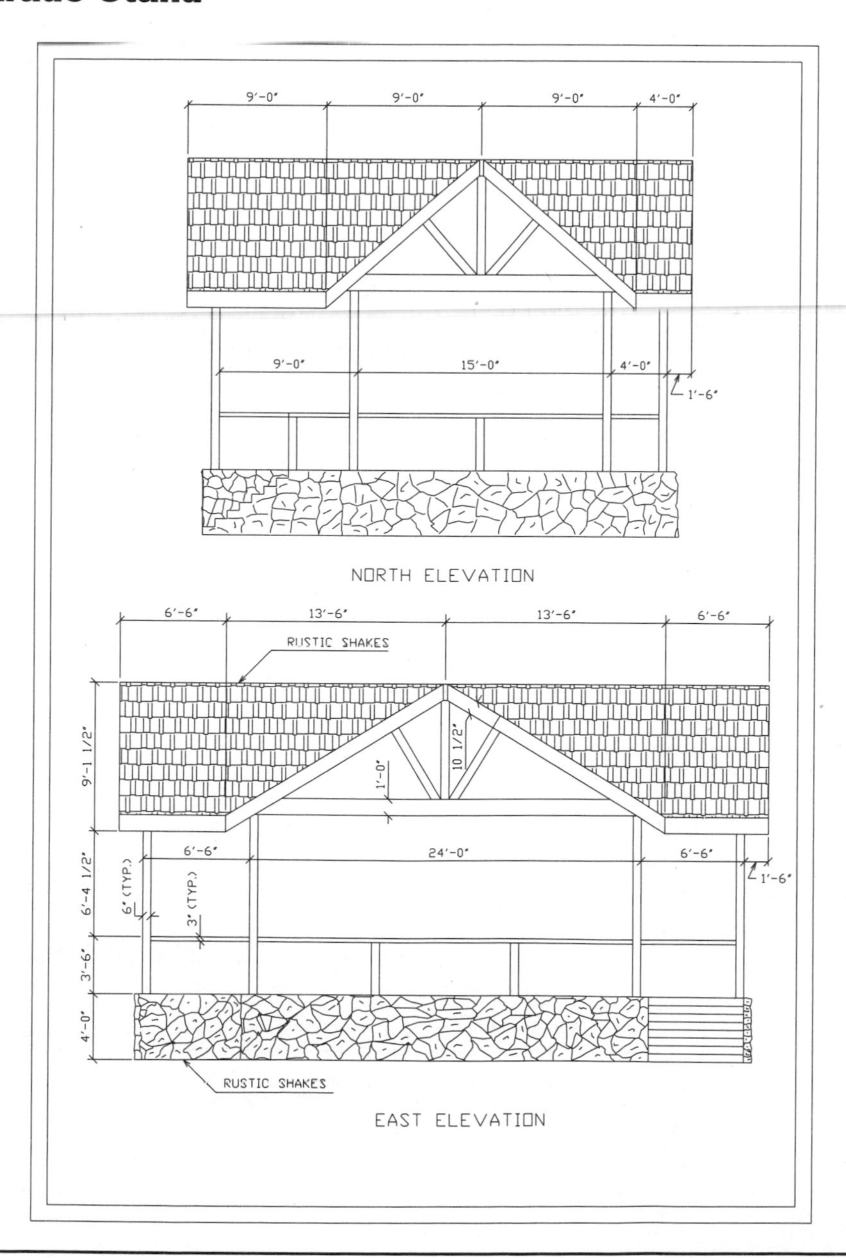

Exercise 12–5

Architectural detail

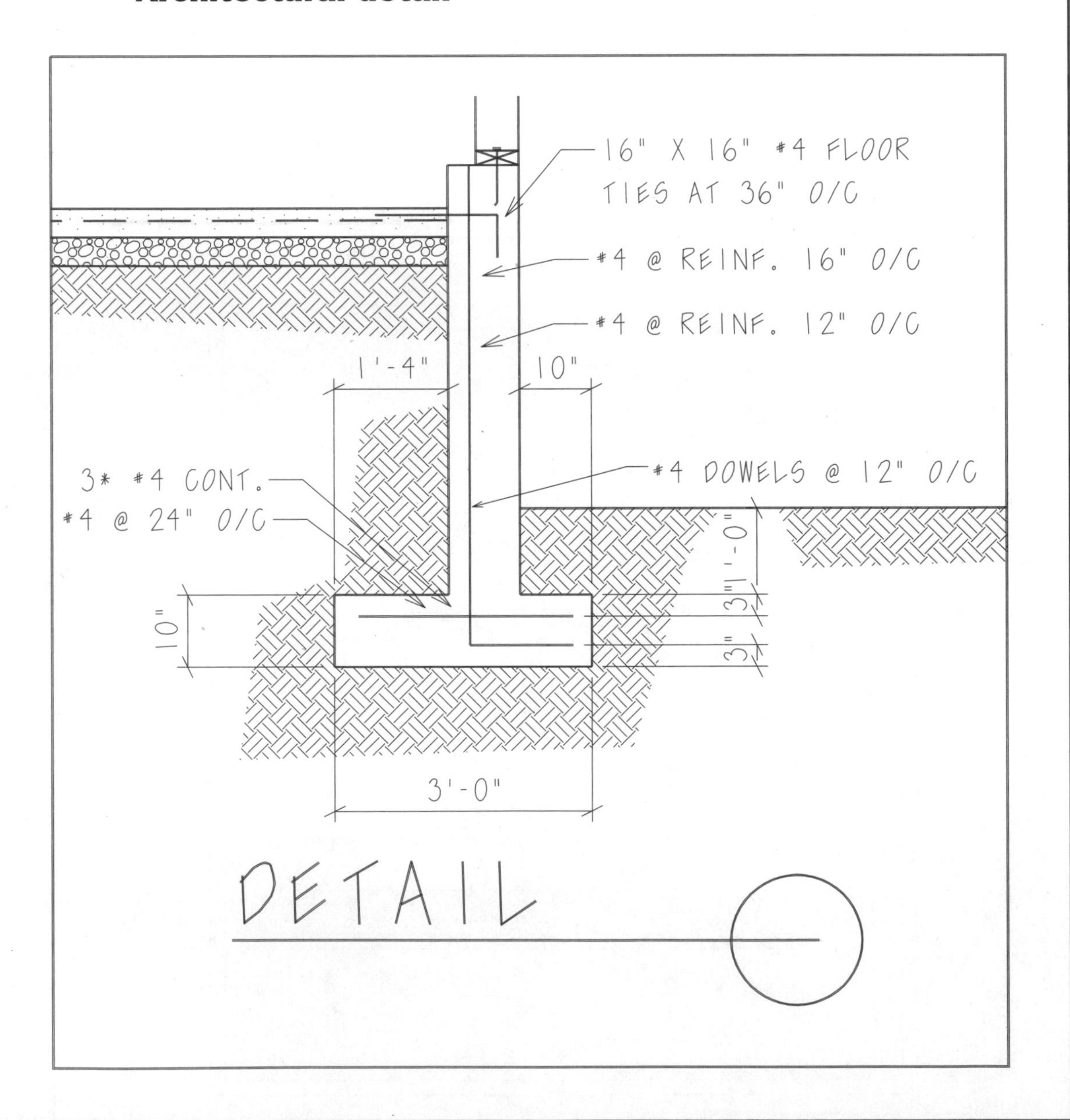

chapter 13

Reference Files

One of the most powerful timesaving features of MicroStation is its ability to view other design files while you are working in your design file. MicroStation lets you display the content of up to 255 other design files while working in your current design file. This function takes place in the form of reference files.

Objectives

After completing this chapter, you will be able to do the following:

- Attach a reference file.
- Move a reference file in the design plane.
- Scale your view of a reference file.
- Rotate your view of a reference file.
- Mirror your view of a reference file.
- Clip off part of your view of a reference file.
- Detach a reference file.
- Reload a reference file.

OVERVIEW OF REFERENCE FILES

When a design file is referenced in your design file, you can view and tentative point snap to all elements in the reference file, but each drawing's data is still stored and maintained in a separate design file. The only information about the referenced design file that becomes a permanent part of your design file is the name of the referenced design file and its directory path.

Elements in the reference file can be copied into your design file. Elements copied from a reference file become part of the current design file. This method is useful if, for example, a design file that was drawn previously contains information that can be used in your current design file.

Reference files may be scaled, moved, rotated, and viewed by levels via tools that are specifically programmed to work with reference files. The reference file behaves as one element when you manipulate it with reference file manipulation tools. The only exception is the regular Copy tool, in which the elements behave as individual elements. All of the manipulations are performed on your view of the referenced file, not the actual file. You cannot edit or modify the contents of a referenced file. If you need to make any changes to it, you actually have to load that file in MicroStation.

When you attach an external reference file, it is permanently attached until you detach it. When you open a design file, MicroStation automatically reloads each reference file; thus, you see the current version of each reference file when you view your design file.

The reference file manipulation tools are available in a tool box and from the Reference Files settings box. Open the Reference Files tool box:

Tools pull-down menu	Select Reference Files.
Key-in window	**dialog toolbox reference** (or **di to ref**) Enter

Micro displays the Reference Files tool box (see Figure 13–1).

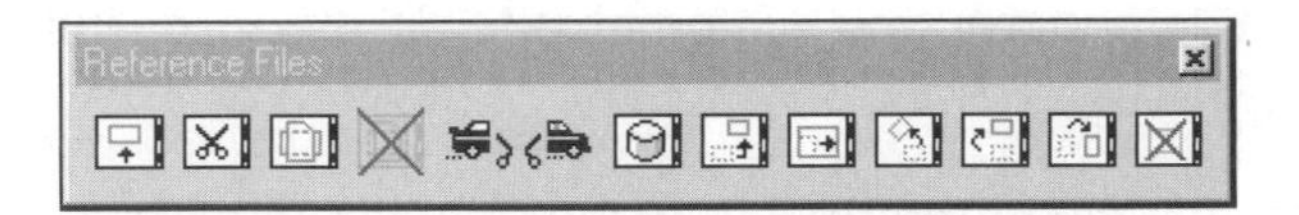

Figure 13–1 The Reference Files tool box

Open the Reference Files settings box:

File pull-down menu	Select Reference.
Key-in window	**dialog reference** (or **di ref**) Enter

MicroStation displays the Reference Files settings box. All reference file manipulation tools are in the settings box Tools pull-down menu (see Figure 13–2).

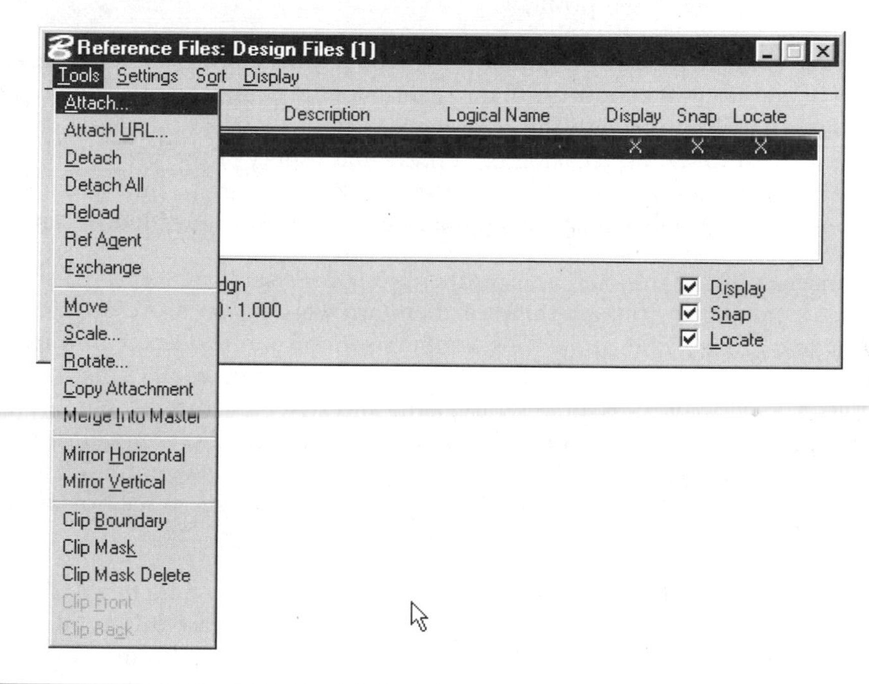

Figure 13–2 The Reference Files settings box

Note: The following reference file tool descriptions will show how the tool is opened from the tool box and from the settings box.

EXAMPLES OF USING REFERENCE FILES

Borders and title blocks are an excellent example of design files that are useful as reference files. The elements that make up the border use considerable space in a file and usually amount to around 40 to 50 KB. If a border and title block are drawn in each design file, it would waste a large amount of space, especially if you multiply 50 KB by 1000 design files. If reference files are used correctly, they can save a lot of disk space.

Accuracy and efficient drawing time are other important design features that are enhanced through the use of reference files. As mentioned earlier, when an addition or change is made to a design file that is serving as a reference file, all the design files that reference it will reflect the modifications. For example, let's say the name of the company is changed. Just change the company name in the title block design file, and all the design files that reference the title block automatically display the new company name the next time they are accessed. (That's much easier than accessing 1000 design files to correct one small detail.) Reference files save time and ensure the drawing accuracy required to produce a professional product.

When combined with the networking capability of MicroStation, external references give the project manager powerful new tools for coping with the realities of file management. The project manager can, by combining drawings through the referencing tools, see instantaneously the work of the various departments or designers working on a particular aspect of the project. If necessary, the manager can overlay a drawing where appropriate, track the progress, and maintain document integrity. At the same time, departments need not lose control over individual designs and details.

Let's look at an example of this function in operation. Let's say that you are a supervisor with three designers reporting to you. All three of them are working on a project, and each one is responsible for one-third of the project. As a supervisor you want to know how much progress each designer has made at the end of each day. Instead of calling up each of the three design files to see the progress, you can create a dummy design file and attach the three design files as reference files. Every day you can call up the dummy design file, and MicroStation will display the latest versions of the reference files attached to your design file. This will make your job a lot easier and give you an opportunity to put together all three pieces of the puzzle to see how they fit into the evolving design.

As mentioned earlier, you can attach up to 255 reference files at any time to a design file (earlier versions defaulted to a maximum of 32). If necessary, you can decrease this number of reference files in the Preference dialog box invoked from the Workspace pull-down menu. The valid range is 16 to 255. MicroStation allocates approximately 0.5 KB of memory for each allowed reference file, regardless of whether an attachment actually exists. Thus, you should set the maximum no higher than necessary. A change in the Preference dialog box is not effective until the next time you start MicroStation.

In addition to attaching a design file as a reference file, you can attach a raster image file as a reference file. Monochrome, continuous-tone (gray-scale), or color images in a variety of supported image formats can be attached.

ATTACHING A REFERENCE FILE

To attach a reference file, invoke the Attach Reference file tool:

Reference Files tool box	Select the Attach Reference File tool (see Figure 13–3).
Reference Files settings box	Tools > Attach...
Key-in window	**reference attach** (or **refe a**) Enter

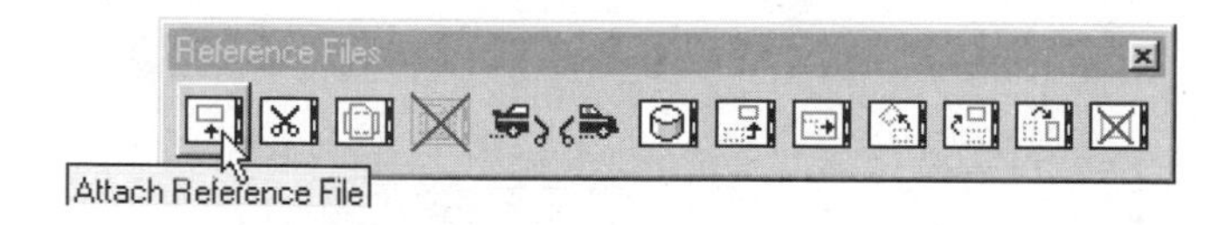

Figure 13–3 Invoking the Attach Reference File tool from the Reference Files tool box

MicroStation displays the Preview Reference dialog box, as shown in Figure 13–4. Select the design file to attach as a reference file from the list box and click the OK button to select it. If you want to make sure you selected the correct design file before selecting it, click the Preview button to display it in the Preview window.

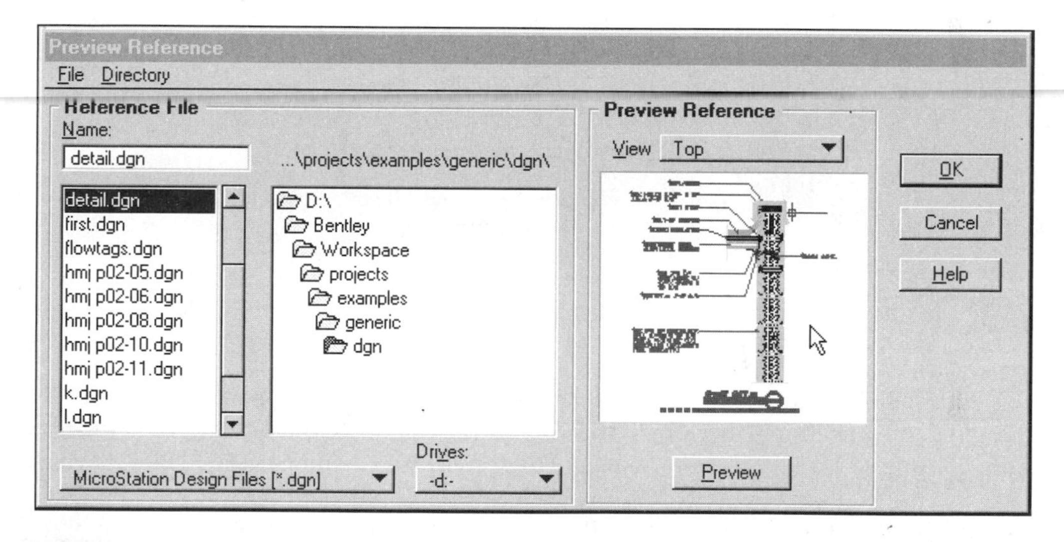

Figure 13–4 Preview Reference dialog box

After you select a design file, MicroStation displays the Attach Reference File dialog box, similar to the one shown in Figure 13–5, where you can enter additional, optional information. The top part of the dialog box displays the name and path of the design file being attached as a reference file. To retain the reference file's full path specification in the attachment information, turn ON the Save Full Path toggle button.

Click in the Logical Name edit field and type in a name. The logical name is a short identifier or nickname with a 22 character maximum that you can use later to identify the reference file when you need to manipulate it. The logical name is optional, unless you attach the same design file more than once to the current design file.

Figure 13–5 Attach Reference File dialog box

The next edit field is for a Description. The Description, which is optional, can serve to describe the purpose of the attachment. If more than two or three reference files are attached, it allows you to identify the purpose of attachment quickly. The Description cannot exceed 40 characters.

Select one of the two options available in the Attachment Mode. By default, the mode is set to Coincident. In Coincident mode, MicroStation attaches the reference file in such a way that the coordinates of the reference file's design plane are aligned with those of the active design file, without any rotation, scaling, or offset.

If the mode is set to Saved View, you can select one of the available Saved Views from the list box. A saved view allows you to display a clipped portion of the design file. You can attach the saved view at a specified scale factor by entering the appropriate factor in the Scale (Master:Ref) edit fields. The Saved View mode is available when the reference file contains saved views. (Saved View manipulation was described in Chapter 3.)

The Scale (Master:Ref) edit fields allow you to specify the scale factor as a ratio of the Master Units of the active design file and the Master Units of the reference file.

Note: You can specify the scale factor only when you attach a saved view of the reference file.

The Nest Depth edit field allows you to control what MicroStation does with reference files that are attached to the file you reference. If the depth is zero, only the file you select is attached to your design. If the depth is one, all files referenced by the file you are attaching will also be attached to your design. If the depth is two, all files referenced by the second level attachments will also be attached to you design. The higher the number the greater the attachment depth.

The Scale Line Styles toggle button controls the scaling of custom line styles in the reference file. If it is ON, custom line style components (for example, dashes) are scaled by the Scale (Master:Ref) factor. If it is OFF, custom line style components are not scaled.

Click the OK button to attach the selected reference file to the current design file, or click the Cancel button to cancel the reference file attachment.

You can reference the same design file more than one time. The only restriction is that you must provide a unique logical name for each attachment. Though it is not required, it is advisable to use the Description field to explain each attachment. That will save other people time in figuring out why the same file is referenced more than once. For example, the same design file might be referenced more than one time when different parts of the referenced design serve as details in your design file.

Figure 13–6 shows a design before and after attaching a reference file (title block).

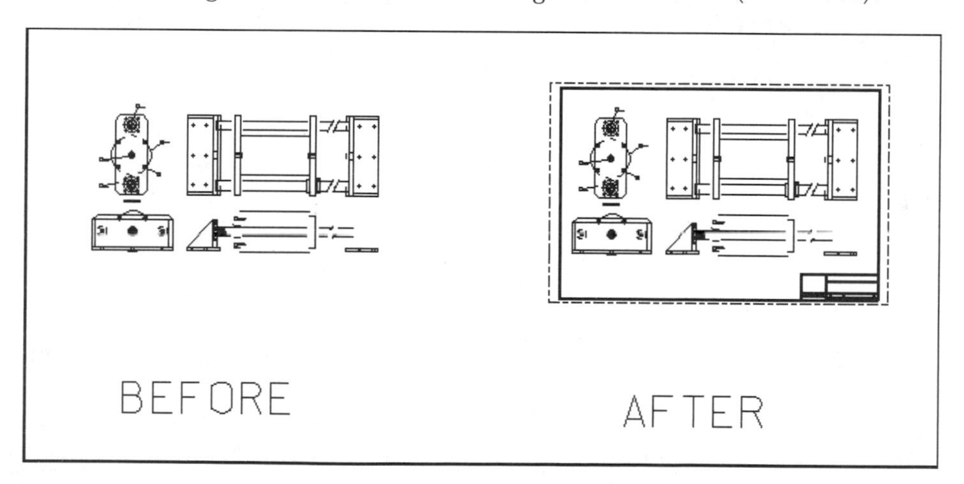

Figure 13–6 Design before and after attaching a reference file

Listing the Attached Reference Files

To list the reference files attached to the current design files, open a Reference File settings box:

Pull-down menu	File > Reference

MicroStation displays the Reference Files settings box, as shown in Figure 13–7. The settings box lists all the reference files attached to the current design file. In addition to listing the reference files, the settings box also provides the logical name of the reference file, and the status of the Display, Snap, and Locate options of the reference files.

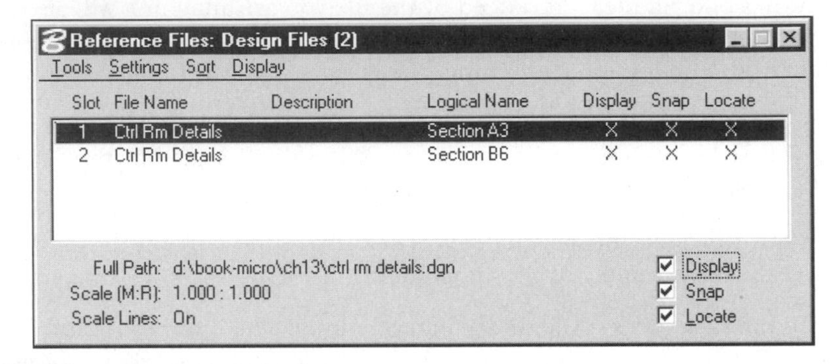

Figure 13–7 Reference Files settings box

The Display toggle button controls the screen display of the specified reference file. If for some reason you don't want to display the reference file but still want to keep it attached, set the Display toggle button to OFF; MicroStation will not display or plot the specified reference file.

The Snap toggle button controls the ability to tentative point snap to reference file elements. If Snap is set to OFF, you can see the reference file elements, but you cannot tentative snap on them. If Snap is ON, you can tentative snap to any element in the specified reference file.

The Locate toggle button controls the ability to locate reference files and copy them into your file via the Copy tool. If Locate is OFF, you cannot identify the elements to use with the Copy tool. If Locate is ON, you can identify the elements for use with the Copy tool.

The default settings for the Display, Snap, and Locate options are set in the Preferences.

If necessary, you can also change the attached file, the option to save its full path, its logical name, and its description. Double-click the reference file line in the Reference Files dialog box and MicroStation displays the Attachment Settings dialog box, shown in Figure 13–8. Make the required changes in the dialog box, then click the OK button to save the changes, or the Cancel button to close the box without making any changes.

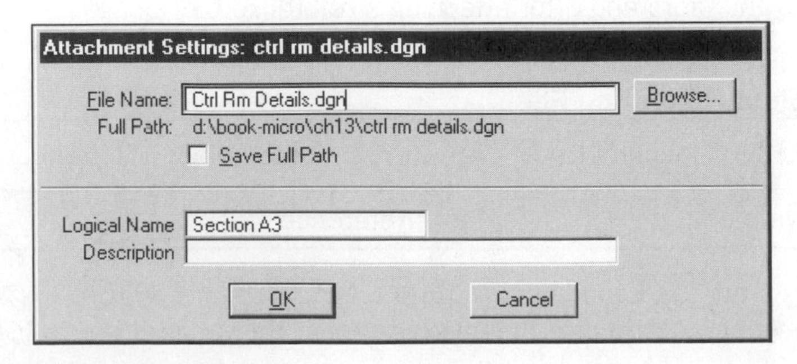

Figure 13–8 Attachment Settings box

REFERENCE FILE MANIPULATIONS

MicroStation provides a set of manipulation tools specifically designed to manipulate reference files. As mentioned earlier, the reference file behaves as one element, and there is no way to drop it. The reference manipulation tools include Move, Scale, Rotate, Mirror Horizontal, Mirror Vertical, Clip Boundary, Clip Mask, and Clip Mask Delete.

Selecting the Reference File

As discussed at the beginning of this chapter, the reference file manipulation tools are available in the Reference Files tool box and in the Tools pull-down menu in the Reference Files settings box. The first step in using any of these tools is to select the reference file to be manipulated.

- If the Reference Files settings box is open, the reference file selected in the list field is the one that will be manipulated when you select a manipulation tool from the tool box or settings box.
- If the Reference Files settings box is closed, you must use the manipulation tools in the tool box, and each command will start by asking you which reference file is to be manipulated. To select a reference file, click on an element in the reference file, then click a second time to select it. If you have the Key-in window open, you can also type the reference file's logical name to select it.

These tools do not remain active after you are finished using them, so you must select the reference file each time you select a tool. For example, if you need to move a reference file a second time, you must select the tool again, and select the reference file (unless it is already selected in the settings box).

Note: The reference file manipulation tool discussions that follow assume you have already selected the file in the Reference Files settings box, and they do not show the reference file selection prompts. If you are using the tool from the tool box and the settings box is closed, you will be prompted to select the reference file to manipulate the tools.

Move Reference File

The Move Reference File tool allows you to move a reference file from one location to another. Before you invoke the tool, highlight the reference file by clicking the Data button in the Reference File settings box.

To move a reference file, invoke the Move Reference file tool:

Reference Files tool box	Select the Move Reference File tool (see Figure 13–9).
Reference Files settings box	Tools > Move
Key-in window	**reference move** (or **refe m**) Enter

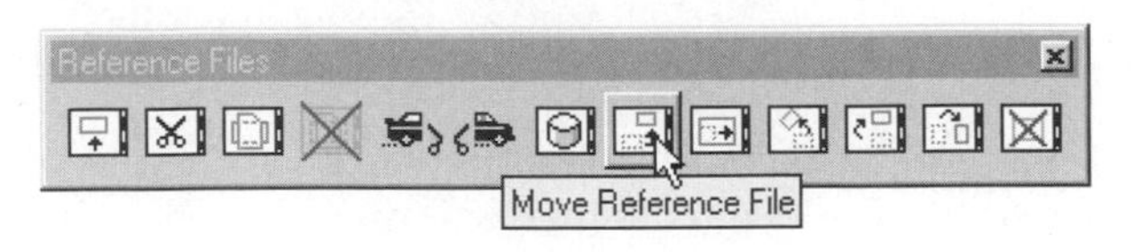

Figure 13–9 Invoking the Move Reference File tool from the Reference Files tool box

MicroStation prompts:

> Move Reference File > Enter point to move from *(Place a data point anywhere on the reference file.)*
>
> Move Reference File > Enter point to move to *(Place a data point where you want to move the reference file in reference to the first data point.)*

After the file is moved, the tool exits. To move another reference file or the same file to another location, select the tool again. Figure 13–10 shows a view before and after moving a reference file border.

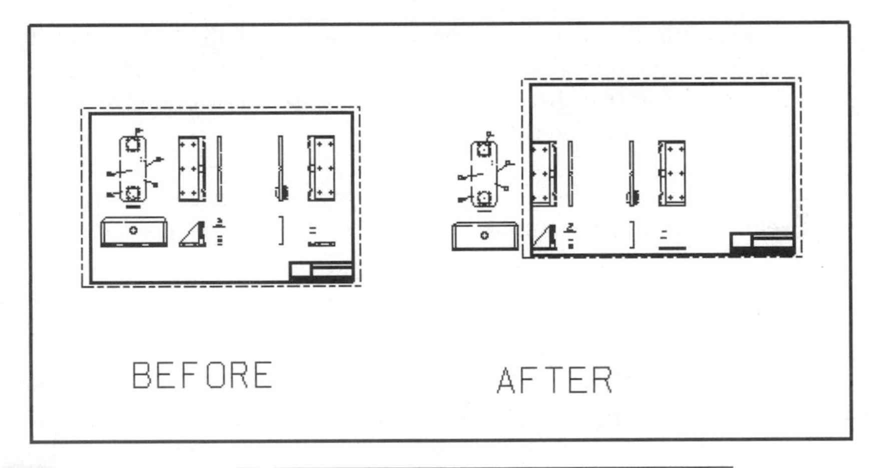

Figure 13–10 The design view before and after moving a reference file

Scale Reference File

The Scale Reference File tool allows you to enlarge or reduce a reference file. Before you invoke the tool, highlight the reference file by clicking the Data button in the Reference File settings box.

To scale a reference file, invoke the Scale Reference file tool:

Reference Files tool box	Select the Scale Reference File tool (see Figure 13–11).
Reference Files settings box	Tools > Scale...
Key-in window	**reference scale** (or **refe s**) Enter

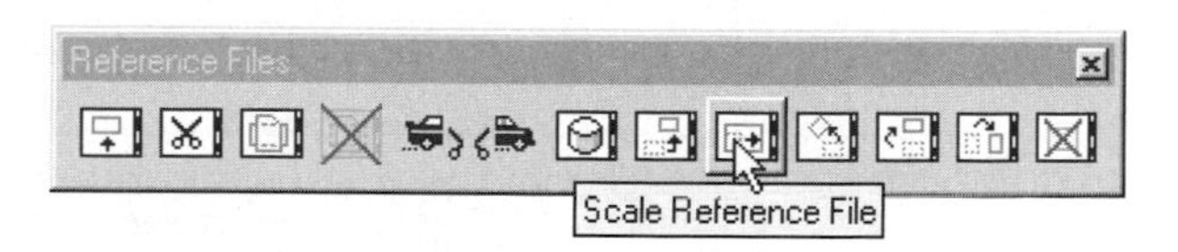

Figure 13–11 Invoking the Scale Reference File tool from the Reference Files tool box

In the Tool Settings Window, specify the scale factor in terms of a ratio of the Master Units of the active design file to the Master Units of the reference file.

MicroStation prompts:

> Scale Reference File > Enter point to scale reference file about *(Place a data point about which the file will scale.)*

For example, if the same working units settings are used in both design files, a ratio of 3:1 scales a reference file up three times; a ratio of 1:5 scales a reference file down five times. If you make a mistake, you can always invoke the Undo tool to undo the last operation.

After the reference file is scaled, the tool exits. To scale another reference file or the same file with a different scale factor, select the tool again.

Note: The scale factor ratio between the active design file and the reference file is *not* cumulative. For instance, if you specify a scale of 3:1 followed by 6:1, the final result will be 6:1. Figure 13–12 shows a view before and after scaling the reference file border.

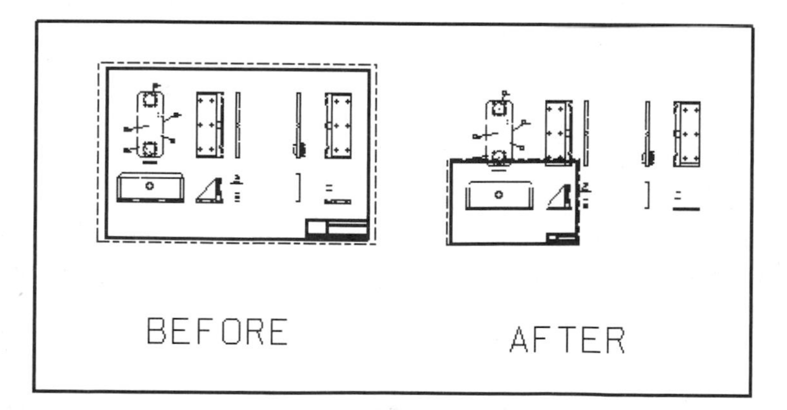

Figure 13–12 The design view before and after scaling a reference file

Rotate Reference File

The Rotate Reference File tool allows you to rotate a reference file to any angle around a pivot point. Before you invoke the tool, highlight the reference file by clicking the Data button in the Reference File settings box.

To rotate a reference file, invoke the Rotate Reference file tool:

Reference Files tool box	Select the Rotate Reference File tool (see Figure 13–13).
Reference Files settings box	Tools > Rotate...
Key-in window	**reference rotate** (or **refe ro**) Enter

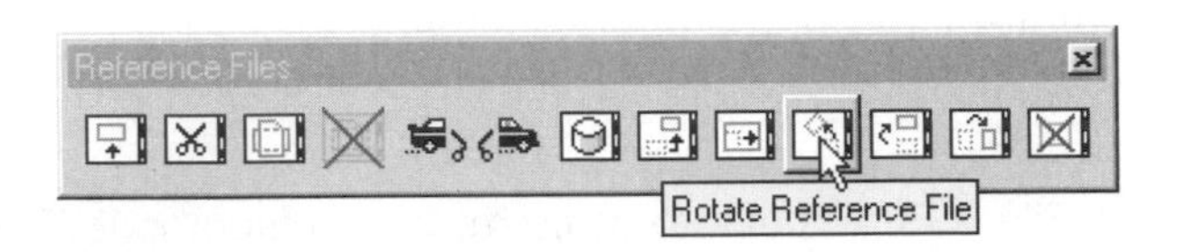

Figure 13–13 Invoking the Rotate Reference File tool from the Reference Files tool box

In the Tool Settings window, specify the Z axis rotation angle.

MicroStation prompts:

Rotate Reference File > Enter point to rotate reference file about *(Place a data point to define a pivot point about which the reference file rotates.)*

For example, an angle of 60 degrees rotates the reference file 60 degrees counterclockwise from its current position around the pivot point. If you make a mistake, you can always invoke the Undo tool to undo the last operation.

After the rotation completes, the tool exits. To rotate another reference file or the same file, select the tool again.

Mirror Horizontal Reference File

The Mirror Horizontal Reference File tool allows you to mirror a reference file about the horizontal (or *X*) axis. Before you invoke the tool, highlight the reference file by clicking the Data button in the Reference File settings box.

To mirror a reference file about the horizontal axis, invoke the Mirror Reference file Horizontal tool:

Reference Files tool box	Select the Mirror Reference File Horizontal tool (see Figure 13–14).
Reference Files settings box	Tools > Mirror Horizontal
Key-in window	**reference mirror horizontal** (or **refe mi h**) Enter

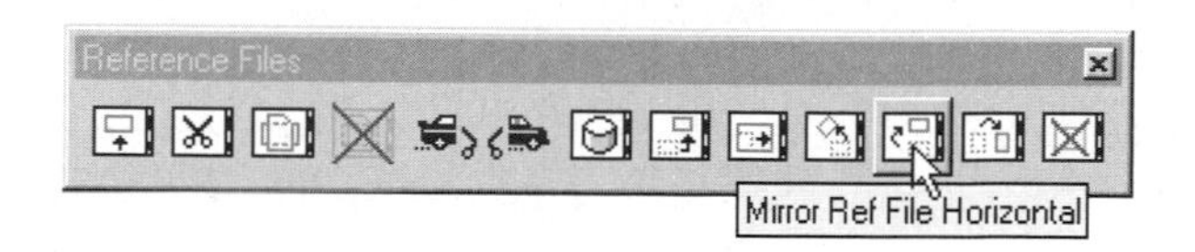

Figure 13–14 Invoking the Mirror Reference File Horizontal tool from the Reference Files tool box

MicroStation prompts:

> Mirror Reference File About Horizontal > Enter point to mirror about *(Place a data point to define the mirror axis.)*

After the mirror is completed, the tool exits. To mirror horizontally, another reference file or the same file, select the tool again.

Mirror Vertical Reference File

The Mirror Vertical Reference File tool allows you to mirror a reference file about the vertical (or *Y*) axis. Before you invoke the tool, highlight the reference file by clicking the Data button in the Reference Files settings box.

To mirror a reference file about the vertical axis, invoke the Mirror Reference file Vertical tool:

Reference Files tool box	Select the Mirror Reference File Vertical tool (see Figure 13–15).
Reference Files settings box	Tools > Mirror Vertical
Key-in window	**reference mirror vertical** (or **refe mi v**) Enter

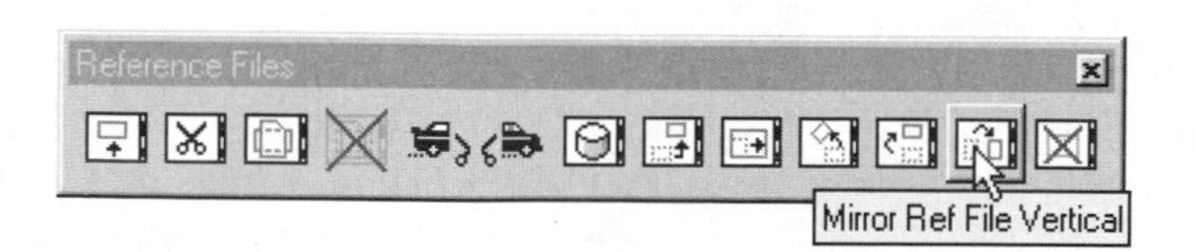

Figure 13–15 Invoking the Mirror Reference File Vertical tool from the Reference Files tool box

MicroStation prompts:

> Mirror Reference File About Vertical > Enter point to mirror about *(Place a data point to define the mirror axis.)*

After the mirror is completed, the tool exits. To mirror vertically another reference file, select the tool again.

Reference Clip Boundary

The Reference Clip Boundary tool lets you display only a desired portion of a reference file. Place a Fence on the reference file to define the clipping boundary. The clipping boundary can have up to 60 vertices, and you can place a circular fence. Nonrectangular clipping boundaries are displayed in a view and plotted only if the Fast Ref Clipping View Attribute is set to OFF for the selected view.

Before you invoke the tool, highlight the reference file by clicking the Data button in the Reference Files settings box.

To clip a boundary, first place a fence, then invoke the Clip Reference File tool:

Reference Files tool box	Select the Clip Reference File tool (see Figure 13–16).
Reference Files settings box	Tools > Clip Boundary
Key-in window	**reference clip boundary** (or **refe c bo**) Enter

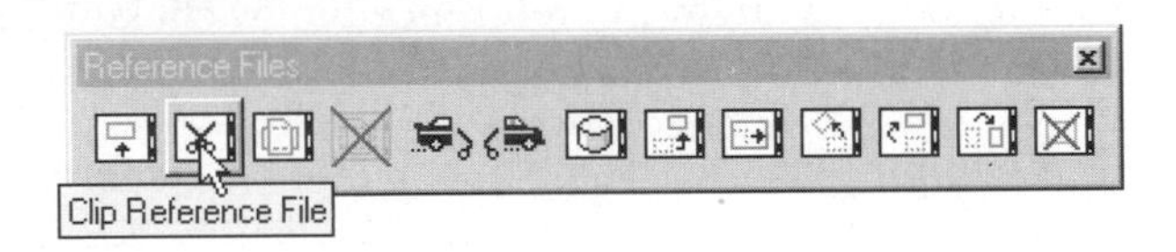

Figure 13–16 Invoking the Clip Reference File tool from the Reference Files tool box

MicroStation displays only the part of the reference file enclosed within the fence. If you make a mistake, you can always invoke the Undo tool to undo the last operation.

Reference Clip Mask

The Reference Clip Mask tool, like the Clip Boundary tool, allows you to display only a portion of a reference file. Clip Boundary displays the part inside a fence, whereas the Clip Mask tool displays the part outside the fence. Place a Fence on the reference file to define the desired clipping boundary. The clipping mask can have up to 60 vertices.

Before you invoke the tool, highlight the reference file by clicking the Data button in the Reference Files settings box.

To clip a mask, first place a fence, and then invoke the Mask Reference File tool:

Reference Files tool box	Select the Mask Reference File tool (see Figure 13–17).
Reference Files settings box	Tools > Clip Mask
Key-in window	**reference clip mask** (or **refe c m**) Enter

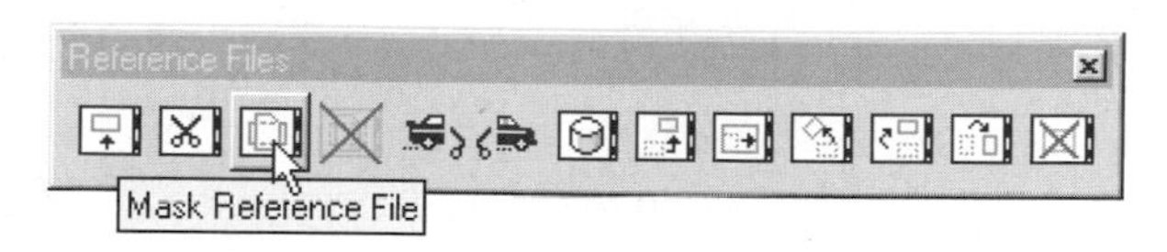

Figure 13–17 Invoking the Mask Reference File tool from the Reference Files tool box

MicroStation displays only the part of the reference file outside the fence. If you make a mistake, you can always invoke the Undo tool to undo the last operation.

Deleting Reference File Clipping Mask(s)

The Delete Clip Mask tool can delete a reference file's clipping masks. Before you invoke the tool, highlight the reference file by clicking the Data button in the Reference Files settings box.

To delete a reference file's clipping mask, invoke the Delete Clip Mask tool:

Reference Files tool box	Select the Delete Clip Mask tool (see Figure 13–18).
Reference Files settings box	Tools > Clip Mask Delete
Key-in window	**reference clip mask delete** (or **refe c m d**) Enter

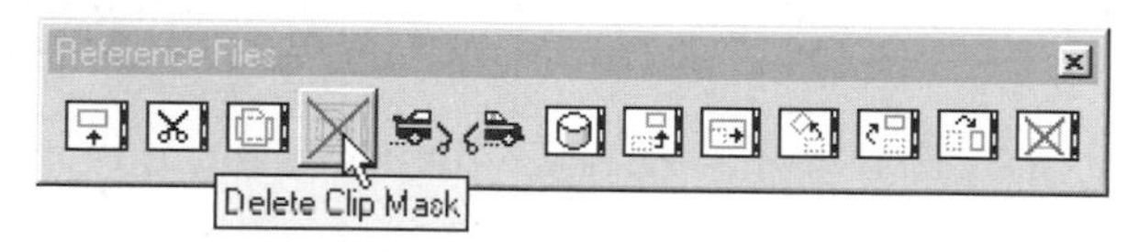

Figure 13–18 Invoking the Delete Clip Mask tool from the Reference Files tool box

MicroStation deletes the selected clip mask.

Detaching the Reference File

When you no longer need a reference file, you can detach it from the current design file. Once the reference file is detached, there is no more link between the reference file and the current design file.

Before you invoke the tool, highlight the reference file you want to detach by clicking the Data button on the file name in the Reference Files settings box.

To detach a reference file, invoke the Detach Reference File tool:

Reference Files tool box	Select the Detach Reference File tool (see Figure 13–19).
Reference Files settings box	Tools > Detach
Key-in window	**reference detach** (or **refe d**) Enter

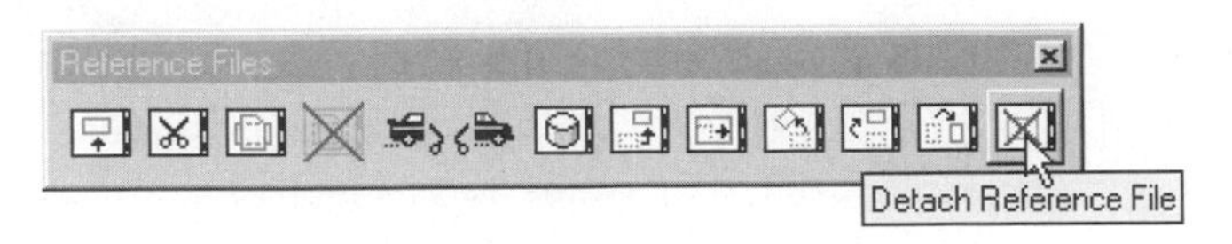

Figure 13–19 Invoking the Detach Reference File tool from the Reference Files tool box

MicroStation displays an alert box to confirm that the selected reference file is to be detached. Click the OK button to detach.

To detach all the attached reference files, invoke the Detach All tool:

Reference Files settings box	Tools > Detach All
Key-in window	**reference detach all** (or **refe d all**) Enter

MicroStation displays an alert box to confirm that all the reference files are to be detached. Click the OK button to detach all the reference files.

Reloading the Reference File

MicroStation automatically loads all reference files attached to a design file only when you first open the design file. To get the latest version of a reference file while your design is open, you must use the Reload tool. The Reload tool is helpful, especially in a network environment, to access the latest version of the reference design file while you are working in a design to which it has been attached.

Before you invoke the tool, highlight the reference file you want to reload by clicking the Data button in the Reference Files settings box.

To reload a reference file, invoke the Reload Reference File tool:

Reference Files tool box	Select the Reload Reference File tool (see Figure 13–20).
Reference Files settings box	Tools > Reload
Key-in window	**reference reload** (or **refe r**) Enter

MicroStation reloads the selected reference file.

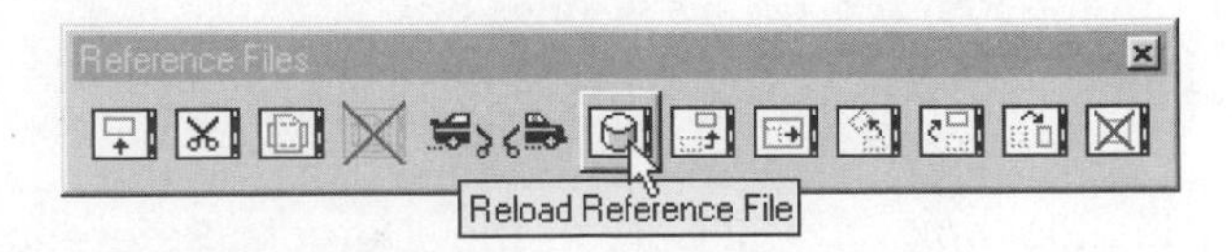

Figure 13–20 Invoking the Reload Reference File tool from the Reference Files tool box

Exchange Reference File

MicroStation provides a tool that allows you to make one of the reference files the active file, so you can make changes to it. When you invoke this tool, the active file is closed and the selected reference file is loaded in MicroStation for editing. Before you invoke the tool, highlight the reference file by clicking the Data button in the Reference Files settings box.

To exchange a reference file for the active design file, invoke the Exchange tool:

Reference Files settings box	Tools > Exchange

MicroStation exchanges the files, and the selected reference file becomes the active design file.

Copy Reference File Attachments

The Copy Attachment allows you to make additional attachments of an existing reference file attachment. Everything about the attachment is copied, although a unique logical name is provided for each copy. You attach it one time and then make copies of the attachment for the other details. Before you invoke the tool, highlight the reference file by clicking the Data button in the Reference Files settings box.

To copy an attachment, invoke the Copy Attachment tool:

Reference Files settings box	Tools > Copy Attachment

MicroStation opens the Copy Attachment dialog box to allow you to enter the Number of Copies you want to make. Enter the number of copies in the edit field, then click the OK button to copy the attachments.

Levels and Level Symbology

In addition to controlling the display of the reference file, you can manipulate the display of specific levels in a reference file. To manipulate the levels, first highlight the reference file you want to manipulate by clicking the Data button in the Reference Files settings box, then invoke the Levels settings box:

Reference Files settings box	Settings > Levels

MicroStation displays the Reference Levels settings box, as shown in Figure 13–21. Select the level numbers you want to turn OFF or ON. Select the appropriate view window number from the View Number options menu. Once the selection is completed, click the Apply button.

Similarly, you can manipulate the Level Symbology (explained in Chapter 14) for a reference file.

Figure 13–21 Reference Levels settings box

REVIEW QUESTIONS

Write your answers in the spaces provided.

1. List at least two benefits of using reference files.

   ___

2. By default, how many reference files can you attach to a design file?

   ___

3. What effect does attaching reference files have on a design file size?

   ___

4. List the tools that are specifically provided to manipulate reference files.

   ___

   ___

5. Give the steps involved in attaching a reference file.

   ___

   ___

6. Explain the difference between the Reference Clip Boundary tool and the Reference Clip Mask tool.

   ___

   ___

7. List the steps involved in detaching a reference file.

   ___

   ___

PROJECT EXERCISE

This project exercise provides step-by-step instructions for creating the design shown in Figure P13–1. The intent is to guide you in using reference files to complete a design.

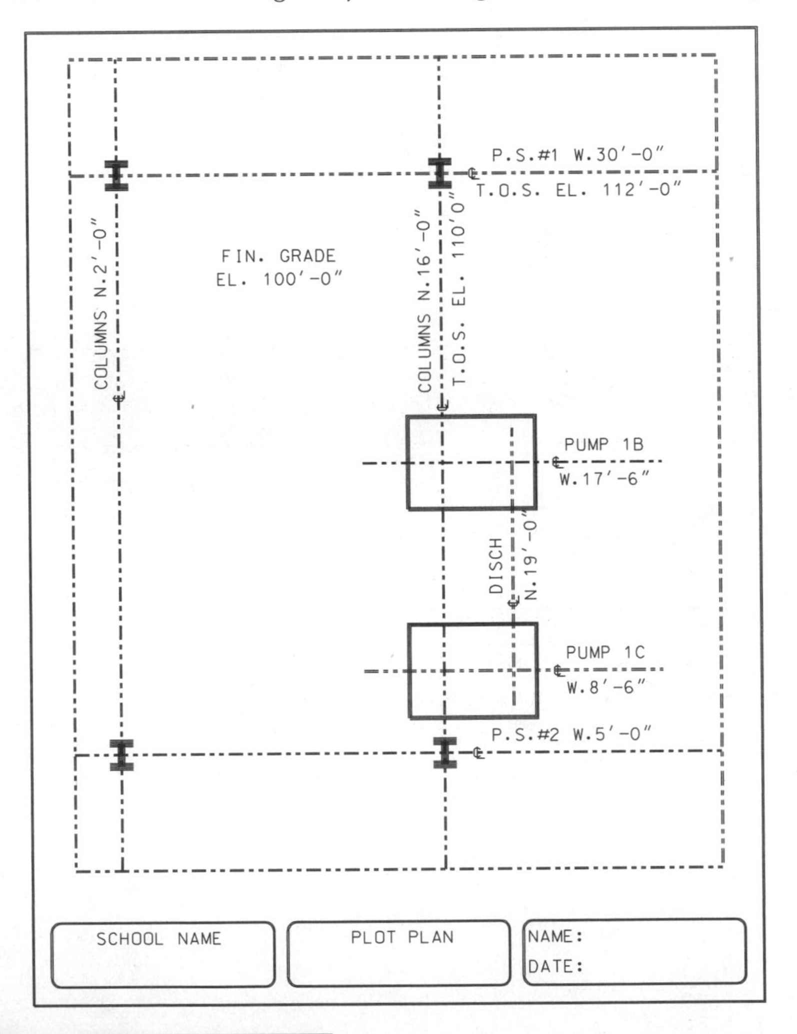

Figure P13–1 Completed project design

This project creates separate design files for the following:

- Pipe rack plot plan
- Pump foundation plot plan

- Border
- Complete design formed from references to the other three design files

Note: As you complete each step in the project procedures, place a check mark by the step to help you keep up with where you are in the project.

Create the Pipe Rack Plot Plan

This procedure creates the design shown in Figure P13–2. The pipe rack is drawn using offsets from the design origin point.

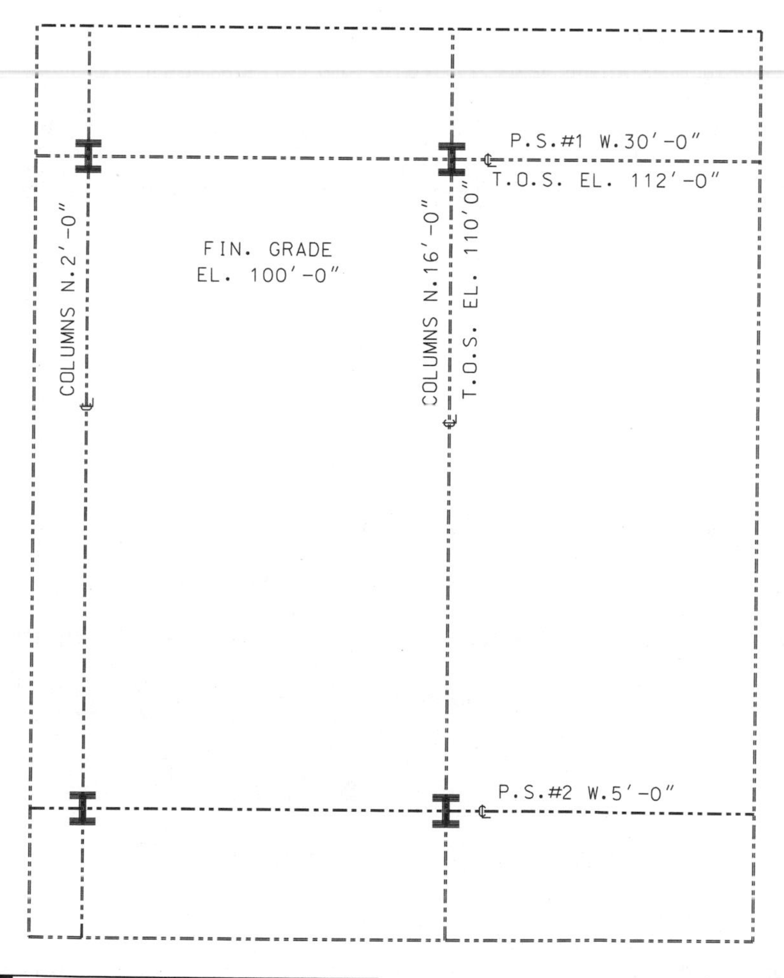

Figure P13–2 Completed pipe rack plot plan

STEP 1: Invoke MicroStation via the normal technique for the operating system on your workstation.

STEP 2: Create a new design file named RACK.DGN using the SEED2D.DGN seed file.

STEP 3: Set the following design parameters:

- Working Units Master Units name to ' (for feet), the Sub Units name to " (for inches), and the resolution ratios to 12 and 8000.
- Grid Master to 0.5, Grid Reference to 2, and Grid Lock to OFF
- Active Level = 2
- Line Weight = 1
- Line Style = 6
- Color = Yellow

STEP 4: Invoke the Save Settings tool from the File pull-down menu to save the design settings.

STEP 5: Draw the following two unconnected lines:

- XY = 2,0 to DL = 0,35
- XY = 0,5 to DL = 28,0

STEP 6: Make parallel copies of the two lines:

- Horizontal line: 25 feet above the original line
- Vertical line: 14 feet to the right of the original line

STEP 7: Draw a block from XY = 0,0 to XY = 28,35, as shown in Figure P13–3.

STEP 8: Set the element attributes to:

- Active Level = 1
- Line Style = 0
- Line Weight = 4
- Color = Blue

STEP 9: Draw a 1-foot by 1-foot I-beam centered on the lower left centerline intersection, as shown in Figure P13–4.

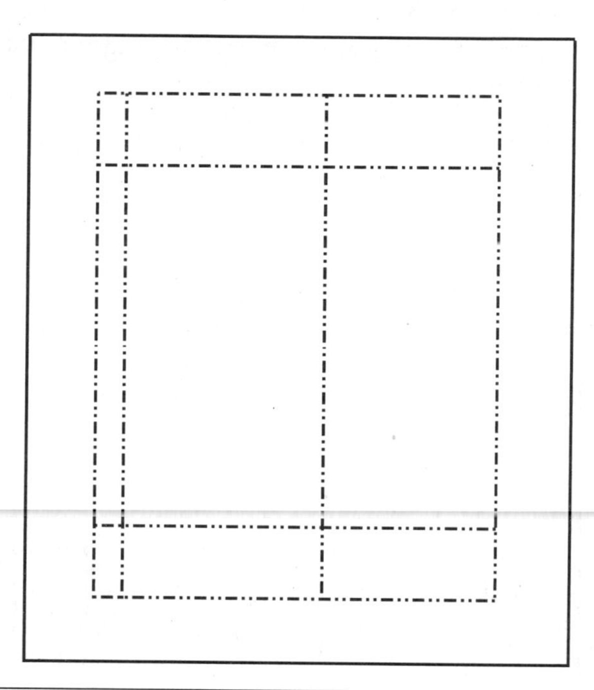

Figure P13–3 Pipe rack centerlines and perimeter

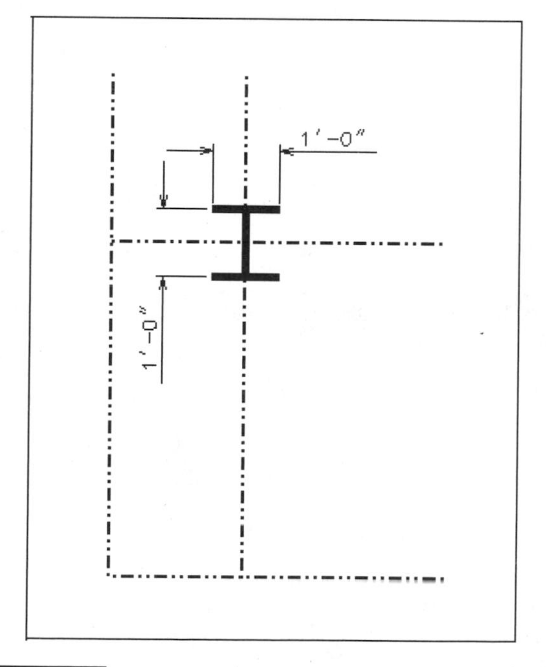

Figure P13–4 Lower left I-beam

STEP 10: Use the Array tool to place a rectangular array with the following settings (see Figure P13–5):

- Rows = 2
- Columns = 2
- Row spacing = 25
- Column spacing = 14

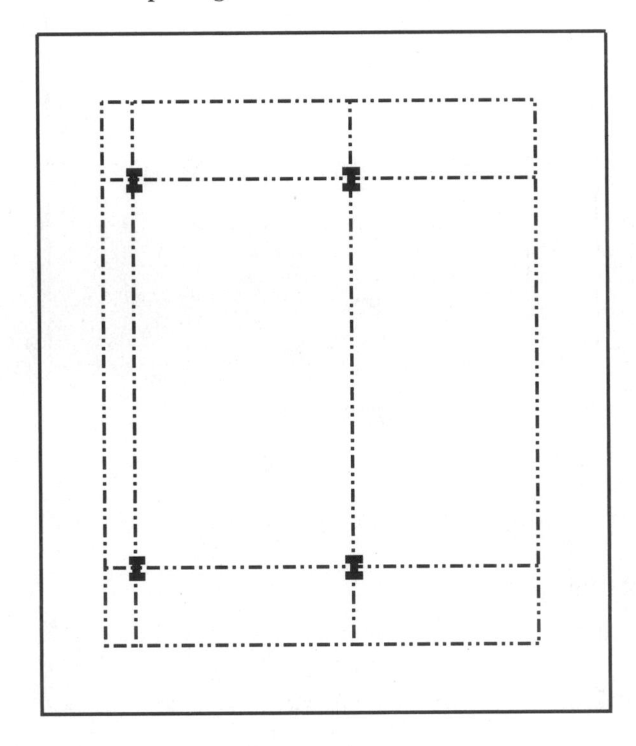

Figure P13–5 Results of the Rectangular Array tool

STEP 11: Set the element attributes to:

- Active Level = 3
- Line Style = 0
- Line Weight = 0
- Color = Blue

STEP 12: Set the text parameters to:

- Font = 3

- Height and Width = 0.125
- Line Spacing = 0.125

Note: The text size is 0.125', so it will plot 0.125" (1/8") high at a plot scale of ¼" = 1'.

STEP 13: Place all text shown in Figure P13–2.

Note: To place the symbol on the centerlines, set the Font to 15 (IGES1001), and place the lowercase letter **q**.

STEP 14: Invoke the Save Settings tool from the File pull-down menu to save the design settings.

Create the Pump Foundations Plot Plan

This procedure creates the design shown in Figure P13–6. The foundations are drawn using offsets from the design origin point.

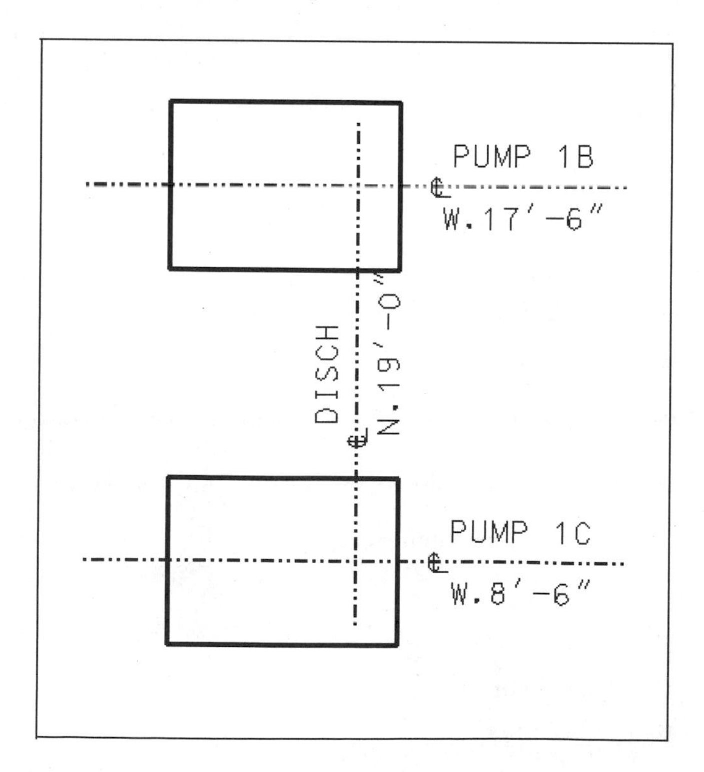

Figure P13-6 Pump foundations plot plan

STEP 1: Create a new design file named PUMPS.DGN, using the SEED2D.DGN seed file.

STEP 2: Set up the design settings as follows:

- Working Units Master Units name to ' (for feet), the Sub Units name to " (for inches), and the resolution ratios to 12 and 8000
- Grid Master to 0.5, Grid Reference to 2, and Grid Lock to OFF
- Active Level = 2
- Line Weight = 1
- Line Style = 6
- Color = Yellow

STEP 3: Invoke the Save Settings tool from the File pull-down menu to save the design settings.

STEP 4: Draw the following two unconnected lines:

- XY = 12.5,8.5 to DL = 13,0
- XY = 19,7 to DL = 0,12

STEP 5: Make a parallel copy of the horizontal line 9' above the original.

STEP 6: Set the element parameters to:

- Active Level = 1
- Line Style = 0
- Line Weight = 2
- Color = Blue

STEP 7: Draw a block from XY = 14.5,6.5 to DL = 5.5,4, as shown in Figure P13–7.

STEP 8: Make a copy of the block exactly 8' above the original.

STEP 9: Set the element attributes to:

- Active Level = 3
- Line Style = 0
- Line Weight = 0
- Color = Blue

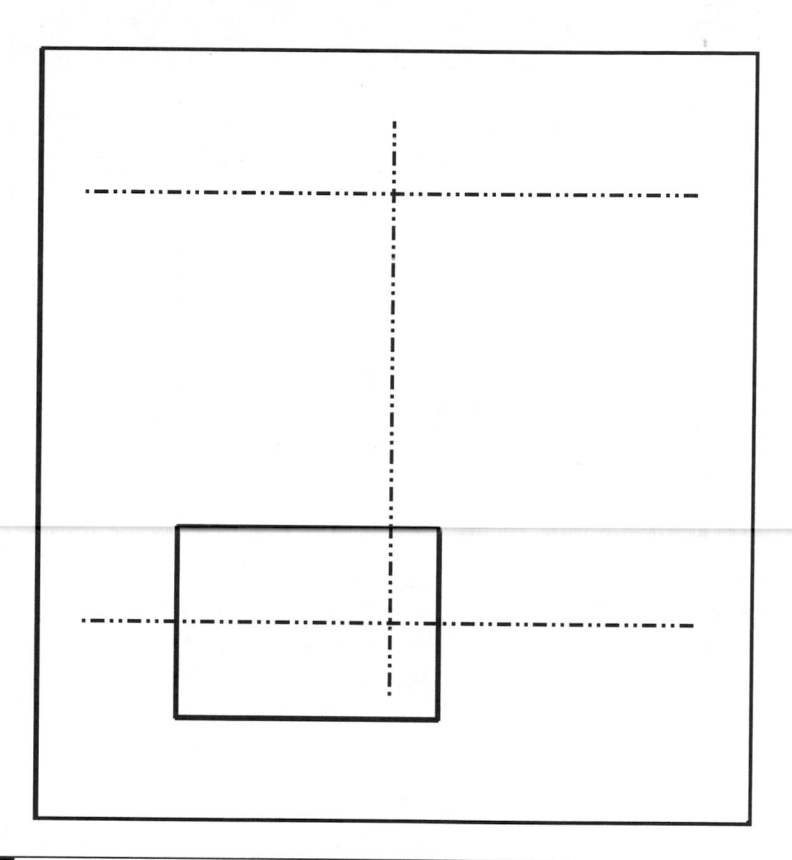

Figure P13–7 Pump centerlines and lower pump foundations

STEP 10: Set the text parameters to:

- Font = 3
- Height and Width = 0.125'
- Line Spacing = 0.125'

STEP 11: Place text as shown in Figure P13–6.

STEP 12: Invoke the Save Settings tool from the File pull-down menu to save the design settings.

Create the Border

This procedure creates the design shown in Figure P13–8.

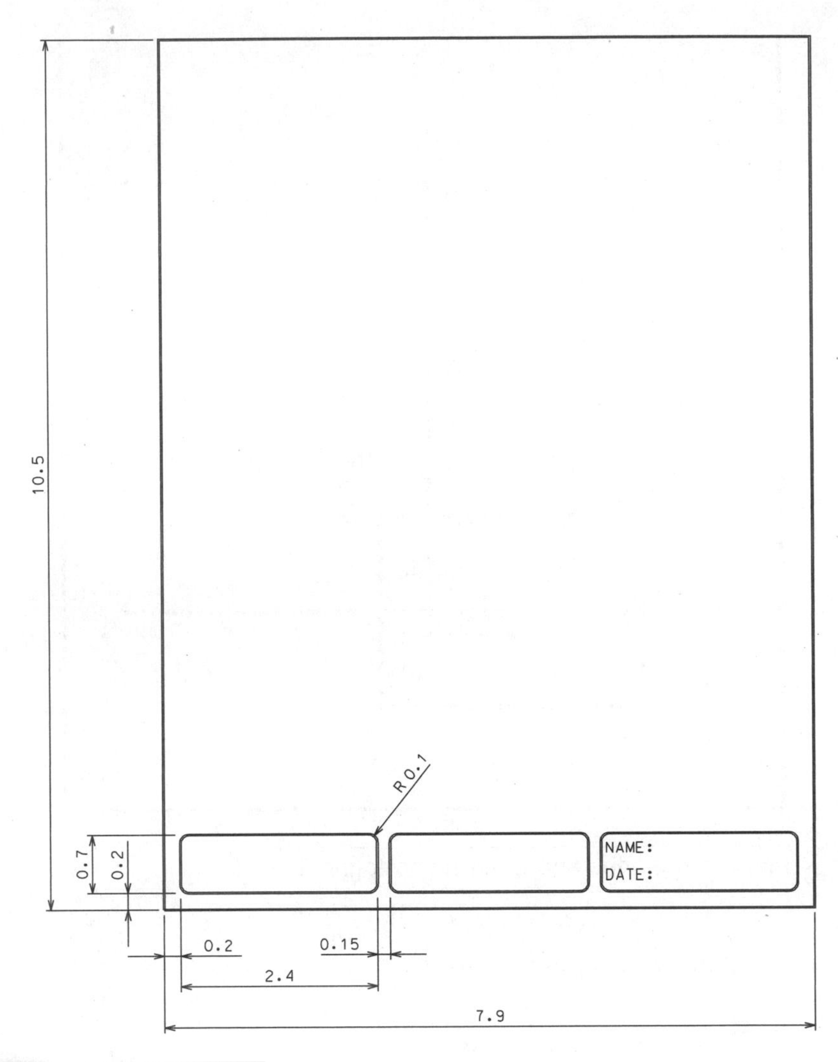

Figure P13–8 Letter-size border

STEP 1: Create a new design file named BORDER.DGN, using the SEED2D.DGN seed file.

STEP 2: Adjust the design settings as follows:

- Working Units Master Units name to IN (for inches), the Sub Units name to TH (for tenths of an inch), and the resolution ratios to 10 and 1000
- Grid Master to 0.1, Grid Reference to 10, and Grid Lock to ON
- Active Level = 1

- Line Weight = 1
- Line Style = 0
- Color = Blue

STEP 3: Invoke the Save Settings tool from the File pull-down menu to save the design settings.

STEP 4: Draw the border and title block using the dimensions shown in Figure P13–8.

STEP 5: Set the following parameters:

- Line Weight = 0
- Text Height and Width = 0.125"
- Text Font = 3

STEP 6: Place the text as shown in Figure P13–8.

STEP 7: Invoke the Fit View tool to fit the view.

STEP 8: Open the Save Views settings box by selecting Saved Views from the Utilities pull-down menu.

STEP 9: In the Saved Views settings box:

- Key-in **border** in the Source Name field.
- Key-in **Letter-size border** in the Source Description field.
- Click the Save button to create a saved view.

STEP 10: Invoke the Save Settings tool from the File pull-down menu to save the design settings.

Create the Complete Design

This procedure creates the composite design by referencing the previous design files (see Figure P13–1).

STEP 1: Create a new design file named CH13.DGN, using the SEED2D.DGN seed file.

STEP 2: Set up the design settings as follows:

- Working Units Master Units name to ' (for feet), the Sub Units name to " (for inches), and the resolution ratios to 12 and 8000
- Grid Master to 0.5, Grid Reference to 2, and Grid Lock to OFF

- Active Level = 1
- Line Weight = 0
- Line Style = 0
- Color = Blue

STEP 3: Open the Reference Settings box by selecting Reference from the File pull-down menu.

STEP 4: Invoke the Attach tool from the Tools pull-down menu in the Reference Files settings box. MicroStation opens the Attach Reference File dialog box.

STEP 5: Select RACK.DGN from the appropriate directory, and click the OK button to attach the design file as a reference file to the current design file. MicroStation displays the Attach Reference File dialog box.

STEP 6: In the Attach Reference file dialog box, make the following settings:

- Key-in **rack** in the Logical Name field.
- Key-in **Pipe rack plot plan** in the Description field.
- Click the OK button to attach the pipe rack design file as a reference file.

STEP 7: Invoke the Fit View All tool to fit the view window.

STEP 8: Similarly, attach the PUMPS.DGN design file as a reference file, keying-in **pumps** for the Logical Name and **Pump foundations** for the Description.

STEP 9: Also attach BORDER.DGN as a reference file. In the Attach Reference File dialog box, set the following settings:

- Key-in **border** in the Logical Name field.
- Key-in **The drawing border** in the Description field.
- Set the Attachment Mode to Saved View.
- Select the border saved view from the Saved Views list box.
- Set the Scale (Master:Ref) ratio to 4 to 1, as shown in Figure P13–9.
- Click the OK button.

Note: The Scale (Master:Ref) ratio scales the border for plotting at 1/4" = 1'. This works because the border design has inches as its Master Units and the active design has feet as its Master Units. The ratio of 4 to 1 means every four feet of active design displays one inch of border reference, or 1" = 4', (divide both sides of the equation by 4 and you get ¼" = 1').

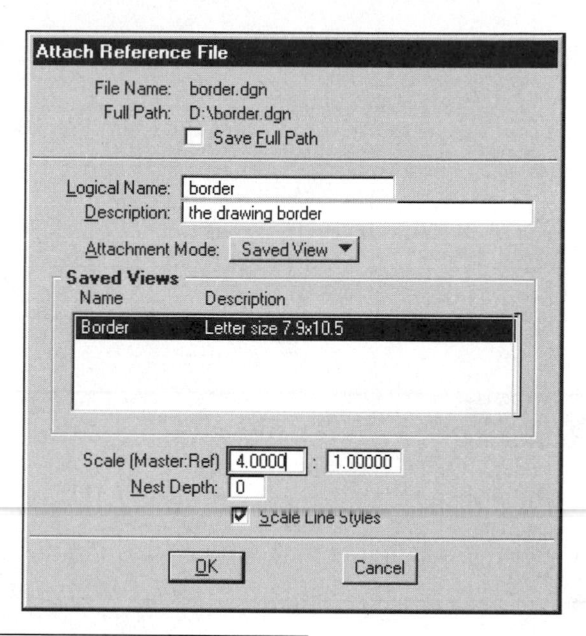

Figure P13–9 Settings for attaching the border

STEP 10: Drag the reference file dynamic outline until all of the pipe rack is inside and close to the top of the outline, then click the Data button to place the border.

STEP 11: If the border is not correctly placed, select it in the Reference Files settings box, then invoke the Move tool from the pull-down menu Tools in the Reference Files settings box.

MicroStation prompts:

Move Reference File > Enter point to move from *(Place a data point somewhere in the design.)*

Move Reference File > Enter point to move to *(Move the drawing pointer the direction and distance the border needs to be moved, then place a second data point to move the reference file.)*

STEP 12: Set the text parameters to:

- Font = 3
- Text Height and Width = 0.125'

STEP 13: Fill in the title block text, as shown in Figure P13–1.

STEP 14: Invoke the Fit View tool to fit the view.

STEP 15: Invoke the Save Settings tool from the File pull-down menu to save the design settings.

DRAWING EXERCISES 13–1 THROUGH 13–5

Use the following table to set up the design files for Exercises 13–1 through 13–5.

SETTING	VALUE
Seed File	SEED2D.DGN
Working Units	10 TH Per IN and 1000 Pos Unit Per Th
Grid	Grid Master = 0.25, Grid Reference = 4

Exercise 13–1

Border

Draw the border shown in the figure. This border is used in the following exercises.

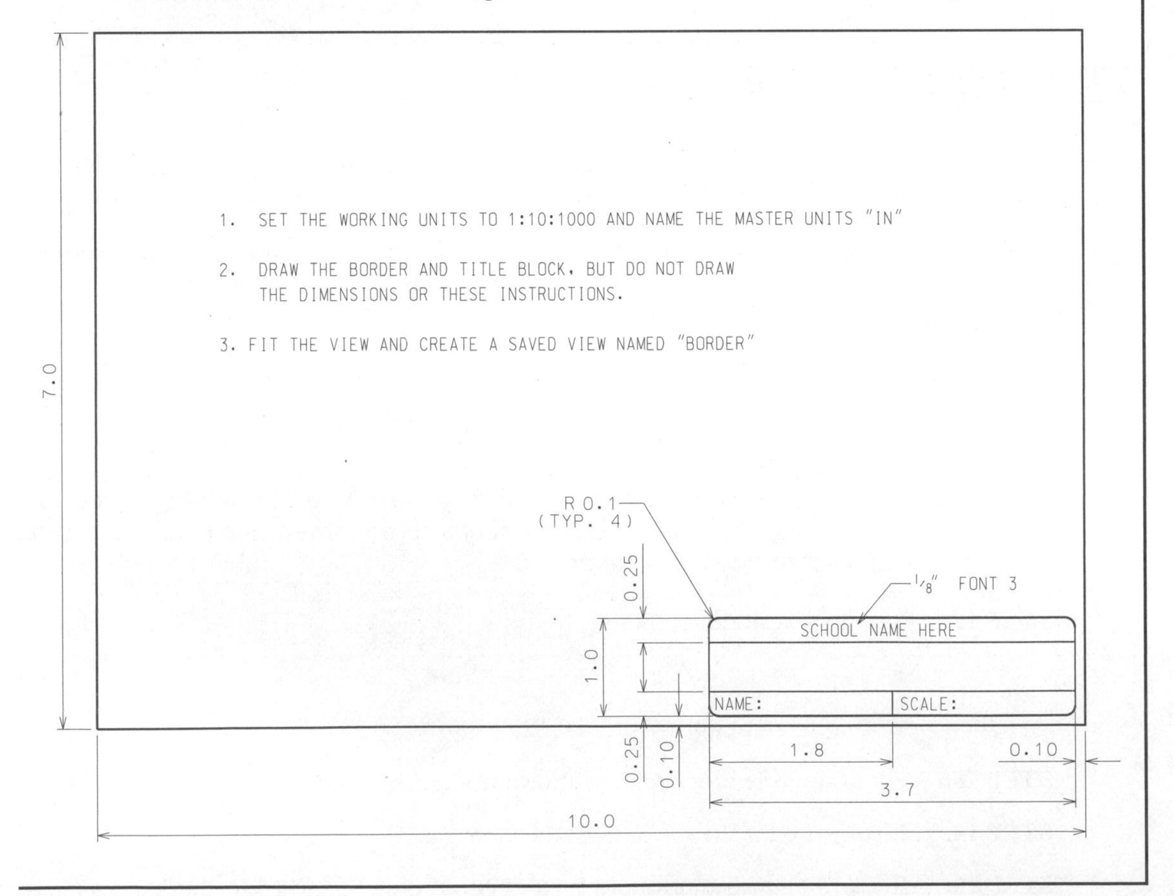

1. SET THE WORKING UNITS TO 1:10:1000 AND NAME THE MASTER UNITS "IN"
2. DRAW THE BORDER AND TITLE BLOCK, BUT DO NOT DRAW THE DIMENSIONS OR THESE INSTRUCTIONS.
3. FIT THE VIEW AND CREATE A SAVED VIEW NAMED "BORDER"
7.0
R 0.1
(TYP. 4)
0.25
1/8" FONT 3
SCHOOL NAME HERE
1.0
NAME:
SCALE:
0.25
0.10
1.8
0.10
3.7
10.0

Exercise 13–2

Create a design file

Reference the border in the new file, then reference the machine part drawings created in Exercises 4–1, 5–2, and 12–3. Clip the boundary of each of the machine parts, scale each one to one-half its true size (Master:ref 0.50000:1.00000), and then move each one inside the border.

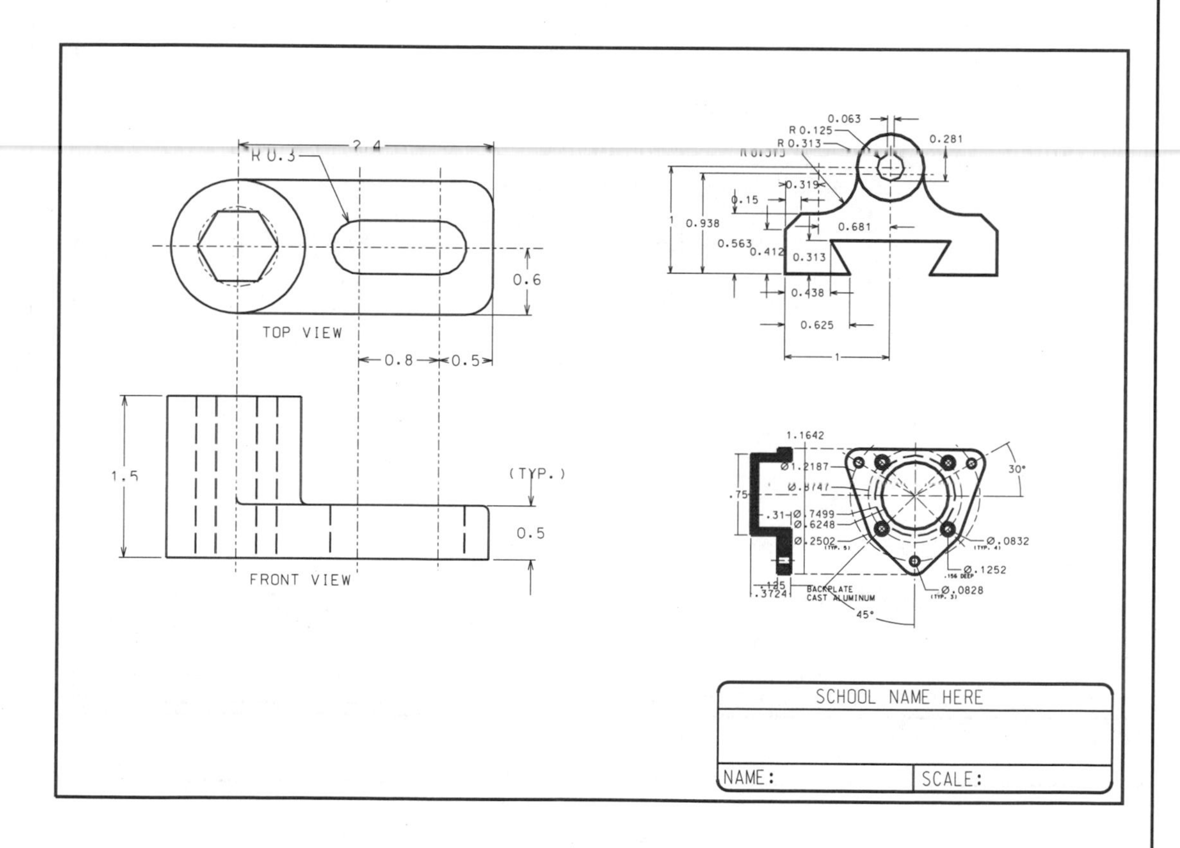

Exercise 13–3

Open the front elevation design drawn in Exercise 7–3. If the design includes a border, delete it. Reference the border design file's "Border" saved view at an attach scale of 1/8" = 1' (Master:ref 4.00000:1.00000).

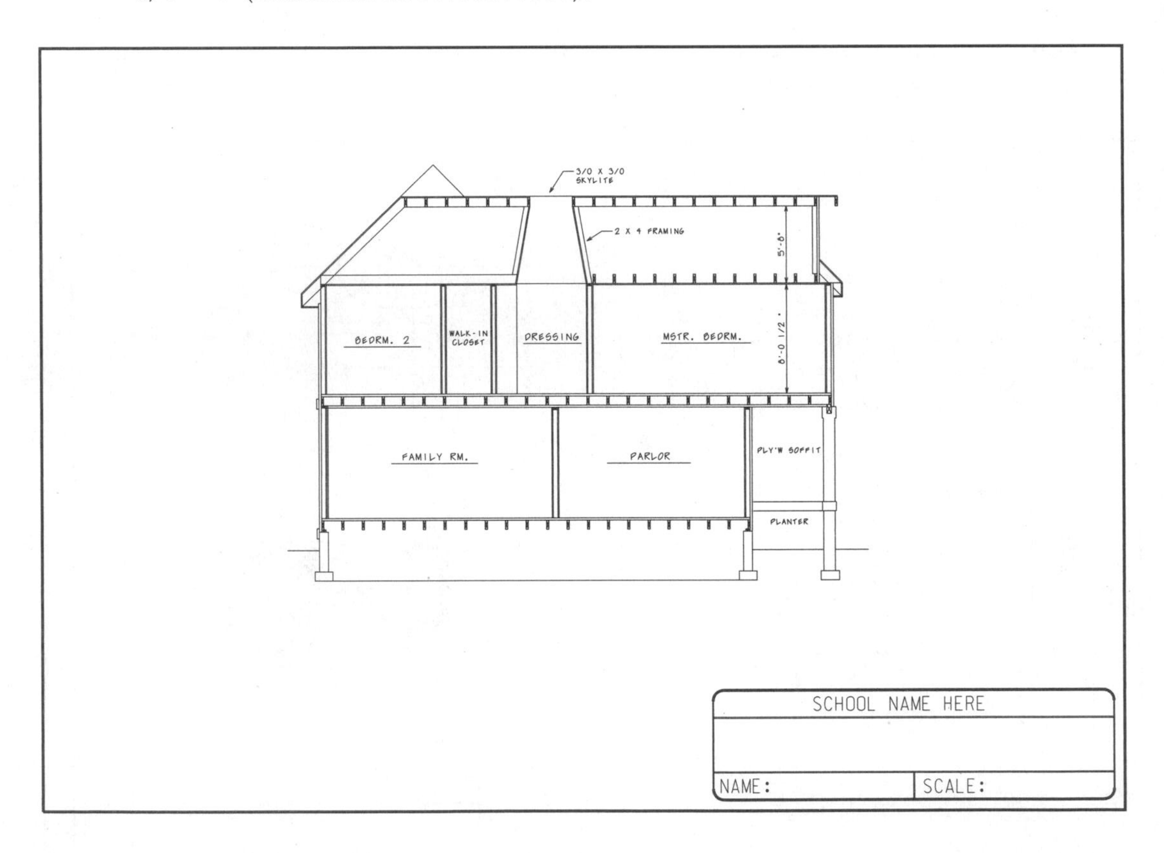

Exercise 13–4

Draw the two parts of a flow diagram shown in Figures Ex13–4a and Ex13–4b. Put each part in a separate design file. Place the ends of the off-page lines at the XY coordinates shown in each figure. The dimension shown in the first figure is intended only to provide a feel for the size of the diagrams. Do not draw the dimension or XY coordinates.

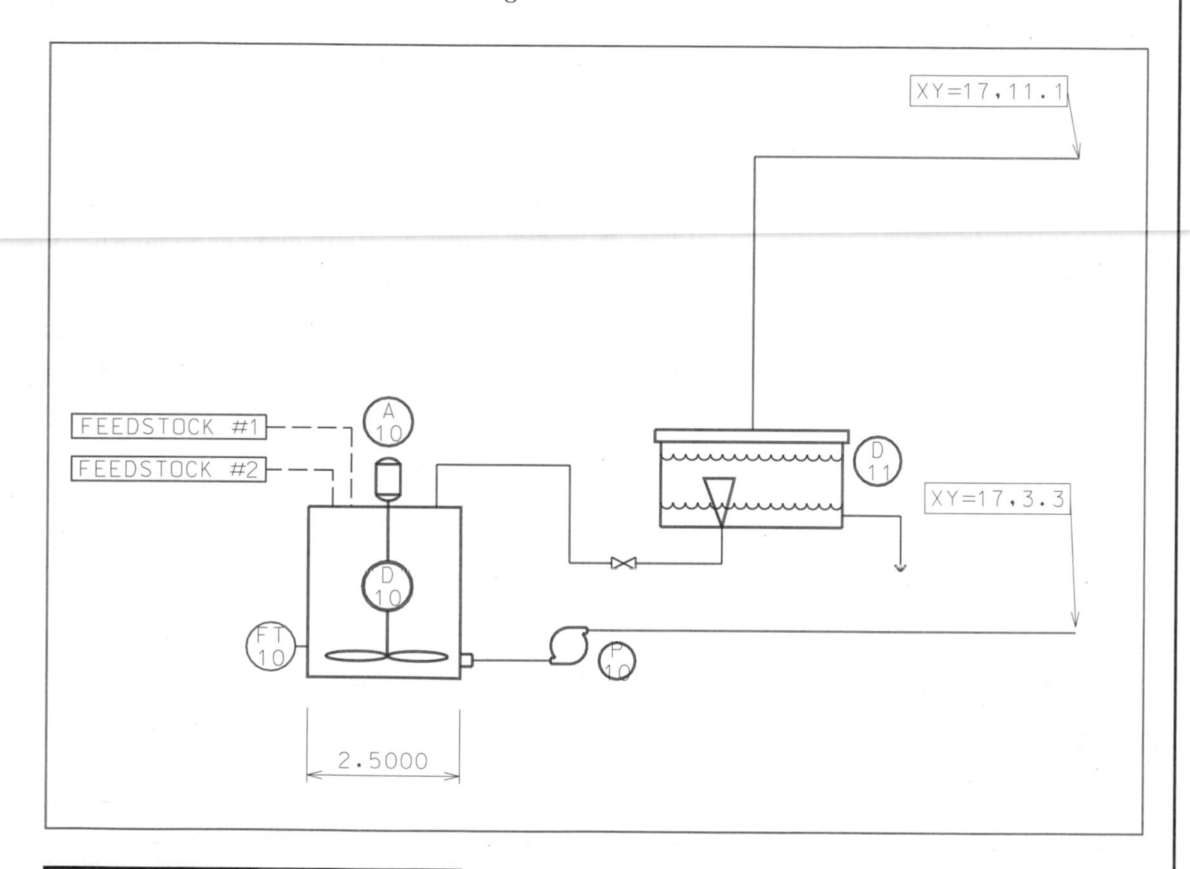

Figure Ex13–4a Flow diagram A.

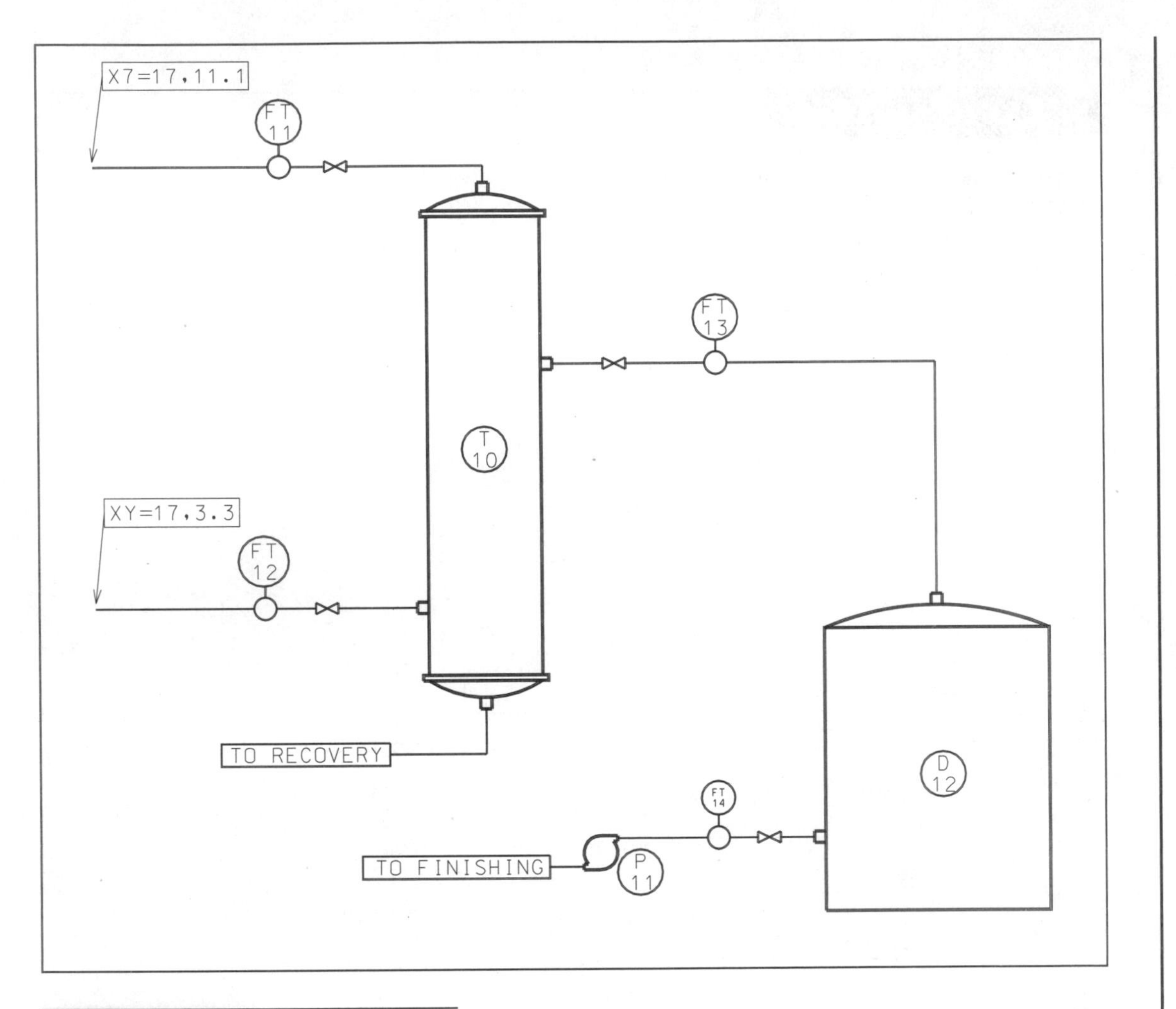

Figure Ex13-4b Flow diagram B.

Exercise 13-5

Completed flow diagram

Create a design file, then reference the two flow diagram parts A and B to create the complete diagram. Reference the border and scale it to encompass the complete flow diagram, as shown in the figure.

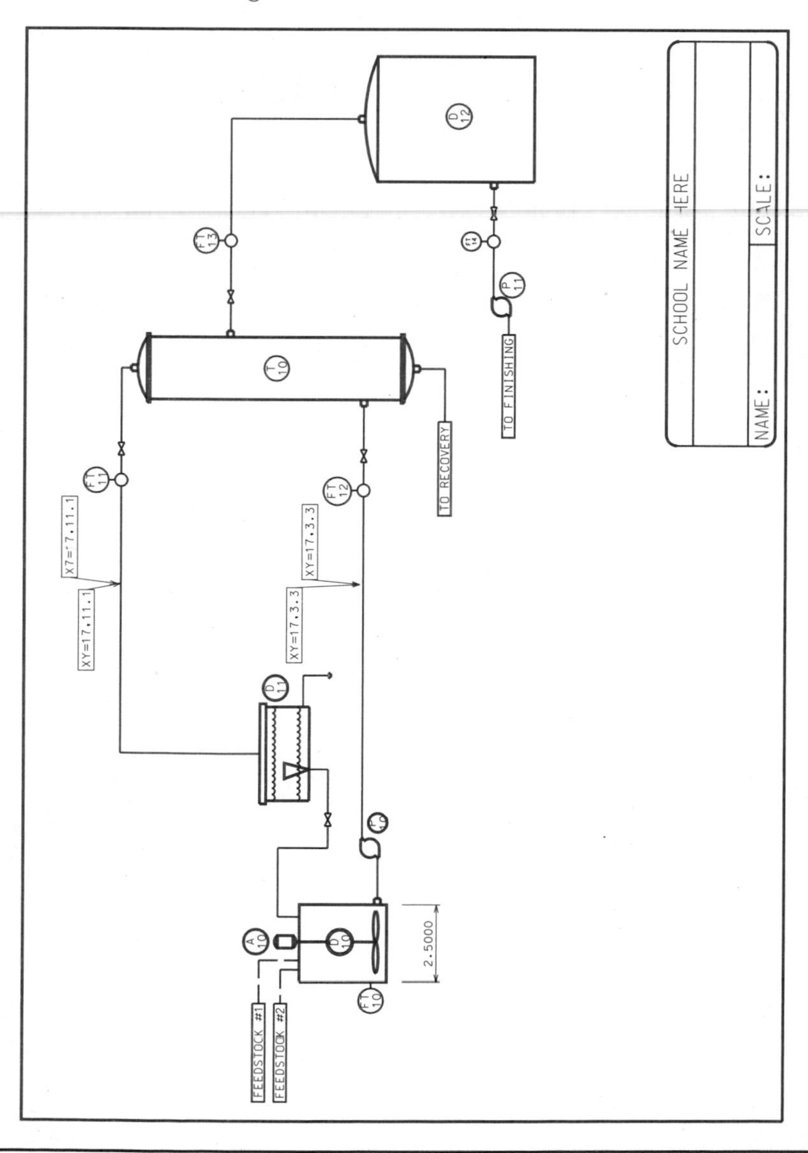

chapter 14

Special Features

MicroStation provides some special features that are used less often than the tools described earlier in this book, but provide added power and versatility. This chapter introduces several such features. For more detailed information on each feature, consult the documentation furnished with MicroStation and the online help.

Objectives

After completing this chapter, you will be able to do the following:

- Create and use graphic groups.
- Use the Duplicate and Merge utilities.
- Select groups of elements according to various selection criteria.
- Define a level symbology to help you determine what level elements are on.
- Use special drafting tools.
- Change the highlight and vector cursor colors.
- Open AutoCAD drawings as MicroStation design files.
- Save MicroStation design files as AutoCAD drawings.
- Import from and export to other CAD graphic formats.
- Save and display graphic images.
- Place text from a glossary of standard terms and add text to the glossary.
- Place flags.

- Use the Image Manager.
- Create dimension-driven designs.
- Use object linking and embedding.

GRAPHIC GROUPS

The Add To Graphic Group tool allows you to group elements together so they act as if they were one element when you apply element manipulation tools with the Graphic Group Lock set to ON. The Drop From Graphic Group tool allows you to drop elements from a graphic group or drop the entire group.

The following MicroStation tools create elements that are part of a graphic group:

- Imported text files of more than 128 lines, or more than 2,048 characters, makes each line of text a separate element, and places all the lines in one graphic group.
- The Place Text Along tool makes each character a separate element and places all the characters as one graphic group.
- The elements of a pattern are part of one graphic group, but the elements that contain the pattern are not part of the group.

Adding Elements to a Graphic Group

To add elements to a graphic group, invoke the Add To Graphic Group tool:

Groups tool box	Select the Add To Graphic Group tool (see Figure 14–1).
Key-in window	**group add** (or **gr a**) Enter

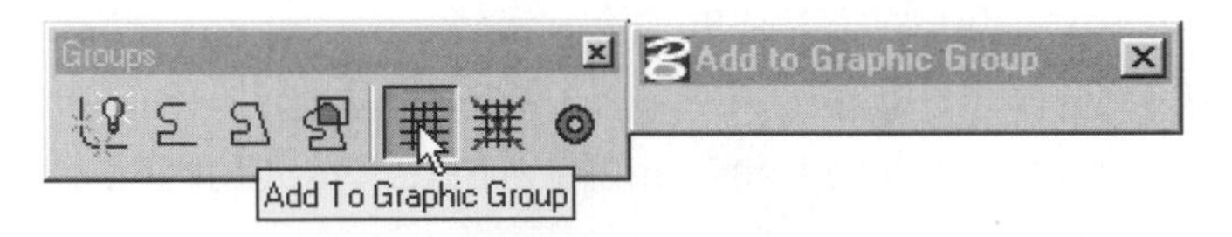

Figure 14–1 Invoking the Add To Graphic Group tool from the Groups tool box

MicroStation prompts:

Add to Graphic Group > Identify element *(Identify the element.)*

If the selected element is not already in a graphic group, a new group is started and MicroStation prompts:

Add to Graphic Group > Add to new group (Accept/Reject) *(Identify the next element to add to the new group, or click the Reset button to reject the selected element.)*

If the selected element is already in a graphic group, elements will be added to that group and MicroStation prompts:

> Add to Graphic Group > Add to existing group (Accept/Reject) *(Identify the first element to be added to the existing graphic group, or click the Reset button to reject the selected group.)*
>
> Add to Graphic Group > Accept/Reject (select next input) *(If more elements are to be added to the new or existing group, select each one. When all elements have been selected, click the Data button in space to accept adding the last selected element.)*

Manipulating Elements in a Graphic Group

The Graphic Group Lock controls the effect the element manipulation tools have on elements in a graphic group. If the lock is OFF, the tools manipulate only the selected element. If the lock is ON, all elements in the graphic group are highlighted and manipulated (even if you cannot see them in the view you are working in). For example, if the graphic group lock is OFF and you select one of the elements in a graphic group to be deleted, only that element is deleted. If the lock is ON, all elements in the graphic group are deleted.

The Graphic Group Lock can be toggled ON and OFF in any of the Lock settings and menus (see Figure 14–2).

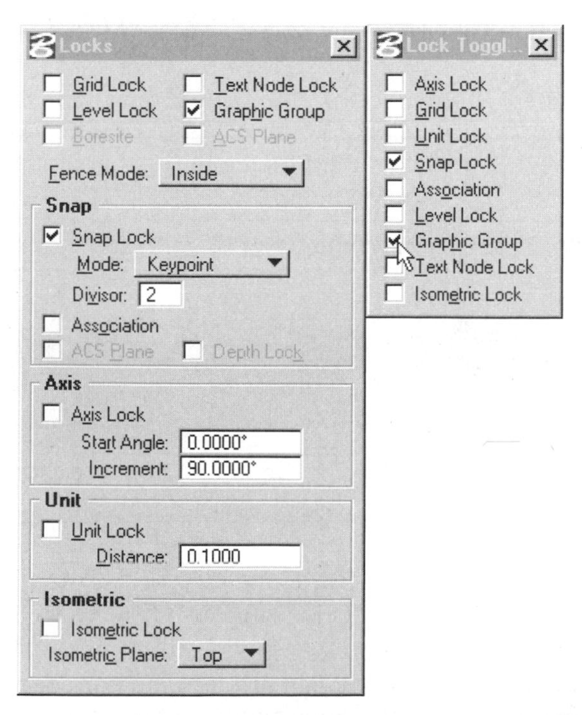

Figure 14–2 Menus and dialog boxes containing the Graphic Group lock

Note: If the Graphic Group Lock is ON, you can determine if the element you select for manipulation is in a group by looking at the element type description in the Status bar. If the element is in a graphic group, the message includes "(GG)."

Copying Elements in a Graphic Group

The Graphic Group Lock affects the way the element copy and fence contents copy tools handle elements in a graphic group.

- If the graphic group lock is set to OFF, the copied elements are not part of any graphic group.
- If the lock is set to ON, the element copy tool copies the entire graphic group, and the copies form a new graphic group.
- If the lock is set to ON, only the graphic group elements in the fence are copied, and the copies become a new, separate graphic group.

Dropping Elements from a Graphic Group

The Drop From Graphic Group tool drops:

- The selected element when the Graphic Group Lock is set to OFF.
- All elements in the group when the Graphic Group Lock is set to ON.

To drop elements from a graphic group, invoke the Drop From Graphic Group tool:

Groups tool box	Select the Drop From Graphic Group tool (see Figure 14–3).
Key-in window	**group drop** (or **gr d**) Enter

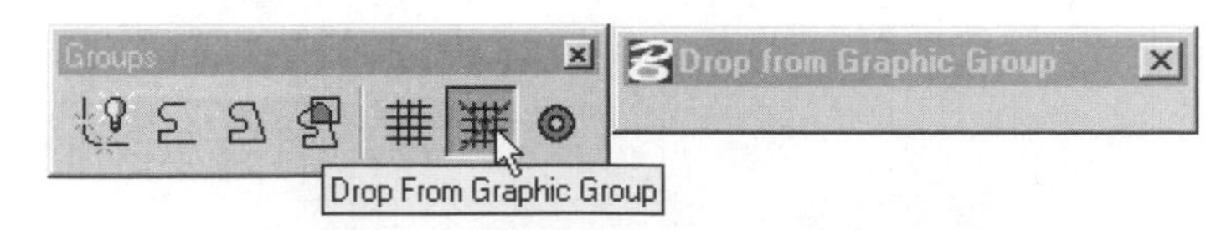

Figure 14–3 Invoking the Drop From Graphic Group tool from the Groups tool box

MicroStation prompts:

Drop From Graphic Group > Identify element *(Identify the element.)*

Drop From Graphic Group > Accept/Reject (Select next input) *(If more elements are to be dropped, select each one. When all elements have been selected, click the Data button in space to accept dropping the last selected element.)*

DUPLICATE ELEMENTS UTILITY

The Duplicate Elements Utility allows you to clean up your design by removing duplicate elements. Duplicate elements are the same type, at the same location, the same size, and on the same level. They can have different symbology (color, style, weight) and still be duplicates. MicroStation users sometimes create duplicate elements by clicking too many times when doing element manipulations. For example, clicking two times without moving the cursor when asked to specify the location of the axis while mirroring the fence contents with the Make Copy button turned on—the second acceptance click puts a copy of the elements on top of the original elements.

To eliminate duplicate elements, invoke the Duplicate Elements Utility:

Pull-down menu	Utilities > MDL Applications

MicroStation displays the MDL dialog box, as shown in Figure 14–4.

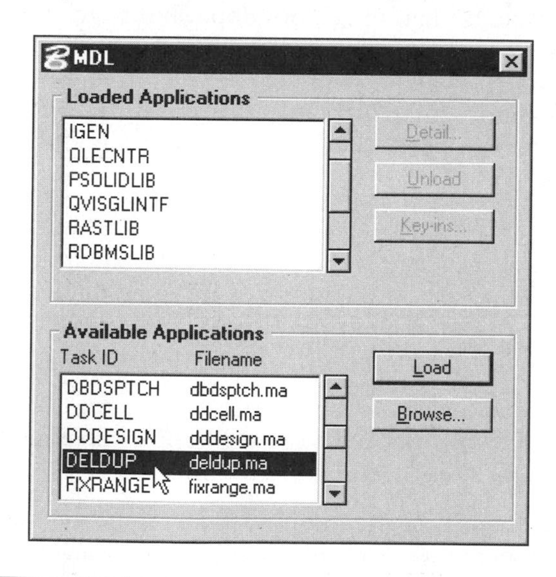

Figure 14–4 The MDL dialog box

In the Available Applications list box, choose the DELDUP option, then click the Load button to open the Duplicate Elements Utility dialog box, as shown in Figure 14–5.

Select the desired options in the dialog box (described below), then click the Apply button to delete the duplicate elements and click the Done button to close the dialog box.

The options on the Duplicate Elements Utility dialog box let you control the way MicroStation eliminates the duplicates:

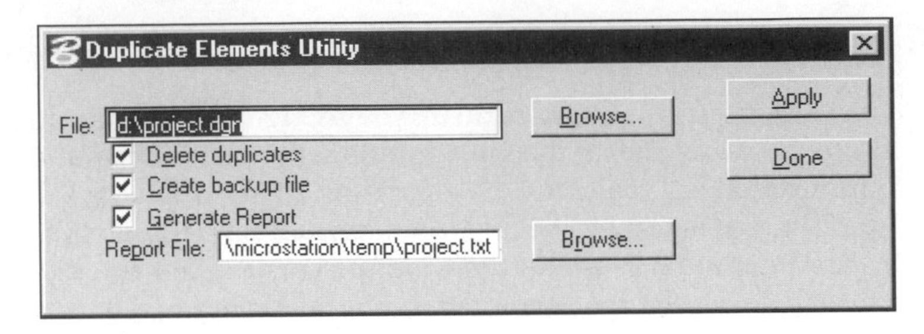

Figure 14–5 The Duplicate Elements Utility dialog box

- The **File** Edit field allows you to enter the specification of the design file you want to remove duplicates from (it does not have to be the active file). You can also click the Browse button next to the field to search for and select the file. Wild cards are allowed in the file name to cause the utility to remove duplicates from several files. For example, enter the file name B-*.dgn to remove duplicates from all design files in the current path whose file name starts with the "B-" characters.

Note: If you enter a wildcard character in the Browser Files edit field, the Browser remains open when you click the OK button, and you must click the Cancel button to close the Browser.

- If the **Delete Duplicates** toggle button is ON, the duplicate elements are actually deleted from the design file. If the button is OFF, the duplicate elements remain in the file, but their properties are set to "NEW, NOT MODIFIED." You can search for elements with this property using the Selection by Attributes feature, discussed later in this chapter.
- If the **Create Backup** toggle button is ON, a copy is made of the design file before the duplicates are removed from the original file. The extension of the copied file will be ".bak." For example; the backup for a design file named "B-1003A.dgn" will be "B-1003A.bak."
- If the **Generate Report** toggle button is on, a report will be generated that describes each original element with duplicates and each duplicate that was removed.
- The **Report File** edit field allows you to enter the name of the report file. The Browse button next to the edit field allows you to browse to a path and enter a report file name.

MERGE UTILITY

The Merge Utility allows you to copy elements from one or more design files or cell libraries into a design file or cell library. The utility runs from the MicroStation Manger dialog box.

Note: If you have a design file open in MicroStation, select File>Close to return to the MicroStation Manager dialog box.

In the MicroStation Manger dialog box, invoke the Merge Utility:

Pull-down menu (MicroStation Manager)	File > Merge...

MicroStation displays the Merge dialog box, as shown in Figure 14–6.

To select the file to be merged into the target file, click the Select... button in the Files to Merge area of the Merge dialog box. MicroStation opens the Select Files to Merge dialog box, as shown in Figure 14–7.

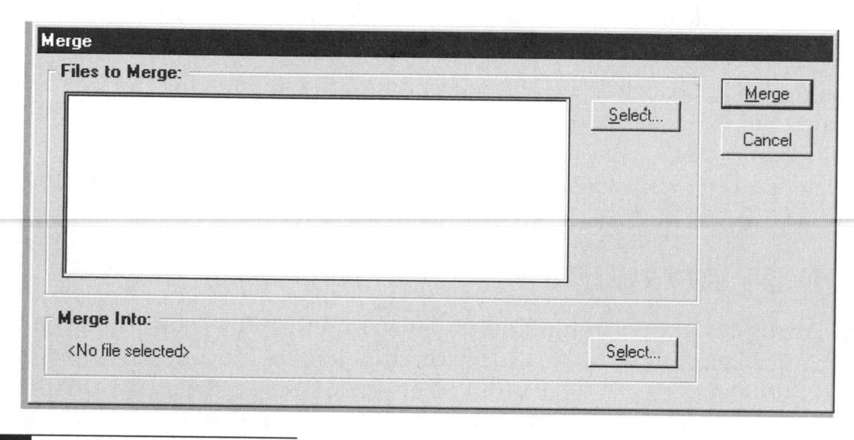

Figure 14–6 The Merge dialog box

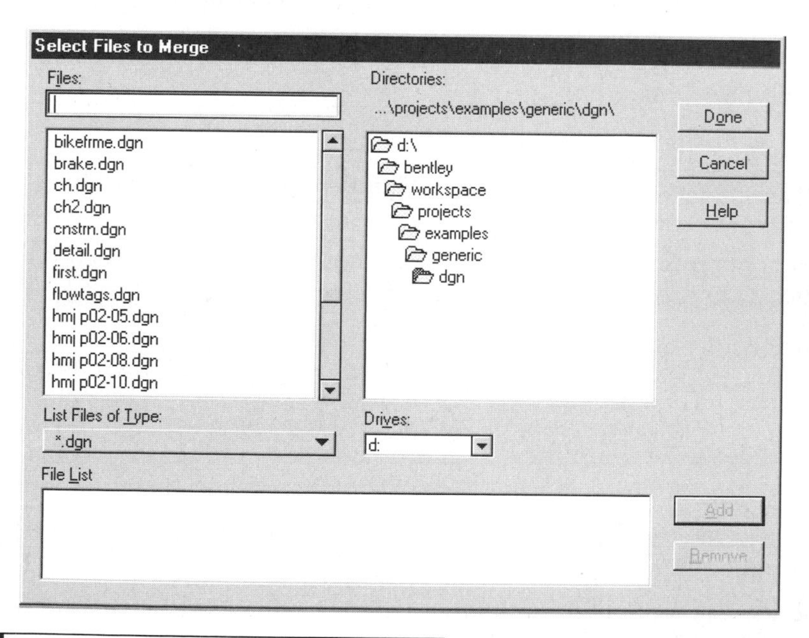

Figure 14–7 The Select Files to Merge dialog box

1. In the upper part of the Select Files to Merge dialog box, find and select a file to merge.
2. Click the Add button near the bottom of the dialog box to add the selected file to the list of files to be merged.
3. Repeat the first two steps for each file that is to be merged into the destination file.
4. When all files are selected, click the Done button to close the dialog box.
5. In the Merge Into area of the Merge dialog box, click the Select... button to open the Select Destination File dialog box.
6. In the Destination File dialog box, select the file into which the previously selected files will be merged, then click the OK button.
7. In the Merge dialog box, click the Merge button to initiate merging the files and close the Merge dialog box.

After the final step, all elements from the files to be merged are copied into the merge to file.

SELECTION BY ATTRIBUTES

The Select By Attributes option in the Edit pull-down menu allows you to limit the selection of elements for manipulation to those that meet certain element attributes. For example, if all red ellipses on level 10 need to be changed to the color green, use the Select By Attributes dialog box to select only those elements, then apply the color change to the selected elements. Handles appear on the selected elements, just as they did with the selection tool described in Chapter 6. You can delete the selected elements, move them, copy them, change their attributes, and apply several other manipulation tools to them.

Select By Attributes Dialog box

Open the Select By Attributes dialog box:

Pull-down menu	Edit > Select By Attributes

The dialog box contains several fields, selection menus, and options to open other dialog boxes (see Figure 14–8).

The dialog box is made up of the following parts:

- Selection by Levels (upper left side of the box)
- Selection by element Types (upper right side of the box)
- Selection by Symbology—color, line style, or line weight (lower left side of the box)
- Selection Mode, which contains three menus that control the way the selection criteria are applied
- The Execute button at the bottom left of the box, which puts the selection criteria currently set in the fields into effect

- The Properties button, which opens the Select By Properties dialog box (see Figure 14–9) where you can choose to select elements by property attributes (such as only filled elements) and classes (such as construction elements).
- The Tags button, which opens the Select By Tags dialog box (see Figure 14–10) where you can limit your selection to elements that contain only certain tags or combinations of tags.

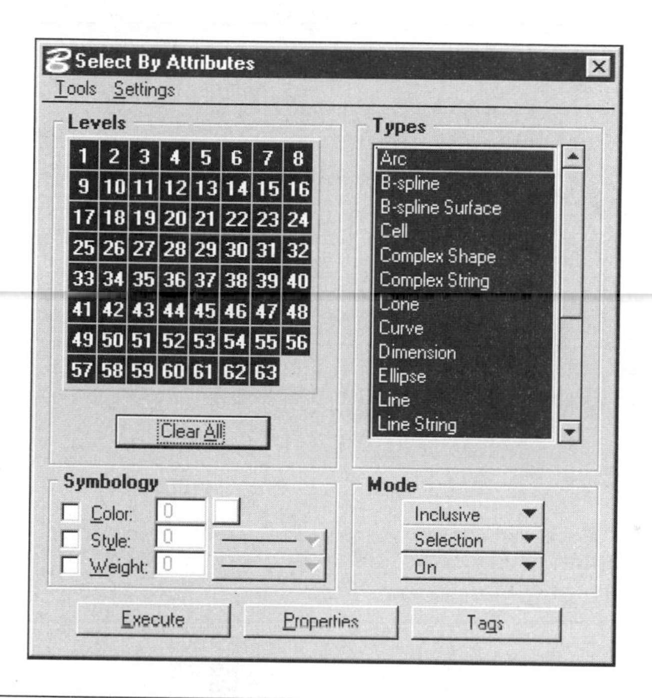

Figure 14–8 Select By Attributes dialog box

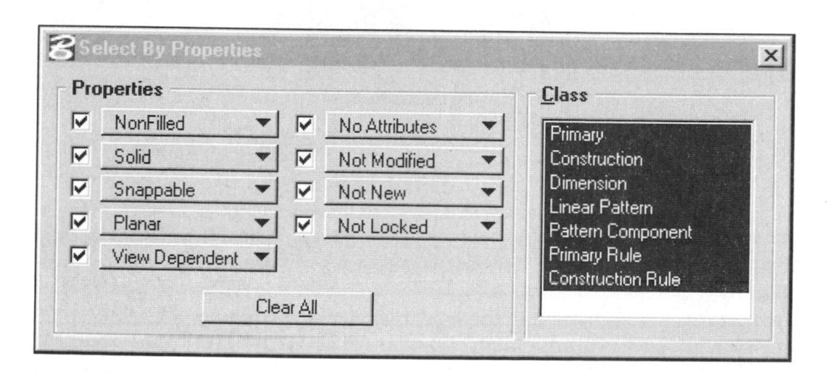

Figure 14–9 Select By Properties dialog box

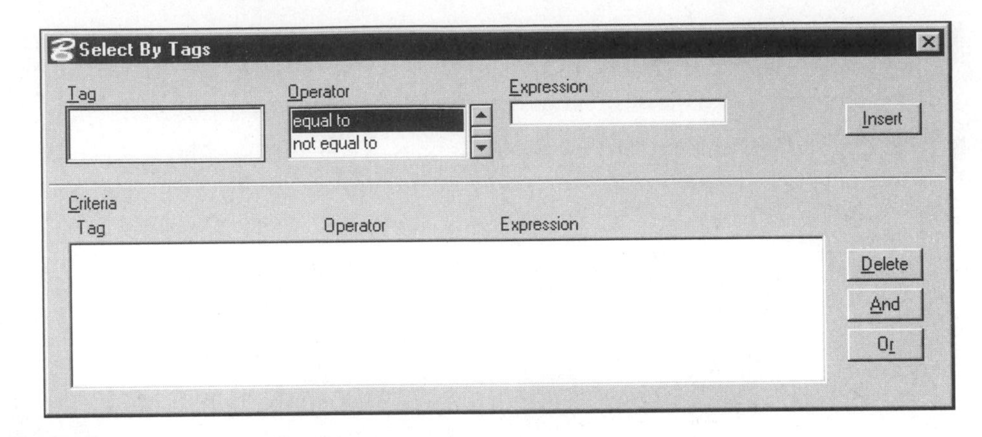

Figure 14–10 Select By Tags dialog box

Note: Your selection criteria can be based on one item in one of the fields or on a combination of items in one or more fields—for example, all elements on level 10 or only ellipses on level 5 that are filled.

Selection by Level

The Levels area contains a grid of level numbers and a toggle button (see Figure 14–11). If a level number is shown with a dark background, it is part of the selection criteria. A light background means the level is not part of the selection criteria.

Figure 14–11 Displaying the Levels field of the Select By Attributes dialog box

To switch the state of a single level, click the Data button on it. To change the state of a group of elements, drag the screen pointer across them while holding down the Data button. For example, if you wish to select elements on levels 10 through 15, set those six levels to a dark background and all other levels to a light background.

The toggle button below the level numbers field will either Clear All levels or Select All levels. The name of the button says what action it is going perform when you click it.

Selection by Type

The Types area contains a list of all element types (see Figure 14–12). Type names shown with a dark background are part of the selection criteria. A light background means that type is not part of the selection criteria.

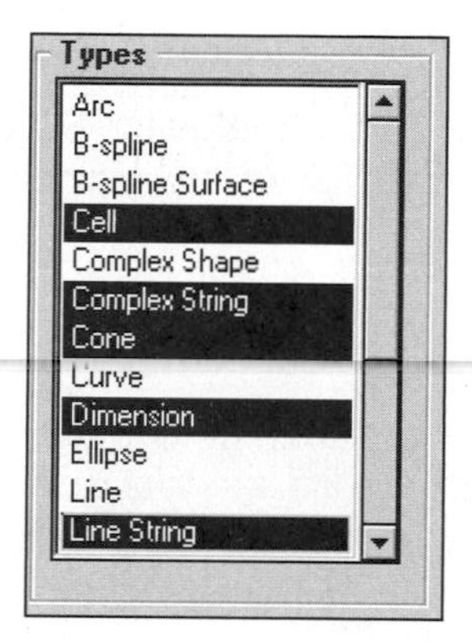

Figure 14–12 Displaying the Types field of the Select By Attributes dialog box

To select a specific type and turn off all others, click the Data button on the desired type. To select additional types, hold down the Ctrl key while clicking the Data button on type names. To select a contiguous group of types, drag the screen pointer across them while holding down the Data button. For example, to select only ellipses and line strings, click the Data button on the Ellipse type, then Ctrl-click the Data button on the Line String type.

Selection by Symbology

The Symbology field allows you to include element color, style, and weight in the selection criteria (see Figure 14–13). To add a symbology item to the selection criteria, turn ON the button to the left of the item's name, then select a value in the menu to the right of the item's name. For example, if the color button is set to ON, elements that are the color shown in the color field are the only ones included in the selection criteria.

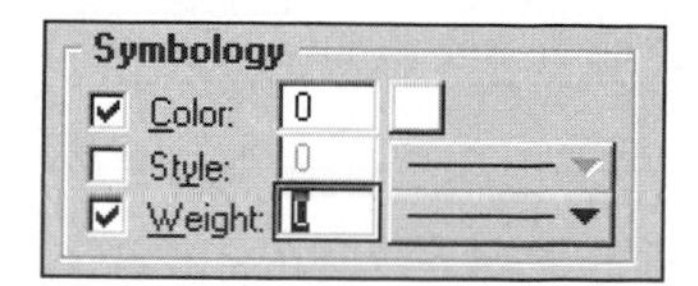

Figure 14–13 Displaying the Symbology field of the Select By Attributes dialog box

Controlling the Selection Mode

The three Mode option menus (see Figure 14–14) control the way the selection criteria are applied when you click the Execute button at the bottom left of the dialog box.

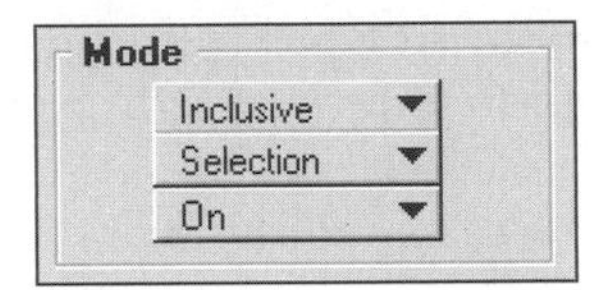

Figure 14–14 Displaying the Mode field of the Select By Attributes dialog box

The first Mode option menu has two options:

- Inclusive—Select only elements that meet the selection criteria.
- Exclusive—Select only elements that do not meet the selection criteria.

For example, if the selection criterion is set to type Ellipse, Inclusive mode causes all ellipses to be selected, and selection in Exclusive mode causes all elements except ellipses to be selected.

The second Mode option menu has three options:

- Selection—Immediately select (place handles on) all elements that meet the selection criteria when the Execute button is pressed.
- Location—Turn on the selection criteria but do not select any elements when the Execute button is pressed. Use the Element Selection tool to select elements that meet the criteria in this mode. Elements that do not meet the criteria cannot be selected.
- Display—Turn on the selection criteria, and make all elements that do not meet the criteria disappear from the view when the Execute button is pressed. No handles are placed on the remaining elements, so the Element Selection tool must be used to select from the elements that still appear in the view.

The third Mode option menu has two options:

- OFF—When the Execute button is pressed, turns off the previously set selection criteria so it has no effect on element selection.
- ON—When the Execute button is pressed, turns on the current selection criteria so it can be used.

Note: If you close the Select By Attributes dialog box with a selection criterion in effect, an Alert window appears prompting you to click OK or Cancel. Click OK to keep the selection criterion in effect, or Cancel to turn off the criterion.

Select By Properties Dialog box

Use the Select By Properties dialog box to include additional element attributes in the selection criteria. Open this box by clicking the Properties button in the Select By Attributes dialog box.

On the left side of the box, select element Properties settings (see Figure 14–15). Each property has an options menu from which you can select the property to be included in the selection criteria. To select a property, turn the toggle button to ON and then select a setting from the Properties option menu. For example, the area property options menu allows you to select one of the two available options, Solid or Hole elements.

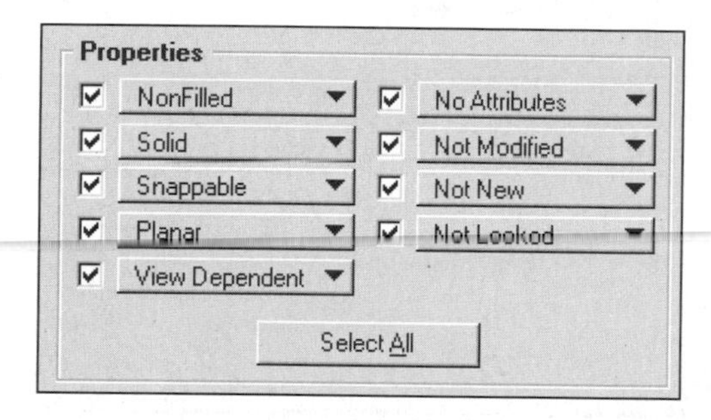

Figure 14–15 Displaying the Properties in the Select By Properties dialog box

The toggle button below the Properties area turns ON or OFF all the property buttons. The name of the button says what will happen next. If it displays Select All, it turns ON all the options in the Properties area; if it displays Clear All, then it turns OFF all the options in the Properties area.

On the right side of the box is a field in which you select what class or classes of elements to include in the selection criteria (see Figure 14–16). If the class is shown with a dark background, it is included. Each name is a toggle switch. Clicking the Data button on the class name adds it to or removes it from the selection criteria. For example, if you want to select only elements in the construction class, set the word Construction in the menu to have a dark background, and set all the others to have a light background.

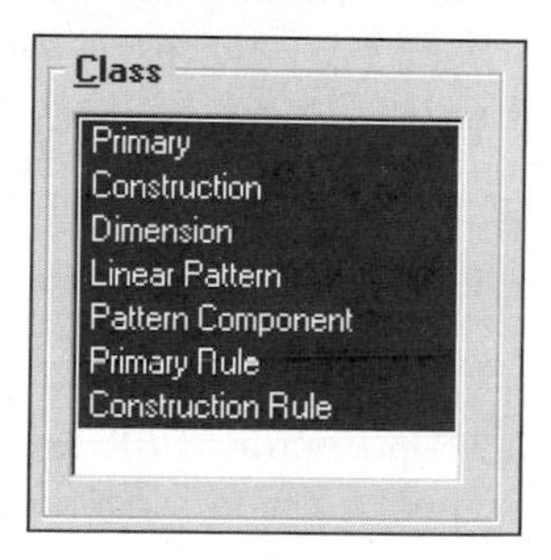

Figure 14–16 Displaying the Class listing in the Select By Properties dialog box

Select By Tags Dialog box

The Select By Tags dialog box (see Figure 14–17) is used to specify criteria based on tag values. If selection criteria based on tag values are specified, elements that do not have attached tags with the specified tag names will not be selected, located, or displayed. To open the Tags dialog box, click the Tags button in the Select By Attributes dialog box.

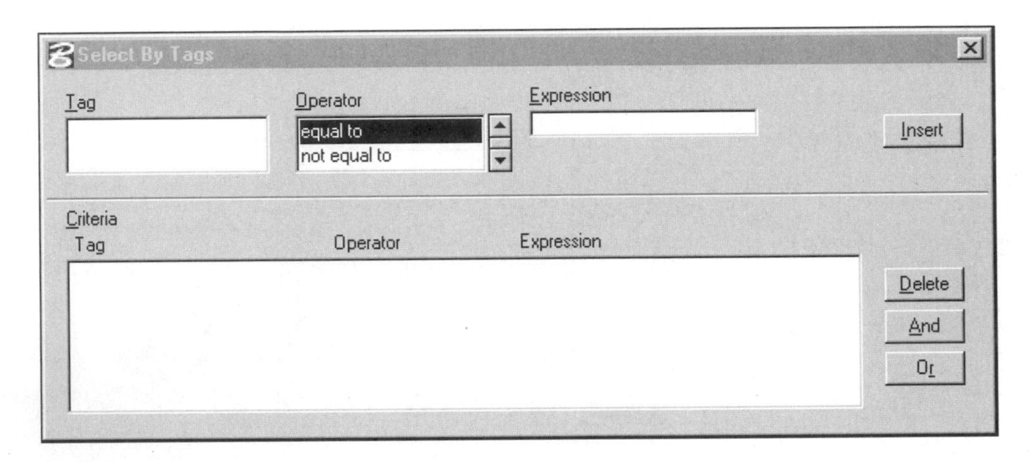

Figure 14–17 Select By Tags dialog box

Putting Selection Criteria into Effect

Once you set all the selection criteria, click the Execute button in the Select By Attributes dialog box to select the elements according to the settings. Handles appear (depending on the settings) on the selected elements. After completing the selection criteria, invoke the appropriate element manipulation tool and follow the prompts.

LEVEL SYMBOLOGY

Designers often make extensive use of the 63 available levels to help organize the various parts of a design. For example, a plot plan may have separate levels for roadways, descriptive text, foundations, utilities, the drawing border, and title block information. An architectural plan might have the walls on one level, the dimensions on another level, electrical information on still another level, and so on. Separating parts of the design by level allows designers to turn on only the part they need to work on and allows them to plot parts of the design separately.

Keeping up with what level everything is on can be confusing. MicroStation's level symbology reduces the confusion by setting unique combinations of display color, weight, and style for each level. When level symbology is set to ON, all elements are displayed using the symbology assigned to the level the elements are on, rather than their true symbology. For example, if level 10 symbology is set to display elements that are red, dashed, and weight 5, all elements on level 10 display with that symbology, no matter what their actual color, weight, and style settings are.

The View Attributes dialog box controls the display of level symbology by turning ON and OFF for selected views.

Note: If you plot a view in which level symbology is set to ON, the elements are plotted with level symbology rather than their true symbology.

The Level Symbology dialog box provides options for creating and modifying level symbology settings. Invoke the Level Symbology settings window:

Pull-down menu	Settings > Level > Symbology...
Key-in window	**dialog levelsymb** (or **di l**) Enter

MicroStation displays the Level Symbology dialog box, as shown in Figure 14–18. The Level Symbology dialog box contains the following fields:

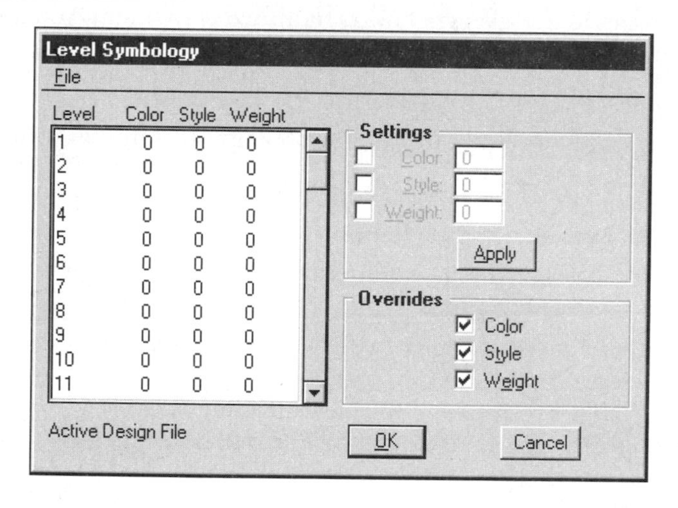

Figure 14–18 Level Symbology dialog box

- Level list box—The list box on the left side of the dialog box shows the current symbology (Color, Style, Weight) settings for each level.
- Settings—The Settings area on the top right of the box is where you set the symbology for the selected level numbers in the Level list box. When one of these buttons is turned ON, a field appears to the right of the button in which you can select the desired setting.
- Overrides—The Overrides area determines which symbology settings are used. If a button is ON (appears depressed, with a dark center), that symbology setting is in use. For example, if the Color button is ON and the Style and Weight buttons are OFF, elements on each level are displayed using the level symbology color, but each element's true style and weight are displayed.

- OK and Cancel buttons—To save the level symbology settings, click the OK button. To discard any changes you made, click the Cancel button. You must click one of the two buttons to close the Level Symbology dialog box before you can do anything else in MicroStation.

Following are the procedures to set up a level symbology or to modify an existing one.

1. Open the Level Symbology dialog box.
2. Turn on the override for each level symbology attribute you want to use.
3. Highlight a level by clicking the Data button on the level line in the Level list box on the left side of the box.
4. Select a Color, Style, and Weight in the Settings area, and then click the Apply button.
5. Repeat procedure steps 3 and 4 for each level.
6. Click the OK button to save the level symbology changes.

Note: Changes to the level symbology setup are permanent. You do not have to invoke Save Settings to keep them.

Following are the procedures to display the level symbology attributes in your design.

1. Open the View Attributes dialog box from the Settings pull-down menu.
2. Turn ON the Level Symbology toggle button.
3. If you want to turn ON level symbology for only one view, set the View Number to the one you want and click the Apply button.
4. If you want the level symbology display to be ON for all open views click the All button.

Note: Any changes you make to the View Attributes are lost when you exit MicroStation, unless you select Save Settings from the File pull-down menu.

CHANGING THE HIGHLIGHT AND POINTER COLOR

The highlight and drawing pointer colors can be changed from the Design File Settings window. These colors are used for the following reasons:

- Highlighting selected elements.
- The drawing pointer when a data point is placed and when an element is manipulated.
- The locate tolerance circle that appears on the pointer during manipulations.

A common case where different colors may be required is when many elements in the design use the same colors as the selection and pointer. In that case, there is no visual indication that an element has been selected, and the pointer can be hard to see.

To change the colors, invoke the Design File Settings dialog box:

Pull-down menu	Settings > Design File..., then select Color from the Category menu (see Figure 14–19).
Key-in window	**mdl load dgnset** (or **md l dgnset**) Enter

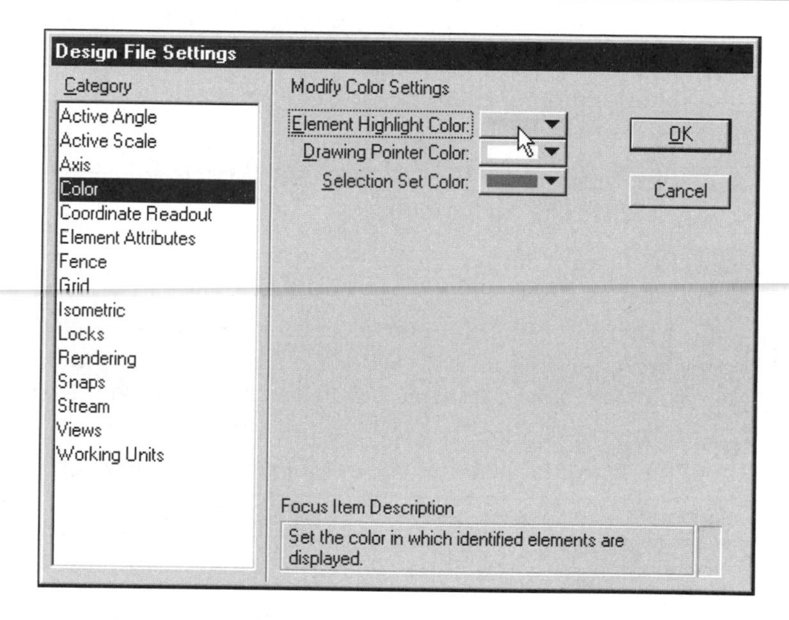

Figure 14–19 Design File Settings dialog box, with Color selected

In the Design File Settings window:

1. Pop up the Element Highlight Color menu and select the desired color for highlighting elements.
2. Pop up the Drawing Pointer Color menu and select the desired color for the drawing pointer.
3. Click the OK button to make the new color settings active.
4. To make the changes permanent in this design file, select Save Settings from the File pull-down menu.

Notes: The color changes apply only to the design file in which they were changed. Each design file has its own highlight and pointer color settings.

Do not set the colors the same as the view window background color. If the color is the same, highlighted elements and the pointer cannot be seen.

IMPORTING AND EXPORTING DRAWINGS IN OTHER FORMATS

MicroStation can import files containing graphic information in several file exchange formats and can export drawings to those formats.

- Several formats can be imported directly in MicroStation from the MicroStation Manager and File Open windows.
- Design files can be exported to several formats from the File Save As window.
- Export and Import options are also provided in the pull-down menu File.

The support of these other formats allows MicroStation users to share design files with clients and vendors using other CAD and graphic applications.

Supported Formats

Table 14–1 describes file exchange formats that MicroStation can open directly and save and that are also available in the Import and Export submenus from the pull-down menu File. Table 14–2 describes file exchange formats that are available only from the Import and Export submenus.

Table 14–1 File Exchange Formats Available for Opening and Saving Files

FORMAT	IMPORT	EXPORT	DESCRIPTION
DWG	Yes	Yes	Native format AutoCAD drawings.
DXF	Yes	Yes	Drawing Interchange Format—developed by Autodesk, Inc., to exchange graphic data among many CAD and graphics applications. DWG and DXF imports are handled identically.
DXB	Yes	Yes	AutoCAD binary DXF files.
CGM	Yes	Yes	Computer Graphics Metafile Format—an ANSI standard for the exchange of picture data between different graphics applications; device and environment independent.
IGES	Yes	Yes	Initial Graphics Exchange Specification Format—a public domain ANSI standard file format that is intended as an international standard for the exchange of product definition data among different CAD applications.

Table 14–2 File Exchange Formats Available in the Import and Export Submenus

FORMAT	IMPORT	EXPORT	DESCRIPTION
Image	Yes	No	Several graphics formats used by text processing and publication graphics packages.
Text	Yes	No	ASCII text files (discussed in Chapter 6).
3D	No	Yes	MicroStation's *three-dimensional* design file format. If the open design file is *3D,* there will be an option to save it as a *two-dimensional (2D)* drawing.

Opening a File Containing Another File Format

The file formats shown in Table 14–1 can be opened directly in MicroStation. The conversion to design file format is done as the file opens.

To open a drawing created in another format from either of the MicroStation Manager or Open Design File dialog boxes, do the following:

1. Pop up the "List Files of Type" menu (see Figure 14–20).
2. Select the format of the drawing to be opened.
3. Follow the usual procedures for opening a file.

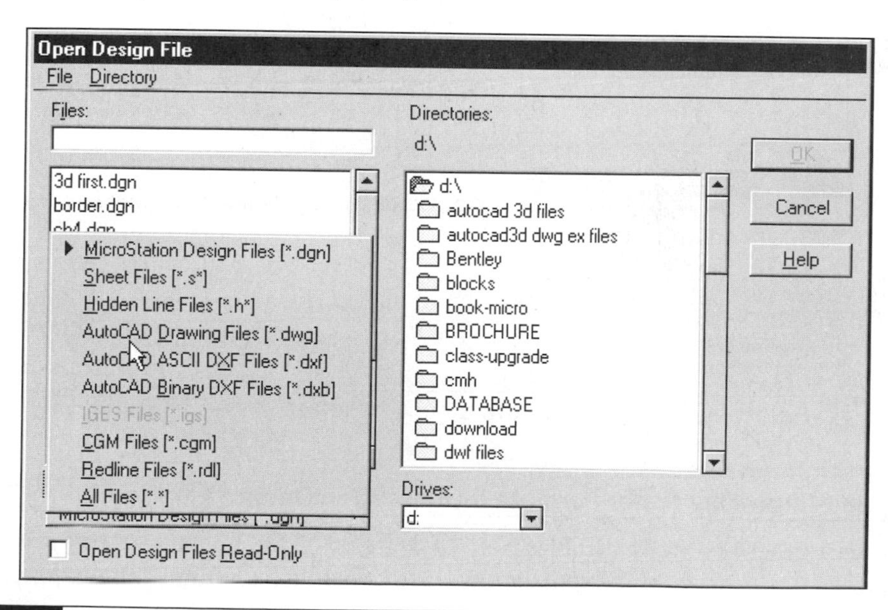

Figure 14–20 Selecting the format of a file to be opened

Saving a Design File That Is in Another File Format

The file formats shown in Table 14–1 can be saved from the File Save As dialog box. The conversion to the other file format is done as the file is saved. To save a design file in another format, follow these steps:

4. Open the Save As dialog box from the File pull-down menu.
5. Pop up the "List Files of Type" menu (see Figure 14–21).
6. Select the format to be used.
7. Supply a directory path and file name.
8. Click the OK button to initiate the conversion and close the dialog box.
9. The contents of the design file are converted to the other format, and the design file remains open in MicroStation.

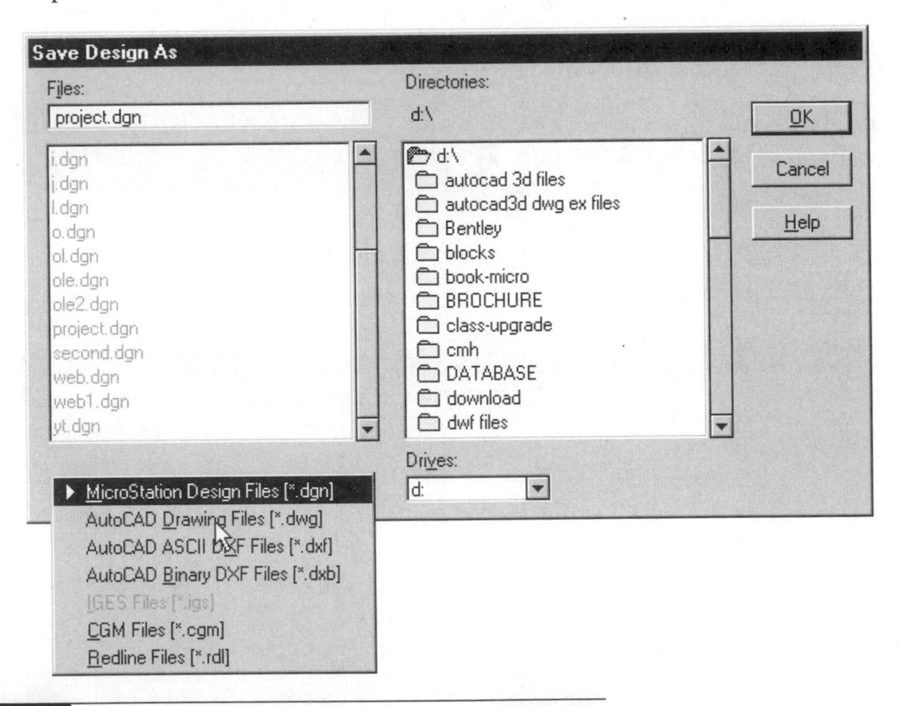

Figure 14–21 Selecting the format to save a design file as

Importing and Exporting Other Formats into an Open Design File

The file formats described in Tables 14–1 and 14–2 are available for import and export from an open design file. The Import and Export options are available from the File pull-down menu.

- The Import option inserts the imported file contents into the open design file. The elements in the imported file are converted to design file elements. Image files remain as images in the design file and are not an element.

- The action of the Export tool varies among the formats. In some cases it opens the Save As window, then follows that window with an Export window containing options to control the way the design file is opened. In other cases, the Export window opens directly.

For detailed information on importing and exporting, refer to the technical documentation furnished with MicroStation.

MANIPULATING IMAGES

In MicroStation, the term "image" refers to graphics files that can be inserted in word processing and graphics processing applications. The pictures of MicroStation windows presented in this textbook are examples of such images. MicroStation provides a set of tools for creating, manipulating, and viewing images under the Utilities pull-down menu, as shown in Figure 14–22. For example, a programmer created an automated drawing procedure that involves several custom dialog boxes; a technical writer is creating a training guide in a Windows-based word processing package and needs pictures of the dialog boxes. The programmer uses the Image Capture option to capture the dialog boxes as bit-mapped image files that the technical writer can insert in the word processing file.

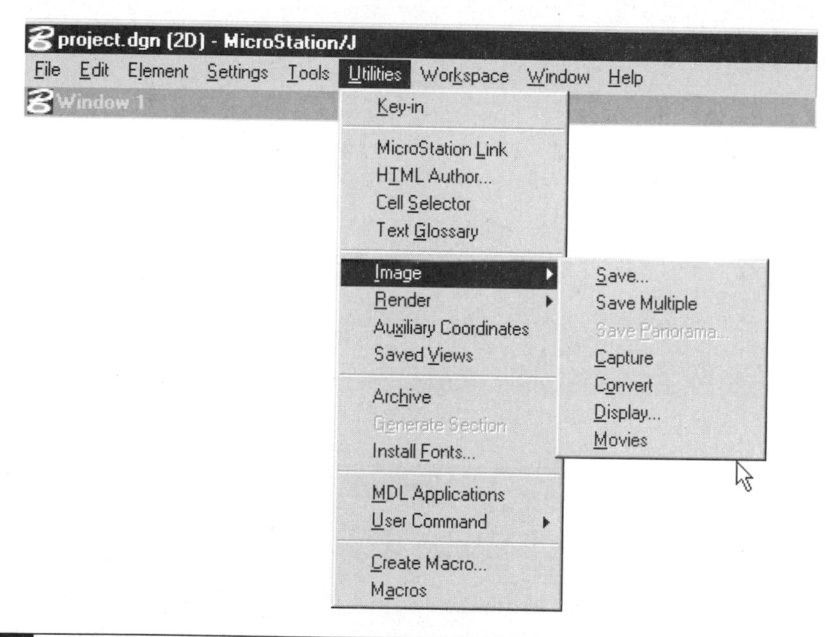

Figure 14–22 Image tools submenu in the Utilities menu

Save

The Save option opens the Save Image dialog box (see Figure 14–23) from which the contents of one of the eight view windows can be saved as an image file using one of several available image formats. To determine the required image format, refer to the documentation furnished with the application for which the image is being created.

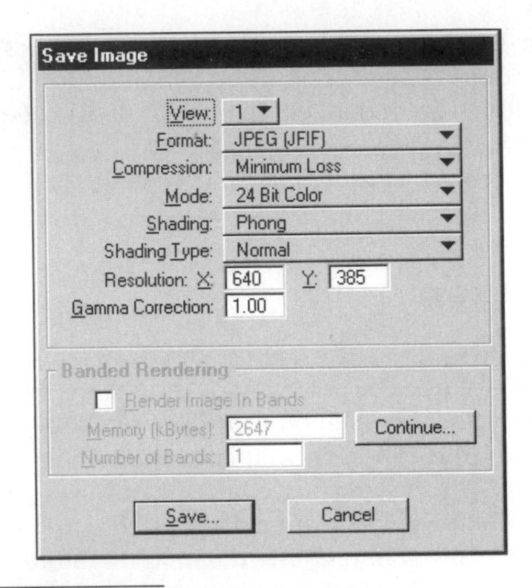

Figure 14–23 Save Image dialog box

Capture

The Capture option opens the Screen Capture dialog box (see Figure 14–24) from which you can select several methods of capturing all or part of the image on the workstation screen. Each capture method opens a Capture Output dialog box in which you can specify a file name and directory path for the captured screen image.

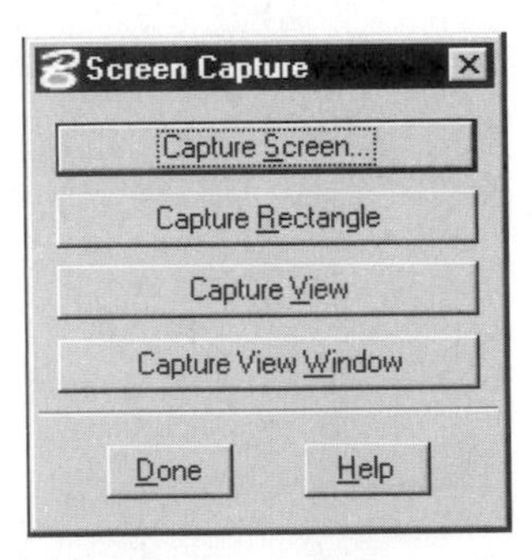

Figure 14–24 Screen Capture dialog box

The Screen Capture methods include the following:

- Capture Screen—Capture the entire screen.
- Capture Rectangle—Capture the contents of a rectangle you define within the MicroStation workspace.
- Capture View—Capture the contents of the view you select. Windows on top of the view are also captured.

- Capture View Window—Capture the contents and window border of the view you select. Windows on top of the view are also captured.

Convert

The Convert option opens the Raster File Conversion dialog box (see Figure 14–25) in which you can select an Input image file and convert it to an Output format file using a file name and path you specify.

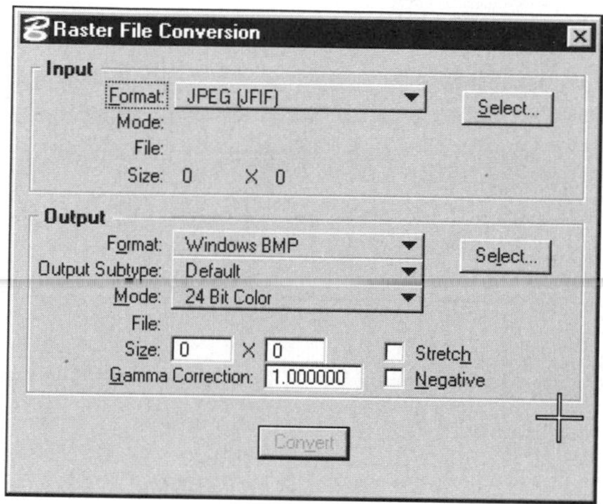

Figure 14–25 Raster File Conversion dialog box

Display

The Display option opens the Display Image dialog box (see Figure 14–26) from which you can select an image file to view in a separate window (see Figure 14–27).

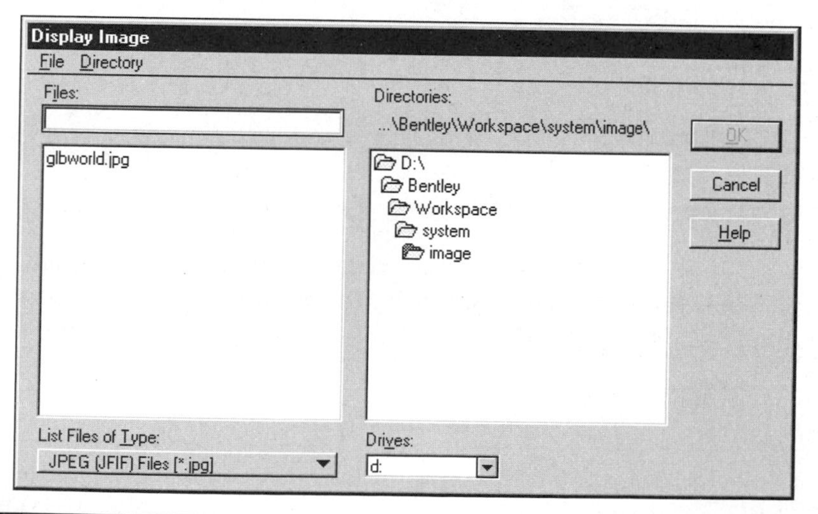

Figure 14–26 Display Image dialog box

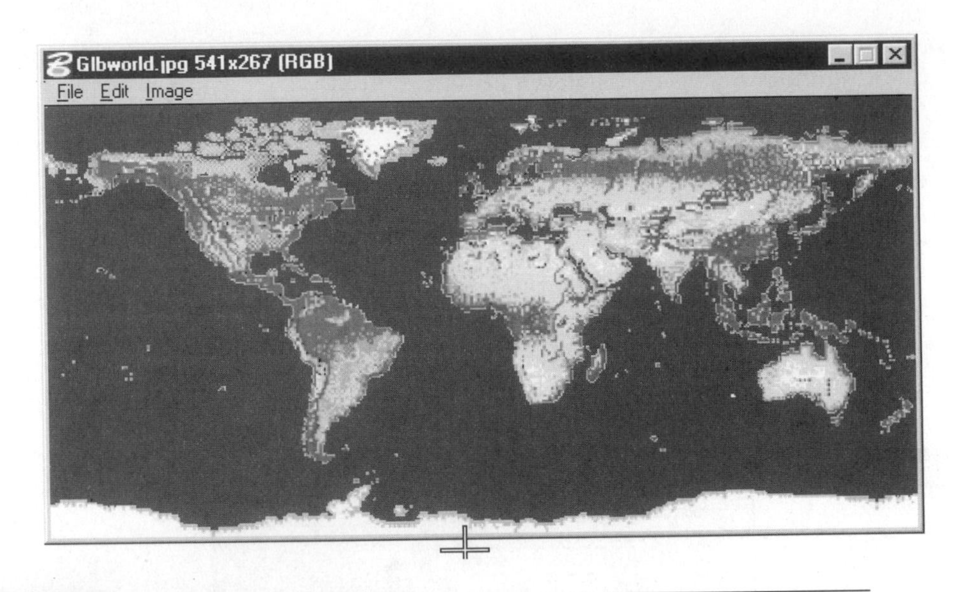

Figure 14–27 Example of displaying an image file in the Display Image window

Movies

The Movies option opens the Movies dialog box (see Figure 14–28) from which an animated sequence file can be selected for viewing in a separate window.

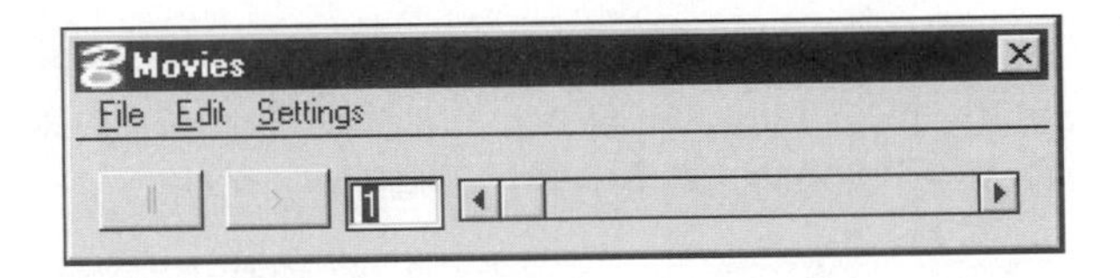

Figure 14–28 Movies dialog box

GLOSSARY TEXT

A large part of the textual annotation added to engineering models consists of standard terms and phrases. MicroStation helps speed up placing such text by providing a text glossary tool. Standard terms and phrases can be selected from the glossary for insertion in a design, and new ones can be added to the glossary.

Glossary Dialog box

Glossary text is viewed, selected, and placed from the Glossary dialog box. Open the Glossary dialog box:

Pull-down menu	Utilities > Text Glossary
Key-in window	**mdl load glossary** (or **md l glossary**) Enter

MicroStation displays a Glossary dialog box, as shown in Figure 14–29. Table 14–3 describes avalable options.

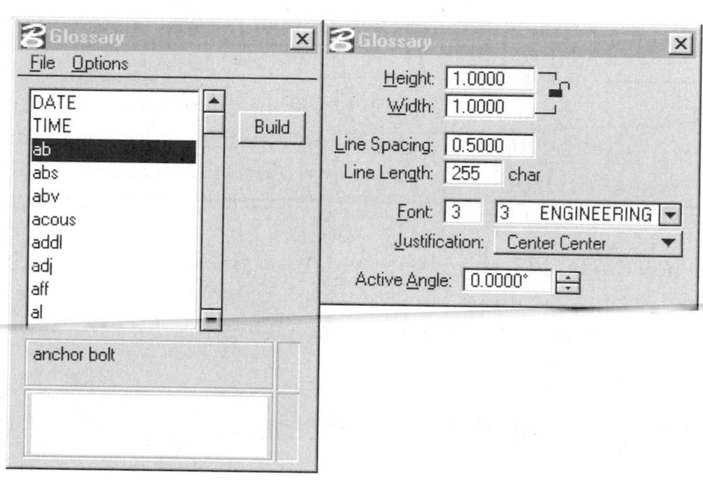

Figure 14–29 Glossary dialog box

Table 14–3 Glossary Dialog box Fields

FIELD	DESCRIPTION
File menu	The File pull-down menu contains an Open option that allows you to select a different glossary text file than the default glossary displayed in the window. The default glossary file path is contained in the MS_GLOSSARY configuration variable.
Options menu	The Options pull-down menu contains the Case submenu for selecting the case to use when placing the selected glossary text in the design: • Default—Place the text as it appears in the Place Text String field. • Uppercase—Make the entire text string uppercase when it is placed. • Lowercase—Make the entire text string lowercase when it is placed.
Alias list	The list box on the left side of the window lists the text aliases. Aliases are short abbreviations that identify the glossary text strings. Click on an alias to select it. In Figure 14–29 the "ab" alias is highlighted.
Associated Text	When an alias is selected, the glossary text associated with the alias appears in the Associated Text field just below the Alias List field. In Figure 14–29 the field contains "anchor bolt," which is the text associated with the "ab" alias highlighted in the Alias list field.

Table 14–3 Glossary Dialog box Fields (continued)

FIELD	DESCRIPTION
Build button	When the Build button is pressed, the text in the Associated Text field is inserted in the Place Text String field at the bottom of the window. Each time the button is clicked, the text associated with the selected alias is appended to the end of the text already in the Place Text String field (separated by a space).
Place Text String	The text string to be placed in the design is constructed in the field at the bottom of the window. Text strings are constructed here by clicking the Build button and by typing in the field.

Note: When the Build button is pressed the first time, the text attribute settings appear in the Tool Settings window. These settings control the appearance of the glossary text when it is placed in the design and can be changed before placing the text.

Constructing Text Strings from Glossary Text

Follow these steps to create and place a text string from the Glossary dialog box.

1. Open the Glossary dialog box as described previously.
2. If the required glossary is in a file other than the MicroStation default file, use the Open option in the Glossary dialog box's File pull-down menu to open the glossary file.
3. Select the alias for the glossary text you need.
4. Click the Build button to insert the selected text in the Place Text String field.
5. Repeat procedure steps 3 and 4 for each piece of glossary text to be inserted.
6. Edit the text in the Place Text String field, if required, to complete constructing the text string.
7. Optionally, change the text attribute settings in the Tool Settings window.
8. Place a data point in the design to place the text.
9. When the text string is placed in the design, the Place Text String field in the Glossary dialog box is cleared. To clear the Place Text String field without placing the text in it, click the Reset button on the pointing device.

Creating Glossary Text

Glossary text is contained in ASCII files that can be edited from word processing applications. For example, in Microsoft Windows, the Notepad is a handy application for editing the glossary.

The default glossary contains a large number of entries, which makes finding specific entries in the Glossary window rather time consuming. A more effective way to use glossaries is to create separate glossary files for each type of drawing. For example, create CIVIL.GLS to hold civil engineering terms and MECH.GLS to hold mechanical engineering terms.

Here is a procedure for creating custom glossary files.

1. Find the EXAMPLE.GLS glossary file and make a copy of it using a different file name.
2. Open the copy in a word processing application.
3. Delete all glossary terms not related to the discipline for which the glossary is being created.
4. Add any required terms that were not in EXAMPLE.GLS.
5. Save the new glossary and close the word processing application.
6. Repeat steps 1 through 5 for each glossary file to be created.

Instructions for creating new glossary text in a glossary file are provided at the beginning of EXAMPLE.GLS, as shown in Figure 14–30.

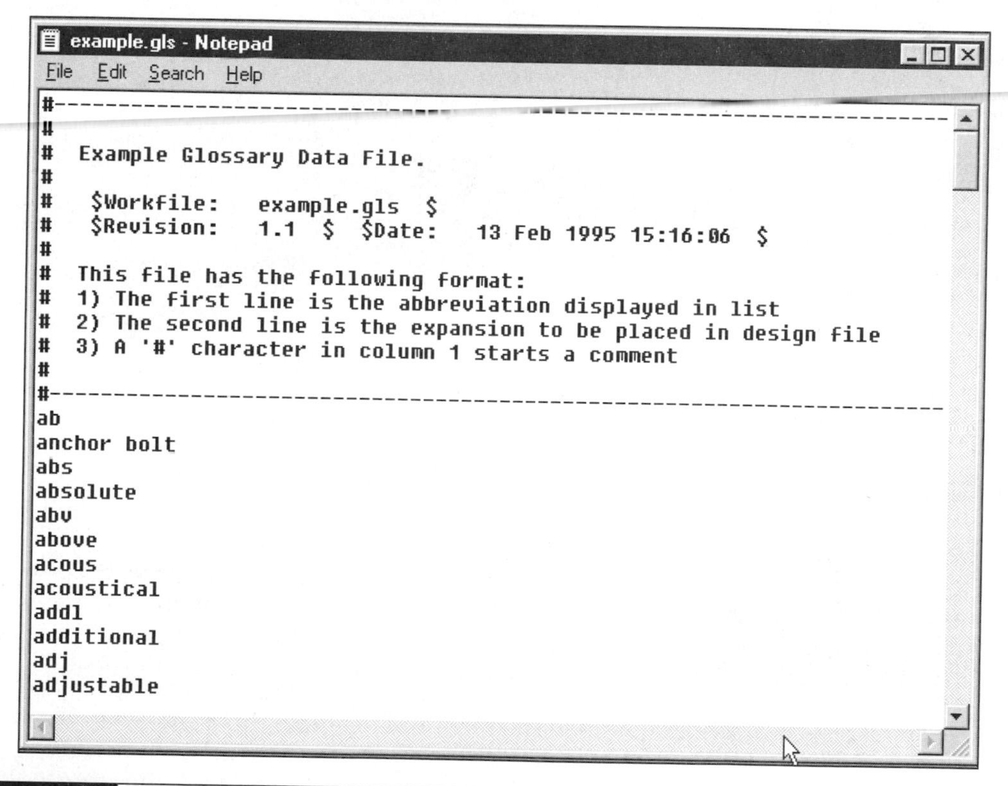
example.gls - Notepad
File Edit Search Help
#
Example Glossary Data File.
#
$Workfile: example.gls $
$Revision: 1.1 $ $Date: 13 Feb 1995 15:16:06 $
#
This file has the following format:
1) The first line is the abbreviation displayed in list
2) The second line is the expansion to be placed in design file
3) A '#' character in column 1 starts a comment
#
ab
anchor bolt
abs
absolute
abv
above
acous
acoustical
addl
additional
adj
adjustable

Figure 14–30 Top lines of EXAMPLE.GLS, the default glossary file

THE ANNOTATION TOOL FRAME

MicroStation provides an Annotation tool frame that contains three tool boxes containing tools that simplify the placement of annotation items such as flags, callout markers, and leaders, and places coordinate values in the design.

Invoke the Annotation tool frame:

Pull-down menu	Tools > Annotation (See Figure 14–31)
Key-in window	**dialog toolbox annotation** (or **di to annotati**) Enter

The Annotation tool frame contains three tool boxes:

- Annotate
- Drafting Tools
- XYZ Text

Note: Like the Main tool frame, the Annotation tool frame displays the last button selected from each tool box in the frame.

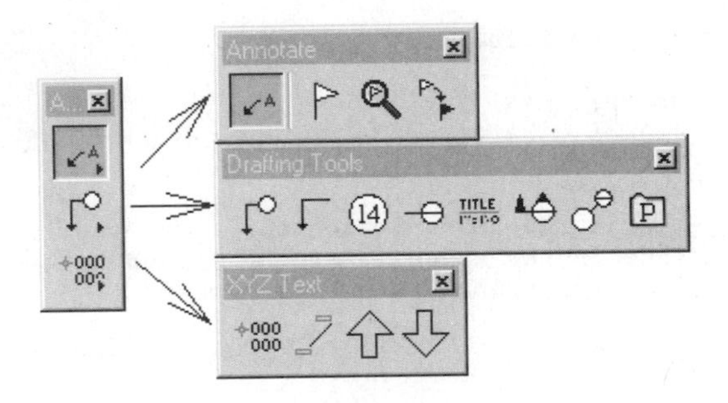

Figure 14–31 The Annotation tool frame and associated tool boxes

The Annotate Tools

The Annotate tools provide a way for the people working on a design file to place design notes in the design file behind flag symbols. The flag holds the text of the note and provides a visual reminder that there is a note at the location of the flag.

The following discussion describes how to place, read, and maintain the flags.

Note: The first tool in the Annotate tool box is Place Note. This is the same tool that is in the Text tool box, and it was discussed in chapter 7.

Place Flag Command

When you select the Place Flag tool, the Tool Settings window will contain several options for controlling the configuration of the flags you place. Table 14–4 explains the options.

Table 14–4 Place Flag Tool Settings

SETTING	EFFECT
Scale	Scales the flag symbol. A number greater than 1 increases the size and a number less than 1 decreases the size.
Level	Controls what level the flag is placed on.
Class	Provides a menu for setting the flag's class to one of the following: • Primary • Construction (if the Construction view attribute is turned OFF for a view, the flag will not be seen in the view or plotted)
Transparent	If ON, the image's background pixels are set to black to make the image appear to be transparent
Image	The name of the bit map file containing the symbol to be used for the flag. See Figure 13–32 for examples of typical flag symbol images and the names of the files containing the images.
Browse	Click this button to open a Browse window in which another image file can be chosen. The default directory for Browse contains a set of image files supplied with MicroStation. The maximum size of an image is 320 x 320 pixels.

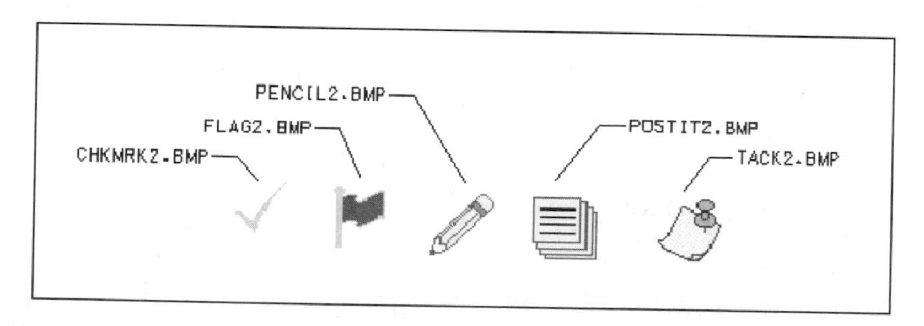

Figure 14–32 Typical flag images and their bit map file names

Invoke the Place Flag command:

Annotate tool box	Select the Place Flag tool, then select the placement options in the Tool Settings window (see Figure 14–33).
Key-in window	**place flag** (or **pl fl**) Enter

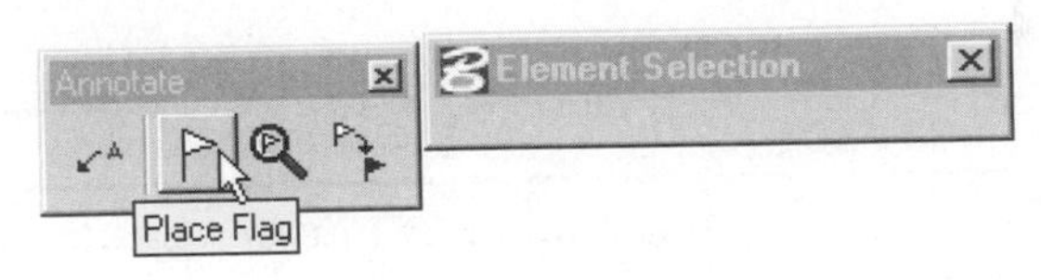

Figure 14–33 Invoking the Place Flag command from the Annotate tool box

MicroStation prompts:

Place Flag > Identify location *(Define the point where the flag is to be placed.)*

After the flag location is defined, the Define Flag Information window opens (see Figure 13–34). Type the flag's message in this window and do one of the following:

- Click OK to place the message with the flag and close the window.
- Click Cancel to close the window without placing text with the flag.

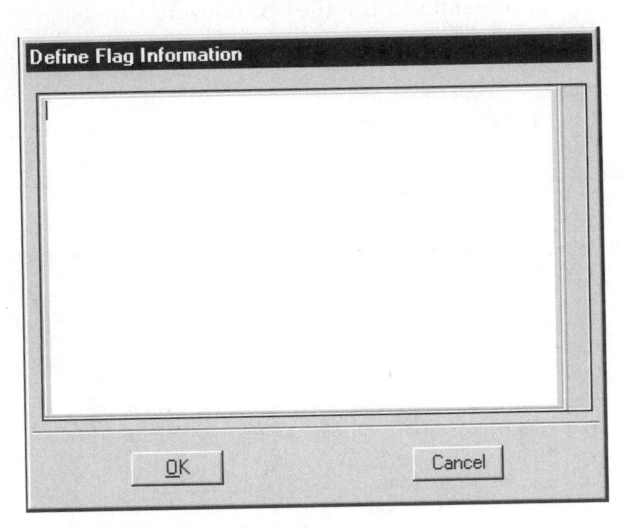

Figure 14–34 Define Flag Information window

Show/Edit Flag Command

The Show/Edit Flag command displays the flag's message and allows editing of the message content.

Invoke the Show/Edit Flag command:

Annotate tool box	Select the Show/Edit Flag tool (see Figure 14–35).
Key-in window	**show flag** (or **sh f**) Enter

Figure 14–35 Invoking the Show/Edit Flag command from the Annotate tool box

MicroStation prompts:

Show/Edit Flag > Select flag *(Click the Data button on the flag to be viewed or edited.)*
Show/Edit Flag > Accept/Reject *(Click the Data button again to accept the selected flag.)*

After the flag is accepted, the Define Flag Information window opens, and the message is ready for editing. After you complete editing the flag text, do one of the following:

- Click OK, to save the changes and close the window.
- Click Cancel, to close the window without saving the changes.

Update Flag Command

The Update Flag command changes the image used for an existing flag's symbol to the image currently set in the Place Flag's tool settings Image field. Before invoking the Update Flag command, invoke the Place Flag command and use the settings box Browse button to find and select the correct image.

Invoke the Update Flag command:

Annotate tool box	Select the Update Flag tool (see Figure 14–36).
Key-in window	**flag update** (or **fl u**) Enter

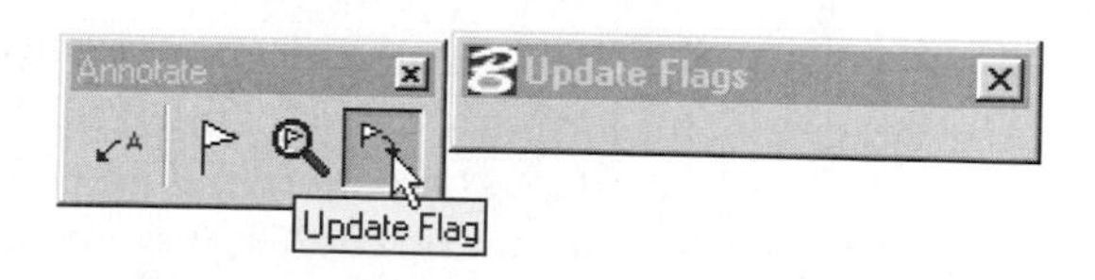

Figure 14–36 Invoking the Update Flag command from the Annotate tool box

MicroStation prompts:

Update Flag > Select flag *(Select the flag to be updated.)*
Update Flag > Accept/Reject *(Click the Data button again to update the flag symbol.)*

The Drafting Tools

The Drafting Tools tool box provides seven tools that simplify placing callout markers and leaders, and a tool to customize the way the elements are placed by the seven commands. Following are brief descriptions of each tool in the toolbox.

Place Callout Leader

Figure 14–37 shows the Place Callout Leader tool and examples of callout leaders placed by the tool. Place callout leaders by specifying the terminator location and the center of the callout symbol. The Tool Settings window provides options for selecting different types of Callout symbols, leader lines, and terminators, and for the text to be placed in the bubble. The text is placed using the active text parameters.

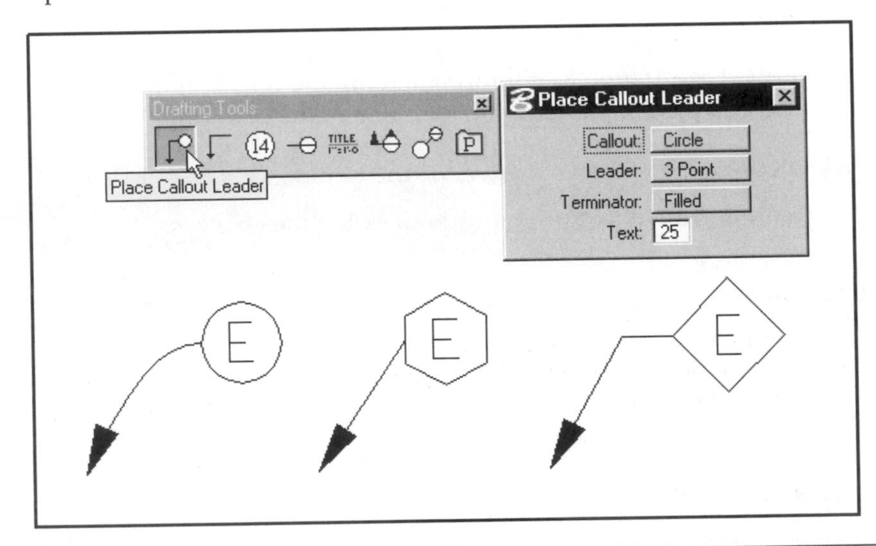

Figure 14–37 The Place Callout Leader tool and examples of callout leaders placed by the tool

Place Leader and Text

Figure 14–38 shows the Place Leader and Text tool and examples of leaders placed by the tool. Place the leaders by specifying the terminator location and the end of the leader. The Tool Settings window provides options for selecting different types of leader lines and terminators, and a place to type a text string to place with the leader. The text is placed using the active text parameters.

Place Callout Bubble

Figure 14–39 shows the Place Callout Bubble tool and examples of bubbles placed by the tool. Place the bubbles by specifying the center of the bubble. The Tool Settings window provides options for selecting different types of leader lines and terminators, and a place to type a text string to place with the leader. The text is placed using the active text parameters.

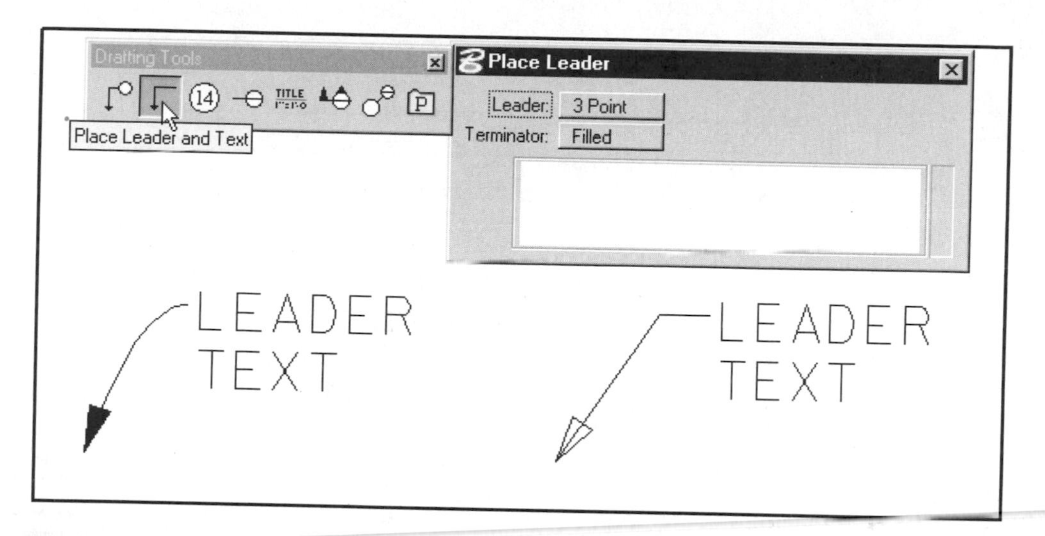

Figure 14–38 The Place Leader and Text tool and examples of leaders placed by the tool

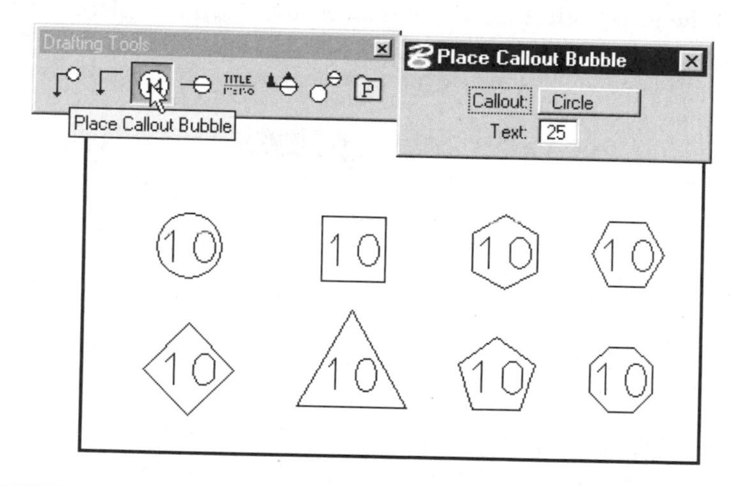

Figure 14–39 The Place Callout Bubble tool and examples of bubbles placed by the tool

Place Selection Marker

Figure 14–40 shows the Place Selection Marker tool and an example of a marker placed by the tool. Place the markers by specifying the left end of the marker line. The Tool Settings window provides options for placing text strings and reference numbers above and below the marker line. The text is placed using the active text parameters.

Figure 14–40 The Place Selection Marker tool and an example of a marker placed by the tool

Place Title Text

Figure 14–41 shows the Place Title Text tool and an example of a title placed by the tool. Place the title text by specifying the left end of the title line. The Tool Settings window provides options for placing text strings above and below the title line. The text is placed using the active text parameters.

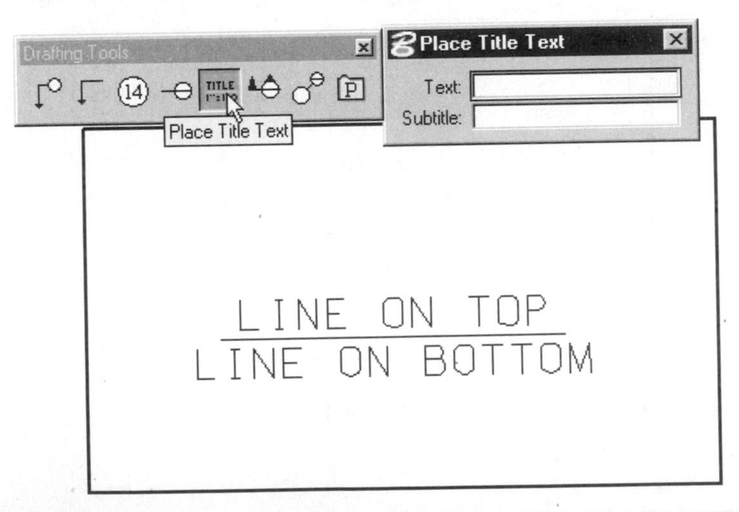

Figure 14–41 The Place Title Text tool and an example of a title placed by the tool

Place Arrow Marker

Figure 14–42 shows the Place Arrow Marker tool and examples of arrow markers placed by the tool. Place the marker by specifying the center of the marker and the direction of the arrow. The Tool

Settings window provides options for placing text strings above (Ref#) and below (Sht#) the marker line, and for placing flipped arrows. The text is placed using the active text parameters.

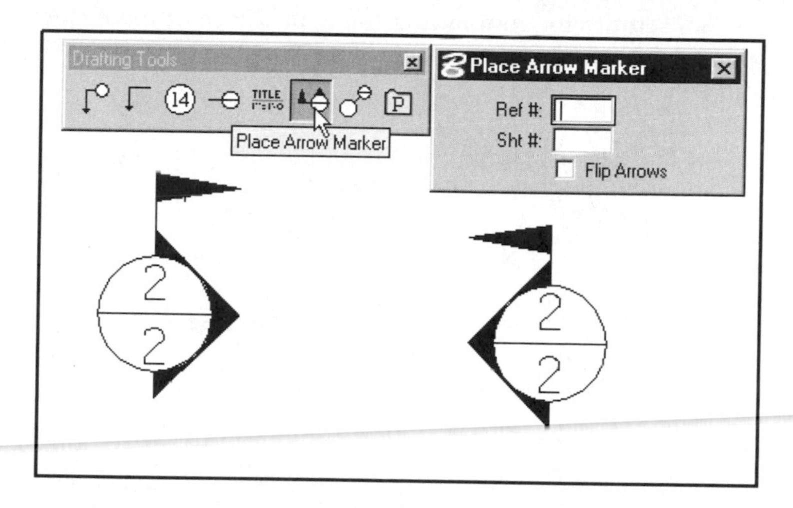

Figure 14–42 The Place Arrow Marker tool and examples of arrow markers placed by the tool

Place Detail Marker

Figure 14–43 shows the Place Detail Marker tool and examples of a detail marker placed by the tool. Place the marker by specifying the center of the detail and edge of the detail circle and the center of the marker. The Tool Settings window provides options for placing text strings above (Ref#) and below (Sht#) the marker line. The text is placed using the active text parameters.

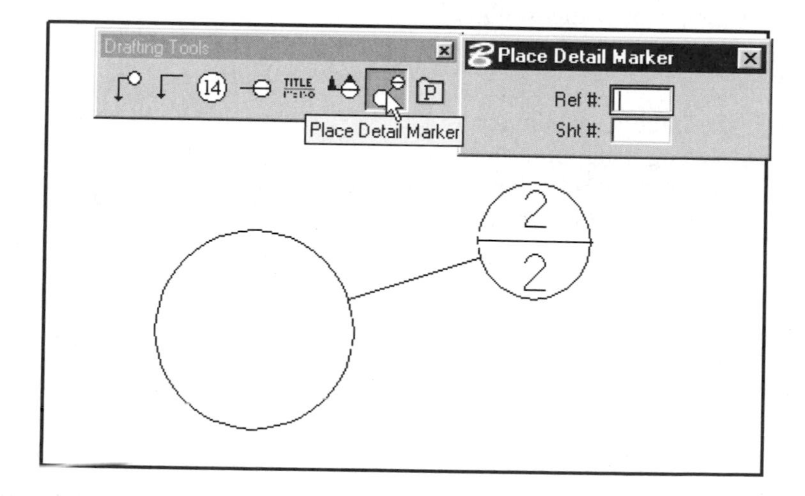

Figure 14–43 The Place Detail Marker tool and an example of a detail marker placed by the tool

Define Properties

Figure 14–44 shows the Drafting Tools Properties dialog box that the Define Properties tool opens. The dialog box provides settings for customizing the appearance of the elements placed by the Drafting Tools. Options are provided to control the size and attributes of the callout bubbles, leaders, bubbles, and text.

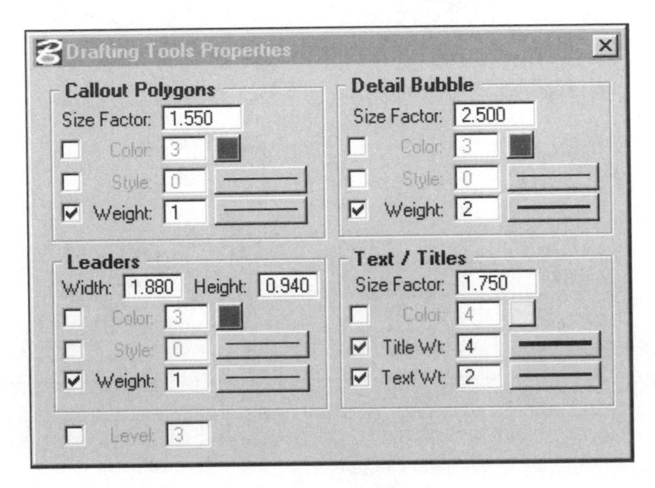

Figure 14–44 The Drafting Tools Properties dialog box

XYZ Text Tool box

The XYZ Text tool box provides tools that allow you to place coordinate labels in the design, export coordinate points to an ASCII file, or import coordinate points from an ASCII file. Following are brief descriptions of each tool in the toolbox.

Label Coordinates

Figure 14–45 shows the Label Coordinates tool and examples of placing coordinate labels. Place coordinate labels by placing data points in the design. The Tool Settings window provides options for controlling the placement of the coordinate labels. The options are described below. The text is placed using the active text parameters.

The Tool Settings window provides the following options for controlling the placement of the coordinate label text:

- The Order menu allows you display the *X* coordinate first or the *Y* coordinate first.
- The Units menu allows you to control the working units format of the coordinates.
- The Accuracy menu allows you to control the accuracy, in decimal places, of the coordinate.
- The Separator menu allows you to have the *X* and *Y* coordinates placed on separate lines, separated by a comma, or separated by a space.

- The View menu is for *3D* files only.
- X, Y, and Z prefix edit fields allow you to change the prefix that is placed before each coordinate value.

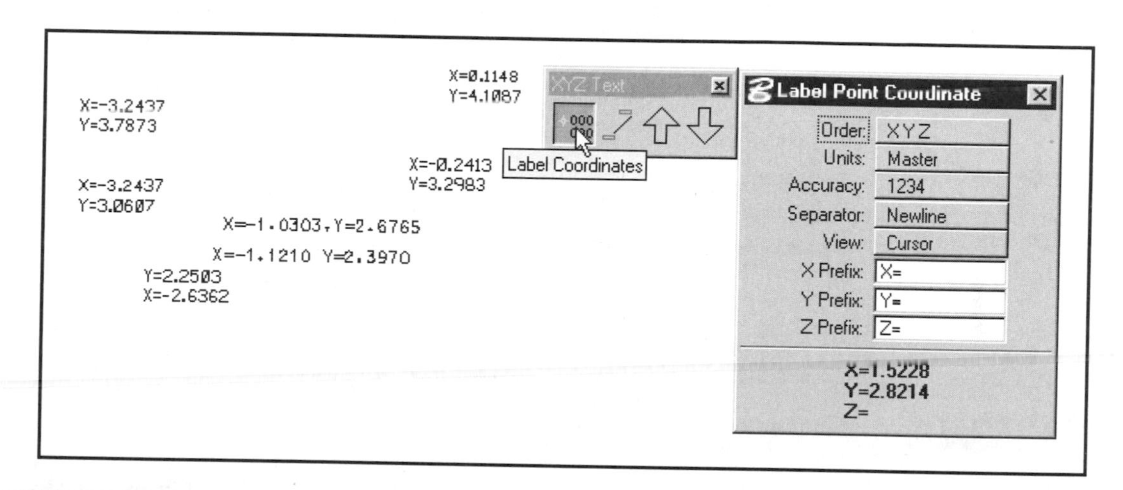

Figure 14–45 The Label Coordinates tool and examples of the labels in a design

Note: The coordinates are placed as multi-line text elements. If the Text Node View Attribute is ON, the Text Node number and cross will be displayed with the coordinate element, and it will plot.

Label Element

Figure 14–46 shows the Label Element tool and examples of placing coordinate labels. The Tool Settings window provides the same coordinate text options as the Coordinate Label tool, and three buttons that control the way elements are selected:

- The Single button allows you to place coordinate labels for a single element.
- The Fence button allows you to place coordinate labels on all elements grouped by a fence.
- The All button allows you to place coordinate labels on all elements in the design. A confirmation box will appear asking you to confirm that you really want to put labels on all elements.

Each time you use this tool you must select one of the control buttons. For example, before you can place a coordinate label for one element, you must click the Single button. The coordinate text is placed using the active text parameters.

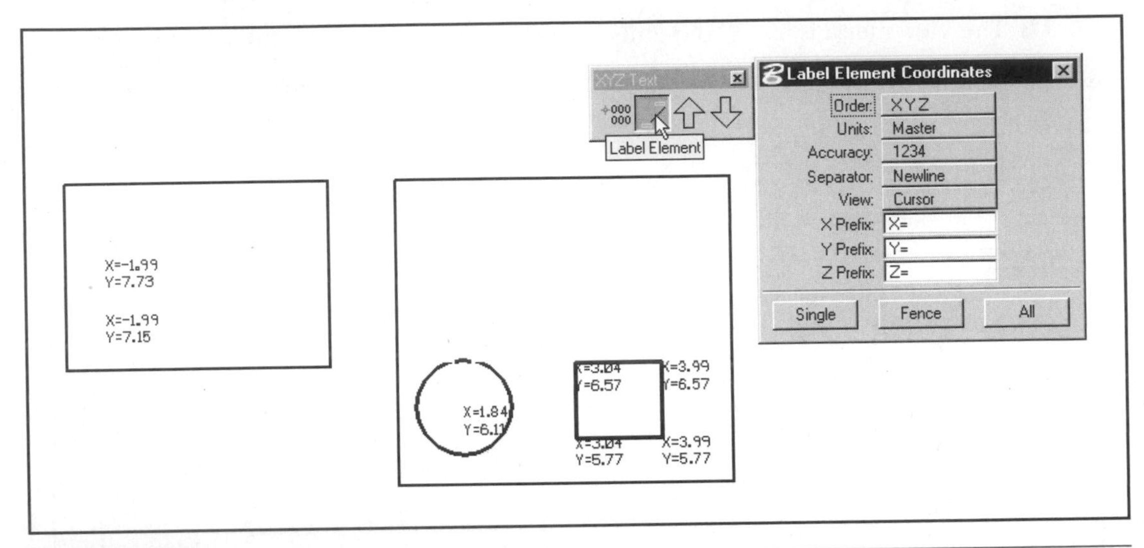

Figure 14–46 The Label Element tool and examples of coordinate labels placed on elements

Note: The coordinate label will be placed at the center of a circle and ellipse, at the center and ends of an arc, and at each vertex of linear elements.

Export Coordinates

Figure 14–47 shows the Export Coordinates tool. This tool allows you to write the coordinates of selected elements to an ASCII file. The first option in the Tool Settings window provides an edit field in which you can type the name of the ASCII file you want the coordinates written to, and a Browse button that opens a file selection dialog box from which you can select the file. You can use an existing or new file. The other options in the Tool Settings window are identical to the Label Element tool settings.

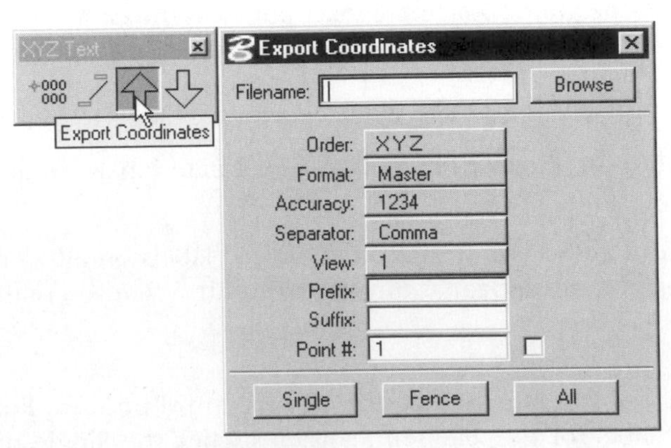

Figure 14–47 The Export Coordinates tool

The buttons at the bottom of the Tool Settings window allow you to export the coordinates for a Single element, the elements grouped by a Fence, or All elements in the design. If you are exporting to an existing file, MicroStation will display the Export File Exists dialog box shown in Figure 14–48. You can elect to append the coordinates to the end of the file or replace the information currently in the file with the new coordinates. Figure 14–49 shows an example of coordinates placed in an export file.

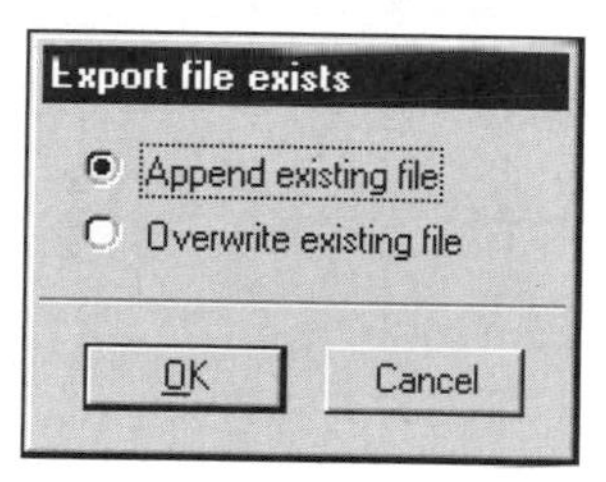

Figure 14–48 The Export File Exists dialog box

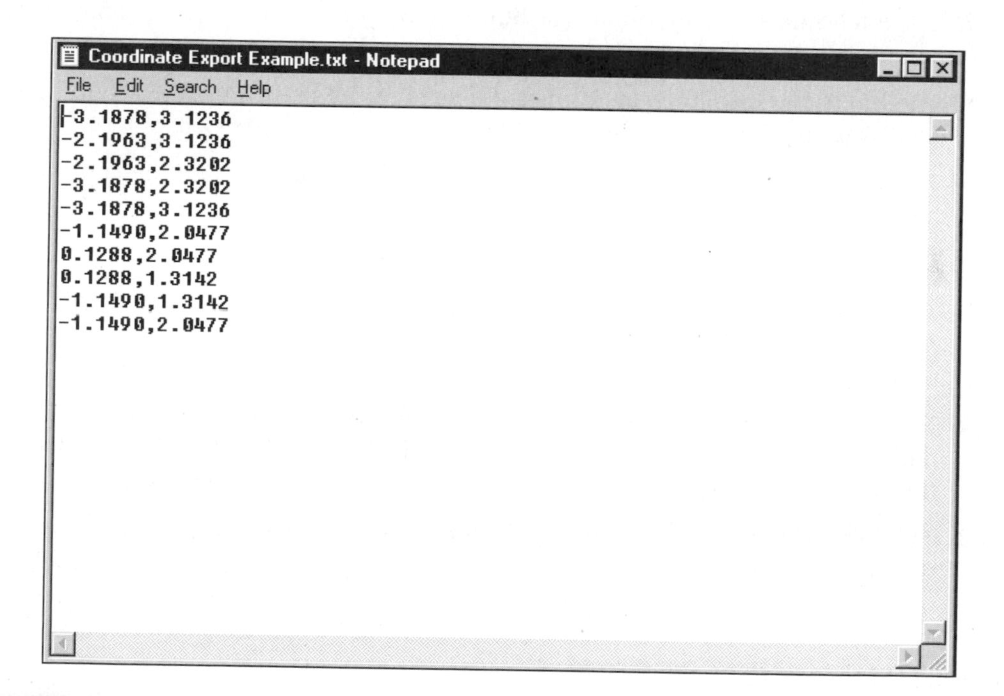

Figure 14–49 Example of exported coordinates

Import Coordinates

Figure 14–50 shows the Import Coordinates tool. This tool allows you to import coordinates from an ASCII file and place them in the active design as a point, text string, or cell.

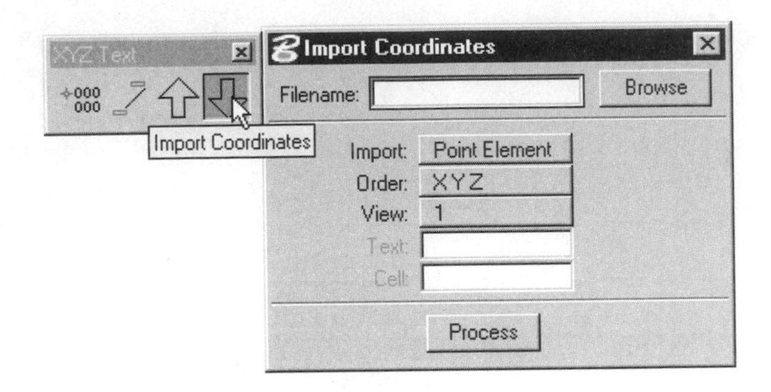

Figure 14–50 The Import Coordinates tool

The associated Tool Settings window provides options to control how the coordinates are imported, and to initiate the importation:

- The FileName edit field and Browse button allow you to either type the name of the file containing the coordinates or browse to it.
- The Import menu allows you to choose what is placed at the imported coordinate location. You can choose to place a Point Element, Text string, or Cell.
- The Order menu allows you to elect to choose the order in which the coordinates are imported (XY or YX).
- The View menu is for *3D* designs.
- The Text edit field is enabled only when the Import menu is set to Text. This field provides a place for you to enter the text string that will be placed at the coordinate locations.
- The Cell edit field is enabled only when the Import menu is set to Cell. This provides a place for you to enter the name of the cell that will be placed at the coordinate locations. The cell library containing the cell must be attached to the design.
- The Process button initiates importing the coordinates from the file whose name is in the Filename field.

DIMENSION-DRIVEN DESIGN

Dimension-driven design provides a set of tools to define constraints on the elements that make up a design. These constraints control the size and shape of the design, so the design can be adjusted for changing requirements by simply entering new dimension values.

Dimension-driven cells are cells defined from constrained designs. The dimensions of such cells can be changed as they are placed.

Example of a Constrained Design

We introduce dimension-driven design by describing the steps required to create the model shown in Figure 14–51. Before we start constructing the model, let's look at the constraints on the model and at the effect of changing a constraint.

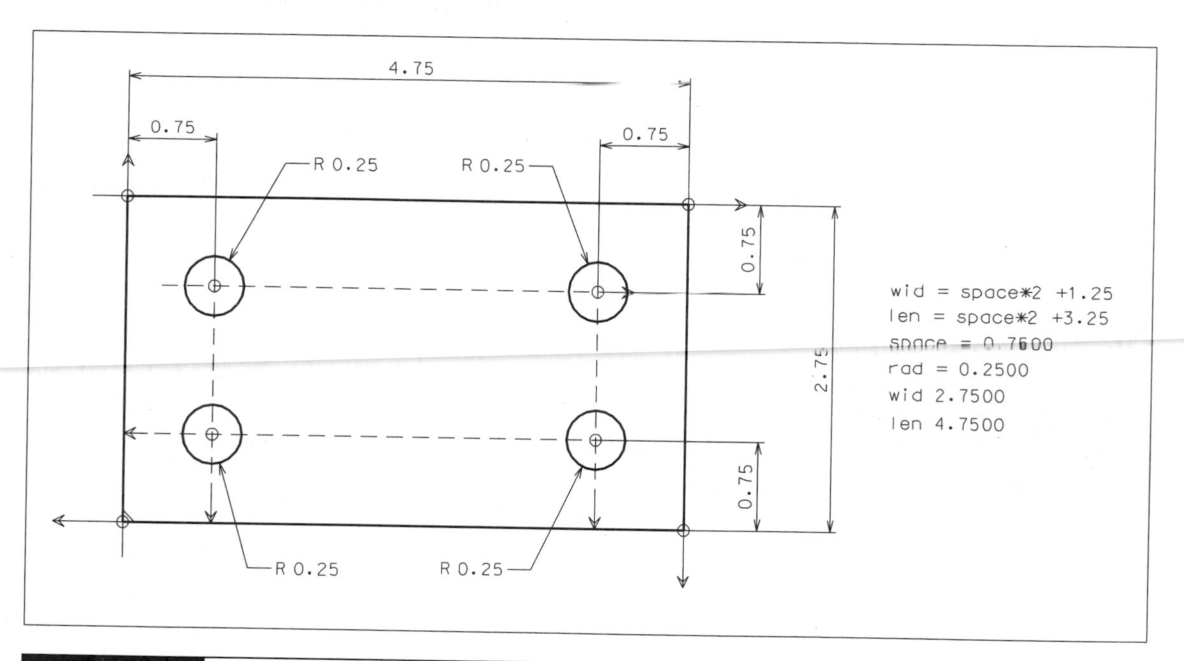

Figure 14–51 Example of a constrained design with the construction elements displayed

The Constraints

The design in the figure contains several "construction" elements that graphically represent the constraints:

- The dashed lines with arrows on one end are the construction lines to which the design is attached (some of the construction lines are covered by the lines and circles of the design).
- The equations and variables to the right of the model define the constraints.
- The Construction View attribute controls the display of construction elements. Figure 14–51 shows the view with Construction lines set to ON.

The following constraints were placed on the design of Figure 14–51:

- The angles of the construction lines are fixed, as indicated by the arrows.
- The construction line intersections are constrained to always be connected, as indicated by the small circles.

- The radius of the circles are set by the “rad = 0.2500” variable.
- The circles are constrained always to be centered on the intersections of the inter-construction lines.
- The space between the center of each circle and the adjacent design edges are set by the “space = 0.7500” variable.
- The overall length of the design is set by the “len = (space * 2) + 3.25” equation. The equation multiplies the circle-center-to-edge space by 2 and adds 3.25 to that total (4.75 = [0.75 * 2] + 3.25).
- The overall width of the design is set by the “wid = (space * 2) + 1.25” equation. The equation multiplies the circle-center-to-edge space by 2 and adds 1.25 to that total (2.75 = [0.75 * 2] + 1.25).

Effect of Changing a Constraint

To illustrate the effect of changing a constraint, we use the Text Edit tool to change the add-on value in the wid equation from 1.25 to 0.75 Master Units (“wid = [space * 2] + 0.75”), then we use the Re-solve Constraints tool to solve the design for the new value. Figure 14–52 shows the design change caused by changing the wid add-on:

- We moved the two rows of circles closer together.
- We decreased the overall width of the design from 2.75 to 2.25 master units.

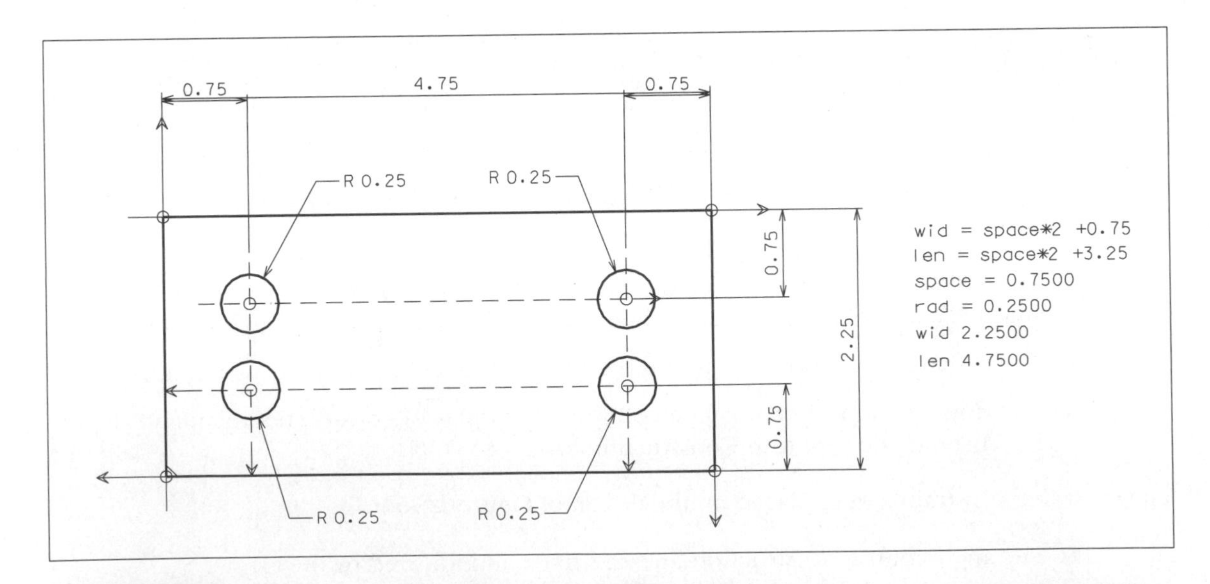

Figure 14–52 Effect of changing the width equation add-on from 2.75 to 2.25 master units

The overall size was reduced because the value we changed is part of the constraint on the overall width.

Dimension-Driven Design Terms

Following are common terms in dimension-driven design.

- **Constraint**—Information that controls how a construction is handled within a model. A constraint can be one of the following types:
- **Location**—Fixes the location of a point in the design plane.
- **Geometric**—Controls the position or orientation of two or more elements relative to each other.
- **Dimensional**—Controlled by a dimension.
- **Algebraic**—Controlled by an equation that expresses a relationship among variables.
- **Construction**—An element, such as a line or circle, on which constraints can be placed to control its relation to other constructions in the model.
- **Well-Constrained**—A set of constructions that is completely defined by constraints and has no redundant constraints. It has what is needed to define it and no more.
- **Underconstrained**—A set of constructions that does not have enough constraints to define completely its geometric shape.
- **Redundant**—A constraint applied to a construction that is already well-constrained. It provides no useful information for the construction.
- **Degrees of Freedom**—A number that sums up a dimension driven cell's ambiguity. The cell is underconstrained.
- **Solve**—Constructing the model from the given set of constraints. Each time a constraint is modified or added the model is solved for the new set of constraints. If the model can be solved for the constraints, the model is updated; if not, an error message is displayed in the Status bar.

Dimension-Driven Design Tools

Dimension-driven tool boxes are available in the DD Design submenu. Invoke the submenu:

Pull-down menu	Select Tools > DD Design
Key-in window	**dialog toolbox dddtools** (or **di to dddt**) Enter

MicroStation displays the Dimension-Driven tool frame. Figure 14–53 shows the tool frame surrounded by the tool boxes that can be opened from it.

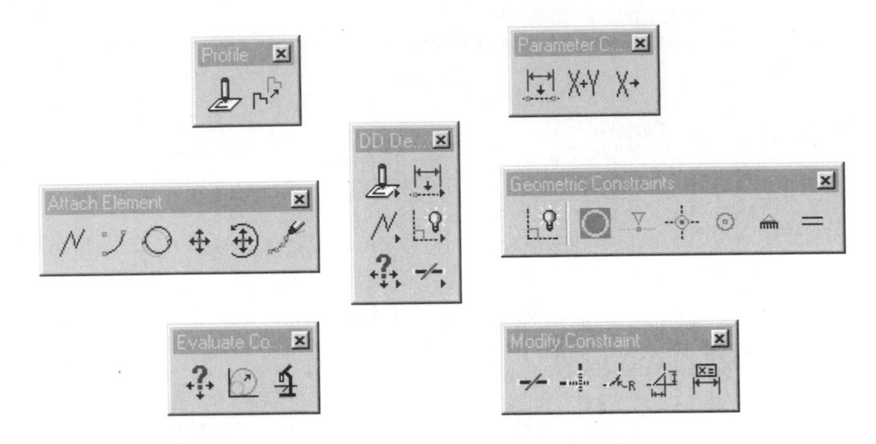

Figure 14–53 Dimension-Driven tool frame and its tool boxes

The tool boxes consist of the following:

- **Profile**—Provides tools for drawing a construction profile and for converting elements to a construction profile.
- **Attach Elements**—Provides tools for attaching elements, such as lines and arcs, to the constrained construction elements.
- **Evaluate Constraints**—Provides tools for solving and obtaining information about the construction.
- **Parameter Constraints**—Provides tools for converting dimensions to constraints and for assigning equations and variables.
- **Geometric Constraints**—Provides tools for placing constraints on the construction elements that are to define the model.
- **Modify Constraint**—Provides tools for modifying constraints

Creating a Dimension-Driven Design

A dimension-driven design requires careful planning before you begin to define the constraints. Numerous tools are available to constrain the design. A good way to be introduced to the method is to walk through the creation of a design. The following discussion shows one way to create the model shown earlier in Figure 14–51.

Draw Construction Elements

Set the line weight to zero, and draw the lines and circles as shown in Figure 14–54. The elements do not have to be drawn to specific dimensions because the constraints to be added later set the dimensions of the design. Draw the elements in the Primary mode. AccuDraw will help you quickly ensure that the lines are drawn horizontally and vertically.

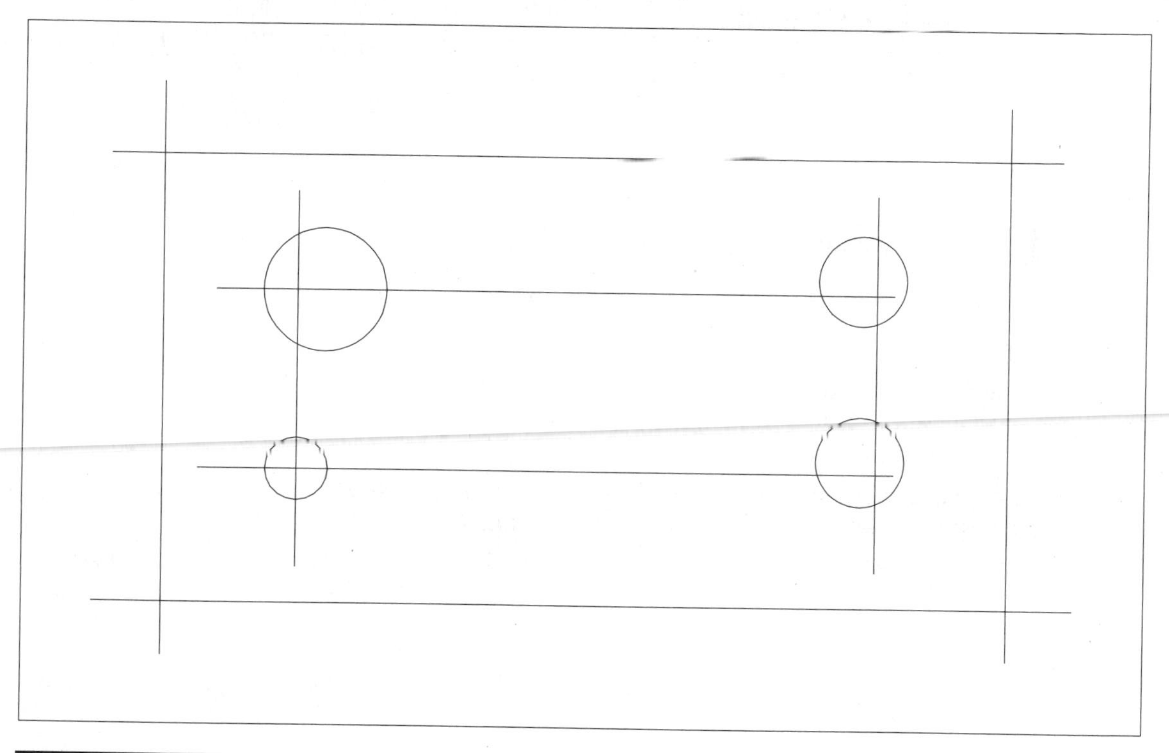

Figure 14–54 Design construction lines and circles

Constrain the Angle of the Lines

The first constraint we are going to put on the design is to force the lines to stay at the angles at which they were drawn. To constrain the angle of the lines shown in Figure 14–54, invoke the Constrain Elements tool:

Geometric Constant tool box	Select the Constrain Elements tool, then in the Tool Settings window, select the Smart Constrain Elements Method (the left-most Method button), click the Settings button to expose the settings, and turn on the Convert to Constructions toggle button (see Figure 14–55).
Key-in window	**constrain element** (or **con e**) Enter

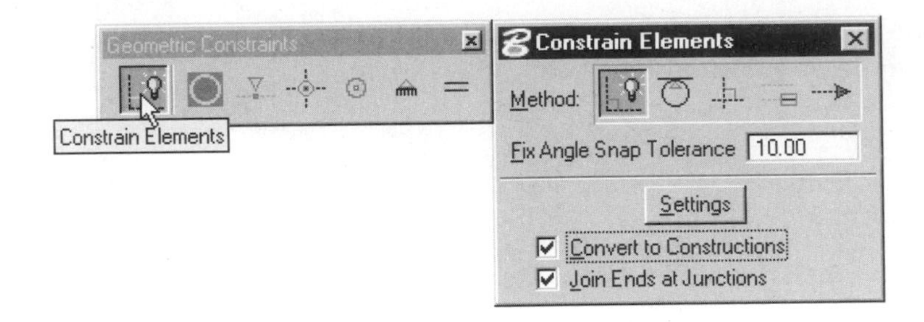

Figure 14–55 Invoking the Constrain Elements tool from the Geometric Constraints tool box

MicroStation prompts:

> Make Smart Constrain Elements > Identify Construction *(Identify one of the lines, and click the Data button again to accept it. The second point does not select another element.)*
>
> Smart Constrain Elements > Identify Second Construction or Same Construction to Fix Angle *(Identify the same line again to fix its angle.)*
>
> Smart Constrain Elements > Accept *(Click the Data button a third time to constrain the angle.)*

An arrow appears at one end of the line, the line's color changes to yellow and its line style to dash, and the line is changed to a construction element. This constraint ensures that the lines always remain at their original rotation angle.

Repeat the steps to constrict the angle of the rest of the elements in the design. Figure 14–56 shows the results of constraining the angles of all the lines.

Note: If the Constructions View Attribute is turned OFF in the view you are working in, the constrained elements disappear from the view.

Constrain Point At Intersection

The Constrain Point At Intersection tool either constrains a point to lie at the intersection of two constructions or forces two constructions to pass through a point. It works with any kind of construction, except points. To constrain the line intersections as shown in Figure 14–56, invoke the Constrain Point At tool:

Geometric Constraints tool box	Select the Constrain Point At tool (see Figure 14–57).
Key-in window	**constrain intersection** (or **con i**) Enter

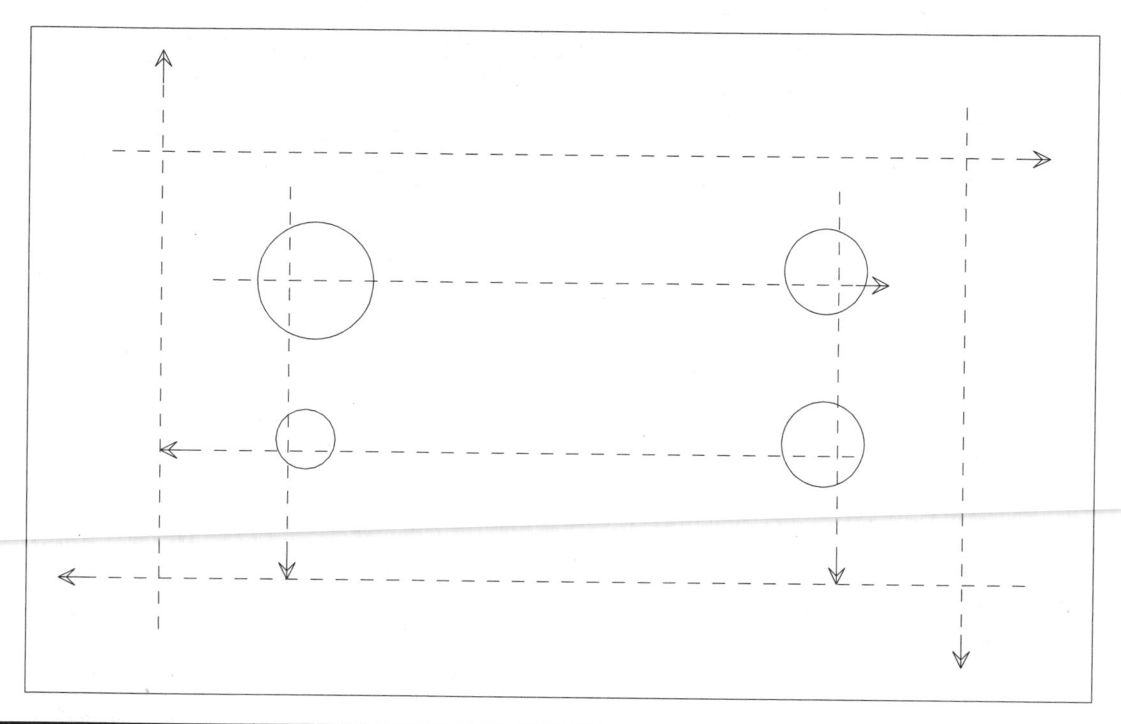

Figure 14–56 Results of constraining the angles of the lines

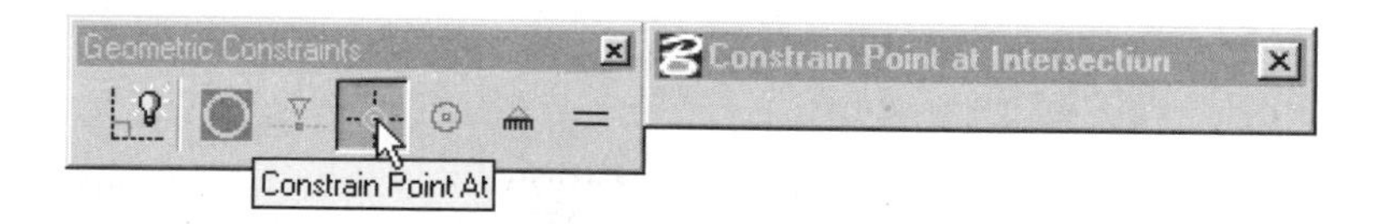

Figure 14–57 Invoking the Constrain Point At tool from the Geometric Constraints tool box

MicroStation prompts:

> Constrain Point at Intersection > Identify Construction *(Identify one of the intersecting lines near the point of intersection, then select the other intersecting line.)*
>
> Constrain Point at Intersection > Identify point or Accept+RESET *(Click the Data button in space, then click the Reset button to constrain the point.)*

A small circle appears at the constrained intersection. This constraint ensures that when one line is modified, the constrained lines always pass through the constraining point.

To constrain the remaining intersections, repeat the procedure for each section of the outer and inner set of lines. Figure 14–58 shows the results of constraining the intersections.

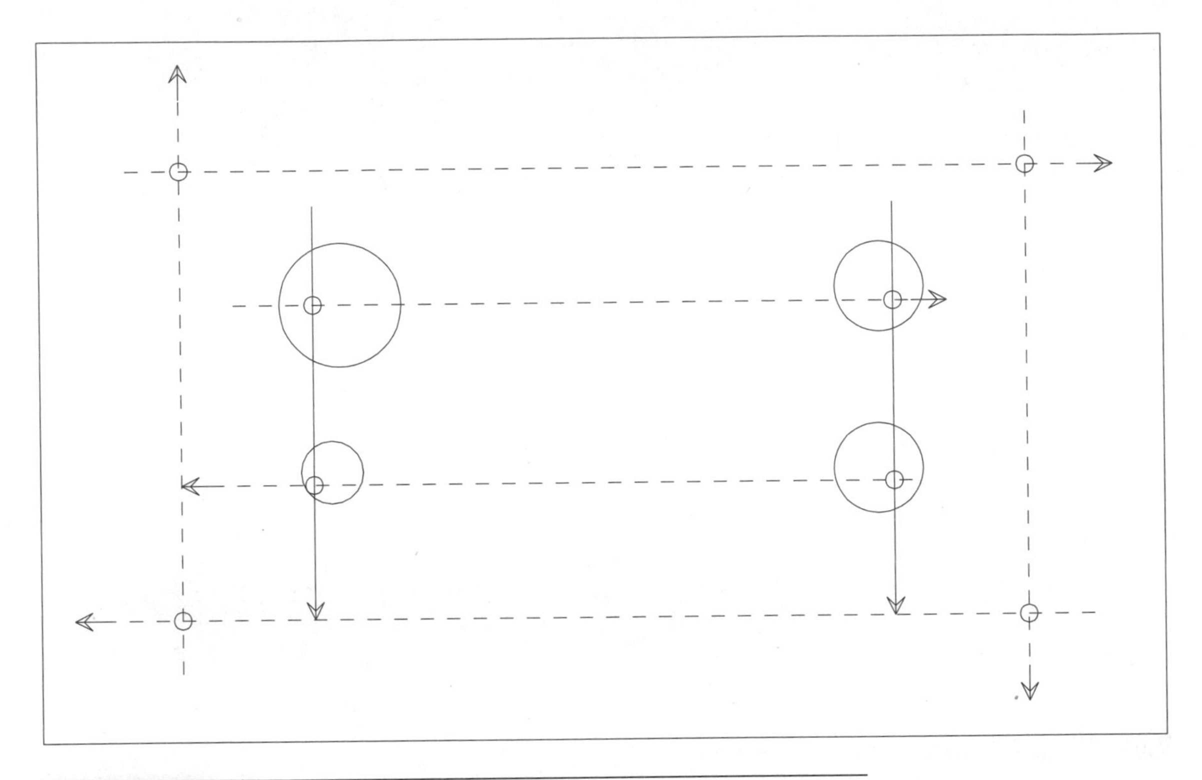

Figure 14–58 Results of constraining the construction intersections

Constrain Two Points to Be Coincident

This tool lets you constrain two points to the same location (coincident), two circles to be concentric (have the same center), or a point to lie at the center of a circle. To constrain the circles to be centered at the intersection points as shown in Figure 14–51, invoke the Constrain Points Coincident tool:

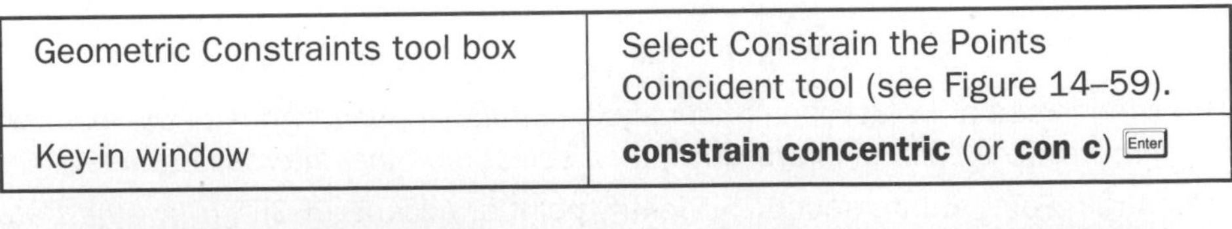

Geometric Constraints tool box	Select Constrain the Points Coincident tool (see Figure 14–59).
Key-in window	**constrain concentric** (or **con c**) Enter

Figure 14–59 Invoking the Constrain Two Points to Be Coincident tool from the Geometric Constraints tool box

MicroStation prompts:

> Constrain Two Points to Be Coincident > Identify Point (or Ellipse) *(Identify the intersection point the circle is to be centered about.)*
>
> Constrain Two Points to Be Coincident > Identify Next Point (or Ellipse) *(Identify the circle.)*
>
> Constrain Two Points to Be Coincident > Accept *(Click the Data button in space to complete the constraint.)*

This constraint causes the circles to change to construction elements that are yellow and dashed. The circles are constrained to stay centered over the intersections when the positions of the intersecting lines are modified.

Repeat the procedure to constrain the other three circles to the intersection points. Figure 14–60 shows the results of constraining the circles to the intersection points.

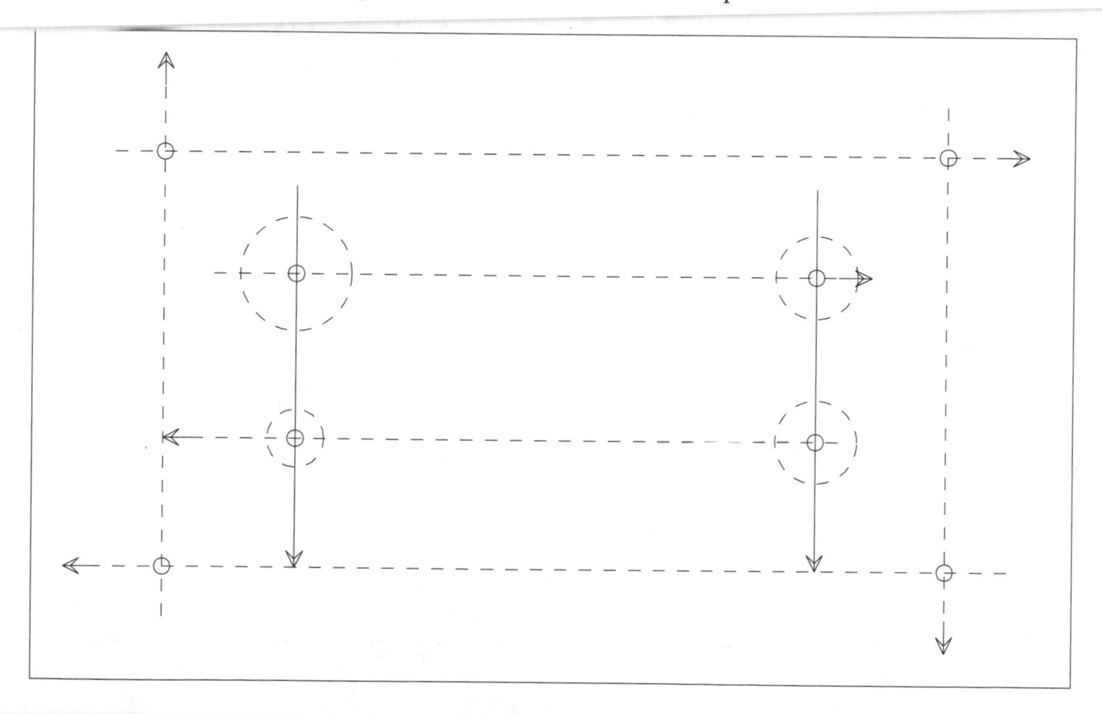

Figure 14–60 Results of constraining the circles

Fix Point at Location

The Fix Point at Location tool enables you to fix the location of a point (or the center of a circle or ellipse) in the design plane. To attach the lower left intersection of the design to its current location in the design plane, invoke the Fix Point tool:

Geometric Constraints tool box	Select the Fix Point tool (see Figure 14–61).
Key-in window	**constrain location** (or **con l**) Enter

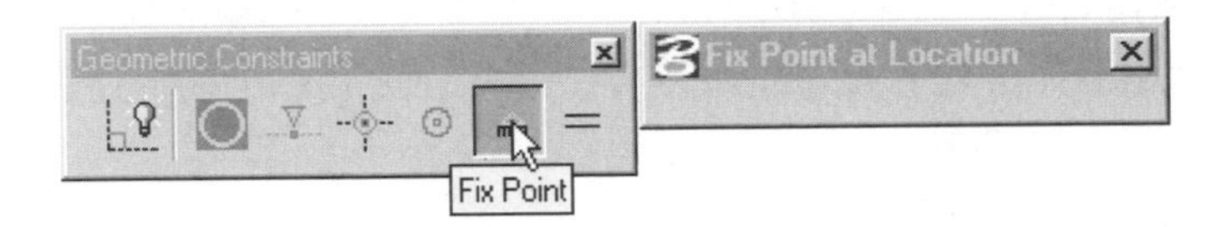

Figure 14–61 Invoking the Fix Point tool from the Geometric Constraints tool box

MicroStation prompts:

Fix Point at Location > Identify Point (or Ellipse) *(Select the lower left intersection point.)*
Fix Point at Location > Accept/Reject *(Click the Data button a second time to accept the point.)*

This tool fixes the design to a location in the design plane, and this point remains fixed when changes are made to the size of the constrained design.

Construct Attached Line-String Or Shape

With the Construct Attached Line-String Or Shape tool you can create a line-string or shape with the vertices attached to construction points, circles, or constraints. It is recommended to set a higher value for the active line weight so that you can easily distinguish it from the construction elements. To connect the shape as shown in Figure 14–51, invoke the Attach Line-String or Shape tool:

Attach Element tool box	Select the Attach Line-String Or Shape tool (see Figure 14–62).
Key-in window	**attach lstring** (or **at ls**) Enter

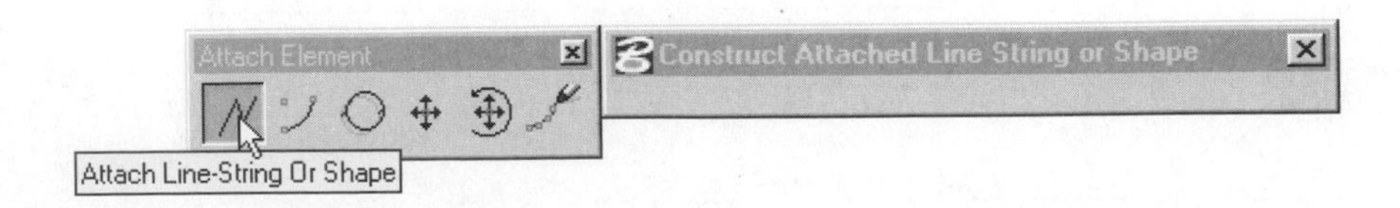

Figure 14–62 Invoking the Construct Attached Line-String Or Shape tool from the Attach Element tool box

MicroStation prompts:

> Construct Attached Line String or Shape > Identify Point or Constraint *(Select one of the constraint points on an intersection of a pair of the outer lines, then select the constraint point at the other end of one of the lines.)*
>
> Construct Attached Line String or Shape > Identify point, or RESET to finish *(Select the other two constraint points on the outer lines, then select the first constraint point again to complete the shape.)*

When the first constraint point is selected again, a closed shape is placed attached to the four points, as shown in Figure 14–63. Changes to the position of the constraint points will change the shape of the attached shape element.

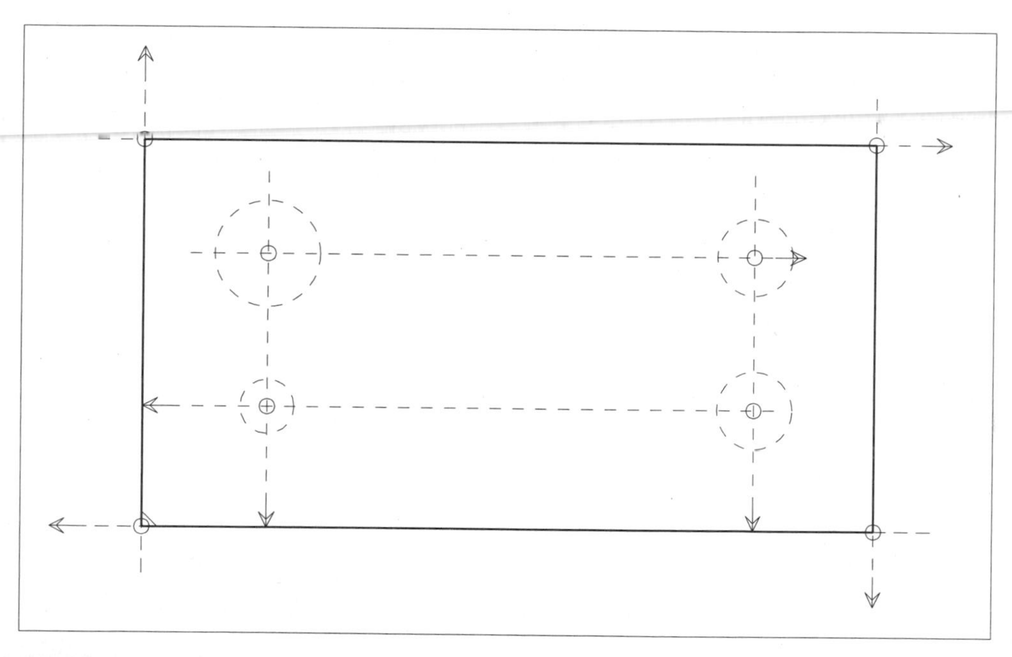

Figure 14–63 Results of drawing the design's outline shape

Construct Attached Ellipse or Circle

The Construct Attached Ellipse or Circle tool enables you to create and attach a circle to a construction circle or point. To attach the design circles to the intersection points as shown in Figure 14–51, invoke the Attach Ellipse tool:

Attach Element tool box	Select the Attach Ellipse tool (see Figure 14–64).
Key-in window	**attach circle** (or **at c**) Enter

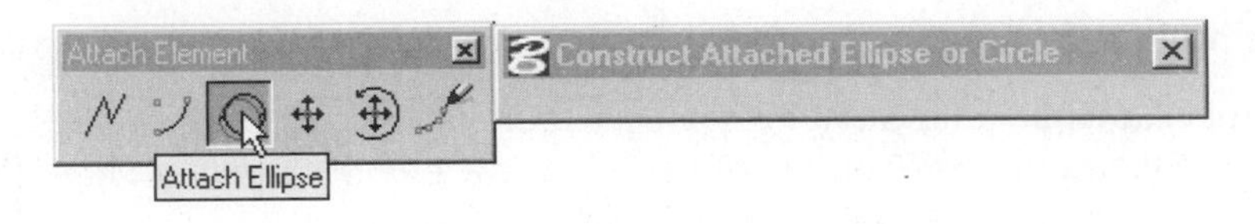

Figure 14–64 Invoking the Construct Attached Ellipse or Circle tool from the Attach Element tool box

MicroStation prompts:

> Construct Attached Ellipse or Circle > Identify Ellipse *(Select one of the construction circles, then click the Data button in space to accept the circle.)*

This attachment causes the design circle always to stay centered on the inner construction line intersections. Repeat the procedure for the other three circles. The results of attaching design circles to each of the construction circles is shown in Figure 14–65.

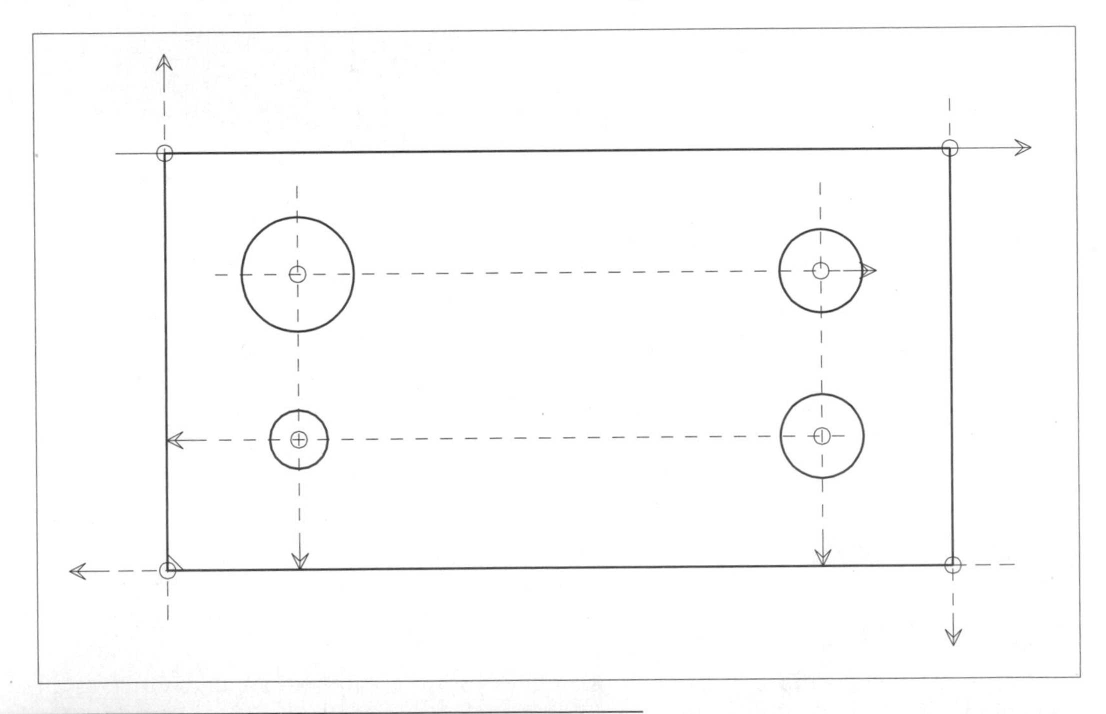

Figure 14–65 Results of drawing the design's circles

Add dimensions to the design, as shown in Figure 14–66. Set the Association lock to ON before placing the dimensions. Associating the dimensions with the design elements allows the dimensions to be used as constraints. When you construct the linear dimensions for the circles, start the dimension on the outside rectangle. If you start the dimension at the circles, an error message will appear.

Also make sure that the dimension extension lines are connected to the design elements, not the constructions. To make sure of that, snap to the starting and ending points until the correct element is highlighted, and then accept it.

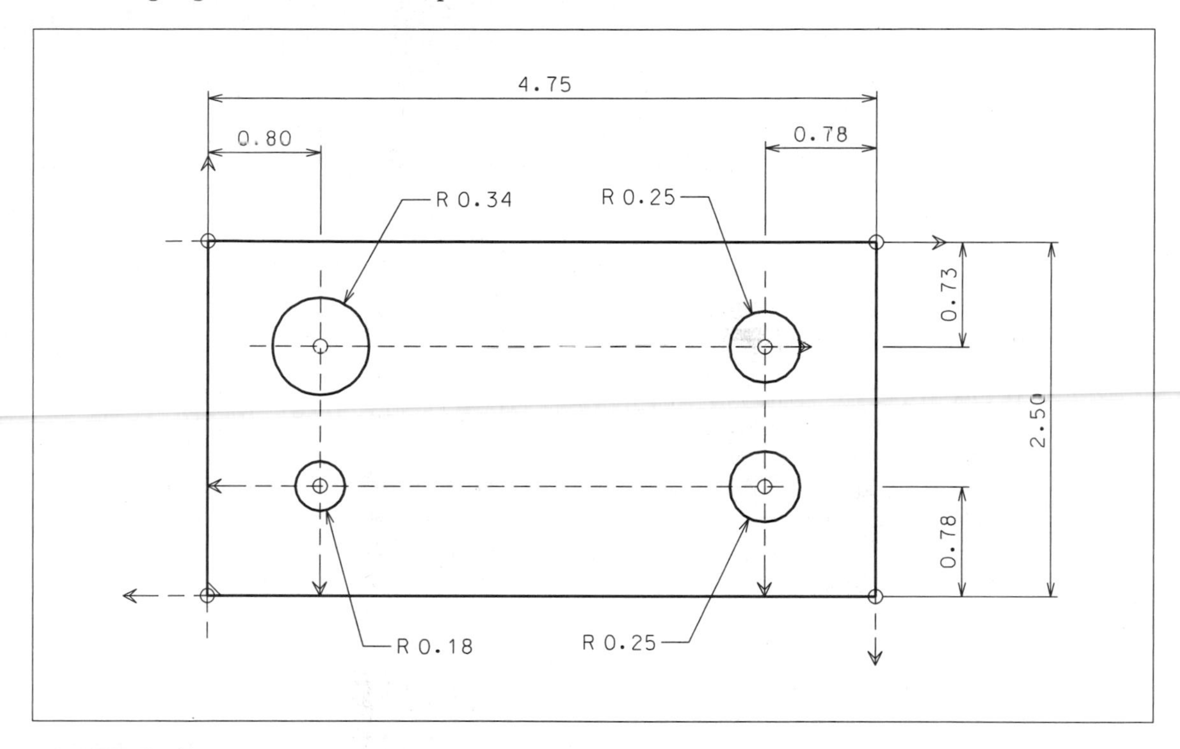

Figure 14–66 Result of dimensioning the design

Note: Ignore the dimension values now. The dimensions will be constrained and their values set later.

Place the following text strings for the equations and variables by using the Place Text at Origin tool, and place each line of text in separate, one-line text elements. Place the text to the right of the design.

wid = (space * 2) + 1.25

len = (space * 2) + 3.25

space = 0.75

rad = 0.25

wid

len

Figure 14–67 shows the design after adding the text.

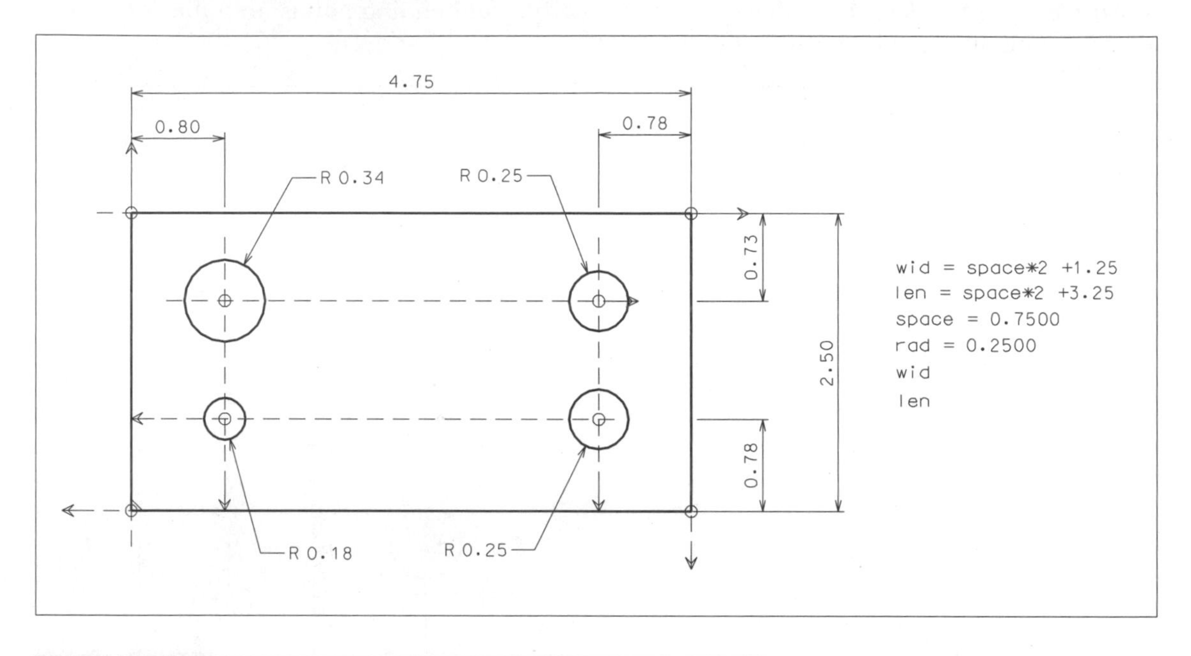

Figure 14–67 Result of adding the text strings to the design

Assign Variable to Dimensional Constraint

With the Assign Variable to Dimensional Constraint tool you can assign a constant or variable to a dimensional constraint. The constant or variable then represents the dimension's value in equations. Assign the "rad" variable as the radius of each circle. Invoke the Assign Variable tool:

Parameter Constraints tool box	Select the Assign Variable tool (see Figure 14–68).
Key-in window	**assign variable** (or **as v**) Enter

Figure 14–68 Invoking the Assign Variable tool from the Parameter Constraints tool box

MicroStation prompts:

> Assign Variable to Dimensional Constraint > Identify variable *(Select the "rad = 0.25" text string.)*Assign Variable to Dimensional Constraint > Identify Constraint *(Select the radial dimension of one of the circles.)*
>
> Assign Variable to Dimensional Constraint > Accept *(Click the Data button in the space to assign the "rad" variable to the circle's dimension and set the circle's radius to the "rad" value.)*

Repeat the procedure for the radial dimensions of the other three circles.

Similarly, constrain the space between each circle center and the adjacent design edges with the Assign Variable. Assign the "Space" variable to each of the four circle-center-to-edge dimensions.

Assign Equation

The Assign Equation tool assigns an algebraic constraint—an equation that expresses a constraint relationship between variables, numerical constants, and built-in functions and constants—to a model. Create the "wid" and "len" equations by invoking the Assign Equation tool:

Parameter Constraints tool box	Select the Assign Equation tool (see Figure 14–69).
Key-in window	**assign equation** (or **as e**) Enter

Figure 14–69 Invoking the Assign equation tool from the Parameter Constraints tool box

MicroStation prompts:

> Assign equation > Identify equation *(Select the "wid = [space * 2] + 1.25" text string.)*
>
> Assign equation > Identify variable, or RESET to finish *(Select the "space = 0.75" text string, select the "wid" text string, then click the Data button in space to complete defining the "wid" equation.)*

Similarly, invoke the Assign equation tool again to create the "len" equation by selecting "len = (space * 2) + 3.25", "space = 0.75", and "len."

Constrain the design's outline dimensions by invoking the Assign Variable tool to assign "len" to the overall horizontal dimension and "wid" to the overall vertical dimension.

Figure 14–70 shows the completed design with the Construction view attribute turned OFF so that only the actual design is displayed.

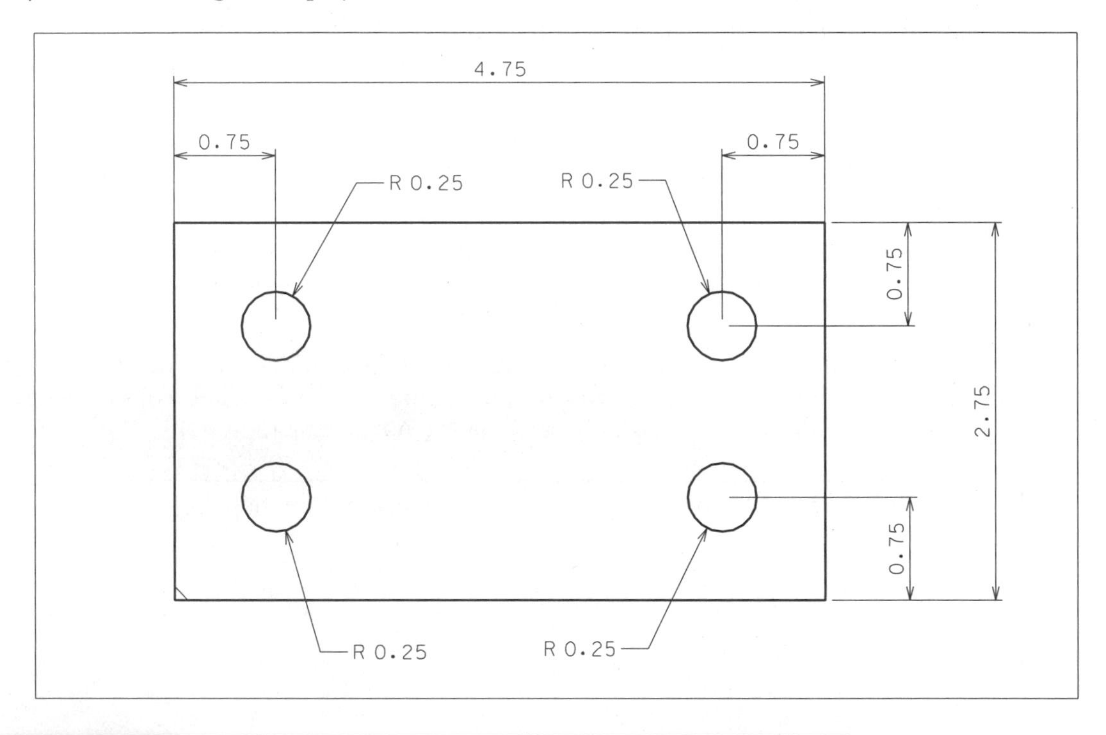

Figure 14–70 Completed design with construction elements turned off

Note: After the design is completed, it may be necessary to invoke the Modify Element tool to adjust the position of some of the dimension elements and to move the text strings.

Modifying a Dimension-Driven Design

Changing one of the variable values and then re-solving the design modifies dimension-driven designs. Constraint equations and variables can be changed by editing the text strings or by invoking the Modify Value tool. For example, the design we just constructed contains four variables:

- The "wid" and "len" variables are set equal to equations that contain constants (1.25 and 3.25). Editing the equations with the Edit Text tool can change the constants. These two constants control the horizontal and vertical space between the circles.

- The "space" variable is set equal to a constant (0.75) that can be changed either by editing the text string via the Edit Text tool or with the Modify Value tool. Changing this variable changes the position of the circles and, because it appears in the two equations, the overall width and length of the design.
- The "len" variable is set equal to a constant (0.25) that can be changed to control the size of the circles.

Modify Value of Dimension or Variable

This tool can edit the value of a dimensional constraint. To change the variable, invoke the Modify Value tool:

Modify Constraint tool box	Select the Modify Value tool (see Figure 14–71).
Key in window	**model edit_dimension** (or **mo e**) Enter

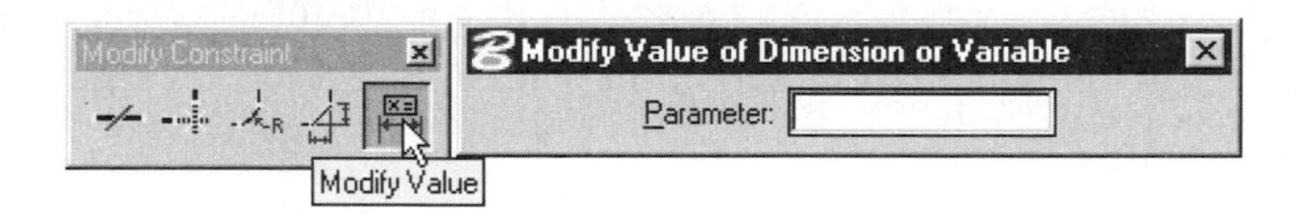

Figure 14–71 Invoking the Modify Value tool in the Modify Constraint tool box

MicroStation prompts:

Modify Value of Dimension or Variable > Identify element *(Select the dimension or variable to be changed.)*

Modify Value of Dimension or Variable > Accept *(Click the Data button in space to accept the selected element.)*

Modify Value of Dimension or Variable > Enter a value *(Enter the new value in the Settings window edit field, then press Enter to re-solve the design for the new value.)*

Re-Solve Constraints

If the Edit Text tool is used to change a variable's value, then the design must be "re-solved" to apply the new constraint value. To re-solve the design, invoke the Re-solve Constraint tool:

Evaluate Constraints tool box	Select the Re-solve Constraints tool (see Figure 14–72).
Key-in window	**update model** (or **up m**) Enter

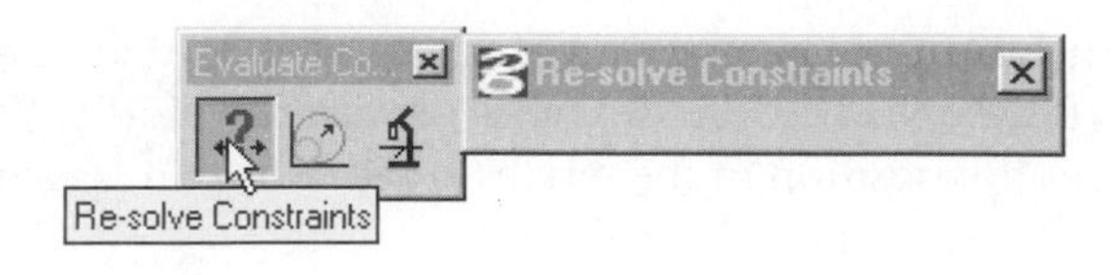

Figure 14–72 Invoking the Re-solve Constraints tool from the Evaluate Constraints tool box

MicroStation prompts:

> Re-solve Constraints > Identify Element *(Select the variable that has been changed or the text of a dimension that is constrained to the variable, then click the Data button in space to initiate re-solving of the design.)*

Creating a Dimension-Driven Cell

Dimension-driven cells are created from dimension-driven designs when the Construction view attribute is set to ON and all of the construction elements of the design are included in the design. When a dimension-driven cell is placed in a design file, its constraints can be changed.

Note: If the Construction view attribute is set to OFF when a cell is created from a dimension-driven design, the cell will *not* be created as a dimension-driven cell.

Object linking and embedding (OLE)

Object linking and embedding (OLE) is a Microsoft Windows feature that combines various application data into one compound document. MicroStation J has client as well as server capability. As a client, MicroStation now permits you to have objects from other Windows applications either embedded in or linked to your design. As a server, MicroStation allows you to send or serve views of design files to other Windows applications by either embedding or linking.

When an object is embedded into a MicroStation design file from an application that supports OLE, it is no longer associated with its source file. If necessary, you can edit the embedded data from inside the MicroStation design by using the original application. But at the same time, this editing does not change the source file.

If, instead, you insert an object as a linked object into a MicroStation design file from an application that supports OLE, the object remains associated with its source file. When you edit a linked object in MicroStation by using the original application, the original file as well as the object inserted in MicroStation changes. Use linking when you want to include the same information in several different design files.

Linked or embedded objects appear on the screen in MicroStation and can be printed or plotted using Windows system drivers.

Let's look at an example of object linking between MicroStation (server) design and Microsoft Word (client). Figure 14–73 shows a design of a desk, a computer, and a chair that contains various text nodes.

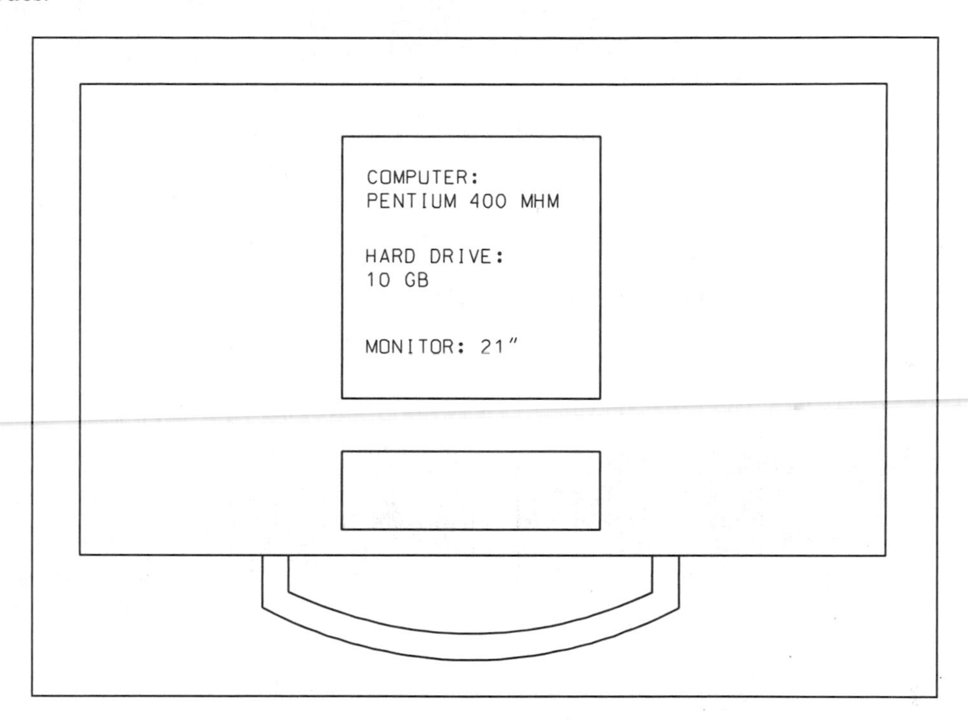

Figure 14–73 Drawing of a desk, a computer, chair, and text nodes

To copy the elements from the view into the clipboard for linking to the Word document, first invoke the OLESERVE MDL application:

Key-in window	**mdl load oleserve** (or **md l oleserve**) Enter

To capture the view objects, invoke oleserve viewcopy tool:

Key-in window	**oleserve viewcopy [view number]** (or **oles v [view number]**) Enter

Specify the view number you want to copy into the clipboard for linking. Invoke the Microsoft Word program, and from the Edit pull-down menu, select Paste Special... to display the Paste Special dialog box, as shown in Figure 14–74. Select the Past link: radio button and select MicroStation View object from the As: list box. Click the OK button to paste the MicroStation View object and close the Paste Special dialog box. MicroStation view objects are inserted into the Word document as shown in Figure 14–75.

Paste Special

Source: MicroStation View
d:\book-micro\ch14\HMJ FIG14-73.dgn!View1

As:

Paste:

Paste link:

MicroStation View Object
Picture

OK

Cancel

Float over text

Display as icon

Result

Inserts the contents of the Clipboard as a picture.

Paste Link creates a shortcut to the source file. Changes to the source file will be reflected in your document.

Figure 14–74 Word's Paste Special dialog box

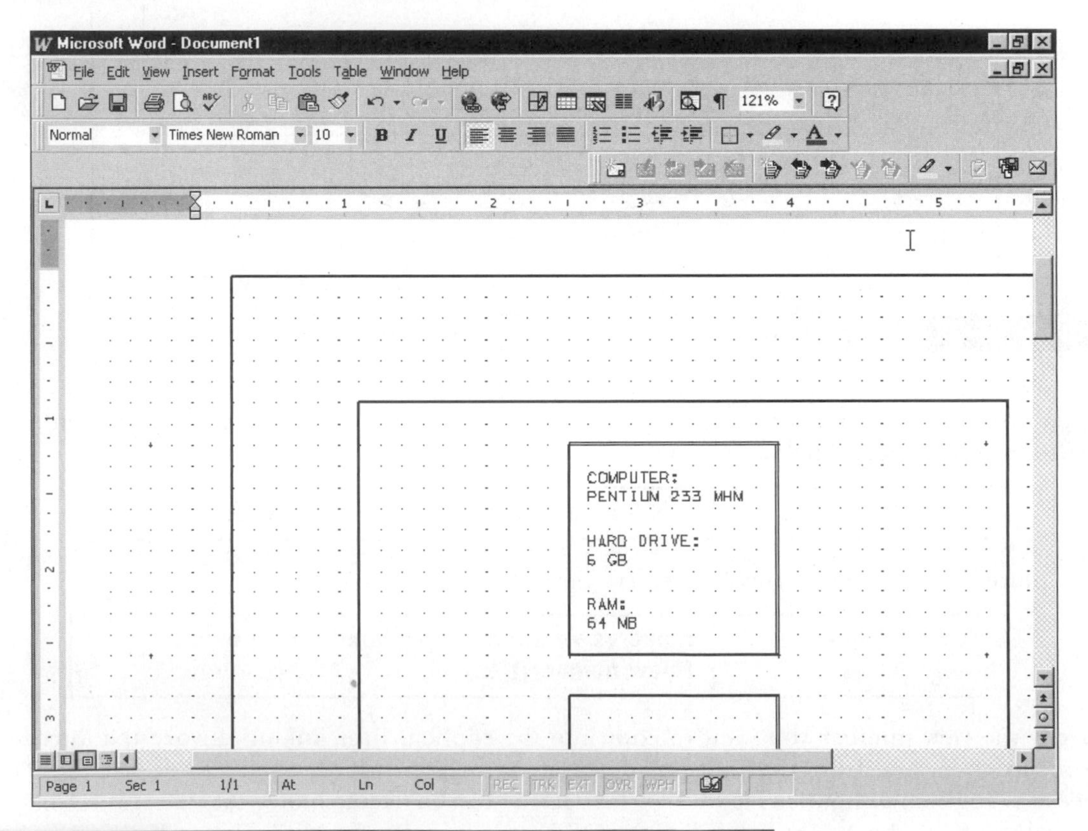

Figure 14–75 Microsoft Word document with MicroStation design

Minimize the Word program and maximize MicroStation (or if the MicroStation program is not open, double-click the design image in the Word document to launch the MicroStation program with the design open). Edit the values of the text nodes in the computer block to 400, 10.0GB, and 21", which represents a Pentium 400 computer with a 10 GB hard drive and a 21" monitor.

Switch back to the Word program and in the Edit pull-down menu, select the Links... option. The Word program will display the Links dialog box shown in Figure 14–76.

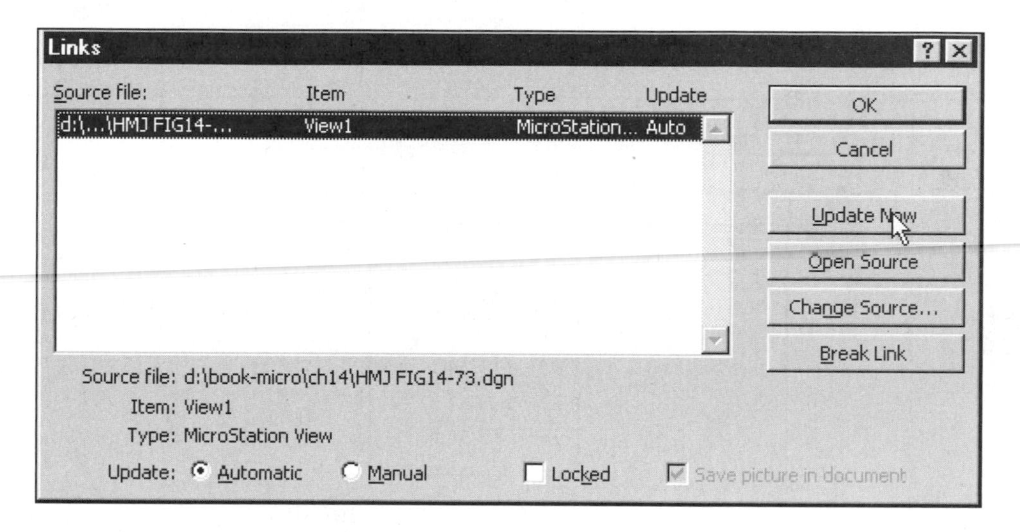

Figure 14–76 Links dialog box

Select the Update Now button and then choose the OK button. The image in the Word document is updated, as shown in Figure 14–77.

In our example, MicroStation J is the server and Microsoft Word is the client.

Alternatively, you can place a linked object into MicroStation, where MicroStation is the client and another application is the server. Let's look an example in which MicroStation J is the client and Excel is the server.

Figure 14–78 shows an Excel spreadsheet containing a table for Area Calculations.

To copy the data into the clipboard for linking to the MicroStation design file, first select all the data, then invoke the Copy tool from Excel program:

Pull-down menu	Edit > Copy

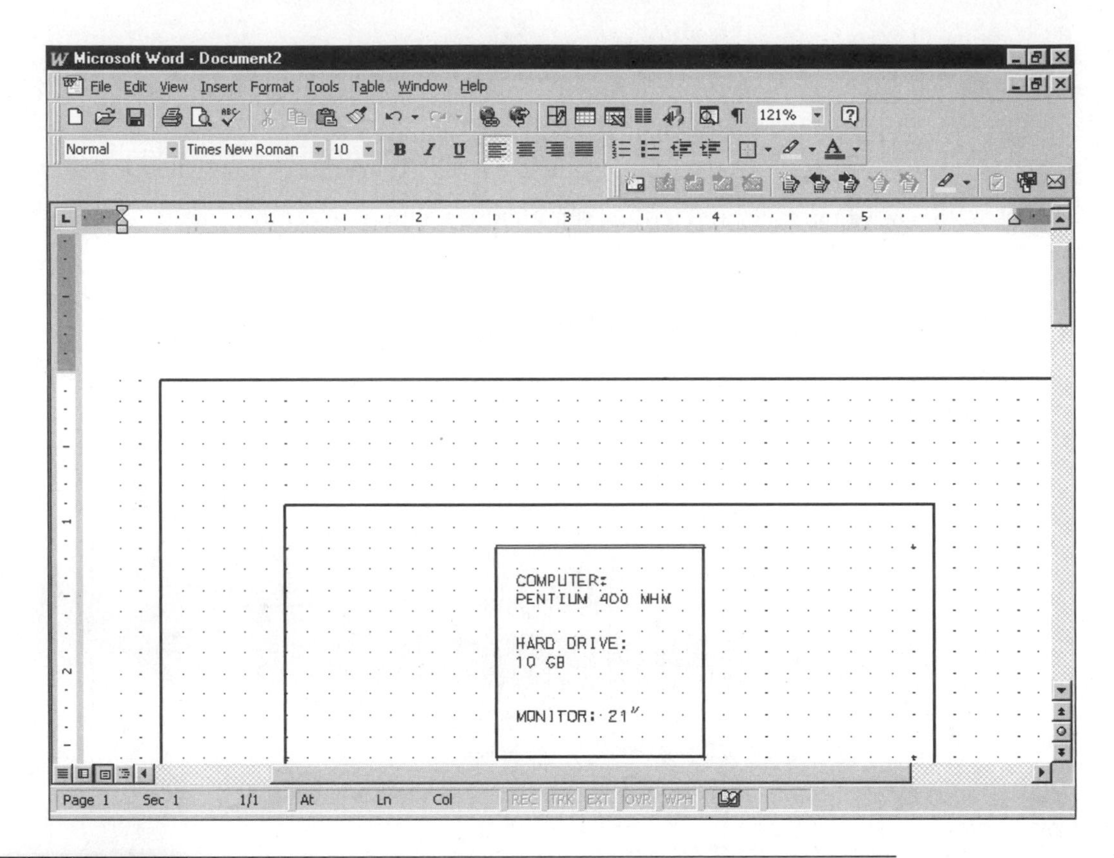

Figure 14–77 Microsoft Word document with updated MicroStation design

The Excel program copies all the selected data to the clipboard. Switch back to the MicroStation program, and invoke the Paste Special tool:

Pull-down menu	Edit > Paste Special...

MicroStation displays the Paste Special dialog box, shown in Figure 14–79. Select Linked Microsoft Excel Worksheet from the Data Type list box and click the OK button.

Enter a data point to define one corner, then enter a second data point to define the diagonally opposite corner. The selected section of the document is linked to your design file.

Note: MicroStation only supports True Type fonts. If the True Type fonts you are using in your source applications are not currently used in MicroStation, you may want to install them. Otherwise, MicroStation substitutes the active font.

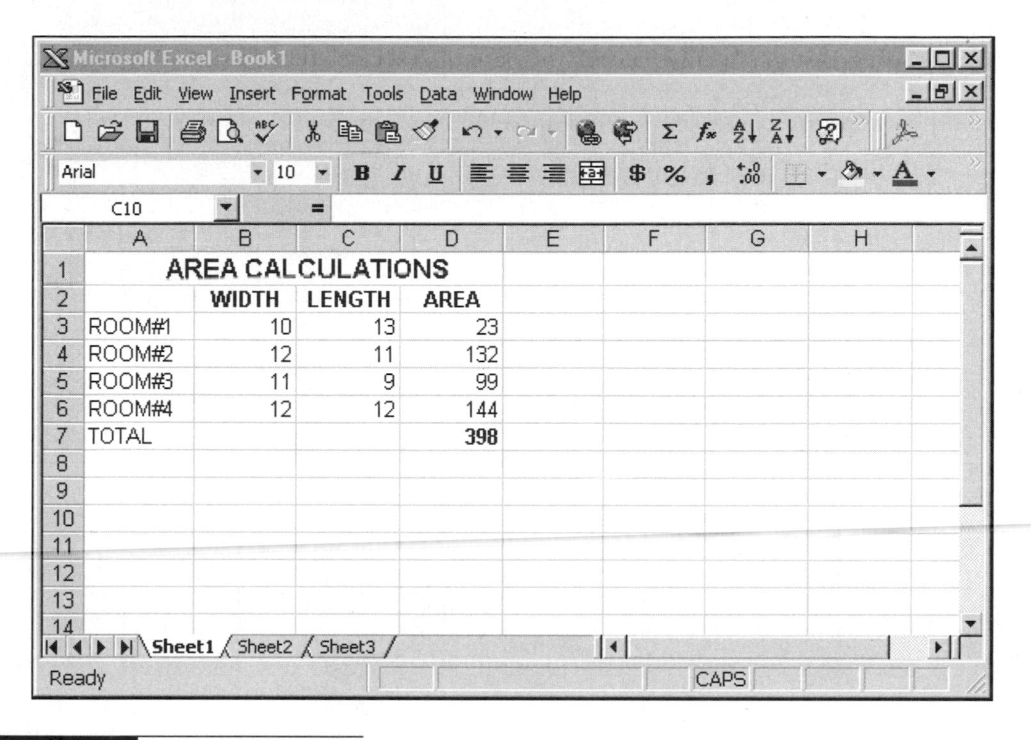

	A	B	C	D
1	AREA CALCULATIONS			
2		WIDTH	LENGTH	AREA
3	ROOM#1	10	13	23
4	ROOM#2	12	11	132
5	ROOM#3	11	9	99
6	ROOM#4	12	12	144
7	TOTAL			**398**

Figure 14–78 Excel spreadsheet

Paste Special

Data Type

Metafile/Picture
Text To Design File
Embedded Microsoft Excel Worksheet
Linked Microsoft Excel Worksheet
Linked Text To Design File
Bitmap to Design File

Paste Cancel

Figure 14–79 MicroStation Paste Special dialog box

The Excel spreadsheet will be linked to the design, as shown in Figure 14–80. MicroStation is now the client and Microsoft Excel is the server.

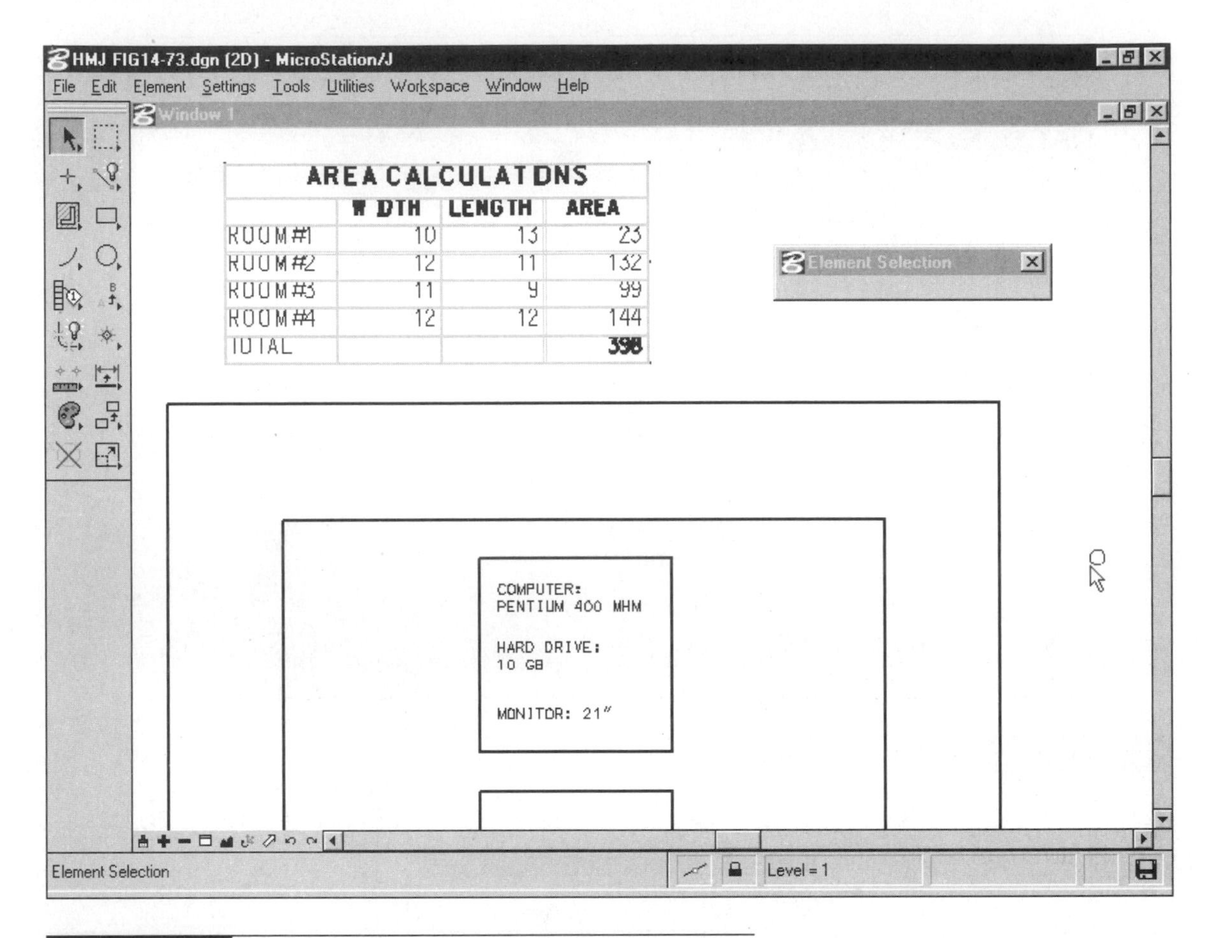

Figure 14–80 MicroStation design with Excel spreadsheet

To edit the spreadsheet, double-click anywhere on the spreadsheet to launch Excel with the spreadsheet document open. Any changes made to the spreadsheet will be reflected in the design. Figure 14–81 shows the changes that were made in the spreadsheet.

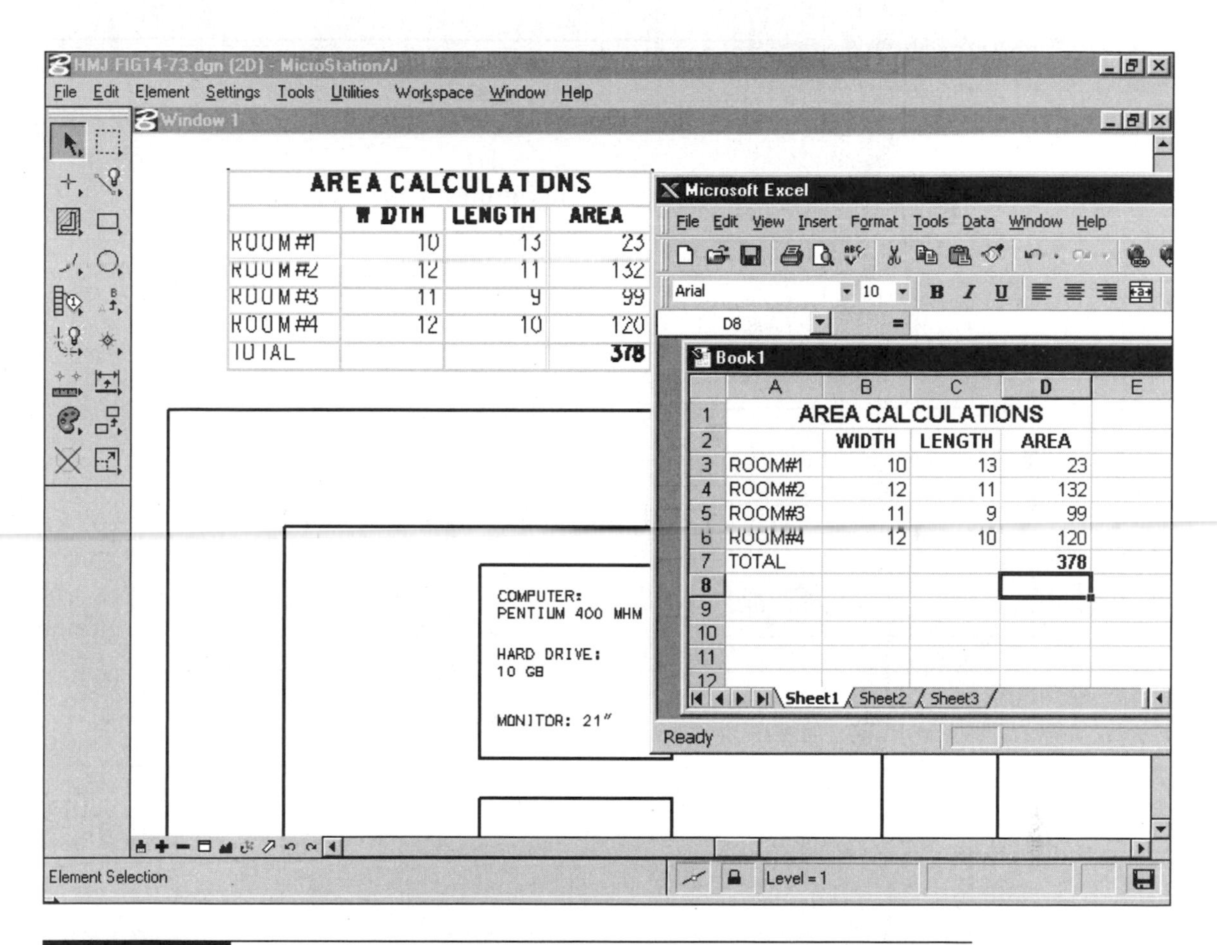

Figure 14–81 MicroStation design with Excel spreadsheet reflecting changes

REVIEW QUESTIONS

Write your answers in the spaces provided.

1. Explain briefly the purpose of creating a graphic group.

___

___

2. List the steps in creating a graphic group.

___

___

3. Explain the purpose of the Graphic Group Lock setting.

___

___

4. Explain briefly the purpose of element selection by element type.

___

___

5. Explain briefly the benefits of setting up a level symbology table.

___

___

6. List the steps involved in setting up a level symbology table.

___

___

7. List the steps involved in converting a MicroStation design file to an AutoCAD drawing file.

___

___

8. Explain briefly the benefits of dimension-driven design.

___

___

chapter 15

Internet Utilities

The Internet is the most important way to convey digital information around the world. You are probably already familiar with the best-known uses of the Internet: e-mail (electronic mail) and the Web (short for "World Wide Web"). E-mail lets users exchange messages and data at very low cost. The Web brings together text, graphics, audio, and movies in an easy-to-use format. Other uses of the Internet include FTP (File Transfer Protocol, for effortless binary-file transfer), Gopher (which presents data in a structured, subdirectory-like format), and Usenet (a collection of more than 10,000 news groups).

MicroStation J allows you to interact with the Internet in several ways. Engineering Links are a set of Web-enabling technologies that simplify management of your MicroStation-based engineering projects. In addition, MicroStation J includes a Web browser, which enables users to directly access data over the Internet without leaving the application. The browsing tools provide everything the user needs for browsing all of the information on the Web.

Objectives

After completing this chapter, you will be able to do the following:

- Launch the Web browser and open files remotely
- Publish MicroStation data to the Internet
- Link Geometry to Internet URLs

LAUNCHING THE WEB BROWSER AND OPENING FILES REMOTELY

Web browsers are applications designed to provide the user with direct, easy access to the diverse content of the World Wide Web. The original Web browser, created at the NCSA (National Center for Supercomputing Applications at the University of Illinois), is called Mosaic. Revolutionary in its time, Mosaic, and the startling capabilities it introduced to the Internet, played a major role in igniting the exuberant explosion in Web usage, which got underway in the early 1990s. Building on their remarkable success, members of the Mosaic development team went on to create Netscape Navigator. Microsoft's Internet Explorer has similar functionality to Netscape Navigator, and also provides access to the diverse content of the World Wide Web.

Highly intuitive in design and function, Web browsers provide a clean, easily navigable window through which to scan pages on the Web. Customizable to suit the user's aesthetics and interests, browsers typically support a number of secondary functions including sending and receiving e-mail, monitoring newsgroups, or even downloading files directly from Web sites.

MicroStation J provides a built-in Web browser that allows users to directly access data over the Internet without leaving their application. MicroStation J's new browsing tools provide everything users need for browsing all of the information on the Web.

To open the Web browser, invoke the MicroStation Link:

Pull-down menu	Utilities > MicroStation Link

By default, MicroStation displays the MicroStation Browser, which looks similar to Figure 15–1. The MicroStation Link allows you to connect to an external browser and bypass the MicroStation Link internal Web browser. The choice of browsers is defined with the variable MS_USEEXTERNALBROWSER. You can set the variable to open Netscape Navigator or Microsoft Explorer instead of the internal browser, if you prefer.

Specify the URL (short for "uniform resource locator") in the Location: edit field in the browser. The URL is the Web site address, which usually follows the format http://www.bentley.com. The URL system allows you to find any resource (a file) on the Internet. Example resources include a text file, a Web page, a program file, an audio or movie clip—in short, anything that you might also find on your computer. The primary difference is that these resources are located on somebody else's computer. A typical URL looks like the examples in Table 15–1.

Table 15–1 Example URLs

Example	Meaning
http://www.bentley.com	Bentley primary Web site
ftp://ftp.bentley.com	Bentley FTP Server

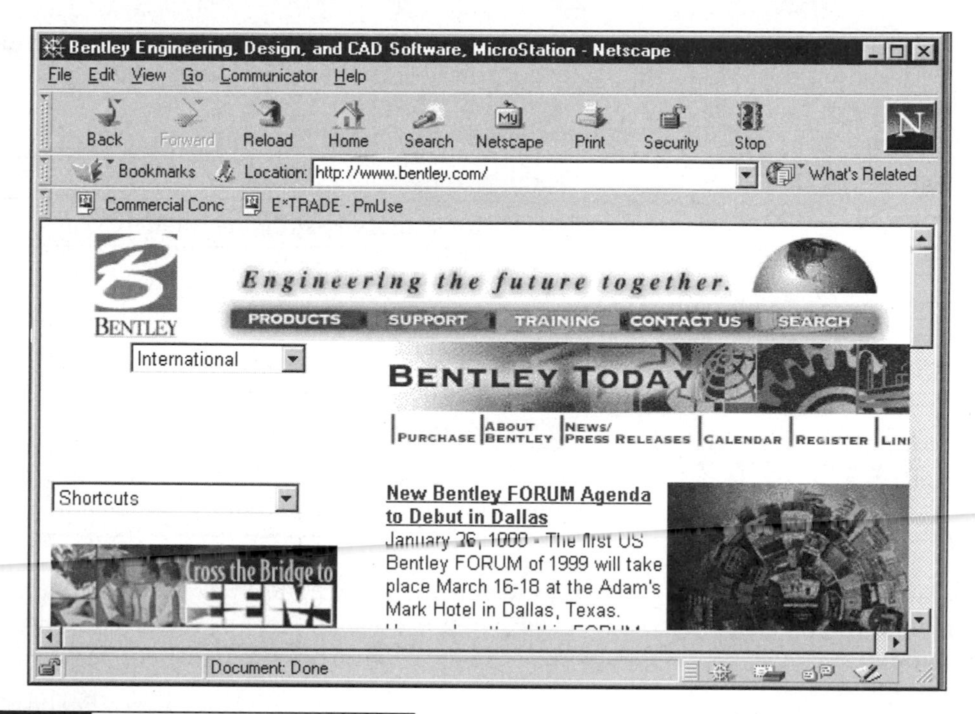

Figure 15–1 MicroStation Web Browser

The “http://” prefix is not required. Most of today’s Web browsers automatically add in this routing prefix, which saves you a few keystrokes. URLs can access several different kinds of resources—such as Web sites, e-mail, news groups—but always take on the following format:

Scheme://netloc

The scheme accesses the specific resource on the Internet, including those listed in Table 15–2.

Table 15–2 URL Prefix Meanings

Scheme Prefix	Meaning
File://	Files on your computer’s hard drive or local network
ftp://	File transfer protocol (downloading files)
http://	Hypertext transfer protocol (Web sites)
mailto://	Electronic mail (e-mail)
news://	Usenet news (news groups)
telnet://	Telnet protocol
gopher://	Gopher protocol

The "://" characters indicate a network address. Table 15–3 lists the formats recommended for specifying URL-style file names in the Location: edit field of the Browser.

Table 15–3 Recommend Formats for Specifying URL-Style File Names

Drawing Location	Template URL
Web site	http://servername/pathname/filename
FTP site	ftp://servername/pathnae/filename
Local file	file:///drive:/pathname/filename File:///drive|/pathname/filename File://\\localPC\pathname\filename File:////localPC/pathname/filename
Network file	file://localhost/drive:/pathname/filename File://localhost/drive|/pathname/filename

Servername is the name of the server, such as www.bentley.com. The *pathname* is the name of the subdirectory or folder name. The *drive:* is the drive letter, such as C: or D:. A *local file* is a file located on your computer. The *localhost* is the name of the network host computer.

MicroStation J allows you to select a URL as a design file location instead of a specific local design file. You can also open files using URLs to open remote settings files, archives, reference files, or cell libraries. Downloaded files from a URL are stored in a directory specified by the configuration variable MS_WEBFILES.

Remote Opening of Design File

To open a design file located at a remote server by selecting a URL, invoke the Open URL tool:

Pull-down menu	File > Open URL...

MicroStation displays a Select Remote Design file dialog box, similar to Figure 15–2.

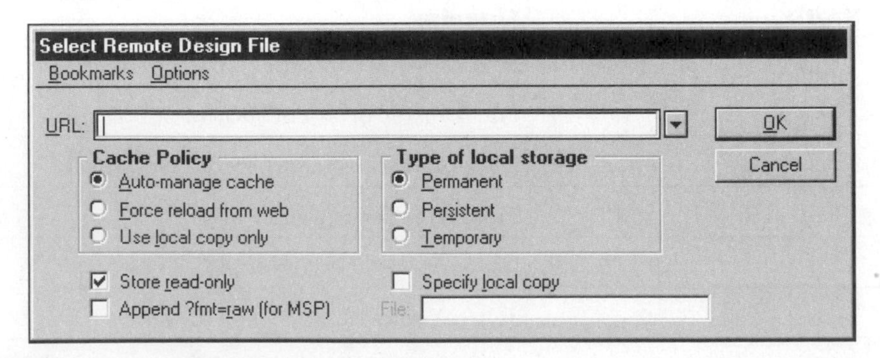

Figure 15–2 Select Remote Design File dialog box

Specify the URL in the URL: edit field. Following are the available options in the dialog box:

Cache Policy

The Cache Policy section of the dialog box allows you to select whether the design files from selected URLs are downloaded or if a previously downloaded local copy is used. Select one of the three following options:

- Auto-manage cache selection—downloads the file from the selected URL only if it is newer than the local version.
- Force reload—reloads from the Web selection and automatically downloads the file from the selected URL.
- Use local copy only selection—uses the local copy and no Internet request is sent.

Type of Local Storage

The Type of Local Storage section of the dialog box allows you to select how the downloaded file is stored. Select one of the three following options:

- Permanent selection—stores the file that is downloaded and it is not deleted.
- Persistent selection—deletes the file only when it exceeds the size requirements. By default, the total permitted size of local copies of remote files is 25MB.
- Temporary selection—deletes the file at the end of the current session.

Store Read Only

The Store Read Only selection stores downloaded files as read-only. This is a useful reminder that the remote file is not being changed and local modifications could lead to confusion.

Append ?fmt=raw

The Append selection appends the string "?fmt=raw" to the end of the selected URL that is submitted for download. This is required only if the selected site is running ModelServer Publisher.

Specify local copy

If necessary, you can specify a different location in the File: edit field for downloaded files from selected URLs rather than the default location.

Remote Opening of Settings File

To open a settings file located at a remote server by selecting a URL, first open the Settings Manager:

Pull-down menu	Settings > Manage

MicroStation displays the Select Settings box. Invoke the Open URL tool:

Pull-down menu (Select Settings)	File > Open URL...

MicroStation displays Select Remote Settings File dialog box, similar to Figure 15–3.

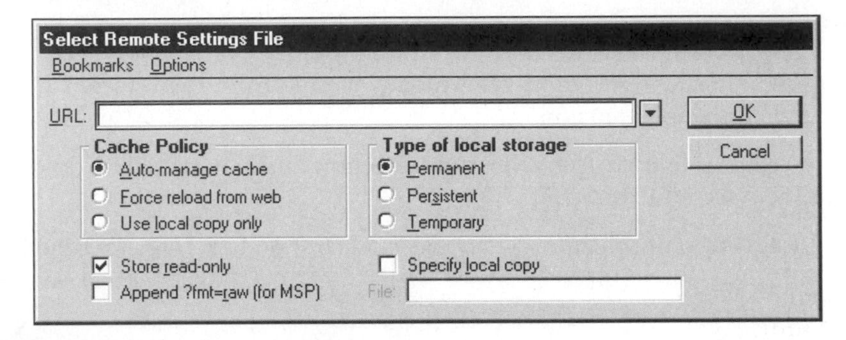

Figure 15–3 Select Remote Settings File dialog box

Specify the URL in the URL: edit field. The options available in the Select Remote Settings file dialog box are the same explained earlier for Select Remote Design File dialog box.

Remote Opening of Archive File

To open an archive file located at a remote server by selecting a URL, first open the Archive Settings box:

Pull-down menu	Utilities > Archive

MicroStation displays the Archive Settings box. Invoke the Open URL tool:

Pull-down menu (Archive Settings)	File > Open URL...

MicroStation displays the Open Remote Archive dialog box, similar to Figure 15–4.

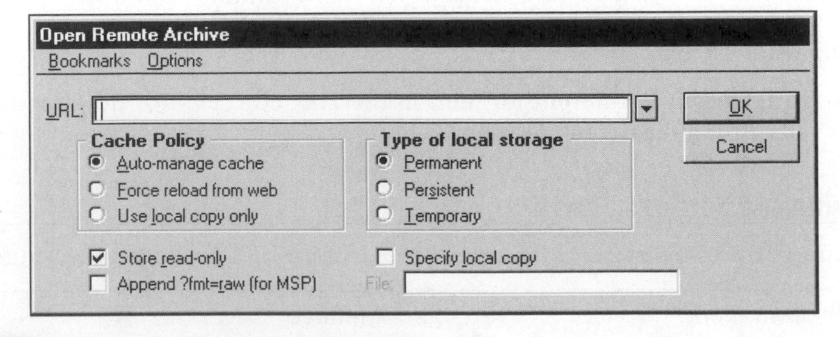

Figure 15–4 Open Remote Archive dialog box

Specify the URL in the URL: edit field. The options available in the Open Remote Archive dialog box are the same as explained earlier for the Select Remote Design File dialog box.

Remote Opening of Reference File

To open a reference file located at a remote server by selecting a URL, first open the Reference Files Settings box:

Pull-down menu	File > Reference

MicroStation displays the Reference Files settings box. Invoke the Open URL tool:

Pull-down menu (Reference Files Settings)	File > Open URL...

MicroStation displays the Select Remote Design File to Attach: dialog box, similar to Figure 15–5.

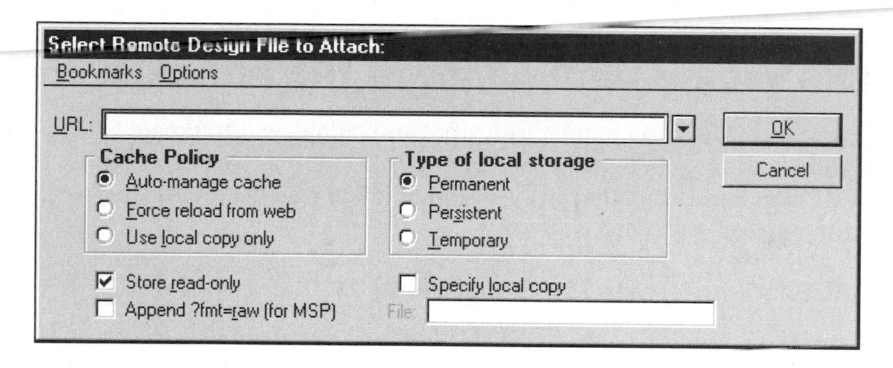

Figure 15–5 Select Remote Design File to Attach: dialog box

Specify the URL in the URL: edit field. The options available in the Select Remote Design File to Attach: dialog box are the same as explained earlier for the Select Remote Design File dialog box.

Remote Opening of Cell Library

To open a cell library located at a remote server by selecting a URL, first open the Reference Files Settings box:

Pull-down menu	Utilities > Cell Selector

MicroStation displays the Cell Selector Settings box. Invoke the Open URL tool:

Pull-down menu (Cell Selector Settings)	File > Load Remote Cell Library

MicroStation displays the Specify URL for Cell Library dialog box, similar to Figure 15–6.

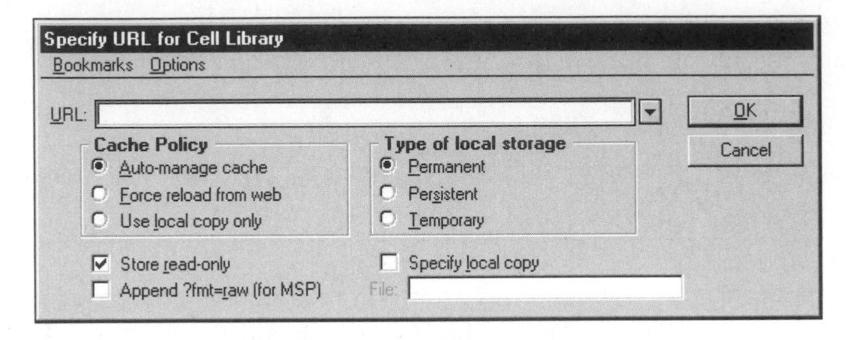

Figure 15–6 Specify URL for Cell Library dialog box

Specify the URL in the URL: edit field. The options available in the Specify URL for Cell Library dialog box are the same as explained earlier for the Select Remote Design File dialog box.

PUBLISHING MICROSTATION DATA TO THE INTERNET

The HTML (HyperText Markup Language) Author tool allows you to create HTML files that can be viewed via MicroStation Link or any external browser (such as Netscape Navigator or Internet Explorer). HTML files can be created from a Design File Saved View, Design File Snapshot, Cell Library, or BASIC Macros.

Creating HTML File from Design File Saved View

The Design File saved view option creates an HTML Web page with the selected view of the design file. To create an HTML file from a design file saved view, first open the HTML Author dialog box:

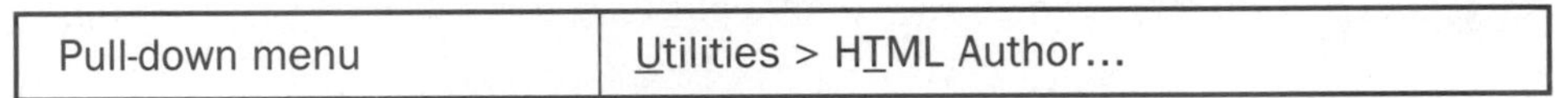

Pull-down menu	Utilities > HTML Author...

MicroStation displays the HTML Author dialog box, similar to Figure 15–7

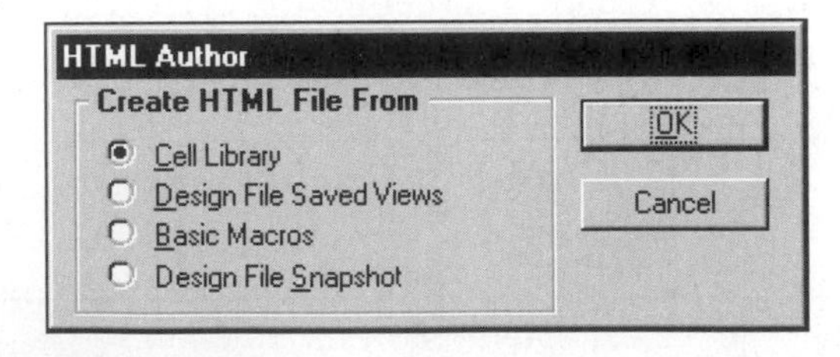

Figure 15–7 HTML Author dialog box

Select the Design File Saved Views radio button from the Create HTML File From section of the dialog box and click the OK button to close the HTML dialog box. The Select Design File dialog box opens. Select the design file from the appropriate directory and click the OK button. MicroStation opens the Design File Walkthrough dialog box, similar to Figure 15–8.

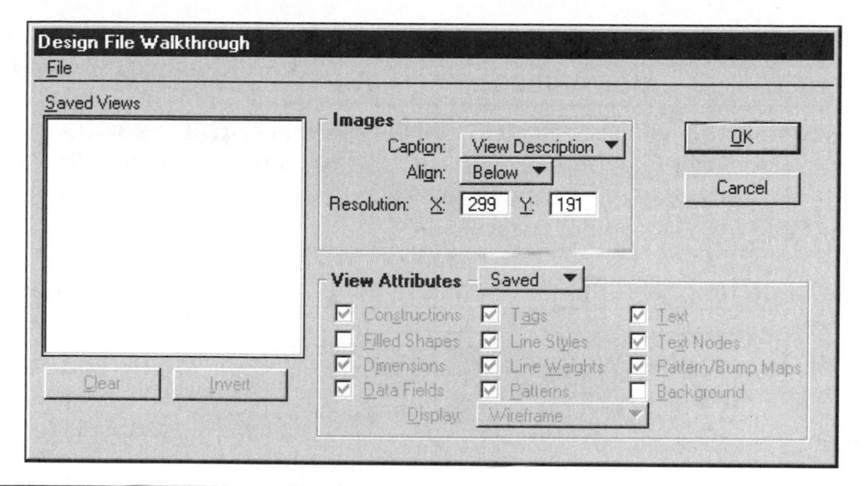

Figure 15-8 Design File Walkthrough dialog box

MicroStation lists all the saved views in the design file in the Saved Views list box. You can select any or all of the saved views to be included in the HTML page.

The Images section of the dialog box sets controls for image display. The Caption: option menu allows you to select whether to display the View Description, View Name, or None as part of the image display on the HTML page. The Align: option menu controls where the text (selected from the Caption option menu) appears with relation to the image on the HTML page. The available options include Below, Left, Right, and Above. The Resolution: edit field sets the resolution of the generated image in pixels.

The View Attributes section of the dialog box lists all the available view attributes and the corresponding settings as saved in the design file. If you need to override the settings, select the Override from View Attributes option menu and make the necessary changes in the View Attributes settings box.

After making the necessary changes, click the OK button to close the Design File Walkthrough dialog box. MicroStation opens the Create HTML File dialog box, similar to Figure 15–9.

Create HTML File
File Name: d:\j.htm
Title: j
Heading: j
OK
Cancel
Design Directory: d:\ Browse...
Image Directory: d:\ Browse...
URLs: Relative Host Name:
Root Directory: d:\ Browse...
Display File In Browser

Figure 15-9 Create HTML File dialog box

Specify the file name of the HTML file being created; a Descriptive title; and a Heading Description (appears top of the graphics) in the File Name: edit field, Title: edit field, and Heading: edit field, respectively. Specify the design (location of the design file) and image directories in the Design Directory: and Image Directory: fields, respectively. Specify if the URLs being used are absolute or relative in the URLs option menu. If it is set to absolute, then specify the host name and root directory in the Host Name: edit field and Root Directory: edit fields. Click the OK button to create the HTML file from Design file saved view.

Creating HTML File from Design File Snapshot

The Design File Snapshot option creates an HTML Web page with a view-only picture of the selected design file. To create an HTML file from a design file snapshot, first open the HTML Author dialog box:

Pull-down menu	Utilities > HTML Author...

MicroStation displays the HTML Author dialog box, similar to Figure 15–7.

Select the Design File Snapshot radio button from the Create HTML File From section of the dialog box and click the OK button to close the HTML dialog box. The Select Design File dialog box opens. Select the design file from the appropriate directory and click the OK button. MicroStation opens the Create HTML File dialog box, similar to Figure 15—10.

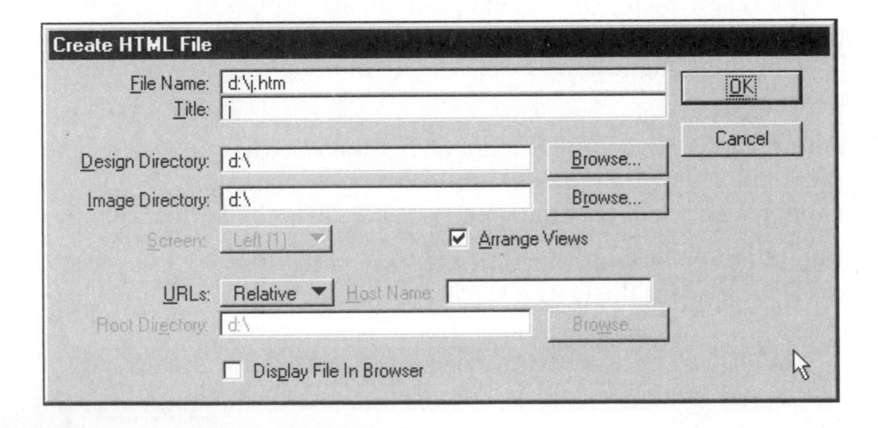

Figure 15–10 Create HTML File dialog box

Specify the file name of the HTML file being created, and the Title in the File Name: edit field and Title: edit fields, respectively. Specify the design (location of the design file) and image directories in the Design Directory: and Image Directory: fields, respectively. Specify if the URLs being used are absolute or relative in the URLs option menu. If it is set to absolute, then specify the host name and root directory in the Host Name: edit field and Root Directory: edit field. Click the OK button to create the HTML file from Design file Snapshot.

Creating HTML File from Cell Library

The Cell Library option creates an HTML Web page from a cell library. To create an HTML file from a cell library, first open the HTML Author dialog box:

Pull-down menu	Utilities > HTML Author...

MicroStation displays the HTML Author dialog box, similar to Figure 15–7.

Select the Cell Library radio button from the Create HTML File From section of the dialog box and click the OK button to close the HTML dialog box. The Select Cell Library to Open dialog box opens. Select the cell library file from the appropriate directory and click the OK button. MicroStation opens the HTML Cell Page dialog box, similar to Figure 15–11.

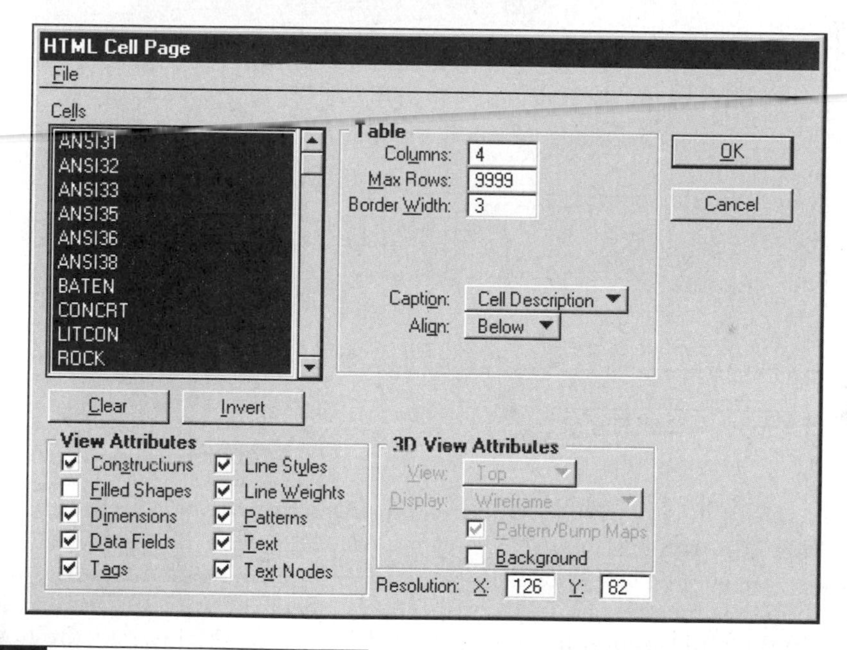

Figure 15–11 HTML Cell Page dialog box

MicroStation lists all the cells in the cell library file in the Cells list box. You can select any or all of the cells to be included in the HTML page.

The Tables section of the dialog box allows you to set the settings (such as the number of columns, maximum number of rows, and border width) for the table to be displayed on an HTML page.

The Caption: option menu allows you to select whether to display the Cell Description, Cell Name, or None as part of the image display on the HTML page. The Align: option menu controls where the text (selected from Caption option menu) appears in relation to the image on the HTML page. The available options include Below, Left, Right, and Above.

The View Attributes section of the dialog box lists all the available view attributes and the corresponding settings as saved in the design file. If necessary, you can make changes in the View Attributes settings.

The 3D View Attributes section of the dialog box sets attributes for rendering generated images including view, display, patterns/bitmaps, and background.

The Resolution: edit fields sets the resolution of the generated image in pixels.

After making the necessary changes, click the OK button to close the HTML Cell Page dialog box. MicroStation opens the Create HTML File dialog box, similar to Figure 15–12.

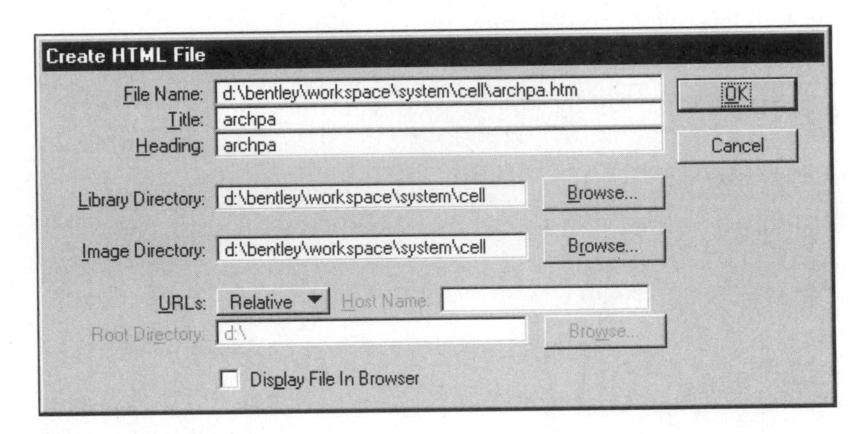

Figure 15–12 Create HTML File dialog box

Specify the file name of the HTML file being created; a Descriptive title; and the Heading Description (appears top of the graphics) in the File Name: edit field, Title: edit field, and Heading: edit field, respectively. Specify the cell library directory location and image directories in the Library Directory: and Image Directory: edit fields, respectively. Specify if the URLs being used are absolute or relative in the URLs option menu. If it is set to absolute, then specify the host name and root directory in the Host Name: edit field and Root Directory: edit field. Click the OK button to create the HTML file from Cell Library.

Creating HTML File from Basic Macros

The Basic Macro HTML Page dialog box is used to create an HTML file that references a directory of MicroStation Basic macros. To create an HTML file from a directory of MicroStation Basic macros, first open the HTML Author dialog box:

Pull-down menu	Utilities > HTML Author...

MicroStation displays the HTML Author dialog box, similar to Figure 15–7.

Select the Basic Macros radio button from the Create HTML File From section of the dialog box and click the OK button to close the HTML dialog box. The Select Basic Macro Directory dialog box opens. Select the directory containing basic macro files and click the OK button. MicroStation opens the Basic Macro HTML Page dialog box, similar to Figure 15–13.

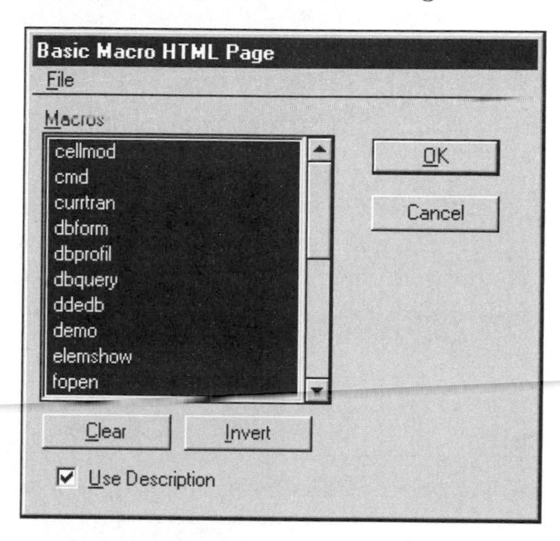

Figure 15–13 Basic Macro HTML Page dialog box

MicroStation lists all the basic macros in the selected directory in the Macros list box. You can select any or all of the basic macros to be included in the HTML page. Click the OK button to complete the selection.

MicroStation opens the Create HTML File dialog box, similar to Figure 15–14.

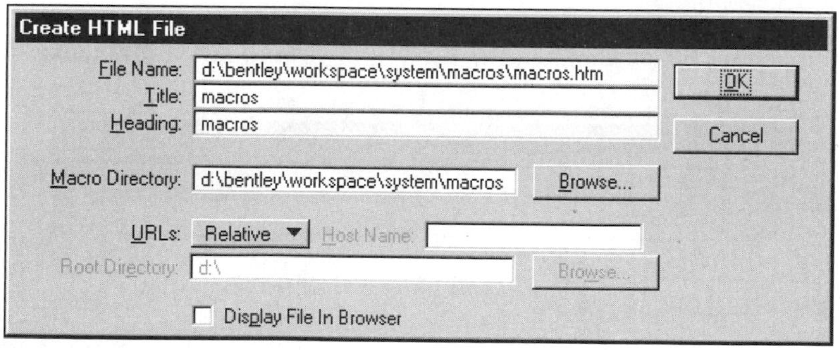

Figure 15–14 Create HTML File dialog box

Specify the file name of the HTML file being created; a Descriptive title; and the Heading Description (appears top of the graphics) in the File Name: edit field, Title: edit field, and Heading: edit field, respectively. Specify the macro directory location in the Macro Directory: edit field. Specify if the URLs being used are absolute or relative in the URLs option menu. If it is set to absolute, then specify the host name and root directory in the Host Name: edit field and Root Directory: edit field. Click the OK button to create the HTML file from Basic Macros.

LINKING GEOMETRY TO INTERNET URLS

MicroStation J allows you to link Internet URLs to geometry. Most browsers highlight a linked entity when the cursor is placed on it and display the link title. Clicking on the linked entity redirects the browser to the linked URL. This feature allows you to connect to component catalogs which can be accessed over the Internet, and, for example, engineers can execute a search over the Web to find a supplier whose components meet their specifications. Having URL links to geometry will also help when someone using the drawing needs more information on that component, since they can just click on the component and it would automatically connect them to the manufacturer's Web page.

Attaching Link to a Geometry

To attach a URL to an element, invoke the Attach Engineering Link tool:

E-Links tool box	Select the Attach Engineering Link tool (see Figure 15–15).

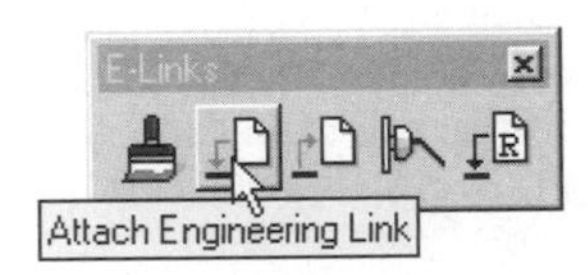

Figure 15–15 Invoking the Attach Engineering Link tool from the E-Links tool box

MicroStation displays the Attach Engineering Link settings box, similar to Figure 15–16.

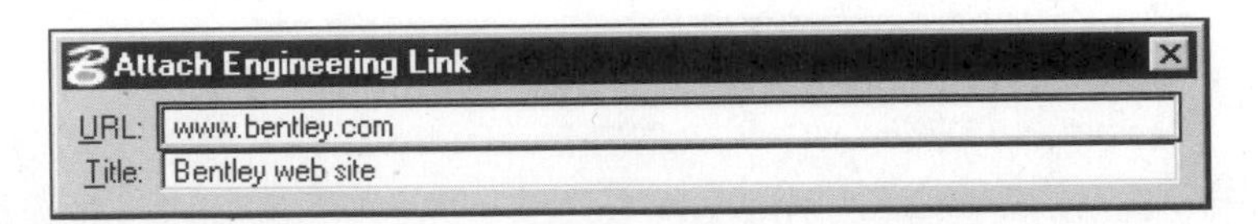

Figure 15–16 Attach Engineering Link settings box

Specify the URL to attach to an element and the title in the URL: and Title: edit boxes.

MicroStation prompts:

Attach Engineering Link > Identify element *(Identify an element to attach.)*

Attach Engineering Link > Accept/Reject *(Place a data point to accept the attachment of the link to the selected element.)*

You can attach multiple elements with the same URL.

Displaying Links

To highlight the elements that have engineering links, invoke the Show Engineering Links tool:

E-Links tool box	Select the Show Engineering Links tool (see Figure 15–17).

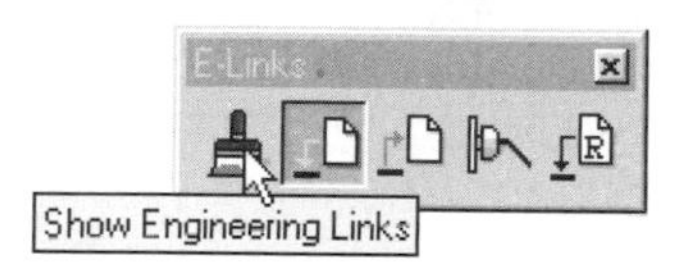

Figure 15–17 Invoking the Show Engineering Links tool from the E-Links tool box

MicroStation highlights the elements that have engineering links.

Connecting to URLs

To connect to the URL attached to an element, first invoke the Follow Engineering Link tool:

E-Links tool box	Select the Follow Engineering Link tool (see Figure 15–18).

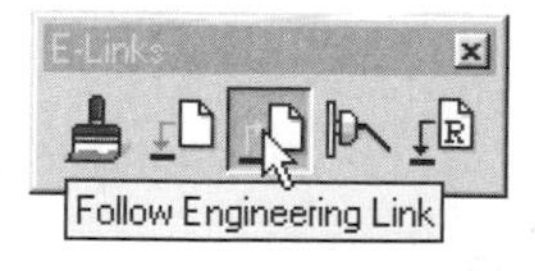

Figure 15–18 Invoking the Follow Engineering Link tool from the E-Links tool box

MicroStation prompts:

Follow Engineering Link > Identify element *(Identify the element to which the URL is attached.)*

Attach Engineering Link > Accept/Reject *(Place a data point to accept the access to the URL.)*

MicroStation opens the default Web browser and connects to the URL.

If the Web browser is not able to access the specified URL, then it will display an error message. Make sure proper connection is established to access the Internet.

chapter 16

Customizing MicroStation

MicroStation is an extremely powerful program off the shelf. But, like many popular engineering and business software programs, it does not automatically do all things for all users. It does, however, permit users to make changes and additions to the core program to suit individual needs and applications. Word processors offer a feature by which you can save a combination of many keystrokes and invoke them at any time with just a few keystrokes. This is known as a *macro*. Database management programs have their own library of user functions that can also be combined and saved as a user-named, custom-designed tool. These programs also allow you to create and save standard blank forms for later use, to be filled out as needed. Utilizing these features to make your own copy of a generic program unique and more powerful for your particular application is known as *customizing*.

Objectives

Customizing MicroStation can include several facets. After completing this chapter, you will be able to do the following:

- Create settings groups and edit existing groups.
- Provide names for levels and group them into user-defined group names.
- Create custom line styles.

- Create and modify workspace components: the project configuration, the user configuration, and the user interface.
- Customize the function keys.
- Install fonts.
- Archive design files and their associated resources.

SETTINGS GROUPS

MicroStation allows you to define a settings group with a user-specified name in three categories: Drawing (default), Scale, and Working Units. Under each settings group of the Drawing category, you can define individual group components. You can set element attributes such as color, weight, line style, level, and class, associating with a primitive tool (such as Place Line, Place Text) as a part of the group component. As a group component you can also save the current multi-line definition and active dimension settings. You can define any number of group components for each settings group.

MicroStation provides an option to save the settings groups and group components to an external file, and by default the file will have an extension .STG. This file follows the same concept as the cell library. Once created, the settings file may be attached to any of your design files and activate one of the settings groups and corresponding group component. Selecting a component does the following:

- All element attributes associated with the component are set as specified in the component definition.
- If a key-in is defined for the component, the corresponding tool is selected, letting you place an element or elements without invoking the tool from a tool box.

Selecting a Settings Group and Corresponding Component

The Select Settings box allows you to select a settings group and corresponding group component from the currently attached settings group file. The settings box also provides an option to attach another settings group file and invoke a settings group and corresponding group component.

To open the Select Settings box, invoke the Manage tool:

Pull-down menu	Settings > Manage

MicroStation displays the Select Settings box, as shown in Figure 16–1. MicroStation displays the name of the currently opened settings group file as part of the title of the settings box. The settings box is divided into two parts. By default, the Group: list box lists the names of the available drawing settings groups. The Component: list box lists the name and type of each component from the selected settings group. To invoke one of the components, select the name of the component, and settings associated with the component are set as specified in the definition. If a key-in is defined for the component, the corresponding tool is selected, letting you place an element(s) without using a specific tool.

Figure 16–1 Select Settings dialog box

If necessary, you can dock the Select Settings box to the top or bottom of the application window. The Select Settings box can also be displayed as a large dialog box. To display the Select Settings box as a large dialog box, invoke the Large Dialog tool:

Pull-down menu (Select Settings box)	Options > Large Dialog

MicroStation displays the Select Settings box as shown in Figure 16–2. The top part lists the names of the available drawing settings groups and bottom part of the settings box lists the name and type of each component from the selected settings group. The Category options menu sets the category for the listing of groups in the Group list box. The Sort options menu sets the manner in which components are sorted in the Component list box By Name or By Type.

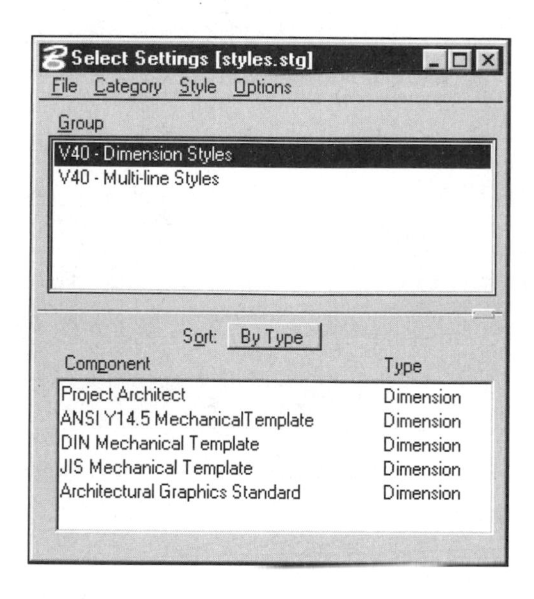

Figure 16–2 Select Settings box (Large Dialog)

Attaching a Settings Group File

To attach a settings group file to a design file, invoke the Open... tool:

Pull-down menu (Select Settings box)	File > Open...

MicroStation displays the Open Existing Settings File dialog box. Select the settings group file (extension .STG) from the Files list box, and click the OK button or press Enter to attach it to the design file. MicroStation lists the available groups and corresponding components.

Modifying a Settings Group

The Edit Settings box allows you to define, modify, and delete settings groups and group components. To open the Edit Settings box, invoke the Edit tool:

Pull-down menu (Select Settings box)	File > Edit

MicroStation displays the Edit Settings box, as shown in Figure 16–3. MicroStation displays the name of the currently opened settings group file as part of the title of the settings box. The settings box is divided into two parts. By default, the top part lists the names of the available drawing settings groups. The bottom part of the settings box lists the name and type of each component from the selected settings group.

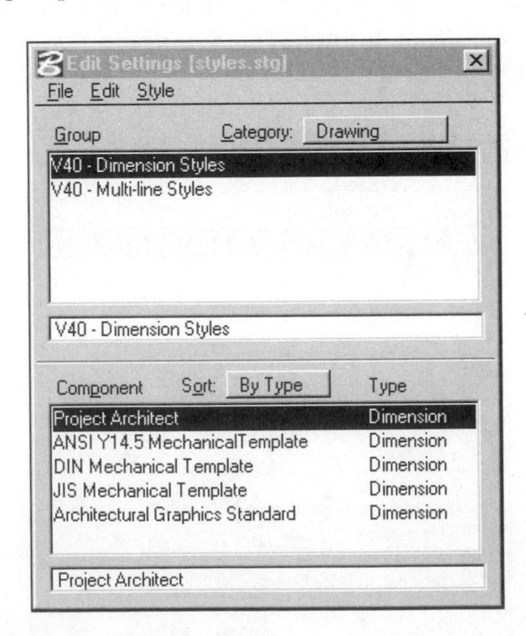

Figure 16–3 Edit Settings box

Creating a Settings Group File

To create a new settings group file, invoke the New... tool:

Pull-down menu (Edit Settings box)	File > New...

MicroStation displays the Create New Settings File dialog box. Key-in the name of the file in the Files: edit field, and click the OK button to create a new settings file.

Creating a Settings Group

To create a new settings group, invoke the Create Group tool:

Pull-down menu (Edit Settings box)	Edit > Create > Group

MicroStation adds a new group, called "Unnamed," to the Group list box. You can change the name from "Unnamed" to any appropriate name in the name edit field located just below the listing of the Group names list box. The maximum number of characters for the group name is 31.

Deleting a Settings Group

To delete a settings group, first select the group to be deleted from the Group list box and invoke the Delete tool:

Pull-down menu (Edit Settings box)	Edit > Delete...

MicroStation opens an alert box to confirm the deletion. Click on the OK button to delete the group. MicroStation deletes the group and all its components.

Creating a New Component

To create a new component, first select the settings group under which you want to create a new component. Select one of the seven component types available:

Pull-down menu (Edit Settings box)	Edit > Create > <component types>

MicroStation adds a new component type "Unnamed," with the listing of the component type, to the Components list box. You can change the name from "Unnamed" to any appropriate name in the name edit field box located just below the listing of the Component names list box. The maximum number of characters for the group name is 31.

Table 16–1 lists the component types and corresponding tools that can be used with each component type.

Table 16–1 Component Types and Corresponding Tools

COMPONENT TYPE	TOOL
Active Point	Points tool box
Area Pattern	Pattern tool box
Cell	Cells tool box
Dimension	Dimension tool box
Linear	Linear Elements tool box Polygons tool box Arcs tool box Ellipses tool box Curves tool box
Multi-line	The key-in **PLACE MLINE CONSTRAINED** corresponds to the Place Multi-line tool
Text	Text tool box

Modify a Component

To modify a component, first select the component from the Component list box, and invoke the Modify tool:

Pull-down menu (Edit Settings box)	Edit > Modify

MicroStation displays the appropriate component settings box.

Make the necessary changes in the settings box. Or instead, click the Match button, and MicroStation matches the settings in the settings box to the selected element. Click the Save button to save the settings and close the settings box. To disregard the changes, click the Close button.

The **Modify Point Component settings box** (see Figure 16–4) allows you to modify the options available related to Construct Point tools. To set the key-in that will be activated automatically when the component is selected, set the key-in toggle button to ON and type the appropriate key-in for the Construct Points tool. Set the appropriate element attributes (level, color, weight, and line style) that will be activated automatically when the component is selected. Select one of the three point types, Zero-Length Line, Cell, or Character, from the Type options menu. If you select the Cell option, set the appropriate settings in the Cell section of the Modify settings box. If you select the Character option, set the appropriate settings in the Character section of the Modify settings box.

The **Modify Area Pattern Component settings box** (see Figure 16–5) allows you to modify the options available related to hatching and patterning tools. To set the key-in that will be activated automatically when the component is selected, set the key-in toggle button to ON and type the appropriate key-in for the hatching or patterning tool. Set the appropriate element attributes (level, color, weight, and line style) that will be activated automatically when the component is selected. Select one of the two pattern types, Pattern or Hatch. If you select the Pattern option, set the appropriate settings in the Pattern section of the Modify settings box. If you select the Hatch option, set the appropriate settings in the Hatch section of the Modify settings box.

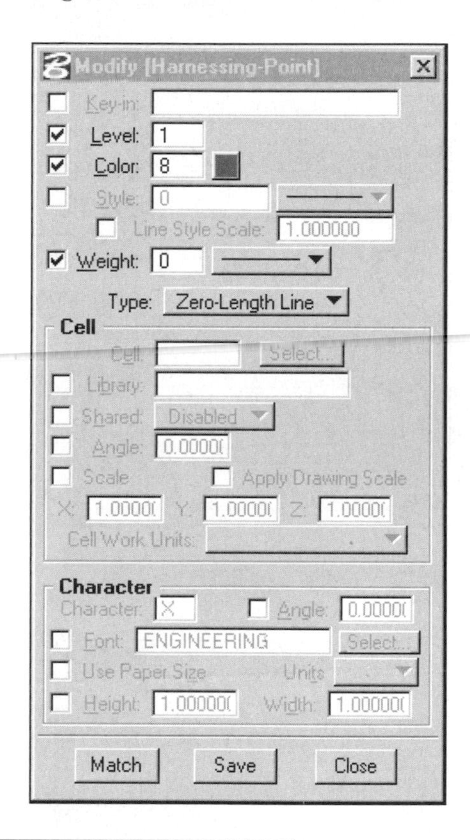

Figure 16–4 Modify Point Component settings box

The **Modify Cell Component settings box** (see Figure 16–6) allows you to modify the options available related to cell placement tools. To set the key-in that will be activated automatically when the component is selected, set the key-in toggle button to ON and type the appropriate key-in for the cell placement tool. Set the appropriate element attributes (level, color, weight, and line style) that will be activated automatically when the component is selected. Key-in the name of the cell library, the cell name, and the appropriate scale factors that will be automatically activated when the component is selected. Select one of the two cell types, Placement or Terminator, from the Type options menu. If the Placement option is selected, then the Cell selection will become the Active Cell when the component is selected in the Select Settings window. If the Terminator option is selected, then the Cell selection will become the Active Line Terminator.

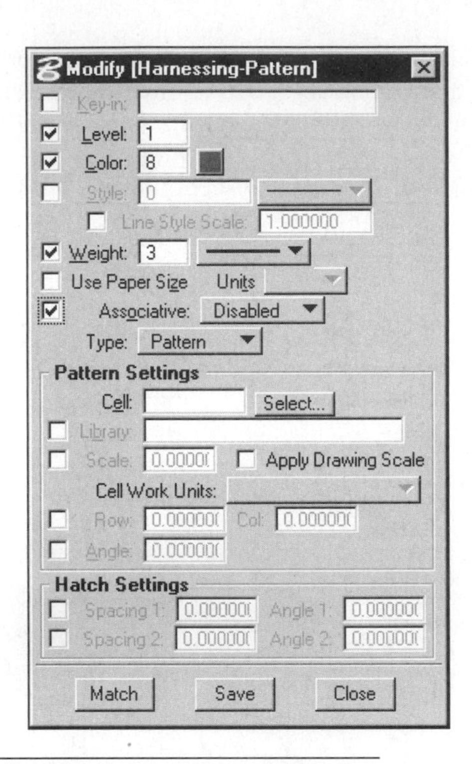

Figure 16-5 Modify Area Pattern Component settings box

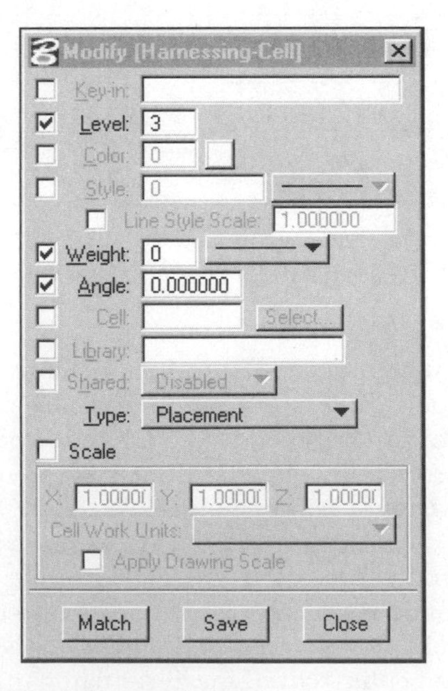

Figure 16-6 Modify Cell Component settings box

Note: The Level, Color, Style, and Weight controls can affect placement of a cell using the component only if the specified cell was created as a point cell.

The **Modify Dimension Component settings box** (see Figure 16–7) allows you to modify the options available related to the Dimension component definition. To set the key-in that will be activated automatically when the component is selected, set the key-in toggle button to ON and type the appropriate key-in for the dimension placement tool. Set the appropriate element attributes (level, color, weight, and line style) that will be activated automatically when the component is selected.

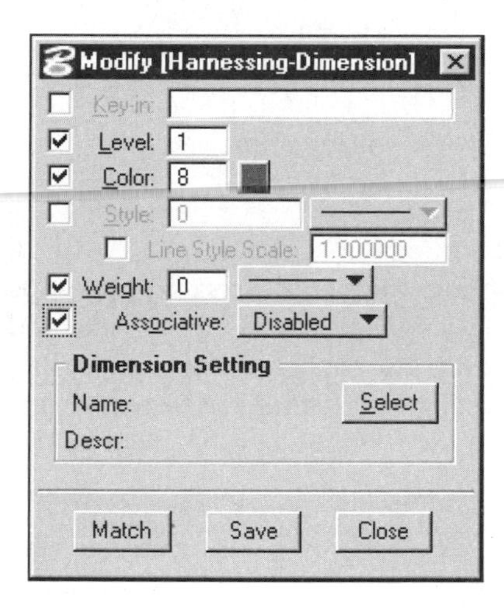

Figure 16–7 Modify Dimension Component settings box

As part of the Component definition, select the dimension style that will set the appropriate dimension settings. Click the Select button to select one of the available dimension styles. MicroStation displays the Select Dimension Definition dialog box listing the available dimension styles. Select the dimension style from the list box and click the OK button. Or, if you want to use the current dimension settings, then first save the dimension settings on a given style name. Invoke the Dimension tool:

Pull-down menu (Edit Settings box)	Style > Dimension

MicroStation displays the Edit Dimension Styles settings box, as shown in Figure 16–8.

Click the Get Active button and MicroStation adds a new dimension style "Dim0." You can change the name from "Dim0" to any appropriate name in the Name edit field, and close the settings box.

Click the Select button from the Modify settings box to select the newly created dimension style from the Select Dimension Definition dialog box.

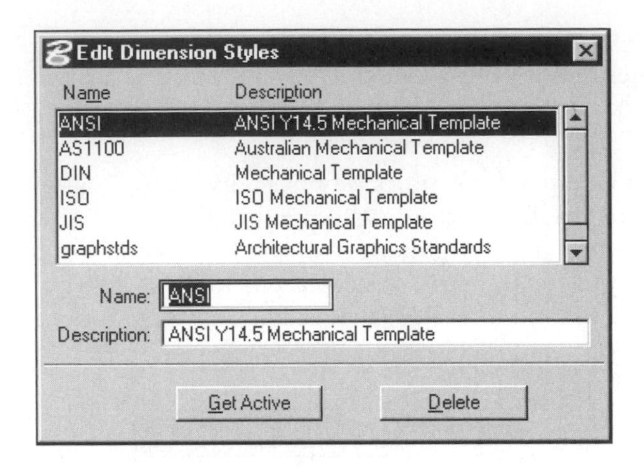

Figure 16–8 Edit Dimension Styles settings box

The **Modify Linear Component settings box** (see Figure 16–9) allows you to modify the options available related to placement of lines, polygons, arcs, circles, ellipses, and curves. To set the key-in that will be activated automatically when the component is selected, set the key-in toggle button to ON and type the appropriate key-in for the Linear tool. Set the appropriate element attributes (level, color, weight, line style, area, and fill) that will be activated automatically when the component is selected.

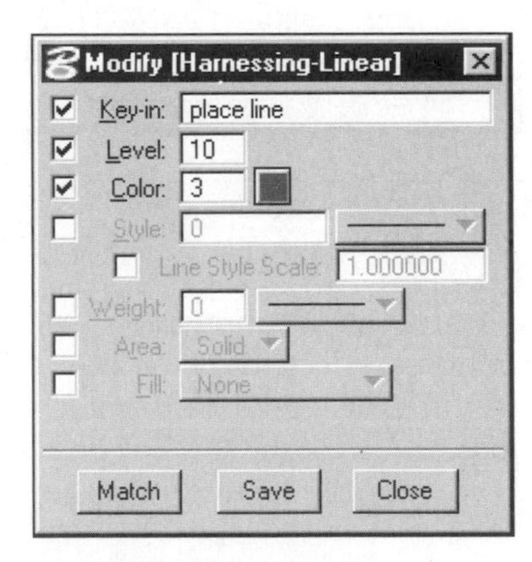

Figure 16–9 Modify Linear Component settings box

The **Modify Multi-line Component settings box** (see Figure 16–10) allows you to modify the options available related to multi-line component definition. To set the key-in that will be activated automatically when the component is selected, set the key-in toggle button to ON and type the appropriate key-in for the multi-line placement tool. Set the appropriate element attributes (level, color, weight, and line style) that will be activated automatically when the component is selected.

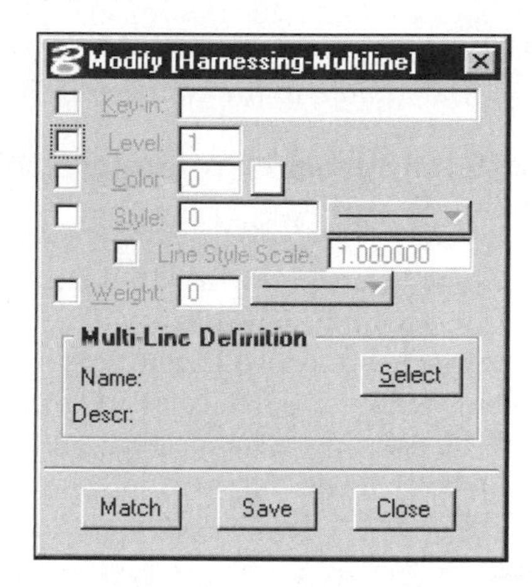

Figure 16–10 Modify Multi-line Component settings box

As part of the component definition, select the multi-line definition that will be set when you invoke the component. Click the Select button to select one of the available multi-line definitions. MicroStation displays the Select Multi-line Definition dialog box listing the available multi-line definitions. Select the multi-line definitions from the list box and click the OK button. Or, if you want to use the current multi-line definition, then first save the definition on a given style name. Invoke the Multi-Line tool:

Pull-down menu (Edit Settings box)	Style > Multi-Line

MicroStation displays the Edit Multi-line Styles settings box, as shown in Figure 16–11.

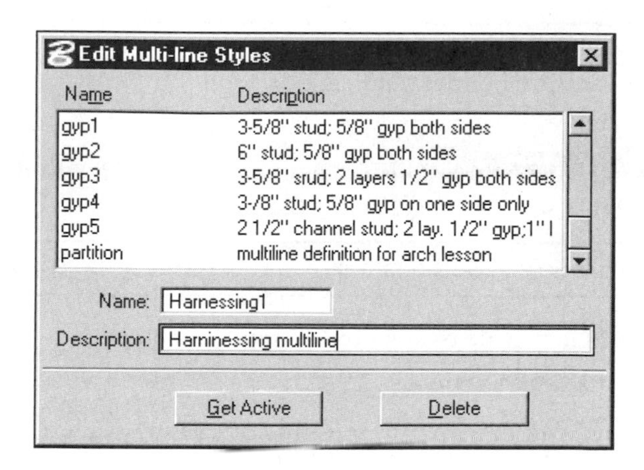

Figure 16–11 Edit Multi-line Styles settings box

Click the Get Active button. MicroStation adds a new multi-line style "Mline0." You can change the name from "Mline0" to any appropriate name in the Name edit field and close the settings box.

Click the Select button from the Modify settings box to select the newly created multi-line style from the Select Multi-line Definition dialog box.

The **Modify Text Component settings box** (see Figure 16–12) allows you to modify the options available related to placement of text. To set the key-in that will be activated automatically when the component is selected, set the key-in toggle button to ON and type the appropriate key-in for the place text tool. Set the appropriate element attributes (level, color, weight, line style, area, and fill) and text attributes (slant angle, line length, line spacing, fraction, vertical, underline, justification, font, size, and intercharacter spacing) that will be activated automatically when the component is selected.

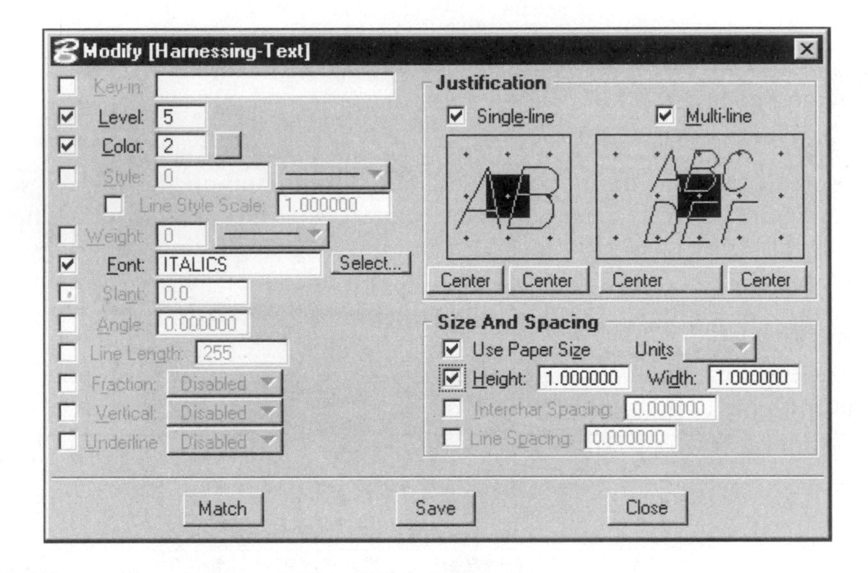

Figure 16–12 Modify Text Component settings box

In addition, you can also specify the relationship between plotting units and design Master Units by selecting one of the available scale settings. To list the available scale settings, select Scale...:

Pull-down menu (Select Settings box)	Category > Scale

MicroStation displays the Select Scale dialog box. To set the scale, double-click the scale settings in the list box and close the dialog box.

You can also set the current Working Units by selecting one of the available Working Units settings. To list the available Working Units settings group, select Working Units...:

Pull-down menu (Select Settings box)	Category > Working Units...

MicroStation displays the Select Working Units dialog box. The dialog box lists the available Working Units settings. To change the Working Units to one of the listed Working Units for the current design, double-click the Working Units setting to make its settings active and close the dialog box. An alert box opens for confirming the adjustment of the Working Units settings.

Importing Dimension and Multi-line Styles from Version 4

To import dimension and multi-line styles created with MicroStation Version 4 as a group, invoke the Import... tool:

Pull-down menu (Edit Settings box)	Style > Import > V4 Styles (w/Names)...

MicroStation displays the Select V4 Style File dialog box. Each style that is selected is converted to a component of the new group and assigned a name identical to its style name.

NAMING LEVELS

In Chapter 3 you were introduced to placing elements on individual levels. Each design file is provided with 63 levels and each level is assigned a number between 1 and 63, inclusive. To make a level active, key-in **LV=<number>** and press Enter. You can turn ON and OFF levels by keying **ON=**<number> and **OFF=**<number>, respectively. Instead of keying a level number, you can assign an alphanumeric "name" that represents a level number and then substitute the "name" for the "number" when using the key-ins. By assigning a name to a level, it is easier to remember what type of elements will be placed on a specific level. Let's say you want to place dimensioning on level 10. Instead of remembering level 10 for dimensioning, assign a name "dim" for level 10. Whenever you want to place dimensioning, make sure you are in the dim level by making it the active level by keying-in **LV=dim** and pressing Enter. It may also be helpful to assign level names if you have a need to "translate" a drawing into or out of MicroStation from or to another CAD system.

In addition to assigning a level name, MicroStation allows you to provide a group name. Under a group name you can assign a set of level names or numbers, and MicroStation allows you to control the display of levels by keying-in a group name for ON and OFF key-in commands. Level names and group names are analogous to the files and folders in Windows operating system.

MicroStation provides an option to save the level name assignments and level group definitions composing the level structure to an external file. You can store any number of definitions to a single file, and by default the file will have an extension .LVL. This file follows the same concept as the cell library. Once created, the level structure file may be attached to any of your design files and activate one of the definitions stored in the file.

Assigning Level Names and Group Names

The Level Manager settings box is used to control level display and level symbology (explained in Chapter 14) for the active design file and attached reference files. In addition, it allows you to open the Level Names dialog box to assign level names and group names. Open the Level Manager settings box:

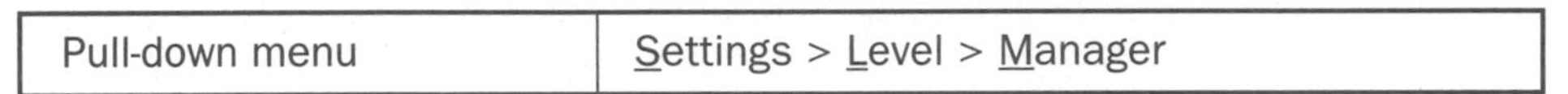

Pull-down menu	Settings > Level > Manager

MicroStation displays the View Levels settings box, as shown in Figure 16–13.

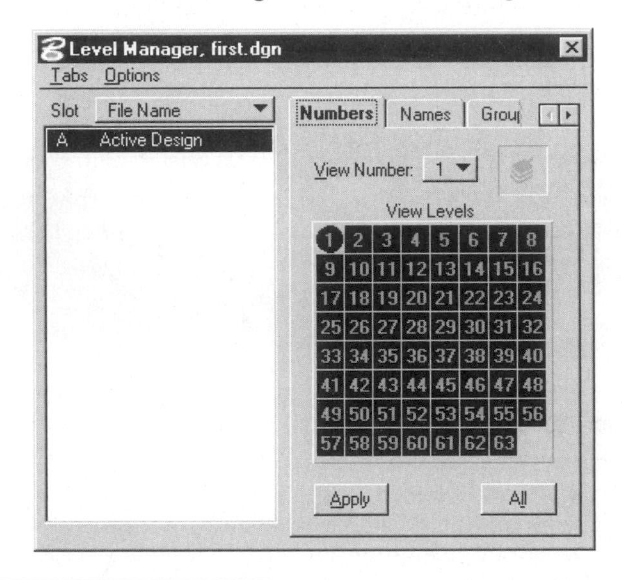

Figure 16–13 Level Manager settings box

The default Level Manager settings box includes the Files list box in addition to four tabs: Names, Numbers, Groups, and Symbology, with the Numbers tab brought to the front. Use the scroll bar in the right corner of the settings box to access the other tabs. The size of the settings box depends on which tab is brought to the front. The scroll bar disappears if all the tabs are visible. You can control the display of the tabs and files list box by selecting appropriate option tool from the Tabs pull-down menu in the Level Manager settings box.

The Slot option menu located just above the Files list box provides three options to choose from: File Name, Logical Name, and Description, by which you can list the reference files attached to the current design file.

The **Files list** box lists the current active design file and reference files attached to the current design file. If you right-click in the list box, a menu appears containing the following three options:

1. Attach Ref—Opens the Preview Reference dialog box from which you can select a design or raster file to attach to the current design file.

2. Detach Ref—Detaches the selected reference file and removes its entry in the list box.
3. Properties—Opens the Design File settings box if the active design file selected; if you have any other file selected, it opens the Reference files dialog box and highlights the selected reference file.

The **Numbers** tab displays the level numbers layout as shown in Figure 16–13. If you right-click in the level map, a menu appears containing the following four options:

1. All On—Sets all the levels to ON. To apply to the current View Number selection, click the Apply button; to apply to all View Windows, click the All button.
2. All Off—Sets all the levels to OFF except the Active Level. To apply to the current View Number selection, click the Apply button; to apply to all View Windows, click the All button.
3. Off By Element—Sets the level to OFF of each selected element on which it is drawn.
4. All Except Element—Sets all levels to OFF except the selected element on which it is drawn.

The Numbers tab also contains the Update Level Usage icon in the upper right corner of the tab. The icon is dimmed when you initially open the Level Manager settings box. When elements are added to the current design file, this icon enables to inform you that the levels in the design file may have changed. Click the Update Level usage icon to update the changes.

The **Names** tab lists the assigned level names in the active design file or for an attached reference file similar to the one shown in Figure 16–14. The active level name is displayed in red. To turn ON or OFF a set of level(s), select all the level name(s) to be included (dragging or Ctrl-clicking selects multiple levels) from the level operations layout display and click the On or the Off button. To make a level active, select a level name and click the Active button.

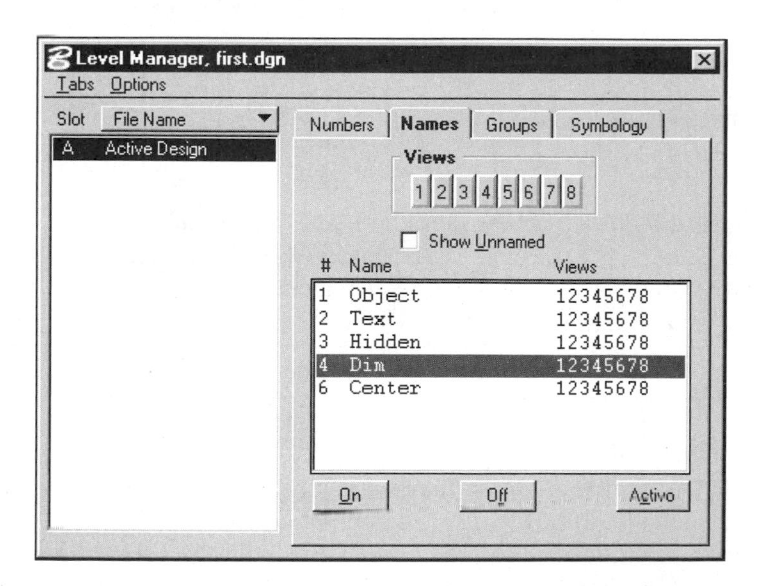

Figure 16–14 Level Manager settings box, with the Names tab selected

The **Groups** tab lists the group names and corresponding levels similar to the one shown in Figure 16–15. The active level name is displayed in red. To turn ON or OFF a specific group, select the group name from the group operations layout display and click the On or the Off button. To turn ON or OFF a set of level(s), select all the level name(s) to be included (dragging or Ctrl-clicking selects multiple levels) from a specific group name and click the On or the Off button. To make a level active, select a level name and click the Active button.

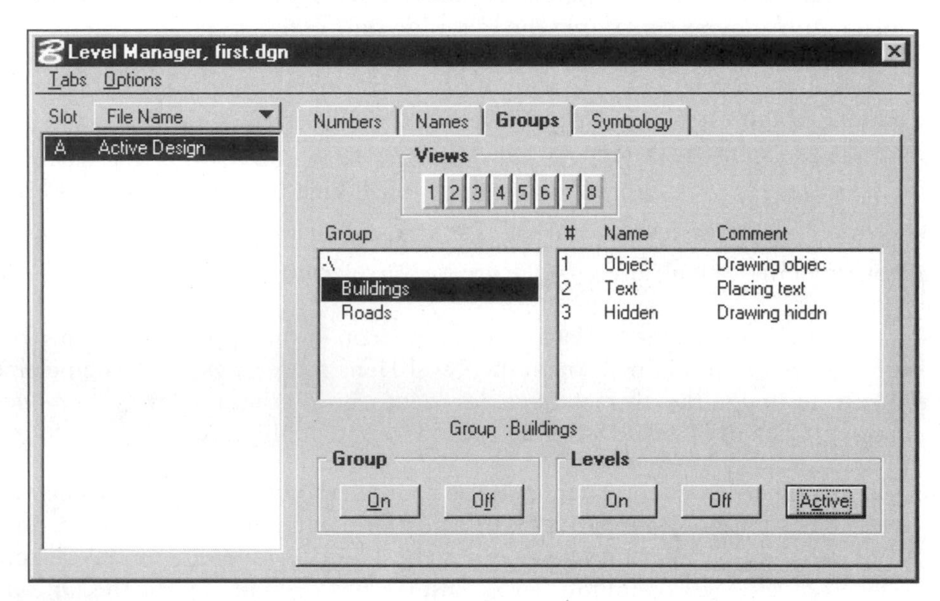

Figure 16–15 Level Manager settings box, with the Groups tab selected

The **Symbology** tab, shown in Figure 16–15, has the same functionality as in the Level Symbology dialog box explained in Chapter 14.

Defining Names

To assign level names and group names, open the Level Names dialog box:

Pull-down menu (Level Manager settings box)	Options > Level Names...
Pull-down menu	Settings > Level > Names...

MicroStation displays the Level Names dialog box, as shown in Figure 16–16.

The Level Names dialog box has two alternate layouts, chosen from the Display menu, as shown in Figure 16–17. The Level Operations layout displays all the level names assigned in the current level structure file. The Level Group operations layout displays all the group names assigned in the current level structure file.

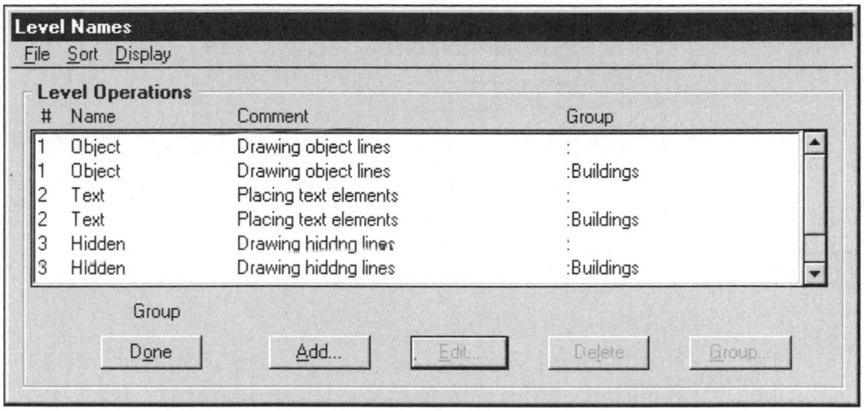

Level Names

File Sort Display

Level Operations

#	Name	Comment	Group
1	Object	Drawing object lines	:
1	Object	Drawing object lines	:Buildings
2	Text	Placing text elements	:
2	Text	Placing text elements	:Buildings
3	Hidden	Drawing hiddng lines	:
3	Hidden	Drawing hiddng lines	:Buildings

Group

Done Add... Edit... Delete Group...

Figure 16–16 Level Names dialog box

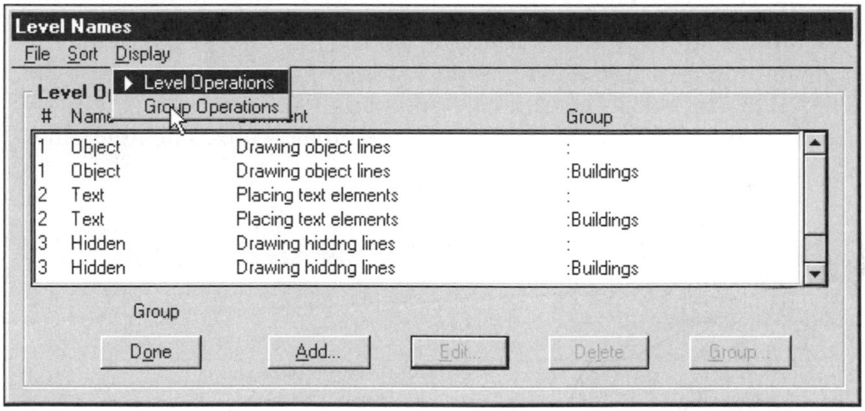

Level Names

File Sort Display

Level Operations
Group Operations

#	Name	Comment	Group
1	Object	Drawing object lines	:
1	Object	Drawing object lines	:Buildings
2	Text	Placing text elements	:
2	Text	Placing text elements	:Buildings
3	Hidden	Drawing hiddng lines	:
3	Hidden	Drawing hiddng lines	:Buildings

Group

Done Add... Edit... Delete Group...

Figure 16–17 The Level Names dialog box has two alternate layouts to choose from the Display menu

The **Level Operations** layout lists the level name, comment, and group affiliation (with path) for each named level, as shown in Figure 16–17. The list is sorted by the criteria specified in the Sort Criteria dialog box (invoked from the Sort...pull-down menu). If a level is a member of multiple level groups, it is listed multiple times—once for each group of which it is a member. To operate on a level, you must first select it. Dragging or Ctrl-clicking selects multiple levels.

To assign a level name, invoke the Level Name dialog box by clicking the Add... button. MicroStation displays the Level Name dialog box, similar to the one shown in Figure 16–18. Key-in the level number in the Number: edit field, the level name in the Name: edit field (maximum valid number of characters is 16), and any comment in the Comment: edit field (maximum valid number of characters is 32).

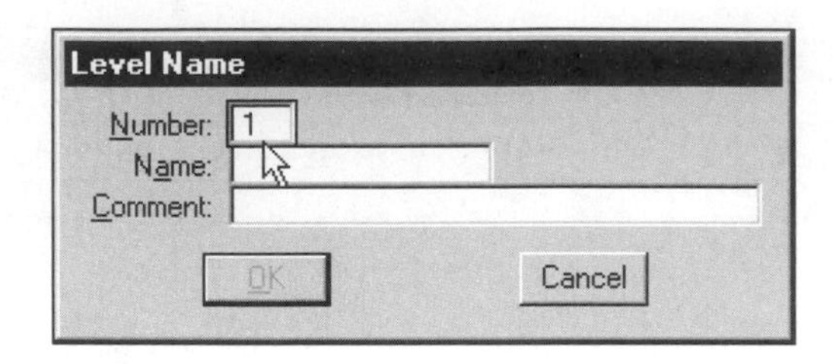

Figure 16–18 Level Name dialog box

To change the level name, the assigned level number, or the comment, first select the level to be changed in the list box, then click the Edit... button. MicroStation displays the Level Name dialog box. Make the necessary changes and click the OK button.

To remove a level name from the list box, first select the level to be removed in the list box, then click the Delete button. MicroStation removes the selected level name assignment from the list box.

The **Group Operations** layout lists the names of the level groups in a tree structure under the group listing, as shown in Figure 16–19. To list the levels assigned to a group, click the group name. MicroStation lists the level numbers, with corresponding level names, including the comments. The list is sorted by criteria specified in the Sort Criteria dialog box. The full path to the selected group is shown below the list box.

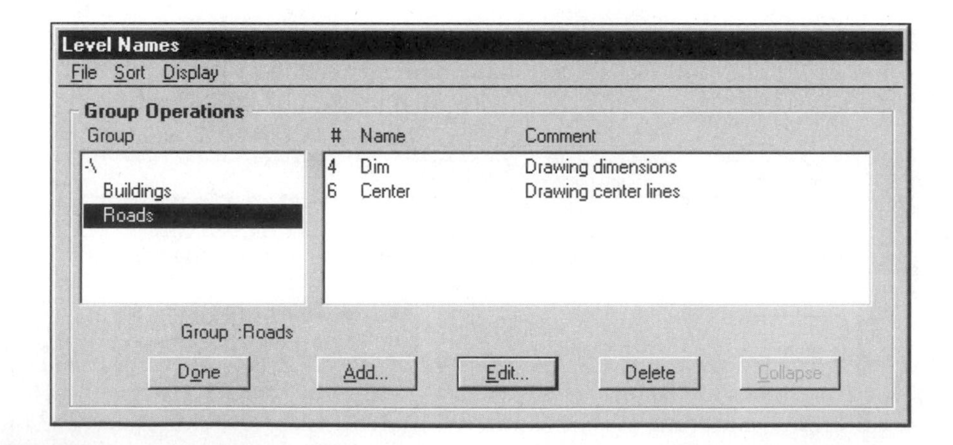

Figure 16–19 The Group operations layout lists names of the level groups in a tree structure

The backslash symbol (\) indicates the "root" of the structure. Each group that has subgroups is indicated with a symbol. The minus sign (–) indicates that the group's subgroups are listed below. The plus sign (+) indicates the group's subgroups are not listed. Double-clicking a group name indicated with a minus or plus sign toggles the listing of its subgroups.

To assign a group or subgroup name, invoke the Level Group dialog box by clicking the Add... button. MicroStation displays the Level Group dialog box, similar to the one shown in Figure 16–20. Key-in the level group name in the Name edit field (maximum valid number of characters is 16).

Figure 16–20 Level Group dialog box

To change the group name, first select the group to be changed in the list box, then click the Edit... button. MicroStation displays the Level Group dialog box. Make the necessary changes and click the OK button.

To remove a group name from the list box, first select the group to be removed in the list box, then click the Delete button. MicroStation removes the selected group name assignment and its subgroups, if any, and discontinues component level name assignment.

To list the selected group's subgroups, click the Collapse button. The Collapse button is disabled (dimmed) if the selected group does not have any subgroups.

Assigning Levels to a Group

To assign a set of level(s) to a group name, first select all the level name(s) to be included (dragging or Ctrl-clicking selects multiple levels) from the Level Operations layout display. Open the Select Target Group dialog box:

Level Names settings box (Level operations display)	Select the Group... button

MicroStation displays the Select Target Group dialog box, as shown in Figure 16–21. The Select Target Group dialog box enables you to copy or move the selected level(s) to a group.

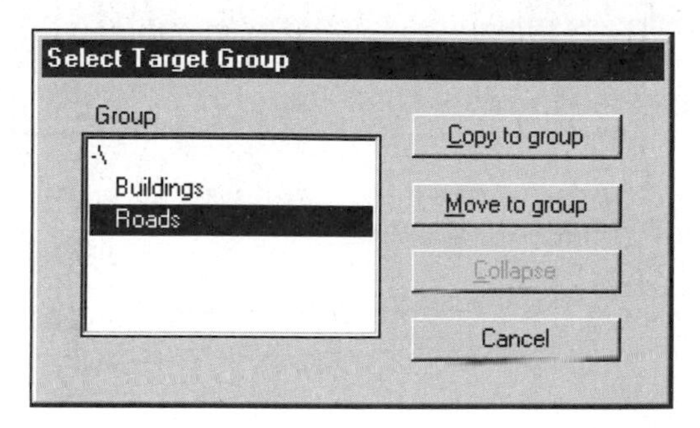

Figure 16–21 Select Target Group dialog box

The dialog box lists the names of level groups in a tree structure, indicated by indenting each successive level. The backslash symbol (\) indicates the "root" of the structure. To specify a group as the destination for the selected level(s), first click the group name. To add the level to the selected group without deleting it from its present group, click the Copy to Group button. If you want to delete from the present group, if any, then click the Move to Group button. The Collapse button toggles the listing of the selected group's subgroups. The Cancel button closes the dialog box without copying or moving the selected level(s).

Saving the Level Structure

To save the current level structure, invoke the Save tool:

Pull-down menu (Level Names settings box)	File > Save...

MicroStation displays the Save Level Structure dialog box. Key-in the file name in the Files: edit field and click the OK button to save the level structure. The default extension for the file name is .LVL.

Note: You can save the level structure for the active design file by invoking the Save Settings tool from the File pull-down menu.

Attaching the Level Structure to the Current Design File

To attach an existing level structure file to the current design file, invoke the Open tool:

Pull-down menu (Level Names settings box)	File > Open...

MicroStation displays the Open Level Structure dialog box. Select the level structure file from the files list box and click the OK button, or press Enter, to attach the level structure to the current design file.

Removing the Level and Group Assignment Names

To remove the level and group assignment names from the list box, invoke the Remove tool:

Pull-down menu (Level Names settings box)	File > Remove...

MicroStation displays an alert box to confirm the request. Click the OK button to remove all the level name assignments from the design file.

To close the Level Names settings box, click the Done button.

LEVEL USAGE IN THE ACTIVE DESIGN

MicroStation lists the level usage by name and number in addition to providing information on the number of elements placed on each level. Open the Level Usage dialog box:

Pull-down menu	Settings > Level > Usage...

MicroStation displays the Level Usage dialog box, as shown in Figure 16–22.

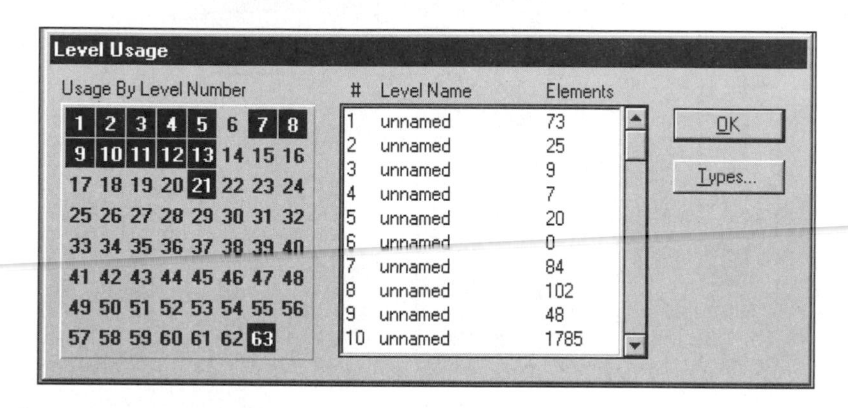

Figure 16–22 Level Usage dialog box

The level numbers with a dark background are the ones that have elements. Click the Types... button to get the listing of element types in the design file.

CUSTOM LINE STYLES

In addition to standard line styles (LC = 0 ... LC = 7), MicroStation allows you to create custom line styles. Custom line styles are created and stored in a style library file (extension .RSC). The style library file delivered with MicroStation is called LSTYLE.RSC. As with the cell library, line styles are created, stored, then recalled as needed. Once a library is created, it is attached to a design file, and all the line styles stored in the library are available. Only one style library can be attached at any time. If no style library is attached, MicroStation's standard line styles (LC = 0 ... LC = 7) will still be available.

Note: If you copy or move your MicroStation design file from one PC to another, be sure to copy the style library as well. Your custom line styles will not appear in your design if the style library is not present.

Activating a Custom Line Style

To browse, activate line styles, and set line style modifiers, open the Line Styles settings box:

Primary tool box	Line Style option menu > Custom (see Figure 16–23)
Key-in window	**linestyle settings** (or **lines s**) Enter

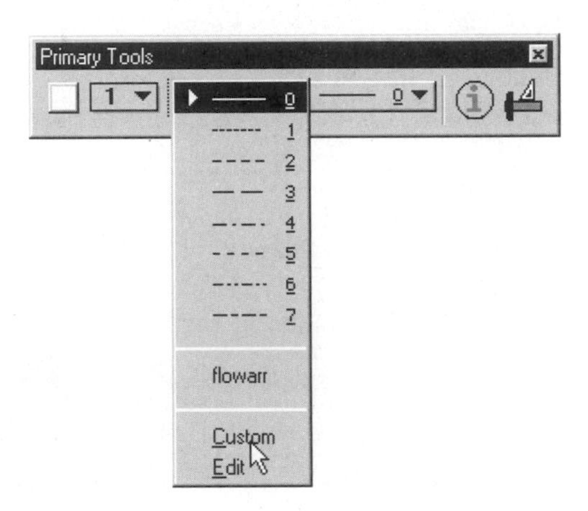

Figure 16–23 Selecting Custom line styles from the Primary Tools box

MicroStation displays the Line Styles settings box, similar to Figure 16–24.

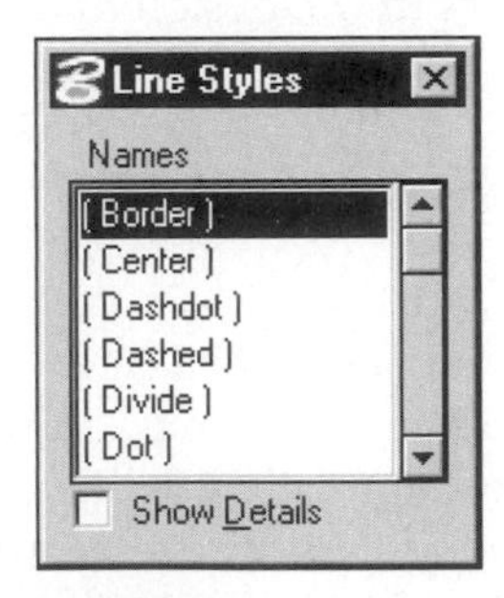

Figure 16–24 Line Styles settings box

The settings box lists the names of line styles from the default line style library LSTYLE.RSC unless it is set for a different line style library. Double-clicking a line style name makes the line style the active line style for element placement. To display all the attributes associated with line styles, turn ON the Show Details toggle button. MicroStation displays a Line Styles settings box similar to Figure 16–25.

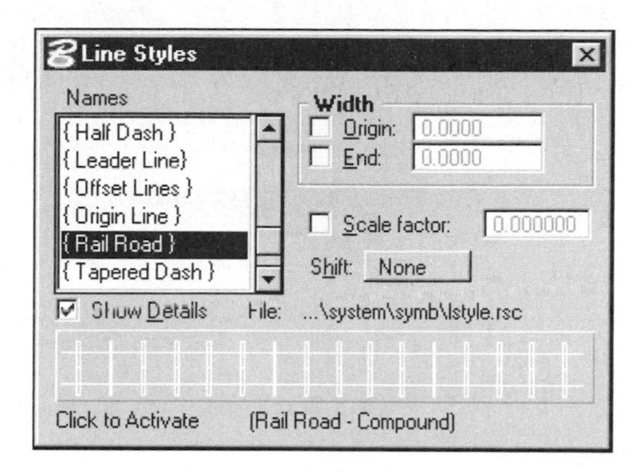

Figure 16–25 Line Styles settings box with additional options displayed

At the bottom of the settings box, MicroStation shows the line style and description of the line style selected in the Names list box with active modifiers applied. Click anywhere on the sample to make the line style the active line style for element placement. If necessary, you can modify the starting and ending width for the strokes as defined in the line style definition. To change the starting width, turn ON the Origin: toggle button and key-in the Width in Master Units in the Origin: edit field. To change the ending width, turn ON the End: toggle button and key-in the Width in Master Units in the End: edit field.

The Scale factor sets the scale to all displayable characteristics (dash length and width, point symbol size) of the active line style. To change the scale factor, turn ON the Scale factor: toggle button and key-in the scale factor in the Scale factor: edit field.

The Shift: option menu sets the distance or fraction by which each stroke pattern in the active line style is shifted or adjusted.

Defining and Modifying Line Styles

To define and modify existing custom line styles, open the Line Style Editor settings box:

Primary tool box	Line Style option menu > Edit (see Figure 16–26)
Key-in window	**linestyle edit** (or **lines e**) Enter

Line style definitions are stored in line style libraries. The settings box lets you open and define or modify line styles. If a line style library is open, its file specification is displayed in the title bar.

MicroStation lists the names of all the available line styles in the Styles list box. Selecting a name causes a sample of the line style to be displayed. In the Components list box, MicroStation lists the types and descriptions of all line styles components. If >> is displayed left of its Type, the component is linked directly to the line style whose name is selected in the Styles list box. Selecting a component causes a sample line with the component to be displayed. To modify a component, you must first select the component and make the necessary changes.

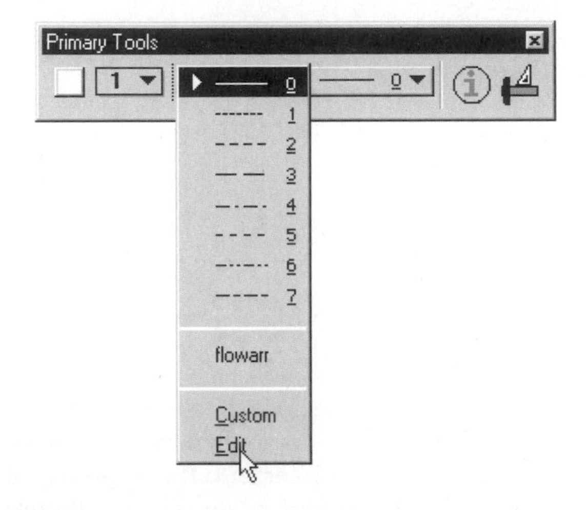

Figure 16–26 Invoking the line style Edit tool from the Primary Tools box

MicroStation displays the Line Style Editor settings box, similar to Figure 16–27.

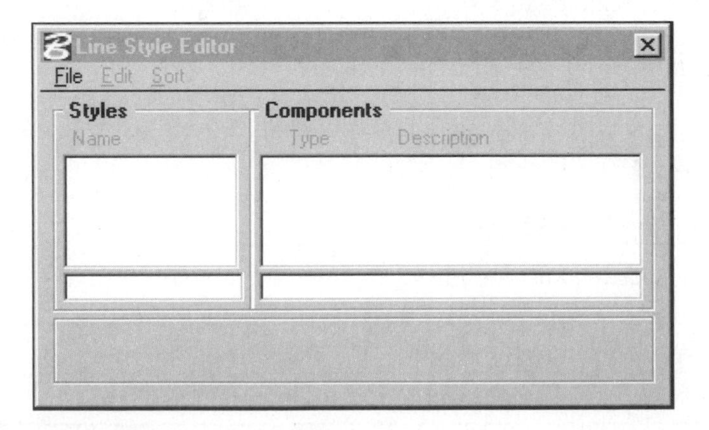

Figure 16–27 Line Style Editor settings box

Attaching an Existing Line Style Library

To attach an existing line style library, invoke the Open tool:

Pull-down menu (Line Style Editor settings box)	File > Open...

MicroStation displays the Open Line Style Library dialog box. Select the line style library file from the Files: list box and either click the OK button or press Enter to attach the library to the current design file.

Creating a New Line Style Library

To create a new line style library, invoke the New tool:

Pull-down menu (Line Style Editor settings box)	File > New...

MicroStation displays the Create Line Style Library dialog box. Key-in the name of the new style library file in the Files: edit field and either click the OK button or press Enter to create and attach the library to the current design file.

Creating a New Line Style

To create a new line style, invoke the Create Name tool:

Pull-down menu (Line Style Editor settings box)	Edit > Create > Name

MicroStation adds a new line style, named "Unnamed," in the Styles list box. Unnamed is automatically selected and is linked to the component selected in the Components list box. If necessary, you can change the name from "Unnamed" to an appropriate line style.

MicroStation provides three tools—Stroke Pattern, Point, and Compound—to customize the line style; they are available in the Create submenu of the Edit pull-down menu.

The **Stroke Pattern** setting provides controls for creating, modifying, and deleting stroke pattern line style components. Open the Stroke Pattern settings:

Pull-down menu (Line Style Editor settings box)	Edit > Create > Stroke Pattern

MicroStation displays the settings for the Stroke Pattern Attributes and Stroke Pattern, as shown in Figure 16–28.

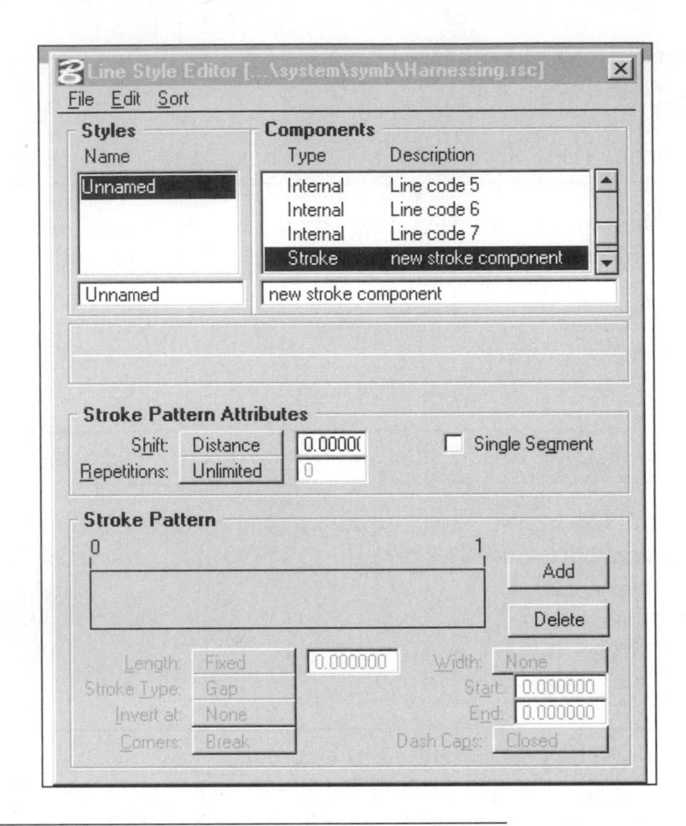

Figure 16-28 Line Style Editor with Stroke Pattern settings

- The Shift: options menu sets the distance or fraction by which the stroke pattern is shifted or adjusted.
- The Repetitions: options menu sets the number of times the stroke pattern is repeated throughout the length of an element or element segment. The number of repetitions needs to be fixed.
- The Single Segment toggle button controls the truncation of the stroke pattern at the end of each element segment.
- The Add button adds a new gap stroke at the end of the stroke pattern. The maximum number of strokes is 32. Each dash stroke is represented by a filled bar. Each gap stroke is represented by an unfilled bar. Click anywhere on the stroke for modification. The stroke is highlighted. Dragging a stroke's handle changes the stroke's length.
- The Length: edit field sets the selected stroke length in Master Units.
- The Stroke Type: options menu allows you to switch a highlighted stroke from gap to dash or vice versa.

- The Corners: options menu controls the behavior of the selected stroke when it extends farther than an element vertex.
- The Width: options menu controls the effect of width settings on the selected dash stroke.
- The Dash Caps: options menu sets the type of end cap on the selected dash stroke when displayed with width.

The **Point** settings provide controls for creating, modifying, and deleting point line style components. Open the Point settings:

Pull-down menu (Line Style Editor settings box)	Edit > Create > Point

MicroStation displays the settings to set the Point settings, as shown in Figure 16–29.

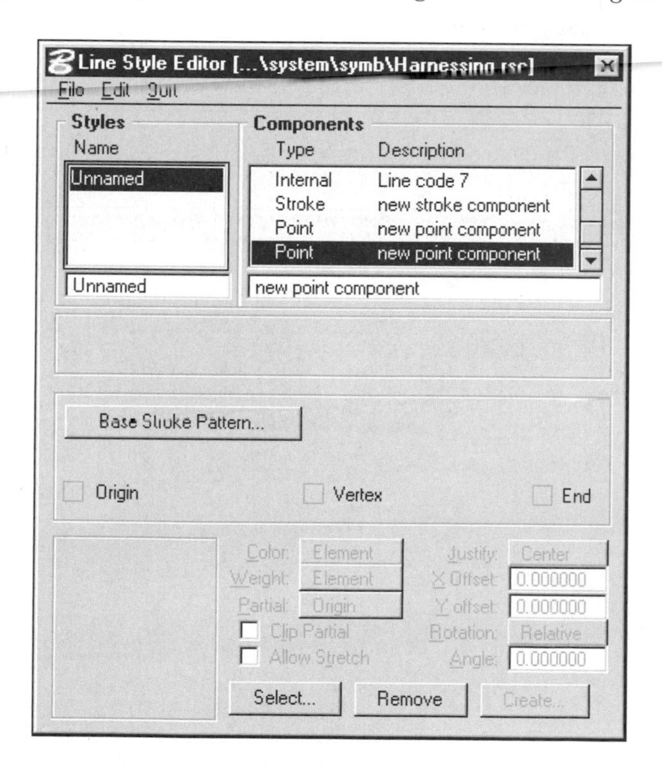

Figure 16-29 Line Style Editor with Point settings

- The Base Stroke Pattern... button opens the Base Stroke Pattern dialog box, from which you can select the stroke pattern component on which the selected point symbol component is based.

- The Origin toggle button selects the origin (first vertex of an element) with which a symbol can be associated.
- The Vertex toggle button selects the internal vertices (of an element) with which a single symbol can be associated.
- The End toggle button selects the end (last vertex of an element) with which a symbol can be associated.

Note: If a symbol is associated with either the origin, the vertex, or the last vertex of an element, the symbol is displayed (below and left of the settings box), and the controls for adjusting the related settings (below and right of the settings box) are enabled.

The **Compound** settings provide controls for creating, modifying, and deleting compound line style components. Open the Compound settings:

Pull-down menu (Line Style Editor settings box)	Edit > Create > Compound

MicroStation displays the settings to set the Compound settings, as shown in Figure 16–30.

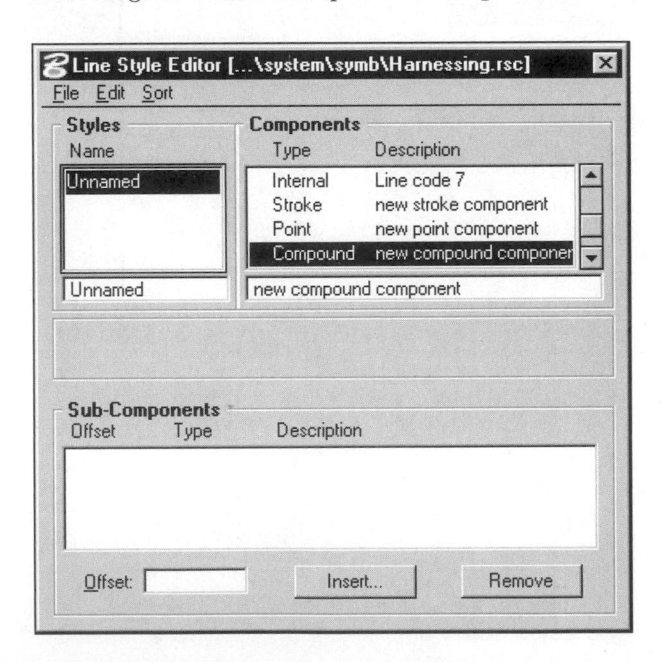

Figure 16–30 Line Style Editor with Compound settings

- The Sub-Components list box lists the offsets, types, and descriptions of all sub-components of the compound component selected in the Components list box.

- The Offset edit field sets the distance, in Master Units measured perpendicular from the work line, by which the selected component is displayed parallel to the work line. If the Offset is set to zero, the selected component is displayed on the work line.
- The Insert... button opens the Select Component dialog box, from which you can select a sub-component to insert in the compound component selected in the Components list box.
- The Remove button removes the sub-component selected in the Sub-Components list box from the compound component selected in the Components list box.

Deleting a Line Style

To delete a line style from the list box, invoke the Delete tool:

Pull-down menu (Line Style Editor settings box)	Edit > Delete

MicroStation deletes the selected line style from the list box.

Linking the Component

To link the selected component from the Components list to the selected line style from the Styles list box, invoke the Link tool:

Pull-down menu (Line Style Editor settings box)	Edit > Link

MicroStation links the selected component to the line style selected in the Styles list box.

Saving the Line Style Library

To save the line styles library, invoke the Save tool:

Pull-down menu (Line Style Editor settings box)	File > Save

MicroStation saves the currently open line style library.

Step-by-Step Procedures for Creating a New Line Style

Following are the step-by-step procedures to a create a line style called FLOWARR, consisting of a 1.5"-long line segment followed by a filled arrowhead (0.1"-wide, 0.125"-long).

STEP 1: Open the Line Style Editor settings box.

STEP 2: Create a new line style library called HARNESS.RSC

STEP 3: Invoke the Name tool under the submenu Create from the Edit pull-down menu. MicroStation adds a new line style, named "Unnamed," in the Styles list box. Change the "Unnamed" to "FLOWARR" in the edit field located under the Name list.

STEP 4: Select the Stroke Pattern option from the submenu Create from the Edit pull-down menu. The description "new stroke component" will be added to the list. Change the description to read "LINE CODE FOR FLOW ARROW" in the edit field located under the Description list.

STEP 5: Select the Link option from the Edit pull-down menu. This establishes a relationship between the new line style name and the new component.

STEP 6: In the Stroke Pattern Attributes section of the settings box, set the Shift Distance to 0.0 and the Repetitions value to Unlimited.

STEP 7: Click the Add button to add a new gap stroke.

STEP 8: Select Fixed in the Length: option menu and key-in 1.5 in the Length: edit field. Set the Stroke Type: option menu to Dash, the Invert at: option menu to None, and the Corners: option menu to Break. The characteristics of the stroke pattern will change to reflect these settings.

STEP 9: Click the Add button to add a new gap stroke at the end of the stroke pattern.

STEP 10: Place a data point anywhere in the new stroke pattern. The stroke is highlighted.

STEP 11: Select Fixed in the Length: option menu and key-in 0.125 in the Length: edit field. Set the Stroke Type: option menu to Dash, the Invert at: option menu to None, the Corners: option menu to Break, the Width: option menu to Full, the Start: Width value to .1, the End Width value to 0.0, and the Dash Caps: option menu to Closed. The characteristics of the stroke pattern will change to reflect these settings (see Figure 16–31).

STEP 12: Save the newly created line style by invoking the Save tool under the File pull-down menu in the Line Style Editor settings box.6

STEP 13: Close the Line Style Editor settings box.

STEP 14: Test the newly created line style by setting the active line style to FLOWARR.

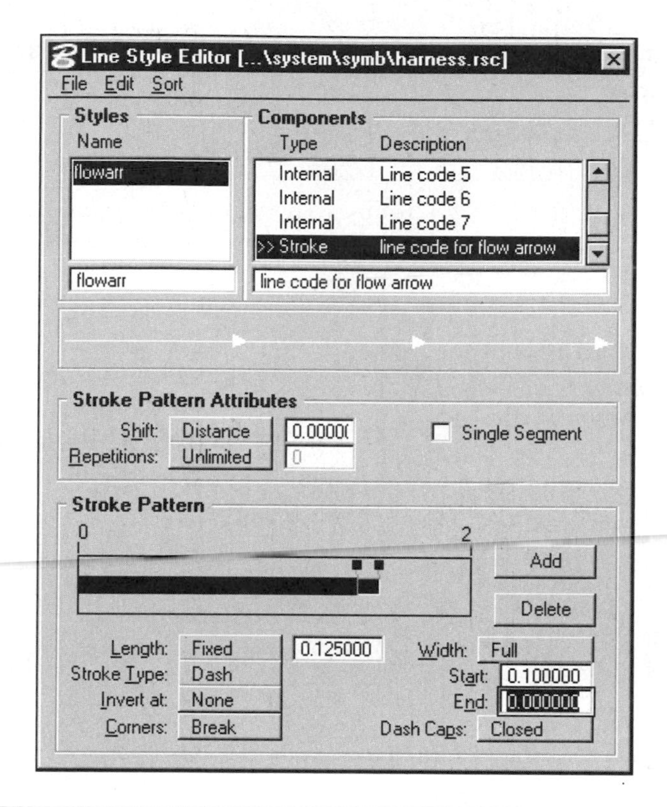

Figure 16–31 Displaying the new line type in the settings box

WORKSPACE

A *workspace* is a customized drafting environment that permits the user to set up MicroStation for specific purposes. You can set up as many workspaces as you need. A workspace consists of "components" and "configuration files" for both the user and the project.

MicroStation comes with a set of workspaces for various disciplines. For example, MicroStation is delivered with a sample "civil" workspace. When the civil workspace is active, the files and tools you need to perform civil engineering design and drafting are available by default.

When the civil workspace is active, tools and tool boxes that are unrelated to that discipline are removed from the interface so that they are out of the way.

Setting Up the Active Workspace

You can select the default workspace from the MicroStation Manager. At the bottom of the MicroStation Manager dialog box (see Figure 16–32), four option menus are provided that allow you to change the major components of the workspace before opening a design file.

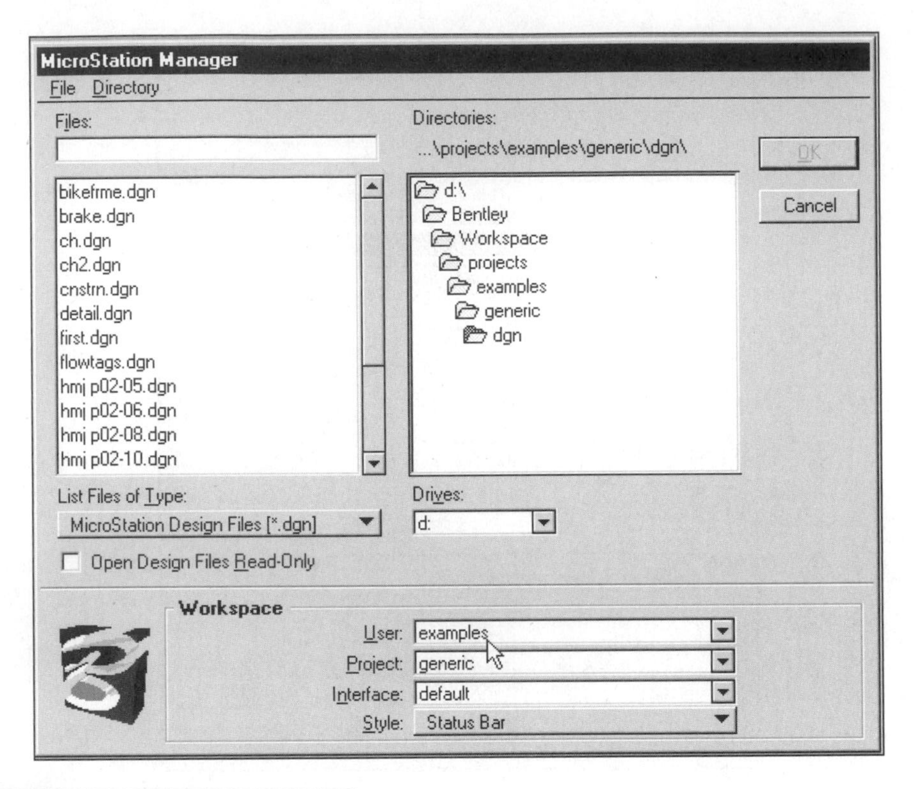

Figure 16–32 MicroStation Manager

User Option Menu

The User option menu in the Workspace section of the dialog box allows you to select the default workspace from the available workspaces. Selecting a workspace from the list reconfigures MicroStation to use that workspace's components. Selecting a workspace also resets the search path to a corresponding folder for loading design files. In addition, MicroStation sets the associated project and user interface.

When MicroStation is started with any workspace as the default workspace, a preference file is created for that workspace, unless one already exists. The settings in the preferences are set either to default settings or to AutoCAD Transition settings, depending on the active user interface component.

To create a new workspace, invoke the New... option from the User option menu. MicroStation displays the Create User Configuration File dialog box, similar to Figure 16–33.

Key-in the name of the new workspace in the Name: edit field and click the OK button. MicroStation displays another dialog box, similar to Figure 16–34. Key-in the description (optional) in the Description: edit field. If necessary, change the components by clicking the appropriate Select...

button for Project: and/or User Interface. Click the OK button to close the dialog box. MicroStation sets the newly created workspace as the default workspace. A workspace can contain only one project and one interface. These components are attached to a workspace. So, to use two different projects with the same interface or two interfaces for one project, you need to make additional workspaces.

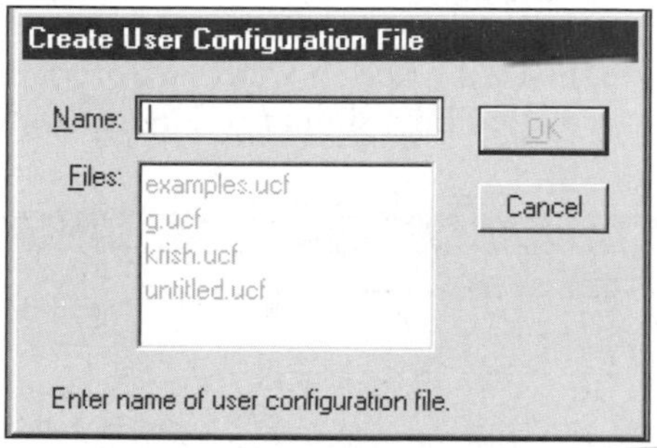

Figure 16–33 Create User Configuration File dialog box

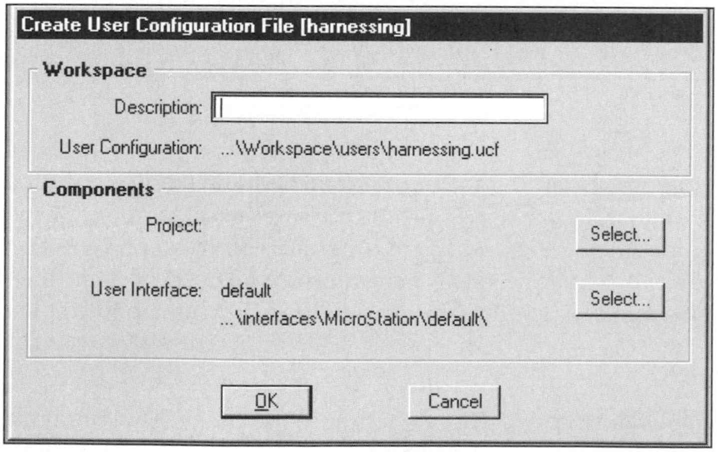

Figure 16–34 Create Workspace (name) dialog box

Project Option Menu

The selection of the project sets the location and names of data files associated with a specific design project. If necessary, you can change the selection of the project from the Project option menu.

You can also create a new Project from the Project: option menu similar to creating a workspace.

User Interface Option Menu

The selection of the user interface sets a specific look and feel of MicroStation's tools and general on-screen operation. If necessary, you can change the selection of the interface from the Interface: option menu. MicroStation comes with discipline-specific interfaces: civil engineering, architecture, mechanical engineering, drafting, and mapping, and it also has interfaces for previous versions (V. 4 and V. 5) and AutoCAD users.

To create a new interface, invoke the New... option from the Interface: option menu. MicroStation displays the Create User Interface dialog box, similar to Figure 16–35.

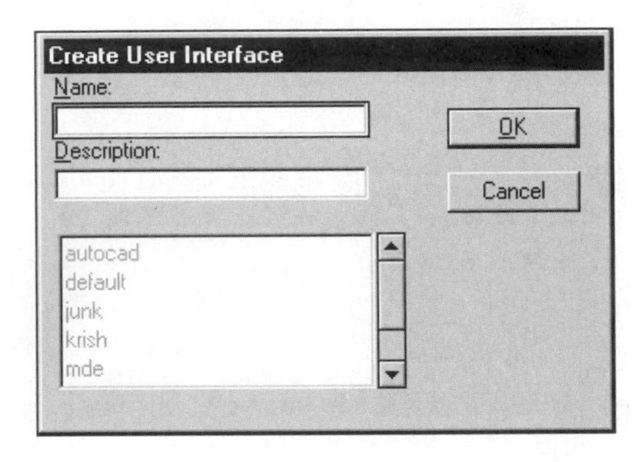

Figure 16–35 Create User Interface dialog box

Key-in the name of the new user interface in the Name: edit field and a description in the Description: edit field (optional). Click the OK button to close the dialog box. MicroStation creates an interface directory under the Bentley\Workspace\interfaces directory and sets the newly created user interface as the default user interface. The new interface takes the default interface as its starting point. Any changes you make while using this new interface is written only to the new interface.

Style Option Menu

The Style: option menu allows you to select whether to use the older Command Window (V. 4 and V. 5) method of communicating with MicroStation or the Status bar. The default selection is the Status bar. Whichever style you select, MicroStation will remember it from session to session.

Setting User Preferences

Preferences are settings that control the way MicroStation operates and the way its tool frames and tool boxes appear on the screen. For example, they affect how MicroStation uses memory on a user's system, how windows are displayed, and how reference files are attached by default. You can change the settings to suit your needs. The user preferences are saved under the same name as the workspace, with the file extension .UCF.

To set the user preferences, invoke the Preferences:

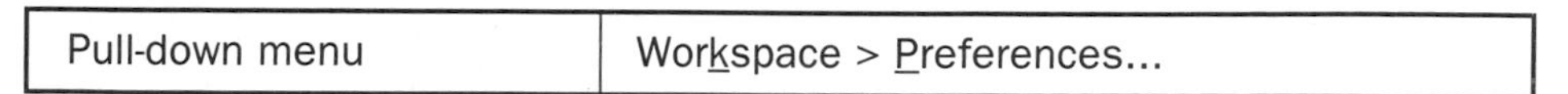

Pull-down menu	Workspace > Preferences...

MicroStation displays the Preferences dialog box, similar to Figure 16–36. MicroStation displays the name of the file under which preference settings are saved as part of the title bar.

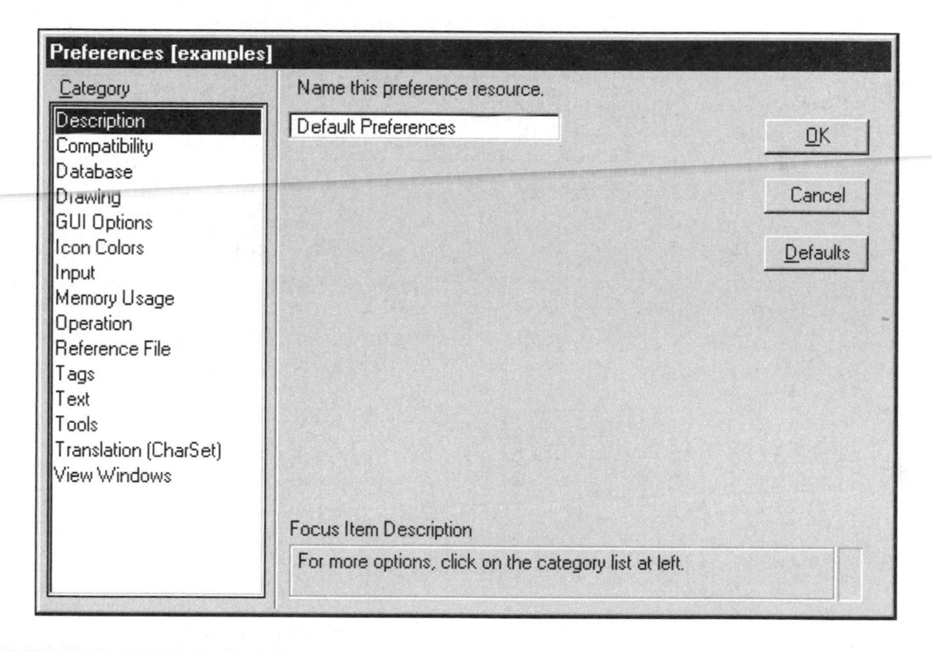

Figure 16–36 Preferences dialog box

User Preferences are divided into categories. The Category list box lists all the available categories. Selecting a category causes the appropriate controls to be displayed to the right of the category list. Each category controls a specific aspect of MicroStation's appearance or operation. Table 16–2 describes briefly all the settings available in the Preferences dialog box.

Click the OK button to save the settings and close the Preferences dialog box.

Table 16–2 Preferences Settings

CATEGORY	PREFERENCE SETTING	DESCRIPTION	DEFAULT
Compatibility	Compatibility: Option menu	Sets the MicroStation version number	5.0+
	Dimensions toggle button	Set to ON to place dimensions elements	ON
	Multi-lines toggle button	Set to ON to place multi-lines	ON
	Shared Cells toggle button	Set to ON to place shared cells	OFF
	Associative Patterning toggle button	Set to ON to enable associative patterning	ON
Database	Block Database Undeletes toggle button	Set to ON, it is impossible to undo the deletion of an element that has a database linkage	OFF
	Use Single AE/MSFORMS Tables toggle button	Set to ON, single AE and MSFORMS tables are maintained	OFF
	Use Database Msline Cache toggle button	Set to ON, a cache of database linkages is maintained	OFF
Drawing	Max. Grid Pts/View edit field	Sets the maximum number of displayable grid points in a view, counted horizontally	90
	Max. Grid Refs/View edit field	Sets the maximum number of displayable grid references (crosses) in a view, counted horizontally	40
	Line Weights...	Sets the display width (in pixels) for each of the 32 line weights	1:1 (2:1 in DOS version)

Table 16–2 Preferences Settings (continued)

CATEGORY	PREFERENCE SETTING	DESCRIPTION	DEFAULT
GUI Options	Dialog Boxes: option menu	Sets the graphical user interface standard that MicroStation emulates or adheres to	Window
	Open Two Application Windows (Windows only) toggle button	Set to ON, MicroStation application windows are opened for use on a two-screen system	OFF
	Dialog Font: option menu	Sets the text size, in points, in dialog and settings boxes	10 pt (DOS); 12 pt (Windows)
	Border Font: option menu	Sets the text size, in points, in window borders	10 pt (DOS); 12 pt (Windows)
Icon Colors	Icon Colors: option menu	Allows you to select the colors for the icons; selection includes Default, Gray Scale, and Custom. Custom selection provides selection of custom color selection	Default
Input	Start in Parse All Mode toggle button	Set to ON, the parsing function is enabled as MicroStation starts	ON
	Disable Drag Operations	Set to ON, MicroStation disregards Data button-up operation when the pointer is in view	OFF
	Ctrl + **Z** to Exit	Set to ON, pressing Ctrl + **Z** exits MicroStation	OFF
Memory Usage	Max Element Cache: edit field	Sets the maximum amount of memory, in KB, reserved to store elements	8000

Table 16–2 Preferences Settings (continued)

CATEGORY	PREFERENCE SETTING	DESCRIPTION	DEFAULT
Memory Usage (continued)	Resource Cache: edit field	Sets the amount of memory, in KB, reserved for resources read from MicroStation resources files and application resource files	24
	Undo Buffer: edit field	Sets the amount of memory, in KB, reserved for recording the possible negation using Undo	256
	Font Cache: edit field	Sets the maximum size, in KB, of the section of memory reserved for data used to display text elements	30
	Conserve Memory toggle button	Set to ON, MicroStation sacrifices speed in some functions in order to use less memory	OFF
	Disable OLE Automation toggle button	Set to ON, OLE Automation Server capability is disabled unless MicroStation is started by an OLE Automation request	OFF
Operation	Locate Tolerance: edit field	Sets the size of the searched area around the pointer for selecting an element	10
	Point Size: option menu	Sets the size of the crosshair pointer	Normal
	Pointer Type: option menu	Controls the alignment of the pointer's crosshairs	Orthogonal
	Display Levels: option menu	Controls whether level Names or Numbers are shown in screen controls	Names

Table 16–2 Preferences Settings (continued)

CATEGORY	PREFERENCE SETTING	DESCRIPTION	DEFAULT
Operation (continued)	Immediately Save Design toggle button	Set to ON, MicroStation does not maintain a backup file and does not have a Save item on the menu File	ON
	Save Settings on Exit toggle button	Set to ON, the File menu's Save Settings item is disabled	OFF
	Compress Design on Exit toggle button	Set to ON, deleted elements are automatically removed from the active design file upon closing	OFF
	Enter into Untitled Design toggle button	Set to ON, when you start MicroStation it automatically creates and opens a design file named "untitled.dgn"	OFF
	Reset Aborts Fence Operations toggle button	Set to ON, resetting during a fence manipulation halts the operation	ON
	Level Lock Applies for Fence Operations toggle button	Set to OFF, fence contents manipulations ignore the Level Lock setting	ON
	Use Semaphore File for Locking toggle button	Set to ON, enables users to specify that a "semaphore" file be used to control write access to files	OFF
Reference File	Locate On When Attached toggle button	Set to ON, the capability to identify elements in a particular reference file is turned ON when the reference file is attached	ON

Table 16–2 Preferences Settings (continued)

CATEGORY	PREFERENCE SETTING	DESCRIPTION	DEFAULT
Reference File (continued)	Snap On When Attached toggle button	Set to ON, the capability to snap to elements in a particular reference file is turned ON when the reference file is attached	ON
	Use Color Table toggle button	Set to OFF, MicroStation ignores any color table attached to a reference file for display purposes	ON
	Use Level Names toggle button	Set to ON, reference file level names are displayed as part of the level information	ON
	Cache When Display Off toggle button	Set to OFF, memory caching of reference files that are not displayed is disabled	OFF
	Reload When Changing Files toggle button	Set to OFF, cached reference files are kept in memory when one design file is closed and another is opened	OFF
	Save Settings to Save Changes toggle button	Set to OFF, the results of reference file manipulations are immediately permanent	OFF
	Ignore Update Sequence toggle button	Set to ON, the Update Sequence menu item is disabled	OFF
	Store Full Path When Attached toggle button	Set to ON, the Save Full Path check box is ON by default	OFF
	Update Self Attachments toggle button	Set to ON, you can modify self-attached reference file elements, and the changes to the master file will be updated in the self-attached reference files	ON

Table 16–2 Preferences Settings (continued)

CATEGORY	PREFERENCE SETTING	DESCRIPTION	DEFAULT
Reference File (continued)	Max Reference: Files edit field	Sets the maximum number of reference files that can be attached to the active design file	255
	Nest Depth: edit field	Sets the number of levels of nested attachments	0
Tags	Prompt on Duplicate Tag Sets toggle button	Set to ON, MicroStation displays an alert box for duplication	ON
	Use Design File Tag Sets By Default toggle button	Set to ON, tag sets in the design file cannot be replaced by tag sets of the same names in cell libraries from which cells are placed	OFF
	Place Tags in Some Graphic Group toggle button	Set to ON, when a tag set is attached to an element all tags in the set become members of the same graphic group	ON
Text	Display Text with Lines Styles toggle button	Set to OFF, displayed with standard solid line style	OFF
	Fit Text by Inserting Space toggle button	Set to OFF, places fitted text by enlarging or shrinking the characters of text	OFF
	Fixed-width Character Spacing toggle button	Set to OFF, spacing between characters is measured from the end of one character to the beginning of the next character	OFF
	Preserve Text Nodes toggle button	Set to ON, text placed as a text node will remain a text node, even if edited down to one line	OFF

Table 16–2 Preferences Settings (continued)

CATEGORY	PREFERENCE SETTING	DESCRIPTION	DEFAULT
Text (continued)	Justify Enter Data Fields Like IGDS toggle button	Set to OFF, the odd space in a center-justified enter data field containing an odd number of extra blank spaces is positioned at the beginning of the enter data field	OFF
	Edit Character: edit field	Sets the text character that denotes each character in an enter data field	—
	Smallest Text: edit field	Sets the size, in pixels, above which text is drawn	4
	Underline Spacing (%) edit field	Sets the distance, as a percentage of the text height, between the baseline and underlining	20
	Degree Display Character edit field	Sets the ASCII character used to display the degree symbol	176
	Text Editor Style option menu	Sets the type of text editing interface	Dialog Box
Tools	Single Click: option menu	Controls how tools are selected with a single click of the Data button	Locked
	Default Tool: option menu	Sets the tool that is automatically selected upon completion of a one-time function	Element selection
	Highlight: option menu	Sets the color with which tools are highlighted to indicate locked selection	Gray
	Layout: option menu	Affects the size of tool boxes	Regular
	Tool Size: option menu	Sets the size of tool icons	Small

Table 16–2 Preferences Settings (continued)

CATEGORY	PREFERENCE SETTING	DESCRIPTION	DEFAULT
Tools (continued)	View Pop-ups: option menu	Sets the method used to open the pop-up menu	Shift -Reset
	Auto-Focus Tool Settings Window toggle button	Set to ON, the input focus automatically moves to the Tool Settings window when a tool with settings is selected	ON
	All Pop-Downs in Tool Setting Window toggle button	Set to ON, Version 5 palettes do not have pop-downs	ON
	Auto-Open Tool Settings Window toggle button	Set to ON, the Tool Settings window opens automatically when MicroStation starts	ON
Translation (CharSet)	Character Translation Table:	Sets the character translation table, which converts external ASCII characters to the form of ASCII used internally by MicroStation	Default
View Windows	Scroll Bars on View Windows toggle button	Set to ON, view windows are displayed with borders, including scroll bars and view control bars	ON
	Black Background → White toggle button	Set to ON, the view background color, if set to black, is displayed in white	OFF
	Tile Like IGDS toggle button	Set to OFF, four views are tiled in the same manner as in IGDS	OFF
	Use Backing Store toggle button	Set to ON, maintains an off-screen copy of each view window so obscured areas can be refreshed instantly when windows are rearranged	ON

Table 16–2 Preferences Settings (continued)

CATEGORY	PREFERENCE SETTING	DESCRIPTION	DEFAULT
View Windows (continued)	Gamma Correction edit field	Affects the brightness of rendered images	1.0
	Direct Drawing (Windows only) option menu	Affects the techniques used to display bitmap in view windows	OFF

Working with Configuration Variables

MicroStation allows you to make necessary changes to the user configuration variables and their respective directory paths from the Configuration Variables dialog box. Invoke the Configuration Variables dialog box:

Pull-down menu	Workspace > Configuration...

MicroStation displays the Configuration dialog box similar to the Figure 16–37.

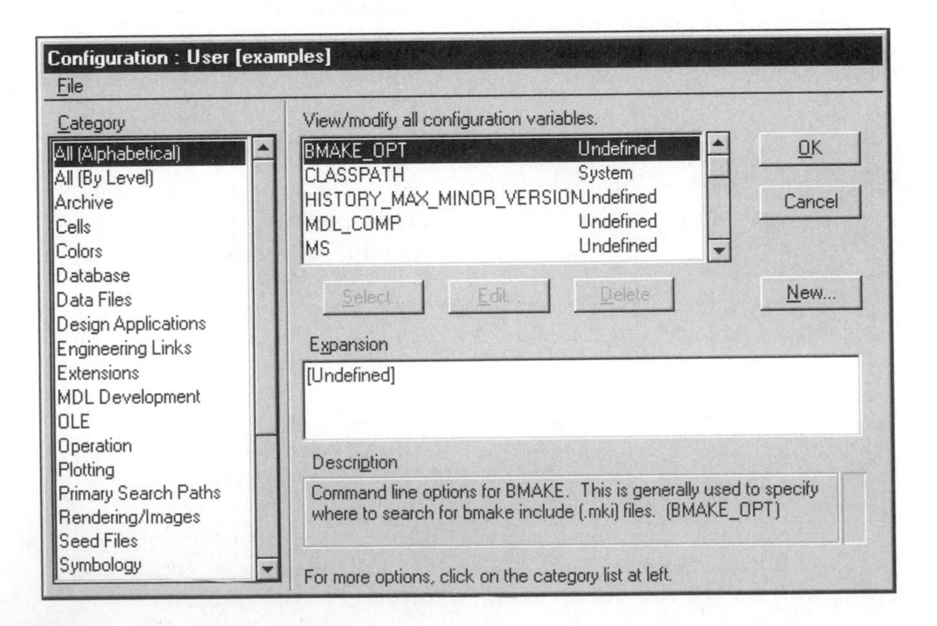

Figure 16–37 Configuration dialog box

The Category list box lists all the available categories. When you select a category, MicroStation lists the controls for settings of that category on the right side of the dialog box. In the Expansion field, the expansion of the variable is shown. In the Description field, a description of the variable and its name are shown.

Use the controls to modify the definition. The procedure varies for the different types of configuration variables.

If a configuration variable needs a path specification, then MicroStation displays a Select Path dialog box, similar to Figure 16–38. Select the appropriate directory and click the Add button. MicroStation adds the path for the selected directory in the Directory List box. You can add multiple directories to the list. Click the Done button to close the dialog box.

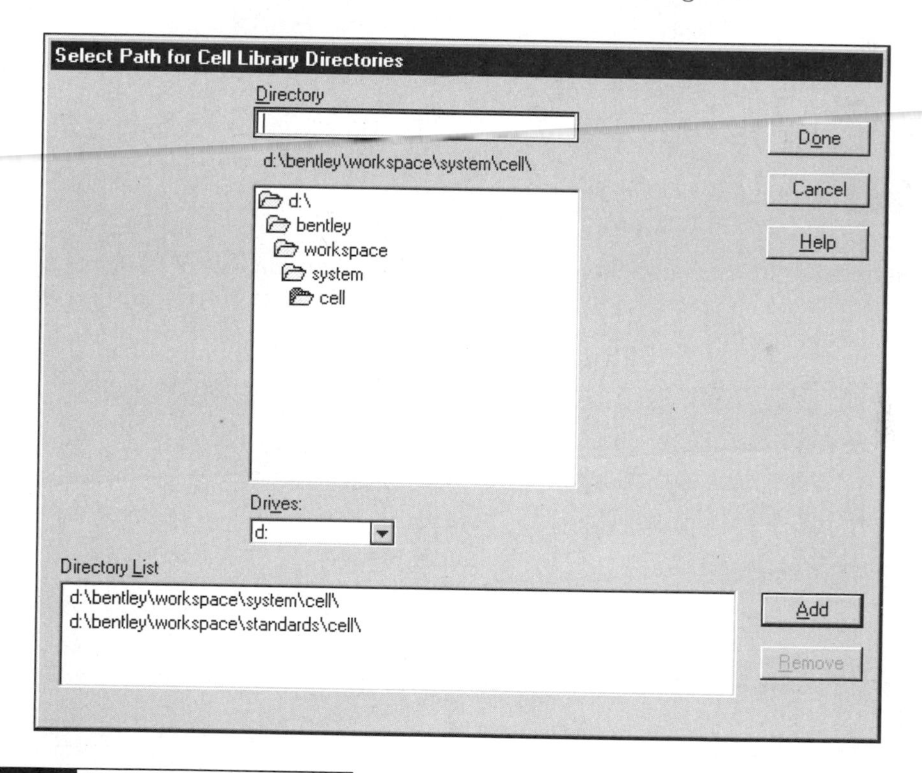

Figure 16–38 Select Path dialog box

If a configuration variable needs a directory specification, then MicroStation displays a Select Directory dialog box, similar to Figure 16–39. Select the appropriate directory and click the OK button to close the dialog box.

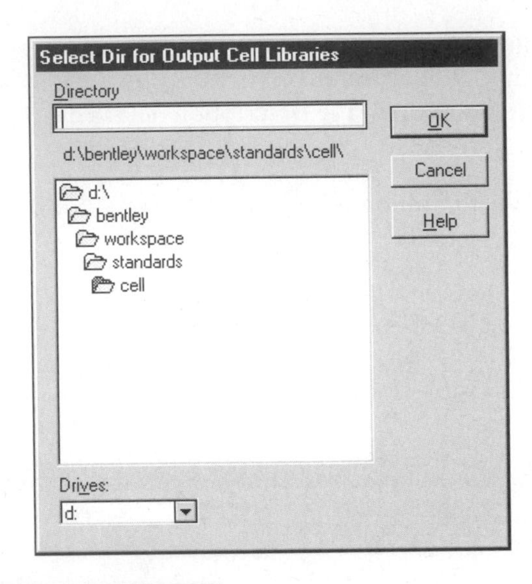

Figure 15–39 Select Directory dialog box

If a configuration variable needs a file name specification, then MicroStation displays a Select File List dialog box, similar to Figure 16–40. Select the desired file from the appropriate directory and click the OK button to close the dialog box.

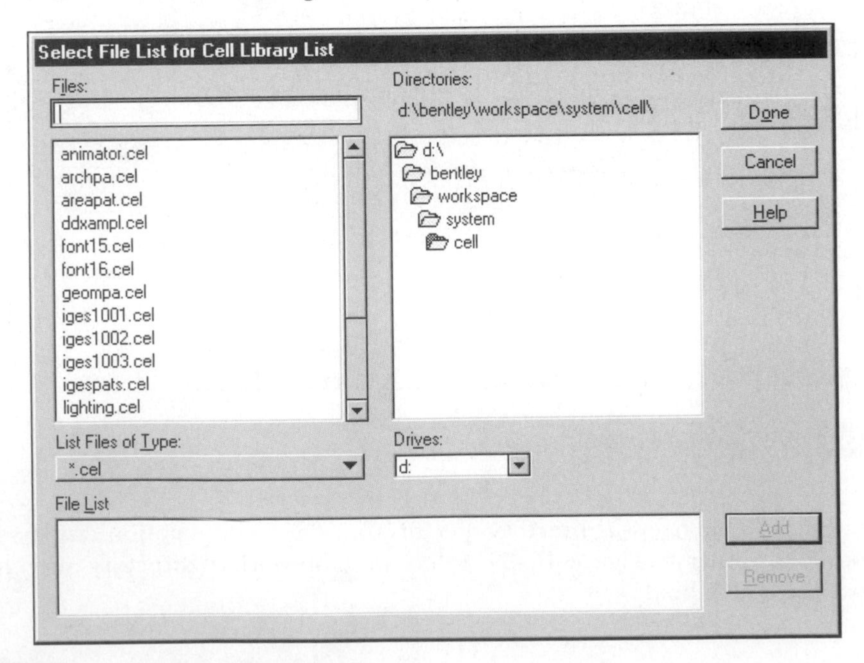

Figure 16–40 Select File List dialog box

If a configuration variable needs a keyword configuration variable, then MicroStation displays an Edit Configuration Variable dialog box, similar to Figure 16–41. Key-in the desired variable in the New Value: edit field. The keyword is also shown in the Expansion field. Click the OK button to close the dialog box.

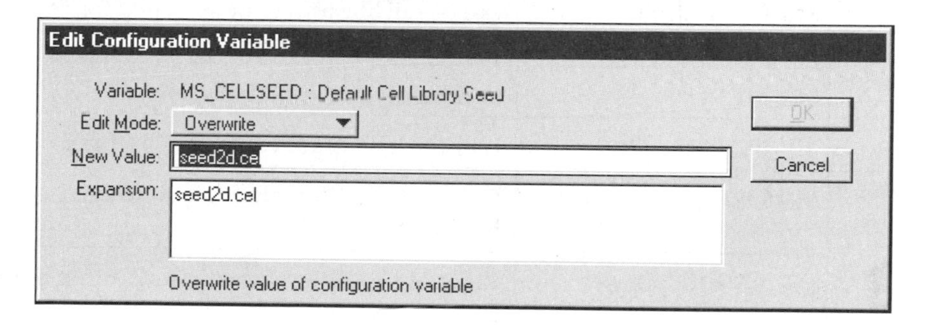

Figure 16–41 Edit Configuration Variable dialog box

Table 16–3 lists all the available configuration variables by category.

Click the OK button to save the configuration variable settings and close the dialog box.

Table 16–3 Available Configuration Variables

CATEGORY	VARIABLE	DESCRIPTION
B/RAS	BRAS	Directory containing the "bras.rsc" resource file
	BRAS_RASTER	Default directory for the Load and Preview dialog boxes when no other raster file is loaded
	BRAS_FILTER	Filter Default filter (e.g., "*.*") for Load and Preview dialog boxes
	BRAS_MAX_SCAN	The maximum number of scan lines a raster file can have
	BRAS_MAX_X	Maximum scan line length (pixels) a raster file can have
	BRAS_MAX_FG_ RUNS	Maximum number of foreground run lengths per scan line in a raster file
	BRAS_BKLOAD_ TICKS	Background Load TicksTime (in MicroStation timer ticks) between background loading time slices
	BRAS_BKLOAD_ SCANS	Number of raster scan lines loaded for each time slice of BRAS fast loading

Table 16–3 Available Configuration Variables (continued)

CATEGORY	VARIABLE	DESCRIPTION
Cells	MS_CELL	DirectorSearch path(s) for cell libraries
	MS_CELLLIST	List of cell libraries to be searched for cells not found in the current library
	MS_MENU	Cell library file containing menu cells
	MS_TUTLIB	LibraryCell library containing tutorial cells
Colors	MS_DEFCTBL	Default Color TableDefault color table if design file has none
	MS_RMENCTBL	Right Menu Color TableDefault menu colors (dialog boxes, borders, etc.) for right screen—specifies a color table (.tbl) file
	MS_LMENCTBL	Color TableDefault menu colors (dialog boxes, borders, etc.) for left screen—specifies a color table (.tbl)file
Database	MS_DBASE	Search path(s) for database files
	MS_SERVER	MDL application to load the database interface software
	MS_DBEXT	The database interface "server" application
	MS_LINKTYPE	User data linkage types recognized by the database interface software (see MS_LINKTYPE)
	MS_TAGREPORTS	Output directory for tag reports
	MS_TAGTEMPLATES	Directory containing tag report templates
Data Files	MS_SETTINGS	Open settings file
	MS_SETTINGSDIR	Directory containing settings files
	MS_LEVELNAMES	Directory containing level structure files
	MS_GLOSSARY	List of files for use with the Glossary settings box (see Utilities>Glossary)
Design Applications	MS_DGNAPPS	List of MDL applications to load automatically when a design file is opened
MDL Development	MS_DBGSOURCE	Location of source code for MDL applications (used by MDL debugger)

Table 16–3 Available Configuration Variables (continued)

CATEGORY	VARIABLE	DESCRIPTION
MDL Development (continued)	MS_MDLTRACE	Additional debugging print statements when debugging MDL applications
	MS_DBGOUT	Output of MDL debugger
	MS_DEBUGFAULT	If set, automatically invokes the debugger when a fault is detected while an MDL application is active
	MS_DEBUG	If set to an integer with bit 1 ON, does not time out
Operation	MS_FKEYMNU	Open function key menu file
	MS_SYSTEM	If set, MicroStation allows the user to escape to the operating system
	MS_APPMEN	Location of application and sidebar menus
	MS_TRAP	Set to "NONE," "MDL," or "ALL" (default)
Plotting	MS_PLTFILES	Directory for plotting output files
	MS_PLTR	Name of plotter configuration file
Primary Search	MS_DEF	Search path(s) for design files
	MS_DESIGNFILTER	File filter for opening and creating design files
	MS_RFDIR	Search path(s) for reference files
	MS_MDLAPPS	Search path(s) for MDL applications displayed in the MDL dialog box
	MS_MDL	Search path(s) for MDL applications or external programs loaded by MDL applications
	MS_RSRCPATH	Search path(s) for resource files loaded by MDL applications
Rendering	MS_MATERIAL	Search path(s) for material palettes
	MS_PATTERN	Search path(s) for pattern maps
	MS_BUMP	Search path(s) for bump maps
	MS_IMAGE	Search path(s) for images
	MS_IMAGEOUT	Directory in which created image files are stored
	MS_SHADOWMAP	Directory where shadow maps will be read from and written to

Table 16–3 Available Configuration Variables (continued)

CATEGORY	VARIABLE	DESCRIPTION
Seed Files	MS_SEEDFILES	Search path(s) for all seed files
	MS_DESIGNSEED	Default seed file
	MS_CELLSEED	Default seed cell library
	MS_SHEETSEED	Seed sheet file drawing composition, DWG import, and IGES import
	MS_TRANSEED	Default seed file for DWG, CGM, and IGES translations
Symbology	MS_SYMBRSRC	List of symbology resource files—last one in list has highest priority
System Environment	MS	The MicroStation root installation directory used by MDL sample "make" files
	MS_CONFIG	Main MicroStation configuration file—sets up all configuration variables
	MS_EDG	Directories used by EDG (not MicroStation)
	MDL_COMP	Command text string to be inserted at the beginning of the command line by the MDL compiler (used to specify where to search for included files)
	RSC_COMP	Text string to be inserted at the beginning of the command line by the resource compiler (used to specify where to search for included files)
	BMAKE_OPT	Command line options for BMAKE (used to search for bmake include (.mki) files)
	MS_ DEBUGMDLHEAP	If set (to the base name of an MDL application or "ALL"), use extended malloc for debugging
Temp and	MS_BACKUP	Default directory for backup files
Backup Files	MS_TMP	Directory for temporary files created and deleted by MicroStation
	MS_SCR	Directory for scratch files created by MicroStation
Translation	MS_CGMIN	CGM Input Directory for CGM translations
	MS_CGMOUT	Output directory for CGM translations
	MS_CGMLOG	Output directory for CGM log files

Table 16–3 Available Configuration Variables (continued)

CATEGORY	VARIABLE	DESCRIPTION
Translation (continued)	MS_CGMTABLES	Directory containing the CGM translation tables
	MS_CGMINSET	Settings file for the CGMIN application
	MS_CGMOUTSET	Settings file for the CGMOUT application
	MS_DWGIN	Input directory for DWG translations
	MS_DWGOUT	Output directory for DWG translations
	MS_DWGLOG	Output directory for DWG log files
	MS_DWGTABLES	Directory containing the DWG translation tables
	MS_DWGINSET	Settings file for the DWGIN application
	MS_DWGOUTSET	Settings file for the DWGOUT application
	MS_IGESIN	Input directory for IGES translations
	MS_IGESOUT	Output directory for IGES translations
	MS_IGESLOG	Output directory for IGES log files
	MS_IGESINSET	Settings file for IGES import
	MS_IGESOUTSET	Settings file for IGES export
User Commands	MS_UCM	Search path(s) for user commands
	MS_INIT	Name of user command to be executed at startup
	MS_EXITUC	Name of user command to be executed at exit
	MS_NEWFILE	Name of user command to be executed when a newfile is opened
	MS_APP	Apps from "TSK" state—search path(s) of applications started from "TSK" statements in user commands
Uncategorized	MS_BANNER	Text file to be displayed in MicroStation's startup banner dialog
	MS_CMDWINDRSC	Command Window resource file—default is used if undefined
	MS_CODESET	AMDL application for handling multi-byte character sets
	MS_DATA	Directory for data files created or used by MicroStation

Table 16–3 Available Configuration Variables (continued)

CATEGORY	VARIABLE	DESCRIPTION
Uncategorized (continued)	MS_DEFCHARTRAN	Default character translation table
	MS_DEMOONLY	If set, MicroStation runs in demonstration mode only
	MS_DGNOUT	Directory containing design files created as a result of "on-the-fly" translation from other file formats
	MS_EXE	Directory containing the MicroStation executable program
	MS_GUIHAND	Identifies auxiliary handlers
	MS_HELPPATH	Path to help files
	MS_INITAPPS	List of initial startup MDL applications
	MS_RIGHTLOGICKB	Right to Left Character—if set, type from right to left (used for foreign language support)
	MS_RSRC	Main MicroStation resource file; typically set to "ustation.rsc"
	MS_RSRVCLRS	Number of colors MicroStation does not use
	MS_TUT_UCMS	Directory containing user commands that drive tutorials
	MS_UNDO	If set, overrides the user preference Undo Buffer
	MS_USERLICENSE	File containing MicroStation license information
	MS_WINDOWMGR	Tells MicroStation which window manager is in use ("MOTIF" or "OPENLOOK"); MicroStation can usually infer this
	WRK_DD_IGDS	Directory for IGDS user commands

Customizing the User Interface

MicroStation allows you to customize any or all of the parts of the active workspace user interface:

- Tool frames and tool boxes
- View border tools
- Pull-down menu

Customization of MicroStation's user interface is stored in user interface modification files. Whenever you create a new user interface, MicroStation creates a folder under the <disk>:\Bentley\Workspace\interfaces\MicroStation\ with the given name of the new user interface. By default, MicroStation places the default user interface file (ustn.r01) in the new folder. If part of the interface is modified, then the file name of the user interface modification file is "ustn" with the file name extension "m" followed by a number from 01 to 99. Each time an interface modification file is saved with the same file name, the number in its suffix is incremented. Therefore, the files might be named "ustn.m01," "ustn.m02," and so on. If the tool box or dialog box modified is provided by an MDL (MicroStation Development Language) application, the file name of the user interface modification file is the file name of the MDL application.

Customizing Tool Boxes

Follow these steps to modify or create a new/modify tool box. Open the Customize settings box:

Pull-down menu	Workspace > Customize

MicroStation displays a Customize settings box similar to Figure 16–42.

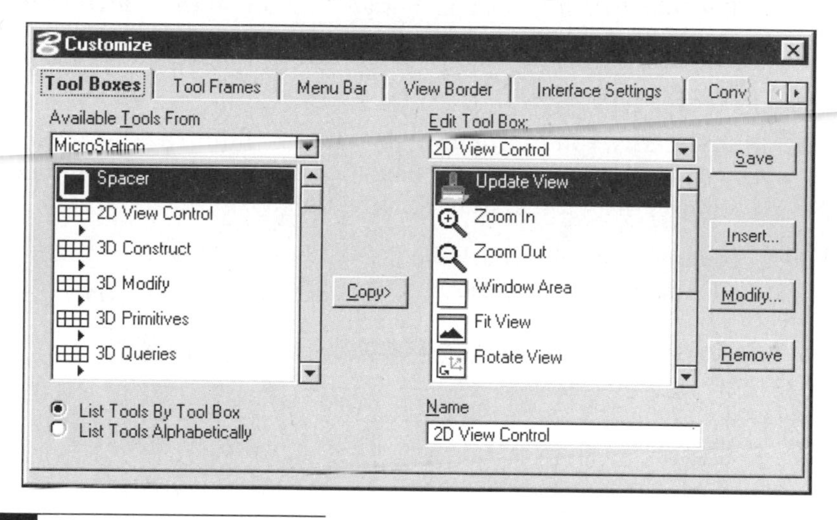

Figure 16–42 Customize settings box

Follow these steps to modify or create a new tool box:

STEP 1: Select the name of the interface to list the available tools from the Available Tools From list box. MicroStation lists the available tool boxes from the selected interface (left side of the settings box) and lists the available tools by default from the 2D View Control in the Edit Tool Box: list box (on the right side of the settings box).

STEP 2: Select one of the available tool boxes to customize from the Edit Tool Box: option menu.

or

Select Create Tool Box... from the Edit Tool Box option menu to create a new tool box. MicroStation displays Create Tool Box dialog box, as shown in Figure 16–43. Key-in the name of the new tool box and click the OK button. MicroStation adds the new tool box to the Edit Tool Box: option menu.

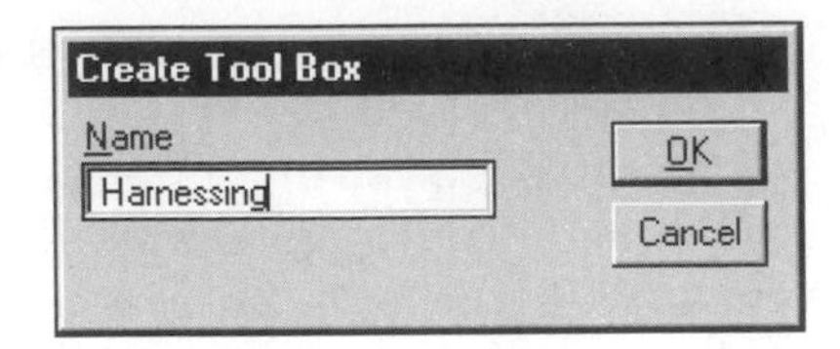

Figure 16–43 Create Tool Box dialog box

STEP 3: From the available tool boxes listed on the Available Tools list box (on the left side of the settings box), double-click the Tool Box name, and MicroStation lists the available tools. Select the tool you want to insert in the Tool Box, and drag and drop it at the appropriate location in the Edit Tool Box: list box.

STEP 4: If necessary, you can rearrange the tools by dragging and dropping in the Edit Tool Box: list box.

STEP 5: If necessary, you can remove the tool from the Tool list box. First select the tool to remove from the list box, then click the Remove button. MicroStation removes the selected tool from the tool box.

STEP 6: To insert a new tool, click the Insert... button. MicroStation displays the Insert Tool dialog box, as shown in Figure 16–44.

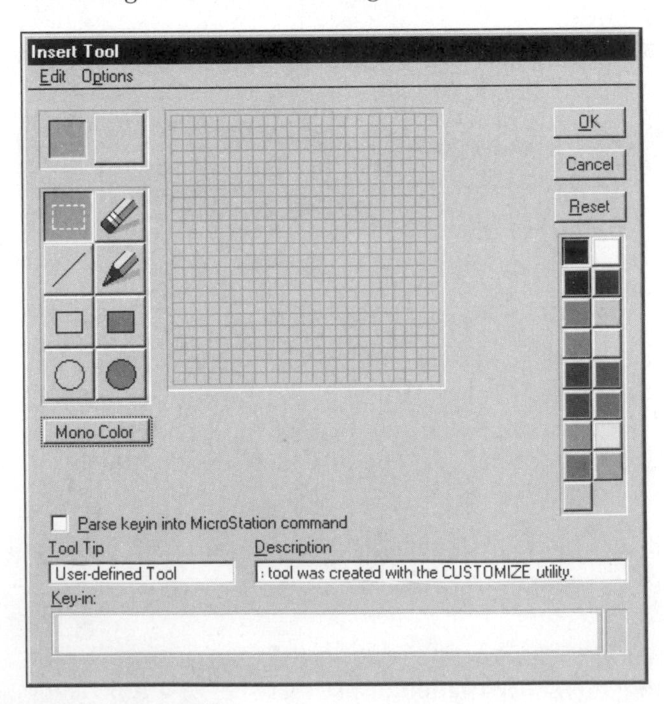

Figure 16–44 Insert Tool dialog box

Use the drawing tools to create the icon for the new tool. Key-in the tool tip text and description for the new tool in the Tool Tip and Description edit fields. Key-in the action strings to be associated with the tool in the Key-in: edit field. If a multiple key-in action string is specified, the key-ins must be separated by semi-colons (;). For example, the following key-in sets the color to red, the line weight to 3, the active level to 3, and invokes the Place Line tool:

CO=RED;WT=3;LV=3;PLACE LINE

Click the OK button to accept the changes and close the dialog box.

STEP 7: To modify a tool, first select the tool to modify from the list box, then click the Modify... button. MicroStation displays a dialog box, as shown in Figure 16–45.

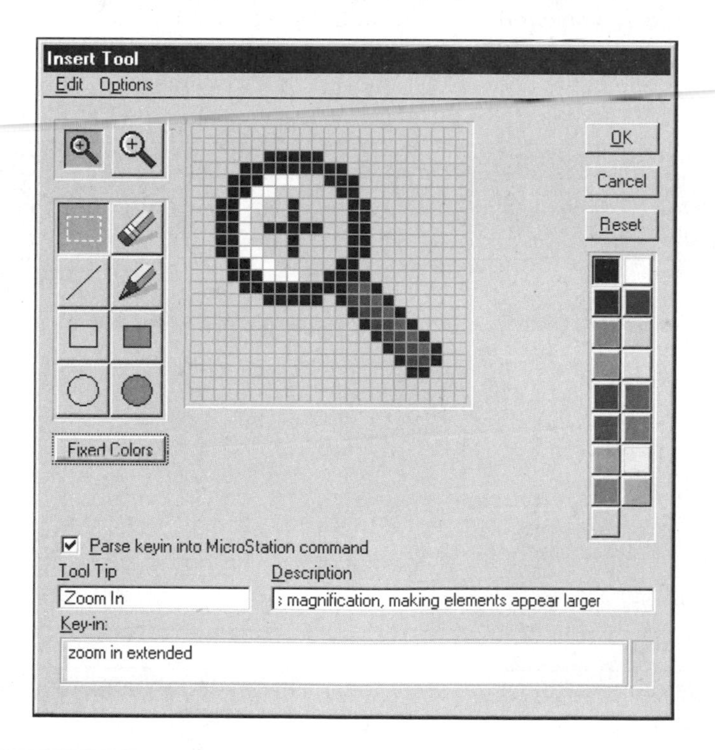

Figure 16–45 Modify tool dialog box

If necessary, use the drawing tools to modify the icon. Make the necessary modifications to the tool tip text, description, and key-ins. Click the OK button to accept the changes and close the dialog box.

STEP 8: Click the Save button to save the Modification/Addition to Tool Boxes. MicroStation displays the customized version of the Tool Box(es).

STEP 9: If no further changes have to be made, close the Customize settings box.

Customizing Tool Frames

The Customize settings box enables you to customize tool frames. Open the Customize settings box:

Pull-down menu	Workspace > Customize

Follow these steps to modify or create a new tool frame:

STEP 1: Select Tool Frames tab in the Customize settings box, and MicroStation lists the available options, similar to Figure 16–46. Select the name of the interface to list the available tools from the Available Tools From list box. MicroStation lists the available tool boxes from the selected interface (on the left side of the settings box), and lists the available tool boxes by default from the 3D Main tool frame in the Edit Tool Frame: list box (on the right side of the settings box).

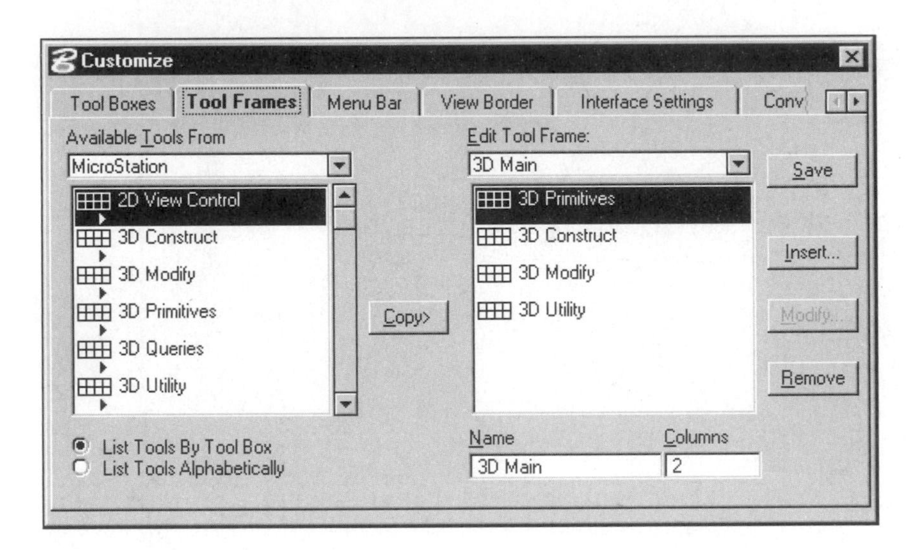

Figure 16–46 Customize settings box, with the Tool Frames tab selected

STEP 2: Select one of the eight available tool frames to customize from the Edit Tool Frame option menu.

or

Select Create Tool Frame… from the Edit Tool Frame: option menu to create a new tool frame. MicroStation displays the Create Tool Frame dialog box, as shown in Figure 16–47. Key-in the name of the new tool frame and click the OK button. MicroStation adds the new tool frame to the Edit Tool Frame: option menu.

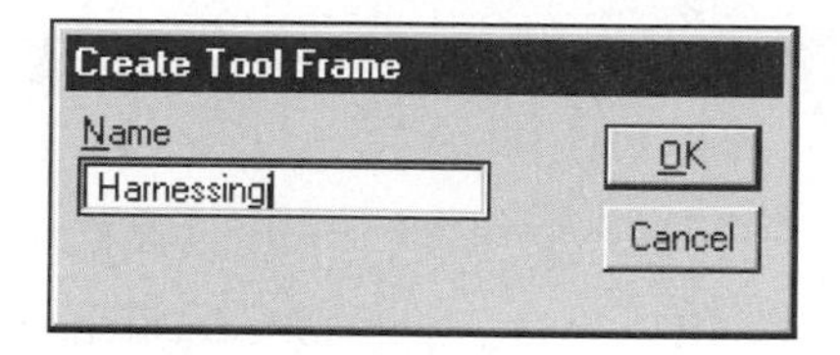

Figure 16–47 Create Tool Frame dialog box

STEP 3: From the available tool boxes listed on the Available Tools list box (on the left side of the settings box), select the tool box you want to insert into the Tool Frame, and drag and drop it to the appropriate location in the Edit Tool Frame: list box.

STEP 4: If necessary, you can rearrange the tool boxes by dragging and dropping in the Edit Tool Frame: list box.

STEP 5: If necessary, you can remove the tool box from the Tool Frame list box. First select the tool box to remove from the list box, then click the Remove button. MicroStation removes the selected tool box from the tool frame.

STEP 6: Click the Save button to save the Modification/Addition to Tool Frames. MicroStation displays the customized version of the Tool Frame(s).

STEP 7: If no further changes need to be made, close the Customize settings box.

Customizing the Pull-down (also called Menu Bar) Menu

The Customize settings box enables you to customize the pull-down menu. Open the Customize settings box:

Pull-down menu	Workspace > Customize

Follow these steps to modify the pull-down menu:

STEP 1: Select the Menu Bar tab in the Customize settings box, and MicroStation lists the available options, similar to Figure 16–48. Select the name of the interface to list the available menus from the Available Menus From list box. MicroStation lists the available menus from the selected interface (on the left side of the settings box), and lists the available menus from the Main menu bar by default in the Menus: list box (on the right side of the settings box). The Available list box is used only for inserting menus and menu items.

STEP 2: To modify a menu name, first select the Menu name in the Menus: list box, then click the Modify... button. MicroStation displays a Modify Menu dialog box similar to Figure 16–49. Make the modification in the Label: edit field and click the OK button.

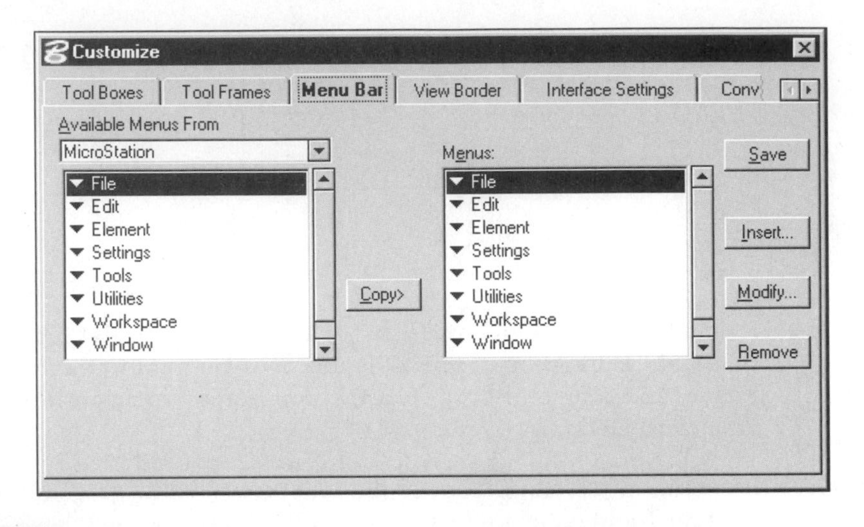

Figure 16–48 Customize Settings box, with the Menu Bar tab selected

Figure 16–49 Modify Menu dialog box

To modify a menu item, first double-click the menu name. MicroStation expands the menu. Select the menu item to modify, then click the Modify... button. MicroStation displays a Modify Menu Item dialog box, similar to Figure 16–50.

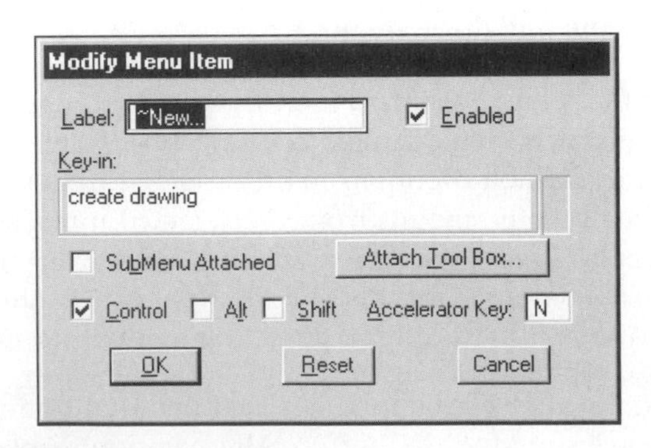

Figure 16–50 Modify Menu Item dialog box

To modify the menu item name, type the new name in the Label: edit field.

Note: Make sure to insert at the appropriate place the tilde (~) character immediately before the character that will be the mnemonic access character—for example E~dit for Edit.

To enable or disable the item in the menu, turn Enabled ON or OFF.

If necessary, make the modifications for any action string associated with the item.

To attach a submenu to the item, turn on SubMenu attached, and click the Attach Tool Box... button. MicroStation displays the Select Tool Box dialog box. Select the appropriate tool box from the list box, and click the OK button to attach the Tool Box to the item.

To assign or modify the keyboard accelerator, select Ctrl alone or with Alt or Shift to indicate the modifier key(s), and type the accelerator key in the Accelerator Key:edit field.

Click the OK button to close the dialog box.

STEP 3: To insert a menu name, first select the entry for the existing menu before which you want to insert the new menu, then click the Insert... button. MicroStation displays an Insert Menu dialog box, similar to Figure 16–51.

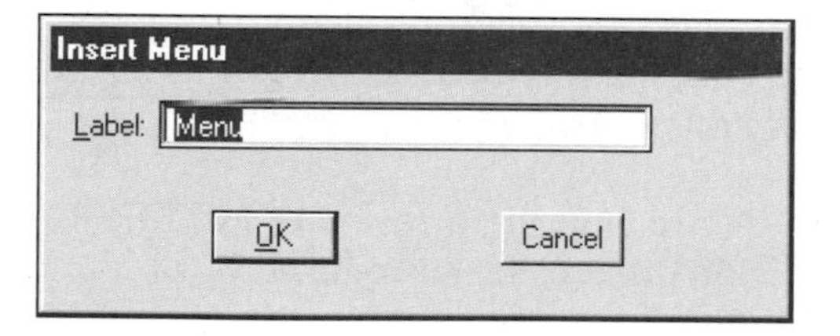

Figure 16–51 Insert Menu dialog box

Key-in the name of the new menu name in the Label: edit field, and click the OK button to close the dialog box.

To insert a menu item, first double-click the menu name. MicroStation expands the menu. Select the entry for the existing menu item before which you want to insert the new menu item, then click the Insert... button. MicroStation displays an Insert Menu Item dialog box, similar to Figure 16–52.

Use the appropriate controls in the Insert Menu Item dialog box to set the menu item, and click the OK button to close the dialog box.

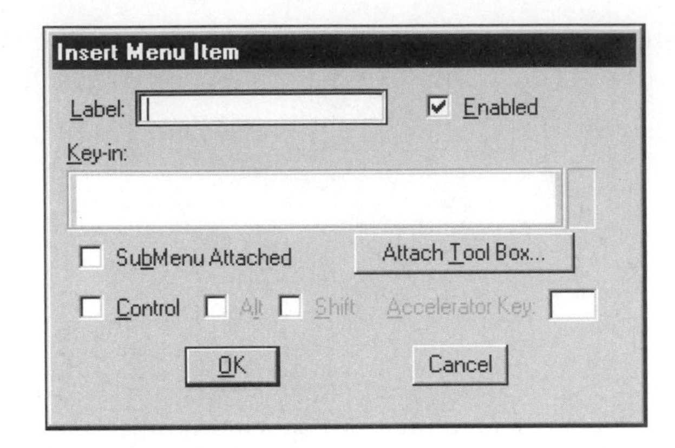

Figure 16–52 Insert Menu Item dialog box

STEP 4: From the available Menu names listed on the Available Menus (on the left side of the settings box), double-click the Menu name. MicroStation lists the available menu items. Select the menu item you want to insert, and drag and drop it at the appropriate location in the Menus: list box.

STEP 5: If necessary, you can rearrange the menu or menu items by dragging and dropping in the Menus: list box.

STEP 6: If necessary, you can remove the menu or menu item from the Menus: list box. First select the menu or menu item to remove from the list box, then click the Remove button. MicroStation removes the selected menu or menu item from the tool box.

STEP 7: Click the Save button to save the Modification/Addition to menu bar. MicroStation displays the customized version of the pull-down menus.

STEP 8: If no further changes have to be made, close the Customize settings box.

Note: To disregard the changes, close the Customize settings box without saving the changes.

Customizing View Border Tools

The Customize settings box can also customize view border tools. Open the Customize settings box:

Pull-down menu	Workspace > Customize

Follow these steps to modify view border tools:

STEP 1: Select the View Border tab in the Customize settings box and MicroStation lists the available options, similar to Figure 16–53. By default, MicroStation lists the available tools in the Available Tools list box (on the left side of the settings box), and lists the Tools from the 2D View Border in the Edit View Border: list box (on the right side of the settings box).

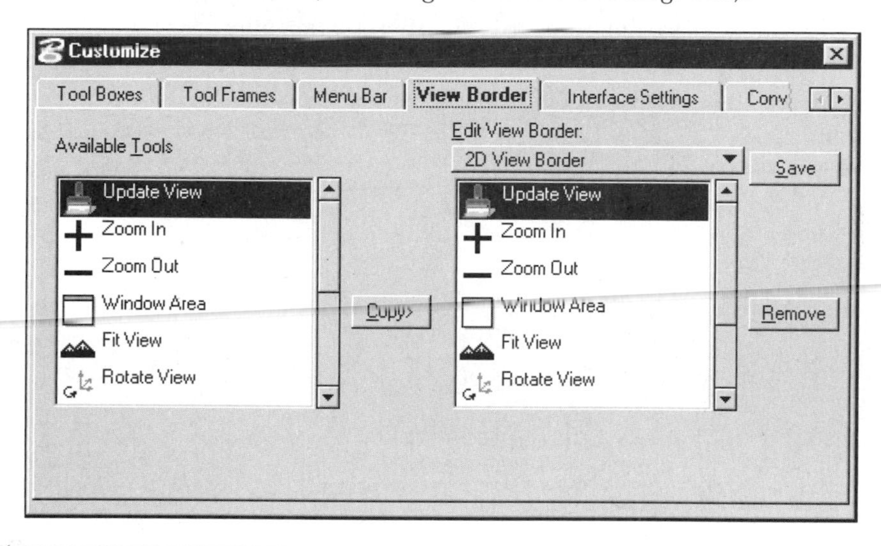

Figure 16–53 Customize Settings box with View Border tab selected

STEP 2: Select one of the two available view borders to customize (2D View Border or 3D View Border) from the Edit View Border: option menu.

STEP 3: From the available tools listed on the Available Tools: list box (on the left side of the settings box), select the tool you want to insert in the View Border, and drag and drop it at the appropriate location in the Edit View Border: list box.

STEP 4: If necessary, you can rearrange the tools by dragging and dropping in the Edit View Border: list box.

STEP 5: If necessary, you can remove a tool from the View Border list box. First select the tool to remove from the list box, then click the Remove button. MicroStation removes the selected tool from the View Border.

STEP 6: Click the Save button to save the Modification/Addition to View Border. MicroStation displays the customized version of the View Border.

STEP 7: If no further changes have to be made, close the Customize settings box.

FUNCTION KEYS

Each personal computer keyboard has a special set of keys called *function keys.* MicroStation has a utility that assigns menu action strings to the various function keys (F1 through F12). Once the assignment is made, then all you have to do is select the function key, and the menu action string is activated.

There are only 12 functions on the keyboard, but MicroStation allows you to make up to 96 function key assignments by combining a function key with the Shift, Alt, and/or Ctrl key. So, for instance, you can assign the F1 with the following combinations:

F1

Shift + F1

Alt + F1

Ctrl + F1

Shift + Alt + F1

Shift + Alt + F1

Alt + Ctrl + F1

Shift + Alt + Ctrl + F1

Function key assignments are stored in an ASCII file, and by default the extension for the function key file is .MNU. This file follows the same concept as the cell library file. Once created, the function key file may be attached to any of your design files to activate the function keys. The default function key menu file is called "funckey.mnu." It is stored under the /Bentley/Worskpace/interfaces/fkeys/ folder.

Creating and Modifying Function Key Definitions

To create and modify function key definitions, open the Function Keys dialog box:

Pull-down menu	Workspace > Function Keys...

MicroStation displays the Function Keys dialog box, as shown in Figure 16–54.

The title bar identifies the file name of the function key menu. The list box lists currently defined function key combinations and their definitions.

The Shortcut Keys section contains controls for selecting a function key definition to change or for creating a new one. Turn ON the appropriate toggle buttons for Ctrl, Alt, and/or Shift modifiers, and select the function key from the Key option menu to change or to create a new definition. Click the Edit... button, and MicroStation opens a Edit Key Definition dialog box similar to Figure 16–55.

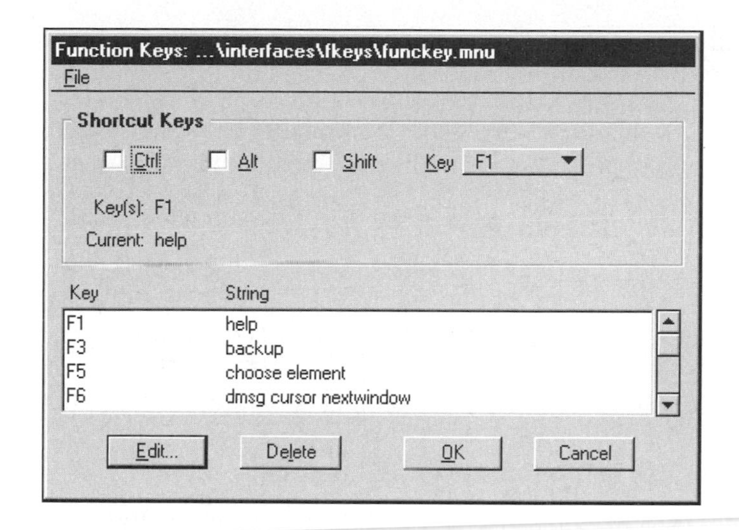

Figure 16–54 Function Keys dialog box

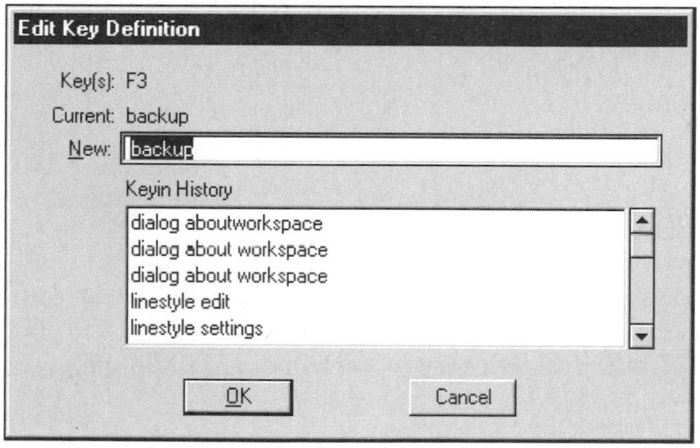

Figure 16–55 Edit Key Definition dialog box

Key-in the action string (for example, place line) in the New: edit field, then click the OK button to accept the function key definition and close the dialog box.

You can also program a function key with multiple action strings (macros) separated by semicolons. For example, the following key-in sets the color to red, the line weight to 2, the active level to 4, and invokes the place line tool:

co=red;wt=2;lv=4;place line

In addition, you can also select the action strings from the Key-in History, as shown in Figure 16–55. Click the OK button to accept the changes and close the Edit Key Definition dialog box.

To delete a function key definition, first select the function key definition from the list box, then click the Delete button. MicroStation deletes the selected function key definition.

Saving the Function Key Definitions

To save the function key definitions, invoke the Save tool:

Pull-down menu (Function Keys dialog box)	File > Save

MicroStation saves the current status of the function key definitions to the function key menu file.

Saving the Function Key Definitions to a Different File

To save the function key definitions to a different menu file, invoke the Save As tool:

Pull-down menu (Function Keys dialog box)	File > Save As...

MicroStation opens the Save Function Key Menu As dialog box. Key-in the name of the menu file in the Files: edit field and click the OK button.

INSTALLING FONTS

Each time a design file is displayed, a font resource file is used that contains font definitions. Provided the font resource file name has not been changed during workspace definitions, MicroStation, by default, uses a resource file called FONT.RSC. MicroStation allows you to add new fonts to the font resource file via the utility called Font Installer.

The Font Installer dialog box helps you to import fonts from different sources into the MicroStation font library; in addition, you can rename and renumber fonts.

To open the Font Installer dialog box, invoke Install Fonts:

Pull-down menu	Utilities > Install Fonts

MicroStation displays the Font Installer dialog box, as shown in Figure 16–56.

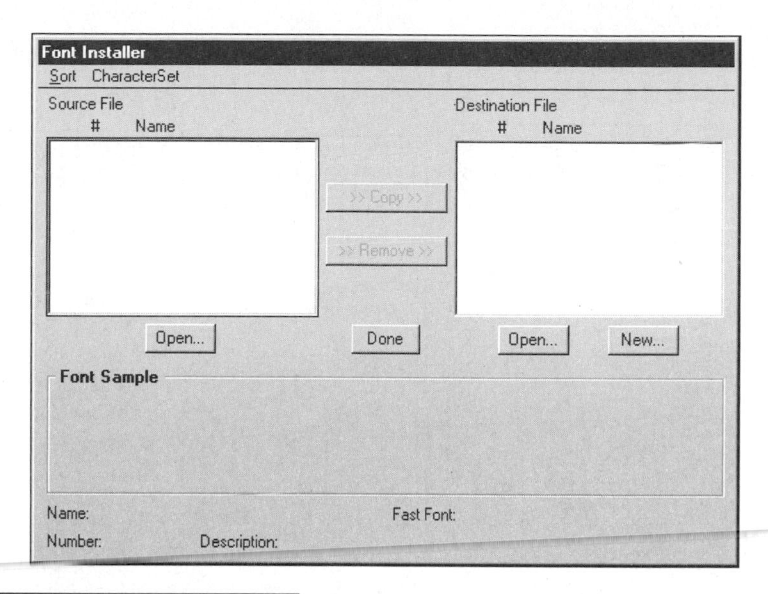

Figure 16–56 Font Installer dialog box

Selecting the Source Font Files

To select the source fonts, click the Open... button located below the Source File list box. MicroStation displays an Open Source Font Files dialog box similar to Figure 16–57.

The Type: option menu sets the file type from one of the following:

> Font Cell Library—MicroStation font cell library, a standard cell library that contains cells that define the characters and symbols in a traditional MicroStation font and the font's attributes (*.cel)
>
> Font Library—MicroStation Version 5 font library (*.rsc)
>
> uSTN V4/IGDS Fontlib—Version 4.1 or earlier (or IGDS) font library
>
> PS Type-1—PostScript Type 1 (uncompressed only) (*.pf*)
>
> Shape Files (AutoCAD)—AutoCAD shape files (*.shx files)
>
> TrueType (*.tf*)

Select the font file from the Files list box and click the Add button. You can select multiple files as the source font files. Click the Remove button to remove the font file from the selected list box. Then click the Done button to accept the selection and close the dialog box. MicroStation lists all the fonts available in the source File list box.

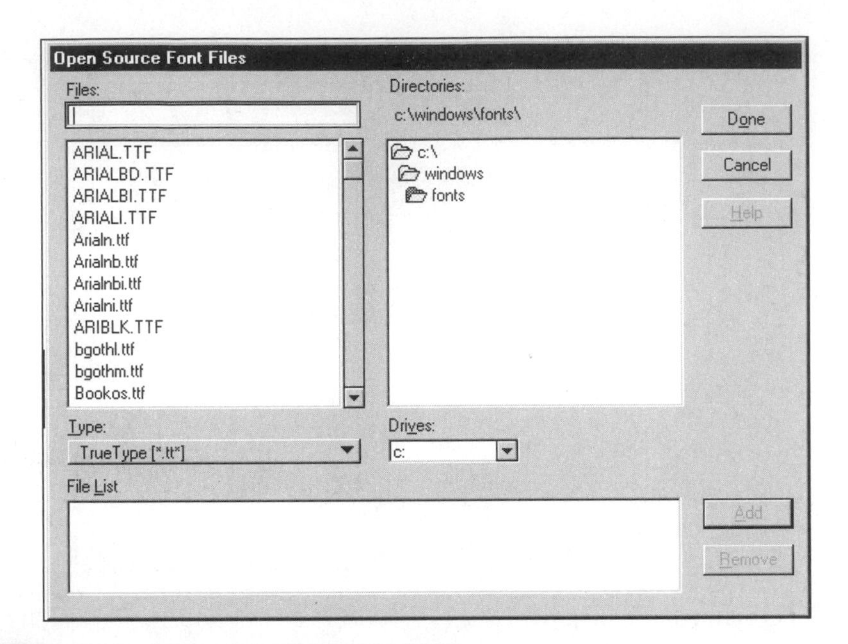

Figure 16–57 Open Source Font Files dialog box

Selecting the Destination Fonts File

To select the destination fonts file, click the Open... button located beneath the Destination File list box. MicroStation opens the Open Font Library dialog box, similar to Figure 16–58. Select the font library from the Files: list box as the destination for the new font insertion, and click the OK button to close the dialog box. MicroStation lists all the available fonts in the Destination File list box.

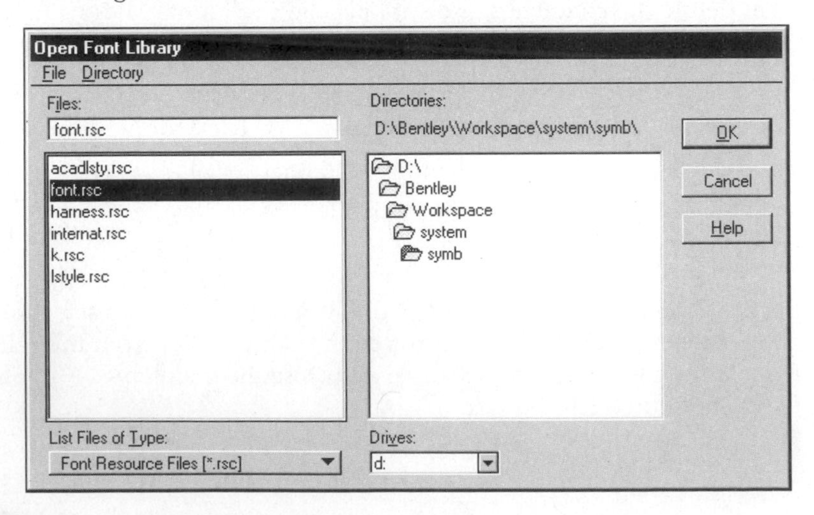

Figure 16–58 Open Font Library dialog box

Importing Fonts

To import the font from the source to the destination file, first select the font in the source file list box, and click the >> Copy >> button. MicroStation copies the selected font into the Destination File list box. To remove the font in the Destination File list box, first select the font to be removed from the list, then click the >> Remove >> button. MicroStation removes the selected font from the Destination File list box.

If necessary, you can change the name of the font, the number of the font, or the description. First select the font in the Destination File list box; MicroStation displays the information in the bottom of the dialog box. Make the necessary changes in the appropriate edit fields.

Click the Done button to close the Font Installer dialog box.

ARCHIVE UTILITY

The Archive utility allows you to select and bundle together the design files you wish to archive, along with all the resources needed to re-create those design files on a different computer system. Following are the classes of resources that can be included in an archive file:

- Design files
- Reference files
- Raster Reference files
- Cell libraries
- Background files (raster images)
- Resource files (line styles, fonts, etc.)
- Material tables
- Material tool boxes
- Materials
- Pattern files
- Bump Map files
- Animation Script Files
- Current workspace files
- All configuration files
- All user interface files

In addition, you can create your own classes to hold additional resource files, such as plotter configuration files, tag templates, and even non-MicroStation files.

Creating a New Archive File

Follow these steps to create a new Archive file:

STEP 1: Invoke the Archive settings box:

Pull-down menu	Utilities > Archive

MicroStation displays an Archive settings box similar to Figure 16–59.

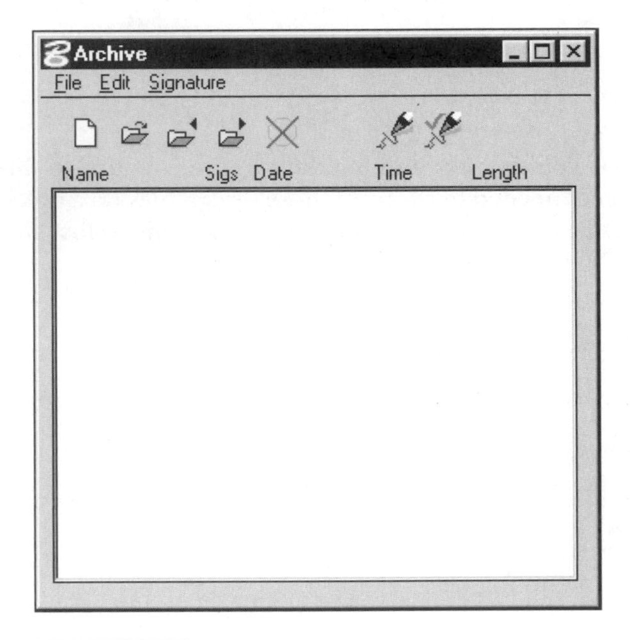

Figure 16–59 Archive settings box

STEP 2: Invoke the New Archive tool:

Pull-down menu (Archive settings box)	File > New...
Archive tool box (Archive settings box)	Select the New Archive tool (see Figure 16–60).

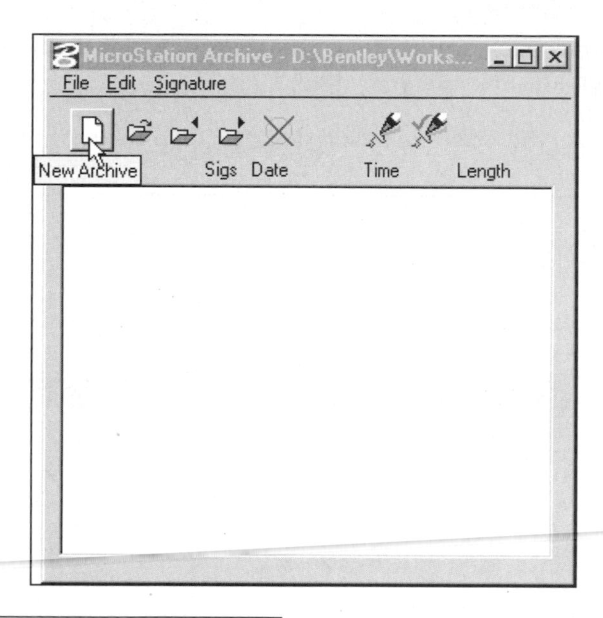

Figure 16–60 Invoking the New Archive tool

MicroStation displays a Create Archive File dialog box, similar to Figure 16–61.

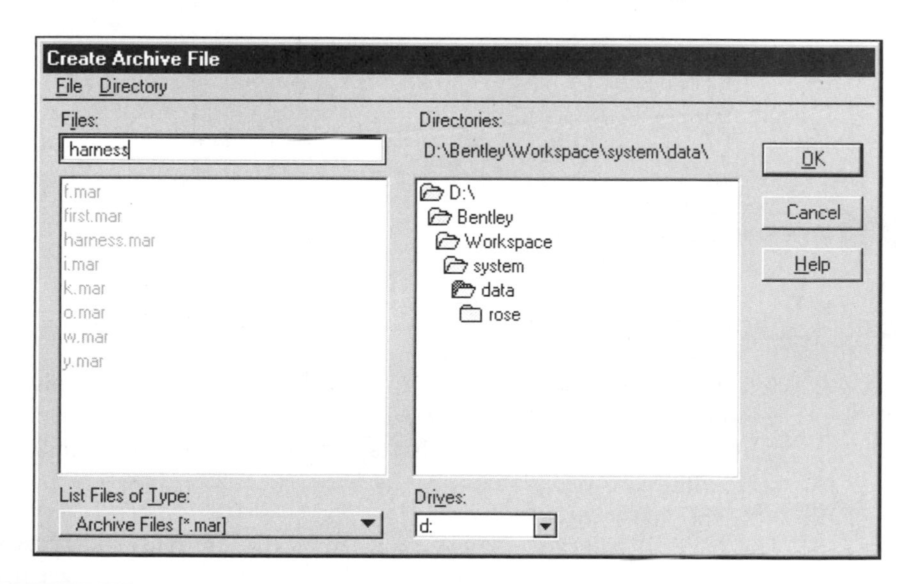

Figure 16–61 Create Archive File dialog box

STEP 3: Specify the Archive File name in the Files: edit field and click the OK button to create a new archive file.

STEP 4: Invoke the Add tool to add design files to the newly created archive file:

Pull-down menu (Archive settings box)	Edit > Add...
Archive tool box (Archive settings box)	Select the Add Files tool (see Figure 16–62).

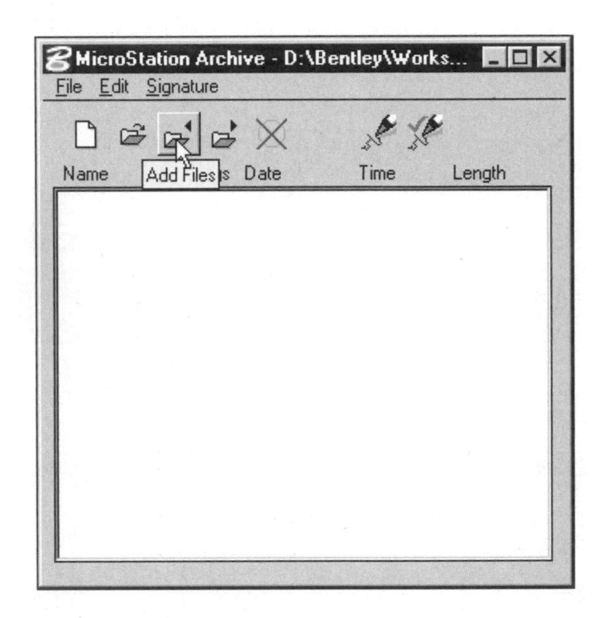

Figure 16–62 Invoking the Add Files tool

MicroStation displays the Select Files To Add dialog box, similar to Figure 16–63.

Select the files to add from the appropriate folders to the archive file. Once the selection is complete, click the Done button to close the Select Files To Add dialog box. MicroStation displays the Add Archive Files dialog box, similar to Figure 16–64.

The Save Directories toggle button stores the file path specification in the archive file. If this path is set to OFF, only the file names are stored in the archive file.

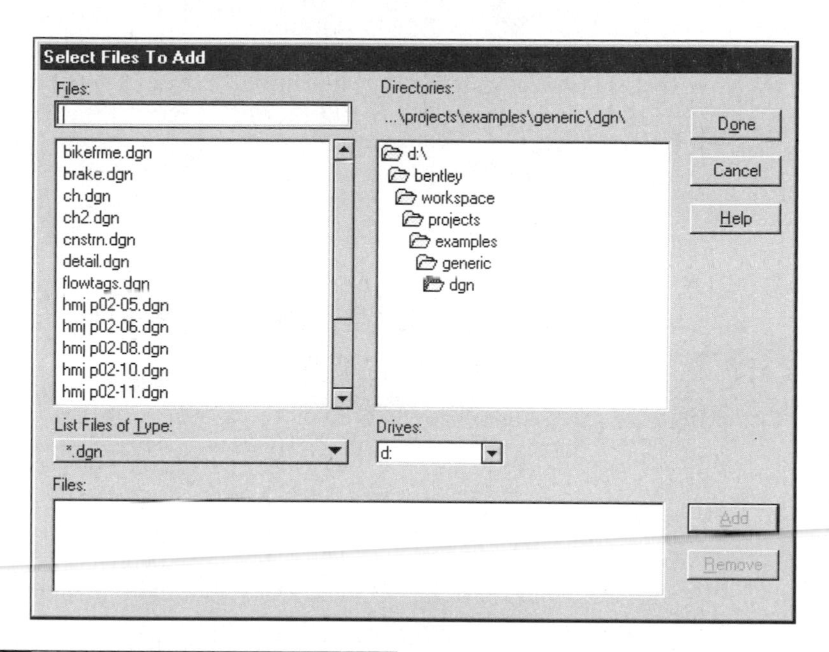

Figure 16–63 Select Files To Add dialog box

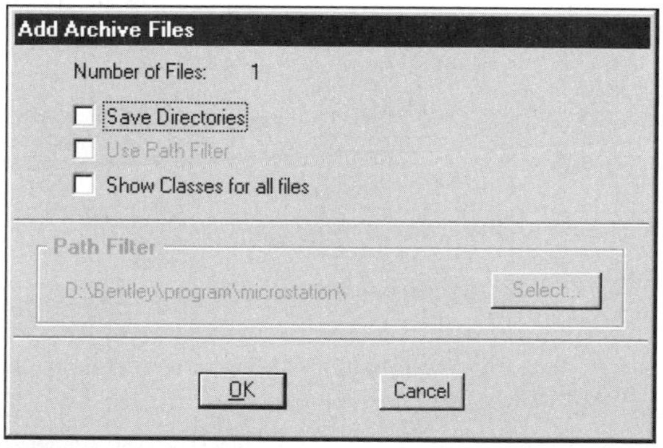

Figure 16–64 Add Archive Files dialog box

The Use Path Filter toggle button sets the filter to remove that part of the path that is unique to a specific system. If it is set to ON, then specify the path by clicking the Select... button in the Path Filer section of the dialog box. MicroStation will archive all the files not defined in the Path Filter section.

The Show Classes for all files toggle button controls the display of the Select Archive Classes dialog box. Select Archive Classes dialog box allows you to select files from various categories of files to be part of the Archive file.

Click the OK button to close the Add Archive Files dialog box. If the Show Classes for all files was set to ON, then MicroStation displays Select Archive Classes dialog box, similar to Figure 16–65.

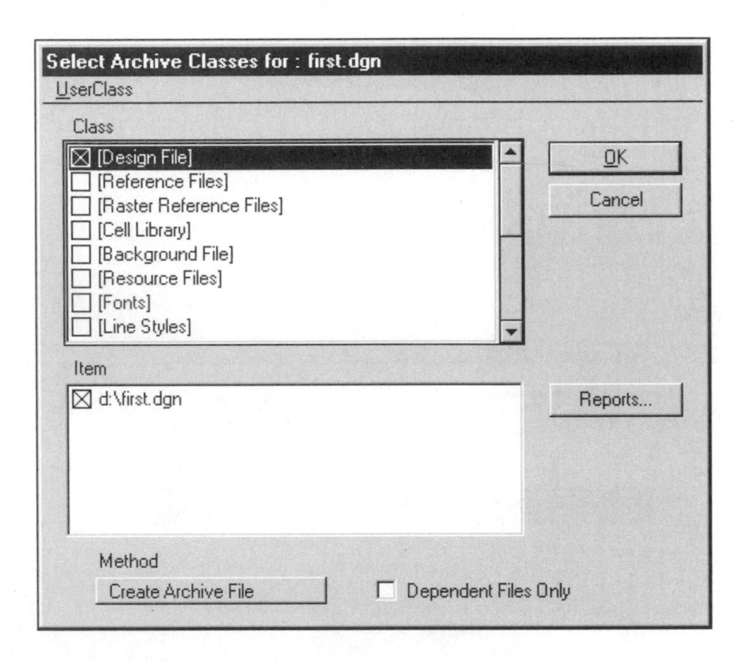

Figure 16–65 Select Archive Classes dialog box

Set the toggle button to ON for specific archive classes to include from the Class list box. MicroStation displays any items specifically called for in the design file that will be shown in the Item list box. By default, they are automatically selected for archiving. Click the related option box to select or deselect each file.

If necessary, click the Reports... button. MicroStation opens the Archive Report dialog box to review the resource file tree used by the current design file. MicroStation provides three different methods by which you can display the report. The methods include Class Summary, Dependency Tree, and File List; you can select one of them from the submenu Style of the File pull-down menu of the dialog box. Figure 16–66 shows the report display in the Dependency Tree format. Click the OK button to close the Archive Report display dialog box.

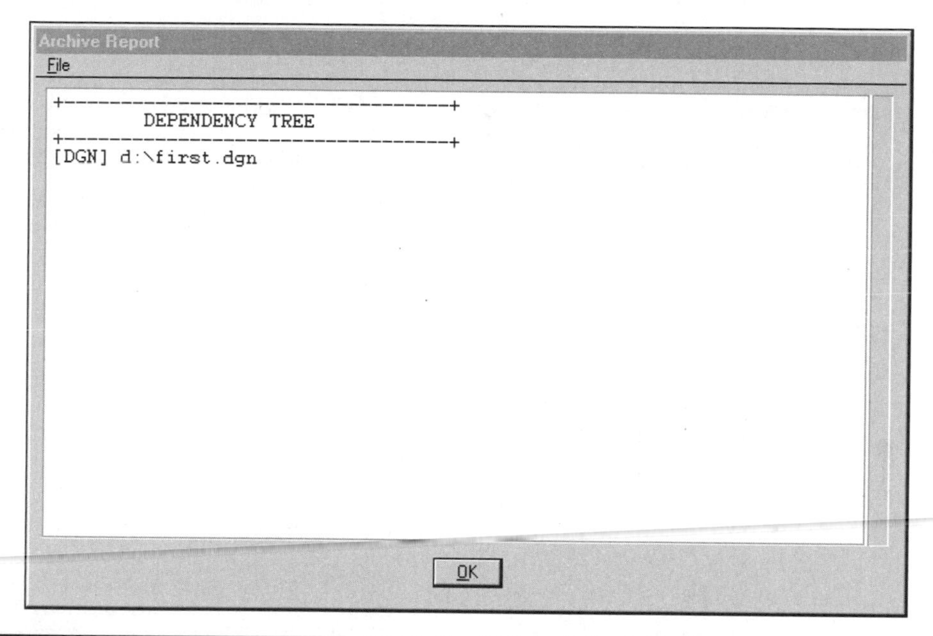

Figure 16–66 Archive Report display in the Dependency Tree format

STEP 5: Click the OK button to create the archive file. A progress indicator is displayed showing the name of the file being processed, as well as the progress of the entire archive process. After creating the Archive file, MicroStation displays the list of the files included in the newly created Archive file in the MicroStation Archive dialog box.

Copying a Design File and Its Resources to a New Destination

Follow these steps to copy a design file and its resources to a new destination:

STEP 1: Invoke the Archive settings box:

Pull-down menu	Utilities > Archive

MicroStation displays an Archive settings box.

STEP 2: Invoke the New Archive tool:

Pull-down menu (Archive settings box)	File > New...
Archive tool box (Archive settings box)	Select the New Archive tool

MicroStation displays a Create Archive File dialog box. Specify the Archive File name in the Files: edit field and click the OK button to create a new archive file.

STEP 3: Invoke the Add tool to add design files to the newly created archive file:

Pull-down menu (Archive settings box)	Edit > Add...
Archive tool box (Archive settings box)	Select the Add Files tool

MicroStation displays the Select Files to Add dialog box. Select the files to add from the appropriate folders to the archive file. Once the selection is complete click the Done button to close the Select Files to Add dialog box. MicroStation displays the Add Archive Files dialog box. The Set Show Classes for all files toggle button should be set to ON. Click the OK button to close the dialog box. MicroStation displays the Select Archive Classes dialog box.

STEP 4: Set the toggle button to ON for specific archive classes to include from the Class list box. MicroStation displays any items specifically called for in the design file that will be shown in the Item list box. By default, they are automatically selected for archiving. Click the related option box to select or deselect each file.

STEP 5: If necessary, click the Reports... button. MicroStation opens the Archive Report dialog box to review the resource file tree used by the current design file.

STEP 6: Select Copy Files from the Method option menu located at the bottom of the Select Archive Classes dialog box, and click the OK button. MicroStation displays a Copy Files dialog box similar to Figure 16–67.

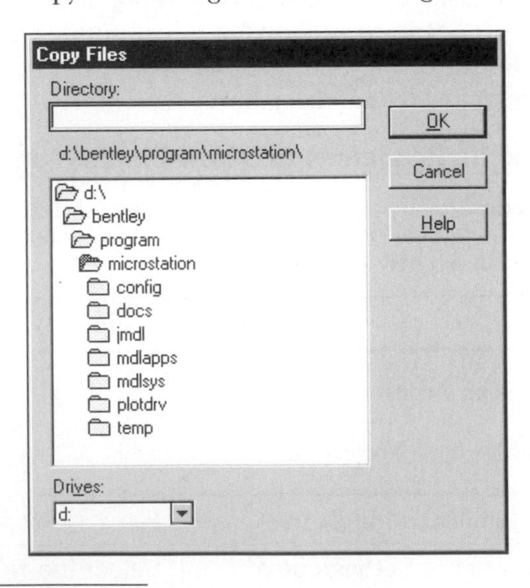

Figure 16–67 Copy Files dialog box

Select a directory to which to copy the selected design files and resource files. Click the OK button to close the dialog box. A progress indicator is displayed.

Extracting Files from an Existing Archive

Follow these steps to extract files from an existing archive:

STEP 1: Invoke the Archive settings box:

Pull-down menu	Utilities > Archive

MicroStation displays an Archive settings box.

STEP 2: Invoke the Open Archive tool:

Pull-down menu (Archive settings box)	File > Open
Archive tool box (Archive settings box)	Select the Open Archive tool (see Figure 15–68).

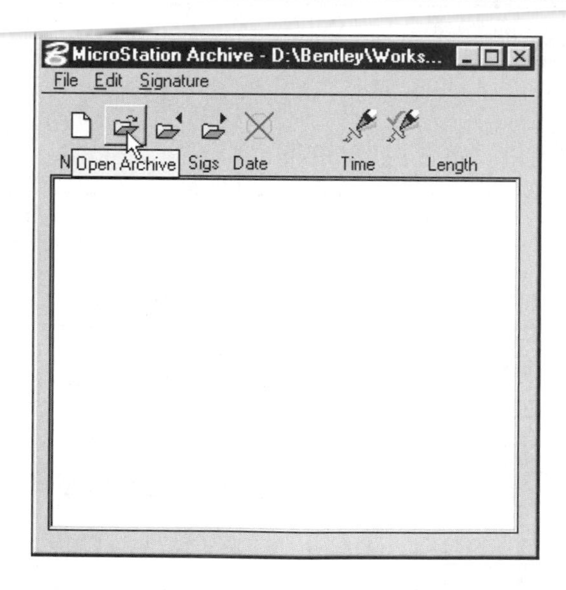

Figure 16–68 Invoking the Open Archive tool

MicroStation displays an Open Archive File dialog box, similar to Figure 16–69.

STEP 3: Select the archive file from which files are to be extracted from the appropriate folder and click the OK button. The files that constitute the archive file are displayed in the Archive File list box.

STEP 4: In the list box, select the files to be extracted. Standard list selection techniques can be used to select more than one file. To select all files in an archive, invoke the Select All item from the Edit pull-down menu.

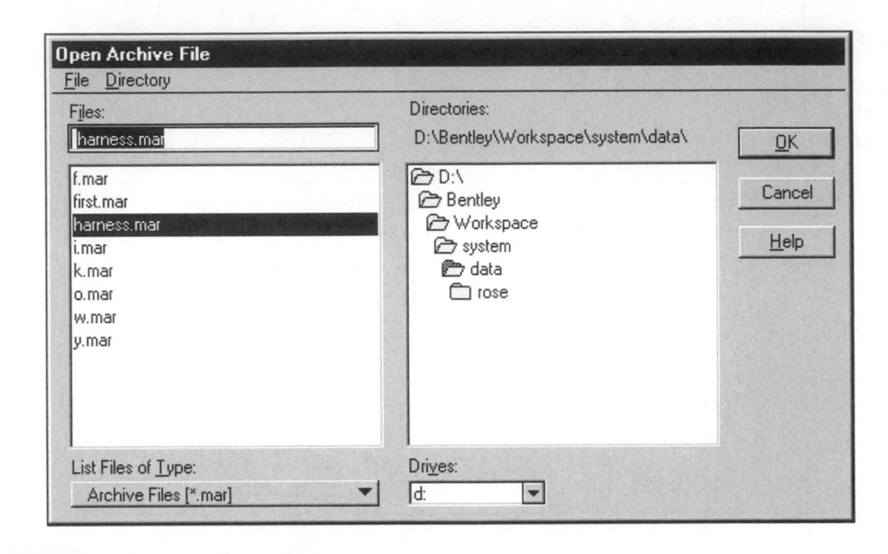

Figure 16–69 Open Archive File dialog box

STEP 5: Invoke the Extract Files tool:

Pull-down menu (Archive settings box)	Edit > Extract...
Archive tool box (Archive settings box)	Select the Extract Files tool (see Figure 15–70).

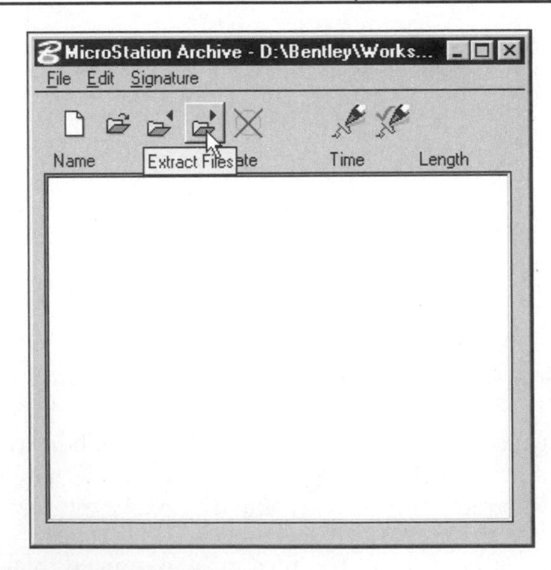

Figure 16–70 Invoking the Extract Files tool

MicroStation displays the Extract Archive Files dialog box, which displays information about the selected files along with destination options.

STEP 6: If necessary, turn ON the toggle button for the Create Stored Directories option. By selecting this option, you direct the Archive utility to use stored directories as the destination path.

STEP 7: If necessary, turn ON the toggle button for the Overwrite Existing Files option. By selecting this option, the extracted file will overwrite an existing file with the same name.

STEP 8: If necessary, turn ON the toggle button for the Preserve Date/Time option. By selecting this option, the date/time of the extracted file will be copied from the compressed file, preserving its original values.

STEP 9: If necessary, change the target directory by clicking the Select... button in the Extract To section.

STEP 10: Click the OK button to begin the file extraction process.

Adding Files to an Existing Archive

Follow these steps to add files to an existing archive:

STEP 1: Invoke the Archive settings box:

Pull-down menu	Utilities > Archive

MicroStation displays an Archive settings box.

STEP 2: Invoke the Open Archive tool:

Pull-down menu (Archive settings box)	File > Open
Archive tool box (Archive settings box)	Select the Open Archive tool

MicroStation displays an Open Archive dialog.

STEP 3: Select the archive file from which files are to be extracted and click the OK button. The files that constitute the archive file are displayed in the Archive File list box.

STEP 4: Invoke the Add Files tool:

Pull-down menu (Archive settings box)	Edit > Add
Archive tool box (Archive settings box)	Select the Add Files tool

MicroStation displays a Select Files To Add dialog box.

STEP 5: Select the file from the appropriate directory and click the Add button. The file will appear in the Files list box. Continue selecting files to archive.

STEP 6: Click the Done button after selecting all the files to be archived. MicroStation displays the Add Archive Files dialog box. Make the necessary changes and click the OK button.

MicroStation adds the selected files to the opened archive file.

SCRIPTS, MACROS, USER COMMANDS

MicroStation has three application software tools that automate often-used command sequences. Depending on the complexity of the operation to be performed, you can choose one of the three application software tools. The tools include scripts, macros, and user commands.

Scripts

A script is the simplest software application in which you create an ASCII file containing the key-in sequence of the MicroStation commands. For example, the following script sets the Active Color, Active Line Weight, and Active Level:

```
active color red
active weight 6
active level 5
```

To load and run a script, key-in the following in the Key-in window: @<script_file> and press Enter. MicroStation executes the command sequence in the order that the script is set.

Note: If the script is not in the current directory, script_file must include the full path to the script.

Macros

Macros are BASIC programs that automate often-used, usually-shortened sequences of operation. Many MicroStation-specific extensions have been added to the BASIC language to customize it for the MicroStation environment. Macros select tools and view controls, send key-ins, manipulate dialog boxes, modify elements, and so on.

For detailed information about creating macros, refer to the MicroStation BASIC Guide that comes with the MicroStation software.

Several sample macros are provided with MicroStation. To load and run a macro, invoke the Macro tool:

Pull-down menu	Utilities > Macro
Key-in window	**Macro** <macro_name> Enter

MicroStation displays a Macros settings box similar to Figure 16–71. Select the macro from the Macro Name list box, click the Run button, follow the prompts, and provide the appropriate responses.

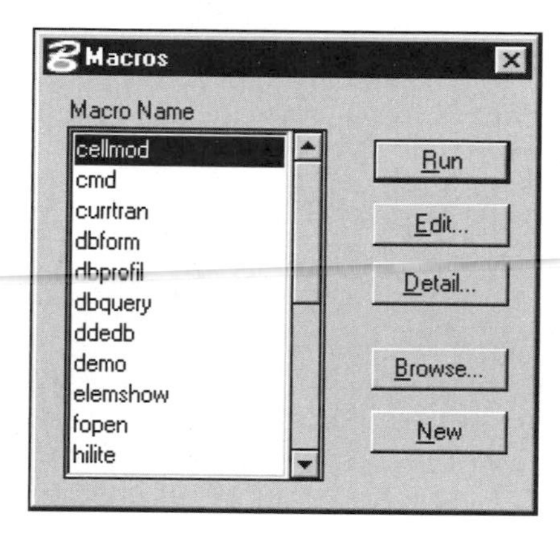

Figure 16–71 Macros settings box

User Commands

A *user command* is an ASCII text file that groups together "statements" that are executed while in a MicroStation design file. A user command is similar to a macro. The coded statements of a user command follow the rules that are part of the MicroStation User Command Language. This language provides an interface consisting of a set of commands and instructions to automate a task or set of tasks. These instructions are arranged in a logical order that imitates the steps that would be followed if you were performing MicroStation tasks manually. The coded statements are performed one after the other until all have been executed or until a condition is met that causes the user command to exit. Refer to the MicroStation User Manual for a detailed description of user commands.

Several sample user commands are provided with MicroStation. To load and run a user command, invoke the tool:

Key-in window	**uc=**<name> Enter
Pull-down menu	Utilities > User command > Run

MicroStation displays a Run User Command dialog box, similar to Figure 16–72.

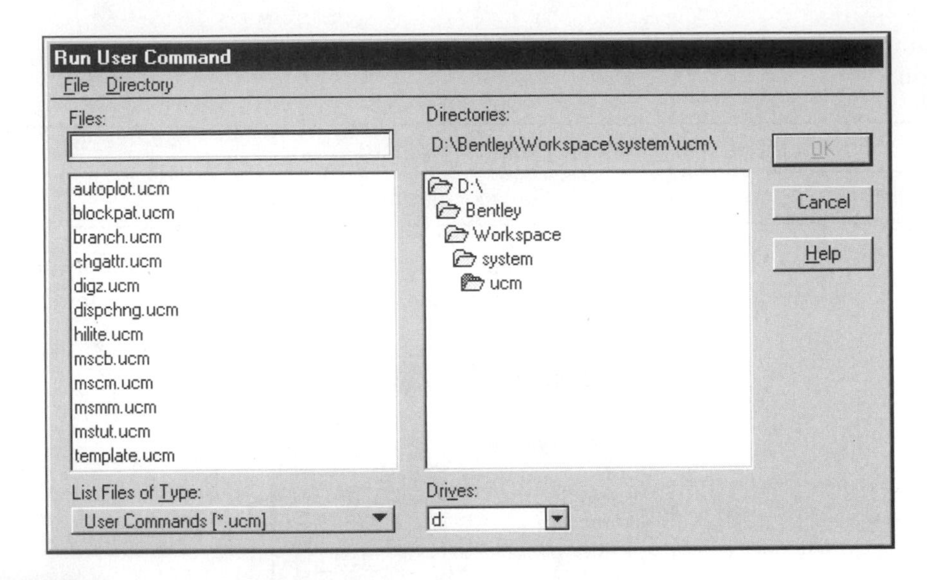

Figure 16–72 Run User Command dialog box

Select the User Command file from the Files list box and click the OK button. MicroStation executes the User Command. Then follow the prompts and provide the appropriate responses.

chapter 17

3D Design and Rendering

WHAT IS 3D?

In *two dimensional* drawings, you work with two axes, *X* and *Y*. In *three dimensional* drawings, in addition to the *X* and *Y* axes, you work on the *Z* axis, as shown in Figure 17–1. Plan views, sections, and elevations represent only two dimensions. Isometric, perspective, and axonometric drawings, on the other hand, represent all three dimensions. For example, to create three views of a cube, the cube is simply drawn as a square with thickness. This is referred to as *extruded 2D.* Only objects that are extrudable can be drawn by this method. Any other views are achieved by simply rotating the viewpoint of the object, just as if you were physically holding the cube. You can also can get an isometric or perspective view by simply changing the viewpoint.

Drawing objects in *3D* provides three major advantages:

- An object can be drawn once and then can be viewed and plotted from any angle.
- A *3D* object holds mathematical information that can be used in engineering analysis, such as finite-element analysis and computer numerical control (CNC) machinery.
- Shading can be added for visualization.

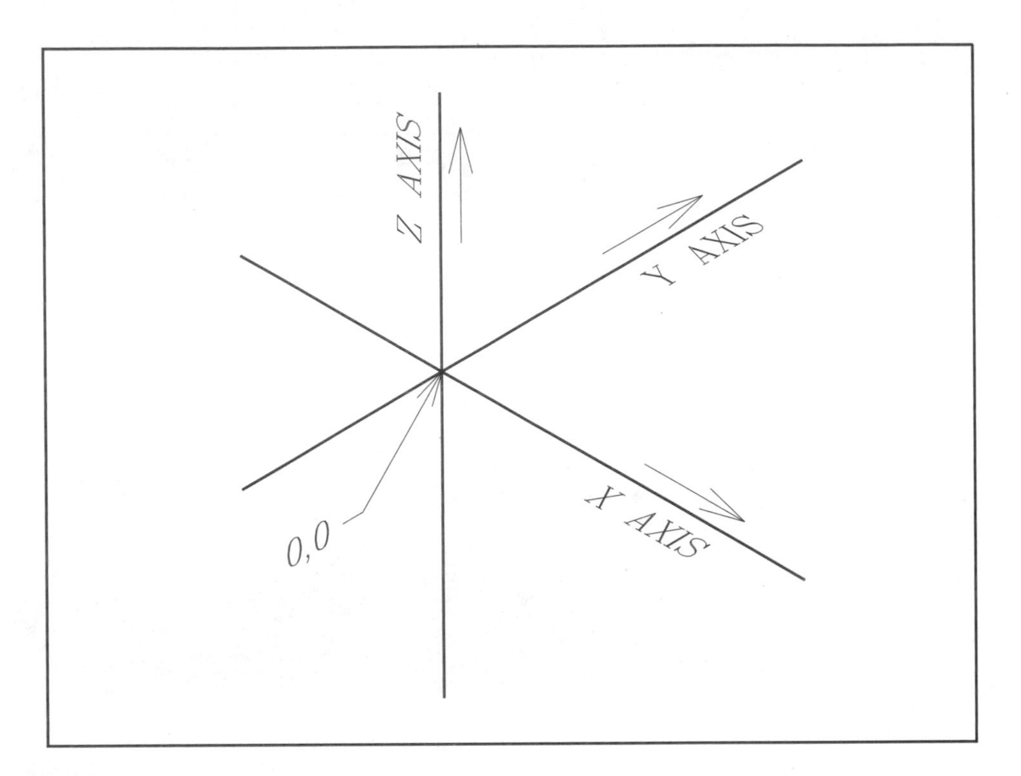

Figure 17–1 *X, Y,* and *Z* axes dor 3D Design

MicroStation provides two types of *3D* modeling: Surface and Solid. Surface modeling defines the edges of a *3D* object in addition to surfaces, whereas solid models are the unambiguous and informationally complete representation of the shape of a physical object. Fundamentally, solid modeling differs from surface modeling in two ways:

- The information is more complete in the solid model.
- The method of construction of the model itself is inherently straightforward.

Using MicroStation's SmartSolids and SmartSurfaces tools you can quickly construct complex *3D* models. With primitive solids or surfaces, you can add finishing touches, such as fillets and chamfers, and you can create a hollow solid with defined wall thickness with the ShellSolid tool.

This chapter provides an overview of the tools and specific commands available for *3D* design.

CREATING A 3D DESIGN FILE

The procedure for creating a new *3D* design file is similar to that for creating a new *2D* design file, except you have to use a seed file that is designed specifically for *3D*. The same holds true for cell libraries. To create a new *3D* design file, invoke the New... tool:

Pull-down menu	File > New...
Key-in window	**create drawing** (or **cr d**) Enter

MicroStation opens the Create Design File dialog box. Click the Select... button, and MicroStation displays a list of seed files available, as shown in Figure 17–2. Select one of the *3D* seed files from the Files list box and click the OK button. Key-in the name for your new 3D design file in the Name: edit field and click the OK button. MicroStation highlights the name of the file you just created in the Files list box of the Create Design dialog box by default. To open the new design file, click the OK button and your screen will look similar to the one shown in Figure 17–3. By default, MicroStation displays four view windows, each set for one of the four standard view orientations. As part of the title of the view window, MicroStation displays the name of the view being displayed.

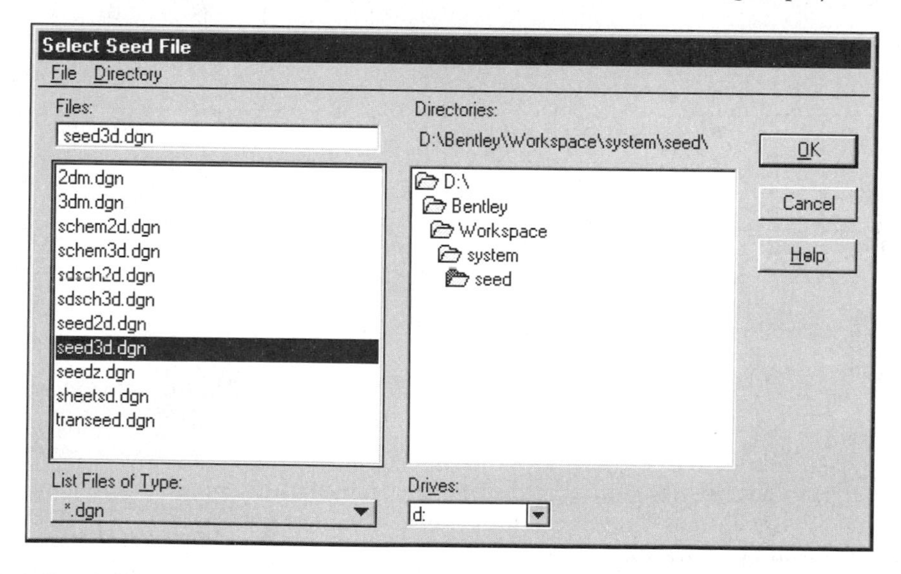

Figure 17–2 Select Seed File dialog box

Note: The text elements you see displayed when you start a new design are construction elements. You can turn off the display of the construction elements by setting the Constructions toggle button to OFF in the View Attributes settings box.

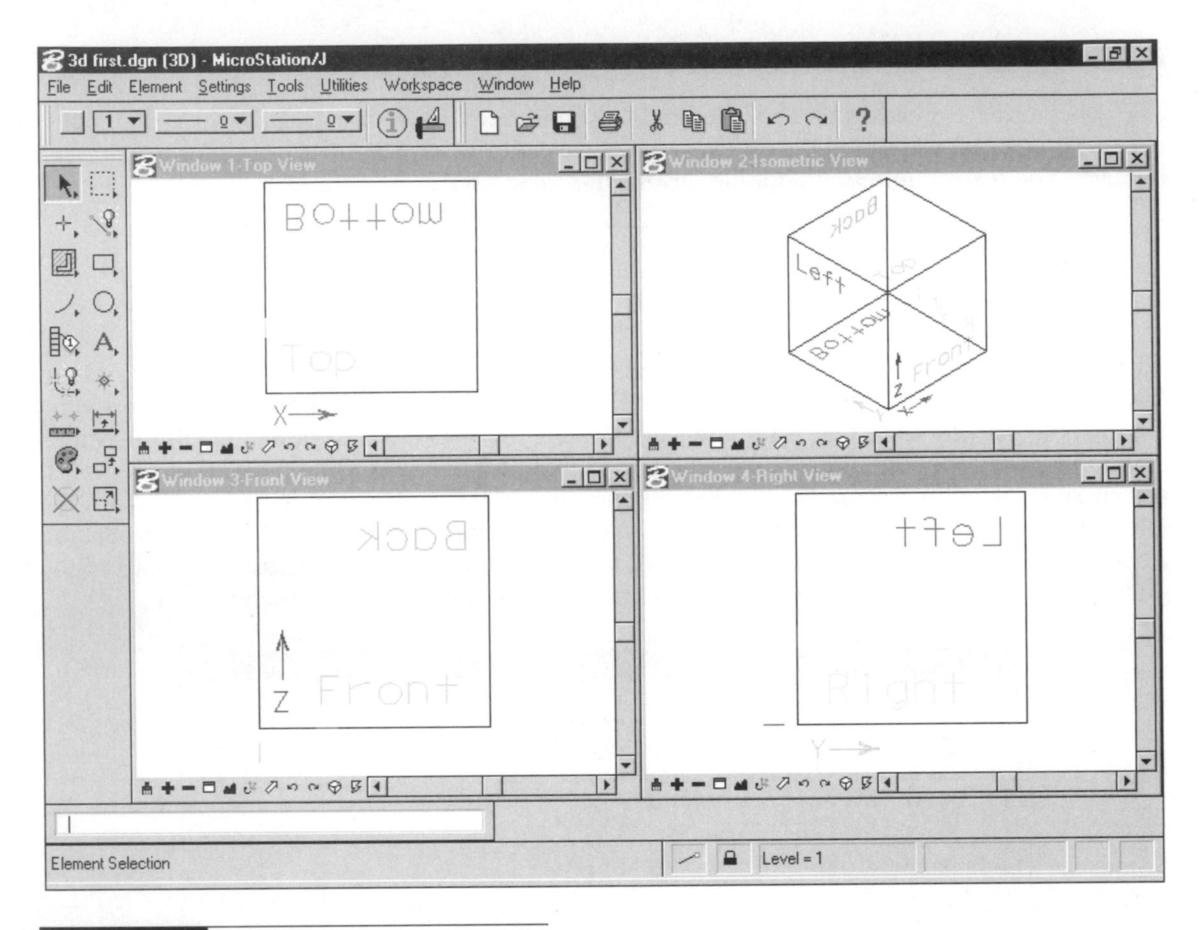

Figure 17–3 MicroStation screen display

VIEW ROTATION

There are seven standard view orientations defined in MicroStation: top, bottom, front, back, right, left, and isometric. You can use one of the four tools available in MicroStation to display the standard view orientation in any of the view windows.

Rotate View

The Rotate View tool displays one of the standard view orientations. In addition, it can rotate the view dynamically. Invoke the Rotate View tool:

View Control bar	Select the Rotate View tool (see Figure 17–4).

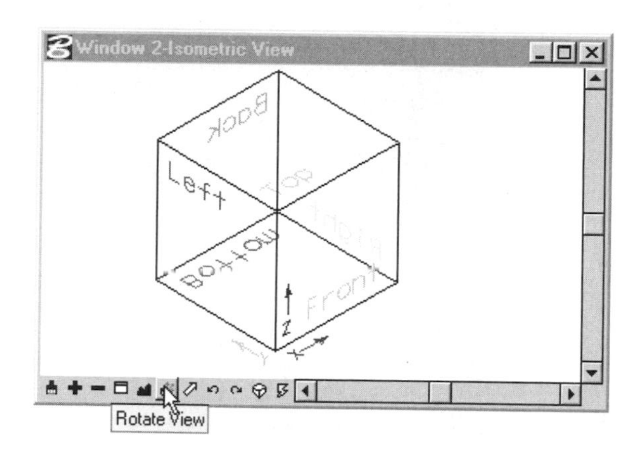

Figure 17–4 Invoking the Rotate View tool from the View Control bar

Select one of the 10 available options in the Tool Settings window. To position the view to one of the standard view orientations, first select the standard view orientation (Top, Bottom, Front, Back, Right, Left, Isometric [Top, Front, or Left] or Isometric [Top, Front, or Right]). Then place a data point in the view window where you want to display the view orientation. Selection of the Dynamic option allows you to position the view at any angle around a data point. The three points allow you to rotate the view by defining three data points (origin, direction of the *X* axis, and a point defining the *Y* axis).

Changing View Rotation from the View Rotation Settings Box

You can also rotate the view by specifying the angle of rotation in the View Rotation settings box. Open the View Rotation settings box:

3D View Control tool box	Select the Change View Rotation tool (see Figure 17–5).
Key-in window	**dialog view rotation** (or **di viewro**) Enter

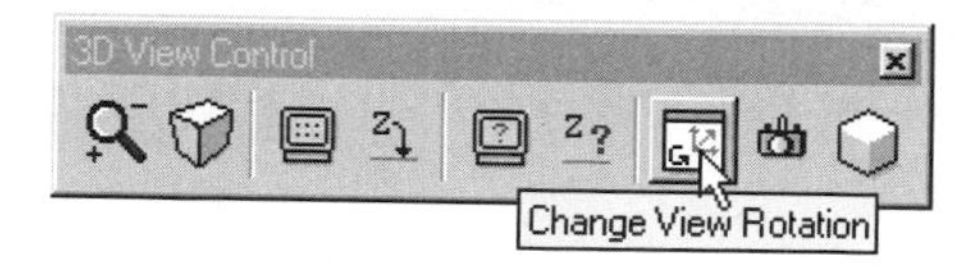

Figure 17–5 Invoking the Change View Rotation tool from the 3D View Control tool box

MicroStation displays the View Rotation settings box, similar to Figure 17–6.

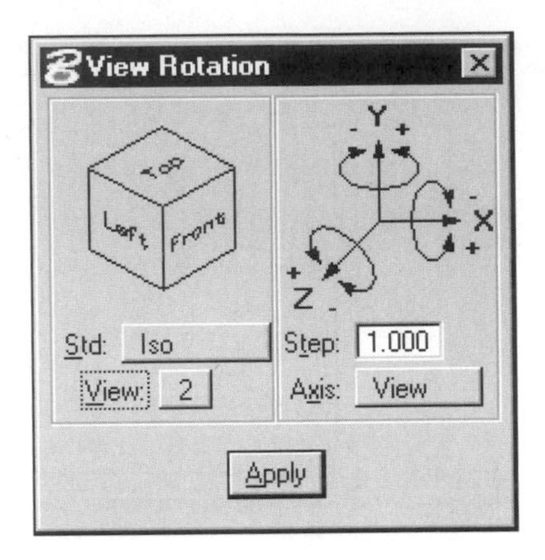

Figure 17-6 View Rotation settings box

The view to be manipulated is selected in the View: option menu. Key-in the rotation increment in degrees in the Step: edit field. Click the "+" control to rotate the view in the positive direction by the Step amount around the specified Axis. Click the "–" to rotate the view in the negative direction by the Step amount around the specified Axis. If you want to reposition the view to one of the standard view orientations, select the view orientation from the Std: option menu. Click the Apply button to rotate the selected view to the specified rotation.

Rotating View by Key-in

You can also rotate the view via one of the three key-in tools:

Key-in window	**vi =** <name of the view> Enter **rotate view absolute =** <xx,yy,zz> Enter or **rotate view relative =** <xx,yy,zz> Enter

In the VI key-in, you can specify either one of the standard view rotations (top, bottom, front, back, right, left, or iso) or the name of the saved view.

In the Rotate View Absolute key-in, xx, yy, and zz are the rotations, in degrees, about the view *X*, *Y*, and *Z* axes (by default, 0 for each).

In the Rotate View Relative key-in, xx, yy, and zz are the relative, counter-clockwise rotations, in degrees, about the view *X*, *Y*, and *Z* axes. This key-in follows what is commonly called the "right-hand-rule." For example, if you key-in a positive *X* rotation and point your right thumb in the view's positive *X* direction, then the way your fingers curl is the direction of the rotation.

DESIGN CUBE

Whenever you start a new *2D* design, you get a design plane—the electronic equivalent of a sheet of paper on a drafting table. The *2D* design plane is a large, flat plane covered with an invisible matrix grid consisting of 4,294,967,296 (2^{32}) positional units (UOR) along the *X* and *Y* axes. In *3D* design, you use that same *XY* plane plus a third-dimension *Z* axis. The *Z* axis is the depth in the direction perpendicular to the *XY* plane. The volume defined by *X*, *Y*, and *Z* is called the *design cube*. Similar to the design plane, the design cube is covered with an invisible matrix grid consisting of 4,294,967,296 (2^{32}) positional units along each of the *X*, *Y*, and *Z* axes. The global origin (0,0,0) is at the very center of the design cube; see Figure 17–7.

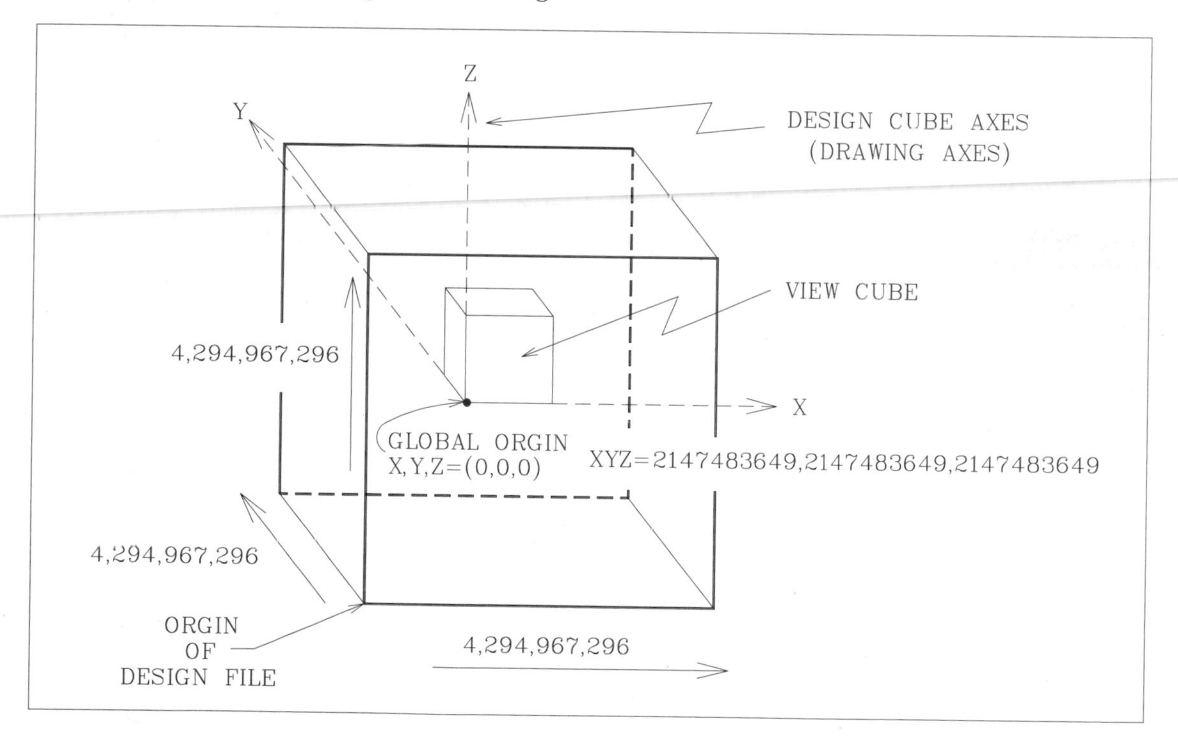

Figure 17–7 Design cube

DISPLAY DEPTH

Display depth enables you to display a portion of your design rather than the entire design. The ability to look at only a portion of the depth comes in handy—especially if the design is complicated. Display depth settings define the front and back clipping planes for elements displayed in a view and is set for each view. Elements not contained in the display depth do not show up on the screen. If you need to work with an element outside the display depth, you must change the display depth to include the element.

Setting Display Depth

To set the display depth, invoke the Set Display Depth tool:

3D View Control tool box	Select the Set Display Depth tool (see Figure 17–8).
Key-in window	**depth display** (or **dep d**) Enter

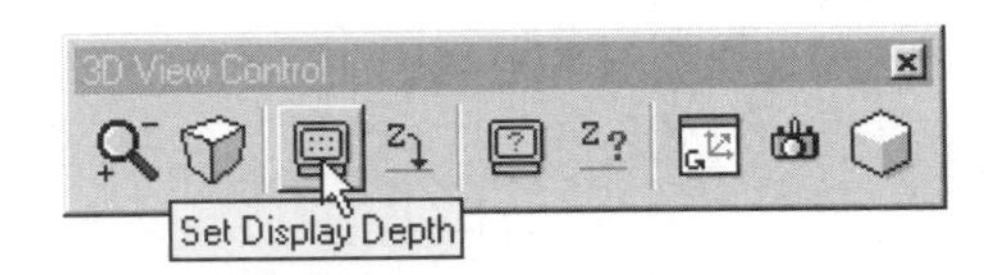

Figure 17–8 Invoking the Set Display Depth tool from the 3D View Control tool box

MicroStation prompts:

> Set Display Depth > Select view for display depth *(Place a data point in a view where you want to set the display depth.)*
>
> Set Display Depth > Define front clipping plane *(Place a data point in any view where you can identify the front clipping plane.)*
>
> Set Display Depth > Define back clipping plane *(Place a data point in any view where you can identify the back clipping plane.)*

You can also set the display depth by keying-in the distances in MU:SU:PU along the view *Z* axis by absolute or relative coordinates. To set the display depth by absolute coordinates, key-in:

Key-in window	**dp =** <front,back> Enter

The <front,back> are the distances in MU:SU:PU along the view *Z* axis from the global origin to the desired front and back clipping planes. MicroStation prompts:

> Set Display Depth > Select view *(Identify the view with a data point to set the display depth.)*

To set the display depth by relative coordinates, key-in:

Key-in window	**dd =** <front,back> Enter

The <front,back> are the distances in MU:SU:PU, and they add the keyed-in values to the current display depth settings. MicroStation prompts:

Set Display Depth > Select view *(Identify the view with a data point to set the display depth.)*

To determine the current setting for display depth, invoke the Show Display Depth tool:

3D View Control tool box	Select the Show Display Depth tool (see Figure 17–9).
Key-in window	**dp = $** (or **dd = $**) Enter

Figure 17–9 Invoking the Show Display Depth Tool from the 3D View Control tool box

MicroStation prompts:

Show Display Depth > Select view *(Place a data point anywhere in the view window.)*

MicroStation displays the current setting of the display depth in the Status bar.

Note: When you are setting up the display depth in all views, you will notice dashed lines indicating the viewing parameters of the selected view. Both the display volume of the view and the active depth plane are dynamically displayed, with different-style dashed lines.

In Figure 17–10, the display depth is set in such a way that only the square box is displayed but not the circles.

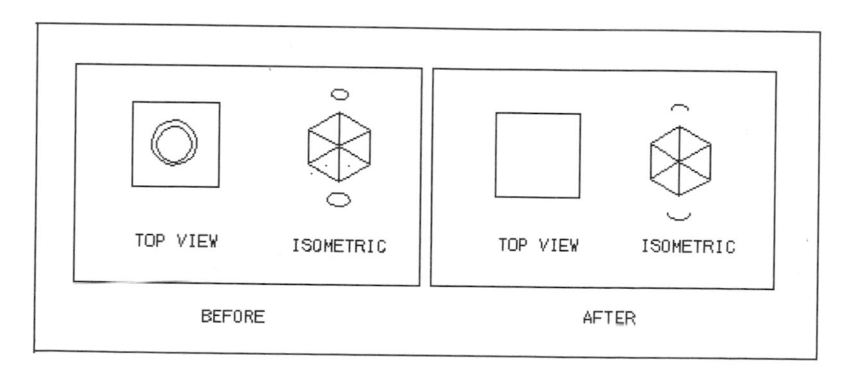

Figure 17–10 Example of setting up the display depth

Fitting Display Depth to Design File Elements

A fast way to display the entire design is to invoke the Fit View tool and select the appropriate view. In *2D* design the Fit tool adjusts the view window to include all elements in the design file. Similarly, in *3D* design the Fit View tool adjusts both the view window and the display depth to include all the elements in the design file. The Fit View tool automatically resets the display depth to the required amount to display the entire design file.

ACTIVE DEPTH

MicroStation has a feature that allows you to place an element in front of or behind the *XY* plane (front and back views), to the left or right of the *YZ* plane (right and left views), and above or below the *XY* plane (top and bottom views). This can be done by setting up the *active depth.* The active depth is a plane, parallel to the screen in each view, where elements will be placed by default. Each view has its own active depth plane, which you can change at any time.

Elements are placed at the active depth if you don't either tentatively snap to an existing element or use a precision input. In the top and bottom views (*XY* plane), the depth value is along its *Z* axis. In the front and back views (*XZ* plane), the depth value is along its *Y* axis. And in the case of right and left views (*YZ* plane), the depth value is along its *X* axis.

Setting Active Depth

To set the active depth, invoke the Set Active Depth tool:

3D View Control tool box	Select the Set Active Depth tool (see Figure 17–11).
Key-in window	**depth active** (or **dep a**) Enter

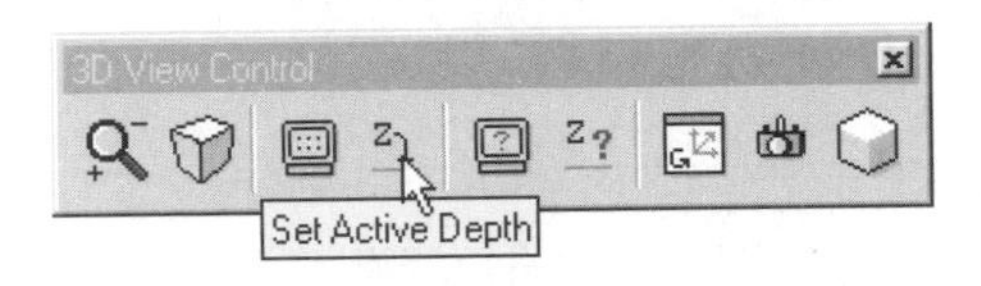

Figure 17–11 Invoking the Set Active Depth tool from the 3D View Control tool box

MicroStation prompts:

Set Active Depth > Select view *(Place a data point in a view where you want to set the active depth.)*

Set Active Depth > Enter active depth point *(Place a data point in a different view where you can identify the location for setting up the active depth.)*

Note: When you are setting up the active depth in all views, you will notice dashed lines indicating the viewing parameters of the selected view. Both the display volume of the view and the active depth plane are dynamically displayed, with different-style dashed lines.

You can also set the active depth by keying-in the distances in MU:SU:PU along the view *Z* axis by absolute or relative coordinates. To set the active depth by absolute coordinates, key-in:

Key-in window	**az** = <depth> Enter

The <depth> is the distance in MU:SU:PU along the view *Z* axis from the global origin to the desired active depth. MicroStation prompts:

Set Active Depth > Select view *(Identify the view with a data point to set the active depth.)*

To set the active depth by relative coordinates, key-in:

Key-in window	**dz** = <depth> Enter

The <depth> is the distance in MU:SU:PU, and it adds the keyed-in values to the current active depth setting. MicroStation prompts:

Set Active Depth > Select view *(Identify the view with a data point to set the active depth.)*

To determine the current setting for active depth, invoke the Show Active Depth tool:

3D View Control tool box	Select the Show Active Depth tool (see Figure 17–12).
Key-in window	**az = $** (or **dz = $**) Enter

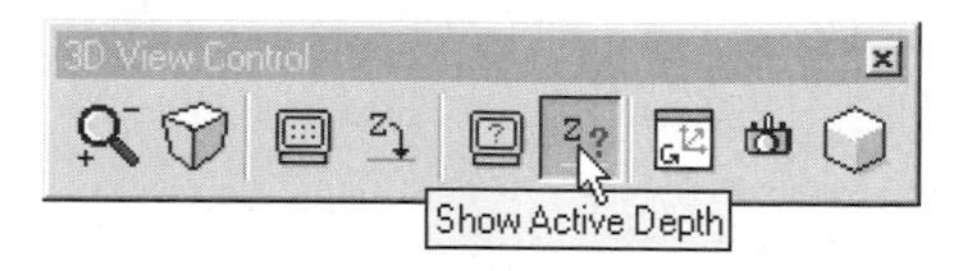

Figure 17–12 Invoking the Show Active Depth tool in the 3D View Control tool box

MicroStation prompts:

Show Active Depth > Select view *(Place a data point anywhere in the view window.)*

MicroStation displays the current setting of the active depth in the Status bar.

Note: Before you place elements, make sure you are working at the appropriate active depth and display depth.

If you set the active depth outside the range of the display depth, then MicroStation displays the following message in the error field:

Active depth set to display depth

The active depth is set to the value closest to the display depth.

Let's say the current display depth is set to 100,450 and you set the active depth to 525. Since MicroStation sets the active depth to the closest value, in this case the active depth is set to 450. Make sure that MicroStation sets the value for the active depth to the value intended. If necessary, change the display depth and then set the active depth.

Boresite Lock

The Boresite lock controls the manipulation of the elements at different depths. If the Boresite lock is set to ON, you can identify or snap to elements at any depth in the view; elements being moved or copied will remain at their original depths. If it is set to OFF, you can identify only those elements at, or very near, the active depth of a view. You can toggle ON or OFF for the Boresite lock from the Lock Toggles settings box, as shown in Figure 17–13.

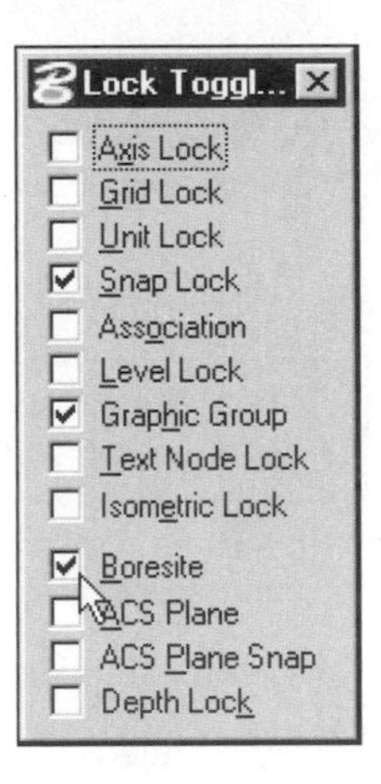

Figure 17–13 Lock Toggles settings box

Note: Tentative points override the Boresite lock. You can tentatively snap to elements at any depth regardless of the Boresite lock setting.

PRECISION INPUTS

When MicroStation prompts for the location of a point, in addition to providing the data point with your pointing device, you can use precision input tools that allow you to place data points precisely. Similar to *2D* placement tools, *3D* tools also allow you to key-in by coordinates. MicroStation provides two types of coordinate systems for *3D* design: The drawing coordinate system and the view coordinate system.

Drawing Coordinate System

The drawing coordinate system is the model coordinate system fixed relative to the design cube, as shown in Figure 17–14. For example, in the TOP view, *X* is to the right, *Y* is up, and *Z* is out of the screen (right-hand rule). In the RIGHT view, *Y* is to the right, *Z* is up, and *X* is out of the screen, and so on. Following are the two key-ins available for the drawing coordinate system:

XY=<X,Y,Z>

DL=<delta_x,delta_y,delta_z>

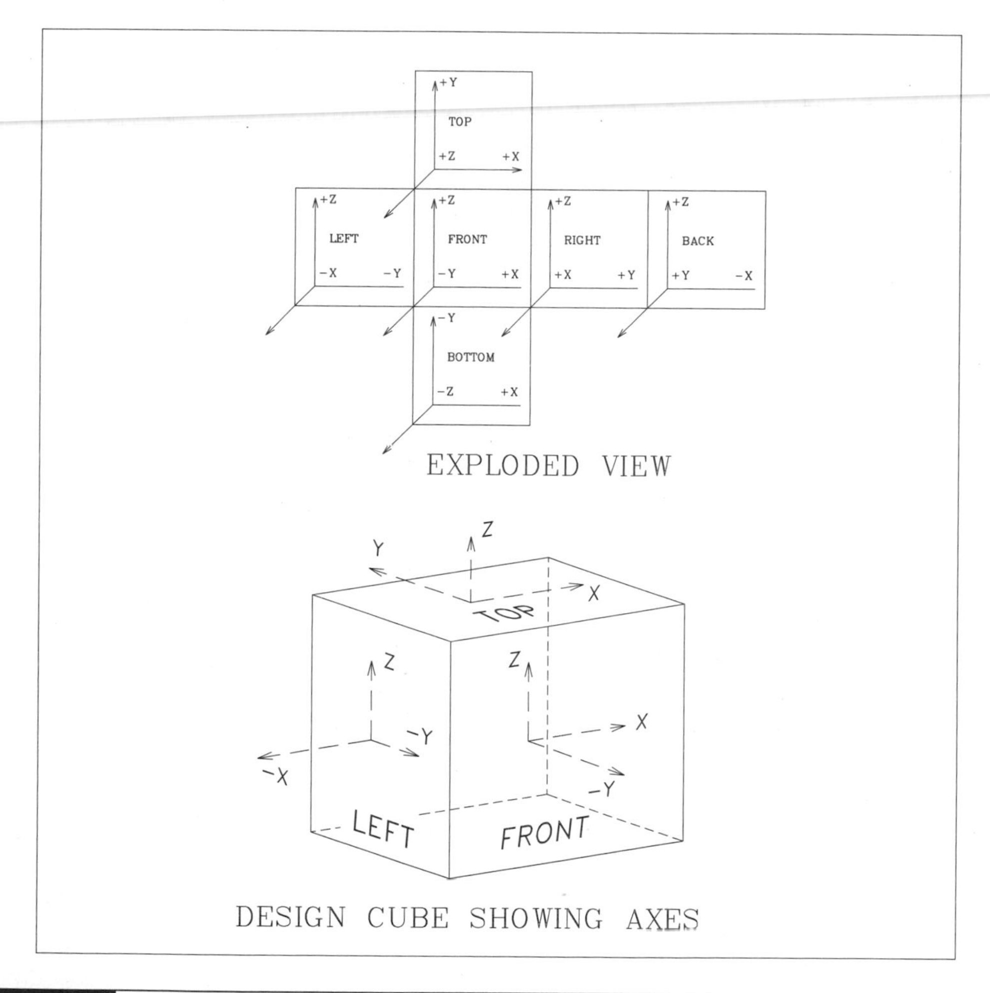

Figure 17–14 Design cube showing the drawing coordinate system

The XY= key-in places a data point measured from the global origin of the drawing coordinate system. The <X,Y,Z> are the *X*, *Y*, and *Z* values of the coordinates. The view being used at the time has no effect on them. The DL= places a data point to a distance along the drawing axes from a previous data (relative) or tentative point. The <delta_x,delta_y,delta_z> are the relative coordinates in the *X*, *Y*, and *Z* axes relative to the previous data point or tentative point.

View Coordinate System

The view coordinate system inputs data relative to the screen, where *X* is to the right, *Y* is up, and *Z* comes directly out from the screen in all views, as shown in Figure 17–15. The view coordinate system is view-dependent, that is, depends on the orientation of the view for their direction. Following are the two key-ins available for the view coordinate system:

DX=<delta_x,delta_y,delta_z>

DI=<distance,direction>

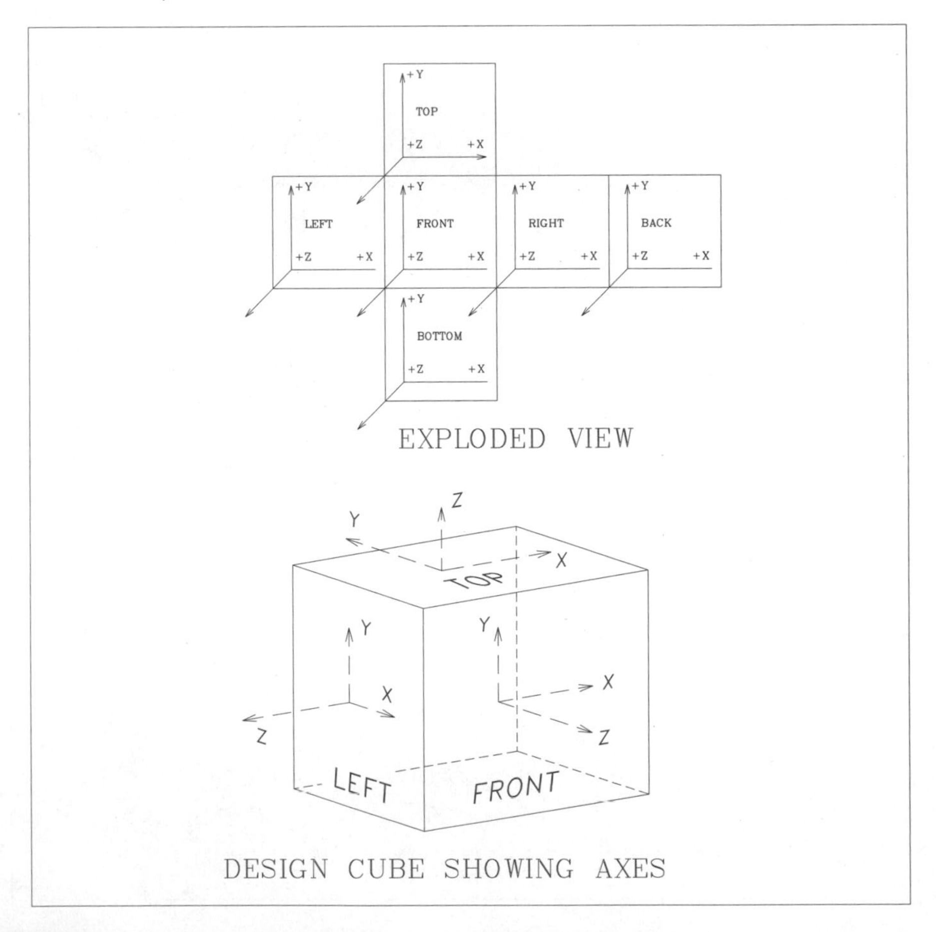

Figure 17–15 Design cube showing the view coordinate system

The DX= key-in places a data point to a distance from the previous data or tentative point (relative) in the same view where the previous point was defined. The <delta_x,delta_y,delta_z> are the relative coordinates in the *X*, *Y*, and *Z* axes relative to the previous data point or tentative point. The DI= key-in places a data point a certain distance and direction from a previous data or tentative point (relative polar) in the same view where the previous point was defined. The <distance,direction> are specified in relation to the last specified position or point. The distance is specified in current working units (MU:SU:PU), and the direction is specified as an angle, in degrees, relative to the *X* axis.

Note: With key-in precision inputs, MicroStation assumes that the view you want to use is the one you last worked in—that is, the view in which the last tentative or data point was placed. The easiest way to make a view current is to place a tentative point and then press the Reset button. Updating a view is also another way to tell MicroStation that the selected view is the view last worked in.

AUXILIARY COORDINATE SYSTEMS (ACS)

MicroStation provides a set of tools to define an infinite number of user-defined coordinate systems called *auxiliary coordinate systems.* An auxiliary coordinate system allows the user to change the location and orientation of the *X*, *Y*, and *Z* axes to reduce the calculations needed to create *3D* objects. You can redefine the origin in your drawing, and establish positive *X* and *Y* axes. New users think of a coordinate system simply as the direction of positive *X* and positive *Y*. But once the directions *X* and *Y* are defined, the direction of *Z* will be defined as well. Thus, the user only has to be concerned with *X* and *Y*. For example, if a sloped roof of a house is drawn in detail using the drawing coordinate system, each end point of each element on the inclined roof plane must be calculated. On the other hand, if the auxiliary coordinate system is set to the same plane as the roof, each object can be drawn as if it were in the plan view. You can define any number of auxiliary coordinate systems, assigning each a user-determined name. But, at any given time, only one auxiliary coordinate system is current with the default system.

MicroStation provides a visual reminder of how the ACS axes are orientated and where the current ACS origin is located. The *X*, *Y*, and *Z* axis directions are displayed using arrows labeled appropriately. The display of the ACS axes is controlled by turning ON/OFF the ACS Triad in the View Attributes settings box, as shown in Figure 17–16.

MicroStation provides you with three types of coordinate systems for defining an ACS: rectangular, cylindrical, and spherical coordinate systems.

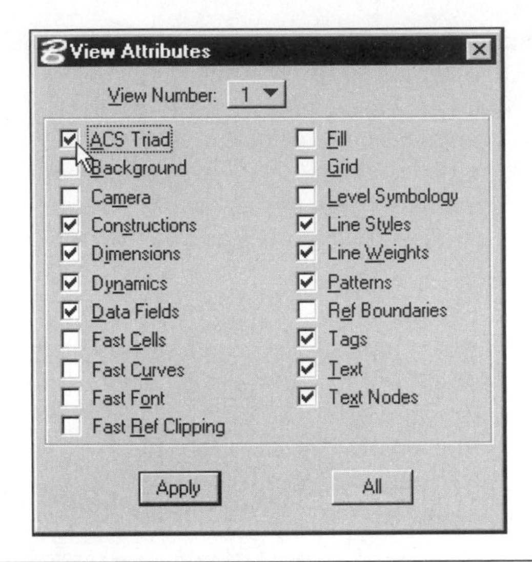

Figure 17-16 View Attributes settings box showing the ACS Triad set to ON

Rectangular Coordinate System

The rectangular coordinate system is the same one that is available for the design cube and is also the default type to define an ACS.

Cylindrical Coordinate System

The cylindrical coordinate system is another *3D* variant of the polar format. It describes a point by its distance from the origin, its angle in the *XY* plane from the *X* axis, and its *Z* value. For example, to specify a point at a distance of 4.5 units from the origin, at an angle of 35 degrees relative to the *X* axis (in the *XY* plane), and with a *Z* coordinate of 7.5 units, you would enter: **4.5,35,7.5**.

Spherical Coordinate System

The spherical coordinate system is another *3D* variant of the polar format. It describes a point by its distance from the current origin, its angle in the *XY* plane, and its angle up from the *XY* plane. For example, to specify a point at a distance of 7 units from the origin, at an angle of 60 degrees from the *X* axis (in the *XY* plane), and at an angle 45 degrees up from the *XY* plane, you would enter: **7,60,45**.

Precision Input Key-in

Similar to the key-ins available for drawing and view coordinates, MicroStation provides key-ins to input the coordinates in reference to the auxiliary coordinate system. Following are the two key-ins available for the auxiliary coordinate system:

AX=<X,Y,Z>

AD=<delta_x,delta_y,delta_z>

The AX= key-in places a data point measured from the ACS origin and is equivalent to the key-in XY=. The <X,Y,Z> are the *X*, *Y*, and *Z* values of the coordinates. The AD= places a data point to a distance along the drawing axes from a previous data (relative) or tentative point and is equivalent to the key-in DL=. The <delta_x,delta_y,delta_z> are the relative coordinates in the *X*, *Y*, and *Z* axes relative to the previous data point or tentative point.

Defining an ACS

MicroStation provides three different tools to define an ACS. The tools are available in the ACS tool box. Before you select one of the three tools, select the coordinate system you wish to use with the new ACS from the Type: option menu in the Tool Settings window. In addition, you can control the ON/OFF toggle buttons for two locks in the Tool Settings window. When the ACS Plane Lock is set to ON, each data point is forced to lie on the active ACS's *XY* plane (Z=0). When the ACS Plane Snap Lock is set to ON, each tentative point is forced to lie on the active ACS's *XY* plane (Z=0).

Defining an ACS by Aligning with an Element

This option lets you define an ACS by identifying an element where the *XY* plane of the ACS is parallel to the plane of the selected planar element. The origin of the ACS is at the point of identification of the element. Upon definition, the ACS becomes the active ACS.

To define an ACS aligned with an element, invoke the Define ACS (aligned with element) tool:

ACS tool box	Select the Define ACS (Aligned with Element) tool (see Figure 17–17).
Key-in window	**define acs element** (or **d a e**) Enter

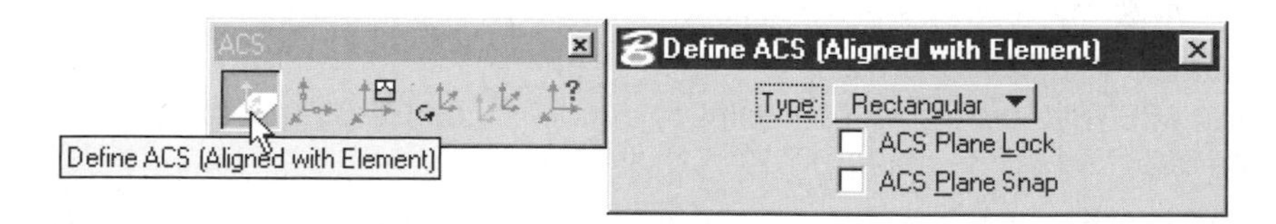

Figure 17–17 Invoking the Define ACS (Aligned with Element) tool from the ACS tool box

MicroStation prompts:

Define ACS (Aligned with Element) > Identify element *(Identify the element with which to align the ACS and define the ACS origin.)*

Define ACS (Aligned with Element) > Accept/Reject (Select next input) *(Place a data point to accept the element for defining an ACS or click the Reject button to cancel the operation.)*

Defining the ACS by Points

This option is the easiest and most used option for controlling the orientation of the ACS. It allows the user to place three data points to define the origin and the directions of the positive *X* and *Y* axes. The origin point acts as a base for the ACS rotation, and when a point is selected to define the direction of the positive *X* axis, the direction of the *Y* axis is limited because it is always perpendicular to the *X* axis. When the *X* and *Y* axes are defined, the *Z* axis is automatically placed perpendicular to the *XY* plane. Upon definition, the ACS becomes the active ACS.

To define an ACS by Points, invoke the Define ACS (By Points) tool:

ACS tool box	Select the Define ACS (By Points) tool (see Figure 17–18).
Key-in window	**define acs points** (or **d a p**) Enter

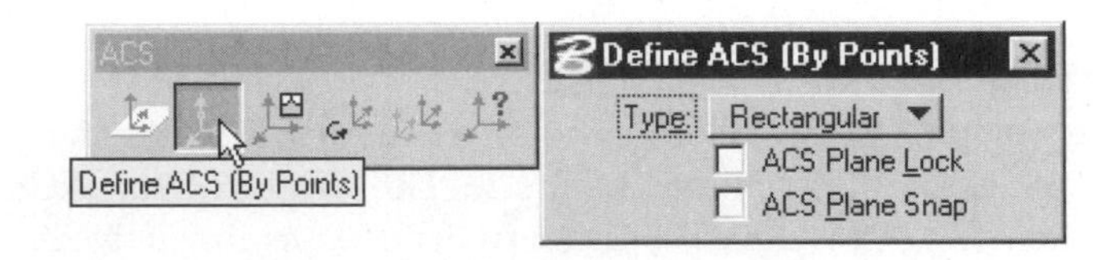

Figure 17–18 Invoking the Define ACS (By Points) tool from the ACS tool box

MicroStation prompts:

Define ACS (By Points) > Enter first point @x axis origin *(Place a data point to define the origin.)*

Define ACS (By Points) > Enter second point on *X* axis *(Place a data point to define the direction of the positive* X *axis, which extends from the origin through this point.)*

Define ACS (By Points) > Enter point to define *Y* axis *(Place a data point to define the direction of the positive* Y *axis.)*

Defining the ACS by Aligning with a View

In this option, the ACS takes the orientation of the selected view. That is, the ACS axes align exactly with those of the view selected. Upon definition, the ACS becomes the active ACS.

To define an ACS by aligning with a view, invoke the Define ACS (Aligned with View) tool:

ACS tool box	Select the Define ACS (Aligned with View) tool (see Figure 17–19).
Key-in window	**define acs view** (or **d a v**) Enter

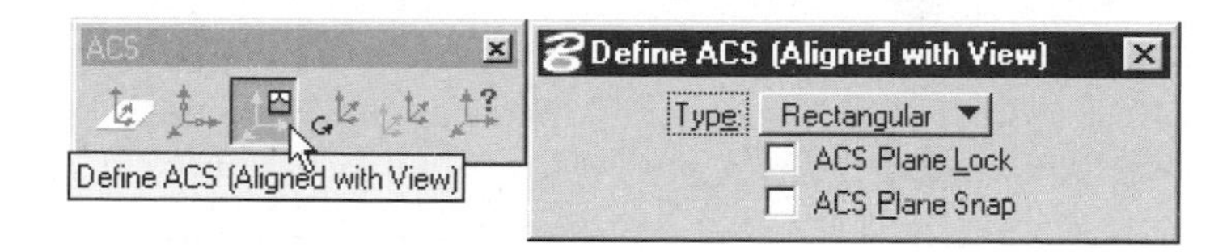

Figure 17–19 Invoking the Define ACS (Aligned with View) tool from the ACS tool box

MicroStation prompts:

> Define ACS (Aligned with View) > Select source view *(Place a data point to select the view with which the ACS is to be aligned and define the ACS origin.)*

Rotating the Active ACS

The Rotate Active ACS tool rotates the Active ACS. The origin of the ACS is not moved. To rotate the active ACS, invoke the Rotate Active ACS tool:

ACS tool box	Select the Rotate Active ACS tool (see Figure 17–20).

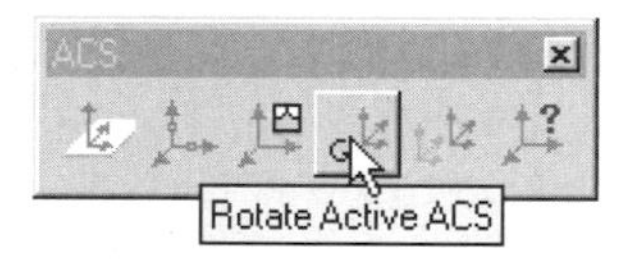

Figure 17–20 Invoking the Rotate Active ACS Tool from the ACS tool box

MicroStation displays the Rotate Active ACS dialog box, as shown in Figure 17–21.

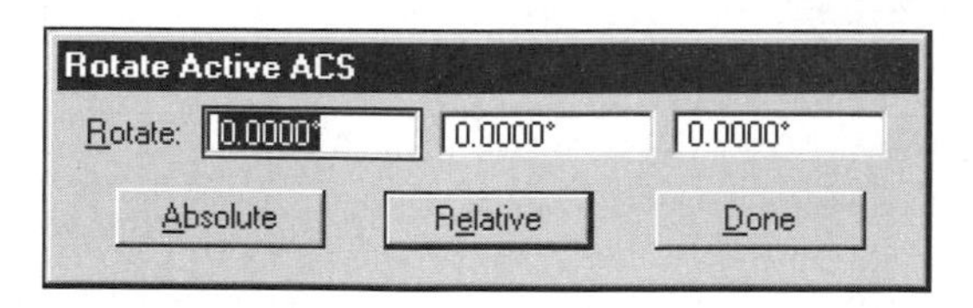

Figure 17–21 Rotate Active ACS dialog box

Key-in the rotation angles, in degrees, from left to right, for the *X*, *Y*, and *Z* axes. Click the Absolute button to rotate the ACS in relation to the unrotated (top) orientation. Click the Relative button to rotate the ACS in relation to the current orientation. When you are finished, click the DONE button to close the Rotate Active ACS dialog box.

Moving the Active ACS

The Move ACS tool allows you to move the origin of the Active ACS, leaving the directions of the *X*, *Y*, and *Z* axes unchanged. To move the ACS, invoke the Move ACS tool:

ACS tool box	Select the Move ACS tool (see Figure 17–22).
Key-in window	**move acs** (or **mov a**) Enter

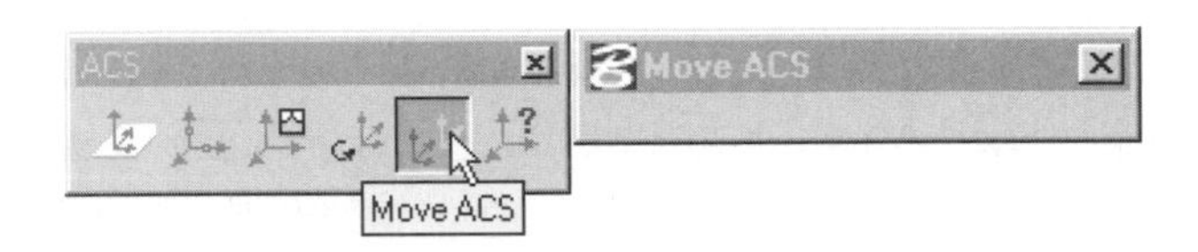

Figure 17–22 Invoking the Move ACS tool from the ACS tool box

MicroStation prompts:

Move ACS > Define origin *(Place a data point to define the new origin.)*

Selecting the Active ACS

The Select ACS tool allows you to identify an ACS for attachment as the active ACS from the saved ACS in each view. To select the ACS, invoke the Select ACS tool:

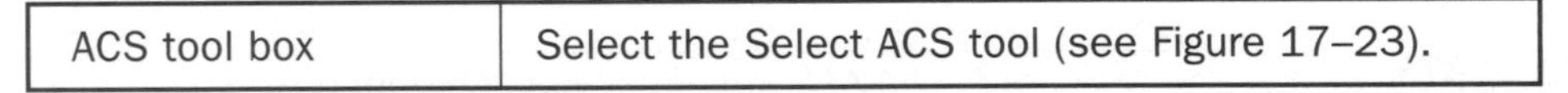

ACS tool box	Select the Select ACS tool (see Figure 17–23).

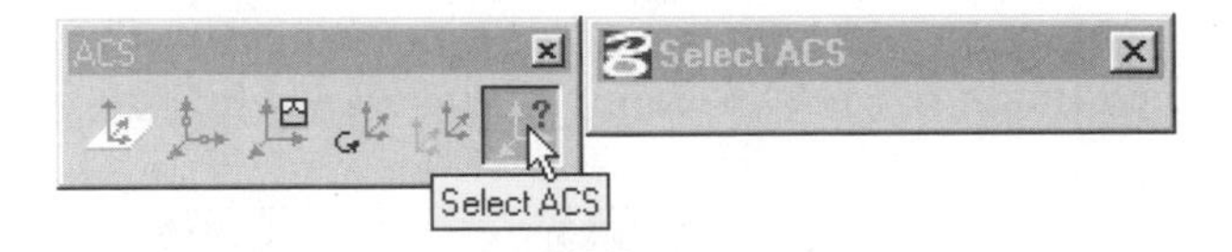

Figure 17–23 Invoking the Select ACS tool from the ACS tool box

MicroStation prompts:

Select ACS > Select auxiliary system @ origin *(Identify the ACS origin from the coordinate triad displayed.)*

Saving an ACS

You can define any number of ACSs in a design file. Of these, only one can be active at any time. Whenever you define an ACS, you can save it for future use. The Auxiliary Coordinate Systems settings box is used to name, save, attach, or delete an ACS.

Open the Auxiliary Coordinate Systems settings box:

Pull-down menu	Utilities > Auxiliary Coordinates

MicroStation displays an Auxiliary Coordinate Systems settings box, similar to Figure 17–24.

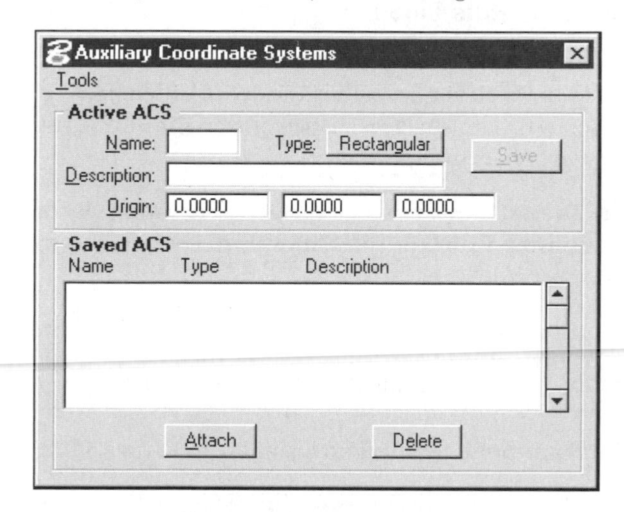

Figure 17–24 Auxiliary Coordinate Systems settings box

Key-in the name for the active ACS in the Name: edit field. The name is limited to six characters. Select the coordinate system you wish to save with the active ACS from the Type: option menu. Key-in the description (optional) of the active ACS in the Description: edit field. The description is limited to 27 characters. If necessary, you can change the origin of the active ACS by keying-in the coordinates in the Origin: edit field. Click the Save button to save the active ACS for future attachment. MicroStation will display the ACS name, type, and description in the Saved ACS list box.

To attach an ACS as an active ACS, select the name of the ACS from the Saved ACS list box and click the Attach button. To delete an ACS, first highlight the ACS in the Saved ACS list box, then click the Delete button.

Note: All the tools that are available in the ACS tool box are also available in the Tools pull-down menu of the Auxiliary Coordinate Systems settings box.

3D PRIMITIVES

MicroStation provides a set of tools to place simple *3D* elements that can become the basic building blocks that make up the model. The primitive tools include slab, sphere, cylinder, cone, torus, and wedge.

Two important settings need to be taken into consideration before the primitives are created that control the way in which solids and surfaces are created and displayed on the screen. The settings include: display method and number of rule lines that represent a surface with a full 360 degree curvature and selection of solids and surfaces.

Display Method and Surface Rule Lines

MicroStation provides two options: wireframe and surfaces, for on-screen display of the solids and surfaces. By default, it is set to wireframe, which is the more efficient mode for working with solids and surfaces in a design session. The surfaces display mode should be used only where the design is to be rendered with an earlier version of MicroStation.

Surface rule lines provide a visual indication of a surface's curvature. By default, it is set to 4—a full cylindrical solid is displayed with 4 surface rules lines. If necessary, you can increase the setting to display with additional surface rule lines.

To change the display method and set the surface rule lines, open the SmartSolid Settings box:

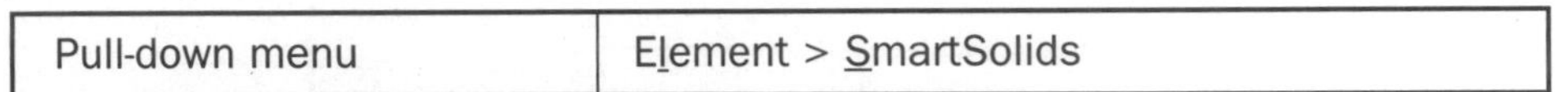

Pull-down menu	Element > SmartSolids

MicroStation displays the SmartSolid Settings box, similar to Figure 17–25.

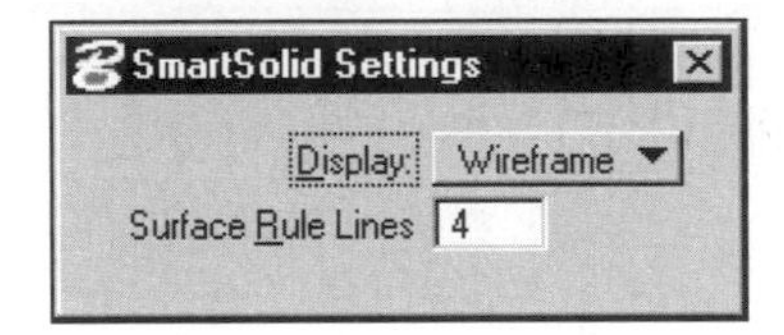

Figure 17–25 SmartSolid Settings box

The Display: drop-down menu allows you to select one of the two display methods. The Surface Rule Lines: edit field allows you to change the required surface rule lines.

Selection of Solids and Surfaces

MicroStation provides three options in the selection of solids and surfaces. By default, surfaces and solids may be identified with a data point anywhere on their surface, not necessarily on an edge line or surface rule line. If necessary, you can select the option that will allow you to select solids and surfaces with a data point on an edge or surface rule line. To change the selection mode, open the Preferences dialog box:

Pull-down menu	Workspace > Preferences...

MicroStation displays the Preferences dialog box, similar to Figure 17–26.

Select Input category from the Category list box. The Locate By Picking Faces: option menu provides the following three selection modes in the selection of solids and surfaces:

- Never—Solids and surfaces can only be identified with a data point on an edge or surface rule line.
- Rendered Views Only—Solids and surfaces rendered with any of the rendering options may be identified with a data point anywhere on their surface.
- Always—Solids and surfaces whether rendered or not may be identified with a data point anywhere on their surface.

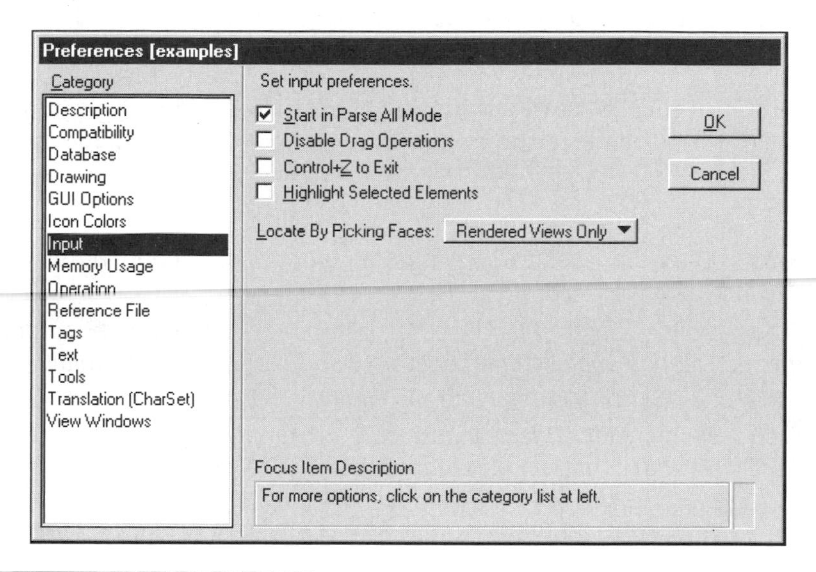

Figure 17–26 Preferences dialog box

Place Slab

The Place Slab tool places a volume of projection with a rectangular cross section. To place a slab, invoke the Place Slab tool:

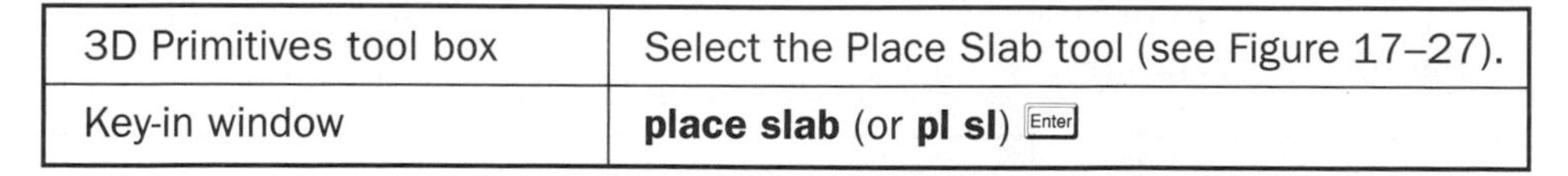

3D Primitives tool box	Select the Place Slab tool (see Figure 17–27).
Key-in window	**place slab** (or **pl sl**) Enter

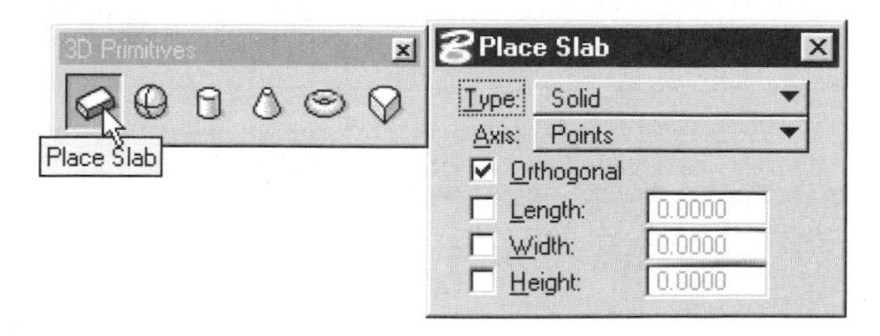

Figure 17–27 Invoking the Place Slab tool from the 3D Primitives tool box

Select the type of surface from the Type: option menu in the Tool Settings window. The surface (not capped) option is considered to be open at the base and top, whereas the solid (capped) option is considered to enclose a volume completely.

From the Axis: option menu in the Tool Settings window, select the direction in which the height is projected relative to the view or design file axes. If set to Screen X, Screen Y, or Screen Z, the height is projected with the selected screen (view) axis. If set to Drawing X, Drawing Y, or Drawing Z, the height is projected with the selected design file axis.

If necessary, turn ON the toggle buttons for Orthogonal, Length, Width, and Height in the Tool Settings window. If Orthogonal is set to ON, the edges are placed orthogonally. If you turn on the constraints for Length, Width, and Height, make sure to key-in appropriate values in the edit fields.

MicroStation prompts:

Place Slab > Enter start point *(Place a data point or key-in coordinates to define the origin.)*

Place Slab > Define Length *(Place a data point or key-in coordinates to define the length and rotation angle. If the Length constraint is set to ON, this data point defines the rotation angle.)*

Place Slab > Define Width *(Place a data point or key-in coordinates to define the width. If the Width constraint is set to ON, this data point accepts the width.)*

Place Slab > Define Height *(Place a data point or key-in coordinates to define the height. If the Height constraint is set to ON, this data point provides the direction.)*

Note: To place a volume of projection with a nonrectangular cross section, use the Extrude tool in the 3D Construct tool box.

Place Sphere

The Place Sphere tool can place a sphere, in which all surface points are equidistant from the center. To place a sphere, invoke the Place Sphere tool:

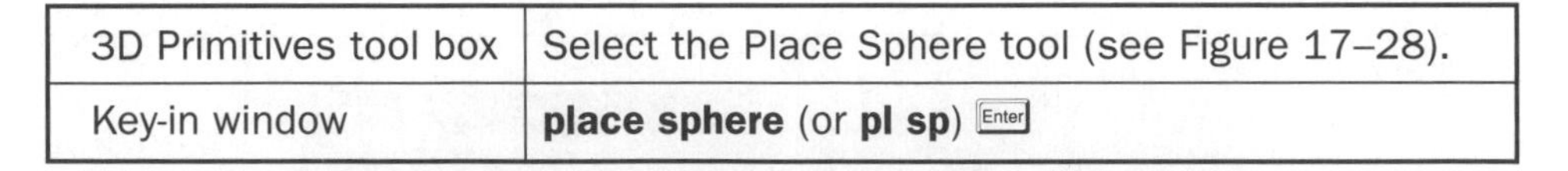

3D Primitives tool box	Select the Place Sphere tool (see Figure 17–28).
Key-in window	**place sphere** (or **pl sp**) Enter

Figure 17–28 Invoking the Place Sphere tool from the 3D Primitives tool box

From the Axis: option menu in the Tool Settings window, select the direction of the sphere's axis relative to the view or design file axes. If set to Screen X, Screen Y, or Screen Z, the sphere's axis is set with the selected screen (view) axis. If set to Drawing X, Drawing Y, or Drawing Z, the sphere's axis is set with the selected design file axis.

If necessary, turn ON the toggle button for Radius constraint, and key-in the radius in the Radius: edit field.

MicroStation prompts:

Place Sphere > Enter center point *(Place a data point or key-in coordinates to define the sphere's center.)*

Place Sphere > Define radius *(Place a data point or key-in coordinates to define the radius. If Radius is set to ON, then the data point accepts the sphere.)*

Note: To place a volume of revolution with a noncircular cross section, use the Extrude tool in the 3D Construct tool box.

Place Cylinder

The Place Cylinder tool places a cylinder of equal radius on each end and similar to an extruded circle. To place a cylinder, invoke the Place Cylinder tool:

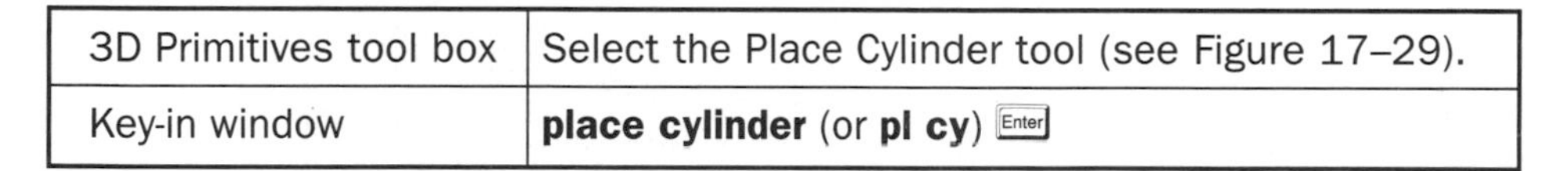

3D Primitives tool box	Select the Place Cylinder tool (see Figure 17–29).
Key-in window	**place cylinder** (or **pl cy**) Enter

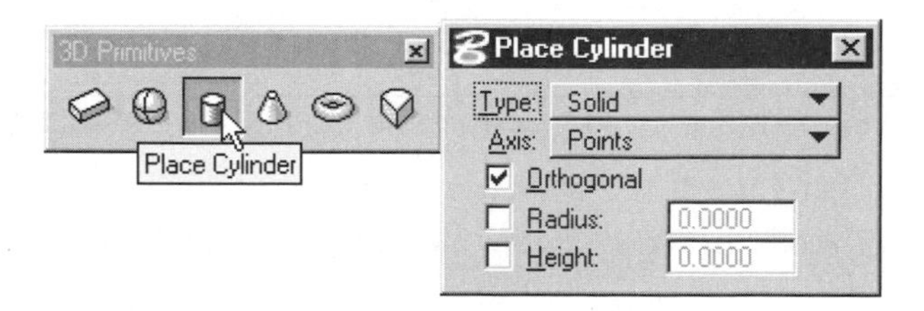

Figure 17–29 Invoking the Place Cylinder tool from the 3D Primitives tool box

Select the type of surface from the Type: option menu in the Tool Settings window.

From the Axis: option menu in the Tool Settings window, select the direction of the cylinder's axis or its height relative to the view or design file axes. If set to Screen X, Screen Y, or Screen Z, the direction of the cylinder's axis or height is set with the selected screen (view) axis. If set to Drawing X, Drawing Y, or Drawing Z, the direction of the cylinder's axis or height is set with the selected design file axis.

If necessary, turn ON the toggle buttons for Orthogonal, Radius, and Height in the Tool Settings window. If Orthogonal is set to ON, the cylinder is a right cylinder. If you turn on the constraints for Radius and Height, make sure to key-in appropriate values in the edit fields.

MicroStation prompts:

> Place Cylinder > Enter center point *(Place a data point or key-in coordinates to define the center of the base.)*
>
> Place Cylinder > Define radius *(Place a data point or key-in coordinates to define the radius. If Radius is set to ON, then the data point accepts the base.)*
>
> Place Cylinder > Define height *(Place a data point or key-in coordinates to define the height. If Height is set to ON, then the data point accepts the cylinder.)*

Place Cone

The Place Cone tool places a cone of unequal radius on each end. To place a cone, invoke the Place Cone tool:

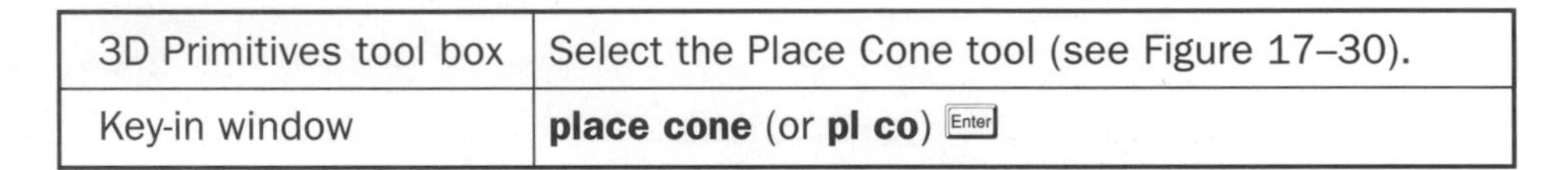

3D Primitives tool box	Select the Place Cone tool (see Figure 17–30).
Key-in window	**place cone** (or **pl co**) Enter

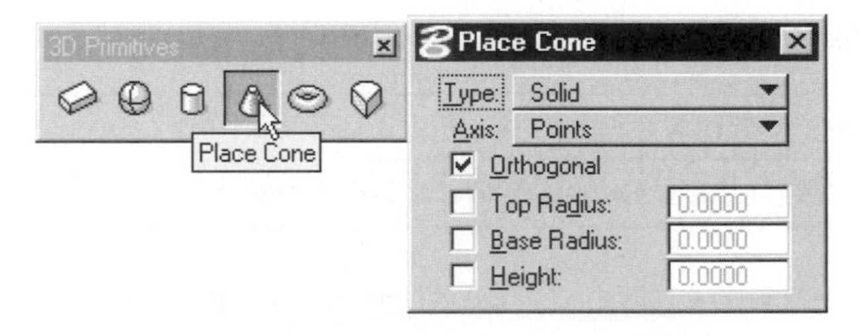

Figure 17–30 Invoking the Place Cone tool from the 3D Primitives tool box

Select the type of surface from the Type: option menu in the Tool Settings window.

From the Axis: option menu in the Tool Settings window, select the direction of the cone's axis or its height relative to the view or design file axes. If set to Screen X, Screen Y, or Screen Z, the direction of the cone's axis or height is set with the selected screen (view) axis. If set to Drawing X, Drawing Y, or Drawing Z, the direction of the cone's axis or height is set with the selected design file axis.

If necessary, turn ON the toggle buttons for Orthogonal, Top Radius, Base Radius, and Height in the Tool Settings window. If Orthogonal is set to ON, the cone is a right cone. If you turn ON the constraints for Top Radius, Base Radius, and Height, make sure to key-in appropriate values in the edit fields.

MicroStation prompts:

Place Cone > Enter center point *(Place a data point or key-in coordinates to define the center of the base.)*

Place Cone > Define radius *(Place a data point or key-in coordinates to define the base radius. If Base Radius is set to ON, then the data point accepts the base.)*

Place Cone > Define height *(Place a data point or key-in coordinates to define the height and top's center. If Height is set to ON, then the data point defines the top's center; if Orthogonal is set to ON, then the data point defines the direction of the height only.)*

Place Cone > Define top radius *(Place a data point or key-in coordinates to define the top radius. If Top Radius is set to ON, then the data point accepts the cone.)*

Place Torus

The Place Torus tool creates a solid or surface with a donut-like shape. To place a torus, invoke the Place Torus tool:

3D Primitives tool box	Select the Place Torus tool (see Figure 17–31).
Key-in window	**place torus** (or **pl to**) Enter

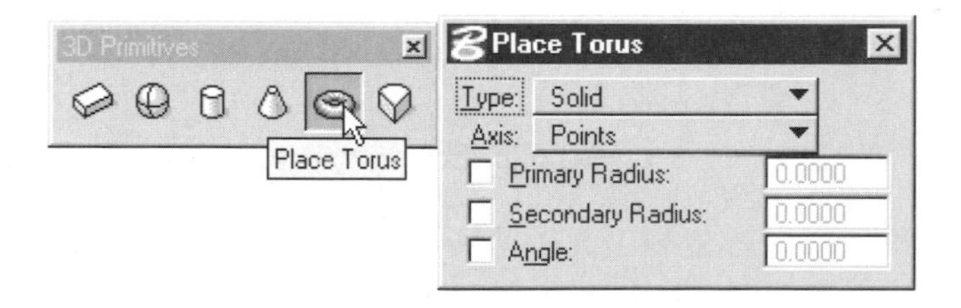

Figure 17–31 Invoking the Place Torus tool from the 3D Primitives tool box

Select the type of surface from the Type: option menu in the Tool Settings window.

Select the direction of the axis of revolution relative to the view or design file axes from the Axis option menu in the Tool Settings window. If set to Screen X, Screen Y, or Screen Z, the axis or revolution is set with the selected screen (view) axis. If set to Drawing X, Drawing Y, or Drawing Z, the axis of revolution is set with the selected design file axis.

If necessary, turn ON the toggle buttons for Primary Radius, Secondary Radius, and Angle in the Tool Settings window. If you turn ON the constraints for Primary Radius, Secondary Radius, and Angle, make sure to key-in appropriate values in the edit fields.

MicroStation prompts:

Place Torus > Enter start point *(Place a data point or key-in coordinates to define the start point.)*

Place Torus > Define center point *(Place a data point or key-in coordinates to define the center point, primary radius, and start angle. If Primary Radius is set to ON, then the data point defines the center and the start angle.)*

Place Torus > Define angle and secondary radius *(Place a data point or key-in coordinates to define the secondary radius and the sweep angle. If Secondary Radius is set to ON, then the data point defines the sweep angle; if Angle is set to ON, then the data point defines the secondary radius; and if both Secondary Radius and Angle are set to ON, then the data point defines the direction of the sweep angle rotation.)*

Place Wedge

The Place Wedge tool creates a wedge—a volume of revolution with a rectangular cross section. To place a wedge, invoke the Place Wedge tool:

3D Primitives tool box	Select the Place Wedge tool (see Figure 17–32).
Key-in window	**place wedge** (or **pl w**) Enter

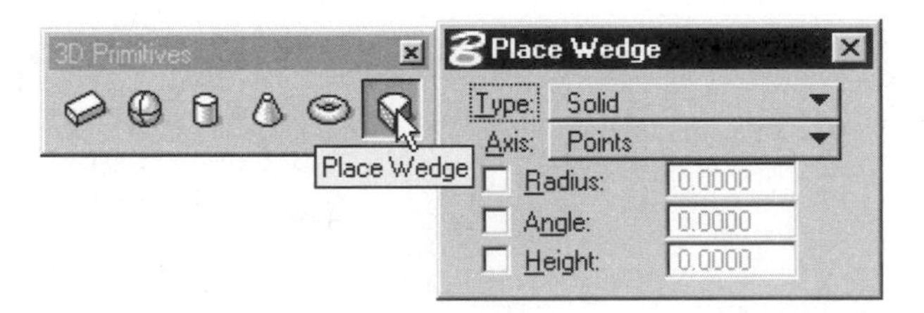

Figure 17–32 Invoking the Place Wedge tool from the 3D Primitives tool box

Select the type of surface from the Type: option menu in the Tool Settings window.

Select the direction of the axis of revolution relative to the view or design file axes from the Axis option menu in the Tool Settings window. If set to Screen X, Screen Y, or Screen Z, the axis of revolution is set with the selected screen (view) axis. If set to Drawing X, Drawing Y, or Drawing Z, the axis of revolution is set with the selected design file axis.

If necessary, turn ON the toggle buttons for Radius, Angle, and Height in the Tool Settings window. If you turn ON the constraints for Radius, Angle, and Height, make sure to key-in appropriate values in the edit fields.

MicroStation prompts:

Place Wedge > Enter start point *(Place a data point or key-in coordinates to define the start point.)*

Place Wedge > Define center point *(Place a data point or key-in coordinates to define the center point and the start angle. If Radius is set to ON, then the data point defines the start angle.)*

Place Wedge > Define angle *(Place a data point or key-in coordinates to define the sweep angle. If Angle is set to ON, then the data point defines the direction of the rotation.)*

Place Wedge > Define Height *(Place a data point or key-in coordinates to define the height. If Height is set to ON, then the data point defines whether the wedge is projected up or down from the start plane.)*

CHANGING THE STATUS—SOLID OR SURFACE

The Convert 3D tool can change the status of an element from surface to solid, or vice versa. To change the status, invoke the Convert 3D tool:

Modify Surfaces tool box	Select the Convert 3D tool (see Figure 17–33).
Key-in window	**change surface cap** (or **chan su c**) Enter

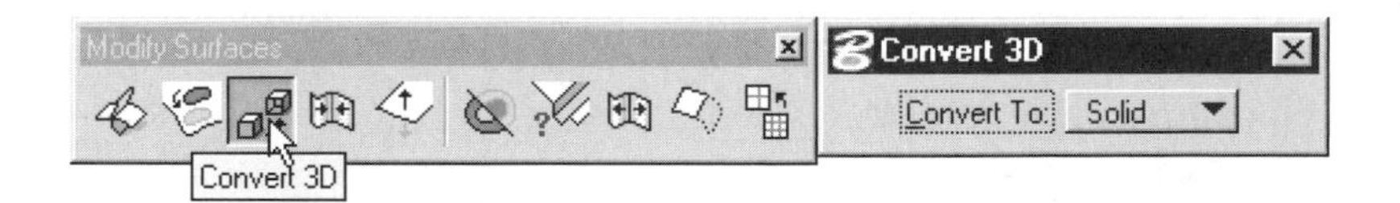

Figure 17–33 Invoking the Convert 3D tool from the Modify Surfaces tool box

Select the type of surface whose status you wish to change from the Convert To: option menu in the Tool Settings window.

MicroStation prompts:

Convert 3D > Identify solid or surface *(Identify the element whose status is to change.)*

Convert 3D > Accept/Reject *(Place a data point to accept the change in status, or click the Reject button to reject the operation.)*

USING ACCUDRAW IN *3D*

AccuDraw *3D* provides the ability to work in a pictorial view rather than the standard orthogonal views. AccuDraw automatically constrains data points to its drawing plane regardless of its orientation to the view.

Open the AccuDraw window:

Primary tool box	Select the Start AccuDraw tool (see Figure 17–34).
Key-in window	**accudraw activate** (or **acc a**) Enter

Figure 17–34 Invoking the Start AccuDraw tool from the Primary tool box

The AccuDraw window opens, either as a floating window or docked at the top of the MicroStation workspace.

In *3D*, when using rectangular coordinates, the AccuDraw window has an additional field for the *Z* axis. For polar coordinates in *3D*, the AccuDraw window has the same two fields as in *2D*.

By using AccuDraw keyboard shortcuts you can rotate the drawing plane axes, making it convenient to draw in an isometric view. For example, it is easy with AccuDraw to place a nonplanar complex chain or complex shape in an isometric view in any direction without reverting to an orthogonal view.

AccuDraw's ability to adhere to the standard view axes while manipulating your drawing in a pictorial view is so important that it maintains the current orientation from tool to tool.

By default, AccuDraw orients the drawing plane to the view axes, similar to working with *2D* design. You can return AccuDraw to this orientation any time the focus is in the AccuDraw window by pressing the **V** key.

To rotate the drawing plane axes to align with the standard top view, focus in the AccuDraw window and press the **T** key. AccuDraw dynamically rotates the compass to indicate the orientation of the drawing plane.

To rotate the drawing plane axes to align with the standard front view, focus in the AccuDraw window and press the **F** key. AccuDraw dynamically rotates the compass to indicate the orientation of the drawing plane.

To rotate the drawing plane axes to align with the standard side (left or right) view, focus in the AccuDraw window and press the **S** key. AccuDraw dynamically rotates the compass to indicate the orientation of the drawing plane.

To rotate the drawing plane axes 90 degrees about an individual axis, focus in the AccuDraw window and press letters **R** and **X** to rotate 90 degrees about the *X* axis, **R** and **Y** to rotate 90 degrees about the *Y* axis, and **R** and **Z** to rotate 90 degrees about the *Z* axis.

To rotate the drawing plane axes interactively, focus in the AccuDraw window and press letters **R** and **A**. Place data points to locate the *X* axis origin, the direction of the *X* axis, and the direction of the *Y* axis.

PROJECTED SURFACES

The Extrude tool creates a unique *3D* object from *2D* elements. Line, line string, arc, ellipse, text, multi-line, complex chain, complex shape, and B-spline curve are the elements that can be projected to a defined distance. Surfaces formed between the original boundary element and its projection are indicated by straight lines connecting the keypoints.

To project a boundary element, invoke the Extrude tool:

3D Construct tool box	Select the Extrude tool (see Figure 17–35).
Key-in window	**construct surface projection** (or **constru s p**) Enter

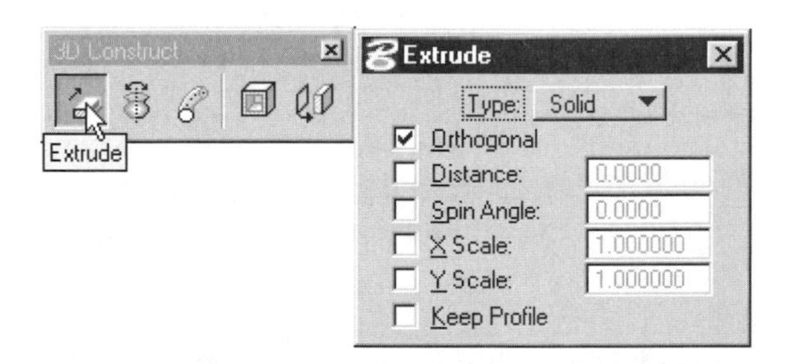

Figure 17–35 Invoking the Extrude tool from the 3D Construct tool box

Select the type of surface from the Type: option menu in the Tool Settings window.

If necessary, you can turn ON the toggle buttons for Orthogonal, Distance, Spin Angle, X Scale, Y Scale, and Keep Profile in the Tool Settings window. If Orthogonal is set to ON, the boundary element is projected orthogonally. If you turn ON the constraints for Distance, Spin Angle, X Scale, and Y Scale, make sure to key-in appropriate values in the edit fields.

MicroStation prompts:

Extrude > Identify profile *(Identify the boundary element.)*

Extrude > Define distance *(Place a data point to define the height. If Distance is set to ON, then the data point provides the direction.)*

EXTRUDE ALONG A PATH

The Extrude Along Path tool creates a tubular surface or solid extrusion along a path. Line, line string, arc, ellipse, text, multi-line, complex chain, complex shape, and B-spline curve are the elements that can be projected along a path. Straight lines connecting the keypoints indicate surfaces formed between the original boundary element and its projection.

To project a boundary element along a path, invoke the Extrude Along Path tool:

3D Construct tool box	Select the Extrude Along Path tool (see Figure 17–36).
Key-in window	**construct extrude along** (or **constru e a**) Enter

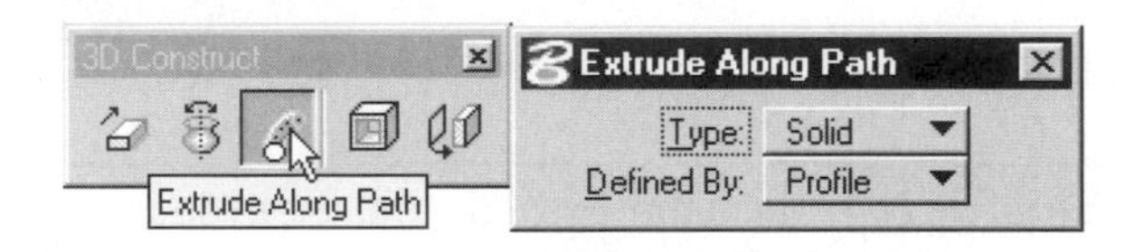

Figure 17–36 Invoking the Extrude Along Path tool from the 3D Construct tool box

Select the type of surface from the Type: option menu in the Tool Settings window.

Select one of the two available options: Circular or Profile from Defined By: option menu. The Circular selection creates a tube with a circular cross-section. You have to specify inside and outside radius. The Profile selection creates a surface by extruding the selected element along the selected path.

MicroStation prompts:

Extrude Along Path > Identify path *(Identify the boundary element.)*

Extrude Along Path > Identify profile or snap to profile at attachment point *(Identify profile or snap to profile at attachment point)*

Extrude Along Path > Accept to create *(Click the Accept button to create the profile.)*

SURFACE OF REVOLUTION

The Construct Revolution tool is used to create a unique *3D* surface or solid of revolution that is generated by rotating a boundary element about an axis of revolution. Line, line string, arc, ellipse, shape, complex chain, complex shape, and B-spline curve are the elements that can be used in creating a *3D* surface or solid. Surfaces created by the boundary element as it is rotated are indicated by arcs connecting the keypoints.

To create a *3D* surface or solid of revolution, invoke the Construct Revolution tool:

3D Construct tool box	Select the Construct Revolution tool (see Figure 17–37).
Key-in window	**construct surface revolution** (or **constru s r**) Enter

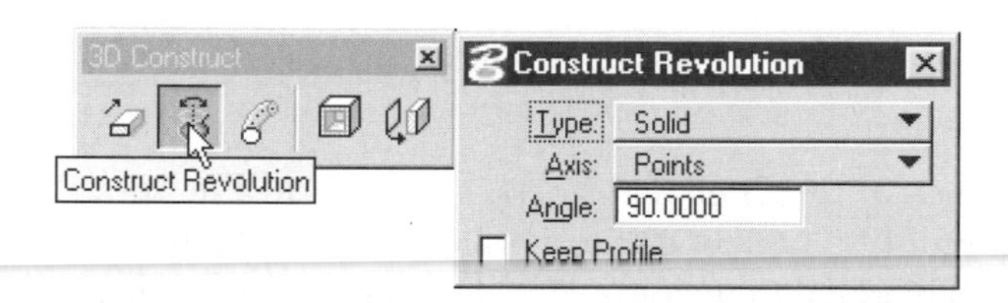

Figure 17–37 Invoking the Construct Revolution tool from the 3D Construct tool box

Select the type of surface from the Type: option menu in the Tool Settings window.

Select the direction of the axis of revolution relative to the view or design file axes from the Axis: option menu in the Tool Settings window. If set to Screen X, Screen Y, or Screen Z, the axis of revolution is set with the selected screen (view) axis. If set to Drawing X, Drawing Y, or Drawing Z, the axis of revolution is set with the selected design file axis.

If necessary, turn ON the toggle buttons for Angle and Keep Profile in the Tool Settings window. If you turn ON the constraint for Angle, make sure to key-in the appropriate value in the edit field.

MicroStation prompts:

Construct Revolution > Identify profile *(Identify the boundary element.)*

Construct Revolution > Define axis of revolution *(Place a data point or key-in coordinates. If Axis is set to Points, this data point defines one point on the axis of revolution and subsequently MicroStation prompts you for a second data point. If not, this data point defines the axis of revolution.)*

Construct Revolution > Accept, continue surface/reset to finish *(Place additional data points to continue, and/or press the Reset button to terminate the sequence.)*

SHELL SOLID

The Shell Solid tool creates a hollowed out solid for one or more selected faces of a defined thickness.

To create a hollowed out solid, invoke the Shell Solid tool:

3D Construct tool box	Select the Shell Solid tool (see Figure 17–38).
Key-in window	**construct shell** (or **constru s**) Enter

Figure 17–38 Invoking the Shell Solid tool from the 3D Construct tool box

Specify the shell thickness in the Shell Thickness: edit field in the Tool Settings window. Set the Shell Outward toggle button to OFF to create a hollowed out solid for one or more selected faces. If set to ON, the material is added to the outside and the original solid defines the inside of the walls.

MicroStation prompts:

> Shell Solid > Identify target solid *(Identify the target solid.)*
>
> Shell Solid > Identify face to open *(Move the screen pointer over the solid, the face nearest the pointer highlights and data point selects the highlighted face)*
>
> Shell Solid > Accept/Reject (select next face) *(Select additional faces, click the Reset button to deselect an incorrect face; to accept the selection of faces, click the Accept button to complete the selection)*

THICKEN TO SOLID

The Thicken To Solid tool is used to add thickness to an existing surface to create a solid. You can specify the thickness by keying-in a distance or graphically.

To add thickness to an existing surface to create a solid, invoke the Thicken To Solid tool:

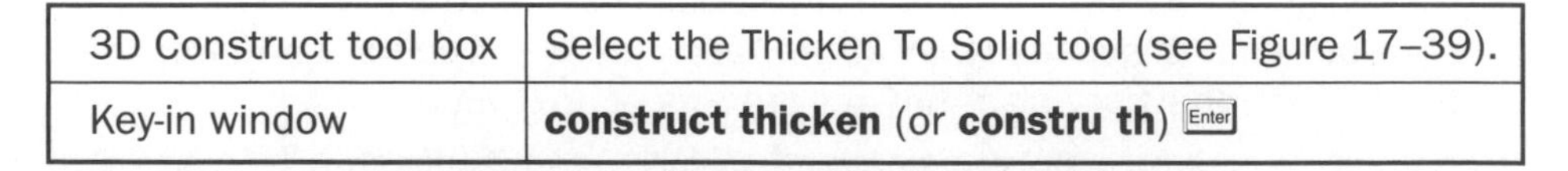

3D Construct tool box	Select the Thicken To Solid tool (see Figure 17–39).
Key-in window	**construct thicken** (or **constru th**) Enter

Figure 17–39 Invoking the Thicken To Solid tool from the 3D Construct tool box

If you need to add the thickness to both sides of the selected surface, set the Add To Both Sides toggle button to ON. To increase the thickness to a specific value, key-in the values in the Thickness: edit field and turn ON the toggle button.

MicroStation prompts:

Thicken to Solid > Identify surface *(Identify the surface.)*

Thicken to Solid > Define thickness *(An arrow is displayed, move the pointer to the side you want to increase the thickness and click the Accept button)*

PLACING *2D* ELEMENTS

Any *2D* elements (such as blocks and circles) that you place with data points without snapping to existing elements will be placed at the active depth of the view. Also, they will be parallel to the screen. Elements that require fewer than three data points to define (such as blocks, circles with radius, circles with diameter/center, and polygons) take their orientation from the view being used. The points determine only their dimensions, not their orientation. Elements that require three or more data points to describe (shapes, circles by edge, ellipses, and rotated blocks) also provide their planar orientation. Once the first three points have been specified, any further points will fall on the same plane.

CREATING COMPOSITE SOLIDS

MicroStation provides three tools that can create a new composite solid by combining two solids by Boolean operations. There are three basic Boolean operations that can be performed in MicroStation:

- Union
- Intersection
- Difference

Union Operation

The union is the process of creating a new composite solid from two solids. The union operation joins the original solids in such a way that there is no duplication of volume. Therefore, the total resulting volume can be equal to or less than the sum of the volumes in the original solids. To create a composite solid with the union operation, invoke the Construct Union tool:

3D Modify tool box	Select the Construct Union tool (see Figure 17–40).
Key-in window	**construct union** (or **constru u**) Enter

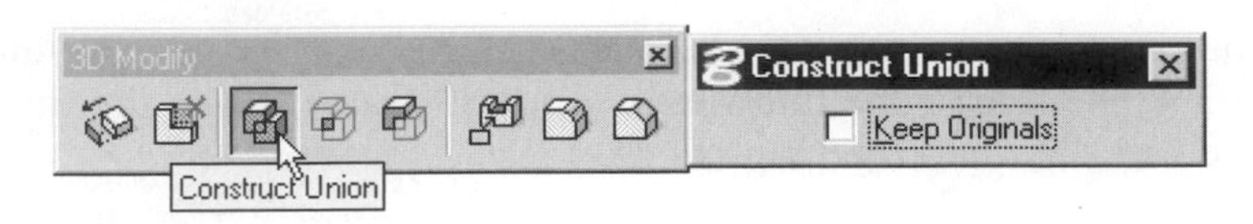

Figure 17–40 Invoking the Construct Union tool from the 3D Construct tool box

To keep the original elements, set the Keep Originals toggle button to ON in the Tool Settings window.

MicroStation prompts:

> Construct Union > Identify first solid *(Identify the first solid element for union.)*
>
> Construct Union > Identify next solid *(Identify the second solid element for union)*
>
> Construct Union > Identify next solid, or data point to finish *(Identify the third element for union or click the data point to accept the union of two selected solids.)*

See Figure 17–41 for an example of creating a composite solid by joining two cylinders with the Construct Union tool.

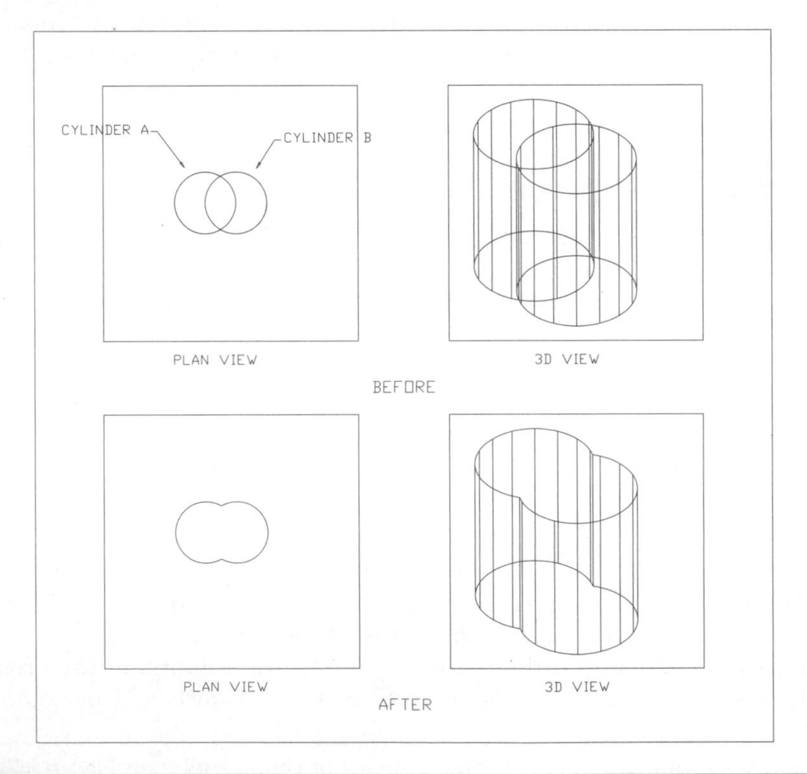

Figure 17–41 Creating a composite solid by joining two cylinders via the Construct Union tool

MicroStation also provides a tool to create a new composite solid from two surfaces. To combine two surfaces by union, invoke the Boolean Surface Union tool (available only by key-in) and select the surfaces to make it into a composite solid. The parts of the solids left are determined by their surface normal orientations. The surface normals can be changed by the Change Surface Normal tool.

Intersection Operation

The intersection is the process of forming a composite solid from only the volume that is common to two solids. To create a composite solid with the intersection operation, invoke the Construct Intersection tool:

3D Modify tool box	Select the Construct Union tool (see Figure 17–42).
Key-in window	**construct intersection** (or **constru i**) Enter

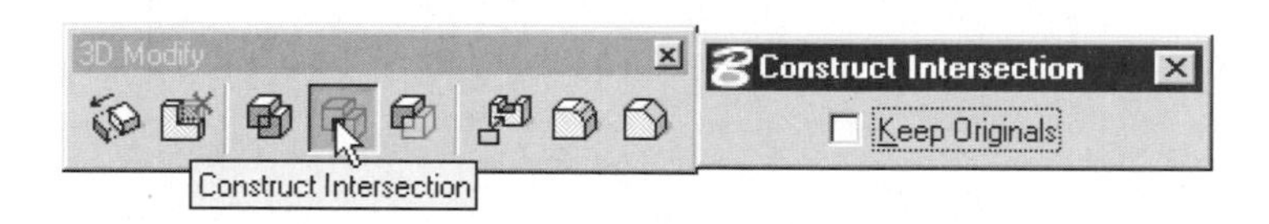

Figure 17–42 Creating a composite solid by joining two cylinders via the Construct Intersection tool

MicroStation prompts:

Construct Intersection > Identify first solid *(Identify the first element for intersection.)*

Construct Intersection > Identify next solid *(Identify the second element for intersection.)*

Construct Intersection > Identify next solid, or data point to finish *(Identify the next element or click the data button to complete the selection)*

See Figure 17–43 for an example of creating a composite solid by joining two cylinders via the Construct Intersection tool.

MicroStation also provides a tool to create a new composite solid from two surfaces. To combine two surfaces by intersection, invoke the Boolean Surface Intersection tool (available only by key-in) and select the surfaces to make it to a composite solid. The parts of the solids left are determined by their surface normal orientations. The surface normals can be changed by the Change Surface Normal tool.

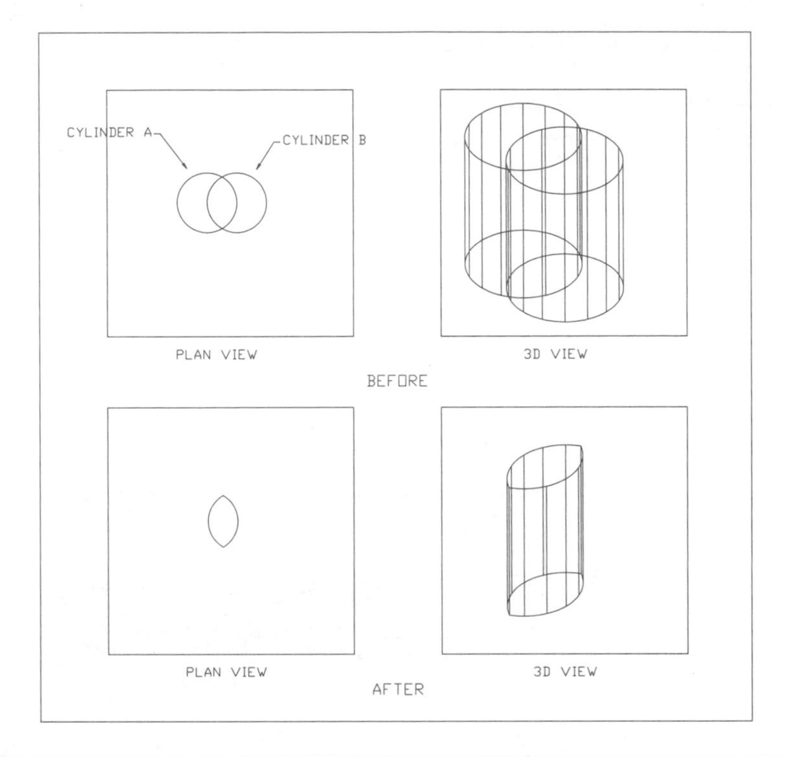

Figure 17–43 Creating a composite solid by intersecting two cylinders with the Construct Intersection tool

Difference Operation

The difference operation is the process of forming a composite solid by starting with a solid and removing from it any volume it has in common with a second object. If the entire volume of the second solid is contained in the first solid, then what is left is the first solid minus the volume of the second solid. However, if only part of the volume of the second solid is contained within the first solid, then only the part that is duplicated in the two solids is subtracted. To create a composite solid with the difference operation, invoke the Construct Difference tool:

3D Modify tool box	Select the Construct Difference tool (see Figure 17–44).
Key-in window	**construct difference** (or **constru d**) Enter

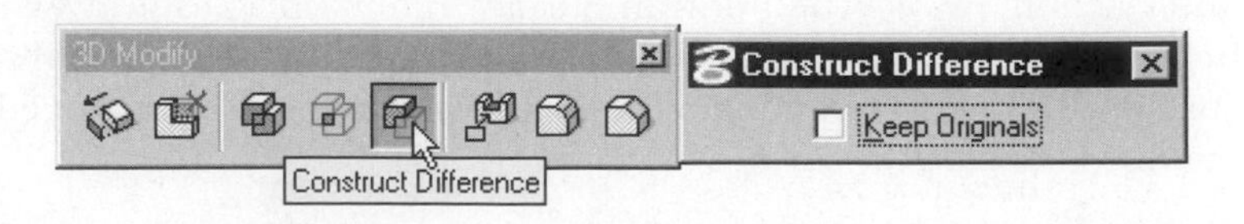

Figure 17–44 Invoking the Construct Difference tool from the 3D Modify tool box.

MicroStation prompts:

> Construct Difference > Identify solid to subtract from *(Identify the first element for the difference operation.)*
>
> Construct Difference > Identify next solid or surface to subtract *(Identify the second solid or surface to subtract.)*
>
> Construct Difference > Identify next solid or surface to subtract, or data point to finish *(Identify the next element to subtract, or click the data button to complete the selection.)*

See Figure 17–45 for an example of creating a composite solid by joining two cylinders via the Construct Difference Between Surfaces tool.

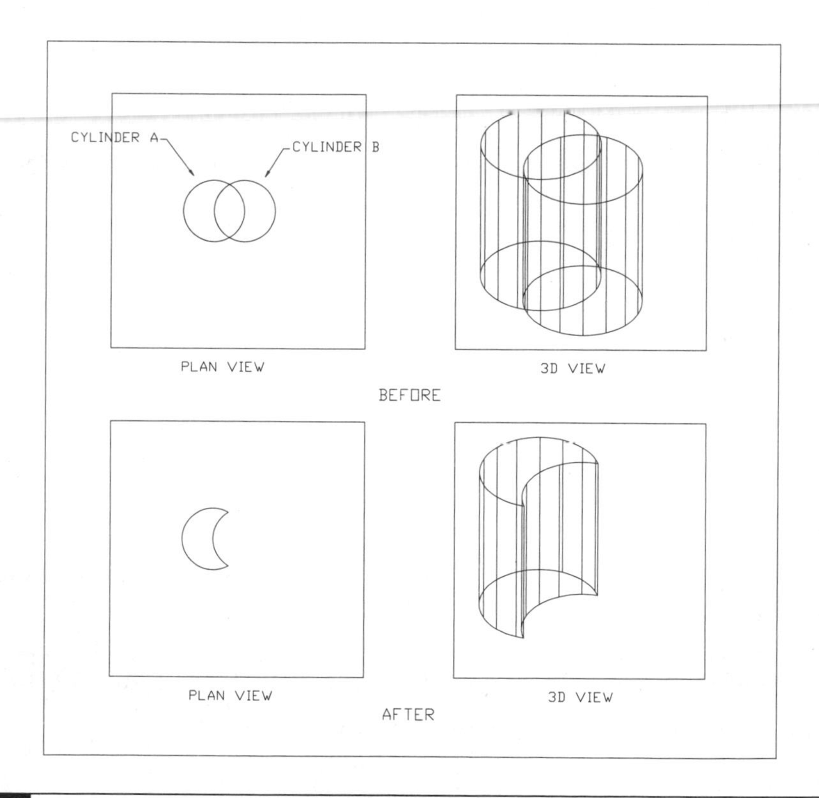

Figure 17–45 Creating a composite solid by subtracting Cylinder B from Cylinder A using the Construct Difference Between Surfaces tool

MicroStation also provides a tool to create a new composite solid from two surfaces. To combine two surfaces by the difference operation, invoke the Boolean Surface Difference tool (available only by key-in) and select the surfaces to make it into a composite solid. The parts of the solids left are determined by their surface normal orientations. The surface normals can be changed by Change Surface Normal tool.

CHANGE NORMAL

The Change Normal Direction tool can change the surface normal direction for a surface. This is useful to control the way the elements are treated while performing the Boolean operations.

To change the surface normal of an element, invoke the Change Normal Direction tool:

Modify Surfaces tool box	Select the Change Normal Direction tool (see Figure 17–46).
Key-in window	**change surface normal** (or **chan n**) Enter

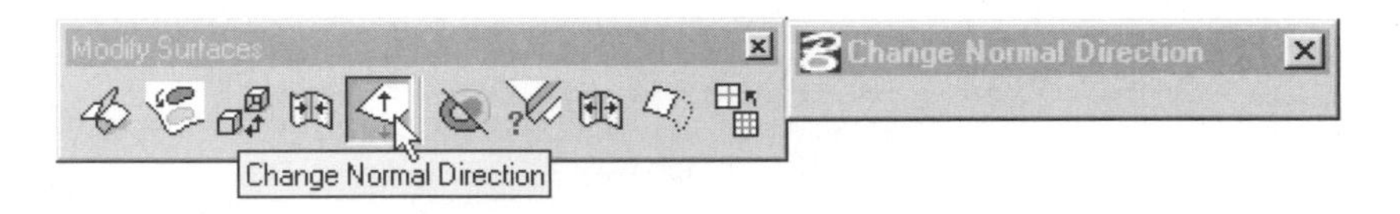

Figure 17–46 Invoking the Change Normal Direction tool from the Modify Surfaces tool box

MicroStation prompts:

Change Surface Normal > Identify element *(Identify the element; surface normals are displayed.)*

Change Surface Normal > Reverse normals, or RESET *(Place a data point to accept the change in the normal direction.)*

MODIFY SOLID

The Modify Solid tool lets you relocate a face of a solid outward (positive) or inward (negative), relative to the center of the solid. To modify a solid, invoke the Modify Solid:

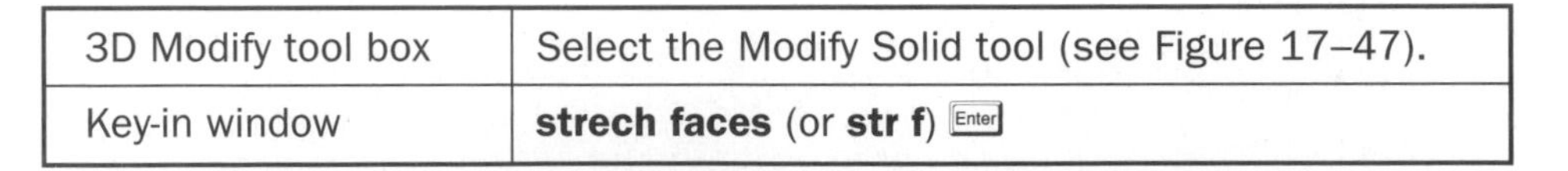

3D Modify tool box	Select the Modify Solid tool (see Figure 17–47).
Key-in window	**strech faces** (or **str f**) Enter

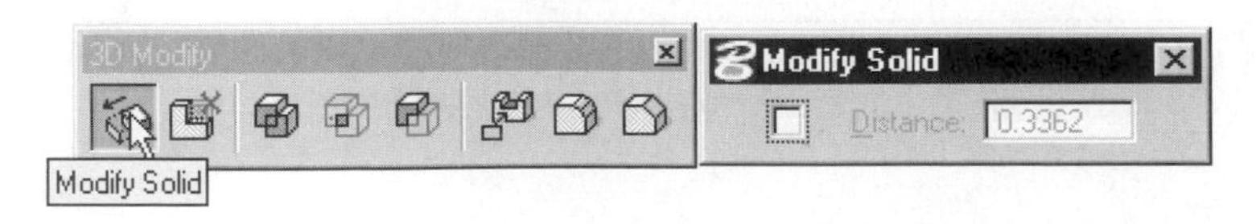

Figure 17–47 Invoking the Modify Solid tool from the 3D Modify tool box

To modify the selected solid face by key-in distance, set the Distance toggle button to ON and key-in the distance in the Distance edit field in the Tool Settings window.

MicroStation prompts:

> Modify Solid > Identify target solid *(Identify the solid to modify.)*
> Modify Solid > Select face to modify *(Select the face to modify.)*
> Modify Solid > Define distance *(Using the arrow as the guide, move the pointer to define the distance dynamically when the Distance toggle button is set to OFF and direction.)*

REMOVE FACES

The Remove Faces tool lets you remove an existing face or a feature from a solid and then close the opening. It can also remove faces that are associated with a cut, a solid that has been added to or subtracted from the original, a shell solid, a fillet, or a chamfer. To remove a face or a feature, invoke the Remove Faces tool:

3D Modify tool box	Select the Remove Faces tool (see Figure 17–48).
Key-in window	**remove faces** (or rem f) Enter

Figure 17–48 Invoking the Remove faces tool from the 3D Modify tool box

Select one of the two available methods to remove faces from the Method: option menu in the Tool Settings window. The Faces selection allows you to remove one or more faces from a selected solid feature. The Logical Groups selection allows you to remove faces that are associated with a cut, a solid that has been added to or subtracted from the original, a shell solid, a fillet, or a chamfer.

MicroStation prompts:

> Remove Faces and Heal > Identify target solid *(Identify the solid to modify.)*
> Remove Faces and Heal > Identify first face to remove *(Select the face to remove.)*
> Remove Faces and Heal > Accept/Reject (select next face) *(Click the Accept button to remove the selected face, if any select additional faces to continue the selection or click the Reject button to cancel the operation.)*

CUT SOLID

The Cut Solid tool lets you split a solid into two or more segments using a cutting profile. Cutting profiles may be open or closed elements. The open element profile must extend to the edge of the solid. To cut a solid, invoke the Cut Solid tool:

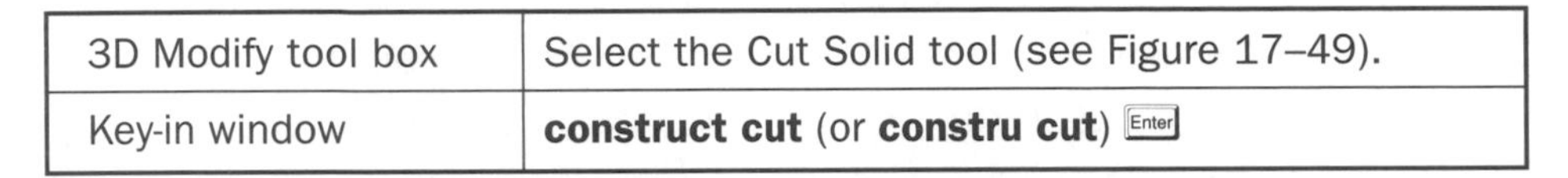

3D Modify tool box	Select the Cut Solid tool (see Figure 17–49).
Key-in window	**construct cut** (or **constru cut**) Enter

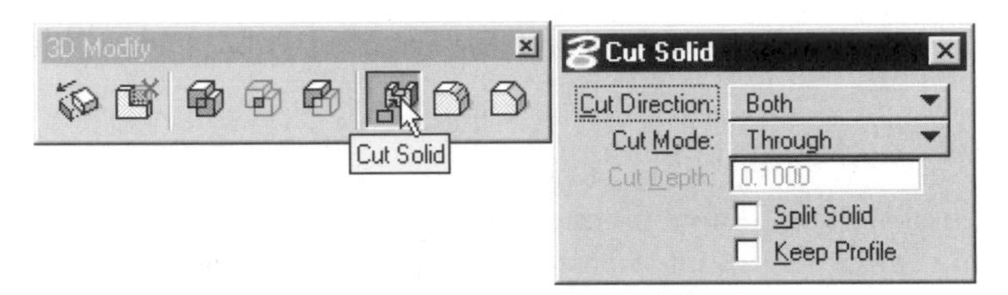

Figure 17–49 Invoking the Cut Solid tool from the 3D Modify tool box

The Cut Direction: option menu in the Tool Settings window sets the direction of the cut relative to the cutting profile's Surface Normal. Available selections include:

- Both—Selects in both directions from the profile's plane.
- Forward—Selects from the forward direction from the profile's plane.
- Back—Selects from the backward direction from the profile's plane.

The Cut Mode: option menu in the Tool Settings window sets the limits of the cut. Available selections include:

- Through—Cuts through all faces of the solid.
- Define Depth—Cuts into the solid a defined distance. Key-in the distance in the Cut Depth: edit field.

The Split Solid: toggle button controls whether the material is removed or not when it is split into segments. When it is set to ON, no material is removed from the solid, and it is split into two or more segments.

The Keep Profile: toggle button controls whether the cutting profile remains in the design. When it is set to ON, the original cutting profile remains in the design.

MicroStation prompts:

Cut Solid > Identify target solid *(Identify the solid.)*
Cut Solid > Identify cutting profile *(Identify the cutting profile.)*

Cut Solid > Accept/Reject *(Click the Accept button to complete the cut or click the Reject button to cancel the operation.)*

CONSTRUCT FILLET

The Fillet Edges tool is used to fillet or round for one or more edges of a solid, projected surface, or a surface of revolution. To construct a fillet, invoke the Fillet Edges tool:

3D Modify tool box	Select the Fillet Edges tool (see Figure 17–50).
Key-in window	**fillet edges** (or **fill e**) Enter

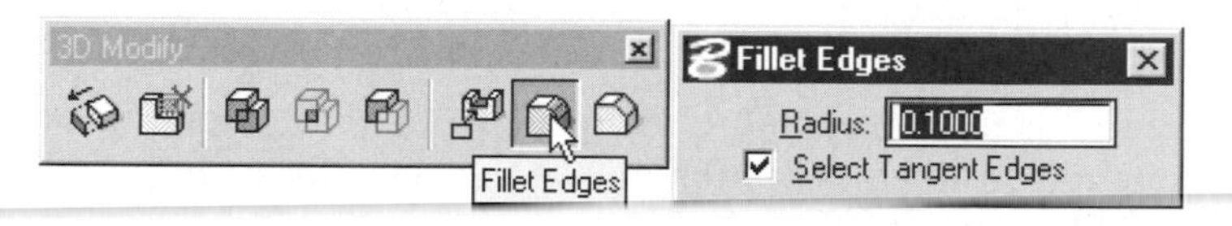

Figure 17–50 Invoking the Fillet Edges tool from the 3D Modify tool box

Key-in the Radius for the fillet in the Radius: edit field in the Tool Settings window. Select the Tangent Edges toggle button; when set to ON, the edges that are tangentially continuous are selected and rounded in one operation.

MicroStation prompts:

Fillet Edges > Identify Edge to Fillet *(Identify the solid.)*

Fillet Edges > Accept/Reject (select next edge) *(Click the Accept button to accept the fillet for the selected edge, and if necessary, select additional edges to fillet or click the Reject button to cancel the operation.)*

The Fillet Surfaces tool helps you construct a *3D* fillet between two surfaces. The fillet is placed by sweeping an arc with a specified radius along the common intersecting curve. The fillet is created in the area pointed to by the surface normals of both surfaces.

To fillet between surfaces, invoke the Construct Fillet Between Surfaces tool:

Fillet Surfaces tool box	Select the Fillet Surfaces tool (see Figure 17–51).
Key-in window	**fillet surfaces** (or **fill su**) Enter

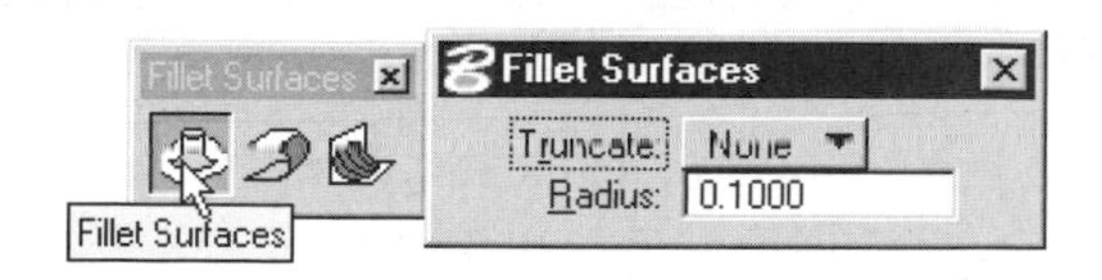

Figure 17–51 Invoking the Fillet Surfaces tool from the Fillet Surfaces tool box

Key-in the radius of the fillet to be drawn in the Radius: edit field.

The Truncate: option menu sets which surface(s) are truncated at the point of tangency with the fillet.

MicroStation prompts:

> Fillet Surfaces > Identify first surface *(Identify the first surface.)*
>
> Fillet Surfaces > Identify second surface *(Identify the second surface.)*
>
> Fillet Surfaces > Accept/Reject *(Click the Data button to accept the fillet, or click the Reject button to cancel the operation.)*

CONSTRUCT CHAMFER

The Chamfer Edges tool is used to chamfer one or more edges of a solid, projected surface, or a surface of revolution. To construct a chamfer, invoke the Chamfer Edges tool:

3D Modify tool box	Select the Chamfer Edges tool (see Figure 17–52).
Key-in window	**chamfer edges** (or **ch e**) Enter

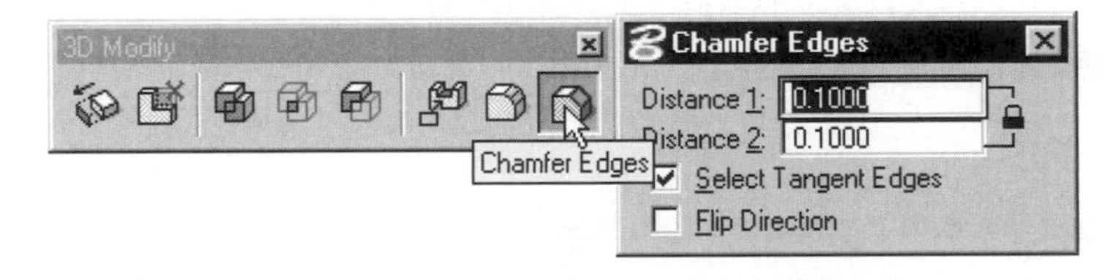

Figure 17–52 Invoking the Chamfer Edges tool from the 3D Modify tool box

Key-in chamfer distances in the Distance 1: and Distance 2: edit fields. The Select Tangent Edges toggle button, when set to ON, makes edges that are tangentially continuous and are selected and rounded in one operation. The Flip Direction toggle button, when set to ON, reverses the direction of the chamfer and sets the values that the faces are trimmed.

MicroStation prompts:

> Chamfer Edges > Identify Edge to Chamfer *(Identify the solid.)*
>
> Fillet Edges > Accept/Reject (select next edge) *(Click the Accept button to accept the chamfer for the selected edge, if necessary, select additional edges to chamfer or click the Reject button to cancel the operation.)*

The Construct Chamfer Between Surfaces tool enables you to construct a *3D* chamfer between two surfaces by a specified length along the common intersection curve. The chamfer is created in the area pointed to by the surface normals of both surfaces.

To chamfer between surfaces, invoke the Construct Chamfer Between Surfaces tool:

Key-in window	**chamfer surface** (or **chamf su**) Enter

The Truncate: option menu sets which surface(s) are truncated at the point of tangency with the chamfer.

Key-in the Chamfer length in the Chamfer Length: edit field.

The Tolerance toggle button sets the override for the system tolerance.

MicroStation prompts:

Construct Chamfer Between Surfaces > Identify first surface *(Identify the first surface.)*

Construct Chamfer Between Surfaces > Accept/Reject *(Click the Data button to accept the first surface selection.)*

Construct Chamfer Between Surfaces > Identify second surface *(Identify the second surface.)*

Construct Chamfer Between Surfaces > Accept/Reject *(Click the Data button to accept the second surface selection.)*

Construct Fillet Between Surfaces > Accept/Reject *(Click the Data button to accept the chamfer, or click the Reject button to cancel the operation.)*

PLACING TEXT

MicroStation provides two options to place text in a *3D* design: (1) placing text (view-dependent) in such a way that it appears planar to the screen in the view in which the data point is placed but rotated in the other views, or (2) placing text (view-independent) in such a way that it appears planar to the screen in all views.

To place text (view-dependent), click the Place Text icon in the Text tool box, select By Origin from the Method: option menu, and follow the prompts. To place text (view-independent), click the Place Text icon in the Text tool box, select View Ind from the Method: option menu, and follow the prompts. Text parameters are set up in the same way as in the *2D* design.

FENCE MANIPULATIONS

Fences are used in a *3D* design in much the same way as in a *2D* design (see Chapter 6). The difference is that a *3D* fence defines a volume. The volume is defined by the fenced area and the display depth of the view in which the fence is placed. The Fence lock options work the same way as in a *2D* design.

CELL CREATION AND PLACEMENT

The procedure for creating and placing cells in *3D* design is the same as in *2D* design (see Chapter 11). Before you create a *3D* cell, make sure the display depth is set to include all the elements to be used in the cell and the origin is defined at an appropriate active depth. If a normal cell was created in the top view and then placed in the front view, it will appear as it did in the top view and rotated in other views. In other words, the normal cell is placed as view-dependent, whereas a point cell when placed will appear planar to the screen in all views. A point cell is placed view-independent.

You can attach a *2D* cell library to a *3D* design file, but the cells will have no depth and will be placed at the active depth of the view in which you are working. However, you cannot create *3D* cells and store them in a *2D* cell library.

Note: You cannot attach a *3D* cell library to a *2D* design file.

DIMENSIONING

The procedure for dimensioning setup and placement in *3D* design is similar to that for *2D* design (see Chapter 9). The main difference is that you have to consider on which plane you want the dimensioning to be located. Before you place dimensions in a *3D* design, make sure the appropriate option is selected from the Alignment: option menu in the Linear Dimension tool box. The view measurement axis measures the projection of the element along the view's horizontal or vertical axis. The true measurement axis measures the actual distance between two points, not the projected distance. And the drawing measurement axis measures the projection of an element along the design cube coordinate system's axis.

RENDERING

Shading, or rendering, can turn your *3D* model into a realistic, eye-catching image. MicroStation's rendering options give you complete control over the appearance of your final images. You can add lights and control lighting in your design and also define the reflective qualities of individual surfaces in your design, making objects appear dull or shiny. You can create the rendered image of your *3D* model entirely within MicroStation. This section provides an overview of the various options available for rendering. Refer to the *MicroStation Reference Guide* for a more detailed description of various options.

Setting Up Cameras

In establishing a viewing position in MicroStation, the assumption you must make is that you are, as it were, looking through a camera to see the image. By default, MicroStation places the camera at a right angle to a view's *XY* plane. If necessary, you can move or reposition the camera to view the model from a different viewing angle.

To enable or disable the default camera setting and make changes to the camera setup, invoke the Camera Settings tool:

3D View Control tool box	Select the Camera Settings tool (see Figure 17–53).

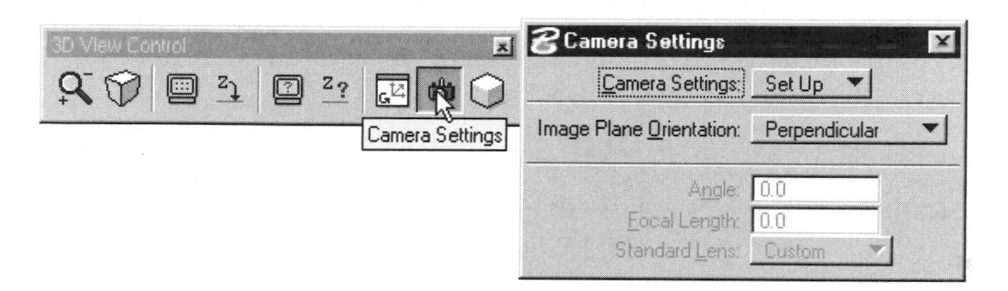

Figure 17–53 Invoking the Camera Settings tool from the 3D View Control tool box

Select one of the available options from the Camera Settings: option menu in the Tool Settings window.

- Turn On—Turns on the camera in a view or views
- Turn Off—Turns off the camera in a view or views
- Set Up—Turns on the camera in a view and sets the camera target and position. The target is the focal point (center) of a camera view. The position is the design cube location from which the model is viewed with the camera. Objects beyond the camera target appear smaller; objects in front of the camera target appear larger and may be outside of the viewing pyramid.
- Move—Moves the camera position
- Target—Moves the target

Select one of the available options from the Image Plane Orientation: option menu in the Tool Settings window:

- Perpendicular—Perpendicular to the camera direction
- Parallel to X axis—Parallel to the view *X* axis; analogous to a bellows camera
- Parallel to Y axis—Parallel to the view *Y* axis; analogous to a bellows camera
- Parallel to Z axis—Parallel to the view *Z* axis; all vertical lines (along the axis) appear parallel

Set the lens angle in degrees and the lens focal length in millimeters in the Angle: and Focal Length: edit fields, respectively.

Select one of the available options from the Standard Lens: option menu if you wish to use the standard lens type commonly used by photographers. MicroStation sets the appropriate lens angle and focal length.

MicroStation prompts depend on the options selected in the Tool Settings window.

Placement of Light Sources

The lighting setup is equally as important as setting the camera angle for producing a high-quality rendered image. MicroStation allows you four types of lighting:

- Ambient lighting
- Flashbulb lighting
- Solar lighting
- Source lighting, including point, spot, and distant

Ambient lighting is a uniform light that surrounds your model. *Flashbulb lighting* is a localized, intense light that appears to emanate from the camera position. *Solar lighting* is sunlight. By defining your location on the earth in latitude, longitude, day, month, and time, you can simulate lighting for most exterior architectural projects. Ambient, flashbulb, and solar lighting are set in the Global settings box invoked from the Rendering submenu of the pull-down Settings menu.

Source lighting is achieved by placing light sources in the form of cells. The MicroStation program comes with three light source cells (point lights—PNTLT, spot lights—SPOTLT, and distant lights—DISTLT) provided in the LIGHTING.CEL cell library.

Point light can be thought of as a ball of light. It radiates beams of light in all directions. Such a light also has more natural characteristics. Its brilliance may be diminished as an object moves away from the source of light. An object that is near a point light will appear brighter; an object that is farther away will appear darker.

Spot lights are very much like the kind of spotlight you might be accustomed to seeing at a theater or auditorium. Spot lights produce a cone of light toward a target that you specify.

Distant light gives off a fairly straight beam of light that radiates in one direction, and its brilliance remains constant so that an object close to the light will receive as much light as a distant object.

Before you place the light cells, adjust the settings in the Source Lighting settings box invoked from the Rendering submenu of the pull-down Settings menu. MicroStation provides various options under the Tool pull-down menu, in the Settings box.

Rendering Methods

MicroStation provides seven different tools to render a view. Depending on the needs and availability of the hardware, you can choose one of the seven tools to render the model.

To render a view, invoke the Render tool:

3D View Control tool box	Select the Render tool (see Figure 17–54).

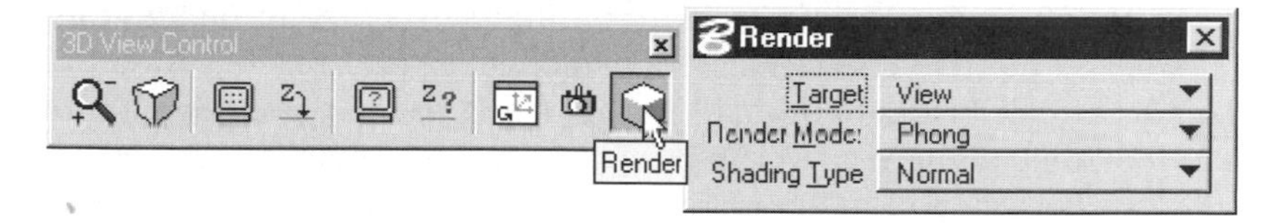

Figure 17–54 Invoking the Render tool from the 3D View Control tool box

Select the type of area or element to be rendered from the Target: option menu in the Tool Settings window. The available options include View, Fence (contents), and Element.

Select one of the available options (Wiremesh, Hidden Line, Filled Hidden Line, Constant, Smooth, or Phong) from the Render Mode: option menu in the Tool Settings window.

- Wiremesh—Similar to the default wireframe display, all elements are transparent and do not obscure other elements.
- Hidden Line—Displays only the element parts that would actually be visible.
- Filled Hidden Line—Identical to a Hidden Line option display except that the polygons are filled with the element color.
- Constant—Displays each element as one or more polygons filled with a single (constant) color. The color is computed once for each polygon, from the element color, material characteristics, and lighting configurations.
- Smooth—Displays the appearance of curved surfaces more realistically than in constant shaded models because polygon color is computed at polygon boundaries and color is blended across polygon interiors.
- Phong—Displays the image after re-computing the color of each pixel. Phong shading is useful for producing high-quality images when speed is not critical.

The Shading Type: option menu sets the rendering method.

- Anti-alias—Displays the image with reduced jagged edges that are particularly noticeable on low-resolution displays. The additional time required for anti-aliasing is especially worthwhile when saving images for presentation, publication, or animated sequences.
- Stereo—Renders a view with a stereo effect that is visible when seen through *3D* (red/blue) glasses. Stereo Phong shading takes twice as long as Phong because two images—one each from the perspective of the right and left eyes—are rendered and combined into one color-coded image.

DRAWING COMPOSITION

One of MicroStation's useful features is the ability to compose multiple views (standard and saved) on a drawing sheet. This will allow you to plot multiple views on one sheet of paper—what-you-see-is-what-you-get (WYSIWYG). The Drawing Composition settings box automates the process of attaching the views of the model. The views are attached as reference files. An attached view in a sheet file can be any standard (top, bottom, right, left, front, back, or isometric), fitted view, or any saved view of a model file. Standard views can be clipped or set to display only certain levels. A view of the model file can be attached in any position at any scale. MicroStation provides a tool that allows you to group a set of views. A group of attached views can be moved, scaled, or detached as one. If necessary, you can remove or add a view to a group. In addition, MicroStation provides a tool that allows you to attach a view by folding an attached view about an orthogonal axis or a line defined by two data points. A folded view is automatically aligned and grouped with the attached view from which it is folded.

Open the Drawing Composition settings box:

Pull-down menu	File > Drawing Composition

MicroStation displays a Drawing Composition settings box, similar to Figure 17–55.

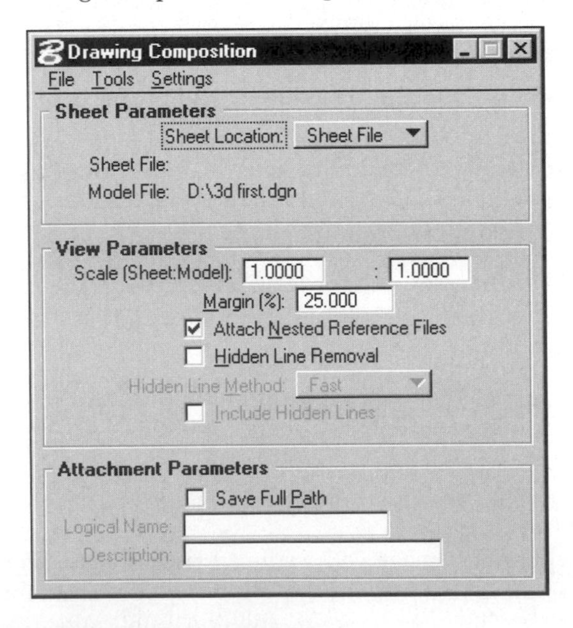

Figure 17–55 Drawing Composition settings box

MicroStation provides two different methods by which you can place the model on the sheet. Method 1 consists of creating or opening a sheet and placing the model (standard views or saved views). In Method 2, you designate one of the view windows in the model design for drawing composition, open an existing border file as a reference file, and then place the model (standard views or saved views).

Method 1

Select Sheet File from the Sheet Location: option menu located in the Sheet Parameters area, as shown in Figure 17–55.

Creating a New Sheet File

To create a new sheet file, open the Create Sheet File dialog box:

Drawing Composition settings box	File > New > Sheet...

MicroStation displays the Create Sheet File dialog box. The default sheet seed file is the SEED.SHT. If necessary, change the seed file by clicking the Seed button. MicroStation displays the Select Sheet File dialog box. Select the appropriate seed file, and click the OK button to close the dialog box. Table 17–1 lists the seed sheet files delivered with the MicroStation program.

Table 17–1 Seed Sheet Files That Come with MicroStation

SEED FILE	SIZE
Seed.sht	11 x 17
Seedah.sht	A (horizontal orientation)
Seedav.sht	A (vertical orientation)
Seedb.sht	B
Seedc.sht	C
Seedd.sht	D
Seede.sht	E
Seedf.sht	F

Key-in the name of the new sheet file in the Files: edit field, and click the OK button. MicroStation displays the name of the newly created sheet file in the Drawing Composition settings box.

Opening an Existing Sheet File

To open an existing sheet file, open the Open Sheet File dialog box:

Drawing Composition settings box	File > Open > Sheet...

MicroStation displays the Open Sheet File dialog box. Select the appropriate sheet file and click the OK button. MicroStation displays the name of the opened sheet file in the Drawing Composition settings box.

Opening a Design File to Place the Views

To open a design file to place the views on the active sheet file, open the Open Model File dialog box:

Drawing Composition settings box	File > Open > Model...

MicroStation displays the Open Design File dialog box. Select the appropriate design file and click the OK button. MicroStation displays the name of the opened design file in the Drawing Composition settings box.

Use one of the tools provided in the Tools pull-down menu in the Drawing Composition settings box to attach the views of the model. Refer to the section on "Attaching Views" for a detailed description of various methods of attaching the views.

Method 2

Create a new design file or open an existing design file. Select Sheet View from the Sheet Location: option menu located in the Sheet Parameters area of the Drawing Composition settings box, as shown in Figure 17–56.

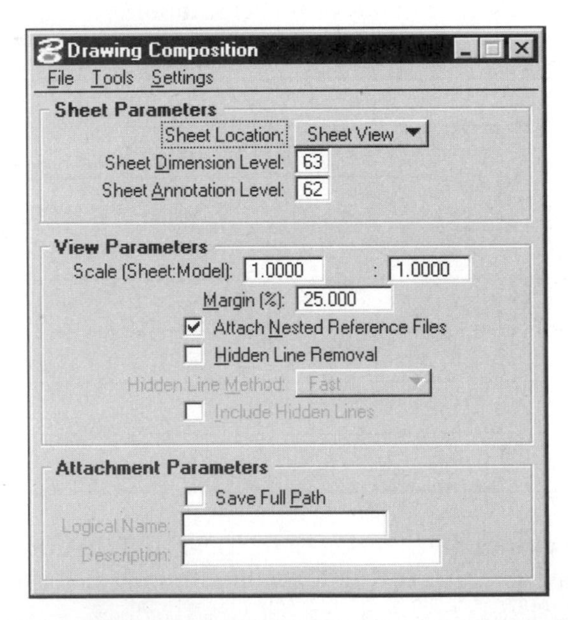

Figure 17–56 Drawing Composition settings box with Sheet View selected

The levels used for text and dimensions in the sheet view are set in the Sheet Parameters section of the settings box. The Sheet Dimension Level: edit field sets the level for dimensions and text that are to be displayed only in one sheet view (level 63 is the default). The Sheet Annotation Level edit field sets the level for any annotations to be displayed in the drawing sheet (level 62 is the default).

Designate a view window as a sheet view:

Drawing Composition settings box	Tools > Open Sheet View > <number of the view>

If necessary, set the Scale in the View Parameters section of the Drawing Composition settings box.

Attaching a Border

Open the Attach Border File dialog box:

Drawing Composition settings box	Tools > Attach Border > Fitted...

MicroStation opens the Attach Border File dialog box. Select the border file and click the OK button.

You can also attach the saved view of a border file by invoking the Saved View... option under the Attach Border menu.

Invoke one of the tools provided in the Tools pull-down menu in the Drawing Composition settings box to attach the views of the model. Refer to the next section on "Attaching Views" for a detailed description of various methods of attaching the views.

Attaching Views

Following are the available methods by which you can attach views of the model onto a sheet.

Attaching a Standard, Fitted View of the Model

To attach one of the standard views (top, bottom, left, right, front, back, isometric, or right isometric), invoke the Attach Standard View tool:

Drawing Composition settings box	Tools > Attach Standard > (select one of the standard views)

MicroStation prompts:

Attach Standard View > Identify view center *(Identify the view center to place the view.)*

MicroStation places the selected standard view. Before you place the view, if necessary, you can change the scale factor in the View Parameters section of the Drawing Composition settings box.

Attaching Saved Views

To attach saved views, invoke the Attach Saved View tool:

Drawing Composition settings box	Tools > Attach Saved View...

MicroStation prompts:

Attach Saved View > Identify view center *(Identify the view center to place the view.)*

MicroStation places the selected saved view. Before you place the view, if necessary, you can change the scale factor in the View Parameters section of the Drawing Composition settings box.

Attaching a Copy of the View

To attach a copy of the view, invoke the Attach the Copy of Existing Attachment tool:

Drawing Composition settings box	Tools > Attach Copy

MicroStation prompts:

Attach the Copy of Existing Attachment > Identify attachment *(Identify an element in the attached view to be copied.)*

Attach the Copy of Existing Attachment > Accept/Copy attachment *(Place a data point where you want to place the copy of the attached view.)*

Attaching a View by Folding it Orthogonally

To attach a view of the model by folding it orthogonally about the edge of an attached view, invoke the Attach Auxiliary By Orthogonal Fold tool:

Drawing Composition settings box	Tools > Attach Folded > Orthogonal

MicroStation prompts:

Attach Auxiliary By Orthogonal Fold > Identify principle attachment *(Identify an element in an attached view from which to fold the new attached view.)*

Attach Auxiliary By Orthogonal Fold > Accept @ fold line *(Identify the edge of the attached view about which the new attached view is to be folded.)*

Attach Auxiliary By Orthogonal Fold > Identify view center *(Place a data point to position the view.)*

Attach Auxiliary By Orthogonal Fold > *(Continue identifying the fold line to place additional views, or click the Reset button to terminate the command sequence.)*

Attaching a View by Folding it from an Attached View

To attach a view of the model by folding it from an attached view about a line defined by two points, invoke the Auxiliary View By Fold Line tool:

Drawing Composition settings box	Tools > Attach Folded > About Line

MicroStation prompts:

> Auxiliary View By Fold Line > Identify principle attachment *(Identify an element in an attached view from which to fold the new attached view.)*
>
> Auxiliary View By Fold Line > Accept at fold line end point *(Place a data point to define one end point of the line about which the new attached view will be folded.)*
>
> Auxiliary View By Fold Line > Identify fold line end point *(Place a data point to define the other end point of the line about which the new attached view will be folded.)*
>
> Auxiliary View By Fold Line > Identify fold line end point *(Continue identifying the fold line to place additional views, or click the Reset button to terminate the command sequence.)*

Additional Tools for Manipulating Views

Following are the additional tools available in the Tools pull-down menu.

Group

The Group option allows you to add to or remove attached view(s) to or from a group.

Clip

The Clip option allows you to clip a portion of the attached view. Refer to Chapter 12 on Reference Files for a detailed explanation on Clipping.

Move

The Move option allows you to move an attached view of the model, a group of views, or all attached views.

Scale

The Scale option allows you to set the scale (Master Scale:Reference Scale) of an attached view of the model, a group of views, or all attached views.

Detach

The Detach option allows you to detach a view of the model, a group of views, or all attached views.

Once you set up all the views, add any annotations and title block information. Invoke the Plot... tool and plot the design.

DRAWING EXERCISES 17–1 THROUGH 17–5

Lay out the objects shown in *3D* form. Create the design to the given dimensions.

Exercise 17–1

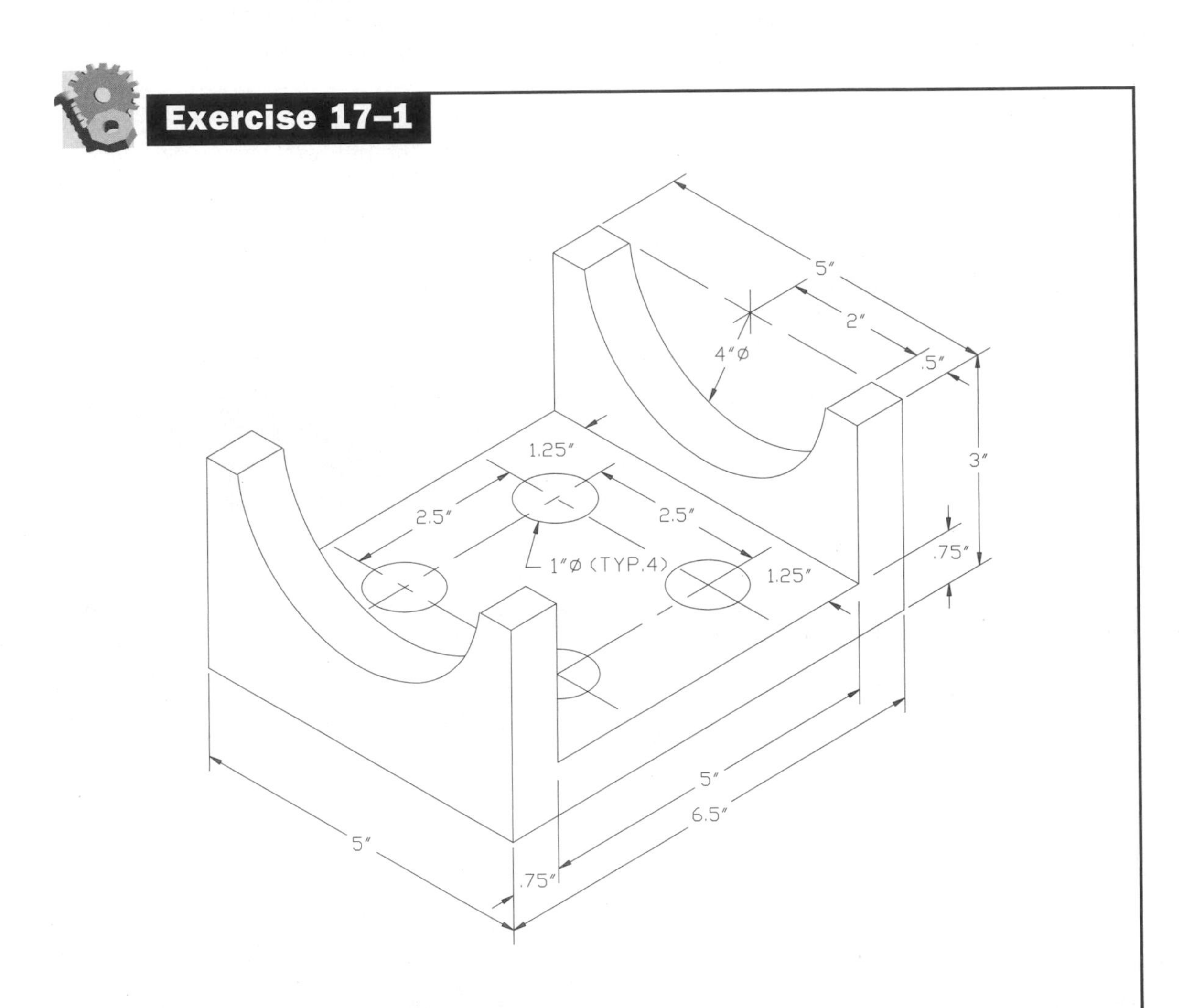

Exercise 17-2

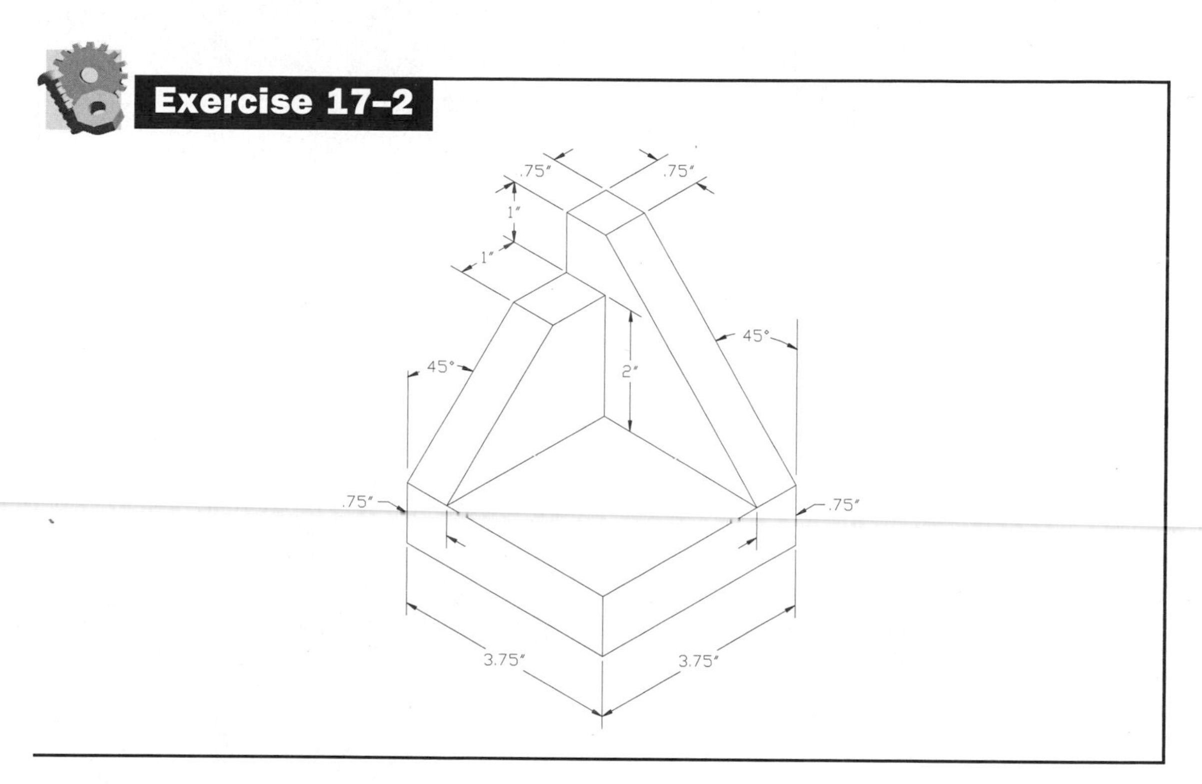

Exercise 17-3

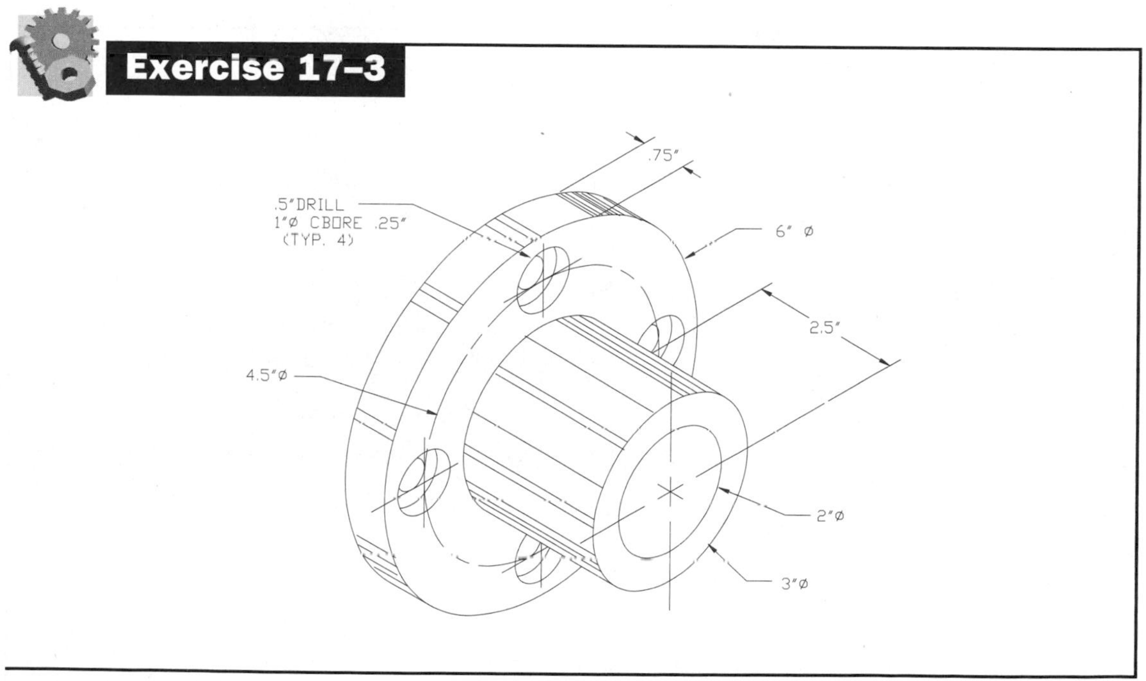

Exercise 17–4

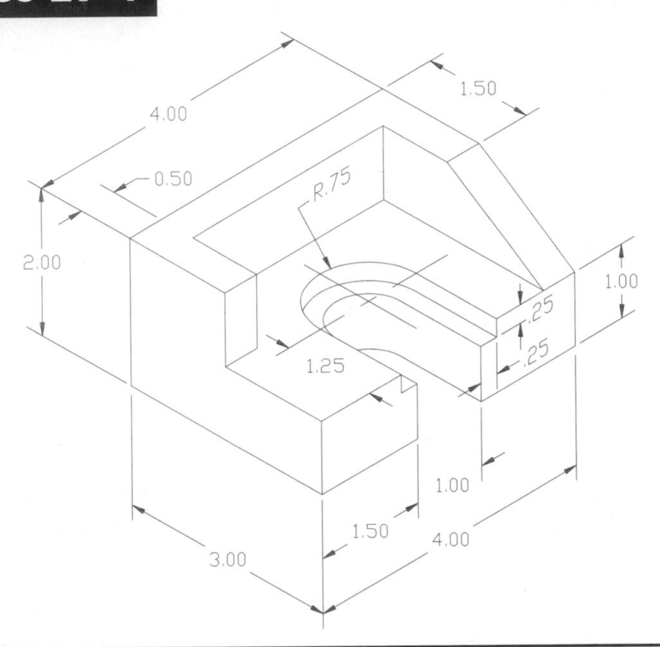

Exercise 17–5

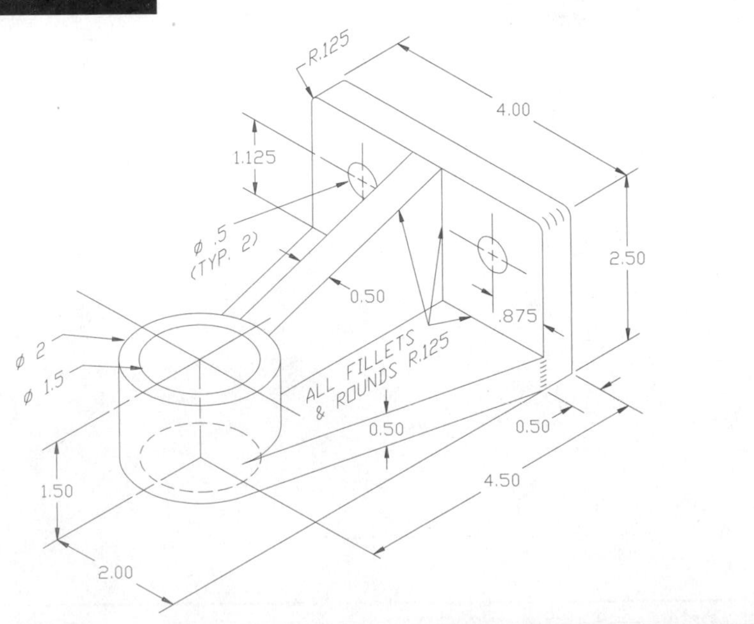

appendix

MicroStation Tool Boxes

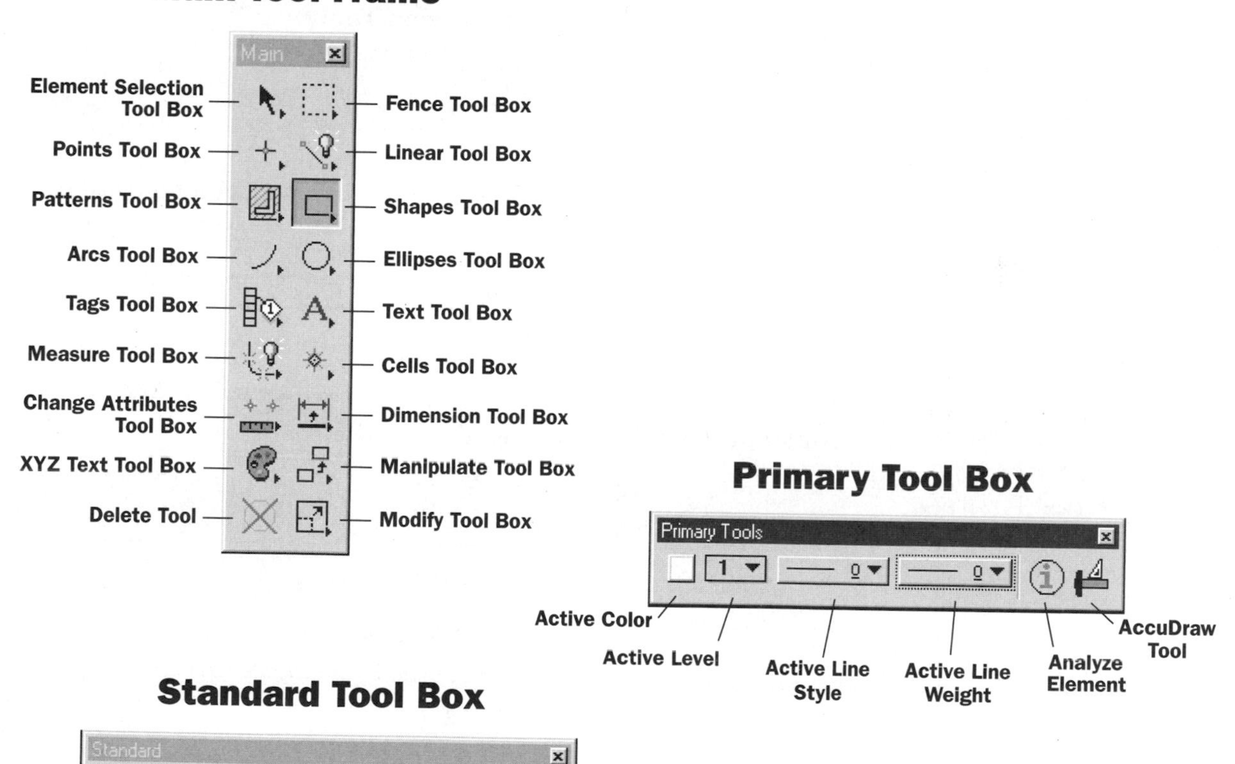

Fence Tool Box

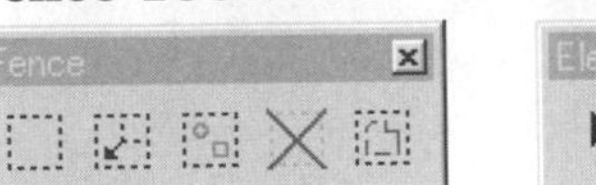

Element Selection Tool Box

Linear Elements Tool Box

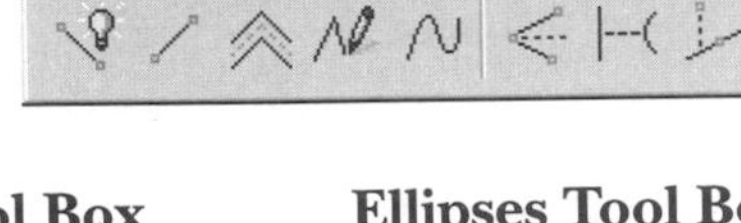

Points Tool Box

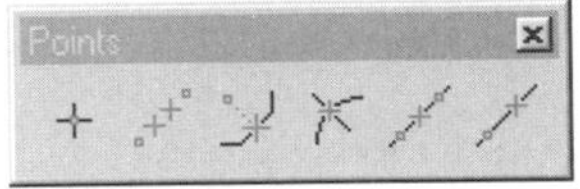

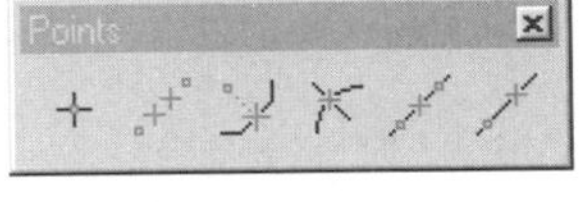

Shapes Tool Box

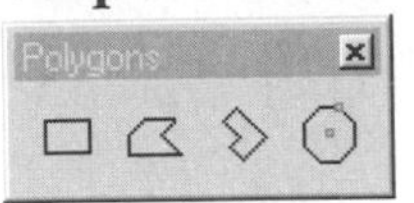

Patterns Tool Box

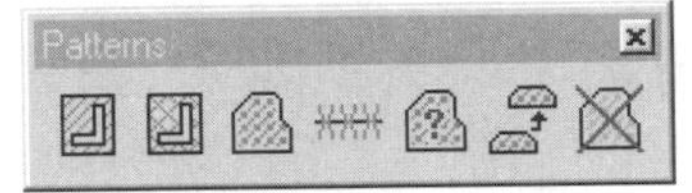

Ellipses Tool Box

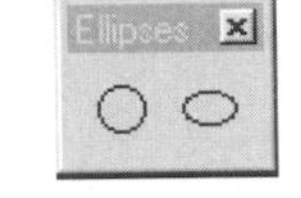

Arcs Tool Box

Text Tool Box

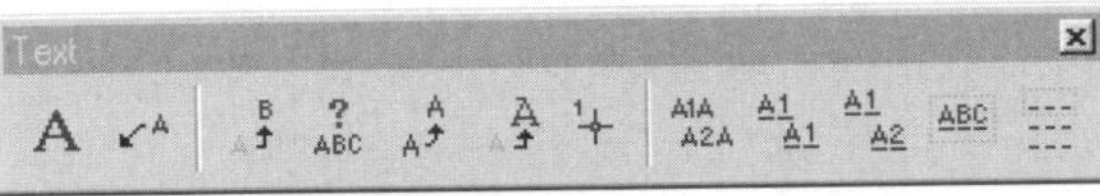

Tags Tool Box

Cells Tool Box

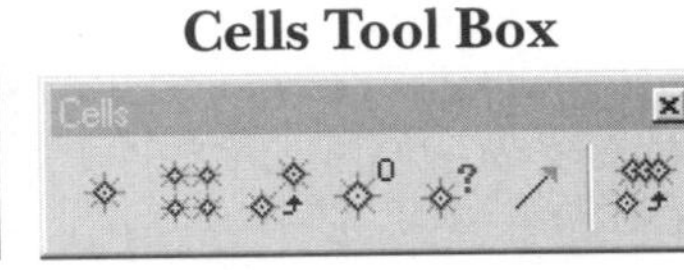

Groups Tool Box

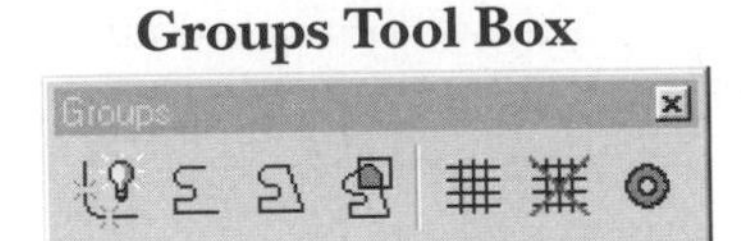

Dimension Tool Box

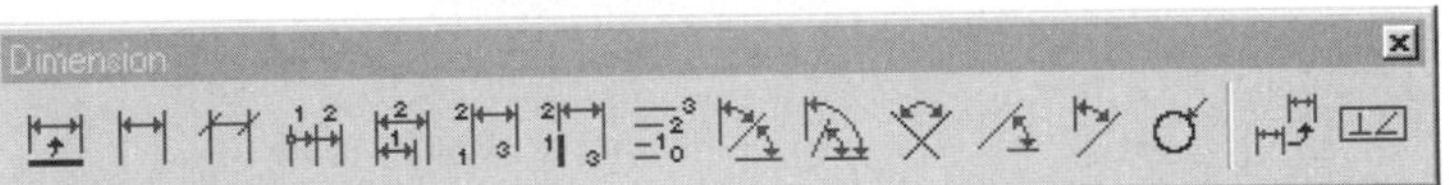

Measure Tool Box

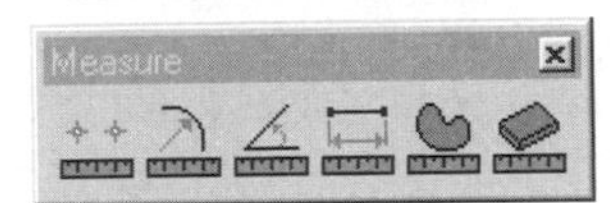

Manipulate Tool Box

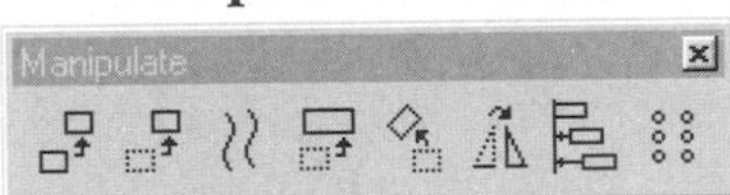

Change Attributes Tool Box

Annotate Tool Frame

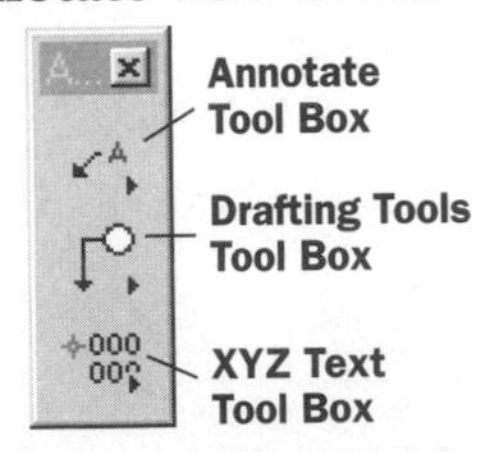

Annotate Tool Box

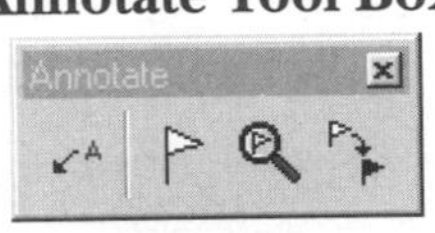

Drafting Tool Box

XYZ Text Tool Box

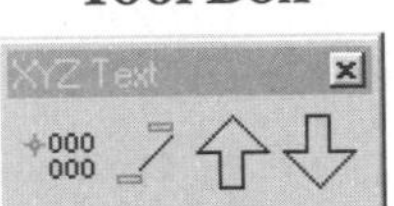

Modify Tool Box

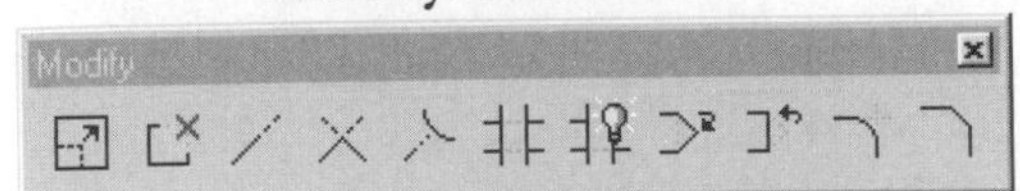

2D View Control Tool Box

3D View Control Tool Box

Rendering Tools Tool Box

appendix B

Key-in Commands

Note: MicroStation is not case-sensitive. That is, key-ins can be uppercase or lowercase, or even a mixture of the two, and MicroStation will still understand the key-ins in the same way.

TOOL NAME	KEY-IN
Add to Graphic Group	GROUP ADD
Attach Active Entity	ATTACH AE
Attach Active Entity to Fence Contents	FENCE ATTACH
Attach Displayable Attributes	ATTACH DA
Attach Reference File	REFERENCE ATTACH (RF=)
Automatic Create Complex Chain	CREATE CHAIN AUTOMATIC
Automatic Create Complex Shape	CREATE SHAPE AUTOMATIC
Automatic Fill in Enter Data Fields	EDIT AUTO
B-spline Polygon Display On/Off	MDL LOAD SPLINES; CHANGE BSPLINE POLYGON
Chamfer	CHAMFER

TOOL NAME	KEY-IN
Change B-spline Surface to Active U-Order	MDL LOAD SPLINES; CHANGE BSPLINE UORDER
Change B-spline Surface to Active U-Rules	MDL LOAD SPLINES; CHANGE BSPLINE URULES
Change B-spline Surface to Active V-Order	MDL LOAD SPLINES; CHANGE BSPLINE VORDER
Change B-spline Surface to Active V-Rules	MDL LOAD SPLINES; CHANGE BSPLINE VRULES
Change B-spline to Active Order	MDL LOAD SPLINES; CHANGE BSPLINE ORDER
Change Element to Active Class	CHANGE CLASS
Change Element to Active Color	CHANGE COLOR
Change Element to Active Level	CHANGE LEVEL
Change Element to Active Line Style	CHANGE STYLE
Change Element to Active Line Weight	CHANGE WEIGHT
Change Element to Active Symbol	CHANGE SYMBOLOGY
Change Fence Contents to Active Color	FENCE CHANGE COLOR
Change Fence Contents to Active Level	FENCE CHANGE LEVEL
Change Fence Contents to Active Style	FENCE CHANGE STYLE
Change Fence Contents to Active Symbology	FENCE CHANGE SYMBOLOGY
Change Fence Contents to Active Weight	FENCE CHANGE WEIGHT
Change Fill	CHANGE FILL
Change Text to Active Attributes	MODIFY TEXT
Circular Fillet (No Truncation)	FILLET NOMODIFY
Circular Fillet and Truncate Both	FILLET MODIFY
Circular Fillet and Truncate Single	FILLET SINGLE
Closed Cross Joint	MDL LOAD CUTTER; JOIN CROSS CLOSED
Closed Tee Joint	MDL LOAD CUTTER; JOIN TEE CLOSED

TOOL NAME	KEY-IN
Complete Cycle Linear Pattern	PATTERN LINE SCALE
Construct Active Point at Distance Along an Element	CONSTRUCT POINT DISTANCE
Construct Active Point at Intersection	CONSTRUCT POINT INTERSECTION
Construct Active Points Between Data Points	CONSTRUCT POINT BETWEEN
Construct Angle Bisector	CONSTRUCT BISECTOR ANGLE
Construct Arc Tangent to Three Elements	CONSTRUCT TANGENT ARC 3
Construct B-spline Curve by Least Squares	MDL LOAD SPLINES; CONSTRUCT BSPLINE CURVE LEAST SQUARE
Construct B-spline Curve by Points	MDL LOAD SPLINES; CONSTRUCT BSPLINE CURVE POINTS
Construct B-spline Curve by Poles	MDL LOAD SPLINES; CONSTRUCT BSPLINE CURVE POLES
Construct B-spline Surface by Cross-Section	MDL LOAD SPINES; CONSTRUCT BSPLINE SURFACE CROSS
Construct B-spline Surface by Edges	MDL LOAD SPLINES; CONSTRUCT BSPLINE SURFACE EDGE
Construct B-spline Surface by Least Squares	MDL LOAD SPLINES; CONSTRUCT BSPLINE SURFACE LEAST SQUARE
Construct B-spline Surface by Points	MDL LOAD SPLINES; CONSTRUCT BSPLINE SURFACE POINTS
Construct B-spline Surface by Poles	MDL LOAD SPLINES; CONSTRUCT BSPLINE SURFACE POLES
Construct B-spline Surface by Skin	MDL LOAD SPLINES; CONSTRUCT BSPLINE SURFACE SKIN
Construct B-spline Surface by Tube	MDL LOAD SPLINES; CONSTRUCT BSPLINE SURFACE TUBE
Construct B-spline Surface of Projection	MDL LOAD SPLINES; CONSTRUCT BSPLINE SURFACE PROJECTION
Construct B-spline Surface of Revolution	MDL LOAD SPLINES; CONSTRUCT BSPLINE SURFACE REVOLUTION
Construct Circle Tangent to Element	CONSTRUCT TANGENT CIRCLE 1

TOOL NAME	KEY-IN
Construct Circle Tangent to Three Elements	CONSTRUCT TANGENT CIRCLE 3
Construct Line at Active Angle from Point (key-in)	CONSTRUCT LINE AA 4
Construct Line at Active Angle from Point	CONSTRUCT LINE AA 3
Construct Line at Active Angle to Point (key-in)	CONSTRUCT LINE AA 2
Construct Line at Active Angle to Point	CONSTRUCT LINE AA 1
Construct Line Bisector	CONSTRUCT BISECTOR LINE
Construct Line Tangent to Two Elements	CONSTRUCT TANGENT BETWEEN
Construct Minimum Distance Line	CONSTRUCT LINE MINIMUM
Construct Perpendicular from Element	CONSTRUCT PERPENDICULAR FROM
Construct Perpendicular to Element	CONSTRUCT PERPENDICULAR TO
Construct Points Along Element	CONSTRUCT POINT ALONG
Construct Surface/Solid of Projection	SURFACE PROJECTION
Construct Surface/Solid of Revolution	SURFACE REVOLUTION
Construct Tangent Arc by Keyed-in Radius	CONSTRUCT TANGENT ARC 1
Construct Tangent from Element	CONSTRUCT TANGENT FROM
Construct Tangent to Circular Element and Perpendicular to Linear Element	CONSTRUCT TANGENT PERPENDICULAR
Construct Tangent to Element	CONSTRUCT TANGENT TO
Convert Element to B-spline (Copy)	MDL LOAD SPLINES; CONSTRUCT BSPLINE CONVERT COPY
Convert Element to B-spline Original	MDL LOAD SPLINES; CONSTRUCT BSPLINE CONVERT ORIGINAL
Copy Fence Content	FENCE COPY
Copy Parallel by Distance	COPY PARALLEL DISTANCE
Copy Parallel by Key-in	COPY PARALLEL KEYIN
Corner Joint	MDL LOAD CUTTER; JOIN CORNER
Create Complex Chain	CREATE CHAIN MANUAL
Create Complex Shape	CREATE SHAPE MANUAL

TOOL NAME	KEY-IN
Crosshatch Element Area	CROSSHATCH
Cut All Component Lines	MDL LOAD CUTTER; CUT ALL
Cut Single Component Line	MDL LOAD CUTTER; CUT SINGLE
Define ACS (Aligned with Element)	DEFINE ACS ELEMENT
Define ACS (Aligned with View)	DEFINE ACS VIEW
Define ACS (By Points)	DEFINE ACS POINTS
Define Active Entity Graphically	DEFINE AE
Define Cell Origin	DEFINE CELL ORIGIN
Define Reference File Back Clipping Plane	REFERENCE CLIP BACK
Define Reference File Clipping Boundary	REFERENCE CLIP BOUNDARY
Define Reference Clipping Mask	REFERENCE CLIP MASK
Define Reference File Front Clipping Plane	REFERENCE CLIP FRONT
Define True North	DEFINE NORTH
Delete Element	DELETE ELEMENT
Delete Fence Contents	FENCE DELETE
Delete Part of Element	DELETE PARTIAL
Delete Vertex	DELETE VERTEX
Detach Database Linkage	DETACH
Detach Database Linkage from Fence Contents	FENCE DETACH
Detach Reference File	REFERENCE DETACH
Dimension Angle Between Lines	DIMENSION ANGLE LINES
Dimension Angle from X-Axis	DIMENSION ANGLE X
Dimension Angle from Y-Axis	DIMENSION ANGLE Y
Dimension Angle Location	DIMENSION ANGLE LOCATION
Dimension Angle Size	DIMENSION ANGLE SIZE
Dimension Arc Location	DIMENSION ARC LOCATION
Dimension Arc Size	DIMENSION ARC SIZE

TOOL NAME	KEY-IN
Dimension Diameter	DIMENSION DIAMETER
Dimension Diameter (Extended Leader)	DIMENSION DIAMETER EXTENDED
Dimension Diameter Parallel	DIMENSION DIAMETER PARALLEL
Dimension Diameter Perpendicular	DIMENSION DIAMETER PERPENDICULAR
Dimension Diameter	DIMENSION DIAMETER
Dimension Element	DIMENSION ELEMENT
Dimension Location	DIMENSION LOCATION SINGLE
Dimension Location (Stacked)	DIMENSION LOCATION STACKED
Dimension Ordinates	DIMENSION ORDINATE
Dimension Radius	DIMENSION RADIUS
Dimension Radius (Extended Leader)	DIMENSION RADIUS EXTENDED
Dimension Size (Custom)	DIMENSION LINEAR
Dimension Size with Arrow	DIMENSION SIZE ARROW
Dimension Size with Strokes	DIMENSION SIZE STROKE
Display Attributes of Text Element	IDENTIFY TEXT
Drop Association	DROP ASSOCIATION
Drop Complex Status	DROP COMPLEX
Drop Complex Status of Fence Contents	FENCE DROP
Drop Dimension	DROP DIMENSION
Drop from Graphic Group	GROUP DROP
Drop Line String/Shape Status	DROP STRING
Drop Text	DROP TEXT
Edit Text	EDIT TEXT
Element Selection	CHOOSE ELEMENT
Extend 2 Elements to Intersection	EXTEND ELEMENT 2
Extend Element to Intersection	EXTEND ELEMENT INTERSECTION
Extend Line	EXTEND LINE DISTANCE

TOOL NAME	KEY-IN
Extend Line By Key-in	EXTEND LINE KEYIN
Extract Bspline Surface Boundary	MDL LOAD SPLINES; EXTRACT BSPLINE SURFACE BOUNDARY
Fence Stretch	FENCE STRETCH
Fill in Single Enter Data Field	EDIT SINGLE
Freeze Element	FREEZE
Freeze Elements in Fence	FENCE FREEZE
Generate Report Table	FENCE REPORT
Global Origin	ACTIVE ORIGIN (GO=)
Group Holes	GROUP HOLES
Hatch Element Area	HATCH
Horizontal Parabola (No Truncation)	PLACE PARABOLA HORIZONTAL NOMODIFY
Horizontal Parabola and Truncate Both	PLACE PARABOLA HORIZONTAL MODIFY
Identify Cell	IDENTIFY CELL
Impose Bspline Surface Boundary	MDL LOAD SPLINES; IMPOSE BSPLINE SURFACE BOUNDARY
Insert Vertex	INSERT VERTEX
Label Line	LABEL LINE
Load Displayable Attributes	LOAD DA
Load Displayable Attributes to Fence Contents	FENCE LOAD
Match Pattern Attributes	ACTIVE PATTERN MATCH
Match Text Attributes	ACTIVE TEXT
Measure Angle Between Lines	MEASURE ANGLE
Measure Area	MEASURE AREA
Measure Area of Element	MEASURE AREA ELEMENT
Measure Distance Along Element	MEASURE DISTANCE ALONG
Measure Distance Between Points	MEASURE DISTANCE POINTS

TOOL NAME	KEY-IN
Measure Minimum Distance Between Elements	MEASURE DISTANCE MINIMUM
Measure Perpendicular Distance From Element	MEASURE DISTANCE PERPENDICULAR
Measure Radius	MEASURE RADIUS
Merged Cross Joint	MDL LOAD CUTTER; JOIN CROSS MERGE
Merged Tee Joint	MDL LOAD CUTTER; JOIN TEE MERGE
Mirror Element About Horizontal (Copy)	MIRROR COPY HORIZONTAL
Mirror Element About Horizontal (Original)	MIRROR ORIGINAL HORIZONTAL
Mirror Element About Line Copy	MIRROR COPY LINE
Mirror Element About Line (Ordinal)	MIRROR ORIGINAL LINE
Mirror Element About Vertical (Copy)	MIRROR COPY VERTICAL
Mirror Element About Vertical (Original)	MIRROR ORIGINAL VERTICAL
Mirror Fence Contents About Horizontal (Copy)	FENCE MIRROR COPY HORIZONTAL
Mirror Fence Contents About Horizontal (Original)	FENCE MIRROR ORIGINAL HORIZONTAL
Mirror Fence Contents About Line (Copy)	FENCE MIRROR COPY LINE
Mirror Fence Contents About Line (Original)	FENCE MIRROR ORIGINAL LINE
Mirror Fence Contents About Vertical (Copy)	FENCE MIRROR COPY VERTICAL
Mirror Fence Contents About Vertical (Original)	FENCE MIRROR ORIGINAL VERTICAL
Mirror Reference File About Horizontal	REFERENCE MIRROR HORIZONTAL
Mirror Fence About Vertical	REFERENCE MIRROR VERTICAL
Modify Arc Angle	MODIFY ARC ANGLE
Modify Arc Axis	MODIFY ARC AXIS
Modify Arc Radius	MODIFY ARC RADIUS
Modify Element	MODIFY ELEMENT
Modify Fence	MODIFY FENCE
Move ACS	MOVE ACS

TOOL NAME	KEY-IN
Move Element	MOVE ELEMENT
Move Fence Block/Shape	MOVE FENCE
Move Fence Contents	FENCE MOVE
Move Reference File	REFERENCE MOVE
Multi-Cycle Segment Linear Pattern	PATTERN LINE MULTIPLE
Open Cross Joint	MDL LOAD CUTTER; JOIN CROSS OPEN
Open Tee Joint	MDL LOAD CUTTER; JOIN TEE OPEN
Pattern Element Area	PATTERN AREA ELEMENT
Pattern Fence Area	PATTERN AREA FENCE
Place Active Cell	PLACE CELL ABSOLUTE
Place Active Cell (Interactive)	PLACE CELL INTERACTIVE ABSOLUTE
Place Active Cell Matrix	MATRIX CELL (CM=)
Place Active Cell Relative	PLACE CELL RELATIVE
Place Active Cell Relative (Interactive)	PLACE CELL INTERACTIVE RELATIVE
Place Active Line Terminator	PLACE TERMINATOR
Place Active Point	PLACE POINT
Place Arc by Center	PLACE ARC CENTER
Place Arc by Edge	PLACE ARC EDGE
Place Arc by Keyed-in Radius	PLACE ARC RADIUS
Place B-spline Curve by Least Squares	MDL LOAD SPLINES; PLACE BSPLINE CURVE LEASTSQUARE
Place B-spline Curve by Points	MDL LOAD SPLINES; PLACE BSPLINE CURVE POINTS
Place B-spline Curve by Poles	MDL LOAD SPLINES; PLACE BSPLINE CURVE POLES
Place B-spline Surface by Least Squares	MDL LOAD SPLINES; PLACE BSPLINE SURFACE LEASTSQUARES
Place B-spline Surface by Points	MDL LOAD SPLINES; PLACE BSPLINE SURFACE POINTS

TOOL NAME	KEY-IN
Place B-spline Surface by Poles	MDL LOAD SPLINES; PLACE BSPLINE SURFACE POLES
Place Block	PLACE BLOCK ORTHOGONAL
Place Center Mark	DIMENSION CENTER MARK
Place Circle by Center	PLACE CIRCLE CENTER
Place Circle by Diameter	PLACE CIRCLE DIAMETER
Place Circle by Edge	PLACE CIRCLE EDGE
Place Circle by Keyed-in Radius	PLACE CIRCLE RADIUS
Place Circumscribed Polygon	PLACE POLYGON CIRCUMSCRIBED
Place Ellipse by Center and Edge	PLACE ELLIPSE CENTER
Place Ellipse by Edge Points	PLACE ELLIPSE EDGE
Place Fence Block	PLACE FENCE BLOCK
Place Fence Shape	PLACE FENCE SHAPE
Place Fitted Text	PLACE TEXT FITTED
Place Fitted View Independent Text	PLACE TEXT VI
Place Half Ellipse	PLACE ELLIPSE HALF
Place Helix	MDL LOAD SPLINES; PLACE HELIX
Place Inscribed Polygon	PLACE POLYGON INSCRIBED
Place Isometric Block	PLACE BLOCK ISOMETRIC
Place Isometric Circle	PLACE CIRCLE ISOMETRIC
Place Line	PLACE LINE
Place Line at Active Angle	PLACE LINE ANGLE
Place Line String	PLACE LSTRING POINT
Place Multi-line	PLACE MLINE
Place Note	PLACE NOTE
Place Orthogonal Shape	PLACE SHAPE ORTHOGONAL
Place Parabola by End Points	MDL LOAD SPLINES; PLACE PARABOLA ENDPOINTS

TOOL NAME	KEY-IN
Place Point Curve	PLACE CURVE POINT
Place Polygon by Edge	PLACE POLYGON EDGE
Place Quarter Ellipse	PLACE ELLIPSE QUARTER
Place Right Cone	PLACE CONE RIGHT
Place Right Cone by Keyed-in Radius	PLACE CONE RADIUS
Place Right Cylinder	PLACE CYLINDER RIGHT
Place Right Cylinder by Keyed-in Radius	PLACE CYLINDER RADIUS
Place Rotated Block	PLACE BLOCK ROTATED
Place Shape	PLACE SHAPE
Place Skewed Cone	PLACE CONE SKEWED
Place Skewed Cylinder	PLACE CYLINDER SKEWED
Place Slab	PLACE SLAB
Place Space Curve	PLACE CURVE SPACE
Place Space Line String	PLACE LSTRING SPACE
Place Sphere	PLACE SPHERE
Place Spiral By End Points	MDL LOAD SPLINES; PLACE SPIRAL ENDPOINTS
Place Spiral By Length	MDL LOAD SPLINES; PLACE SPIRAL LENGTH
Place Spiral by Sweep Angle	MDL LOAD SPLINES; PLACE SPIRAL ANGLE
Place Stream Curve	PLACE CURVE STREAM
Place Stream Line String	PLACE LSTRING STREAM
Place Text	PLACE TEXT
Place Text Above Element	PLACE TEXT ABOVE
Place Text Along Element	PLACE TEXT ALONG
Place Text Below Element	PLACE TEXT BELOW
Place Text Node	PLACE NODE

TOOL NAME	KEY-IN
Place Text On Element	PLACE TEXT ON
Place View Independent Text	PLACE TEXT VI
Place View Independent Text Node	PLACE NODE VIEW
Polar Array	ARRAY POLAR
Polar Array Fence Contents	FENCE ARRAY POLAR
Project Active Point Onto Element	CONSTRUCT POINT PROJECT
Rectangular Array	ARRAY RECTANGULAR
Rectangular Array Fence Contents	FENCE ARRAY RECTANGULAR
Reload Reference File	REFERENCE RELOAD
Replace Cell	REPLACE CELL
Review Database Attributes of Element	REVIEW
Rotate ACS Absolute	ROTATE ACS ABSOLUTE
Rotate ACS Relative	ROTATE ACS RELATIVE
Rotate Element Active Angle Copy	ROTATE COPY
Rotate Element Active Angle Original	ROTATE ORIGINAL
Rotate Fence Contents by Active Angle (Copy)	FENCE ROTATE COPY
Rotate Fence Contents by Active Angle (Original)	FENCE ROTATE ORIGINAL
Rotate Reference File	REFERENCE ROTATE
Scale Element (Copy)	SCALE COPY
Scale Element (Original)	SCALE ORIGINAL
Scale Fence Contents (Copy)	FENCE SCALE COPY
Scale Fence Contents (Original)	FENCE SCALE ORIGINAL
Scale Reference File	REFERENCE SCALE
Select ACS	ATTACH ACS
Select and Place Cell	SELECT CELL ABSOLUTE
Select and Place Cell (Relative)	SELECT CELL RELATIVE

TOOL NAME	KEY-IN
Set Active Depth	DEPTH ACTIVE
Show Active Depth	SHOW DEPTH ACTIVE
Show Active Entity	SHOW AE
Show Linkage Mode	ACTIVE LINKAGE
Show Pattern Attributes	SHOW PATTERN
Single Cycle Segment Linear Pattern	PATTERN LINE SINGLE
Spin Element (Copy)	SPIN COPY
Spin Element (Original)	SPIN ORIGINAL
Spin Fence Contents (Copy)	FENCE SPIN COPY
Spin Fence Contents (Original)	FENCE SPIN ORIGINAL
Symmetric Parabola (No Truncation)	PLACE PARABOLA NOMODIFY
Symmetric Parabola and Truncate Both	PLACE PARABOLA MODIFY
Thaw Element	THAW
Thaw Elements in Fence	FENCE THAW
Truncated Cycle Linear Pattern	PATTERN LINE ELEMENT
Uncut Component Lines	MDL LOAD CUTTER; UNCUT

appendix C

Alternate Key-ins

Note: MicroStation is not case-sensitive. That is, key-ins can be uppercase or lowercase, or even a mixture of the two, and MicroStation will still understand the key-ins in the same way.

AA = ACTIVE ANGLE	set active angle
AC = ACTIVE CELL	set active cell; place absolute
AD = POINT ACSDELTA	data point—delta ACS
AE = ACTIVE ENTITY	define active entity
AM = ATTACH MENU	activate menu
AP = ACTIVE PATTERN CELL	set active pattern cell
AR = ACTIVE RCELL	set active cell; place relative
AS = ACTIVE SCALE	set active scale factors
AT = TUTORIAL	activate tutorial
AX = POINT ACSABSOLUTE	data point absolute ACS
AZ = ACTIVE ZDEPTH ABSOLUTE	set active depth

CC = CREATE CELL	create cell
CD = DELETE CELL	delete cell from cell library
CM = MATRIX CELL	place active cell matrix
CO = ACTIVE COLOR	set active color
CR = RENAME CELL	rename cell
CT = ATTACH COLORTABLE	attach color table
DA = ACTIVE DATYPE	set active displayable attribute type
DB = ACTIVE DATABASE	attach control file to design file
DD = SET DDEPTH RELATIVE	set display depth (relative)
DF = SHOW FONT	open Fonts settings box
DI = POINT DISTANCE	data point—distance, direction
DL = POINT DELTA	data point—delta coordinates
DP = DEPTH DISPLAY	set display depth
DR = TYPE	display text file
DS = SEARCH	specify fence filter
DV = VIEW	delete saved view
DX = POINT VDELTA	data point—delta view coordinates
DZ = ZDEPTH RELATIVE	set active depth (relative)
EL = ELEMENT LIST	create element list file
FF = FENCE FILE	copy fence contents to design file
FI = FIND	set database row as active entity
FT = ACTIVE FONT	set active font
GO = ACTIVE ORIGIN	Global Origin
GR = ACTIVE GRIDREF	set grid reference spacing
GU = ACTIVE GRIDUNIT	set horizontal grid spacing
KY = ACTIVE KEYPNT	set Snap divisor
LC = ACTIVE STYLE	set active line style
LD = DIMENSION LEVEL	set dimension level

LL = ACTIVE LINE LENGTH	set active text line length
LS = ACTIVE LINE SPACE	set active text node line spacing
LT = ACTIVE TERMINATOR	set active terminator
LV = ACTIVE LEVEL	set active level
NN = ACTIVE NODE	set active text node number
OF = SET LEVELS <level list> OFF	set level display off
ON = SET LEVELS <level list> ON	set level display on
OX = ACTIVE INDEX	retrieve user command index
PA = ACTIVE PATTERN ANGLE	set active pattern angle
PD = ACTIVE PATTERN DELTA	set active pattern delta (distance)
PS = ACTIVE PATTERN SCALE	set active pattern scale
PT = ACTIVE POINT	set active point
PX = DELETE ACS	delete ACS
RA = ACTIVE REVIEW	set attribute review selection criteria
RC = ATTACH LIBRARY	open cell library
RD = NEWFILE	open design file
RF = REFERENCE ATTACH	attach reference file
RS = ACTIVE REPORT	name report table
RV = ROTATE VIEW	rotate view (relative)
RX = ATTACH ACS	select ACS
SD = ACTIVE STREAM DELTA	set active stream delta
SF = FENCE SEPARATE	move fence contents to design file
ST = ACTIVE STREAM TOLERANCE	set active stream tolerance
SV = SAVE VIEW	save view
SX = SAVE ACS	save auxiliary coordinate system
TB = ACTIVE TAB	set tab spacing for importing text
TH = ACTIVE TXHEIGHT	set active text height
TI = ACTIVE TAG	set copy and increment value

TS = ACTIVE TSCALE	set active terminator scale
TV = DIMENSION TOLERANCE	set dimension tolerance limits
TW = ACTIVE TXWIDTH	set active text width
TX = ACTIVE TXSIZE	set active text size (height/width)
UC = USERCOMMAND	activate user command
UCC = UCC	compile user command
UCI = UCI	user command index
UR = ACTIVE UNITROUND	set unit distance
VI = VIEW	attach named view
WO = WINDOW ORIGIN	Window Orgin
WT = ACTIVE WEIGHT	set active line weight
XD = EXCHANGEFILE	open design file; keep view config.
XS = ACTIVE XSCALE	set active X scale
XY = POINT ABSOLUTE	data point absolute coordinates
YS = ACTIVE YSCALE	set active Y scale
ZS = ACTIVE ZSCALE	set active Z scale

appendix

Primitive Commands

Note: MicroStation is not case-sensitive. That is, key-ins can be uppercase or lowercase, or even a mixture of the two, and MicroStation will still understand the key-ins in the same way.

NAME OF THE COMMAND	PRIMITIVE COMMAND
ACTIVE ANGLE PT2	/ACTAN2
ACTIVE ANGLE PT3	/ACTAN3
ACTIVE CAPMODE OFF	/WOCMDE
ACTIVE CAPMODE ON	/CAPMDE
ACTIVE SCALE DISTANCE	/ACTSCA
ACTIVE TNJ CB	/TJST#
ACTIVE TNJ CC	/TJST7
ACTIVE TNJ CT	/TJST6
ACTIVE TNJ LB	/TJST2

NAME OF THE COMMAND	PRIMITIVE COMMAND
ACTIVE TNJ LC	/TJST]
ACTIVE TNJ LMB	/TJST5
ACTIVE TNJ LMC	/TJST4
ACTIVE TNJ LMT	/TJST3
ACTIVE TNJ LT	/TJSTO
ACTIVE TNJ RB	/TJST14
ACTIVE TNJ RC	/TJST13
ACTIVE TNJ RMB	/TJSTl 1
ACTIVE TNJ RMC	/TJST 1 0
ACTIVE TNJ RMT	/TJST9
ACTIVE TNJ RT	/TJSTl 2
ACTIVE TXHEIGHT PT2	/TXTHGT
ACTIVE TXJ CB	/TXJS8
ACTIVE TXJ CC	/TXJS7
ACTIVE TXJ CT	/TXJS6
ACTIVE TXJ LB	/TXJSZ
ACTIVE TXJ LC	/TXJSl
ACTIVE TXJ LT	/TXJSO
ACTIVE TXJ RB	/TXJS14
ACTIVE TXJ RC	/TXJS13
ACTIVE TXJ RT	/TXJSlz
ACTIVE TXWIDTH PT2	/TXTWDT
ALIGN	/ALIGN
ATTACH AE	/ATCPTO
CHANGE COLOR	/CELECR
CHANGE STYLE	/CELELS
CHANGE SYMBOLOGY	/CELESY

NAME OF THE COMMAND	PRIMITIVE COMMAND
CHANGE WEIGHT	/CELEWT
CONSTRUCT BISECTOR ANGLE	/ANGBIS
CONSTRUCT BISECTOR LINE	/PERBIS
CONSTRUCT LINE AA 1	/CNSAA1
CONSTRUCT LINE AA 2	/CNSAA2
CONSTRUCT LINE AA 3	/CNSAA3
CONSTRUCT LINE AA 4	/CNSAA4
CONSTRUCT LINE MINIMUM	/MDL2EL
CONSTRUCT POINT	/CNSINT
CONSTRUCT POINTALONG	/NPAE
CONSTRUCT POINT BETWEEN	/NPNTS
CONSTRUCT POINT DISTANCE	/PPAE
CONSTRUCT POINT PROJECT	/PRJPNT
CONSTRUCT TANGENT	/LNTNNR
CONSTRUCT TANGENT ARC 1	/PTARCC
CONSTRUCT TANGENT ARC 3	/ATN3EL
CONSTRUCT TANGENT BETWEEN	/LTZELP
CONSTRUCT TANGENT CIRCLE 1	/CTNlEL
CONSTRUCT TANGENT CIRCLE 3	/CTN3EL
CONSTRUCT TANGENT FROM	/PTFROM
CONSTRUCT TANGENT TO	/PITO
COPY ED	/EDCOPY
COPYELEMENT	/CPELE
COPY PARALLEL DISTANCE	/CPYPP
COPY PARALLEL KEYIN	/CPYPK
COPY VIEW	/COPY
CREATE CHAIN MANUAL	/CONNST

NAME OF THE COMMAND	PRIMITIVE COMMAND
CREATE SHAPE MANUAL	/CPXSHP
DEFINE ACS ELEMENT	/AUXELE
DEFINE ACS POINTS	/AUX3PT
DEFINE ACS VIEW	/AUXVW
DEFINE AE	/DEFPTO
DEFINE CELL ORIGIN	/DOCELL
DELETE ELEMENT	/DLELEM
DELETE PARTIAL	/DLPELE
DELETE VERTEX	/DVERTX
DEPTH ACTIVE PRIMITIVE	/ADEPTH
DEPTH DISPLAY PRIMITIVE	/DDEPTH
DIMENSION ANGLE LINES	/ANGLIN
DIMENSION ANGLE LOCATION	/PITLOC
DIMENSION ANGLE SIZE	/PTSIZ
DIMENSION ARC LOCATION	/ARCLOC
DIMENSION ARC SIZE	/ARCSIZ
DIMENSION AXIS DRAWING	/ACTAXD
DIMENSION AXIS TRUE	/ACTAXP
DIMENSION AXIS VIEW	/ACTAXV
DIMENSION DIMAETER PARALLEL	/DIAPAR
DIMENSION DIAMETER PERPENDICULAR	/DIAPER
DIMENSION DIAMETER POINT	/DIACIR
DIMENSION FILE ACTIVE	/MEAACT
DIMENSION FILE REFERENCE	/MEAREF
DIMENSION JUSTIFICATION CENTER	/JUSC
DIMENSION JUSTIFICATION LEFT	/JUSL
DIMENSION JUSTIFICATION RIGHT	/JUSR

NAME OF THE COMMAND	PRIMITIVE COMMAND
DIMENSION LOCATION SINGLE	/LOCSNG
DIMENSION LOCATION STACKED	/LOCSTK
DIMENSION PLACEMENT AUTO	/ADMAUT
DIMENSION PLACEMENT MANUAL	/PADMAU
DIMENSION RADIUS POINT	/RADRAD
DIMENSION SIZE ARROW	/SIZARW
DIMENSION SIZE STROKE	/SIZOBL
DIMENSION UNITS DEGREES	/UNITDG
DIMENSION UNITS LENGTH	/UNITLN
DIMENSION WITNESS OFF	/WITLOF
DIMENSION WITNESS ON	/WITLON
DROP COMPLEX	/DRCMPX
EDIT AUTO	/EDAUTO
EDIT SINGLE	/EDSING
EDIT TEXT	/EDTEXT
EXTEND ELEMENT 2	/EXLIN2
EXTEND ELEMENT INTERSECTION	/EXLNIN
EXTEND LINE DISTANCE	/EXLIN
EXTEND LINE KEYIN	/EXLINK
FENCE ATTACH	/AAEFCN
FENCE CHANGE STYLE	/CFNCLS
FENCE CHANGE SYMBOLOGY	/CFNCSY
FENCE COPY	/CPFNCC
FENCE DELETE	/DLFNCC
FENCE DETACH	/RATFCN
FENCE LOCATE	/FNCLOC
FENCE MIRROR COPY HORIZONTAL	/MHCPFC

NAME OF THE COMMAND	PRIMITIVE COMMAND
FENCE MIRROR COPY LINE	/MLCPFC
FENCE MIRROR COPY VERTICAL	/MVCPFC
FENCE MIRROR ORIGINAL	/MFVERT
FENCE MIRROR ORIGINAL HORIZONTAL	/MFHRIZ
FENCE MIRROR ORIGINAL LINE	/MFLINE
FENCE MOVE	/MVFNCC
FENCE REPORT	/RPTACT
FENCE ROTATE COPY	/RTCPFC
FENCE ROTATE ORIGINAL	/RFNCC
FENCE SCALE COPY	/SCCPFC
FENCE SCALE ORIGINAL	/SCFNCC
FENCE TRANSFORM	/TRSFCC
FENCE WSET ADD	/ADWSFN
FENCE WSET COPY	/ADWSFC
FILE DESIGN	/FILDGN
FILLET MODIFY	/PFILTM
FILLET NOMODIFY	/PFILTN
FILLET SINGLE	/FILTRM
FIT ACTIVE	/FIT1
GROUP ADD	/ADDGG
GROUP DROP	/DRFGG
IDENTIFY CELL	/IDCELL
IDENTIFY TEXT	/TXNODA
INCREMENT ED	/CIDATA
INCREMENT TEXT	/CITEXT
INSERT VERTEX	/IVERTX
JUSTIFY CENTER	/EDCJST

NAME OF THE COMMAND	PRIMITIVE COMMAND
JUSTIFY LEFT	/EDLJST
JUSTIFY RIGHT	/EDRJST
LABEL LINE	/LABLN
LABEL LINE	/LBLINE
LOCELE	/LOCELE
LOCK ACS [OFF|ON|TOGGLE]	/CPLOCK
LOCK ANGLE [OFF|ON|TOGGLE]	/ANGLLK
LOCK AXIS [OFF|ON|TOGGLE]	/AXLKFF
LOCK BORESITE [OFF|ON|TOGGLE]	/BORSIT
LOCK FENCE CLIP	/CLIP
LOCK FENCE INSIDE	/INSIDE
LOCK FENCE OVERLAP	/OVRLAP
LOCK GGROUP [OFF|ON|TOGGLE]	/GGLOCK
LOCK GRID [OFF|ON|TOGGLE]	/GRIDLK
LOCK SCALE [OFF|ON|TOGGLE]	/SCALLK
LOCK SNAP KEYPOINT	/KEYSNP
LOCK SNAP [OFF|ON]	/SNPOFF
LOCK SNAP PROJECT	/SNAPLK
LOCK TEXTNODE [OFF|ON]	/TXTNLK
LOCK UNIT [OFF|ON]	/UNITLK
MEASURE ANGLE	/LINANG
MEASURE AREA	/AREAPT
MEASURE AREA ELEMENT	/AREAEL
MEASURE DISTANCE ALONG	/MDAE
MEASURE DISTANCE PERPENDICULAR	/PRPND
MEASURE DISTANCE POINTS	/PERIM
MEASURE RADIUS	/RADIUS

NAME OF THE COMMAND	PRIMITIVE COMMAND
MIRROR COPY HORIZONTAL	/MHCPEL
MIRROR COPY LINE	/MLCPEL
MIRROR COPY VERTICAL	/MVCPEL
MIRROR ORIGINAL HORIZONTAL	/MEHRIZ
MIRROR ORIGINAL LINE	/MELINE
MIRROR ORIGINAL VERTICAL	/MEVERT
MODIFY ARC ANGLE	/MDARCA
MODIFY ARC AXIS	/MDARCX
MODIFY ARC RADIUS	/MDARCR
MODIFY ELEMENT	/MDELE
MODIFY FENCE	/MDFNC
MODIFY TEXT	/TXNODC
MOVE ACS	/AUXORC
MOVE ELEMENT	/MVELEM
MOVE FENCE	/MVFNC
NULL	/NULCMD
PLACE ARC CENTER	/PARCC
PLACE ARC EDGE	/PARCE
PLACE ARC RADIUS	/PARCR
PLACE ARC TANGENT	/PTARCC
PLACE BLOCK ORTHOGONAL	/PBLOCK
PLACE BLOCK ROTATED	/PRBLOC
PLACE CELL ABSOLUTE	/PACELL
PLACE CELL ABSOLUTE TMATRX	/PACMTX
PLACE CELL RELATIVE	/PACELR
PLACE CELL RELATIVE TMATRX	/PACRMX
PLACE CIRCLE CENTER	/PCIRC

NAME OF THE COMMAND	PRIMITIVE COMMAND
PLACE CIRCLE DIAMETER	/PCIRD
PLACE CIRCLE EDGE	/PCIRE
PLACE CIRCLE RADIUS	/PCIRR
PLACE CONE RADIUS	/PRCONR
PLACE CONE RIGHT	/PRCONE
PLACE CONE SKEWED	/PCONE
PLACE CURVE POINT	/PPTCRV
PLACE CURVE SPACE	/PSPCRV
PLACE CURVE STREAM	/PSTCRV
PLACE CYLINDER RADIUS	/PRCYLR
PLACE CYLINDER RIGHT	/PRCYL
PLACE CYLINDER SKEWED	/PCYLIN
PLACE ELLIPSE CENTER	/PELL1
PLACE ELLIPSE EDGE	/PELL2
PLACE ELLIPSE HALF	/PPELL1
PLACE ELLIPSE QUARTER	/PPELL2
PLACE FENCE BLOCK	/PFENCB
PLACE FENCE SHAPE	/PFENCE
PLACE LINE	/PLINE
PLACE LINE ANGLE	/PLINAA
PLACE LSTRING POINT	/PPTLST
PLACE LSTRING SPACE	/PSPLST
PLACE LSTRING STREAM	/PSTLST
PLACE NODE	/PTEXTN
PLACE NODE TMATRX	/PTNMTX
PLACE NOTE VIEW	/PVITXN
PLACE PARABOLA HORIZONTAL MODIFY	/PPARMD

NAME OF THE COMMAND	PRIMITIVE COMMAND
PLACE PARABOLA HORIZONTAL NOMODIFY	/PPARNM
PLACE POINT	/PLPNT
PLACE POINT STRING	/PDPTST
PLACE POINT STRING DISJOINT	/PCPTST
PLACE SHAPE	/PSHAPE
PLACE SHAPE ORTHOGONAL	/POSHAP
PLACE TERMINATOR	/PTERM
PLACE TEXT	/PTEXT
PLACE TEXT ABOVE	/PTXTA
PLACE TEXT ALONG	/PTAE
PLACE TEXT BELOW	/PTXTB
PLACE TEXT FITTED	/PFTEXT
PLACE TEXT FVI	/PVIFTX
PLACE TEXT ON	/PTOE
PLACE TEXT TMATRX	/PTXMTX
PLACE TEXT VI	/PVITXT
REFERENCE CLIP BACK	/RFCBCK
REFERENCE CLIP BOUNDARY	/RFCBND
REFERENCE CLIP FRONT	/RFCFRO
REFERENCE DETACH	/RFDTCH
REFERENCE DISPLAY OFF	/RFDISO
REFERENCE DISPLAY ON	/RFDIS1
REFERENCE LEVELS OFF	/RFLEVO
REFERENCE LEVELS ON	/RFLEV1
REFERENCE LOCATE OFF	/RFLOCO
REFERENCE LOCATE ON	/RFLOC1
REFERENCE MOVE	/RFMOVE

NAME OF THE COMMAND	PRIMITIVE COMMAND
REFERENCE ROTATE	/RFROT
REFERENCE SCALE	/RFSCAL
REFERENCE SNAP OFF	/RFSNPO
REFERENCE SNAP ON	/RFSNP1
REPLACE CELL	/RPCELL
REVIEW	/RVWATR
ROTATE 3PTS	/VIEWPL
ROTATE COPY	/RTCPEL
ROTATE ORIGINAL	/ROTELE
ROTATE VIEW POINTS	/VIEWPL
ROTATE VMATRX	/VMATRX
SCALE COPY	/SCCPEL
SCALE ORIGINAL	/SCAELE
SELECT CELL ABSOLUTE	/PSCELL
SELECT CELL ABSOLUTE MATRX	/PSCMTX
SELECT CELL RELATIVE	/PSCELR
SELECT CELL RELATIVE TMATRX	/PSCRMX
SET CONSTRUCT [OFF|ON|TOGGLE]	/CONST1
SET CURVES [FAST|SLOW|OFF|ON|TOGGLE]	/FCURV1
SET DELETE [OFF|PN|TOGGLE]	/DLENSW
SET DIMENSION [OFF|PN|TOGGLE]	/DIMEN1
SET DYNAMIC [FAST|SLOW|OFF|ON|TOGGLE]	/DRAG
SET ED [OFF|ON|TOGGLE]	/UNDLI1
SSET FONT [FAST|SLOW|OFF|ON|TOGGLE]	/FFONT1
SET GRID [OFF|ON|TOGGLE]	/GRID1
SET NODES [OFF|ON|TOGGLE]	/TXNOD1
SET PATTERN [OFF|ON|TOGGLE]	/PATRN1

NAME OF THE COMMAND	PRIMITIVE COMMAND
SET TEXT [OFF|ON|TOGGLE]	/TEEXT1
SET TPMODE ACSDELTA	/AXDTEN
SET TPMODE ACSLOCATE	/AXXTEN
SET TPMODE DELTA	/MDELTA
SET TPMODE DISTANCE	/MANGL2
SET TPMODE LOCATE	/LOCATE
SET TPMODE VDELTA	/MDLTVW
SET WEIGHT [OFF|ON]	/DWGHT1
SHOW HEADER	/HEADER
SURFACE PROJECTION	/PRJELE
SURFACE REVOLUTION	/SURREV
SWAP SCREEN	/SWAP
TRANSFORM	/TRSELE
UPDATE1...UPDATE8	/UPDATE
UPDATE BOTH	/UPDBTH
UPDATE LEFT	/UPDAT2
UPDATE RIGHT	/UPDAT1
UPDATE VIEW	/UPDATV
VIEW OFF	/VIEWOF
WINDOW AREA	/WINDA1
WINDOW CENTER	/WINDC1
WINDOW VOLUME	/WINVOL
WSET ADD	/ADWSEL
WSET COPY	/ADWSEC
WSET DROP	/WSDROP
ZOOM IN 2	/HALF1
ZOOM OUT 2	/DOUBL1

appendix

Seed Files

NAME OF THE SEED FILE	WORKING UNITS			VIEW ATTRIBUTES	TEXT PARAMETERS
	MU	SU	RESOLUTION		
BENTLEY/WORKSPACE/ SYSTEM/SEED/SCHEM2D.DGN	MM	TH	1MM=100TH 1TH=1000PU	SEE NOTE	TX=0.5000 LS=0.0500
BENTLEY/WORKSPACE/ SYSTEM/SEED/SCHEM3D.DGN	MM	TH	1MM=100TH 1TH=1000PU	SEE NOTE GRID DISP. OFF	TX=0.0100 LS=0.0500
BENTLEY/WORKSPACE/ SYSTEM/ SEED/SDSCH2D.DGN	IN	TH	1IN=10TH 1TH=1000PU	SEE NOTE GRID DISP. OFF	TX=6.0000 LS=0.5000
BENTLEY/WORKSPACE/ SYSTEM/SEED/SDSCH3D.DGN	IN	TH	1MM=100TH 1TH=1000PU	SEE NOTE GRID DISP. OFF	TX=0.1250 LS=0.5000
BENTLEY/WORKSPACE/ SYSTEM/SEED/SEED2D.DGN	MU	SU	1MU=10TH 1SU=1000PU	SEE NOTE	TX=1.0000 LS=0.5000
BENTLEY/WORKSPACE/ SYSTEM/SEED/SEED3D.DGN	MU	SU	1MU=10TH 1SU=1000PU	SEE NOTE GRID DISP. OFF	TX=0.1250 LS=0.5000
BENTLEY/WORKSPACE/ SYSTEM/SEED/SEEDZ.DGN	MU	SU	1MU=10TH 1SU=1000PU	SEE NOTE GRID DISP. OFF	TX=0.1250 LS=0.5000
BENTLEY/WORKSPACE/ SYSTEM/SEED/TRANSEED.DGN	MU	SU	1MU=10TH 1SU=1000PU	SEE NOTE	TX=1.0000 LS=0.5000
BENTLEY/WORKSPACE/ARCH/ SEED/ARCHSEED.DGN	'	"	1'=12" 1"=8000PU	SEE NOTE FILL DISP. ON	TX=1.0000 LS=0.5000

NAME OF THE SEED FILE	WORKING UNITS			VIEW	TEXT
	MU	SU	RESOLUTION	ATTRIBUTES	PARAMETERS
BENTLEY/WORKSPACE / ARCH/SEED/SDARCH2D.DGN	'	"	1'=12"	SEE NOTE	TX=0:9000
BENTLEY/WORKSPACE / ARCH/SEED/SDARCH3D.DGN	'	"	1'=12" 1"=8000PU	SEE NOTE GRID DISP. OFF	TX=1.0000 LS=0.3000
BENTLEY/WORKSPACE / CIVIL/ SEED/CIV2D.DGN	FT	TH	1'=10TH 1TH=100PU	SEE NOTE	TX=1:5000 LS=5:0000
BENTLEY/WORKSPACE / CIVIL/SEED/CIV3D.DGN	FT	TH	1'=10TH 1TH=100PU	SEE NOTE GRID DISP. OFF	TX=25.4 LS=25.4
BENTLEY/WORKSPACE / MAPPING/SEED/MAP2D.DGN	FT	TH	1'=10TH 1TH=100PU	SEE NOTE	TX=2:5.00 LS=5:0.00
BENTLEY/WORKSPACE / MAPPING/SEED/MAP3D.DGN	FT	TH	1'=10TH 1TH=100PU	SEE NOTE	TX=2:5.00 LS=5:0.00
BENTLEY/WORKSPACE / MAPPING/SEED/ SDMAP2D.DGN	FT	TH	1'=10TH 1TH=100PU	SEE NOTE GRID DISP. OFF	TX=1.0000 LS=0.5000
BENTLEY/WORKSPACE / MAPPING/SEED/ SDMAP3D.DGN	FT	TH	1'=10TH 1TH=100PU	SEE NOTE GRID DISP. OFF	TX=1.000 LS=5:000
BENTLEY/WORKSPACE / MAPPING/SEED/ SDMAPM2D.DGN	M	MM	1M=1000MM 1MM=10PU	SEE NOTE GRID DISP. OFF	TX=2.0000 LS=3.0002
BENTLEY/WORKSPACE / MAPPING/SEED/ SDMAPM3D.DGN	M	MM	1M=1000MM 1MM=10PU	SEE NOTE GRID DISP. OFF	TX=6.0000 LS=3.0001
BENTLEY/WORKSPACE / MECHDRAF/SEED/ MECHDET.DGN	IN	TH	1IN=1000TH 1TH=254PU	SEE NOTE GRID & NODES DISP. OFF	TX=0.0047 LS=0.0295
BENTLEY/WORKSPACE / MECHDRAF/SEED/ MECHDETM.DGN	MM		1MM=1000 1=100PU	SEE NOTE GRID DISP. OFF	TX=0.0300 LS=0.0150
BENTLEY/WORKSPACE / MECHDRAF/SEED/ MECHLAY.DGN	IN	TH	1IN=1000TH 1TH=254PU	SEE NOTE GRID & NODES DISP. OFF	TX=0.0047 LS=0.0295

NAME OF THE SEED FILE	WORKING UNITS			VIEW ATTRIBUTES	TEXT PARAMETERS
	MU	SU	RESOLUTION		
BENTLEY/WORKSPACE / MECHDRAF/SEED/ MECHLAYM.DGN	MM		1MM=1000 1=100PU	SEE NOTE GRID DISP. OFF	TX=0.0300 LS=0.0150
BENTLEY/WORKSPACE / MECHDRAF/SEED/ SDMECH2D.DGN	IN	TH	1IN=1000TH 1TH=254PU	SEE NOTE GRID DISP. OFF	TX=2.0000 LS=0.5000
BENTLEY/WORKSPACE / MECHDRAF/SEED/ SDMECH3D.DGN	IN	TH	1IN=1000 1TH=254PU	SEE NOTE GRID DISP. OFF	TX=0.1000 LS=0.0295
BENTLEY/WORKSPACE / MECHDRAF/SEED/ SDMENG2D.DGN	MM	SU	1MM=1000SU 1SU=100PU	SEE NOTE GRID DISP. OFF	TX=2.0000 LS=0.0500
BENTLEY/WORKSPACE / MECHDRAF/SEED/ SDMENG3D.DGN	MM	SU	1MM=1000SU 1SU=100PU	SEE NOTE GRID DISP. OFF	TX=0.0100 LS=0.0500

Note: Following are the Display Attributes turned on by default unless noted differently in the table:

CONSTRUCTIONS	LINE STYLES
DATA FIELDS	LINE WEIGHTS
DIMENSIONS	PATTERNS
DYNAMICS	TEXT
GRID	TEXT NODES

Index

B

C

D

E

F

G

H

I

J

K

L

N

O

P

Q

R

S

T

✓ 1¢

U

V

W

X

Z